5th International Conference on New Forming Technology (ICNFT 2018)

MATEC Web of Conferences Volume 190 (2018)

Bremen, Germany
18 - 21 September 2018

Editors:

F. Vollertsen
T.A. Dean
Y. Qin
S.J. Yuan

ISBN: 978-1-5108-7177-9

Printed from e-media with permission by:

Curran Associates, Inc.
57 Morehouse Lane
Red Hook, NY 12571

proceedings.com

Some format issues inherent in the e-media version may also appear in this print version.

Printed by Curran Associates, Inc. (2018)

For additional information, please contact EDP Sciences – Web of Conferences at the address below.

EDP Sciences – Web of Conferences
17, Avenue du Hoggar
Parc d'Activité de Courtabœuf
BP 112
F-91944 Les Ulis Cedex A
France

Phone: +33 (0) 1 69 18 75 75
Fax: +33 (0) 1 69 28 84 91

contact-edps@webofconferences.org

Additional copies of this publication are available from:

Curran Associates, Inc.
57 Morehouse Lane
Red Hook, NY 12571 USA
Phone: 845-758-0400
Fax: 845-758-2633
Email: curran@proceedings.com
Web: www.proceedings.com

TABLE OF CONTENTS

Preface

Worldwide there is a high demand for formed parts, for example in automotive, aerospace and electronics industries as well as for telecommunication and medical purposes. Forming offers a great potential concerning resource conservation and environmental friendliness. Further improving these aspects is in actual focus as well as reducing costs and qualifying forming processes for new applications.

Since 2004 the International Conference on New Forming Technology (ICNFT) is an inspiring forum for researchers and professional practitioners to discuss aspects of leading edge novel forming technologies. Former ICNFT conferences were held in China in 2004, in Germany in 2007, again in China in 2012 and in the UK 2015 respectively.

The 5th International Conference on New Forming Technology (ICNFT2018) is held September 18-21, 2018 in Bremen, Germany. Delegates from 15 countries submitted 135 abstracts to the conference and 81 full papers were accepted by the International Scientific Committee and published in the peer reviewed conference proceedings. There are two keynote sessions for 8 plenary lectures and 21 topic sessions for more than 88 oral and poster presentations. These cover fields of forging, extrusion, bulk forming, drawing, sheet forming/hot-stamping, rolling/roll forming, hydro- and tube forming, micro-forming, powder technology, materials and characterization, tooling and heat treatment, design and modelling, and advanced manufacturing processes. The contents deal with new/enhanced forming processes for both conventional and new materials. They well reflect the state of the art of the current forming technology and give a strong indication to the trend of developing new forming technology.

Special sessions are offered for processes and materials including a plenary Keynote session dedicated to Prof. Frank Vollertsen´s 60th birthday, for micro cold forming with the final colloquium of the DFG funded Collaborative Research Center "Micro Cold Forming" (SFB 747) and for dry metal forming with intermediate colloquium of the DFG funded Priority Program "Dry Metal Forming" (SPP 1676).

A guided tour to Bremen's University of Excellence and its research institutes are part of the conference. The 5th ICNFT gives you the opportunity to exchange thoughts and opinions on technological trends and future challenges as well as international collaboration perspectives in the wonderful surrounding of Bremen. Participation from industry in this conference will particularly enhance the industrial relevance of the research in developing future forming technology.

In this context, we like to thank all members of the International Scientific Committee as well as the Program Committee for their support in attracting contributions and reviewing papers according to CIRP regulations. Furthermore, special thanks shall be extended to the International Academy for Production Engineering (CIRP), the German Research Foundation (DFG), Arbeitsgemeinschaft Umformtechnik (AGU), Wissenschaftliche Gesellschaft Lasertechnik e.V. (WLT) and Bremer Institut für angewandte Strahltechnik (BIAS) for taking over sponsorship for this event. Finally, we would like to thank all those who have contributed to the 5th ICNFT 2018 "behind the curtain". Of these the Local Organizing Committee, which has assembled all the papers for the proceedings and cared for all registration affairs as well as the many little but important things associated with such a large event should especially be acknowledged.

F. Vollertsen
Chairman of the 5th ICNFT
July 2018

Conference Chair

F. Vollertsen, Germany

Conference Co-Chairs

T.A. Dean, UK
Y. Qin, UK
S.J. Yuan, China

Program Committee

T.A. Dean, UK
J. Lin, UK
Y. Qin, UK
F. Vollertsen, Germany
S.J. Yuan, China

Organizing Committee

S. Friedrich, Germany
T. Seefeld, Germany
C. Steffens, Germany

Scientific Committee

J.M. Allwood, UK
B. Awiszus, Germany
N. Bay, Denmark
B.-A. Behrens, Germany
N. Ben Khalifa, Germany
R.B. Bergmann, Germany
G. Bissacco, Denmark
M. Bredol, Germany
A. Brosius, Germany
S. Bruschi, Italy
B. Clausen, Germany
L. Cser, Hungary
T. A. Dean, UK
J.R. Duflou, Belgium
A. Fischer, Germany
A. Fortunato, Italy
L. Fratini, Italy
M. Graf, Germany
P. Groche, Germany
G. Habedank, Germany
T. Hanemann, Germany
G. Hirt, Germany
W. Homberg, Germany
B. Kaftanoglu, Turkey
O. Kessler, Germany
F. Klocke, Germany
V. Kräusel, Germany
B. Kuhfuß, Germany
D. Landgrebe, Germany
B. Lauwers, Belgium
B. Lenczowski, Germany
J. Lin, UK
G. Liu, China
H. J. Maier, Germany
P. Martins, Portugal
A. Mehner, Germany
M. Merklein, Germany
H. Palkowski, Germany
K. Partes, Germany
S. Rosiwal, Germany
Y. Qin, UK
M. Schaper, Germany
C. Schenck, Germany
A. Schmidt, Germany
M. Schmidt, Germany
R. Schmitt, Germany
A. Schulz, Germany
V. Schulze, Germany
T. Seefeld, Germany
T. Shimizu, Japan
A.E. Tekkaya, Germany
H. Toenshoff, Germany
G. Tovar, Germany
E. Uhlmann, Germany
F. Vollertsen, Germany
M. Yang, Japan
S.J. Yuan, China
M. Zäh, Germany

The 5th ICNFT is sponsored by

Deutsche Forschungsgemeinschaft
German Research Foundation

Arbeitsgemeinschaft Umformtechnik

Bremer Institut für angewandte Strahltechnik

SFB 747

SPP 1676

edp sciences

MATEC Web of Conferences

Statement of Peer review

In submitting conference proceedings to MATEC Web of Conferences, I certify to the Publisher that I adhere to the **Policy on Publishing Integrity** of the journal in order to safeguard good scientific practice in publishing.

1. All articles have been subjected to peer review administered by the proceedings editors.
2. Reviews have been conducted by expert referees, who have been requested to provide unbiased and constructive comments aimed, whenever possible, at improving the work.
3. Proceedings editors have taken all reasonable steps to ensure the quality of the materials they publish and their decision to accept or reject a paper for publication has been based only on the merits of the work and the relevance to the journal.

Title, date and place of the conference

5th International Conference on New Forming Technology (ICNFT2018)
September 18-21, 2018 in Bremen, Germany

Proceedings editor(s):

F. Vollertsen, T.A. Dean, Y. Qin, S.J. Yuan

Date and editor's signature July 16, 2018

A Review of Progressive and Compound Forming of Bulk Microparts by Using Sheet Metals

M.W. Fu[1,*], *J.Y.* Zheng[1], and *B. Meng*[2]
[1]Department of Mechanical Engineering, The Hong Kong Polytechnic University, Hung Hom, Kowloon, Hong Kong
[2]School of Mechanical Engineering and Automation, Beihang University, Beijing 100191, P.R. China

Abstract: In the last decade, the concept of progressive microforming has emerged and developed gradually, which is considered as an efficient and promising method to fabricate the micro-scaled part. Micro-cylinder parts, micro-flanged part, and multi-flanged microparts are representative micro bulk parts by the progressive microforming system using sheet metal. In these cases, many efforts focus on the forming process, such as microblanking and microextrusion. Meanwhile, the quality of the fabricated parts also attracts attention. In this paper, an intensive review on the development of progressive microforming technologies and the formed parts is presented, and the influence of size effect to dimensional accuracy, material flow, geometrical feature, and fracture is also discussed.
Keywords: Micro forming, Deformation, Sheet metal

1. Introduction

Due to the increasing demand of product miniaturization in microelectronics, biomedicine, consumer electronics and other industrial clusters [1-4], microforming technologies are gaining more attention and development effort. Microforming is defined as the production of micro-scale parts or structures with at least two dimensions in the sub-millimeter range [2]. This process is considered as a promising micro-manufacturing method due to its good productivity, high material usage, near-net-shape characteristic and the good mechanical properties of the deformed parts [1].

The experience of manufacturing macro-scaled parts cannot be applied directly in the fabrication of microparts due to size effect [3-6]. Fu and Chan [7] reviewed the state-of-the-art microforming technologies and had a panorama of the research in the last two decades on size effect related deformation behaviors, and discussed some research issues from the implementation of mass production. However, some critical issues such as dimensional accuracy of parts, final part damage control and positioning in production processes cripple the application of microforming in mass production.

To solve these problems and meet the increasing demands, the concept of progressive and compound microforming emerged and developed gradually. **Fig. 1** shows some progressive microformed parts. Progressive microforming is one of the state-of-the-art microforming processes [8], which is a forming technology to fabricate micro-scale parts by using progressive forming process. In each progressive forming operation, one or more distinct die is used and after the specific processing sequence, a complete part is fabricated from metal strips.

Regarding this process, Hirota [9] firstly proposed a new method to fabricate micro-pin parts by extruding a metal sheet in the thickness direction. Ghassemali et al. [10, 11] adopted this process to produce axisymmetric parts with micro-pins and investigated the phenomena of variations of material flow and microstructure evolution due to the size effect. According to this novel method, further researches have been conducted. Fu and Chan [12, 13] investigated the feasibility of fabricating the bulk cylinders and flanged parts via directly using sheet metal in meso- and micro-scales, and the characteristics of final parts, as well as the production process, were examined. They also discussed the relationship between grain size and deformation behaviors. Based on the prior studies, Meng et al. [14-16] conducted further and extensive studies of micro progressive forming. In these cases, the cylinder, double-flanged and variable-thickness flanged microparts were manufactured by progressive microforming, respectively. They further investigated the micromechanical damage and deformation behavior in progressive microforming and developed the systematic knowledge for micro-formed part design.

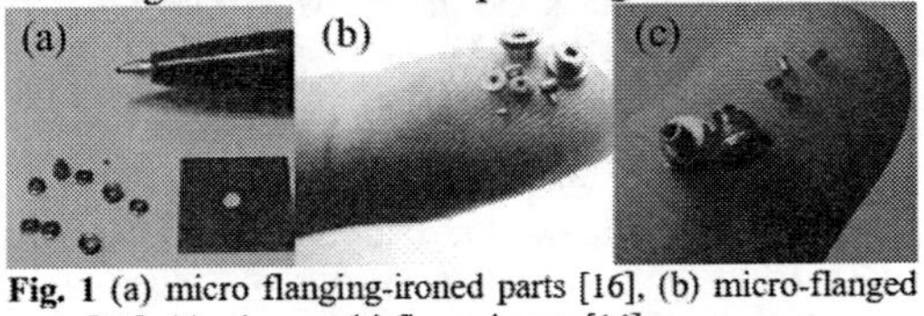

Fig. 1 (a) micro flanging-ironed parts [16], (b) micro-flanged parts [13], (c) micro multi-flanged parts [14].

*Corresponding author: mmmwfu@polyu.edu.hk

Base on the shapes of bulk parts, three types of progressive microforming parts can be classified according to the previous research: micro-cylinder parts, micro-flanged/ironed parts and multi-flanged microparts.

Although the researches on progressive microforming processes have been conducted for many years and several modeling methodologies have been established, there are still some problems that impede the application of the progressive microforming, and it still contains many unknowns behind size effects. In this paper, the studies on the processes of progressive microforming and fabricated microparts are reviewed comprehensively. To provide a summary of the progressive microforming technologies and progressive tooling and parts design. The mechanism of how the size effect affects the deformation behavior and characteristics of microparts is discussed, including the evolution of microstructure, fracture behavior, dimensional accuracy, and undesirable geometry, etc.

2. Progressive formed parts

2.1 Micro-cylinder parts

Cylinder parts can be fabricated via cutting metal wire or blanking metal sheets, and the latter is proved as a more considerable method. In this process, both microblanking and micropunching are shearing operations with the advantage of low cost and high productivity over the micromachining processes. Microblanking is employed to produce bulk metal parts from sheet metals, while micropunching is a simple process for fabrication of microholes [8].

Among the prior arts, Xu et al. [17] conducted micro-blanking experiments via using brass foil to investigate the size effect on deformation behavior. Fu and Chan [12, 13] conducted experiments by designing two scales of progressive forming systems to fabricate meso-/micro-cylinder parts via microblanking process and focused on the characteristic of formed parts. Meng et al. [14-16] extensively investigated the material deformation behavior and the fracture effect of shearing in micro-cylinder parts forming processes by using the same method. In addition, the influence of forming conditions on deformation behavior, fractured surface quality and dimensional accuracy in shearing operation were systematically investigated.

2.1.1 Evolution of microstructure

Meng et al. [15] conducted research on the blanked part. The microstructure of the micro-cylinder part can be divided into four zones (I-IV) according to the variation of grain sizes and stress states, and the stress states during the shearing process are also presented via FE simulation, as shown in **Fig. 2**. Among these zones, the less-deformed zone, Zone I is defined as the dead zone by Ghassemali et al. [18]. The highly elongated and rotated grains close to the shear bands can be seen in zone IV. When the elongated grains extend to the center section, zone II is compressed to be formed. The arched shape is associated with the different flow velocities between the center and sides. Zone III is the gradient distribution zone. The material in zone II is subjected to the tensile stress in the horizontal direction, which is induced by the restriction near the edge of moving punch. The tensile stress in zone II transmits the punching force to the punch edge to separate the material from the sheet metal. Furthermore, the bandwidth of zone II is related to the original grain size. As the fine-grained material is used, the grains are distributed evenly in zones II and III, thus the compressive band is not obvious. The stress state of zone I is under triaxial compressive stress, which is attributed to the fact that zone II hinders the material in zone I to flow to both sides. The grain size of zone III increases to the rollover gradually. The stress state of zone III is the same as that in zone II, but the tensile stress is lower, resulting in the gradient distribution of grain size.

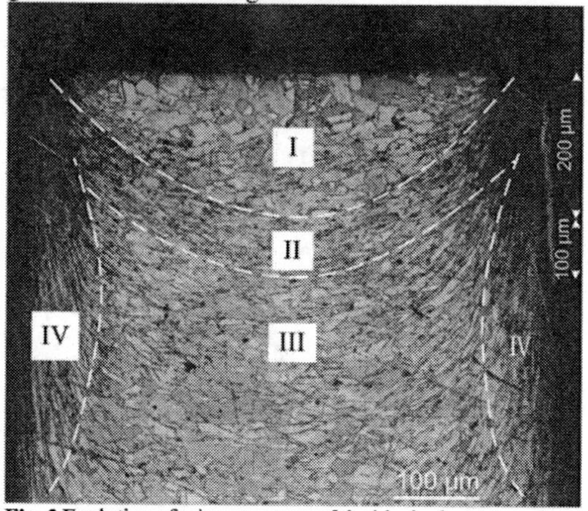

Fig. 2 Evolution of microstructure of the blanked part (zone I, II, III, IV) [15].

2.1.2 Deformation load

Deformation load represents the energy needed during the forming process, which should be as small as possible, so it could be used as one of the criteria in part design. Fu and Chan [12] indicated there were three stages of the shearing process in blanking operation which is shown in **Fig. 3**. In the first stage, punch presses the workpiece to deform elastically and then plastically, resulting in strain hardening and the increase of deformation load. In the second stage, the material is further pushed forward into the die by shearing. The length of the shear band decreases with the punch penetration, causing the decrease of deformation load. In the third stage, the deformation exceeds its limit and fracture took place. This leads to a sudden decrease of deformation load. The remaining load after fracture is to overcome the frictional force at the tooling-workpiece interface.

Meng et al. [15] investigated the interactive effect of grain size, the lubricated condition and the velocities on the micro shearing process. It is revealed the scatter of shearing load increases with grain size using mechanical oil or grease lubricant. The deformation load is reduced by using grease lubricant, but its scatter increases. The lubricant with low viscosity has a limited role in the shearing process, and the high viscosity lubricant is more effective.

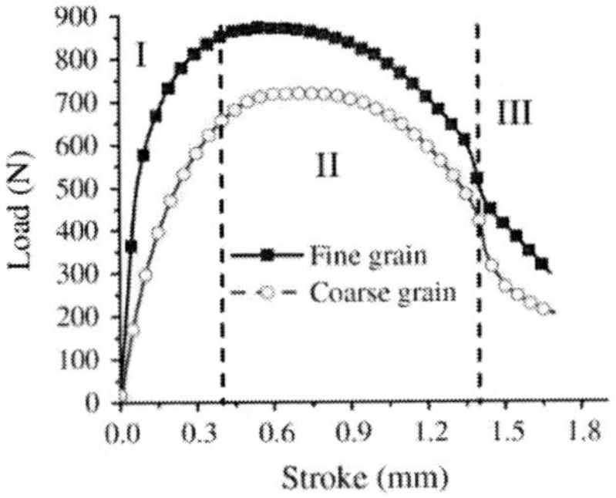

Fig. 3 Load–stroke curves in the blanking operations for the materials with different grain sizes [12].

2.1.3 Ultimate shear stress

The ultimate shear strength is obtained based on the following equation [19]:

$$\sigma_s = \frac{F_{max}}{\pi d l} \quad (1)$$

where F_{max}, d and l are the maximum shearing force, punch diameter and the thickness of sheet material, respectively. Xu et al. [17] indicated the ultimate shearing strength increases with the decrease of the foil thickness. Fu and Chan [13] got the same conclusion by using two scales of forming systems. In the researches of Xu et al. [17] and Meng et al. [15], the ratio of blanking clearance to grain size c/g (**Fig. 4**) becomes one of the main factors to influence the deformation behavior of micro punching/blanking, and there exists a minimum peak value when c/g≈1, as shown in **Fig. 5**. When c/g>1, the punch-die clearance has the width of several grains, and the shear band involves a large number of grain boundaries. When c/g<1, the deformation of grain matrix takes leading position in the shearing band, and the grain boundary sliding is still dominant in the thickness direction. When c/g≈1, the clearance can only accommodate one grain. In this condition, the grain boundary sliding and coordination in the clearance width direction reduces to a minimum level, meanwhile the grain matrix deformation reaches the lowest magnitude. So the ratio of punch-die clearance to grain size leads to the lowest shearing stress when c/g≈1.

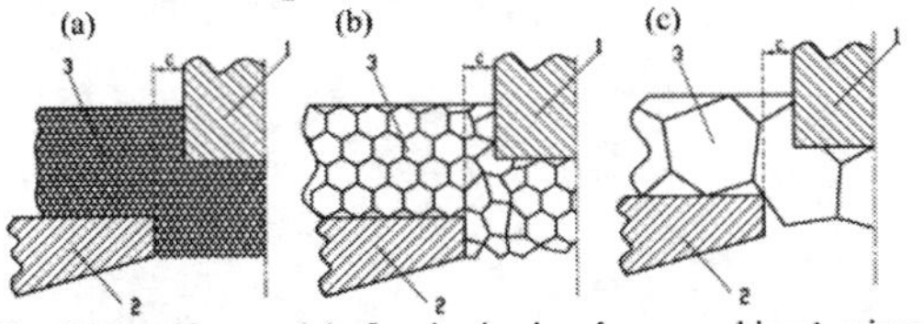

Fig. 4 Size effect model of grain size in micro-punching 1-micro punch; 2-micro die; 3-metal foil:(a) c>g; (b) c≈g and (c) c<g [17].

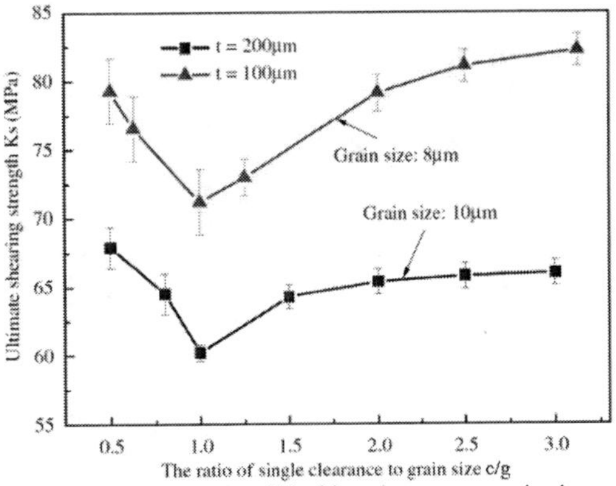

Fig. 5 Effects of the ratio of blanking clearance to grain size on ultimate shearing strength [17].

2.1.4 Dimensions accuracy of microparts

Fu and Chan [13] revealed that the length of blanked cylinder decreases with the increase of grain size, which is attributed to the aggravated lateral material flow when using the coarse-grained material, as shown in **Fig. 7**. When the grain size is large compared to the blanked cylinder, the individual grain properties become significant and make the deformation inhomogeneous, which further results in the irregular shape of the blanked cylinder. In the subsequent researches, Meng et al. [14-16] fabricated different micro-cylinder parts, as shown in **Fig. 7**, and the results supported the prior conclusion. It is found that the length of the blanked cylinder decrease but the diameter increase by grease lubricant, and the billet diameter reached the maximum when the punch-die clearance is close to grain size, as shown in **Fig. 8**.

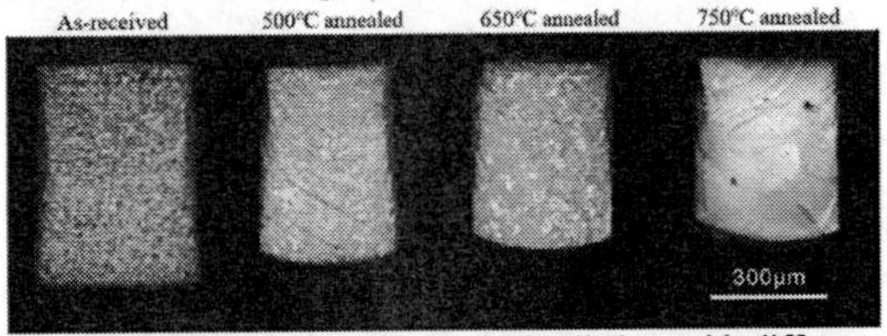

Fig. 6 Microstructures of the blanked cylinders with different material states of the micro-scaled case [13].

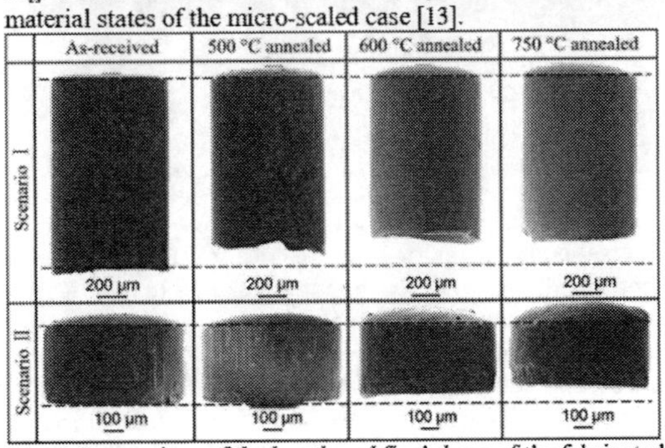

Fig. 7 Comparison of the length and final shape of the fabricated micro-pins [16].

https://doi.org/10.1051/matecconf/201819001001

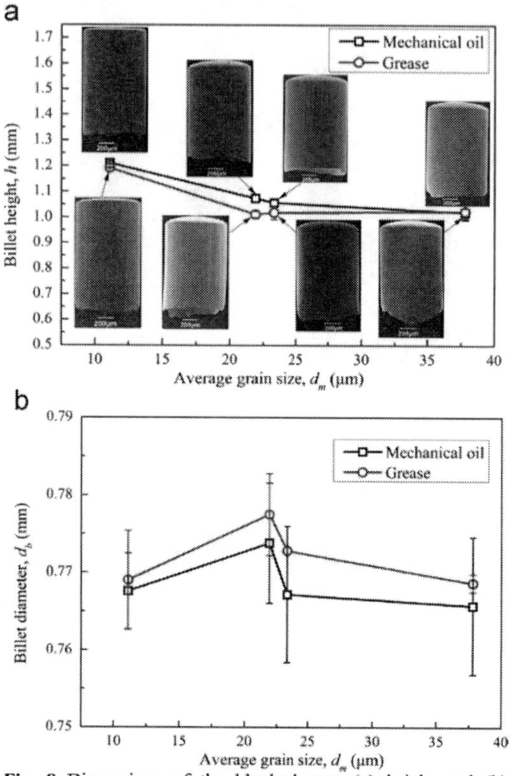

Fig. 8 Dimensions of the blanked part: (a) height and (b) diameter [15].

2.1.5 Fracture

The fracture mechanism based on the initiation, growth, and coalescence of microvoids, which can only occur at the grain boundaries or inclusions, results in the rough surface finish. The irregular shape of the blanked microbillet and the inclined fracture surface could be found as shown in **Fig. 9**.

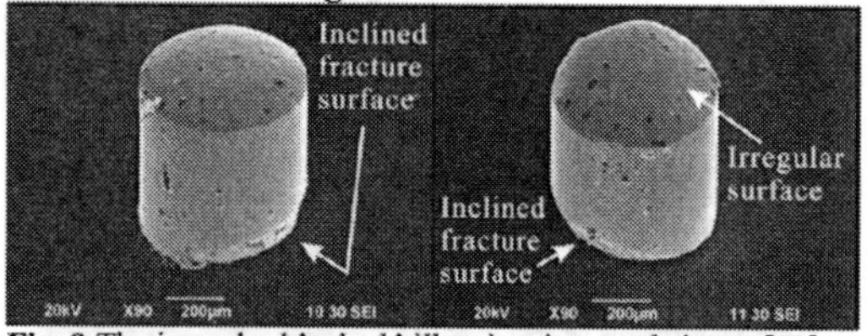

Fig. 9 The irregular blanked billets in micro-scaled case [13].

Xu et al. [17] indicated that the fracture mechanism of micro-blanking significantly changes from fracture mode with ductile dimples as shown in **Fig. 10** (a) and (b) to ductile fracture with slip separation as shown in **Fig. 10** (c) with the decrease of foil thickness. Meanwhile, the number of the dimple is decreased with the brass foil thickness. The irregularities occur which can strongly affect the fracture with the decrease of blanking clearance, as shown in **Fig. 11**. However, the fracture mechanism does not significantly change with ductile dimples with the decrease of blanking clearance, which indicates that blanking clearance has no effect on fracture of microblanking. In addition, grain size has no effect on fracture mechanism of micro-blanking, as shown in **Fig. 12**.

Meng et al. [15] found that there are no obvious dimples on the blanked surface of the micro-cylinder part, especially for the coarse-grained material. In addition, the high velocity worsens the surface finish. The surface breakages on the blanked part appear under the high shear velocity due to the inhomogeneous deformation rate throughout the shear band, as shown in **Fig. 13**.

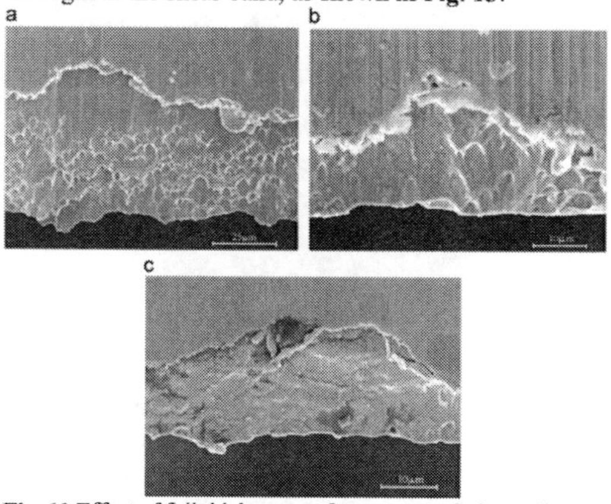

Fig. 10 Effect of foil thickness on fracture morphology of cross-section. (a) t=200μm; (b) t=100 μm and(c) t=60 μm [17].

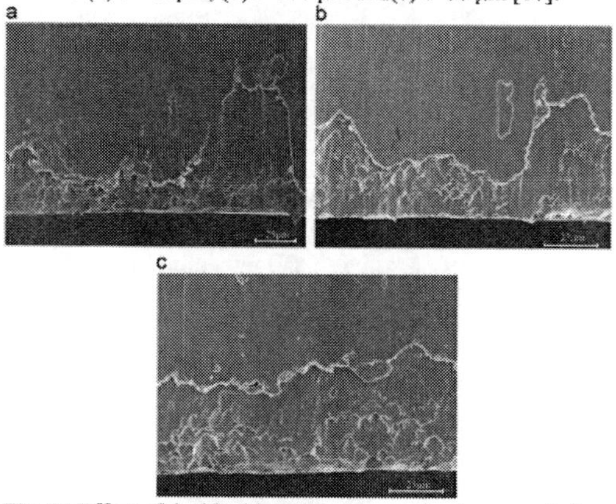

Fig. 11 Effect of the blanking clearance on fracture morphology. (a) c/t=2.5%; (b) c/t=5% and(c) c/t=10% [17].

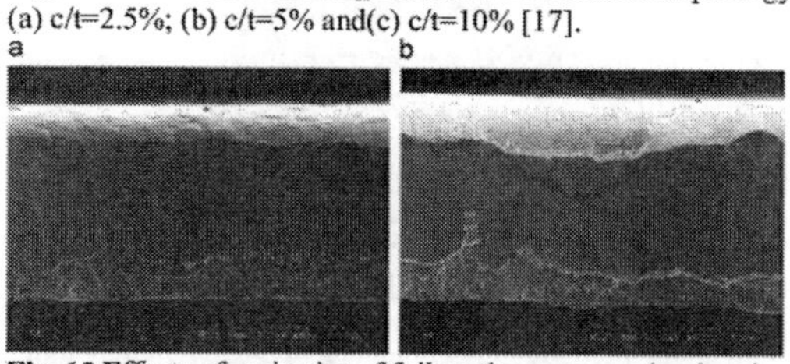

Fig. 12 Effects of grain size of foil on the cross-section in micro blanking. (a) d=8.2 mm and (b) d=98.2 mm [17].

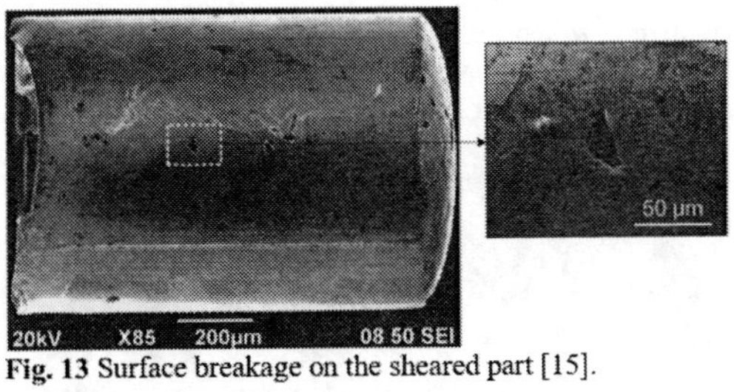

Fig. 13 Surface breakage on the sheared part [15].

2.1.6 Rollover of sheet

Fu and Chan [12] revealed the rollover size is large and the rollover surface becomes rough with the increase of grain size, which could be related to the large hardening exponent and attributed to the fact that strain incompatibility among grains makes the surface grains move normal to the surface to accomplish deformation, as shown in **Fig. 14**.

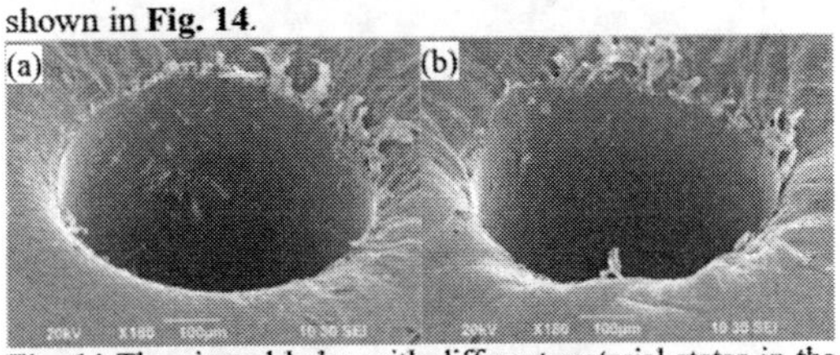

Fig. 14 The pierced holes with different material states in the micro-scaled case (a) 500°C annealed (b) 750°C annealed [12].

2.1.7 Undesirable geometries

Meng et al. [15, 16] conducted further and extensive study of the micro-cylinder part forming under interactive effects of multi-factors, including grain size, lubricated condition, shear velocity and die dimensions. In this research, the micro-cylinder part was divided into different zones, as shown in **Fig. 15**. The experiment results with different grain sizes, dimensions of the part, lubricated conditions and shear velocities are shown in **Fig. 16** to **Fig. 18**. It is revealed that the forming quality deteriorates with the increase of grain size and shear velocity. In addition, a high ratio of thickness to diameter of the micro-cylinder part and high viscosity lubricant leads to the good surface finish.

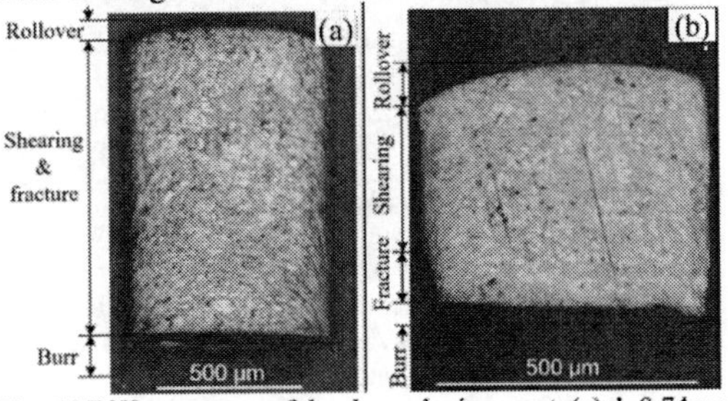

Fig. 15 Different zones of the sheared micro-part: (a) d=0.74mm, l=1.52mm; (b) d=0.6mm, l=0.4mm [16].

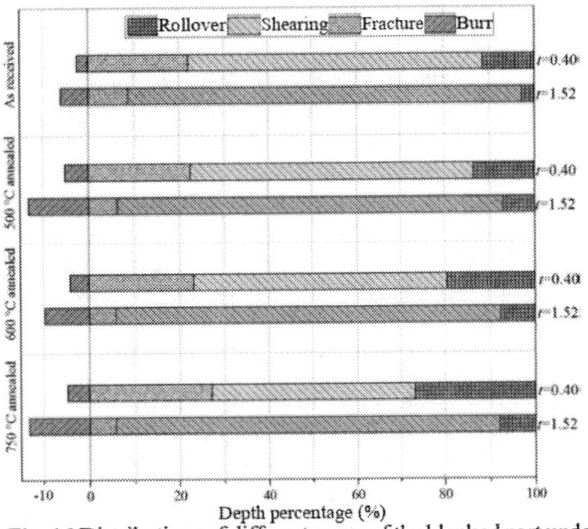

Fig. 16 Distributions of different zones of the blanked part under diverse material conditions [16].

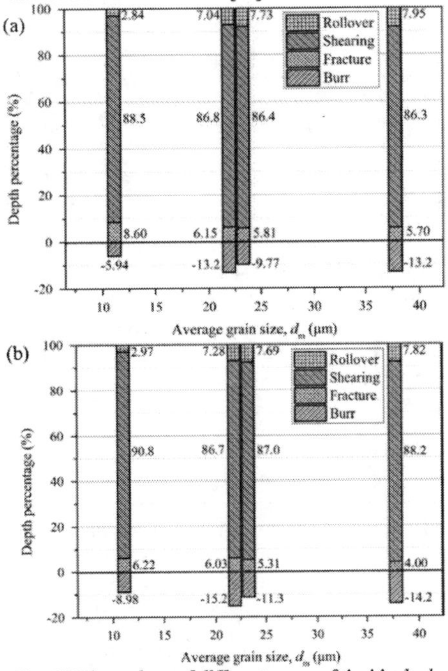

Fig. 17 Dimensions of different zones of the blanked part under different lubricated conditions: (a) mechanical oil (b) grease [15].

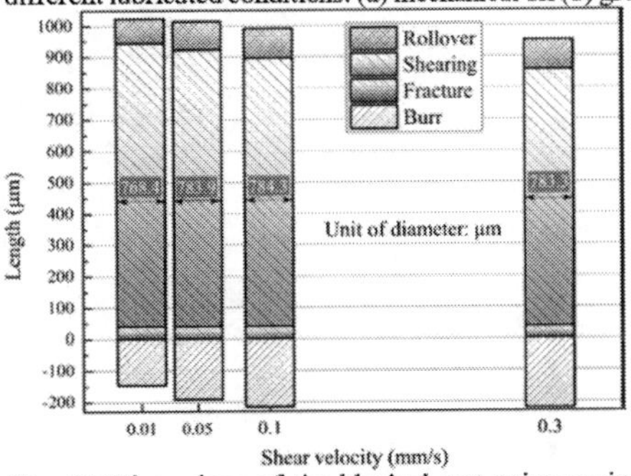

Fig. 18 Dimensions of the blanked part using various shear velocities [15].

2.2 Micro-flanged/ironed parts

Micro-flanged/ironed parts are fabricated by extrusion and blanking processes base on guide hole after punching process. Forward extrusion is defined as a plastic deformation process in which billet material is compressed in the die and the material flows in the same direction as the movement of punch [8]. Microextrusion is widely investigated in terms of size effect, constraint condition, and tooling-workpiece interfacial friction.

Fu and Chan [12, 13] used the progressive forming system to fabricate the flanged parts via the progressive punching, extrusion, and blanking. In the second operation, the workpiece was positioned based on the guide hole, and the material was pushed to flow towards die cavity. Then the flanged part was blanked from the sheet. Pure copper sheets with the thicknesses of 1.5mm as meso-scale and 0.8mm as micro-scale are selected as testing material. Various grain sizes were obtained via heat treatment. The forming process and the flanged part are shown in **Fig. 19**.

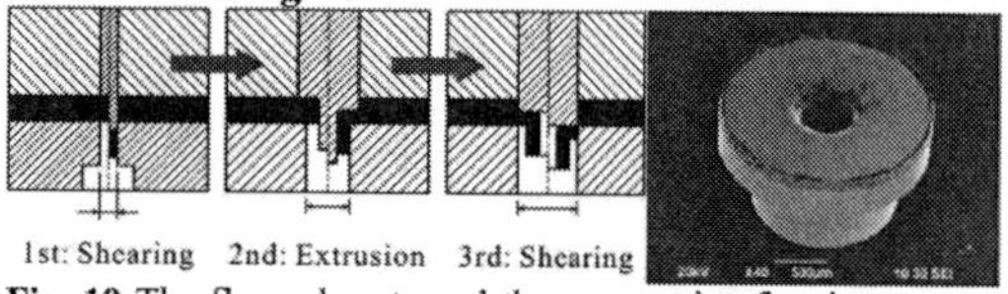

Fig. 19 The flanged parts and the progressive forming system [13].

Meng et al. [16] developed a promising and efficient progressive microforming method to fabricate micro-scaled hollow geometries with higher flange and variable thickness by using an integrated hole flanging-ironing process, as shown in **Fig. 20**. The pure copper sheet metal with the thickness of 0.4mm was chosen as the experimental material, and the sheets were annealed to obtain different grain sizes. The progressive micro forming system consists of three forming steps, viz. shearing, hole flanging-ironing, and blanking. A microhole was punched and a micro-pin part was blanked out after the first operation. And then the fine pre-pierced hole is used for positioning and a long flange is formed by micro-scaled hole flanging-ironing process. The last operation was used to blank out the micropart from the workpiece.

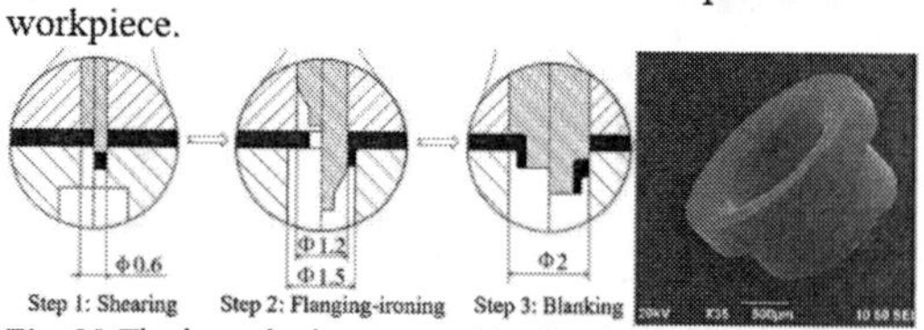

Fig. 20 The ironed micropart and its forming process [16].

2.2.1 Evolution of microstructure

Fu and Chan [13] investigated shear bands connecting the punch and die edges forming after the second operation in progressive microforming, as shown in **Fig. 21**. The flowing material is constrained by the surrounding material outside the deformation zone. The shear band divides the material flow into two directions. The inner material flows towards the die cavity to contribute the length of the extrudate, while the outer material flows laterally. In the second stage, the material in between the punch and die edges are subject to shearing deformation. The length of shear band decreases with the increase of punch stroke.

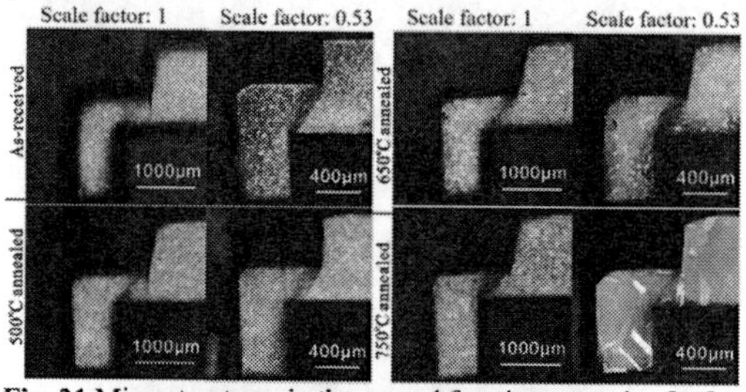

Fig. 21 Microstructures in the second forming operation [13].

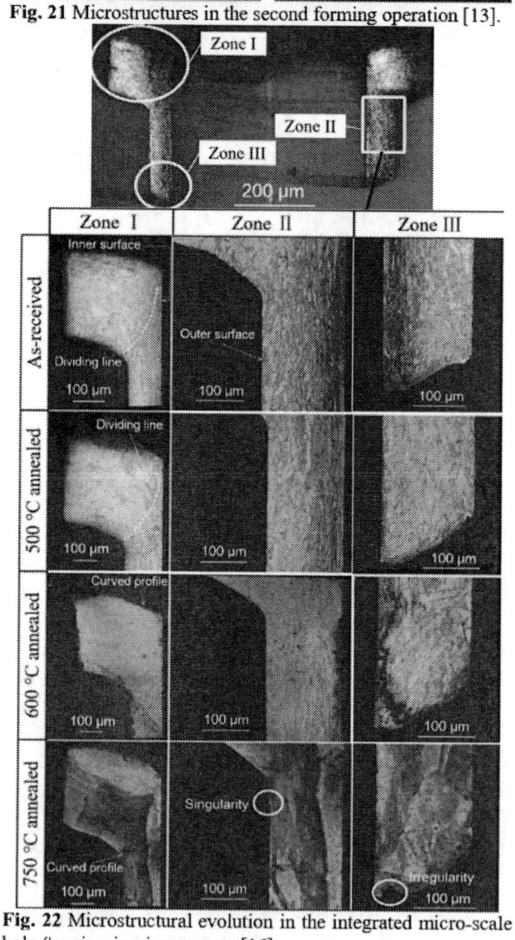

Fig. 22 Microstructural evolution in the integrated micro-scale hole flanging-ironing process [16].

In the study of Meng et al. [16], the evolution of microstructure was observed, as shown in **Fig. 22**. It is discovered that the hole flanging-ironing process produces a flanged feature with a curved profile followed by a straight wall, and the short curved profile and the desirable boring straightness are achieved by using the fine-grained material. zone I is the junction between the blanked part and the flanged edge, and the demarcation lines were observed to distinguish the bending region and the zone of edge stretching and ironing deformation with fine-grained materials. The inhomogeneous microstructure distributions are induced by the formation of dead zone [18]. However, the separatrix becomes blurred for coarse-grained materials. This phenomenon is caused by the decrease of the grain number involved in the deformation zone when the grain size is equivalent to the geometrical dimension of the workpiece. Singularities were observed in zone II at the outer profile of the flanged edge using the coarse-grained material. This could be attributed to the loss of the grain number and enhanced anisotropic properties of individual grains. It is found from zone III that the generated irregularities at the flanged tip are intensified with the increasing grain size.

2.2.2 Deformation pressure

A typical load-stroke curve of the microextrusion process is shown in **Fig. 23**, in which the difference between two curves was influenced by rollover caused by the last operation. It can be seen that the deformation load decrease with the increase of grain size.

The variation of the deformation load in the micro-scaled hole flanging-ironing process and the corresponding deformed shapes are presented in **Fig. 24**. Five deformation stages including elastic deformation, deflexion, flow forming, ironing, and sliding are identified in the micro-scaled hole flanging-ironing process, and the deformation load increases with the decrease of grain size.

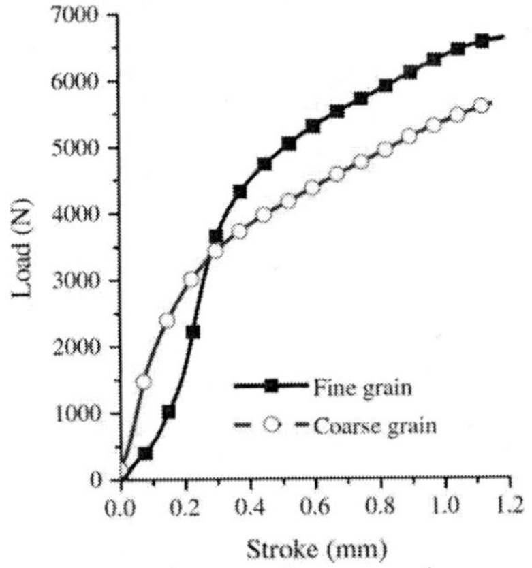

Fig. 23 The load-stroke curve in an extrusion process with different grain sizes [12].

2.2.3 Dimension accuracy

Meng et al. [16] examined the dimensional accuracy of height and tapering angle between inner and outer surfaces of the flanged micropart as shown in **Fig. 25**. It is revealed that the length of the flanged micropart is reduced with the increasing grain size, while both the tapering angle and its scatter present an opposite tendency, which could be explained by the coupled model of free surface roughening and open-closed lubricant pockets.

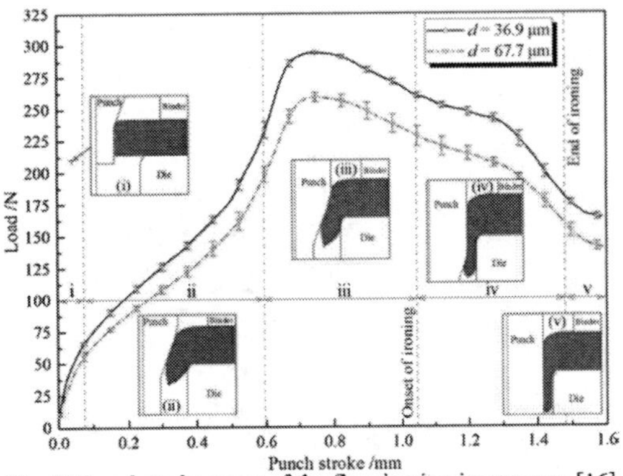

Fig. 24 Load-stroke curve of the flanging-ironing process [16].

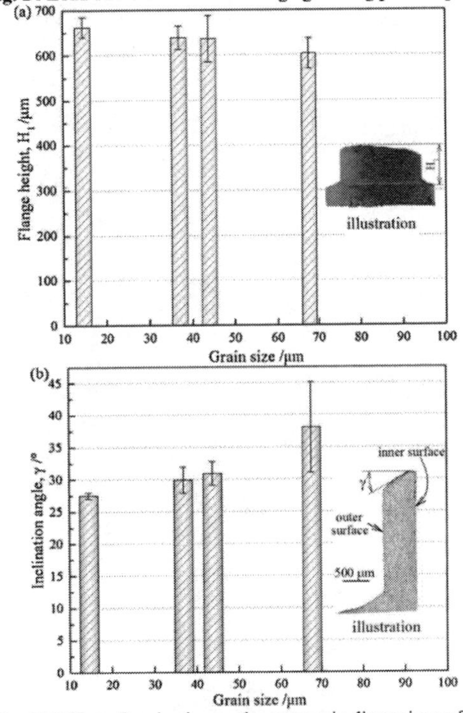

Fig. 25 Effect of grain size on the geometric dimensions of the flanged part: (a) flange height, and (b) tapering angle between inner and outer surfaces [16].

2.2.4 Fracture

Shearing and fracture regions could be found on the blanked surface of the micro-scaled part, as shown in **Fig. 26**. It can be seen that the parabolically shaped microvoids occurred on the fracture surfaces and the number of microvoids decreases with the increase of grain size. But the size of fracture surface does not have a significant change in the decreasing number of grains over the thickness of the workpiece.

Fracture and shearing surface were observed on the blanked part after the third operation. In the flanged ironing micropart, it is found that the flanged microparts with the fine-grained materials have the desirable forming

https://doi.org/10.1051/matecconf/201819001001

quality. The surface integrity of the formed part worsens with the increasing grain size, and the irregularity occurred. Meanwhile, the surface integrity of the junction between the flanged edge and the non-deformed zone also changes with the grain size. The surface of the coarse-grained material becomes rough.

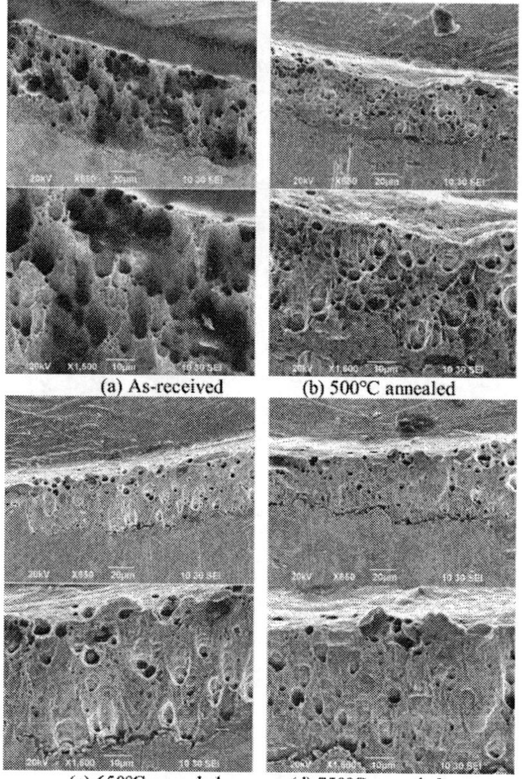
(a) As-received (b) 500°C annealed
(c) 650°C annealed (d) 750°C annealed

Fig. 26 The fracture surfaces of the micro-scaled parts with different material states [13].

2.3 Multi-flanged parts

The multi-flanged part was fabricated by progressive microforming with micropunching, two microextrusions, and microblanking in the study of Meng et al. [14], as illustrated in **Fig. 27**. This study aimed at the effect of grain size on material flow, geometrical features, dimensional accuracy and fracture behavior in microforming. The validity of different fracture criteria in microforming was also discussed.

2.3.1 Evolution of microstructure

The shear bands extending from punch edge to die orifice occurred in two continuous extrusion operation, as shown in **Fig. 28**. The grains in the shear bands undergo plastic deformation, and the effective strain is highly localized in this region. The width of the shear bands decreases with the increase of grain size. The shear band involves a large number of grains to participate in the plastic deformation using the fine-grained material, resulting in more severe deformation in this region.

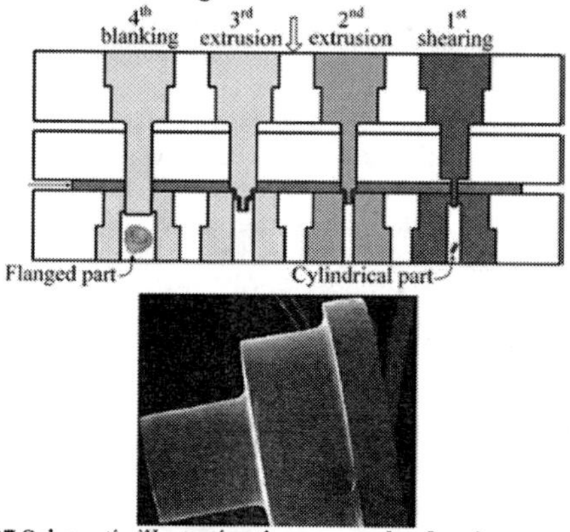

Fig. 27 Schematic illustration the progressive forming system of a multi-flanged part [14].

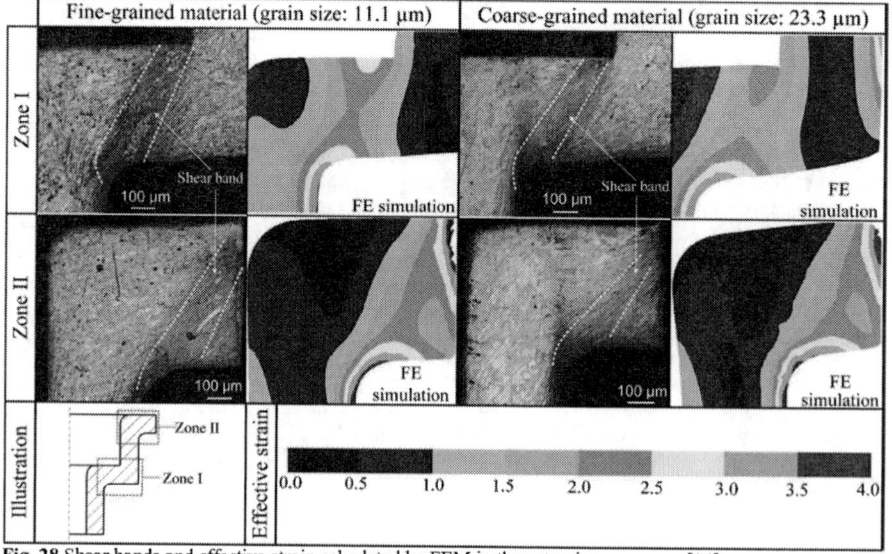

Fig. 28 Shear bands and effective strain calculated by FEM in the extrusion processes [14].

2.3.2 Geometrical features

The irregular geometric defects of the multi-flanged part are identified to exist in the three regions, as illustrated in **Fig. 29**. It is observed that the undesirable deformation deteriorates with the increase of grain size. The serious inclination of the flanged feature, irregular shape as well as the obvious burr is found using the coarse-grained material. This is due to the fact that the size and orientation of individual grains affect material flow behavior significantly when the characteristic dimension is in the order of grain size. In addition, the material stiffness and strength are reduced for the material annealed at high temperature, which contributes to the further deformation of the formed features.

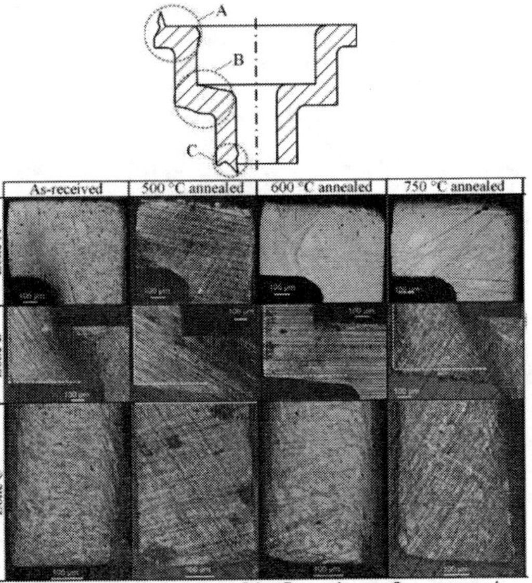

Fig. 29 Geometrical features of the flanged part after progressive deformation [14].

2.3.3 Dimensional accuracy

The dimensional variation of the two-level flanged parts is shown in **Fig. 30**. It is found that the diameters of the flanged part have the same geometries for both the fine- and coarse-grained materials, in contrast to the length of the extrudate shows a decreased tendency with the increase of grain size. This is caused by the fact that the deformation along the diametrical direction was restricted by tooling, while the material along the length direction underwent free deformation.

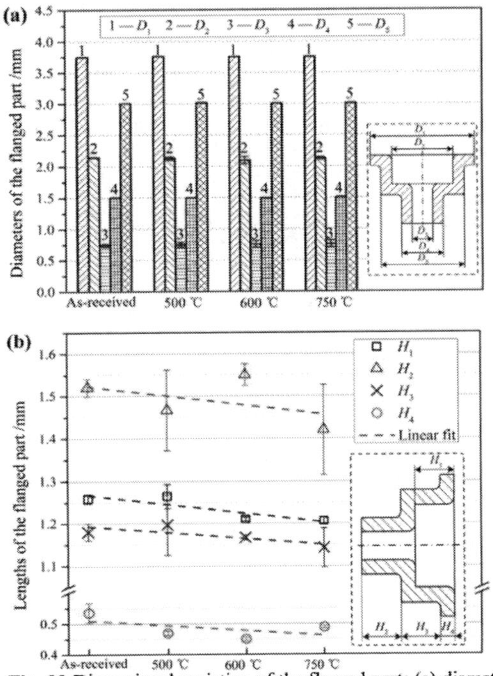

Fig. 30 Dimensional variation of the flanged part: (a) diameter and (b) length [14].

3. Conclusion

Progressive forming is proven to be a feasible and efficient microforming technology for production of micro-scaled parts. However, the occurrence of size effect while miniaturization of the forming system and parts results in some issues. In this paper, the state-of-the-art researches and the latest development of progressive microforming systems are reviewed comprehensively.

Dead zones of the deformed parts the shear band can be observed on the cross-section, which reveals that the material flow toward inner and outer from the shear band in the progressive microforming process, and less deformed area occurs far from the shearing area. Hardness test on the cross section shows that the largely deformed area has high hardness compared to the dead zones, and the hardness distribution supports the conclusion.

The deformation load in progressive microforming decreases with the increase of grain size, both in microblanking and microextrusion. This has a good agreement with the Hall–Petch relation. In addition, high viscosity lubricant can reduce the deformation load.

There is a minimum ultimate shear stress in the microblanking process when the grain size is equal to punch-die clearance, since the grain boundary sliding and coordination, as well as the grain matrix deformation, reach the lowest magnitude. The ultimate shear stress also increases with the decrease of the thickness of the metal sheet, because with the decrease of grains involved in shear deformation, slip and coordination of grains are difficult, leading to large shearing resistance.

The height of micro-cylinder parts, micro-flanged/ironed parts and multi-flanged part decreases with the increase of the grain size, but tapering angle of flanging-ironed parts has an opposite tendency. The height of micro-cylinder parts decreases and the diameter of them increases with the high viscosity lubricant. There is a maximum diameter when the grain size is equal to punch-die clearance.

The forming quality in terms of geometric dimension and surface finish deteriorates with the increase of grain size and the miniaturized thickness of sheet metal, which is related to the limited formability of few grains involved in the shear deformation.

Fracture mechanism changes from ductile dimples to ductile fracture with slip separation while scaling down to micro scale. The irregularities occurred with the decrease of punch-die clearance, but the fracture mechanism has no significant change. High shearing velocity worsens the surface finish and leads to surface breakage. Meanwhile, the number of microvoids decreases and the undesirable forming quality appears with the increase of grain size.

References

1. U. Engel, R. Eckstein, Microforming - from basic research to its realization, Journal of Materials Processing Technology 125 (2002) 35-44.
2. M. Geiger, M. Kleiner, R. Eckstein, N. Tiesler, U. Engel, Microforming, Cirp Ann-Manuf Techn 50(2) (2001) 445-462.
3. F. Vollertsen, Z. Hu, H.S. Niehoff, C. Theiler, State of the art in micro forming and investigations into micro deep drawing, Journal of Materials Processing Technology 151(1-3) (2004) 70-79.
4. F. Vollertsen, H.S. Niehoff, Z. Hu, State of the art in micro forming, Int J Mach Tool Manu 46(11) (2006) 1172-1179.
5. A. Messner, U. Engel, R. Kals, F. Vollertsen, Size Effect in the Fe-Simulation of Micro-Forming Processes, Journal of Materials Processing Technology 45(1-4) (1994) 371-376.
6. F. Vollertsen, D. Biermann, H.N. Hansen, I.S. Jawahir, K. Kuzman, Size effects in manufacturing of metallic components, Cirp Ann-Manuf Techn 58(2) (2009) 566-587.
7. M.W. Fu, W.L. Chan, A review on the state-of-the-art microforming technologies, Int J Adv Manuf Tech 67(9-12) (2013) 2411-2437.
8. M.W. Fu, W.L. Chan, Micro-scaled products development via microforming: deformation behaviors, processes, tooling and its realization, Springer, Berlin Google Scholar, 2014.
9. K. Hirota, Fabrication of micro-billet by sheet extrusion, Journal of Materials Processing Technology 191(1-3) (2007) 283-287.
10. E. Ghassemali, A.E.W. Jarfors, M.J. Tan, S.C.V. Lim, On the microstructure of micro-pins manufactured by a novel progressive microforming process, Int J Mater Form 6(1) (2013) 65-74.
11. E. Ghassemali, M.J. Tan, A.E.W. Jarfors, S.C.V. Lim, Progressive microforming process: towards the mass production of micro-parts using sheet metal, Int J Adv Manuf Tech 66(5-8) (2013) 611-621.
12. W.L. Chan, M.W. Fu, Meso-scaled progressive forming of bulk cylindrical and flanged parts using sheet metal, Materials & Design 43 (2013) 249-257.
13. M.W. Fu, W.L. Chan, Micro-scaled progressive forming of bulk micropart via directly using sheet metals, Materials & Design 49 (2013) 774-783.
14. B. Meng, M.W. Fu, C.M. Fu, K.S. Chen, Ductile fracture and deformation behavior in progressive microforming, Materials & Design 83 (2015) 14-25.
15. B. Meng, M.W. Fu, C.M. Fu, J.L. Wang, Multivariable analysis of micro shearing process customized for progressive forming of micro-parts, Int J Mech Sci 93 (2015) 191-203.
16. B. Meng, M.W. Fu, S.Q. Shi, Deformation characteristic and geometrical size effect in continuous manufacturing of cylindrical and variable-thickness flanged microparts, Journal of Materials Processing Technology 252 (2018) 546-558.
17. J. Xu, B. Guo, C.J. Wang, D.B. Shan, Blanking clearance and grain size effects on micro deformation behavior and fracture in micro-blanking of brass foil, Int J Mach Tool Manu 60 (2012) 27-34.
18. E. Ghassemali, A. Jarfors, M. Tan, S. Lim, Dead-zone formation and micro-pin properties in progressive microforming process, ICTP 2011, 10th International Conference on Technology of Plasticity, Steel Research International: Special Edition, Aachen, DE, Sep 25-29, 2011, Wiley-VCH Verlagsgesellschaft, 2011, pp. 1014-1019.
19. M. Geiger, F. Vollertsen, R. Kals, Fundamentals on the manufacturing of sheet metal microparts, Cirp Ann-Manuf Techn 45(1) (1996) 277-282.

Applications of the Gurson's model in sheet metal forming

Dorel Banabic[1,*] and *Abdolvahed* Kami[2]

[1]Manufacturing Engineering Department, Technical University of Cluj Napoca, Memorandumului 28, Cluj Napoca, Romania
[2]Mechanical Engineering Department, Semnan University, Semnan, Iran

Abstract. Recent advances in the modelling of metals encompass modelling of metals structural inhomogeneity, damage, porosity, twinning/untwining and non-local and second order effects. This presentation is focused on modelling the void growth in ductile fractures. The growth and coalescence of microscopic voids are the main mechanisms in ductile fracture of bulk metallic parts. In sheet metals, ductile fracture is preceded by necking during which existing voids do not have significant growth. However, necking is highly sensitive to plastic flow direction which in turn is sensitive to the presence of voids. Also, under biaxial strain loading, the final fracture in the necking region is still controlled by void growth; hence an accurate fracture prediction is crucial for crash simulations. Finally, in super-plastic sheet forming, void growth and coalescence may precede or accompany necking. Therefore, there is as increasing interest in modelling of voids in the sheet metals. As an application, we show how the predictions of some forming limit curves (FLCs) can be affected by accurate simulation of voids growth.

Keywords: Ductile fracture, Sheet metal, Forming limit curve

1 Introduction

Fracture in ductile metallic materials is due to evolution of tensile load-induced voids. During plastic deformation, new voids will nucleate around inclusions and second phase particles. These voids grow along with the initial voids in the material structure and finally coalesce. As a result, micro-cracks will be formed and material fracture will occur. This procedure is schematically illustrated in Fig. 1.

The first works on the voids growth were conducted by McClintock [1] and Rice and Tracey [2]. They studied the voids growth in a rigid-plastic infinite medium. Later, Gurson [3] proposed a new model for porous ductile materials which predicted both the growth of such voids and their effect on the yield criterion. The Gurson's model was modified by considering models for void nucleation and coalescence [4-6]. This modified model is known as Gurson-Tvergaard-Needleman (GTN) model which has been widely adopted in the engineering community during last 30 years [7].

The sheet metal exhibit a significant anisotropy due to the rolling process used in their production procedure. So, to apply Gurson type models on sheet metal, it is important to consider plastic anisotropy. Benzerga and Besson [8] extended Gurson's model for a material matrix which follows Hill's orthotropic criterion [9]. Later, Monchiet et al. [10, 11] and Keralavarma and Benzerga [12, 13] extended the anisotropic Gurson model to spheroidal voids. Recently, Morin et al. [14] developed a Gurson type model for general ellipsoidal voids in an anisotropic Hill matrix. Stewart and Cazacu [15] extended Gurson's model for anisotropic porous aggregates which display tension–compression asymmetry (e.g. metals with hexagonal crystal structure) and contain spherical voids. The material matrix behaviour can be defined by the anisotropic Cazacu et al. [16] yield criterion [7].

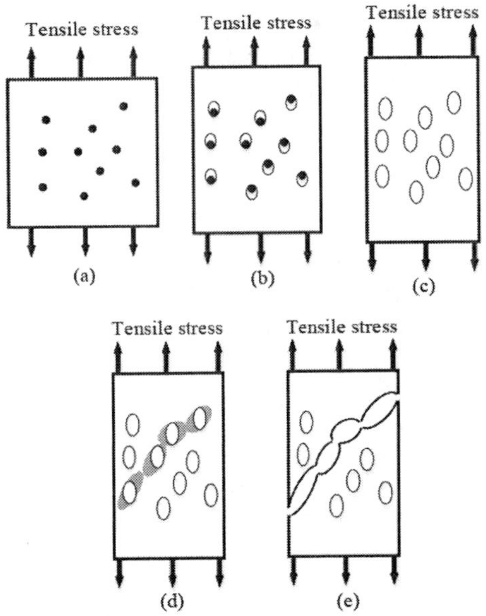

Fig. 1. Evolution of the voids and consequent fracture of sheet metal under tensile loading, (a) initial material with second phase particles and inclusions, (b) nucleation of new voids, (c) voids growth, (d) onset of necking between voids and (f) voids coalescence and sheet metal fracture.

* Corresponding author: banabic@tcm.utcluj.ro

In contrary to bulk components where ductile failure is clearly due to the void evolution, sheet metals fail mainly through necking. The voids can influence the normal to the yield criterion well before necking, which affects the prediction of forming limits. Also, voids provide the final rupture mechanism in the necking region, especially under biaxial strain conditions [7].

Assuming that voids are spherical and remain so during deformation, the main problem is to include GTN in anisotropy of material matrix. Most of studies have been focused on Hill'48 quadratic anisotropy (e.g. [8, 17]). Liao et al. [18] derived an approximate potential close to the original Gurson formulation for anisotropic sheets containing thickness holes. Wang et al. [19] modified Liao et al. model by using average anisotropy parameter in the constitutive equations. Chen and Dong [20] extended the GTN model to characterize the material matrix through Hill quadratic and Barlat-Lian 3-component [21] expressions of the equivalent stress.

In this paper, to explore the role of voids in sheet metal formation, the implementation of a simple anisotropic GTN model was implemented in ABAQUS. Then, the results of the numerical simulation of a deep drawing process were presented by GTN model. Finally, the results of numerical simulation of Nakazima tests with GTN model and its application in calculation of FLCs of single and sandwich sheets were provided.

2 GTN model

As mentioned in the previous section, different extensions of the original GTN model [4-6] have been presented considering the anisotropic behaviour of the material matrix. A common one is the extension which uses Hill'48 yield criterion to describe material matrix yielding. In this section, a brief description of this GTN extension will be presented. In the next section, its implementation in ABAQUS/Explicit code will be described.

The potential function of this model is as follows [17]:

$$\Phi=\left(\frac{\bar{\sigma}}{Y}\right)^2+q_1 f^*\left[2\cosh\left(-q_2\frac{3p}{2Y}\right)-\frac{q_3}{q_1}f^*\right]-1 \quad (1)$$

where $\bar{\sigma}$ is the Hill'48 equivalent stress, Y is the yield stress of the matrix material, p is hydrostatic pressure, q_1, q_2 and q_3 are adjustment parameters and f^* is a quantity, which depends on the current value of void volume fraction. This quantity is defined as Eq. (2).

$$f^*=\begin{cases} f, & \text{if} \quad f\le f_c, \\ f_c+\dfrac{f_f^*-f_c}{f_f-f_c}(f-f_c), & \text{if } f_c<f<f_f, \\ f_f^*, & \text{if} \quad f\ge f_f \end{cases} \quad (2)$$

All admissible stress states of the sheet metal are defined by the $\Phi\le 0$ inequality. Accurately, when $\Phi<0$ defines the elastic states and $\Phi=0$ indicates to elastoplastic states. Accordingly, the flow rule associated to the GTN potential can be expressed in the following form:

$$\dot{\varepsilon}^{(p)}=\dot{\lambda}\frac{\partial\Phi}{\partial\sigma}, \qquad \text{where}\begin{cases}\dot{\lambda}=0, & \text{if } \Phi<0,\\ \dot{\lambda}\ge 0, & \text{if } \Phi=0.\end{cases} \quad (3)$$

where, $\dot{\varepsilon}^{(p)}$ is the plastic part of the strain rate tensor.

The change of the void volume fraction f is caused by the growth of the initial void volume fraction f_0 ($\dot{f}^{(g)}$) and the nucleation of new voids at the limits of the second phase particles and inclusions ($\dot{f}^{(n)}$):

$$\dot{f}=\dot{f}^{(g)}+\dot{f}^{(n)} \quad (4)$$

where

$$\dot{f}^{(g)}=(1-f)\left[\dot{\varepsilon}_{11}^{(p)}+\dot{\varepsilon}_{22}^{(p)}+\dot{\varepsilon}_{33}^{(p)}\right] \quad (5)$$

and

$$\dot{f}^{(n)}=\frac{f_N}{S_N\sqrt{2\pi}}\exp\left[-\frac{1}{2}\left(\frac{\bar{\varepsilon}^{(p)}-\bar{\varepsilon}_N}{S_N}\right)^2\right]. \quad (6)$$

The quantities denoted as S_N, $\bar{\varepsilon}_N$ and f_N in Eq. (6) are also material constants that must be determined by identification.

The plastic strain be subdivided into volumetric and deviatoric parts. So, it is possible to derive the following formuations for the flow rule:

$$\dot{\varepsilon}^{(p,dev)}=\dot{\lambda}\frac{\partial\Phi}{\partial\bar{\sigma}}, \qquad \dot{\varepsilon}^{(p,vol)}=-\dot{\lambda}\frac{\partial\Phi}{\partial p}, \quad (7)$$

So, one can easily conclude that:

$$\dot{\varepsilon}^{(p,dev)}\frac{\partial\Phi}{\partial p}+\dot{\varepsilon}^{(p,vol)}\frac{\partial\Phi}{\partial\bar{\sigma}}=0. \quad (8)$$

Eq. (8) will be used as a consistency condition in calculations of the state variables appearing in implementation of the GTN model.

3 Implementation of GTN model

The GTN damage model has been comibined with Abaqus/Explicit to perform simualtions of deformation. To this end, the damage model has been implemented in a VUMAT routine. Abaqus/Explicit divides nonlinear processes simulations into small $[t,t+\Delta t]$ time increments. The VUMAT routine recieves the components of the stress tensor ${}^t\sigma_{ij}$, the equivalent plastic strain ${}^t\bar{\varepsilon}^{(p)}$ and the void volume fraction ${}^tf$ at the reference state (time t) and evaluates these parameters at the current state (time $t+\Delta t$) i.e. VUMAT evaluates ${}^{t+\Delta t}\sigma_{ij}$, ${}^{t+\Delta t}\bar{\varepsilon}^{(p)}$ and ${}^{t+\Delta t}f$.

The ${}^{t+\Delta t}\sigma_{ij}$ componenets can be expressed as [17]:

$$ {}^{t+\Delta t}\sigma_{ij} = {}^{t+\Delta t}\sigma_{ij}^{(e)} - \frac{E}{1+\nu} \left[{}^{t+\Delta t}_{t}\varepsilon_{ij}^{(p)} + \frac{\nu}{1-2\nu}\, {}^{t+\Delta t}_{t}\varepsilon_{\ell\ell}^{(p)}\delta_{ij} \right], \tag{9} $$

where

$$ {}^{t+\Delta t}\sigma_{ij}^{(e)} = {}^{t}\sigma_{ij} + \frac{E}{1+\nu}\left[{}^{t+\Delta t}_{t}\varepsilon_{ij} + \frac{\nu}{1-2\nu}\, {}^{t+\Delta t}_{t}\varepsilon_{\ell\ell}\delta_{ij} \right] \tag{10} $$

and

$$ {}^{t+\Delta t}_{t}\varepsilon_{ij}^{(p)} = \int_{t}^{t+\Delta t\tau} \dot{\varepsilon}_{ij}^{(p)}\,\mathrm{d}\tau. \tag{11} $$

The ${}^{t+\Delta t}\sigma_{ij}^{(e)}$ are componenets of a stress tensor that would occur in the current configuration if the incremental logarithmic strain were purely elastic. At first ${}^{t+\Delta t}\sigma_{ij}^{(e)}$ will be evaluated using Eq. (10) and then the corresponding pressure and equivalent stress will be calculated, respectively:

$$ {}^{t+\Delta t}p^{(e)} = -\frac{1}{3}\,{}^{t+\Delta t}\sigma_{\ell\ell}^{(e)}, \tag{12} $$

$$ {}^{t+\Delta t}\bar{\sigma}^{(e)} = \sqrt{{}^{t+\Delta t}\sigma_{ij}^{(e)} P_{ijk\ell}\, {}^{t+\Delta t}\sigma_{k\ell}^{(e)}}. \tag{13} $$

By placing the ${}^{t+\Delta t}p^{(e)}$ and ${}^{t+\Delta t}\bar{\sigma}^{(e)}$, it is possible to evaluate ${}^{t+\Delta t}\Phi^{(e)}$ and to decide whether the increment of the logarithmic strain is elastic or elastoplastic. The value of ${}^{t+\Delta t}\Phi^{(e)}$ is equal to [17]:

$$ {}^{t+\Delta t}\Phi^{(e)} = \left[\frac{{}^{t+\Delta t}\bar{\sigma}^{(e)}}{{}^{t}Y}\right]^2 + q_1\,{}^{t}f^{*} \left[2\cosh\left(-q_2\frac{3\,{}^{t+\Delta t}p^{(e)}}{2\,{}^{t}Y}\right) - q_1\,{}^{t}f^{*}\right] - 1, \tag{14} $$

where

$$ {}^{t}Y = Y\left[{}^{t}\bar{\varepsilon}^{(p)}\right] \tag{15} $$

and

$$ {}^{t}f^{*} = \begin{cases} {}^{t}f, & \text{if } {}^{t}f \le f_C, \\ f_C + \dfrac{f_F^{*} - f_C}{f_F - f_C}\left({}^{t}f - f_C\right), & \text{if } f_C < {}^{t}f < f_F \\ f_F^{*}, & \text{if } {}^{t}f \ge f_F \end{cases} \tag{16} $$

are parameters associated to the reference configuration. If ${}^{t+\Delta t}\Phi^{(e)} \le 0$, the incremental logarithmic strain ${}^{t+\Delta t}_{t}\varepsilon$ is purely elastic and the parameters of the current configuration can be set as follows: ${}^{t+\Delta t}\sigma_{ij} = {}^{t+\Delta t}\sigma_{ij}^{(e)}$, ${}^{t+\Delta t}\bar{\varepsilon}^{(p)} = {}^{t}\bar{\varepsilon}^{(p)}$, and ${}^{t+\Delta t}f = {}^{t}f$. On the other hand, if ${}^{t+\Delta t}\Phi^{(e)} > 0$, the material evolves through elastoplastic states during the time interval $\left[t, t+\Delta t\right]$. In such a case, the plastic terms on the right-hand side of Eq. (9) are different from zero. These terms will be evaluated using a backward Euler approximation of Eq. (11):

$$ {}^{t+\Delta t}_{t}\varepsilon_{ij}^{(p)} \approx {}^{t+\Delta t}\dot{\varepsilon}_{ij}^{(p)}\Delta t. \tag{17} $$

Eq. (10) can be rewritten in the normalized form in terms of the plastic strain increments [17]:

$$ {}^{t+\Delta t}\hat{\sigma}_{ij} + \frac{E}{{}^{t+\Delta t}Y}\left[\frac{{}^{t+\Delta t}_{t}\varepsilon^{(p,dev)}}{(1+\nu)\,{}^{t+\Delta t}\hat{\bar{\sigma}}} P_{ijk\ell}\,{}^{t+\Delta t}\hat{\sigma}_{k\ell} + \frac{{}^{t+\Delta t}_{t}\varepsilon^{(p,vol)}}{3(1-2\nu)}\delta_{ij} \right] \approx {}^{t+\Delta t}\hat{\sigma}_{ij}^{(e)}, \tag{18} $$

where

$$ {}^{t+\Delta t}Y = Y\left[{}^{t+\Delta t}\bar{\varepsilon}^{(p)}\right], \tag{19} $$

$$ {}^{t+\Delta t}\hat{\sigma}_{ij}^{(e)} = \frac{1}{{}^{t+\Delta t}Y}\,{}^{t+\Delta t}\sigma_{ij}^{(e)}, \tag{20} $$

$$ {}^{t+\Delta t}\hat{\sigma}_{ij} = \frac{1}{{}^{t+\Delta t}Y}\,{}^{t+\Delta t}\sigma_{ij}, \tag{21} $$

$$ {}^{t+\Delta t}\hat{\bar{\sigma}} = \frac{1}{{}^{t+\Delta t}Y}\,{}^{t+\Delta t}\bar{\sigma} = \sqrt{{}^{t+\Delta t}\hat{\sigma}_{ij} P_{ijk\ell}\,{}^{t+\Delta t}\hat{\sigma}_{k\ell}}. \tag{22} $$

And ${}^{t+\Delta t}_{t}\varepsilon^{(p,dev)}$ and ${}^{t+\Delta t}_{t}\varepsilon^{(p,vol)}$ are the deviatoric and volumetric parts of plastic strain incerement (see Eq. (7)), respectively.

In the case of an elastoplastic evolution of the material, Eq. (18) is accompanied by the following constraints:

- First consistency condition, ${}^{t+\Delta t}\Phi = 0$ [17]:

$$ {}^{t+\Delta t}\hat{\bar{\sigma}}^2 + q_1\,{}^{t+\Delta t}f^{*}\left[2\cosh\left(-\frac{3}{2}q_2\,{}^{t+\Delta t}\hat{p}\right) - q_1\,{}^{t+\Delta t}f^{*}\right] = 1, \tag{23} $$

where

$$ {}^{t+\Delta t}f^{*} = \begin{cases} {}^{t+\Delta t}f, & \text{if } {}^{t+\Delta t}f \le f_C, \\ f_C + \dfrac{f_F^{*} - f_C}{f_F - f_C}\left({}^{t+\Delta t}f - f_C\right), & \text{if } f_C < {}^{t+\Delta t}f < f_F, \\ f_F^{*}, & \text{if } {}^{t+\Delta t}f \ge f_F, \end{cases} \tag{24} $$

and

$$ {}^{t+\Delta t}\hat{p} = -\frac{1}{3}\,{}^{t+\Delta t}\hat{\sigma}_{\ell\ell} \tag{25} $$

- Incremental form of the second consistency condition [17]:

$$ {}^{t+\Delta t}\hat{\bar{\sigma}}\,{}^{t+\Delta t}_{t}\varepsilon^{(p,vol)} - \frac{3}{2}q_1 q_2\,{}^{t+\Delta t}f^{*}\sinh\left(-\frac{3}{2}q_2\,{}^{t+\Delta t}\hat{p}\right)\,{}^{t+\Delta t}_{t}\varepsilon^{(p,dev)} \approx 0 \tag{26} $$

The Eqs (18), (23) and (26) are a nonlinear set of equations having ${}^{t+\Delta t}\hat{\sigma}_{ij}$, ${}^{t+\Delta t}\bar{\varepsilon}^{(p)}$ and ${}^{t+\Delta t}f$ as unknowns. The VUMAT routine solves this set in a numerical manner, using a forward finite difference Newton scheme combined with a line search strategy. The reference state parameters ${}^{t}\hat{\sigma}_{ij} = {}^{t}\sigma_{ij} / {}^{t}Y$, ${}^{t}\bar{\varepsilon}^{(p)}$ and ${}^{t}f$ define the start point of the solution procedure.

 https://doi.org/10.1051/matecconf/201819001002

4 Results and discussions

The GTN model was used in ABAQUS/Explicit code to predict fracture of AA6016-T4 sheet in a square cup drawing simulation. Furthermore, the model was utilized for construction of FLCs of single AA6016-T4 sheet and metal/polymer/metal sandwiches.

4.1 Prediction of fracture in deep drawing

The GTN model was used for prediction of fracture in square cup drawing of 1-mm thick AA6016-T4 sheet. The mechanical properties of the sheet along with its Lankford's coefficients were obtained by conducting tensile experiments in 0°, 45° and 90° directions with respect to the rolling direction. The Lankford's coefficients obtained from these experiments are $r_0 = 0.5529$, $r_{45} = 0.4091$ and $r_{90} = 0.5497$ [22]. The hardening behaviour of AA6016-T4 sheet can be described by Swift's law:

$$Y = 525.8(0.0113 + \bar{\varepsilon}^P)^{0.27} \tag{27}$$

Fig. 2 illustrates the fracture of the metallic sheet during the square cup drawing tests performed with a blank holding force of 10 kN. This figure also presents a comparison between the predictions of the GTN model with the experiments.

As Fig. 2 shows, the fracture path is almost the same in both images. In FE simulation with ABAQUS/Explicit and GTN model, the cup depth at the onset of fracture was predicted as 17.90 mm which is in a good agreement with the depth of 18.70 mm in the experimental specimen.

4.2 FLC construction for single sheets

The GTN model was also used for construction of FLC of AA6016-T4 sheet. For this purpose, FE simulations of Nakazima tests were conducted using ABAQUS/Explicit and GTN model. The simulations were performed on different notched specimens with different widths. These simulations were allowed to continue until the fracture emergence. Then, the major and minor limit strains were measured at five paths normal to the fracture line. The Bragard's method [23, 24] was utilized to find the limit strains at the onset of necking. The accuracy of the simulation results was validated by their comparison with the experimental tests of Nakazima tests.

Fig. 3 compares GTN-predicted FLC with the experimental FLC. As the figure depicts, GTN model provides good predictions for the limit strains in almost all areas of FLC. This indicates that the anisotropic GTN-Hill'48 model has the capability of predicting limit strains for almost all forming paths.

For better understanding of GTN model capability on prediction of limit strains, FLC of AA6016-T4 sheet was calculated using M-K and MMFC models; as shown in Fig. 3. Although M-K and MMFC models provided good predictions for the left branch and middle section of FLC, they failed to predict the accurate limit strains on the right branch of AA6016-T4 sheet. On the other hand, GTN model presented good predictions for both the left and right branches of the FLC. This observation indicate that although GTN model has more computational cost, it is more powerful than M-K and MMFC models in prediction of FLC of AA6016-T4 sheet [17].

(a)

(b)

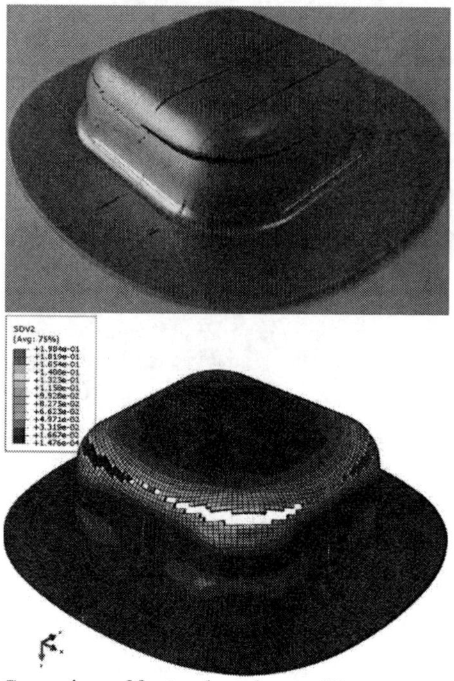

Fig. 2. Comparison of fractured specimens of the square cup drawing (a) experiment (b) FE simulation [22].

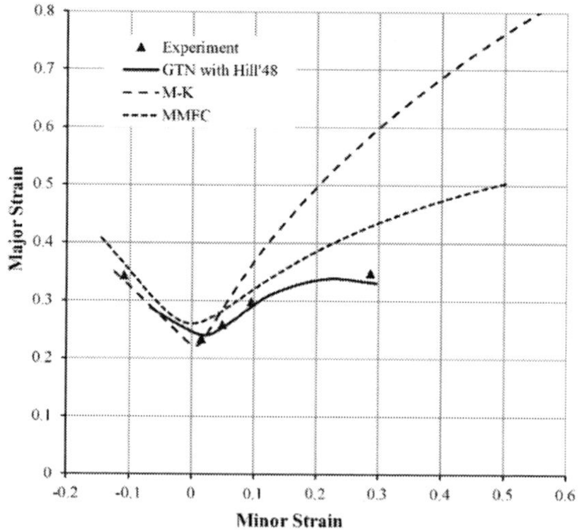

Fig. 3. Comparison of M-K, MMFC models with GTN models in terms of predicting experimental FLCs [17].

4.3 FLC construction for sandwich sheets

The FLC of a Bondal sandwich sheet was calculated using GTN model. The sandwich sheet has DC06 skins and a viscoelastic core layer, where the skins and the core layer are 0.6 mm and 0.05 mm thick, respectively. Mechanical properties of Bondal sheet layers and the GTN model parameters of DC06 skins are listed in Table 1.

Fig. 4 depicts the FLCs obtained by GTN and M-K models. This figure reveals that, GTN model predicts the strains in the middle and right side of the experimental FLC with good accuracy. But, it fails to predict the limit strains on the left side of the experimental FLC. On the other hand, as Fig. 4 suggests, the FLC calculated by the M-K model coincides with the experimental forming limit strains in only two points. This poor prediction of M-K model might be originated from the fact that the value of thickness defect in M-K model is adjusted based on the limit strain associated with the plane strain point of the experimental FLC. However, from Fig. 4, one may conclude that GTN model provides more accurate FLCs compared to those of M-K model, as in the case of AA6016-T4 sheet (see also Fig. 3).

In order to study the effect of thickness and material of sandwich sheet layers on their formability, the FLCs of two sets of sandwiches were calculated using GTN model. Configurations of these sandwich sheets are presented in Table 2; where top skin refers to the skin sheet in contact with punch (in the Nakazima tests).

In Table 2, each sandwich sheet is designated as a four-character code in which the characters (from left to right) represent top skin material, bottom skin material, top skin thickness and bottom skin thickness, respectively. Furthermore, A, S, 1 and 2 stand for aluminum AA5754 alloy, mild-steel, thickness of 0.3 mm and thickness of 0.6 mm, respectively. All sandwich sheets in Table 2 are equipped with a polymeric core layer of 0.6 mm thick. Mechanical properties and GTN parameters of the skins of the sandwiches are presented in Table 3.

Table 1. Mechanical properties of Bondal sheet layers and GTN model parameters of DC06 skins [25, 26].

Material	Young Modulus, [GPa]	Yield stress, [MPa]	Tensile strength, [MPa]	Lankford coefficient		
				r_0	r_{45}	r_{90}
DC06	210	152	279	2.027	1.751	2.467
Polymer	8.8	15	61	-	-	-
-	GTN parameters					
-	f_0	f_N	S_N	$\bar{\varepsilon}_N^p$	f_C	f_F
DC06	0.0005	0.0008	0.1	0.3	0.0219	0.1677

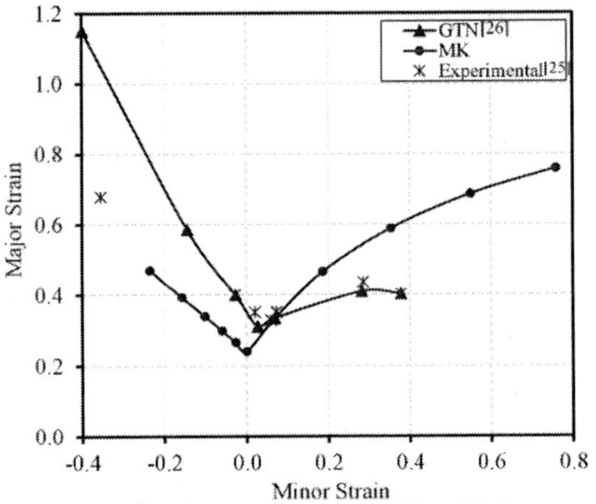

Fig. 4. Comparison between M-K and GTN models in prediction of experimental FLC of the Bondal sandwich sheet [25, 26]

Table 2. Thickness and material of the metallic layers comprising the sandwich sheets.

Sandwich sheet	Sheet name	Material-thickness (mm) of top skin	Material-thickness (mm) of bottom skin
Set #1	AA11	A-0.3	A-0.3
	SS11	S-0.3	S-0.3
	AS11	A-0.3	S-0.3
	SA11	S-0.3	A-0.3
Set #2	AS12	A-0.3	S-0.6
	SA12	S-0.3	A-0.6
	AS21	A-0.6	S-0.3
	SA21	S-0.6	A-0.3

Table 3. Mechanical properties and GTN parameters of the skins for the sandwiches categorized in set #1 and set #2 [27].

Material	Young Modulus, [GPa]	Yield stress, [MPa]	Ultimate strength, [MPa]	Lankford coefficient		
				r_0	r_{45}	r_{90}
Mild-steel	198	148	389	2.20	1.90	1.60
AA5754	70	100	230	0.87	0.76	0.71
-	GTN parameters					
-	f_0	f_N	S_N	$\bar{\varepsilon}_N^p$	f_C	f_F
Mild-steel	0.001	0.039	0.1	0.21	0.0601	0.1810
AA5754	0.001	0.034	0.1	0.32	0.0028	0.0977

Total thickness of all sandwich sheets categorized in set #1 was 1.2 mm, with their skin thickness ratio being equal to one. These sandwich sheets are considered to evaluate the effect of layers material arrangement on the formability of the sandwiches. FLCs of the set #1 sandwiches were calculated by numerical simulation of Nakazima tests using GTN model, as shown in Fig. 5. Clearly, SS11 had the highest forming limit, while the lowest forming limit belonged to SA11. The figure implies that, the alteration of layers material has a significant effect on the FLC. So that, by replacing the bottom skin of AA11 sheet with mild-steel (AS11), its formability will be improved. On the other hand, when the top skin of AA11 is substituted by mild-steel (SA11), a significant reduction will be observed in formability of the sandwich sheet. As in case of AA11 sheet, the minimum limit strain on the FLC is equal to 0.138, which is 26.09% lower than that of AS11 (0.174) and 33.33% higher than that of SA11 (0.092). The results illustrated in Fig. 4 are compatible with the reports of Sokolova et al. [28].

FLCs of the sandwich sheets in set #2 were also calculated through GTN model. Categorized in set #2 were the sandwich sheets composed of AS and SA with skin thickness ratio of 2 (or 0.5). FLCs of these sheets are also demonstrated in Fig. 6. This figure indicates that, formability of AS sheets is higher than that of SA sheets, which is in agreement with the results in Fig. 5. Moreover, Fig. 6 shows that AS12 and AS21 have almost the same formability (this holds true also for SA12 and SA21 sheets), implying that, in particular configurations of the sandwich sheet (e.g. AS), overall thickness of the sheet (rather than thickness ratio of the layers) is the dominant formability-controlling parameter. It should be noted that, this conclusion is just based on the results obtained at thickness ratio of 2 (or 0.5), and different results may be observed at higher thickness ratios [29].

5 Conclusions

Implementation of anisotropic GTN model with Hill'48 yield criterion and its application in FE simulation of square cup drawing and Nakazima tests were addressed in this work. Based on the results, the following conclusions can be drawn:

1- The results showed that GTN model is capable of correct fracture prediction in the square cup drawing tests. The drawing depth at the onset of fracture predicted by GTN model is 17.90 mm, which was in a good agreement with the depth of 18.70 mm obtained in the experimental cup drawing.

2- GTN model allows for computation of limit strains on the FLC of AA6016-T4 aluminium sheet with good accuracy. The limit strains predicted by GTN model were more accurate than those of M–K and MMFC models.

3- Once the bottom skin of AA-type sheets was replaced by mild-steel (AS-type sheet), its formability was improved. Contrarily, replacement of AA-type sheets top skin with mild-steel will result in formability decrease. When GTN model served as the basis, minimum limit strain of the FLC of AA11 sheet was found to be 0.138, i.e. 26.09% lower than FLC of AS11 (0.174) and 33.33% higher than that of SA11 (0.092).

4- In case of AS12, AS21, SA12 and SA21 sheets, the arrangement of skin material was found as the dominant formability-controlling factor rather than the skin thicknesses. Moreover, for sandwich sheets of the same type (i.e. AS-type or SA-type), sheet formability was found to be controlled by overall thickness of the sandwich rather than thickness ratio of its layers.

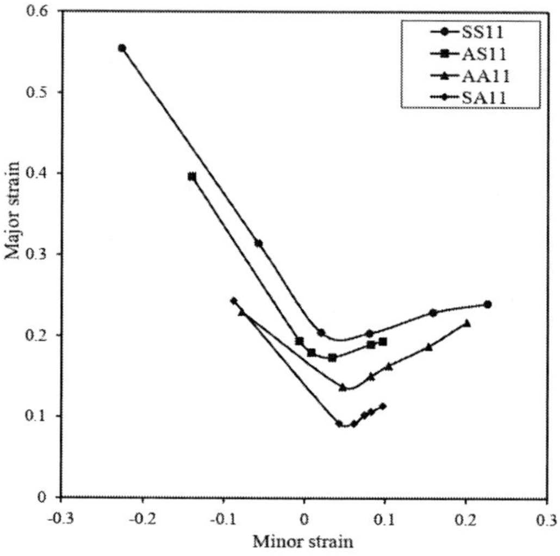

Fig. 5. The FLCs of the sandwich sheets of set #1, calculated by GTN model [29].

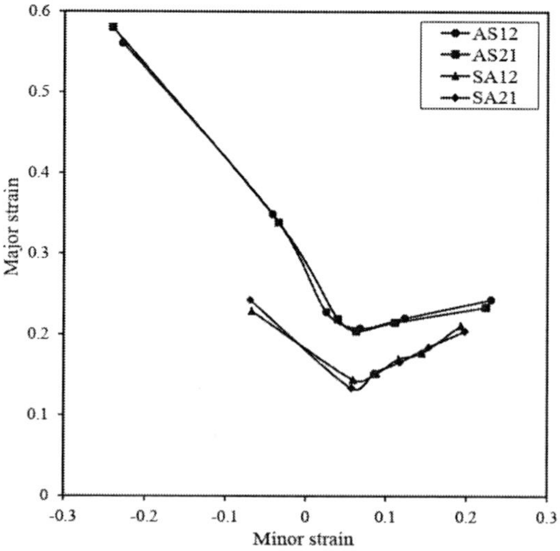

Fig. 6. FLCs of the sandwich sheets of set #2, calculated by GTN model [29].

References

* Corresponding author: banabic@tcm.utcluj.ro

1. F.A. McClintock, J. Appl. Mech. **35**, 363-371 (1968).
2. J.R. Rice and D.M. Tracey, J. Mech. Phys. Solids **17**, 201-217 (1969).
3. A.L. Gurson, J. Eng. Mater. Technol. **99**, 2-15 (1977).
4. V. Tvergaard, Int J Fracture **17**, 389-407 (1981).
5. V. Tvergaard, Int J Fracture **18**, 237-252 (1982).
6. V. Tvergaard and A. Needleman, Acta Metall. **32**, 157-169 (1984).
7. D. Banabic, *Multiscale Modelling in Sheet Metal Forming*. Springer, 2016.
8. A.A. Benzerga and J. Besson, Eur. J. Mech. A. Solids **20**, 397-434 (2001).
9. R. Hill, Proc R Soc Lond A-Math Phys Eng Sci **193**, 281-297 (1948).
10. V. Monchiet, C. Gruescu, E. Charkaluk, and D. Kondo, Comptes Rendus Mécanique **334**, 431-439 (2006).
11. V. Monchiet, O. Cazacu, E. Charkaluk, and D. Kondo, Int. J. Plast. **24**, 1158-1189 (2008).
12. S.M. Keralavarma and A.A. Benzerga, Comptes Rendus Mécanique **336**, 685-692 (2008).
13. S.M. Keralavarma and A.A. Benzerga, J. Mech. Phys. Solids **58**, 874-901 (2010).
14. L. Morin, J.-B. Leblond, and D. Kondo, Int J Solids Struct. **77**, 86-101 (2015).
15. J.B. Stewart and O. Cazacu, Int J Solids Struct. **48**, 357-373 (2011).
16. O. Cazacu, B. Plunkett, and F. Barlat, Int. J. Plast. **22**, 1171-1194 (2006).
17. A. Kami, B.M. Dariani, A. Sadough Vanini, D.S. Comsa, and D. Banabic, J. Mater. Process. Technol. **216**, 472-483 (2015).
18. K.C. Liao, J. Pan, and S.C. Tang, Mech. Mater. **26**, 213-226 (1997).
19. D.A. Wang, J. Pan, and S.D. Liu, Int J Damage Mech **13**, 7-33 (2004).
20. Z. Chen and X. Dong, Comp Mater Sci **44**, 1013-1021 (2009).
21. F. Barlat and K. Lian, Int. J. Plast. **5**, 51-66 (1989).
22. A. Kami, B. Mollaei Dariani, A. Sadough Vanini, D.-S. Comsa, and D. Banabic, Proc Rom Acad Ser A **15**, 300-309 (2014).
23. A. Bragard, Baret, J.C., Bonnarens, H., Rapp. Cent. Rech. Metall. **33**, (1972).
24. R. D'Haeyer, Bragard, A., Rap. Cent. Rech. Metall. **42**, (1975).
25. A. Kami, B.M. Dariani, D.S. Comsa, D. Banabic, A.S. Vanini, and M. Liewald, Proc Rom Acad Ser A **18**, 281–290 (2017).
26. A. Kami, B. Mollaei Dariani, D.S. Comsa, D. Banabic, A. Sadough Vanini, and M. Liewald, Rom. J. Techn. Sci. Appl. Mechanics **61**, 248-264 (2016).
27. M. Brunet, S. Mguil, and F. Morestin, J. Mater. Process. Technol. **80-81**, 40-46 (1998).
28. O. Sokolova, A. Carradó, and H. Palkowski, Adv. Mater. Res. **137**, 81-128 (2010).
29. A. Kami, K. Chung, and D. Banabic, Rom. J. Techn. Sci. Appl. Mechanics **62**, 30–39 (2017).

An integrated precise engineering for micro forming

Ming YANG[1,*], *Tetsuhide* Shimizu[1], Jun Hu[1], and *Tomomi* Shiratori[2]

[1]Tokyo Metropolitan University, Faculty of system design, 191-0065 Tokyo, Japan

[2]Komatsuseiki kosakujo Co., , Japan

Abstract. An integrated precise engineering was proposed and developed to push the micro metal forming technology to higher level for manufacturing micro metallic devices in this study. A textured Diamond Liked Carbon (DLC) coating was applied to micro die for improving tribological performance. High-energy assisted processes, such as die-embedded localize heating and ultrasonic vibration systems were developed to improve formability of micro materials and accuracy of processes. The results show that .the integrated precise engineering solved the issues due to the size effects, and improved very much the tribological properties, formability of work materials and accuracy of product. Furthermore, the integrated precise engineering was applied in micro forming to manufacturing a micro metallic pump which realized commercialization.

Keywords: Micro forming, Textured DLC coating, High-energy assisted process, fine grain material, micro metallic pump

1 Introduction

Metal forming in sub-millimeter or micrometer scales becomes a key technology for the audio-video or information technological devices due to miniaturization of structure parts. Furthermore, fabrication of micro devices, such as Bio-MEMS (Micro Electrical mechanical Systems) or micro equipment for medical application based on metal forming has been attended attractive as well due to metals have the properties of ductility, conductivity and bio-compatibility [1,2]. However, many issues, such as accuracy of machinery, scattering in properties of thin materials, fabrication of tools/dies, assembly of micro components, still exit in metal forming in micro and meso scales. Researches on formability, flow resistance of thin foil materials in micro and meso scales were performed previously. Not only the flow stress, but also accuracy of the products is influenced by size effects [3,4]. Springback properties of thin foils in various thickness and grain sizes change due to its size effect [5-8]. As the size is scaling down, the accuracy of production and forming limit of workpiece usually become worse because of occurrence of the size effects [9]. In general, the dimensional accuracy needs to be increased according to down-sizing of product. It is important to develop the micro forming processed with an integrated precise engineering to improve formability of the work material and accuracy of product. [10].

The authors have been working on a novel micro metal forming system in which micro parts or units are fabricated by micro press forming and automatic assembly processes in a progressive die, and developed several processes, such as a high-energy assisted micro forming, which combines heating and vibration to improve formability of the material and accuracy of products by using servo-press system with a resistance heating system and vibration system embedded in the dies. Furthermore, a textured Diamond Liked Carbon (DLC) coating system was applied to micro forming die in order to improve tribological properties in the processes. Micro deep drawing and micro forging was carried out to evaluate the performance of the high-energy assistance and effect on formability of the material and accuracy of the products.

In this study, progresses in development of the die surface treatment, high-energy assistant forming process, and application of the integrated precise engineering to manufacturing the metallic devices are introduced.

2 Integrated precise engineering for micro forming

Most significant issues in micro forming due to the size effects should be overcome to achieve the targets. In this chapter, the authors will introduce some approaches to overcome the issues due to the size effects: development of the die surface treatment using textured DLC coating to improve tribological properties and its evaluation by a bending test; development of a localize contact heating system with die coating by controlling

* Corresponding author: yang@tmu.ac.jp

electrical resistance of the film coated, and an die embedded ultrasonic vibration system for the high-energy assistant forming processes in order to improve formability of the work materials and accuracy of products; and application of the integrated precise engineering to manufacturing a micro metallic pump.

2.1 Micro-texturing of DLC thin film coatings for surface treatment of micro dies

Due to the low effect of lubricant in microscale dimensions, the tribological issues in micro forming confronts with further severe problems to enhance the performance of micro forming dies [11]. In particular, tool life of the micro-dies has not been well satisfied the industrial requirement. Against these backgrounds, a number of applications of hard film coatings on micro forming die substrates have been gradually increasing. Particularly, diamond like carbon (DLC) films has been received a great attention for its excellent tribological performance, such as low friction coefficient and high wear resistance under dry friction [12].

The authors have been focusing on the tribological performance of micro-textured DLC films under dry sliding friction. In view of the application to the microforming dies, tribological properties were investigated under the contact state as operated in microforming process. Firstly, the micro-textured DLC films were fabricated by masking with metallic mesh during the ionized physical vapor deposition (I-PVD). Then, the tribological properties of this micro-textured DLC films were evaluated by ball-on-disk type friction tests. Based on this evaluation, the effect of the texture size on the dry friction and wear performance were discussed. It has been experimentally proven that surface texturing works to improve the friction characteristics by promoting discharge of abrasion powder at the contact interface [13,14].

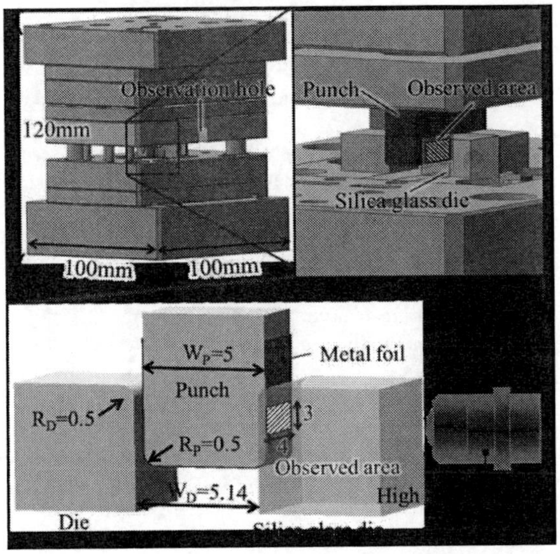

Fig.1. Schematic of in-situ observation of micro bending test

To verify the tribological effect in micro forming processes, a micro bending test with in-situ observation system was performed by using the textured DLC coating on the die surface. Fig. 1 shows the outline of the bending die and the method for observing the sliding surface in situ during processing. The bending die is a compact die with each side about 100 mm, and the die dimensions are: die shoulder radius R_D punch shoulder radius R_P = 0.5 mm, punch width W_P = 5 mm, die width W_D = 5.14 mm. By making one of the bending die (8 mm × 8 mm × 8 mm) with quartz glass, it was possible to observe the sliding interface where the workpiece and die contact with each other by descent of the punch from the back of the die through the glass die. We developed a system capable of recording and observing the observable area of the glass die shoulder end portion, 4 mm × 3 mm, at the maximum magnification of 750 times by a high speed camera with a zoom lens. The prepared metal mask was fixed along the shape of the quartz glass die surface, and DLC film was coated on the micro die to surface texturing. Same conditions are utilized for the coating process as same as the previous study [13]. Fig. 2 shows the optical image of the fabricated texture shape and the AFM image on the die shoulder. Texture width of about 45 μm, texture interval of about 20 μm, texture height of about 1.1 μm, and even a glass die shoulder portion with a shoulder radius of 500 μm succeeded in creating the surface having a uniform texture structure.

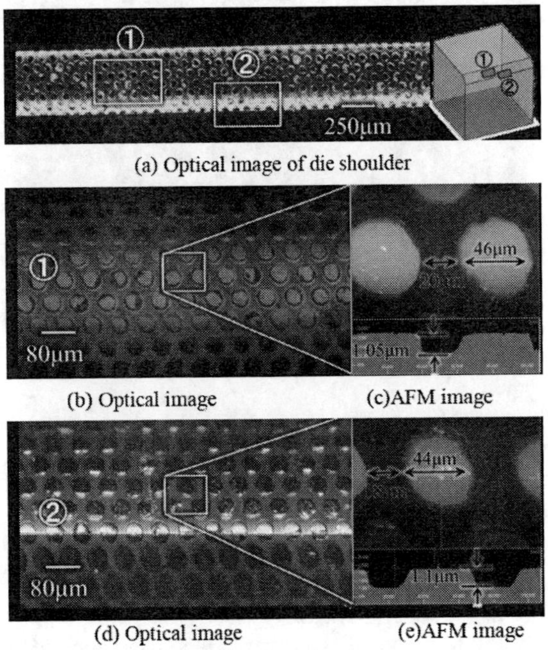

Fig.2. Optical and AFM images of textured DLC on micro die

In order to verify the practicality of this device, DLC was formed on the quartz glass die surface and in-situ observation of the peeling process of the DLC film during micro bending was performed. A stainless steel foil material (JIS: SUS 304 H) having a thickness of

50 μm was used as a test material and processed at a punch descending speed of 1 mm/s under no lubrication. In-situ observation images of the sliding interface between the micro die and the workpiece at the 5th, 10th, and 20th processing times in the micro press bending test are shown in Fig 3. The DLC coating without texture, called was the flat DLC (FDLC), also applied for the comparison. In this case, the DLC film at the end of the die shoulder was peeled off by the fifth time of forming, and scratch-like damage occurred in a part of the DLC film by the tenth. Furthermore, the DLC film was peeled off in the whole area of the observation range by the 20th, so the test was stopped. On the other hand, in the die in which the textured DLC (TDLC) was coated, although partial peeling of the DLC film was observed at the end of the die shoulder, remarkable peeling as in the case of the die with the FDLC coated could be confirmed even at the 20th processing was completed without damages even after processing 100 times. The effect of the texture size on the dry friction and wear performance is confirmed in the micro forming.

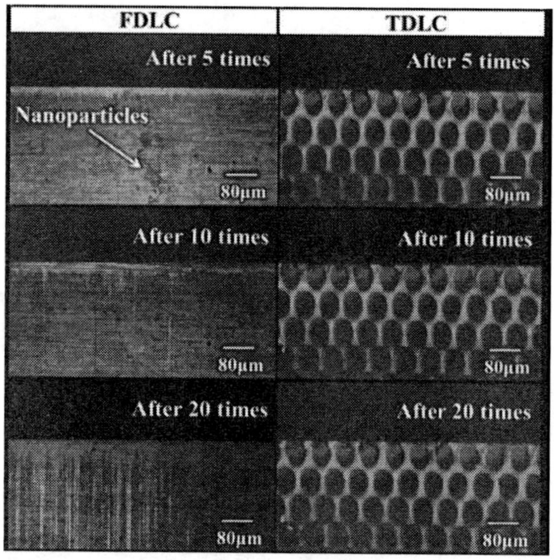

Fig. 3. In-situ observation images during progressive micro bending test

2.2 Developing high-energy assisted forming processes

Improvement of forming limit of the work material and accuracy of product is required for micro forming processes as discussed previously. It is known that heating work material during the process can improve ductility of workpiece and make the deformation more uniform in micro scale [15]. The authors developed a die embedded system for localized heating or other high-energy assistance during the process to achieve micro forming with higher forming limit and accuracy [16]. In next sections, author will introduce development of two high-energy assisted forming processes: a localized heating system by using a die with an electrical resistance controlled surface coating, and an ultrasonic vibration assistance with die-embedded vibrator.

2.2.1 Localized Heating system

The desktop servo-press system with a resistance heating system embedded dies was developed by the authors [17]. The heat-assisted micro forming system with resistance heating was designed to satisfy low energy consumption and high forming accuracy by heating only the workpiece material and limiting the heat transferring to tools. Two contact probes embedded in the dies to keep contact of probes to workpiece and keep heating during the forming process. Because of down-sizing in dimensions of workpiece the targeted temperatures of several hundred Celsius degrees could be achieved in a few seconds and the consumed energies were limited to a small amount, and deformation of the workpiece becomes more ductile and uniform in the heat-assisted micro deep drawing process. However, several issues in temperature distribution and conductivity properties of workpiece need to be solved for application in real manufacturing processes. Here, temperature distribution is dependent on the positioning of the electrode and shape of workpieces. The variety of the electrical resistivity in the working materials also influences the process conditions [18].

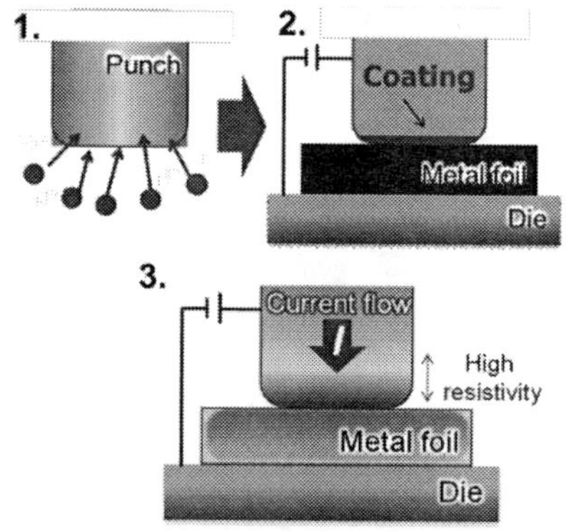

Fig. 4. Conception of Surface modification for heating

In order to increase the versatility of the heating system, a novel heating system is developed by using die surface as the heat source. The authors select proper coating system for performing both of mechanical function for improving tribological properties and electrical functions for heating workpiece locally. The die with the particular coating, which has higher electrical resistivity, was applied as the die embedded heating system. Fig. 4 shows the conception of the heating system. Since the surface layer of the die has higher electrical resistivity and becomes a heating source during the current flowing from upper die through specimen to lower die, this method can achieve the heating of work materials independent from its resistivity and its geometry. In this study, the surface coating with AlCrSiN films of thickness 0.5, 1.0 μm were deposited on upper die and bottom die for compression tests and micro forging tests. Pure titanium, pure copper, and stainless steel

(JIS:SUS304 with ultra-fine grain size) rods ϕ 1 mm × 2 mm were prepared for compression tests and SUS304 UFGSS of thickness 0.2 mm was prepared for forging tests. Diameter of upper die for compression test is 10 mm and that for forging test is 1 mm. The range of flew current density which was defined as the current intensity divided by the initial cross-sectional area was 0-64 A/mm^2.

Fig. 5 shows resultant temperature rising for various current densities during the compression test. As results, higher temperature of work materials was obtained for the die coated AlCrSiN films than that without surface treatment. In particular for the Ti rod under 64 A/mm^2, there was a temperature difference of 400 °C between the die with and without surface treatment. Forming load also decreased by increasing the temperature of work materials by using the surface treated dies. As for the forging tests, depth of forged parts increased by the proposed method. Furthermore, micro forging test also shows that a precise contour shape can obtain by using this heating system, which is very difficult by using standard forging in room temperature because of its high strength and low ductility due to the ultra-fine grain structure.

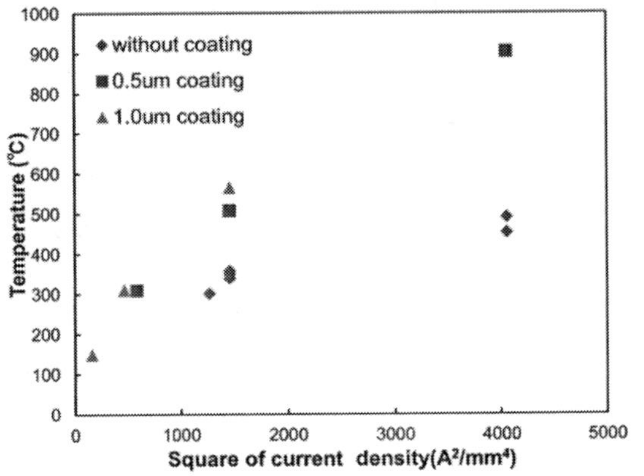

Fig. 5. Comparison of temperature rising with and without surface coating

2.2.2 Vibration assisted micro forming

Another high-energy assisted forming is of using high frequency vibration. Since the amplitude of the vibration becomes relatively larger as dimension of the material decreases, and the effect also increases relatively in the case of micro forming as compared with macro forming. Fig. 6 shows topography image of the Phosphor bronze C5191 foil surface before and after vibration assisted micro forming process. Surface roughness Ra was reduced from 102 nm to 21 nm. In this case, static punch load was 305.7 MPa and vibration frequency and amplitude were 1 kHz and 4.7 μm, respectively. It is found the surface asperities are deformed to very smooth with almost no asperity remained, and as a result, the surface roughness became much lower. What should be noticed is that the nominal static stress applied on the surface was far smaller than the yield stress of Phosphor bronze C5191. This

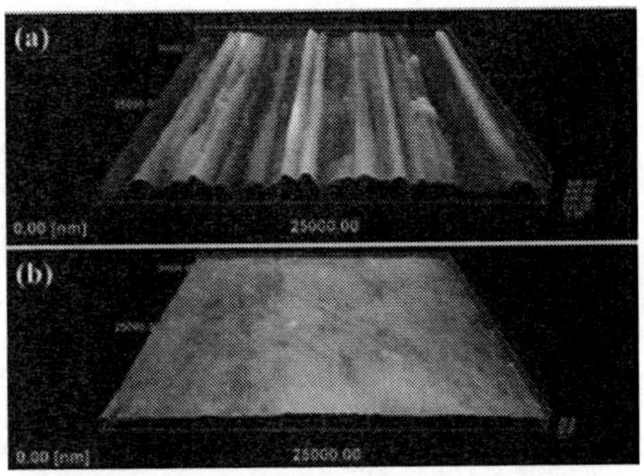

Fig. 6. Topography of the Phosphor bronze C5191 foil surface (a) before (Ra =102 nm) and (b) after vibration assisted forming process with frequency of 1 kHz (Ra=21nm)

indicates that the presence of vibration induced plastic deformation of the surface layer significantly [19].

Different frequency and vibration amplitude give different strain on the thickness direction with different energy density. As a result, the deformation of asperity and surface finishing results are different. Fig. 7 shows the surface reduction obtained at different amplitudes and frequency and without vibration. It is found that

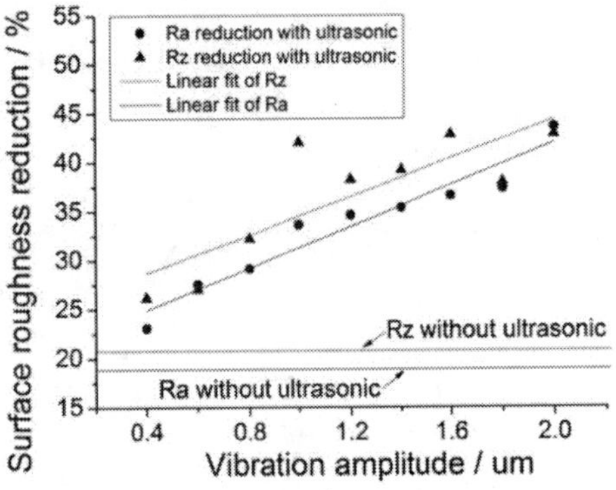

(a)

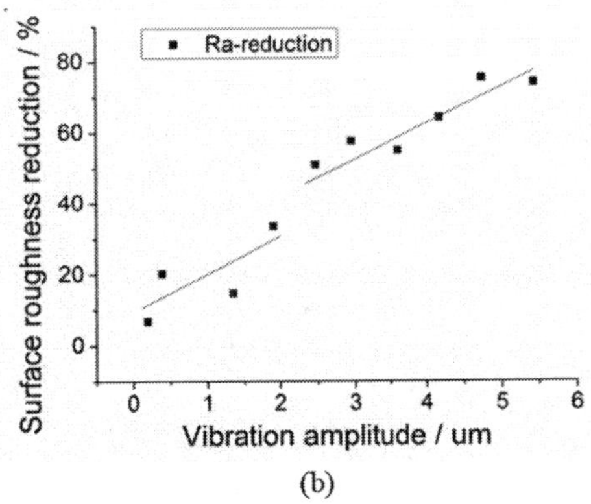

(b)

Fig. 7. Relation between surface roughness reduction and vibration amplitude for (a) vibration frequency 100 kHz and (b) vibration frequency 1 kHz reduction of the roughness increases with increase of the amplitude for both frequencies. Three different effects work during the vibration assisted forming process: (1.) friction reduction between the tools and workpiece due to varied contact, (2.) impact force while the departed punch retouches the workpiece with high acceleration, (3.) material softening due to high frequent elastic waving. In this case, vibration with high frequency accompanied material softening effect and as a result, the reduction ratio was higher than that with the low frequency for similar amplitude by comparing Fig. 7(a) and (b) for amplitude smaller than 2.0 μm. Furthermore, the reduction ratio jumped up higher for the amplitude larger than 2.5 μm (Fig. 7(b)). It is because that while the amplitude became larger than 2.5 μm, the punch departs from contact and retouches the workpiece within one vibration cycle, and as a result, an impact force added on the surface of workpiece [20].

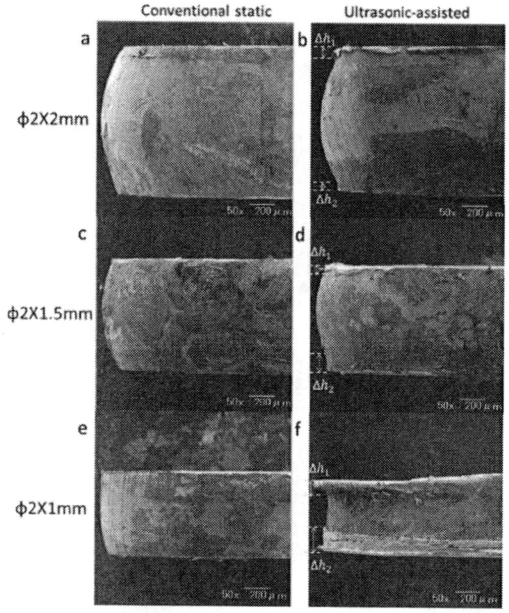

Fig. 8. SEM images of typical specimens in different original height after conventional static and ultrasonic-assisted compression

A vibrator with natural frequency of 60 kHz, was designed for micro forming Punch is connected on the vibrator and the vibration is further amplified by the punch with staircase structure, and maximum amplitude of 3.2 μm can be achieved on the top surface of the punch. A compression test is performed by the vibration system. The SEM images of typical deformed specimens in different original height after conventional static and ultrasonic-assisted compression were shown in Fig. 8. In the conventional static compressions without ultrasonic vibration in Fig. 8(a), (c) and (e), all 3 specimens had a significant barrelling shape, which indicated a relative high friction value at the top and bottom interface. However, when the compressions were conducted with ultrasonic assistance, both top and bottom surfaces had more expansion and even anti-barrelling shapes in depth of Δh_1 and Δh_2 were formed around the corners in every specimen, as shown in Fig. 8(b), (d) and (f). It can be seen that the deformation of surface layer becomes more significant for the workpieces with smaller height. This phenomenon involves dynamic impact effect, which is caused while the punch detaches the workpiece in the case of the amplitude larger than springback of workpiece [21].

2.3 Manufacturing of micro metallic device with fine grained material

For manufacturing micro metallic devices, the authors integrated the precise engineering technologies including the material development and the die fabrication. A micro pump was developed by using precise piercing and diffusion bonding processes with ultra-fine grain stainless steel foil (SUS304) [22]. Fig. 9 shows the Electron Back Scatter Diffraction (EBSD) results of thickness 10 μm foil. This grain phase almost changed from general austenitic phase to strain-induced-martensitic phase. This could be regarded as "full-martensitic stainless steel". This material can achieve diffusion bonding at a low temperature of 973 K [23].

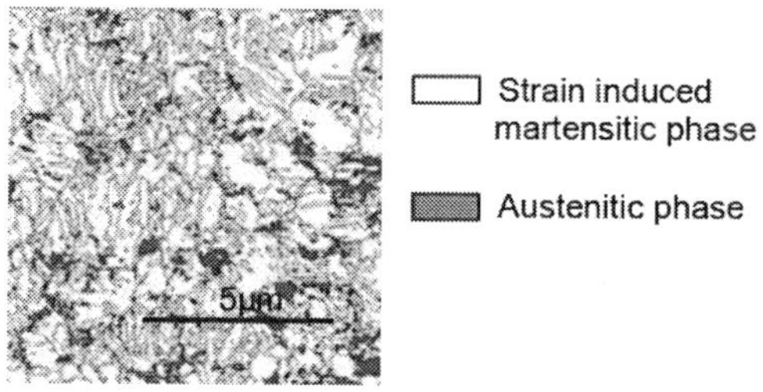

Fig. 9. Grain condition of fine-grained SUS304 observed by SEM with EBSD

Furthermore, the average grain size decreased to 1.5 μm, which brings in improvement of the deformation uniformity and the production quality in the metal forming processes. Moreover, in order to achieve a diffusion bonding under the low stress, surface

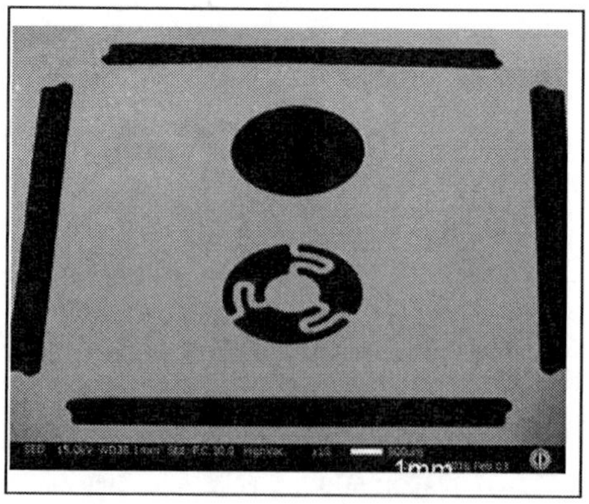

Fig. 10. SEM image of pierced 0.010mm foil

https://doi.org/10.1051/matecconf/201819001003

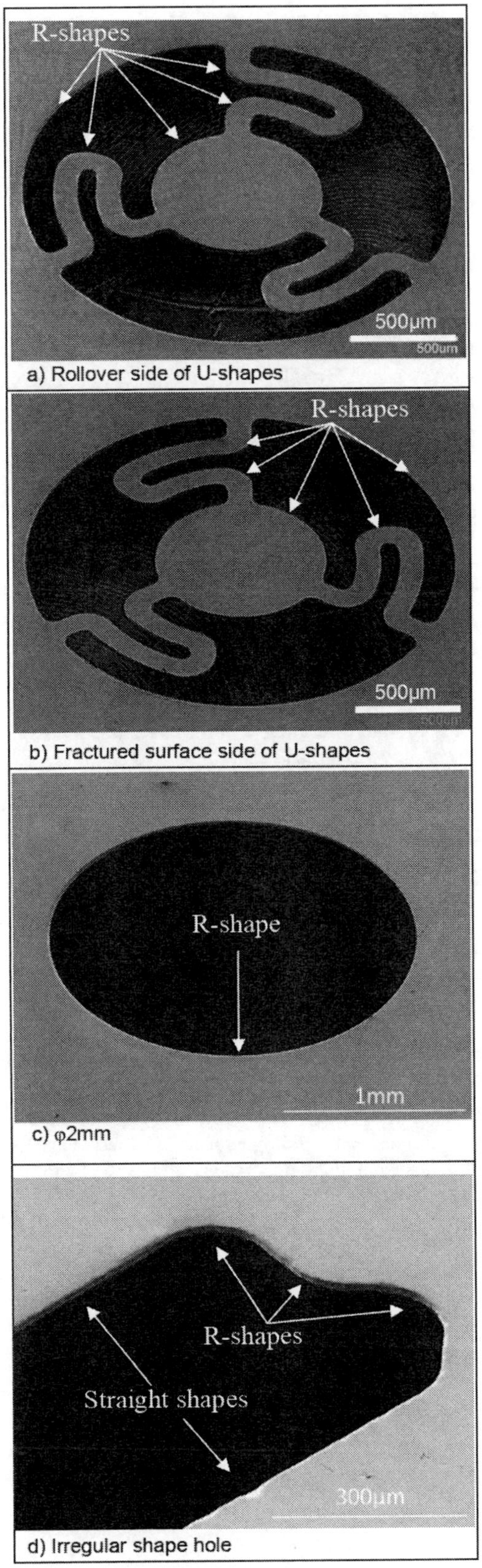

Fig. 11. SEM image of enlarged pierced holes

roughness was also improved. The punch for the piercing process was cut by penetrating it into the die to fabricate zero clearance between the punch and the die in order to guaranttee the dimension accuracy and the edge shape. Fig. 10 shows the SEM image of pierced 0.010 mm stainless SUS304 foil. Fig. 11 shows details of pierced shapes and enlarged images. The shapes of pierced holes and irregular shapes are uniform in sheared surface and there are almost no burrs. Finally eight components of the micro pump were joined by diffusion bonding at 973 K. Fig. 12 shows the photo image of the pump and its components. The pump is driven by the piezo actuator, and the diaphragm is deformed accordingly. The flow rate is regulated by two valves allocated inlet and outlet of the pump, which linked to the vertical movement of the diaphragm. Dimensional accuracy of the valves determine the pump performance since it affects leakage property of pump significantly. Performance of the pump shows that the pierced metal foil maintains its flatness enough to stabilize the performance of metal micro-pump, such as flow rate stability and repeatability [24].

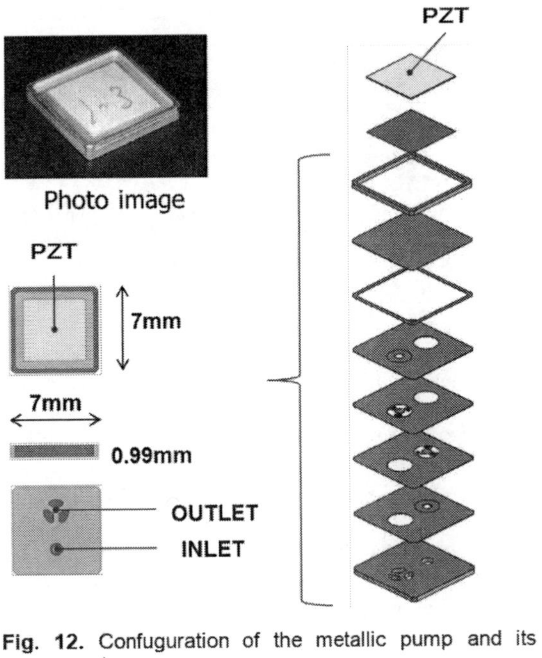

Fig. 12. Confuguration of the metallic pump and its components

3 Conclusions

An integrated precise engineering is proposed and developed for micro forming in this study. The integrated approach can not only improve formability of the work material, but also accuracy of product. The main findings are summarized below.

- The die surface was treated using textured DLC and its effect was evaluated by a micro bending test. The tribological properties were improved by stabilizing friction during the sliding between the workpiece and die and reducing wear rate of the die.

- A localize contact heating system with die coating by controlling electrical resistance of the film effectively reduced flow stress and improved ductility of the work material during micro forming processes.
- A die embedded ultrasonic vibration system improves formability of the work materials and especially enhance deformation in the surface layer.
- The integrated precise engineering was applied to manufacture a micro metallic pump which realized commercialization.

References

1. Geigerl M, Kleine M, Eckstein R, Tieslerl N, Engel U. Microforming. CIRP Ann 2001; 50(2): 445-462.
2. Engel U, Eckstein R. Microforming-from basic research to its realization. J Mater Process Tech (2002); 125-126: 35-44.
3. Kals TA, Eckstein R. Miniaturization in sheet metal working. J Mater Process Tech, 103: (2000), 95-101.
4. Suzuki K, Matsuki Y, Masaki K, Sato M, Kuroda M. Tensile and microbend tests of pure aluminum foils with different thicknesses. Mater Sci Eng A (2009), 513–514: 77-82.
5. Diehl A, Engel U, Merklein M. Size effects in bending processes applied to metal foils. Prod Eng Res Devel, (2010); 4: 47-56.
6. Li HZ, Dong XH, Shen Y, Diehl A, Hagenah H, Engel U, Merklein M. Size effect on springback behavior due to plastic strain gradient hardening in microbending process of pure aluminum foils. Mater Sci Eng A (2010); 527: 4497-4504.
7. Li HZ, Dong XH, Wang Q, Shen Y, Diehlc A, Hagenah H, Engel U, Merklein M. Determination of material intrinsic length and strain gradient hardening in microbending process. Int J Solids Struct (2011); 48: 163–174.
8. Aoyama T, Shimizu T, Zheng Q, Yang M. Effect of Heating on Springback in Heat Assisted Microbending. Adv Mater Res (2014); 939: 409-414.
9. Vollertsen F, Biermann D, Hansen HN, et al. Size effect in manufacturing of metallic components. CIRP Ann 2009; 58: 566-87.
10. M. YANG, "Fabrication of MEMS Using Micro Metal Forming Process", Proc. ICNFT, (2004), pp.135-140.
11. Engel, U., Tribology in Microforming. Wear, 260(2006), 265-273.
12. Grill, A., 1997. Tribology of diamond like carbon and related materials: an updated review. Surface and Coating Technology, 94–95(1997), 507–513.
13. Tetsuhide Shimizu, Tai Kakegawa, Ming Yang, Micro-texturing of DLC thin film coatings and its tribological performance under dry sliding friction for microforming operation, Procedia Engineering 81(2014), pp.1884 – 1889.
14. T. Kakegawa, T. Shimizu, Y. Sagisawa, M. Yang, Surface Texturing on Micro-Die and In-situ Observation of Its Dry Friction and Wear Behavior, J. of JSTP, 56-657(2016), 891-896.
15. B. Eichenhueller,E. Egerer,U. Engel, Microforming at elevated temperature-forming and material behavior,The International Journal of Advanced Manufacturing Technology,33(2007), 119-124.
16. M. Yang, T. Shimizu, Novel processes and related technologies for micro metal forming, Int. J. Materials and Product Technology, Vol. 54, Nos. 1/2/3 (2017), 32-44.
17. H. Tanabe, M. YANG, Design and Evaluation of Heat Assisted Microforming System, Steel Research International, Special Edition (2011), 1020-1024.
18. Qiu Zheng, Tetsuhide Shimizu, Ming Yang, Numerical Analysis of Temperature Distribution and Its Optimization for Thin Foils in Micro Deep Drawing Assisted by Resistance Heating, Steel Research International, 85 (2015), 1-8.
19. Yang Bai, Kenji Nishikawa, Ming Yang, Metal Surface Modification with Vibration-Aided Micro-Forging, Materials Transactions, Vol. 53, No. 3 (2012) pp. 489-494.
20. Yang Bai, Ming Yang, Investigation on mechanism of metal foil surface finishing with vibration-assisted micro-forging, J. of Materials Processing Technology 213 (2013) , pp.330– 336.
21. Hu, J., Shimizu, T., Yang, M., "Impact effect of superimposed ultrasonic vibration on material characteristics in compression tests." Procedia Engineering, (2017).
22. T Shiratori, Y. Suzuki, S. Nakano, M. Yang, and T. Komatsu, Effects of grain size on the sheared surface in micropiercing of austenitic stainless steel SUS304, Manufacturing Rev., 2-9(2015), 1-6.
23. M. Katoh et al., "Reduction of diffusion bonding temperature with recrystallization at austenitic stainless steel," Tetsu-to-hagane, 102 (2016), 34-39.
24. T. Shiratori, S. Nakano, M. Katoh, Y. Suzuki, T. Aihara, M. Yang, Development of metal MEMS manufacturing technologies using pierced metal foil and diffusion bonding process at low temperature, Proc. 4M/IWMF, (2016).

Simulation of thermal behaviours and powder flow for direct laser metal deposition process

Quanren Zeng[1], *Yankang* Tian[1], *Zhenhai* Xu[1, 2] and *Yi* Qin[1,*]

[1] Department of Design, Manufacture and Engineering Management, University of Strathclyde, G1 1XQ, Glasgow, UK
[2] National Key Laboratory for Precision Hot Processing of Metals & School of Materials Science and Engineering, Harbin Institute of Technology, Harbin 150001, China

Abstract. Laser engineering net-shaping (LENS), based on directed energy deposition (DED), is one of the popular AM technologies for producing fully dense complex metal structural components directly from laser metal deposition without using dies or tooling and hence greatly reduces the lead-time and production cost. However, many factors, such as powder-related and laser-related manufacturing parameters, will affect the final quality of components produced by LENS process, especially the powder flow distribution and thermal history at the substrate. The powder concentration normally determines the density and strength of deposited components; while the thermal behaviours of melt pool mainly determines the cooling rate, residual stress and consequent cracks in deposited components. Trial and errors method is obviously too expensive to afford for diverse applications of different metal materials and various manufacturing input parameters. Numerical simulation of the LENS process will be an effective means to identify reasonable manufacturing parameter sets for producing high quality crack-free components. In this paper, the laser metal powder deposition process of LENS is reported. The gas-powder flow distribution below the deposition nozzle is obtained via CFD simulation. The thermal behaviours of substrate and as-deposited layer/track during the LENS process are investigated by using FEM analysis. Temperature field distributions caused by the moving laser beam and the resultant melt pool on the substrate, are simulated and compared. The research offers a more accurate and practical thermal behaviour model for LENS process, which could be applied to further investigation of the interactions between laser, melt pool and powder particles; it will be particularly useful for manufacturing key components which has more demanding requirement on the components' functional performance.

Keywords: Additive manufacturing, Powder, Simulation

1 Introduction

Laser engineering net-shaping (LENS), based on a kind of directed energy deposition (DED), is one of the popular AM technologies for producing complex metal structural components production. It could be used to fabricate complex, fully-dense metal components from CAD files directly without using dies, tooling or further machining, which hence greatly reduces the lead-time and production cost. Although LENS is a promising additive manufacturing process, it has not been widely used until recently. One of the main reasons is the high temperature and cooling rate during the LENS process will produce large residual stresses which may cause detrimental cracks and geometrical distortion to the deposited components. Besides, the range of deposited component size, imperfect surface quality and high cost of metal powder also limited LENS applications. Many factors, such as the laser-related and powder-related input manufacturing parameters will affect the final quality of components produced by LENS process, especially the thermal history of laser metal-powder deposition process mainly determining the cooling rate, residual stress and consequent cracks on the surface of deposited components. Trial and errors method is obviously too expensive to afford for diverse applications of different metal materials and various manufacturing input parameters. Numerical simulation of the LENS process will be an effective means to identify reasonable manufacturing parameter sets for producing high quality crack-free components.

Many researches have already tried to investigate the LENS or direct laser metal deposition (LMD) process numerically or experimentally [1-4]. Peyre et al [5] developed a three-step analytical and numerical approach to predict the shapes of manufactured structures and thermal loadings induced by the LMD process using multi-physics COMSOL. This approach takes into account the moving interface during metal deposition which allows the conductivity front to move simultaneously with the moving laser source and hence could accurately represent the LMD process. Liu et al. [6] proposed a FE model to predict thermal stress and

* Corresponding author: qin.yi@strath.ac.uk

https://doi.org/10.1051/matecconf/201819002001

deformations generated in the multi-layer laser metal deposition process. The thermal distribution, thermal stress field, geometry deformation and effect of deposition parameters on residual stress and deflections are explained. Verma et al [7] managed to simulate the thermal effect of a moving heat source on the SS316 and Ni substrates for LMD using Ansys Parametric Design Language. Manvatkar et al [8] developed a three-dimensional, transient, heat transfer, and fluid flow model for the laser assisted multilayer additive manufacturing process with coaxially fed austenitic stainless steel powder. Johnson et al [9] simulated the thermo-mechanical history and 3D microstructure evolution of 304L stainless steel tubes manufactured using LENS. Stender et al [10] proposed the process modelling of the LENS process in which thermal mechanical finite element simulation was focused.

Although many modelling and experimental researches relating to the LENS/LMD process have been done, less of them considered the metal powder dynamics effect on the final thermal mechanical behaviors during the LENS. In this paper, the laser metal powder deposition process of LENS is reported. The gas-powder flow distribution below the deposition nozzle is obtained via CFD simulation. The thermal behaviours of substrate and as-deposited layer/track during the LENS process are investigated by using FEM analysis. Temperature field distributions caused by the moving laser beam and the resultant melt pool on the substrate, are simulated and compared. The research offers a more accurate and practical thermal behaviour model for LENS process, which could be applied to further investigation the interactions between laser, melt pool and powder particles; it will be particularly useful for manufacturing key components which has more demanding requirement on the components' functional performance.

1.1 Physical phenomena involved in LENS process

For the LENS process, as shown in Fig.1, a high-power laser is used to melt metal powder supplied

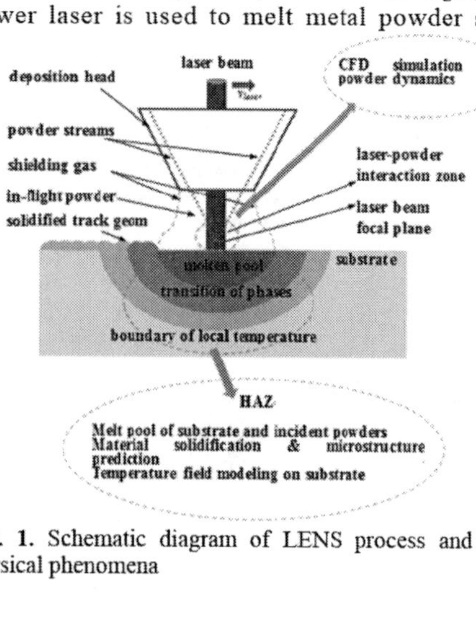

Fig. 1. Schematic diagram of LENS process and involved physical phenomena

coaxially to the focus of the laser beam through a deposition head in LENS. The laser beam typically travels through the centre of the laser deposition head and is focused to a small spot by one or more lenses. The laser head moves horizontally to scan the layer area to be deposited and then will be moved up vertically as each layer is completed. Metal powders are delivered and distributed around the circumference of the head by using a pressurized carrier gas. An inert gas is also used to shield the melt pool from atmospheric oxygen for better control of properties, and to promote layer to layer adhesion by providing better surface.

1.2 Interactions of sub-processes in LENS

To better understand the physic phenomena of LENS process, it is necessary to use a combination of modelling methods to simulate different sub-processes at the corresponding stages in the LENS process first. As seen in Fig.2, there are at least 4 sub-processes to model and simulate for the whole LENS process: (1) powder dynamics; (2) melt pool temperature field formation; (3) solidified track with microstructure evolution; (4) machining (if needed). Linking variables, such as particle mass concentration (PMC), velocity, temperature field of melt pool, cooling rate and residual stress, are transferred in-between different sub-processes for the holistic analysis.

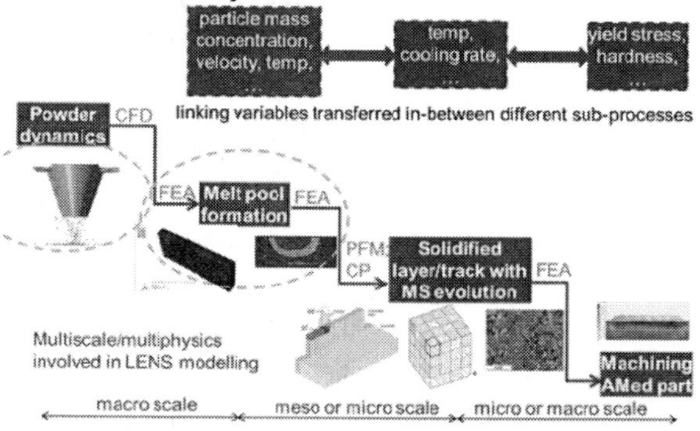

Fig. 2. Numerical realization of whole LENS process by linking different sub-processes

As for the powder dynamics, CFD modelling of the gas-powder flow and powder distribution is key to the density and strength of deposited components. As for the melt pool formation, the moving temperature field on the substrate caused by moving laser heat source is critical to determine the cooling rate, residual stress and consequent cracks in deposited components.

To study the influence of gas-powder flow and thermal behaviour during LENS process on the deposited components and its mechanical properties, accurate modelling, analysis and optimisation of these corresponding sub-processes are indispensable.

2 Modelling gas-powder flow in LENS

Generally, the gas-powder flows in the nozzles' passages and within the interaction zone between the nozzles' tips and the substrate are not a simple one-phase turbulent flow. In fact, this flow can be characterized as a two-phase flow, in which the primary phase is the turbulent inert gas and the secondary phase consists of the metal powder particles. The behavior of particles in a turbulent flow depends on the properties of both the particles and the flow. Turbulent dispersion of both the particles and the carrier gas can be handled by the concept of eddy diffusion energy in some range of the particle size distribution. The interaction of the particle and gas could be solved by two-way coupling the discrete and continuous phase until the solutions in both phases achieve a kind of stable balance. To realize the simulation and analyse powder flow behaviours involved in the powder feeding system of the LENS deposition head, especially the turbulence phenomenon for the gas-powder flow, a CFD numerical model has been developed based on the real nozzle setup shown in Fig.3(a).

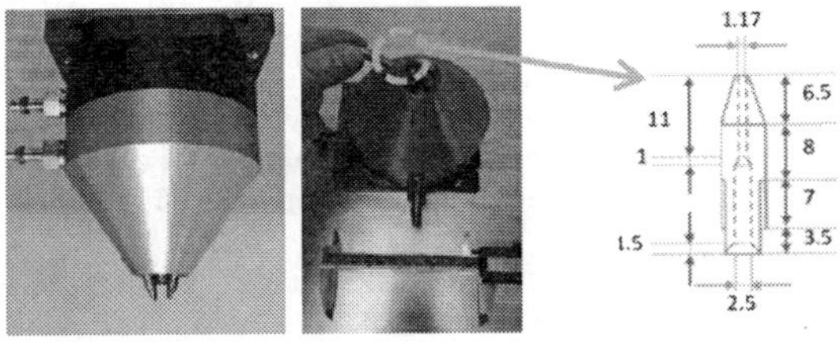

(a) laser deposition head of LENS system (b) size of nozzles

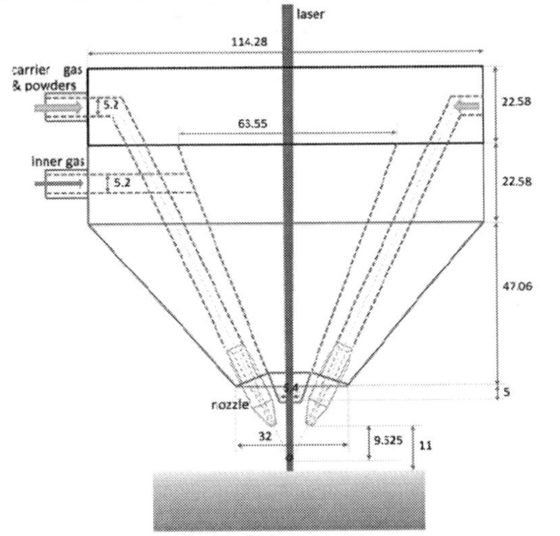

(c) measured geometrical size of LENS laser deposition head

Fig. 3. The laser deposition head with its real geometry of internal passages for CFD simulation

As shown in Fig3(a), the typical construction of a laser deposition head is a conical metal structure with 4 copper powder nozzles facing to a focal point and 1 main central nozzle for gas and laser passing through coaxially. The real geometry of the copper nozzle with internal passages and ducts are given in Fig.3(b). Fig.3(c) shows the schematic diagram of how the laser deposition head deliver powders with assistance of gas and gives detailed geometry with internal passages and ducts are. The gas-powder mixture is firstly injected into the inlets of laser deposition head and reaches a fully developed flow while travelling through the passages and ducts inside the deposition head and nozzles. Once the gas-powder flow is sprayed out from the 4 nozzles' tips, 4 gas-powder mixed jets will firstly flying without interactive actions and then intersect with each other to form a small circular focused spot which show the maximus powder mass concentration of the gas-powder flow.

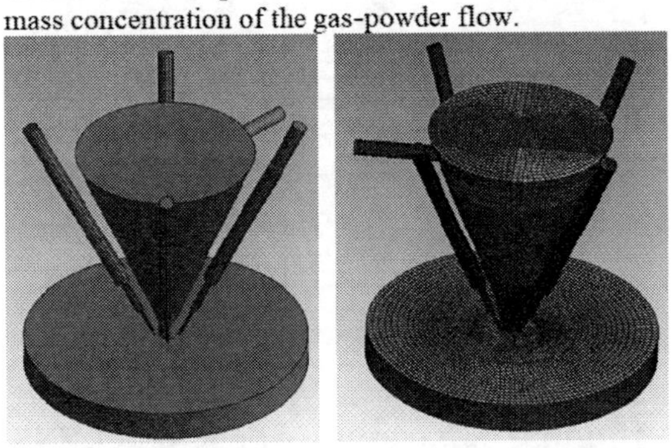

(a) extracted volume for CFD (b) discretized domain for CFD

Fig. 4. Extracted closed volume of LENS deposition head for CFD simulation & its meshes

For CFD simulation, only the closed volume of the laser deposition head is the boundary condition within which the fluid flow will be analysed. Fig. 4(a) shows the extracted geometry for CFD simulation of the powder dynamics. The external area shown in Fig.4(a) is where the powders ejected from the internal nozzle and in-flight until contact with the top surface of the substrate. Fig.4(b) discretizes the CFD domain with different grades of meshes. Near the nozzle's tip, much more refined meshes were used to consider the improvement of overall calculation speed and the required accuracy at some local critical areas with complicated geometry (such as the nozzle tip and focus plane). Some assumptions are taken for modelling of the gas-powder flow: (1) gas-powder mixture is treated as a steady-state turbulent flow with a constant velocity and pressure distribution at the inlets in which k-ε turbulent model is adopted; (2) the powder particles volume fraction is less than 10% and the one-way coupled discrete phase modelling (DPM) is adopted; (3) the heat transfer by laser radiation is temporarily neglected; (4) the particle size is assumed to be spherical and with an average diameter of 100um. For the CFD simulation, a 25L/min argon jet is supplied and ejected from the inner gas nozzle to protect the laser optics from the rebounding particles and to shield the melt pool. The TiAl powders were delivered to the melt pool by using carrier gas (argon) via 4 nozzles which are spaced equally around the inner gas nozzle. The total flow rate of the carrier gas for the 4 nozzles is 5 L/min. These 4 nozzles are installed around the main inner gas nozzle and point to the melt pool which could make the carrier gas-powder jets be trapped to the melt pool. The diameter of the nozzle tip is 1.17mm. Each of the nozzles is titled 65 degree from the substrate surface. The initial conditions for CFD simulation of gas-powder flow are given and listed in Table 1.

Table 1. Input parameters from CFD simulation of gas-powder flow

Inputs for CFD	Value
average diameter of powders	100um
powder flow rate	100g/h
powder flow velocity	0.118 m/s
carrier gas flow rate	8.33×10^{-5} m^3/s
carrier gas flow velocity	0.588 m/s
pressure of carrier gas at inlet	8 bar
pressure of inner gas at inlet	2 bar
inner gas flow rate	6.67×10^{-5} m^3/s
inner gas flow velocity	0.471m/s
nozzle outer diameter	22.56mm
material density	3910kg/m^3
carrier gas density	1.67 kg/m^3
stand-off	9.525mm

3 Thermal behaviour modelling in LENS

3.1. Laser heat source caused temperature field on the substrate

A 3D FE model for simulation of the thermal transfer when laser heat source moving on the substrate is established with user subroutine DFlux in ABAQUS. The transient heat conduction equation between the moving laser heat source and the substrate could be expressed as follow in the 3D Cartesian coordinate system :

$$\frac{\partial}{\partial x}\left(k\frac{\partial T}{\partial x}\right)+\frac{\partial}{\partial y}\left(k\frac{\partial T}{\partial y}\right)+\frac{\partial}{\partial z}\left(k\frac{\partial T}{\partial z}\right)=\rho_s C_s\frac{\partial T}{\partial t} \quad (1)$$

where, k is the thermal conductivity; ρ_s is the substrate material density; C_s is the specific heat; T is the temperature and t is time variables. The typical boundary conditions are represented as

$$q_s = \underbrace{k_n\frac{\partial T}{\partial n}}_{\text{Conduction}} + \underbrace{h_c(T-T_R)}_{\text{Convection}} + \underbrace{\sigma_{S-B}\varepsilon_c(T^4-T_R^4)}_{\text{Radiation}} \quad (2)$$

where n refers to the direction vertical to the substrate surface; k_n is the thermal conductivity vertical to the substrate surface; h_c is the surface heat transfer coefficient; ε_c is emissivity constant; $\sigma_{S\text{-}B}$ is Stefan–Boltzmann constant; T_R is the reference temperature, respectively. The 1st term on the right-hand side in Eq. [4] represents heat energy being transferred to the substrate due to thermal conduction from the surface whose unit normal is n. The 2nd and the 3rd terms on the right-hand side refer to convection and radiation from the surface open to air. For laser beam heat source which follows Gaussian distribution, the heat flux, q_s, transferring from laser beam to the substrate surface could be expressed as:

$$q_s=\frac{\eta P d}{\pi\cdot r_{lb}^2}\exp\left(-\frac{d(x^2+y^2)}{r_{lb}^2}\right) \quad (3)$$

where P is the laser power; η is the thermal absorption coefficient of the laser beam, r_{lb} the radius of the laser beam spot; d is the beam distribution parameter, respectively. The values of d, r_{lb}, and η are considered as 3.0, 0.26 mm, and 0.28, respectively. It is worthy to mention that the high value of d(>2.0) allows the distribution of the applied heat flux to follow a high peak with a steep descent within a small focused area, which is typical for the type of laser used in the LENS machine.

The effective beam radius r_{lb} could be obtained by measuring several melt pool radii that are produced by the LENS system's laser on a SS316 substrate without any powder material. The absorption coefficient, η, of the laser beam is a complex function of substrate temperature, incident surface quality, and shielding atmosphere. To avoid complexity, an average value of η is estimated following Bramson's equation as [3,11]

$$\eta=0.365\left(\frac{R}{\lambda}\right)^{1/2}-0.067\left(\frac{R}{\lambda}\right)+0.006\left(\frac{R}{\lambda}\right)^{3/2} \quad (4)$$

where R is the temperature-dependent electrical resistivity of the material and λ is the wavelength of the laser beam, the latter being equal to 1.067um in the present case.

During the LENS processing, the powders travel through the passages within the laser deposition head and are then ejected out from the nozzles; they will fly to the focus of the laser beam and finally fall into the melt pool if not fully being melt when flying via the focus of laser beam. It is too complicate to track the real-time position and temperature of the individual powders. It is a reasonable assumption that most of the incident powders will be trapped by the molten pool at melting points and some of them may be renounced by the substrate due to large momentum and keeps the room temperature.

In this paper, three-dimensional 8-node heat transfer brick element (DC3D8 in ABAQUS) with temperature as the nodal degree of freedom is used for modelling the moving laser caused thermal behaviours on the substrate. The transient heat transfer calculations are performed using a uniform time-step and a number of very small time increments within each time-step. For the moving laser beam of a moving speed 5m/s and spot diameter of 0.26mm, the simulated temperature field around the laser spot on the substrate are obtained in Section 4. The realization and the application of the heat input to the heat transfer elements in the substrate are realized through an ABAQUS user subroutine DFlux.

4 Simulation & Results

4.1 Gas-powder flow

As mentioned in Fig.2, there are many output variables from CFD simulation, such as particle velocity, particle mass concentration (PMC), DPM number of collision and trajectory of powders after being ejected from the nozzles. PMC is the most important variable

https://doi.org/10.1051/matecconf/201819002001

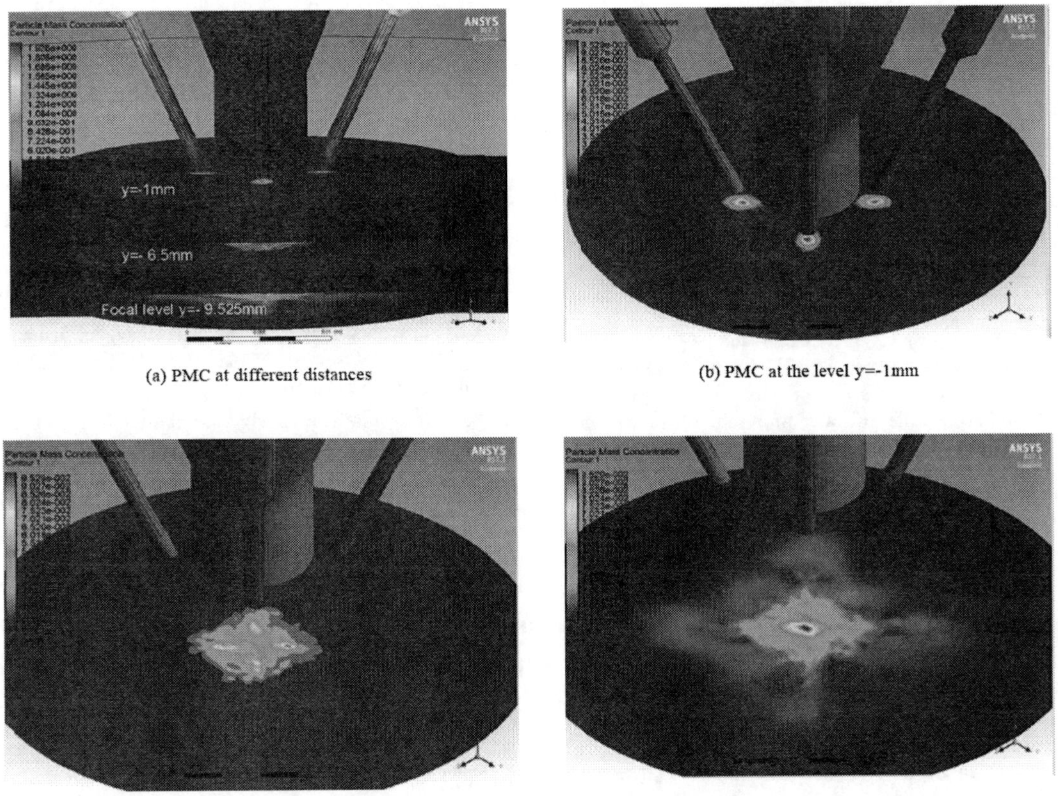

(a) PMC at different distances

(b) PMC at the level y=-1mm

(c) PMC at the level y=-6.5mm

(d)PMC at the focus y=-9.525mm

Fig. 5. PMC at different locations away from the tips of nozzle

determining the density of particles on a specified plane or in a specified volume. It could give the straightforward impression where the particles are more likely to gather.

Fig.5(a) shows the simulated result of variation of particle mass concentration along the stand-off distance of LENS. After ejecting out from the nozzles, there is a clear trend that the 4 jets of particles are travelling to and concentrating on a point of intersection. At the level of y=-1mm (Fig.5(b)), the particle mass concentration of the 4 jets are separate and discrete. When the jets travels to the level of y=-6.5mm (Fig.5(c)), the boundary of 4 jets is becoming blurred. When the 4 jets travels to the level of y=-9.525mm (Fig.5(d)), the highest particle mass concentration of the 4 jets seem to converge at the middle of the domain. If the focal point of a laser beam is just at this location, then the highest laser power heat could be employed to melt the densest cloud of particles. Matching the laser beam focal point and the position of the convergent highest particle mass concentration is extremely important for full fusion of the deposited substrate and particles and consequently generation of crack-free and high-quality deposited tracks. The Fig.5(d) shows a larger influencing area of particle mass concentration than that of Fig.5(b) and (c). It is mainly because that it is near the focal point and some particles will collide and even bounce back once comes into contact with the substrate.

4.2 Thermal behaviors

The laser-related processing parameters and substrate geometry are influential factors for the quality of component produced by LENS process. For the thermal behaviour modelling and simulation, the substrate material is stainless steel (SS316) and its geometry is simplified to be a rectangular block with a size dimension 150mm×60mm×15mm. The bottom and the left side of the workpiece are applied with fixed boundary condition. Substrate material SS316L is assumed to be isotropic and follows Johnson-cook plastic criterion. No material phase transformation is considered. The laser power is 1000W, laser moving velocity 50mm/s. DFlux subroutine is used for simulating the temperature field on the substrate caused by the moving laser beam heat source and the heat flux coming from a laser beam is assumed to follow Gaussian distribution

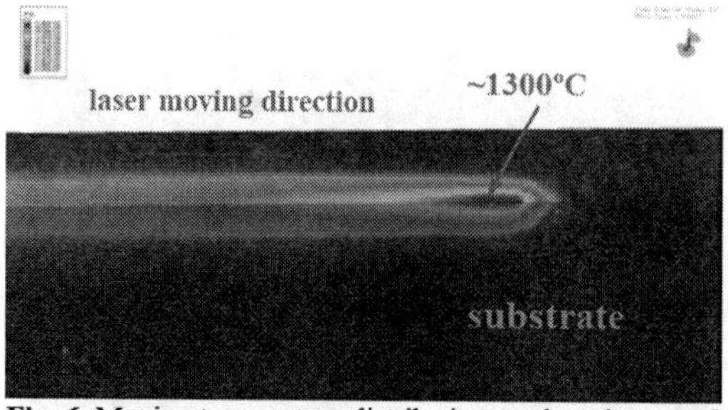

Fig. 6. Moving temperature distribution on the substrate after the LENS process achieving quasi-stable state (*t*=0.89s)

As shown in Fig.6, a transient temperature field caused by the input moving laser beam is obtained via simulation. The temperature field has a steep gradient distribution at the side of laser moving ahead and a comparatively gradual distribution at the side after laser moving. For the situation, the maximum temperature near the laser spot is around 1300 °C. If the material's melting point is reached, then the area will be the location of melt pool.

5 Conclusions

The laser metal powder deposition process of LENS is investigated and analysed. Numerical simulations of the key sub-processes within the LENS process are carried out. The gas-powder flow distribution below the deposition nozzle is obtained via CFD simulation. The thermal behaviours of substrate during the LENS process are investigated by using FEM analysis. Temperature field distributions caused by the moving laser beam and the resultant melt pool on the substrate are obtained. This research could offer a more accurate and practical thermal behaviour model for LENS process, which could be applied to further investigation the interactions between laser, melt pool and powder particles; it will be particularly useful for manufacturing key components with LENS process which has more demanding requirement on the components' functional performance.

Acknowledgement

The authors would like to acknowledge the support from EU H2020 project MMTECH (GA 633776) for the research reported in this paper.

References

1. H. Qi, J. Mazumder, H. Ki, J. App. Phys. **100**, 024903 (2006); doi: 10.1063/1.2209807
2. F. Bruckner, D. Lepski, E. Beyer, J. Therm Spray Tech., **16**, (2008). DOI: 10.1007/s11666-007-9026-7
3. V.D. Manvatkar, A.A. Gokhale, G. Jagan Reddy, A. Venkataramana, A. De, Metallur. & Mater. Trans. A, **42A** (2011).
4. L.E. Murr, et al., J. Mater. Sci. Technol., **28** (2012).
5. P. Peyre, P. Aubry, R. Fabbro, R. Neveu, and A Longuet, J. Phys. D. Appl. Phys., **41** (2008).
6. H. Liu, T.E. Sparks, F.W. Liou, and D. M. Dietrich, *Proceedings of 24th Solid Freeform Fabrication Symposium*, Austin, TX, August 12-14, (2013)
7. V. Verma, M. Shukla, *Proceedings of the 15th International RAPDASA conference*, Stellenbosch, South Africa, (2014).
8. V. Manvatkar, A. De, T. DebRoy, *Journal of Applied Physics* **116**, *124905* (2014) (https://doi.org/10.1063/1.4896751)
9. K.L. Johnson, T.M. Rodgers, O.D. Underwood, J.D. Madison, K.R. Ford, S.R. Whetten, D.J. Dagel, J.E. Bishoo, Computational Mechanics, Published online Dec (2017) (https://doi.org/10.1007/s00466-017-1516-y)
10. M.E. Stender, L.L. Beghini, M.G. Veilleux, S.R, Subia, J,D. Sugar, *Proceeding of the ASME 2017 Presure Vessels and Piping Conference (PVP2017-65992)*, Waikoloa, Hawaii, USA (2017)
11. V.D. Manvatkar, A. De, A.A. Gokhale, G. Jagan Reddy, *Proceeding of the Materials Science and Technology (MS&T 2011)*, Columbus, Ohio, USA (2011).

Laser edge forming to increase the bending radius in hemming

Peer Woizeschke[1,*], *Lukas* Heinrich[1], *Philip* Eichner[2,3]

[1]BIAS - Bremer Institut für angewandte Strahltechnik GmbH, Klagenfurter Str. 5, 28359 Bremen, Germany
[2]University of Bremen, Klagenfurter Str. 5, 28359 Bremen, Germany
[3]Present address: VEMAG Maschinenbau GmbH Verden, Weserstraße 32, 27283 Verden (Aller), Germany

Abstract. Hemming can be used, for example, to join dissimilar materials which cannot be welded by conventional laser or arc based processes. A critical value in hemming is the ratio of outer sheet bending radius and outer sheet thickness because cracks can occur in outer sheets with low ductility due to a small bending radius. In this study, a novel approach is introduced for increasing the bending radius by laser based edge forming. The laser beam melts the edge of the inner sheet prior to the hemming process. In laser edge forming, the edge thickness increases due to the surface tension of the melt. The edge cross-section has typically a nearly circular shape after solidification. Two laser beam irradiation strategies were investigated. In case of the first strategy, a defocused laser beam was applied. The second strategy combined a focused laser beam resulting in the formation of a vapor capillary (keyhole) with sinusoidal beam oscillation transversal to the sheet edge. The results show that the position of the cylindrical edge cross-section according to the sheet plane could be affected by the applied irradiation strategy. The needed laser power was significantly lower when using a focused laser beam due to the deep penetration effect. The potential of the laser edge forming technique for crack-free hemming processes was demonstrated with 1 mm thick aluminum alloy sheets by carrying out stripe tests.

Keywords: Joining, Metal Forming, Laser

1 Introduction

Hemming can be applied as process to increase the stiffness of a sheet edge or to eliminate the sharpness of an edge on the one hand. On the other hand, hemming can be applied for joining or assembling two sheets by folding an outer sheet around an inner sheet. The hemming joining technique has potential to replace partially the spot welding in the production of hollow sections for car body structures in case of certain (also dissimilar) material combinations, see Figure 1a and 1b. A main challenge in hemming is to avoid damages in form of cracks in the outer sheet being non-acceptable in part manufacturing [1]. Especially in case of materials with low ductility, the bending of the outer sheet can result in critical tensile strains [2]. In order to enhance the bending ability, for example in case of aluminum alloys, process and tool design modifications were analyzed by experiments and simulations and summarized by trend-lines [3] and a new flanging operation to deliver additional metal into the bending zone [4] as well as an electromagnetic forming (EMF) technology [2] were developed. In case of high-strength steels with low ductility, crack-free hemming was achieved by hemming using a punch with stopper [5]. In doing so, the zone of the outer sheet with high tensile stress is additionally compressed. A main parameter regarding the critical bending conditions is the ratio of the bending radius of the outer sheet and its sheet thickness. For the application of hemming in car manufacturing, a

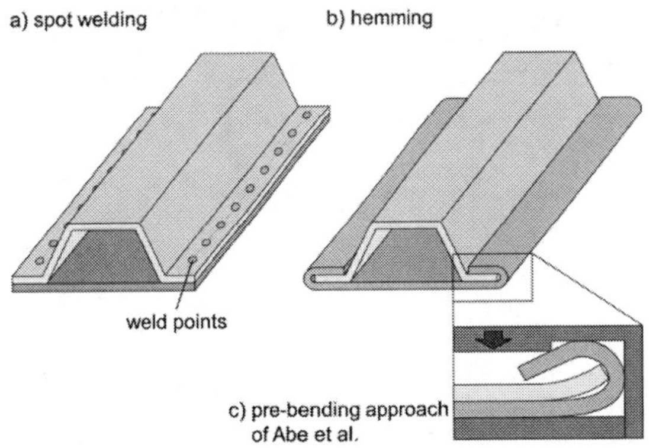

Fig. 1. (a) Sketch of a typical spot welded hollow section; (b) sketch of a hemmed variant of the hollow section and (c) schematic description of the approach of Abe et al. according to [7] for using a pre-bent inner sheet to increase the bending radius of the outer sheet.

* Corresponding author: woizeschke@bias.de

concept was invented [6] in which a groove is inserted on the inside of the outer sheet to improve the bending process. Basically, the bending radius is determined in conventional hemming processes by the inner sheet thickness. In this regard, Abe et al. presented an approach at which a pre-bending of the inner sheet was used to increase the bending radius of the outer sheet [7], see sketch in Figure 1c. The process enabled the crack-free production of ultra-high strength steel hollow parts by 3-stage hemming [7]. In case of classical hemming of high-strength steel sheets cracks occurred [5].

The results of Abe et al. motivated the work presented in this paper. The investigated novel concept is based on appropriately modifying the edge of the inner sheet by laser melting. Laser beam edge melting might be applicable to form advantageously the edge of the inner sheet. Vollertsen et al. developed the "free form heading" laser melting process for wires/rods to reduce the number of steps of forming processes [8]. A laser beam melts the rod end during the process to produce preforms with a drop-like shape and increased diameter due to the surface tension of the molten material. Woizeschke et al. transferred the process approach to sheet edges and demonstrated the fundamental feasibility of producing cylindrical forms along sheet or foil edges by laser melting processes [9]. In principle, different laser beam irradiation strategies are applicable [9]. Figure 2 demonstrates the laser edge forming by using a defocused laser spot which is guided on the sheet surface and moved along the sheet edge [9].

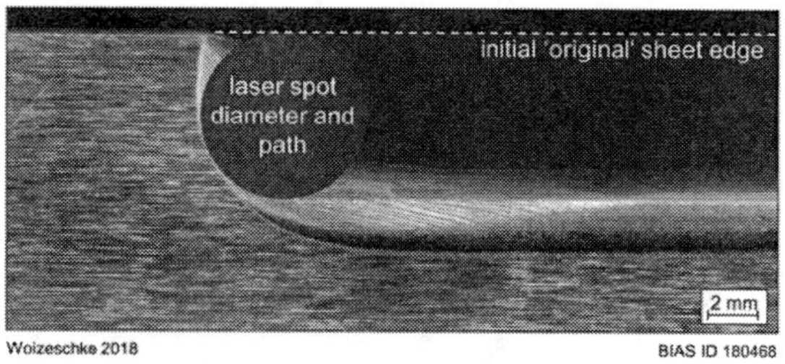

Fig. 2. Example for laser edge forming by using a defocused laser spot according to [9].

In this study, laser based edge forming shall be investigated to enable the precise increase of the bending radius in hemming to produce crack-free joints in case of materials with low ductility. Two strategies of laser edge forming, a heat conduction based variant with a defocused large laser spot on the workpiece (named "defocused" strategy in this study) and a deep penetration based variant with a focused laser beam on the workpiece superimposed by a spatial transversal beam oscillation (named "oscillation" strategy), shall be analyzed regarding advantages and disadvantages with respect to the process behaviors and the resulted edge forms.

2 Materials and methods

2.1. Material

The experiments of this study were carried out with sheets of the aluminum alloy EN AW 6082 (AlSi1MgMn) in the heat treatment condition T4. The base material thickness was 1 mm. The specimens with the dimension of 160 mm x 100 mm were shear cut and the edges were milled to generate reproducible conditions. The milled edges are called 'original' or 'conventional' edges in contrast to the 'laser formed' edges in this study.

2.2 Laser setup

A disk laser (wavelength of 1030 nm; beam quality of 8 mm mrad; TruDisk 8002) was used as laser source for the melting processes. A gantry system (company Föhrenbach) was equipped with a laser processing optic to perform the laser edge forming. The laser beam was guided to the laser head by a fiber with a core diameter of 200 µm. The focal lengths of collimation as well as focusing were 200 mm and, thus, the resulting laser focal diameter was 0.2 mm. A scanning system (DC-scanner ILV-0205; ILV) for spatial beam modulation was mounted on the laser head to enable the beam oscillation transversal to the main process direction which was parallel oriented to the sheet edge. The scanning system allowed the application of sinusoidal beam oscillation. The maximum amplitude depended on the oscillation frequency. Argon was applied as shielding gas for the melt pool during the laser melting processes.

2.3 Irradiation strategies

Two irradiation strategies were carried out to produce suitable edge forms for hemming processes. In case of the "defocused" irradiation strategy, a defocused laser beam with a comparatively large laser spot diameter on the workpiece was used to melt the sheet edge without any additional beam oscillation. In case of the "oscillation" irradiation strategy, the main movement of the laser head was superimposed by the transversal beam oscillation. Thus, the laser beam focus, which was positioned on the workpiece surface at this strategy, followed a sinusoidal path along the edge. The process parameters of the experiments in this study are listed in Table 1.

2.4 Measuring methods

The thickness of the laser based thickened/formed sheet edge was measured at seven positions along the edge in 20 mm intervals by using a digital caliper. The starts and the ends of the edge are not considered in this study. The averages and standard deviations in the analysis are based on 70 samples per strategy.
The metallographic analysis was performed by cross-sectional polishes of the laser formed edges. The polishes were electrolytically etched according to the etching parameters which are given in Table 2. A sequence of abrasives up to grit 1200 was used before etching.

https://doi.org/10.1051/matecconf/201819002002

Micrographs of the prepared polishes were taken by a microscope under polarized light.

Table 1. Process parameters for both irradiation strategies.

process parameter	irradiation strategy "defocused"	irradiation strategy "oscillation"
main process direction	parallel to the sheet edge	parallel to the sheet edge
superimposed laser beam movement	-	sinusoidal oscillation transversal to the main process direction
focal plane	+16 mm above the workpiece	on the workpiece surface
spot diameter on the workpiece	6.1 mm	0.2 mm
shielding gas	argon	
opening diameter of the shielding gas nozzle	28 mm	
shielding gas flow	20 l/min	30 l/min
laser output power	5.5 kW	1.1 kW
main process speed	0.8 m/min	0.8 m/min
oscillation frequency	-	100 Hz
oscillation width	-	6.3 mm
initial waiting time	-	0.2 s

Table 2. Etching parameters of the cross-section polishes.

electrolyte	aqueous solution of 2-percent tetrafluoroboric acid HBF_4-solution: 950 ml H_2O + 38 ml HBF_4
voltage	25 V
duration	60 s
flow rate	10 ml/min

2.5 Feasibility test

In order to demonstrate the feasibility of hemming with laser formed inner sheet edges, simple stripe tests have been carried out with 1 mm thick aluminum sheets. The outer sheet stripe was manually bent around the inner sheet (with and without laser formed edge). After bending the outer sheet was hemmed by pressing it on the base sheet (next to the formed edge). The surfaces of the outer sheets were analyzed with respect to cracks or other defects.

3 Results

3.1 Process observations

Figure 3 shows snapshots of the laser forming for both strategies during the laser melting processes. The focused oscillating laser beam in case of the "oscillation" strategy generated a highly dynamic process behavior with significant spatter formation (see Figure 4b), whereas no spatters were observed during the process of the "defocused" strategy with a defocused non-oscillated laser beam.

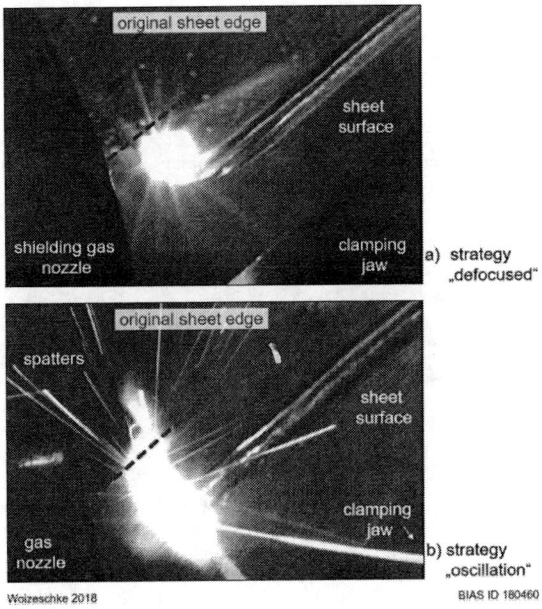

Fig. 3. Snapshots during the laser forming processes for (a) the "defocused" strategy and (b) the "oscillation" strategy (red broken lines: original sheet edges before laser melting).

3.2 Characteristics of the laser formed edges

The laser formed edges for both strategies are shown by front and side views in Figure 4 and Figure 5. In both cases, no adhering spatters on the sheet surfaces were observed. The measured values of the edge form thickness (diameter) along the edge are plotted for the two strategies in Figure 6 and Figure 7. The averages of the edge form thickness vary along the edges by less than 0.1 mm. The average value of the "defocused" strategy was determined to 3.44 ± 0.07 mm. In case of the "oscillation" strategy the average is 2.98 ± 0.08 mm. Both strategies resulted in edge forms with approximately circular cross-section. The center of the form cross-section was below the sheet plane in case of the "defocused" strategy considering that the laser beam was guided on the upper sheet surface, whereas the center was positioned above the sheet plane in case of the "oscillation" strategy, see Figure 8a and Figure 9a. The micrographs of the cross-sections show grain coarsening within the laser formed edges. In contrast to the defocused variant, the laser forming with the focused oscillating laser beam produced an area with significantly smaller grains, see Figure 9c. The red dotted lines show a difference in the fusion line orientation.

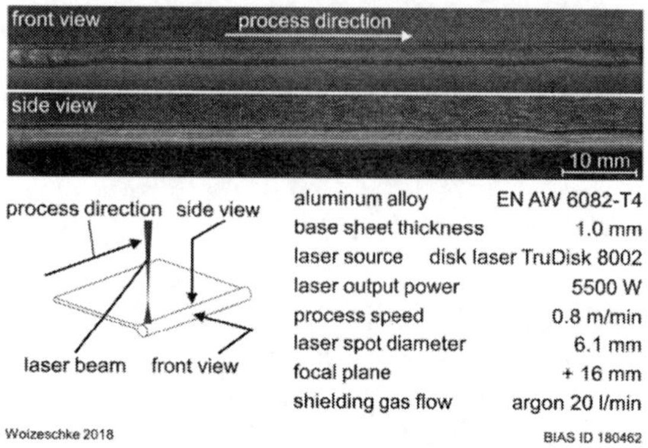

Fig. 4. Front and side view on an exemplary laser formed edge for the "defocused" irradiation strategy.

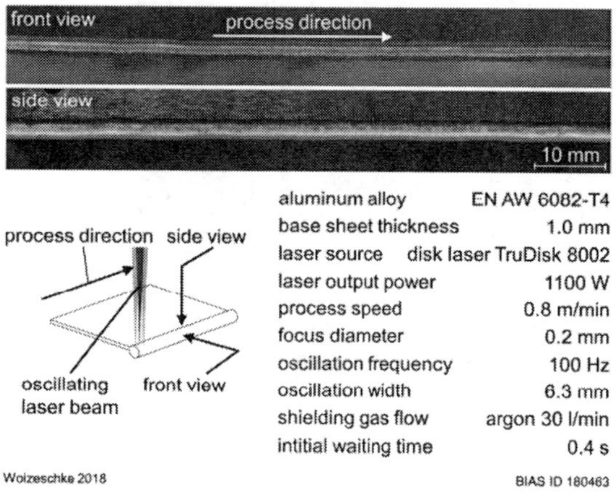

Fig. 5. Front and side view on an exemplary laser formed edge for the "oscillation" irradiation strategy.

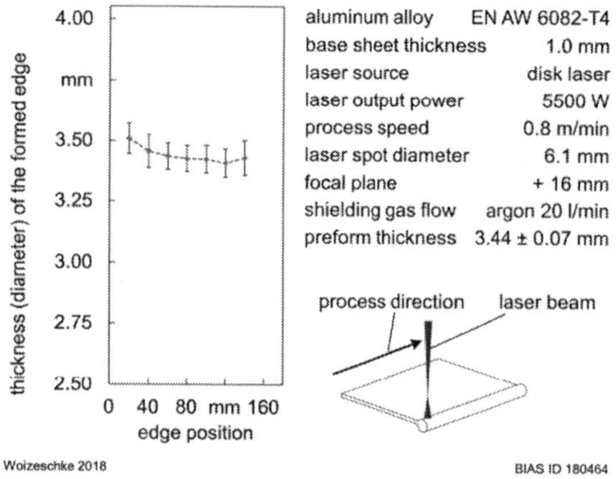

Fig. 6. Measured thickness (diameter) values of the edge forms at the seven measuring positions along the edge for the "defocused" strategy (values are based on 70 samples each).

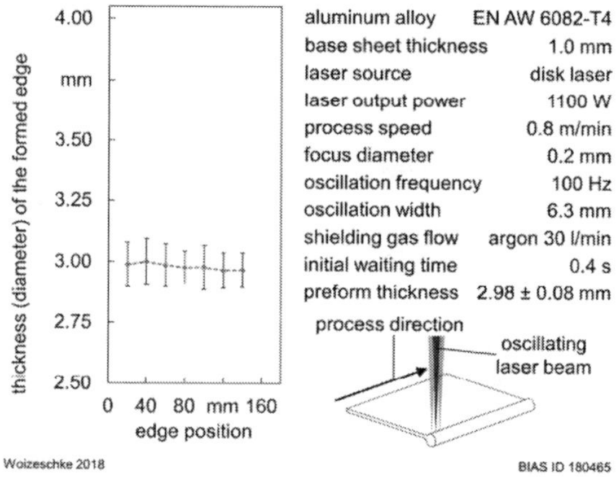

Fig. 7. Measured thickness (diameter) values of the edge forms at the seven measuring positions along the edge for the "oscillation" strategy (values are based on 70 samples each).

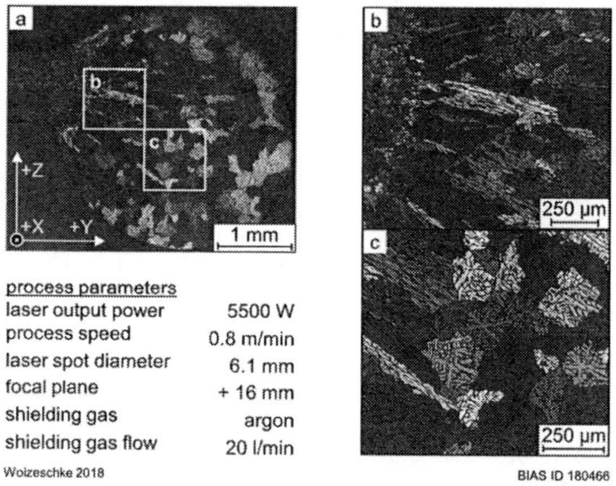

Fig. 8. Exemplary cross-section micrograph (electrolytically etched polish) of a laser formed edge produced by the "defocused" irradiation strategy (red dotted line: fusion line).

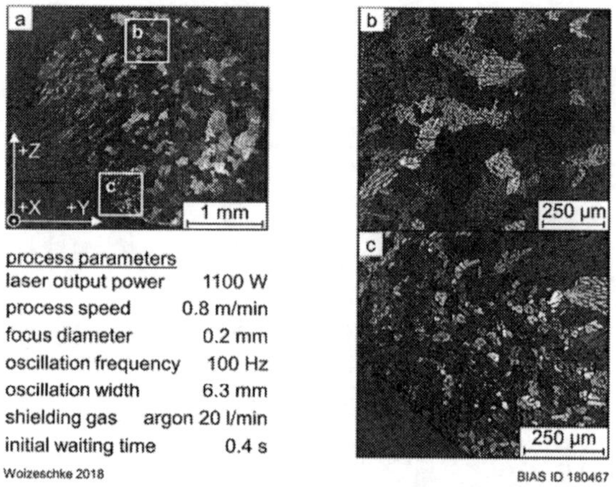

Fig. 9. Exemplary cross-section micrograph (electrolytically etched polish) of a laser formed edge produced by the "oscillation" irradiation strategy (red dotted line: fusion line).

3.3 Potential demonstrating forming test

In order to demonstrate the effect of the increased edge radius of the inner sheet on the bending behavior of the outer sheet in hemming, the effect of a laser formed edge is compared to the behavior in case of a conventional milled sheet edge by sheet forming experiments, see Figure 10. The non-manipulated conventional milled edge of the inner sheet resulted in cracks on the outer surface of the outer sheet when the sheet was completely bent around the inner sheet (Figure 10a). In case of the laser formed edges, crack formation was not observed (Figure 10b).

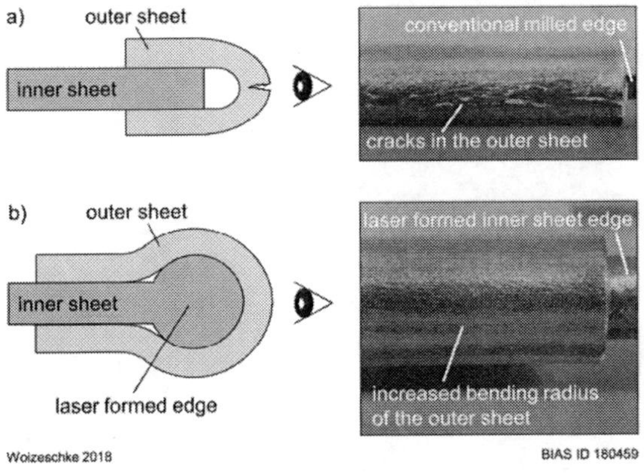

Fig. 10. Sheet forming tests (a) with a non-manipulated conventional milled edge at the inner sheet and (b) with a cylindrically laser formed edge at the inner sheet.

4 Discussion

Both irradiation strategies could produce cylindrical shaped edge forms at the edges of 1 mm thick aluminum sheets. The significantly different process behaviors are caused by the different modes of the applied laser material processes. The comparatively calm process behavior without spatter formation with the defocused laser spot resembles a heat conduction welding process. The laser energy is absorbed on the upper surface. In case of the "oscillation" strategy, the focused laser beam resulted in keyhole formation in the workpiece. The keyhole, a vapor capillary, is formed during laser material processes when the power density reaches a threshold, for example approximately $5x10^6$ W/cm² in case of aluminum alloys and solid-state laser sources with wave lengths of 1 µm. The keyhole formation caused, on the one hand, the high process dynamic in the molten pool with spatter formation in case of beam oscillation. On the other hand, the keyhole formation is responsible for the significantly higher process efficiency of the "oscillation" strategy in comparison with the "defocused" strategy. The needed laser output power for the laser edge forming was five times higher in case of the "defocused" irradiation strategy than in case of beam oscillation with focused laser beam. In typical keyhole (deep penetration) laser welding of aluminum alloys, the absorption can be increased to values over 90 % by the formation of the vapor capillary due to multiple reflections within the keyhole, see [10]. In contrast, the (one-time Fresnel) absorption of aluminum alloys in laser processes like heat conduction welding or laser edge forming with the "defocused" strategy is typically in the range of 5 % to 30 % depending on the surface conditions, the substrate temperature and the alloy composition.

In addition to the process efficiency, the results demonstrate that the irradiation strategy can be used to affect or adjust the orientation of the cylindrical edge form with respect to the plane of the inner sheet. A visible difference between the two irradiation strategies is the orientation of the fusion line within the cross-sections, see Figure 8 and Figure 9. This is probably a main reason for the different edge form position.

The suitability and the advantage of such laser formed edges for hemming processes have been demonstrated by sheet forming experiments, see Figure 10. The cylindrical forms which can be produced by non-contact fast laser melting processes enable the reproducible and precise adjustment of the bending radius by creating a form-locked connection between the outer sheet and the laser formed edge of the inner sheet.

5 Conclusion

This study can be concluded as follows:

- Different laser irradiation strategies are suitable to produce reproducible cylindrical forms at aluminum sheet edges with tolerances below 0.1 mm
- The choice of the irradiation strategy affects significantly the process efficiency of the laser edge forming process due to different absorption mechanisms
- The irradiation strategy can be used to affect the position of the edge form with respect to the plane of the inner sheet
- Observed typical cracking of the outer sheets in case of a simple hemming test with conventional milled edges at the inner sheets can be avoided successfully by laser edge forming due to the adjustable increased bending radius.

Acknowledgements

The authors are grateful for the financial support by the Senatorin für Wissenschaft, Gesundheit und Verbraucherschutz of the country of Bremen.

The 'BIAS ID' numbers are part of the figures and allow the traceability of the results.

References

1. N. Le Maoût, S. Thuillier, P. Y. Manach, Engineering Fracture Mechanics **76** (2009) 1202-1214
2. P. Jimbert, I. Eguia, I. Perez, M. A. Gutierrez, I. Hurtado, Journal of Materials Processing Technology **211** (2011) 916-924
3. H. Livatyali, A. Müderrisoğlu, M. A. Ahmetoğlu, N. Akgerman, G. L. Kinzel, T. Altan, Journal of Materials Processing Technology **98** (2000) 41-52
4. S. F. Golovashchenko, Journal of Materials Engineering and Performance **14** (2005) 508-515
5. Z. Hamedon, K. Mori, Y. Abe, Procedia Engineering **81** (2014) 2074-2079
6. M. Blechschmidt, M. Liewald, S. Hoenle, T. Altermann, J. Hesse, patent, DE 10 2015 000 981 B4 (2015)
7. Y. Abe, K. Mori, K. Nakagawa, Procedia Engineering **207** (2017) 974-979
8. F. Vollertsen, R. Walther, CIRP Annals – Manufacturing Technology **57** (2008) 291-294
9. P. Woizeschke, H. Brüning, F. Vollertsen, Laser Magazin **5** (2015) 22-24
10. Y. Kawahito, N. Matsumoto, Y. Abe, S. Katayama, Journal of Materials Processing Technology **211** (2011) 1563-1568

Deburring and rounding of sheet metals and functionally relevant edges of parts using laser radiation

Judith Kumstel[1,*], and *Sven* Lüken[1]

[1] Fraunhofer ILT - Institute for Laser Technology, 52074 Aachen, Germany

Abstract. The manufacturing of sheet metal parts or component edges often leads to burrs. Due to different requirements regarding the appearance, haptics, tribology et cetera the edges have to be deburred or rounded in a subsequent process. Especially for complex geometries, automated post-processing has been difficult up to now. Laser remelting is one possible solution. It is fully automated and suitable for most 3D geometries. During laser remelting the laser beam is focused on the surface, thus a thin surface layer up to 100 µm is melted. The surface tension in the liquid state is smoothing out the surface and the edge resolidifies rounded. The radius of the rounding is adjustable by the chosen process parameters, especially laser power P_L and feed speed v_{feed}. Furthermore the resulting roughness and the gloss level are controllable via the process parameters. Typical processing speeds are 100 mm edge length per second. First studies with industry partners showed promising results. Laser remelting of push belts for the torque transmission in CVTs lead to an increase of fatigue strength by up to 200% which enables a higher torque.

Keywords: Deburring, Laser beam machining (LBM), Surface modification

1 Introduction

When cutting or separating sheet metal parts, it is often not possible to produce edges free of burrs and defects. Burrs and similar edge defects impair the mechanical properties with negative consequences for processability and loadability (edge formability, fatigue properties). Furthermore they considerably increase the risk of injury and can endanger the functionality of components (e.g. damage to cable insulation, wear).

Although there are many different processes for deburring and edge processing, these have significant shortcomings depending on the task at hand (e.g. contamination of the components and therefore the need for an additional cleaning step, no selective processing and thus damaging of areas that have not to be processed, not suitable for the processing of free-form edges). This sometimes results in high production costs and/or the achievable component quality is limited.

A rather new and very unexplored approach for deburring and edge processing is the laser remelting. Laser remelting of sheet metal is a flexible process and is based on remelting the edge using laser radiation. In the liquid state, the roughness of the edge can flow out due to the surface tension and is smoothed. Burrs and microcracks can also be melted down in this way. In addition, the molten edge solidifies rounded.

A comprehensive investigation of laser deburring of steel materials is shown in [1]. However the investigation is limited to fine burrs with the aim of producing sharp-edged, burr-free edges on small components. Although there is already a first industrial implementation of laser deburring in the series production of sheet metal rings for CVT car transmissions, the potential of laser edge processing is only just beginning to be exploited [2, 3]. Systematic investigation of the quantitative relationships between the process parameters of laser edge processing and the expected property profiles have not yet been carried out. Also, the application limits of the method have not been identified. This paper examines the influence of the process parameters of laser remelting on the edge rounding.

A similar process based on the same basic principle but developed for other applications and dimensions is the laser rod end melting. Laser rod end melting results in a rounding of the end of cylindrical rods for rods with a maximum diameter of 1 mm [4]. After the generation of the spherical geometry at the rod end in the master forming stage, this part can be used as a preform for a micro cold forming operation in a subsequent process stage for the production of metallic micro parts such as screws or springs [5].

2 Basics of laser remelting

During laser remelting the laser beam (round or rectangular) is focused on the surface and guided over the surface with the feed speed v_{feed}, thus a thin surface layer is melted (see fig. 1). The surface tension in the liquid state is smoothing out the surface and the edge resolidifies rounded. No material is removed but rather relocated while molten. For metals, diode-pumped solid-state lasers in CW mode are used. The thickness of the remelted layer is in the range of 50- 400 µm.

* Corresponding author: judith.kumstel@ilt.fraunhofer.de

https://doi.org/10.1051/matecconf/201819002003

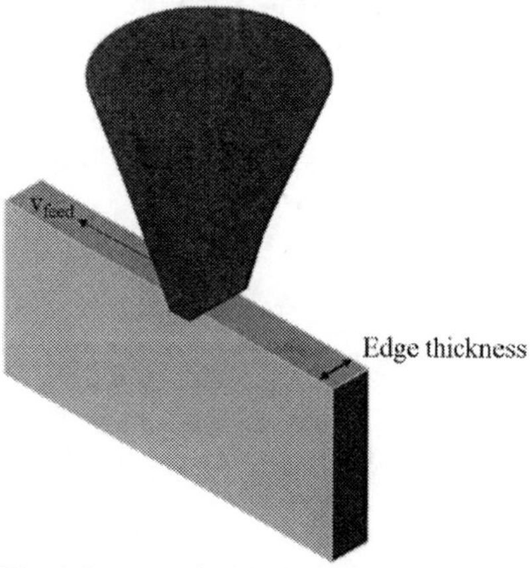

Fig. 1. Process principle for laser edge rounding with round laser beam.

The process parameters (e.g. laser power, feed speed and laser beam shape and size) have to be adapted to the initial roughness and the treated type of material. Furthermore the level of edge rounding can be adjusted by a suitable selection of the process parameters and reaches from simple deburring up to reinforcing the edge (see fig. 2).

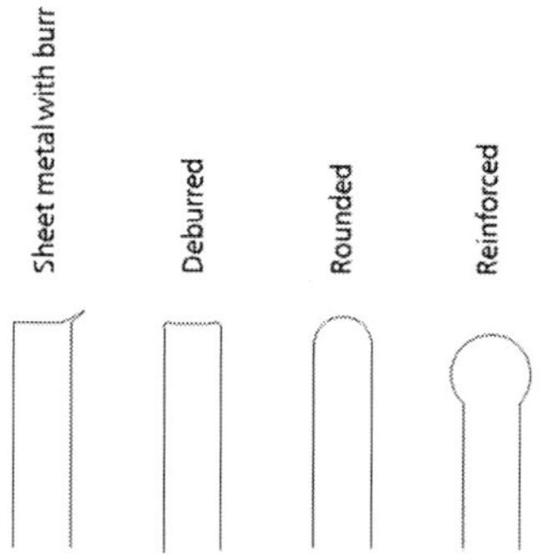

Fig. 2. Possible shapes of the sheet metal edges.

Laser remelting is carried out under inert gas atmosphere to avoid oxidation.

3 Experimental setup & approach

To investigate the influence of the laser beam shape and size on the edge rounding, an experimental setup is developed with which different beam geometries can be realized quickly and cost-effectively (fig. 3). At first, a rectangular top-hat intermediate focus is generated by a micro lens array. The micro lens array rearranges the collimated circular Gaussian beam into a rectangular top-hat beam. An adjustable aperture in the following intermediate focus then crops the beam into the desired shape.

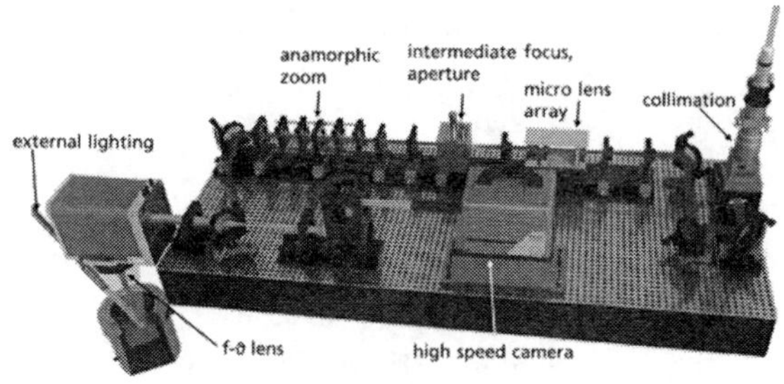

Fig. 3. Experimental Setup

This approach allows a flexible and cost efficient realization of different shaped laser beams, because the manufacturing of different apertures is very quick and cheap. The cut intermediate focus is then projected onto the surface of the sheet metal via a complex optical system, a laser scanner and an f-ϑ lens. For the variation of the aspect ratio of the beam an anamorphic zoom telescope is integrated. The used laser source is a disc laser from Trumpf with a maximum output laser power of P_L=1 kW.

All the investigations are performed on sheet metals of the austenitic stainless steel 1.4301 with an edge thickness of 1 mm (fig. 4, right).

The approach for the investigation of the cause-and-effect relationship between the shape of the laser beam, the process parameters and the resulting rounding of the sheet metal edge is as follows:

- Sheet metal edges are laser remelted under variation of the shape and size of the laser beam, the laser intensity I and the feed speed v_{feed}. The varied process parameters are listed in table 1, the tested shapes of the laser beam are shown in figure 4, left.

Table 1. Varied process parameters.

Feed Speed v_{feed} [mm/s]	Intensity I [W/mm²]
25, 50, 100	500-2200

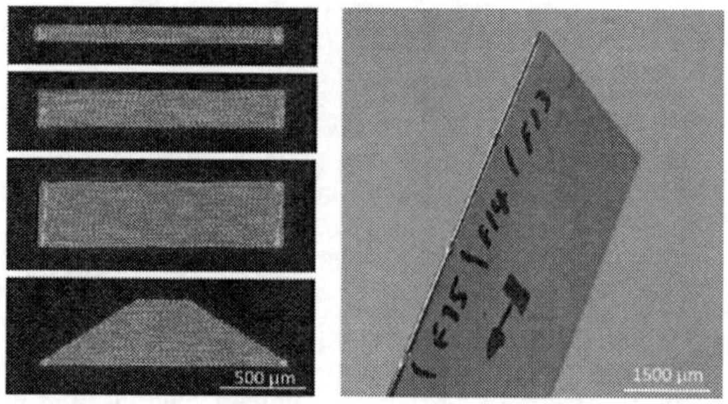

Fig. 4. Left: Overview of the tested laser beam shapes, right: Sample with rounded edges using different process parameters.

The naming of the shapes is as follows: A line shaped laser beam with a length of 1500 µm and a width in scanning direction of 100 µm is named "line 1500x100 µm²".

To make statistical statements, each parameter combination is tested three times.

- The laser remelted sheet metal edges are analyzed concerning the achieved rounding with a white light interferometer. By shooting at different heights, the rounded edge can be precisely displayed with the white light interferometer (transversal resolution: 0.1 nm, lateral resolution: ~100 nm), so that the radius can be determined.

- To get an impression of the surface quality after laser remelting, light microscopy photos of the edges are taken (fig. 5, right).

- To measure the depth of remelting, cross sections are prepared (fig. 5, left).

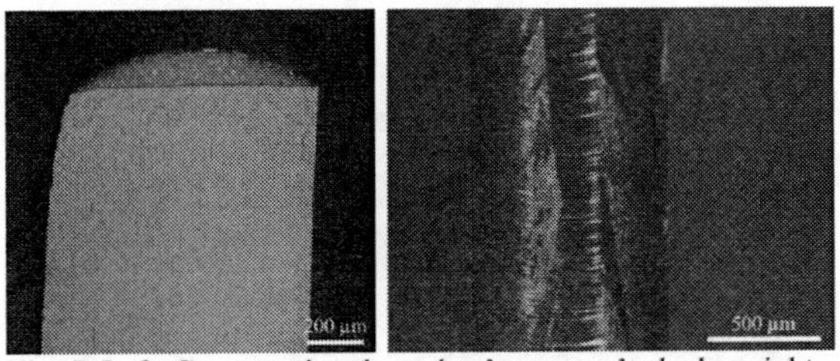

Fig. 5. Left: Cross section through a laser remelted edge, right: light microscopy of the top of the remelted edge (process parameters: line 1500x100 µm², v_{feed}=25 mm/s, I=2000 W/mm²).

4 Results

In a first step the influence of feed speed v_{feed} and intensity I on the resulting radius of the remelted edges are analysed for the line shaped laser beam of 1500x100 µm² (fig. 6).

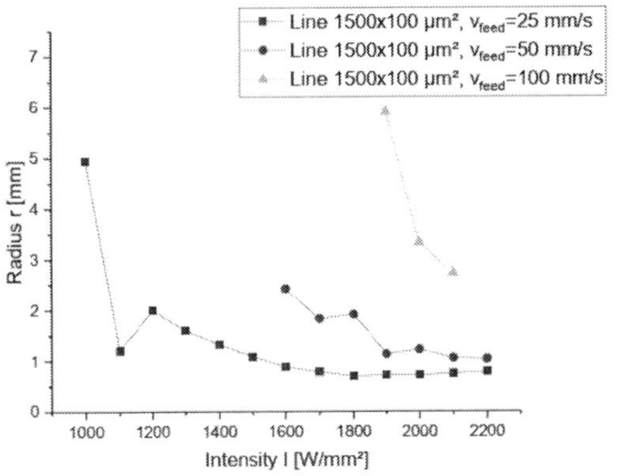

Fig. 6. Average radius r of the laser remelted edge at different feed speeds v_{feed} and intensities I for the line shaped laser beam 1500x100 µm².

The reproducibility of the results is very high, the standard deviation is between 0.01 and 0.04 mm. Due to their small sizes the error bars in the graphs are not visible. The achievable radius increases for a faster scanning. A reason for this behavior could be the smaller remelting depth for faster feed speeds and same intensities, because the shape of the molten pool differs depending on the feed speed (fig. 7).

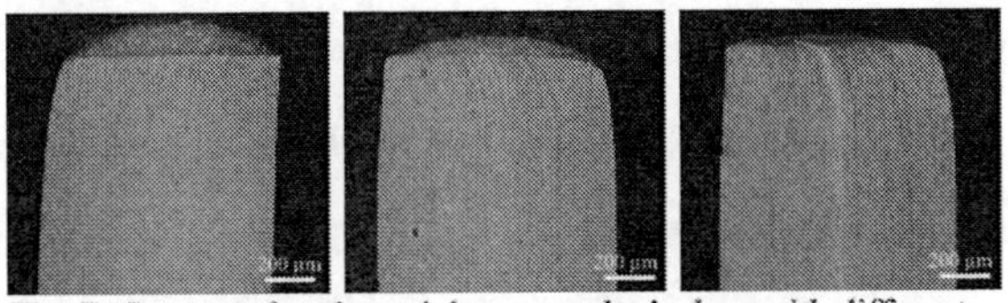

Fig. 7. Cross section through laser remelted edges with different feed speeds v_{feed}=25 mm/s (left), v_{feed}=50 mm/s (middle), v_{feed}=100 mm/s (right) (process parameters: line 1500x100 µm, I=2000 W/mm²).

For the considered laser beam geometry and feed speed, it is not possible to increase the radius further by reducing the intensity, as the process becomes unstable and the sheet metal edge is no longer remelted completely.

For v_{feed}=25 mm/s the influence of the line width on the rounding of the laser remelted edge has been analysed (fig. 8). For this the length of the line shaped laser beam has been kept constant at 1500 µm, only the width in scanning direction has been varied from 100 µm to 500 µm.

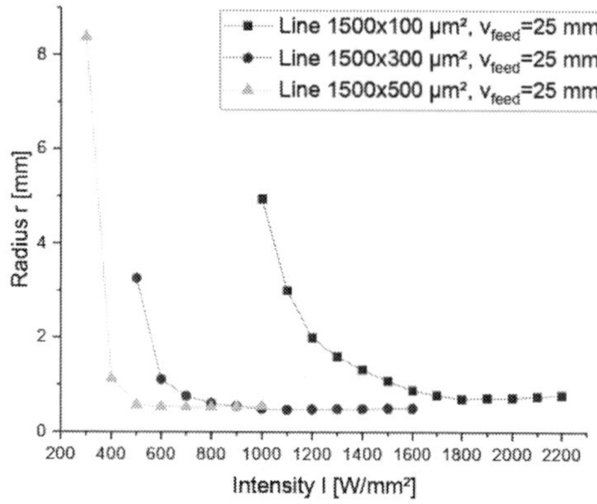

Fig. 8. Average radius r of the laser remelted edge for different line widths and intensities I at a (constant) feed speed of v_{feed}=25 mm/s.

It seems that the minimal achievable radius is independent of the laser beam shape. For higher intensities each graph becomes saturated with r≈0.55 mm. At a sheet thickness of 1 mm, this corresponds to half the sheet thickness and thus to a spherical bulge of the edge. Accordingly, it is not possible to set a rain force by laser remelting for the investigated process parameters.

Outside the saturated area of the curves the following relationship applies: At same intensities larger radii are possible the smaller the line width is. This correlation also leads to the assumption that a smaller remelting depth leads to a larger radius of the remelted sheet metal edge. For this reason, the remelting depth has been determined for various process parameters that lead to different radii (fig. 9).

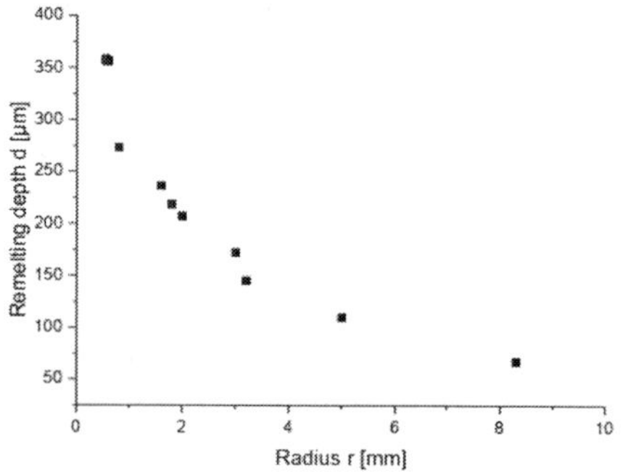

Fig. 9. Remelting depth d in dependency of the radius r of the laser remelted sheet metal edge for different process parameters.

Now the question arises what the maximum possible radius that can be achieved by laser remelting is. It is assumed that the main influencing factors on the rounding of the thin sheet metal edge are the surface tension and mass conservation. To achieve smaller bulges in the middle of the edge and thus a bigger radius of the rounded edge, a smaller remelting depth at the edge region is necessary due to the conservation of mass.

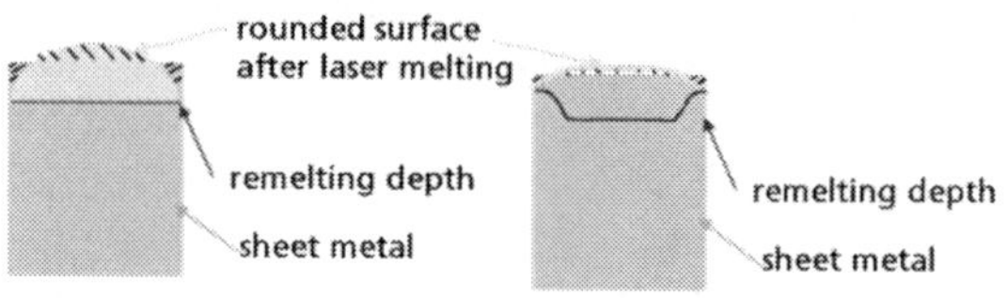

Fig. 10. Assumed schematics of the mechanisms of the bulge formation during laser remelting.

Therefore, in the next step a laser beam with reduced energy in the edge region is formed by using apertures (fig. 4, left, bottom). The results of edge rounding using the cut line are compared with those obtained with the line (fig. 10). Both laser beams have the same length with 1500 µm and also the same width of 500 µm in scan direction.

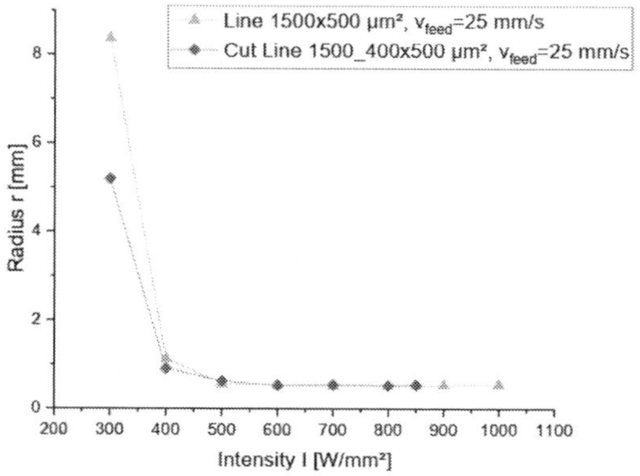

Fig. 11. Average radius r of the laser remelted edge for different shaped lines and different intensities I for the feed speeds v_{feed}=25 mm/s.

It seems that the assumption of reducing the intensity in the edge regions leads to a smaller bulge, e.g. a bigger radius is not true. For small intensities the regular line shape leads to a bigger radius than the cut line.

5 First industrial implementations

Laser remelting of push belts for the torque transmission in CVTs lead to an increase of the fatigue strength by up to 200% in comparison to the tumbling method used previously [2, 3]. This enables a higher torque. The pushbelt (comprising around 400 elements and 18 rings) transmits the torque between two pairs of conical pulleys. By changing the distance between the pulleys the transmission ratio can be changed continuously.

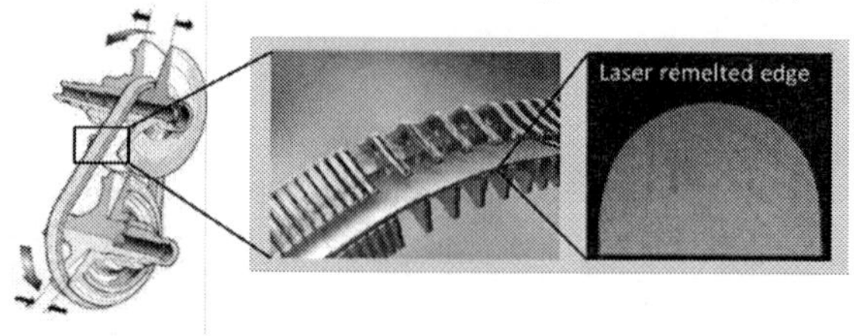

Fig. 12. Laser remelted push belts.

6 Summary & Outlook

On the austenitic stainless steel 1.4301 with an edge thickness of 1 mm basic correlations between process parameters and the resulting radius of the remelted sheet edge have been investigated. The most important relations are:

- Radius r increases with decreased laser power/intensity I and remelting depth d up to the point at which the process gets unstable
- Radius r increases for smaller line widths
- Radius r increases for faster feed speed v_{feed}.

Above a certain intensity, which depends on the line width and the feed speed, it is not possible to further reduce the radius of the remelted edge. Regardless of the line width in feed direction, the minimum achievable radius is constant with r≈0.55 mm.

The maximum radius that can be achieved is limited by the instability of the remelting process due to small intensities or remelting depths (r≈8 mm). It seems that larger radii are possible the smaller the line width in feed direction. Therefore one solution to further increase the radius of the remelted edge is to decrease the line width in feed direction. For the realization of smaller line widths either a new optic with significantly reduced power losses is required, or apertures have to be used which crop the beam even more. However the smaller the line width should be, the more laser power has to be cut out in the intermediate focus. In this case a laser source with a higher maximum output power is required.

The remelting of the sheet edges shows a high reproducibility. The investigations were carried out three times. The standard deviation on the resulting radii is always between 0.01 and 0.04 mm.

References

1. T. Schmidt-Sandte, „Laserstrahlbasierte Entgratverfahren für feinwerktechnische Anwendungen“, Dissertation, Technische Universität Braunschweig, (2003)
2. Presentation of T. Kiedrowski, BOSCH, at LaP2014 in Aachen, Germany
3. Presentation of J. Vermeer, BOSCH, at AKL14 in Aachen, Germany
4. H. Brüning, F. Vollertsen, “Energy efficiency in laser rod end melting”, LIM 2015
5. H. Brüning, F. Vollertsen, “Surface accuracy achieved by upsetting of preforms generated by laser rod end melting”, MATEC Web of Conferences 21, 09006 (2015)

Incremental magnetic pulse welding of dissimilar sheet metals

Verena Psyk[*], *Maik* Linnemann, *Christian* Scheffler, *Petr* Kurka, and *Dirk* Landgrebe

Fraunhofer Institute for Machine Tools and Forming Technology IWU, Reichenhainer Strasse 88, 09112 Chemnitz, Germany

Abstract. Magnetic pulse welding is a solid state welding process using pulsed magnetic fields resulting from a sudden discharge of a capacitor battery through a tool coil in order to cause a high-speed collision of two metallic components, thus producing an impact-welded joint. The joint is formed at room temperature. Consequently, temperature-induced problems are avoided and this technology enables the use of material combinations, which are usually considered to be non-weldable. The extension of the typically linear weld seam can easily reach several hundred millimetres in length, but only a few millimetres in width. If a larger connected area is required, incremental or sequential magnetic pulse welding is a promising alternative. Here, the inductor is moved relative to the joining partners after the first weld sequence and then another welding process is initiated. Thus, the welded area is extended gradually by arranging multiple adjacent weld seams. This paper demonstrates the feasibility of incremental magnetic pulse welding. Furthermore, the influence of important process parameters on the component quality is investigated and evaluated in terms of geometry and micrographic analysis. Moreover, the suitability of different mechanical testing methods is discussed for determining the strength of the individual weld seams.

Keywords: Manufacturing process, Welding, Magnetic pulse welding

1 Motivation and principle of incremental magnetic pulse welding

Industrial manufacturing is more and more influenced by social, ecological, and health aspects. Politics supports this trend by establishing corresponding national and international laws and agreements such as the 2020 Climate & Energy Package [1] or the Agenda 2030 [2]. Saving energy and resources as well as reducing emissions are essential aspects of this development. Regarding product design and manufacturing, this specifically means that one major focus has to be placed on reducing weight by consistently implementing lightweight design concepts e.g. multi-material design [3].

When choosing product materials, technological properties and weight reduction potential, but also availability and costs of specific materials are decisive. Furthermore, it is essential to consider the readiness of cost-efficient joining technologies suitable for the specific material or material combination. Conventional joining technologies, i.e. usually thermal ones such as gas metal arc welding or resistance spot welding often reach their limits when it comes to multi-material combinations. [4]

Magnetic pulse welding (sometimes also called electromagnetic pulse welding) is a technology offering a high potential, especially for joining dissimilar materials, including material combinations which are usually considered to be non-weldable. Feasibility has been proven for material combinations aimed at e.g. car body manufacturing such as aluminium and steel [5, 6], including stainless steel [7] and press hardening steel (see Fig. 1). In addition, feasibility was verified for material combinations which are of special interest for applications related to electrical engineering or heat transfer and conduction such as aluminium and copper [8]. Fig. 1 shows examples of joints produced by magnetic pulse welding of similar and dissimilar sheet metals at Fraunhofer IWU.

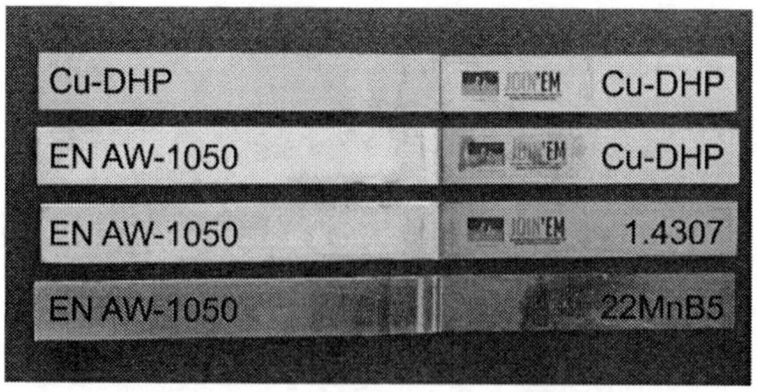

Fig. 1. Joints produced by magnetic pulse welding of similar and dissimilar materials

[*] Corresponding author: verena.psyk@iwu.fraunhofer.de

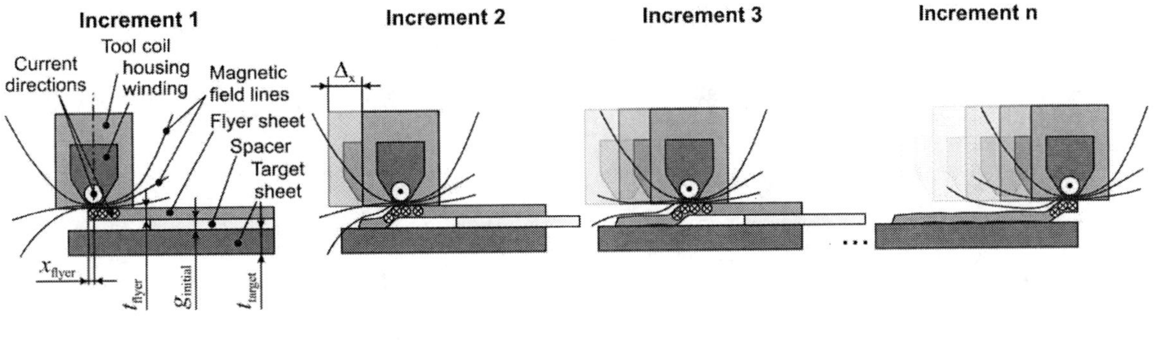

Fig. 2. Principle of incremental magnetic pulse welding

The process of magnetic pulse welding was initially suggested and patented by Lysenko et al. in 1970 [9]. This technology is an impact-based welding technique which features some similarities with the well-known explosive welding or cladding techniques. However, magnetic pulse welding is significantly less critical with regard to safety, which makes it much easier to implement it into an industrial environment.

In order to perform the process, the flyer – i.e. the accelerated joining partner – and the target – i.e. the static joining partner – are positioned with a defined initial gap between them and the flyer is accelerated by electromagnetic forming [10]. This means that a capacitor battery is charged to an application-adapted charging energy and discharged through a so-called tool coil or inductor. As a consequence, damped sinusoidal current flows through the tool coil and induces an according pulsed magnetic field, which in turn induces a time-dependant current in the workpiece – i.e. in the flyer sheet – which is directed opposed to the coil current. Lorentz forces occur due to the interactions of the currents and the magnetic field, which accelerate the flyer to velocities of up to several hundred meters per second within a few microseconds. The flyer movement is directed away from the tool coil and towards the target. After overcoming the initial gap, flyer and target are involved in a high-speed collision. If the collision parameters (impact velocity and impact angle) lie within a process window, which is specific for the material combination, an impact welded joint is generated [11]. In contrast to classic welding techniques, magnetic pulse welding forms the joint at room temperature without significant heating of the parts, thus avoiding temperature-induced problems such as thermal softening or the formation of continuous intermetallic or oxidic phases deteriorating the weld quality.

Due to the setups typically used for the process, the weld seam is usually of linear shape either along the circumference of tubular components, or along the edges of joint sheet metal parts. The extension of such weld seams can easily reach several hundred millimetres in length, but their width is limited to a few millimetres.

However, in some applications it is necessary to have a larger connected area of the joining partners, e.g. in order to achieve better electrical or heat conducting properties. In these cases incremental magnetic pulse welding is a promising alternative. In this process variant the tool coil is moved relatively to the joining partners after the first magnetic pulse welding sequence and then another welding process is initiated (see Fig. 2). In doing so, the extension of the joint area can be extended gradually by arranging multiple adjacent weld seams. Feasibility of the principle extension of an electromagnetic forming process by applying an incremental or sequential approach has been proven for a structuring process in [12], for a simple 3D-shaping process in [13] and for a more complex 3D-shaping process of large sheet metals in [14]. For tube welding processes, first investigations were carried out regarding incremental magnetic pulse welding in [15], but up to now, the transfer to magnetic pulse welding of sheet metal has not been studied.

2 Procedure of analysing incremental magnetic pulse welding

2.1 Reference of single sequence magnetic pulse welding

In this analysis the reference used was a conventional magnetic pulse welded joint consisting of one single weld sequence. In particular, in this reference a one millimetre thick flyer sheet made of EN AW-1050 was welded to a two millimetre thick target sheet made of Cu-DHP. Suitable process parameters guaranteeing high weld quality were chosen based on the results of a detailed study quantifying the influence of important adjustable process parameters on quality criteria of the resulting weld such as transferable force, electrical and thermal conductivity, and width of the weld seam [11]. Specifically, the following values were chosen:

- Capacitor charging energy E=30 kJ
- Initial gap between flyer and target $g_{initial}$=1 mm (see Fig. 2)
- Relative position of tool coil and flyer x_{flyer}=2 mm (see Fig. 2)

2.2 Parameters of sequential welding

In the sequential welding tests, for the first weld sequence the same process parameters were applied as in the single sequence reference. Then further weld sequences were added using the same capacitor charging energy. This paper specifically investigates and evaluates the influences of different numbers of

https://doi.org/10.1051/matecconf/201819002004

sequences *n*, on the one hand, and different relative movements of coil and joining partners Δx, on the other hand, on the component quality in terms of geometry, micrographic analysis, and mechanical strength.

Parts with n=2 sequences, n=3 sequences, and n=5 sequences were examined. The considered relative movements of Δx=9 mm, Δx=13.5 mm, and Δx=18 mm were selected, taking into consideration the cross section geometry of a typical single sequence weld. Fig. 3 shows that for these kinds of welds, the contact area of the two sheets is approximately 9 mm followed by the slope geometry of the flyer, which extends approximately for another 9 mm. This means that the gap between flyer and target is similar to the initial gap width between flyer and target in the first weld sequence at a distance of 18 mm to the flyer edge.

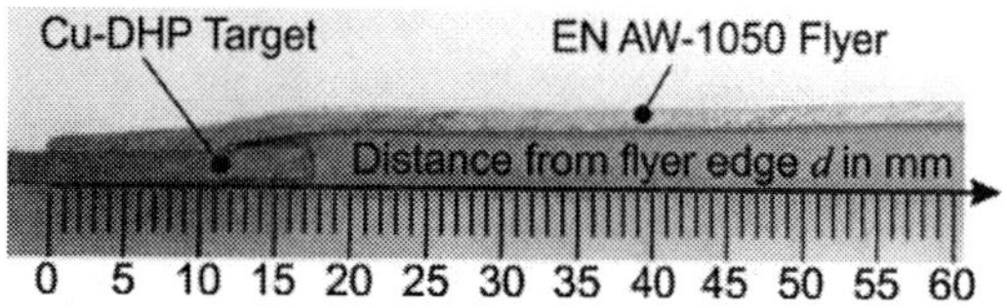

Fig. 3. Cross section of a typical joint produced by magnetic pulse welding

3 Characterisation and evaluation of the joining result

3.1 Thickness of the welded specimens

A significantly simplified approach for estimating the overall thickness of the welded part is adding the nominal thicknesses of the joining partners. In the example regarded here, this leads to a value of 3 mm for the total width. However, it is well known that both – flyer and target – can undergo significant changes in thickness due to the impact and the corresponding severe plastic deformation of the joining partners, which seems to be an indispensable prerequisite for magnetic pulse welding. This in turn influences the thickness of the part produced by magnetic pulse welding [8]. The deformation and the resulting thickness of flyer and target significantly depend on density and strength ratio of the joining partners. As exemplarily shown in Fig. 4, the impact of a copper flyer onto an aluminium target causes significantly more deformation of the target than the impact of an aluminium flyer onto a copper target.

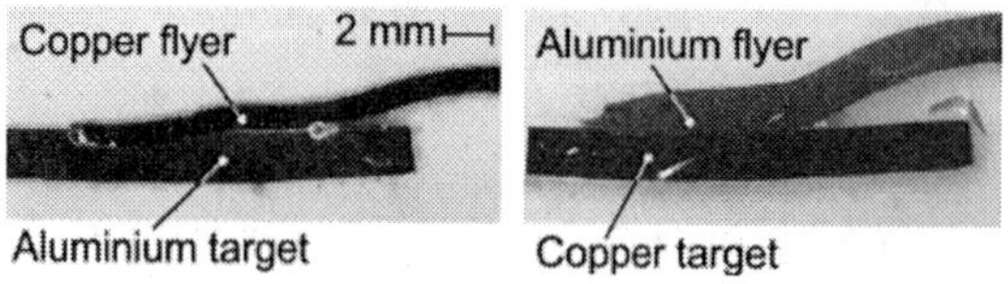

Fig. 4. Resulting cross sections of magnetic pulse welding of a copper flyer to an aluminium target (left) and vice versa (right)

Considering e.g. the application of the process in manufacturing of components related to heat transfer, essential aspects include the contact properties. Therefore, the flatness of the contact surfaces used to introduce the heat, which is closely related to the thickness distribution is important. Thus, the thickness distributions have been examined regarding different sheets produced by sequential magnetic pulse welding and a reference specimen manufactured by single sequence magnetic pulse welding. As shown in Fig. 5, each single weld sequence can easily be recognized in the thickness plots, as it causes a local thickness minimum. Consequently, at least one surface of the resulting part produced by sequential welding features more or less distinctive waviness. This waviness is augmented by an increase of the coil movement relative to the position of the joining partners in-between two subsequent joining sequences.

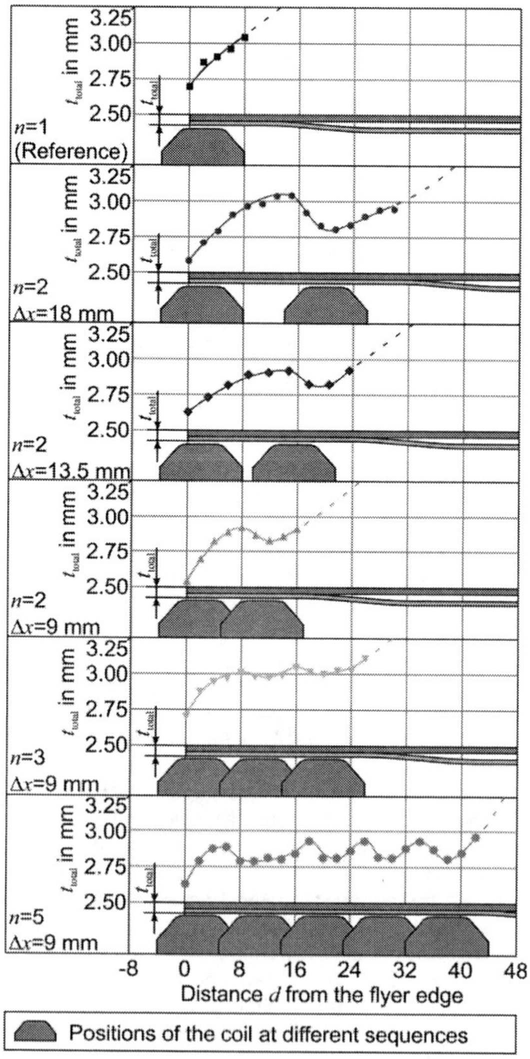

Fig. 5. Coil positions and corresponding thickness distributions of specimens produced by single sequence and incremental magnetic pulse welding

3.2 Microstructural investigations of the welded specimens

Another important evaluation criterion for joints produced by magnetic pulse welding is the extension of the welded area, which is directly related to the conductivity of the weld and the force, which can be transferred via the weld [11]. Obviously, it is desirable to maximise the extension of the welded area. Therefore, influences of the different parameters on the welded areas in sequential magnetic pulse welding were analysed. Welded areas are frequently characterised by a more or less distinctive wavy structure of the interface between the two joining partners. Sometimes they even feature vortexes, while non-welded sections feature a gap or a shattered structure (see Fig. 6).

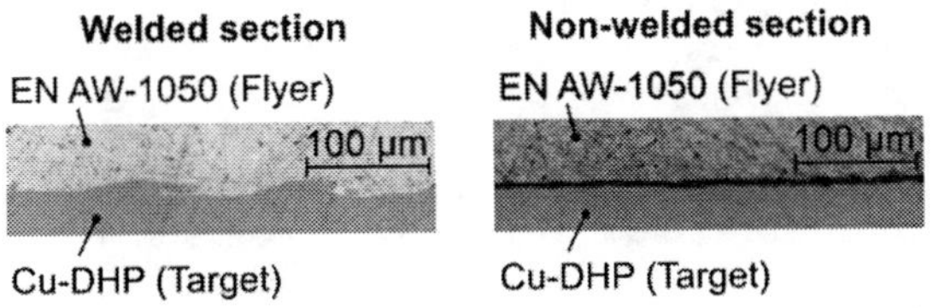

Fig. 6. Exemplary micrographs of welded and non-welded sections

As shown in Fig. 7, all specimens produced by incremental magnetic pulse welding feature multiple welded areas. As expected, the number of welded areas and the total length of the welded area l_{total}, i.e. the sum of the individual lengths of all welded areas, rise with an increasing number of sequences. The total weld lengths indicated in Fig. 7 are mean values calculated from all experiments performed with the parameter set, while the local distributions of the weld extensions represent one specific test case and consequently may slightly deviate from the mean value.

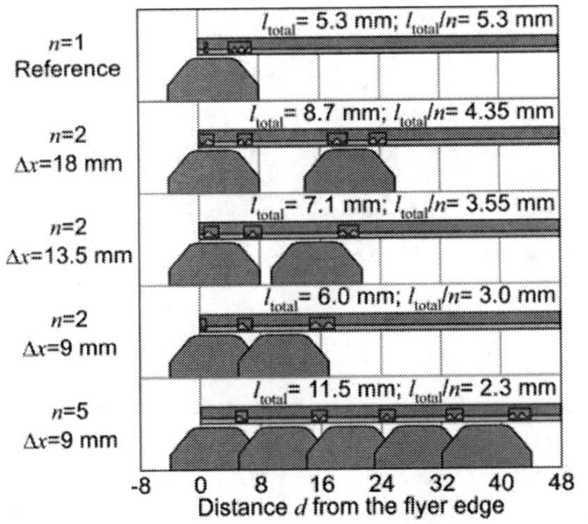

Fig. 7. Coil positions and corresponding welded areas of specimens produced by sequential magnetic pulse welding

The direct comparison with the positions of the coil during the different weld sequences allows attributing the individual welded sections to the different weld sequences. It is most likely that all welding areas at a distance $d \leq 10$ mm from the flyer edge can be attributed to the first weld sequence. The differences in the welding area for this sequence can probably be attributed to slight inaccuracies in the manual positioning of the specimens in relation to the tool coil and in the flatness of the parts resulting in slight deviations of x_{flyer} and $g_{initial}$.

For the specimens welded with two sequences all further welded areas can be attributed to the second weld sequence. For the specimen with five welding sequences, the weld area at a distance of approx. 25 mm from the flyer edge can be attributed to the third welding sequence, while the welded areas at a distance of $d \approx 34$ mm and $d \approx 42$ mm can be attributed to the fourth and fifth welding sequence, respectively.

This shows that in the ideal case, two welded areas at both sides of the coil result from one single welding sequence. This is the case for the first welding sequence in nearly all specimens. In contrast, most of the further welding sequences lead to only one single welded area. The reason for this probably lies in the fact that there is only a very small distance between flyer and target in the area close to the preceding welding sequences. This means that especially for processes with small relative movement of coil and joining partners in-between two subsequent joining sequences, the available acceleration distance for the flyer is very small at the side facing the preceding welding sequence. Therefore, the collision velocity necessary for magnetic pulse welding cannot be reached as it is in the range of 300 m/s for this specific material combination, [8]. If the relative movement of tool coil and workpiece is sufficiently long so that the initial distance between flyer and target is approximately the same as in the first welding sequence, the acceleration is more effective. Consequently, the collision conditions are more beneficial so that two welding areas result again from one welding sequence. In the considered examples this is the case for a relative movement of tool coil and joining partners of Δx=18 mm. These correlations explain to some extent the tendency of an increasing total weld length per welding sequence l_{total}/n with increasing relative movement Δx, which is illustrated in Fig. 7. Here mean values are represented which were calculated from all experiments performed with the specific parameter set.

3.3 Discussion of mechanical characterisation methods for specimens produced by sequential magnetic pulse welding

3.3.1 Lap shear test

The lap shear test is a frequently applied mechanical test for characterising the mechanical quality of joints produced by single sequence magnetic pulse welding, see e.g. [5, 7, 16, 17, 18]. For this purpose, either simple rectangular or waisted specimens are used, similar to those used in a conventional tensile test, but featuring the joint produced by magnetic pulse welding in the centre of the specimen, and the test is carried out in a standard testing machine.

As already shown e.g. in [11] for single sequence welds, the strength of a joint produced by magnetic pulse

https://doi.org/10.1051/matecconf/201819002004

welding can exceed that of the base material, causing failure in the base material and not in the joint. Only low quality welds fail in the weld area, so that the failure type can be regarded as an evaluation criterion for a high-quality joint. In Fig. 8 the difference is shown based on a copper aluminium joint failing in the joint and in the flyer base material, respectively. In this sample the flyer is the weaker base material. The reason for this is that the copper sheet features remarkably smaller thickness compared to the aluminium base material. Therefore, the aluminium sheet can transfer higher forces although its material strength is lower compared to the copper material.

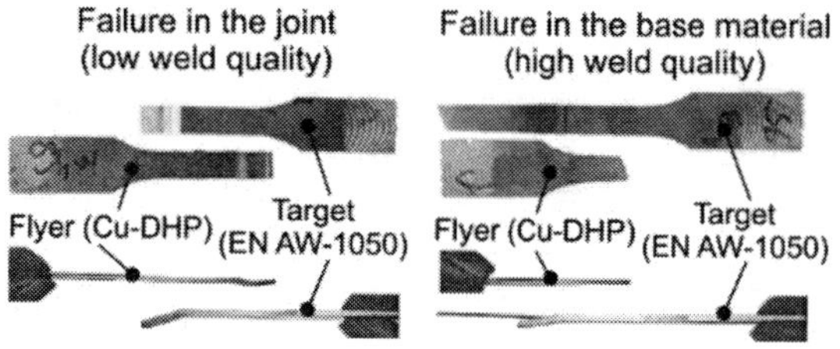

Fig. 8. Failure cases in lap shear tests of joints produced by single sequence magnetic pulse welding

However, the lap shear test can only provide an overall evaluation of the complete welding zone. Thus, the informative value is expected to be limited with regard to testing of parts produced by incremental magnetic pulse welding. Differentiated conclusions regarding the contribution of the individual weld sequences to the overall transferable force will not be possible because in this test setup the complete welded area is always loaded, since the forces are applied at the two opposing ends of the weld zone. Thus, even if only one weld sequence is of high quality, failure will occur in the base material, and minor quality of the other welds cannot be detected at all.

3.3.2 Peel test

In contrast to the lap-shear test, forces are applied perpendicular to the welded zone at only one end of the welding zone in the peel test. Theoretically, this should allow testing one weld sequence after the other and drawing conclusions regarding the quality of each individual weld seam. Therefore, the end of welded sheets were bent 90° and peel tests were performed on the prepared specimens in a conventional testing machine (see Fig. 9).

However, the regarded welded sheets feature significantly different material strength and stiffness. The bending stiffness of the copper sheets is approx. 1.8 to 2.8 times as high as that of the aluminium sheet due to the higher Young's modulus and thickness in this joining example. As shown above, the thickness of the joining zone is remarkably higher than that of the base materials, which also implies that the stiffness is even higher here. At the same time the quasistatic yield strength of the copper target is approx. 1.8 times as high and the quasistatic tensile strength is even 2.3 times as high as the corresponding values of the aluminium flyer. As a consequence, no remarkable elastic or plastic deformations occur in the copper base material or in the deformation zone, while the aluminium base material is strongly bent. Finally, bending failure occurs in the aluminium base material close to the welding zone. Some qualitative deformation stages of the testing process are shown in Fig. 9. They clearly show that this kind of test is not suitable for joints produced by magnetic pulse welding – neither single sequence nor multiple sequence welds – of different materials or different sheet thicknesses. However, the test might lead to meaningful results in cases of welds with similar sheet metal parts.

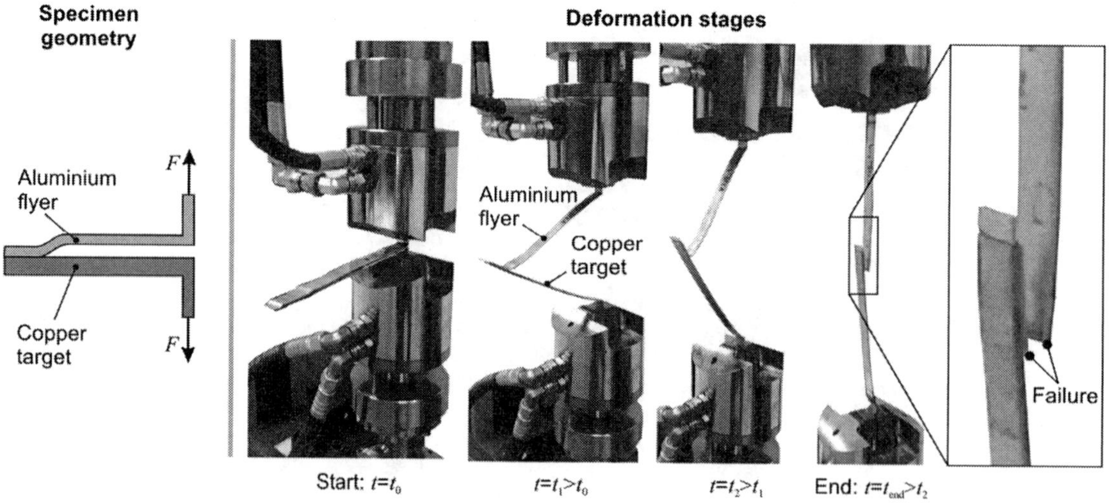

Fig. 9. Specimen geometry and deformation stages in peel tests

3.3.3 Chisel test

In spot welding, in addition to the peel test the so-called chisel test is another established way for characterising the joint quality by applying the test force on one side of the joint [19]. Thus, a similar test was tried here. Fig. 10 shows setup, specimen geometry, representative deformation stages and a typical tested specimen. Tests were carried out on six reference parts produced by single sequence welding and six sequentially welded specimens.

These tests used only specimens with a number of sequences of n=3 and a relative movement of Δx=9 mm. As known from the investigations of the microstructure, these specimens provide relatively short weld lengths; consequently the expected transferable force is relatively low. This means that choosing this parameter set will allow a worst-case estimation of the weld strength. During the tests, the force and the displacement of the moving chisel were recorded, and the failure type was analysed after the test.

Similar to the lap shear tests, also for these tests occurring failure cases can be differentiated between failure in the joint (i.e. detachment of the joint) and failure in the weaker base material (i.e. the aluminium flyer material). Fig. 10 shows an exemplary sequentially welded specimen featuring both failure types for different weld sequences. In case of failure in the joint, the two sheets are separated without any major damage. Only slight particle residues from the aluminium flyer can be detected on the copper sheet. In case of failure in the base material, the adherence of the two dissimilar material sheets to each other is so high that no separation of the materials occurs. Instead, shear failure of the aluminium base material occurs, although the chisel used in these tests does not feature a sharp, but a rounded edge with a radius of 1 mm.

In the reference case using welding with one single magnetic pulse welding sequence, all tested specimens failed in the aluminium base material. This proves that the set of process parameters resulting in high quality welds in terms of lap shear strength, electrical conductivity of the joint, and length of the welded area [11] also leads to high quality in terms of the acting loads during the chisel test.

In case of the specimens produced by incremental magnetic pulse welding it is important to consider that the weld sequences were tested in reversed order due to the setup, specifically the orientation of the specimen. This means that the welding sequence that was produced first was the last one to be tested, while the welding sequence that was produced last was the first one to be tested. In most cases the welding sequence tested first already failed in the base material. In this case further testing of the other welding was not possible any longer, because no suitable force application area could be created for the chisel in order to test the further sequences. For two specimens, the first and second tested welding sequences failed in the joint. However, here the welding sequence that was tested last (i.e. the sequence, which was welded first) failed in the base material. The thickness measurement of these specific specimens has shown that the thickness values in the area of the weld sequences failing in the joint are slightly (i.e. up to two tenths of a millimetre) higher compared to most of the other specimens. This suggests that shortcomings in terms of weld quality can also be detected via the thickness distribution.

Fig. 11 shows three distinctive areas featured by force-displacement measurements during the chisel test of the specimens produced by single sequence magnetic pulse welding. An initial slight increase of the force up to approx. 200-400 N is followed by a much steeper rise up to approx. 2500 N, before a similarly steep drop of the force occurs. This principle shape corresponds to the deformation of the specimen. This deformation is characterised by elastic-plastic bending of the free flyer end at the beginning of the test. The increasing slope of the curve characterises the transition from plastic bending to shearing. Finally, necking of the aluminium flyer causes the force to drop again until failure occurs.

As expected, force-displacement measurements taken for specimens produced by incremental magnetic pulse welding with material failure in the weld sequence, which were tested first, have the same principle shape. The forces measured during elastic-plastic bending are on a similar level as the reference specimen, but the force achieved during shearing of the aluminium is approx. 30% lower than in case of the single sequence welding.

For specimens featuring base material failure in the first weld sequences and joint failure in the other weld sequences, distinctive drops of the force values can be detected in the measurement curve. These can be

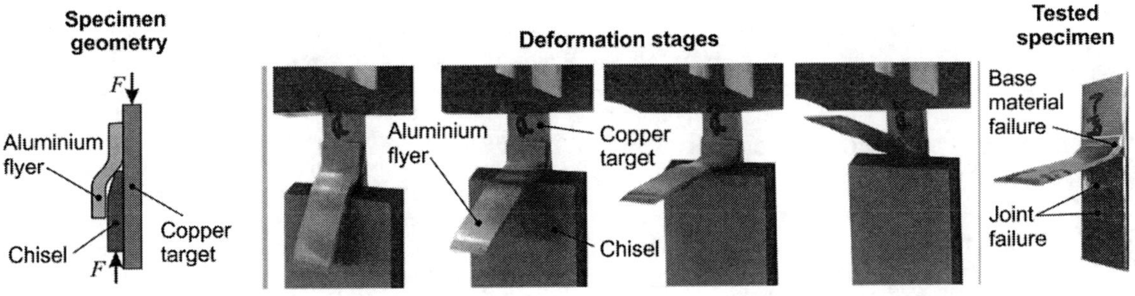

Fig. 10. Specimen geometry and deformation stages in chisel tests

attributed to the detachment of the weld. However, the force peaks related to failure in the joint do not exceed values of approx. 400-450 N. As this is one magnitude lower compared to the maximum force values measured during the tests, it can be concluded that such low quality joints do no significantly contribute to the overall strength of the incremental weld. In contrast, the measured forces during shearing of the aluminium base material in the first weld sequence was on the same level as the single sequence weld, proving that the subsequent welding processes do not deteriorate the quality of the earlier sequences.

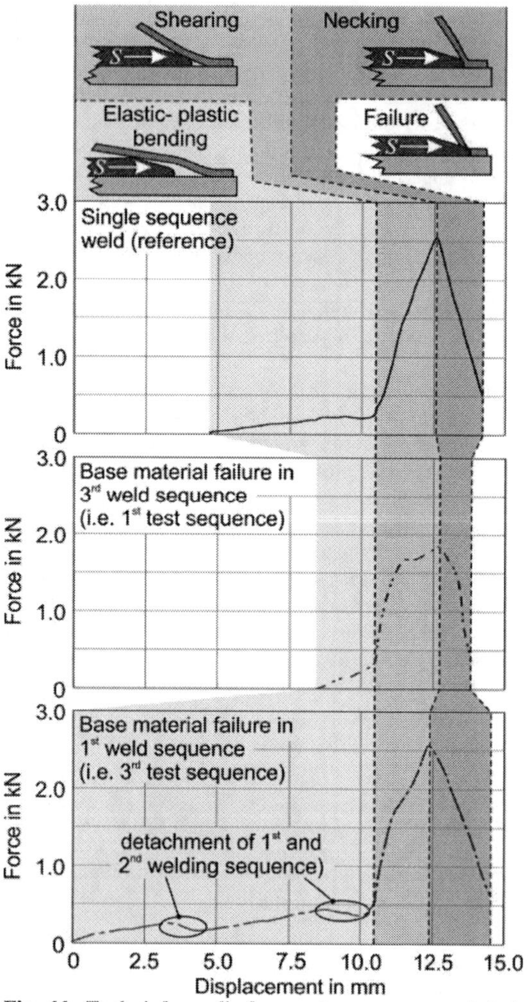

Fig. 11. Typical force displacement curve measured during chisel tests

4 Summary and conclusions regarding the process design

Magnetic pulse welding offers important advantages for manufacturing dissimilar material joints, which cannot be realised by conventional techniques, but the width of the typically linear weld seam is limited to a few millimetres. Larger joint areas can be realised by incremental magnetic pulse welding. Here, multiple weld seams are arranged next to each other by performing several weld sequences and moving tool coil and joining partners relative to each other in-between the sequences.

The feasibility of this new process variant was proven and a basic analysis of the process was carried out for copper aluminium joints. In addition to the typical process parameters known from single sequence magnetic pulse welding, additional parameters significantly influence the resulting joint quality in the incremental approach. These are specifically the relative displacement of tool coil and joining partners between two subsequent sequences and the number of weld sequences necessary to join a defined area.

Measurements of the thickness distribution have shown that decreasing relative movement of the tool coil to the joining partners decreases thickness variations and consequently waviness of the part's surface. However, for short relative movement of tool coil and workpiece the conditions for welding are less favourable since the gap width between flyer and target in direct proximity to the preceding weld seam hardly allows for reaching the flyer velocity required for welding. Therefore, the extension of the weld is only small and due to the direct relation to the corresponding transferable force, the mechanical strength is also low. In extreme cases magnetic pulse welding will not be possible any longer. Thus, a compromise has to be found depending on the requirements related to flatness of the part, on the one hand, and required mechanical strength of the weld, on the other hand.

If high mechanical strength is required, the relative movement must be long enough to guarantee an initial gap width similar to that of a high quality single sequence weld for all weld sequences. In the specific case regarded here, this means a relative movement of at least 18 mm.

If flatness of the part is of major importance as it is the case e.g. in heat transfer elements, the relative movement should be as short as possible. However, this obviously leads to an increase of the number of weld sequences required to connect a specific area, implying higher effort and cost in manufacturing. For the specific case regarded here, a relative movement of Δx=9 mm seems to be a lower boundary. Mechanical load tests for this value have shown that high weld quality characterised by failure in the base material can basically still be achieved for all sequences of a part produced by incremental magnetic pulse welding. However, the transferable force of the subsequent weld sequences and the robustness of the process tend to be lower compared to single sequence magnetic pulse welding. On the other hand, variations of the thickness are relatively low.

If both high strength and good flatness of the part is required, a two-stage process might be a solution. In this case incremental magnetic pulse welding must be carried out with long relative movement of the coil and the joining partners in order to optimise the weld quality of the individual weld sequences, while accepting that the flatness of the part has to be improved in a second manufacturing step. The improvement might be achieved by levelling buckles via electromagnetic forming steps in-between the individual weld sequences. For this operation the same tool coil already applied for the magnetic pulse welding process might be used again, so that no additional equipment is required. However, this procedure will probably not lead to an ideally flat surface. Therefore, milling of the welded surface can be expected to be the best option for flattening it, if geometrical requirements are very high.

The presented results were achieved within the project JOINing of copper to aluminium by ElectroMagnetic fields – "JOIN'EM". This project is funded by the European Union within the framework of the Horizon 2020 research and innovation program under Grant Agreement No. 677660.

References

1. https://ec.europa.eu/clima/policies/strategies/2020_en#tab-0-1 (30.01.2018)
2. United Nations (Editor), Transforming our world: The 2030 agenda for sustainable development, A/RES/70/1, online available https://sustainabledevelopment.un.org/content/documents/21252030%20Agenda%20for%20Sustainable%20Development%20web.pdf (30.01.2018).
3. M. Goede, M. Stehlin, L. Rafflenbeul, G. Kopp, E. Beeh, Super Light Car—lightweight construction thanks to a multi-material design and function integration. Eur. Transp. Res. Rev. **1**:5–10 (2009).
4. A. C. Serrenho, J. B. Norman, J. M. Allwood, The impact of reducing car weight on global emissions: the future fleet in Great Britain. Phil. Trans. R. Soc. A **375**: 20160364 (2017).
5. T. Aizawa, M. Kashani, K. Okagawa, Application of magnetic pulse welding for aluminium alloys and SPCC steel sheet joints. Welding Journal, **86**:119-124, 2007.
6. V. Psyk, T. Lieber, P. Kurka, W.-G. Drossel: Electromagnetic joining of hybrid tubes for hydroforming. Procedia CIRP 23 (2014) 1 – 6.
7. S. D. Kore, P. P. Date, S. V. Kulkarni, Electromagnetic impact welding of aluminium to stainless steel sheets. J. Mater. Process. Tech. **208**:486-493, 2008.
8. V. Psyk, C. Scheffler, M. Linnemann, D. Landgrebe, Manufacturing of hybrid aluminum copper joints by electromagnetic pulse welding– an analysis of important process parameters. AIP Conference Proceedings **1896**, 110001 (2017).
9. D. N. Lysenko, V. V. Ermolaev, A. A. Dudin, Method of Pressure Welding, US-patent 3,520,049, 1970.
10. V. Psyk, D. Risch, B. L. Kinsey, A. E. Tekkaya, M. Kleiner, Electromagnetic Forming – A Review. J. Mater. Process. Tech. **211**:787–829 (2011).
11. V. Psyk, C. Scheffler, M. Linnemann, D. Landgrebe, Process analysis for magnetic pulse welding of similar and dissimilar material sheet metal joints. Procedia Engineering **207**:353-358 (2017).
12. V. Psyk, P. Kurka, S. Kimme, M. Werner, D. Landgrebe, A. Ebert, M. Schwarzendahl, Structuring by electromagnetic forming and by forming with an elastomer punch as a tool for component optimisation regarding mechanical stiffness and acoustic performance. Manufacturing Rev. **2**:23, 2015.
13. X. H. Cui, J. H. Mo, J. Li, J. Zhao, Y. Zhu, L. Huang, Z. W. Li und K. Zhong, Electromagnetic Incremental Forming (EMIF): A Novel Alumimum Alloy Sheet and Tube Forming Technology, J. Mater. Process. Tech. **214**:409-427, 2014.
14. M. Linnemann, C. Scheffler, P. Kurka, V. Psyk, D. Landgrebe, Numerische und experimentelle Untersuchung von inkrementellen, elektromagnetischen Umformvorgängen. Zeitschrift für wirtschaftlichen Fabrikbetrieb – ZWF **112**:454-458 (2017).
15. H. Yua, Z. Fan, C. Li; Magnetic pulse cladding of aluminum alloy on mild steel tube; Journal of Materials Processing Technology **214**:141–150 (2014).
16. T. Aizawa, M. Kashani, K. Okagawa, Application of magnetic pulse welding for aluminium alloys and SPCC steel sheet joints. Welding in the world **49**: 212-222 (2005).
17. S.D. Kore, P.P. Date, S.V. Kulkarni, Effect of process parameters on electromagnetic impact welding of aluminum sheets. International Journal of Impact Engineering **34**:1327–1341 (2007).
18. I, Kwee, V, Psyk, K, Faes, Effect of the Welding Parameters on the Structural and Mechanical Properties of Aluminium and Copper Sheet Joints by Electromagnetic Pulse Welding. World Journal of Engineering and Technology, 2016, 4, 538-561.
19. N. n. ISO 10447:2015 Resistance welding – Testing of welds – Peel and chisel testing of resistance spot and projection welds.

Investigation of microstructure and hardness of a rib geometry produced by metal forming and wire-arc additive manufacturing

Markus Hirtler[1,*], *Angelika* Jedynak[1], *Benjamin* Sydow[1], *Alexander* Sviridov[1], and *Markus* Bambach[1]

[1]Brandenburg University of Technology Cottbus-Senftenberg, Chair of Mechanical Design and Manufacturing, 03046 Cottbus, Germany

Abstract Within the scope of consumer-oriented production, individuality and cost-effectiveness are two essential aspects, which can barely be met by traditional manufacturing technologies. Conventional metal forming techniques are suitable for large batch sizes. If variants or individualized components have to be formed, the unit costs rise due to the inevitable tooling costs. For such applications, additive manufacturing (AM) processes, which do not require tooling, are more suitable. Due to the low production rates and limited build space of AM machines, the manufacturing costs are highly dependent on part size and batch size. Hence, a combination of both manufacturing technologies i.e. conventional metal forming and additive manufacturing seems expedient for a number of applications. The current study develops a process chain combining forming and additive manufacturing. First, a semi-finished product is formed with forming tools of reduced complexity and then finished by additive manufacturing. This research investigates the addition of features using AlSi12 created by Wire Arc Additive Manufacturing (WAAM) on formed EN-AW 6082 preforms. By forming, the strength of the material was increased, while this effect was partly reduced by the heat input of the WAAM process.

Keywords: Additive Manufacturing, Metal forming, Microstructure

1 Introduction

The demand for individualized products is a significant driver for improvements of manufacturing processes. The costs for conventional fabrication, e.g. by closed-die forging, rises with increasing complexity and decreasing batch-size [1]. A short cycle time is beneficial for die-based forming processes, making them economical for large-scale production. For the production of customized parts or small batches, the tooling costs are unacceptably high. Additive manufacturing (AM) methods do not rely on dedicated tools, and hence, allow for the production of single parts or small batch sizes irrespective of the complexity. Currently available powder-based AM processes show a low deposition rate of 0.3 kg/h [2] for Selective Laser Melting (SLM) and 3.0 kg/h [3] for Laser Metal Deposition (LMD). A more promising technology is Wire Arc Additive Manufacturing (WAAM) with higher deposition rates of up to 10 kg/h [4], good power efficiency and low investment costs [5].

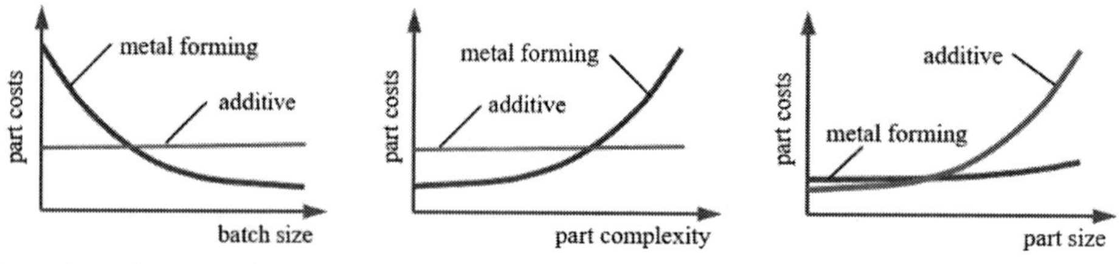

Fig. 1. Comparison of part costs for metal forming and additive processes regarding batch size, part complexity and part size.

* Corresponding author: markus.hirtler@b-tu.de

https://doi.org/10.1051/matecconf/201819002005

However, resolution and surface roughness of SLM are better than WAAM [6]. With both methods, nearly full density can be achieved [7]. The dependence of part costs on part size, part complexity and batch size are illustrated schematically in Fig. 1. Instead of using a single technology for manufacturing the desired part shape, a combination of processes could be beneficial for manufacturing smaller batch sizes or product variants. One reasonable synergy can be obtained by combining wire arc additive manufacturing and metal forming, yielding two possible process routes, which are visualized in Fig. 2. First, an additive manufacturing process, e.g., WAAM can be used to generate a pre-shaped semi-finished part. The semi-finished part will then be formed by using a single forming tool to achieve the final contour.

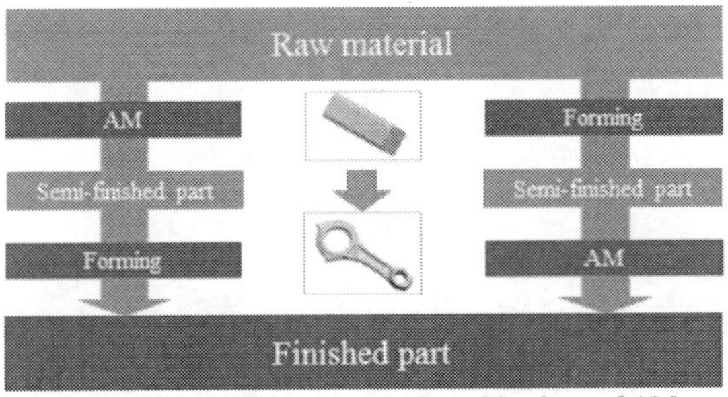

Fig. 2. The two possible routes and combinations of AM and metal forming processes.

Behrens et al. investigated deposited claddings on a shaft which was formed subsequently [8]. The hybrid part was made of a 42CrMo4 bar cladded with the high-alloyed steels 1.8401 and 1.4718. The forming processes applied were forging and cross-wedge-rolling (CWR). It was found that the cladded part remains undamaged after forging and CWR causes a significant reduction of pores.

Another process route was examined by Zhang et al. where SLM was used to fabricate Ti6Al4V preforms for forming [9]. The microstructure was improved and anisotropy was minimized after forming and subsequent heat treatment. A reduction of porosity was obtained by high deformation rates and a high degree of deformation, which also leads to higher fatigue strength. Previous work of Bambach and Sizova [10] focused on the forming of AM-preforms made of Ti6Al4V and Inconel 718 [11]. A good formability of both materials was observed in the 3D printed state, with lower activation energy for hot forming. Williams et al. used WAAM to produce a wing spar from aluminum, which was generated on sheet metal, but without strength-enhancing forming [4]. Martina et al. used WAAM to produce Ti6Al4V parts and realized local forming by rolling during the AM process [12]. They found that the local forming leads to grain refinement of the primary β-grains, a thickness reduction of the α-phase lamellae and a change from a columnar to an equiaxed microstructure.

The second process route utilizes a conventionally pre-formed part, which misses features of the final workpiece. These features, such as cooling or reinforcement ribs or other structural or functional elements are then added via AM.

Bambach et al. [13] showed that AM can be used to create local reinforcements on sheet metal parts. Extensive work on additively manufactured features which are added to formed parts have been performed by Merklein et al. [14–17]. While most existing work focusses on additive manufacturing of features on pre-formed sheet metal parts, this work examines the WAAM of stiffening ribs created by forging. Forging of features such as ribs with large aspect ratios is problematic, especially on hydraulic presses with their long contact times between tool and workpiece. A possibility to overcome limitations of the forging process is to decrease the complexity of the forged shape and to add the respective feature by WAAM.

2 Methodology

The geometry used for the current study is presented in Fig. 3 and represents a forged T-section with a length of 120 mm. The width of the rib is 20 mm. In the forging process, a rib height of 50 mm can be achieved. Higher ribs cause a much higher tool wear and increase the probability of forging defects. Using WAAM, the height of the bridge was expanded to a total of 95 mm. As reference, T6-heat treated EN-AW 6082 bulk material was used which was milled to a T-section geometry with a cross section of 120 mm x 50 mm x 20 mm. On the rib of this section, AlSi12 was added using WAAM.

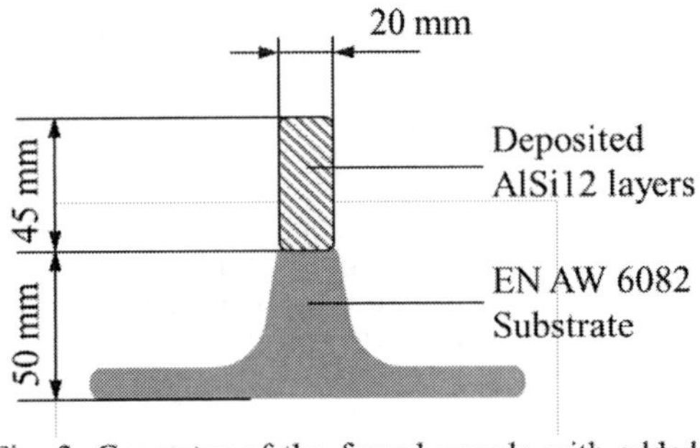

Fig. 3. Geometry of the forged sample with added AM structure.

The aluminum alloy EN-AW 6082 was used as substrate material and AlSi12 was used as a filler metal for the demonstration of the combined forming and AM process. The chemical composition of both materials is presented in Table 1 and Table 2, respectively.

Table 1. Chemical composition of EN-AW 6082 in wt%. [18]

Si	Mg	Mn	Fe	Cr	Zn	Cu	Ti
0.80	0.66	0.41	0.43	0.02	0.06	0.05	0.01

Fig. 4. Processing stages of samples. Left: forged, middle: machined, right: deposited AM layers.

Table 2. Chemical composition of AlSi12 in wt%. [19]

Si	Fe	Cu	Zn	Mn	Mg	Ti
12.0	<0.6	<0.3	<0.2	<0.15	<0.1	<0.15

The temperature of the samples was continuously monitored during the forming and the additive manufacturing process with thermocouples. The position of the thermocouple during the AM process is marked in Fig. 6. The maximum recorded temperature during the deposition of the AM layers was 460 °C and this value occurs during the deposition of the first layers. For the deposition of the subsequent layers, an average temperature between 270 °C and 325 °C was recorded.
From the sample geometry, test specimens were extracted for hardness, porosity and microstructure measurements at three distinct positions (substrate material, joining zone and additive layer).

2.1 Production of samples

The three stages (forged, machined, deposited AM layers) of a specimen are illustrated in Fig. 4. After the forging process and the heat treatment the top side was milled to obtain a flat top, which improves the welding process.

2.1.1 Forming and heat treatment

The setup for the forming process can be seen in Fig. 5. A round bar of EN-AW 6082 with a diameter of 75 mm was used as billet. After the forming process the true plastic strain in the middle of the part was between 0.5 and 1 and ~2.6 at the edge of the part.

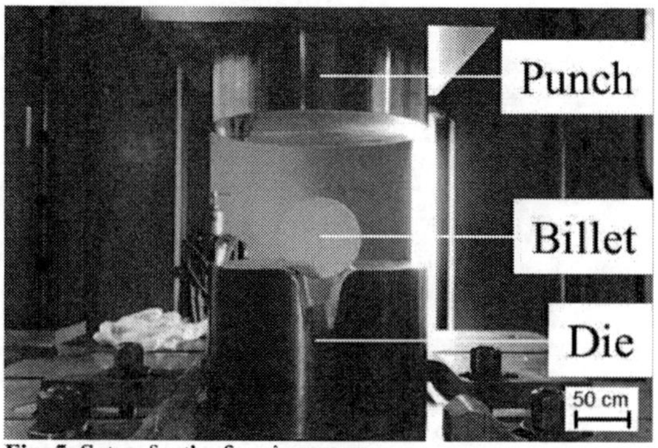

Fig. 5. Setup for the forming process.

The forming parameters are listed in Table 3.

Table 3. Forming parameters.

Punch speed	40 mm/s
Starting temperature	500 °C
Max. Force	3167 kN
Lubrication	Boron nitride

After forming a subsequent heat treatment including solution annealing at 530±10 °C for 2 hours, quenching in water and a subsequent aging at 165±5 °C for 10 hours was performed.

2.1.2 Wire+arc additive manufacturing

A six-axis FANUC® robot with the welding power source TPS/i 500 by Fronius® was used to deposit the layers on the substrate material. The wire had a diameter of 1.2 mm. The shielding gas was pure Argon with a gas flow rate of 15 l/min. The applied WAAM process parameters are presented in Table 4. The seams were deposited in longitudinal direction next to each other. Because of the thicker material base, the forged parts were preheated up to 190 °C while the thinner reference material was held at 65° C from the beginning of the welding process.

Table 4. Process parameters for WAAM.

Layer	1st	2nd	3rd	$\geq 4^{th}$
Wire feed rate [m/min]	7.1	6.3	6.1	6.0
Current [A]	140	105	96	91
Voltage [V]	15.0	14.0	13.7	13.6

2.2 Investigations

The hardness, microstructure and porosity were analyzed to access the impact of the process combination on the mechanical properties. Fig. 6 shows the locations of the extracted specimens, separated into three zones: (1) AM zone, (2) interface zone and (3) substrate zone. For each zone, one or more micro sections were cut from the specimens. These micro sections were extracted not only from the middle but also from the edge. For microscopic analysis, the samples were etched with a Keller reagent and examined by means of a light optical microscope. Porosity was determined with the software ATLAS by TESCAN. The hardness was measured via two methods i.e. the Vickers hardness test and the Ultrasonic Contact Impedance (UCI) with a BAQ UT100-V device both under conditions of HV 0.5.

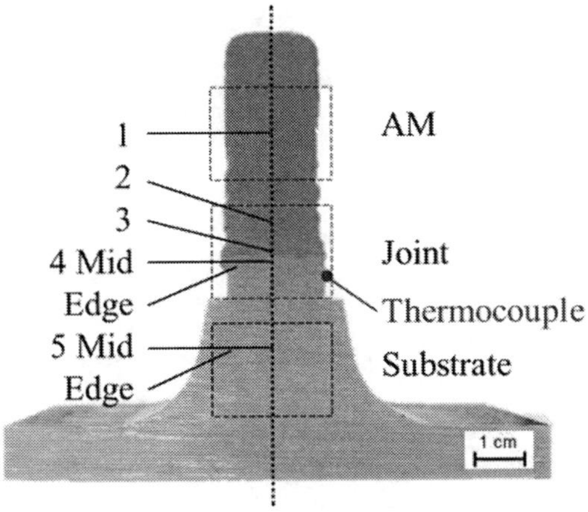

Fig. 6. Positions of specimen extraction for metallography. Pos. 1: AM layers; Pos. 2-4: interface zone with Pos. 3 as interlace area and Pos. 5: substrate. Position of thermocouple for temperature measurement

3 Results

3.1 Microstructure/Porosity

The AM layers show the development of columnar dendritic grains (see Fig. 7). The orientation of the grains is similar for formed & welded (FW) and milled & welded parts (MW) at Pos. 1. Formation of highly oriented columnar dendrites is an indication of a low temperature gradient. A similar structure for welded EN AW 6082 was also identified by Babu et al. [20]. The increasing part height leads to reduced heat conduction and the repetitive heating of previously deposited layers to additional continuous thermal energy input.

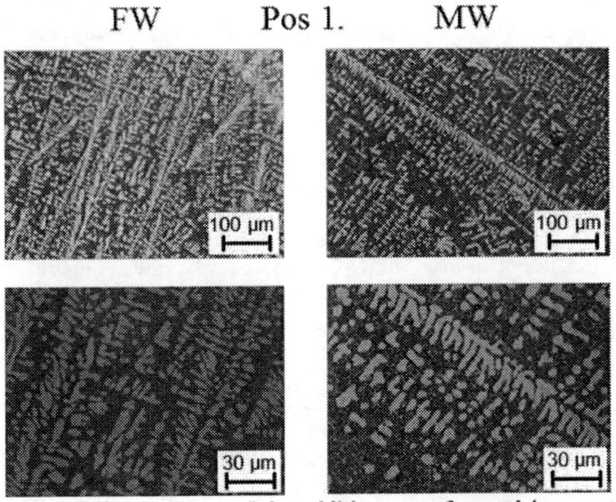

Fig. 7. Microstructure of the additive manufactured layers at Pos. 1. FW – formed and welded, MW – milled and welded.

Fig. 8 shows Pos. 2 slightly above the bonding zone. Compared to Pos. 1 it can be seen that the dendritic structure is less oriented. The crystal growth reduces the amount of elongated grains caused by the repeated heating during the welding process.

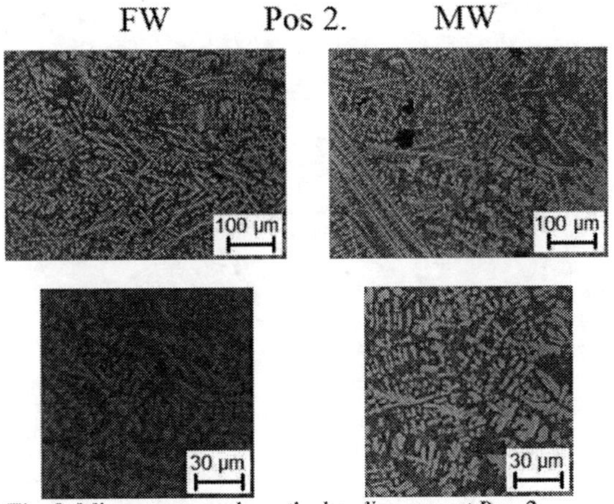

Fig. 8. Microstructure above the bonding zone at Pos. 2.

The mid bonding zone (Pos. 3) is illustrated in Fig. 9. The direction of grain growth of the primary phase is oriented along the largest temperature gradient, which induces the formation of a narrow columnar zone with mainly α-Al grains and eutectic Al-Si-structures lying in-between at the boundary layer. The top area of the T-section was not governed by the forming process as indicated by the coarse grain structure. As mentioned the FW part was preheated for the AM process. The thinner MW part caused a high cooling rate during the deposition of the first AM layers. This results in a globular grain formation.

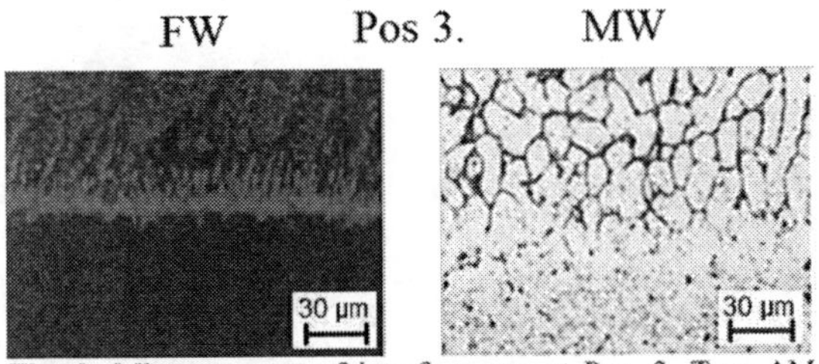

Fig. 9. Microstructure of interface area at Pos. 3. Top: AM layer, bottom: substrate.

In Fig. 10 the microstructure of the edge and mid area for a FW part is displayed. The elongation caused by forming can be seen in the edge area of the part while the mid position is less deformed. The microstructure of the MW part consists of fine grains, which is also reported by Kumar et al. [21].

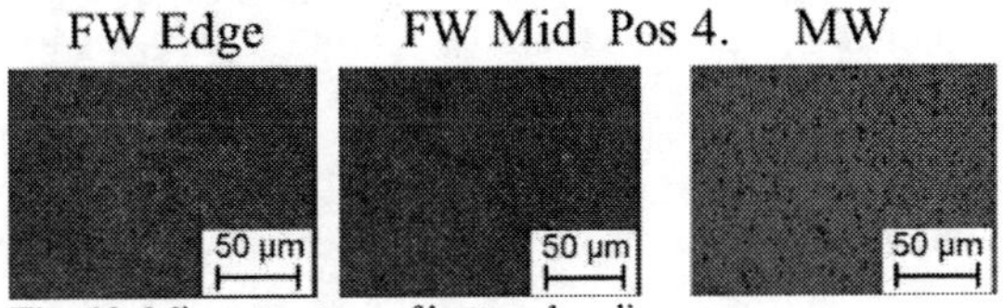

Fig. 10. Microstructure of bottom bonding zone

The microstructure of the substrate is illustrated in Fig. 11. The substrate was less affected by the heat input. This shows that the formed structured remains in the same condition if the location of the heat input is not in close vicinity. Thus, mechanical properties such as tensile strength should remain unaffected.

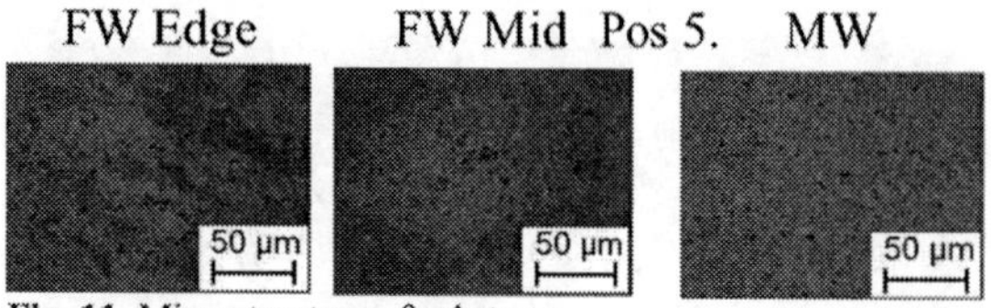

Fig. 11. Microstructure of substrate.

The results for the porosity are presented in Table **5**. It can be seen that the obtained porosity shows no difference between FW and MW in the pure AM or substrate zone. For the interface area the porosity reaches about 1.0 %. In general, the welding of aluminum is difficult due to take-up of hydrogen, as molten aluminum has a much higher solubility than solidified aluminum. The maximum value of 1.2 was measured directly in the interface zone. The porosity during deposition of subsequent layers is below 1.0 %. This value is the limit for additive manufacturing processes according to VDI3405 [22]. There were no hot cracks detected in the samples.

Table 5. Porosity for FW and MW in %.

Area	Pos.	FW	MW
AM	1	0.6	0.6
	2	1.2	1.2
Joint	3	1.0	<0.1
	4	<0.1	<0.1
Substrate	5	<0.1	<0.1

3.2 Hardness

The obtained hardness values are shown in Fig. 12. The hardness of the non-forged base material was 40 ± 2 HV. After the transition area, the hardness increased by the fine eutectic structure. In the AM zone, the value stabilized at 55 ± 4 HV, which is in good agreement with the norm values of 52-56 HV (50-55 HB) [23; 24]. The hardness in the AM zone of the forged parts was nearly the same i.e. 61 ± 5 HV. However, the first two AM layers have higher values up to 120 HV which could have been caused by a formation of fine eutectic phase due to a locally higher solidification rate [25]. On the other side, the hardness was only 70 HV in the transition area. At around 10 mm distance to the transition area, the hardness level increased from 70 to 80 HV. A similar effect was also reported by Benoit et al. [26] in their arc welding experiments. This is probably triggered near to the heat-affected zone in the first layers. In the lower forged material the hardness has a range of 57-70 HV, which can also be impacted by the heat effect during the deposition of the AM material.

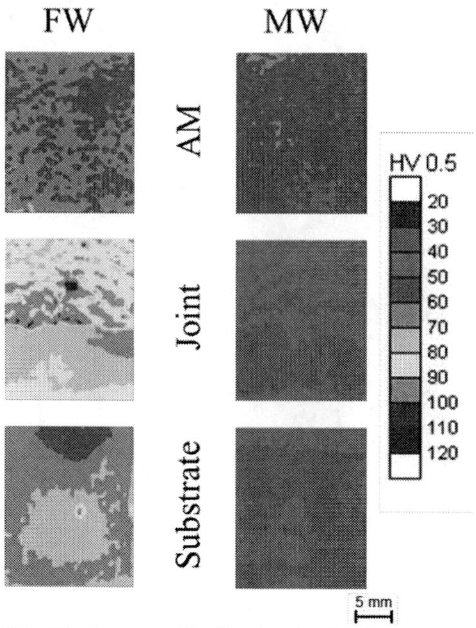

Fig. 12. Hardness distribution for the three areas.

4 Conclusion

It was shown that the additive deposition of layers with a WAAM process on a pre-formed Aluminum component is feasible. The forging process induced a higher hardness in the base material while the hardness of the milled material dropped. Further investigations of the material bonding especially in the 1st AM material layer are needed to enhance the process chain. This demands an improvement of the AM process, as a reduction of the hardness since large hardness gradients may diminish the part strength. The porosity with a maximum of 1.2 % also degrades the mechanical properties, but is in the standard range of welding processes. In further studies, an accurate temperature control and optimized process parameters have to be established. While dissimilar aluminum alloys were used in this study, attempts with EN-AW6082 wire for WAAM will be made to avoid differences in material properties. Other additive manufacturing methods such as laser metal deposition will also be examined for the suitability for a process combination.

The authors thank the European Regional Development Fund (ger. Europäischer Fond für Regionale Entwicklung, EFRE) under the project number 85002924 for the support.

5 References

1 D. Bauer, K. Borchers, T. Burkert, et al., *Handlungsfelder Additive Fertigungsverfahren* (Verein Deutscher Ingenieure, Düsseldorf, 2016).

2 S. Bremen, D. Buchbinder, W. Meiners, et al., LTJ, **6**, 24–28 (2011).

3 K. Shah, I. u. Haq, A. Khan, et al., Materials & Design (1980-2015), 531–538 (2014).

4 S. W. Williams, F. Martina, A. C. Addison, et al., Materials Science and Technology, **7**, 641–647 (2015).

5 O. Yilmaz, A. A. Ugla, Proceedings of the Institution of Mechanical Engineers, Part B: Journal of Engineering Manufacture, **10**, 1781–1798 (2016).

6 A. Gebhardt, *Generative Fertigungsverfahren* (Hanser, München, 2016).

7 Alphons Anandaraj Antonysamy, *Microstructure, Texture and Mechanical Property Evolution during Additive Manufacturing of Ti6Al4V Alloy for Aerospace Application* (Manchester).

8 B.-A. Behrens, L. Overmeyer, A. Barroi, et al., Prod. Eng. Res. Devel., **6**, 585–591 (2013).

9 Q. ZHANG, Z.-l. LIANG, M. CAO, et al., Transactions of Nonferrous Metals Society of China, **5**, 1036–1042 (2017).

10 I. Sizova, M. Bambach, Procedia Engineering, 1170–1175 (2017).

11 M. Bambach, I. Sizova, F. Silze, et al., Journal of Alloys and Compounds, 278–287 (2018).

12 F. Martina, P. A. Colegrove, S. W. Williams, et al., Metall and Mat Trans A, **12**, 6103–6118 (2015).

13 M. Bambach, A. Sviridov, A. Weisheit, et al., Metals, **12**, 113 (2017).

14 M. Merklein, P. Dubjella, A. Schaub, et al., *Interaction of Additive Manufacturing and Forming* (Nürnberg).

15 M. Merklein, D. Junker, A. Schaub, et al., Physics Procedia, 549–559 (2016).

16 A. Schaub, B. Ahuja, L. Butzhammer, et al., Physics Procedia, 797–807 (2016).

17 J. Domblesky, F. F. Kraft, Journal of Materials Processing Technology, **1-3**, 82–86 (2007).

18 Estral Forme d`Alluminio estruse a Misura, *Inspection certificate EN AW 6082 - Doc. Nr. 119669* (2017).

19 MIGAL, *Datenblatt - AlSi12*, http://www.migweld.de/produkte/aluminium-schweissdraht/ml-4047-alsi12/.

20 N. K. Babu, M. K. Talari, D. Pan, et al., Materials Chemistry and Physics, **2**, 543–551 (2012).

21 R. Kumar, U. Dilthey, D. K. Dwivedi, et al., Materials & Design, **2**, 306–313 (2009).

22 VDI, *VDI 3405 Additive Fertigungsverfahren - Grundlagen, Begriffe, Verfahrensbeschreibungen* (Beuth, Berlin, 2014).

23 DIN, *Metallische Werkstoffe – Umwertung von Härtewerten (ISO 18265:2013)* (Beuth, Berlin, 2014).

24 Metallgießerei Chemnitz GmbH, *Übersicht über die gebräuchlichen Gußlegierungen und ihre Eigenschaften*, http://www.metallgiesserei.biz/wordpress/wp-content/uploads/2010/11/Legierungstabelle-DIN-EN-1706-Aluminium-2014.pdf.

25 G. P. Dinda, A. K. Dasgupta, S. Bhattacharya, et al., Metall and Mat Trans A, **5**, 2233–2242 (2013).

26 A. Benoit, P. Paillard, T. Baudin, et al., Science and Technology of Welding and Joining, **1**, 75–81 (2014).

Energy of Laser Induced Shockwaves

Tobias Czotscher[1,*] and *Tim* Wünderlich[1]

[1]BIAS GmbH – Bremer Institut für angewandte Strahltechnik GmbH, Klagenfurter Straße 5, 28359 Bremen, Germany

Abstract. High throughput experimentation is a possibility to develop new materials in a short time in order to meet the demands of efficient characterisation of compositions. Thus, fundamentals of a new hardness measurement method are investigated based on laser-induced shockwaves. In this study, plasma is created with a nanosecond pulsed TEA CO_2 laser on top of an indenter. Further interactions of the plasma with the high intensity laser beam result in a shockwave. The pressure of the shockwave is used to push an indenter inside a material surface. So far, the energy transfer of the shockwave on indenters is not fully understood. Therefore, pendulum experiments are conducted to calculate how much energy can be transferred from the shockwave into the indenter. For these experiments, a bob, which geometry is equal to the indenter geometry, is connected to a thread pendulum and maximum deflection angles are recorded with a high-speed camera. Under standard conditions and the assumption of a spherical expansion of the shockwave, the experiments show that with a 6 J pulse energy a shockwave energy of up to 9 μJ can be used for indentation tests.

Keywords: Micro forming, Energy efficiency, Quality control

1 Introduction

Conventional material developments are based on expensive experimental investigations of different material properties. To meet the demands of efficient identification of compositions, which pursue the objective to realizing a specific performance profile of a material, high-throughput experimentation in materials science has been recognized as a new scientific approach to generate knowledge [1]. New techniques must operate at least at the same accuracy level as existing methods. One example is the hardness measurement. The conventional indentation and evaluation process usually takes several seconds. A new method was presented in [2] and is based on laser induced shockwaves to establish a new high throughput measurement method. When energy density of a CO_2 laser pulse exceeds a critical point, a plasma is induced and results in a shockwave [3]. The shockwave pressure is used to push an indenter inside a material. The induced depth and diameter in the material is measured and can be correlated with the material hardness. Previous experiments showed that a shift in focal position below the indenter increases the created indentations [2]. In addition, it was shown that although the plasma temperature increases up to 19000 K [4] above the indenter, no significant heating of the tested sheet underneath the indenter is measured during the indentation process [5].

In the context of high throughput, lasers offer high flexibility and high dynamic. Laser systems are used as measuring devices since the 1960s. At that time, Brech and Cross first introduced material emission analysis with a ruby laser [6]. Most laser-induced shockwave studies concern the analysis and understanding of surface hardening effects with wavelengths in the range of 1064 nm [7]. In the region of this wavelength and shorter, the metal is mainly directly ablated to create a high temperature plasma. Fabbro et al. showed that by using a confined and for the wavelength transmissive medium above the surface, the acting pressure on the metal surface is increased [7]. However, direct ablation leads to a change in surface conditions of the formed or indented material, which is not advantageous, also in terms of high throughput. Instead, pulsed CO_2 lasers are used for high-speed processing of thin metals, such as forming or cutting processes, because pre-treatment [8] and ablation layers are not necessary [9], because almost all the irradiation of the CO_2 laser wavelength is absorbed by the instantly formed plasma [10]. Barchukov et al. showed for pulsed CO_2 laser that the plasma initiates approximately 5 mm above the surface [11], but plasma formation can be influenced by the shift in focal position of the laser beam [2]. According to Marpaung et al., pulsed CO_2 laser-induced plasma can be described in two characteristic plasma regions [12]. One part is the primary plasma, which is found above the surface. The primary plasma is the initial energy source for the shockwave. The other part is the secondary plasma, which expands from the primary plasma. Vollertsen et al. observed that for forming operations of thin aluminium sheets with a thickness of 50 μm with 5.3 J pulse energy average pressures of

* Corresponding author: czotscher@bias.de

4.5 MPa are possible [13]. Further experiments showed that the pressure decreases regressively with increasing distance [14] and that the plasma height correlates with the maximum pressure [15]. For micro forming operations, Veenaas et al. showed that CO_2-lasers are a reliable tool, where the maximum shockwave pressure was found 9 mm above the surface on Aluminium rods [8].

However, the energy transfer of the shockwave on the indenters is not fully understood and analysed so far. In addition, high pressure deviations are observed in some studies [13]. Defined laser-induced shockwaves would enable high throughput hardness measurement with a measurement time for the hardness indentation and procedure of less than one second. Therefore, pendulum experiments are conducted to calculate how much energy can be transferred from the shockwave into the indenter. If no friction occurs, the transferred energy is equal to the deflection angle. The goal of this paper is to understand the energy transfer from laser induced shockwaves on indenters.

2 Experimental set-ups

The experiments are conducted with a pulsed TEA CO_2 laser from SLCR. The laser has a pulse duration of 100 ns and a maximum pulse energy of 6 J. A focusing mirror with a focal length of 200 mm is used, resulting in a 2 mm x 2 mm laser spot. Two different pendulum experimental set-ups were applied to capture the bob movement (as shown in Fig. 1 a and Fig. 1 b). For both set-ups the same camera and camera settings are used. Additionally, the influence of focal position f_p (as shown in Fig. 1 c), of pulse energy and of the positioning of the laser on the bob was studied.

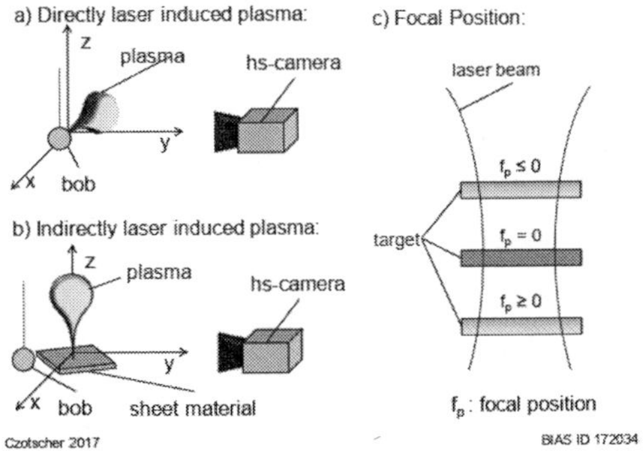

Fig. 1. Schematic experimental setup (directly a) and indirectly b)) and the influence of focal position c).

The motion of the bob was recorded with the high-speed camera Phantom VEO 410L from Vision Control. The high-speed camera Phantom v5.1 captured the deflection angle of the bob in both experiments with an array of 1280 x 800 pixels, a framerate of 1000 fps and an exposure time of 200 µs. For the first setup, the laser beam is directly irradiated on a hardened 100Cr6 steel bob with a diameter of 4 mm. The bob possesses a through hole of 1 mm in diameter to fix the indenter with a thread pendulum. The mass of the bob is 0.241 g. The induced shockwave results in a movement of the bob, which is recorded with a high-speed camera. The schematic setup is found in Fig. 2. In the second setup, the shockwave is induced on a separate target and the expanding shockwave induces a movement on the bob (see Fig. 3). The maximum deflection angle of the bob is measured. The bob was positioned at different positions relative to the laser beam. For each position, five measurements were conducted. Moreover, the standard deviation was calculated for these measured values. After each created plasma, several seconds were waited until the next laser shot was given on another part of the sample to minimize interaction effects between the already created plasma and the new plasma.

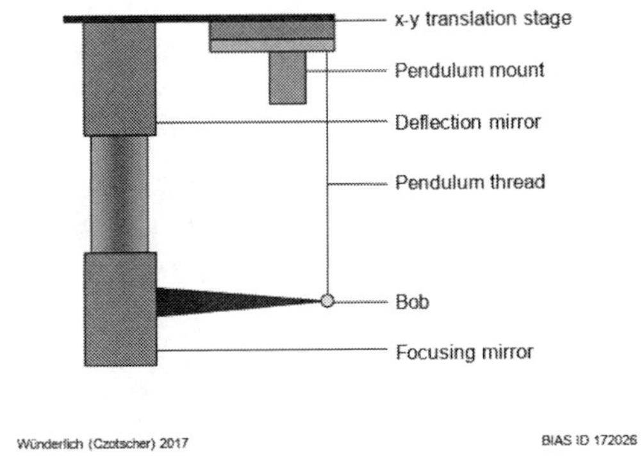

Fig. 2. Pendulum setup with laser beam, which is directly irradiated on indenter.

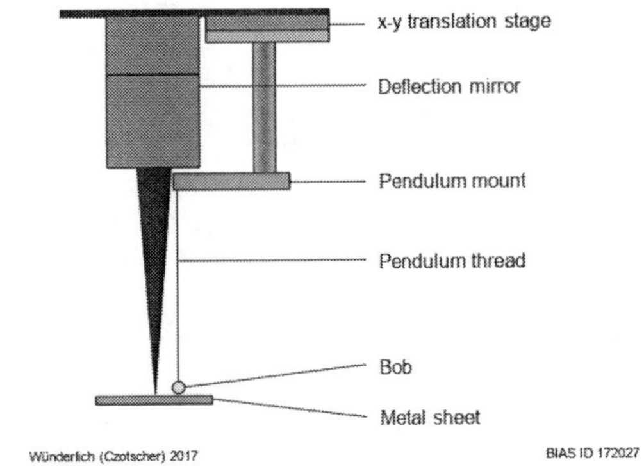

Fig. 3. Pendulum setup with laser beam, which is irradiated on external target.

3 Method

3.1 Shockwave expansion model

Idealisations of a real pendulum are made to calculate the transferred energy from the kinetic energy of the shockwave on the bob: The pendulum consists of a point mass, which is connected to an inextensible thread.

Motion only occurs in two dimensions. The kinetic energy from the shockwave is transferred without loss into potential energy. As a result, the energy transferred into the bob E_p can be calculated by the maximum deflection angle φ:

$$E_p = m_k \cdot g \cdot l_p \cdot (1\text{-}\cos\varphi) \qquad (1)$$

Where m_k is the mass of the bob, g the gravity and l_p is the length of the thread. When the expansion radius of the shockwave r_s, is known and half width of the laser spot $d_s = 1$ mm is taken as the initial point of the shockwave, the transferred energy on the bob E_p and the radius of the bob r_k can be used to determine the maximum possible shockwave energy E_s. Previous research done by Kawaguchi analysed the formation and expansion of TEA CO_2 laser-induced shockwaves [16]. They showed that the upper part of the expanding shockwave is spheroidal. The central part of the expanding plasma is cylindrical, whereas in the lower regions deviations in shape occurred. Additionally, Walter et al. showed that for larger distances from the shockwave initiation that the shockwave form is spherical [17]. The shockwave that is formed along the plasma transforms into a spheroidal shockwave, which is illustrated in Fig. 4.

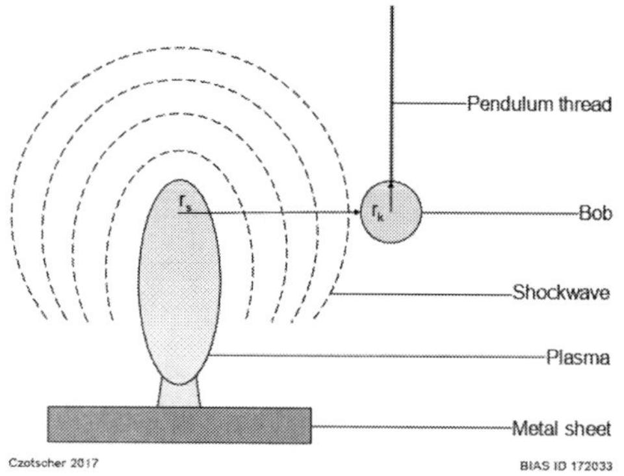

Fig. 4. Expansion of shockwave according to [15].

Accordingly, the part of shockwave energy that is transferred on the bob is relative to its area. The bob area A_b can be described as:

$$A_b = \pi \cdot r_k^2 \qquad (2)$$

By considering for the upper part a spherical and for the lower part a cylindrical expansion of the shockwave, the total surface area of the shockwave A_s can be calculated as follows:

$$A_s = 2 \cdot \pi \cdot (r_s - d_s) \cdot h + 2 \cdot \pi \cdot (r_s - d_s)^2 + \pi \cdot (r_s - d_s)^2 \qquad (3)$$

The height of the cylinder h is taken from Veenaas et al. experiments with TEA CO_2 laser that analysed the pressure at different heights along the expanding plasma [8]. By cross-multiplication, the shockwave energy can be calculated as follows:

$$E_s = E_p \cdot [2 \cdot (r_s - d_s) \cdot h + 3 \cdot (r_s - d_s)^2] / (\pi \cdot r_k^2) \qquad (4)$$

4 Results

4.1 Direct laser irradiation

Shockwaves induced with different pulse energies directly on the bob show a positive influence of the transferred energy on the bob. However, the differences are small and between 5.6 J and 6.0 J no significant changes are observed (as shown in Fig. 5). The influence on shift in focal position is demonstrated in Fig. 6. Additional to the focus position, the focus is shifted $f_p = -10$ mm below and $f_p = +10$ mm above the target. The deviation increases if the focal position is shifted. The transferred energy increases with the shift in focal position fp = -10 mm. The transferred energy in dependence of the positioning of the laser beam on the bob is shown in Fig. 7. The laser beam is positioned $z = +1$ mm above the centre of the bob and $z = -1$ mm below the centre. The transferred energy decreases when the laser is not positioned concentrically on the bob. The transferred energy decreases more for $z = +1$ mm compared to $z = -1$ mm and the concentrically position. The deviation increases when the laser beam is below the bob ($z = -1$ mm).

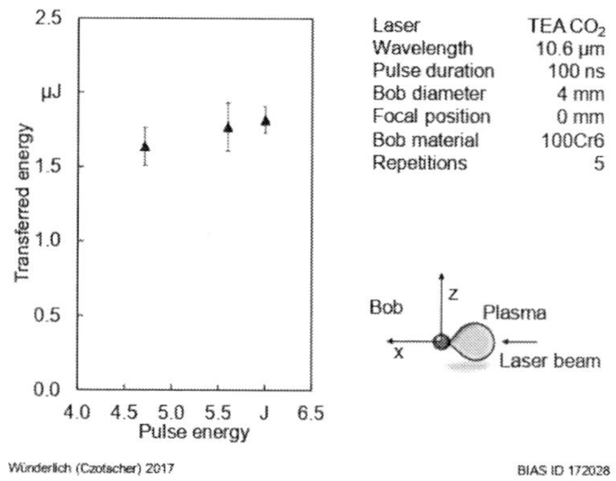

Fig. 5. Correlation between pulse energy and transferred potential energy on bob.

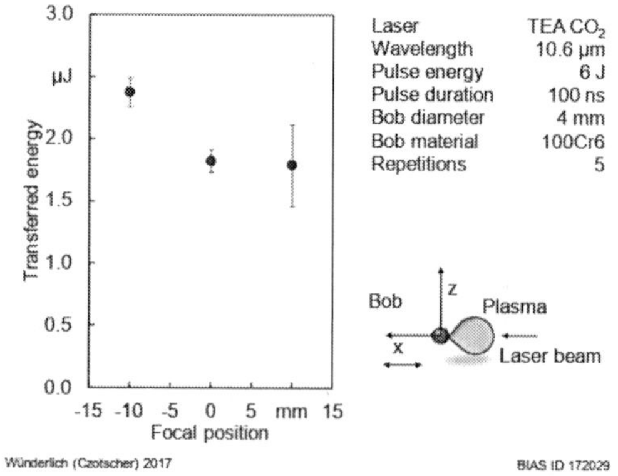

Fig. 6. Influence of focal position on transferred energy on bob.

https://doi.org/10.1051/matecconf/201819002006

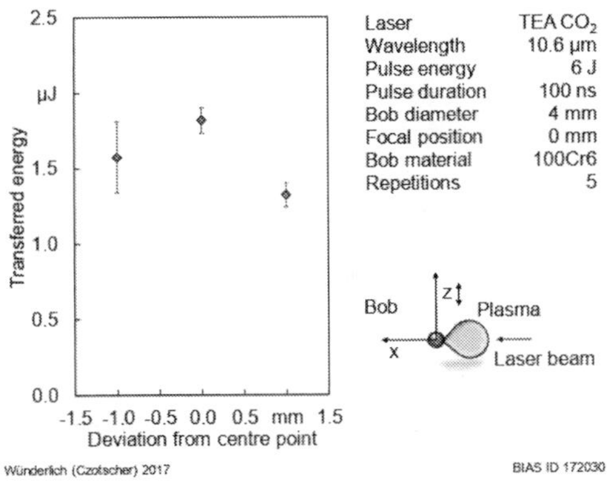

Fig. 7. Influence of positioning of the laser beam on bob.

4.2 Laser irradiation on external target

Fig. 8 shows the calculated transferred energy at different pendulum positions relative to the induced shockwave centre. The centreline shows the estimated centre of the shockwave mentioned in [2]. Values measured closer to the surface show that the transferred energy is higher compared to larger distances. In this case, the energy decreases exponentially. At the X-distance of 17.8 mm and higher, the transferred energy does not change significantly in z-direction.

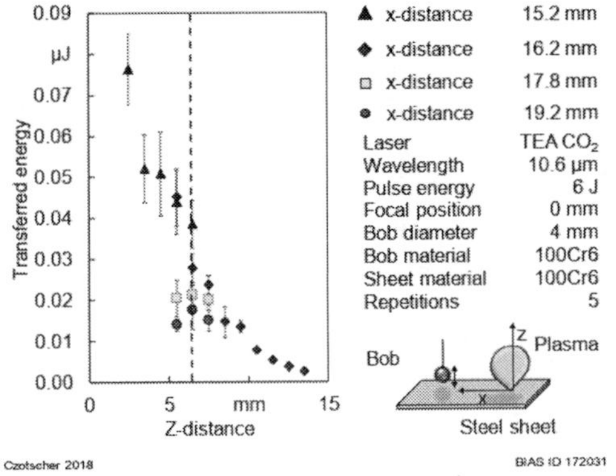

Fig. 8. Transferred energy measured at different Z-distances from induced shockwave.

With Eq. 4 and the calculated transferred energy in Fig. 8, the shockwave energy can be calculated as shown in Fig. 9. No significant changes are observed between the distance of 6.5 mm and 7.5 mm. The average values are between 5 µJ and 9 µJ. When at the distance of 6.5 mm, the bob is moved closer the shockwave, the transferred energy on the bob increases exponentially (see Fig. 10). By extrapolation and considering half of the laser beam spot width of 1 mm, a maximum transferred energy of E_p =1.8 µJ is estimated.

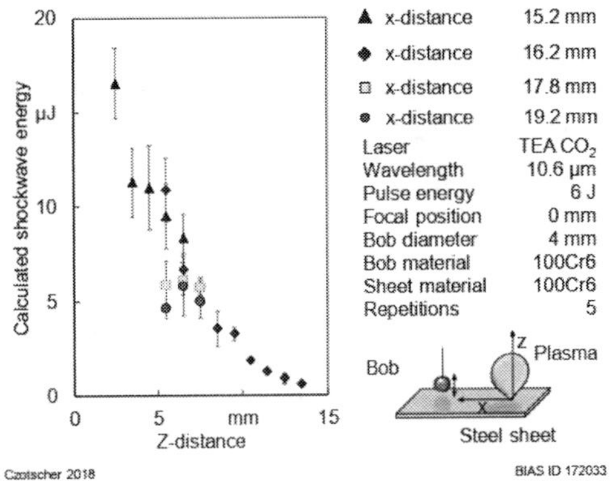

Fig. 9. Calculated shockwave energy with Eq. 4 for different Z-distances from induced shockwave.

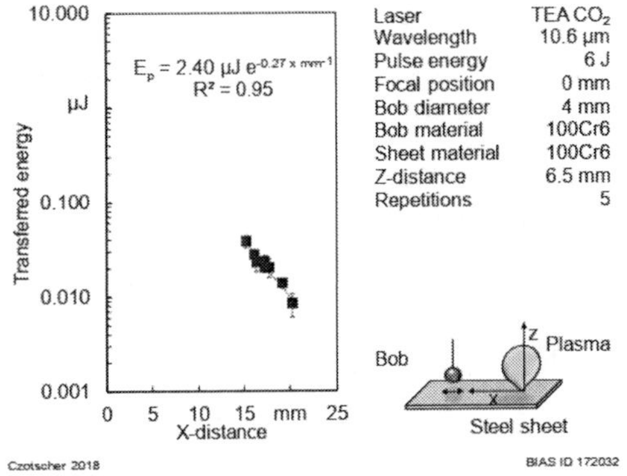

Fig. 10. Transferred energy measured at different X-distances from indirectly induced shockwave.

5 Discussion

For direct interaction of the laser beam with the bob, the deviation of transferred energy decreases when the laser beam is focused directly on the bob (see Fig. 6 and Fig. 7). However, the transferred energy increases if the focal position is shifted behind the bob (f_p = -10 mm). The increase can be explained by a larger focal area, which emits more electrons out of the bob surface. These electrons absorb further irradiation of the laser beam. Moreover, when the focal position of the laser is shifted below the workpiece, as shown in [2], the created plasma is closer to the material surface. Consequently, higher forces act on the bob. The positive effect of more emitted electrons due to a larger spot area also explains why a shift in focal position above the bob surface of f_p = 10 mm does not lead to a significant change in transferred energy compared to the position in focus (f_p = 0 mm). However, as plasma observations showed in [2], the plasma form becomes more instable the further the focal position of the laser is shifted. This also increases the deviation in transferred energy. Additionally, if the laser beam is not positioned concentrically on the probe, it negatively affects the transferred energy and increases the deviation (see

Fig. 7). In Fig. 8, it is observed for indirect induced shockwaves that the transferred kinetic energy increases when the pendulum with the bob is placed closer to the surface. This effect can be explained as following: The reflection of the shockwave on the target surfaces leads to a superposition of the induced shockwave and the reflected part. This increases the measured deflection angle of the pendulum and therefore, the transferred energy. This effect was not observed in [8], when the plasma was induced on a small surface (Aluminium rod). When the bob is moved further up a plateau is found, which is in the region of estimated expansion centre of the shockwave [15]. Again, a decrease in transferred energy can be found by moving the pendulum even further up. These results can be explained by the measurements obtained in Fig. 7. The expanding shockwave does not push the bob concentrically and thus less energy is transferred.

All in all, for direct interaction of the bob with the laser beam, 1.8 µJ transferred energy is measured for the position in focus. For indirect induced energy, also 1.8 µJ are calculated by extrapolation (see Fig. 10). By assuming a spherical expansion of the shockwave and using Eq. 4, in total 6 J pulse energy are transformed into 9 µJ kinetic shockwave energy. This shockwave energy is sufficient to create indentations in metallic materials as shown in [2]. It was shown that with a 4 mm indentation ball, indentation diameter of up to 210 µm were created on aluminium alloy materials.

6 Conclusions

- A TEA CO_2 laser pulse with an energy of 6 J induced on 100Cr6 can be transferred in a shockwave energy of up to 9 µJ. This energy is sufficient to create an indentation diameter in aluminium alloy with a 4 mm indenter of 210 µm.
- For directly and indirectly laser-induced shockwaves, the calculated initial shockwave energy is the same. However, the creation of the shockwave on large targets, such as metal sheets, leads to a superposition of the laser-induced shockwave and the reflected shockwave on the target surface. This leads to an overpressure in the surface region and to relatively higher transferred energies on the bob in this area.

Acknowledgement

Financial support of the subproject D02 "Laser induced hardness measurements" of the Collaborative Research Centre SFB1232 by the German Research Foundation (DFG) is gratefully acknowledged.

References

1. R. Potyrailo, K. Rajan, K. Stoewe, I. Takeuchi, B. Chisholm, H. Lam, ACS Comb. Sci. **13**, 579-633 (2011).
2. T. Czotscher, *36th International Congress on Applications of Lasers & Electro-Optics*, (ICALEO, Atlanta, 2017)
3. J.D. O'Keefe, C.H. Skeen, C.M. York, J. Appl. Phys. **44**, 4622-4626 (1973)
4. L.J. Radziemski, D.A. Cremers, T.M. Niemczyk, Spectrochim. Acta Part B At. Spectrosc. 40, 517–525 (1985)
5. T. Czotscher, F. Vollertsen, *Proceedings of the 8th International Conference on High Speed Forming* (ICHSF, Columbus, OH, 2018) (to be published)
6. F. Brech, L. Cross, Appl. Spectrosc. **16**, 59-64 (1963)
7. R. Fabbro, J. Fournier, P. Ballard, D. Devaux, J. Virmont, J. Appl. Phys. **68**, 775–784 (1990)
8. S. Veenaas, H. Wielage, F. Vollertsen, Prod. Eng. **8**, 283-290 (2013)
9. A. Miziolek, V. Palleschi, I. Schechter, *Cambridge University Press* (Cambridge, 2006)
10. H.W. Bergmann, H. Hügel, Strahltechnik Band **6**, 35-44 (1998)
11. A.I. Barchukov, F.V. Bunkin, V.V. Konov, A.A. Lyubin, Sov. Phys. **39**, 42–45 (1974)
12. A.M. Marpaung, R. Hedwig, M. Pardede, T.J. Lie, M.O. Tjia, K. Kagawa, H. Kurniawan, Spectrochim. acta, Part B At. Spectrosc. **55**, 1591-1599 (2000).
13. F. Vollertsen, H. Schulze Niehoff, H. Wielage, Prod. Eng. **3**, 1-8 (2009)
14. T. Czotscher, F. Vollertsen, *Proceedings of the 9th International WLT-Conference on Lasers in Manufacturing*, 1-7 (LiM, Munich, 2017)
15. T. Czotscher, S. Veenaas, F. Vollertsen, J. Technol. Plast. **42**, 1-7 (2017)
16. K. Kawaguchi, K. Yonezawa, M. Nakajima, K. Horioka, Plasma Fusion Res. Rapid Commun. **10**, 3–4 (2015)
17. D. Walter, A. Michalowski, R. Gauch, F. Dausinger, *Proceedings of the 4th International WLT-Conference on Lasers in Manufacturing*, 557-562 (LiM, Munich, 2007)

Electrohydraulic incremental bulk metal forming

Lasse Langstädtler[1,3*], *Holger* Pegel[1,3], *Marius* Herrmann[1,3], *Christian* Schenck[1,2,3] and *Bernd* Kuhfuss[1,2,3]

[1] Bremen Institute for Mechanical Engineering (bime), Badgasteiner Straße 1, 28359 Bremen, Germany
[2] MAPEX - Center for Materials and Processing
[3] University of Bremen

Abstract. Electrohydraulic forming is a working media based high speed technique that is usually applied for sheet metal processing. In this process a shock wave acts as a flexible punch that transmits the punching force in a very short period of time. This force is usually used to accelerate the workpiece towards the passive tool. In Contrast to sheet forming, the electrohydraulic method is still not adapted to bulk forming. Although, the exchange of a mechanical rigid punch by a shockwave with a flexible shape enables special advantages especially if parts with millimeter dimensions or smaller are to be processed. But in case of deep and small cavities with a high aspect ratio are to be filled, the forming energy is not transferable within one shock wave. To overcome these obstacles the incremental electrohydraulic forming is introduced. As an example, the electrohydraulic extrusion of cylindrical samples (aluminum Al99.5) with an initial diameter of 1.5 mm was performed with a series of consecutive shock waves.

Keywords: Micro forming, Extrusion, Aluminum

Nomenclature

α_e	extrusion angle
C	capacitance
d_b	bulge diameter
d_d	guiding channel diameter
d_e	extrusion channel diameter
d_{eS}	extruded sample diameter
d_s	initial sample diameter
d_w	wire diameter
Δe	step size
e	extrusion height
E_C	loading energy of the capacitor
E_R	relative energy
E_{st}	pulse energy
E_V	vaporization energy
ΣE	total energy
h	bulge forming height
h_R	relative bulge forming height
I	discharge current
l_d	guiding channel length of the die
l_e	extrusion channel length of the die
l_s	initial sample length
l_w	wire length
s_0	sheet metal thickness
t	time
t_R	current rise time
U_0	loading voltage

1 Introduction

To develop new metallic materials, time and cost intensive iterative procedures are required. Therefore, high-throughput technologies, as already used for medicine research, are suggested to decrease development time and reduce the costs [1]. Processes like single droplet solidification [2] provide rapidly generated and modified small material samples. Though, to characterize the new materials is still stated to be a fundamental problem [3].

One approach to characterize the material behavior of these micro-samples is to deform the samples and compare the required forces and the resulting strains to already characterized materials. Due to the low size and the spherical shape of the samples a suitable solution is a bulk forming operation. At conventional bulk forming of micro samples, the force transmission from punch to micro sample is an ambitious challenge. Especially, if the extrusion dies feature high aspect ratio between channel diameter and length. Here, as the mechanical punch also needs to provide the high aspect ratio, it tends to brake. In addition, close tolerances are required, as the samples material tends to flow into the gap between punch and die. Furthermore, when multiple bulk forming stages are aimed while pushing the sample completely through the die, the conventional punch is not flexible to undergo these stages within a single tool, Figure 1.

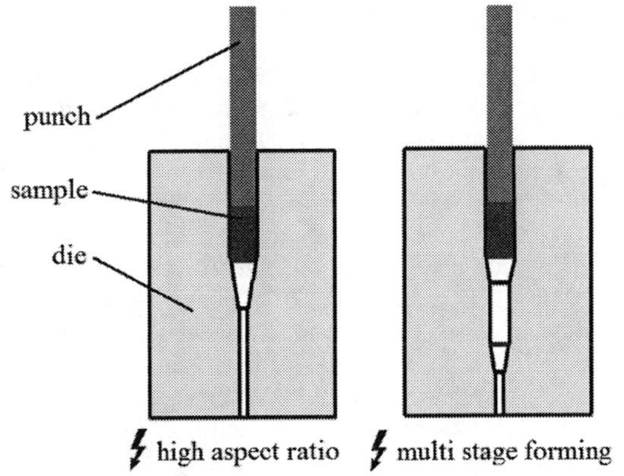

Fig. 1. Motivation for electrohydraulic bulk forming.

* Corresponding author: langstaedtler@bime.de

Consequently, a new bulk forming procedure is introduced as incremental electrohydraulic bulk forming. This high-speed process offers several advantages for micro forming like high deformation rates and a contactless force transmission [4, 5]. Here the idea is to characterize the micro samples while they are incrementally formed into deep dies with different forming stages that cause defined forming-load conditions. The increments thereby should be quite small to sample the stress-strain curve with high resolution. Comparing measured deformations with simulation results enables the determination of characteristic values that describe the material behavior. In the process shock waves caused by an electric spark or a wire explosion in a working media filled pressure chamber are an approach to act as flexible punches. The energy E_V is required to vaporize the wire and to build the shock wave. The electrohydraulic punch transmits the punching force in a few microseconds with high pressures up to several GPa [6].

By the action of force, the workpiece is pushed towards and in the passive tool, e.g. an extrusion die. In the die the workpiece material is deformed. Conventionally, the process energy is usually provided completely by a single shock wave. When deep and small cavities with a high aspect ratio are used for extrusion, multiple shock waves are required for an incremental extrusion, as the energy cannot be transferred within one single step. With the proposed new incremental method with flexible punch, extrusion in multi-stage dies including both compression and expansion operations are enabled.

In order to extrude micro samples by the action of shock waves, the energy transmission of the shock waves needs to be controlled. This was investigated by sheet metal bulge forming tests. Afterwards, by applying consecutive shock waves, the incremental bulk forming was investigated for the extrusion of micro samples. Here, the use of cylindrical samples should enable a repeatable testing due to its constant diameters as they vary for spherical microscopic samples [7]. The shock waves were varied in the amount of energy which was stepwise applied and the influence on the extrusion into the depth of the die was investigated.

2 Experimental set-up

The shock waves were generated by short circuiting an LC-resonator over a vaporizing aluminum wire. The wire had a diameter of $d_w = 0.3$ mm and a length of $l_w = 20$ mm and was placed in a pressure chamber filled with distilled water. The water was renewed after each vaporization due to pollution of small aluminum particles that stay in the water after vaporization. The resonator had a capacitance of $C = 100$ µF and the loading voltage U_0, and in consequence the loading energy $E_C = 0.5 \cdot C \cdot U_0^2$. The loading voltage was varied from $U_0 = 0.5$ kV to $U_0 = 2.7$ kV for sheet forming and from $U_0 = 1.7$ kV to $U_0 = 2.7$ kV for bulk forming.

2.1 Electrohydraulic sheet metal bulging

Sheet bulge forming (bulge diameter $d_b = 15$ mm, sheet thickness $s_0 = 1$ mm, sheet dimension 50x50 mm², sheet material AlMg3) tests were performed to investigate the possibility to dose the shock wave, Figure 2. The relative energy E_R, the quotient of the loading energy E_C and the vaporization energy E_V, was varied from $E_R < 1$ (partial vaporization) to $E_R > 1$ (increased energy deposition), Equation 1.

$$E_R = E_C / E_V \tag{1}$$

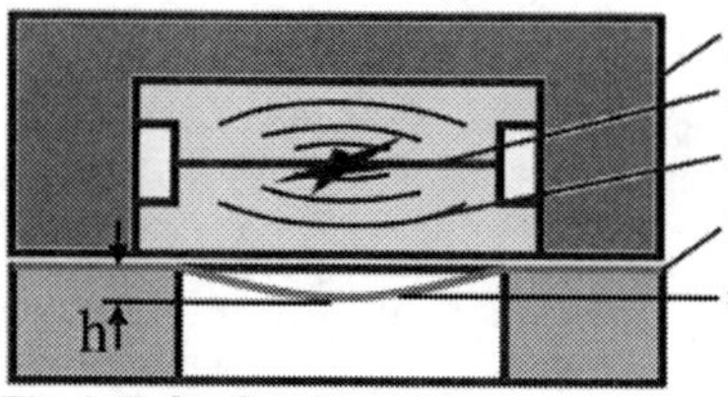

Fig. 2. Bulge forming set-up.

To vaporize an aluminum wire an amount of 14 MJ per kilograms is required [8]. E_V can be calculated theoretically. In this set-up, for three different wire diameters of $d_w = 0.3$ mm, $d_w = 0.4$ mm and $d_w = 0.5$ mm, the vaporization energies are $E_{V0.3} \approx 50$ J, $E_{V0.4} \approx 100$ J and $E_{V0.5} \approx 150$ J. For each relative energy the characteristic discharge current I and the forming height h were measured. The relative forming height h_R was calculated with equation 2.

$$h_R = h(E_R) / h(E_R = 1) \tag{2}$$

2.2 Electrohydraulic bulk forming

Cylindrical aluminum (Al99.5) samples with an initial diameter of $d_s = 1.5$ mm and an initial length of $l_s \approx 3.0$ mm were electrohydraulically extruded to a diameter of $d_e = 1.28$ mm, Figure 3a). Both end faces were initially flat. The length of the extrusion channel was $l_e = 6$ mm. The extrusion die, made of unhardened St355 steel without reinforcement, had a diameter of $d_d = 1.5$ mm and a length of $l_d = 5.2$ mm in the first part of the die. The extrusion angle was $\alpha_e = 128°$. The extrusion was performed stepwise by transferring different amounts of pulse energy E_{st} ($E_{st} = 162$ J, $E_{st} = 288$ J and $E_{st} = 382$ J) while keeping E_{st} constant for the consecutive shock waves of one extrusion. The total energy ΣE was calculated with Equation 3 and the current and voltage curve were measured, Figure 3b).

$$\Sigma E = \Sigma E_{st} \tag{3}$$

The depth e was measured with a laser line-triangulation scanning set-up [5]. The sample was completely plunged in the extrusion channel of the die if e = 5.2 mm was reached. The difference between each step was calculated as step size Δe with Equation 4.

$$\Delta e = e_{i+1} - e_i \tag{4}$$

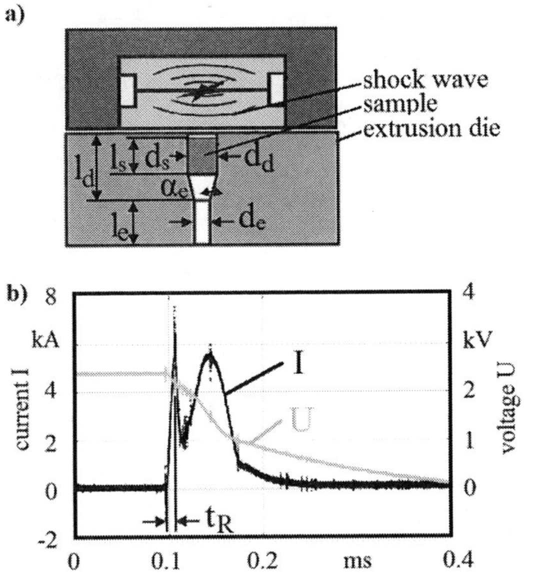

Fig. 3. Experimental set-up: a) pressure chamber with extrusion die, b) typical discharge current I and voltage U for $U_0 = 2.4$ kV.

3 Experimental results

3.1 Electrohydraulic sheet metal bulging

The relative forming height h_R increased with the relative energy E_R, Figure 4a). A forming was even reached for $E_R < 1$. Each E_R resulted in a characteristic discharge current chart, Figure 4b). With $E_R < 0.3$ the wire did not vaporize and the electrical circuit kept closed which resulted in an oscillation of the current without forming of the sheet metal. With $0.3 < E_R < 1$ the circuit broke by melting or vaporizing of the wire but the energy was not high enough to ignite a plasma channel. Hence, the capacitor was not completely discharged. However, the discharge resulted in a force that was high enough to deform the sheet metal. For $E_R > 1$ the wire was vaporized completely and the adjunctive plasma kept the circuit closed until the capacitor was completely discharged. Further increasing E_R decreased the peak rise time t_R as the wire vaporized faster. Thus, by varying E_R, the shock wave was adjustable and the current charts qualitatively characterized the kind and strength of the discharged energy. The wire diameter d_w did not influence the resulting relative forming heights h_R. Consequently, any chemical reactions of the vaporized aluminum did not contribute to the forming. However, the absolute forming steps decreased with decreasing d_W.

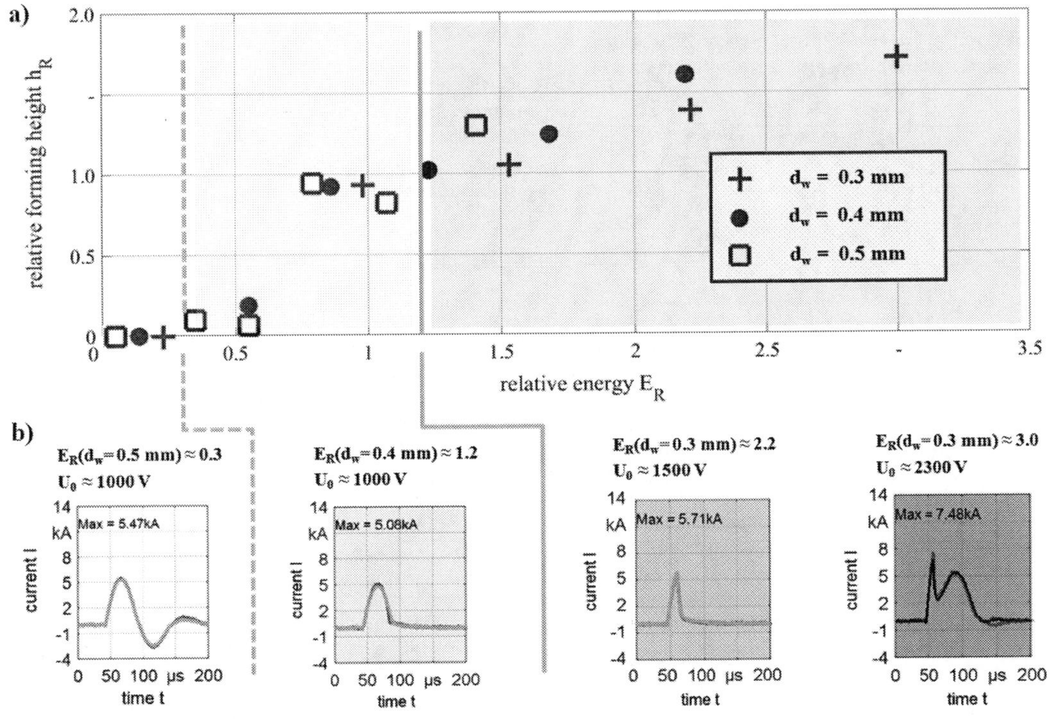

Fig. 4. Sheet metal bulging with different wire diameters d_w: a) relative forming height h_R as a function of the relative energy E_R, b) example of characteristic discharge currents I for different E_R.

https://doi.org/10.1051/matecconf/201819003001

3.2 Electrohydraulic bulk forming

The extrusion started at $e \approx 2.0$ mm due to the length of the samples. Increasing ΣE resulted in an increase of the extrusion depth e, Figure 5. With the shown set-up it was possible to repeatedly generate shock waves that were gentle enough to incrementally extrude micro samples. For $E_{st} = 162$ J the extrusion stopped after a small deformation. However, a complete extrusion with $E_{st} = 288$ J and $E_{st} = 382$ J was possible.

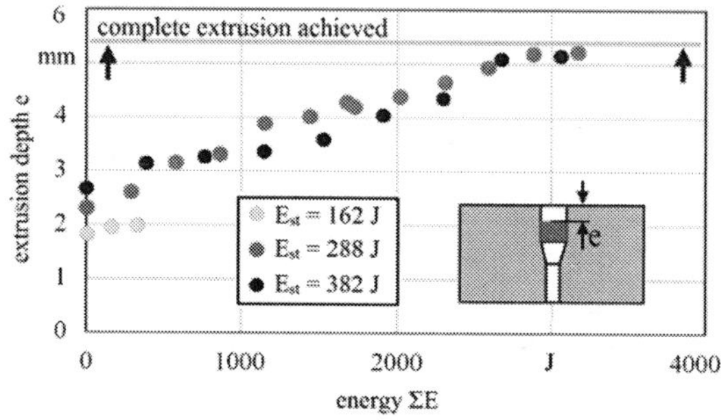

Fig. 5. Extrusion depth e for incremental extrusion.

Concerning the geometry of the samples, a convex deformation of the end face in the direction of extrusion (B) and an increase of the length of the sample was recognized, Figure 6. This is comparable to conventional extrusion. The other end face (A) in contrast exhibits a concave deformation. Thus, the hydraulic punch features a flexible shape. Both shapes of the end faces could be explained by friction between the sample and the die that reduced the axial material flow relatively to the center of the sample.

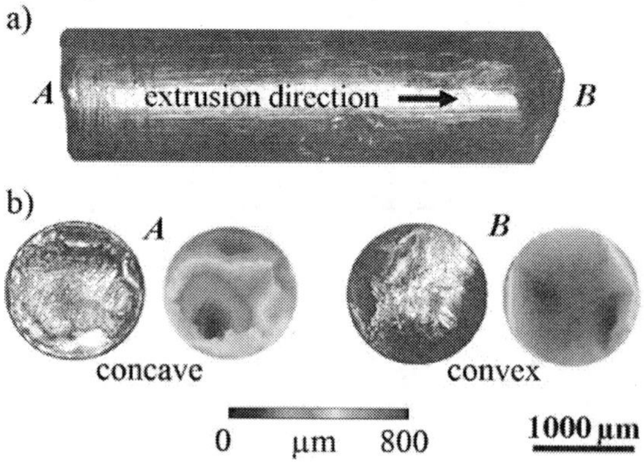

Fig. 6. Incrementally extruded Al99.5 cylindrical sample ($E_{st} = 382$ J): a) side view, b) concave and convex face of the sample.

In Figure 7, the subsequent step size Δe is plotted as a function of the actual extrusion height e. The step size decreased with an increase of the extrusion depth. Though, until now it is not clear which effects are mainly to be addressed causing the spreading of the achieved results.

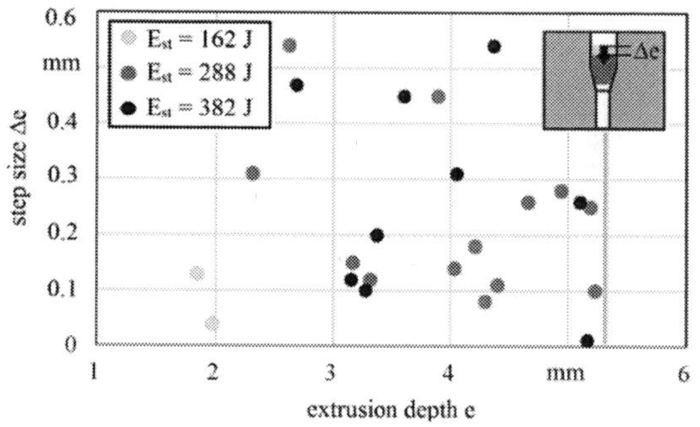

Fig. 7. Step size Δe as function of the actual extrusion depth e.

Comparing both, the diameter of the extrusion channel of the die $d_e \approx 1.28$ mm and the diameter of the extruded sample $d_{eS} \approx 1.27$ mm, a very good fit was reached for both energies, although the extrusion die was neither hardened nor reinforced. There was no need of pre-strains, which are conventionally required for extrusion dies, to compensate the elastic spring back of the sample after extrusion. In contrast to conventional extrusion operations, the samples diameter was nearly the same (even lower) than the dies diameter $d_{eS} < d_e$ and in addition, this was reached with a very simple die which had a low stiffness. This can be explained by the very high deformation rate that results in an increased samples material plasticity.

As the die as well is loaded by the shock wave, it possibly tends to deform. Hence, a further explanation for this good replication is an elastic deformation of the die which can lead to a reduction of the diameter which operates like a pre-strain. This is underlined by the fact, that different energies had the same influence on the increase of extrusion depth which indicates a self-regulation feature of the process.

4 Conclusion

In this contribution the potential of electrohydraulic forming was analyzed to fully extrude micro samples within deep dies with the purpose to establish a new method for high throughput materials testing. Therefore, two fundamental questions had to be answered: i) How and in which range can the shock wave that result in an action of force be dosed to realize an incremental forming, and ii) can the shock wave be used as a flexible punch in bulk metal forming, especially extrusion? The first question was examined with sheet metal bulge testing with the following conclusions:

- The action of force could be controlled by the relative energy E_R that is the proportion of supplied energy and vaporization energy of the wire which itself is a function of the wire diameter d_w.
- With values of $E_R < 1$ small forming was achievable although the wire did not completely vaporize or melt - this enabled very small deformation steps.

- The relative energy could be monitored by the discharge current whereat the kind of force generation was imprinted in the current curve progression.

The second question was analyzed with electrohydraulic extrusion experiments with the following results:

- Bulk metal micro-forming and especially full extrusion was possible with the means of electrohydraulic forming.
- Even full extrusion within a deep cavity -with an aspect ratio of 5- could be realized.
- Sufficient small forming increments with a step size of 100 μm and below were attained.
- The shock wave easily adapted to changing shapes of the workpiece end face.
- Due to high pressures for bulk forming according to the investigated step sizes extrusion was realized with relative energies of $E_R > 5$.

These results state the principle feasibility of the proposed high throughput materials testing approach. Additionally, three aspects inherently follow from these findings and will be subject of future work:

- Multistage full extrusion in one single die that is out of question with a mechanical punch will be enabled with the electrohydraulic technique.
- Even full lateral and backward extrusion should be feasible by using a shock wave instead of a mechanical punch.
- The tool design and manufacturing will be strongly simplified if the special demands of high speed forming are taken into account. Hardening and reinforcing of the tools may not be mandatory as shown by the extrusion experiments.

Acknowledgments

The authors would like to gratefully acknowledge the financial support of subproject D04 within the Collaborative Research Center SFB 1232 "Farbige Zustände" by the German Research Foundation (DFG).

References

1. L. Mädler, 4th International conference on Nanomanufacturing Bremen (2014)
2. N. Ellendt, N. Ciftci, C. Goodreau, V. Uhlenwinkel, L. Mädler, IOP Conference Series: Materials Science and Engineering 117 (1) (2016)
3. A. Beinhauer, C. Heinzel, O. Riemer, Fachbeiträge 8. Kolloquium Mikroproduktion, Editor: F. Vollertsen, C. Hopmann, V. Schulze, J. Wulfsberg, BIAS Verlag (2017)
4. L. Langstädtler, L. Schönemann, C. Schenck, B. Kuhfuss, ASME Journal of Micro- and Nano-Manufacturing, Volume **4**, Issue 2 (2016)
5. L. Langstädtler, M. Herrmann, C. Schenck, B. Kuhfuss, 8th Conference on Tribology in Manufacturing Processes & Joining by Plastic Deformation, Denmark (2018)
6. S.F. Golovashchenko, Project report - Electrohydraulic Forming of Near-Net Shape Automotive Panels (2013)
7. L. Langstädtler, H. Pegel, M. Herrmann, C. Schenck, D. Stöbener, J.F. Westerkamp, A. Fischer, B. Kuhfuss, International Conference on High Speed Forming, Columbus - Ohio, (2018)
8. M. Löffler, H.A. Wieland, International Conference on Pulse Power Applications, E.16/1-7, (2001)

Bulk Metal Forming of Additively Manufactured Elements

Thomas Papke[1*], *Daniel* Junker[1], *Michael* Schmidt[2], *Tobias* Kolb[2], and *Marion* Merklein[1]

[1]Friedrich-Alexander-Universität Erlangen-Nürnberg, Institute of Manufacturing Technology, Egerlandstr. 13, 91058 Erlangen, Germany
[2]Friedrich-Alexander-Universität Erlangen-Nürnberg, Institute of Photonic Technologies, Konrad-Zuse-Str. 3, 91052 Erlangen, Germany

Abstract. Current trends in the industry go towards individualised parts in high volume production. Facing the high amount of different parts, efficient processes with a high flexibility are necessary. Within this context, laser-based additive manufacturing of metal parts become more important. These processes enable the production of individual geometries using a variety of metal materials. However, main challenges of this technology are rough surfaces, limited geometric precision and varying material properties. Therefore machining after the additive manufacturing process is a common approach for improving surface quality and geometric accuracy. However, the machining process affects the grain structure at the edges of the part. Using forming instead leads to a deformation of the grains but does not create new interfaces. Furthermore, the material properties can be improved due to work hardening and fibre flow. Within this work, a bulk metal forming process of an additively manufactured element of stainless steel 316L will be analysed. The aim of the investigation is a fundamental understanding of the forming behaviour and the material flow of an additively built metal part compared to the conventionally manufactured material. Therefore mechanical properties of both materials will be identified by upsetting tests. Additionally, fibre flow and Vickers hardness before and after the forming operation are analysed. It is detected that the forming behaviour of the additively manufactured material is strongly influenced by the layer-wise building process.
Keywords: Additive Manufacturing, Cold Forming, Steel

1 Introduction

A current trend in the industry is an increasing variety and the individualization of products on costumers desires. This leads to shorter product life cycles and therefore the necessity of highly flexible processes to face the quick changes of the product development. Within this context, laser-based additive manufacturing processes have become more important in the last few decades and changed from a prototyping technology to a production technology [1]. Additive manufacturing, in general, uses the principle that every three dimensional part can be divided up into infinitesimal small layers. As each layer describes the cross section of the part at this specific plane, the part also can be produced by joining these layers one after another [2]. However, the assumption of infinitesimal layers is idealised for real manufacturing processes. The layer thickness is influenced by process parameters like e.g. the size of the powder or the recoating process.

Not just design parts made by plastics can be produced by additive manufacturing but also functional metal parts for example by selective laser beam melting (LBM). Within this process, a thin layer of metal powder is applied onto a building plate by using a recoater. Subsequently, a laser beam melts the powder to manufacture the cross section before the building plate is lowered and the next layer of powder is applied. To prevent the molten metal from oxidation during the process the building plate is placed in a building chamber that is flooded with an inert gas like nitrogen or argon [3]. The advantages of the additive manufacturing technology are the high geometrical freedom for designing components and the high material efficiency as only the needed material for the part is used and remaining powder can be saved for the next building process. The main disadvantages are a poor geometrical accuracy due to the layer-wise production and a rough surface caused by powder particles sticking to the part. Furthermore, the mechanical properties highly depend on process parameters, homogeneity of the powder material, machine maturity and instabilities during the process, therefore, the reproducibility of parts built by LBM is still a challenge. For this reason, the mechanical properties have a high deviation between additively manufactured parts compared to components build by a forging process [4]. The main advantage of forging processes is the formation of a fibre flow and therefore relatively high and constant mechanical properties as well as a high geometrical accuracy for each manufactured part which is favourable for dynamically stressed parts.

Current research within the Collaborative Research Centre 814 "Additive Manufacturing" investigates the combination of additive manufacturing and sheet metal forming. Therefore, functional elements are added onto a sheet metal by LMB and this hybrid part is formed by deep drawing [5] or bending [6] afterwards whereas the

* Corresponding author: thomas.papke@fau.de

forming operation is primarily applied to the sheet metal. The aim is a fundamental knowledge of interactions combining these two technologies and to design process chains with a high degree of flexibility and the productivity of forming processes.

The objective of the investigation presented in this work is to analyse the forming behaviour of additively manufactured elements within a bulk metal forming operation. A possible application for this approach would be a hybrid semi-finished product that gets the final shape in a combined sheet metal and bulk metal forming process. In order to determine the forming behaviour of additively manufactured elements by LBM, an upsetting test is conducted as a simple bulk metal forming operation. Using this test the forming behaviour as well as the mechanical properties can be analysed. The material stainless steel 316L (1.4404) is chosen. This steel is common in additive manufacturing and can also be used for cold forging operations. To evaluate the forming of additively manufactured elements they are compared with conventionally manufactured specimens. Additionally, the microstructure and the resulting work hardening after forming are analysed.

2 Experimental Setup

2.1. Laser Beam Melting

At this point, the manufacturing of high-quality parts with 316L is possible with commercially available machine parameter sets for this kind of stainless steel. However, the properties of metal parts manufactured with laser beam melting depend strongly on the powder material used [7]. Hence, the material quality is an important objective in the research of additive manufacturing. The particle size distribution, the chemical composition and the humidity of the powder affect its usage. Therefore an analysis of the used powder is conducted. The humidity is measured with a Karl-Fischer Coulometer (KF 860 Thermoprep) and determined the mass of the contained water is 145.8 µg inside a 7.19 g sample, so the content is below 0.002 %. Fig. 1 shows a scanning electron microscope (SEM) image of the used powder.

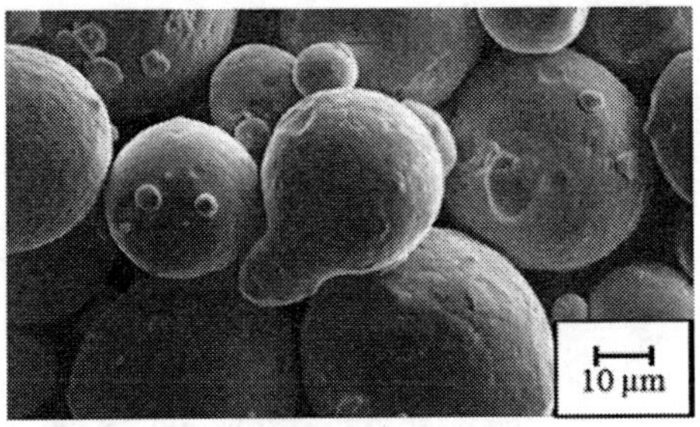

Fig. 1. SEM image of particle shape

The chemical composition is determined with an inductively coupled plasma-optical emission spectroscopy (ICP-OES) and by a combustion analyser at an external testing laboratory. The analysed chemical composition is shown in Table 1. Furthermore, the particle size distribution of 25 µm to 45 µm is measured by dynamic image analysis (Camsizer Retsch Technologies).

Table 1. Chemical composition analysis of the used powder by plasma-optical emission spectroscopy in wt-%

C	Cr	Fe	Mn	Mo	Ni	P	S	Si
0.017	18.4	Rest	0.970	2.16	12.7	<0.020	0.007	0.620

The test specimens for the bulk forming operation have been produced with a M2 machine of the company Concept Laser GmbH. The machine is equipped with a 400 Watt laser and the models are sliced in 25 µm layers. The scan parameters use a hull-core-island strategy according to [8]. To guarantee an isotropic microstructure the scanning orientation is switched for each layer. For a simple comparison, the volumetric energy input E_v for the different regions is calculated by dividing laser power P by the multiplication of scan velocity v, hatch distance h and the layer height l [3] as shown in equation (1):

$$E_v = P \,/\, (v\,h\,l) \tag{1}$$

The parameter set of the hull is defined with $E_v = 89.4$ J/mm³ and for the core area it is defined with a value of $E_v = 86.6$ J/mm³. The hull-core strategy is often applied by various machine manufacturers since it decreases the manufacturing time as the core can be molten with less accuracy only every second layer. The small layer height of 25 µm is chosen in order to produce parts with a relative density higher than 99.9 %. To get a constant quality of the additively manufactured material cylinders with a diameter of 50 mm and a height of 55 mm are produced. The test specimens are extracted from these cylinders.

To analyse the strength and the forming behaviour of the additively and the conventionally manufactured material 316L, cylinders with a diameter of 10 mm and a height of 15 mm are tested. First, bars with a diameter of 11 mm are extracted from the additively manufactured cylinder in the longitudinal direction by electrical discharge machining. Second, to reduce surface effects of the electrical discharge machining the final geometry is realized by turning. For the specimen of conventional material, hexagonal bars are taken as raw material and the samples are also manufactured by electrical discharge machining and finished by turning.

2.2 Forming Test

To investigate the forming behaviour of additively manufactured material an unidirectional forming is realised by an upsetting test. It allows the identification of mechanical properties and the analysis of the material flow under compression load.

https://doi.org/10.1051/matecconf/201819003002

The upsetting tests, schematically sketched in Fig. 2, are conducted on a universal testing machine LFEM 300 from the company Walter+Bai with a maximum force of 300 kN. Due to the high load during compression testing, a carbide tool is used. For constant low friction conditions, a Teflon film is put in between the specimen and the punches. The tests are conducted in accordance with DIN 50106 [9] for upsetting tests with a compression velocity of 5 mm/min and the specimen is compressed to 50 % of the initial height or until the set force limit of 200 kN is reached.

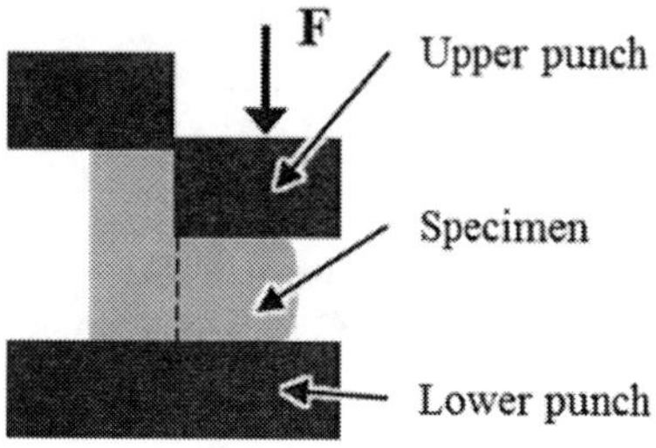

Fig. 2. Schematic principle of upsetting test

To analyse the forming behaviour the stress-strain curve can be determined from the machine data. For avoiding deviations by deflection of the machine frame an elastic correction curve of the machine is recorded prior to the tests. Additionally, the tests are recorded by CCD-Cameras of an ARAMIS System for an in-situ analysis of the forming behaviour.

2.3 Metallographic Analysis

In order to investigate the microstructure of additively manufactured and conventionally manufactured material cross sections are analysed after the upsetting tests. After polishing the specimens are etched with V2A-etchant for 20 s to visualize the microstructure.

2.4 Hardness Measurement

For a comprehensive investigation of mechanical properties, the hardness of the specimens is measured before and after the forming process. The measurements are conducted with the Fischerscope HM2000 from the Helmut-Fischer GmbH that allows a high-resolution measurement of the Vickers Hardness (HV0.5). To investigate the influence of work hardening on the hardness, a grid of measurement points is used. The distance between the grid points is 0.3 mm in vertical and horizontal direction. The hardness before the forming operation is measured at several points and identified as the initial hardness.

3 Results and Discussion

3.1 Upsetting Test

To analyse the mechanical properties stress-strain curves are calculated from machine data recorded during the upsetting tests. Fig. 3 shows the curves, representing mean values, of conventionally and additively manufactured material exemplarily.

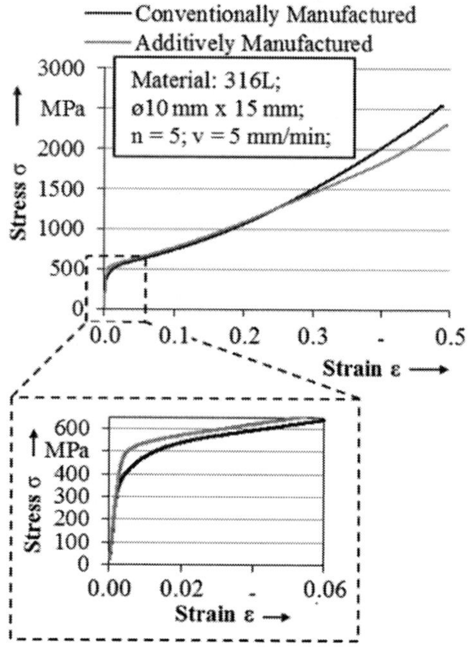

Fig. 3. Stress-strain-curves for conventionally and additively manufactured material calculated from machine data

In the beginning, the elastic behaviour of both materials is similar. In the further evolution of the curve a difference of the yield strength, which represents the beginning of plastic forming, is analysed. The additively manufactured specimen has a yield strength (508.7 ± 2.4 MPa) which is 25 % higher than that of the conventional material of 408.8 ± 1.1 MPa. Afterwards, the conventional material has a higher work hardening until a strain of 0.02 before the curves are progressing parallel, although the stress of the additively manufactured material is slightly higher (Fig. 3). At a stress value of about 1000 MPa and a strain of about 0.2 the work hardening of the additively manufactured specimen is decreasing and at about 1300 MPa the stress-strain curve crosses that of the conventional material at a strain of about 0.25. Until the crossing of the curves the materials' properties are reproducible. One reason for the decreasing stress could be shearing effects inside the additively manufactured specimen that ends in the presented distortion after the upsetting test.

To support the explanation of the flow behaviour Fig. 4 gives an overview of the change in shape of the specimens for strains from 0.1 to 0.4 and after the forming test. For strain 0.1 the shape of the specimen for both materials can be described as cylindrical. However, for strain 0.2 a beginning distortion of the specimens' outer shape orthogonal to the forming direction can be determined for the additively manufactured material. In contrast, the shape of the conventionally manufactured material remains cylindrical. This could explain the deviation of the flow curves, which begins at this strain.

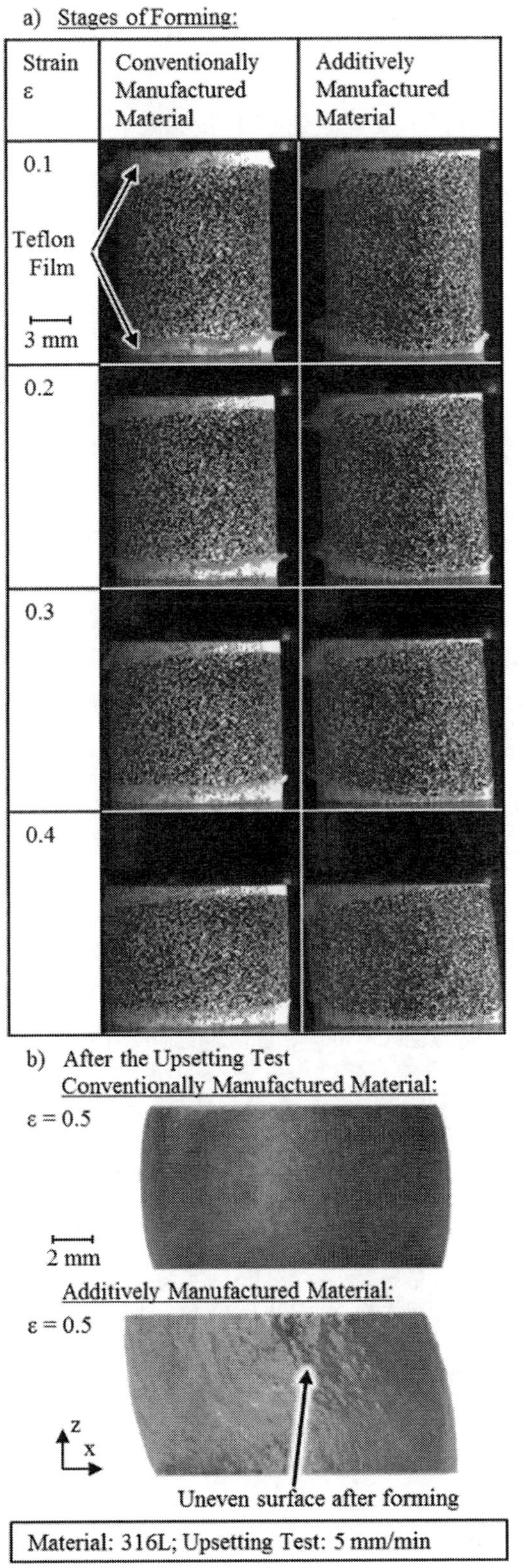

Fig. 4. Shape of specimens for conventionally and additively manufactured material for a) subsequent stages of forming and b) after the upsetting test

The difference in the shapes increases for higher strain values as 0.3 and 0.4 until the final shape after the forming process is reached. As shown in Fig. 4 b), the specimens of the upsetting tests of the conventionally manufactured material result in a symmetrical cylinder as it is common for this test. Contrary to this, for the upsetting tests of the additively manufactured material a distortion of the specimen is observed for each of the five repetitions, exemplarily shown in Fig. 4 b). Furthermore, an uneven surface of the additively manufactured specimen results from the free forming operation. That indicates a different forming behaviour of the additively manufactured material in contrast to the conventional material.

3.2 Metallographic Analysis of Flow Behaviour

It is assumed that the inhomogeneous forming behaviour is caused by the specimen's microstructure, therefore the initial state of the microstructure of conventionally and additively manufactured material is analysed by polished and etched cross sections, shown in Fig. 5.

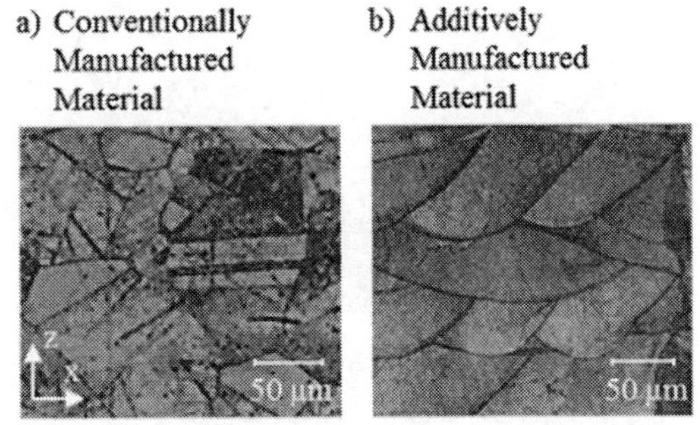

Fig. 5. Analysis of the microstructure before forming of a) conventionally and b) additively manufactured material

For the conventional material, a small grained structure can be identified compared to the additively manufactured material. Within the additively manufactured part, the microstructure is dominated by the shape of welding beads representing the single layers, which correspond to the scan track melt pools. In the additively manufactured structure, just a few beads are grown over several adjacent layers.

To analyse the fibre flow due to the upsetting test, the specimens are analysed after the forming process. Fig. 6 shows the fibre flow for the formed conventionally and additively manufactured material.

The fibre flow of the conventionally manufactured material shows the typical shape after an upsetting specimen. Analysing the change of the direction of the fibres relative to the punch direction (z-direction), the microstructure in the cross section can be divided into highly and hardly formed areas. Along the diagonals, slip bands caused by the forming during the upsetting test are identified. However, the fibre flow of the additively manufactured material does not show a similar orientation. Due to the etching of the cross section, the layers caused by the additive manufacturing process are visible. The separation of the microstructure in highly and hardly formed areas as it was identified for the conventionally manufactured material is not possible. Based on the orientation of the layers after the upsetting test, a distortion of the upper side of the specimen relative to the lower side along the x-axis can be seen.

a) Conventionally Manufactured Material:

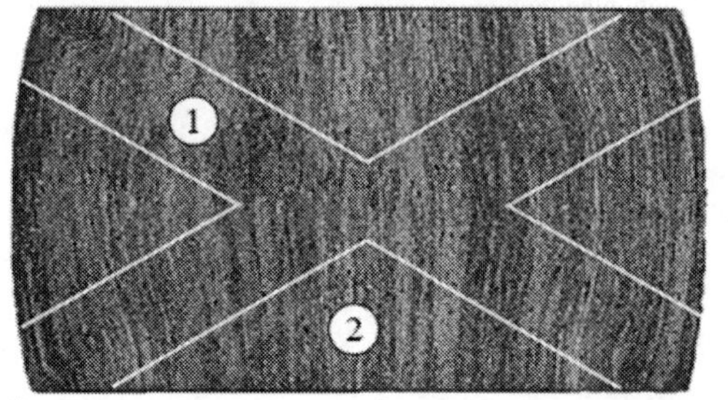

① Highly Formed Area ② Hardly Formed Area

b) Additively Manufactured Material:

Direction of Distortion

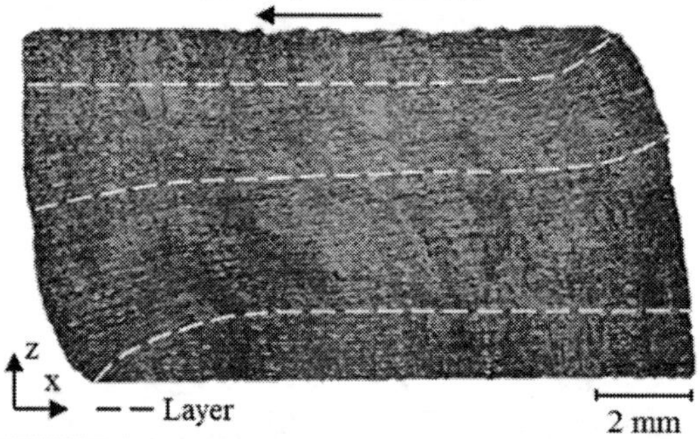

Material: 316L; Etching: V2A-Etchant, t = 20 s

Fig. 6. a) Conventionally and b) additively manufactured etched cross sections of specimens after upsetting tests

An explanation that can be assumed regarding the cross section shown in Fig. 6 b) could be the slide of the layers onto each other due to the forming process. Commonly known for forming is the formation of slip bands under high loads, on which the material slides during deformation [10]. The initiation of slip bands could start at the corners of the specimen in Fig. 6 a), which is indicated due to the diagonal alignment of the fibres. As LBM in principle is a welding process, a heat affected zone results for each layer and causing brittle phases [11]. Thus a possible explanation may be that while a force causes the forming of the specimen the weak areas of the layers begin to deform whereas the brittle phases within the heat affected zones have higher yield strength and resist the deformation. Reaching the maximum formability, the weak areas start to slip onto the brittle layers.

3.3 Hardness Measurement

For a comprehensive investigation of the mechanical properties, the Vickers hardness (HV0.5) is measured before and after the forming process. For the initial state, the mean hardness value of the conventionally manufactured material is 245.45 ± 16.35 HV0.5 compared to 257.30 ± 15.49 HV0.5 of the additively manufactured material. Therefore, no significant difference can be measured that could explain the difference of 100 MPa between the yield strengths.

A hardness mapping for both materials after the upsetting test is shown in Fig. 7. The measuring grid for specimens of both materials after forming is set to 0.3 mm.

a) Conventionally Manufactured Material:

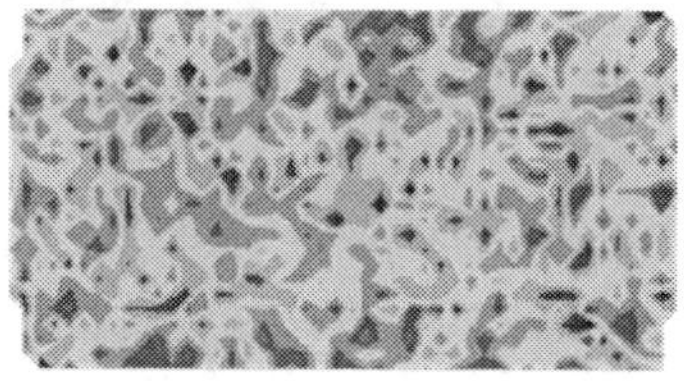

b) Additively Manufactured Material:

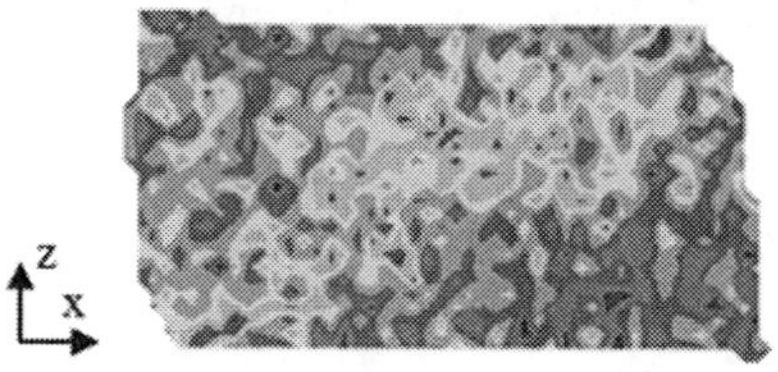

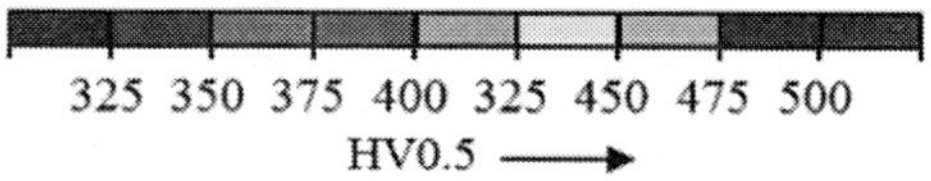

Fig. 7. Mapping of hardness values for a) conventionally manufactured material and b) additively manufactured material specimens after upsetting test

The analysis of the conventionally manufactured specimen depicted in Fig. 7 a), shows work hardening with a minimum value of 350 HV0.5 within the whole cross section, which is about 100 HV0.5 higher than that measured for the unformed sample. Higher values up to 560 HV0.5 are mainly measured within the highly formed areas. In comparison, the work hardening of the additively manufactured specimen up to more than 500 HV0.5 takes place along the diagonal slip band, whereas the hardness in the adjacent material only increases about 50 HV0.5 to a value of about 300 HV0.5. However, for the additively manufactured material high hardness values are located along the diagonal direction, which is also indicated by the distortion of the layers along the x-axis in Fig. 6 b). The increase in hardness could explain the distortion of the upper side of the specimen relative to the lower side.

4 Conclusion

Regarding the results of the presented investigations, it can be concluded that the layer-wise structure of additively manufactured material influences the forming behaviour. It is obvious that forming of additively manufactured material differs significantly from forming conventionally manufactured material what could be verified by stress-strain curves and the analysis of the microstructure after forming. Although no significant differences in the hardness of the initial materials can be measured, the layer-wise structure caused by the additive manufacturing process has a huge effect on the material flow during forming.

Concluding the presented results, forming of additively manufactured material can be used like conventional material until the maximum strain is lower than 0.2. For higher strains, the forming behaviour becomes significantly different. An analysis of the specimen geometry during the upsetting test shows distortion of the outer shape orthogonal to the forming direction. In contrast, the shape of conventionally manufactured specimen remains symmetrical cylindric due to the forming operation, which is typical for the upsetting test. The change in the forming behaviour of the additively manufactured material could be indicated in the stress-strain-curve as well. For stresses higher than 1300 MPa the work hardening decreases. Therefore, it could be concluded, that for higher deformation the layer-wise structure of the additively manufactured material has to be taken into account. It seems that every layer has to be considered separately and that the heat affected zone appears as a frontier between the layers. To prove this hypothesis, further investigations are necessary for example by analysing the microstructure by a selective electron microscope and electron backscatter diffraction.

5 Summary and Outlook

As the individualization of products is a current trend in industry additive manufacturing is getting more interesting for production. One disadvantage of this technology is a poor reproducibility of the mechanical properties compared to those realized by forming processes. Therefore, bulk metal forming of additively manufactured material is investigated within this work.

To analyse the forming behaviour upsetting tests for conventionally manufactured material and additively manufactured material have been carried out. During these tests, a distortion of the additively manufactured material was observed. Furthermore, the stress-strain curves and the material properties such as the Vickers hardness depending on the material flow are determined also. The results show higher yield strength for additively manufactured material compared to the conventionally manufactured material. Afterwards, the flow behaviour is analysed by visualization of the fibre flow after the forming process by investigating etched cross sections of the specimens. It was claimed that the layers of the additively manufactured material show a distortion orthogonal to the forming direction, which could be caused by brittle phases between the layers.

For a comprehensive investigation of the mechanical properties and visualization of the forming zone, a mapping of the Vickers hardness is conducted for both materials after the upsetting test. As expected, the conventionally manufactured material shows higher hardness values in highly formed zones, which are confirmed by the fibre flow. However, the additively manufactured material shows high hardness values in the diagonal direction of the cross section. Therefore, it can be seen that the hardness distribution for both materials after the forming process is different. This result could support the hypothesis that the forming behaviour is influenced by the layer-wise structure.

Further investigation will deal with the analysis of phases and the mechanical properties of the single layers caused by laser additive manufacturing. For fundamental investigation of forming additively manufactured material, the layer-wise structure has to be taken into account. Furthermore, the influence of a forming process on the orientation of the layers should be investigated by bulk metal forming processes.

This research work has been founded by the German Research Foundation (DFG) within the scope of the Collaborative Research Centre 814 "Additive Manufacturing" subproject B5.

References

1. M.Schmidt, M. Merklein, D. Bourell, D. Dimitrov, T. Hausotte, K. Wegener, L. Overmeyer, F. Vollertsen, G.N. Levy, CIRP Annals **66/2** 561-583 (2017)
2. A. Gebhardt, *Generative Fertigungsverfahren* **4** (2013)
3. W. Meiners, *Direktes selektives Laser Sintern einkomponentiger metallischer Werkstoffe* (1999)
4. B. Ahuja, A. Schaub, D. Junker, M. Karg, F. Tenner, R. Plettke, M. Merklein, M. Schmidt, *Proc. 15th Annual Conference of RAPDASA*, 30-42 (2014)
5. A. Schaub, J. Degner, A. Bhrigu, M. Schmidt, M. Merklein, *Int. Conf. on Comp. Manuf.*, (2016)
6. M. Merklein, P. Dubjella, A. Schaub, L. Butzhammer, M. Schmidt *6th Int. Conf. on Add. Tech.* 309-316 (2016)
7. A. Cooke, J. Slotwinski, *NISTIR 7873* (2012)
8. S. Kumar, J.-P. Kruth, Advanced Engineering Material **10.8** 750-753 (2008)
9. DIN50106:2016-11 *Testing of metallic materials – Compression test at room temperature* (2016)
10. K. Lange, *Umformtechnik* (1984)
11. G. Schulze, *Die Metallurgie des Schweißens* (2010)

Generation of residual stresses in rotary swaging process

Dhia Charni[1,*], *Svetlana* Ishkina[2,3], *Jérémy* Epp[1,4], *Marius* Herrrmann[2,3], *Christian* Schenck[2,3,4], *Hans-Werner* Zoch[1,3,4], and *Bernd* Kuhfuss[2,3,4]

[1]Leibniz Institute for Materials Engineering - IWT, Badgasteiner Str. 3, 28359 Bremen, Germany
[2]Bremen Institute for Mechanical Engineering - bime, Badgasteiner Str. 1, 28359 Bremen, Germany
[3]University of Bremen
[4]MAPEX Center for Materials and Processing

Abstract. Infeed rotary swaging is an established incremental cold forming production technique for axisymmetric workpieces. Among others, work hardening as well as near net shape forming are advantages of this production technique. Due to the incremental open die forging process, the rotary swaging induces a complex material flow history which is dependent on the process control. This material flow history influences the material modifications such as work hardening as well as residual stresses. In this study, the properties of steel tubes were investigated after rotary swaging using varying parameters by experimental and simulation analysis. In particular, the influence of lubrication with different feeding velocities was analysed. The workpiece quality, the hardness and the residual stresses were characterized in detail. After rotary swaging, an influence of the process parameters on the geometrical and surface quality could be observed. The workpieces showed significant work hardening which was higher at the surface and process dependent, while below 100 µm, this increased hardness was observed over the complete wall thickness independently of the process parameters. The residual stress state was highly fluctuating at the surface and was in tensile for all conditions. The results showed that the process parameters influenced the properties near the surface, while a few hundred micrometers below the surface, the workpiece properties seemed to be driven only by the total deformation.

Keywords. Cold forming, Microstructure, Residual stress.

1. Introduction

Infeed rotary swaging is a non-cutting chipless cold forming process and has an important field of application in the automotive industry. Commonly it is used to produce axisymmetric components, such as axes, steering spindles and gear shafts [1]. Thereby the cross section is incrementally reduced by repeated radial strokes of oscillating dies [2]. During the process, the swaging unit is rotating and the base jaws are passing the cylinder rollers which induces simultaneously radial short strokes of the dies, as shown in Figure 1. The forming process has several positive effects on the produced workpieces like undisturbed fibre flow as well as improved material properties like increased tensile strength and hardness due to work hardening [1]. Furthermore, especially for the production of hollow parts, an adjustable wall thickness production is possible to yield an optimal use of material resources for lightweight applications [3].

Due to the incremental open die forging process, rotary swaging takes place with a complex material flow which is dependent on the process control. This can be shown by the neutral plane shape and position which in general cases is located in the forging zone [4]. The neutral plane is defined as the plane in the workpiece in which the material is only deformed radially whereas alongside the neutral plane, the material is deformed radially and axially and flows away from the neutral plane. Lahoti and Altan [5] studied the influence of process parameters on the location of the neutral plane and concluded amongst others that the neutral plane shifted towards the preform (undeformed workpiece part) when the friction increased or when back push or front pull was applied. Consequently, more material flowed in the forward direction [4, 5]. In another work, Wu et al [6] analysed the influence of several process parameters on the neutral plane that appeared to be not a planar cross section. Moumi found out that even its shape and its position changed during one single stroke [4].

This non-steady material flow has also been investigated by Liu. He showed that the material flow was fastest at the beginning of the stroke and then started to slow down to the end of this stroke [7]. This non steady material flow influenced the complex material modification such as work hardening as well as residual stresses [8].

Due to the use of components produced by rotary swaging under static and dynamic loads it was necessary to investigate the mechanisms of residual stress generation in order to be able to predict and control the

*Corresponding author: charni@iwt-bremen.de

mechanical properties [9]. Ameli and Movahhedy investigated the residual stress distribution after rotary swaging process and found that these were significantly influenced by the incremental degree of the resulted deformation. It was found for tubular components that on the outer surface mainly tensile and on the inner surface mainly compressive residual stresses were present [10].

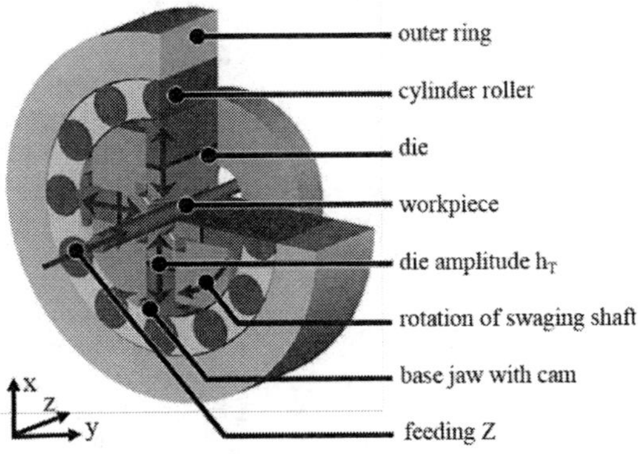

Fig. 1. Schematic operation of rotary swaging machine.

Since the final product shape can be achieved by different material flow histories, rotary swaging is a particularly suitable forming process to generate various residual stress and work hardening distributions. To achieve controlled material modifications, it is necessary to understand the interdependency between the material flow, which is controlled by the process parameters and the resulting material modifications - in particular the residual stress state. As a consequence of the induced work hardening and residual stress distribution, it is expected that various static and dynamic mechanical properties can be achieved without modification of the component geometry itself but just by tuning the rotary swaging process parameters.

For this qualitative study, infeed rotary swaging with different process conditions was performed. The local work hardening and microstructural evolution was characterized. Additionally, residual stress distributions generated by this process were investigated. The residual stresses were measured and broad variations were observed. Based on these results, a first assessment of the influence of the rotary swaging process control on the resulting material modifications could be achieved.

2. Experimental setup

2.1 Rotary swaging

The material used for this study is cold drawn E235JR steel tubes of $l_0 = 300$ mm initial length, $d_0 = 20$ mm initial diameter and $s_0 = 2$ mm initial wall thickness. The tubes exhibited a zinc phosphate coating of about 1 µm thickness as usually applied for corrosion protection or friction reduction to decrease die wear. Single pass rotary swaging experiments were carried out on these tubes, using a rotating shaft at 300 rpm, reducing the diameter down to $d_1 = 15$ mm. The stroke following angle that is defined as the angle of rotation around the z-axis between two successive strokes was about 51°. The used dies were conventional powder-metallurgical ASP®2023 made of 1.2379 steel with a die angle $\alpha = 10°$ and a calibration length $l_{cal} = 20$ mm. The reduction zone of these dies featured a spray-coated tungsten carbide layer (see Figure 2) which increased the effective friction between the dies and the workpiece to reduce the axial reaction force of the preform [2]. For all experiments the stroke height of the dies was set to $h_T = 1$ mm and the stroke frequency to $f_{st} = 37.5$ Hz. The tubes were fed with a linear direct drive into the swaging unit with four different feed velocities ($v_f = 500$ mm/min, 1000 mm/min, 1500 mm/min and 2000 mm/min) and two different lubrication conditions (with and without lubrication), see Table 1.

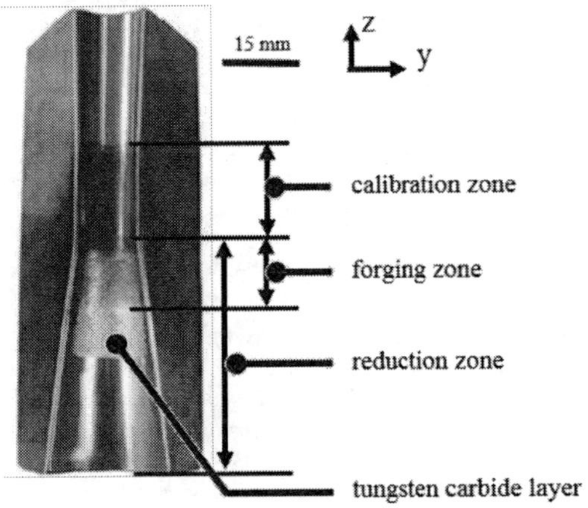

Fig. 2. Typical swaging die (one out of four).

Table 1. Design of experiments.

Sample	Lubrication condition	Feed velocity (mm/min)	Feed per stroke (mm/stroke)
Init	Initial condition before rotary swaging		
L500	With lubricant	500	0.22
L1000		1000	0.44
L1500		1500	0.66
L2000		2000	0.88
D500	No lubricant (dry)	500	0.22
D1000		1000	0.44
D1500		1500	0.66
D2000		2000	0.88

From the stroke frequency and the feed velocity, a feed per stroke v_{st} can be deduced. Thus the lowest $v_f = 500$ mm/min matches $v_{st} = 0.22$ mm/stroke and the highest $v_f = 2000$ mm/min matches $v_{st} = 0.88$ mm/stroke. The lubricant used was a mineral oil Condocut KNR 22. Before each dry rotary swaging the dies and workpieces were cleaned thoroughly using ethanol. In addition, the samples were mounted in a clamping device at the swaging machine. They were clamped with a moderate tension force allowing the samples to rotate to minimize the torsion of the tubes that can be induced by the rotating dies upon stroke contact.

2.3 Residual stress measurement

The residual stresses were measured on selected samples (Init, L500, L2000, D500 and D2000) by X-Ray diffraction (XRD) using a GE inspection technologies diffractometer Type ETA 3003 with vanadium filtered Cr-Kα radiation. The {211} diffraction peak of α-iron was evaluated using the conventional $\sin^2\psi$-method [12]. Axial and tangential residual stresses were evaluated along a path of 20 mm on the outer surface of the tubes with a step size of 1 mm starting at an arbitrary position (Figure 3). The beam diameter was 1 mm. The full width at half maximum (FWHM) of the diffraction peaks was also evaluated in order to obtain additional information on the microstructure and the microstresses. In cold forming processes of non-hardened steel, the FWHM correlates well with the hardness and allows a qualitative evaluation of its distribution [13]. Residual stress and FWHM depth profiles were also measured at one position on each sample. For this, the material removal was done by electrolytic etching using a solution of H_3PO_4 and H_2SO_4.

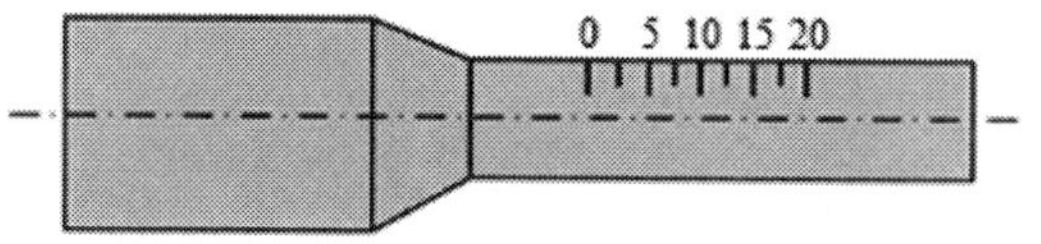

Fig. 3. Sketch of a swaged sample showing the positions of surface residual stress measurements.

2.4 Metallographic examination and hardness measurement

The microstructure of the samples was investigated after standard metallographic preparation. After cutting transverse sections from the different tubes, these were embedded, ground and polished. An optical microscope was used to examine the effect of the process on the microstructure at the outer wall surface after etching using a 3%-HNO_3 alcoholic solution. Additionally, the hardness was measured using Vickers indentations with a load of 1 N for depth profiles analysis on the cross section and 5 N for measurements directly at the outer surface. All results were calculated from the average of 3 measurements.

3 Results and discussion

3.1 Workpiece geometry and quality

The workpiece geometry was analysed after the rotary swaging experiments. The final diameter of all formed samples exhibited almost the same value (deviation less than 100 µm). Hence, differences were recognized in the final length of the tubes. For higher feed velocity the length difference was increased, also for forming without lubricant ("dry") longer workpieces were achieved, see Figure 4. Furthermore, the wall thickness is smaller than for the workpieces formed by lower feed velocity or with lower friction conditions during forming., which was seen in cross sections. This means that for these conditions, the material flow in axial direction is higher.

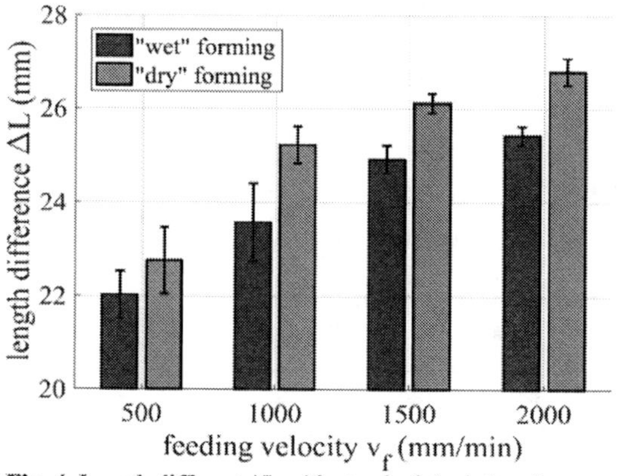

Fig. 4. Length different ΔL with standard deviation for wet and dry forming with different feed velocities.

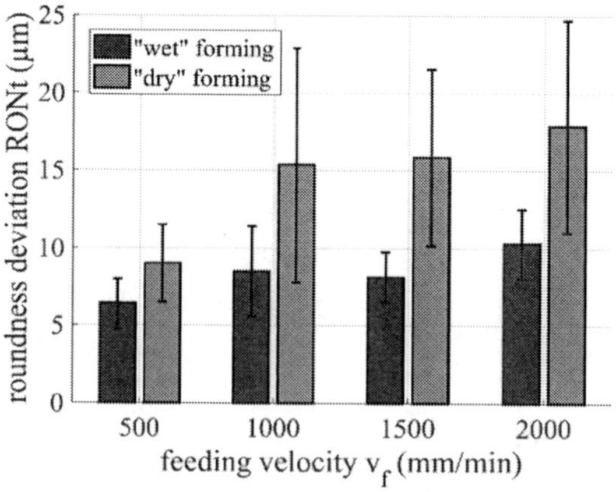

Fig. 5. Roundness deviation RONt with standard deviation for wet and dry forming with different feed velocities.

Additionally, the roundness deviation was measured, see Figure 5. It could be observed that with higher feed velocity the roundness deviation increased. Furthermore, dry forming compared to forming with lubricant led to higher roundness deviation as well as higher scatter.

https://doi.org/10.1051/matecconf/201819004001

To examine the workpiece quality surface roughness measurements were carried out. For the forming with different feeding velocities it was recognized, that with increasing v_f also the surface roughness Sa increased, see Figure 6. Due to the faster infeed of the workpiece the surface was deformed by less strokes and the deformation per stroke was higher. Thus a higher surface roughness Sa could be explained. This was also observed for dry forming. The higher roundness deviation and the surface roughness as well as the higher standard deviation of these properties during dry forming could be explained by the higher amount of the produced abraded particles which adhere on the surface of the workpiece and the dies due to the absent flushing of the forging zone by the lubricant [14].

The different number of strokes that formed the workpiece led to different surface developments. This can be observed, as illustrated in Figure 7, showing the surfaces of wet and dry formed workpieces with v_f = 500 mm/min (L500 and D500) and v_f = 2000 mm/min (L2000 and D2000). The surface after forming with lubricant using the lowest feed velocity showed no obvious surface pattern, see Figure 7 a). For this forming with lubricant and the lowest feeding velocity, which represents the lowest effective friction condition, the zinc phosphate coating on the workpiece can no longer be recognized on the L500 samples, see blank surface in Figure 7 a). This shows that the effective friction was low and allowed the dies to slide on the surface of the tube during contact. For the lubricated forming with the highest feeding velocity a fine texture and still some deposits of the workpiece coating (dark spots) can be recognized due to less strokes on the surface, see Figure 7.b).

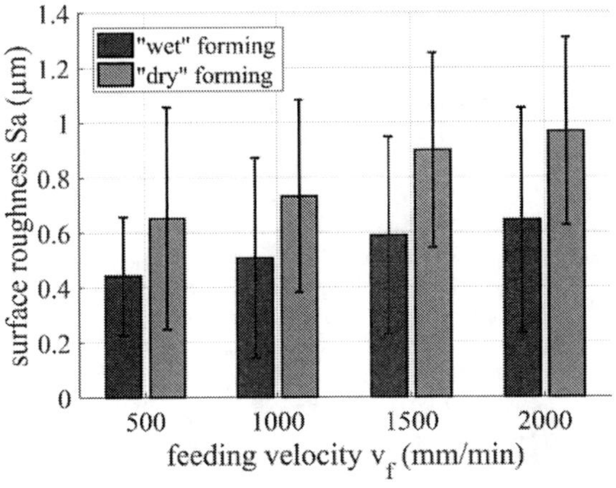

Fig. 6. Surface roughness Sa with standard deviation for wet and dry forming with different feed velocities.

However, dry forming produces a texture at the surface leading to pronounced helical patterns, see Figure 7 d). This can be explained by the rotation of the dies around the workpiece between the strokes. Due to the absent lubricant and to the high feed velocity, the effective friction is very high. Furthermore, dry processing highly increased the process forces and the contact pressures [15]. Therefore, the adhesion was increased, preventing the dies from sliding on the tube which led to a slight torsion of the tube while the dies were closed. Thus, the helical texture was more pronounced and less removing of the workpiece coating was recognized.

Moreover, it was noticed that during the dry rotary swaging with high velocities, the D2000 samples were rotating much more than the D500 samples, while the lubricated samples were almost not rotating during the process.

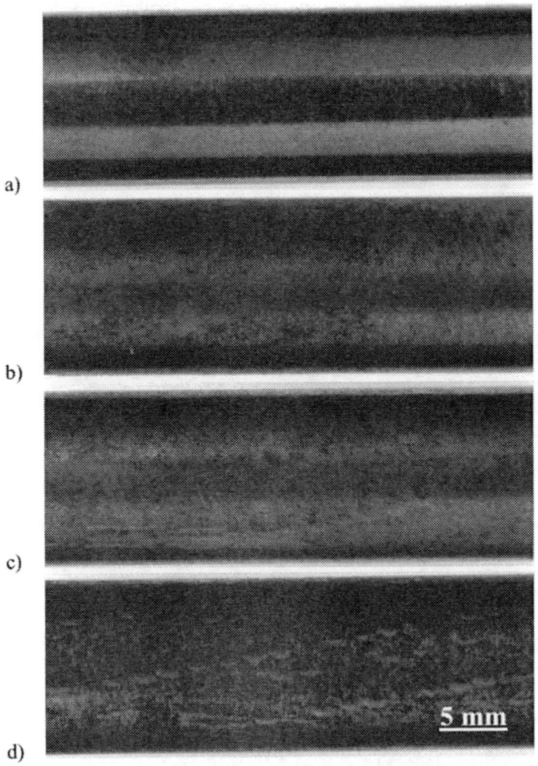

Fig. 7. Photos of the tube surfaces; a) L500, b) D500, c) L2000, d) D2000.

The observed surface patterns can also be identified on the roundness profiles of L500 and D2000 samples as the samples always exhibit an octagon-like shape. On the roundness profile of the lubricated formed tube around every 45 ° an edge on the surface of about 3 µm can be seen, Figure 8 a). This shape is more remarkably noticed on the workpiece processed with high effective friction with edges on the surface of about 6 µm, see Figure 8 b). This supports the theory of the sliding of the dies on the surface of the workpiece occurring during low effective friction rotary swaging, so that the sliding on the octagon shape is profiling round edges.

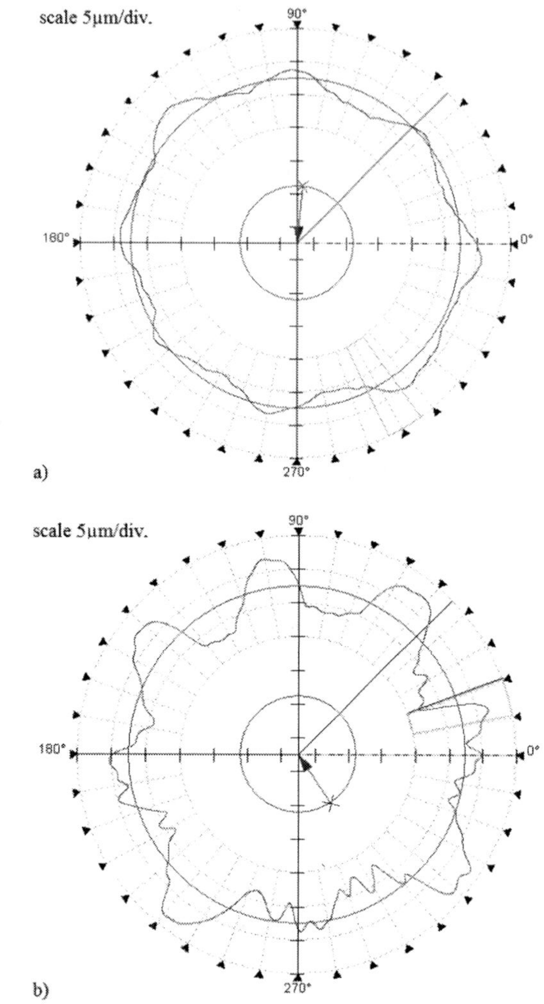

Fig. 8. Roundness profile: a) L500, b) D2000

3.2 Simulated material flow

The simulation enabled the investigation of the material flow in axial direction (forward and reverse) as well as in radial direction. The velocity magnitude and direction of the workpiece nodes during one stroke of the dies in rotary swaging process is shown in Figure 9, with feed of the samples in positive z-direction. It represents the moment of tools closing at 0.74 ms after the first contact of the workpiece with tools and 0.71 ms before the tools totally closed. The arrows show the direction and the magnitude is presented by the length of the arrows as well as the colour of the workpiece. The top section, see Figure 9 a), is illustrating the lower friction between the workpiece and the die while Figure 9 b) represents the high friction for the friction pair "coated E235JR and 1.2379 steel". It can be observed, that the higher the friction coefficient is, the higher become the velocity of the nodes in the feeding direction (z-direction, forward), see Figure 9. It means that the higher the friction is, the more material flows forwards and less backwards. For lower friction coefficient, see Figure 9 the material tends to flow backwards and thus, a back-flow of the material takes place.

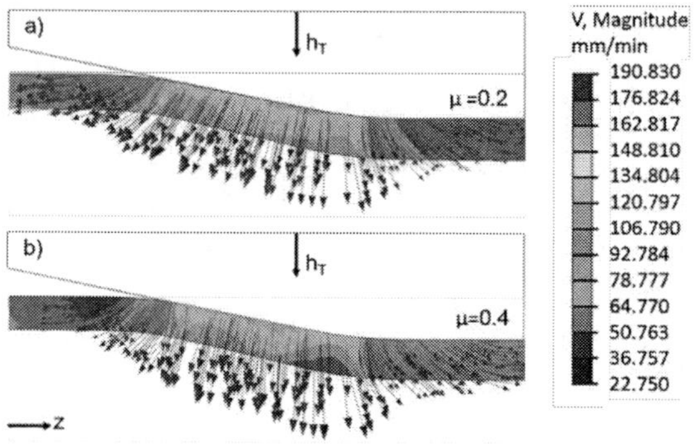

Fig. 9. Material velocity distribution of tubes during rotary swaging process at v_f = 2000 mm/min for two different friction coefficients $\mu = 0.2$ and $\mu = 0.4$.

3.3 Microstructure and hardness

The surface near microstructure of the different conditions can be seen at high magnification in Figure 10. On the lubricated samples L500 and L2000, a layer of axially deformed grains was formed within the first 30-50 µm. This could be explained by the sliding of the dies on the workpiece and thus the additional deformation near the surface. On the other hand, the dry processed samples D500 and D2000 showed only slightly deformed grains near the surface. Additionally, the micrographs show that for higher feed velocities, the microstructure tended to be less axially deformed near the surface than for lower feed rate.

The effect of the rotary swaging on the Vickers hardness depth profile is shown in Figure 11. In initial conditions, a non-homogeneous hardness distribution was present through the thickness with hardness values close to 150 HV0.1 and a decrease to one minimum of 119 HV0.1 at half the thickness. This was due to the prior cold drawing process, leading to a gradient in hardness over the entire thickness, as already observed in [16]. After rotary swaging, the measured hardness distribution was strongly increased for all process conditions and reached more or less homogeneous hardness values around 195 HV0.1. The hardness increase could be attributed to the work hardening due to the plastic deformation during the rotary swaging process. However, no clear effect of the process parameters on the hardness distribution could be identified from these results. These hardness depth profiles were completed by surface hardness measurements, see Figure 12. It can be seen that the hardness values were higher than in deeper layers at the cross section. Here again, no clear difference between the process conditions could be identified. Based on these results, it could be assumed that in this case, the work hardening effect of the process is independent of the investigated process parameters.

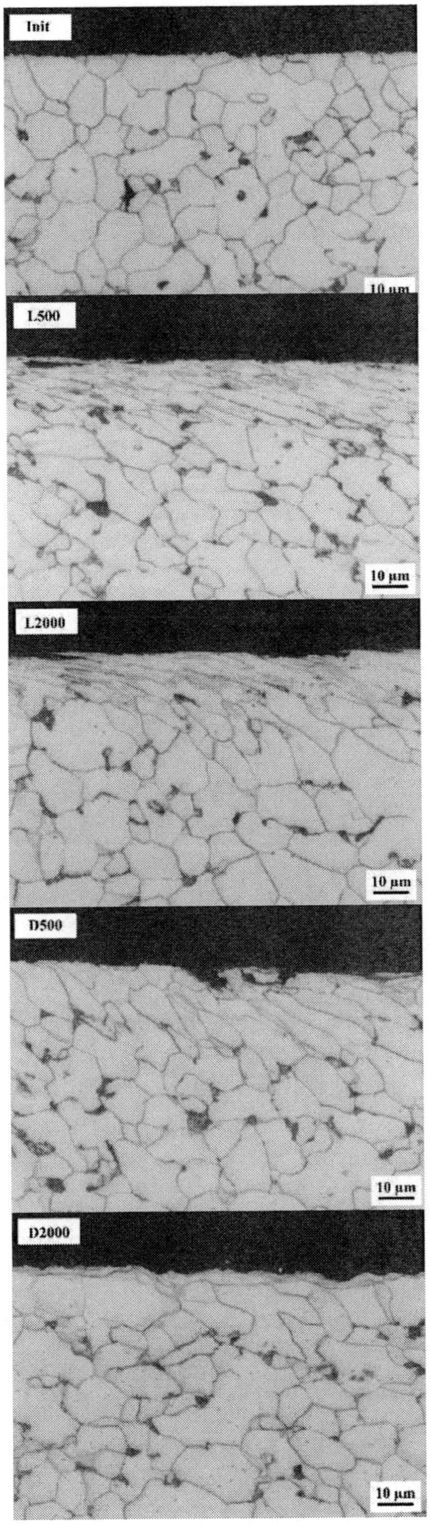

Fig. 10. Optical micrographs of longitudinal section of different samples.

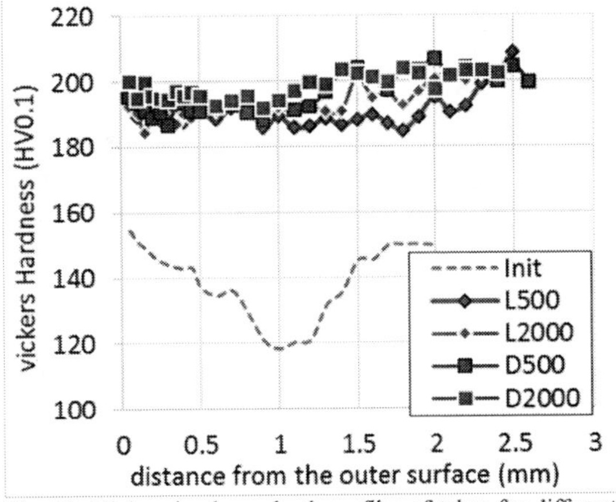

Fig. 11. Vickers hardness depth profiles of tubes for different process conditions compared to the initial state.

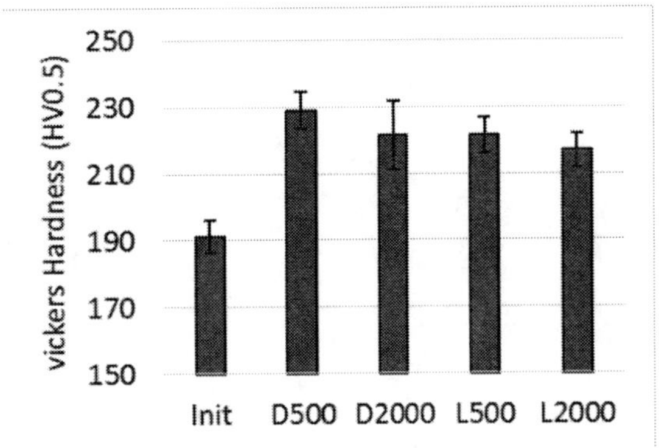

Fig. 12. Vickers hardness measured at the outer surface of tubes for different process conditions compared to the initial state.

3.4 Residual stress

On each sample, the residual stresses were measured at the surface along the axial direction of the tube. Figure 12 and 13 show the evolution of the surface residual stresses in axial and tangential directions along the axial position of the tube. In initial condition steady low compressive residual stresses in axial and slight tensile residual stresses in tangential directions, are present. After the rotary swaging process, all samples exhibit mainly tensile residual stresses in axial direction which are systematically shifted to lower tensile or slight compression in tangential directions. All measured residual stress distributions are highly fluctuating along the measured path, what can be attributed to the influence of the local deformation caused by incremental forming of the dies illustrated in the surface patterns (see Figure 7). The measured FWHM distributions are presented in Figure 14 and also exhibit variations for all processed samples. Compared to the initial state before rotary swaging which showed homogeneous low FWHM values, all processed samples exhibited strongly increased values due to work hardening as already observed with the hardness measurements.

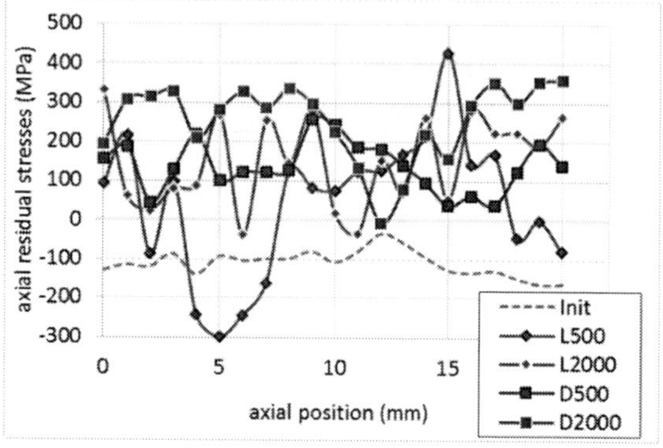

Fig. 12. Axial residual stress distribution measured along the axial direction of the tubes.

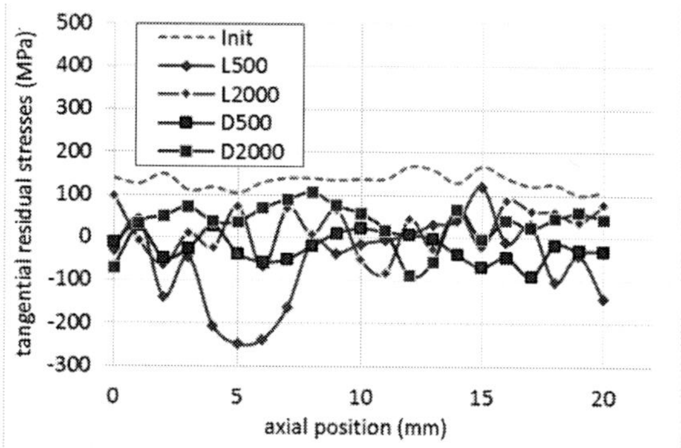

Fig. 13. Tangential residual stress distribution measured along the axial direction of the tubes.

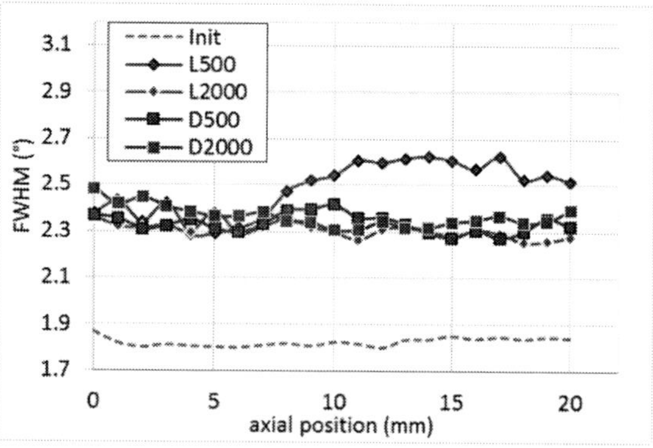

Fig. 14. FWHM distribution measured along the axial direction.

A comparison of the average values of residual stresses and FWHM measured for all samples is shown in Figure 15. Statistically, the L500 sample provided the highest FWHM values. These FWHM which reflects the local hardness correlates well with the disturbed layer under the surface found in the metallography results. The presence of a strongly deformed microstructure layer led to high FWHM by strong plastic deformation. This effect is not as pronounced for the sample L2000, while the micrographs also show a highly deformed layer. This might be explained by the fact that the surface near state is locally changing and the regions of metallographic investigations and of the XRD measurements are not exactly comparable.

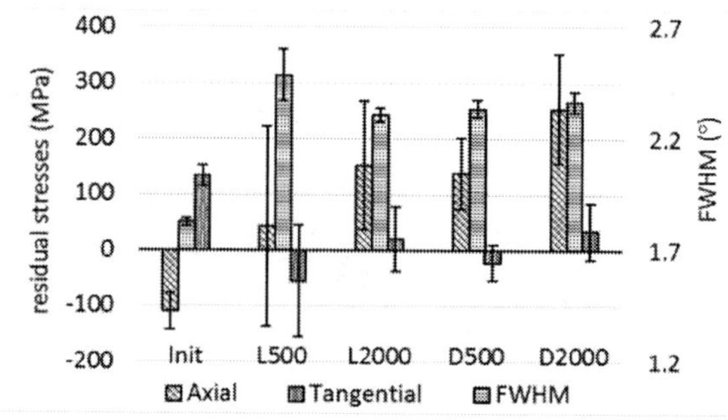

Fig. 15. Comparison of average residual stresses and FWHM at the surface.

Concerning the average residual stresses, a slight trend could be identified by the average plot in Figure 15: the high feed rate seems to induce higher tensile residual stresses than the low feed rate and the process performed without lubrication leads to the same trend. One possible explanation for this behaviour is the inhomogeneous deformation along the depth of the tube wall caused by external forces, in particular in the forging zone as observed in the simulation [17]. These stresses are result of the elastic response of the material to inhomogeneous plastic deformation.

Additionally, to the surface measurements, the axial residual stress and the FWHM were measured in the depth of previously investigated samples at a representative position in regions featuring low residual stress fluctuation. The results are shown in Figure 16 and 17. It can be observed that the initial condition was almost stress free below the surface and that the FWHM was increased in the first 50 μm while in deeper regions, slightly decreasing values were present. After rotary swaging, the residual stress state was thoroughly changed in the whole evaluated depth region. Mainly tensile residual stresses were found with decreasing values in the first 50 μm, followed by increasing tensile stresses in further depth. In 200 μm depth, the four conditions produced high tensile residual stresses that all converged to similar values. The evolution of FWHM after rotary swaging followed the same behaviour for all samples and in a depth of 50 μm no pronounced differences could be observed anymore. From these results it seems that for the considered process, the rotary swaging process parameters influenced the surface near work hardening and the residual stress state in the first 200 μm, while in deeper depth the effect of the swaging conditions became less important. The differences at the very surface can be explained by the presence of a variously deformed layer. Additional influence of the heat generated by plastic deformation during rotary swaging is expected to play an important role in the generated residual stresses and local work hardening. The absence of extensive lubrication that flushes away the abraded particles can also influences the thermal impact of the process.

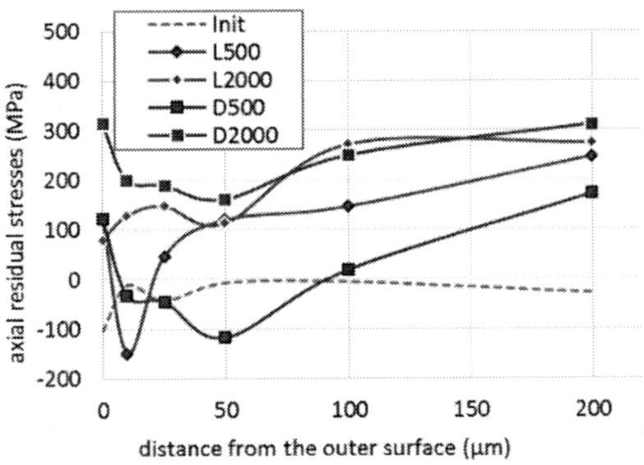

Fig. 16. Axial residual stress depth profile of the tubes.

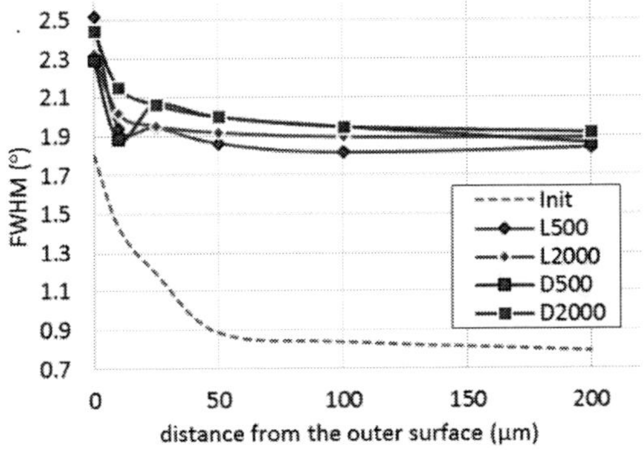

Fig. 17. FWHM depth profile of the tubes.

4. Conclusion

In this study, workpiece quality, material flow, the residual stress and the microstructure of steel formed with different rotary swaging parameters have been investigated. The results show that the workpiece quality is highly influenced by the feeding velocity of the process as well as the lubrication condition. Thus with higher feed and higher friction a higher roundness deviation and surface roughness was produced. Furthermore, with higher friction coefficient (dry process) more material flow occurred in axial direction which is illustrated by the experimentally measured elongation of the tubes as well as by simulation. This increased axial material flow correlates in this case of rotary swaging to a slightly higher tensile residual stress in axial direction at the investigated positions. The increase of hardness due to work hardening is pronounced at the very surface and then almost constant over the entire wall thickness. It could be shown that in the investigated parameter range, the hardness increase is almost independent on the used process parameters. The microstructure is mainly influenced near the surface by the friction condition. The obtained results can be explained by two simplified scenarios: Low effective friction lubricated rotary swaging process involves sliding of the dies on the workpiece surface. Additionally, this leads to an increased axial back flow of the material. On the other hand, dry rotary swaging process with high effective friction involves slight torsion of the workpiece with minimal peeling and leads to more axial forward flow of the material.

The investigations of the residual stress revealed that a fluctuation along the axial position results due to complexity of the interaction between the dies and the workpiece as well as the complex material flow history. The axial residual stress is mainly tensile in the first few hundreds of microns. An effect of thermo-mechanical loading during the process is expected to induce differences in the resulting residual stress distribution. The results show that the very surface is strongly influenced by the process parameters, while in higher depth the properties are comparable for all conditions. This has to be confirmed in future investigations. Moreover, for better understanding of the material modifications induced by rotary swaging process, it is important to consider the thermomechanical effects with a larger range of parameters. Additionally, the influence of further process parameters on the material modifications needs to be investigated, especially the overall deformation. Finally, the influence of workpiece rotation freedom will help to control and understand the effect of the torsion induced by the rotating dies.

Acknowledgements

The authors would like to thank the German Research Foundation (DFG Deutsche Forschungsgemeinschaft) for funding this work within the sub-project (ZO 140-23-1 and KU 1389/16-1) "Control of component properties in rotary swaging process" of the priority program SPP 2013 "The utilization of residual stresses induced by metal forming".

References

1. E. Rauschnabel, V. Schmidt, J Mater. Process. Tech., **35**, 371-383 (1992)
2. B. Kuhfuss, E. Moumi, *Incremental Forming. In: Micro Metal Forming*, Edition F. Vollertsen, Springer Verlag (2013)
3. Kuhfuss, B., *Hollow drive shafts-a contribution towards weight and cost reduction in auto motive construction*, Automotive Technology International, Sterling Publications Limited, London (1997)
4. G.D. Lahoti, T. Altan, J. Eng. In., 98, 265–71 (1976)
5. E. Moumi, S. Ishkina, B. Kuhfuss, T. Hochrainer, A. Struss, M. Hunkel, Proc. Eng., **81**, 2342 – 2347 (2014)
6. Y. Wu, X. Dong, Q. Yu, Int. J. Mech. Sci., **93**, 102–110 (2015).
7. Y. Liu, M. Herrmann, C. Schenck, B. Kuhfuss, AIP Conf. Proc., **1896**, 080013 (2017)
8. R. Paukert, Stofffluss beim Radialumformen – Teil 1, DRAHT, 35, 108-111 (1984)
9. B. Scholtes, , *Eigenspannungen in mechanisch randschichtverformten Werkstoffzuständen:*

Ursachen, Ermittlung und Bewertung, Oberursel, DGM-Informationsges (1991)

10. A. Ameli, M. R. Movahhedy, Int. J. Adv. Manuf Tech. , **33**, 7-17 (2017)
11. M. Herrmann, C. Schenck, B. Kuhfuss, DMOAJ FMT (to be published)
12. J. Epp, , G. Huebschen, I. Altpeter, R. Tschuncky, H.-G., Herrmann, *X-ray diffraction techniques for material characterization: Materials Characterization Using Nondestructive Evaluation (NDE) Methods*, (Eds), Woodhead publishing, 81-124 (2016)
13. J. Epp, H. Surm, T. Hirsch, F. Hoffmann, J. Mater. Proc. Tech., **211**, 637-643 *(2011)*
14. *M. H*errmann, C. Schenck, B. Kuhfuss, Key Eng. Mat., **651-653**, 1042-1047 (2015)
15. H. Hetzner, *Systematische Entwicklung amorpher Kohlenstoffschichten unter Berücksichtigung der Anforderungen der Blechmassivumformung*, dissert., University of Erlangen (2014)
16. J. Dong, J. Epp, A. Rocha, R. Nunes, H.-W. Zoch, Met. Mat. Trans. A, **47A**, 877-888 (2016)
17. Z. Wang, B. Gong, *Residual Stress in the Forming of Materials,* ASM Handbook of Residual Stress and Deformation of Steel, ASM International, USA, 141–149 (2002).

The influence of microstructure deformation on the corrosion resistance of cold formed stainless steel

Ingmar Bösing[1,4,*], *Marius* Herrmann[2,4], *Ilya* Bobrov[3,4], *Jorg* Thöming[1,4,5], *Bernd* Kuhfuss[2,4,5], *Jérémy* Epp[3,4,5] and *Michael* Baune[1,4]

[1] UFT, Center for Environmental Research and Sustainable Technology, Leobener Strasse 6, 28359 Bremen, Germany
[2] bime, Bremen Institute for Mechanical Engineering, Badgasteiner Str. 1, 28359 Bremen, Germany
[3] Leibniz Institute for Materials Engineering IWT, Badgasteiner Str. 3, 28359 Bremen, Germany
[4] University of Bremen
[5] MAPEX Center for Materials and Processing

Abstract. Rotary swaging is an incremental cold forming process to produce axisymmetric workpieces from rods and tubes. The process also induces changes of the microstructure of the material depending on the process parameters. This in turn influences the mechanical properties like hardness as well as the electrochemical properties. As a result of changed electrochemical properties, the passivity of the material and thus the corrosion behavior changes. In order to investigate the influence of rotary swaging on the corrosion behavior of the stainless steel AISI304, deformed micro parts are electrochemically analyzed. The measurements reveal a dependency of corrosion rate and impedance on both feed velocity and final diameter of the rotary swaging process. A higher feed velocity decreases the corrosion rate and increases the impedance indicating a better resistance against corrosion as a side effect of rotary swaging.
Keywords: Cold forming, Stainless steel, Corrosion

1 Introduction

Rotary swaging is an incremental cold forming process to produce axisymmetric workpieces from rods and tubes. In the macro-range it has an important field of application in the automotive industry for example to produce axes, steering spindles and gear shafts [1]. It is also suitable for micro manufacturing to produce micro parts [2]. During infeed rotary swaging the workpiece is axially fed into the swaging unit and the diameter is reduced incrementally by a radially oscillating motion of three tools [3]. The swaging unit is rotating and the base jaws with the cam are passing the cylinder roller and induces the simultaneously radially stroke of the tools (Figure 1). The process shows several advantages of the produced parts like undisturbed fiber flow as well as improved mechanical properties [1]. Due to cold working, the tensile strength as well as the hardness increases during the process. Furthermore, the formability and also the microstructure of the material is influenced. This influence on the material properties depends on the process parameters like deformation degree or feed velocity as shown for the material AISI 304 [4]. Processes like rotary swaging, which cause mechanical deformation also on the microscale, can induce a transformation of this austenitic stainless steel into martensite. This change of microstructure influences the mechanical behavior of the produced workpieces [4]. The lower the infeed velocity during rotary swaging the higher the hardness compared to the initial state [5]. Thus more martensite is generated during the forming with lower feeding velocity [6]. Furthermore, the microstructure is significantly influenced by the deformation degree insofar as a texture is generated, grain size becomes smaller and the elongated grains are formed in the longitudinal direction [7].

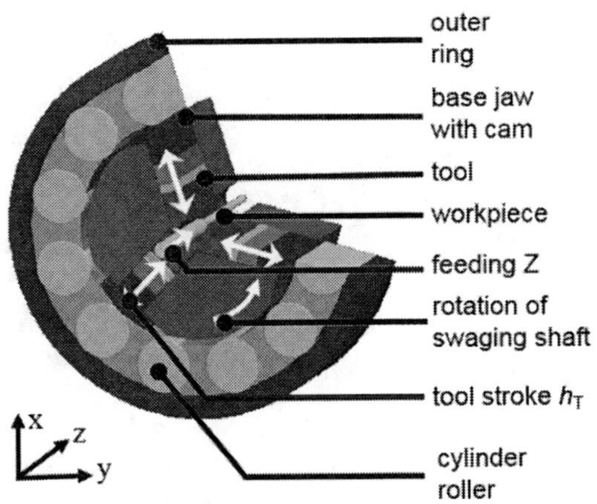

Fig. 1: Principle of infeed rotary swaging.

* Corresponding author: ingmar.boesing@uni-bremen.de

The changes in microstructure might affect not only the mechanical properties but also the electrochemical properties such as corrosion and the formation of protective passive layers. Stainless steel shows the ability to spontaneously form a protective oxide layer on the material surface. This passive layer is just a few nanometers thick and protects the steel against corrosion [8].

The thickness, composition and stability – and thus the protective effect of the passive layer depend on the composition of the alloy (e.g. a high content of chromium leads to a more protective layer [9, 10]), the environmental conditions and the microstructure of the steel [11].

The microstructure on different scales influences the formation of the passive layer and can accelerate or decelerate corrosion processes. It could be observed that a smaller grain size leads to a more stable passive layer in passivating environment [12]. But the higher amount of lattice defects associated with small grains can also lead to a faster dissolution under aggressive conditions like acidic media [13]. Not only the grain size but also the crystal structure can have an impact on the corrosion behavior. Martensite for example shows a higher electrochemical activity and thus a faster dissolution by corrosive reactions than austenitic phase [14].

Electrochemical measurements are a common and a well-known tool to study passivity and corrosion. Corrosion is an electrochemical phenomenon since the oxidative conversion of metals is accompanied by an electron transfer and thus a measurable current. The corrosion reactions can be forced by applying a potential between the investigated sample and a counter electrode. By measuring the flowing current during the reaction it is possible to compare different samples with respect to their resistance against corrosion. A lower anodic current indicates lower dissolution by either a more noble composition or a better passivation.

By rotary swaging the grain size of the material decreases while the martensite content in stainless steel increases [5, 7]. Both phenomena influences the passivity of the material. A smaller grain size leads to a better passivation while a higher martensite content leads to a faster dissolution. By choosing the optimal process parameters it should be possible not just to get improved mechanical properties like a higher hardness but also a better passivation and resistance against corrosion.

In this paper it will be shown that the martensite content depends on the feed velocity while the grain size depends on the final diameter. The additional electrochemical measurements reveal the dependency of both material parameters and thus process parameters on the corrosion behavior.

2 Experimental set up

2.1 Rotary swaging

Rotary swaging experiments were conducted with annealed workpieces of material AISI 304 (X5CrNi18-10). The workpieces were fed with a direct linear motion drive with different feed velocities of $v_f = 1$ mm/s, 5 mm/s, 7.5 mm/s and 10 mm/s into the swaging head. The stroke height of the tools ($h_T = 0.1$ mm) and the stroke frequency ($f_{st} = 102$ Hz) were kept constant. A lubricant (Shell Macron 01 F-2) was applied to all forming operations. The initial diameter of the workpieces was $d_0 = 1$ mm. Two different forming operations were carried out. Thus two tool sets of the material 1.2379 (X155CrVMo12) were used, one with a nominal diameter of $d_{nom1} = 0.75$ mm and second with a nominal diameter of $d_{nom2} = 0.46$ mm. Both tool sets had the same settings with a tool angle $\alpha = 10°$ and a calibration length $l_{cal} = 20$ mm (Figure 2). Workpieces were deformed in one stage with the first tool set to a final diameter of $d_1 = 0.8$ mm. An additional set of workpieces were deformed in two stages first to 0.8 mm and finally to $d_1 = 0.5$ mm with the second tool set. When removing the workpiece from the swaging unit, the tools were opened to avoid further impact on the specimen.

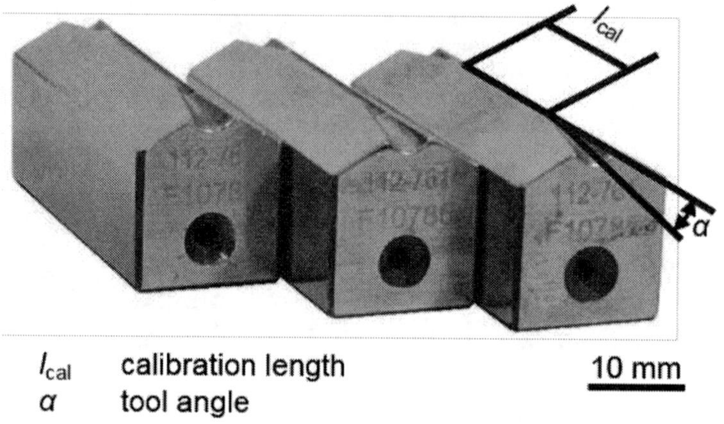

Fig. 2. Toolset for rotary swaging.

 https://doi.org/10.1051/matecconf/201819004002

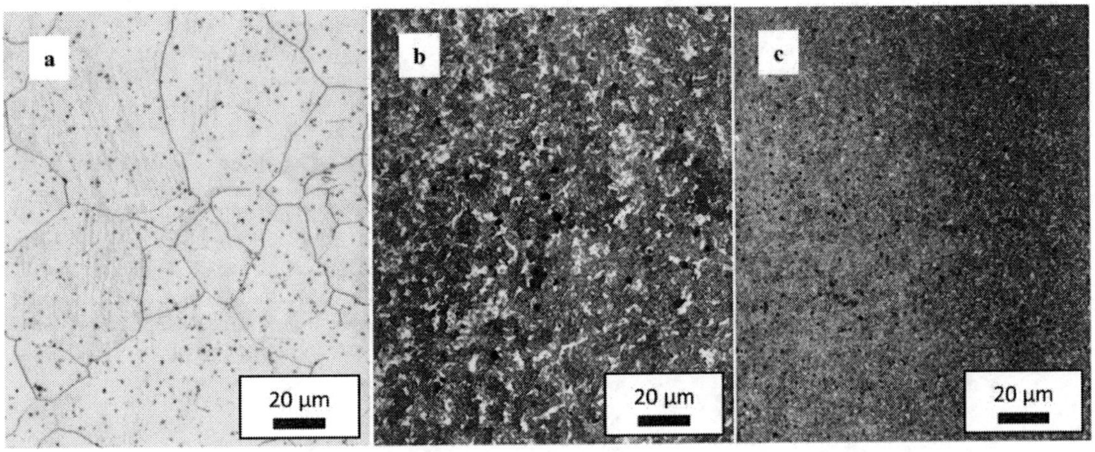

Fig. 3. Microstructure (cross-section polish) of (a) annealed material, (b) deformed material with final diameter $d_1 = 0.8$ mm ($v = 1.0$ mm/s), (c) deformed material with final diameter $d_1 = 0.5$ mm ($v_f = 1.0$ mm).

2.2 XRD measurements

For this study quantitative phase analysis [15] was performed using Cr-radiation ($K_{\alpha 1,\, \alpha 2}$), produced by long fine focus tube 12*0.4 (Panalytical). Diffracted radiation was detected by a position sensitive detector "Miostar 2" (Photron-X, Ottobrunn, Germany), equipped with vanadium filter. The measurements were performed on the outer surface of the wire, using a fox-mini-lens (IFG, Berlin, Germany), with incident beam diameter of 200 μm. All the diffraction profiles were obtained by varying 2θ from 60° to 164° with a resolution of 0.05°.

The collected XRD data was refined by Rietveld method [16] using the software TOPAS 4.2 (Bruker AXS, Karlsruhe, Germany). In this method, the whole measured pattern is refined with a calculated pattern, taking into account the instrumental contributions and the crystal structure of the present phases, including microstructural features like crystallite size and microstrains. The calculated criterion residuals weighted profile (weighted R-factor; R_{wp}) gives a reliable information about the fit quality. Thereby, the smallest R_{wp} value represents the best refinement. It is calculated by:

$$R_{wp} = \sqrt{\frac{\sum U_i (Y_i obs - Y_i calc)^2}{\sum (Y_i calc)^2}} \qquad (1)$$

In this equation $Y_i obs$ is the measured and $Y_i calc$ is the calculated intensity at each 2θ position i [16].

2.3. Electrochemical measurements

All electrochemical measurements were carried out in a standard three electrode cell using a Metrohm Autolab potentiostat with the PGSTAT204 and the FRA32 modules. All experiments were performed at room temperature. The counter electrode was a platin electrode and the reference electrode was an Ag/AgCl electrode, while the investigated material acts as the working electrode. The samples were embedded in a PTFE (Teflon) holder. All potentials in this paper are referred to the Ag/AgCl electrode.

The electrochemical measurements, potentiodynamic polarization scans and electrochemical impedance spectroscopy, were done in phosphate buffered saline (PBS) containing 0.2 M NaCl + 0.1 M phosphate buffer solution [17] (pH = 7.5). Before each measurement the working electrodes were wet grinded with SiC grinding paper up to 2000 grid, degreased in ethanol and cleaned in an ultrasonic bath containing deionized water.

The polarization scans were performed in PBS with a scan velocity of 1 mV/s. Each polarization scan started from -0.8 V and ended at 1.0 V. Before every measurement the open circuit potential (OCP) was measured for 300 seconds.

The electrochemical impedance spectroscopy (EIS) measurements were performed at OCP which was measured for 1800 seconds. The EIS measurements were performed in the frequency range between 10^5 and 0.1 Hz with an AC amplitude of ±10 mV_{rms}.

3 Results and discussion

3.1 Microstructure analysis

The microstructure of the annealed (initial state) material and two rotary swaged samples are shown in Figure 3. The annealed material is recognizable as austenitic steel with a grain size of 36.4 μm. By rotary swaging the grain size of the material was reduced significantly. The workpieces that were deformed in one stage to the final diameter $d_1 = 0.8$ mm showed a grain size of 2.2 μm. With an additional deformation stage and a final diameter of $d_1 = 0.5$ mm the grain size continued to be decreased down to 0.94 μm.

Tab 1: Weighted dislocation density and refined martensite content in dependence of final diameter and feed velocity.

Sample	Average dislocation density, cm^{-2}	Martensite content, mass-%
As received	1.17×10^{9}	0
$d_1 = 0.8$ mm		
$v_f = 10$ mm/s	1.81×10^{10}	21
$v_f = 7.5$ mm/s	1.82×10^{10}	27
$v_f = 5$ mm/s	1.61×10^{10}	35
$v_f = 1$ mm/s	1.59×10^{10}	47
$d_1 = 0.5$ mm		
$v_f = 5$ mm/s	1.56×10^{10}	55
$v_f = 1$ mm/s	1.14×10^{10}	80

A dependency of the grain size and the feed velocity could not be observed. Other studies came to the same result [5].

3.2 XRD Measurements

XRD was used to study the phase transformations taking place during rotary swaging. Figure 4 shows the XRD patterns of the annealed and deformed samples. In these diagrams γ represents peaks of the austenitic phase and α represents martensite peaks. The labels of each scan describes final diameter of the wire and the feed velocity used during rotary swaging of the investigated samples. Figure 3 illustrates the evolution of martensite peaks and corresponding evolution of austenite peaks depending on the process parameters. The increase of the α-peak is correlated with the generation of deformation-induced martensite during the process. Moreover, a broadening of the diffraction peaks can be observed.

The values of R_{wp} for the refinements of studied samples were in range of 3.2 to 3.9 which correspond to a high fit quality. Phase composition was refined taking into account two existing phases: initial austenite (FCC, face-centered cubic) and deformation-induced martensite (BCC, body-centered cubic). Also the refinement was performed taking into account the texture with preferred orientation in <011> direction, resulting from rotary swaging process. The degree of preferred orientation correction was refined by March-Dollase approach. The precision of the quantitative analysis of phase contents is estimated to be around ± 5 mass-% in the present case. Microstructural parameters, such as microstrain (e_0) and volume weighted crystalline block size (D), also obtained by this refinement were used to calculate the dislocation density by:

$$\rho = \sqrt{\frac{3k \times (e_0)^2}{(D)^2 \times 2.48 \times 10^{-21}}} \quad (2)$$

with k a structure depending factor, which takes values (calculated) of 16.1 for FCC materials and 14.4 for BCC materials [18]. The dislocation density was calculated for each of both existing phases. The average was calculated taking into account the determined respective phase contents. The results of these evaluations are shown in Table 1.

It shows, that dislocation density values for initial state were an order of magnitude smaller than for the wires after rotary swaging. The dislocation density values for samples subjected the rotary swaging with varied process parameters were comparable, but there seems to be no dependency on the process parameters and it has to be kept in mind that the measurements were performed on the outer surface and therefore, differences along the radius might take place. On the other hand, the process parameters had a significant influence on the martensite content. An increasing feed velocity led to a decreasing formation of martensite, while a higher diameter reduction increased the amount of martensite.

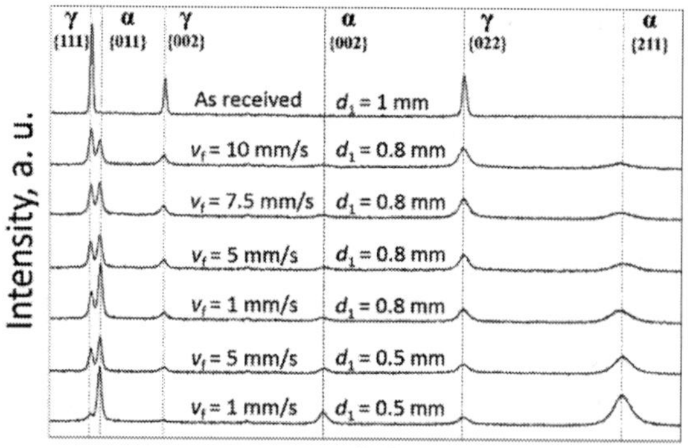

Fig. 4. : XRD-scans for rotary swaged wires by different feed velocities v_f and for initial wire (no RS).

3.3 Electrochemical measurements

3.3.1 Corrosion rate

To determine the corrosion potential and the corrosion current density of the deformed samples and the initial state material polarization scans in PBS solution were performed. During the polarization scan a potential between the sample and the counter electrode was increased using a scan velocity of 1 mV/s and the current between the sample and the counter electrode was measured. The reference electrode served as a potential reference during the measurement (for more detailed description of the reference electrode the reader is referred to [19]). The applied potential caused corrosion on the surface of the workpiece and the measured current was directly linked to the converted mass via Faraday's law. So a high anodic current in areas of dissolution reactions (e.g $Fe \rightarrow Fe^{2+}$) indicates a fast corrosion process. Thus a

https://doi.org/10.1051/matecconf/201819004002

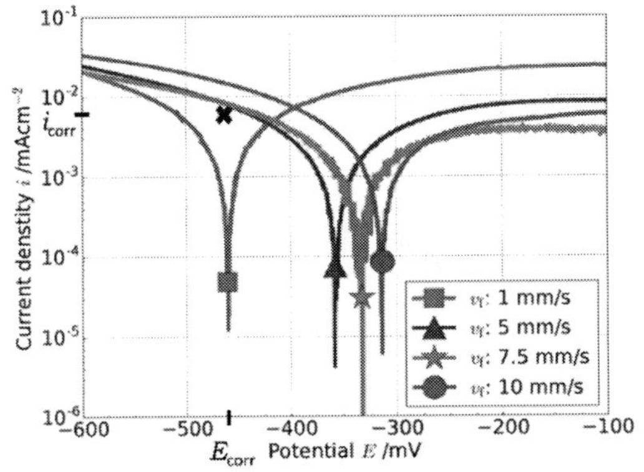

Fig. 5. Polarization scan in PBS. Final diameter d_1 = 0.8; variable feed velocity; scan velocity 1 mV/s; all potentials referred to Ag/AgCl electrode; the corrosion current density i_{corr} and the corrosion potential E_{corr} of sample v_f = 1 mm/s are shown as an example.

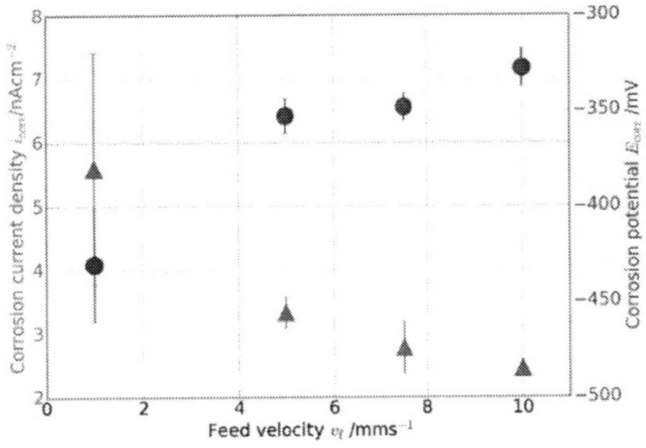

Fig. 6. Dependency of corrosion current density i_{corr} and corrosion potential E_{corr} from feed velocity for: final diameter d_1 = 0.8 mm.

low corrosion current density i_{corr} and high corrosion potential E_{corr} indicate good corrosion resistance.

The aim of the polarization scans was to uncover a correlation between the process parameters feed velocity and final diameter and the electrochemical values corrosion potential and corrosion current density. In addition, the corrosion rate, which describes the dissolution by corrosion, can be calculated from the polarization scans.

Figure 5 shows the polarization scans of the deformed material with d_1 = 0.8 mm. The absolute value of the current density $|i|$ is plotted against the potential. At the beginning of the measurements a cathodic current related to the reduction of oxygen can be seen. While shifting to higher (less negative) potentials the cathodic current decreases. At the corrosion potential E_{corr} the cathodic current is equal to the anodic current and the total current is zero. For the determination of the corrosion current density the linear courses of the anodic and cathodic currents are extended and the intersection of both lines indicates the value of the corrosion current (Tafel plot). At higher potentials the anodic current, which is related to the oxidation of the material, increases.

The lowest corrosion potential and highest corrosion current density is assigned to the sample with the lowest feed velocity (v_f = 1 mm/s) during the forming process.

With increasing feed velocity the corrosion potential increases while the corrosion current density decreases (Table 2). Both of these characteristics indicate higher corrosion resistance.

Along with an increasing feed velocity the martensite content declines. It is known that martensite shows higher electrochemical activity than austenite [14, 20]. This explains the lower corrosion current density and the higher corrosion potential with increasing feed velocity, which is obvious when corrosion current density and corrosion potential are plotted against feed velocity. Figure 6 shows this for a final diameter of 0.8 mm, but the same trends were found for a final diameter of 0.5 mm (not plotted).

While E_{corr} and i_{corr} are clearly a function of feed velocity v_f the dependence on the final diameter is not that clear. The smaller final diameter (due to a second deformation step) yields a higher martensite content and a smaller grain size compared to the samples with larger final diameter and the untreated material. On the one hand a higher martensite content leads to a higher

Table 2. Corrosion potential and corrosion current density obtained by potentiodynamic polarization scans in PBS.

Sample		Corrosion Potential E_{corr} /mV	Corrosion current density i_{corr} /nAcm^{-2}	Corrosion rate CR /µma^{-1}
d = 0.8 mm	v_f = 10 mm/s	-327.7	2.45	2.56
	v_f = 7.5 mm/s	-384.8	2.78	2.90
	v_f = 5 mm/s	-352.9	3.32	3.47
	v_f = 1 mm/s	-443.9	5.59	5.84
d = 0.5 mm	v_f = 10 mm/s	-343.5	2.20	2.30
	v_f = 5 mm/s	-451.4	3.15	3.29
	v_f = 1 mm/s	-475.1	9.40	9.82
As received		-390.5	4.38	4.57

electrochemical activity while on the other hand a smaller grain size leads to a better passivation of stainless steel.

So the differences between samples treated with the same feed velocity but different final diameter shrink with increasing feed velocity (Table 2). Comparing sample $d_1 = 0.8$ mm and $v_f = 10.0$ mm/s with sample $d_1 = 0.5$ mm and $v_f = 10.0$ mm/s both corrosion current densities are nearly the same (2.45 nA and 2.20 nA respectively). The smaller grain size of two times deformed samples ($d_1 = 0.5$ mm) seems to compensates the higher martensite content resulting in nearly the same corrosion behavior.

The corrosion rate is the material loss per time due to corrosion under the given conditions. It is a function of the corrosion current density and is given in micrometers per year.

The corrosion rate CR is given by

$$CR = \frac{i_{corr} M t}{n F} \tag{3}$$

with the corrosion current density i_{corr}, the molar mass M the time t, the charge number n which indicates the number of electrons exchanged in the dissolution reaction and F is the Faraday constant. The quotient M/n is sometimes referred to as equivalent weight. The different values of the corrosion rate are summarized in Table 2. According to Equation 3 i_{corr} is the only variable parameter so the corrosion rate correlates linear with the corrosion current density. Due to the significant smaller grain size compared to the initial state material the deformed samples with low martensite content ($v_f = 5$ mm/s – $v_f = 10$ mm/s) show even a lower corrosion rate than the pure austenitic as received material.

The polarization scans reveal a dependency of the corrosion potential and the corrosion rate on the feed velocity (Figure 5). The corrosion potential increases with feed velocity v_f, while the corrosion rate decreases with v_f. So an increasing feed velocity improves the corrosion behavior.

3.3.2 Corrosion resistance

To underline the findings of the polarization scans and to investigate the corrosion resistance of the samples electrochemical impedance spectroscopy (EIS) measurements were performed in PBS solution. During the EIS measurements an AC voltage with decreasing frequency (10 kHz – 0.1 Hz) was applied between the sample and a counter electrode. While the measurement the AC resistance – the impedance – as well as the phase shift was measured. The impedance of the electrochemical system can be interpreted as a resistance against corrosion.

In Figure 7 the real part of the impedance (Z') is plotted against the imaginary part of the impedance (-Z''). With decreasing frequency the absolute impedance increases. Figure 6 reveals the same relationship that was already observed in the polarization scans.

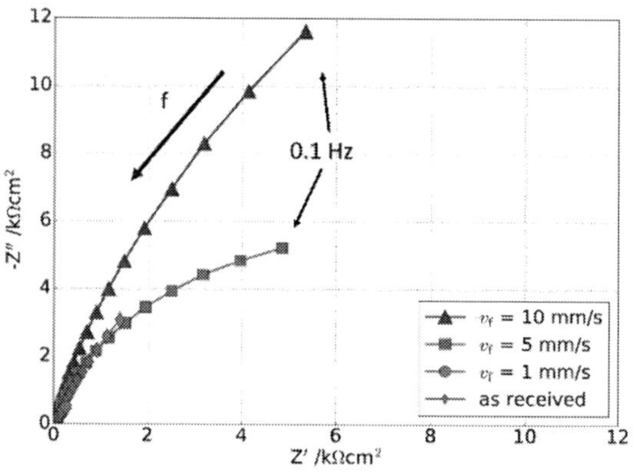

Fig. 7. Nyquist plot of the rotary swaged and initial material recorded at OCP in PBS solution.

The impedance, so the resistance against corrosion, increases with increasing feed velocity and thus decreasing martensite content.

The differences between the impedance values for one and two deformation steps also shrink with increasing feed velocity. At a feed velocity of $v_f = 10$ mm/s the impedance values of both samples are nearly the same. The higher martensite content appears to compensate for the smaller grain size and nearly the same corrosion behavior can be observed. It can also be seen that it is possible to increase the impedance of the deformed material compared to the initial material. The smaller grain size improves the corrosion resistance ($v_f = 10$ mm/s and $v_f = 5$ mm/s compared to as received) but with increasing martensite content ($v_f = 1$ mm/s) the impedance gets smaller compared to the initial material (Table 3).

The EIS measurements confirm the same relationship like the polarization scans between corrosion resistance and the feed velocity. A higher feed velocity leads to a better passivation and a better corrosion resistance visible through a high impedance (Figure 7) and a low corrosion current density (Figure 5).

Table 3. Impedance values measured in PBS for one ($d_1 = 0.8$ mm) and two ($d_1 = 0.5$ mm) deformation steps.

Sample		Impedance Z /MΩ
$d_1 = 0.8$ mm	$v_f = 10$ mm/s	5.78
	$v_f = 5$ mm/s	5.31
	$v_f = 1$ mm/s	4.17
$d_1 = 0.5$ mm	$v_f = 10$ mm/s	5.90
	$v_f = 5$ mm/s	4.46
	$v_f = 1$ mm/s	2.01
As received		4.06

4 Conclusion

In this study workpieces deformed by rotary swaging were investigated by analyzing the microstructure and the electrochemical properties. The grain size of AISI 304 decreases with every deformation stage and strain-induced martensite is built during the process. It could be shown that the martensite content increases with increasing deformation while an increasing feed velocity can decrease the amount of formed martensite. On the other hand, no significant impact of the feed velocity on the grain size nor the dislocation density could be observed.

The martensite content is directly linked to the corrosion behavior. A high martensite content leads to faster dissolution measurable by a higher corrosion current density and corrosion rate as well as by a low impedance. With increasing feed velocity the corrosion current density (and thus the corrosion rate) decrease while the impedance of the workpiece increase, which can be attributed to the decreasing amount of built martensite.

The grain size also has a significant effect on the corrosion behavior namely a smaller grain size leading to a lower corrosion current density and a higher impedance. Both indicates a better resistance against corrosion.

By choosing a high feed velocity it is therefore possible to achieve a better protection against corrosion compared to the initial material, even despite a higher martensite content, due to the significant reduction of the grain size. These effects were visible in a lower corrosion rate and a higher impedance of the deformed material with feed velocities from $v_f = 5$ mm/s to $v_f = 10$ mm/s.

The results confirmed that it is not just possible to influence the mechanical and geometrical properties of stainless steel AISI 304 by rotary swaging but also the resulting electrochemical properties. The choice of the right process parameters can lead to an increased hardness as well as to an increased corrosion resistance.

Besides martensite content and grain size other factors can also influence the corrosion behavior or the passivation. For example microstructural properties like texture [21], grain size distribution [22] and coincident site lattice boundaries (by suppression of sensitization through chromium depletion) [23, 24] can play an important role. Further investigations on the influence of rotary swaging on these microstructure properties and thus the properties of the passive layer also seem to be very promising and will be in focus of future investigations.

Financial support of subproject D03 "Electrochemical High Throughput Characterization of Metallic Micro Samples" and D01 "Qualification of material conditions with mechanical and physical measuring methods" of the Collaborative Research Center SFB 1232 "Farbige Zustände" by the German Research Foundation (DFG) is gratefully acknowledged.
The authors would also like to thank the German Research Foundation for funding this work within the sub-project "Control of component properties in rotary swaging process" of the priority program SPP 2013 "The utilization of residual stresses induced by metal forming".
Further acknowledgements to Georg Pesch for fruitful scientific discussion and proof reading of this paper.

References

1. E. Rauschnabel and V. Schmidt, J. Mater. Process. Technol., vol. **35**, no. 3–4, pp. 371–383 (1992)
2. B. Kuhfuss, E. Moumi, and V. Piwek, Microsyst. Technol., vol. **14**, no. 12, pp. 1995–2000 (2008)
3. B. Kuhfuss; E. Moumi: Incremental Forming. In: *Micro Metal Forming*, Hrsg. F. Vollertsen, Springer Verlag, ISBN-13: 9783642309151 (2013)
4. S. K. Ghosh, P. Mallick, and P. P. Chattopadhyay, J. Iron Steel Res. Int., vol. **19**, no. 4, pp. 63–68 (2012)
5. B. Köhler, B. Clausen, H.-W. Zoch, *8. Kolloquium Mikroproduktion*, (BIAS Verlag, 91-98, 2017)
6. E. Moumi; B. Kuhfuss, Properties Of Alloy 204 Micro Parts Processed by Rotary Swaging, nanoMan (2014)
7. S. J. Lim, H. J. Choi, and C. H. Lee, J. Mater. Process. Technol., vol. **209**, no. 1, pp. 283–288 (2009)
8. J. Kruger, Int. Mater. Rev., vol. **33**, no. 1, pp. 113–130 (1988)
9. K. Sieradzki and R. C. Newman, J. Electrochem. Soc., vol. **133**, no. 9, pp. 1979–1980 (1986)
10. D. E. Williams, R. C. Newman, Q. Song, and R. G. Kelly, Nature, vol. **350**, no. 6315, pp. 216–219 (1991)
11. P. Schmuki, J. Solid State Electrochem., vol. **6**, no. 3, pp. 145–164 (2002)
12. K. D. Ralston and N. Birbilis, Corrosion, vol. **66**, no. 7, pp. 1–4 (2010)
13. L. Liu, Y. Li, and F. Wang, J. Mater. Sci. Technol., vol. **26**, no. 1, pp. 1–14 (2010)
14. C. C. Xu and G. Hu, Anti-Corrosion Methods Mater., vol. **51**, no. 6, pp. 381–388 (2004)
15. J. Epp, XRD methods for materials Characterization in *Mat. Charact. Using Nondestr. Evaluation (NDE) Meth.*, Woodhead publishing pp. 81–124 (2016)
16. G. Will, *Powder diffraction: The rietveld method and the two stage method to determine and refine crystal structures from powder diffraction data* (Springer Science & Buisness Media, 2006)
17. I. Bösing, J. Thöming, and M. Baune, Electrolyte Composition for Distinguishing Corrosion Mechanisms in Steel Alloy Screening, vol. **2017**, 2017
18. G. K. Williamson and R. E. Smallman, Philos. Mag., vol. **1**, no. 1, pp. 34–46 (1956)
19. A. Bard and L. Faulkner, *Electrochemical Methods: Fundamentals and Applications* (New York: Wiley, 2001)
20. S. Sunada, N. Nakamura, H. Kawase, H. Notoya, S. Sanuki, and K. Arai, Nippon Kinzoku

Gakkaishi/Journal Japan Inst. Met., vol. **55**, no. 6, pp. 660–666 (1991)

21. B. R. Kumar, R. Singh, B. Mahato, P. K. De, N. R. Bandyopadhyay, and D. K. Bhattacharya, Mater. Charact., vol. **54**, no. 2, pp. 141–147 (2005)
22. S. Gollapudi, Corros. Sci., vol. **62**, pp. 90–94, 2012
23. H. Y. Bi, H. Kokawa, Z. J. Wang, M. Shimada, and Y. S. Sato, Scr. Mater., vol. **49**, no. 3, pp. 219–223, 2003
24. M. Shimada, H. Kokawa, Z. J. Wang, Y. S. Sato, and I. Karibe, Acta Mater., vol. **50**, no. 9, pp. 2331–2341 (2002)

MATEC Web of Conferences **190**, 04003 (2018) https://doi.org/10.1051/matecconf/201819004003
ICNFT 2018

Axial and Radial Material Flow Analysis in Infeed Rotary Swaging of Tubes

Yang Liu[1,*], *Marius* Herrmann[1], *Christian* Schenck[1,2], and *Bernd* Kuhfuss[1,2]

[1]bime, Bremen Institute of Mechanical Engineering, University of Bremen, 28359, Bremen, Germany

[2]MAPEX Center for Materials and Processing

Abstract. In rotary swaging – an incremental cold forming production technique to reduce the diameter of axisymmetric parts – the material flow can be assumed to be predominantly axial and radial. The actual ratio of this axial and radial flow influences the mechanical properties and especially in tube forming the final geometry. It is known that during mandrel free infeed rotary swaging of tubes the wall thickness changes. The change is depending on the process parameters like incremental and cumulated strain. Hence, the ratio of axial and radial material flow changes. Consequently, the analysis of the wall thickness of rotary swaged tubes enables fundamental insight how to control the material flow direction. In this study, the infeed rotary swaging process of steel tubes with different wall thicknesses from 3 mm to 7 mm and rods were investigated with FEM under two feeding velocities. The axial and radial material flow and the resulting geometry were studied by the relative wall thickness. It could be seen that the relative wall thickness was affected by the feeding velocity as well as the initial wall thickness. The findings of the simulation were validated by rotary swaging experiments.

Keywords: Metal forming, Cold forming, Modeling

Nomenclature		
d_0	initial diameter	mm
d_1	final diameter	mm
f_{st}	stroke frequency	Hz
h_T	stroke height	mm
l_0	initial length	mm
l_1	final length	mm
l_h	drilling length	mm
R_n	radial neutral boundary	-
R_{nr}	position of R_n	mm
R_{np}	radial relative position of R_{nr}	%
s_0	initial wall thickness	mm
s_1	final wall thickness	mm
$s_{1\text{-fem}}$	final wall thickness of FE model	mm
$s_{1\text{-exp}}$	final wall thickness of experiment	mm
s_n	wall thickness that enables $\Delta s_p = 0$	mm
v_f	feeding velocity	mm/min
y_0	original mesh height	mm
y_1	final mesh height	mm
z_0	original mesh length	mm
z_1	final mesh length	mm
I	reduction zone	-
II	calibration zone	-
III	exit zone	-
α	die angle	°
Δs_p	change of wall thickness	%
Δy_p	change of mesh height	%
Δz_p	change of mesh length	%
δ	relative error of FE model	%
μ	friction coefficient	-

1 Introduction

Rotary swaging is an incremental cold forging process. Usually it is used for reducing the cross section of rods and tubular workpieces, or for achieving certain wall thicknesses of tubes [1]. This technology can improve the material strength as well as the surface condition with reduced roughness values. In addition, this process can produce advantageous inner profiles for tubes [3]. Nowadays, rotary swaging is widely used in automobile and textile industry [4]. The main part of a rotary swaging machine is the swaging head, wherein two or more dies are concentrically arranged. During the process, the dies rotate with the driving shaft. Simultaneously, the dies strike repeatedly towards the workpiece in the center. The forming is achieved either by axially feeding of the workpiece into the swaging

*Corresponding author: liu@bime.de

head (infeed rotary swaging) or by radially feeding of the dies (plunge rotary swaging).

The principle of mandrel free infeed rotary swaging is shown in Figure 1. According to the geometry, the die is distinguished into three main zones: reduction zone (I), calibration zone (II) and exit zone (III). During the process, the plastic deformation of the workpiece appears mainly in the reduction zone, which features the die angle α. The main material flow is initiated in this zone. In the calibration zone and exit zone the geometry of the workpiece is predominantly maintained. The main parameters of the process are feeding velocity (v_f), striking frequency (f_{st}) and stroke height (h_T). The geometry of the tubes is given by the initial diameter (d_0) and the initial wall thickness (s_0). After the deformation a final diameter (d_1) and final wall thickness (s_1) are achieved. When the calibration zone is fully filled during the process, a quasi-constant condition is established. Thus, the same amount of material is deformed with the same conditions with every further stroke. In this situation, a stable phase is assumed [5].

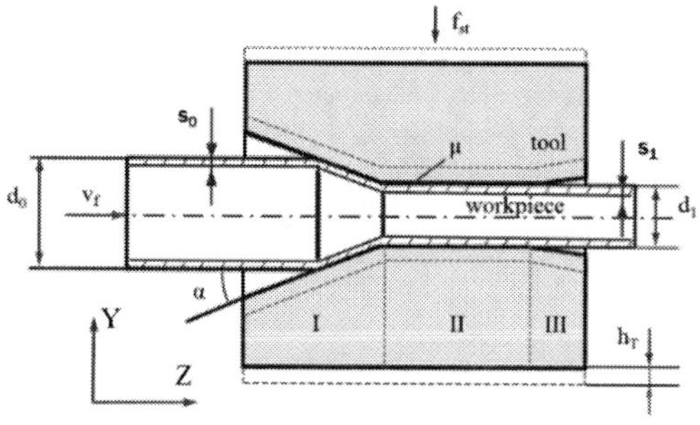

Fig. 1. Principle of mandrel free tube infeed rotary swaging.

In former research, rotary swaging has been widely investigated by Finite Element Method (FEM) and experiments. Rong et al. [6] studied the radial material flow in pure magnesium rods with simulations. They found that the radial strain distribution was not notably affected by different values of v_f. Moumi et al. [7] studied the material flow by 2D axisymmetric simulations and the means of the axial neutral plain –the zone where the actual axial velocity equals Zero. They identified that the neutral plane was changing in shape and position during one stroke. Gan et al. [8] and Abdulstaar et al. [9] investigated the microstructures and tensile properties of commercial pure magnesium and pure aluminum in rotary swaging experiments. They found that both materials were gradually refined and the tensile yield strength was increased due to material flow. Ishkina et al. realized asymmetric strokes and used flat dies to change the radial material flow. They produced hexagonal wires and found that the shear strain distribution could be affected by these measures. In consequence, the formability of AISI304 in subsequent cold extrusion operations was improved [2, 10].

According to former research [11], tubes with d_0 = 8 mm and s_0 = 0.4 mm developed a thicker s_1 after forming. For tubes with s_0 higher than a half of d_1, s_1 must become definitely smaller than s_0. Hence, the radial material flow must change significantly depending on the initial wall thickness s_0. Due to the constancy of material volume, when the cross section of the workpiece is reduced in the process, the material flows both in axial and radial directions. Additionally, the reducing of the cross section of the workpiece is finished with many increments. That makes the material flow complex, because the same part of the workpiece experiences different load conditions in different zones of the dies. Beneath the final geometry of the product the material flow also changes mechanical properties. Consequently, a profound knowledge of the material and its dependency on processing parameters is needed for improved controllability and further optimization.

In this paper, mandrel free infeed rotary swaging process of steel alloy (1.0308) tubes was studied with varying wall thickness and feeding velocity values by simulations and experiments. The data from simulation and experiment were represented as "fem" and "exp" for short.

2 Methods

In FE model, the process was realized in the general finite element software ABAQUS (version 6.13). A 2D axisymmetric explicit model was used based on [12] (Figure 2) with following assumptions: isotropic material; negligible thermal influence; constant friction condition between the contact pair. Coulomb friction was assumed despite the expected high normal forces as it was used and discussed in former literature [6, 7, 12 and 13]. The processing parameters in this FE model are shown in Table 1.

Table 1 Parameters of the experimental settings.

Name	Value	Unit
d_0	20	mm
d_1	15	mm
f_{st}	37.5	Hz
h_T	1	mm
v_{f500}	500	mm/min
v_{f2000}	2000	mm/min

In the model, the material was steel alloy (1.0308) whose plastic properties were from literature [15], the values of s_0 were 3, 4, 5, 6, 6.5 and 7 mm. A square shaped CAX4RT mesh was used on the workpiece. The

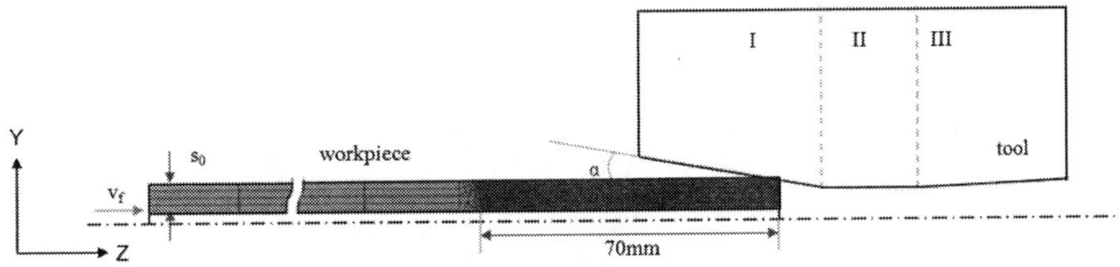

Fig. 2. FE model in ABAQUS 6.14, $s_0 = 7$ mm [12].

length of the workpiece with fine mesh was 70 mm. This section that was involved in the deformation was meshed according to the initial wall thicknesses and radius. Re-meshing was disabled. In this FE model, for thicker tubes with s_0 higher than 5 mm more radial meshes were used as a common measure to avoid hourglass phenomenon [14]. Consequently, different mesh sizes were deployed. The length for each square was: $s_0/20$ for $3 \text{ mm} \leq s_0 \leq 5$ mm, $s_0/30$ for $6 \text{ mm} \leq s_0 \leq 7$ mm, $s_0/40$ for the rod [12]. According to literature the friction coefficient value of lubricated contact pair steel - steel is around 0.1 [16], so $\mu = 0.1$ was used in this study. After the simulations were finished, the values of $s_{1\text{-fem}}$ were measured at a point 35 mm from the end of undeformed material (A-A section in Figure 3), where the material was supposed to have experienced the whole deformation process after the stable phase was established.

In a detailed view, the shape change of mesh elements in axial and radial directions was analyzed as a measure of axial and radial deformation at A-A section. In Figure 3 the shape change of a mesh at the outer surface is illustrated. In this study, the shape changes of meshes from the outer surface to the inner surface were investigated.

As different mesh sizes used in the FE model, the changes of mesh shape in axial and radial direction were normalized by the initial dimensions:

$$\Delta z_p = 100\% * (z_1 - z_0) / z_0 \tag{1}$$

$$\Delta y_p = 100\% * (y_1 - y_0) / y_0 \tag{2}$$

Δz_p is the axial deformation, due to the features of rotary swaging, the workpiece should always be elongated, so the values of Δz_p should always be positive. But according to former research the radial deformation for the meshes could be stretched or compressed, which means that the values of Δy_p could be either positive or negative. It is supposed there might be an area where the value of Δy_p is zero, which could be named as radial neutral boundary (R_n) whose location could be marked as R_{nr}. The radial relative position according to s_1 could be calculated as R_{np} as equation (3), which represents the radial ratio of R_{np} from the outer surface.

$$R_{np} = 100\% * (d_1/2 - R_{nr}) / s_1 \tag{3}$$

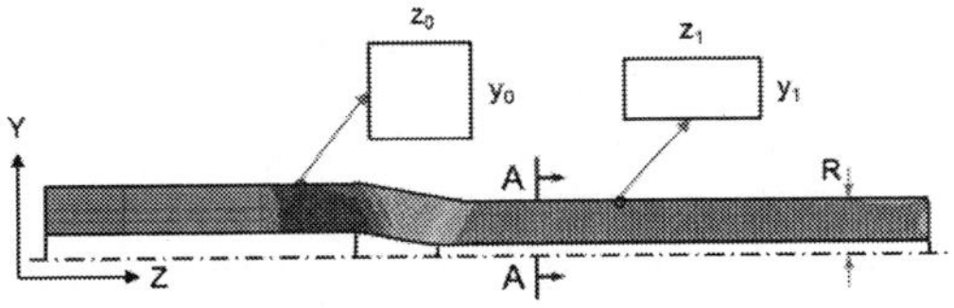

Fig. 3. Shapes of mesh before and after deformation (A-A section for measuring $s_{1\text{-fem}}$).

In the experiments, the tubes were manufactured from rods ($d_0 = 20$ mm and $l_0 = 300$ mm) by drilling to ensure equal starting conditions (Figure. 4). The length of drilling was $l_h = 80$ mm. For each s_0 three workpieces were prepared. Conventional swaging dies with tungsten carbide layer in the reduction zone were used [5], the contact pair was lubricated with oil (Condocut KNR 22). During the process the workpieces were fed into the center of the dies with a length of 70 mm. Then they were pulled back. After the process, the geometry of the workpiece was changed as shown in Figure. 4.

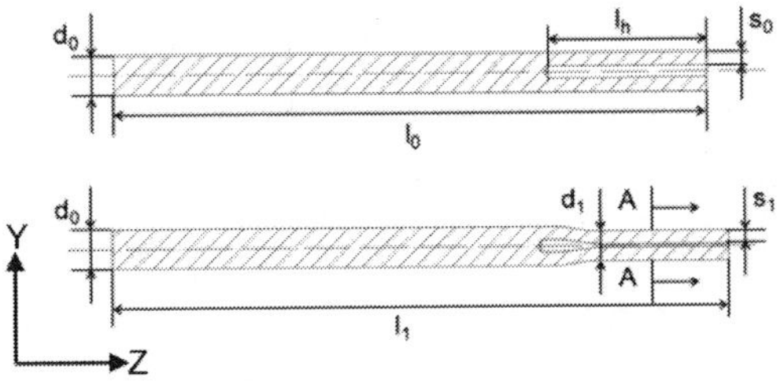

Fig. 4. Section view of the workpiece before and after the process, $s_0 = 7$ mm. (A-A section for measuring $s_{1\text{-exp}}$).

After the experiments the values of $s_{1\text{-exp}}$ were measured at A-A section. The physical probes were measured by preparing metallographic specimens. A laser microscope was used to take five times enlarged photos of the cross sections. Based on the micrographs the values of s_1 were measured eight times for each of the three samples and the average value and variance were calculated. For the validation of the FE model, the values of relative error were calculated by the differences between the s1 values from simulation and experiment divided by the value from experiment (equation (4)).

$$\delta = 100\% * (s_{1\text{-fem}} - s_{1\text{-exp}}) / s_{1\text{-exp}} \quad (4)$$

The percentage of the wall thickness development (Δs_p) was calculated by (equations (5)).

$$\Delta s_p = 100\% * (s_1 - s_0) / s_0 \quad (5)$$

3 Results and discussion

No matter which s_0 was used, the length of the workpiece increased. So, the axial material flow is discussed firstly. The resulting values in axial direction Δz_p for v_{f500} are shown along the radius (Y-axis) in Figure 5. These values of Δz_p for both rod and tubes could be treated as constant values regardless of their locations in radial direction. Thus, the axial material flow is homogeneous. The mean values of Δz_p are shown in Figure 6. Δz_p increases almost linearly with higher value of s_0; the influence of the feeding velocity v_f is low and vanishes with higher s_0. The reason for increasing Δz_p along with s_0 could be the increasing radial resistance due to more material to be deformed.

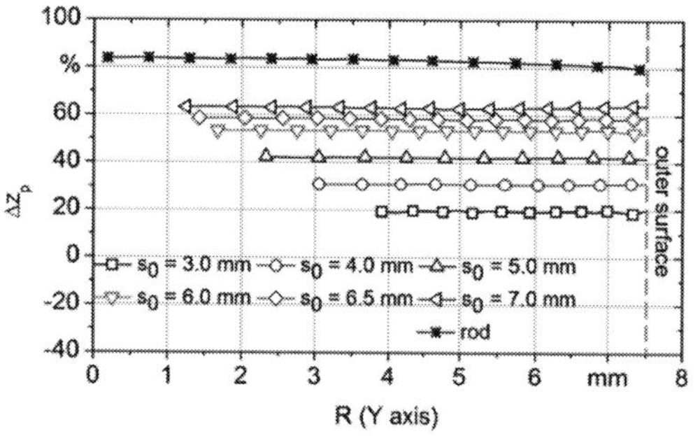

Fig. 5. Axial elongation Δz_p of meshes for different wall thicknesses of tubes and rod for the feeding velocity v_{f500} at the position A – A.

With the same value of s_0, higher v_f led to an earlier contact between the dies and the workpiece during each stroke. Also, the contact length in the reduction zone was shorter and its center was shifted to the undeformed part; in the same time, due to an earlier contact the distance between the calibration zone and the deformed workpiece was longer, thus the friction resistance worked later compared to the situation with lower v_f. Due to this reasons, the material flow towards axial direction was easier for higher v_f.

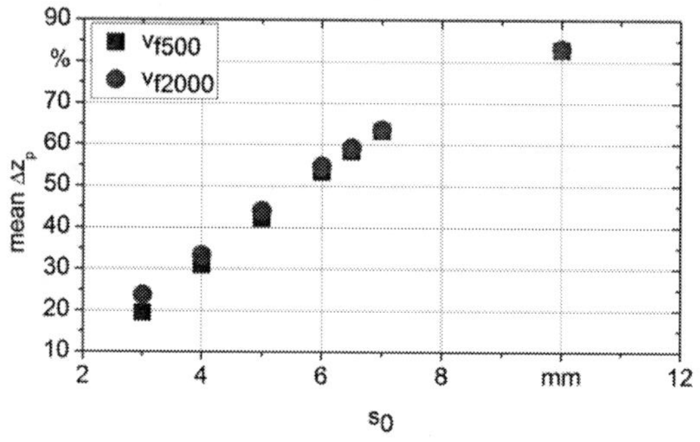

Fig. 6. Mean values of Δz_p of tubes with different s_0 for v_{f500} and v_{f2000}.

Figure 7 shows the values of Δy_p over the radius. In contrast to the axial direction, the material flow in radial direction was not homogeneous with the exception of rods. For tubes, the radial material flow depended on s_0. For s_0 equaled 3 and 4 mm, the values of Δy_p were always positive, which means over the whole radius the material was radially stretched. For tubes with higher values of s_0, the values of Δy_p changed from positive to negative from inner surface to outer surface. In this case, there was a radial neutral boundary (R_n) that distinguished the radial material flow into local radial compression ($\Delta y_p < 0$) and local radial stretching ($\Delta y_p > 0$).

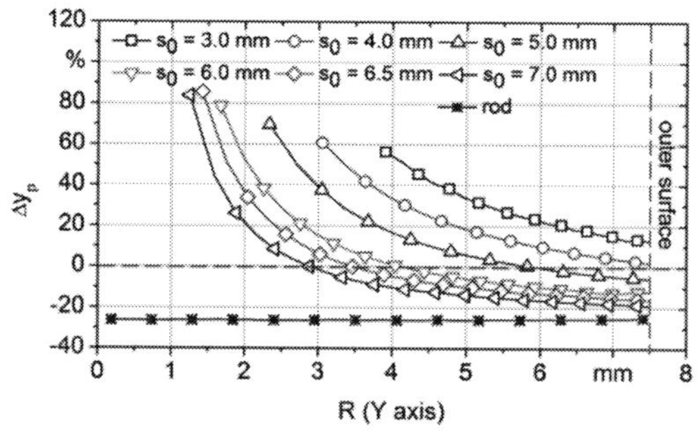

Fig. 7. Radial shape changes Δy_p along the radius for v_{f500}.

The ratios of R_n (R_{np}) in radial direction (Y axis) for both v_f with s_0 from 5 mm to 7 mm were determined and shown in Figure 8. The relative locations of this R_{np} shifted towards the central axis with bigger s_0. Thus in thicker tubes more radial material had the similar behavior to that of rods as expected. For both v_f the location of R_n was almost linearly dependent on s_0; with the same value of s_0 the location of R_n from higher v_f was always nearer to the inner surface. So the effect of increasing v_f was forcing more material flow to axial direction. The effect of v_f in changing radial material flow could help to control the inner profile, for example by using two v_f during the process might create stepped inner structures.

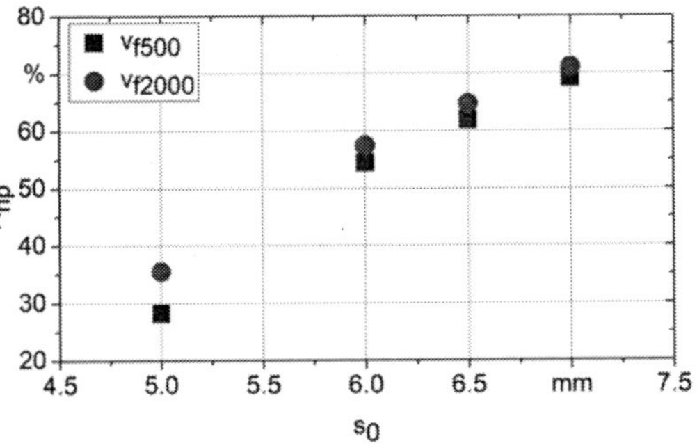

Fig. 8. Ratio of R_{np} according to s_1 for both v_f with different s_0

According to former discussion, increasing the wall thickness affected the material flow more than increasing the feeding velocity. Nevertheless, the trend could be derived that the more material to be deformed in one stroke, the harder for material to flow in radial direction. This should be caused by higher resistance against deformation in radial direction or a lower resistance in axial direction. The axial elongation of material could be treated as to be linearly dependent on s_0.

For the experimental results, two photographs of tubes (s_0 equals 3 mm and 7 mm) deformed with v_{f500} are shown in Figure 9. The values of s_1 measured from experiments and FE model are shown in Figure 10, the value of variance of each measurement of experiment is in terms of error bars. In Figure 10 it could be found that the values of simulation is always higher than those of experiment. The over-valuation of s_1 of FE model could be caused by no-ideal material flow in circumferential direction in 2D-axisymmetric model or by a discrepancy of material property values. Also a simulation in an extended research shown by increasing the friction coefficient from 0.1 to 0.2 the values of s_1 were reduced, and a former research shown dry rotary swaging of tubes resulted in larger elongation and thinner walls [17]. Hence, the friction coefficient value might be underestimated in the present study. Nevertheless, the maximum value of δ was 6.7% for s_0 equaled 4 mm with v_{f500} while the average values were 4.44% and 3.30% for v_{f500} and v_{f2000}. Thus a good agreement between experiment and FE model was achieved and so the FE model was validated.

To view the changes of wall thickness based on both FE model and experiment, the values of Δs_p were shown in Figure 11. According to Figure 11 the thickening of the wall (positive values of Δs_p) decreased linearly with increasing s_0 up to an initial wall thickness of about $s_0 = 6$ mm. Above this threshold the wall thickness decreased by the rotary swaging process ($\Delta s_p < 0$). Also, with a processing parameter dependent value of s_0 a not changing wall thickness was observed, which agreed with the simulation results. In addition, it could be verified that higher v_f led to a lower value of s_1, thus increasing v_f increased the resistance for material flow towards radial direction and forced more material flow towards axial direction. This result agreed with former experimental research of tubes ($s_0 = 2$ mm) [17].

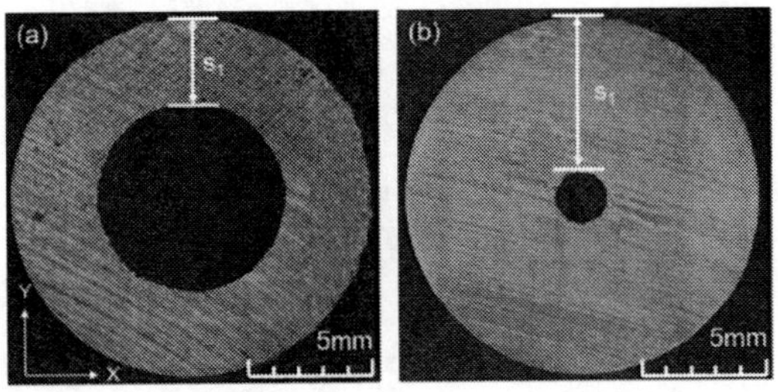

Fig. 9. Cross section of deformed tubes of v_{f500} (position A-A); (a) $s_0 = 3$ mm, $s_1 = 3.63$ mm; (b) $s_0 = 7$ mm, $s_1 = 6.50$mm.

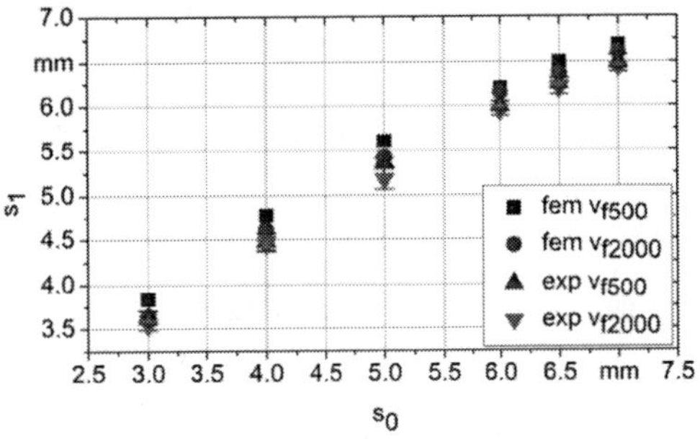

Fig. 10. Comparison of simulations and experiments.

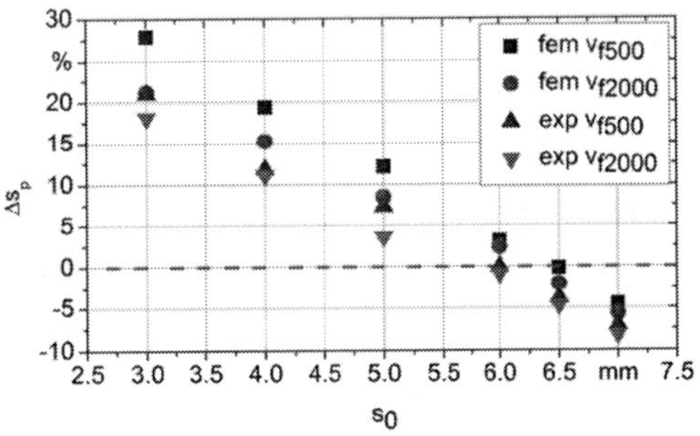

Fig. 11. Relative change of the wall thickness Δs_p at position A-A as a function of the initial wall thickness s_0. Positive values indicate an increase.

5 Conclusions and outlooks

In this paper the axial and radial material flow of steel tubes in mandrel free infeed rotary swaging process were investigated. A 2D-axisymmetric FEM model was built, the shape changes of the deformed meshes were calculated. The experimental validation was carried out by comparing the wall thickness change after the process. Several conclusions could be drawn:

- 2D-axisymetric model is reliable to study the radial and axial material flow in mandrel free tube rotary swaging process;
- axial elongation of the material in mandrel free tube infeed rotary swaging process is homogeneous over the radial position;
- increase of the wall thickness increases the radial material flow more than increasing the feeding velocity;
- the material flow in radial direction near the inner surface of tubes increases with higher initial wall thickness;
- for tubes with a certain wall thickness, there is a radial neutral boundary (R_n) where the material is neither compressed nor stretched, the location of R_n is nearer to the outer surface with lower v_f.

Based on the conclusions, the next research should be focused on the validation of radial neutral boundary, and if it affects the mechanical properties that are known to be affected by the actual material flow. Also, more friction conditions should be considered. Other processing parameters like stroke height or frequency are also needed to figure out their influences, which could help to control the material flow. The influence of feeding velocity should be investigated due to the potential to change the wall thickness.

The authors gratefully acknowledge the support by the German Research Foundation for a subproject within the priority program 2013, as well as the support by China Scholarship Council (CSC).

7 References

[1] T. Altan, S.-I. Oh, G. Gegel, American Society for Metals **1983**, 353 (1983).
[2] S. Ishkina, C. Schenck, B. Kuhfuss, E. Moumi, K. Tobeck, Int. J. Precis. Eng. Manuf **18,** 1035 (2017)
[3] S. J. Lim, H. J. Choi, K. H. Na, C. H. Lee, Solid State Phenomena **124.** 1645 (2007).
[4] Q. Zhang, K. Jin, D. Mu, P. Ma, J. Tian. Procedia Engineering **81**, 2336 (2014).
[5] M. Herrmann, C. Schenck, B. Kuhfuss, Procedia CIRP **40**, 654 (2016).
[6] R. Li, Z.-R. Nie, T.Y. Zuo, Transactions of Nonferrous Metals Society of China **16**, 1015 (2006).
[7] E. Moumi, S. Ishkina, B. Kuhfuss, T. Hochrainer, A. Struss, M. Hunkel, Procedia Engineering **81**, 2342 (2014).
[8] W. Gan, Y. Huang, R. Wang, G. Wang, A. Srinivasan, H.-G. Brokmeier, N. Schell, K. Kainer, N. Hort, MATER DESIGN **63**, 83 (2014).
[9] M. A. Abdulstaar, E. A. El-Danaf, N. S. Waluyo, L. Wagner, Mater. Sci. Eng., A **565**, 351 (2013).
[10] S. Ishkina, B. Kuhfuss, C. Schenck. Key Engineering Materials **651**, 627 (2015).
[11] Y. Li, J. Huang, G. Huang, W. Wang, J. Chen, Z. Zeng. MATER DESIGN **56,** 822 (2014).
[12] Y. Liu, M. Herrmann, C. Schenck, B. Kuhfuss, AIP Conference Proceedings **1896** (2017).
[13] X. Tan, Tribology International **35**, 385 (2002).
[14] E. Q. Sun, In MSC Software Corporation' s 2006 Americas Virtual Product Development Conference: Evolution to Enterprise Simulation, 17 (2006)
[15] E. Doege, H. Meyer-Nolkemper, I. Saeed, *Fließkurvenatlas metallischer Werkstoffe*, Hanser Verlag, Wien (1986).
[16] H. Czichos, K.-H. Habig, J.-P. Celis, *Tribologie-Handbuch: Tribometrie, Tribomaterialien, Tribotechnik*, Vieweg + Teubner, Wiesbaden (2010).
[17] M. Herrmann, B. Kuhfuss, C. Schenck, Key Engineering Materials **651**, 1042 (2015).

Influence of a modified drawing process on the resulting residual stress state of cold drawn wire

Markus Baumann[1*], *Alexander* Graf[1], *René* Selbmann[2], *Katrin* Brömmelhoff[3], *Verena* Kräusel[1], *Dirk* Landgrebe[1], and *Markus* Bergmann[2]

[1]Professorship for Forming and Joining, Chemnitz University of Technology, 09107 Chemnitz, Germany
[2]Fraunhofer Institute for Machine Tools and Forming Technology, 09126 Chemnitz, Germany
[3]Institute for Materials Science and Technology - Metallic Materials, Technical University Berlin, 10587 Berlin, Germany

Abstract. Torsion bars are used in automotive engineering as well as in other industrial applications. Such elements are produced by bending cold drawn wires. In conventional drawing processes tensile residual stresses occur near the surface of the wire. Small bending radii, which are required in limited assembly spaces, result in component failure due to reduced formability. Additional operations such as heat treatment or shot peening are necessary to influence the residual stress of the wire and to improve the dynamic stability of the torsion bar. The aim of the research is to reduce tensile residual stresses near the surface of the wire in order to eliminate process steps and to enhance formability. Therefore, a forming technology is developed by using a modified drawing die geometry on the basis of gradation extrusion. Finite element simulation is used to investigate the influences of element geometry, number of elements and process modification on the resulting residual stresses after wire drawing of a steel alloy. The results are evaluated and compared with the conventional wire drawing process. Furthermore, the requirements for the design of an experimental test device will be outlined as well as the measurement of the residual stresses by using X-ray diffraction.

Keywords. Residual stress, Cold forming, Finite-Element-Method

1 Introduction

The present study investigates the effect of using a modified drawing die geometry for cold drawn wires to influence the residual stresses of the wire. The aim is the reduction of tensile residual stresses near the surface.

Wire drawing is a method of bulk metal forming. The diameter is reduced by successive passes through a set of conical dies. Residual stresses are generated due to the inhomogeneous plastic deformation [1]. These stresses influence the mechanical behaviour of the wires and their durability and have an important effect on the functional properties of the wire for further forming operations or applications [2]. In general, residual stresses in the macro and micro range affect the strength properties. The superposition principle can lead to macro residual stresses and external load stresses exceeding the yield point of the material prematurely. However, in case of technical application this principle could be used to absorb a local treatment by elastic deformation because of reversal of the sign of the residual stress in comparison to the load stress [3]. Scholtes [4] stated that the vibration resistance caused by the development of residual compressive stresses could be improved by using shot peening or deep rolling. These additional processes influence the residual stresses after wire drawing. With regard to the optimization of the fatigue strength, residual compressive stresses are more suitable than tensile residual stresses due to reduced stress intensity. Llorca et al. [5] demonstrated that tensile residual stresses on the wire surface reduce the fatigue strength. In order to optimize the fatigue strength, the profile of the residual stresses should be customized to the load distribution since the external forces, torques and the residual stress profile of the wire get to a stress condition which determines the fatigue behaviour under vibration stress.

Tekkaya [6] described and compared how the residual stress was influenced by changing the percentage reduction of the wire diameter in one conical die. The final state of residual stresses is influenced by numerous parameters in the wire drawing process. The present study was intended to investigate the effect of modified drawing geometries of the die in the first section of drawing. Therefore the finite element method (FEM) was used. He et al. [2] investigated the influence of texture-based material models for simulated residual stresses after wire drawing. The computed distribution of the residual stresses over the cross section was evaluated using experimental specimens measured by X-ray

* Corresponding author: markus.baumann@mb.tu-chemnitz.de

https://doi.org/10.1051/matecconf/201819004004

diffraction. The measured and simulated residual stresses showed radial, tangential and axial residual compressive stresses in the wire core. When examining the cross section, it becomes apparent that the tangential and axial residual stresses are transformed into tensile stresses in the area from the wire core to the surface of the wire. The radial residual stresses were zero. Eutectoid steel (0.07 % C) was used as the investigated material. The level of the residual stresses can be influenced as described in [7]. The increase of the residual stresses was achieved by minimizing the drawing steps. Överstam [8] investigated the influence of production tolerances and die geometry on the residual stresses. In [9] and [10] the modification of the approach angle and form of the die was determined and compared to the resulting residual stresses.

The influence of changing the die geometry using extrusion was investigated in [11]. The authors describe the principle of severe plastic deformation (SPD) for the process of gradation extrusion. To achieve an ultrafine grain size of below one micrometer in the peripheral layer near the surface, repeated deformation of the material has to be realized while changing the forming direction. The design of the die includes different sections providing the specific forming conditions. An aluminium alloy was used in the test series [12].

2 Materials and method

The aim is to apply the experiences from gradation extrusion to wire drawing. Therefore the main results of gradation extrusion are presented first. Afterwards the methods of the FE simulations and the modified die geometries are described.

2.1 Gradation extrusion

Gradation extrusion (GE), a process which combines common impact extrusion with severe plastic deformation, is used to create property graded materials (Fig. 1). The processed material exhibits an increased strength in the near-surface layers and a relatively ductile core with lower strength [11].

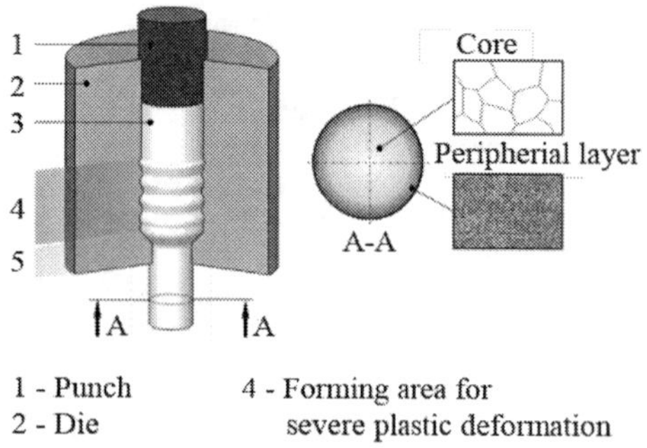

Fig. 1. Principle of gradation extrusion [12]

The aim of the GE process is to provide a severe plastic deformation process which has a high potential for applications in industrial manufacturing. It is intended to be integrated into process chains and to render the advantageous properties of the material accessible for various fields of applications. Therefore the process was designed to be material efficient, productive and applicable on standard machinery. Ideally, the forming process can combine both the microstructure modification and the generation of the geometry. In order to meet these aims, the basic approach consists in integrating severe plastic deformation into a common extrusion process and focusing the severe plastic deformation of the material to a local area with the highest load bearing requirements, which is generally the lateral area of a part. The name gradation extrusion refers to the major characteristic effect of the method: the generation of a graded microstructure. The processed workpiece possesses an ultrafine grained microstructure in the lateral area which gradually becomes coarser towards the middle or core of the workpiece where the grain size corresponds to the initial coarse or normal microstructure. The distribution of mechanical properties basically corresponds to the microstructure gradation [13].

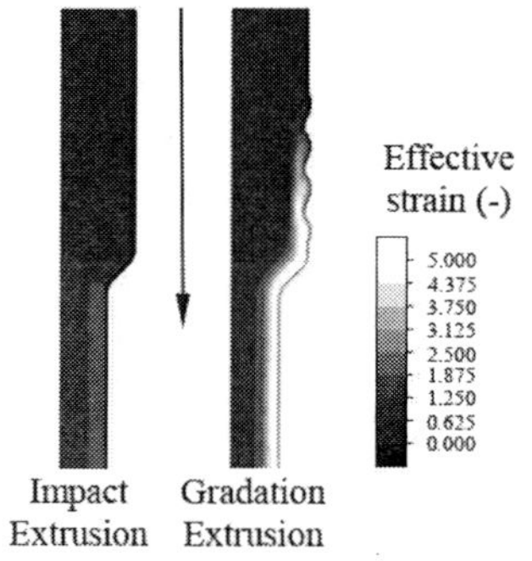

Fig. 2. Distribution of effective strain determined with FE simulation [14]

Fig. 2 shows the distribution of effective strain obtained from the FE simulation representing the steady state of material flow. A pronounced gradient of effective strain was found, ranging from the interior region to the surface. This gradient increases with every forming element in axial direction. The final cross-sectional reduction leads to an additional introduction of strain, resulting in a final gradient ranging approximately from 1.1 (interior) to 4.9 (surface). The process basically fulfils the requirements for severe plastic deformation processes (SPD processes), since it includes large effective strains, load path changes, high hydrostatic pressure and temperatures below recrystallization temperature. These results show that processing by gradation extrusion has a high potential for grain refinement by SPD, which leads to the assumption that distinct microstructural gradients may occur [15]. These previous studies demonstrate that special geometries influence the microstructure and the effective strain. The results should be transferred to the wire drawing process.

https://doi.org/10.1051/matecconf/201819004004

2.2 Finite element simulation

The investigations described in this paper were conducted using the finite element program ABAQUS 6.14. The forming process was simulated with an explicit solver. To calculate the residual stresses of the wire after forming, the tool was removed and an implicit simulation of the wire followed. Using a defined node set over the cross section in the middle of the wire the axial, tangential and radial residual stresses were determined. Due to the axial symmetry of the wire drawing process, an axially symmetric two-dimensional FE model was set up (Fig. 3). The initial diameter of the wire was 12 mm, and resulted after forming in a diameter of 10.8 mm, which means a effective strain for the process of $\varphi = 0.2107$. The conventional die has a reduction die angle of 12°. In the simulation the workpiece consists of 4200 elements with an element size of the mesh, which is 0.1 x 0.5 mm^2 in the peripheral region and 0.5 x 0.5 mm^2 in the core. The results of the FE simulations depend on this size of the mesh. Due to the expected high degree of deformation, the near surface layer of the wire was modelled using adaptive mesh refinement. In order to minimize calculation time, the die is defined as a rigid body in this studies. Further, elastic-plastic properties of the die require investigations such as mechanical characterization and experimental tests to validate the obtained results.

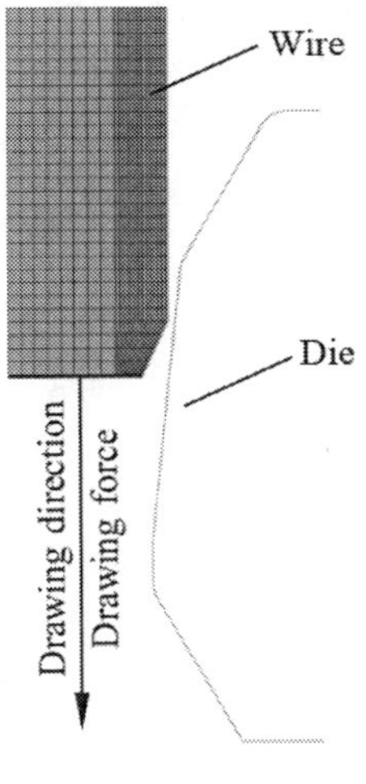

Fig. 3. Axially symmetric two-dimensional FE model for simulations of wire drawing

The austenitic steel X5CrNi18-10/1.4301 was defined for the simulations. The flow curve was established by using Hensel-Spittel approach (1).

$$k_f = A \cdot e^{m_1 \cdot \vartheta} \cdot \vartheta^{m_9} \cdot \varphi^{m_2} \cdot e^{\frac{m_4}{\varphi} + m_7 \cdot \varphi} \cdot (1+\varphi)^{m_5 \cdot \vartheta} \cdot \dot{\varphi}^{m_3 + m_8 \cdot \vartheta} \quad (1)$$

A = 1188.736, $\vartheta = 20$; $m_1 = -0.00173$; $m_2 = 0.29211$; $m_3 = 0.00456$; $m_4 = 0.00021$; m_5-$m_9 = 0$; $\dot{\varphi} = 0.0018$ [16]

Young's modulus, E, was set to 198.7 GPa and Poisson's ratio, υ, was set to 0.3. The material for these investigations was assumed to be isotropic and elastic-plastic. The influences of the temperature and the strain rate were at first not considered in the modelling, despite the fact that they would influence the flow rate. Further studies are planned using special material models as described in [2] and elastic-plastic modelling of the die. The material models will consider the temperature behaviour, the microstructure, the strain rate and texture of the material. The investigation described in this present paper focused on the fundamental relationships between the residual stresses in the wire after drawing regarding the die geometry by using the defined mesh size of the wire and friction behaviour to the die. The friction between the wire and the die was modelled according to Coulomb's law with a friction coefficient of $\mu = 0.03$ [2]. The drawing direction was determined by a boundary condition at the bottom of the wire (Fig. 3). In this context, certain displacement was defined at the 29 respective nodes. The drawing force is evaluated as a sum using these nodes. The penalty formulation was defined as a contact constraint to the die.

2.3 Modified die geometry

The developments of different die geometries are based on a conventional die for wire drawing with an reduction die angle of 12°. The wire diameter of each investigated die geometry is reduced from 12 mm to 10.8 mm. The cylindrical part, also referred to as bearing length, is implemented with the same length of 1.64 mm at each geometry variant. Influences of different bearing geometries are investigated by Överstam [8]. To identify the correlation between the drawing die geometry and resulting residual stresses a single forming element was realized. Three different die geometries are investigated with two variants (V) each. Fig. 4 presents a comparison between the convex, concave, step (each V1 and V2) and the conventional geometry.

The convex geometry V1 should cause deformation in the diameter of a conventionally formed die from 12 mm to 11.30 mm with a reduction die angle of 12° and further severe deformation down to 10.8 mm by using a forming angle of the element of 60°. The size of the forming element is 0.25 mm (convex V1). V2 of this convex geometry exhibits a forming angle of 15° of the element by the same size of 0.25 mm, which implies an initial diameter of 11.63 mm before the forming element. The geometry concave V1 is implemented with the same size, angle of the forming element and end position as the forming element in geometry convex V1. Further forming to a diameter of 10.8 mm is done with a conventional reduction die angle of 12°. At the geometry concave V2, the forming element is positioned in front of the bearing length at the end of the conventional forming. The forming element size is 0.25 mm with a forming angle of 60°.

 https://doi.org/10.1051/matecconf/201819004004

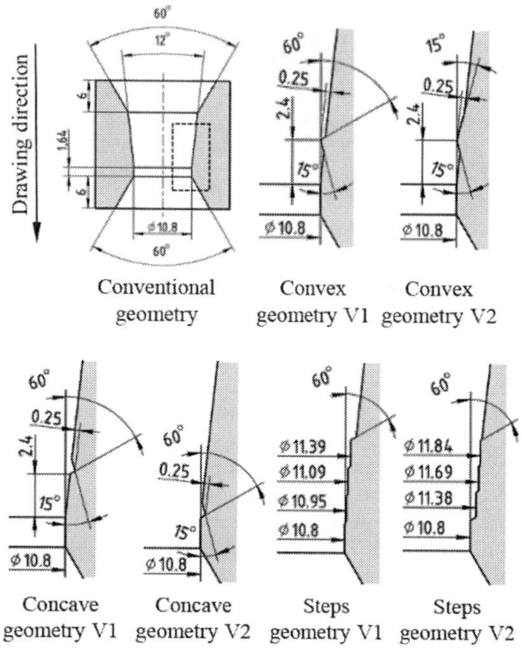

Fig. 4. Die geometries

The geometries step V1 and V2 comprise forming of the wire in four stages. The overall effective strain of the process ($\varphi = 0.2107$) is divided into step V1 with stage 1 entailing approx. half the overall effective strain ($\varphi = 0.105$), in stage 2 the effective strain value is 50% of the effective strain of stage 1 ($\varphi = 0.0525$), and stage 3 and 4 each comprise one half of the remaining effective strain ($\varphi = 0.0266$). In step V2 the stage geometry is inverted (0.0266, 0.0266, 0.0525, 0.105). The distance between each stage is 1.5 mm.

R = 0.1 mm applies to the active forming radii of each forming element of the different geometries and variants. The geometry for gradation extrusion presented by Bergmann [14] was not designed. Contrary as shown in Fig. 1, the contour of the forming element will not be filled in due to the tensile forces in wire drawing. The first geometrical element in the drawing direction causes an overall severe reduction of the wire diameter, or induces cracking of the wire.

3 Results of FE simulation

In order to identify the effect on residual stresses after drawing by the differently designed die geometries, the axial, radial and tangential stresses are evaluated. The following figures present the results. The values of the stresses are analysed at nodes in radial direction in the middle of the drawn wire. It has to be considered that the results of the residual stresses depend on the defined size of the mesh, the material model and further named boundary conditions.

Fig. 5 shows the residual stresses which occur when applying a conventional die geometry in wire drawing. The stress distribution of the axial and tangential directions exhibits compressive residual stresses in the core of the wire and transforms into tensile residual stresses on the surface. No radial residual stresses are detected on the surface of the wire. These residual stress distributions are typical for conventional wire drawing processes [1]. The axial and tangential tensile residual stresses on the surface of the wire are critical for further forming operations such as bending of the wire, as applied for producing torsion bars.

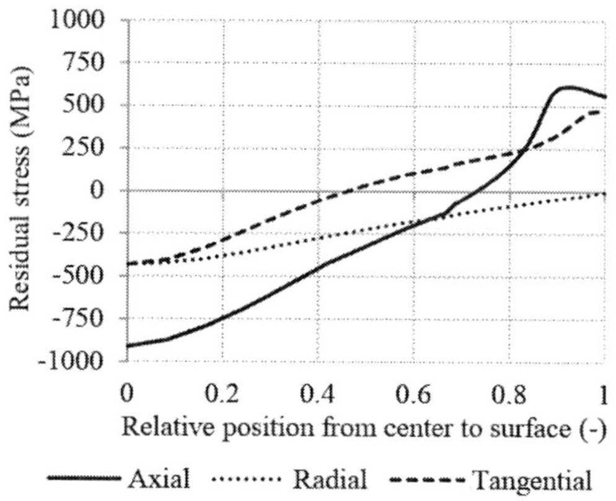

Fig. 5. Results of residual stress for conventional geometry

The die geometry for the first drawing step was changed (Fig. 4) in order to influence the residual stress distribution of the wire. Fig. 6 compares the results of the FE simulation for convex geometries. The radial stresses of the two geometrical variants show approximately the same values and flows. Large deviations regarding axial and tangential residual stress occur between convex V1 and V2.

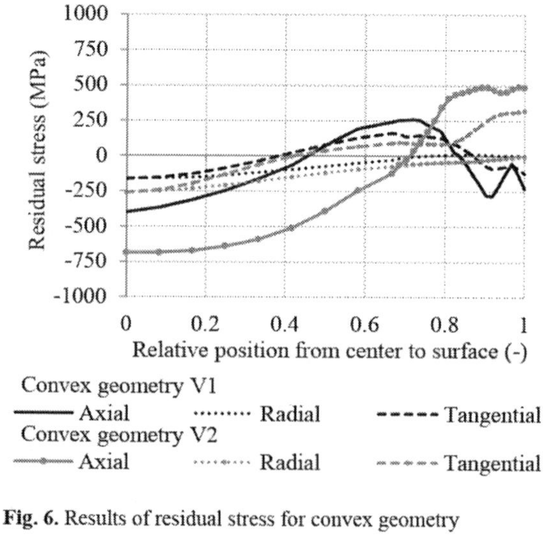

Fig. 6. Results of residual stress for convex geometry

By using a convex element with an angle of 60° and a size of 0.25 mm of the forming element as it is the case in convex V1, the residual stresses on the surface change to compressive axial and tangential residual stresses compared to forming with the conventional die geometry. The forming element has an influence on the overall stress distribution. The compressive stresses in the core are halved and zero crossing occurs much earlier compared to the stress flow of the residual stress of wire forming due to conventional die geometry. In convex V2, which implies an angle of 15° of the forming element instead of 60° in convex V1, the stresses in the core are reduced, but the tensile residual stresses at the surface are not minimized to zero or compressive stresses, compared to the conventional geometry shown in Fig. 5. The analysis of the material flow in the FE simulations showed that the material does not flow back in wire drawing after shaping the convex forming element. The convex geometry affects the forming of the wire in the same manner as a die with large reduction die angle.

Fig. 7 presents the results of the residual stress distribution by forming the wire by conventional geometry with an additional concave forming element. Concave V1, with a concave forming element in the position of the geometry convex V1 shows approximately the same stress distribution as in forming with a conventional die. Only a small forming effect was detected in the FE simulation analysis. Moving the forming element down to the end of the reduction zone of the conventional die geometry before the bearing starts, which is realized in the geometry concave V2, there is a large influence on the stress distribution of axial, radial and tangential residual stresses compared to results of conventional geometry. The compressive stresses in the core are halved compared to the conventional geometry and the zero crossing occurs much earlier. In addition, there is a severe decline from the tensile residual stresses to compressive residual stresses in the stress distribution to the surface of the wire, followed by an increase to zero and drop back to the range of compressive stresses.

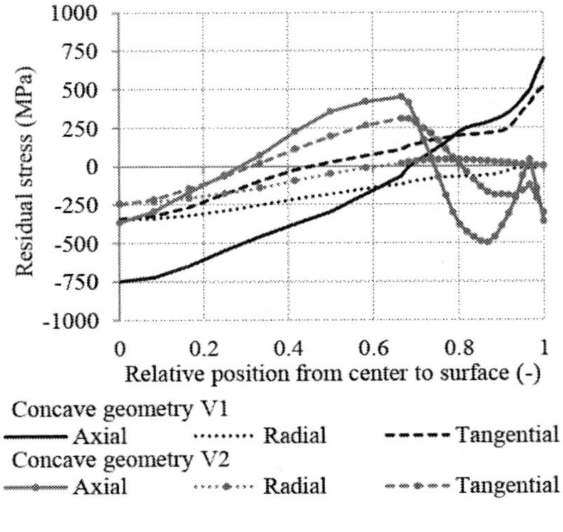

Fig. 7. Results of residual stress for concave geometry

The most significant influence on the residual stress distribution is achieved with the forming by step geometry. Fig. 8 demonstrates that over a large area of the cross section the stress distribution is in the range of compressive stresses.

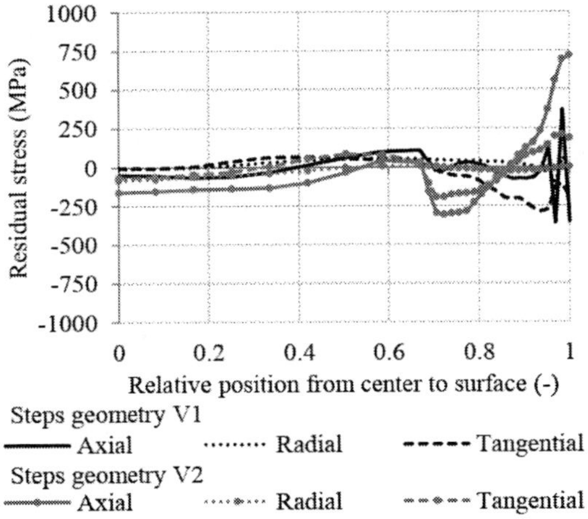

Fig. 8. Results of residual stress for steps geometry

Forming with the geometry step V2, which means a large strain rate occurs in the last stage, causes tensile residual stresses on the surface of the wire. When using the geometry step V1, which exhibits the largest strain rate in the first stage, compressive tangential stresses and varying axial stresses near the surface are implied. The differences occur due to multiple change of forming direction compared to the convex and concave geometries which is recognizable in the higher effective strain.

Because the axial and tangential tensile residual stresses at the near surface layer of the wire are critical for further forming operations, such as bending of the wire, these results of the distribution are compared in separate charts. Fig. 9 demonstrates the axial residual stresses of the different geometry variants compared up to 30 % under the surface of the wire. In Fig. 10 the tangential residual stresses are compared dependent on the different geometry variants.

Considering the defined mesh size, the material model and further named determinations of the FE simulations, Fig. 9 and Fig. 10 demonstrate that the aim of the studies, to generate compressive axial and tangential residual stresses in the near surface layer, is achieved with the geometries convex V1, concave V2 and step V1. On the basis of these results the aim for further studies is to combine and investigate more geometry variants to deduce tendencies of the individual geometry parameters. The aim of the presented studies was to investigate whether geometrical changes on the forming surface of the die influence the residual stresses of the wire in one drawing step.

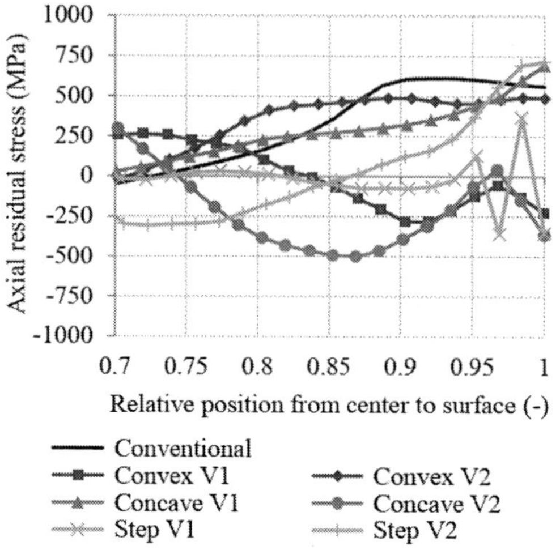

Fig. 9. Comparison of the axial residual stresses of the different die geometry variants

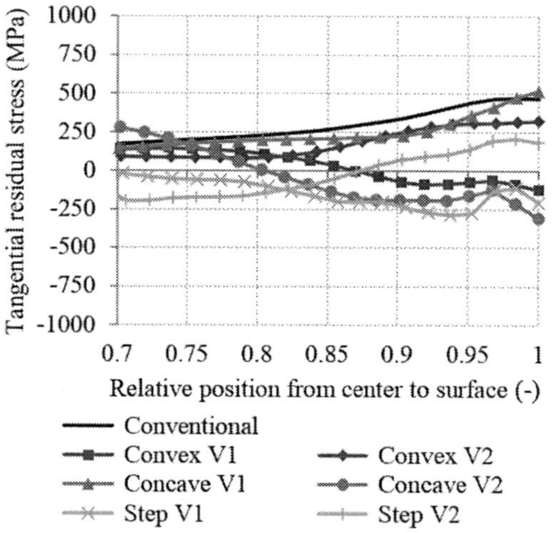

Fig. 10. Comparison of the tangential residual stresses of the different die geometry variants

To analyse the distribution of effective strain over the cross section and especially in the near surface layer the values are plotted by using the same node set as the residual stresses are evaluated. In Fig. 11 the effective strain curves, which are the results by forming the wire with the defined mesh size, are shown. The effective strain value in the core of all geometrical variants is $\varphi = 0.21$. Nearly from the half of the wire diameter up to the surface the values of the effective strains are different. The maximum effective strain is reached by forming the wire with the step geometry. The absolute values of the effective strain depend on the relative position from the center to the surface. They are compared in Table 1.

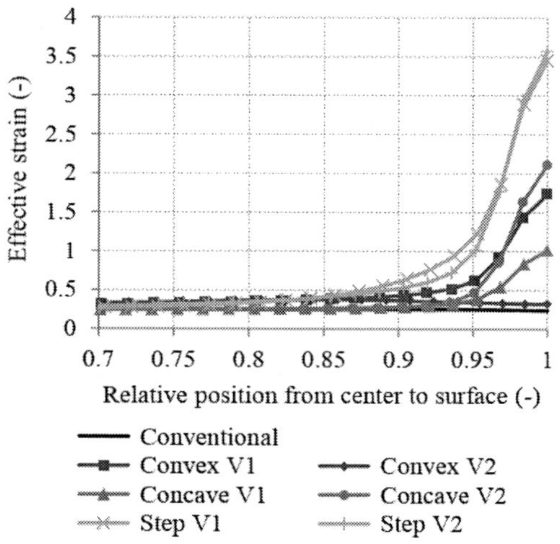

Fig. 11. Comparison of the effective strain at the near surface layer of the different geometry variants

Table 1. Effective strain dependent on relative position from the center to the surface

	Effective strain φ dependent on the relative position from the center to the surface (-)			
Geometrical variant	**0.7**	**0.8**	**0.9**	**1**
Conventional	0.24	0.25	0.26	0.24
Convex V1	0.31	0.36	0.43	1.74
Convex V2	0.33	0.37	0.36	0.32
Concave V1	0.24	0.26	0.28	1.02
Concave V2	0.24	0.26	0.28	2.11
Step V1	0.27	0.34	0.64	3.45
Step V2	0.29	0.35	0.53	3.57

Because of the multiple change of the forming direction the effective strain in the geometry variants step V1 and V2 is larger in the observed surface layers compared to the other geometrical variants. The variants with compressive residual stresses in the near surface layer (convex V1, concave V2 and step V1) show higher effective strain on the surface layer than the variants with tensile residual stresses at the surface layer. An exception is the geometry variant step V2. Further investigations are necessary to identify dependencies and tendencies of geometrical parameter influences.

Fig. 12 presents the results of average drawing forces values as function of the forming die geometry. Material modelling and the size of the defined mesh is important for the absolute values of the drawing force. However, it is possible to draw a relative comparison of the forces as function of the die geometry. The angle of the elements of the convex geometries (V1: 60°, V2: 15°) has a significant impact on the resulting drawing force. Compared to the conventional die geometry, the geometry convex V1 showed an increase of approx. 70% generated by the severe deformation of the forming element.

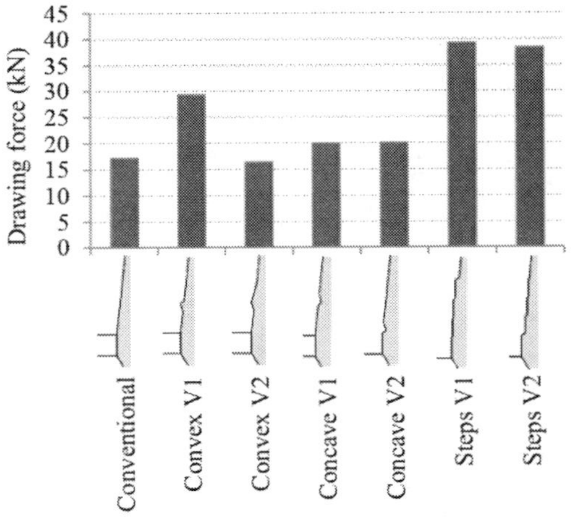

Fig. 12. Results average value of drawing force

The influence of the concave elements on the drawing force is not very significant. The forming elements of the step geometries have the greatest effect on the residual stress flow, but also the drawing force increases considerably because of the multiple change of forming direction due the step geometry. In addition the effective strain in the near surface layer is much higher.

4 Conclusions and Outlook

The aim of this investigations is to assess die geometry variants and to obtain geometries which reduce tensile residual stresses near the surface of the wire. The results of the FE simulation indicate that there are significant and different influences on the stress distribution of the residual stresses when using specific forming elements for wire drawing. It has to be considered that the results depend on the defined mesh size, the material model and the named boundary conditions. Regarding these conditions the pursued effect is achieved by using one convex forming element with a reduction angle of 60° and an element size of 0.25 mm. A concave forming geometry at the end of the forming could also lead to compressive residual stresses in the surface of the wire. The results of the axial and tangential residual stresses near the surface of the wire are compared in Fig. 9 and Fig. 10. The stepped forming geometries generate the lowest residual stress values in the core of the wire compared to the other geometries. However, more FE simulations with different die geometries and parameter variations are planned to deduce tendencies of the geometrical parameters on the resulting residual stresses of the wire. In these studies the limitations due to manufacturing constrictions such like radii or element sizes will be regarded.

In order to realize the planned further investigations, specific material models are required for the FE simulation. Isotropic hardening was assumed in these studies. The wire temperature, the forming ratio, the high strain rate because of drawing speed up to 100 m/min and information about the texture of the microstructure have to be considered for wire drawing. In addition, experimental tests are essential to enable a validation of the simulations by a residual stress analysis by means of X-ray and neutron diffraction. Laboratory X-ray diffraction will be used to analyse the surface residual stresses and the residual stresses up to a depth of 500 µm after different steps of electrochemical removal. Higher depths beyond 500 µm up to the middle of the wire will be analysed by neutron diffraction. The analysis will yield the macro and micro residual stresses and their distributions over the wire diameter as well as information about the textures which are also important for the material modelling.

For the experimental work a test device shall be developed which will be implemented in a high speed testing machine. The die shall be interchangeable to use the test device for all drawing steps. Investigations of the behaviour of wear of the forming elements are part of advanced experimental studies. Using experimentally determined residual stresses by neutron diffraction and data from the experimental tests, the FE model can be validated by comparing numerical computations and experimental results [1]. As described, Neutron diffraction is necessary to determine the residual stresses in deeper layers of the wire.

The industrial wire drawing operations consist of more than one drawing step. FE simulations with four or five steps have to be computed and the residual stresses have to be evaluated.

In comparison to gradation extrusion, wire drawing is a different process due to the tensile forces in the forming direction. After forming by the convex forming element, the wire material does not flow back to the die. The planned experimental work will comprise the influence of the forming technology using specific forming elements in the drawing die.

The support provided by the priority program –SPP 2013–, which is financed by German Research Foundation, is gratefully acknowledged.

References

[1] J. M. Atienza, J. Ruiz-Hervias, M. L. Martinez-Perez, F. J. Mompean, M. Garcia-Hernandez, M. Elices: Residual stresses in cold drawn pearlitic rods. *Sript. Mat.*, **52**, no. 12, pp. 1223–1228 (2005)
[2] S. He, A. Van Bael, S. Y. Li, P. Van Houtte, F. Mei, A. Sarban: Residual stress determination in cold drawn steel wire by FEM simulation and X-ray diffraction. *Mat. Sci. and Eng.*, **346**, no. 1–2, pp. 101–107 (2003)
[3] V. Hauck: Structural and Residual Stress Analysis by Nondestructive Methods, pp. 235 (1997), no. 1–2, pp. 101–107 (2003)

[4] B. Scholtes: Eigenspannungen in mechanisch randschichtverformten Werkstoffzuständen: Ursachen, Ermittlung und Bewertung. *DGM Informationsgesellschaft* (1991)
[5] J. Llorca, V. Sánchez-Gálvez: Fatique limit and fatique life prediction in high strength cold drawn eutetoid steel wires. *Fatique Fract. Eng. Mater. Struct*, **12**, no. 1, pp. 31–45 (1989)
[6] A. E. Tekkaya: Ermittlung von Eigenspannungen in der Kaltmassivumformung. *Berichte aus der Umformtechnik der Universität Stuttgart* (1986)
[7] D. J. Celentano, M. Palacios, E. L. Rojas, M. A. Cruchaga, A. A. Artigas, A. E. Monsalve: Simulation and experimental validation of multiple-step wire drawing processes. *Finite Elem. Anal. Des.*, **45**, no. 3, pp. 163–180 (2009)
[8] H. Överstam: The influence of bearing geometry on the residual stress state in cold drawn wire, analysed by the FEM. *Journal of Materials Processing Technology*, **171**, no. 3, pp. 446–450 (2006)
[9] J. M. Atienza, M. Elices: Influence of residual stresses in the tensile test of cold drawn wires. *Mat. and Struc.*, **36**, no. 262, pp. 548–552 (2003)
[10] Ch. Siva Naga Teja, N. Guru Murty, P. Satish Reddy: FE Analysis of Wire Drawing Process with different die contours. *Int. journal of scie. eng. and adv. tech.*, **4**, no. 2, pp. 134–143 (2016)
[11] R. Neugebauer, A. Sterzing, R. Selbmann, R. Zachäus, M. Bergmann: Gradation extrusion Severe plastic deformation with defined gradient. *Mat.-wiss. u. Werkst.*, **43**, no. 7, pp 582–588 (2012)
[12] M. Bergmann, A. Sterzing, D. Landgrebe: Influencing the gradient of material properties by gradation extrusion. *Appl. Mech. and Mat.*, **794**, pp. 166–173 (2015)
[13] R. Neugebauer, M. Bergmann: Local Severe Plastic Deformation by Modified Impact Extrusion Process. *Steel Research International - Special Edition: 14th International Conference on Metal Forming*, pp. 471–474 (2012)
[14] M. Bergmann: Verfahren zur Herstellung gradiert hochgradig plastisch umgeformter Werkstoffe. *Verlag wissenschaftlicher Scripten* (2013)
[15] D. Landgrebe, A. Sterzing, N. Schubert, M. Bergmann: Influence of die geometry on performance in gradation extrusion using numerical simulation and analytical calculation. *CIRP Annals-Man. Tech.*, **65**, pp. 269–272 (2016)
[16] M. Spittel, A. Spittel: Datasheet X5CrNi18–10 1.4301, *Metal Forming Data of Ferrous Alloys deformation behaviour*, **2C1**, SpringerMat. (2009)

Numerical modelling of T-pipe joints forming

*Jarosław*Bartnicki[1,]

[1]Lublin University of Technology, 36 Nadbystrzycka, 20-618 Lublin, Poland

Abstract. This paper presents results of numerical calculations of T-pipe joints (tee) forming process for a wide scope of industrial applications. FEM analyses were conducted for the chosen process technological parameters and tools designs. The kinematics of metal flow in the area of the formed flanges was analysed. Distributions of stresses, strains and damage criterion during forming were determined. Calculated values of forces and moments acting on tools and workpieces allow for designing of tools geometry for experimental verification of the proposed forming process.

Keywords: FEM, design optimization, cold forming

1 Introduction

The solutions for different liquid, sand, gases transportations are very different. Many of them concern applications of different advanced polymer technologies but for high pressure pipe ducts steel materials are commonly applied. In this case it is extremely important to preserve good sealing of designed pipelines. Different welding technologies are normally used for connections making but especially difficult case among them is T-pipe joints (tee) forming and welding with another pipe structures.

For a wide scope of application connections made with T-pipe joints are realized by drilling the wholes inside one pipe and its welding into another. For increasing mechanical characteristics and life time of these connections flanging technology is applied. This solution is favourable for pressure distribution and makes the welding processes easier for making and verifying – Fig. 1. The proposed solution is also less material consuming comparing with traditional one and it can be also applied in existing installations using portable devices for drilling and tee making. Typical applications areas are in food processing, pharmaceutical, pulp and paper industry, stainless steel piping systems, water treatment, shipbuilding and conventional and nuclear submarines[1].

The proposed solution for bigger pipe installations must be equipped with special machine for tee making. This aggregate consist of drilling section for elliptic hole making, and special flanging head. Combining rotations and axial movement of this element results in flanges (collars) creation. For smaller pipes tee making can be realized by means of portable tools because this process does not require application of bigger values of forces and torques.

The observations of this problematic and numerical verification of material flow and technological parameters are fundamental for planned new devices for T-pipe joints development. The most important for new solution designing is tools geometry and load parameters knowledge combined with final parts dimensions accuracy [2-4].

Fig. 1. The example of traditional (left) and formed by flanging T-pipe joint (right) [1]

2 Numerical modelling

Numerical calculations of proposed scope of flanging processes were realized using Deform 3D software. In numerical models were applied pipes with external diameter Ø150 and Ø300 mm and created collars with diameters Ø50 and Ø100 mm.

2.1 Numerical models

In numerical simulations the pipes with the mentioned above two different external diameters and two collars were considered. As the material model weldable fine-grain structural steel for pressure vessels P355NH was used [2]. For calculations in the cold metal

*Corresponding author: j.bartnicki@pollub.pl

https://doi.org/10.1051/matecconf/201819004005

forming conditions all tools and workpieces have the same initial temperature T = 20°C.The constant friction model was assumed for all cases with friction factor m = 0.25, as the processes are realized with good lubrication.

Some calculations were realized with workpiece mounted inside the jaws with diameter corresponding to the pipe dimensions. After verification of state of stress and strain, the workpiece were limited to the halfpipes fixed by boundary conditions at the external edges. The workpiece was meshed by tetra - type elements with refinement of forming zone by three times smaller ones with minimum 5 elements on wall thickness as assumption. This solution permits to realize simulation faster.

For the proper circular collar shape forming the initial holes inside the pipes was designed as elliptic calculating geometrically created collar. For example, the collar Ø50 mm formed into the pipe Ø150 mm required initial elliptic whole 30 x 37 mm. The examples of worked out models of the analysed cases are presented in Fig. 2.The first model consists of formed pipe with initial elliptic whole, outer pipe as mounting for workpiece and two bars – tools which flange the collar. In the second, developed model physical fixation is replaced by boundary conditions with fixed edges (red nodes in Fig 2).

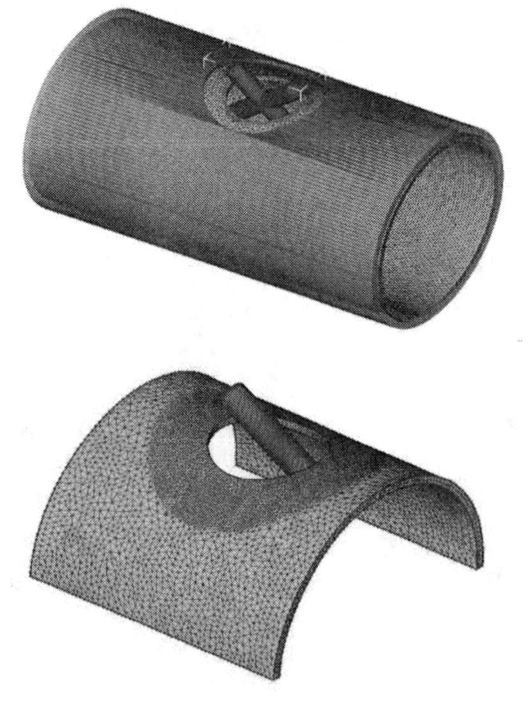

Fig.2. Worked out numerical models for calculations with physical pipe stabilization (above) and with boundary conditions with fixed displacement (below)

For time calculation optimization in later realized simulation the box with fine mesh was applied. These modifications help to make simulation faster. All these changes were made after state of stress verifications with the first model without any changes.

2.2 Feed rates and tools configurations

The technological values crucial for the analysed cases were feed rates f written in mm of working tools movement per rotations. This parameter is strictly combined with dimensional accuracy and effective time of realized collar tee connections.

Analysing accessible solution [1] it was assumed to apply solution with 1, 2, 3 and 4 rotating and axially moving tools. The forming tools – bars have the same diameter Ø10 mm and specially rounded ends. According to the chosen option, the feed rates applied in calculations were 0.5; 1.0 and 2 mm per one rotation.

Another very important technological parameter in this forming process was tools – bars inclination angle α. The values for this parameter were taken within the range of α = (35÷60)°, which results in collar flanging mechanism and final joint edge shape. This parameter also changes final strain distribution. Smaller inclination angle and feed rates lead to obtaining bigger values of strain with significant part of redundant strains.

3 Simulations results

3.1 States of effective stress and strain

During numerical simulations realized with the presented assumptions and technological parameters it was possible to verify distributions of stress and strain during collars flanging. In Fig. 3 and Fig. 4 the results of calculations for Ø50 collar forming into Ø150 mm pipe are presented.

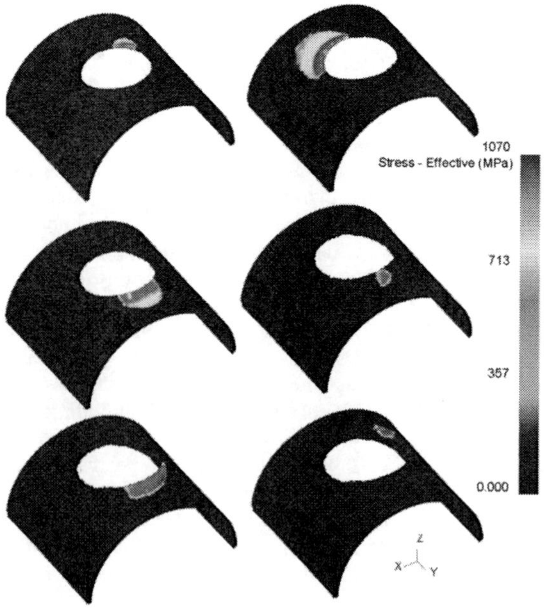

Fig.3. Progression of shape and distribution of effective stress during Ø50 collar forming into Ø150 mm pipe

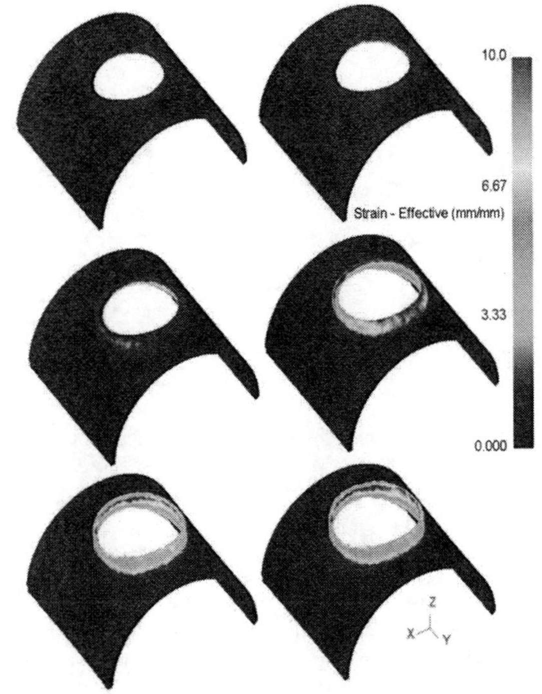

Fig.4. Progression of shape and distribution of effective strain during Ø50 collar forming into Ø150 mm pipe

As it is shown in Fig. 3 the distribution of effective stress is really local and maximum values of this parameter exceed 1000MPa. In zones of contact between tools and workpiece material is moved towards the direction of axial movement of tools. Presented in Fig. 4 effective strain distributions show real range of material deformation during flanging process. Values of this parameter are slightly different according to number of moving tools but the distributions are depended on different feed rates applications Bigger values of feed rate f = 1 and 2 mm per rotation provide to obtaining effective strain values reaching 12÷14. Normally this information is important considering fine grain material structures into plastically formed zones. However this observation must be verified in experimental tests in real flanging collars device [5].

3.2 Wall thickness and damage criterion distributions

The wall thickness distributions in longitudinal and cross sections of chosen case are shown in Fig. 5.

Presented results show very favourable distribution of wall thickness. In all analysed zones of collar forming there are no significant decreases of this parameter. Very similar situation is observed in real case of collar forming by two rotating bars in steel P355NH pipe which is shown in Fig. 6. This information is very important for the future scope of use of T-pipe joint in welding of high pressure pipe ducts with very elevated mechanical characteristics.

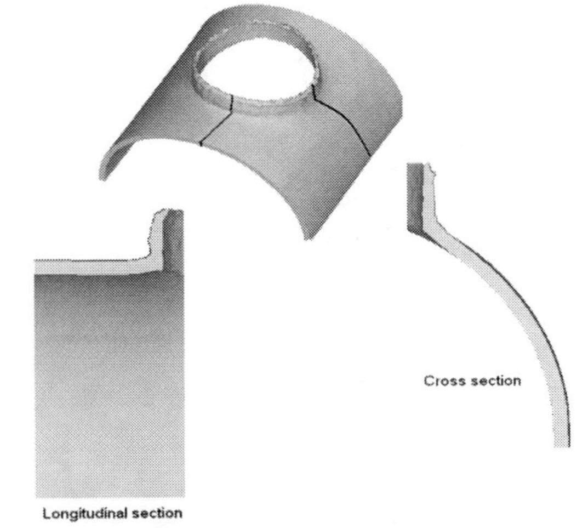

Fig.5. Wall thickness shapes - calculation results during Ø50 collar formed into Ø150 mm pipe

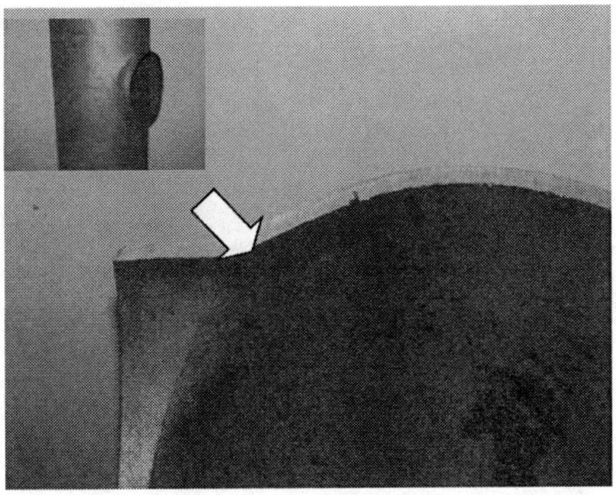

Fig.6. Wall thickness shape in cross section – real part view during Ø50 collar formed into Ø150 mm pipe

The differences between calculated and obtained in experiment distributions of collar wall thickness are not significant. The application of pipe with initial wall thickness 4.00 mm results in minimal calculated value 3.48 mm and measured in experiment 3.56 mm. During the process observation it was stated that the friction conditions are very important for this values distribution. The application of old pipes with rusted inner surface in experiments results in bigger wall thinning during flange forming. The application of parts from construction steel with analogical dimensions and process conditions results in final collar wall thickness equal 3.16 mm. This shows the importance of friction phenomena reduction during process realization. In the further research this value will be determined in experiments.

The distributions of damage criterion by Cockroft – Latham were also analysed during calculations. One chosen distribution of this parameter at the final stage of forming is presented in Fig. 7. The obtained results confirmed that flanged material should not be damaged

 https://doi.org/10.1051/matecconf/201819004005

by internal defects especially by cracking. In the case of lower feed rate values as f = 0.5 mm/rot. C-L criterion integral reaches 0.8 comparing with 0.3÷0.4 in case of bigger analysed feed rates.

3.3 Load parameters

Load parameters as axial forming force and torque were also analysed during simulations. Values of these parameters for the extreme conditions of forming are presented in Fig. 8 and Fig. 9.

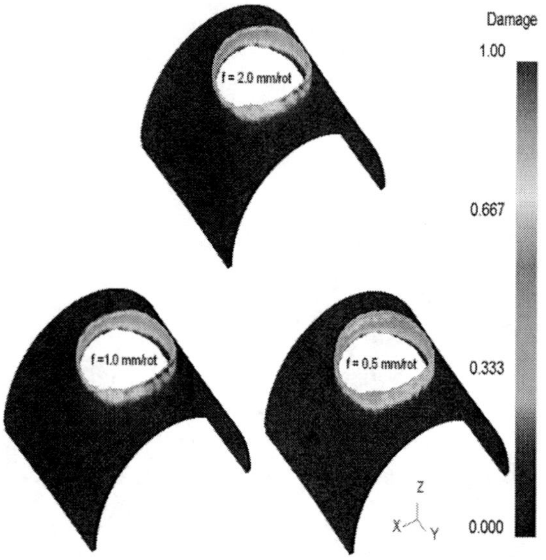

Fig.7. Cockroft-Latham damage criterion values for different feed rates - calculation results during Ø50 collar forming into Ø150 mm pipe

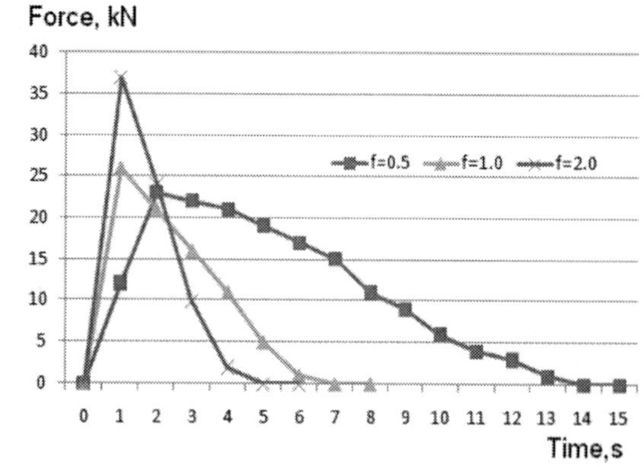

Fig.8. Forming forces during collar forming – distributions with different feed rates

Numerical simulations results give information about values of forming forces and torques which is necessary in constrains for the future device designing. The knowledge of basic technological parameters and its influencing each other is very important for the process designing.

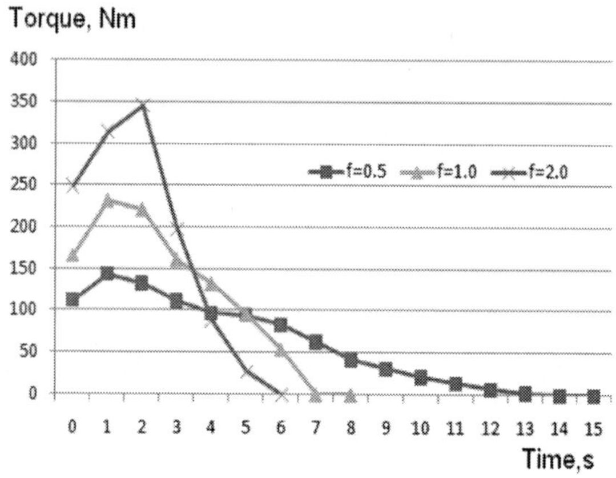

Fig.9. Torques during collar forming – distributions with different feed rates

4 Conclusions

Worked out numerical calculations of T-pipe joints forming show practical possibility of this process application. Regarding the existing solutions for this type of parts manufacturing it should be interesting to develop special device for this activity dedicated to the hydraulic press or for handy operated devices for smaller pipe diameters.

Some results concerning especially feed rate f and inclination angle α and initial elliptic hole dimensions will be very useful for this process recognizing. The observed significant torque values reduction depending on applied feed rates will be very useful for handy operated portable device. In this case, the crucial limitation is pipe fixation during collar forming in existing installation. The device body stiffness is also very important because it will decide about whole dimensions accuracy and tools positions stability.

All results must be verified in experiments, It is especially important in the case of microstructure, hardness distribution and eventually internal defects very difficult to forecast by numerical calculations.

Another very important factor are friction conditions determining parts wall thickness distributions important for certification procedures of these T-joints.

Finally, all these issues will decide on the scope of the proposed solution applicability and safety rules for users.

References

1. https://t-drill.com/technologies/tee-forming/
2. Montanari L, Cristino VA, Silva MB, Martins PAF Int J Adv Manuf Technol 69(5) (2013)
3. Thipprakmas S, Phanitwong W Int J Adv Manuf Technol 61(5) (2012)
4. Cao TT, Lu B, Ou H, Long H, Chen J Int J Mach Tools Manuf 110(2016)
5. Yang C, Wen T, Liu LT, Wang H J Mat Process Technol 214(2014)

Push-bending Process and Forming Quality of Stainless-steel Tubes

Chen Guo-qing[1,*], *Yang Jian*[1], *Li* Hong-xiang[1], *Fu* Xue-song[1], *Tang* Rui[2], *Zhou* Wen-long[1]
[1]Key Laboratory of Solidification Control and Digital Preparation Technology (Liaoning Province), School of Materials Science and Engineering, Dalian University of Technology, Dalian116085, China
[2]Science and Technology on Reactor Fuel and Materials Laboratory, Nuclear Power Institute of China, Chengdu 610213, China

Abstract. The push-bending process of stainless-steel tubes was investigated by using the experimental method and finite element (FE) analysis based on the commercial software MSC.MARC. The effects of the processing parameters and the lubricating condition on the forming quality of as-obtained ASTM304 tubes were discussed. The results demonstrated that the experiment results were consistent with the simulation both in deformation behaviour and the stress distribution during the push-bending process, which verified the reliability of the established FE-model. The results also show that the maximum residual stress concentrates in the middle of tubes, the residual strain on both the external side increases firstly, then decrease, and increase again with the increase of the location angel. After the push-bending, the interior wall thickness increases while the exterior wall thickness decreases. When the relative bending radius is larger, smaller variation of wall thickness will be obtained. With the increase of the friction coefficient, the wall thickness thinning rate will decrease. With the increase of the pushing speed, the wall thickness thinning rate decrease a little while the thickening rate increases slightly. When push-bending of the large radius tubes, a good lubricating condition can reduce the wall thickening problem at the intrados, which will ensure a better forming quality.
Keyword. Bending, Quality control, Simulation

1 INTRODUCTION

As important lightweight structures and liquid conveying or heat exchanging components, metallic tubular parts are widely used in the fields of aeronautics and aerospace, automobile, oil and chemical industries, etc. It has great advantages of efficiency, cost, and quality when tubular productions are made by plastic forming technologies. Among them, bending is most commonly used [1].

The studies of tube bending mainly focus on the deformation behaviour during the process. The researches on deformation vary from wrinkling instability, cracking, cross section deformation to spring back phenomenon. Yang [2-4] did some works on the wrinkling activity from the effects of parameter to the prediction model. Megharbel [5] introduced a method to analysis bending tubes with different section shapes, and built equations to predict the moment for forming tube to a specific radius of curvature. There are also a lot of studies on spring back. The predicting equation [6-7], the mechanism [8], the time-dependent behaviour [9-10], the effects of different parameters [11] are all included in the study of spring back.

However, most of studies are focus on the rotary bending only. There are few results on the push- bending process. In this paper, a FE model is established to analysis the push-bending process. The model is verified with the experiment from the aspects of the wall thickness and the residual stress. The model is also used to study the contact behaviour during the push-bending. The effects of some parameters on wall thickness are mentioned.

2 Research Method

2.1 Experimental method

The experiments are conducted using hydraulic machine which provides a force of 3150KN, as shown in Fig.1. The mechanism is composed of sleeve, bending die and punch. The stainless tubes of AISI304 are used in the experiment with length l of 250mm. The outer diameter d of the tubes is 34mm, and the thickness t is 4mm. The radius R after the push-bending process is set as 136mm, and the bending angle θ is 90°. During the experiment, engine oil is used as lubricant. The wall thickness of the tube at different locations after bending is detected by ultrasonic generator and oscilloscope. X ray diffraction (XRD) is used to detect the residual stress along extrados of the tube.

2.2 FE method

The finite element model is established using the commercial finite element software MSC.MARC. The element size we used in the simulation is 2mm*2mm. Considering the symmetry character of the tube and the bending process, the model is simplified as a half model. The mechanical properties of the material are shown in

*Corresponding author: gqchen@dlut.edu.cn

https://doi.org/10.1051/matecconf/201819004006

Tab.1. The solid shell 182 is used in the model due to the consideration of hourglass and reducing integral function. Coulomb friction law is employed with the friction factor of 0.08. The whole process is divided into two parts, the loading process and the unloading process. During the loading process, the punch is set with a certain velocity.

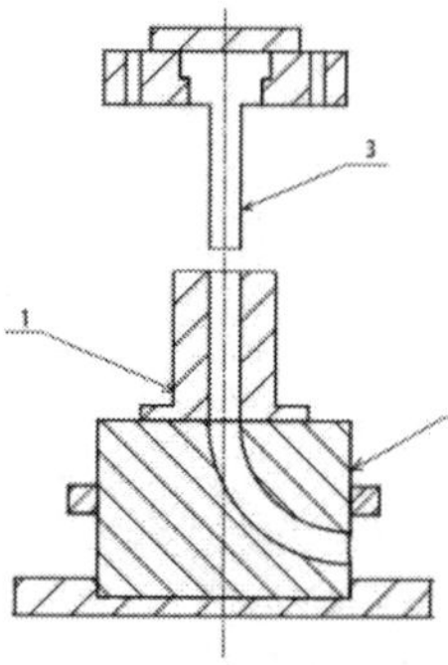

(a)

(b)

Fig. 1. The equipment of push-bending: (a) The Assembly diagram: 1: sleeve; 2: bending die; 3: punch; (b) The real figure.

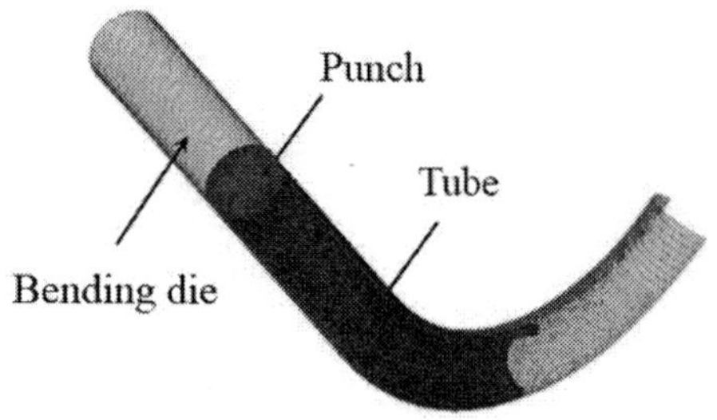

Fig.2. Finite element model of push-bending.

Table 1. Mechanical properties of AISI 304.

Density, ρ(kg/m3)	7930
Young's modulus, .E(GPa)	193
Poisson's ratio, ν	0.3
Tensile strength, σb(MPa)	694.72
Yield strength, σs(MPa)	242.31

2.3 Comparison of the methods

In order to verify the reliability of the finite element model, the results of the experiment and FE simulation are compared in two aspects: the residual stress and the wall thickness.

The results of the wall thickening/thinning of both the experiment and simulation are compared as shown in Fig.3. With the increase of the angle where the points are located, the exterior wall thickness decreases first and then increase. However, the interior wall thickness changes in an opposite rule. The results of both the experiment and the simulation show the same rule. The maximum wall thickness error of experiment and simulation is 3.38% of the exterior wall thickening, and 6.20% of the interior wall thinning.

Fig.4 (a) shows the bent tube in experiment and simulation. The tubes are in good agreement in the shape, and there is no obvious pints on the surface. Fig.4 (b) is the comparative results of the residual stress at extrados. In the loading process, the exterior wall is in the role of tensile stress. When in the unloading process, there will be compressive stress at the extrados. With the increase of the angle where the test points are located, the residual compressive stress increases first and then decrease, and increases again in the location with a larger angle. The gap between simulation results and experiment results may be caused by the residual stress in the original production process and the error of X-ray diffraction method.

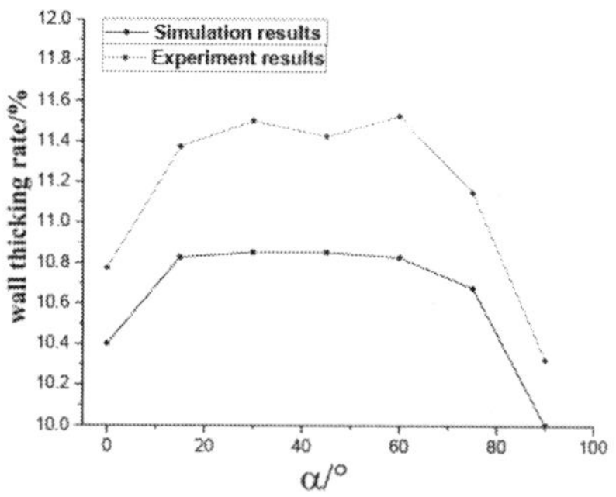

(a)

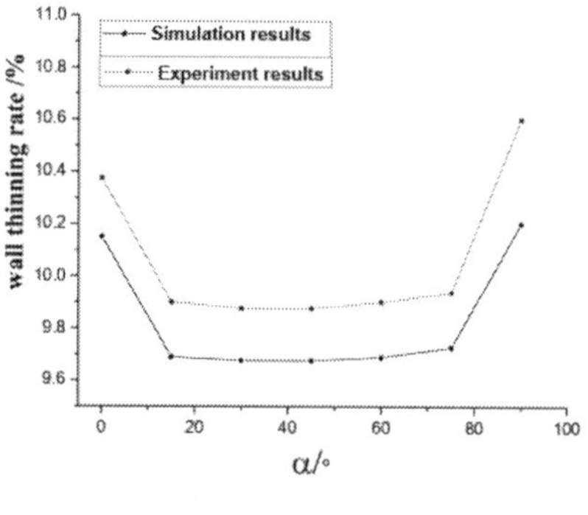

(b)

 https://doi.org/10.1051/matecconf/201819004006

Fig.3. Wall thickening/thinning rate distribution of bent tubes: (a) wall thickening rate at the extrados; (b) wall thinning rate at the intrados.

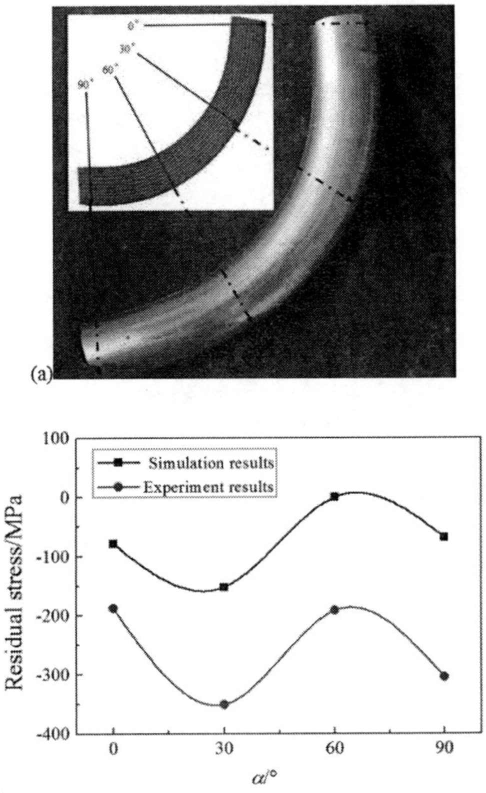

(a)

(b)

Fig.4. The residual stress distribution of the bent tube: (a) Point position; (b) The residual stress.

3 Result and Discussion

During the bending process, the tube and the die are in contact and have a relative motion which produces deformation in tube. The exterior wall thickness decreases due to the tensile stress, while the interior wall thickness increases due to the compressive stress. The FE model is used to study the contact behaviour during the process, and the wall thickness along with the change of the coefficient of friction, relative bending radius and the bending speed.

3.1 Contact behaviour during the push bending

Fig.5 shows the contact status during push-bending. It is found that the tube was not always in contact with the inner wall of the die closely. The yellow region in the figure means in which place the tube and the die was in contact. Along with the bending process, more and more sections had a deformation and this cause gaps between the die and the tube. As shown in Fig. 5(b) (c), an arc area appeared in which the tube and the die were in contact. At the meantime, a characteristic region at the top and the bottom of the tube was always in contact with the die until the unloading process.

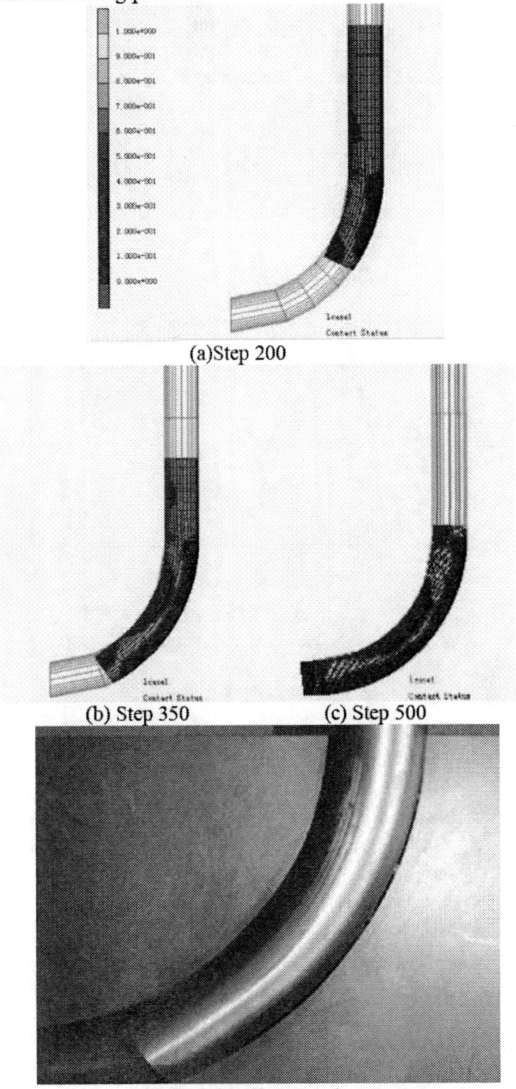

(a)Step 200

(b) Step 350 (c) Step 500

(d) Push-bending experiment

Fig.5. Contact status of different time steps :(a) Step 200; (b) Step 350; (c) Step 500.

3.2 The effects of friction coefficient

In the push-bending process, the exterior wall thickness decreases, and the interior wall thickness increases. Usually, the wall thinning rate and the wall thickening rate, as shown in Eq. (1) and Eq. (2), are used to express the two different wall thickness changes above.

Wall thinning rate at the extrados:

$$I_{ti} = t_{min}/t*100\% \quad (1)$$

Wall thickening rate at the intrados:

$$I_{ta} = t_{max}/t*100\% \quad (2)$$

Where t is the original wall thickness of the tube. t_{min} is the wall thinning at the extrados. t_{max} is the wall thickening at the intrados.

Fig.6 (a) (b) show the effects of friction coefficient on the wall thickness at the extrados and intrados separately. With the increase of friction coefficient, the wall thinning rate decreased at the extrados, and the thickening rate increased at the intrados. During bending, the metal flow occurs at the exterior wall because of the axial tension stress. The increase of friction coefficient can hinder the metal flow and reduce the thinning rate. At the meanwhile, the increasing friction coefficient hinders the metal flow at the interior wall, intensifying the wall thickening rate.

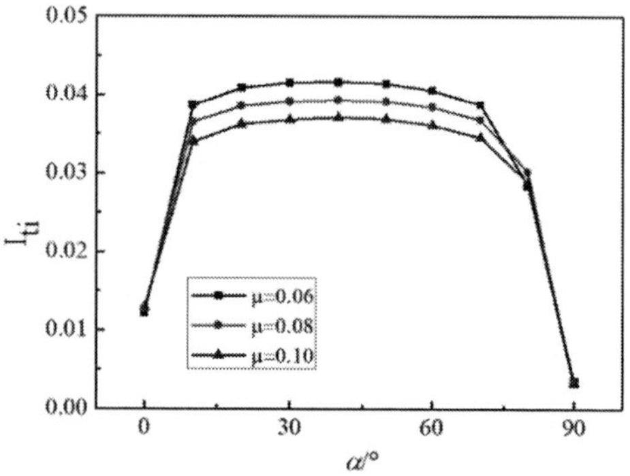

(a) Thinning rate

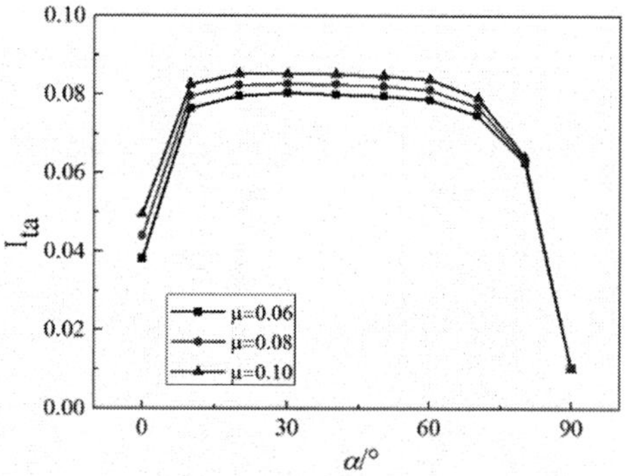

(b) Thickening rate

Fig.6. Effects of friction coefficient on thickness of tube

3.3 The effects of relative bending radius

Fig.7 (a) (b) show the effects of relative bending radius on the wall thickness. When the relative bending radius R/D is 3, 3.5, 4, the maximum value of the thinning rate was 4.64%, 4.37, 4.12%, and the maximum thickening rate was 10.79%, 9.17, and 8.04%. Along with the increases of the relative bending radius, both the wall thinning rate and the wall thickening rate decreased. When the relative bending radius is too small, it will be hardly to get a bent tube with high quality. Also, the wall thickening rate was significantly higher than the thinning rate. The tube is more likely to wrinkle than to rupture.

3.4 The effects of pushing speed

Fig.8 shows the thinning rate and the thickening rate with the increase of the pushing speed. The maximum value of the thinning rate was 4.19%, 4.17%, 4.16, and the maximum value of the thickening rate was 8.01%, 8.03%, 8.04%. Both the interior wall and the exterior wall changed a little. When the pushing speed increased from 6mm/s to 10mm/s, the thinning rate decreased by 0.03%, and the thickening rate increased by 0.03%. The effects of the pushing speed were very little.

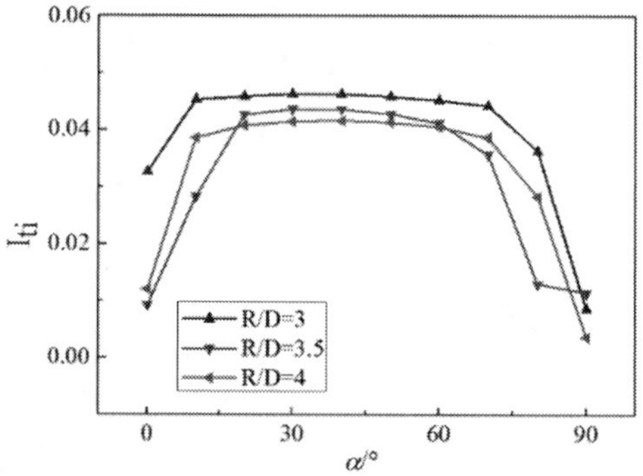

(a)

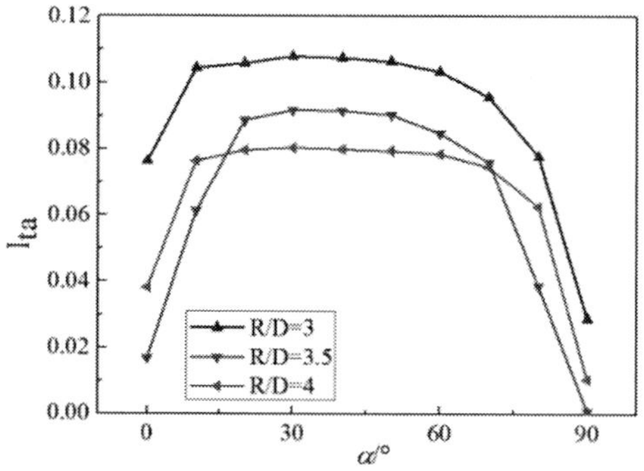

(b)

Fig.7. Effects of relative bending radius on thickness of tube: (a) thinning rate; (b) thickening rate.

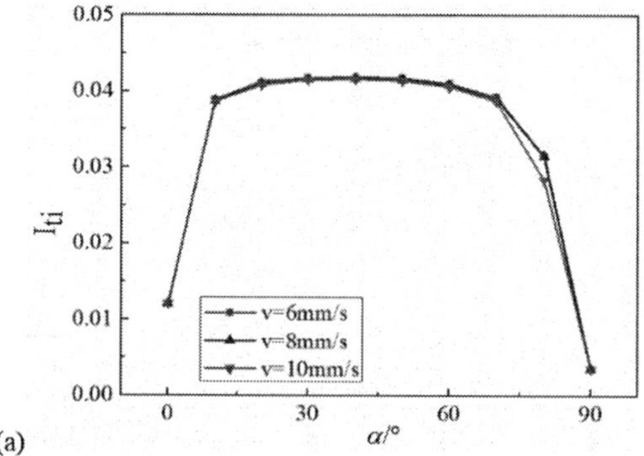

(a)

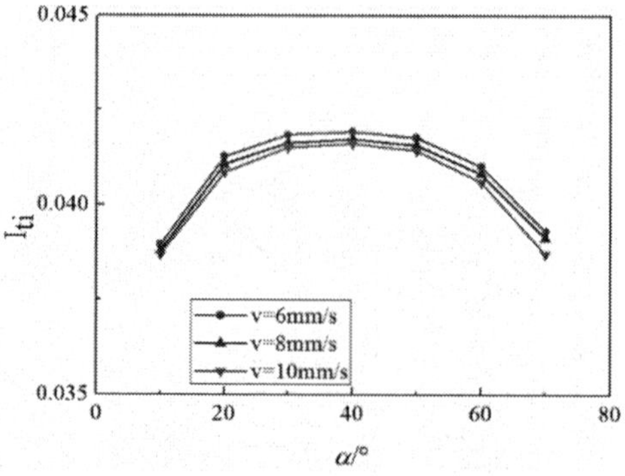

(b)

https://doi.org/10.1051/matecconf/201819004006

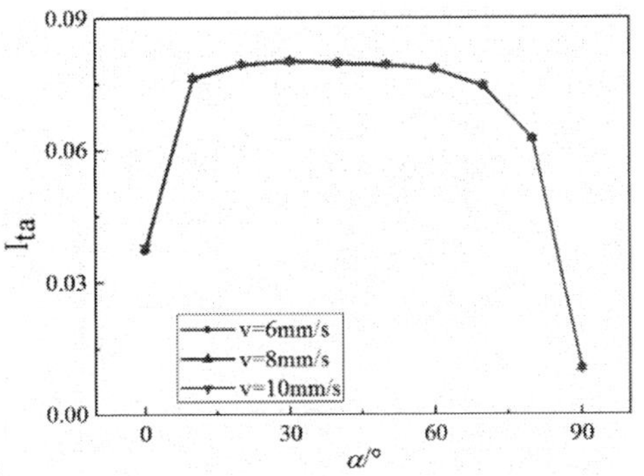

(c)

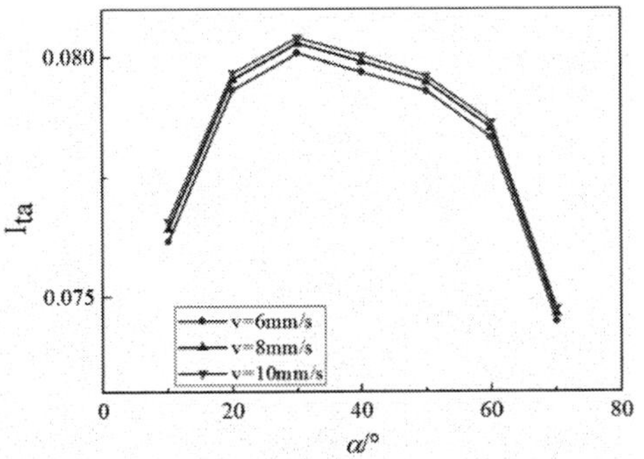

(d)

Fig.8. Effects of pushing speed on the wall thickness: (a) thinning rate; (b) thinning rate in the middle; (c) thickening rate; (d) thickening rate in the middle.

5 Conclusion

The FE model was established of push-bending to study the deformation behaviour, and results were compared with the experimental results. The contact status of different steps showed where the tube and the dies are in contact closely and where are not. For the push-bending, the wall thickness changes with the technological parameters. With the increase of friction coefficient or bending speed, the thinning rate decreased and the thickening rate increase. When bending speed changed from 6mm/s to 10mm/s, the wall thickness changed only a little. Both the thinning rate and thickening rate increased when a smaller relative bending radius was used. As a result, to push-bend a large radius tubes, a good lubricating condition which will help to ensure a better forming quality.

This work was supported by the International Science & Technology Cooperation Program of China [No. 2015DFR60370] and the Fundamental Research Funds for the Central Universities [No.DUT17GF209].

References

1. Wen T. On a new concept of rotary draw bend-die adaptable for bending tubes with multiple outer diameters under non-mandrel condition. J MATER PROCESS TECH. **214(2)**,311-317(2004)
2. Yang H, Lin Y. Wrinkling analysis for forming limit of tube bending processes. J MATER PROCESS TECH. **152(3)**,363-369(2004)
3. Li H, Yang H, Zhan M, Gu RJ. The interactive effects of wrinkling and other defects in thin-walled tube NC bending process. J MATER PROCESS TECH. **s187-188(3)**,502-507(2007)
4. Li H, Yang H, Zhan M, Kou YL. Deformation behaviours of thin-walled tube in rotary draw bending under push assistant loading conditions. J MATER PROCESS TECH.**210(1)**,143-158(2010)
5. El Megharbel A, El Nasser GA, El Domiaty A. Bending of tube and section made of strain-hardening materials. J MATER PROCESS TECH. **203(1-3)**,372-380(2008)
6. Al-Qureshi HA. Elastic-plastic analysis of tube bending. International Journal of Machine Tools and Manufacture. **39(1)**,87-104(1999)
7. Al-Qureshi HA, Russo A. Spring-back and residual stresses in bending of thin-walled aluminium tubes. MATER DESIGN. **23(2)**,217-222(2002)
8. Zhan M, Yang H, Huang L, Gu R. Springback analysis of numerical control bending of thin-walled tube using numerical-analytic method. J MATER PROCESS TECH. **177(1-3)**, 197-201(2006)
9. Daxin E YL. Springback and time-dependent springback of 1Cr18Ni9Ti stainless steel tubes under bending. Materials and Design.**31(3)**, 1256-1261 (2010)
10. Lim H, Lee MG, Sung JH, Kim JH, Wagoner RH. Time-dependent spring back of advanced high strength steels. INT J PLASTICITY. **29(1)**, 42-59(2012)
11. Jiang ZQ, Yang H, Zhan M, Xu XD, Li GJ. Coupling effects of material properties and the bending angle on the spring back angle of a titanium alloy tube during numerically controlled bending. MATER DESIGN. **31(4)**, 2001-2010(2010)

Effect on the mechanical properties of sheet metals after the use of drawbeads in deep drawing

Harald Schmid[1,*] and *Marion* Merklein[1]

[1]Institute of Manufacturing Technology, Friedrich-Alexander-Universität Erlangen-Nürnberg, Egerlandstraße 11-13, 91058 Erlangen, Germany

Abstract. The use of modern and more complex materials leads to a need for detailed knowledge and controlling of the material flow. Next to the blank holder force or the lubrication system, drawbeads are nowadays one common way to bring serial forming processes under control. Although drawbeads are already well investigated regarding their retention forces, the effect on the resulting mechanical properties after a drawbead passage is not analysed in detail yet. This work will show the influence of a common drawbead geometry used in forming processes on typical mechanical properties of sheet metal. Therefore, two different sheet metal materials are preloaded in a modified strip drawing test with industrial used drawbead geometry. In testing, three different pressure levels between 2.5 MPa and 7.5 MPa and three levels of drawing speed between 10 mm/s and 50 mm/s are combined to nine variations which will be examined. Afterwards, specimens are cut out by a laser cutting machine of the drawn strips. Those already preloaded and deformed specimens are then tested under standard conditions with the use of an optical measurement system. The results like tensile strength or elongation at fracture are compared to the initial state and each other and discussed with consideration of pressure and speed levels to work out the main effects. Those results are a contribution for the optimization of deep drawing simulations of parts including drawbeads or to evaluate the influence of drawbeads on forming processes to specify or even expand process limitations.

Keywords: Deep drawing, Hardening, Sheet metal

1 Introduction

Deep drawn parts are more complex in geometry and material these days than in the past [1]. This requires the accurate knowledge and prediction of the material flow to control it and to achieve satisfying product quality. While modifications of the blank holder force or the lubrication system are options to control forming processes, drawbeads are also a popular way to improve the material flow. Drawbeads are already examined in detail regarding retention forces but not when considering modified mechanical properties of the sheet metal after a drawbead passage. This causes the necessity for a more detailed investigation of the impact of drawbeads on the used sheet material itself. At the moment, a lot of conventional steel materials in the automotive industry are replaced with high strength steel for the reason of light weight design, according to Mohrbacher et al. [2]. Because of the high formability and their high strength, the sheet thickness and therefore the overall weight can be reduced, what is also described by Bleck and Ratte in [3].

In the past, different authors have already investigated drawbeads and their effects on the forming process. Liu et al. could already successfully predict in [4] the forming and springback behaviour of sheet material undergoing multiple bending-unbending cycles, as they appear in a drawbead. A FE simulation combined with a nonlinear isotropic / kinematic hardening model shows almost the same results compared to the conducted experiments. According to their findings, the predicted results are highly influenced by the used hardening model and the results present a decreasing accuracy with more bending-unbending cycles compared to the experimental data. That's another reason, why sheet metal hardening after drawbeads should be examined. In [5], the hardening of sheet metal was examined for two different levels of blank holder pressure and in three layers over the thickness. It could be demonstrated, that a drawbead leads to different hardening levels in the sheet layers and in drawing direction after the passage. Also, results from a modified strip drawing test with a drawbead geometry are compared to results from a deep drawing process. In addition, in [6] the influence of drawbeads on material hardening was also investigated on the upper and lower side in drawing direction. Additionally, the measured strain development was correlated with the hardening behaviour and an overall hardening as well as a rising accumulated strain was described.

Halkaci et al. show in [7], that a deep drawing process is changed with drawbeads and improves the limiting drawing ratio by up to 10 %. This also indicates a shift in mechanical properties for the use of drawbeads. Samuel develops a numerical model in [8] to calculate forces and stresses during a drawbead passage and compares these findings with his experimental data. A new material model is used, which is also based on the strain hardening exponent n from tensile tests. The numerical results are found in a good agreement with the experimental data in between 8% deviation. It is also observed, that the drawbead geometry and friction

* Corresponding author: harald.schmid@fau.de

coefficient highly affect the retention and pressure forces for the calculations in a process with drawbeads. Looking on the used materials DC04 and DP800, also some investigations have been conducted regarding the hardening behaviour. Silvestre et al. [9] investigated different hardening behaviour of different steel grades. It was shown, that mild steels like DC04 are presenting a cyclic hardening behaviour when they are loaded and reverse loaded again and again. In comparison, the advanced high strength steels like the investigated DP800 show cyclic softening behaviour. Also, the mild steel was increasing its yield strength with every cycle, while the advanced high strength steels showed a constant yield strength value with every cycle. This can be seen as a saturation and a stable microstructure, according to [9]. Finally, the two used steel grades already show a different hardening behaviour, what obviously also has effects when considering the sheet metal after a drawbead passage. Additionally, Schreijäg [10] examined in 2013, that macroscopic forming conditions of DC04 mainly depend on the anisotropy of the texture of the used sheet metal. Connecting this with the deep drawing process through drawbeads, it could be possible to achieve different mechanical properties after a drawbead passage, depending on the rolling direction.

All presented findings can be interpreted as a note for significant changes in mechanical properties after a drawbead passage. Such a shift can lead to the incorrect determination of material models and poor numerical calculations what finally leads to bad forming results. This makes further and more detailed investigations of drawbead processed parts necessary. Especially standardized material characterization techniques seem to be suitable to compare results of preloaded parts to their initial state. Within these investigations, the uniaxial tensile test is used to investigate the influence of drawbeads on sheet metal parts. The tensile test is very useful, as it is an international standardized material test. Also, it is possible to determine well known and meaningful parameters like the tensile strength TS, the yield strength YS, the uniform strain ε_u or the elongation at break ε_f. At the moment, there are insufficient investigations about the influence of different process parameters on the material characteristics. This testing will help to determine the behaviour of different sheet metal materials under varying conditions in the drawbead passage and to improve the deep drawing process itself.

2 Materials

Because of their use in the automotive industry, conventional steel and a high-strength steel will be examined after a drawbead passage. Therefore, DC04, a kind of mild steel, is used very commonly in deep drawing processes. Also, an advanced high strength steel HCT780X, also called DP800, will be observed as a modern lightweight-material. It is often used in the automotive industry in the last years. The mild steel DC04 offers a high formability in forming processes and is used here in an initial thickness of t_0 = 1.0 mm. The initial tensile strength is TS = 314.4 MPa, the yield strength YS = 162.1 MPa, the uniform elongation ε_u = 25.9 % and the fracture strain ε_f = 44.2 %. DP800 presents a significant higher tensile strength of TS = 817.9 MPa and the yield strength YS = 516.1 MPa. The uniform elongation of DP800 is about the half of DC04 and amounts with ε_u = 12.9 % while the fracture strain is about ε_f = 17.4 %. The strain hardening exponent of DP800 is about n_{DP800} = 0.27 while DC04 has an exponent of n_{DC04} = 0.16, what is about the half. These material parameters in the initial state will be used in this work to quantify the changes of the material parameters after a drawbead.

3 Experimental setup

First of all, the sheet metal strips with a size of 700 x 50 mm² need to pass the drawbead passage where they are preloaded. Therefore, a modified strip drawing test is used, that is also described in [6] and shown in Figure 1. The used drawbead geometry is built according to the Numisheet 2008 benchmark and has a peak radius of 6.5 mm and a height of 3.6 mm with running-in and running–out radii of 2.5 mm. The drawbead tool is built into a strip drawing test machine, as can also be seen in Figure 1. Also, the schematic process is illustrated. After being lubricated with deep-drawing oil Multidraw KTL N 16 (Zeller + Gmelin GmbH), the metal strips are subsequently drawn through the drawbead. This is done with varying velocities of v = 10 mm/s, 30 mm/s and 50 mm/s while the blank holder pressure is varied between p_N = 2.5 MPa, 5.0 MPa and 7.5 MPa. During this process, the velocity v and the normal pressure p_N are controlled automatically and kept constant. After that, A50 tensile test specimens are laser cut out of the preloaded and deformed metal strips. Therefore, the strips have to be fixed on the laser machine bed because of their deformation to cut the correct specimen geometry. The preloading as well as the cut out is executed in 0° to the rolling direction. Specimens in 45° or 90° to the rolling direction cannot be extracted after preloading because of the strips' geometric boundaries.

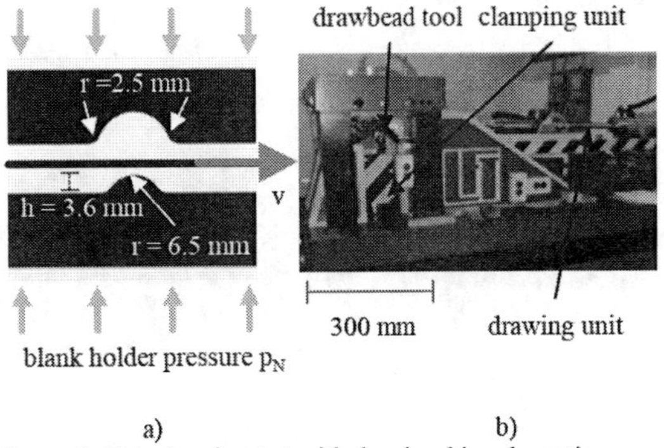

Figure 1. Strip drawing test with drawbead in schematic process illustration (a) and used test bed (b)

Finally, five specimens were tensile tested for every variation (n = 5) on a universal testing machine Z100 (Zwick GmbH & Co. KG) based on DIN EN ISO 6892 and by using the optical measurement system ARAMIS (GOM GmbH). The evaluation of the elongation was averaged for every specimen over the measurement area with a length of l_0 = 50 mm. The process and methodology is also presented in Figure 2.

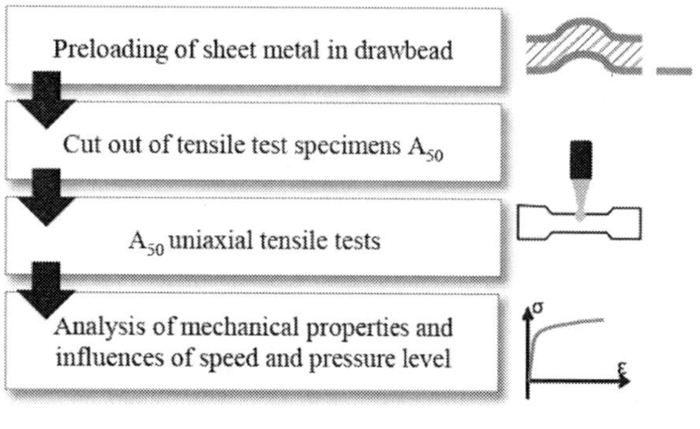

Figure 2. Methodology used for tensile test examination

Because of their importance for the deep drawing process, the influence of the blank holder pressure p_N and the velocity v in a drawbead are examined. The results are pointed out in the next chapter.

4 Results

After drawing the metal strips through the drawbead, every metal strip has a different thickness depending on the material and the varied parameters in this process. This is considered in the tensile tests, which are performed for the analysis of the stresses. In Figure 3, the thickness results for the preloaded strips are presented. In general, the thinning is more obvious for DC04 compared to DP800. Also, the increase in pressure leads to thinner sheet metal strips after a drawbead. The effect of the velocity on the sheet thickness t_1 at constant pressure p_N is in between the standard deviation and therefore not significant. For the highest pressure of p_N = 7.5 MPa, the material DP800 has a thickness of t_1 = 0.95 mm and for DC04 of t_1 = 0.9 mm. According to the thickness reduction results in Figure 3, changes already become visible and trends can be indicated.

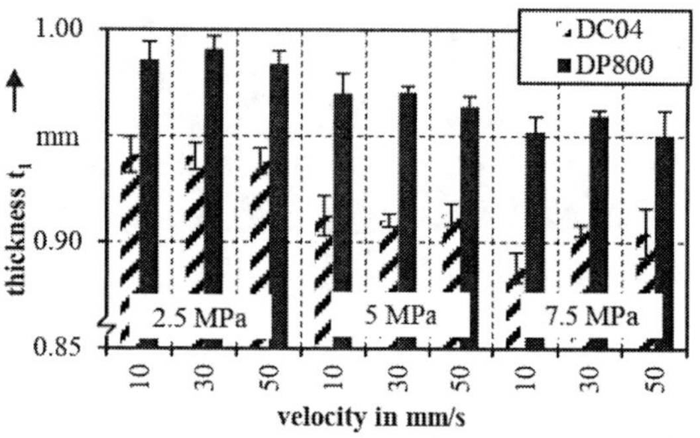

Figure 3. Strip thickness after drawbead for tested variations

For more detailed results, further experimental investigations under varying parameters have been conducted. The influence of the velocity and the blank holder pressure in the strip drawing test with drawbead is analysed in the tensile test setup. To get an impression and overlook of the changes in the material characteristics, a true stress-true strain diagram of both materials is drawn in Figure 4. It shows the curves for DC04 and DP800 in the initial state and after a drawbead preloading for the test variation of v = 30 mm/s and the blank holder pressure p_N = 5 MPa.

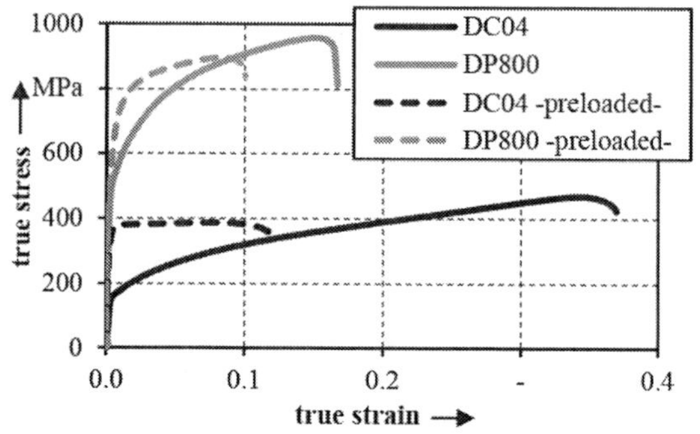

Figure 4. True Stress- true strain diagram of initial and preloaded DC04 and DP800 (p_N = 5 MPa, v = 30 mm/s)

It can be observed, that for both materials a significant change is visible. In Figure 4, the fracture strain is clearly smaller for the preloaded materials. Also, the maximum of the true stress changes significantly. These curves also indicate the need for further, detailed comparison of the standard material parameters to quantify the occurred changes. In Figure 5, the changes of a varying velocity and a constant blank holder pressure of p_N = 5 MPa are shown for a better comparison between the different variations and the two materials.

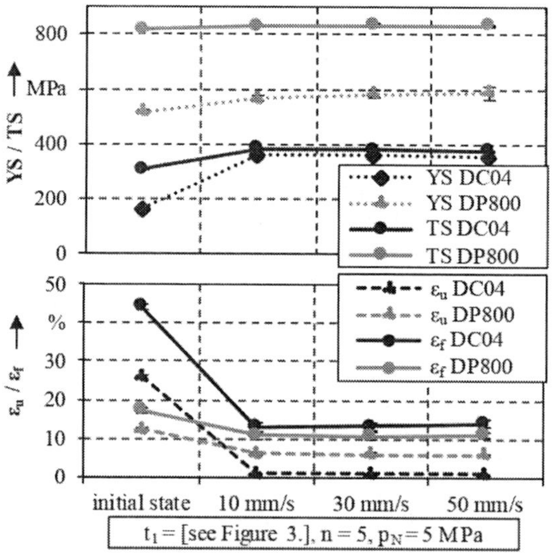

Figure 5. Material parameters after drawbead passage for the variation of the drawing velocity v and constant pressure p_N

The tensile strength TS for DC04, in black, is increased by about 100 MPa, while the yield strength YS is increased by nearly 200 MPa. The uniform elongation ε_u decreases to only 1 % while the fracture strain also decreases from 45 % to about 15 %. For DP800 in grey, there is only a slight increase of the tensile strength TS and yield strength YS visible. The fracture strain is divided in half, what can also be said for the uniform elongation. For both materials, there is no big difference in between the curves when varying the velocity. The difference for the materials is in between the standard deviation, so the variation of the velocity in between v = 10 – 50 mm/s most likely does not have an impact on the material itself. At least the parameters do not change significantly compared to the other preloaded specimens.

Analysing the variation of the blank holder pressure p_N with a constant velocity of v = 50 mm/s, the results are more manifold. As it can be seen in Figure 6 for DC04, there is a visible difference between the varied blank holder pressure levels. The tensile strength rises with more pressure what also takes effect for YS. This can also be seen in Figure 7. The uniform elongation and the fracture strain are decreasing with higher pressure levels. The values for 5 MPa and 7.5 MPa are only demonstrating small changes compared to the gradient from 2.5 MPa to 5 MPa. Considering the diagram in Figure 6 for DP800, the same findings can be observed. Although the tensile strength does not increase in the same amount compared to DC04, a higher pressure level also provides higher tensile strength levels. Looking on the strain in the preloaded specimens, a decrease is visible. By rising the pressure level from 2.5 MPa to 5 MPa and further on to 7.5 MPa, the fracture strain and the uniform elongation are clearly minimized.

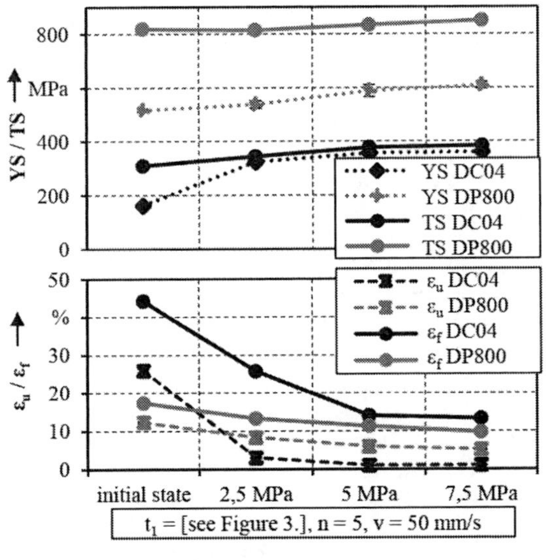

Figure 6. Material parameters after drawbead passage for the variation of the blank holder pressure p_N and the constant drawing velocity v

Compared to each other, DC04 shows a higher increase of the tensile strength while DP800 does not have a high rise. The uniform elongation is also minimized roughly for DC04 while value for DP800 does not decrease in the same amount. Because the two tested materials show a different behaviour in the tensile tests, it is necessary to analyse them by using specific values and to compare them respecting the initial state of the parameters.

5 Discussion

For the comparison and detailed discussion of the conducted tensile tests, the tensile strength and the fracture strain are chosen to be investigated in detail. All in all, the parameters of the variations have a standard deviation of about maximum 7 % compared to the average value. This deviation was examined with a number of specimens of n = 5. The tensile strength TS provides a good overview of the maximum tolerable stresses and is therefore a parameter describing maximum tolerable loads. The fracture strain is used as a value for the ultimate plastic strain to failure, what is a descriptive value. It is the parameter describing the material failure regarding the strain and not only the stresses.

The analysed parameters are set in relation to their initial state and the increase or decrease is given in percent. Also, the standard deviation is calculated with five specimens tested and shown in Figure 7. For a better overlook, the results for both materials are drawn into one diagram considering the variation of the tensile strength TS in Figure 7 and the fracture strain ε_f in Figure 8. In general, the tensile strength for DC04 is increased nearly up to 25 % for the variation p_N = 7.5 MPa and v = 50 mm/s, while for the same setting the fracture strain is decreased by over 70 %.

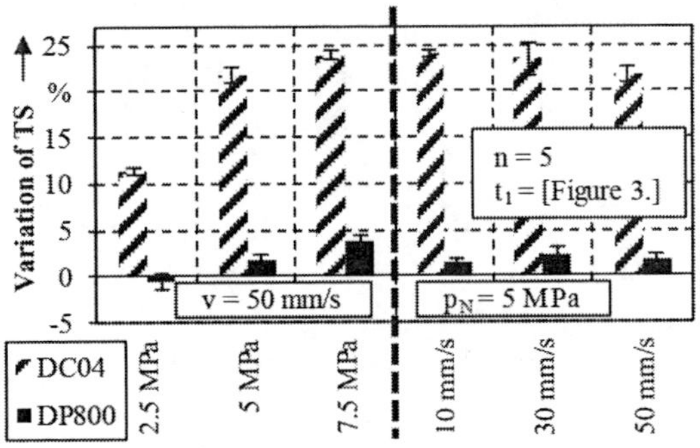

Figure 7. Modification of the tensile strength TS in tensile tests after preloading in a drawbead passage

On the first hand, it is visible that the increase in tensile strength TS is way higher for DC04 compared to DP800. As said before, the tensile strength for DC04 increases nearly up to 25 %, while DP800's rises only by about 4 %. One variation of DP800 (p_N = 2.5 MPa, v = 50 mm/s) even shows a very small decrease. In practical use, this can likely be seen as no influence on TS at all.

It is visible, that with an increase of the normal pressure p_N in the strip drawing test, the value for the tensile strength TS after preloading is increasing. For the velocity of v = 50 mm/s, the influence of the normal pressure p_N increases the level of TS from about 11 % at p_N = 2.5 MPa over to about 21 % at p_N = 5 MPa up to about 25 % at a blank holder pressure of 7.5 MPa. The small difference for the tensile strength TS between p_N = 5 and 7.5 MPa indicates that the material is saturated by plastic deformation. Regarding the DC04 results for varying the velocity, there is a very small decrease in the yield strength what could also be called a trend. The standard deviations of TS with the influence of the velocity are overlapping, what means there is no significant difference.

When looking at DP800 in Figure 7, the increase of TS is only 4 % and somewhat lower compared to DC04. Indeed, one can also detect an increase in the tensile strength TS for a rising of the pressure p_N, but only up to 4 % and not to 25 % like for DC04. For the variation of the velocity v, there are no significant differences in the development of TS for DP800. For Figure 7, it can be summarised for the investigated boundaries, which a higher blank holder pressure p_N leads to higher values for TS with maximum up to 25 % for DC04 and 4 % for DP800. The velocity between v = 10 – 50 mm/s tends to have no influence, because TS does not show any significant shift for these variations. DP800 all in all shows a lower level of residual formability considering the tensile strength compared to DC04. After considering the changes in the applicable stresses, the effect of the drawbead on the maximum bearable strain in the tensile test will be considered. In Figure 8, the influence of the parameters v and p_N on the fracture strain ε_f is given.

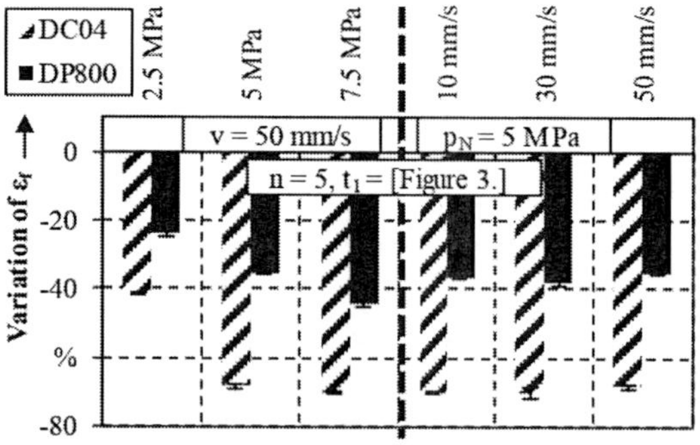

Figure 8. Percentage modification of the fracture strain in tensile tests after preloading in a drawbead passage

For DC04, the fracture strain decreases with an increasing blank holder pressure. Starting with a drop of the fracture strain of 40 % at p_N = 2.5 MPa, it drops over 70 % for the pressures of p_N = 5 MPa and 7.5 MPa. Similar to the tensile strength, there is no significant change in the values between 5 MPa and 7.5 MPa anymore. For varying the strip velocity under a constant blank holder pressure of p_N = 5 MPa, the fracture strain drops for all variations by about 70 %.

When looking at DP800 as a representative for advanced high strength steels, the drop for the fracture strain compared to the initial state is lower than for DC04. For p_N = 2.5 MPa, the fracture strain is lowered by about 20 % compared to the initial specimens. For p_N = 5 MPa and 7.5 MPa the reduction is 38 % and 42 %. Like DC04, DP800 also does not show any compelling changes for the fracture strain when modifying the strip velocity.

Altogether, it is remarkable that DC04 shows higher modifications for the tensile strength and the fracture strain after drawing the sheet metal through a drawbead geometry. One possible explanation could be the microstructure evolution. According to Nesterova et al. [11], DP800's microstructure evolution of the softer ferrite matrix, is mainly influenced by the hard phase (martensite). This process, of course, is not happening for single phase steels. According to them, DP800 also shows much larger work hardening rates at lower plastic strains, which are correlating pretty well with local internal stresses. Connecting this with the analysed material parameters, it can be a reason for the lower strain values compared to DC04.

Soares et al. [12] also examined the strain hardening behaviour of dual phase steels. He figured out, there is a two-staged strain hardening behaviour for a DP800 steel. Up to a logarithmic strain of 0.025, the strain hardening exponent n is increasing rapidly. According to their research, this is due to a lot of mobile dislocations, which are introduced to ferrite by martensitic transformation in the production process. Afterwards, the exponent decreases slightly and comes to a constant value. As a summary, the authors of [12] remark, that this first stage is the most important for the whole forming process and responsible for the fast increase in work hardening of DP steels. This could also be an explanation for the different behaviour of the two steel grades represented by DC04 and DP800.

For the change of the blank holder pressure, a modification could be proved, while the velocity in the investigated area does have ignorable influences on the material parameters. Only for DC04 a small change for the yield strength in Figure 5 can be seen. The higher modifications for DC04 could possibly arise from its higher ability to deform what also can be seen in the initial fracture strain of nearly ε_f = 45 %. DP800, also a cold rolled steel with two different phases, has a lower initial fracture strain. The hardening exponent of DP800, which is about n_{DP800} = 0.27 and about the double of n_{DC04} = 0.16, also leads to a higher work hardening and consequently to lower strain levels. Those differences and also a varying texture of the two materials can explain the changes that can be seen.

6 Summary and Outlook

First of all, strip drawing tests with a drawbead geometry were performed to replicate stresses and strains in a deep drawing process. The loads were applied to DC04 and DP800 sheet metal in a drawbead passage. The drawing velocity was varied between v = 10 – 50 mm/s and the blank holder pressure between p_N = 2.5 – 7.5 MPa. Afterwards, the deformed sheet material was clamped on to the machine bed and the tensile test specimens A_{50} were cut out by a laser cutting machine. In the next step, those specimens were examined with a conventional uniaxial tensile test using an optical measurement system. The preloading process, which is also shown schematically in Figure 1, obviously significantly changes material properties of both sheet metal materials.

The most distinctive findings are:

- The tensile strength TS is increased by about 25 % for DC04 and only by 4 % for DP800.
- The fracture strain ε_f is decreased by over 70 % for DC04 and by 40 % for DP800.
- Increasing the blank holder pressure between p_N = 2.5 – 7.5 MPa increases TS in general and decreases ε_f.
- Varying the strip velocity between v = 10 – 50 mm/s has no significant influence on the tested properties.

It is shown in tensile tests, that drawbeads change the material properties in deep drawing processes. Depending on the material, the tensile strength rises while the fracture strain is decreased. For deep drawing parts, where preloaded metal is part of the final geometry, this needs to be taken into account definitely. Especially for numerical calculations of those parts, it is very important. In the future, it would also be interesting to investigate different drawbead geometries or other materials like aluminium. It is therefore also a need to examine material characteristics for other stress states after preloading like plane strain, shearing or biaxial tests and consider these results in calculations.

For the support in the research project EFB 08/114 (AiF 18328N) the authors would like to thank the European Research Association for Sheet Metal Working e.V. (EFB) as well as the German Federation of Industrial Research Associations „Otto von Guericke“ e.V. (AiF).
The authors would also like to show their gratitude to Ms. Ines Becker, who participated in this experimental study within the scope of her bachelor’s thesis.

References

1. M. Barthau, M. Liewald, H. Christian, J. Phys.: Conf. Ser., **896**, 12040, (2017) DOI: 10.1088/1742-6596/896/1/012040
2. H. Mohrbacher, S. Jansto, and F. Siciliano, *International Symposium on Niobium Microalloyed Sheet Steel for Automotive Application, (*2006)
3. W. Bleck and E. Ratte, *Fünftes Industriekolloquium SFB 362 „Fertigen in Feinblech“. Abschlusskolloquium Werkstoffe – Verfahren – Konzepte*, 35–40, (2005)
4. S.-D. Liu, S. Jiang, M. Garnett, *21st Biennial Congress of the International Deep Drawing research Group,* **21**, 203–212, (2000)
5. H. Schmid, S. Suttner, M. Merklein, *7. WGP-Jahreskongress Aachen*, 37–42, (2017)
6. H. Schmid, S. Suttner, M. Merklein, J. Phys.: Conf. Ser., **896**, 12010, (2017) DOI: 10.1088/1742-6596/896/1/012010
7. H. Selcuk Halkaci, M. Turkoz, M. Dilmec, Journal of Materials Processing Technology, **214**, 1638–1646, (2014) DOI: 10.1016/j.jmatprotec.2014.03.008
8. M. Samuel, Journal of Materials Processing Technology, **122**, 94–103, (2002) DOI: 10.1016/S0924-0136(01)01233-X
9. E. Silvestre, J. Mendiguren, L. Galdos, E. Sáenz de Argandoña, International Journal of Mechanical Sciences, **101-102**, 10–20, (2015) DOI: 10.1016/j.ijmecsci.2015.07.013
10. S. Schreijäg, *Microstructure and mechanical behavior of deep drawing DC04 steel at different length scales* (KIT Scientific Publishing, Karlsruhe, (2013)
11. E. V. Nesterova, S. Bouvier, B. Bacroix, Materials Characterization, **100**, 152–162, (2015) DOI: 10.1016/j.matchar.2014.11.031
12. G. Corrêa Soares, B. Mendonça Gonzalez, L. de Arruda Santos, Materials Science and Engineering: A, **684**, 577–585, (2017) DOI: 10.1016/j.msea.2016.12.094

Development of a novel high straining backward extrusion process

Qiang Wang[1,]*, *Zhimin* Zhang[1], *Xubin* Li[1], and *Huifang* Zhang [1]

[1]College of Materials Science and Engineering, North University of China, 030051 Taiyuan, China

Abstract. In this study, a new method of backward extrusion is proposed. In this new process, a punch with a movable mandrel was designed. A hollow billet was firstly backward extruded and subsequently upset with the use of the punch after the mandrel returned. The extrusion and upsetting processes were successively executed in order for a higher effective strain to be imposed and a fibrous tissue flow direction to be controlled. In order for the capability of this process to be investigated, experimental and finite element (FE) methods were used. The effective strain of the final part prepared by both the conventional and the new process were compared along the bottom radial and wall axial direction respectively. In the results, it is shown that the plastic strain applied through the processed sample was approximately higher in twice the value of the sample processed via conventional backward extrusion. Consequently, this may improve the mechanical properties and anisotropy of the final products. The difference of the UTS and the TYS between radial and tangential at the bottom was less than 3%.This new process has proven to be promising for parts with a central hole at the bottom production in order for the parts low performance to be improved.
Keywords. Extrusion, Strain, Aluminium

1 Introduction

Nowadays, the extrusion process has a great important role in manufacturing industries. Between various conventional deformation processing methods, the backward extrusion (BE) method has become one of the most promising manufacturing processes in usage, due to its material savings and high surface quality[1 , 2]. Moreover, backward extrusion has been used in steel, aluminum and magnesium alloy forming for the production of hollow-shape symmetric and cylindrical products [3]. Due to the evolution of the microstructures and mechanical properties of the deformed materials being directly related to the amount of plastic deformation, the phenomenon associated with strain development must be understood [4]. This is critical, considering the fact that most existing backward extrusion procedures suffer from a smaller deformation percentage of the imposed effective strain within the processed parts[5,6]. This may consequently lead to an unfavorable microstructure inhomogeneity and poor overall mechanical performance of the products [7].

Recently, extensive studies have been performed for various new techniques, based on the extrusion process, for imposing a higher effective strain and a better strain homogeneity. The conventional direct extrusion process with a high ratio in one/two steps [8, 9] and the hydrostatic extrusion with an extra-high ratio [10] were studied. The multiple direct extrusion (MDE) process was earlier proposed [11] as a new technique for grain refinement. A repetitive indirect extrusion technique, known as accumulative back extrusion (ABE), was introduced as a novel extrusion technique for the first time in 2009[12]. ABE is a kind of SPD process in which back extrusion (BE) and compression are cyclically executed. Certain studies are presented regarding this method [13-17]. A novel cyclic forward-backward extrusion (CFBE) method, consisting of the forward-backward extrusion followed by constrained back-pressing, was earlier proposed for UFG materials [18, 19] production. A new backward extrusion utilizing a small diameter billet was also recently described as a technique for grain refinement [20]. A novel low temperature indirect extrusion process utilizing artificial cooling was devised, and the effects of artificial cooling during indirect extrusion on the microstructure and tensile properties of an extruded commercial AZ31 alloy were studied [21].

In this study, a new backward extrusion method for a hollow billet using the punch with the movable mandrel to be produced, named Expanding-Reducing Extrusion (ERE), including conventional indirect extrusion followed by upsetting, is proposed. The description of this novel high-straining process, emphasizing the effective strain, metal flow behavior and the experimental results for evaluation of the potential of both grain refinement and mechanical properties improvement, is presented.

* Corresponding author: wangqiang@nuc.edu.cn

2. Principle of the new backward extrusion process

Schematic illustrations of the conventional and new backward extrusion methods are shown in Figs. 1 and 2. During the conventional backward extrusion process, the solid billet was initially placed in the matrix and combined with the pressure of the punch, the material was compressed and flow was initiated into the gap between the punch and the matrix. The principles of the new backward extrusion process are shown in Fig. 2 where the equipment used, contained an innovative punch with a movable mandrel. The billet, in the form of a hollow cylinder, was placed into the die cavity (Fig. 2a). Following the mandrel insertion into the billet, the billet was pressed by the punch in order for the material to be back-extruded into the gap, between the punch and the die (Fig. 2b). Upon the return of the mandrel, the bottom of the extruded part was upset with the use of the same punch for the diameter to be reduced and the fibrous tissue flow direction (Fig. 2c) to be controlled.

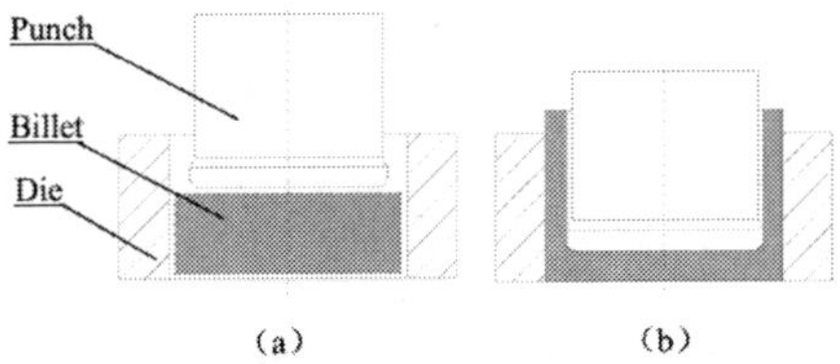

Fig.1. Schematic view of the conventional backward extrusion

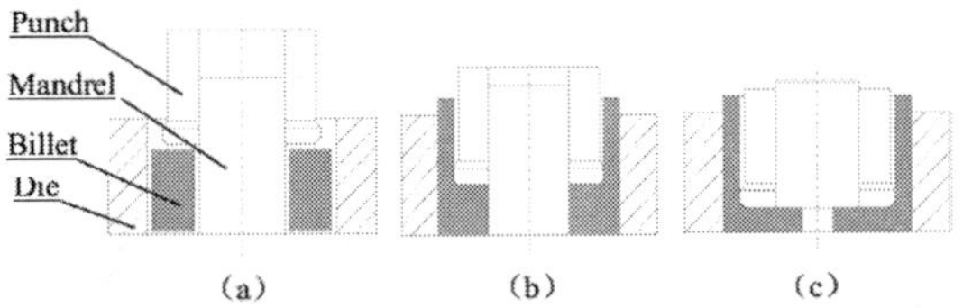

Fig.2. Schematic view of the new method of backward extrusion

3 Experimental and FE procedure

3.1 Simulation

The DEFORM 3D-6.1 software was used for FE simulation. The employed material in this study was 7A04 aluminum alloy (Al-6.15wt%Zn-2.63wt%Mg-1.6wt%Cu). The size of the billet and mold was predetermined as shown in table 1. The true stress–strain data of the 7A04 alloy, based on the results of the compression test, were inserted into the FE software. The material model was defined as elastic–plastic. All die parts were considered as rigid bodies. The billet was defined as a deformable object during analysis. A Tetrahedral element type was used, and the mesh sensitivity test was executed. The suitable mesh number was reported to be 50000. The selected friction type was the shear friction type, and due to the nature of the process, the friction factor (m) was selected to be 0.3. The simulation type was the Lagrangian Incremental. Global re-meshing was chosen, and the type of interference depth selected was relative and its value selected to be 0.7. The conjugate-Gradient was chosen as the solver of the simulation.

3.2 Experimental procedure

The dimensions of all objects in the experiment matched the dimensions of the simulated objects perfectly, in order for the results of the simulation to be validated. The deformation processing was executed in the temperature of 470°C, whereas the temperature error was controlled within a divergence of°C±5 The forming mold was pre-heated to a matching temperature with the billet. In order for the friction between aluminum specimens and tools to be reduced, the contact surfaces were lubricated beforehand with oil-based graphite. The equipment utilized was a hydraulic oil press machine of 30MN in capacity and the average forming speed of a ram was 10mm/s.

The products processed were solution treated at 470°C for 2h and then subsequently aged at 125°C in an air furnace. Specimens were cut from the bottom and wall and machined into tensile specimens of 5mm gauge diameter and 25mm gauge length. Tensile testing was carried out on an Instron-10kN material test machine at speed of 1 mm/min. Microstructure examination was performed in an optical microscope (OM). The grain size was determined using a linear intercept method from a large number of non-overlapping measurements.

Table 1. Computational condition in finite element (unit: mm)

	size			Billet size	
	I.D	O.D	Hole diameter	Diameter	Length
Conventional method	485	515	0	510	80
New method	485	515	280	O.D510/ I.D400	190

4 Results and discussion

4.1 Equivalent strain analysis

The effective strain distribution of the samples, prepared by both the conventional and the novel method of backward extrusion, are shown in Fig.3, respectively. The observed effective strain contours display the difference between the effective strain of the traditional and the new backward extrusion to be very high. Furthermore it can be observed from Fig.3, that the equivalent strain of the sample prepared by traditional backward extrusion from the solid billet is extremely low, particularly at the bottom of the component. While relative to employing expanding-reducing extrusion from the hollow billet, which makes the deformations become large clearly. The equivalent strain of the bottom and the wall was increased. This was interpreted that the deformation increased with the use of the novel method of backward extrusion, instead of the conventional method of backward extrusion.

https://doi.org/10.1051/matecconf/201819006001

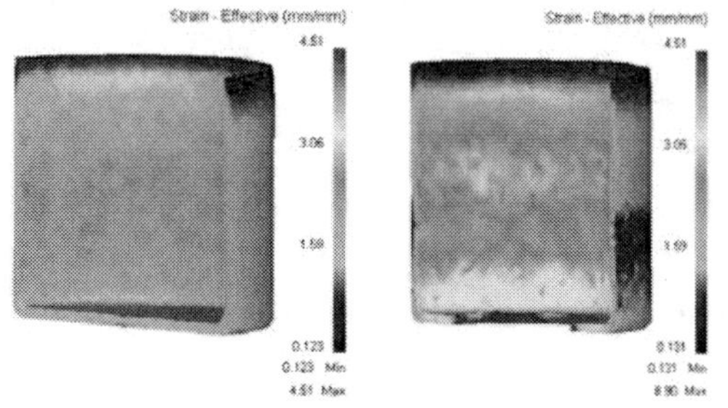

Fig. 3. Distribution of effective strain for the sample processed by (a) conventional and (b) new backward extrusion

The effective strain value of 11 points was selected to be used for the bottom length in the middle node of thickness, whereas the 11 points location was the same in both methods and the distance for each point application method was 10mm from the center hole surface to the exterior area, respectively. Shown in Fig.4 , are the quantitative values of the effective strain through the bottom length of the produced part, by both described processes. Clearly observed from the graph, is the effective strain obtained from the bottom length of the part produced from the novel method of backward extrusion, having at least twice the value of the effective strain obtained from any point of the part manufactured from the traditional method of backward extrusion, although the variation range of the effective strain was relatively high. Spreading smoothly towards the part interior, the effective strain in the case of the novel backward extrusion process was gradually increased in significance. The effective strain in the novel process was even higher, reaching approximately three times the strain of the part produced from the conventional method.

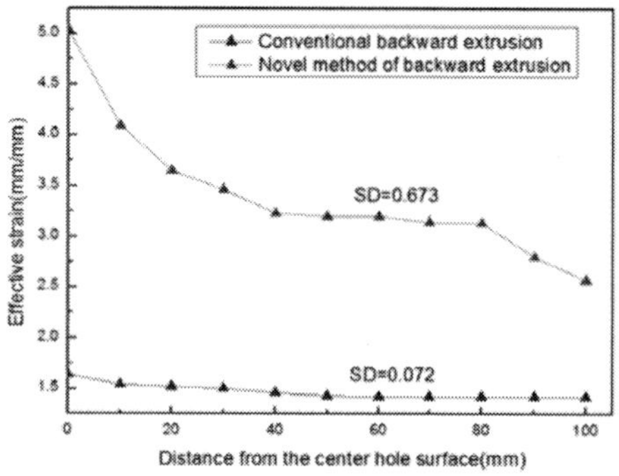

Fig. 4. Comparison of effective strains along the bottom length of processed products

The quantitative values of the effective strain along the wall length of the produced parts by both methods are compared, as shown in Fig.5. The effective strain value of 12 points was selected to be used along the wall length in the middle node of thickness, whereas the application location of the 12 points value was the same in both methods and the distance of each two points application was 30mm from the bottom to the top of the part. The Standard Deviation value of effective strain for the conventional and the novel method of backward extrusion was 0.3 and 0.28 respectively. It was interpreted that the homogeneous degree of effective strain in the novel backward extrusion is relatively higher. The effective strain of the novel method of backward extrusion through the wall length of the part was higher by 1.5 times the effective strain at any point of the parts having been manufactured by the traditional method of backward extrusion.

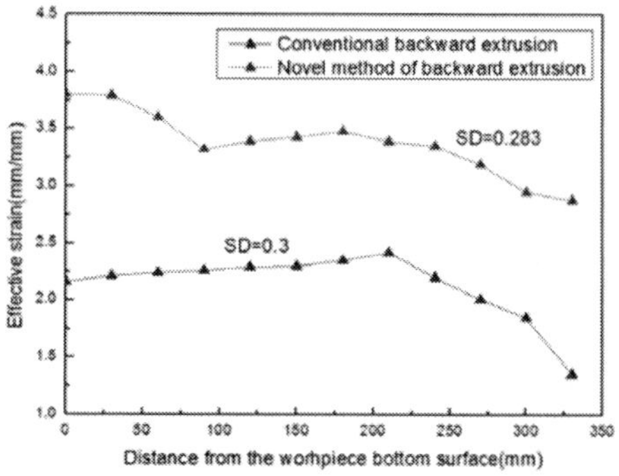

Fig. 5. Comparison of effective strains along the wall length of processed products

4.2 Microstructure and mechanical properties

The 7A04 alloy cylinder parts were extruded successfully without cracks or discontinuities according to the present technique. The prepared initial hollow billet and the extruded and upset processed sample are both shown in Fig. 6(a, b and c), accordingly.

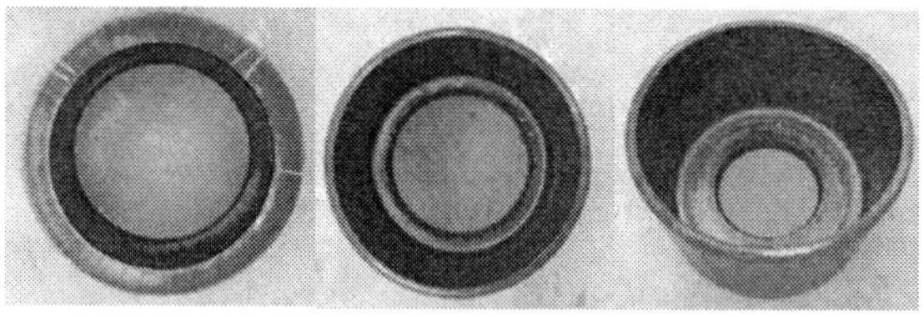

Fig.6. (a) Initial hollow billet, (b) Extrusion sample, (c) Extrusion sample with upsetting

Fig.7 presents the mechanical properties of 7A04 alloy part extruded by new process at different positions and directions. The UTS and the TYS were over 560MPa and 480MPa respectively, the elongation was above 10% in both bottom and wall positions. The mechanical properties reach a peak at wall axial direction, but the difference from the bottom was very small. The difference of the UTS and the TYS between radial and tangential in the bottom was less than 3%. It is noted that the mechanical properties anisotropy was obviously reduced due to the uniformity of deformation. For comparative purposes, the mechanical properties of 7A04 alloy part extruded by conventional process also were tested. The UTS and the TYS were about 540MPa and 460MPa respectively, the elongation was about 8%. It can be seen that mechanical properties were

https://doi.org/10.1051/matecconf/201819006001

remarkably improved for the 7A04 alloy component when it is extruded by the new process.

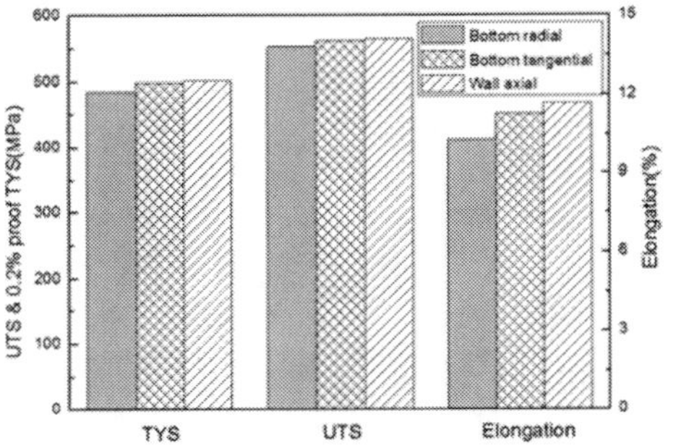

Fig.7. Mechanical properties of 7A04 alloy part obtained by new method of backward extrusion

Fig.8 shows the microstructure of the transverse sections of the extruded part of 7A04 alloy in T6 condition. It is obvious that the extrusion specimens have recrystallized and the grains refine greatly. The calculation shows that the average grain size (7.1μm) of wall specimen was less than that (7.5μm) of bottom specimen. It is obvious that grain refinement by dynamic recrystallization is very effective in alloy. At the same time, a large amount of fine and uniformly distributed second phase particles can be observed both bottom and wall positions.

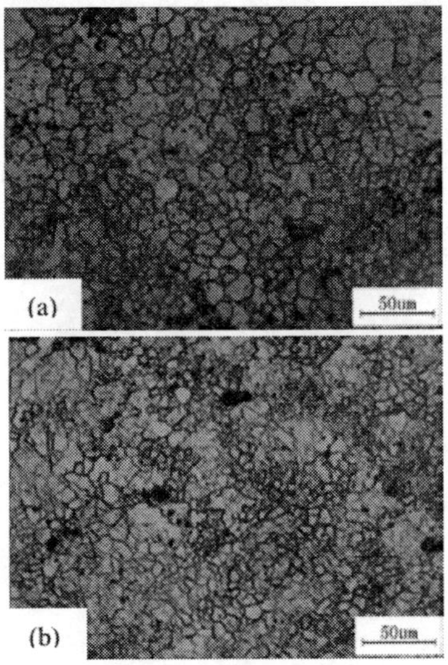

Fig.8. Microstructure of 7A04 alloy part extruded in (a) bottom and (b) wall position

The novel method of backward extrusion can apply a higher plastic strain compared to the strain of the traditional method, especially at the bottom of the component. A large body of researchers observed that the microstructure and formation of small sized grains corresponded to effective strain [22] amount. A higher plastic strain causes to the grain size of the microstructure to decrease and consequently enhance mechanical properties [23]. Therefore, this new method proves to be promising for the production of cylinder parts with higher strength compared to parts produced from the conventional technique.

5 Conclusions

In this study, the traditional and new backward extrusion processes were simulated for the 7A04 aluminum alloy products, with the use of the DEFORM-3D numerical simulation software. The effective strain diagram for both kinds of forming processes was analyzed and the effective strain values of the product are compared. The results showed that the novel method of backward extrusion can impose a higher effective strain. The effective strain of the workpieces in the novel method of backward extrusion were at least 1.5 times in value compared to the corresponding workpieces of the conventional processing method that reached 2 times in value, the strain at the bottom of the workpieces. Correspondingly, the grain of the final products was finer and the mechanical properties of the products improved. This new expanding-reducing extrusion (ERE) method proved to be extremely promising for higher strength parts production comparing to the conventional extrusion method.

The authors are grateful for the research support of the Program for the Top Young and Middle-aged Innovative Talents of Higher Learning Institutions of Shanxi.

References

1. D.Hinz, A.Kirchner, D.Brown, B.M.Ma, O. Gutfleisch, J. Mater. Process. Tech **135**, 358-365 (2003)
2. M.M.Moshksar, R.Ebrahimi, Int. J. Mech. Sci **40**, 1247-1263 (1998)
3. B.Barišić, Z.Car, M.Ikonić, METABK **47**, 313-316 (2008)
4. Hyoung Seop Kim,Min Hong Seo,Sun lg Hong, J. Mater. Process. Tech **113**, 622-626 (2001)
5. Roberto B.Figueiredo, Paulo R.Cetlin,Terence G.Langdon, Acta. Mater **55**, 4769-4779 (2007)
6. Megumi Kawasaki, Roberto B.Figueiredo,Terence G.Langdon, J. Mater. Sci **47**, 7719-7725(2012)
7. F.P.E.Dunne, Int J Plasticity **14**,413-433 (1998)
8. H.K. Lin,J.C.Huang, Mater.Trans **43**, 2424-2432 (2002)
9. Yongjun Chen, Qu Dong Wang, Jianguo Peng, Chun Quan Zhai, Mater. Sci. Forum **503–504**, 865-870 (2006)
10. Y. M. Hwang, M. T. Yang, Key. Eng. Mater **233**,311-316 (2003)
11. L.Zaharia,R.Chelariu,R.Comaneci, Mater.Sci.Eng.A **55**, 293–299 (2012)
12. S.M. Fatemi-Varzaneh, A. Zarei-Hanzaki, Mater. Sci. Eng. A **504**,104-106 (2009)
13. H. Alihosseinia, G. Faraji, A.F. Dizaji, K. Dehghani, Mater. Charact **68**,14-21(2012)

14. G.Faraji, H.Jafarzadeh, H.J.Jeong, M.M.Mashhadi, H.S.Kim, Mater. Design **35**, 251-258 (2012)
15. N. Haghdadi, A. Zarei-Hanzaki, D. Abou-Ras, M.H. Maghsoudi, A. Ghorbani, M. Kawasaki, Mater. Sci .Eng. A **595**, 179-187 (2014)
16. S.M.Fatemi-Varzaneh, A.Zarei-Hanzaki, S.Izadi, J. Mater. Sci **46**,1937–1944 (2010)
17. G.Faraji, M.M.Mashhadi, H.S.Kim, Mater. Sci. Eng. A **528**, 4312–4317(2011)
18. H.Alihosseini, M.AsleZaeem, K.Dehghani, Mater. Lett **68**, 204-208(2012)
19. H. Alihosseini, M. Asle Zaeem, K. Dehghani, H.A. Shivaee, Mater. Lett **74**, 147–150(2012)
20. V. Shatermashhadi, B. Manafi, K. Abrinia, G. Faraji, M. Sanei, Mater. Design **62**, 361–366 (2014)
21. Beomcheol Kim, Chan Ho Park, Ha Sik Kim, Bong Sun You and Sung Soo Park, Scripta. Mater **76**, 21-24 (2014)
22. G.Faraji, A.Babaei, M.M.Mashhadi, K.Abrinia, Mater. Lett **77**, 82-86 (2012)
23. S.M.Fatemi-Varzaneh, A.Zarei-Hanzaki, Mater. Sci. Eng. A **504**, 104-109 (2009)

Investigation on precision and performance for hot gas forming of thin-walled components of Ti_2AlNb-based alloy

Xueyan Jiao[1,2], *Zhiqiang* Liu[1,2], *Yong* Wu[3], and *Gang* Liu[1,2 *]

[1]State Key Laboratory of Advanced Welding and Joining, Harbin Institute of Technology, Harbin, 150001, China
[2]School of Materials Science and Engineering, Harbin Institute of Technology, Harbin, 150001, China
[3]College of Mechanical and Electrical Engineering, Nanjing University of Aeronautics and Astronautics, Nanjing, 210016, China

Abstract. Ti_2AlNb-based alloys have received considerable attention as potential materials to replace the nickel alloy at 600-750 °C, depending on their advantages of high specific strength, good corrosion and oxidation resistance. To realize the precision and performance control for Ti_2AlNb-based alloy thin-walled components, the microstructure evolution was analyzed for setting up the unified viscoplastic constitutive equations based on the physical variables and simulating the forming process coupled between the deformation and the microstructure evolution. Through the finite element model with coupling of microstructure and mechanical parameters, the microstructure evolution and shape fabricating can be predicted at the same time, to provide the basis for the process parameters optimization and performance control. With the reasonable process parameters for hot gas forming of Ti_2AlNb thin-walled components, the forming precision and performance can be controlled effectively.

Keywords: hydroforming, performance, precision

1 Introduction

With the rapid development of aerospace industry, the speed of aircrafts has been dramatically improved, resulting in higher requirement on heat-resistant and lightweight materials. In this case, the Ti_2AlNb-based alloys, combing high specific strength, good oxidation resistance and sufficient creep resistance at elevated temperatures, have great potential application in aerospace field [1-3]. However, the superior high-temperature performance and the intrinsic brittleness have brought difficulties in the deformation of Ti_2AlNb-based alloys, especially for the complex components with irregular section. Moreover, as multi-phase alloys, the Ti_2AlNb-based alloys have multiple phase transformations and compositions, which can result in abundant microstructure morphologies. Boehlert [4] has pointed out that the mechanical properties of the Ti_2AlNb-based alloys were very sensitive to the microstructures, which was determined by the thermal processing and heat treatment history. To achieve the accurate prediction of deformation behavior and microstructure evolution simultaneously, the unified viscoplastic constitutive equations based on the microstructure evolution mechanism were established in this paper.

Developed on the tube and pipe hydroforming process, hot gas forming is considered as one of novel forming processes to produce complicated shaped components of metal materials at high temperature [5]. Compared with the traditional forming technology, the hot gas forming can obtain higher inner-surface accuracy in the closed components with a complex cross-section. In recent years, hot gas forming has been widely applied to manufacturing parts in various fields, such as automobile, aircraft and aerospace, and ship building industry, due to the increasing demands for lightweight parts [6].

In this paper, the simulations of hot gas forming process for the Ti-22Al-24Nb-0.5Mo quadrate tubes were performed using unified viscoplastic constitutive equations. The corresponding hot gas forming process of Ti-22Al-24Nb-0.5Mo quadrate tubes was also conducted to verify the simulation accuracy. The effect of forming temperature on wall thickness distribution, micro-structural evolution and mechanical property was investigated.

2 Unified viscoplastic constitutive equations for Ti-22Al-24Nb-0.5Mo alloy

The Ti-22Al-24Nb-0.5Mo alloy was tensile tested at 910~1040°C with strain rates of 0.1~0.0004s^{-1} and the deformed samples with different strains were investigated by electron backscattered diffraction (EBSD)

* Corresponding author: gliu@hit.edu.cn

 https://doi.org/10.1051/matecconf/201819007001

to gain the microstructural information of grain size, phase volume fraction, recrystallization, and globularization. During the hot deformation, the microstructure evolves with the increasing strain, in turn, the mechanical behavior will be also affected by the microstructural evolution. When the Ti-22Al-24Nb-0.5Mo alloy was deformed at 950 °C, the obvious softening phenomenon occurred after the peak stress, which is caused by damage and softening mechanism such as dynamic recovery, dynamic recrystallization and globularization. Besides, the temperature-rise induced by plastic deformation also caused the material softening. The softening phenomenon also occurred in this alloy when it was deformed at temperature above 970 °C with high strain rate, while the strain hardening phenomenon presented with low strain rate because of grain growth during the deformation. The constitutive equation listed in Eq.(1), which considered the microstructure evolutions, including grain size, recrystallization, phase volume fraction, damage, globularization and temperature rise, was established to predict the microstructure evolution and stress variation under a complicated deformation condition. The specific meaning of variables and calculation of parameters have been reported in reference [7]. The genetic algorithm toolbox in the MATLAB software was used to optimize the material parameters in the constitutive model.

$$\begin{cases}
\dot{\varepsilon}_{p,O}=\left(\dfrac{\sigma/(1-D)-H-\sigma_{0,O}}{\lambda K_{B2/\beta}(1-\zeta f_O\omega)}\right)^n(\bar{d})^{-\mu} \\
\dot{\varepsilon}_{p,B2/\beta}=\left(\dfrac{\sigma/(1-D)-H-\sigma_{0,B2/\beta}}{K_{B2/\beta}(1-\zeta f_O\omega)}\right)^n(\bar{d})^{-\mu} \\
H=B\bar{\rho}^{0.5} \\
f_{B2/\beta}=1-\psi_1\exp(\psi_2(T_{T,B2/\beta}-T)) \\
f_O=\psi_3\exp(\psi_4(T_{T,O}-T)) \\
f_{\alpha_2}=1-f_O-f_{B2/\beta} \\
\dot{\varepsilon}_p=f_{B2/\beta}\dot{\varepsilon}_{B2/\beta}+f_O\dot{\varepsilon}_O \\
\dot{S}=\dfrac{q_1[x\bar{\rho}-\bar{\rho}_c(1-S)](1-S)^{q_2}\dot{\varepsilon}_p^{q_3}}{\bar{d}} \\
\bar{\rho}_c=q_4\dot{\varepsilon}_p^{q_5} \\
\dot{x}=q_6(1-x)\bar{\rho} \\
\dot{\bar{\rho}}=A\dot{\varepsilon}_p^{\delta}\cdot|\dot{\varepsilon}_p|-k_1\bar{\rho}\dot{\varepsilon}_p^{\delta_1}\cdot|\dot{\varepsilon}_p|-k_2\bar{\rho}^{\delta_2}-[k_3\bar{\rho}/(1-S)]\dot{S} \\
\dot{\bar{d}}=\alpha_1\bar{d}^{-\gamma_1}+\alpha_2\dot{\varepsilon}_p\bar{d}^{-\gamma_2}-\alpha_3\dot{S}^{\gamma_3}\bar{d}^{\gamma_4} \\
\dot{\omega}=\chi_1(1-\omega)+(1-\omega)\chi_2\dot{\varepsilon}_p^{\chi_3} \\
\dot{T}=\dfrac{\eta\sigma}{C_v d_n}|\dot{\varepsilon}_p|-\lambda'\Delta T \\
\dot{D}=\eta_1(1-D)\dot{\varepsilon}_p^{d_1}+\eta_2\dot{\varepsilon}_p^{d_2}\exp(\eta_3\varepsilon_p)-\eta_4 D \\
\sigma=E(1-D)(1-\zeta f_O\omega)(\varepsilon_T-\varepsilon_p)
\end{cases} \tag{1}$$

3 Numerical simulation for hot gas form -ing of quadrate tube

The established unified viscoplastic constitutive model based on microstructure internal state variables was embedded in the commercial FEM software Abaqus by the subroutine VUMAT. The hot gas forming of quadrate tube was simulated with the constitutive model coupled of microstructure and mechanical parameters. Microstructure evolution and shape fabricating were predicted by the constitutive model at the same time. The simulation was respectively performed at 950 °C, 970 °C and 990 °C with forming pressure of 15 MPa. The simulated process for hot gas forming of Ti-22Al-24Nb-0.5Mo quadrate tube is shown in Fig. 1.

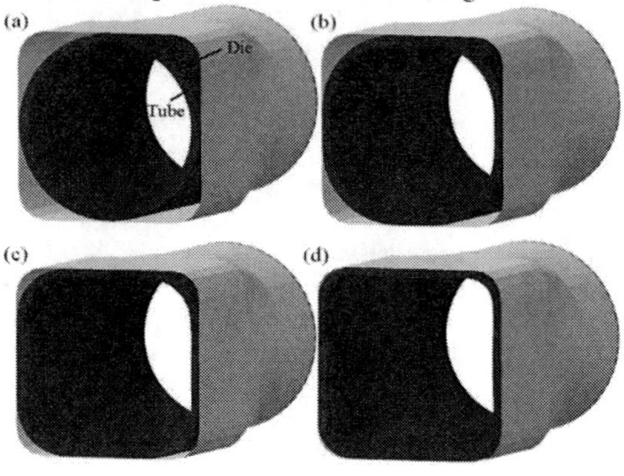

Fig. 1. The simulated process for hot gas forming of Ti-22Al-24Nb-0.5Mo quadrate tube, a) initial status, b-c) intermediate status, d) finished status.

Fig. 2 shows the thickness distribution and the thinning ratio distribution along the quadrate section of tubes. As shown in Fig. 2a, the reducing of wall thickness concentrates on the transition region (TR) between straight region (SR) and round region (RR) due to the effect of friction. As illustrated in Fig. 2b, the severe thinning occurs in the transition region for the tube formed at 950 °C, while the relatively uniform thickness can be achieved in the tube formed at 970 °C and 990 °C.

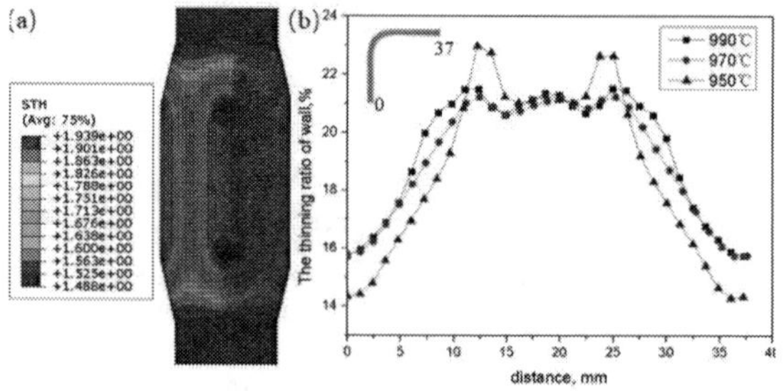

Fig. 2. The simulated results of a) thickness distribution at 970°C, b) the thinning ratio distribution along quadrate section of the tubes.

Fig. 3 displays the simulated distributions of strain, stress and microstructural variables for the quadrate tube formed at 970 °C. The strain and stress distributions show the analogical trend of that the higher value concentrates on the round region and transition region, as shown in Fig. 3a-b. In accordance with the strain distribution, the distribution of relative dislocation exhibits larger value in the round region and transition region. Besides, the zone with larger strain has smaller relative grain size due to dynamic recovery and dynamic recrystallization, while the undeformed zone shows bigger relative grain size due to the single thermal effect.

Fig. 4 demonstrates the simulated damage distribution for the tubes formed at 950-990 °C. It is found that the severe damage distributes on the round region and transition region, where has performed larger deformation. The damage increases with the increasing strain resulting from the nucleation and growth of defects, such as micro-voids and micro-cracks during deformation. In addition, the damage also increases with the decreasing forming temperature. Since the damage can be partially restored at high temperature, slighter damage is obtained at 990 °C and 970 °C. As indicated by the simulation results, the distribution of damage and thinning ratio are similar for the tubes formed at 990 °C and 970 °C, which are much lower than those at 950 °C. Considering the wall thickness, microstructure damage and energy consumption, the forming temperature of 970 °C is suggested as the optimum deformation temperature.

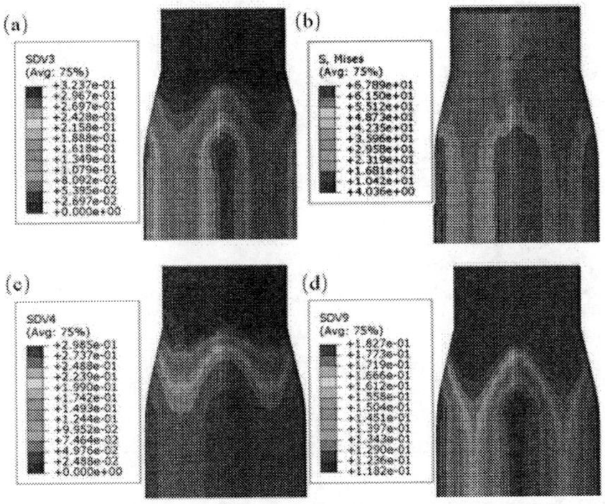

Fig. 3. The simulated distributions of a) strain, b) stress, c) relative dislocation density and d) relative grain size.

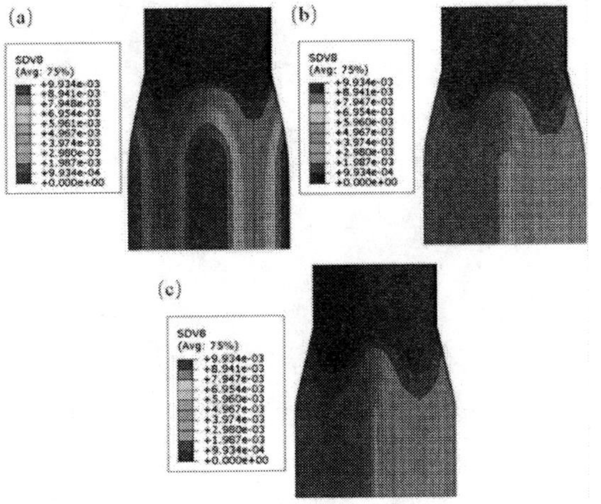

Fig. 4. The simulated results of damage distribution at a) 950 °C, b) 970 °C and c) 990 °C.

4 Hot gas forming experiments

The initial Ti-22Al-24Nb-0.5Mo tube billets with a diameter of 40 mm and thickness of 2.0 mm were applied in the experiments. From the microstructure of experimental Ti-22Al-24Nb-0.5Mo alloy shown in Fig. 5, it is noted that the initial microstructure was consisted of equiaxed primary α_2 phase (dark contrast), lamellar O phase (gray contrast) and the matrix of β/B2 phase (bright contrast). The hot gas forming process was conducted at 950-990 °C with forming pressure of 15 MPa on a self-designed high pressure gas forming platform. The quadrate tube with radius of 6 mm in round corner was shown in Fig. 6. Tensile specimens with gauge length of 15 mm and width of 3 mm were cut from the straight region of the quadrate tubes.

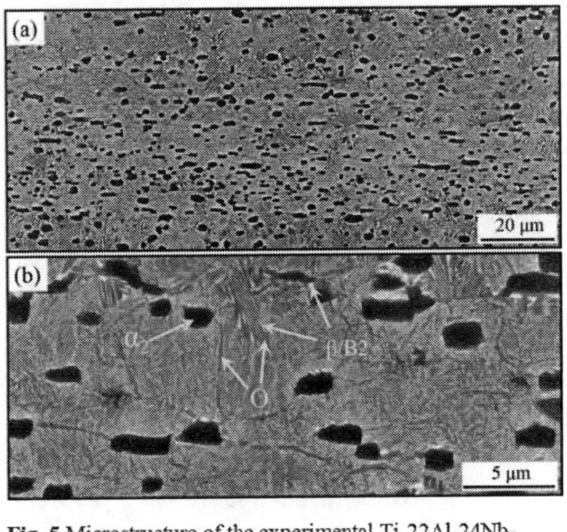

Fig. 5 Microstructure of the experimental Ti-22Al-24Nb-0.5Mo alloy, a) low magnification image, b) high magnification image.

https://doi.org/10.1051/matecconf/201819007001

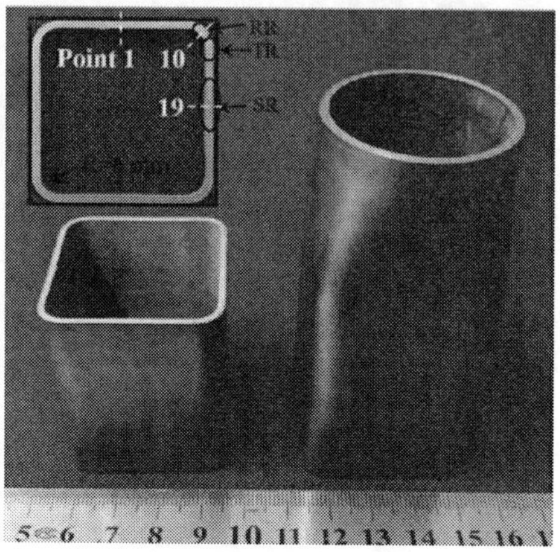

Fig. 6 Hot gas formed quadrate tube.

5 Mechanical properties and precision of the components

5.1 Microstructure and mechanical properties

Fig. 7 displays the SEM microstructures of the Ti-22Al-24Nb-0.5Mo quadrate tubes formed at different temperatures. Compared to the initial microstructure shown in Fig. 5, the O phase was significantly coarsened at the forming temperature of 950 °C and dissolved at and above 970 °C. Furthermore, the volume fraction of coarsened O phase for the tube formed at 950 °C was determined to be 19.3 %, which would play a key role on the mechanical property. Besides, the matrix β/B2 grains have slightly grown up at 990 °C due to the partial dissolution of equiaxed α_2 phase, which contributes to pinning the β/B2 grain boundaries.

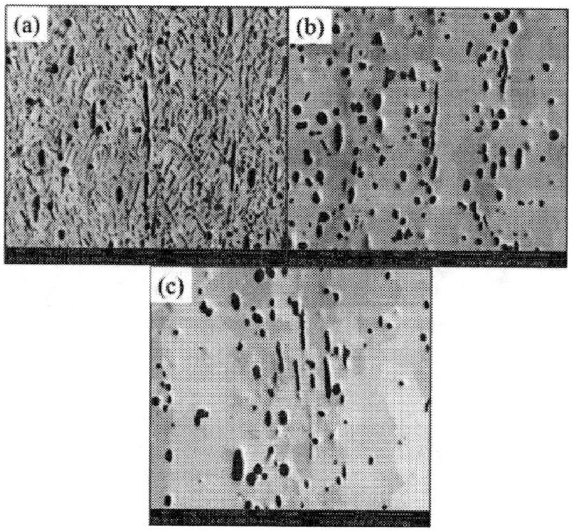

Fig. 7. Cross-section microstructure of the quadrate tubes formed at different temperatures, a) 950 °C, b) 970 °C, c) 990 °C.

5.2 Mechanical properties

Fig. 8 shows the tensile properties at 750 °C for the Ti-22Al-24Nb-0.5Mo quadrate tubes. It is noted that the tensile strength increased but the elongation-to-failure decreased with the increasing of forming temperature. The tensile strength of quadrate tubes formed at 950 °C, 970 °C and 990 °C was 793 MPa, 906 MPa, and 921 MPa, respectively. This trend also confirmed the damage results achieved in simulations, demonstrating that the microstructural damage is harmful to the strength. Due to the existing of O phase (Fig. 7a), the quadrate tube formed at 950 °C exhibited a high elongation-to-failure of 14.9 %. It has been reported that the O phase contributed to the improvement of ductility by dispersing operative dislocations [8].As the forming temperature increases, the elongation-to-failure decreases due to the reduced volume fraction of O phase. When the forming temperature increased to 990 °C, the elongation-to-failure decreased to 2.8 %, indicating that the larger grain size of β/B2 phase was detrimental to the ductility.

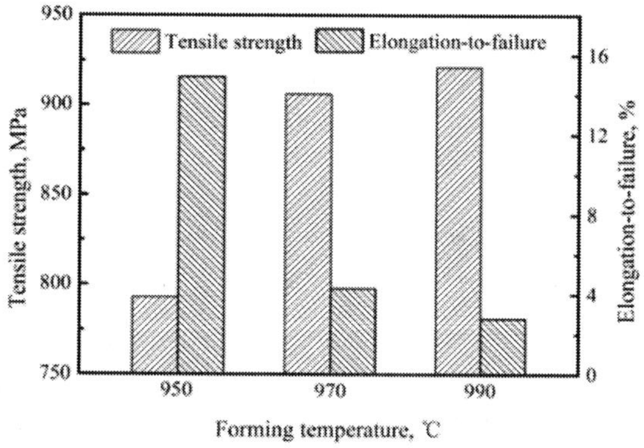

Fig. 8 Tensile properties at 750 °C for the Ti-22Al-24Nb-0.5Mo quadrate tubes.

5.3 Precision of the component

Table 1 lists the thickness thinning ratio along the cross-section of quadrate tubes formed at different temperatures. The point number is marked in white front, as shown in Fig. 6. Apparently, the maximum thickness thinning occurred in the transition region between straight region and round region, which also verified the thickness distribution achieved in simulation. Due to the friction on the straight region, the transition region primely satisfies plastic yielding condition and produces compressive strain through the thickness [9]. From the comparison of thickness thinning ratio for the quadrate tubes formed at different temperatures, it is noted that the biggest fluctuation of wall thickness appeared in the tube formed at 950 °C due to the significant strain softening. However, a relatively stable thinning ratio of 15 %-19 % was achieved in the quadrate tubes formed at 970 °C and 990 °C. Based on the simulation results, the suggested forming temperature is 970 °C, which is optimum to gain the qualified component at low energy consumption.

Table 1. Thickness thinning ratio along the cross-section of quadrate tubes formed at different temperatures.

Point number	Thinning ratio (%) for tube formed at		
	950 °C	970 °C	990 °C
1	16.37	16.58	15.75
2	17.04	15.61	16.50
3	18.54	16.68	16.95
4	18.15	16.68	17.75
5	20.00	17.46	18.50
6	19.63	18.44	18.50
7	16.22	17.95	18.50
8	14.41	17.12	15.60
9	13.43	16.10	16.15
10	13.89	16.93	16.00
11	13.50	16.98	15.25
12	13.67	16.83	16.00
13	16.41	16.68	17.50
14	17.95	18.78	17.25
15	18.86	18.15	17.05
16	17.83	17.95	16.90
17	17.10	16.83	15.40
18	16.65	15.71	14.97
19	16.63	15.85	15.20

6 Conclusions

The finite element model of hot gas forming was established based on the unified viscoplastic constitutive model with coupling of microstructure and mechanical parameters. The microstructure evolution and shape fabricating for the hot gas forming of Ti-22Al-24Nb-0.5Mo quadrate tube was predicted by simulation. Besides, the corresponding hot gas forming of Ti-22Al-24Nb-0.5Mo quadrate tube was also conducted at 950-990 °C. The results on wall thickness distribution, microstructure evolution and mechanical properties were summarized as follows:

(1) The unified viscoplastic constitutive equations can effectively predict the wall thickness distribution and microstructural evolution for the Ti-22Al-24Nb-0.5Mo quadrate tube. The forming temperature of 970 °C is suggested to gain the qualified component at low energy consumption.

(2) The maximum thinning occurred in the transition section of the quadrate tube due to the effect of friction. The biggest fluctuation of wall thickness appeared in the quadrate tube formed at 950 °C resulting from the significant strain softening. The forming temperature of 970 °C and 990 °C is suggested to achieve the uniform thickness for the Ti-22Al-24Nb-0.5Mo quadrate tubes.

(3) During the gas forming, the initial O phase was coarsened at 950 °C and dissolved at 970 °C. The tube formed at 950 °C exhibited high elongation-to-failure of 14.9 % at 750 °C due to the existing of ductile O phase. As the forming temperature increased to 970 °C and 990 °C, the elongation-to-failure decreased to 4.3 % and 2.8 %, and strength increased to 906 MPa and 921 MPa, respectively.

This work was financially supported by National High-level Personnel of Special Support Program (No. W02020239).

References

1. D. Banerjee, A.K. Gogia, T.K. Nandy, V.A. Joshi, Acta Metall. **36**, 871 (1988)
2. S.R. Dey, S. Roy, S. Suwas, J.J. Fundenberger, R.K. Ray, Intermetallics. **18**, 1122 (2010)
3. J. Kumpfert, Adv. Eng. Mater. **3**, 851 (2001)
4. C.J. Boehlert C.J., Mater. Sci. Eng., A. **267**, 82 (1999)
5. G. Liu, Y. Wu, D. Wang, S. Yuan, Int. J. Adv. Manuf. Tech. **81**, 1809 (2015)
6. K.Wang, G. Liu, J. Zhao, J. Wang, S. Yuan, Mater. Des, **91**, 269 (2016)
7. Y. Wu, High temperature unified viscoplastic constitutive model and high pressure gas bulging of Ti-22Al-24.5Nb-0.5Mo sheet, Harbin Institute of Technology, 100 (2017)
8. Y. Wu, L. Zhen, D.Z. Yang, M.S. Kim, S.K. Hwang, Y. Umakoshi, Intermetallics, **12**, 43 (2004)
9. G. Liu, S. Yuan, B. Teng, J. Mater. Process. Tech. **177**, 688 (2006)

Influence of heat treatment on the microstructure and mechanical properties of TC17 Titanium Alloy

Wang Min[1], Huang Darong[1], Fan Ya[1], Nin YongQuan[1], Guo Hongzhen[1],Yao Zekun[1]

(1. Northwestern Polytechnical University School of Materials Science and Engineering, xi'an 710072, China)

Abstract. The influence of heat treatment on the microstructure and mechanical properties of TC17 titanium alloy was investigated by changing holding time before forging and aging temperature after β-forging. A better heat treatment mechanism was chosen according to its working requirements. The results showed that the holding time and aging temperature had a significant influence on microstructures and mechanical properties of TC17 alloy. With the increase of holding time before forging in the single β phase region, the original β grain size enlarge, the strength and fracture toughness decrease, and the plasticity increases. The size of lamellar α-phase bundles enlarge with the rise of aging temperature after β-forging, the orientated relationship becomes simple, and the content of β-transformation increases. And the plasticity and fracture toughness of the alloy increase, the strength and hardness decreases. The fracture toughness is better under the heat treatment mechanism at 873℃/2h +918℃/64min before forging, 800℃/4h,WC+660℃ /8h AC after forging, with the high and low cycle fatigue properties meet the technical requirements.

Key Words: Heat treatment, Forging, Microstructure

1. Introduction

Titanium alloys have been widely used in aircraft, aircraft engines and other fields owing to their high strength and toughness, relatively low density, excellent creep and corrosion resistance[1-2]. TC17 (Ti-5Al-2Sn-2Zr-4Mo-4Cr) of near β-titanium alloy belongs to medium temperature and high strength titanium alloy. The properties of TC17 correspond with Ti-6A1-6V-2Sn, the abilities of plasticity and fracture toughness are better than Ti-6A1-6V-2Sn and the creep property is higher than Ti-6A1-4V. TC17 alloy has been attracted considerable attention in the fabrication of high load-bearing components, such as fan plate of aircraft engine and compressor disk. TC17 alloy includes α stable element Al and isomorphous β element Mo and eutectic β element Cr、Fe and Si. The tensile strength and yield strength can be enhanced by strengthening α and β phases [3].

It is difficult for titanium alloys to obtain stable and consistent workpiece in production, and it is easy to produce various macroscopic and microcosmic defects, which will affect the mechanical properties of the workpieces. The fracture toughness decreases with the increase of strength of high strength titanium alloy, and the strengthening of alloy properties is constrained by both plasticity and toughness at a certain strength level. Along with the application of damage tolerance design idea in titanium alloy, the fracture toughness of titanium alloy has been attracted more and more attention and research. The fine lamellar structure can be obtained by using β forging, which can improve the fracture toughness of the alloy greatly [4-5]. Titanium alloy β forging has a series of advantages, such as good plasticity, low deformation resistance, high production efficiency and long service life of the forging equipment [6]. TC17 titanium alloy is taken as the research subject in this paper, studying the effect of heat treatment on the microstructure and mechanical properties of TC17 titanium alloy by changing holding time before forging and aging temperature after β-forging, and optimal heat treatment system has been selected according to the service requirements of TC17 titanium alloy, which provides experimental and theoretical basis for actual practice.

2. Materials and procedures

The material used in this investigation was a TC17 forging bar provided by a factory, 250mm in diameter. The measured chemical compositions(wt%) of the bar are as follows: 5.53%Al, 2.23%Sn, 2.29%Zr, 4.30%Mo, 4.08%Cr, 0.12%Fe, 0.01%C, Ti balance. The β transus temperature is 893℃ as determined with the metallographic observation method, as follows: test samples of TC17 titanium alloy were heated at a series of temperatures with 5℃ intervals between tests around the expected β transus temperature for 30 min and then water cooling(WC); the β transus temperature is taken about a middle temperature between the temperatures on the basis of the appearance and disappearance of α_p grains. Figure 1 shows the original microstructure of the bar before forging. It could be observed that the original microstructure is typical equiaxed structure, which consists of equiaxed α phase and β transformation structure; The equiaxed α phase is uniformly distributed in the β matrix and contains a small amount of short rod-like α phase.

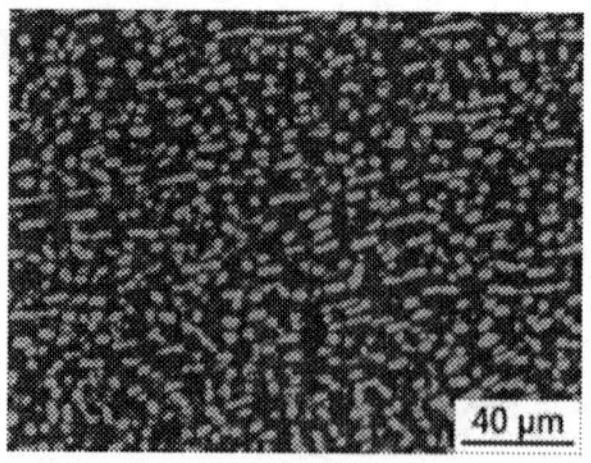

Fig.1 The initial microstructure of TC17 alloy bar

 https://doi.org/10.1051/matecconf/201819007002

The pre-heating time of titanium alloy billet is calculated by the formula "0.8min/mm×minimum section size(mm)". With reference to the actual production, three heating specification were selected: 873°C/2h+918°C/15min (specification 1) 、 873°C/2h+918°C/64min (specification 2) 、 918°C/120min (specification 3) . Isothermal forging of TC17 titanium alloy specimens in β region were carried on the 630t strain rate controllable oil press. The forging temperature and deformation degree are the same and the strain rate is different according to the specimen size, air cooling(AC) after forging. The solid solution temperature after forging was 800°C and three kinds of different aging temperatures were employed: 620°C, 640°C, 660°C.The specific heat treatment scheme as shown in Table1. The specimens were heat treated in a SX2-10-13 chamber electric furnace; microstructure was observed by Leica DMI3000M metallographic microscope; the room temperature tensile test were conducted using the ENST-1196 stretcher; the fracture toughness (T-L) values were obtained using the INSTRON-1251 model test machine. The properties of high and low cycle fatigue were carried out on the MTS810 electro-hydraulic servo fatigue testing machine.

Tab.1 Subsequent heat-treatment processes after forging

Heat treatment	Heat treatment system
I	800°C/4h，WC+620°C/8h，AC
II	800°C/4h，WC+640°C/8h，AC
III	800°C/4h，WC+660°C/8h，AC

3. Results and discussion

3.1The effect of holding time before forging on microstructure

The microstructure of the TC17 titanium alloy heated by different heating specifications and carried out isothermal forging are presented in Fig2. The parameters of forging are: temperature is 918 °C，deformation degree is 60%, strain rate is 7.63 × $10^{-3}s^{-1}$， slow cooling after forging and heat treatment by system II. Under different heating specifications，the microstructure of TC17 titanium alloy after isothermal forging is typical basket-weave structure. The primary equiaxed α phase disappears completely; the precipitated lamellar α phase in β grain boundary is broken remarkably; lamellar and secondary α phase interwoven distribution in intracrystalline, the ratio of length to width is 10 to 1. The original β grain is elongated along the deformation orientation and the ratio of length to width is 2 to 1, a small amount of fine recrystallized grain appears in grain boundary. Under the three heating specifications, the growth degree of the β grain is different due to different holding time at 918°C. The holding time increases gradually from specification I to specification III, and the initial grain size in microstructure increases gradually, but specification III changed little compared to specification II. Owing to the size of the original β grain is large, so the deformation of grain is inhomogeneous in the forging process. As a result, the number of recrystal grains formed nearby the grain boundaries differ greatly. In addition, after the billet was heated at 918°C/2h, forging and heat treatment, more secondary α phases appeared in the β intragranular of the alloy, which precipitated directly from the grain boundaries and grew in the crystal. The lamellar α phase is longer in the crystal, this phenomenon may be on account of the interruption of heating for a certain period during the forging process and the decrease of the die temperature，the deformation degree of the billet increases in the two phase zone.

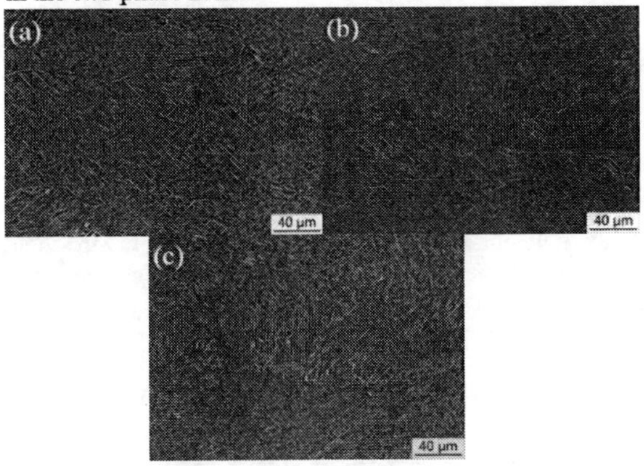

Fig.2 Microstructure of TC17 alloy under different heating specifications (7.63×$10^{-3}s^{-1}$)

(a) 918°C/15min; (b) 873°C/2h+918°C/64min;

(c) 918°C/2h

The microstructure of the TC17 titanium alloy heated by different heating specifications and carried out isothermal forging are presented in Fig3. The parameters of forging are: temperature is 918 °C，deformation degree is 60%, strain rate is 3.05 × $10^{-3}s^{-1}$， slow cooling after forging and heat treatment by system II. Under different heating specifications, the microstructure of TC17titanium alloy after isothermal forging is typical basket-weave structure. The lamellar α phase breaks obviously at the grain boundary, lamellar and secondary α phase interwoven distribution in intracrystalline, the ratio of length to width is 10 to 1. The original β grain is elongated along the deformation orientation and the ratio of length to width is 2 to 1, a small amount of fine recrystallized grain appears in grain boundary. The holding time of billets in the β single phase field increases from specification I to specification II, the original β grain size increases, and the number of recrystallized grain decreases near the grain boundary. It could be that the larger grain boundary area and more deformation storage energy of the billet under the heating condition of 918°C /15min, and it is easier to satisfy the energy condition of recrystallization. Compared with Fig.2， the strain rate

https://doi.org/10.1051/matecconf/201819007002

decreases and the deformation time increases, the grain size of the recrystallized grains enlargement. In addition, the primitive β grains have more time to coordinate deformation, and the grain deformation inhomogeneity reduces.

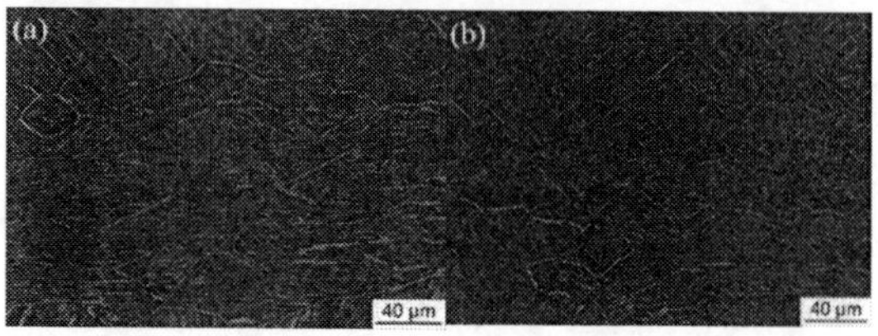

Fig.3 Microstructure of TC17 alloy under different heating specifications (3.05 × $10^{-3}s^{-1}$)

(a) 918℃/15min; (b) 873℃/2h+918℃/64min;

3.2 The effect of holding time before forging on mechanical properties

The tensile properties at room temperature of the TC17 titanium alloy heated by different heating specifications and carried out isothermal forging are presented in Tab.2 The parameters of forging are: temperature is 918°C, deformation degree is 60%, strain rate is 7.63 × 10^{-3}s-1, slow cooling after forging and heat treatment by system II.

It is found from the data in table 2 that only the yield strength of specification I exceeds the technical standard, and the other properties of the alloy satisfy the technical requirements. The change of heating scheme will have obvious influence on the properties of TC17 titanium alloy. From specification I to specification III, the heat holding time in β single phase filed increases gradually, the strength of the alloy decreases and the plasticity increases gradually.

The fracture toughness at room temperature of the TC17 titanium alloy heated by different heating specifications and carried out isothermal forging are presented in Table3. The parameters of forging are: temperature is 918°C, deformation degree is 60%, strain rate is 3.05 × 10^{-3}s-1, slow cooling after forging and heat treatment by system II. The fracture toughness of the alloy after specification I is slightly higher than that of others, the test results of heating specifications II and III are basically same. The fracture toughness of the alloy under the three specifications all reach the technical standard. The fracture toughness of TC17 titanium alloy decreases with the increase of the holding time in β single phase filed, which may be due to the size of the primitive β grains enlarge, the grain boundary area decreases, the grain boundary of the crack propagation path decreases and the energy consumption decreases correspondingly. So the fracture toughness decreases slightly.

Table 2 The room-temperature tensile properties of TC17 alloy at different heating modes

specification	Tensile strength σ_b/MPa		Yield strength $\sigma_{0.2}$/MPa		Elongation δ/%		Area reduction Ψ/%	
	Test	average	Test	average	Test	average	Test	average
1	1302	1299	1242	1238	11.0	10.25	15	16
	1296		1234		9.5		17	
2	1219	1217.5	1168	1168	12.0	12.0	23	20
	1216		1168		12.0		17	
3	1188	1184.5	1136	1134	12.0	13.0	20	21
	1181		1132		14.0		22	
technical standard	≥1120		1020~1193		≥5		≥10	

Table 3 The fracture toughness of TC17 alloy at different heating modes

specification	K_{IC}/ MPa·$m^{1/2}$
1	64.9
2	62.7
3	62.6
technical standard	≥55

3.3 The effect of heat treatment after forging on microstructure

The microstructure of TC17 titanium alloy carried out isothermal forging are presented in Fig4. The parameters of β-forging temperature is 918 °C, deformation degree is 60%, strain rate is 7.63 × $10^{-3}s^{-1}$, and heat treatment by different system. The microstructure of all specimens are: typically basket-weave structure. The primary β grains are elongated and deformed along the direction of metal flow obviously, secondary lamellar α distributes in the β matrix. When the plastic deformation occurs, the grain boundary of α phase is fragmented obviously, and there are more fine recrystallized β grains at the grain boundaries. The stacking fault energy of the β phase of BCC structure is high, the intracrystalline dislocation reduces easily at high temperature, the piling-up of dislocation degree of the grain boundary is relatively larger, the distortion storage required for the recrystallization are easy to satisfy. Furthermore, the dislocation density of crystal boundary is large, so it is easy to occur recrystallization. Therefore, recrystallization grains are formed easily near the grain boundaries during the heat treatment. The near β titanium alloy after solid solution and quench, ω phase and metastable β phase are formed easily in the structure owing to the fast cooling, which is unstable in thermodynamics and will resolve if heated. The subsequent aging process causes the ω phase and metastable β phase transforms into diffuse lamellar α phase[7]. The microstructure, as shown in Fig.4, illustrates that with the increase of aging temperature, the concentration size of secondary α phase in the intracrystalline tend to increase, the distribute orientation became simple, the content of β transus increased.

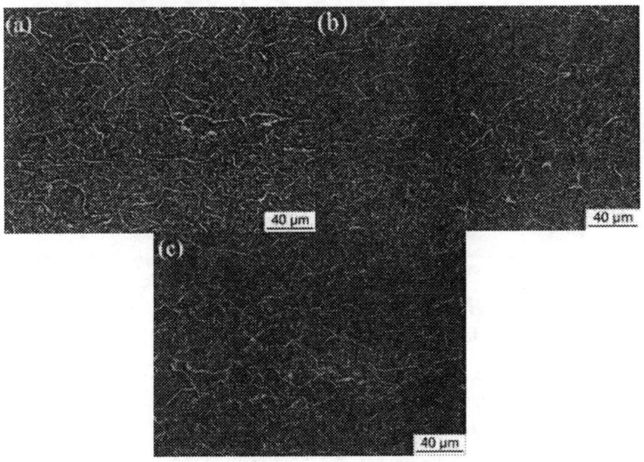

Fig.4 The microstructure of TC17 alloy at different heat-treatments($7.63\times10^{-3}s^{-1}$)

(a) 800℃/4h, WC+620℃/8h, AC; (b) 800℃/4hWC+640℃/8h, AC; (c) 800℃/4h, WC+660℃/8h, AC

When the strain rate decreased to $3.05\times10^{-3}s^{-1}$, other parameters of β-forging are same, the microstructure of specimens shows in Fig.5. The microstructure of all specimens are typically basket-weave structure, the primary β grains are elongated along the direction of specimen deform and the ratio of length to width is more than 2:1. The secondary α phase precipitate and intertwine distribute in the crystal, the grain boundary α phase is broken obviously, as the strain rate decreases, the forging time increases from 122s to 305s, and the distortion energy produced by deformation could has more time to release through recovery. In addition, the density of pile-up in the grain boundary decreases, the condition of recrystallization nucleation cannot be fully satisfied, so the number of recrystallized grains decreases. After solid solution, the alloy precipitated ω phase and metastable β phase, which will transform into secondary α phase during the subsequent aging process. Fig.5 verifies the results that with the increasing of aging temperature, the distribution of the secondary α phase in the crystal tended to be simple, and the content of the β transus increased. There is a part of coarse primary β grains in the alloy under the condition of II as showed in Fig.3(b), which caused by the longer heating time.

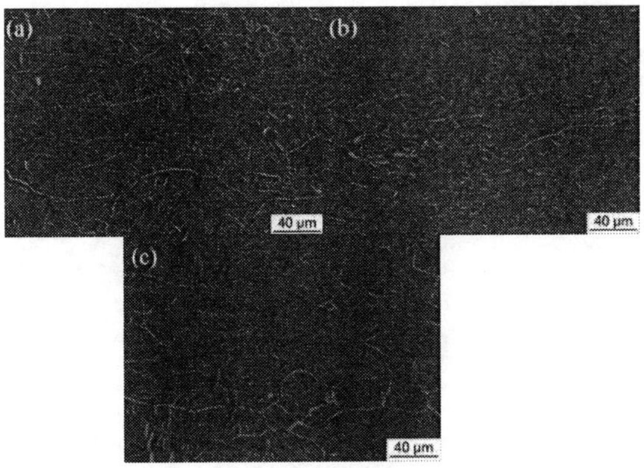

Fig.5 The microstructure of TC17 alloy at different heat-treatments($3.05\times10^{-3}s^{-1}$)

(a) 800℃/4h, WC+620℃/8h, AC; (b) 800℃/4h, WC+640℃/8h, AC; (c) 800℃/4h, WC+660℃/8h, AC

3.4 The effect of heat treatment after forging on mechanical properties

The tensile properties at room temperature after different heat treatments are shown in Tab.4. The tensile strength, ductility and shrinkage section of the TC17 alloy after the different heat treatments satisfy the technical standards. Under the system III, The yield strength is within the range of technical standards, the value of system I and system II is higher than the technical standards. It can be concluded from Tab.4, with the increase of aging temperature, the strength of the alloy decreases gradually, and the plasticity increases. When the aging temperature is rising from 620 ℃ (I) to 640℃ (II), the strength and plasticity of alloy only change a little, less than 1.1% and 5.1%, respectively. When the aging temperature is rising from 640℃ to 660 ℃ (III), the strength and plasticity of the alloy increase greatly, exceeding 7.9% and 31.7%, respectively. When the aging temperature increases, the size of secondary α phase cluster in crystal increases and distribute orientation becomes simple. The length of slip band in internal extends, the effective slip length of dislocation increases, and the extent of the direction variety decreases when the crack passes through the α cluster. So the strength reduces and the plasticity improves. What is more, with the increase of the β transus, the β matrix, BCC, has more slip flat than the α phase, and the critical stress required by the sliding surface is also lower. The atomic dense degree of BCC is lower and the cavity is easy to induce plastic deformation under shear stress. In the meantime, it also helps to reduce strength and enhance plasticity. Combining with the microstructure, it is found that the change of microstructure is more remarkable when the aging temperature increases from 640℃ to 660 ℃ than from 620℃ to 640 ℃. Therefore the change degree of the strength and plasticity is greater.

Tab.4 The room-temperature tensile properties of TC17 alloy at different heat-treatments

Heat treatment	Tensile strength σ_b/MPa		Yield strength $\sigma_{0.2}$/MPa		Elongation δ/%		Area reduction Ψ/%	
	Test	average	Test	average	Test	average	Test	average
I	1298	1296.5	1260	1259.5	11.5	10.5	22	19.5
	1295		1259		9.5		17	
II	1283	1282.5	1250	1250	11.5	11	23	20.5
	1282		1250		10.5		18	
III	1173	1176	1155	1151.5	16.0	15.0	31	27
	1179		1148		14.0		23	
technical standard	≥1120		1020-1193		≥5		≥10	

The fracture toughness results of TC17 alloy under different heat treatment are shown in Tab.5. The fracture toughness of TC17 alloy improves with the increase of aging temperature. The influence of different heat treatment on fracture toughness of TC17 is remarkable, and it has obvious regularity. The fracture toughness of the alloy meets the technical requirements only under the system III. The equiaxed α phase has stronger ability to resist crack initiation, and it is beneficial to improve fracture toughness of titanium alloy. Most of the cracks expand along the α/β interface or through α clusters in the Widmanstatten or basket-weave microstructure. Because of the different orientations of α phase cluster in β crystal, after the crack extending to the cluster boundary, it will be hindered by another α cluster if continue to expand, then have to change directions and produce branches. Thus, crack encountering clusters in different directions will cause the crack blunting and stress relaxation at its tip. It is necessary to change the direction of expansion frequently, making the path circuitous and increasing the total length of the branch and the crack. Thus, more energy need to be absorbed and fracture toughness is enhanced[8-9]. In Tab.5, fracture toughness enhances with the increase of aging temperature, closely relating to the number and size of secondary α phases and their clusters in the alloy.

Tab.5 The fracture toughness of TC17 at different heat treatments

Heat treatment	fracture toughness K_{IC}/ MPa·m$^{1/2}$
I	45.6
II	53.8
III	61.6
technical standard	≥55

Analyse the room temperature mechanical properties of alloy under three heat treatment systems, TC17 titanium alloy has preferable comprehensive mechanical properties during the heat treatment system III, that is 800°C/4h, WC+660°C/8h and air cooling. Therefore, at the strain rate of 3.05 x 10^{-3} s^{-1}, the system III can be taken as the better heat treatment system.

4. Conclusion

1. The effect of heat treatment on microstructure of TC17 alloy is mainly reflected in the difference of secondary intracrystalline α phase and β transus matrix, with the increasing of aging temperature, the size of intracrystalline secondary lamellar α phase cluster increases, orientation relationship becomes simple, and the content of β transformed structure increases.

2. The effect of aging temperature on the mechanical properties of TC17 alloy shows strong regularity, with the increasing of aging temperature, the strength of the alloy reduces, plasticity and fracture toughness increase, high cycle fatigue performance is not sensitive to the aging temperature.

3. In conclusion, TC17 under the experimental conditions of this paper, using heat treatment of 873℃/2h +918℃ /64min before forging, AC 800°C/4h,WC+660°C/8h,AC afterβ-forging, tensile properties at room temperature, fracture toughness, high and low cycle fatigue performance all meet technical requirements. The mechanical properties matches well, so heat treatment of 800°C/4h,WC+660°C/8h,AC is the better heat treatment method.

Reference

[1] Li C, Zhu M, Wang N, et al. Application of titanium alloy in airplane[J]. Chinese Journal of Rare Metals, 2009, 33(1):84-91.

[2] Leyens C,Peters M. Titanium and titanium alloys [M]. ley-VCH Verlag GmbH & Co.KGaA, 2003.

[3] Briggs R D. Tough, high-strength titanium alloys; methods of heat

treating titanium alloys: US, US8262819[P]. 2012.

[4]Yongqiang Zhang, Hongzhen Guo, Rui Liu. Microstructure and mechanical properties of isothermal forging of TC18 alloy in beta phase region [J]. Rare metal materials and EngineeringA, 2013 ,42(3):634-638.

[5] Ezugwu E 0, Bonney J, Yamane Y. An overview of the machinability of aeroengine alloys [J]. Journal of Materials Processing Technology-2003, 134 (2): 233-253.

[6] R.C. Picu, A. Majorell. Mechanical behavior of Ti-6Al-4V at high and moderate temperatures-Part II: constitutive modeling[J]. Materials Science and Engineering A, 2002, 326: 306~316.

[7] Julien Da Costa Teixeira, Benoît Appolaire, Elisabeth Aeby-Gautier, et al. Transformation kinetics and microstructures of Ti17 titanium alloy during continuous cooling [J]. Mater. Sci. Eng. A, 2007, 448: 135-145.

[8] Xiaohui Shi, Weidong Zeng, Qingyang Zhao. The effects of lamellar features on the fracture toughness of Ti-17 titanium alloy[J]. Materials Science & Engineering A, 2015, 636: 543-550.

[9] RICHARDS N L. Quantitative evaluation of fracture toughness-microstructural relationships in alpha-beta titanium alloys[J]. JMEPEG, 2004, 13: 218 -225.

https://doi.org/10.1051/matecconf/201819007003

Hot stretch forming of titanium sheet by resistance heating

Xiaoqiang Li[1*], *Hongrui* Dong[1], *Guiqiang* Guo[1], and *Dongsheng* Li[1]

[1]School of Mechanical Engineering & Automation, Beihang University, 100191 Beijing, China

Abstract. Titanium alloys are important aerospace materials due to excellent comprehensive performance, but it is difficult to form at room temperature. Electrically assisted hot stretch forming is used to form the titanium panels of aircraft. This paper studies the effect of temperature, strain, stress relaxation time and cooling rate on forming accuracy by stress relaxation test, tensile test and forming experiment. It is shown that for the influence of temperature on forming accuracy, there is no large difference among 773K, 873K, 973K. And with increase of strain, the forming accuracy becomes lower. For the influence of stress relaxation time, there is no large difference among 0min, 15mins, 30mins. And with decrease of cooling rate, the forming accuracy is improved significantly. It's concluded that when the temperature reaches a high level, temperature and stress relaxation time have no great influence on forming accuracy. And the forming accuracy is mainly influenced by strain and cooling rate. Electrically assisted hot stretch forming is an effective method to form titanium alloy. And the part with high forming accuracy can be obtained by adopting high temperature, low strain, proper stress relaxation time and low cooling rate.

Keyword: Hot stretch forming, stress relaxation, cooling rate

1 Introduction

With the growing use of composites in aircraft transport, titanium alloys are widely used in aerospace due to their high strength to weight ratio, perfect compatibility with composite materials and excellent corrosion resistance.[1] These components are typical extrude shape, which used to be made with aluminum and be formed by stretching at room temperature.[2] But titanium alloys are difficult to form due to their poor formability and large springback at room temperature. In order to improve the formability and decrease the springback, hot forming is used to form the titanium panels of aircraft. However, traditional heating method by furnace has the shortcoming of low efficiency and high consumption and requires complex equipment. So in order to improve heating efficiency and simplify experimental equipment, many other assisted heating methods are put forward, such as laser and electricity.

Among all titanium alloys, Ti-6.5Al-2Zr-1Mo-1V is a kind of near α titanium with high Al equivalent. It is used to manufacture bulkhead, panel and other important structural parts due to its attractive properties such as high specific strength, thermal stability and welding ability.[3][4] For the manufacturing of bulkhead, a new process of electrically assisted hot stretch forming is put forward, which combines traditional stretch forming with hot titanium forming technique and improves formability and forming accuracy by self-resistance heating. This technique saves material and machining time, which are two serious cost issues for today's aircraft manufacturers. And this process consists of four stages: heating sheet, stretching forming, stress relaxation and cooling, and the corresponding process parameters at each stage are temperature, strain, stress relaxation time and cooling rate respectively, as shown in Fig.1. Because bulkheads are usually slender members with curvature and forming quality of these components directly affects the assembly precision, the aerodynamic shape and the service life of aircraft, bulkheads with high quality need to be obtained.[5]

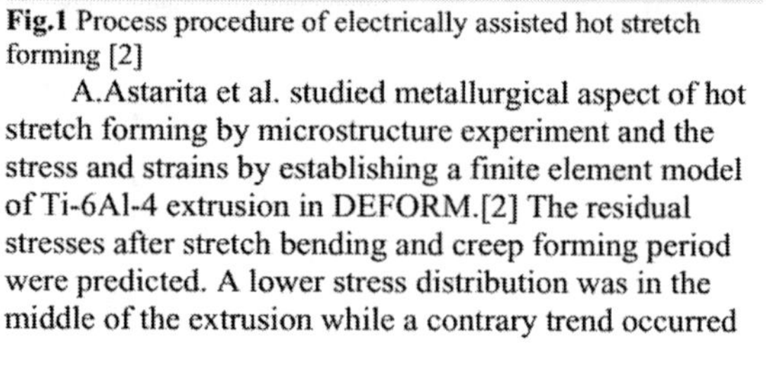

Fig.1 Process procedure of electrically assisted hot stretch forming [2]

A.Astarita et al. studied metallurgical aspect of hot stretch forming by microstructure experiment and the stress and strains by establishing a finite element model of Ti-6Al-4 extrusion in DEFORM.[2] The residual stresses after stretch bending and creep forming period were predicted. A lower stress distribution was in the middle of the extrusion while a contrary trend occurred

* Corresponding author: lixiaoqiang@buaa.edu.cn

https://doi.org/10.1051/matecconf/201819007003

on the side because of the effect of the clamps. And the stress was reduced and became more homogeneous after the creep forming period. Deng et al. created a hot stretch forming model in ABAQUS to research the effect of dwell time on spring-back of Ti-6Al-4V and OT4 extrusion during creep forming period. The results indicated that the spring-back decreased greatly with the increase of dwell time.[6] Above previous researches didn't consider the effect of strain, which has great effect on residual stress and springback. And different stages interact. For example, stress relaxation exists in the process of stretching forming. So it's necessary to study the effects of these parameters on forming accuracy and determine which parameter is the main effect factor.

This paper studies the stress relaxation and hot stretch behaviour of Ti-6.5Al-2Zr-1Mo-1V and the effects of temperature(773K, 873K, 973K), strain(0%, 2%, 4%), stress relaxation time(0min, 15mins, 30mins) and cooling rate(80K/min, 120K/min, 160K/min) on forming accuracy.

2 Experiment

2.1 Test material and designed parts

The test material is Ti-6.5Al-2Zr-1Mo-1V and its chemical composition is listed in table 1. The as-received alloy was annealed after hot and cold rolling into 1mm-thick sheet. And the geometry of stress relaxation specimen is as shown in Fig.2

Table.1 The chemical composition of Ti-6.5Al-2Zr-1Mo-1V alloy (%)

Al	Zr	Mo	V	Si	Fe	Ti
5.5-7.0	1.5-2.5	0.5-2.0	0.8-2.5	≤0.15	≤0.25	Bal

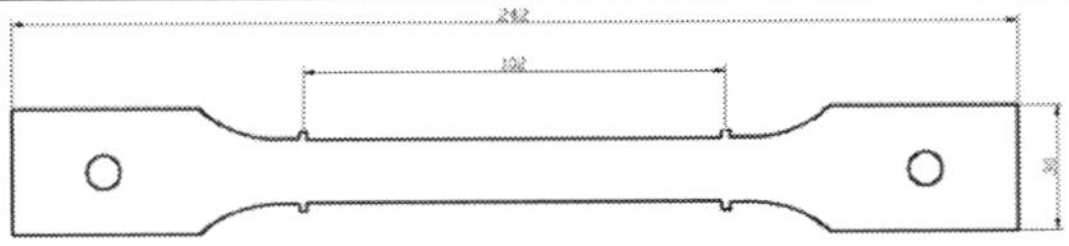

Fig.2 Geometry of stress relaxation specimen

Based on the existing die and equipment, the designed part and die are as shown Fig.3. The material of forming sheet is Ti-6.5Al-2Zr-1Mo-1V, and the material of die is self-designed lime cement plate, which can withstand a very high temperature. The corresponding dimensions are as shown in Fig.3.(a). And the diagrams of geometrical position before forming and after forming are as shown in Fig.3.(b) and Fig.3.(c) respectively.

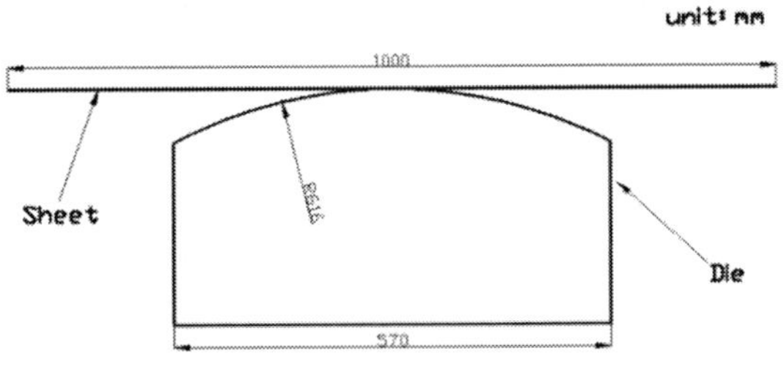

(a)

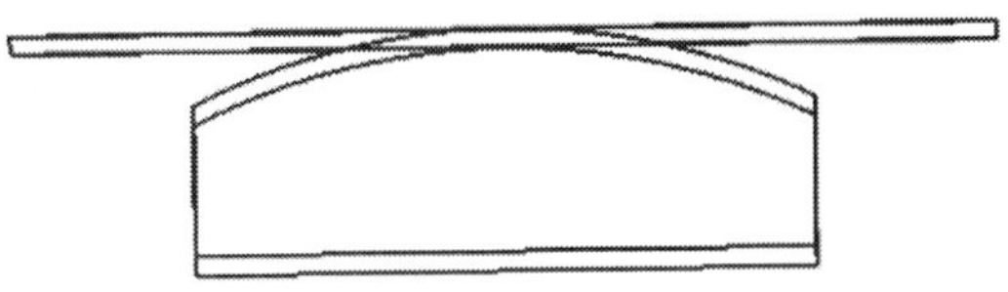

(b)

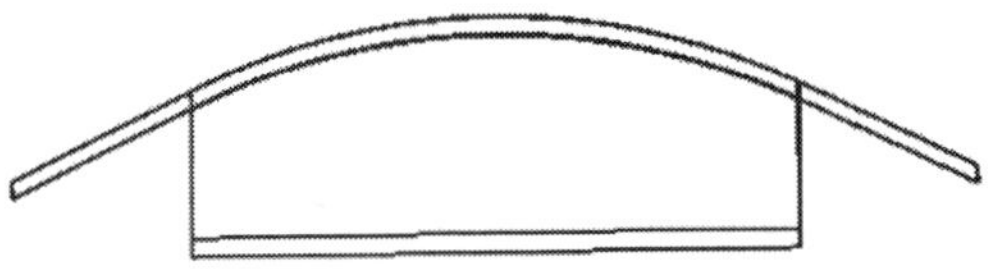

(c)

Fig.3. Schematic of designed part and die
(a).The geometry of part and die (b).The diagram of geometrical position before forming (c). The diagram of geometrical position after forming

Because the effect of strain will be studied and the stress does not always have positive correlation with strain under high temperature, it is need to study tensile behaviour simply. The geometry of tensile specimen is as shown in Fig.4.

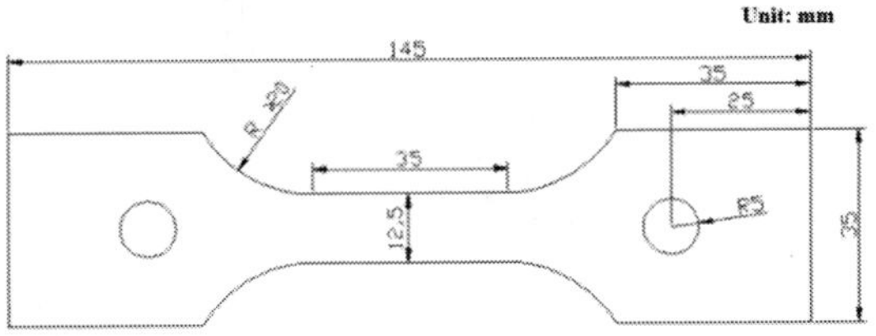

Fig.4 Geometry of tensile specimen

2.2 Experiment equipment and procedure

The stress relaxation experiments were conducted by using a Zwick/Roell Z100 electric universal test machine equipped with a heating furnace. The furnace offered a constant temperature zone with 200mm length and the possibility to control the atmosphere temperature with ± 3K precision by three independent heating zones. The designed test matrix about stress relaxation is as shown in table.2. Firstly, the specimen was heated to the desired temperature and then kept warm to make temperature uniform. Secondly, stretched the specimen to the pre strain 10%. Finally, fixed the clamp and recorded time-stress curve for 60mins.

Table.2 The designed test matrix about stress relaxation

Temperature	Pre-strain	Stress relaxation time
773K	10%	60 mins
873K	10%	60 mins
973K	10%	60 mins

In order to study the effects of temperature, strain, stress relaxation time and cooling rate on forming accuracy, the designed experiment matrix is as shown in table 3.

Table.3 Experiment matrix about electrically assisted hot stretch forming

	Temperature (K)	Strain	Stress relaxation time (min)	Cooling rate (K/min)
Effect of temperature	773	2%	15	120
	873	2%	15	120
	973	2%	15	120
Effect of strain	873	0%	15	120

https://doi.org/10.1051/matecconf/201819007003

	873	2%	15	120
	873	4%	15	120
Effect of stress relaxation time	873	2%	0	120
	873	2%	15	120
	873	2%	30	120
Effect of cooling rate	873	2%	15	80
	873	2%	15	120
	873	2%	15	160

The experiments were made with 50 tons stretching machine and an incubator, in which there is a die, as shown in Fig.5. Sheet was heated by self-resistance. The temperature was measured by thermocouple and controlled by closed loop control system.

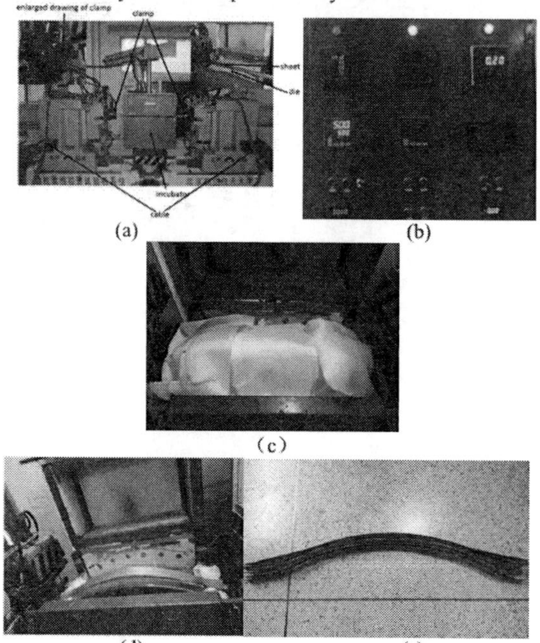

(a) (b) (c) (d) (e)

Fig.5. Diagram of stretching machine
(a). Stretching machine (b). Temperature control system (c). Incubator with insulation cotton (d). Picture in forming (e). Formed part

Firstly, the specimen was put into incubator to control temperature by closed-loop control of current and the blank was heated by electricity current to temperature 773K, 873K, 973K. During forming process, the temperatures remains constant. Secondly, the blank was stretched by the lifting of die to strain 0%, 2%, 4%. Thirdly, the stress was relaxed by fixing incubator and clamp for 0 min, 15mins, 30mins. Finally, the blank cooled at the cooling rate of 80K/min, 120K/min, 160K/min.

After the specimens were formed, the profiles of formed specimens were measured with three coordinate measuring machine, as shown in Fig.6.

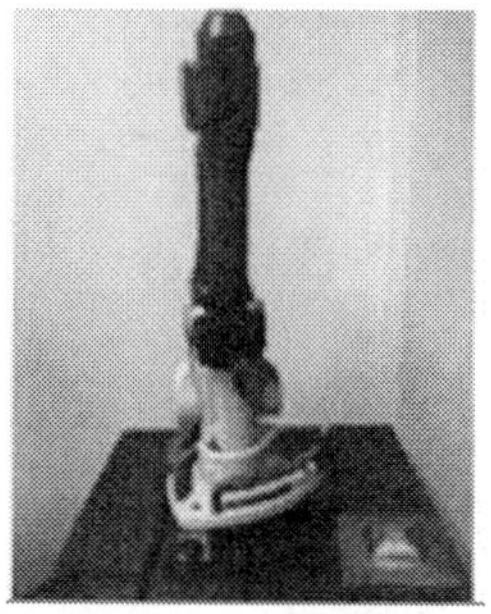

Fig.6 Diagram of three coordinate measuring machine

The tensile experiment was conducted with the same equipment as stress relaxation experiment. The designed test matrix is as shown in table.4. The specimen was heated to temperature 873K and then stretched until it cracked.

Table.4 Experiment matrix about tensile experiment

Temperature	Strain rate
873K	0.001/s

3 Result and discussion

3.1 Result and discussion of stress relaxation test

Fig.7 is recorded time-stress curve of Ti-6.5Al-2Zr-1Mo-1V at 773K, 873K, 973K in stress relaxation tests. As shown in Fig.7, the stress relaxation curve can be divided into three parts. At first, stress declines rapidly. Subsequently, the stress relaxation goes into the transition stage. Finally, the stress relaxation goes into an apparent stable stage. It can be concluded that the increase of temperature is beneficial to accelerate the decrease of stress, which is because high temperature is beneficial to accelerate dynamic recovery and dynamic recrystallization. In order to quantify stress relaxation efficiency under different experimental condition, stress relaxation ratio (SRS) is introduced, which is expressed as Eq.1.[7]

$$\Delta = \frac{\sigma}{\sigma_0} \quad (1)$$

where σ is instantaneous stress value and σ_0 is initial stress value.

Ultimate SRS values of 773K, 873K, 973K are 50.8%, 13.4%. 2% respectively. It indicates that with increase of temperature, the residual stress decreases and elastic strain can be transformed into plastic strain more thoroughly, which is helpful to reduce springback.

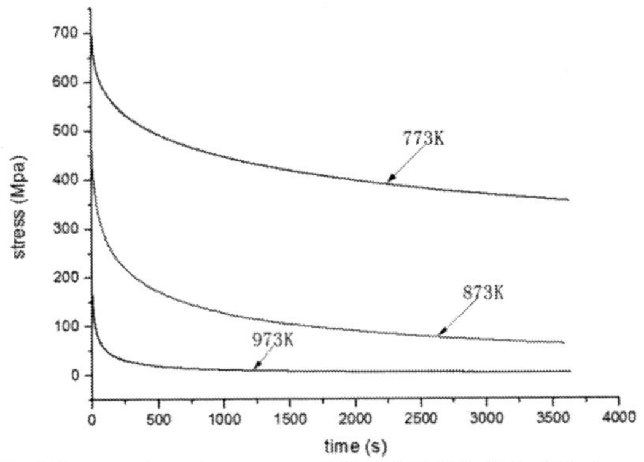

Fig.7 Stress relaxation curves of Ti-6.5Al-2Zr-1Mo-1V at 773K, 873K, 973K

3.2 Result and discussion of tensile test

Fig.8 is recorded true stress-true strain curve of Ti-6.5Al-2Zr-1Mo-1V at 873K in the process of hot tensile test. As shown in Fig.8, when strain reaches 2% and 4%, the material has got into plastic stage. And stress value of strain 4% is larger than stress value of strain 2%, which indicates that hardening effect is still dominant when the forming temperature is 873K. So for the initial stress value in the process of stress relaxation, the value of 4% is larger than the value of 2%.

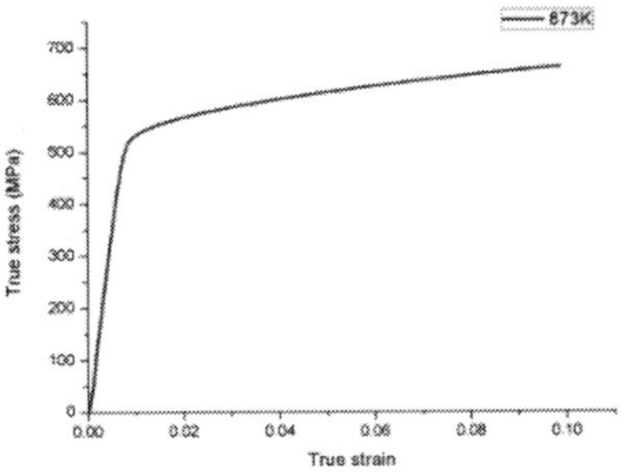

Fig.8. True stress-true strain curve of Ti-6.5Al-2Zr-1Mo-1V at 873K in the process of hot tensile test

3.3 Result and discussion of forming experiment

The profiles of formed part under different conditions are shown in Fig.9. In Fig.9, the first parameter 873 indicates temperature 873K, the second parameter 2 indicates strain 2%, the third parameter 15 indicates stress relaxation time 15mins, the last one 120 indicates cooling rate 120K/min, and they are expressed as 873-2-15-120.

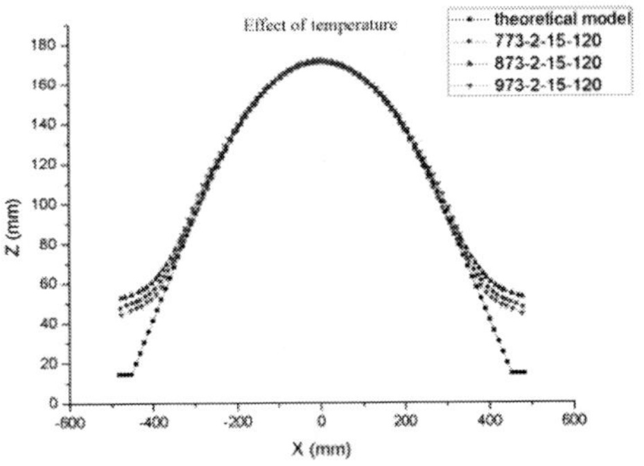

(a)

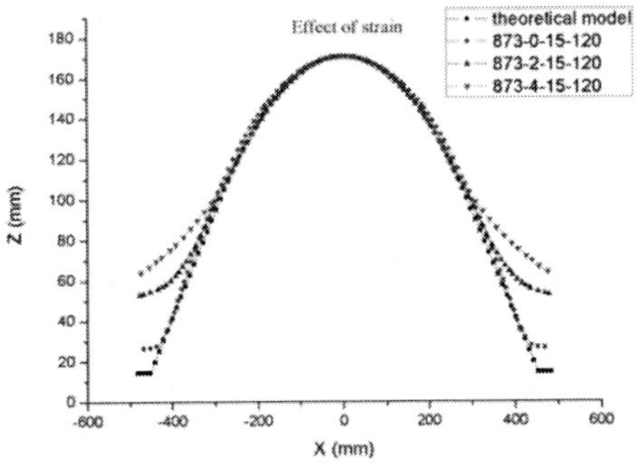

(b)

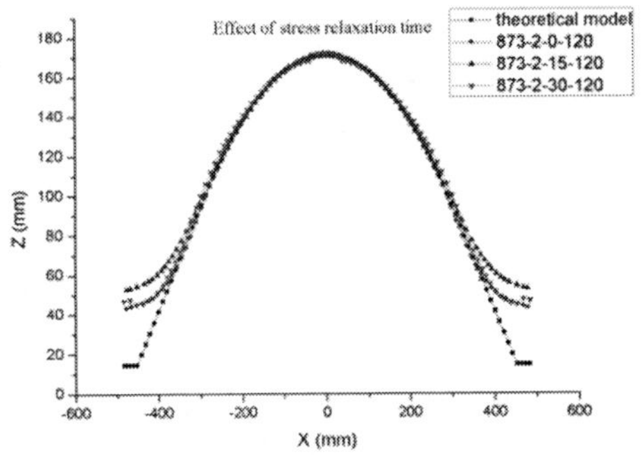

(c)

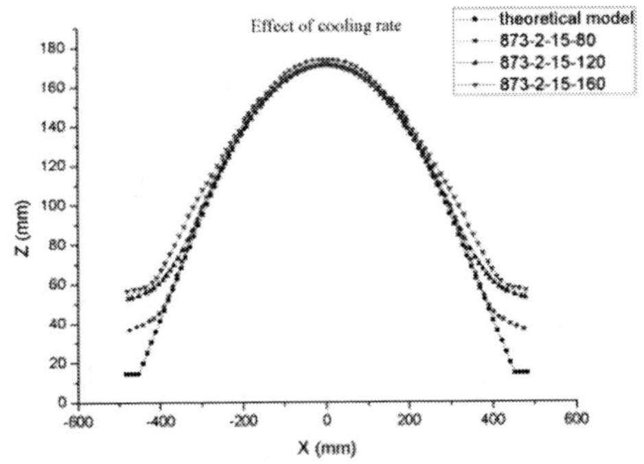

(d)

Fig.9. The profiles of formed part under different conditions
(a) The profiles under different temperature

(b) The profiles under different strain
(c) The profiles under different stress relaxation time
(d) The profiles under different cooling rate

For the effect of temperature, as shown in Fig.9(a), when stress relaxation time and cooling rate are proper, there is not large difference among 773K, 873K, 973K. It is concluded that when the temperature reaches a high level, it will have no great influence on forming accuracy of arc section. But for the whole section, forming accuracy of 973K is the highest, which is because with increase of temperature, the residual stress decreases.

For the effect of strain, as shown in Fig.9(b), with increase of strain, both of the forming accuracies of arc section and the whole section decrease. This is because with increase of strain, the corresponding stress increases after stretching, and hardening effect is dominant, which results in higher residual stress after stress relaxation and larger springback after unloading.

For the effect of stress relaxation time, as shown in Fig.9(c), when temperature, strain and cooling rate are proper, there is not large difference among 0min, 15mins, 30mins. It should have been that stress relaxation time had the largest effect on forming accuracy, which is because stress relaxation time effects the residual stress greatly, but that's not the case. This is because in the process of heating, stretching, relaxation stress and cooling, all of them effect the residual stress level. When the temperature reaches 873K and the cooling rate is controlled properly, the residual stress is close for different stress relaxation time.

For the effect of cooling rate, as shown in Fig.9(d), with increase of cooling rate, the forming accuracy decreases, which is because residual stress increases with increase of cooling rate.

4 Conclusion

1. With increase of temperature, the residual stress decreases and elastic strain can be transformed into plastic strain more thoroughly, which is helpful to reduce springback.
2. When the sheet is formed under high temperature, besides in the process of stress relaxation, there is still stress relaxation behaviour in the process of forming and cooling rate, which has great influence on the forming accuracy.
3. When the temperature reaches a high level, temperature and stress relaxation time have no great influence on forming accuracy. And when the temperature reaches a high level, the forming accuracy increases significantly with decrease of strain and cooling rate.
4. Electrically assisted hot stretch forming is an effective method to form titanium alloy which is difficult to form under room temperature. And the part with high forming accuracy can be obtained by adopting high temperature(＞773K)， low strain, proper stress relaxation time and low cooling rate.

References

1. M. Peters, C. Leyens, *Titanium and Titanium Alloys - Fundamentals and Applications* (2003)
2. A. Astarita, E. Armentani, E. Ceretti, L. Giorleo， P. Mastrilli, Key. Eng. Mater, **554-557**,647-656 (2003)
3. P. Yao, P. Li, C. Li, K. Xue, G. Gan, Chin. J. Rare. Met, **39,** 967-974 (2015)
4. W. Cheng, H. Yang, X. Fan, Z. Sun, Trans. Nonferrous Met. Soc. China, **21**. 1963-1969 (2011)
5. G. Guo, D. Li, X. Li, et al, Int. J. Adv. Manuf. Technol, **1–2**,1-13 (2017).
6. T. Deng, D. Li,, X. Li, P. Ding, K. Zhao, P. I. Mech. Eng. B-J. Eng, **230**, 505-516 (2016)
7. Y. Liu, J. Zhu, Mech Mater, **40**, 792–795 (2008)

In-situ measurement technologies for flow behaviour and contact status of fibre-reinforced thermoplastics during forming with a six degree of freedom press

Birk Wonnenberg[1,*], *Franz* Dietrich[1], and *Klaus* Dröder[1]

[1]Institute of Machine Tools and Production Technology, Technische Universität Braunschweig, 38106 Braunschweig, Germany

Abstract. The paper presents investigations on a forming process, which is implemented in a forming press based on a Stewart platform. In contrast to common forming techniques, this buildup offers not only one but six degrees of freedom. This is of particular interest when it comes to the forming of materials that show significant anisotropic behaviour such as fibre-reinforced plastics. Therefore, an experimental setup is presented to record characteristic variables during the forming process of fibre-reinforced thermoplastics. The contact state is of particular interest for this kind of forming process because it changes continuously in shape and position as the forming process progresses. For this purpose, temperatures at different places in the tool are recorded to provide information about the flow velocity of the material and the contact state between tool and workpiece. This allows the determination of the exact time and position of the contact between material and forming tool as well as the duration of this contact. The results are compared with optical measurements analysed by image processing algorithm and process forces measured by load cells.

Keywords: Measurement, Press, Flow

1 Introduction

Forming processes with more than one degree of freedom in tool path have the ability to generate local material characteristics. An example for a machine, which can generate these forming processes and used in the experiments, is given in [1, 2]. This machine is based on a Stewart platform, which has six degrees of freedom within the tool movement path. These additional degrees of freedom in the pressing movement provide a broad range of feasible process parameters. For instance, the tool geometry can be manufactured in shapes depending on the tool path, which result in different forming processes and material properties.

The anisotropic behaviour of certain materials is of particular interest for forming with a tool path more than one degree of freedom. It is proposed, that with the additional degree of freedom in the tool path a material specific forming process can generate beneficial properties in anisotropic materials. To generate these properties an understanding of the process is necessary and therefore a measurement technique to observe the process.

The scope of the paper is a measurement technique to detect the flow behaviour and contact state during forming. In particular, an in-situ measurement technique for hot forming of fibre-reinforced thermoplastic to detect the contact state between the material and the tool. The contact state has significant impact on the heat distribution and therefore on the viscosity of the material, which can be derived from the measurement.

To develop and test the measurement techniques, a simplified process within an experimental setup is needed. In the forming process, the material is pressed in a shape similar to a plate. The simplicity of this geometry facilitates process observation. The forming process is depicted in the progress from left to right in figure 1. In this process, the upper tool moves towards the material, pressing it onto the lower tool. During the process, the upper tool rotates around the tool centre point, so that a rolling movement is generated. This tool movement leads to a material flow, which should be observed in the forming process.

In the following sections, the material properties and sensor concepts in other scientific works are presented, followed by a description of the experiments.

2 Material behaviour in the forming process and a proposal for a measurement technique

In this section, the material properties are described. From this description, observable properties are derived and candidates for the measurement techniques are evaluated regarding the applicability in the experiments.

* Corresponding author: b.wonnenberg@tu-braunschweig.de

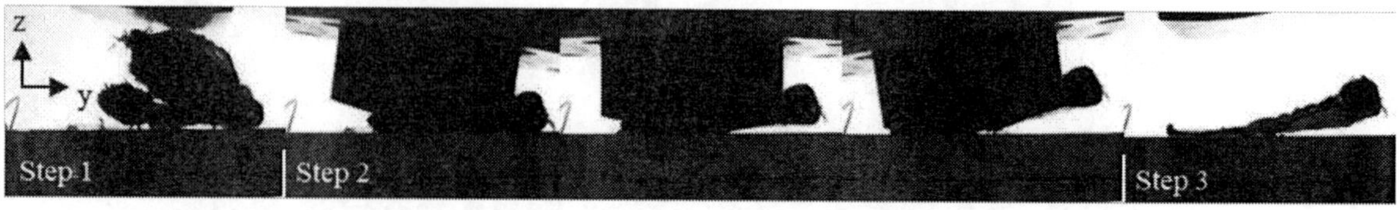

Fig. 1. Lateral view on the forming process in the progress from insertion of the material to finish (left to right)

The developed measurement technique used in the experiments shall help the understanding of the process. This understanding can provide a basis for an empirical description of the process. For the experiments, as example for an anisotropic material glass-fibre-reinforced thermoplastic with a polypropylene matrix reinforced with a chopped fibre content of 30% was used.

In order to observe the process in the experiments, an understanding of the material is important; an overview is given in [3]. The flow behaviour and the material properties are directly related to the material temperature. Therefore, the heat change is of great importance processing the material. The material flow within forming tools is described in figure 2 [3]. While pressing, the material flow velocity has a gradient towards the middle of the cavity. As a consequence, warm material is transported from the middle to the surface. This process is dependent on the viscosity of the material, which is correlated with the temperature.

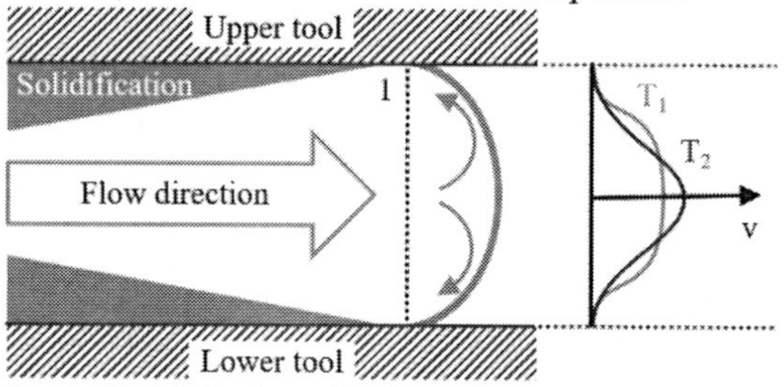

Fig. 2. Flow behaviour of molten thermoplastic in a forming process; (left) feeding current, flow front and solidification between the forming tools; (right) flow velocity profile at (1) in respect to temperature ($T_1 < T_2$) [3]

From the information given in [3] the flow behaviour of the material in the forming process can be derived from the flow front and the heat flux to the tool. The work cited shows that it is possible to detect the flow front by presence of the material in a certain area optically, but not the contact state. The contact zone is not clearly defined in an optical perspective because of small gaps between the tool and the material. However, the contact zone can be determined by the heat flux to the tool. In a warm state, the matrix is melted and the material is soft and permeable. The heat flux from the material to the tool is relative low without pressure. If the forming process presses the material, it becomes solid and the heat flux rises.

With the description of the material, a measurement concept can be created. A starting point for the concept creation is the evaluation of other research work, which concerns the detection of material and the flow behaviour within forming processes.

In [4] closed-loop control of product properties in metal forming are reviewed by Polyblank et al. and an overview of sensors to observe the forming process are presented. From this overview thermocouples, load cells and a camera were chosen to observe the process, due to its simplicity of use. [4]

Apart from the use in other research work, the utilization of a camera in [5] to observe a forming process is a promising approach. Shimizu et al. [5] used a forming tool with transparent flank to observe the processes. In this work, the wear due to the contact between tool and sheet is examined. This setup from [5] could be the approach for the next attempt, if the experimental setup in this paper proves to be insufficient.

Hamedon et al. [6] used a borescope to observe the forming process, for instance sheet metal forming. The borescope was used in the process within the forming tool. However, in [6] this optical measurement was used to detect the edge of a bended metal sheet with a high contrast to the surrounding and a free line of sight within the forming tool.

3 Experimental setup and experimental procedure

In this section, the experimental setup with the forming press, tool and sensors are developed and evaluated with a preliminary test. In the experiments, a material flow in a plain open forming tool was examined with regard to material state. The material state can be derived from the contact conditions between the material and the tool surface. While, the contact condition can be determined by the heat flux from the material into the tool. This shall reveal that the material state within the process can be determined by thermal sensors.

3.1 Experimental setup

The six degrees of freedom press used for the experiments is a hydraulically actuated Stewart platform. [1] The press is designed with a minimal cylindrical workspace of 100 mm in diameter and 50mm in height and can rotate the tool 15° around the x and y-axis as well as 5° around the z-axis. The machine is designed to press with 50 kN in z-axis and 5 kN in x- and y-axis. Load cells are installed on each limp to record forces in the process.

The press was equipped with a tool, which was prepared with thermocouples. The positions of the thermocouples are based on the work of Yoneyama and Tozawa [7]. The thermocouples in [7] are placed in two different depth below the tool surface and measure the

* Corresponding author: b.wonnenberg@tu-braunschweig.de

temperature within the tool. With the tool temperature in two depth, the heat flux can be calculated. In addition, the surface temperature of the tool can be calculated with an inverse heat conduction.

The thermocouples are placed on and under the surface of the lower tool. Four sensors are inserted in holes from below. These holes vary in depth relatively to the upper surface of the lower tool. Sensors 3_2 and 4_2 are 2 mm and sensors 3_3 and 4_3 are 4 mm below the surface. Sensors 1_1 to 4_1 are placed on the tool surface, in top view sensor 3_1 is in line with sensors 3_2 and 3_3, whereas sensor 4_1 is in line with sensors 4_2 and 4_3. Sensors 1_1 and 2_1 are on the insertion area of the material. In figure 3, the position of the sensors is shown, sensor 2_1 is in the centre of the lower tool, sensor 1_1 is 25 mm to the right side and sensors 3_1, 3_2 and 3_3 are 5 mm to the left side, seen from the perspective of the camera. Sensors 4_1, 4_2 and 4_3 are 10 mm to the left side from the tool centre.

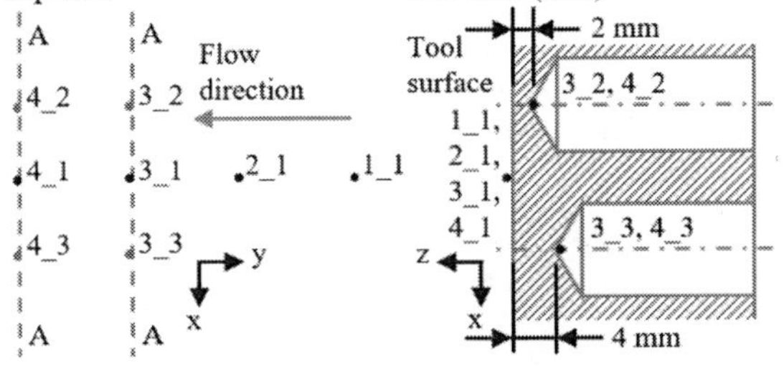

Fig. 3. Position of thermocouples; (left) top view with the flow direction of the material while pressing and position of sensors 1_1 to 4_3; (right) side view with the boreholes with 2 mm remaining material thickness above sensors 3_2 and 4_2 and 4 mm above sensors 3_3 and 4_3

Figure 4 shows the set up for a preliminary test of the sensors within a forming tool. The sensors were tested by applying ethylene-vinyl acetate (EVA) with a temperature of 160 °C on the tool surface. The EVA was applied in one step to get a synchronous heat input.

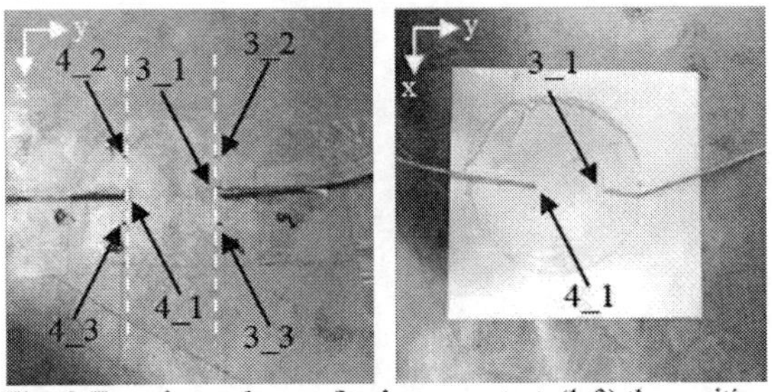

Fig. 4. Experimental setup for the sensor test, (left) the position of the sensors 3_1 to 3_3 and 4_1 to 4_3 on the lower tool; (right) distribution of EVA after testing and removing from the tool surface

3.2 Experimental procedure

The experiment procedure was subdivided into a heating phase of the material, an insertion of the material into the press and a pressing phase. In the first phase, the material was melted in a convection oven at 220 °C for 10 min while the tools were kept at room temperature. In the second phase, the material was inserted into the press and on the surface of the lower tool. For the last phase, a pressing process was chosen with three forming steps. The first step is a vertical movement (z-axis) of the upper tool towards the material with a constant angle. In a second step, a rolling movement rotates the tool (A-rotation axis) and in a third step the tool is lifted again. The other axis were kept constant without displacement over the process. The movement is displayed in figure 5.

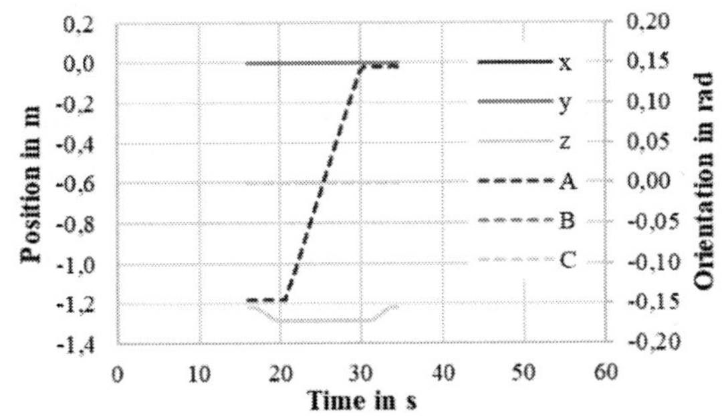

Fig. 5. Motion path of the forming tool calculated from internal position sensor readings of the press

In order to obtain diverse observations of the material flow, the experiments were separated into two cases. The cases vary according to the insertion position of the material into the process. The insertion position influences the forming results. In case A the material is placed on the centre of the forming tool during insertion. In case B the position of the material on the forming tool is shifted further to the right side. The positions have a huge impact on the flow behaviour in the forming process, which can be easily seen in figure 6. Both figures show the process after the first forming step, in which the upper tool moves towards the lower tool in z-axis. On the left side of figure 6, the material is pressed towards the left side and fills the gap between upper and lower tool entirely. On the right side of figure 6, which shows the case with the insertion position shifted to the right side, the material fills only three quarters of the same gap. After the first forming step in case A sensors 1_1 to 4_1 and in case B sensors 1_1 to 3_1 are in contact with the material.

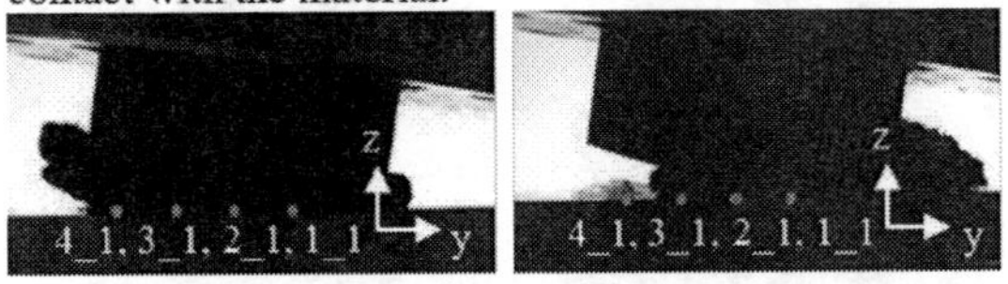

Fig. 6. Side view on the material distribution in case A (left) and case B (right) after the first forming step (tool movement in z-axis); Sensors 1_1 to 4_1 are marked with points

3.3 Image processing

The process in the open tool was observed with a camera to detect the flow front of the material. In order to measure the positon of the flow front image processing algorithm were used to search for the furthest right and left edge of the material. The results from the image processing are displayed in figure 7. An edge detection filters the tool and material surface. By shape recognition, the upper and lower tool are detected and removed from the image. The remainder is the material in the forming process, coloured in blue in figure 7. The farthest left (1) and right (2) edge are detected and the positions are plotted in relation to the lower tool centre.

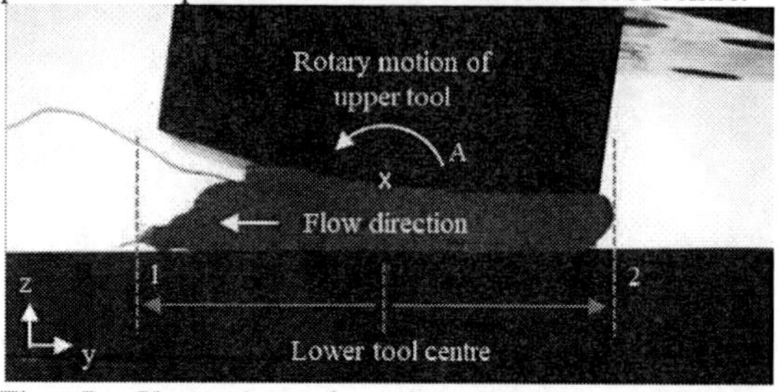

Fig. 7. Observation of a hot forming process with superimposed rolling motion (A) in the open die (moulding compound is coloured in blue after the detection in the image processing); (1) is the farthest left and (2) the farthest right edge of the material measured from the lower tool centre

4 Results

This section describes the results of the preliminary test and the measurements of case A and B. The temperature, force and the results from the image processing are displayed in figure 9. Distinctive points in the graphs are marked by numbers to simplify the discussion of the results. The results show a correspondence between heat distribution within the tool and the material distribution on the surface of the tool, measured by thermocouples and a camera with image processing.

4.1 Preliminary test

In figure 8, the temperature profiles of the preliminary test are displayed. With the application of EVA on the surface of the tool, the temperature increases at each sensor. At (1) the profile of sensors 3_2, 3_3, 4_2 and 4_3 rise nearly instantaneously after the increase at sensors 3_1 and 4_1 but with a smaller gradient. Sensors 3_1 and 4_1 are attached to the surface of the lower tool and thus reach the highest values at (3), 98 and 78 °C. The peak values of sensors 3_2, 3_3, 4_2 and 4_3 are 25, 24, 26 and 23 °C. 10 s after the application at (2) sensors 3_2, 3_3, 4_2 and 4_3 show the respectively same spread as before the application.

4.2 Results from case A and B

The results of the experiments are displayed in figure 9. Figure 9 displays six diagrams and is separated in two columns for cases A and B. Diagrams a) and b) show the thermal measurement of cases A and B respectively. Diagrams c) and d) show the forces measured in cases A and B. Diagrams e) and f) show results from the image processing with the location and velocity of the flow front in cases A and B.

The time scale of the following diagrams are adjusted and synchronised, so that at 20 s the first forming step is finished and at 30 s the second forming step is finished.

For case A, diagram a) shows the temperature profile. The initial temperature for all sensors is 19 °C. Sensors 1_1, 2_1 and 3_1 detect a first increase (1a) of the temperature with the insertion of the material. The second increase (4a) is caused by the first forming step, when the material is compacted over sensors 1_1, 2_1 and 3_1. Sensor 4_1 was not covered by the material during insertion, but the first increase of the temperature is induced by the first forming step, which can be explained by the left side of figure 6. Material is pushed onto the section where sensor 4_1 is attached which leads to the temperature increase that is measured by sensor 4_1. The sensors under the surface of the tool (3_2, 3_3, 4_2, 4_3) detect the first temperature increase directly after the first forming step.

The measured peak temperature in case A are in order 2_1, 1_1, 3_1, 4_1 for the surface sensors, starting at 56 °C for sensor 2_1 (4a), 65 °C for sensor 1_1 (5a), 77 °C for sensor 3_1 (6a) to 117 °C for sensor 4_1 (7a). Under the surface, sensors 3_2, 3_3, 4_2 and 4_3 have almost equal peak values of 26 °C at (3a), only sensor 4_2 detects 25 °C.

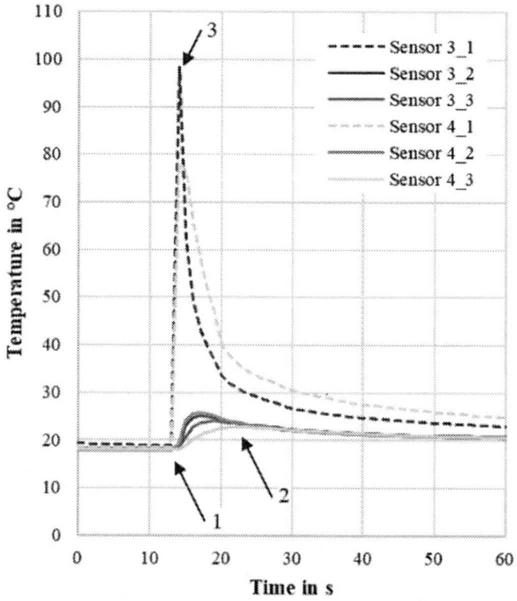

Fig. 8. Temperature profile of the sensors tested with EVA heated to 160 °C; at (1) the measured temperature of sensors 3_1 and 4_1 rise to a peak temperature of 98 and 78 °C; between (1) and (2) sensors 3_2, 3_3, 4_2 and 4_3 have a peak at 25, 24, 26 and 23 °C

In case B the measured temperatures are presented in diagram b) of figure 9. The initial temperature for all sensors is 20 °C. Sensors 1_1 and 2_1 detect the first temperature increase with the insertion of the material (1b) and the second after the first forming step (5b). In case B sensor 3_1 measures a temperature increase after the first forming step (2b). Sensor 4_1 detects the first rise at the end of the second forming step (3b).

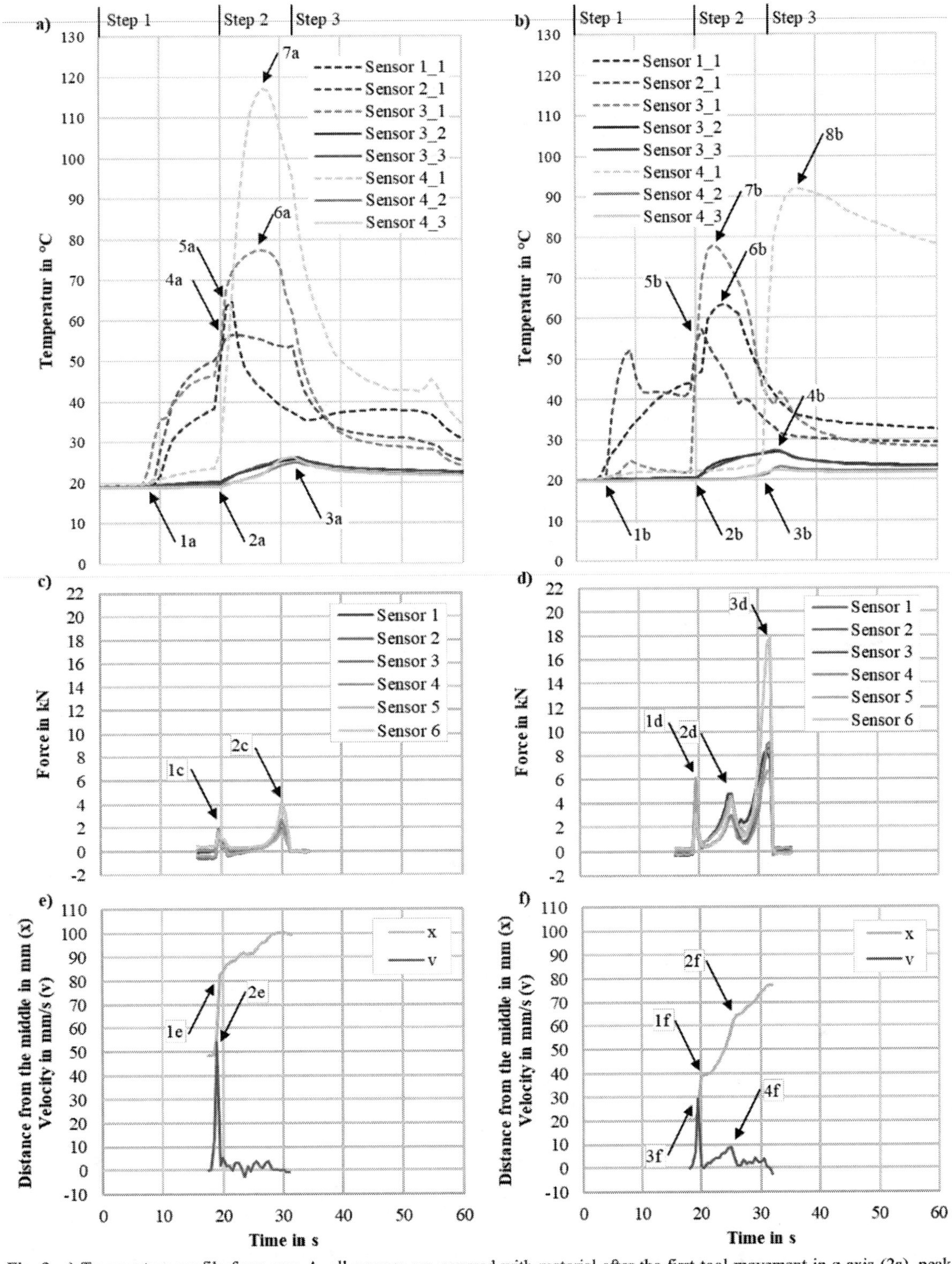

Fig. 9. a) Temperature profile from case A, all sensors are covered with material after the first tool movement in z-axis (2a), peak temperature of sensors 1_1 to 4_3: 65, 56, 77, 26, 26, 117, 25, 26 °C (3a to 7a); **b)** Temperature profile from case B, only sensors 1_1 to 3_3 are covered with material after the first tool movement in z-axis (2b), peak temperature of sensors 1_1 to 4_3: 63, 57, 78, 27, 27, 92, 23, 22 °C (4b to 8b); **c)** Case A, force signals from load cells above the upper tool; **d)** Case B, force signals from load cells above the upper tool; **e)** Movement of the flow front to the left side measured optically from the centre of the lower tool in case A; **f)** Movement of the flow front to the left side measured optically from the centre of the lower tool in case B

Similar to sensor 3_1, sensors 3_2 and 3_3 measure the first increase of the temperature under the surface after the first forming step (2b). After the second forming step sensors 4_2 and 4_3 detect a temperature rise concurrent with the surface sensor 4_1 at (3b). The measured temperature increases after the first step, which matches with the right side of figure 6. With the first step, material is pushed onto sensor 1_1 to 3_1, while the section with sensor 4_1 is still without contact to the material.

The peak temperature in case B are in order 2_1, 1_1, 3_1, 4_1 for the surface sensors, were sensor 2_1 has a peak at 57 °C (5b), sensor 1_1 at 63 °C (6b), sensor 3_1 at 78 °C (7b) and sensor 4_1 has a peak at 92 °C (8b). Under the surface, sensors 3_2 and 3_3 show almost equal temperatures at 27 °C (4b) and 4_2 and 4_3 have lower peak values at 23 °C and 22 °C (4b).

In parts c) and d) in figure 9 the force signals from the load cells are displayed. Two force peaks were measured at the end of the first (1c, 1d) and second forming step (2c, 3d). Case B exhibits a third force peak in the middle the second forming step (2d). This is due to a movement of the upper tool, which was adjusted with a small step in z-direction towards the lower tool to achieve a constant material thickness.

Cases A and B were observed with a camera. With an image processing, the flow front in these experiments was measured optically. The results of this measurement are presented in the following section.

In case A, the left edge of the material located at about 50 mm left from the centre of the lower tool after the insertion. With the first forming step the left flow front moved to 88 mm (1e) with a peak velocity of about 50 mm/s (2e). Due to the second forming step the flow front moved to 100 mm.

In case B, the material was only 20 mm left from the lower tool centre after insertion. The variant position of the material after insertion was part of the experimental design. After the first forming step, the flow front moved to 39 mm (1f) with a peak velocity of about 30 mm/s (3f). In the middle of second forming step, another peak in the velocity of about 10 mm/s (2f, 4f) was detected. After the second forming step, the flow front was 77 mm left from the lower tool centre.

5 Discussion

The discussion sets the results of the measurements into context with each other. Furthermore, the results are compared to literature.

The preliminary test showed that thermocouples on and under the surface are able to detect increases of the surface temperature at the same time. The peak temperature between a sensor on the surface and under the surface differ between 52 to 75 °C. In [7] the surface temperature was calculated by using the inverse heat conduction problem using two thermocouples in two different depth. In the experiments, conducted in [7], the specimen was made out of aluminium and had a higher initial temperature. The outcome is a steep temperature gradient in the tool. In this paper, the signal difference, 3 °C, in preliminary test is too little, in respect to the measurement inaccuracy, ±1,5 °C form the thermocouple type K class 1 [8] and ±0,5 °C form the measuring instrument, to calculate a surface temperature. The same applies to the case A and case B, see figure 9 (3a, 3b, 4b). To calculate the surface temperature of the tool surface or the heat flux the measurement accuracy has to be higher.

The contact state between material and tool in the forming process can be detected by thermocouples. Even sensors under the surface of the tool can detected a temperature increase, if warm material is pressed on the tool surface during the forming process. Material, which only lays on the surface and has no contact to the upper tool, does interfere with thermocouples on the surface but not with those under the surface. This is due to the pressure dependent thermal conductance between the material and the tool. In figure 9 diagram d) displays the amount of pressing force used in case B. At (2d) the pressing force has a peak, which has no influence on the temperature profile of sensors under the surface (3_2 to 4_3) between (2b) and (3b, 4b). This indicates that, after a compaction a contact between material and tool is enough for a detectable heat flow, while the amount of pressing force has no significant influence on the detection of the temperature increase or absolute value.

In both cases A and B a highest temperature peak was measured with surface sensor 4_1 (7a, 8b), which is further away from the insertion area than sensors 1_1 to 3_1. This can be explained by the flow behaviour of the material, see figure 2. While pressing warm material flows from the core to the surface. Although the surface temperature is higher, the measured tool temperature at this point is colder (3b) than close to the insertion area (4b). This leads to the conclusion that measuring the temperature under the tool surface with thermocouples cannot give significant data about the material temperature without knowing the distribution of the material and the heat distribution within the tool at this point. The material distribution on the other hand can be determined with thermocouples under the tool surface.

In figure 9 diagram c) and d) display the profile of the process force measured by the load cells. The load cells show peak values at specific points in the pressing process. At the end of the first (1c, 1d) and second step (2c, 3d) a peak was detected. For itself, the measurement is not significant enough to interpret the flow behaviour, as the reaction force is influenced by other parameters, for instance the material viscosity. However, these values can be compared to simulations and therefore should be recorded for further studies.

With the help of figure 9 diagrams e) and f) it can be shown, that the insertion position of the material has a huge impact on the material flow during the forming process with a six-degree of freedom tool path. The initial positions differ 30 mm from each other. This difference leads to an almost filled gap between the upper and the lower tool in case A after the first forming step (1e). In contrast to this, the gap between the forming tools in case B is only filled after the second forming step due to the different material flow (2f). The progress of the material flow in cases A and B corresponds with the temperature profiles of the thermocouples under the surface. Therefore, it is possible to detect the flow front in the process with the thermocouples.

With the presented measurement technique, the forming process and the flow of the material can be observed. The observations show that the results of the forming process depend strongly on the flow behaviour of the material and therefore on the location of the material insertion.

6 Conclusion

Hot forming with a six degree of freedom forming press provides a broader range of process parameters than linear pressing movements. For instance, the tool motion path can induce shearing in the material or localised pressure conditions. Furthermore, the forming tools can be manufactured with less constraints concerning the geometrical freedom. The combined influence of the tool geometry and the motion path on the forming process can direct the flow of the material. The forming result depends on the insertion position of the material and the specific material flow in each procedure.

To determine the material state and the flow direction in the forming process, it is proposed to use the flow front and the contact state of the material in certain areas. To detect the flow front and the contact state measurement techniques are necessary and an experimental setup is developed for these techniques. The purpose measuring techniques are tested concerning the detection of the flow front and contact state of the material within the tool.

The basis of the experimental set up was a six degree of freedom forming press. The press was equipped with an open tool to observe the process optically. In addition, the tool was equipped with thermocouples on and under the surface. With the help of these thermocouples, the heat from the hot forming of fibre-reinforced thermoplastic was measured. The experiment is split into two cases with different insertion places of the material and thus different flow behaviour to demonstrate the usability of the measurement technique.

Thermocouples and load cells are utilized to observe the process state. In contrast to the presented experiments, an optical observation in closed tools is difficult realize, as long as the tool is not transparent.

With the presence of warm material pressed on the surface of the tool an increase of the temperature can be detected with thermocouples. For this purpose, the thermocouples can be positioned on or under the tool surface. The load cell measurements allow the recording of the process force and the detection of specific forming steps in the process. The presence of the material in a certain area can be captured optically with a camera. With image processing, the flow front can be detected and the flow velocity can be calculated.

Thus, in this state, the measuring technique with thermocouples is not suitable for closed loop control of the process, yet. The acquired information is a punctual measurement with a signal that only has two states: contact and no contact. For continuous control of the process this measurement techniques must be enhanced in resolution and significance of the signal, which will be the subject of further research.

The distance between thermocouples and the surface of the tool has little effect on the measured temperature. In this process with these parameters, it is not possible to calculate neither the heat flux nor a reliable estimation of the surface temperature. Therefore, the depth of two thermocouples must differ more widely or a higher measurement accuracy is needed for this experimental setup.

As a concluding remark, this measurement technique allows to detect the flow front of a warm material within the forming tool. The information can be used as a reference for simulations or for the observation of the forming process in production for quality reasons. The information from a punctual measurement has a lower resolution than an optical measurement, but it can be utilized in closed tools and give an information whether the material is in direct contact or not, which is difficult to determine optically.

This work was supported by the German Research Foundation (DFG DR 973/4-1)

References

1. J. Hesselbach, B.-A. Behrens, F. Dietrich, S. Rathmann, J. Poelmeyer, Prod. Eng. Res. Devel., *Flexible forming with hexapods*, **1** 429–436 (2007)
2. F. Dietrich, *Non - linear modelling of hydraulically actuated production machines using optimised experiments* (Vulkan, Essen, 2014)
3. A. Radtke, *Steifigkeitsberechnung von diskontinuierlich faserverstärkten Thermoplasten auf der Basis von Faserorientierungs- und Faserlängenverteilungen* (Fraunhofer IRB-Verl., Stuttgart, 2009)
4. J.A. Polyblank, J.M. Allwood, S.R. Duncan, J. Mater. Process. Technol., *Closed-loop control of product properties in metal forming: A review and prospectus*, **214** 2333–2348 (2014)
5. T. Shimizu, T. Kakegawa, M. Yang, Y. Qin, T.A. Dean, J. Lin, S.J. Yuan, F. Vollertsen, MATEC web conf., *Development of in-situ observation system of dynamic contact interface between dies and materials during microforming operation*, **21** 9008 (2015)
6. Z. Hamedon, K. Mori, Y. Abe, J. Mater. Process. Technol., *In-situ measurement of three-dimensional deformation behaviour of sheet and tools during stamping using borescope*, **214** 945–950 (2014)
7. T. Yoneyama, Y. Tozawa, CIRP Ann. Manuf. Technol., *Direct Measurement of Stress and Heat between Work and Tool in Metal Forming*, **39** 219–222 (1990)
8. DIN EN 60584-1, *Thermoelemente - Teil 1: Thermospannungen und Grenzabweichungen* (IEC 60584-1:2013), (2014)

Feasibility study of a novel hot stamping process for Ti6Al4V alloy

Mateusz Kopec[1, 3], *Kehuan Wang*[1], *Yaoqi Wang*[2], *Liliang Wang*[1*], *and Jianguo Lin*[1]

[1] Department of Mechanical Engineering, Imperial College London, London SW7 2AZ, UK
[2] Beijing Aeronautical Manufacturing Technology Research Institute,Beijing 100024, China
[3] Institute of Fundamental Technological Research Polish Academy of Sciences, 02-106 Warsaw, Pawinskiego 5B, Poland

Abstract. To investigate the feasibility of a novel hot stamping process for the Ti6Al4V titanium alloy using low temperature forming tools, mechanical properties of the material were studied using hot tensile tests at a temperature range of 600 - 900°C with a constant strain rate of $1s^{-1}$. Hot stamping tests were carried out to verify the feasibility of this technology and identify the forming window for the material. Results show that when the deformation temperature was lower than 700°C, the amount of elongation was less than 20%, and it also had little change with the temperature. However, when the temperature was higher than 700°C, a good ductility of the material can be achieved. During the forming tests, parts failed at lower temperatures (600°C) due to the limited formability and also failed at higher temperatures (950°C) due to the phase transformation. The post-form hardness firstly decreased with the temperature increasing due to recovery and then increased due to the phase transformation. Qualified parts were formed successfully between temperatures of 750 - 850°C, which indicates that this new technology has a great potential in forming titanium alloys sheet components.

Key words: titanium, hot stamping, metal forming

1 Introduction

In order to meet fuel consumption targets for the aviation sector, increasing demands for low density and high strength materials were observed. Due to applications of such lightweight materials in aircraft vehicles, a considerable weight reduction could be achieved [1]. Therefore, titanium alloys are found in aircraft applications because of the combination of weight, strength, corrosion resistance and temperature stability [2]. However, titanium alloys are considered as difficult-to-form materials thus the forming of complex-shaped components is time, energy and cost intensive. Only a few conventional techniques could enable aircraft companies to produce qualified complex-shaped components. Superplastic forming is definitely one of the leading titanium forming techniques. Other known techniques including superplastic forming with diffusion bonding, hot stretch forming, creep forming, hot gas-pressure forming or isothermal hot forming [2]. It is unquestionable that components formed by these techniques have a high quality, however these components are achieved under conditions of a very high temperature, slow strain rate and simultaneous heating of tools and sheet during the process. As a result, the low efficiency of the forming process and simultaneous high production cost are observed. Recent development of forming technologies such as the solution heat treatment, forming and in-die quenching (HFQ) [3], the Quick-plastic forming [4], and the hot stamping using rapid heating [5] could lead to the increase of productivity and become an alternative to conventionally used techniques. Hamedon et al. [6] proposed a hot stamping process to form titanium components from sheet metal utilizing cold dies and resistance heating. This solution led to the reduction of the tool wear and is more efficient and economical in comparison to conventionally used isothermal hot forming techniques. However, this technology was validated by forming a simple U-shaped component under resistance heating condition.

In order to further study the feasibility of hot stamping process for titanium alloys under furnace heating condition, the hot stamping tests of Ti6Al4V titanium alloy utilizing low temperature forming tools (used at

* Corresponding author: liliang.wang@imperial.ac.uk

room temperature) and a hot blank was carried out. To investigate the formability of Ti6Al4V alloy, the uniaxial tensile tests were performed at temperatures ranging from 600°C to 900°C with constant strain rate of 1 s^{-1}. Ti6Al4V components were successfully formed at temperatures ranging from of 750°C to 850°C. Microstructure of formed components was characterized by Scanning Electron Microscope (SEM).

It was found that this new technology has a great potential in forming titanium alloys sheet components and a qualified part could be formed successfully. Proposed hot stamping process to form components from sheet metal using cold dies, promises to be a more efficient and economical process than the traditionally used isothermal hot forming techniques [6]. Moreover the post-form mechanical properties and microstructure can be tailored by the adjustment of heating temperature and application of low temperature forming.

2 Experimental details

2.1 Uniaxial tensile test and microstructure examination

Ti6Al4V titanium alloy sheet with initial thickness of 1.5 mm is used throughout the research. A Gleeble 3800 thermo-mechanical testing machine was used to characterize the properties of material through uniaxial tensile test at elevated temperatures. Dog-bone shaped specimens with a gauge length of 46 mm, a width of 12 mm and a thickness of 1.5 mm were used. During the tests, flat samples were heated up with the heating rate of 2°C/s to the testing temperatures ranging from 600°C to 900°C and deformed. After tensile tests, samples were air quenched immediately. Hardness of the samples after hot forming tests was measured using Vickers hardness tester. Hardness tests were carried out at room temperature by applying 10kgf load on indenter for 10 seconds. Microstructures were examined via scanning electron microscopy (SEM) using backscattered electron (BSE) mode. Metallographic specimens were prepared by standard mechanical polishing method.

2.2 Hot forming test

The forming tests on the titanium alloy were conducted using a 250 tonne Instron press (Fig.1). Rectangular samples of Ti6Al4V with dimension of 90 x 8 x 1.5 mm were first heated in the furnace to the target temperatures ranging from 600°C to 950°C.Temperature of the samples was monitored using a thermocouple wire attached to them. After reaching the target temperature, the sample was quickly placed in the tool for 10 seconds, which was stamped at a speed of 10 mm/s when the press was activated. The formed part was held in the low temperature die after forming to quench it to room temperature.

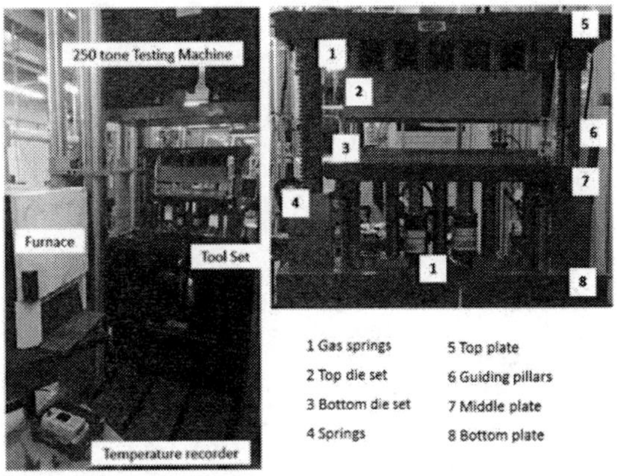

Fig. 1. Testing setup.

3 Results & discussion

3.1 Effect of heating temperature and strain rate on mechanical properties of Ti6Al4V alloy

In order to determine the best forming condition, tensile tests were conducted at temperatures ranging from 600°C to 900°C with a constant strain rate of 1s^{-1} (Fig.2). Simultaneous increase of elongation with the temperature was observed. Strain hardening could be found during deformation at temperatures ranging from 600°C to 700°C, but only slight change in elongation was found. However, over the temperature of 700°C, material softening and subsequent increase of elongation from 20% to 60% at 900°C were observed (Fig.3). High temperatures enhance dislocation mobility resulting in ductility increase during deformation. It was concluded that the investigated titanium alloy has a good formability over the temperature of 700°C.

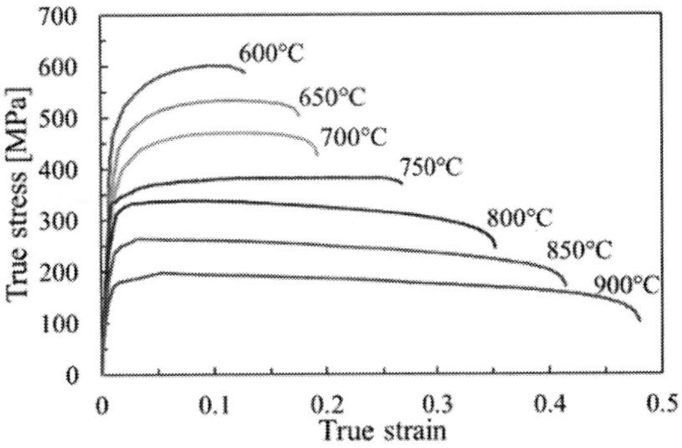

Fig. 2. Representative flow stress - strain curves of Ti6Al4V titanium samples tested under the range of 600 - 900°C.

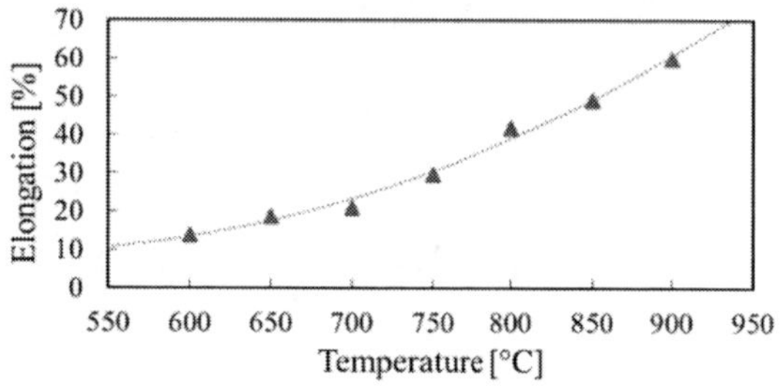

Fig. 3. Variations of elongation of Ti6Al4V titanium alloy with heating temperature obtained by tensile test.

Satisfactory elongation of investigated material (>30%) was observed during testing at temperatures ranging from 750°C to 900°C. In order to investigate the effect of strain rate on material property, tensile tests were carried out at temperatures of 750°C - 850°C, strain rates of $0.1s^{-1}$ - $5s^{-1}$ and the same heating rate of 2°C/s. Evolution of elongation with the different strain rates was presented in Fig.4. It could be observed that the elongation increased with the increasing temperature from 0.2 - 0.35 at 750°C to 0.5 - 0.6 at 850°C and decreasing strain rate from 0.2 at strain rate of $5s^{-1}$ to 0.35 at strain rate of $0.1s^{-1}$ (at testing temperature of 750°C). The same tendency is observed for temperatures of 800°C and 850°C.

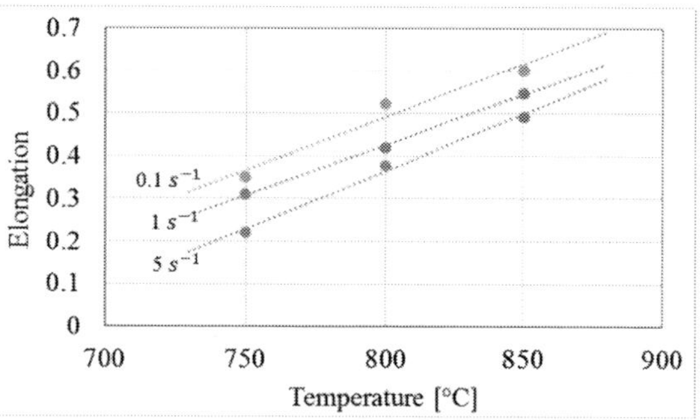

Fig. 4. Elongation evolution of Ti6Al4V titanium samples tested under different strain rates at temperatures ranging from 750 - 850°C.

3.2 Hot forming of Ti6Al4V

During the hot stamping of titanium alloys, three main stages may be distinguished: heating, transferring, and forming and quenching (Fig.5*)*. The effective control of temperature at each stage is required since the temperature affects not only the formability of the material during deformation but also the post-form properties of the formed component. In order to monitor the temperature evolution during the whole process, a thermocouple wire was attached to the specimen. At the first stage, titanium sample was heated to the targeted temperature and then soaked within 180 s, until the temperature became stable. Then, titanium sample was placed in the tool, stamped by low temperature dies and quenched to room temperature within tool.

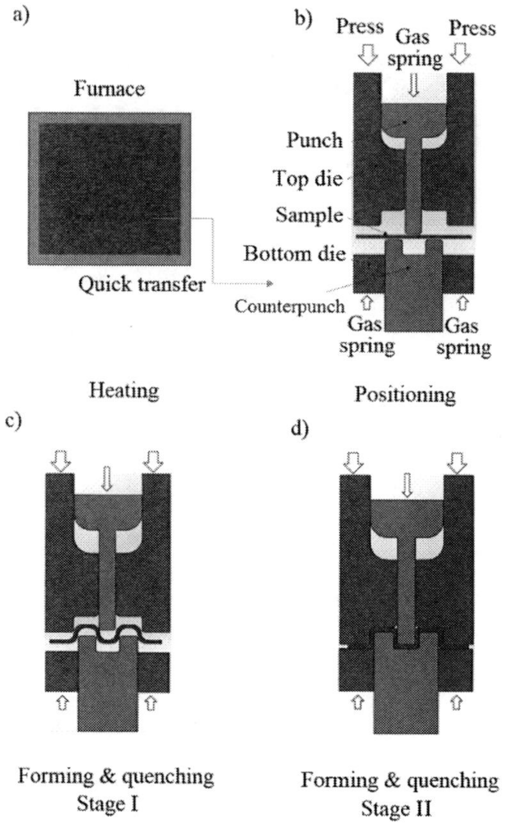

Fig. 5. Schema of the forming process during heating of the sample (a), at the positioning stage (b), during forming (c) and at the final stage (d).

Ti6Al4V components formed at temperatures ranging from 600°C to 950°C on hydraulic press utilizing low temperature forming tools are shown in Fig.6a-e. However, parts were successfully formed only in the temperature range of 750°C to 850°C. Macroscopic observations reveal that sample formed at 600°C has cracks on the surface. The occurrence of cracks is probably caused by the limited ductility of the material (Fig.7a). Evident cracks were also observed in feature formed at 900°C as shown in Fig.7*b*. As the heating temperature reached 950°C, the forming also failed resulting in the fracture of the sample into pieces (Fig.6e). It was found that during the uniaxial tensile test, elongation increased with the temperature increasing. However, the uniaxial tensile tests were performed at isothermal conditions. During actual forming, there is a considerable drop of temperature during the transferring and the forming. Because the microstructure and mechanical properties of material are strongly temperature dependent, such a drop of temperature led to decrease of formability.

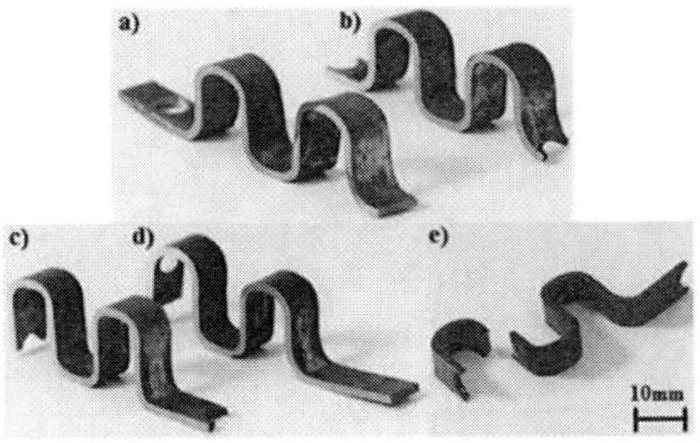

Fig. 6. View of parts formed at 600°C (a), 750°C (b), 850°C (c), 900°C (d), and 950°C (e).

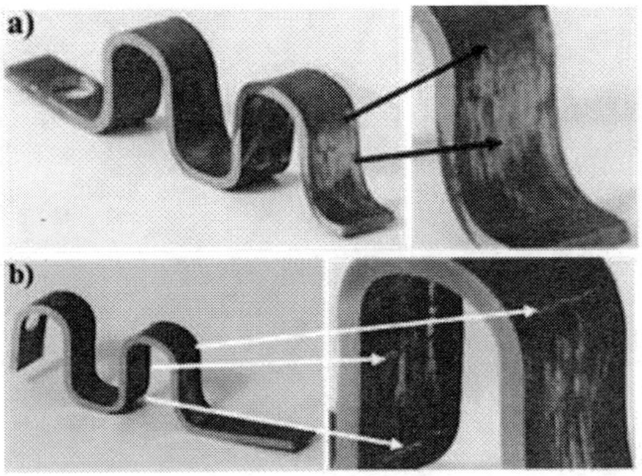

Fig. 7. View of cracks of parts formed at 600°C (a) and 900°C (b).

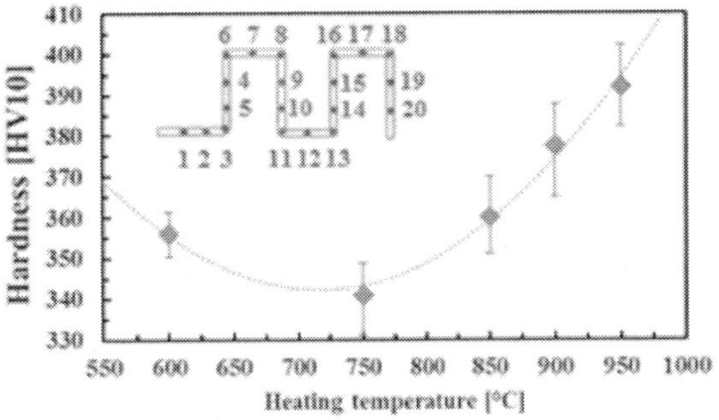

Fig. 8. Effect of heating temperature on post-form strength of formed components.

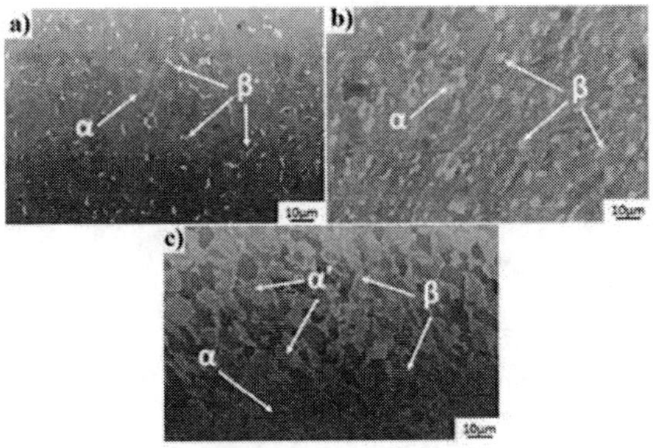

Fig. 9. Microstructures of the Ti6Al4V alloy formed at 600°C (a), 850°C (b), and 950°C (c).

3.4 Effect of hot stamping conditions on post-form properties of the formed components

In order to investigate the effect of heating temperature on the post-form hardness and microstructure of the as-formed part, hardness tests and SEM observations were conducted on shaped components. Hardness measurements were performed in the longitudinal direction of formed parts as it shown in Fig.8. For each part, 20 indentations were used for calculation of an average hardness value. It could be observed that the hardness values firstly decreased from 356HV10 to 341HV10 at 750°C, and then increased to 392HV10 at 950°C (Fig.8). It was concluded that during forming at temperatures up to 950°C, phase transformation occurred and the content of β-phase increased significantly as shown in Fig.9. As the sample was cooled within low temperature dies, transformed beta phase and secondary α‘ phase formed, which improved the hardness of the material at the room temperature (Fig.8). The fraction of transformed beta and α‘ phase increased with the temperature (Fig.9c), resulting in the simultaneous hardness increase [7, 8]. The increasing fraction of transformed beta and α‘ phase formed during the forming process may also decrease the formability of the material, which resulted in the forming failure at temperatures ranging from 900 to 950°C [9].

4 Conclusions

In this paper, an innovative forming technology to shape Ti6Al4V titanium alloy utilizing low temperature forming tools and a hot blank was studied through uniaxial tensile test and hot stamping process. The uniaxial tensile properties of Ti6Al4V alloy at temperatures from 600°C to 900°C with the strain rate of $1s^{-1}$ have been obtained. The results of tensile tests showed that the investigated material has a good formability over the testing temperature of 700°C. It was observed that during deformation at temperatures from 600°C to 700°C, work hardening occurred. As the temperature increase to 750°C, the hardening rate decreased and the flow stress curve became flat due to recovery. With the temperature increase to 900°C, an obvious material softening during the tensile test occurred. Ti6Al4Vcomponents were successfully formed at temperatures ranging from 750°C to 850°C. The hardness of formed parts firstly decreased from 357HV10 at 600°C to 341HV10 at 750°C due to recovery and then increased to 392HV10 at 950°C as a result of phase transformation. It was found that through the adjustment of the heating temperature, the post-form mechanical properties can be tailored, which indicates that this new technology has a great potential in forming titanium alloys sheet components.

The strong support from the Aviation Industry Corporation of China (AVIC) Beijing Aeronautical Manufacturing Technology Research Institute (BAMTRI) for this funded research is much appreciated. The research was performed at the AVIC Centre for Structural Design and Manufacture at Imperial College London.

References

1. M. Peters, J. Kumpfert, C. Ward, C. Leyen, Adv. Eng. Mater. **5** , 419 (2003)
2. S. Semiation, e. al., *Metalworking: Sheet Forming - Forming of titanium and titanium alloys* (ASM International), **14b**, 656 (2006)
3. L. Wang, T. Dean, J. Lin, *Proceedings of the 3rd International Conference on Advanced High Strength Steel and Press Hardening (ICHSU2016)*, W. Scient. 289 (2017)
4. P. Bariani, S. Bruschi, A. Ghiotti, F. Michieletto, CIRP Annals, **62**, 251 (2013)
5. T. Maeno, K. Mori, R. Yachi, CIRP Annals, **66**, 269 (2017)
6. Z. Hamedon, K. Mori, T. Maeno, Y. Yamashita, NMSTU, **5**, 12 (2013)
7. S. Jadhav e. al, IOP Conf. Ser.: Mater. Sci. Eng. **201**, 1 (2017)
8. T. Furuhara, B. Poorganji, H. Abe, T. Maki, JOM, **59,** 64 (2007)
9. A. Rollett, F. Humphreys, G. Rohrer, M. Hatherly, *Recrystallization and related annealing phenomena* (Elsevier, 2004).

Numerical study of the induction heating of aluminium sheets for hot stamping

Yankang Tian[1], *Libo* Wang[2], *Gerald* Anyasodor[1] and *Yi* Qin[1,*]

[1]Centre for Precision Manufacturing, Department of DMEM, The University of Strathclyde, Glasgow, G1 1XJ, UK
[2]School of Manufacturing Sciences and Engineering, Sichuan University, Chengdu, Sichuan, China

* Corresponding author: qin.yi@strath.ac.uk

Abstract. Induction heating is one of the most popular metal heating technologies because of its high heating rate and high energy efficiency. This method is suitable for heating workpieces/blanks in different shapes, sizes and materials. Induction heating of metal sheets has been investigated by various research organizations and industrial companies. However, information concerning the induction heating of aluminium blanks is limited. Further, investigations were required by industries to address the issues relating to the uneven temperature distributions developed in the metal sheets so that an optimized design could be developed to help the enhancement of the technology. Aiming at the study of the induction heating process for hot stamping, especially the temperature distribution developed in the aluminium sheets, this paper presents in-depth analysis of induction heating using 3D FE simulations, involving uses of DEFORM and COMSOL. Different coil arrangements, level of powers, frequencies, cycle times, etc. have been modelled and simulated to examine their effects on the heating efficiency and developed temperature profiles in the Aluminium sheets. It is revealed from the simulations that design features in the induction coils like shapes of cross-sections and angles of coil corners affect the uniformity of the developed temperatures in the metal sheets. Heating with an optimized combination of the coil design and the power setting could help to achieve higher heating rates and temperature uniformity. Nevertheless, the application could be constrained by some practical factors.

Keywords: Hot stamping, Temperature, Simulation

1 Introduction

In a hot stamping process, a blank is heated up to a desired temperature before being transferred to the press and then formed. During the last decade, due to dramatic development of the automotive industry, demand for hot stamping of high strength material has been increasing, which also raises new requirements for the heating process. In order to enable the hot stamping process for various applications, several metal-blanks heating technologies have been developed and used in industry.

Convection furnace heating is the most commonly used heating method due to its simplicity and ease for handling metal sheets of different materials, shapes, and sizes. Moreover, the temperature of the blank can be controlled and held very easily, and it is suitable for mass production. Nevertheless, a convection heating process usually has a very low heating rate and requires a longer cycle time. A furnace system which needs to be warmed up before production usually has low energy efficiency.

Resistance heating is another popular metal-blank heating method with high heating rates, high energy efficiency, and it is cost effective. The process may require the blanks to have regular shapes and even thicknesses. Temperature uniformity of the workpiece depends largely on the geometry of the blanks and where the electrodes are applied.

A method based on induction heating may have a good balance between process flexibility and energy efficiency [1]–[3]. A wide range of materials can be heated up by the induction heating method while a fast heating rate can be achieved. Metal sheets in different shapes and sizes can be heated by induction heating. This method is energy efficient since the heat is generated from inside the workpiece. Induction heating has already been widely used for surface heat treatment minimizing heat treatment distortion and obtaining favourable compressive residual stresses [1], [2], [4].

 https://doi.org/10.1051/matecconf/201819008002

Induction heating uses an alternating electromagnetic field to heat the electrically conductive materials without physical contacts [5], [6]. The material is heated up due to Faraday's law. Temperature distribution in the workpiece and the heating rate induced by induction heating are determined mainly by the workpiece material, coil design, power of the alternating current and other electrical parameters. Induction heating of metal sheets has been investigated by various research institutions and industrial companies. However, information concerning the induction heating of Al-alloy blanks for hot stamping is limited. Further, investigations were required by industries to address the issues relating to the uneven temperature distributions developed in the metal sheets so that an optimized design could be developed to help enhancing the technology.

A series of numerical simulations have, therefore, been conducted to study the induction heating process with reference to different kinds of setups and heating parameters. In this paper, the models developed will be described firstly, followed by presenting the numeric results. The focus is on the factors that influence inhomogeneous temperature distributions in the blanks and possible methods for reducing the temperature differences in the blanks. Several recommendations on deploying induction heating for aluminium-sheet hot stamping applications will also be given.

2 Models for the numerical study of induction heating

Induction heating of Al-alloy blanks was studied using 3D simulations. The numerical studies involved simulating heating processes with different coil shapes/sizes, positioning, blank sizes, power parameters, as well as motions of the blanks. This section describes material properties and set-ups for the numerical models.

2.1 Material properties

Table 1 Material Properties

Parts	Material	Specific Heat Capacity $(J/kg \cdot k)$	Emissivity	Electrical Resistivity $(\Omega \cdot m)$
Blank	Al-alloy	935	0.75	3.8e-05
Coil	Copper	386	1	1.67e-05

Table 2 Temperature-dependent behaviour of Al-alloy

Al-alloy sheet					
T (°c)	Thermal Expansion	T (°C)	Thermal Conductivity $(W/(m \cdot K))$	T (°C)	Mass Density (kg/m^3)
20	2.18e-05	50	280	20	2697.5
200	2.18e-05	100	220	250	2697.5
250	2.21e-05	150	200	300	2687.5
300	2.26e-05	200	200	350	2677.5
350	2.32e-05	250	200	400	2667.5
400	2.37e-05	300	200	450	2657.5
450	2.42e-05	600	200	500	2647.5
500	2.48e-05			550	2637.5
550	2.51e-05			600	2632.5
600	2.56e-05			700	2632.5

Table 1 gives the material properties of the workpiece and coil used in the simulations of induction heating. Thermal expansion, thermal conductivity and mass density of aluminium alloy were considered as temperature-dependent properties and are listed in Table 2. Material properties were kept to be consistent in all simulations.

2.2 Design of induction coil

The designs of the induction coils in this study involved design of the cross-section shapes and arrangement of coils used for induction heating.

Cross-section shape of an induction coil is a very important design aspect. It needs to be considered carefully during the design of an induction heating process. Due to the electrical resistance of the coil material, heat is generated when high electric current is passing through the coil. If no cooling is deployed, the temperature in the coil will increase continuously. Circulated water cooling is usually adopted for transferring the generated heat away and hence, reducing the temperature in the coil.

When an alternative current is applied on a cylindrical conductor, the highest current density is observed at the inner side of the conductor [7]. This phenomenon should be considered when choosing the suitable cross-section shape for the induction coil. For induction heating of a cylindrical surface, cylindrical coil design can take the advantage of such phenomenon. While for induction heating of a metal sheet, a coil with rectangular cross-section can be placed closer to the specimen and generate more even eddy current and heating [7].

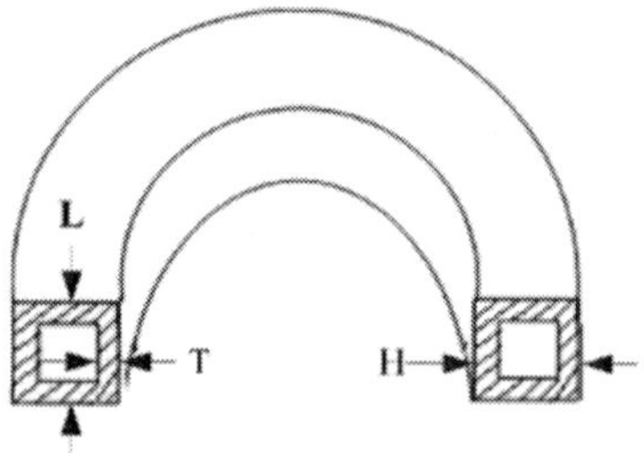

Figure 1 Cross-section of induction coil

The induction coil with hollow rectangular cross-section is chosen in this study for induction heating of shaped aluminium-alloy blanks. As shown in Figure 1, the length L, height H and wall thickness T are three key-parameters for the coil design. Cold water flows through the channel inside the coil for cooling. Size of the channel influences the flow rate of cooling water.

Wall thickness of the coil was chosen to be 1.2 times of the effective skin-depth δ. The effective skin-depth for copper coil was $\delta = 1.53$ mm and $\delta = 0.78$ mm respectively, when the applied power frequencies were 1800 Hz and 7000 Hz. The design wall thickness T was chosen as 2.0 mm. Based on the requirement for heating power, design length L and height H were both chosen to be 8 mm.

Two different coil arrangements, namely longitudinal flux induction heating (LFH) and transverse flux induction heating (TFH), as shown in Figure 2, were considered for induction heating of shaped Al-alloy sheets. Performance of these two designs was evaluated by using coupled electromagnetic-thermal simulations.

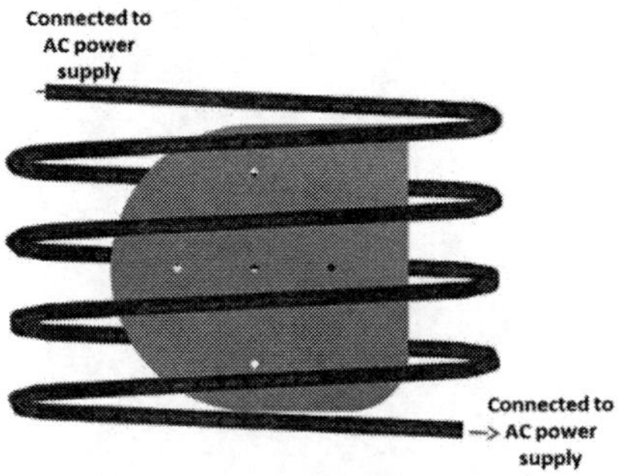

Longitudinal flux induction heating (LFH)

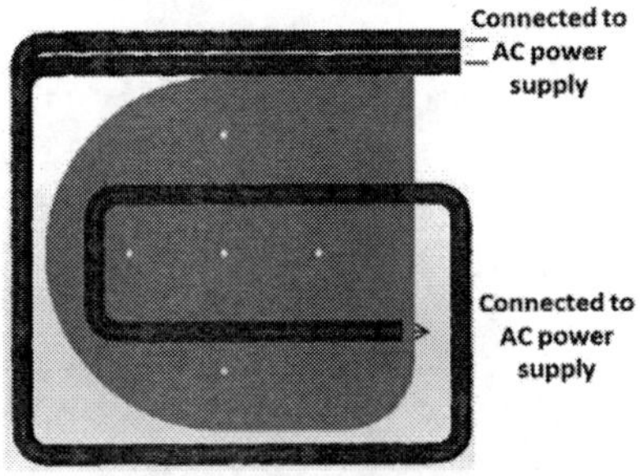

Transverse flux induction heating (TFH)

Figure 2 Two coil arrangements

2.3 Induction heating of blanks in different sizes

Induction heating of Al-alloy blanks in different sizes and shapes were studied numerically. Figure 3 shows the geometry of the shaped Al-alloy blank used for induction heating. Shape of the blank is not regular and it has several holes in it.

Induction heating of Al-alloy blanks of 400 mm × 600 mm × 2 mm was also studied with FE simulations. Two different induction heating methods were considered. One was to use a stationary coil and a moving blank which meets the workpiece from one end firstly. Coils with different shapes of corners were used in the simulations to study the effect of coil shapes. These simulations were conducted with Finite Element (FE) software DEFORM. This software can be used to solve coupled electromagnetic – thermal problems. Figure 4 shows the engineering drawing used in these models for simulations.

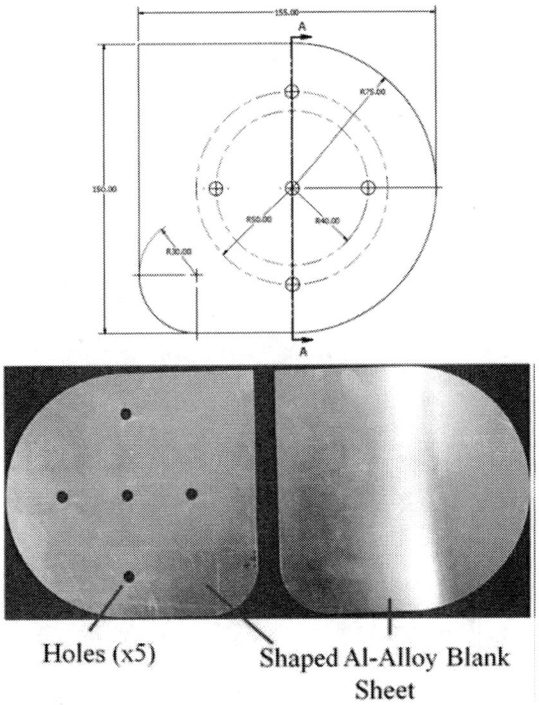

Figure 3 Geometry of small-sized Al-alloy blank

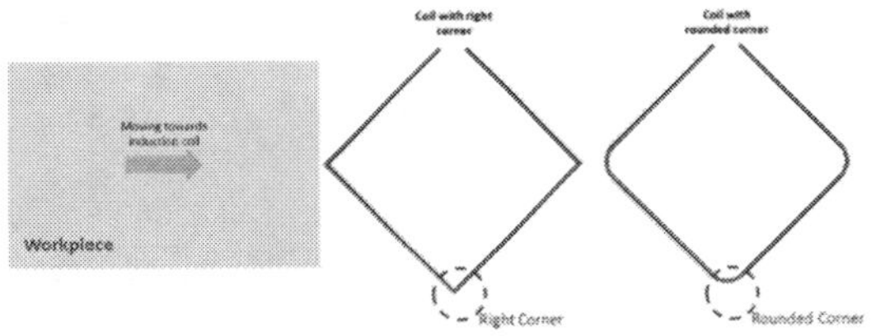

Figure 4 Model of induction heating with motion on the blank

After the investigation into shape effect of the coil, an improved design of the induction coil was developed and the heating performance of this design was also evaluated. Figure 5 shows the geometry of an improved induction coil design.

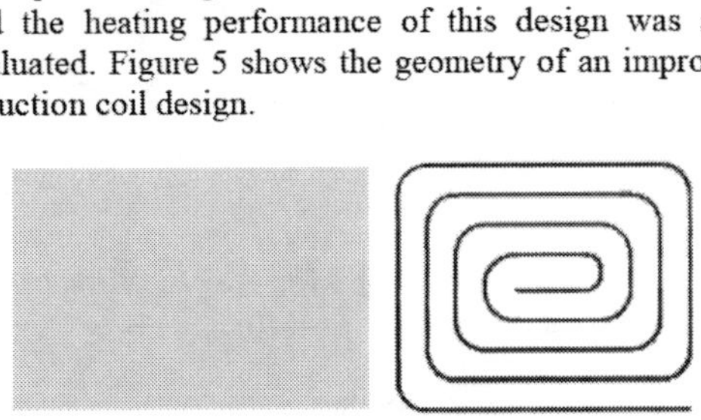

Figure 5 Induction heating with alternative coil design

The 2nd method used stationary blank and arrangement of induction coils which cover the surfaced

https://doi.org/10.1051/matecconf/201819008002

area of the workpiece. Parametric simulations of different power frequencies were conducted using FE software COMSOL Multiphysics. AC/DC module of the software was used to study induction heating process. Eddy currents and heat transfer were coupled in this model. 3D model presented at Figure 6 illustrates the arrangement of coils and workpiece used for the parametric simulations. The workpiece was placed in the middle between the upper and lower coil sets. Table 3 lists the details of conducted simulations in this study.

Figure 6 Model of induction heating with coil arrangement

Table 3 List of details of conducted simulations

No.	Blank Size	Coil Feature	Current Density (A/mm^2)	Current Frequency (Hz)	Blank Motion (mm/s)
1	Small	Transverse (LFH)	54	5000	N/A
2	Small	Longitudinal (TFH)	54	5000	N/A
3	Large	TFH, right corner	14	2000	50
4	Large	TFH, rounded corner	14	2000	50
5	Large	TFH, rounded corner	14	2000	50
6	Large	TFH, rounded corner	27	5000,10000,15000	N/A

3 Results

3.1 Induction heating of small Al-alloy blanks (Simulation No. 1 and No. 2)

Figure 7 shows the temperature in the Al-alloy sheet in a transverse flux induction heating. TFH induction heating is highly efficient. Within 8 seconds after the heating started, the maximum nodal temperature reached 525.0 °C. However, severe edge effect was observed. From the beginning of the heating process, temperature at the centre of the sheet was always lower than the temperature at the edge. Higher current density occurred near the edge of the workpiece. Temperature rises at the edge area was joule heating caused by the eddy current while the temperature increment at the sheet centre was contributed by weak induction heating and thermal conduction. Temperature difference in the workpiece induced by the edge effect reaches as high as 470.0 °C at 7 second.

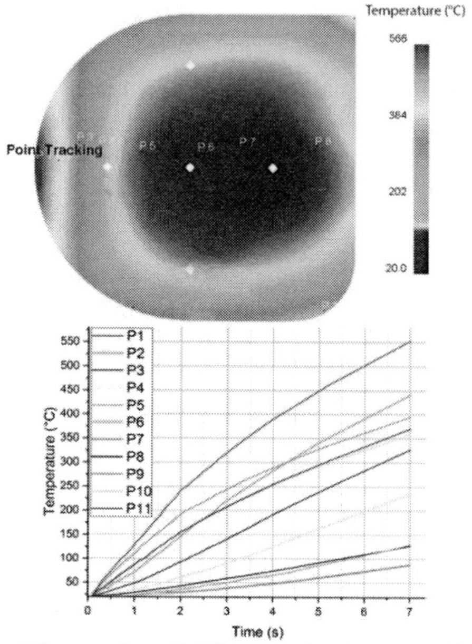

Figure 7 Temperature distribution using transverse flux induction heating in 8 seconds (simulation No. 1)

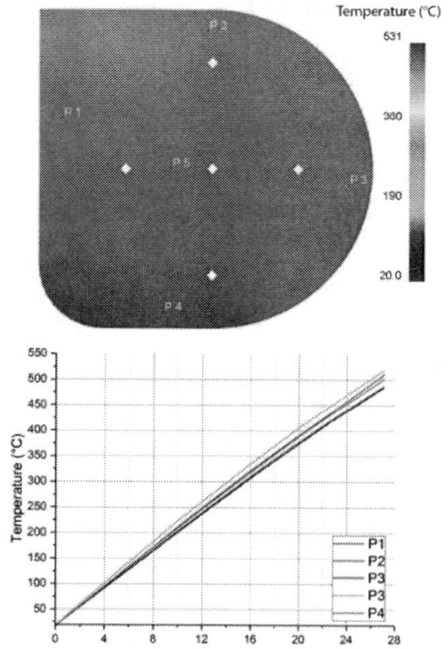

Figure 8 Comparison of induction heating with LFH and TFH methods

A similar simulation process was applied to the longitudinal induction heating model and the final numerical result of the temperature distribution is plotted in Figure 8. This result shows that within 27 seconds after the power was applied, part of the workpiece reached 525.0 °C. Without seeing severe edge effect, temperature was distributed more evenly across the whole workpiece and the highest temperature occurs at the round corner. The greatest temperature difference in the workpiece is about 50.0 °C . However, heating efficiency with this coil design is relatively low.

3.2 Induction heating of large Al-alloy blanks with motion (Simulation No. 3, No. 4, and No. 5)

Figure 9 shows the change of temperature in the Al-alloy sheet when it passed through the induction coil. The sheet was designed to move continuously but the simulation calculated the generated heat based on the relative position of coil and workpiece at each time step. However, the numerical results can still reveal uneven temperature distribution on the workpiece induced by edge effect. Area with relatively high temperature in the sheet had a similar shape of the coil and moved in a direction opposite to the movement of the sheet. The highest temperature was observed at the long edges, especially at each corner of the sheet. High heating rate occurred in these areas were on the account of edge effect where the coil stretched outside the sheet and led to high eddy current density. Temperature at the centre of the sheet was generally lower.

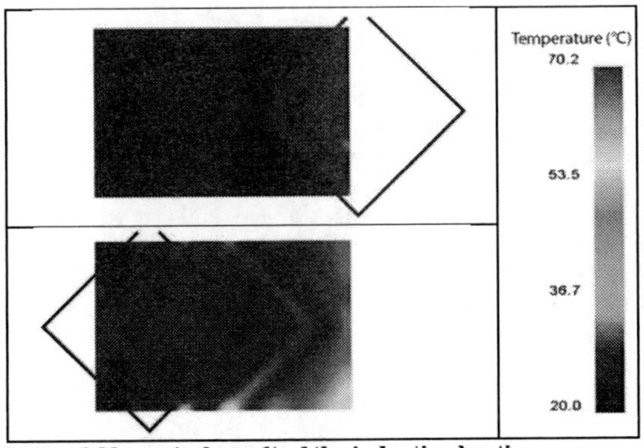

Figure 9 Numerical result of the inductive heating process with coil with right corner (Simulation No. 3)

Figure 10 presents the numerical results of induction heating which used coil with rounded corners. Although the induction edge effect in the Al-alloy sheet could still be clearly observed, more of the area in the centre of the sheet was heated up by this coil design. Coil with rounded corners has the quality to produce more even temperature distribution in the workpiece in an induction heating process.

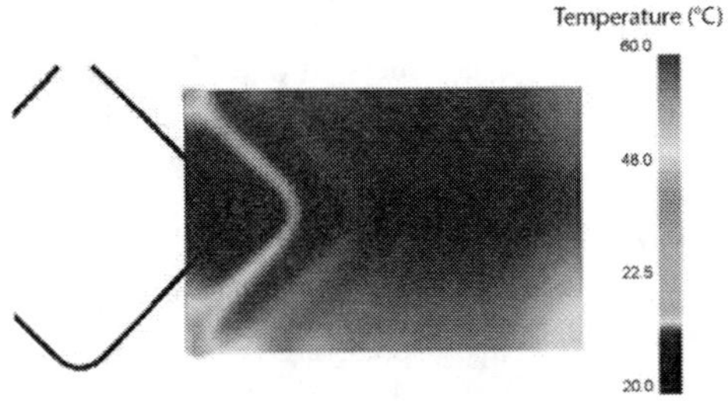

Figure 10 Temperature distribution of the inductive heating process with coil with rounded corner (Simulation No. 4)

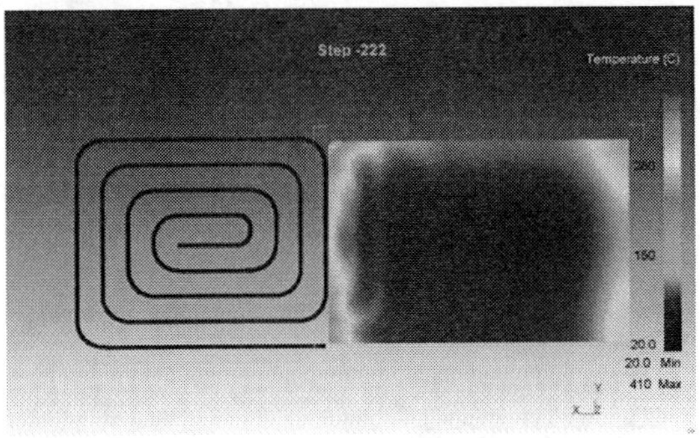

Figure 11 Simulation result of the inductive heating process with optimized induction coil design (Simulation No. 5)

After the study of cross-sections of the induction coils, coil designs and edge effect, an alternative induction coil design was proposed. Figure 11 presents the results of the numerical study of this coil design. The results show that the proposed coil design has the potential to produce more even temperature distribution pattern in the workpiece. Data reading from the simulation result also suggests that temperature in the AL-sheet can increase by as high as 45.0 °C in one second.

3.3 Induction heating of large Al-alloy blanks with array of coils (Simulation No. 6)

The second induction heating method adopted in the numerical study was use of coil arrays. For heating Al-alloy blank of 400.0 mm × 600.0 mm × 2.0 mm , a total of 8 induction coils were used.

In order to study the effect of alternating current frequencies on the temperature development of induction heating, the induction coils were energized by alternating currents of different frequencies. Temperature distributions on the workpiece are plotted at Figure 12.

From the results of different current frequencies, edge effect can be clearly observed. All these simulations were conducted using the current density of 27.0 A/mm^2. As the frequency of the power supply increased, the edge effect became more obvious. For all the simulations, the edges of the workpiece were firstly heated up while the centre of the blank was still relatively cool. At 40 s after the heating processes started, the hottest part of the blank had a temperature above 525.0 °C but the coolest part was still below 150.0 °C. The highest temperature on the blank is severely affected by the supply frequency. The temperature difference in the workpieces increases with increasing current frequency.

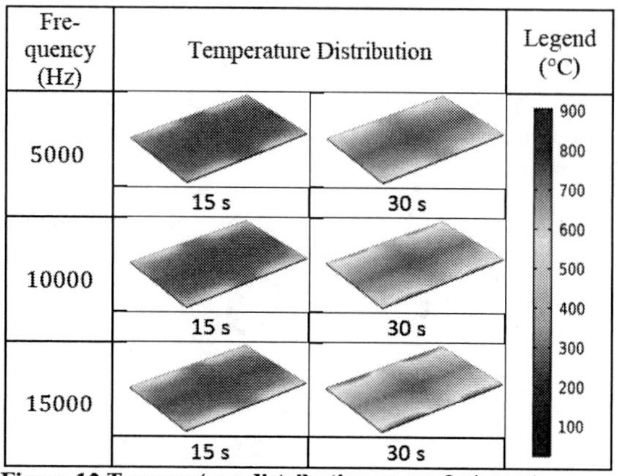

Figure 12 Temperature distribution on workpiece at different current frequencies

4 Conclusions

This study aims to examine the induction heating of aluminium blank for hot stamping applications. Numerical simulations were conducted to investigate effects of the coil designs and power parameters on the induction heating processes for Al-alloy blanks. Aspects like coil design selection, edge effect in induction heating, corner shape of the coil, blank movement, and current properties, etc. were considered and studied. Main conclusions from this study are as follows:

- Both TFH and LFH induction heating process were studied numerically. The TFH heating process can achieve high heating rates, being caused by the induction edge effect. The edge effect also causes the metal sheet to be heated up unevenly. LFH heating method can reduce the edge effect. However, this method is only feasible for small blanks in practise.
- Coils with sharp corners and rounded corners were both used for induction heating study for moving blank and the numerical results show that the coil with rounded corners is likely to result in more uniform heating.
- An alternative coil design was proposed for the heating of large Al-alloy sheets and a numerical simulation showed its potential to produce quick and relatively more uniform heating.
- Induction heating with coil arrays was also studied. Parametric simulations were conducted for different supplied current frequencies. The results suggest that, to obtain relatively more even temperature distribution on Aluminium sheets with thickness of 2 mm, current with low alternating frequency is more applicable.

To adopt induction heating of aluminium blanks in hot stamping production, in practice, special attention should be paid to the selection of the coil design, suitable heating method, blank geometry and its movement, and power control. Only with "optimal" combinations of these factors/parameters, the aluminium blank could be heated up to acceptable temperatures and distributions with desired heating rates, which would help to achieve high-quality hot stamping.

5 References

[1] L. Jakubovičová, G. Andrej, K. Peter, and S. Milan, "Optimization of the induction heating process in order to achieve uniform surface temperature," *Procedia Eng.*, vol. 136, pp. 125–131, 2016.

[2] S. Klonk, "Numerical modelling of induction heating for complex geometrical parts," 2014.

[3] M. W. Kennedy, S. Akhtar, J. A. Bakken, and R. E. Aune, "Review of Classical Design Methods As Applied To Aluminum Billet Heating With Induction Coils," pp. 707–722, 2011.

[4] B. J. Yang, A. Hattiangadi, W. Z. Li, G. F. Zhou, and T. E. McGreevy, "Simulation of steel microstructure evolution during induction heating," *Mater. Sci. Eng. A*, vol. 527, no. 12, pp. 2978–2984, 2010.

[5] A. Bermúdez, D. Gómez, M. C. Muñiz, and P. Salgado, "Transient numerical simulation of a thermoelectrical problem in cylindrical induction heating furnaces," *Adv. Comput. Math.*, vol. 26, no. 1–3, pp. 39–62, 2007.

[6] D. Istardi and A. Triwinarko, "Induction Heating Process Design Using COMSOL ® Multiphysics Software," vol. 9, no. 2, pp. 327–334, 2011.

[7] S. Zinn and S. L. Semiatin, "Elements of Induction Heating," vol. 1, pp. 9–18, 1988.

Research on the Compound Process of Hot Stamping and Heat Treatment for Duralumin Alloy Fairing with Large Aspect Ratio: Experiment and modeling

Lingzhi Xu[1],*Lihua* Zhan[1,2,3,*], *Minghui* Huang[1], *Liang* Guo[1],*Ziyao* Ma[1]
[1]School of Mechanical and Electrical Engineering, Central South University, Changsha 410083, China
2State Key Laboratory of High-Performance Complex Manufacturing, Central South University, Changsha 410083, China
[3]Nonferrous Metal Oriented Advanced Structural Materials and Manufacturing Cooperative Innovation Center, Central South University, Changsha, 410083, China

Abstract. The compound process of hot stamping and solution quenching heat treatment (HFQ®) was utilized to fabricate hard aluminium fairing of high aspect ratio. A simulation model was established by MSC.Marc to optimize the shape and size of the blamk as well as the effects of process parameters, namely, initial temperature, stamping speed, blank holder force (BHF), on extreme final temperature and thickness distribution of formed component were investigated. The reliability of finite element simulation was proved by the hot stamping experiments with self-designed moulds under the optimize process parameters achieved by simulation results. The results indicated that the optimized blank is well deformed without flange wrinkle and broken round; the maximum cooling rate in quenching can inhibite the precipitation when the initial temperature ranges from 350 to 450℃; BHF effects dominated on thickness distribution and 10KN is the most reasonable value under the studied condition while the optimum stamping speed is 100mm•s^{-1}. Compared with the thickness of the stamping trials and numerical simulation, the deviation is less than 3.6%. Furthermore, The desired mechanical property is achieved without significant inhomogeneity.

Key word: Aluminium, Finite element method, Hot stamping, Manufacturing process

Introduction

Nowadays, lightweighting is one of the most importance development directions on the aerospace industries. Therefore, the lightweight and high strength characteristics of duralumin alloy, represented by the Al-Cu-Mg alloy in 2XXX series, determine its application prospects in aerospace and other industries[1].The duralumin alloy is mainly characterized by low density, high specific strength and good heat resistance, and its mechanical properties can be significantly strengthened by solution and aging. Such wonderful performance has making those aluminum alloy one of the best materials for manufacturing load bearing structures in aerospace vehicles.

Stamping is a typical sheet metal forming process. In such process, the sheet blank undergoes a large tension-compression stress due to punching into the deep die, resulting in a great plastic deformation[2].Stamping process is widely used in manufacturing thin-walled components in cylindrical, box, staircase, cone, sphere and other complex shapes. However, the components of aerospace vehicles are not only have complex structures, but also require high forming accuracy and good mechanical performance. It is very difficult to achieve desired shape and performance requirement by the tradition cold forming process through single forming, some of which are unable to fabricate by cold forming, especially for aluminum alloys. For example, fairing, an important structural on a launch vehicle[3], is a complicated structure with streamlined curved surfaces which currently manufacturing by cold forming process. But the production efficiency and the finished product rate are relatively low, makes it difficult to meet the mass production needs. Some studies have shown that the duralumin alloy will produce dynamic recovery and dynamic recrystallization softening during the deformation under the elevated temperature, which will weaken or eliminate the work hardening caused during the deformation process[4]. Therefore, the hot forming can improve the plasticity of the material, reduce the flow stress and overcome the forming defects of duralumin alloy at room temperature.

* Corresponding author: yjs-cast@csu.edu.cn

 https://doi.org/10.1051/matecconf/201819008003

According to recent researchs, many advanced forming processes such as hot stamping, superplastic forming, differential temperature drawing, etc.[5-6] show unique advantages for forming complex thin-walled components. DMFinch et al.[7] conducted theoretical and experimental studies on warm forming of aluminum alloy sheets. The results show that the temperature and blank holder force have important influence on the forming of aluminum alloy box and tube. Under the studied temperature, the formability of 2020-T4, 7075-T6 and other aluminum alloy sheets have been greatly improved. MJPainter et al.[8] found that the stamping limit of Al-2Mg aluminum alloy cylindrical parts increases with the increase of forming temperature and stamping speed, which means that the formability of Al-2Mg aluminum alloy cylindrical parts increases greatly compared with normal temperature. Zhenghua Guo[9] has conducted the warm forming experimental on the 5182 aluminum cylindrical shape (diameter 40mm), and found that when forming temperature increase from 100 ℃ to 250 ℃ without lubrication, the ultimate stamping limit increase from 4mm to 10.5mm. And with the Grafo LN-1169 lubricant, the ultimate stamping limit could increase to 15.5 mm. Above explained that the formability of the 5182 aluminum alloy sheet increased dramatically with increasing temperature and greater lubrication. Lei Fu[10-12] utilized HFQ process[13] to AA6111 aluminum alloy, but the properties of formed components has not been studied. Wenke Li [14] studied the hot stamping process on fabricating U-shaped parts of 6061 aluminum alloy through the comparison between finite element simulation and hot stamping experiment, and found that the hot stamping process can significantly improve the forming quality under the premise of basically no weakened performance. For the 2XXX series of duralumin alloy, there is few research on the performance control of thermoplastic formation process. Therefore, in this paper, 2A12 aluminum alloy thin-walled fairing with large aspect ratio that widely used in aerospace vehicle are taken as the object of this study to conduct a feasibility study on the compound process of hot stamping and quenching heat treatment, which known as HFQ® [15]. This process are considered to be an integrated high-efficiency and short-flow method to achieve the shape and performance of the component. The material were heated to 500℃ and hold in the solution temperature for 40min, which in the state of low deformation resistance and good plasticity, and then be quickly put into the mould for heat stamping. The mould is equipped with sink, which can introduce cooling water to realize the rapid quenching of components. After quenching, the component is pressured in specific shape and be further aging heat treatment. The studied compound process is more simple than the traditional cold forming process that consists of multiple forming, intermediate annealing and solution-quenching-aging. Moreover, this process effectively prevent the loss of forming accuracy in the subsequent high temperature solution-quenching due to the quenching residual stress. The research work is of great significance for improving the precision and efficiency of the forming of duralumin alloy components and expanding its application in aerospace and other fields.

1 The establishment of simulation model

It has been shown in the existing literature that the ultimate aspect ratio of a qualified part formed by cold drawing is 0.64. In this paper, the studied duralumin alloy thin-walled fairing fabricated by hot stamping has an aspect ratio of 1.2, and consists mainly of arcuate surfaces, including the die radius of R10, the punch radius of R20 and the circular arc of R90. The geometry is modeled through the 3D software Pro / E which is shown in Fig. 1, and then imported into MSC.Marc software to build the whole simulation model.

Fig. 2 shows the finite element model followed by the punch, blank holder die, blank and die from top to bottom. A certain amount of sheet metal flange were left for blank holder process. In order to improve the efficiency of simulation, double-cavity symmetrical pressing method was used. A 1/4 model was established and the number of set-up units was 3690, a grid re-division was performed every five time steps, and the sheet was meshed using 3D-Shell No. 75.

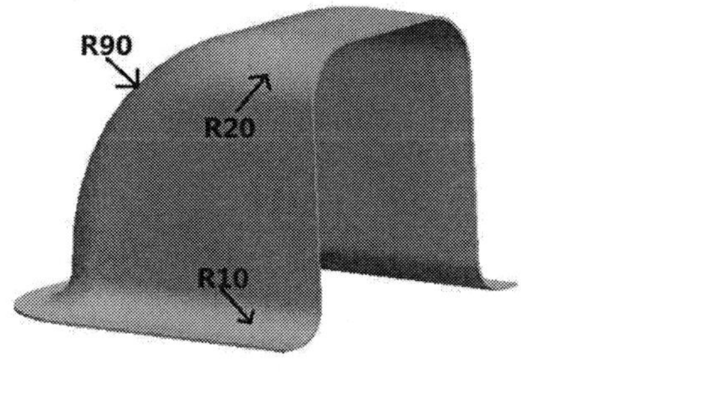

Fig.1. Three-dimensional modeling of the fairing

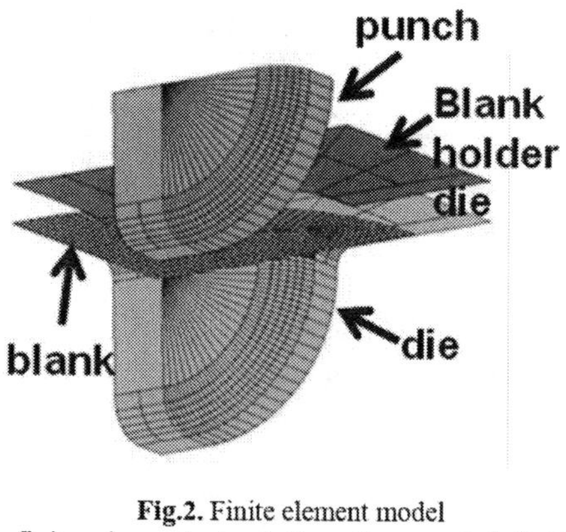

Fig.2. Finite element model

In finite element simulation, the exact definition of material properties has a vital impact on the validity and correctness of simulation results. The elastic-plastic isotropic model is adopted, which has a density of 2.78E-9t•mm^{-3}, an elastic modulus of 72GPa and a

 https://doi.org/10.1051/matecconf/201819008003

Poisson's ratio of 0.31. Since the numerical simulation of hot stamping is thermo-mechanical coupling, it is necessary to set the thermal physical parameters of the material, namely, the thermal conductivity is 117.2W•m^{-1}•K^{-1}, the specific heat capacities at different temperatures and the average coefficient of linear expansion as shown in Table 1.

Table 1. The thermal physical parameters of 2A12 duralumin alloy

Temperature T/℃	specific heat c/J·(Kg·K)$^{-1}$	temperature interval T/℃	average coefficient of linear expansion $a_l/10^{-6}·K^{-1}$
100	921	20～100	22.7
200	1047	20～200	23.8
300	1130	20～300	24.7
400	1172	20～400	—

The true stress-strain relationship at different strain rates and temperatures during the hot stamping process is described by the Z-parameter thermo-hydro-mechanical constitutive model (1), taking into account the effects of temperature changes.

$$\begin{cases}\sigma=\frac{1}{\alpha}\ln\left\{\left(\frac{Z}{A}\right)^{1/n}+\left[\left(\frac{Z}{A}\right)^{2/n}+1\right]^{1/2}\right\} \\ Z=\exp\left(\frac{Q}{RT}\right)\end{cases} \qquad (1)$$

$\alpha = 0.026 - 0.018\varepsilon - 0.028\varepsilon^2 + 0.081\varepsilon^3$,

$n = 9.31 - 28.25\varepsilon + 83.07\varepsilon^2 - 84.38\varepsilon^3$,

$Q = 230.67 - 214.59\varepsilon + 347.42\varepsilon^2 - 289.14\varepsilon^3$,

$\ln A = 34.78 - 9.58\varepsilon - 114.13\varepsilon^2 + 201.77\varepsilon^3$。

According to the definition principle of contact body, the sheet is defined as deformable contact body, and the punch, die and blank holder are defined as rigid contact body, with the friction coefficient of 0.1. The heat transfer coefficient between the sheet metal and air can be ignored in simulation for its much smaller than that between the sheet metal and the mould during the hot stamping process. In addition, only the influence of the gap on the heat transfer coefficient is considered to simplify the calculation. The contact heat transfer coefficient between the sheet metal and the mould is set to 0.7kW • (m2 • ℃)$^{-1}$ and the mould initial temperature is set to a constant temperature 25 ° C.

2 Simulation results and analysis

For the hot stamping process, the shape and size of the blank is an important factor affecting the forming accuracy and performance of the component. The shape of the blank has a significant effect on the flow of the sheet material during forming. A large blank will not only waste material, but also increase the chances of wrinkling and rupture; while a small blank will cause excessive sheeting thinning, leads to the loss of forming accuracy. Therefore, a reasonable blank shape and size can help eliminate the defects of forming process.

The results of the 1/4 model simulation show that the rectangular corner has a significant impediment to the flow of the sheet material during the forming of the rectangular blank, resulting in the uneven distribution of the part thickness and the wrinkling at the flange. But with rounded corners as R90's blank shape, the fracture of the forming part is improved, and the flow of sheet material was more reasonable. The final shape of the smallest optimized blank size is 170mm × 150mm, whose specific size shown in Fig. 3.

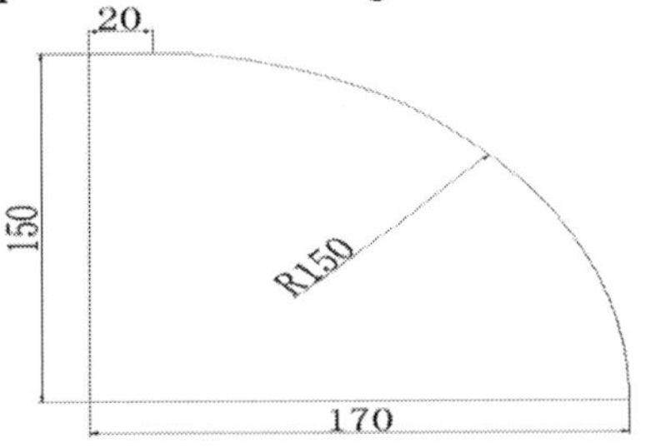

Fig.3. Specific size of quarter rounded blank（mm）

Based on the above blank, a simulation analysis under the die gap of 1.3mm, the die radius of 10mm, the forming temperature of 450 ℃, the stamping speed of 100mm • s^{-1}, the blank holder force of 10KN, the dwell time of 6s, the friction coefficient of 0.1 was carried out as an example to obtain the overall thickness distribution of the formed component, shown in Fig. 4.The maximum thinning ratio is usually used as an important index to evaluate the forming quality of the formed component, and the value is generally not more than 30% [16]. As for the studied 2A12 aluminum alloy sheet with the thickness of 1.2mm, the minimum thickness of the formed component required to be thicker than 0.84mm.

It can be seen from Fig. 4 (b) that under the above simulation paremeters, the minimum thickness of formed component is 0.867mm, and the maximum thinning rate is calculated to be 27.7%, that is, no rupture failure occurs. As the component is completely symmetrical, the date of thickness of cross-section along the transverse and longitudinal path were obtained and the thickness-expansion distance curve is shown in Fig.4(b).

As can be seen from the Fig.4(b), the transverse and longitudinal expansion distances increase from 150mm × 170mm to 174.8mm × 210.6mm, with an increase of 16.5% and 23.9%, respectively, and the thickness of the point on the transverse path is generally lower than that of the longitudinal one. The thickness of the selected points on the transverse section gradually increases with the expansion distance, and the maximum value is 1.49 mm at the flange due to the superposition of the flanges. The thickness of the selected points on the longitudinal section firstly decreases gradually with the expansion distance and reach a minimum distance of 74mm from the center point of 0.93mm, whose rate of thinning was 22.5%, then gradually increases and also reaches the maximum at the flange of 1.51mm. The above simulation results explains that 2A12 aluminum alloy fairing forming by hot stamping is a plastic forming process of local thinning and flange thickening.

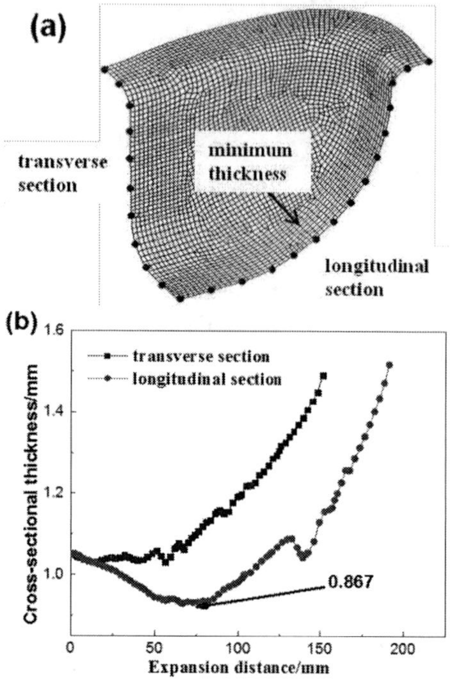

(a) Overall thickness distribution of the formed component
(b) The thickness distribution of each section

Fig.4. Overall thickness distribution of the formed component and the thickness distribution of each section

2.1 Effect of Initial Temperature on Forming

The forming temperature influences significantly on the plasticity of 2A12 hard aluminum alloy. With the increases of forming temperature, the forming limit is higher and the plasticity is better during thermoforming. However, the deformation resistance is decreases simultaneously, which leads to the increased risk of rupture of the component. The extreme final temperature and the maximum thinning rate of the formed component are shown in Fig.5 (a) and (b) respectively with the change of initial forming temperature of 350 ℃, 400 ℃, 450 ℃, while the drawing speed is 100mm • s^{-1}, blank holder force is 10KN, die gap is 1.3mm and dwell time is 6s remain unchanged.

As can be seen from Fig. 5 (a), the minimum final temperature of the sheet is 307.6 ℃, 355.7 ℃ and 402.6 ℃, respectively, and the maximum cooling rate is 42.4 ℃•s^{-1} and 44.3 ℃•s^{-1}, 47.4 ℃•s^{-1}, respectively. It is reported that the maximum cooling rate of the sheet is higher than 38 ℃• s^{-1}, which can inhibit the change of percipitation [17]. Fig. 5 (b) shows that the maximum thinning rate of the sheet increases slightly with the increasement of the forming temperature. While the forming temperature increases from 350 ℃to 450 ℃, the maximum thinning rate increases from 26.83% to 28.42%, with a difference of 1.59%, indicating that the change of forming temperature within 350 ℃ ~ 450 ℃ has little effect on the maximum thinning rate of the hot stamping process of the studied fairing. Taking into account the loss of temperature in the sheet transfer from the heating furnace to the mould, and the higher the minimum final temperature, the faster the cooling speed in the quenching stage, the more able to improve the quality of the formed component, so the suitable forming temperature is 450 ℃.

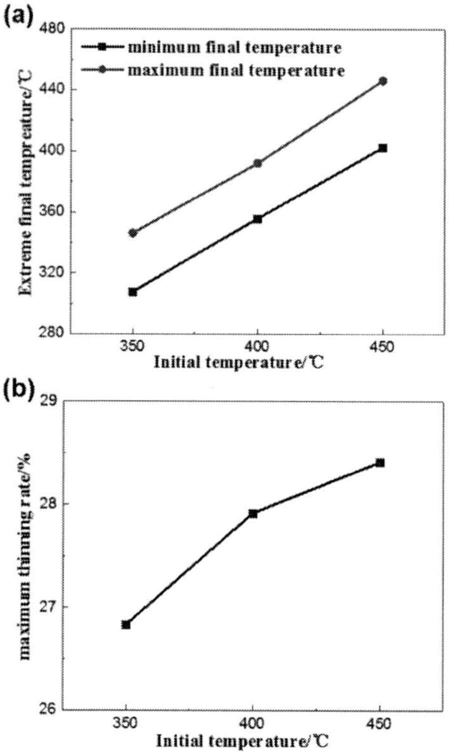

(a) The extreme final temperature with the change of initial forming temperature (b) The maximum thinning rate with the change of initial forming temperature

Fig.5. Effect of different initial temperature on the extreme final temperature and the maximum thinning rate

2.2 Effect of BHF on Forming

The selection of blank holder force is very important for the stamping forming process, unreasonable blank holder force will lead to wrinkling or rupture.The extreme final temperature and the maximum thinning rate of the formed component are respectively shown in Fig.6 (a) and (b) with the change of blank holder force of 6KN, 8KN, 10KN, 12KN, while the stamping speed is 100mm • s^{-1}, the initial forming temperature is 450 ℃, die gap is 1.3mm and dwell time is 6s remain unchanged.

As can be seen from Fig.6 (a), the extreme final temperature hardly changes with the blank holder force. This is because only the gap factor is taken into account in the definition of contact model to simplify the calculation, result in the same extreme final temperature under different blank holder force. Fig.6(b) indicate that the maximum thinning rate of the formed component increases as the blank holder force increases. When the blank holder force is 6KN, 8KN and 10KN, the

https://doi.org/10.1051/matecconf/201819008003

maximum thinning rates were 25.91%, 27.08% and 28.42%, respectively. However, the maximum thinning rate exceed 30% when the blank holder force was 12KN, which indicated the occurrence of rupture failure occurred. On the contrary, in order to avoid the wrinkling problems caused by small blank holder force, 10KN should be a reasonable value.

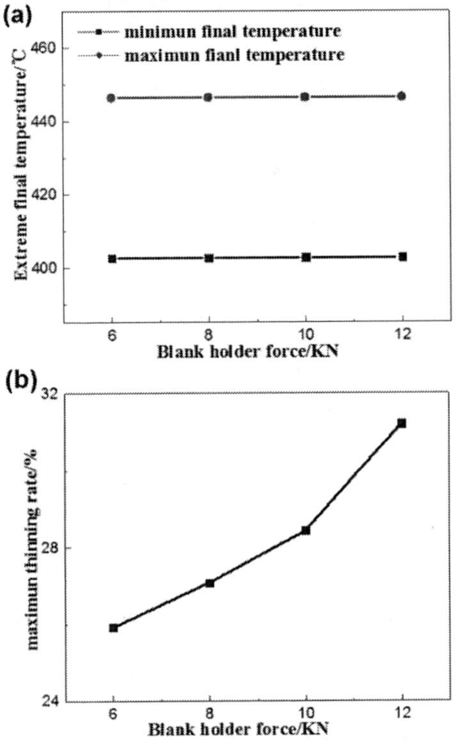

(a) The extreme final temperature with the change of BHF
(b) Tthe maximum thinning rate with the change of BHF

Fig.6. Effect of different BHF on the extreme final temperature and the maximum thinning rate

2.3 Effect of stamping speed on Forming

Stamping speed is as well an important process parameter in hot stamping for it not only determines the strain rate of sheet material, but also affects the distribution of temperature field of the sheet, thus affects its plasticity. The extreme final temperature and the maximum thinning rate of the component are respectively shown in Fig.7 (a) and (b) with the change of stamping speed of 60mm • s^{-1},100mm • s^{-1},140mm • s^{-1},180mm • s^{-1}, while the blank holding force is 10KN, the initial forming temperature is 450 ℃, die gap is 1.3mm and dwell time is 6s remain unchanged.

As can be seen from Fig. 7 (a), as the stamping speed increases, the extreme final temperature of the fairing increases and the difference between maximum and minimum temperatures decreases. When the stamping speed is 60mm • s^{-1} and 180mm • s^{-1}, the minimum temperatures are 362.96 ℃ and 421.21 ℃ respectively, the maximum temperatures are 434.8 ℃ and 448.92 ℃ respectively. This is because the contact time between the sheet and the mould is shortened with the increases of stamping speed, so the temperature distribution of the sheet is more even. From Figure 7 (b), it can be seen that the maximum thinning rate increases with the increasement of stamping speed, and when the stamping speed is over 100mm • s^{-1}, the rupture failure occurred. So for the hot stamping of larger aspect ratio of complex duralumin alloy component, stamping speed is needed to control within a certain range. In summary, combined with the extreme final temperature and thickness changes of the components, the process requirements are satisfied when the stamping speed is 100mm • s^{-1}.

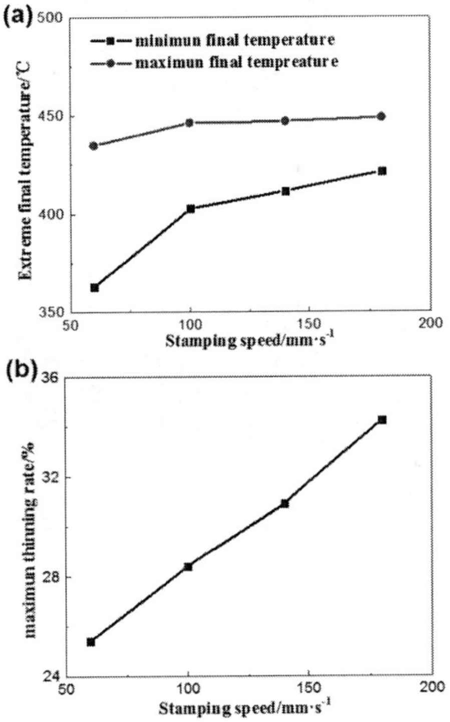

(a) The extreme final temperature with the change of drawing speed (b) The maximum thinning rate with the change of drawing speed

Fig.7. Effect of different drawing speed on the extreme final temperature and the maximum thinning rate

3 Experimental research and analysis

3.1 Experimental Research on the Compound Process of 2A12 DurAlumin Alloy

A self-design mould is developed, which shown in Fig.8, to verify the accuracy of the finite element model. And then the hot stamping of fairing experiment was carried out under the parameters that obtained by previous simulation as follows: forming initial temperature of 450 ℃, mould initial temperature of 25 ℃, blank holder force of 10KN, stamping speed of 100mm • s^{-1}, die gap of 1.3mm and dwell time of 6s.

https://doi.org/10.1051/matecconf/201819008003

The hot stamping process of the aluminum alloy sheet is shown in Fig. 9, the hot stamping mould was firstly installed on a hydraulic press, and then the blank was heated in a heat furnace to a given temperature for a certain time before transferred to the mould and implement the hot stamping. After quenching, the pressure was discharged and the formed component was taken out, as shown in Fig.10, which contains two fairing members.

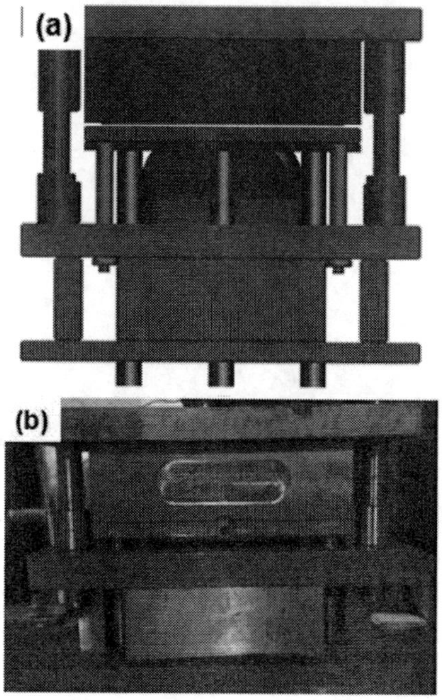

（a）3-D map of the mould （b）Physical map of the mould

Fig.8. Structure of the self-design mould

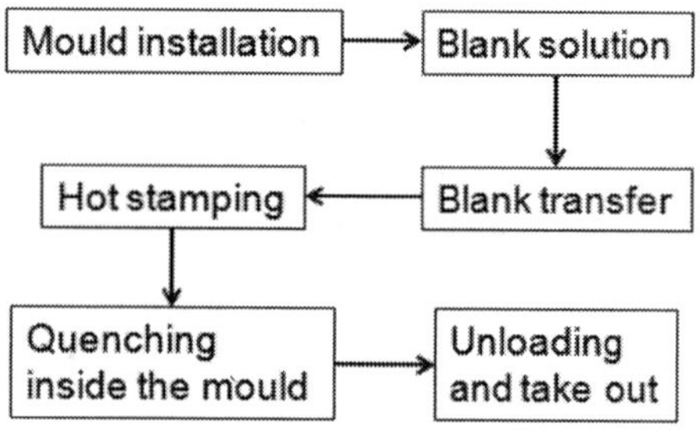

Fig.9. The process of hot drawing

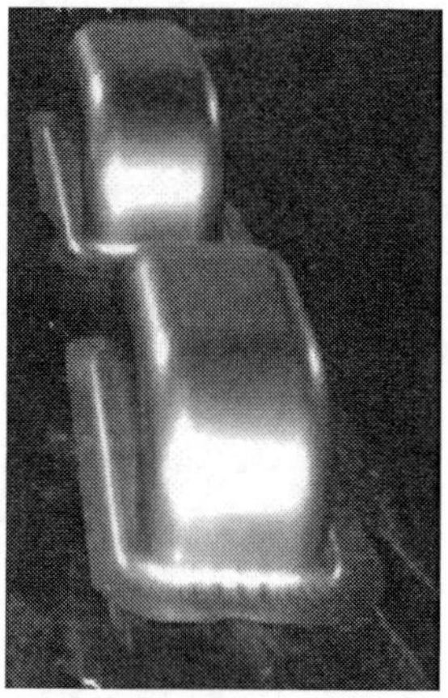

Fig.10. The formed fairing fabricated by hot stamping

3.2 Comparison and Analysis of experiment and simulation of formed fairing thickness

In order to obtain the thickness distribution of the fairing, the formed component was cut from the transverse and longitudinal symmetry plane, as shown in Fig.11.

The cross-sectional thickness of the above 1/4 part along the horizontal and vertical paths were measured by micrometer to compared with the finite element simulation results under the same process parameter. Comparison result in Fig.12 shows that the simulation values of the thickness agrees well with the experimental values along the expansion path, and the maximum deviation does not exceed 3.6%, which also proves the reliability of the finite element simulation.

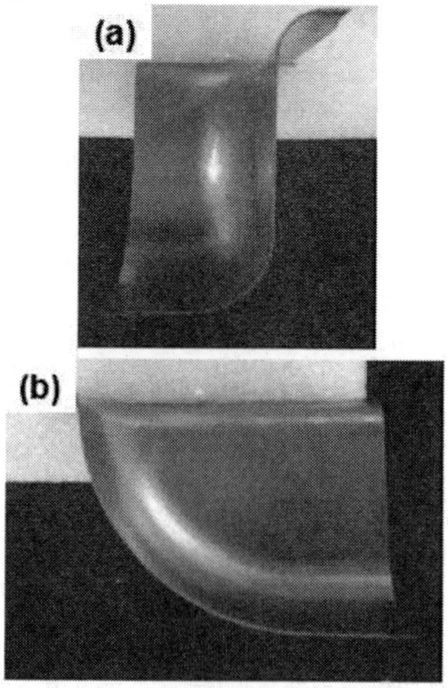

Fig.11. The quarter formed fairing(transverse and longitudinal cross-section)

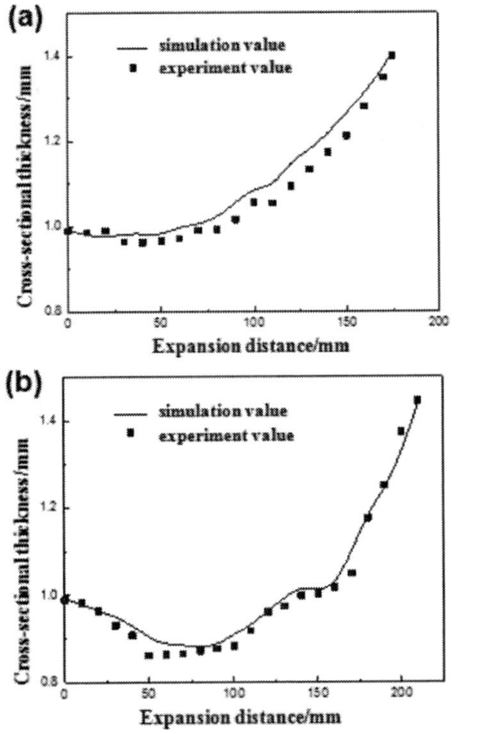

(a) Horizontal cross-section (b) Vertical cross-section

Fig.12. The thickness value of experiment and simulation on horizontal and vertical cross-section

4 Mechanical properties of formed fairing

In order to test the preformance of the fairings formed by the studied process, the components were subjected to natural aging treatment (for 4 days or more at room temperature) after the compound process, and then room temperature tensile test and micro-hardness test were carried out to obtain its tensile strength, yield strength, elongation and micro-hardness. The tensile test specimens with a parallel length of 25 mm and a width of 10 mm were taken from the corresponding positions of the components by wire cutting and the size is shown in .

4.1 Tensile mechanical properties of the formed fairing

The tensile strength, yield strength and elongation results at room temperature of the sidewall and the bottom of the fairing are listed in Table 2. As can be seen from the table, there is a certain difference in the strength and plasticity between the sidewalls and the bottom of the formed fairing, but the deviation is small. In addition, comparison of the properties of the hot stamping parts with the properties of the sheet after trandition solution-quenching-natural aging heat-treated shows that the hot stamping parts can still meet the strength and plasticity requirements.

Table 2. Tensile mechanical properties of the sidewall and the bottom of the formed fairing

Type	Tensile strength /MPa	Yield strength /MPa	Elongation/%
2A12-T4 sheet	474	327	11.8
Sidewall of the formed fairing	478	325	12.3
bottom of the formed fairing	468	318	10.2

4.2 Micro-hardness of the formed fairing

Vickers micro-hardness test was carried out for 4 position, namely the flange(A), sidewall(B), bottom(C) and rounded(D), of the formed fairing as shown in Fig.13. Three fairing parts, labeled # 1, # 2, # 3, were tested for average value, and the results are shown in Table 3.

It can be found from the table that there is a certain difference of micro-hardness between the 4 positions. According to the simulation result above, those 4 parts of the sheet suffered from different temperature and stress state during the whole hot stamping stage, result in the differences in micro-hardness. But the maximum deviation of the average micro-hardness value of the various position is within 8%, which meet the quality requirements of the parts. Additionally, the micro-hardness of all the three fairing in the same position have the maximum deviation of only 1.9%, indicating that the experiment has good repeatability.

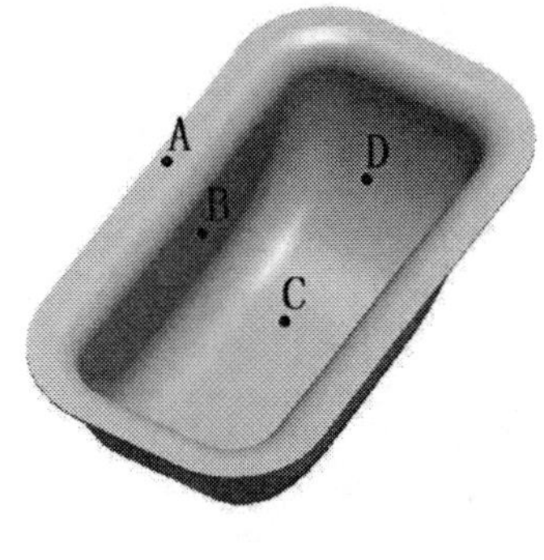

Fig.13. Position of A, B, C and D

https://doi.org/10.1051/matecconf/201819008003

Table 3. Micro-hardness of the formed fairing in A, B, C and D(HV)

Position	#1	#2	#3	Average valus
A	134.3	133.7	136.2	134.7
B	132.5	133.1	133.6	133.1
C	128.8	129.9	130.2	129.6
D	124.6	124.1	124.3	124.3

Conclusion

(1) A finite element model was established by using MARC software to optimize the shape of blank and to study the influences of initial forming temperature, blank holder force and stamping speed on temperature and thickness distribution systematically during the fairing hot stamping process, which help to determine the process parameters on the subsequent experiments.
(2) The reliability of finite element simulation was proved by the hot stamping experiments with self-designed moulds. Compared with the component thickness of the stamping trial and numerical simulation, the deviation was less than 3.6%.
(3) Mechanical properties and micro-hardness at room temperature of the formed fairing were tested, the results show that the hot stamping fairing can meet the strength and plasticity requirements, as well has good repeatability.

Acknowledgements

This research was supported by the National Natural Science Foundation of Hunan Province (Grant No. 13JJ2009)

References

1. J.Z. Cao, Z.T. Wang. L. A. F. Tech, **4**, 1-12, 2013
2. X.F. Xu, G.Q. Tong. J. Plast. E **16**, 64-68, 2009
3. B.F. Song. *Aerospace Technology Introduction*. National Defense Industry Press. 2006
4. H.J. McQueen, J.J. Jonas. Trea. Mater. Sci. Tech. **6**, 394, 1971
5. P.F. Variani, S Bruschi, A Ghiotti, et al. CIRP A-M Tech. **62**, 251-254, 2013
6. S. Yoshihara, H. Nishimura, H. Yamamoto, et al. J. M. P. T. **142**, 609-613, 2003
7. D.M. Finch, S.P. Wilson, J.E. Dom. Trans. ASM. **36**, 254-289, 1946
8. M.J. Painter, R. Pearce. A. American Society for Testing and Materials, Formability of Metallic Materials-2000 AD, Special Technical Publications. 105-118, 1982
9. Z.H. Guo, Z.G. Li, K.J. Ouyang, X.H. Dong. Developing R. China Metal. Equi. Manu. Tech. **38**, 80-82, 2003
10. L. Fu, B.Y. Wang, Q.L. Meng, et al. J. CSU(Sci. Tech.) **3**, 936-941, 2013
11. J. Zhou, B.Y. Wang, J.G. Lin, L. Fu. ARCH. CIV. MECH. ENG. **13**, 401-411, 2013
12. J. Zhou, B.Y. Wang, J.G. Lin, L. Fu. Trans. Nonferrous Met. Soc. China. **24**, 2014
13. M.S. Mohamed, A.D. Foster, J. Lin, D.S. Balint, T.A. Dean. INT J MACH TOOL MANU. **53**, 27-38, 2012
14. W.K. Li, L.H. Zhan, J. Zhao. CHIN J NONFERROUS MET. **26**, 1159-1166. 2016
15. A.D. Foster, T.A. Dean and J Lin, Priority Date: 19 Sept. 2008, Process for forming aluminium alloy sheet component, International Publication No.: WO2010/032002 A1
16. A.D. Foster, M. Mohamed, J. Lin, et al. Steel Res Int, **79**, 133-140, 2008
17. H.Y. Li, J. Bin, X.F. Wang, Y. Tang. CHIN J NONFERROUS MET. **21**, 2068-2074, 2011

Investigation of diffusion behavior of carburized sheet metal in hot stamping

Alexander Horn[1,*] and *Marion* Merklein[1]

[1]Institute of Manufacturing Technology, Friedrich-Alexander-Universität Erlangen-Nürnberg, Egerlandstraße 13, 91058 Erlangen, Germany

Abstract. Today's manufacturing of structural car body parts faces several challenges, like forming accuracy and passenger safety. Besides these two requirements, lightweight design plays an important role. One possibility to fulfill these partially rivalling demands is the application of hot stamped parts. The combination of hot forming and in die quenching reduces not only springback, but also results in tensile strengths of up to 1500 MPa. This makes a simultaneous reduction of sheet thickness and therefore weight reduction possible. Further development enabled a tailored adjustment of mechanical properties, for example by applying different cooling conditions along the parts. One of the biggest issues of these state of the art processes is the formation of a transition zone due to heat transfer. A promising approach to adjust the mechanical properties with a minimized transition zone is the carburization of sheet metal. Therefore, the parts are coated with graphite, heat treated and subsequently quenched. In this work, the time variant process of carbon diffusion is investigated. Sheets with two different thicknesses are carburized and quenched. The resulting mechanical properties are analyzed using uniaxial tensile tests and microhardness measurements. The results are correlated with the carbon content measured by EDX-analysis.

Keywords: Hot stamping, Heat treatment, Carburization

1 Introduction

Ecological aspects as well as increasing safety standards are two reasons for growing importance of lightweight construction in automotive industries [1]. There are different strategies to improve lightweight design in today's manufacturing of structural car body parts. One of them is the substitution of commonly used steel grades by either light-weight materials such as aluminum or materials with higher strength [2]. By using high or highest strength materials, sheet thickness can be reduced, which leads to a weight reduction of the components. Since high strength materials have only limited formability and high process forces are required, temperature assisted forming processes are necessary [3]. Hot stamping of ultra-high strength materials has developed to a state of the art process for manufacturing lightweight, but safety-relevant components.

There are two different process routes in hot stamping. In direct hot stamping, the sheets are austenitized, transferred to the press and then subsequently formed and in-die quenched. Tensile strength of up to 1500 MPa for the commonly used boron manganese steel 22MnB5 can be achieved. These high strengths are a result of a fully martensitic structure which is developed when a cooling rate of at least 27 K/s is ensured during in-die quenching and forming [4]. Contrary to direct hot stamping nearly complete preforming is done before austenitization in the indirect hot stamping process. The parts are quenched and calibrated in the press subsequently to the heat treatment [5]. An overview of the process steps in direct and indirect hot stamping can be seen in Figure 1. Since decarburization and oxidation occur during austenitization, only coated sheet material is used for hot stamping. In direct hot stamping, Al-Si coatings are commonly used [6]. Due to the limited formability of Al-Si at room temperature, other coatings, such as zinc, are preferred in indirect hot forming [4]. Furthermore, Zn-coatings offer cathodic corrosion protection [7].

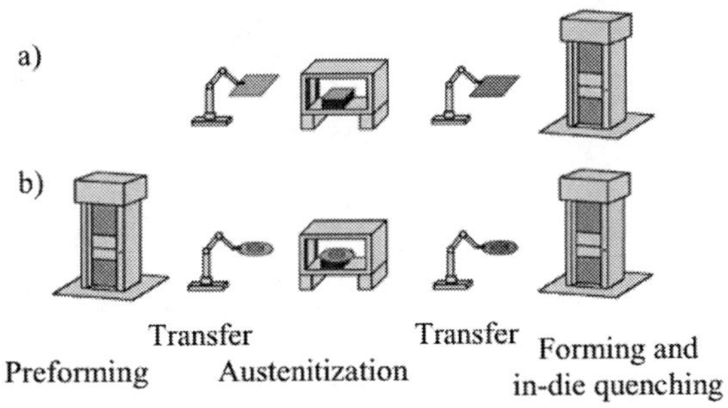

Fig 1. Overview of the process steps in direct (a) and indirect (b) hot stamping according to [8]

* Corresponding author: alexander.horn@fau.de

https://doi.org/10.1051/matecconf/201819008004

One challenge of hot stamped components is a low residual elongation. Especially in crash relevant structures high energy absorbance is desired. An example for this are b-pillars, where the bottom of the part should have increased ductility for better energy absorbance, while the upper part should retain its structural integrity for passenger safety. By adapting different steps of the hot stamping process, it is possible to manually adjust the resulting mechanical properties. Strategies for tailoring of mechanical properties in hot stamping are partial hot stamping, partial austenitization, annealing and tailored products. By preventing a full austenitization of particular areas, or decreasing the heat transfer between tool and sheet below the critical value, it is possible to retain the original ferritic-pearlitic microstructure besides the martensitic structure [9]. One challenge of these state of the art process variants is the formation of a transition zone due to heat transfer between areas of different temperatures. According to Feuser [10], manufacturing of parts with a transition zone less than 30 mm is not feasible.

A method for minimizing this transition zone, while still being able to adjust the mechanical properties, is tailored carburization [11]. In this process, the sheet metal is locally coated with graphite and subsequently carburized. Areas, where carburization is not desired, are masked with boron-nitride. During the heat treatment in the oven at elevated temperatures, the carbon atoms diffuse into the base material. The locally increased carbon content results in higher strength in these areas after quenching. Existing process understanding from case hardening cannot be transferred directly due to different dimensions and objectives. Since carburization is a time variant process and depends on temperature, relevant process parameters have to be analyzed. Aim of this work is to investigate the influence of sheet thickness and several heat treatment parameters on mechanical properties, such as tensile strength or Vickers hardness and carbon content after carburization. Depending on these influencing factors, a process window for carburization of sheet metal will be identified. The process window identified within this work will be the basis for the application of tailored carburization. Furthermore, the applicability of energy-dispersive X-ray spectroscopy to the analysis of carbon content is investigated.

2 Material and Methodology

A hot rolled, uncoated complex phase steel CP-W® 800 is used in this work. The investigated sheet thicknesses are $t_{01} = 1.6$ mm and $t_{02} = 2.5$ mm. Compared to conventional hot stamping steel 22MnB5, CP-W® 800 has a lower carbon content. The difference between both alloys is 0.11 wt%. This enables the possibility of further carburization. The chemical composition of CP-W® 800 can be seen in Table 1.

The carbon source for carburization is a graphite foil of type 90/10 from KERAFOL Keramische Folien GmbH with a thickness of 0.15 mm. The bonding between base material and graphite is realized by application of the adhesive COT Resbond 931C (Polytec PT GmbH). During air curing, the ceramic binder evaporates, so that the adhesive consists of 99% graphite after hardening. To limit oxidation and decarburization during austenitization, the specimens are wrapped into heat treatment foil with a thickness of 0.051 mm from Schröder Industrieöfen GmbH.

Table 1. Chemical composition of CP-W® 800 in wt% [12]

C max.	Si max.	Mn max.	P max.	S max.
0.14	1.00	2.20	0.080	0.015
Al total	**Ti + Nb max.**	**Cr + Mo max.**	**V max.**	**B max.**
0.015 – 2.0	0.25	1.00	0.20	0.005

Besides the thickness of the steel sheets, heat treatment temperature and holding time are investigated as influencing factors. The different specifications are shown below in Table 2. Subsequently to the carburization in the oven, the specimens are quenched in water.

Table 2. Investigated influencing factors

Investigated parameter	Variation
Sheet thickness t_0 in mm	1.6, 2.5
Heat treatment temperature T_{oven} in °C	900, 950
Holding time t_{hold} in min	5, 60, 180, 360

After carburization and quenching, tensile tests are conducted using a universal testing machine Zwick Z100 in combination with a digital image correlation system Aramis (GOM GmbH). Afterwards, small segments are separated from the tested samples and embedded in epoxy resin for microhardness measurements in a Fischerscope HM1000.

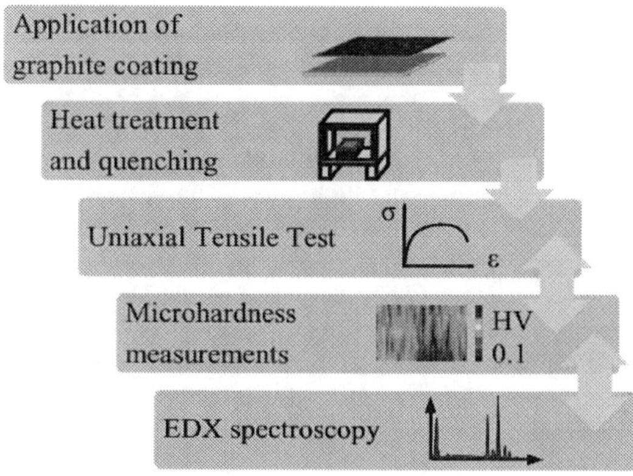

Fig. 2. Used methodology in this work.

The Vickers hardness HV 0.1 is measured on the middle plane of the samples' cross section as well as along the sheet thickness. Afterwards, carbon content of

https://doi.org/10.1051/matecconf/201819008004

the carburized samples is measured via energy dispersive X-ray spectroscopy (EDX). The measurements are carried out with a scanning electron microscope of type ZEISS MERLIN GEMINI II, which is equipped with an OXFORD X-MaxN Silicon Drift Detector 50 for EDX analysis. The results are correlated with mechanical properties from tensile tests and microhardness measurements. An overview of the used methodology is given in Figure 2.

3 Results

3.1 Tensile strength

Figures 3 and 4 show the resulting tensile strength after carburization at 900 °C and 950°C in comparison with uncoated specimens. It can be seen in Figure 3, that an increase in tensile strength by carburization is achievable. After a heat treatment of five minutes at 900 °C, the values are below the initial state, with the best results for a sheet thickness of 1.6 mm. After one hour, the tensile strength of the specimens with a thickness of 1.6 mm is slightly increased to (787.2 ± 4.9) MPa. For the samples with t_{02} = 2.5 mm, the values are decreased, independent of potential carburization. After three hours of heat treatment, major differences can be seen. While tensile strength is further decreased for specimens without graphite coating, carburization results in a significant increase in tensile strength. The maximum of tensile strength of (1258.9 ± 2.0) MPa is reached for specimens with a sheet thickness of 1.6 mm. This is significantly higher compared to the initial state with (840.4 ± 1.0) MPa and the specimens with t_{02} = 2.5 mm. Doubling the dwell time to six hours results in a decrease of tensile strength of carburized samples to (855.4 ± 23.5) MPa for t_{01} = 1.6 mm and (809.9 ± 21.8) MPa for t_{02} = 2.5 mm, respectively. This can be explained by decarburization and continuous grain growth as simultaneous effects to carburization. The tensile strength of specimens without graphite coating remains on the same level as before.

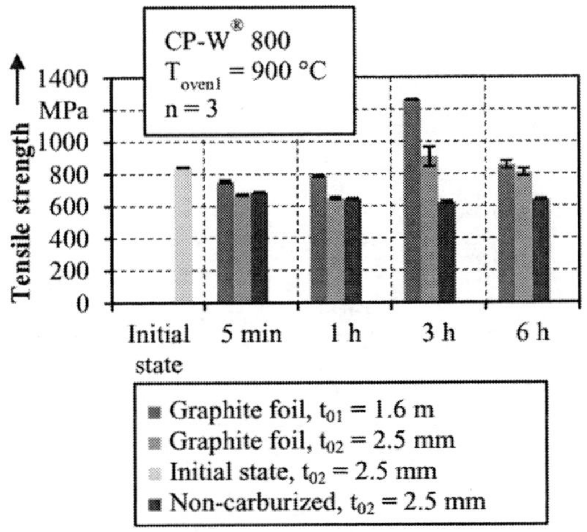

Fig. 3. Tensile strength of coated and uncoated specimens after a heat treatment at 900 °C

Figure 4 shows the resulting tensile strength after a heat treatment at 950 °C and subsequent quenching. The trend of tensile strength is similar to Figure 3, with a difference only after five minutes of heat treatment. Contrary to Figure 3, the carburized samples with a sheet thickness of 2.5 mm show the highest tensile strength with (726.7 ± 25.0) MPa. Considering the standard devitiation, the difference to samples with a thickness of 1.6 mm is not significant. Except for the samples heat treated for five minutes at 900 °C, the investigated specimens with a sheet thickness of 1.6 mm achieve higher values of tensile strength compared to those with a thickness of 2.5 mm. With decreased sheet thickness, the diffusion path of carbon atoms is decreased as well. As a result, the strength gradient over the sheet thickness is smaller compared to t_{02} = 2.5 mm. Therefore, overall tensile strength of the samples is increased.

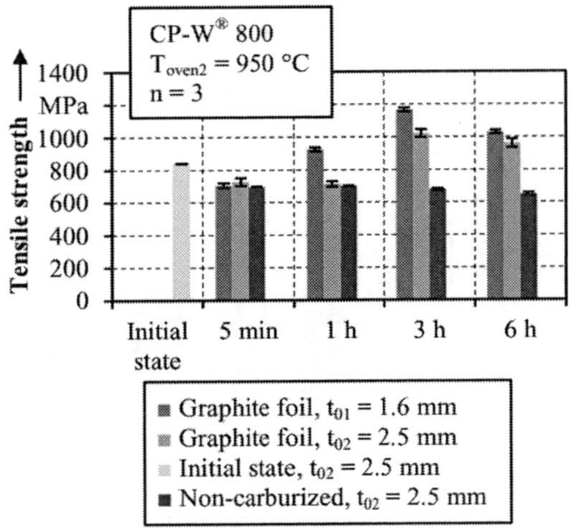

Fig. 4. Tensile strength of coated and uncoated specimens after a heat treatment at 950 °C

Comparing the absolute values of tensile strength in Figure 3 and Figure 4, further differences are visible. It can be seen, that a heat treatment at 950 °C results in higher tensile strength than a heat treatment at 900 °C. An exception can be seen for the maximum in Figure 4 with (1167.2 ± 12.8) MPa for three hours and t_{01} = 1.6 mm. This is about 90 MPa lower compared to the maximum in Figure 3. Basically, carburization is improved at higher temperatures. This explains the elevated values for tensile strength after a heat treatment at 950 °C. Furthermore, not only carburization, but also decarburization is reinforced when temperature is increased. Parallel ongoing decarburization and grain growth are an explanation that the overall maximum tensile strength is reached at a heat treatment temperature of 900 °C instead of 950 °C.

3.2 Vickers hardness

Figure 5 shows the Vickers hardness HV 0.1 after carburization between five minutes and six hours at 900 °C and 950 °C in comparison with the initial state before heat treatment. The measurements are conducted

https://doi.org/10.1051/matecconf/201819008004

in the middle plane of the samples' cross section. Furthermore, the distribution of Vickers hardness HV 0.1 along the sheet thickness is shown in Figure 7 for selected samples.

Figure 5 corresponds well with the findings from tensile testing. Carburization shorter than three hours results in lower Vickers hardness compared to the initial state with (353.4 ± 2.1) HV 0.1. The only exception is a heat treatment of one hour at 950 °C, as already seen in Figure 4. For a sheet thickness of 2.5 mm, a significant improvement of Vickers hardness up to (406.2 ± 9.4) HV 0.1 can be achieved only for six hours of carburization at a temperature of 950 °C. Again, the best results can be seen for a sheet thickness of 1.6 mm. The maximum is (474.4 ± 4.2) HV 0.1 for carburization parameters of three hours and 950 °C. After six hours, the hardness is only around 8.5 HV 0.1 below the maximum. For 900 °C, the difference between three and six hours is 24.6 HV 0.1 with (450.4 ± 3.6) HV 0.1 as the higher value after three hours. Contrary to Figure 3 and Figure 4, the overall maximum is achieved for three hours at 950 °C instead of 900 °C.

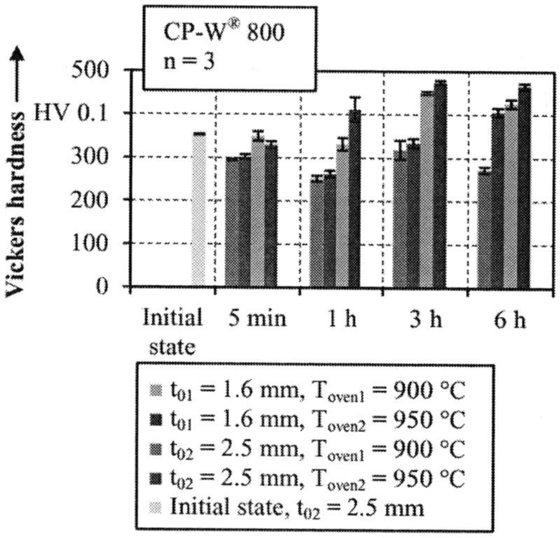

Fig. 5. Vickers hardness of carburized samples in comparison with initial state

For further investigation, the distribution of Vickers hardness along the sheet thickness is analyzed. The location of the measuring points is shown in Figure 6, while the resulting values can be seen in Figure 7. The distance between the measuring points is (97.8 ± 0.4) µm for samples with a thickness of 1.6 mm and (112.0 ± 0.02) µm for specimens with a thickness of 2.5 mm.

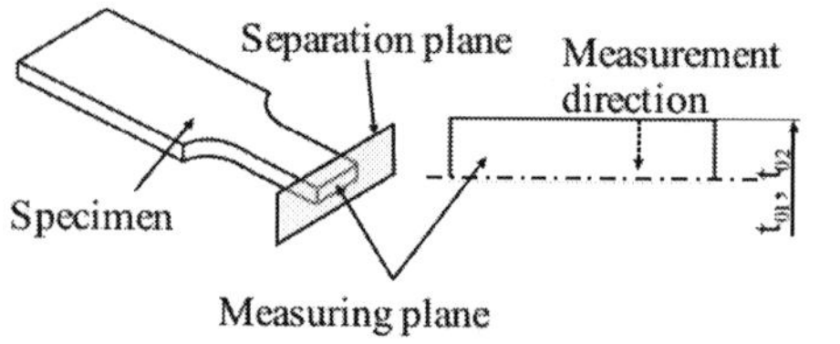

Fig. 6. Location of measuring points for the distribution of Vickers hardness along the sheet thickness.

Figure 7 shows the influence of carburization time. After three hours at 950 °C, an almost homogeneous hardness distribution with values ranging from (459.6 ± 0.04) HV 0.1 to (472.7 ± 9.5) HV 0.1 can be seen. As already mentioned, decarburization and grain growth are two parallel ongoing effects. This is shown for samples with a heat treatment time of six hours. The hardness is decreased to (366.6 ± 6.6) HV 0.1 for a thickness of 1.6 mm and (356.0 ± 7.8) HV 0.1 for samples with t_{02} = 2.5 mm, respectively. Vickers hardness increases with growing distance from the surface. From the surface to a distance of around 0.2 mm, samples carburized for six hours at 950 °C show high accordance in their hardness, independent of the sheet thickness. From there to the mid-section, increased Vickers hardness can be seen for samples with a thickness of 1.6 mm. Due to the lower thickness, diffusion of carbon has proceeded into deeper regions of the material with regard to the overall thickness. Furthermore, the distribution of hardness is more homogeneous in the sheet plane as well. This can be seen by increased standard deviation for specimens with a thickness of 2.5 mm.

Figure 7 also shows an explanation for the different mechanical properties for specimens with carburization temperatures of 900 and 950 °C after a holding time of three hours. As already mentioned, after carburization of three hours at 950 °C, a very homogenous hardness distribution is achieved. Contrary to that, a heat treatment temperature of 900 °C results in a clear hardness gradient. In the middle of these samples, Vickers hardness is around 12.6 to 20.2 HV 0.1 below the values of specimens with a heat treatment temperature of 950 °C. This can be explained with a decreased diffusion coefficient due to lower temperature. Apart from the midsection, samples carburized at 900 °C show up to 82.3 HV 0.1 higher values for Vickers hardness. Besides decarburization, which is especially present in the surface near region, grain growth due to elevated temperatures is a possible explanation [13], which starts around 950 °C [14].

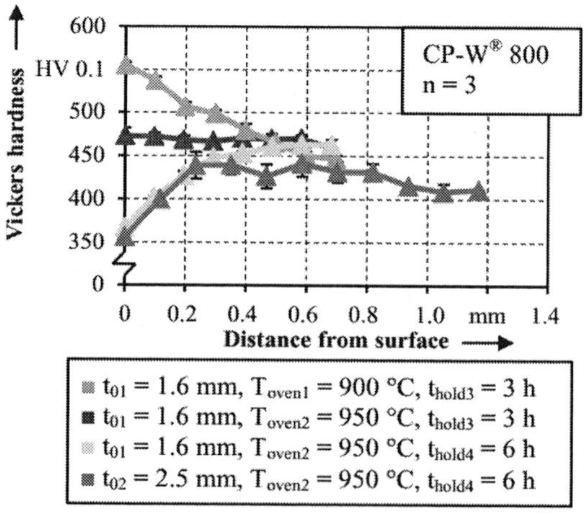

Fig 7. Distribution of Vickers hardness along the sheet thickness of selected samples with a sheet thickness of 1.6 and 2.5 mm after carburization of three and six hours at 900 and 950 °C

Calculating the average Vickers hardness along the sheet thickness, highest values are achieved for sheets with a thickness of 1.6 mm and carburization of three hours at 900 °C with (492.4 ± 19.1) HV 0.1, followed by three hours at 950 °C with (468.7 ± 5.3) HV 0.1 and six hours at 950 °C with (435.4 ± 16.9) HV 0.1. The lowest average Vickers hardness along the sheet thickness can be seen for samples with a thickness of 2.5 mm and heat treatment parameters of six hours and 950 °C with (419.5 ± 16.2) HV 0.1. This corresponds well with the results from tensile testing.

3.3 EDX analysis

Figures 8 and 9 show the carbon content of four different samples measured with EDX spectroscopy. The carbon content of samples carburized at 900 °C is illustrated in Figure 8, while the measurement data for specimens heat treated at 950 °C can be seen in Figure 9. The values were measured via line measurement along the sheet thickness. For the presentation of the results the measured values were averaged over a range of 100 µm. This was done near the surface and in the mid-section of the samples.

The carbon content of samples with a sheet thickness of 1.6 mm carburized for three hours at 900 °C is depicted in Figure 8. The results are similar to the results from Vickers hardness measurements. In the mid-section, the carbon content is lower compared to the values in the surface-near area of the specimens. Considering the standard deviation, the effect is not significant. In addition to the statistical error, displayed by the standard deviation, the measurement data contains an overall absolute measurement error due to low accuracy of energy-dispersive X-ray spectroscopy on light elements such as carbon [15]. Therefore, only a qualitative analysis of carbon content is done within this work.

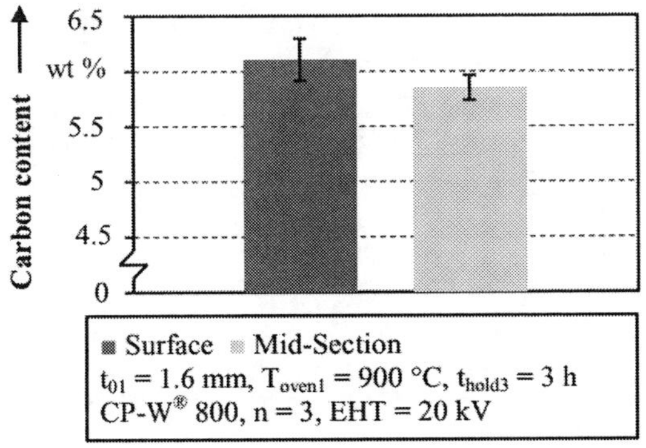

Fig 8. Carbon content of sample with a sheet thickness of 1.6 mm carburized for three hours at 900 °C measured via EDX spectroscopy

The carbon content of samples carburized at 950 °C can be seen in Figure 9. Regarding the specimens with a holding time of three hours and a thickness of 1.6 mm, a very homogeneous distribution of carbon content can be seen. There are only small differences between the mid-section and near-surface areas. A comparable behavior was already shown for the distribution of Vickers hardness HV 0.1 in Figure 7. The standard deviation in the mid-section is increased compared to the surface-near area, which might be a sign for inhomogeneous carburization along the sheet plane. Since possible contaminations of the vacuum chamber have an influence on the measurements, a comparison of the results of EDX spectroscopy amongst each other is not possible.

The results for samples with a thickness of 2.5 mm carburized for six hours at 950 °C correspond well with the findings from Figure 7. Comparing the carbon content in the surface-near area to the values in the mid-section, a clear difference can be seen. The weight percentage of carbon in the mid-section is significantly higher in comparison with the weight percentage of carbon near the surface. This supports the assumption, that decarburization takes place simultaneously to the desired carburization process.

The carbon content of samples with a sheet thickness of 1.6 mm and carburization parameters of six hours and 950 °C can be seen in Figure 9, too. The results are contrary to the findings in Figure 7. The Vickers hardness measurements showed that the hardness is decreased near the surface. Decarburization was stated as possible reason for the gradient. The carbon content measured via energy-dispersive X-ray spectroscopy shows a different behavior. The weight percentage of carbon is increased in the surface-near area and lower in the mid-section of the samples.

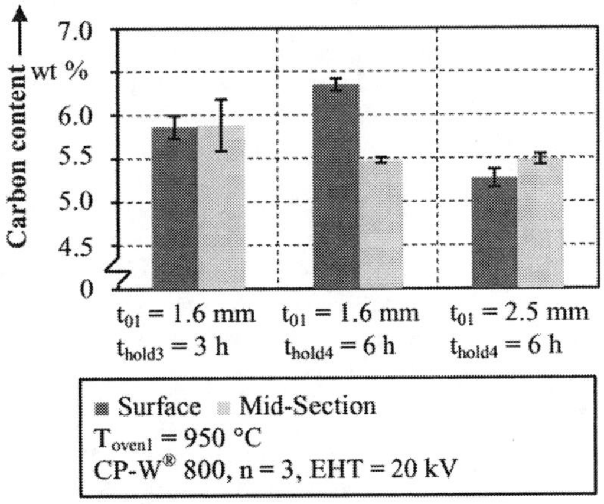

Fig 9. Carbon content of samples carburized at 950 °C measured via EDX spectroscopy

On basis of the investigations carried out within this work, EDX is not applicable for neither quantitative nor qualitative analysis of carbon content. In order to verify the findings in Figures 8 and 9, or as a possible alternative with increased accuracy, wavelength-dispersive X-ray spectroscopy (WDX) has to be taken into consideration. Since WDX spectroscopy faces the same challenges in measuring light elements such as carbon, further investigations have to be carried out as well.

4 Conclusion

Within this work, different carburization parameters such as temperature and holding time as well as the influence of sheet thickness have been investigated. Therefore, the samples were coated with graphite foil, heat treated and subsequently quenched. It was shown, that an improvement of mechanical properties of sheet metal is achievable by carburization. The tensile tests indicated, that sheet thickness has an significant influence on mechanical properties after carburization. Samples with a thickness of 1.6 mm showed higher tensile strengths compared to specimens with a thickness of 2.5 mm. Furthermore, it was revealed, that negative effects such as grain growth and decarburization have an influence as well. This was especially apparent for samples carburized longer than three hours. Since temperature and dwell time have a positive as well as a negative effect on the resulting mechanical properties, the process window is limited to three hours for both investigated temperatures. An extended process window includes carburization times of up to six hours for a temperature of 950 °C. This process window is applicable for both investigated sheet thicknesses. Considering an industrial application of the process, the identified process window is not feasible for production lines with cycle times of few seconds, but is an alternative for more flexible productions of high strength crash relevant parts.

The qualitative results from EDX spectroscopy were partly in accordance to the findings from other investigations such as Vickers hardness. Taking into account an overall error due to the limitations of the measurement method, a quantitative analysis of carbon content was not possible. As a possible improvement in the determination of carbon content, the application of wavelength-dispersive X-ray spectroscopy has to be investigated. Although it offers increased accuracy, WDX faces the same challenges in measuring light elements.

Since an applicable process window was identified within this work, further research has to focus on the detailed investigation of the transition zone between carburized and non-carburized areas.

The authors would like to thank the German Research Foundation (DFG) for funding the project "Tailored Carburization: Gradierung mechanischer Eigenschaften in der Blechebene und über die Blechdicke" (DFG ME 2043/51-1). Moreover, the authors acknowledge the investigations of Yang Yang within the framework of his master thesis at the Institute of Manufacturing Technology (LFT).

References

1. A. Turetta, S. Bruschi, A. Ghiotti, JMPT **117**, pp. 396-400 (2006)
2. M. Skrikerud, J. Friberg, "Press hardening, Process Design and Tool Systems" *Proceedings on the 4th International Conference on Hot Sheet Metal Forming of High-Performance Steel - CHS2*, edited by M. Oldenburg, B.Prakash and K. Steinhoff, (Luleå 2013), pp. 423-430.
3. R. Neugebauer, T. Altan, M. Geiger, M. Kleiner, A. Sterzing, CIRP Annals **55** (2), pp. 793-816 (2006)
4. H. Karbasian, A.E. Tekkaya, JMPT **210** (15), pp. 2103-2118 (2010)
5. S. Bruschi, A. Ghiotti, Comprehensive Materials Processing **3**, pp. 27-54 (2014)
6. L. Pelcastre, J. Hardell, A. Rolland, B. Prakash, JMPT **228**, pp. 117-124 (2016)
7. C. W. Lee, D. W. Fan, R. Sohn, S. J. Lee, R. Cooman, Metallurgical and Materials Transactions A **43A**, pp. 5122-5127 (2012)
8. C. Schäffner, "Konduktive Erwärmung für die Warmblechumformung", Dissertation, Universität Hannover, 2009
9. M. Merklein, M. Wieland, M. Lechner, S. Bruschi, A. Ghiotti, JMPT **228**, pp. 11-24 (2016).
10. P. Feuser, "Ein Ansatz zur Herstellung von pressgehärteten Karosseriekomponenten mit maßgeschneiderten mechanischen Eigenschaften: Temperierte Umformwerkzeuge, Prozessfenster, Prozesssimulation und funktionale Untersuchung", Dissertation, Friedrich-Alexander-Universität Erlangen-Nürnberg, 2012.
11. B.F. Rolfe, A. Abdollahpoor, Y. Wang, M. Pereira, H. Kong, D. Fabijanic, K. Hu, R. Han, K. Gao, A. Liu, L. Pan, Y. Bi, Z.J. Wang, Y.S. Zhang, "Local Patchworking of Sheet Properties Combining Carburization and Hot Stamping" in *Advanced High Strength Steel and Press Hardening*, WORLD SCIENTIFIC 2017, pp. 346-351
12. N.N.: Complex phase steels CP-W® and CP-K®. thyssenkrupp, 2017
13. T. Svec, M. Merklein, "Prozessfenster zur Verarbeitung alternativer Stahlgüten im Presshärtprozess" in *Proceedings of the 7th Erlanger Workshop Warmblechumformung*, edited by M. Merklein (Erlangen, 2012), pp. 55-70
14. F. Hippenstiel, *Tailored Solutions in Microalloyed Engineering Steels for Power Transmission Industry*, Material Science Forum **539-543**, pp. 4131-4136 (2007)
15. J. J. Friel, *X-ray and Image Analysis in Electron Microscopy*, Princeton Gamma-Tech (1994)

Characterisation of the contact pressure dependent interfacial heat transfer coefficient for a hot stamping process following a data driven approach

Xiaochuan Liu[1], *Yang* Zheng[1], *Omer* El Fakir[1], *Rajab* Said[2] and *LiLiang* Wang[1,*]

[1]Department of Mechanical Engineering, Imperial College London, London, SW7 2AZ, UK
[2]ESI Group, Paris, 75015, France

Abstract. In the present research, the contact pressure characteristics of a hot stamping process was revealed by data mining an existing data base of verified FE simulation results, collected by a cloud-based finite element (FE) simulation platform. Following a data driven approach, a dedicated, novel testing scheme for a hot stamping process was developed and applied to determine the effect of contact pressure on the interfacial heat transfer coefficient (IHTC) between an AA7075 aluminium alloy and cast-iron tools. In this novel testing scheme, only one single IHTC test was required which has improved testing efficiency significantly.

Keywords: Hot stamping, Aluminium, Interfacial heat transfer coefficient

1 Introduction

Aluminium alloys have been widely applied in the sheet hot stamping industry, and their heat transfer characteristics in such processes require a more thorough understanding. The interfacial heat transfer coefficient (IHTC) is one of the most important heat transfer characteristics that determines the cooling rate of a component in a hot stamping process. When the cooling rate is above the critical value, the full mechanical strength of the formed component can be retained after the artificial ageing process [1]. Therefore, the IHTC should be determined and then implemented in finite element (FE) simulations to precisely simulate the temperature evolution and cooling rate of the component during the hot stamping process.

Great efforts have been made to determine the IHTC at different contact pressures, which is one of the most important influential factors that affects the IHTC. In the study of Chang et al. [2], the IHTC between a boron steel and a tool steel was determined at different contact pressures of 0.006, 0.04, 1, 5, 15 and 25 MPa, and found to increase exponentially with increasing contact pressure from approximately 0.3 to 4.6 kW/m^{2}K. A linearly increasing trend of the IHTC with contact pressure was obtained by applying different contact pressures of 4, 8, 15 and 24 MPa, in the study of Caron et al [3]. Another eight different contact pressures were selected from 0 to 24 MPa and applied to investigate the effect of contact pressure on the IHTC between an aluminium alloy and a cast-iron steel by Liu et al. [4]. It was found that the IHTC increased logarithmically with the increasing contact pressure from approximately 0.8 to 15.1 kW/m^{2}K. Larger contact pressures of 30 and 40 MPa were used to determine the IHTC values for an titanium alloy in hot stamping processes in the study of Zhang et al. [5].

In previous studies, different static contact pressure conditions were selected based on experience to determine the IHTC values, which would not satisfy the exact contact pressure conditions encountered in hot stamping processes, leading to a decrease in testing efficiency. In addition, the applied contact pressure may change over extremely short time periods in hot stamping processes, resulting in sudden changes in the IHTC. In the present research, a variable contact pressure trend was revealed by data mining an existing data base of verified FE simulation results to represent the exact contact pressure characteristics of hot stamping processes. Following a data driven approach, a dedicated, novel test scheme was developed and applied to determine the effect of contact pressure on the IHTC between a hot AA7075 aluminium alloy specimen and cold cast-iron tools in a single IHTC test, significantly improving testing efficiency.

2 Development of a novel testing scheme for the characterisation of the contact pressure dependent IHTC

The contact pressure distribution in hot stamping processes was generated by data mining an existing data base of verified hot stamping simulation results run under different forming conditions, collected by a cloud-based finite element (FE) simulation platform Smart Forming [6]. As shown in Fig. 1, the probability of contact pressure for all elements of the tools as a function of progression

* Corresponding author: liliang.wang@imperial.ac.uk

https://doi.org/10.1051/matecconf/201819008005

of hot stamping process could be analysed to develop a novel testing scheme for the characterization of the contact pressure dependent IHTC to ensure that the exact contact pressure conditions encountered in hot stamping processes would be satisfied in the IHTC tests. The x-axis shown in Fig. 1 represented the normalized time periods of a hot stamping process rather than the actual time of a hot stamping process, leading to a dimensionless unit.

During the first time-period from 0 to 0.4 progression of hot stamping process, the contact area of the blank increased slowly from a small value and the contact pressure distribution was concentrated in the contact pressure range I from 0.4 to 0.8 MPa in terms of probability. This indicated that the blank was being compressed between the blankholder and tools during the first time-period, resulting in a small blank area being contacted with the tools at small contact pressures. Therefore, the contact pressure range I determined the IHTC during the blankholding stage. During the second time-period from 0.4 to 0.75 progression of hot stamping process, the blank contact area increased rapidly. Meanwhile, the contact pressure range II from 2 to 7 MPa started to increase and then had the largest probability when the progression of hot stamping process was larger than 0.5. It was deduced that larger contact pressures were applied to deform the blank since the progression of hot stamping process of 0.4. Therefore, the contact pressure range II determined the heat transfer between the blank and tools during the initial forming stage, i.e. the second time-period. During the third time-period from 0.7 to 1 progression of hot stamping process, the blank contact area increased dramatically to 100%, indicating that the blank was continually deformed and subsequently quenched by the tools. The contact pressure range III from 12 to 25 MPa had the largest probability during this time period, dominating the IHTC during the late forming and the subsequent initial quenching stages.

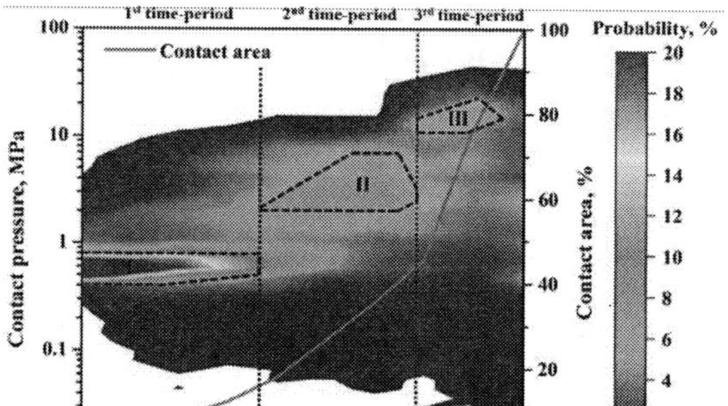

Fig. 1. The contact pressure distribution in the hot stamping processes.

Due to the contact pressure distribution being dominated by three contact pressure ranges, their average values of 0.6, 4.5 and 18.5 MPa were the most representative contact pressures in the hot stamping processes. In order to represent the contact pressure characteristics of hot stamping processes, a variable contact pressure trend, suddenly increasing from 0.6 to 4.5 MPa at 0.4 progression of hot stamping process and finally to 18.5 MPa at 0.75 progression of hot stamping process, could be used as a dedicated novel efficient testing scheme to determine the effect of contact pressure on the IHTC in a single IHTC test. Compared with previous studies, the novel test scheme developed by using the data driven approach for the characterisation of the contact pressure dependent IHTC would significantly improve testing efficiency.

3 Determination of the IHTC using the dedicated novel testing scheme

In the present research, an IHTC test facility was applied to simulate hot stamping processes [4]. A 120 x 10 x 2 mm^3 AA7075 specimen was screwed onto two blankholders, as shown in Fig. 2 (a), and heated by using direct resistance heating. Once the target temperature of 470°C was reached, the specimen was compressed by cold cast-iron punch and die at the pre-defined variable contact pressure, as shown in Fig. 3. A pair of thermocouple was embedded mid-thickness at the centre of the specimen to record its temperature histories during compression.

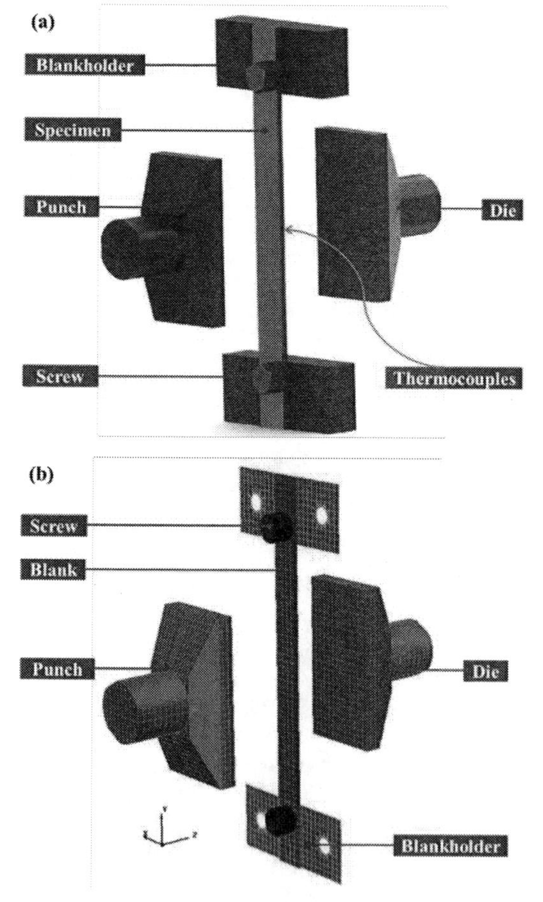

Fig. 2. (a) Overall schematic structure of the IHTC test facility; (b) The FE model of the IHTC test in PAM-STAMP.

 https://doi.org/10.1051/matecconf/201819008005

A FE model was built up in PAM-STAMP to simulate the IHTC tests, as shown in Fig. 2 (b), which was comprised of four stages. In the first gravity and holding stages, the specimen was located on the blankholders and fixed by the screws. In the subsequent stamping and quenching stages, the punch moved towards the specimen and compressed it against the die. A constant IHTC value was implemented between the specimen and tools to obtain the temperature evolution of the specimen during the first time-period, eliminating the effect of contact pressure on the IHTC. The simulated temperature evolution was then compared with the experimental results measured from the same location on the specimen to determine the IHTC value during the first time-period. Subsequently, different constant IHTC values were implemented in the second and third time-periods respectively to obtain the temperature evolution of the specimen after the first time-period. The best agreement between the experimental and simulated temperature evolutions identified the IHTC value during each time-period. Main simulation processing parameters were shown in Table 1.

Table 1. Main simulation processing parameters.

Parameter	Value
Initial specimen temperature	470 °C
Initial tool temperature	25 °C
Punch moving speed	400 mm/s
Element size	2 mm
Element number of specimen	240
Element number of tool	325
Air convection coefficient	20 W/m^{2}K

4 Results and discussions

As shown in Fig. 3, the simulated temperature evolution of the specimen at a constant IHTC value of 2 kW/m^{2}K agreed well with the experimental results during the first time-period from 0 to 0.4 seconds. When the contact pressure suddenly increased from 0.6 to 4.5 MPa at 0.4 seconds, the temperature decreasing rate was significantly enhanced, resulting in a lager IHTC value of 7 kW/m^{2}K during the second time-period. When the contact pressure increased to 18.5 MPa at 0.75 seconds, the temperature decreasing rate was increased again. Due to the good agreement between the experimental and simulated temperature evolutions during the third time-period being achieved, the IHTC was considered to be increased to 11 kW/m^{2}K.

Therefore, the IHTC values at three different contact pressures were determined in a single IHTC test by using the dedicated novel test scheme, significantly improving testing efficiency. In addition, the determined IHTC values could represent the heat transfer between the blank and tools during the blankholding, forming and quenching stages.

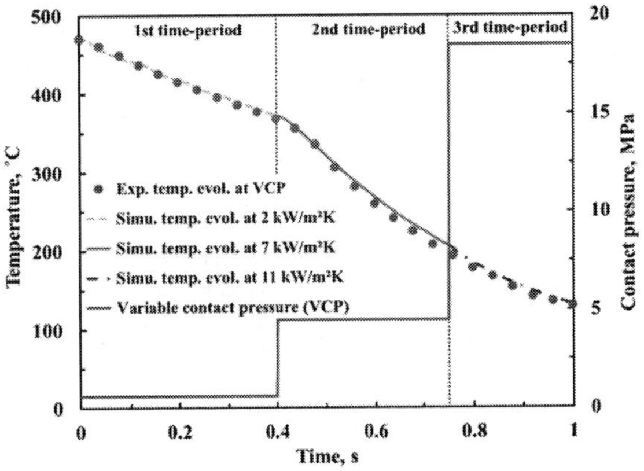

Fig. 3. The experimental and simulated temperature evolutions under variable contact pressure conditions.

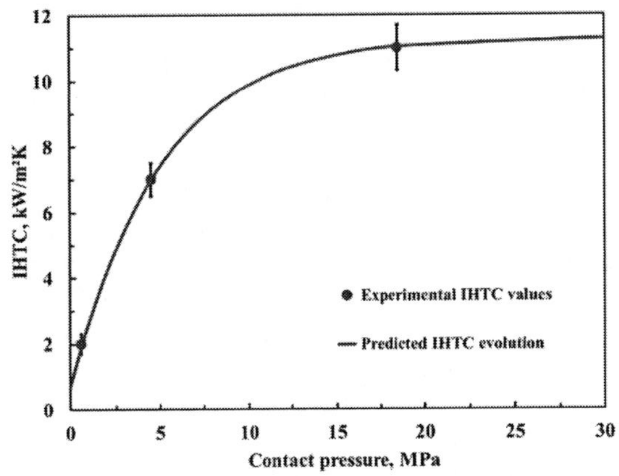

Fig. 4. The experimental and predicted IHTC evolutions with contact pressure.

As shown in Fig. 4, the IHTC increased dramatically from 2 to 7 kW/m^{2}K when the contact pressure increased from 0.6 to 4.5 MPa, followed by a continuous increase to approximately 11 kW/m^{2}K at 18.5 MPa. When the contact pressure was larger than 18.5 MPa, a plateau of the IHTC was observed. This was caused by the evolution of the real contact area between the specimen and tools with contact pressure. Due to the asperities on the contacting surfaces, the real contact area was less than the apparent contact area initially. The asperities on the hot specimen were deformed plastically and elastically by the cold tools during compression, resulting in the increasing real contact area [7] and thus higher IHTC values. When the real contact area approached the apparent contact area, the IHTC reached its peak value and thus the plateau was observed.

The determined IHTC values were applied to the calibration of an experimentally verified IHTC model, which was developed in the study of Liu et al. [4], as shown in Eqs. (1-5).

$$h = h_a + h_c \tag{1}$$

The overall IHTC h is a sum of two independent parts, the null-pressure IHTC h_a, representing the heat transfer across the air gap as a constant value of 0.7 kW/m^{2}K, and the solid-contact IHTC h_c, representing the heat transfer between the contacting solids when the contact pressure is applied.

$$h_c = \alpha \cdot K_{st} \cdot N_P / R_{st} \tag{2}$$

where α is a model parameter, K_{st} is the harmonic mean thermal conductivity of the contacting solids, R_{st} is the root mean square of surface roughness of the contacting solids, and N_P is a pressure dependent parameter, which captures the effect of contact pressure on the IHTC.

$$K_{st} = 2 \cdot k_s k_t / (k_s + k_t) \tag{3}$$

where k_s and k_t are the thermal conductivities of the specimen and tools respectively.

$$R_{st} = \sqrt{R_s^2 + R_t^2} \tag{4}$$

where R_s and R_t are the surface roughnesses of the specimen and tools respectively.

$$N_P = 1 - \exp(-\lambda \cdot P / \sigma_U) \tag{5}$$

where λ is a model parameter, P is the applied contact pressure, and σ_U is the ultimate strength of the specimen at the elevated temperature. Table 2 shows the material properties and calibrated parameters of the IHTC model.

Table 2. Material properties and model parameters of the IHTC model.

h_a	k_s	σ_U	R_s
0.7 kW/m^{2}K	0.14 kW/mK	21 MPa	3.9e-7 nm
R_t	k_t	α	λ
1.8e-7 nm	0.0352 kW/mK	8.1e-5	4.2

The IHTC evolution as a function of contact pressure was therefore predicted by the IHTC model, which could be implemented in the FE simulations to precisely simulate the temperature evolutions and cooling rates of components during hot stamping processes. This would be beneficial to the optimization of the hot stamping processing parameters and tool design to ensure that the critical cooling rates of the formed components were achieved. In addition, an excessive contact pressure could be prevented from being applied to reduce the tool wear and thus increase the tool life.

6 Conclusions

In the present research, a variable contact pressure trend was determined by data mining an existing data base of verified FE simulation results, collected by the cloud-based finite element (FE) simulation platform Smart Forming, in order to satisfy the contact pressure characteristics of hot stamping processes. Following the data driven approach, a dedicated, novel testing scheme was generated and applied to determine the IHTC values between a hot AA7075 aluminium alloy specimen and cold cast-iron tools, which were subsequently applied to the calibration of the experimentally verified IHTC model to predict the IHTC evolution as a function of contact pressure. Therefore, the effect of contact pressure on the IHTC was obtained in a single IHTC test by using the dedicated novel testing scheme, which has significantly improved testing efficiency.

The authors thank the European Commission for their support on the H2020 project 'Low Cost Materials Processing Technologies for Mass Production of Lightweight Vehicles (LoCoMaTech)', Grant No: H2020-NMBP-GV-2016(723517). This research was also supported by the China Scholarship Council (Grant CSC no. 201608060261): A non-profit institution enabling talented Chinese students to undertake a PhD programme overseas.

References

1. M. Raugei, O. El Fakir, L. Wang, J. Lin, D. Morrey, J. Clean. Prod. **83**, 80 (2014)
2. Y. Chang, X. Tang, K. Zhao, P. Hu, Y. Wu, J. Mater. Process. Technol. **228**, 25 (2016)
3. E. Caron, K. J. Daun, M. A. Wells, Int. J. Heat Mass Transf. **71**, 396 (2014)
4. X. Liu, K. Ji, O. El Fakir, H. Fang, M. M. Gharbi, L. Wang, J. Mater. Process. Technol. **247**, 158 (2017)
5. Z. Zhang, X. Li, Y. Zhao, X. Li Metall. Mater. Trans. B Process Metall. Mater. Process. Sci. **45**, 1192 (2014)
6. Smartforming, http://smartforming.com/ (2016)
7. B. Buchner, M. Buchner, B. Buchmayr, Tribol. Int. **42**, 897 (2009)

New Device and Technology for High Speed Hydroforming

Matthias Hermes[1,*], *Aynur* Keskin[1], and *Christoph* Berlinger[1]

[1]Fachgebiet für Umformtechnik und Fügetechnik, Fachhochschule Südwestfalen, 59872 Meschede, Deutschland

Abstract. In the article a new production device is presented that combines the methodologies of high speed forming with conventional hydroforming. The result is a robust production system that can be easy controlled by the use of standard industrial components. The new system can be used as well for high automated production systems for mass production of small tubular and sheet metal parts as for the flexible manufacturing of workpieces with larger dimensions. The device is in principle a hydraulic driven high speed pressure generator that can be mounted directly on a hydroforming tool. Moreover, an additional drive for the axial feed of the tube is installed. In the paper the idea, the design of the device and the technical data is explained. Furthermore, the results of hydroforming experiments are presented and an overview on the specifications and the advantages of the new process is given. The strain rate is measured for an example workpiece and compared to the theoretical value. A first view on the forming limits is a further important topic of the paper. For the investigated copper tube, the forming limits are extended compared to a quasi-static hydroforming test.

Keywords: Hydroforming, Machine, Process

1 Introduction

Hydroforming is an important production technology since many years. Complex thin walled lightweight components for example for automotive applications can be produced in one process step on special production devices [1]. For smaller parts this technology is often inefficient because of high machine costs and the relatively high process time.

In recent years high speed forming methods were developed to produce sheet metal or tubular parts. Advantages of the high forming speed are among other things the extension of process limits and the production of workpieces with high geometrical complexity and accuracy in one process step [2, 3]. Very well-known is the electromagnetic forming process that can be used for forming operations on tubular parts and sheet metal parts [4, 5]. For bigger workpieces the energy of the electromagnetic devices is not sufficient. High speed forming technology then is driven by explosive materials or the detonation of gas mixtures. The problem of all high speed methods is the difficult control and automation of the process by standard control systems. Moreover, tools and production system components are not robust enough for industrial application and bigger production lots.

Obviously better control possibilities seem to have the plunger based systems because of the more conventional drive technology [6]. This technology is robust and components can be adjusted by conventional mechanical engineers. The problem of these processes is the limit in the production of complex large-surface parts because of the short pressure pulse. In the state of the art large surface parts are produced by quasi-static processes like deep drawing or conventional hydroforming. Thus a good innovation is the combination of the methodologies of high speed forming with conventional hydroforming. An interesting possible approach is the combination of quasi-static high pressure sheet metal hydroforming and pneumo-mechanical high speed forming on one and the same high speed forming machine [7].

In the following paper a new process and production device is shown that goes a similar way. It combines the methodologies of high speed forming with conventional tube hydroforming (THF). The result is a robust production system that is easy to control by standard industrial components. The new system can be used as well for highly automated production systems for mass production of smaller tubular and sheet metal parts as for the flexible manufacturing of workpieces with larger dimensions. The device is in principle a hydraulic driven high speed pressure generator that can be mounted directly on the hydroforming tool. When the idea was created a realistic approach was made how fast a hydraulic driven system could be. The idea was to fill the gap somewhere between conventional forming technology and the explosive forming or electromagnetic forming. To get a technological relevance the forming pressure should be as high as a conventional hydroforming process. In Fig. 1 (acc. to [8]) the planned range of forming speed and forming pressure for the new device is shown.

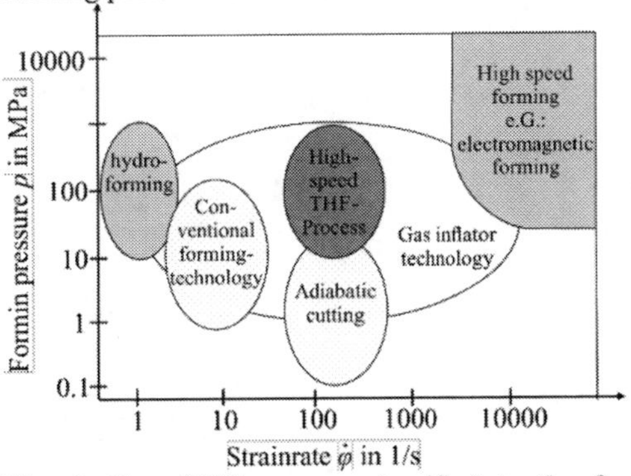

Fig. 1 New THF process classified to the forming speed according to [8]

* Corresponding author: hermes.matthias@fh-swf.de

 https://doi.org/10.1051/matecconf/201819009001

2 Analysis of the process idea

To develop a new device that fits to the process specifications of tube hydroforming (THF) the principle parameters has to be clarified. The THF-process has different parameters that are of interest. In Fig. 1 the different parameters are shown.

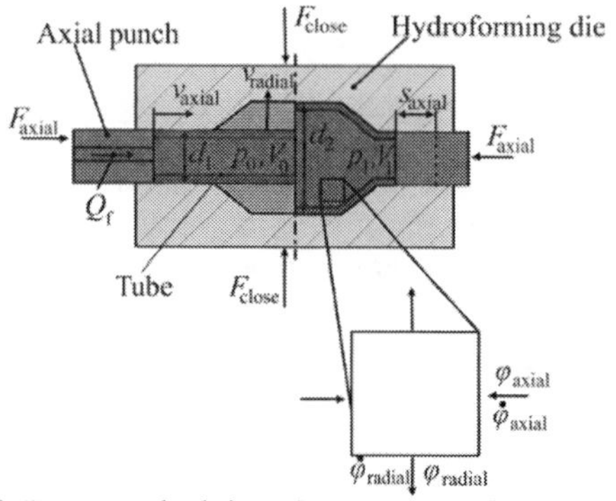

Fig. 2. Process principle and parameters of conventional and high speed THF process

The initial parameter d_1, the final diameter d_2 and the superposed axial feed s_{axial} of the tube are of interest for quasi static THF process. Moreover, the filling pressure p_0 and the forming pressure p_1 have to be defined for a good working process. For high speed processes the parameters v_{axial} und v_{radial} are of additional interest because they define the forming speed during the process. The radial forming speed is influenced by the flow rate Q_f. To reach extended forming limits and big diameters d_2 without cracks it is important to superpose an axial feed and efforts a high speed drive at the axial punch and a fast control system. Then it is possible to move the feed system with a high axial speed v_{axial}. Based on these thoughts an interesting process variant can be designed. For explanation in Fig. 2 a hypothetic THF process with low and with high speed (HSTHF) is shown.

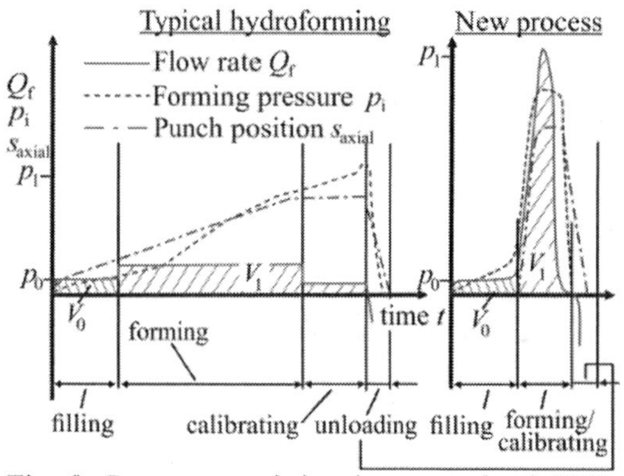

Fig. 3. Parameter variation in conventional THF and potential of HSTHF.

On the left side a typical hydroforming process diagram is shown. A relatively low flowrate fills the tube to V_0 and then V_1 is reached by further filling with a high pressure system. Simultaneously the axial feed pushes the tube into the die and the forming zone. The last optional step is calibrating the workpiece with high pressure and a small flow rate of fluid.

The diagram on the right shows a HSTHF process with superposed axial feed. The filling of V_0 is a quasi-static process, too. The second step shows the filling of V_1 by a high pressure pulse. The volume V_1 is filled during this pulse. This can be reached by a very high flow rate realized by a special high pressure pump that has a high speed drive. Moreover, the volume $V_2 - V_1$ can be defined exactly by the pump and an overloading in principle is not necessary because the calibration step and the main forming step is superposed in one small time step. Moreover, it is thinkable that by the high short pressure pulse shockwaves are superposed with the simple pumping of the fluid. This can influence the forming process as well. Additionally, the axial feed of the tube is realized by a high speed drive and can be started during the same time step or earlier. Compared to conventional THF there are some differences that occurs an interesting potential:

1. The short pressure pulse and the high flow rate make high forming speeds possible. So the typical extended forming limits because of high strain rates are thinkable. Moreover, the superposed axial feed makes it possible to reduce the thinning of the tube wall thickness and enhance the forming limits further.
2. The high forming speed achieves a high radial speed v_{radial} and so a high accuracy and surface quality can be achieved. Moreover, complex geometries with small radii can be formed and calibrated.
3. The short process time has the potential to use the process in mass production and can be coupled in process chains with other fast production processes with short circle times.
4. The closing force F_{close} can be reduced by the short pressure pulse and the defined volume of fluid $V_2 - V_1$. That allows locking the die for example by mass inertia.
5. The defined volume of forming fluid that is pumped into the workpiece allows producing big surface parts, because there is more than a short pressure pulse like in other high speed forming methods.
6. Sealing is often a problem at fluid based forming processes. Because of the high flow rate small leakage means no eminent reduction of the pressure maximum because of the dependency of pressure and flow rate.
7. The short pressure pulse gives the potential of shock waves superposed with the high flowrate that can enhance the forming process.

These potentials have to be investigated and achieved by an experimental device that was developed at the laboratory. The device is designed very flexible, so it can realize a big range of parameter combinations and process curves according to Fig. 2. The design and the specifications are shown in chapter 2.

2 Concept and design of the device

The list of requirements for the device includes the following main points:

- Realization of a high flow rate,
- Realization of a high and short pressure pulse,
- Realization of a high speed drive for axial feeding of the tube,
- the use of conventional hydraulic components and a conventional PLC,
- and a design that makes a flexible adjusting of all parameters and further developments possible

Considering these requirements, a flexible testing device for HSHF was designed [9]. The functional scheme of the new device is shown in Fig. 3.
The high flow rate is based on at least one hydraulic accumulator that is loaded with a standard hydraulic pump with relatively small flow rate. In the realized device two accumulators were used. The accumulator characteristic can be adjusted by a N_2 bottle and a high pressure gas valve. In a steel block a cartridge seat valve NG 46 is mounted. This size is selected to reduce pressure losses and to reach a high flow rate in the primary side of the device.

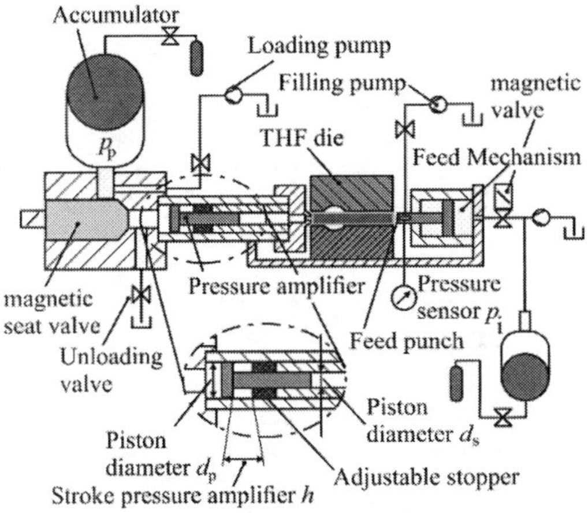

Fig. 4. Functional scheme of the forming device

To start the process the seat valve is opened by the control system (Siemens S7) and hydraulic fluid runs into a pressure amplifier. The pressure pulse pushes the piston onto an adjustable stopper that defines the stroke. By that the volume of forming fluid is pumped on the secondary side of the system into the THF-die. The pressure amplifier raises the pressure of the conventional hydraulic pressure to the forming pressure and separates the hydraulic oil from the water based forming fluid. The flow rate is of course reduced. The transformation ratio of the amplifier can be adjusted by exchangeable piston/sleeve inserts. The primary pressure pump (loading pump) is a conventional hydraulic pump that defines the pressure in the low pressure side of the pressure amplifier (p_p = 30 MPa). At the moment two variants of amplifiers are possible to mount in the system. The first delivers a forming pressure of 250 MPa but has because of a smaller bore on the secondary side a lower flow rate. The second amplifier has a bigger flow rate and a maximum pressure of 140 MPa. The primary flow rate is defined by the accumulators. They have each the potential of 30 l/s.
On the other side of the THF-die an axial feed system is installed. This has a hydraulic drive system with an accumulator. By that the axial feed can be driven also with higher speed values. The start signal for the installed magnetic valve is controlled by the S7 also. The start time can be adjusted in steps of 0.05 sec. So the valve can be opened simultaneously, before or after the process start of the forming process.
The feed punch is prepared for the filling system that fills the secondary system and the tube with the forming fluid. The filling pump has enough pressure to push the pressure amplifier in the start position. Moreover, it is possible to pre-form the workpiece quasi-static and to use the system for conventional THF (p_{max} = 360 MPa). The device and its control system allow the investigation of a wide range of different HSTHF process variants in addition to the variants shown in Fig. 2. The design of the device is very flexible and is continuously extendable. In Table 1 actual specifications are listed.

Table 1. Specifications of the devive (status quo).

Specification	Potential
Filling pump pressure $p_{0,max}$	0…360 MPa
Primary pressure max $p_{p,max}$	30 Mpa
Secondary pressure (forming pressure, depends on plunger diameter) max. $p_{i,max}$	250 (140) MPa
Primary flow rate $Q_{p,max}$	approx. 60 l/s
Secondary flow rate max. $Q_{f,max}$	approx. 7,5 (11,5) l/s
Force of the axial feed punch max. $F_{axial, max}$	7,8 kN
Axial feed speed max $v_{axial,max}$	approx. 3,5 m/s

In Fig. 4 the experimental setup is shown in reality. The red painted high pressure device is coupled with flexible pipes and control cables. This makes it possible to mount the device on other production systems for example stamping dies or special tube forming machines. The device is continuously developed. An important aim of the project is getting knowledge for future industrial machines.

https://doi.org/10.1051/matecconf/201819009001

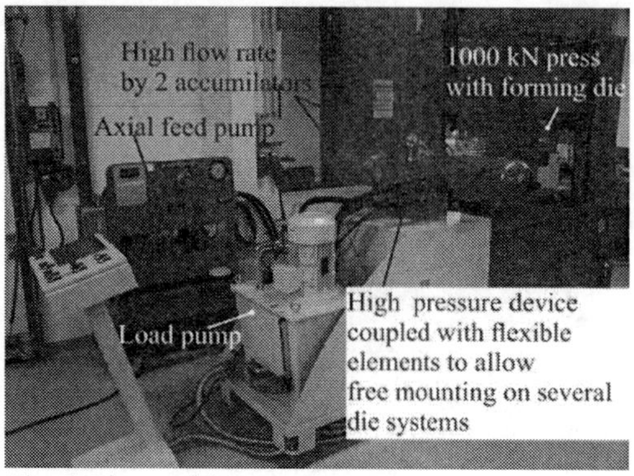

Fig. 5. Photo of the experimental setup

3 First results

To get more information about the process potential the theoretical values were investigated by experiments.
In Fig 5 the pressure rise of the 140-MPa-amplifier is shown. It can be said that the pressure amplifier generates the expected pressure values. The rise is more interesting with view on the forming speed. At the current experiments the rise is generated in time steps smaller than 20 ms. For future developments this should be shorter. The reason lies in the opening time of the seat valve of 40 ms. The aim is to reach 1 or 2 ms by changing the seat valve to a faster one. In the described concept in Fig. 2 the pressure peak is short and falls after reaching the maximum. This wasn't tested till now. In principle it is possible by unloading the primary pressure by the unloading valve (Fig. 3).

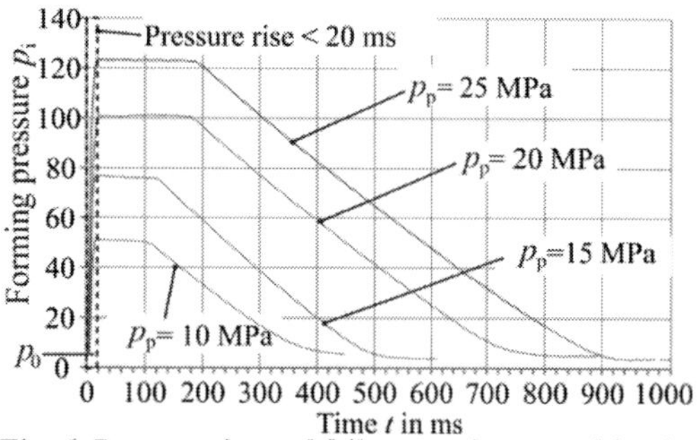

Fig. 6. Pressure rise and fall at actual status of the device.

More interesting is the radial and axial forming speed v_{axial} and v_{radial}. Till now some investigations regarding v_{rad} were done. The theoretical value can be calculated by the time t_f that is needed to fill the workpiece geometrie with the volume $V_1 + V_2$ during the forming process. This can be calculated with the theoretical maximum flow rate of the device:

$$t_f = \frac{V_1+V_2}{Q_{f,max}} \quad (1)$$

The average strainrate assuming a pure radial forming can be estimated with

$$\dot{\varphi} \approx \frac{ln\frac{d_2}{d_1}}{t_f}, \quad (2)$$

where d_1 und d_2 depend of the die design. To prove this theory experiments were made with a die that was mounted into the device. A copper tube d_1 = 22 x 1 mm was used as specimen. Special probes where developed to measure by contact the time between starting and the end of the expansion. The experimental setup and the result are shown in Fig. 6.

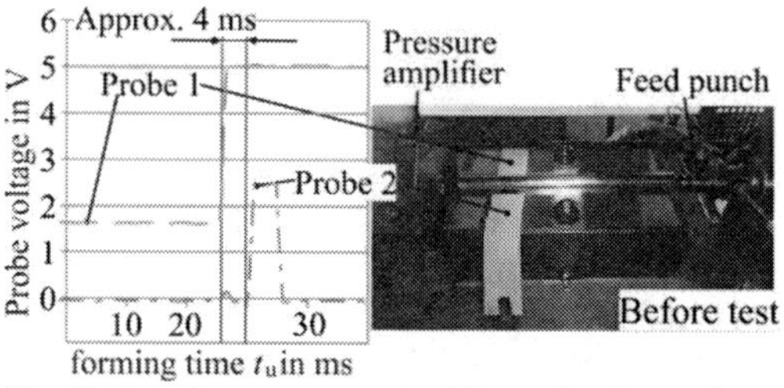

Fig. 7. Speed measurement with conductive probes

The probes where made of thin copper sheets and coated with isolation paint. The probes where positioned between the die halves and conducted to a fast measurement system that determines the time of contact between the tube wall and the probe 1 and probe 2 by a voltage signal. The time between the contacts is about 4 ms with maximum loading of the device and using the 250 MPa pressure amplifiers. With (1) and (2) a strain rate of approx. 70 s^{-1} could be calculated. This fits very well to the theoretical value. It has to be said that this is not very fast for ha high speed process but even faster than quasi static processes. The influence of additional axial feed on the forming speed is not yet tested. The 140-MPa-amplifier can approximately deliver the double forming speed. Interesting is that the relatively slow seat valve does not influence the process. In future the forming speed is planned be raised by further development of the device. A second interesting experiment is shown in Fig. 7.
Here the influence of the axial feed combined with the high forming speed was tested. The results are shown at the example of copper tubes (d_1 = 22 x 1 mm, material R250 HH) that were expanded to d_2 = 33 mm into a forming die. For lubrication industrial petroleum jelly was used. For comparison a tests under quasi static forming were done and failed by a typical tangential crack in the tube wall. The next test group were made under dynamic forming and shows no crack but a stronger thinning. Moreover, in some areas on the surface the material localizes and the stadium before cracking could be detected. The thinning was analyzed by cutting the tube in halve. The tubes were 9 mm shorter so the forming comes not only out of the wall thickness. The second test where superposed by an axial feed with s_{axial} = 17 mm. The workpieces has a little less thinning in the expanded area. So even in this early state of the investigations it can be said that the axial feed has a positive effect on the HSTHF process.

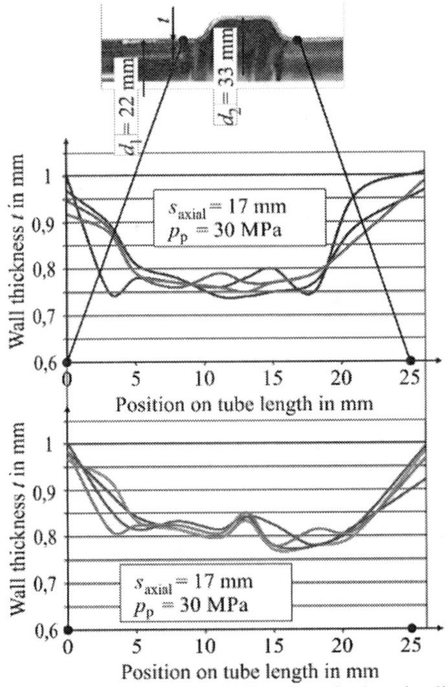

Fig. 8. Wall thickness deviation over the longitudinal axis

4 Conclusion and outlook

First step of the work was the design of a flexible high speed hydroforming device that uses conventional hydraulic and control components. The technical function of the system was tested and showed a good performance. First aim of the investigations with the device concentrated on the function of the machine. Input output relation of the parameters fit to the requirements. Investigations regarding the strain rates showed that parameters can be controlled with a good precision. Further tests were made with copper tubes. They are expanded from 22 mm on 33 mm, showed enhanced process limits in comparison to the quasi static THF process. Moreover, it could be seen that the installed high speed axial feed system reduces thinning of the wall thickness.

The device was designed to make future upgrading easy possible. Aim of the next work is to extend values of forming speed, flow rate and pressure pulse. The Pressure pulse should be shortened.

Furthermore, aim of future process investigations is the variation of workpiece materials. Moreover, the forming limits, the variation of the parameters forming time and the effect of axial feed will be investigated.

At the moment a FE-model is developed to investigate process mechanisms in detail.

Further activities in the field of applied research are the integration of the device in coupled processes and in special forming machines to analyze the possibilities of process automation. Furthermore, the superposition of other production operations for example high speed blanking will be investigated.

The project is supported by transfluid Maschinenbau GmbH, Schmallenberg, Germany. The company has helped to build the device and supports the experiments with raw material and tools.

References

1. R. Neugebauer, *Hydro-Umformung*. ISBN-13 978-3-540-21171-6, Springer 2007
2. X. Hu, G.S. Daehn, *Acta mater.* **44**, 1021 (1996)
3. V. Balanethiram, X. Hu, M. Altynova, G.S. Daehn,. *Journal of Materials Processing Technology*, **45**, 595, (1994)
4. C. Beerwald, *Grundlagen der Prozessauslegung und –gestaltung bei der elektromagnetischen Umformung*. Dissertation, Dortmund (2004)
5. V. Psyk, D. Risch, B. L. Kinsey, A. E. Tekkaya, M. Kleiner, *Journal of Materials Processing Technology*, **211**, 5, 787, (2011),
6. W. Homberg, E. Djakow, O. Akst, *ICHSF 2012*, Dortmund, Germany, 2012
7. E. Djakow, W. Homberg, D. Tabakajew, *ICHSF 2016*, Dortmund, Germany, 2016
8. H. Hoffmann, R. Neugebauer, G. Spur, *Handbuch Umformen*. ISBN: 978-3-446-42778-5, Hanser 2012
9. M. Hermes, M. Böhmer, *European patent application*, EP 17400044, (2017)

Springback prediction of CP2 titanium sheets in hydroforming with membrane diaphragm process with finite element method

Cahit Sertac Aydogan[1,*], *Hasan Ali* Hatipoglu[1], and *Omer* Keles[2]

[1]Tusas Aerospace Industries (TAI), Akıncı, Ankara, Turkey
[2]Gazi university, Department of Mechanical Engineering, Maltepe, Ankara, Turkey

Abstract. This study aims to predict and to examine the springback of CP2 Titanium sheets in hydroforming with membrane diaphragm process, which is the branch of hydroforming process. The Hydroforming with Membrane Diaphragm was used for experimental investigation, in which specified CP2 titanium sheets were bent with various radii and angles on a specifically designed die. Springback results were then used to validate the finite element model constructed previously. Results show that the numerical model of this study can be used to predict the approximate springback values. Thus, the scrap quantities of the sheet metal titanium products can be reduced by making some modifications on the die, such as the springback compensation based on the obtained approximate springback values.

Keywords: Hydroforming, Springback, Finite element method (FEM)

1 Introduction

The Hydroforming with membrane diaphragm is a type of forming process. In this technique, a chamber filled by the fluid is covered by an elastic diaphragm. The surface of the diaphragm behaves as a comprehensive punch that can fit any shape of a rigid die by the action of fluid pressure (Fig. 1). The advantages of the process are that there is less springback value of the sheet, strain distribution is more even compare to other methods because of the smooth effect of fluid pressure and lastly, high quality parts with less damage to the sheet. The disadvantages are that high capacity press is needed to withstand the high pressures (up to 80 MPa) in the chamber; this method is suitable for producing a small number of parts. Finally, design precautions must be taken to avoid the tearing or wrinkling of parts because final geometry of the part is more challenging than in conventional stamping. So, qualified skilled workforce is needed.

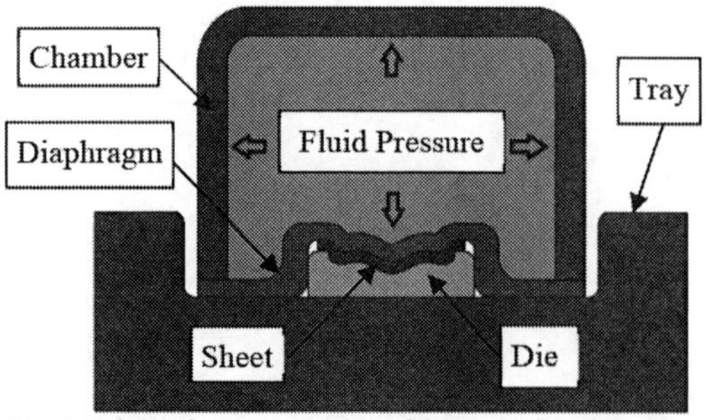

Fig. 1. Schematic representation of the Hydroforming with Membrane Diaphragm.

Some studies are found in literature related to this process. Hatipoglu et al. [1] studied about hydroforming with membrane diaphragm process, which was modeled by finite element method. Three experiments were conducted which is bending of a straight and contoured flange specimen and bulging of a circular specimen made of Aluminum 2024 T3. By these experiments, the influences of bend radius, blank thickness, flange length and orientation of the rolling direction of the specimen were examined. Experimental and numerical results were compared with each other.

Kulkarni and Prabhakar [2] aimed at developing a predictive finite element technique for the springback of aluminum 2024-T3 and 2024-O. Experiments were implemented to various bend angles and part thicknesses. Then, results were compared to finite element model findings. Palaniswamy et al. [3] studied about the springback reduction in the flexforming process by using optimized blank dimensions. The flexforming process is similar to the forming operations in stamping but is carried out in a fluid form cell. In the fluid form cell, a press ram with the oil and a flexible rubber diaphragm replaces the solid punch in a conventional press. Besides, sensitivity analysis was performed to demonstrate the influence of the friction coefficients and the initial blank dimensions for the cone shaped parts.

Dirikolu and Akdemir [4] examined the effect of parameters for the simulation of flexible forming process, which is another name of the hydroforming with membrane diaphragm, such as rubber hardness, blank material type, contact friction and die design. In another study, Yamashita et al. [5] investigated the applicability

* Corresponding author: caydogan@tai.com.tr

of the numerical simulation for the Maslennikov's sheet metal drawing technique. The effect of the forming parameters was investigated, such as dimension and hardness of the rubber ring, friction coefficients and mechanical properties of the sheet.

Aim of the paper is to investigate the numerical modeling of the Hydroforming with membrane Diaphragm Process mainly focusing on the springback problem. Experimental work was done for the validation of the numerical model. Firstly, the experimental setup was submitted and the experimental results were discussed. Then, the numerical model was presented and the numerical results were compared to the experimental results.

2 Experimental setup

Blank dimensions were 60 mm width, 100 mm length and 0.5 mm thickness for the experiments. CP2 Titanium was used as the work piece material. The blanks were located on a die, which was specifically designed for the experimental purpose using locator pins. The rolling direction was parallel to the bending line. Experiments were repeated three times for each condition and the average of the obtained results taken. Experiments were carried out at 60 MPa for the working pressure. All measurements were carried out by Mitutoyo 187-907 protractor with 5" precision. Experiments were conducted on Fluid cell press from Quintus technologies, which have 1.2 m width, 3 m length and 0.27 m height rectangular forming trays (Fig. 2).

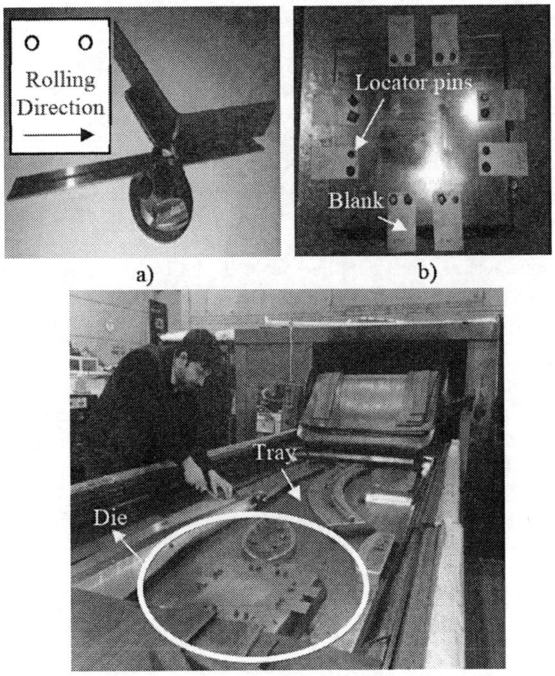

Fig. 2. (a) Protractor and scheme of the rolling direction. (b) Specifically designed die with locator pins and 8 specimens. (c) Fluid cell press from quintus technologies with the specifically designed die.

2.1 Die design

The die was designed bilaterally which has 0°, 60°, 70°, 80°, 90°, 100°, 110°, 120° bending angles and has 2 and 5 mm radii. AISI 1040 steel was used as the die material. Suitability of the mechanical properties and easy presence was the main criteria to choose this material. The die was modelled by using Catia V5-R20 software programme (Fig. 3).

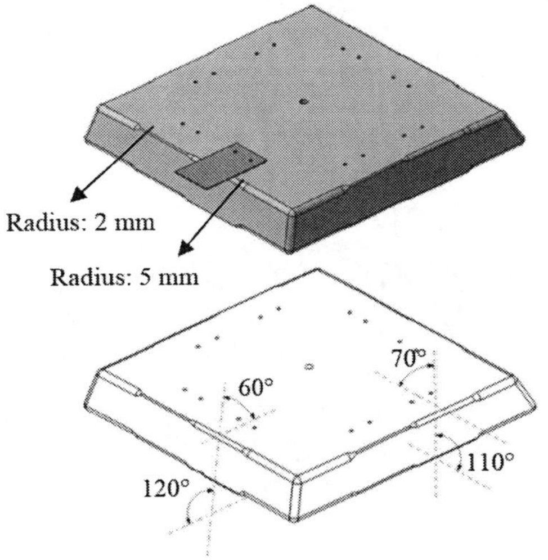

Fig. 3. Cad model of the die

3 Numerical modelling

Numerical models of physical experiments were prepared by using Dynaform software and solved by Ls-Dyna solver. Fig. 4 shows the three bodies in the model: the die (rigid), the blank (plastically deformable) and the diaphragm (elastically deformable). The springback phenomenon can be described as a combination of two stages. First stage is the forming and second stage is the unloading, in which work piece elastically recovers itself. Explicit scheme was used for the loading (forming) stage and implicit scheme was used for the unloading (springback) as suggested in Noels et al. [6].

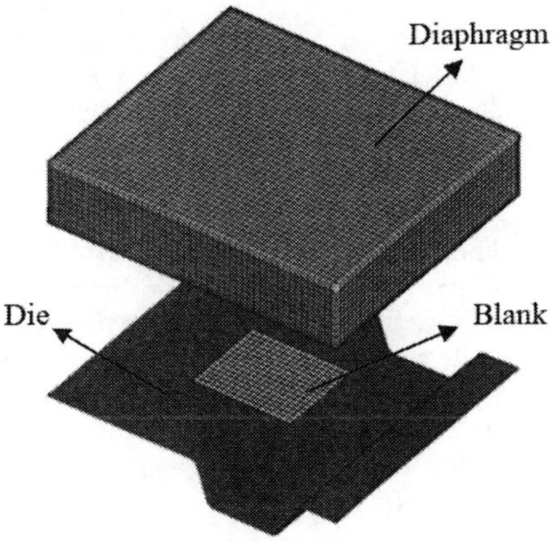

Fig. 4. Parts in the numerical model.

https://doi.org/10.1051/matecconf/201819009002

3.1 Mesh structure and element type

The blank was divided into 2x2 mm fully integrated shell element using 5 integration points through the thickness. The mesh was then refined to 0.5x0.5 mm elements at critical areas such as the bending zone by using adaptivity algorithm. This algorithm reduces the computation time drastically. The die was divided into meshes, which depend on mesh parameter to maximum size (4 mm), minimum size (0.1 mm), chordal deviation (0.1 mm) and angle change for adaptivity (5°). The diaphragm was divided into 2x2x2 mm constant stress solid elements (Fig. 5).

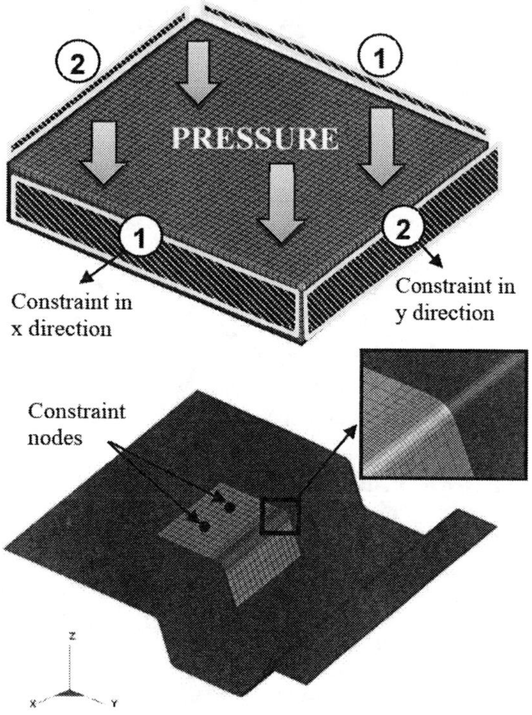

Fig. 5. Mesh structure and boundary conditions of the system.

3.2 Boundary conditions

Applied pressure destabilizes the system by forcing diaphragm to flow in the x and y directions. Therefore, the edge nodes of the diaphragm were constrained in the x and y directions to prevent any motion. For the blank, the two nodes were constrained in all directions to simulate locator pins (Fig. 5).

Loading curve (pressure vs time curve) is essential in slow-dynamic-explicit analysis. The trend of the curve should minimize the dynamic effects in the process [7]. In this case, the slope of the curve in the first region is low where the diaphragm is moving and enclosing the die, while the slope of the curve in the second region is high where the diaphragm is more stationary (Fig. 6).

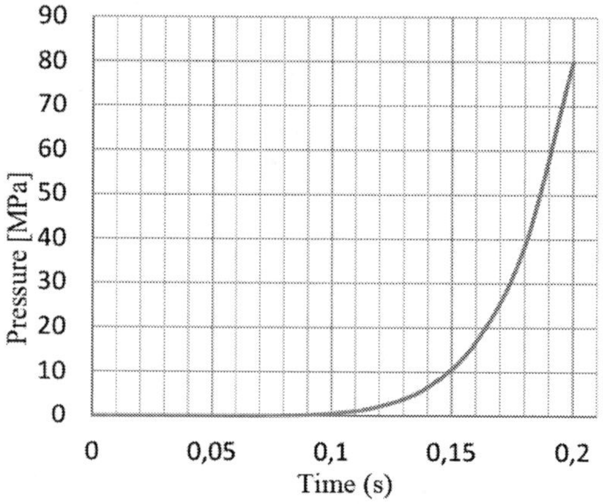

Fig. 6. The loading curve used in the analysis.

3.3 Contact features and friction coefficient

Definition of the contact algorithm between surfaces (diaphragm-die, diaphragm-blank, die-blank) is important to obtain precise and correct solutions in sheet metal forming applications. Penalty method was used for the contact algorithm and Coulomb model was used for friction. Friction coefficient between blank (Titanium) and die (Steel) was accepted as 0.365, according to Adamus [8]. A special scaling factor was applied for the contact stiffness between diaphragm and the die, also between the diaphragm and the sheet in order to compensate the bulk modulus differences between these materials [1].

3.4 Material characterization

The materials in the process were CP2 Titanium for the blank, rubber for the diaphragm and AISI 1040 steel for the die. Mechanical properties of these materials were obtained from uniaxial tensile (ASTM E 8M for CP2 Titanium and ASTM D 412 for rubber). These mechanical properties were supplied from TAI (Turkish Aerospace Industry). Hollomon law was used for the flow curve of CP2, which is implemented by the following equation:

$$\sigma = K \times \varepsilon^n \tag{1}$$

Where K is the strength coefficient, n is the hardening exponent, σ is the effective stress, ε is the effective strain. Barlat 3-parameter anisotropic function was implemented for the yield surface. Mooney-Rivlin material model was used for the rubber. The strain energy density function is given in Mooney-Rivlin by:

$$W = \sum_{k+m=1}^{n} C_{km} \times (I_1 - 3)^k + (I_2 - 3)^m + \frac{1}{2} K \times (I_3 - 1)^2 \tag{2}$$

I_1, I_2 and I_3 are the strain invariants. K is the bulk modulus and $I_3 = 1$ for incompressible materials. Two-parameter Mooney-Rivlin model (n=1 in Eq. (2)) was used. The Mooney-Rivlin constants (C_{01}, C_{10}) were obtained from experiments [9]. Lastly, the die was accepted as rigid

https://doi.org/10.1051/matecconf/201819009002

body. The necessary mechanical data used for the material models is summarized in Table 1 (supplied by TAI).

Table 1. Material data for CP2 titanium, AISI 1040 and rubber.

Materials			
Properties	CP2	Rubber	AISI 1040
Density, ρ [t / mm³]	4.52E-09	1.00E-09	7.80E-09
Elastic modulus, E [MPa]	111500	6	207000
Poisson's ratio, ν	0.31	0.49	0.34
Yield strength, σ_0 [MPa]	452		
Anisotropy value, R_0	0.8		
Anisotropy value, R_{45}	2.03		
Anisotropy value, R_{90}	2.43		
Strength coefficient, K [MPa]	742		
Hardening exponent, n	0.105		
Mooney-Rivlin constant, C_{10} [MPa]		0.8	
Mooney-Rivlin constant, C_{01} [MPa]		0.2	

3.5 Energy controls and units

Increasing the speed of process allow the analysis to take less computer time. However, virtual inertial effect resulting from the high process speed may lead to erroneous solutions, since the sheet metal bending process is quasi-static. For this reason, the speed of the process should be adjusted not to exceed a certain limit [10]. Generally, Energy files should be examined whether the kinetic energy exceeded 10% of internal energy or not. As seen from figure 7, the kinetic energy does not exceed 10% of internal energy, so the acceptable quasi-static solution can be obtained. The unit system used was mm-s-t-N at the simulations.

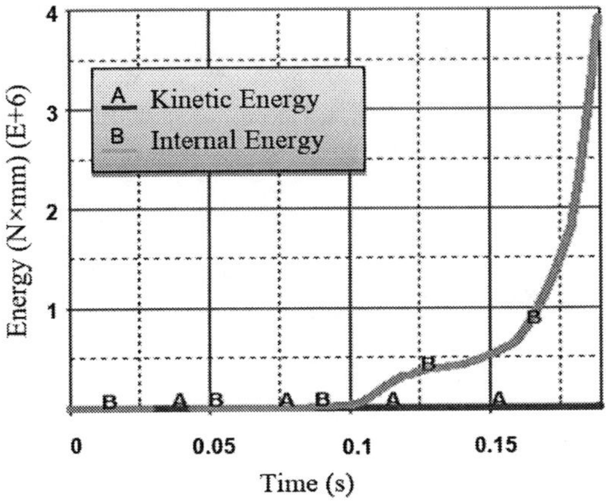

Fig. 7. Kinetic-internal energy control.

4 Numerical and experimental results

The numerical springback (N.S) and experimental springback (E.S) results were evaluated by measuring the springback angles, which is defined as the angle deviation from bending angles. Results were obtained for all the variation of bend angles and radii. Fig. 8 shows a sample result including part geometry and plastic strain distribution after springback.

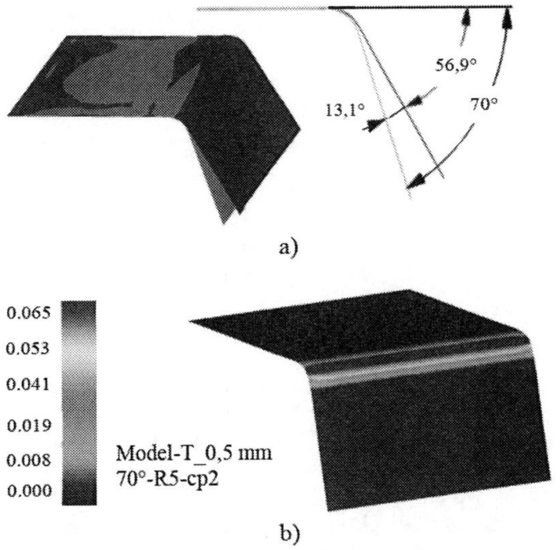

Fig. 8. (a) Part after springback (b) Plastic strain distribution.

Numerical springback and experimental springback results, which were obtained by finite element analyses and experiments, are given in Table 2.

Table 2. Numerical springback (N.S) and experimental springback (E.S) results for the bending specimen.

	Bend Radius			
	R:2 mm		*R:5 mm*	
Bend Angle [°]	*N.S [°]*	*E.S [°]*	*N.S [°]*	*E.S [°]*
60	5.7	4.5	5.2	4.7
70	5.8	6.7	13.1	11.3
80	5.5	6.5	10.6	12.7
90	6.6	6.0	11.8	12.1
100	7.4	8.2	15.5	16.2
110	8.8	9.9	16.5	18.8
120	9.2	13.7	18.7	22.7

In addition, graph of results is given in Fig. 9 for each bend radius and angle. As seen in the graph, the trend line of numerical results is compatible with the experimental results.

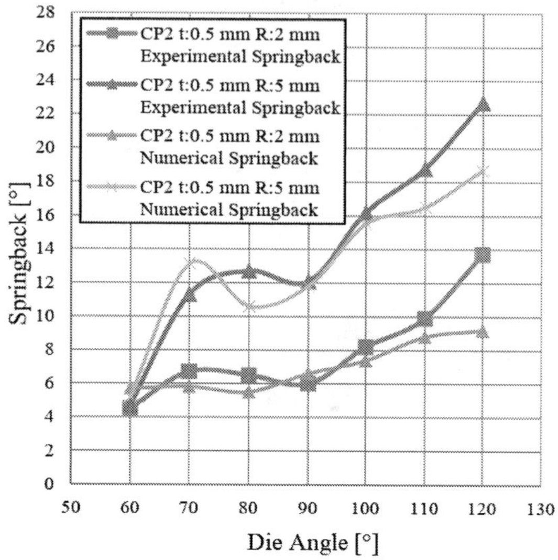

Fig. 9. Graph of result in Table 2.

It was hypothesized that the difference in experimental and numerical springback results can be due to a mismatch in the rate of forming of the parts on the hydroforming with membrane diaphragm and the strain rate at which material properties were conducted. In addition, springback is also influenced by many other parameters such as material thickness, die radius, ambient conditions and the speed of process, which is an important factor in creating dynamic effects and it affects the accuracy of the finite element analysis.

Furthermore, results show that the deviation of springback values between 70° and 90° bend angles are lower than the other intervals. This phenomenon could be explained by the diaphragm contact mechanics affecting on the blank material. For the acute bend angles, the diaphragm is in full contact with the blank, whereas for the obtuse bend angles, there is a region where the diaphragm loses its contact with the blank (Fig. 10). The gap observed is filled by the diaphragm reaching the maximum pressure. Consequently, this inconsistent contact between diaphragm and blank causes this unexpected trend shown in Fig. 9.

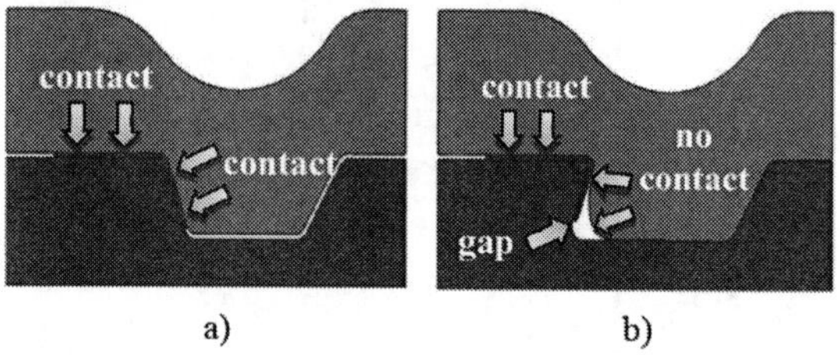

Fig. 10. The diaphragm contact mechanics (a) for the acute bend angles (b) for the obtuse bend angles.

5 Conclusion

Finite element analysis was used to simulate the Hydroforming with Membrane Diaphragm process and to estimate the springback angles. Principally, the springback phenomenon can be minimized with suitable die modifications but cannot be eliminated. According to this study, the numerical and experimental springback results show good agreement. So, the springback values obtained from the numerical model can be used for die modifications in order to compensate springback and to get proper titanium productions.

The second outcome of this study is that the increment of springback angles does not follow a regular trend between 0° and 120° bend angles in this manufacturing process based on the numerical and experimental results. It is thought that this phenomenon occurs due to the nature of diaphragm contact mechanics.

This work was supported by TUSAŞ Aerospace Industries (TAI), Ankara, Turkey.

References

1. H.A. Hatipoglu, A.E. Tekkaya, Key Eng. Materials, **344**, 469-476 (2007)
2. P. Kulkarni, S. Prabhakar, *Proceedings of 4th European Ls-Dyna Users Conference*, 27-34 (2003)
3. H. Palaniswamy, G. Ngaile, T. Altan, J. Mater. Process. Technol, **146**, 28-34 (2004)
4. M.H. Dirikolu, E. Akademir, J. Mater. Process. Technol, **148**, 346-381 (2004)
5. M. Yamashita, T. Hattori, N. Nishimura, J. Mater. Process. Technol, **187**, 192-196 (2007)
6. L. Noels, L. Stainier, J. Ponthot, J. Comp. and App. Math, **168**, 331-339 (2004)
7. L.M. Kutt, A.B. Pifko, J.A. Nardiello, J.M. Papazian, Computers & Structures, **66**, 1-17 (1998)
8. J. Adamus, Int. J. of Applied Mechanics and Engineering, **11** , 727-733 (2006)
9. M. Mooney, J. Appl. Phys, **11**, 582-597 (1940)
10. O.K. Demir, Technical paper. http://www.figes.com.tr/ls-dyna/, 2018 (Accessed 02.01.2018).

Investigation of a Modified Hydrodynamic Deep Drawing Assisted by Radial Pressure with Inward Flowing Liquid Process

Maziar Khademi[1,*], *Milad* Sadegh yazdi[1], *Mohammad* Bakhshi-Jooybari[1] and *Hamid* Gorji[1]

[1]Metal Forming Research Center, Department of Mechanical Engineering, Babol Noshirvani University of Technology, Babol, 48187-1167, Iran

Abstract. Hydrodynamic Deep Drawing (HDDRP), the combination of hydroforming and conventional deep drawing, accommodates the advantages of the two processes. A technique, called HDDRP with inward flowing liquid, has been introduced based on the idea of insertion of radial pressure around the blank rim. The radial pressure created on the blank edge, can increase the drawing ratio. Thus, increasing the radial pressure to an amount greater than the cavity pressure, and independent control of these pressures is the basic idea of this research for forming cylindrical parts. To perform the experiments, two independent pumps were used to provide the two pressures independently. The pressure supply system and the die set were designed in a way that provides simultaneous control of the pressures throughout the process. Then, the effects of radial pressure paths on thickness distribution of cylindrical St13 cups were investigated. In addition, a comparison between HDDRP and HDDRP with inward flowing liquid processes has been performed experimentally. Results indicated that using a higher radial pressure than the cavity pressure and controlling their values at any moment of the process enhances the thickness distribution of the formed part in all regions.

Keywords: Hydroforming, Sheet metal, Steel

1 Introduction

Of the various metal forming processes, hydroforming is one of the non-traditional ones. This technique uses a pressurized fluid or gas in place of a conventional solid tooling like punch, die, etc. for manufacturing purpose. So, liquid is used as the medium of energy transfer to form the workpiece. Since 1990s hydroforming has been widely adopted for making metal sheet components in many industrial fields, especially in the automotive, aerospace, electronics and kitchen industries [1-3].

Production of sound parts by using hydroforming, mainly depends on different important parameters such as control of liquid pressure, blank holding force, exact design of tooling and process capability [4, 5] .

By development of technologies, different hydroforming methods have now been proposed such as standard hydroforming [6, 7], hydromechanical deep drawing [8], radial hydromechanical deep drawing (hydro-rim) [9], hydrodynamic deep drawing assisted by radial pressure [10] and hydrodynamic deep drawing process assisted by radial pressure with inward flowing liquid [11].

Kandil [7] carried out an experimentally investigated the effect of process parameters such as sheet material properties, punch geometry, punch load, initial pressure for cylindrical cup forming. The results showed that by using hydroforming deep drawing process compared with conventional deep drawing process, higher limiting drawing ratio can be achieved and uniform strain distribution along the cup wall were obtained.

Bagherzadeh et al. [8] studied the application of hydro-mechanical deep drawing process on laminated sheets. Their analysis includes 3D finite element (FE) simulation with implementing Fortran based code for accurate modelling of non-uniform oil pressure distribution. To verify FE results, some experiments were conducted on the laminated aluminium/steel sheet. Their results showed that higher limiting drawing ratio (LDR) and lower thinning in low formable aluminium sheet would be achieved in Al/St lay-up than St/Al lay-up.

Tirosh et al. [12] exerted external fluid pressure on the blank rim in Hydro-rim process to reduce radial stress and to decrease the interfacial friction. They reported that the most important benefit of this method is that if changes in the material properties can be controlled, by controlling the process speed, higher drawing ratios can be reached.

Gorji et al. [10] studied forming of copper and St14 conical–cylindrical cups in the HDDRP process. The effect of pressure path on the thickness distribution and drawing ratio of the sheet was also investigated. They

* Corresponding author: Maziar.khademi@gmail.com
m.khademi@stu.nit.ac.ir

 https://doi.org/10.1051/matecconf/201819009003

illustrated that for the pressure path with a certain maximum amount, the part was formed with appropriate thickness reduction. They reported an LDR of 9.76 in forming pure copper conical cups. The HDDRP process has been developed by Wang et al. [11], called HDDRP with inward flowing liquid. The key point in this process is that a radial pressure higher than the cavity pressure will be achieved if the liquid flees inward from the rim cavity to the die cavity. They demonstrated that this mechanism can decrease the forming force and increase the LDR, significantly.

In this paper, the HDDRP assisted by radial pressure with inward flowing liquid has been further modified. The effects of different parameters have been investigated by the simulation method in the previous work [13]. In this process, the mechanism of the implementation of the hydraulic circuit and its repeatability are examined. Also, the mechanism of the formation of higher radial pressure and the method of simultaneously controlling the cavity and radial pressures was studied. Eventually, the effect of higher radial pressure on thickness distribution has been examined experimentally.

2 Experimental Procedure

The schematic of the pressure supply in the modified HDDRP with inward flowing liquid and the dimensioned geometry of the die set are illustrated in Fig. 1.

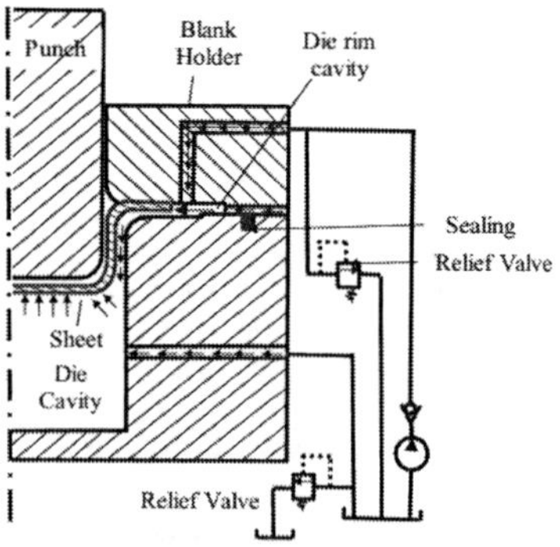

Fig 1-a: Scheme of the die set and pressure unit.

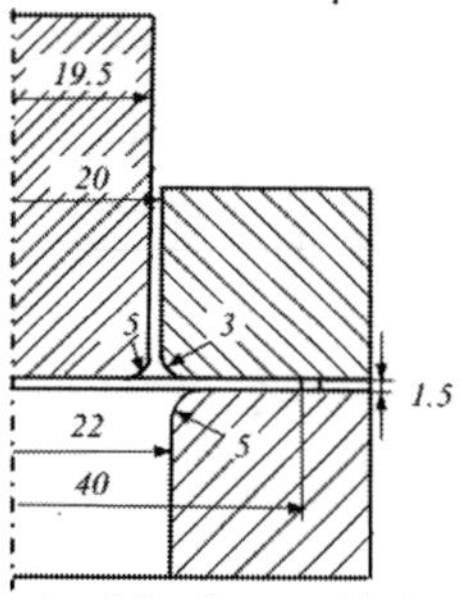

Fig. 1-b. Schematic of the die set with the geometrical dimensions (mm).

Fig. 2 shows the components of assembled die set for the hydroforming of cylindrical cups. A Denison and Mayes Group universal testing machine with a capacity of 600 kN, was used to form St13 circular blanks. For the supply of liquid pressure, a variable flow rate hydraulic pump with the maximum pressure of 50MPa and a constant flow rate hydraulic pump with a maximum pressure of 100MPa were used. Fig. 3 illustrates the typical pressure path used in the experiments for the cavity and radial pressures. In the hydroforming process, a small pre-bulging can be created on the sheet bottom to improve the drawing process. In this research, 1 MPa pre-bulging pressure was applied (path OA). After the pre-bulge stage, the pressure was linearly increased (path AB) to a maximum value (path BC). The slope of this path depends on different parameters such as the punch velocity, sheet thickness and punch profile bottom [14]. The punch speed is about 20 mm/s. The pressure path variation was controlled by a computer system connected to a digital manometer.

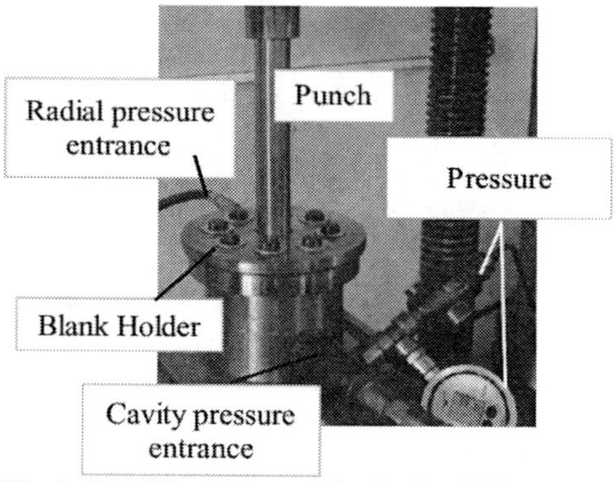

Fig. 2. HDDRP with inward flowing liquid die set.

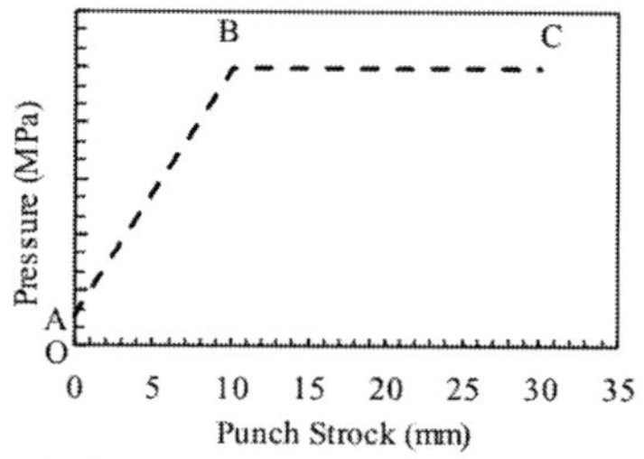

Fig. 3 Typical pressure path in this study.

2.1 Material properties

Table 1 shows material properties obtained from uniaxial tensile tests, where specimens were cut at different orientations to the rolling direction (0°, 45°, and 90°) according to ASTM-A370 standard. The sheet diameter and thickness were 80 mm and 1.5 mm, respectively.

Table 1. Mechanical properties of St13 sheets [13].

Parameter	Value
Poisson's ratio	0.32
Young's modulus, (GPa)	210
Strain hardening exponent	0.29
Harding coefficient, (MPa)	539
Yield stress, (MPa)	158
Anisotropy value (r_0)	0.972
Anisotropy value (r_{45})	1.1761
Anisotropy value (r_{90})	1.361

2.2 Die and hydraulic circuit design

In designing the die of this process, the most important goal was to create a radial pressure higher than the cavity pressure. For this purpose, according to Fig. 1-a, the die should be sealed to prevent oil leakage and pressure drop. Therefore, as shown in Fig. 1, an O-ring was used for sealing in the die using the standard mentioned in [16] to prevent fluid leakage. Moreover, because of high pressure and leakage prevention, the blank holder was fastened to the die by eight high strength screws (Fig. 6).

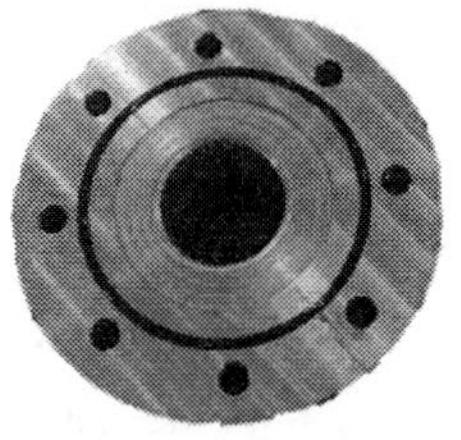

Fig. 6 The O-rings used for sealing the die.

The second most important point in creating higher and controllable radial pressure at any moment is that the radial pressure be created independently by an additional pump (Fig. 1). Therefore, the pressure from the second pump should be transferred in a way to around the blank rim. For this purpose, as shown in Fig. 7, the conventional balnkholder is redesigned. Inside the blank holder, a path was embedded to flow liquid from the pump toward the gap between the die and the holder.

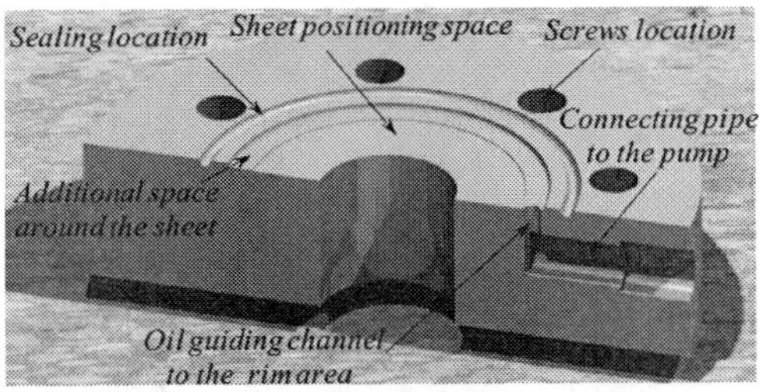

Fig. 7 The different sections of blank holder.

The hydraulic circuit of this process is shown in Fig. 8. As can be seen, this process uses two independent pumps to create pressure. The cavity pressure path is also controlled at any moment through a digital manometer connected to a computer. Also, the radial pressure path is measured and controlled by the analog manometer in the pump at any time. Two relief valves were attached to the die cavity and the blank rim cavity to control the maximum pressure values of the cavity and radial pressures. After the pressure in the cavity and rim area reached the maximum values, the relief valves were opened, and the process continued with the constant pressure. Utilization of two independent pumps provides the possibility of controlling the radial and cavity pressure paths, simultaneously[13].

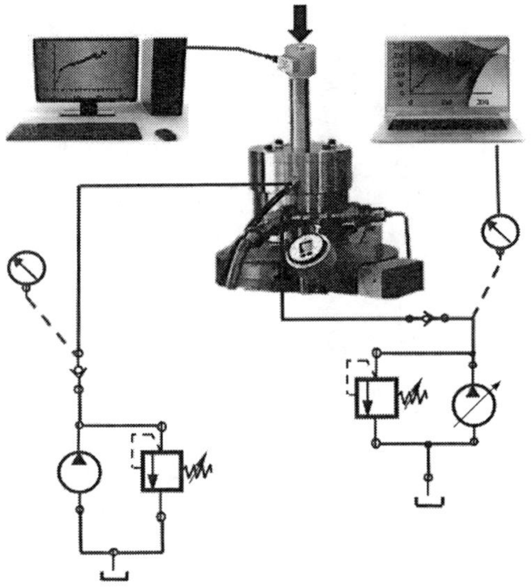

Fig. 8 Hydraulic circuit of modified HDDRP with inward flowing liquid.

3 Process simulation

For the process simulations, the commercial software, ABAQUS 6.13 /Explicit, was used. Fig. 4 shows the FE model of the die set. By considering the symmetry of the process, only one-quarter of the sheet and the die set is modelled [15]. The blank was modelled as 3D deformable with C3D8R element. The die components were considered rigid and R3D4 element type was assigned to these parts. The thickness distribution of the formed cylindrical cup is the most important parameter investigated in this paper. So, five elements were used in the thickness direction to measure precisely the thickness variation. The anisotropy property for St13 sheet was simulated based on the parameters shown in Table 1. Thus, the Hill yield criterion was used to define anisotropic yielding [13]. The friction coefficient was assumed to be 0.14 between the sheet and punch surface but for other surfaces, it was considered 0.04 [14]. The contact algorithm between the sheet and die components, was modeled as penalty method based on Coulomb law. During the process the punch could only move in the z-direction parallel to its central axis and the other die components were fully constrained. In HDDRP with inward flowing liquid, the liquid pressure distribution under the blank flange area is different from that in the conventional Hydroforming deep drawing (Fig. 5). Based on the fluid model proposed by Wang et al. [3] the non-uniform distribution of the fluid pressure under the blank flange area was considered and embedded into the FEM model by a suitable subroutine *VDLOAD based on Visual Fortran code and implemented in ABAQUS.

https://doi.org/10.1051/matecconf/201819009003

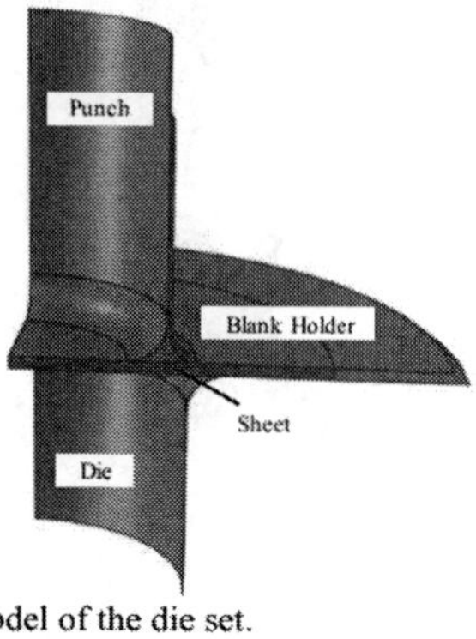

Fig. 4. FE model of the die set.

The pressure distribution in the r-direction can be obtained by Eq. (1) where the boundary conditions can defined as $P = P_R$ at $r = R$, where R is the instant radius of the sheet flange and $P = P_C$ at $r = R_1$, that R_1 is the die entrance radius [11].

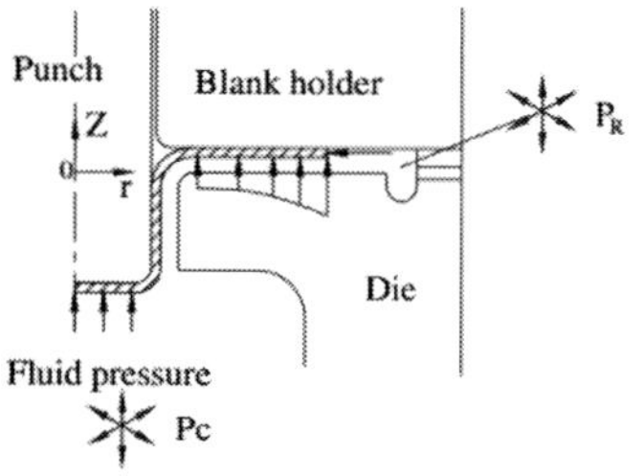

Fig. 5 Pressure distribution in the flange area[11].

$$P_r = P_R - (P_R - P_C) * \frac{\ln(r/R)}{\ln(R_1/R)} \quad (1)$$

Since increasing the radial pressure to a value higher than the cavity pressure depends on the parameters such as die design, pump power and oil properties and leakage, it should be determined that how much the radial pressure can be increased. So, to illustrate this issue and to more closely examine the process, the Fluent software was used.

4 Results and Discussions

The results of the simulations showed that for forming an St13 cylindrical part in HDDRP with inward flowing liquid, the maximum pressure of 20 MPa is required at least. In order to determine the maximum possible radial pressure, based on the specifications of the pump and the die, the Fluent software was used. The simulation was performed based on the geometric dimensions of Fig. 1-b. The results of this investigation are shown in Figure 9. As can be seen, according to the dimensions of the die and the minimum forming pressure of the cavity as input data, radial pressure up to 65 MPa can be created.

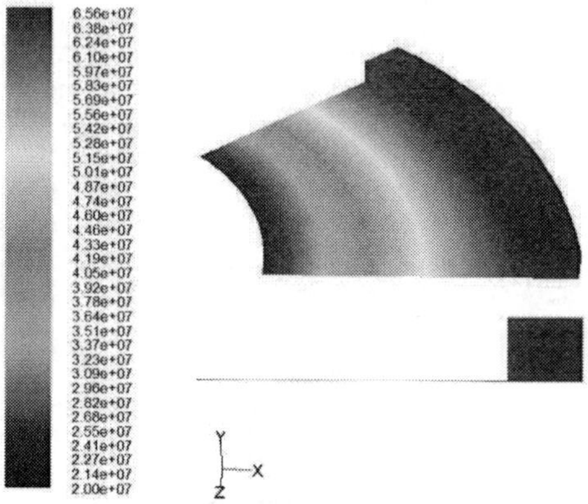

Fig. 9 Fluid Analysis of process with Fluent software, Pressure distribution diagram in the flange area, (MPa).

By Increasing the size of the gap between the punch and die : the part does not contact to die properly so it does not fully form. Also, reducing the size of the gap causes the part contact with die become increases and the wall thickness decreases.

An experimental test was carried out at the cavity pressure of 30MPa and the radial pressure of 57MPa. The 57MPa pressure was the maximum radial pressure that obtained in the experimental state. Because of not reaching the pump to its maximum power the difference with the maximum value of the experimental results and the Fluent software was existed. Figure 10 shows the thickness distribution obtained from experiment and simulations at the same punch stroke for St13 material. As it is shown in the figures, there is a good agreement between the simulation and experimental results.

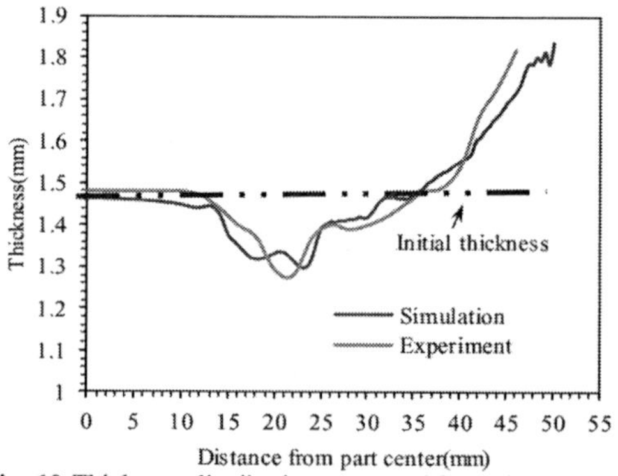

Fig. 10 Thickness distribution curves of formed St13 parts: radial pressures, 57MPa; cavity pressure, 30 MPa; Punch stroke, 29.5 mm

To ensure the repeatability of the test, this test was performed again (radial pressure (55MPa) is approximately equal to the first time (57MPa)).

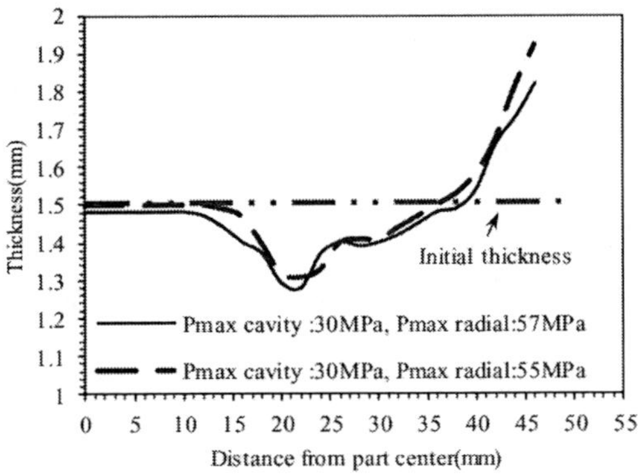

Fig. 11 Thickness distribution curves of formed St13 parts with same pressures; cavity pressure, 30 MPa; radial pressures, 57 and 55MPa Punch stroke, 29.5 mm.

The maximum radial pressure depends to the specifications of pressure supply system, the die geometry and initial blank thickness. Also, the effect of the radial pressure depends on the parameters such as the gap between the blank holder and the die, the gap between the punch and the die, the thickness of the sheet, and so on. Fig.12 shows the effect of the gap value variation between punch and die on part thickness distribution. The effect of gap variation between the punch and the die shows before value of 1mm, the thickness distribution improves but more than 1 mm (for the gap values of 1.5 to 2mm), the workpiece was not formed successfully in the end of wall area as shown in Fig13."

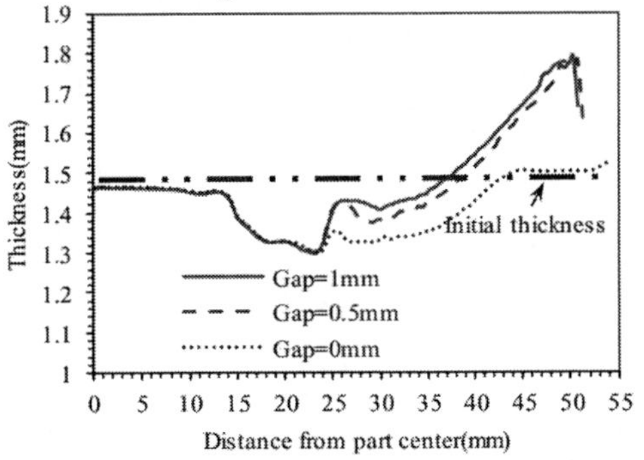

Fig. 12 Thickness distribution curves of formed St13 parts for different gap: radial pressures, 57MPa; cavity pressure, 30 MPa; Punch stroke, 29.5 mm

By Increasing the size of the gap between the punch and die, the part does not contact the die properly. Therefore, it does not fully form. Also, reducing the size of the gap increases the part contact with die. Therefore, the wall thickness decreases.

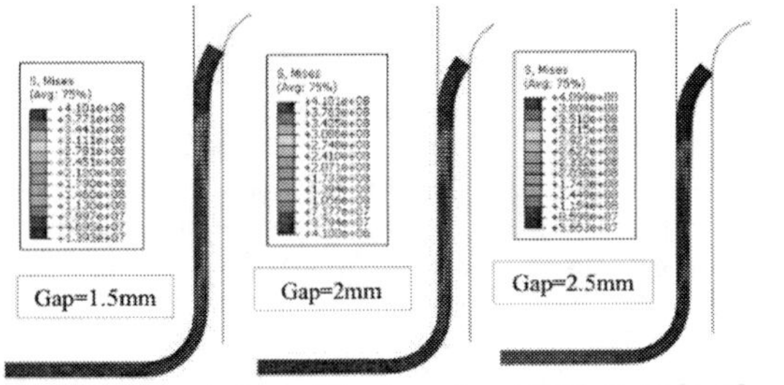

Fig. 13 Sectional view of the formed part St13 at stroke of 29.5 mm at maximum cavity and radial pressures of 30 and 57 MPa, respectively.

Figure 14 shows the formed parts obtained from experiments at a typical punch stroke for St13. The thickness of the formed parts was measured after cutting, the results of which are shown in the following figure.

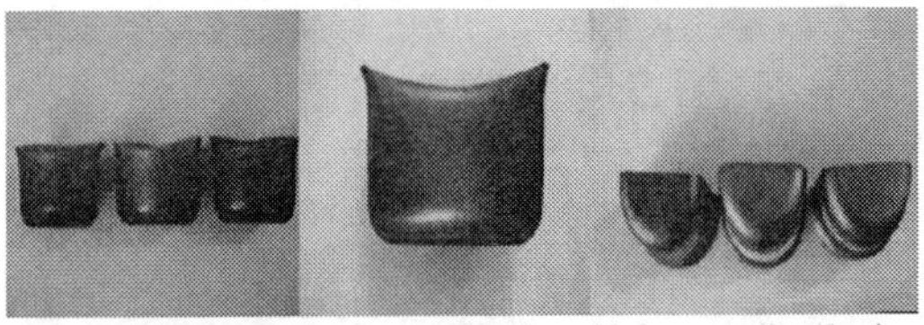

Fig. 14 Formed part sample for thickness distribution measuring at punch stroke 29.5 mm.

Finally, to compare the thickness distribution of the formed part in the case where the radial pressure exists, and in the absence of radial pressure, (or in the case radial pressure is almost equal to the cavity pressure) another experiment test is carried. To ensure repeatability, this test has also been repeated two times as shown in fig 15(cavity pressure (34MPa) is approximately equal to the first time (35MPa)).

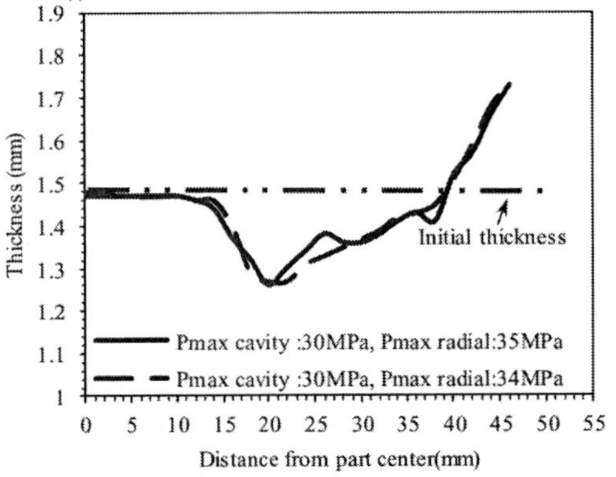

Fig. 15 Thickness distribution curves of formed St13 parts with same pressures; cavity pressure, 30 MPa; radial pressure, 35 MPa, Punch stroke, 29.5 mm

The experimental result of the effect of increasing radial pressure on the thickness distribution curve is shown in fig.16. Radial pressure paths with maximum values of 34 and 57 MPa were used, while the cavity pressure path was kept constant at the maximum of 30 MPa. By increasing the radial pressure, thinning decreases along the formed part. Because of contact inner surfaces of the part with the corner radius of the punch, the maximum tensile stresses occur in this region. Consequently, maximum thinning occurs in punch radius region. By increasing the radial pressure, tensile stresses occurred in this area decrease. So, by increasing the radial pressure, thinning decreases along the formed part. In other words, radial pressure variation has a significant effect on the formed cup thinning. This agrees with the finding of the previous studies [17,18].

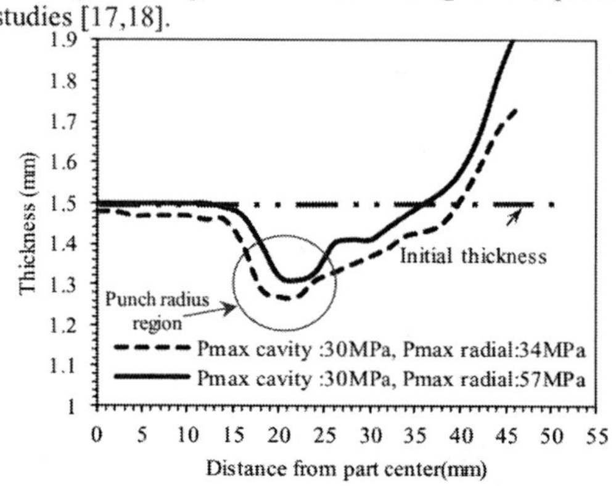

Fig. 16 Thickness distribution curves of formed St13 parts with same pressures; cavity pressure, 30 MPa; radial pressures, 57 and 55MPa Punch stroke, 29.5 mm

5 Conclusion

In this study modified Hydrodynamic deep drawing assisted by radial pressure with inward flowing liquid was investigated.

1. In the experimental section, two independent pumps have been used to provide cavity and radial pressures and the pressure supply system and die set was designed to provides simultaneous change and control of the radial and cavity pressures throughout any time during the process.

2. Results showed that by redesigning the conventional die set and pressure supply system, higher radial pressure can be achieved.

3. It was experimentally shown that by using higher radial pressure than the cavity pressure and controlling their values at any moment improves the thickness distribution of the formed part in all regions.

4. The simulation results by the Fluent Software also showed that the amount of radial pressure, according to the geometry of the die increased up to 3 times of maximum cavity pressure.

References

1. M. Bakhshi-Jooybari, A. Gorji, and M. Elyasi, *Developments in sheet hydroforming for complex industrial parts*, in Metal Forming-Process, Tools, Design, InTech. (2012)
2. S. M. Kulkarni, V. V. Ruiwale, M. S. Jadhao, A. A. Kadam, S. R. & Kale, *A review on Hydroforming Processes,* Ijcet. (2017).
3. H. Hu, J. F. Wang, K. T. Fan, T. Y. Chen, & Wang, S. Y., *Development of sheet hydroforming for making an automobile fuel tank.* Proc. Inst. Mech. Eng. B, **229**(4): p. 654-663, (2015).
4. G. Palumbo, S. H. Zhang, L. Tricarico, C. Xu, & L. X. Zhou, *Numerical/experimental investigations for enhancing the sheet hydroforming process.* Int J Mach Tool Manu, **46**(11): p. 1212-1221, (2006).
5. B. Modi, and D.R. Kumar, *Development of a hydroforming setup for deep drawing of square cups with variable blank holding force technique.* Int. J. Adv. Manuf. Technol. , **66**(5-8): p. 1159-1169, (2013).
6. S. Thiruvarudchelvan, and W. Lewis, *A note on hydroforming with constant fluid pressure.* J Mater Process Tec, **88**(1-3): p. 51-56, (1999).
7. A. Kandil, *An experimental study of hydroforming deep drawing.* J Mater Process Tec, **134**(1): p. 70-80, (2003).
8. S. Bagherzadeh, M. Mirnia, and B.M. Dariani, *Numerical and experimental investigations of hydro-mechanical deep drawing process of laminated aluminum/steel sheets.* JMP, **18**: p. 131-140, (2015).
9. S.-H. Zhang and J. Danckert, *Development of hydro-mechanical deep drawing.* J Mater Process Tec, **83**(1-3): p. 14-25, (1998).
10. A. Gorji, H. Alavi-Hashemi, M. Bakhshi-Jooybari, S. Nourouzi, and S.J. Hosseinipour, *Investigation of hydrodynamic deep drawing for conical–cylindrical cups.* Int. J. Adv. Manuf. Technol., **56**(9-12): p. 915, (2011).
11. H. Wang, L. Gao, and M. Chen, *Hydrodynamic deep drawing process assisted by radial pressure with inward flowing liquid,* Int. J. Mech. Sci., **53**(9): p. 793-799, (2011).
12. J. Tirosh, A. Shirizly, D. Ben-David, and S. Stanger, *Hydro-rim deep-drawing processes of hardening and rate-sensitive materials.* Int. J. Mech. Sci., **42**(6): p. 1049-1067, (2000).
13. M.S. Yazdi, M. Bakhshi-Jooybari, H. Gorji, M. Shakeri, and M. Khademi, *Investigation of Forming Cylindrical Parts in a Modified Hydrodynamic Deep Drawing Assisted by Radial Pressure With Inward Flowing Liquid.* J. of Manuf. Sci. and Eng., **140**(3): p. 031007, (2018).
14. M. Salahshoor, , A. Gorji, and M. Bakhshi-Jooybari, *The study of forming concave-bottom cylindrical parts in hydroforming process.* Int. J. Adv. Manuf. Technol., **79**(5-8): p. 1139-1151, (2015).
15. M. Khademi, A. Gorji, M. Bakhshi, and M.S. Yazdi, *Investigation of Wrinkling in Hydrodynamic Deep Drawing Assisted by Radial Pressure With Inward Flowing Liquid.* Pro. Eng, **183**: p. 65-70, (2017).
16. J. Whitlock, *The Seal Man's O-Ring Handbook, EPM.* Inc., Stockbridge, GA, (2004).
17. E. Onder, A. E. Tekkaya, *Numerical Simulation of Various Cross Sectional Workpieces Using Conventional Deep Drawing and Hydroforming Technologies*, Int. J. Mach. Tools Manuf., **48**(5) : p. 532–542, (2008).
18. M. Hosseinzade, H. Mostajeran, M. Bakhshi-Jooybari, A. H. Gorji, S. Nourouzi, and S. J. Hosseinipour, *Novel Combined Standard Hydromechanical Sheet Hydroforming Process*, Proc. Inst. Mech. Eng., Part B, **224**(3) : pp. 447–457, (2010).

Investigation into the micro deep drawing capabilities of a specially engineered refined aluminium alloy

A. Dhal[1], *S.K.* Panigrahi[1, *], and *M.S.* Shunmugam[1]

[1]Indian Institute of Technology Madras, Chennai, Tamil Nadu, India - 600036

Abstract: During miniaturisation, size of the part comes close to grain size of the material. There is an overall decrease in the total number grains undergoing deformation and most of these are surface grains. Therefore, microscale deformation is marked by abnormal stress-strain response which limits the manufacturing capabilities of microforming. Two distinct phenomena responsible for this are: (i) dominance of single crystal deformation behaviour, and (ii) increased strain localisation due to incompatibly between surface and core grains during deformation. The present work attempts to neutralise these effects by increasing the number of grains in the deformation zone. This has been achieved by engineering refined microstructure in the materials. To develop the refined microstructure, cryorolling followed by controlled annealing treatment has been employed. Microscale deformation behaviour and microforming capabilities of the refined material have been compared with its coarse-grained counterpart by analysing their tensile curves and by post-mortem study of micro deep drawn components over a wide range of sample thicknesses. Material with fully recrystallised, equiaxed, strain-free refined microstructure is found to have the best strain hardening response both in micro and macro deformation domains. This property is also reflected in the micro deep drawing capabilities of the same material.

Keywords: Micro forming, Nano structure, Deep drawing

1 Introduction

In the past decade, there has been a visible trend in miniaturisation particularly in the fields of consumer electronics, biomedical devices, avionics etc. The demand of micro components commonly used in such devices is rising in an unprecedented rate. To cater this rising demand, there is a need to escalate the manufacturing process to mass production scale. Currently popular micro manufacturing processes such as photolithography (LIGA) and micro machining are not designed for mass production of complex 3D microparts. On the other hand, microforming can be used for mass production of such intricate products. It also has economic and ecological advantages in terms of assembly-free near net shape production, minimal material wastage, good product accuracy and repeatability, and automation capabilities [1].

Despite its prospects, microforming remains unexploited in micro-components industries. It is because effective material deformation in microscale remains a big challenge. At micro-scale, the deformation behaviour does not follow the conventional trends. This has been attributed to an overall decrease in the total number grains undergoing deformation and an increase in the share of surface grains compared to the core grains [2]. Since surface grains have free unconstrained exteriors, their grain boundaries have limited dislocation accumulation capabilities compared to the core grains. Therefore, incompatibility exists between core and surface grains for dislocation movement, resulting in an increase of strain localisation at their interface. This results in premature necking in the material during microscale deformation and reduced sheet metal forming capabilities [3].

Apart from the challenges associated with the change in material behaviour in micro scale, there are additional hurdles related to the tool development and material handling in case of microforming. The press tool for microforming should have close dimensional tolerances for mating parts, and controlled positioning as well as misalignment errors among them. Unlike macro press tools, these factors are highly critical in nature in case of microtools. Material handling issues such as adhesion of the samples and easy retrieval of the formed component from the die should be considered. These aspects need to be considered while design and development of microforming tools and systems [4].

The total number of grains in the deformation zone can be increased by refining the microstructure to micrometer or sub-micrometer domain. Therefore, use of samples with refined microstructure instead of the conventional coarse-grained microstructure provides a prospective alternative to tackle the size effect related challenges. Significant amount of research has been devoted to study the influence of grain size effect on the microscale deformation behaviour [5-6]. However, to the best of authors' knowledge, there is no evidence of the application of material with refined microstructures in the form of thin sheet for microscale sheet metal operations. In case of sheet metal forming, a plane strain deformation is desirable as deformation along the sheet thickness results in early failure of the component. Therefore, strain localisation in the material is severely detrimental to the forming capabilities of the sheet metal. For successful microscale sheet metal forming, this creates an additional hurdle as microforming is intrinsically susceptible to premature strain localisation. To improve the microformability, thin sheets used for

* Corresponding author: skpanigrahi@iitm.ac.in

https://doi.org/10.1051/matecconf/201819010001

microforming should possess refined microstructure and have good strain hardening ability.

Cryorolling followed by controlled annealing is an efficient process to develop thin sheets with refined microstructure [7]. During cryorolling, dynamic recovery is supressed resulting in preservation of high amount of dislocation density in the material in form of tangles and clusters. Such material has extremely low ductility, poor strain hardening potential and minimal forming capabilities. Controlled annealing of cryorolled sheets progressively results in development of strain free, equiaxed refined grains [8]. At the initial stage of annealing, the randomly intertwined dislocations get detangled, annihilated and rearranged into ordered, low-angled, sub cells. In the subsequent stages of annealing, defect-free, sub-micrometric, high-angled equiaxed grains are recrystallised in the material. Such grains eventually replace the low angled sub cells developed in the initial stages of annealing. However, further annealing results in grain growth and may result in partial and complete loss of the refined microstructure [8].

The present work proposes an experimental methodology to assess the microforming capabilities of cryorolled sheets and understand the relationship between the microstructure and microformability at various stages of annealing. Such investigation is beneficial for optimising the annealing strategies to engineer desirable refined microstructure for improved microformability. In the present work, samples having three distinctively different microstructure have been designed: (i) partially recrystallised refined microstructure (PRR) with large amount of underdeveloped low angled dislocation cells, (ii) completely recrystallised refined microstructure (CRR) with fully developed equiaxed, high angled recrystallised fine grains, (iii) coarse grained microstructure (CG) with fully grown and very coarse grains. To obtain these distinctive microstructures, three different thermo-mechanical processing routes have been established. First, the micro-scale deformation behaviour of these three different samples have been established by analysing and comparing the tensile stress-strain curves of progressive thinned miniatured samples. Then the micro deep drawing capabilities of the three samples was examined by performing micro deep drawing of the materials, followed by post-mortem analysis of the cross section of the formed micro-cups. These results were correlated with the microstructure of the samples obtained by electron microscopy technique.

2 Material and methods

The material used for the present investigation is aluminium AA5083 alloy (Al-93.5%, Mg-4.5%, Mn-0.4%, Si-0.3%, Fe-0.2%), which is very popular in sheet metal forming industries because of its good ductility and corrosion resistance. The material has been received in form of 10 mm thick plates. These plates were first machined into 50×50 mm square samples and then annealed at a temperature of 450°C for 2 hours in an enclosed furnace followed by water quenching to dissolve any solute completely into the matrix and soften the material for cryorolling. The annealed samples were cryorolled by using a 2-high rolling mill (110 mm roll diameter) operating at 8 rpm. Cryorolling method involved dipping of the base samples in liquid nitrogen bath for 15 minutes followed by its rolling and re-dipping the rolled specimen in the liquid nitrogen bath. This process was continued for a few numbers of passes until the required reduction has been obtained. The final thickness of the cryorolled specimens was approximately 0.9 mm. These specimens were polished using fine emery paper (grit size = 10 μm) to obtain samples of various thicknesses (0.1, 0.3, 0.5 and 0.7 mm). To obtain samples corresponding to three distinctive microstructures, three different types of annealing treatments were employed. PRR material was obtained after annealing cryorolled samples at 200°C for 30 minutes followed by water quenching. CRR material was similarly obtained after annealing cryorolled sheets at 300°C for 30 minutes followed by water quenching. Finally, to obtain CG material, the cryorolled samples were annealed at 450°C for 12 hours in the furnace followed by water quenching. The selection of these annealing conditions is based on the authors' previous work which investigated the microstructure evolution and grain growth kinetics of the same material at various annealing parameters [8]. The microstructures of the PRR, CRR and CG material were visualised by electron back scattered diffraction (EBSD) technique using a FEI Inspect F50 scanning electron microscope operating at 30 kV. To capture the intricate microstructural details of the PRR samples, high resolution micrographs were captured in a Philips CM12 transmission electron microscope (TEM). The samples for the both the electron microscopy processes were prepared using standard techniques (details available in authors' previous work [8]).

Tensile tests were carried for materials with all the three microstructural conditions using a high precision Instron 5984 micro mechanical testing machine equipped with pneumatic grippers. To remove surface roughness and irregularities introduced on the samples in previous stages of polishing, the tensile samples were re-polished using a super-fine emery paper (grit size = 1 μm), followed by electrolytic polishing. A 1:9 acid-alcohol solution of $HClO_3$+CH_3OH was used for this purpose. To verify the consistency in the sheet thickness, the cross section of the samples was observed under a microscope. Image of one the sample has been shown in fig. 1. Miniature sub-standard specimens (length = 3 mm, width = 1 mm) were used for carrying out the tensile test. The sub-standard dimensions were chosen to capture the micro-deformation behaviour of the material more precisely. The deformation speed of the tensile test was fixed at 0.3 mm/min. Three tests were performed at each condition for the sake of repeatability.

A custom designed, indigenously developed micro deep-drawing setup was used to evaluate the micro deep drawing capabilities. The tolerances of the press tool components have been controlled well within 5 μm. An

https://doi.org/10.1051/matecconf/201819010001

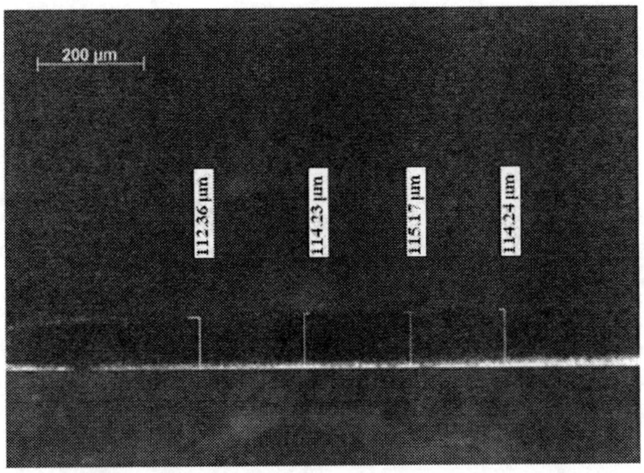

Fig.1. Cross sectional view of the sample with a thickness of 114 µm ± 2.8 µm

inverted configuration of micro deep-drawing has been adopted to prevent deflections and improve stability during forming. The diameters of punch and die used in the micro deep drawing tool are 1.5 mm and 2.0 mm respectively. A corner radius of 0.5 mm has been provided in both the punch and die. The photograph of the press tool has been shown in fig. 1. The samples used for micro deep-drawing were first obtained by punching circular blanks (blank diameter = 3 mm, blank thickness 0.1 and 0.3 mm) from the PRR, CRR and CG sheets. The samples were then drawn up to a height of 1 mm in the micro press tool at a forming speed of 0.3 mm/min. The load-displacement characteristics were recorded using a 2 kN load cell (0.5% accuracy) and 20 nm resolution linear encoder. To perform the post mortem analysis of the formed cup, the components (only with thickness of 0.1 and 0.3 mm) were embedded in a hardened resin mould, sliced into half to expose their cross section. The polished cross section was visualised under Zeiss STEMI 2000CS stereo microscope and the thickness of the section was measured from the centre to sample edge at an interval of 0.05 mm. The average value was computed by analysing at least 5 samples corresponding to each processing condition. The appearance of the top surface of the micro cups was also observed under the stereo microscope.

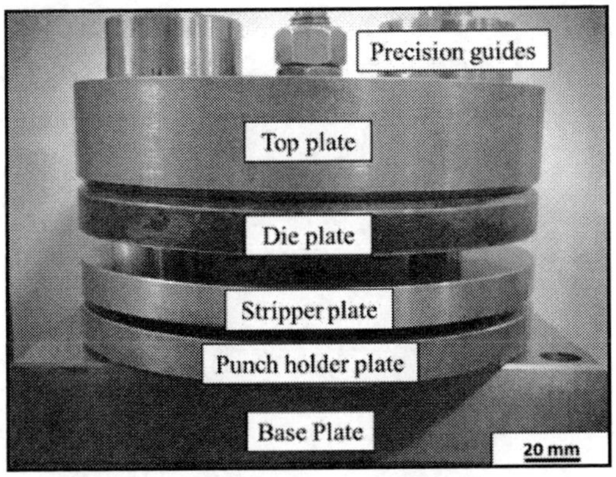

Fig.2. Photograph of the micro deep drawing press tool used in the present investigation

3 Results and discussions

3.1 Microstructural study

The inverse pole figures of PRR, CRR and CG materials as captured by the EBSD technique has been shown in fig. 3a-3c and the TEM micrograph of the PRR material have been shown in fig. 3d.

EBSD inverse pole figures reveal significant differences between the PRR, CRR and CG materials in terms of their grain size, morphology and orientation of the grains. Analysis of the EBSD pole figure (fig. 3a) reveals presence of a composite microstructure consisting of both equiaxed and elongated grains. TEM micrograph (fig. 3d) reveals the internal features of the elongated grains as clusters of relatively low angled equiaxed grains (represented as dashed yellow line). Residual dislocations are also found in form of polygonised dislocation walls (red arrows) inside these low angled grains. Dislocation density analysis presented in the authors' earlier paper [8] showed that after annealing at 200°C, there is an 18.3% drop in dislocation density of the material compared to the cryorolled state.

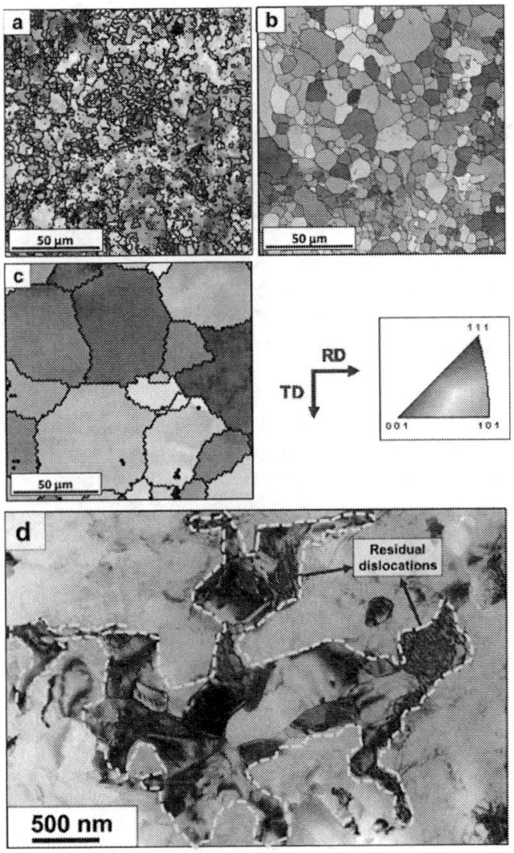

Fig.3. EBSD inverse pole figure of (a) PRR material (b) CRR material (c) CG material (d) High resolution TEM image of the PRR material

This indicates that annealing at 200°C results in annihilation and organisation of dislocation tangles into hierarchical structures comprising of low angled, equiaxed grains embedded inside elongated grains. The average grain size of the selective equiaxed grains is found to be 0.29±0.12 μm. EBSD pole figure of the sample annealed at 300°C (CRR sample) shows completely recrystallised microstructure, with well defined, equiaxed, high angled grains. Decrease in dislocation density is 86.7% compared to as-cryorolled condition [8]. Abnormal grain growth results in coexistence of very fine and moderately coarse grains. The average grain size of the refined grains is found to be 3.7±1.3 μm. Finally, the EBSD micrograph of the CG material obtained after annealing at 450°C for 12 hours reveals fully coarse equiaxed grains. The dislocation density is negligibly low for this material. The average grain size is found to be 88.3±23.6 μm.

3.2 Tensile stress-strain analysis

Tensile stress-strain curves for all three different materials corresponding to samples of different thicknesses (0.1, 0.3, 0.5 and 0.7 mm) have been compiled in fig 4. To account for the scatter in the stress-strain response, three samples were tested in each condition. The mean values and standard deviations of both yield and ultimate strength have been plotted in fig. 5.

Two substantial differences can be seen in the stress-strain response of PRR material. First, compared to CG and CRR materials, PRR material has greater yield strength. Second, the ductility and strain hardening potential of PRR material is found to be significantly lower than that of CG and CRR materials. There is a direct correlation between the mechanical properties such as yield strength, ductility and strain hardening capability with their corresponding microstructure. During cryorolling, material has gone through high degree of strain hardening. PRR material obtained after annealing at 200°C for 30 minutes, undergoes some level of dislocation annihilation. However, it is still having a fair amount of stored dislocations (in form of low angled dislocation cells and polygonised dislocation boundaries) which obstruction the free motion of dislocation, thereby increasing the material's yield strength. Due to its nanometric grain sizes, the volume fraction of grain boundaries and triple junctions in the PRR material is much greater compared to CRR and CG materials. Dislocations travelling from three different directions often get intertwined together and tend to pile up at the triple junctions. Therefore, due to frequent interruptions by triple junctions and residual dislocations in the PRR material, the movement of the newly nucleated dislocations obtained during tensile deformations gets obstructed frequently along the grain boundaries. This results in faster strain localisation and premature fracture of the material, thereby reducing its uniform plasticity/strain hardening potential. As the poor strain hardening is the intrinsic property of the PRR material, this trend is consistently visible throughout all the deformation thicknesses. In microscale, due to very fragile nature of the specimen, the samples of 0.1 mm thickness fail even faster than the samples of other thicknesses.

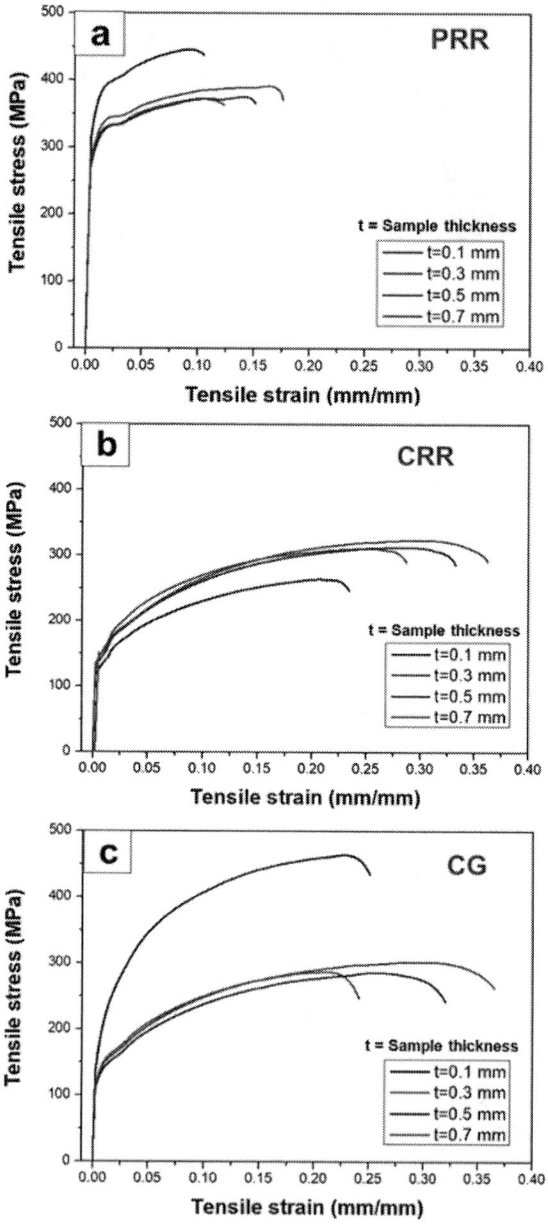

Fig.4. Stress-strain curves for: (a) PRR material (b) CRR material (c) CG material

On the other hand, in case of both CRR and CG material, the ductility and the strain hardening ability of the material is dramatically improved. However, this is achieved by a decrease in both yield and ultimate strength of the materials. The improvement in its strain hardening potential can be associated with the microstructures of the CRR and CG materials.

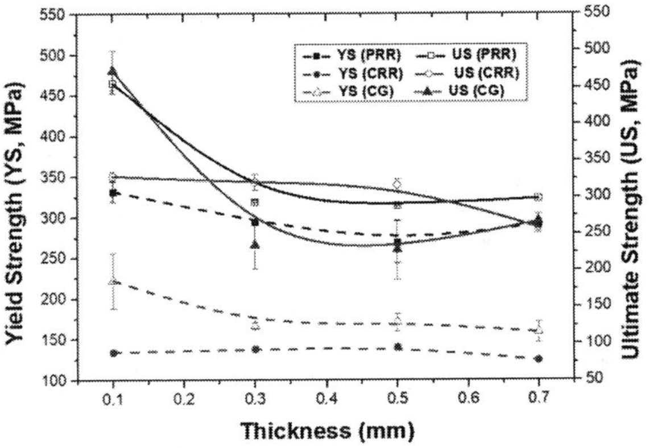

Fig.5. Mean and standard deviations of yield and ultimate strength for various samples

These materials have fully recrystallised and equiaxed grains devoid of excessive residual dislocations. In such microstructures, dislocation motion is relatively easy and free from excessive obstruction by triple junctions.

However, two major deviations in the stress-strain curves are observed between the CG and CRR materials. First, in CG material, with miniaturisation, the ductility is progressively degraded. Second, at microscale (thickness of 0.1 mm), there is a sudden upsurge in the stress-strain response. This could be attributed to the size effect during microscale deformation. Since the sample thickness of 100 μm is very close to the average grain size of the CG material 88.5 μm, there is a dominance of single grain deformation mechanism. The large grain is highly resistant towards plastic deformation, which may result in the relatively smaller grains in its vicinity to deform to greater extent. Therefore, the strain distribution is non-uniform and inconsistent leading to poor formability.

3.3 Evaluation of micro deep drawing capabilities

Microforming capability is the ability of the material to undergo the required plastic deformation without failure. In micro deep drawing operation, since the thickness of the sheet is much smaller, failure occurs mostly along the sample thickness. In the present work, the cross-sectional thickness of the deformed micro components (micro cups) has been measured for samples of PRR, CRR and CG microstructure with an initial sheet thickness of 0.1 mm and 0.3 mm. The average results compiled from 5 samples corresponding to each processing condition have been shown in fig. 6.

Since the strain hardening capability of the PRR material is lower compared to CRR and CG materials, it undergoes faster strain localisation. As a result, the material undergoes abrupt thinning in both the cases for PRR sheet. Thinning is prominent along the curvature of the cup. CRR material shows consistently good forming behaviour marked by its good resistance to thinning in both cases. CG material shows good forming behaviour in case of sample thickness of 0.3 mm. However, when

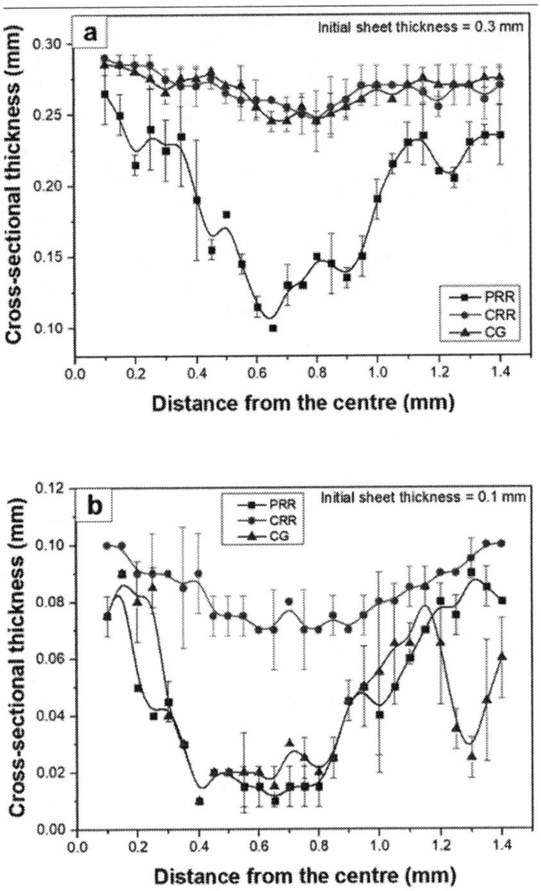

Fig.6. Variation of cross-section thickness of the micro components for an initial sheet thickness of (a) 0.1 mm (b) 0.3 mm

sample of 0.1 mm is formed, it shows very erratic distribution of thickness along the deformation zone. This can be attributed to the prominence of single grain deformation which causes incontinent strain distribution and an increase in process scatter. The micro deep drawing capabilities as assessed from the post deformation cross-sectional thickness correlate well with the uniaxial tensile behaviour of the materials.

The appearance of the micro cups has been presented in fig. 7. The micro cups drawn from PRR and CRR materials have visually much smoother surface. On other hand, micro cup drawn from CG material exhibits inferior surface with pronounced undulations.

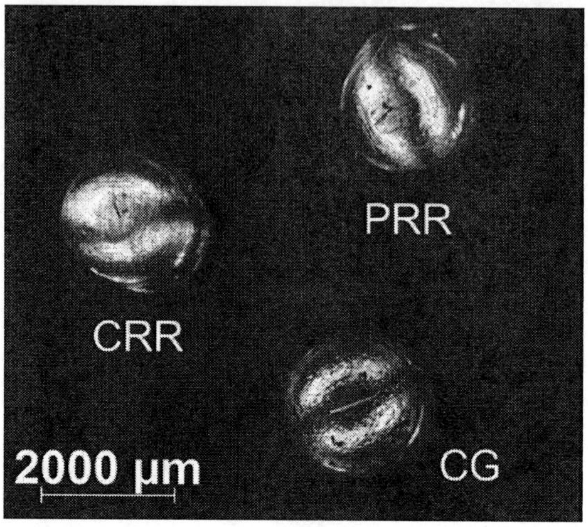

Fig.7. Macrographs of the micro cups drawn from PRR, CRR and CG material

4 Conclusions

The microscale deformation behaviour and micro deep drawing capabilities of three different material with varying microstructures (PRR, CRR and CG) have been investigated by tensile testing and post-mortem cross-sectional analysis of the deformed micro cups. The PRR material was found to have inferior strain hardening tendency resulting in accelerated strain localisation. This was reflected in its poor micro deep drawing performance. The behaviour has been attributed to two distinctive microstructural features: (i) the intertangling of the dislocation introduced during deformation with the pre-existing residual dislocation in the grain boundaries and (ii) frequent obstruction of the dislocation movement by closely packed triple junctions. The CRR and CG material were found to have significantly greater strain hardening potential. However, when the sample thickness is reduced to 100 µm, dominance of single grain deformation prevails in the CG material, which results in inconsistent and non-uniform deformation behaviour. CRR material shows consistently good strain hardening behaviour over various deformation scales. This is also reflected in its micro deep drawing behaviour, which shows stable and uniform sheet thickness throughout its cross-section. For CG material, the surface also appears visually inferior as marked by an undulated and rippled surface. On the other hand, for PRR and CRR materials, surfaces appear relatively smoother.

It is clear from the present investigation that refined microstructures have good microforming capabilities at various deformation domains (0.7 mm to 0.1 mm). However, in case of microscale sheet metal forming, the microforming capability is highly sensitive to plastic instability. Therefore, refined microstructure with good strain hardening capabilities is desirable. Cryorolling introduces a large amount of dislocation, which requires a certain degree of annealing to relieve. Refined microstructure with dislocation free, equiaxed grains are most suitable for microscale sheet metal forming.

This work was financially supported by Aeronautical Research and Development Board, India (ARDB/01/2031831/M/I). The electron microscopy facility has been funded by FIST grant, Department of Science and Technology, India (SR/FST/ET11-059/2012 (G)).

We acknowledge the contributions of Mr. D. Ramamoorthy, Applied Engineering & Technology (P) Ltd, Chennai, in the design and development of micro deep drawing tool used in the present investigation.

References

1. M. Gieger, M. Kleiner, R. Eckstein, N. Tiesler, U. Engel, CIRP Ann. 50(2), 445-462 (2001)
2. H. Justinger, G. Hirt, J. Mat. Proc. Tech. 209, 2111-2121 (2009)
3. W.L. Chan, M.W. Fu, J. Lu, Mater. Des. 32(1), 198-206 (2011)
4. F. Vollertsen, Z. Hu, H. Schulze Niehoff, C. Theiler, J. Mat. Proc. Tech. 151, 70-79 (2004)
5. M.W. Fu, B. Yang, W.L. Chan, J. Mat. Proc. Tech. 213, 101-110 (2013)
6. J.T. Gau, S. Teegala, K.M. Huang, T.J. Hsiao, B.T. Lin, J. Mat. Proc. Tech. 15, 298-305 (2013)
7. Y. Wang, M. Chen, F. Zhou, E. Ma, Nature 419, 912-915 (2002)
8. A. Dhal, S.K. Panigrahi, M.S. Shunmugam, J. Alloy Chem. 726, 1205-1219 (2017)

Micro formed holographic security features in steel

Eckart Uhlmann[1,2], *Stephanie* Schulze, née Frenzel[1*], *Julian* Polte[2], *Christoph* Hein[2], *Johannes* Matthiesen[3] and *Manfred* Böge[4]

[1] Institute for Machine Tools and Factory Management IWF, Technische Universität Berlin, Pascalstr. 8 – 9, 10587 Berlin, Germany

[2] Fraunhofer Institute for Production Systems and Design Technology IPK, Pascalstr. 8 – 9, 10587 Berlin, Germany

[3] HoloDimensions GmbH, In de Butz 4, 22967 Tremsbüttel, Germany

[4] gebr. Böge Metallveredelungs GmbH, Kurt–A. –Körber–Chaussee 27 – 31, 21033 Hamburg, Germany

Abstract. Counterfeiting, with a global trade volume of approximately 450 billion euro a year compromises business results of companies in all industrial segments. At worst, product plagiarism causes severe damages to the individual brand reputation due to product defects or liability issues. Therefore, product integration of inseparable and unique security features is essential for a preserving market share in all manufacturing branches. This paper presents a novel approach regarding a replication technology for the manufacturing of holographic security features. Using a microstructured aluminium substrate with holographic properties as a base material, a forming die is manufactured by a combination of physical vapour deposition (PVD) and galvanic coating processes. Furthermore, a process adapted hardness progression between individual layers was created. By the use of this die, the forming of a holographic microstructure into high strength aluminium alloy (3.3547) and spring steel (1.1248) could be demonstrated within a preliminary study.

Keywords: micro-forming, stamping, hard-elastic layer

1. Introduction

Holographic labeles are accepted security features for large-scale consumer goods in the form of adhesive tags. However, these labels exhibit the disadvantage of easy removal and imitation. Other proven security solutions as data matrix codes as well as RFID labels do not show significant advantages and include eligible costs that can easily exceed the selling price of a product. Evolved and improved security features therefore have to meet requirements at both, counterfeit protection and economic viability. Holographic structures are outstandingly suitable for the use as security features because of the machine readability on the one hand and an opportunity for the customer to perform a visual authenticity check on the other hand. In order to use holographic structures as a security feature, however, unresolved questions still need to be answered. On the one hand, current research and development is targeting the cost-efficient production of unique and function-adapted holographic structures. On the other hand, the development of technologies for the indelible integration of holographic structures into products is an aim of current scientific and technological development [4].

Furthermore, the challenging process chains for the manufacturing of holographic structures by ultraprecision machining or lithographic techniques causes a reliable and cost-efficient technology for the application of holograms on product surfaces. These are needed to increase the acceptance of holograms as security features. Hence, the replication of microstructures and holograms by embossing is a promising approach. The replication of microstructures by embossing is an established manufacturing process in the field of plastics processing. In particular, hot-embossing is established for the moulding of microstructures and nanostructures into plastic substrates. However and due to the process forces and the effect on the wear of the stamping die, the replication of microstructures into metallic substrates has not been accepted. For this reasons, design and evaluation of a process-adapted stamping die are the subject of the investigation presented.

2. Development of a stamping die with a process-adapted hardness profile

The general requirements for a process adapted forming die are a microstructured hard material layer on a hard elastic substratum. The main difficulty in manufacturing a mask following this structure persist in the implementation of the microstructures. This is because hard material layers are not accessible to established methods such as lithography or ultraprecision machining.

*Corresponding author: schulze@iwf.tu-berlin.de

 https://doi.org/10.1051/matecconf/201819010002

For this reason, a completely new approach based on the use of a holographic aluminium shim is presented. The holographic aluminium shim was manufactured by HoloDimensions GmbH by lithography and roll-to-roll replication technology. Starting from this aluminium shim, the layer structure is performed by means of physical vapour deposition process (PVD). As shown in figure 1, the layer systems consists of a titanium nitride (TiN) and a copper (Cu) layer. The titanium nitride layer represents the hard material layer, the surface and also ensures the transmission of the microstructure into the stamping die. The copper layer forms the adhesion promoter for the subsequent chemical and galvanic nickel (Ni) plating processes.

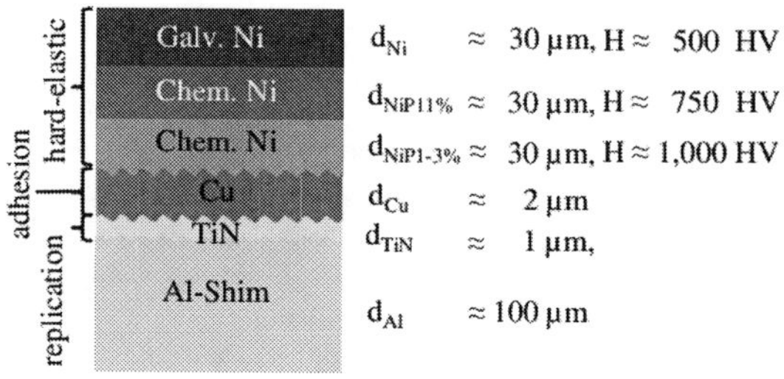

Fig. 1. Layer system consisting of Al-TiN-Cu-NiP-Ni.

In order to implement the previously presented layer system as replication of the holographic aluminium shim, the substrate has to be conditioned for the coating process. This includes a three-stage ultrasonic assisted cleaning process using acetone, isopropanol and deionized water. The deposition of ultra-hard and stoichiometric TiN-layers on the aluminium substrate is performed with an optimised parameter set based on a Ni target, a sputtering power P = 600 W, the use of the radio frequency mode (RF) in addition with argon (Ar) and nitrogen (N) process gases. Furthermore, a PVD coating machine tool Z400, LEYBOLD HERAEUS AG, Cologne was used.

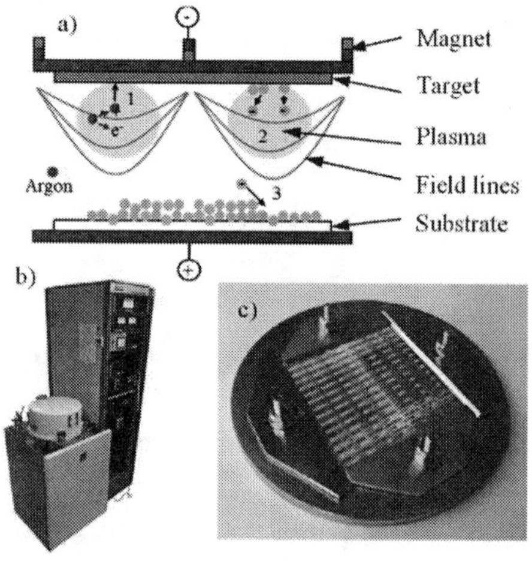

Fig. 2. Schematic representation of the PVD coating process (a), Coating machine tool LH Z400 (b), Substrate holder (c)

In figure 2, the schematic representation of the PVD coating process, the machine tool and the clamping system for the fixture of the aluminium shim are presented.

By the use of optical inspection with a microscope Axio Image 2, CARL ZEISS MICROSCOPY GMBH, Jena, Germany and a magnification between 10x ≤ f ≤ 100x, the replication accuracy of the aluminium microstructure in the TiN layer has been assessed (Fig. 3). As a result of the investigation it is shown that the replication of the holographic structure could be done while providing excellent replication quality.

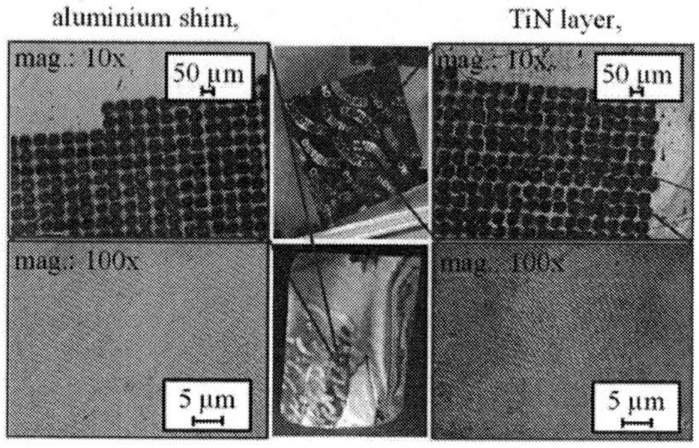

Fig. 3. Comparison of the replication accuracy of hologram structures of the aluminium shim in comparison to the TiN replicate.

Additionally, the replicated holographic structure was evaluated with the method of white-light interferometry and a NewView 5010 ZYGO CORP., Midfield, USA measurement device. The results are shown in Figure 4 exemplarily.

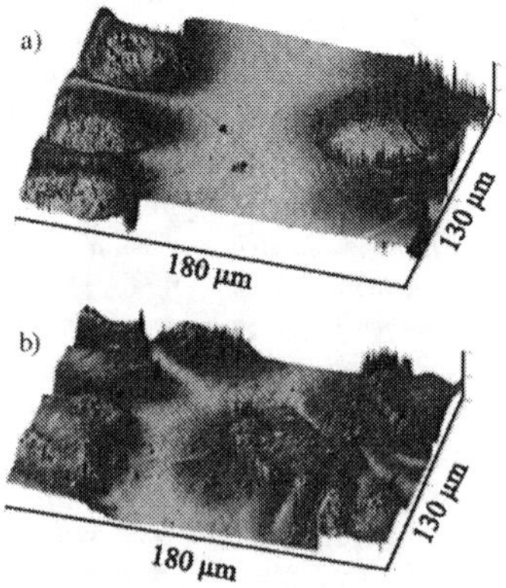

Fig. 4. Comparison of the replication accuracy by means of WLI of a) Al-Shim and b) TiN-replicate.

As shown in figure 4, no significant rounding of sharp-edges was detected. During replication by sputtering, the TiN-replicate can be considered as a negative of the holographic aluminium shim. The height peaks occurring in the 3D-profile are artefacts that are based on process-related measurement errors and can therefore be neglected. It confirms the already obtained impression that the positive form of the TiN-replicate is the exact negative form of the aluminium shim. Even structures in the nanometer range are transferred in full. Furthermore, the PVD coated Cu-layer is cathodically degreased and etched to enable an adhesion of the nickel phosphorus (NiP) layer. Afterwards the NiP is chemically deposited without an externally applied current. The first NiP-Layer has a phosphorus content of $1\% \leq n_{Ph} \leq 3\ \%$ and the second NiP-Layer applied has a phosphorus content of $10.5\ \% \leq n_{Ph} \leq 12\ \%$. The layer composite concludes with an electrolytic nickel plating (pure nickel). Once there is sustainable layer system, the aluminium shim is chemically dissolved.

After all, the stamping die is composed of a base unit (fig. 5) and the holographic stamping foil. The base unit of the stamping die consists of the adapter and the holographic carrier plate. The material used for the composition of the carrier is X153CrMoV12 steel (1.2379). In order to apply the holographic stamping foil to the carrier plate, cohesive bonding techniques was used.

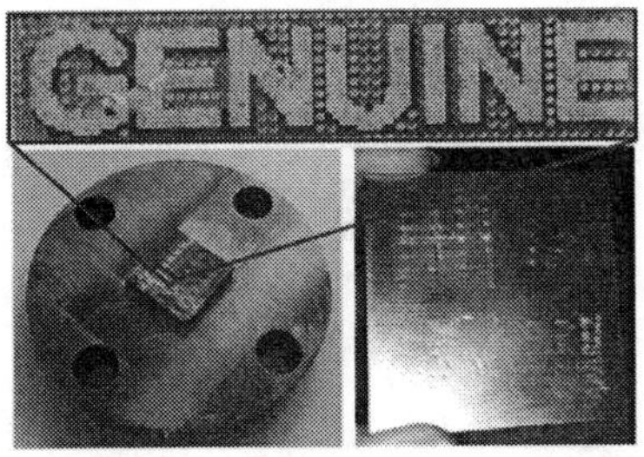

Fig. 5. Embossing result in spring steel strip (1.1248).

For the implementation of the power and path controlled embossing process the materials testing machine tool Z150, ZWICK GMBH, Ulm, Germany shown in figure 6 was used.

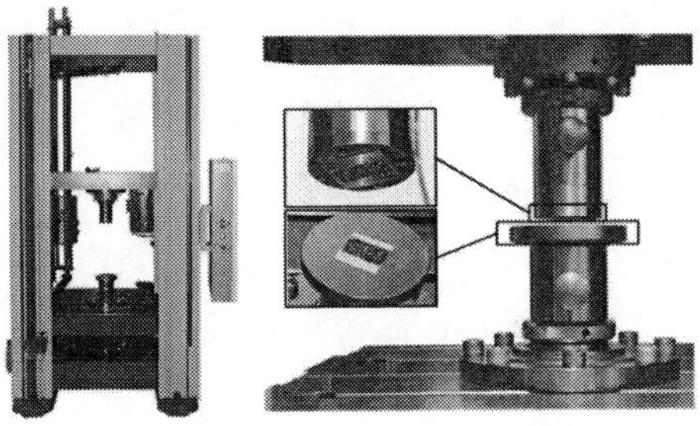

Fig. 6. Stamping test setup on a materials testing machine tool Z150, ZWICK GMBH, Ulm, Germany.

3. Forming of holographic security features

For the determination of the forming force F_F, the following formula can be used [2]:

$$F_F = 2 \cdot k_{fo} \cdot s / \eta_F$$

with

$$k_{fo} \approx 2 \cdot R_{p0.2}. \qquad (1)$$

For the forming of high-strength aluminum alloy AMP 8000 with a forming efficiency $\eta_F \approx 0.8$, a die surface $S = 400\ mm^2$, the 0.2 % proof stress of the alloy and a yield strength $R_{p0.2} \approx 120\ N/mm^2$, a forming force $50\ kN \leq F_F \leq 150\ kN$ was predefined.

As shown in figure 7 even at low forming forces F_F a hologram can be identified visually. An increasing forming force F_F leads to a reinforced visibility in succession with an improved transmission of the microstructures. An investigation of the processing result by atomic force microscopy (AFM) confirms the result of the visual assessment. The transmission of the majority of the holographic structure is possible even at low forming forces F_F. However, the holographic effect and the transmission of the height profile increases with an equally rising forming force F_F.

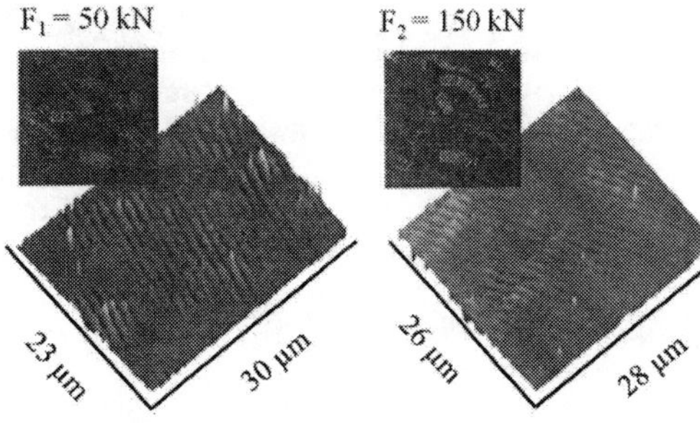

Fig. 7. Comparison of the embossing results in AMP 8000 with different forming forces F_F measured by atomic force microscopy (AFM) and an AFM machine tool N8 NEOS, BRUKER CORP., Billerica, USA.

Nevertheless, the TiN-stamp is subject to signs of wear whose initial stage can be observed in Fig. 8 after embossing no. 50 in the form of chipping in spring steel strip (1.1248).

Beyond that, even with a few forming cycles initial signs of wear are traceable. These show up as breakouts of the TiN-layer and chiping. The breakouts are evidently caused by the elastic movement or compression of the TiN-layer under forming forces $F_F \leq 150$ kN.

https://doi.org/10.1051/matecconf/201819010002

The hard-elastic structure of the embossing foil prevents breakage at high forming forces F_F [3]. As a primary reason for the increased process reliability, the reduction of the compressive stress load by their derivation into the hard-elastic layer structure is presumed [5, 6].

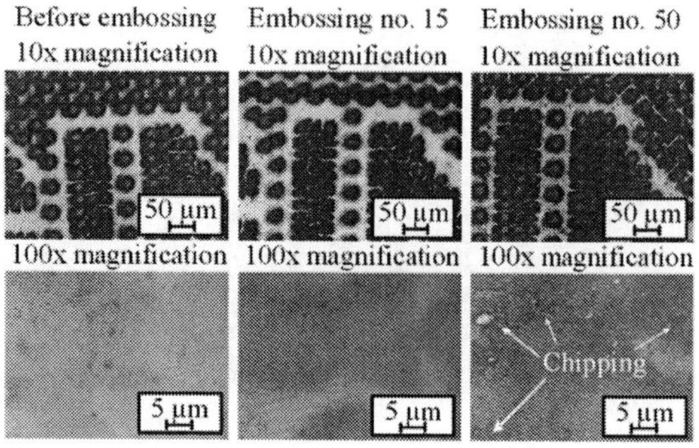

Fig. 8. Comparison of wear behaviour of the holographic TiN die

In addition, WLI-images show that the structural plateau edges of the imaged hologram are reduced in terms of profiling (figure 8), but due to direction of force application the plane surface sections nearly unaffected by wear.

Furthermore, it was discovered that features with an increasingly smaller structural size tend to have a lower wear tilt. Under the assumption of a difference of the surface roughness Ra between substrate holographic substructures, this fact might preserve the integrity of the holographic structure. The chipping of the ultra-hard TiN-layer, which can be identified as white spots in figures 8 and 9, is evidently caused by the elastic movement and the compression of the whole forming die under the effect of the forming force F_F. However, the hard-elastic structure of the forming die prevents the breakage of the whole surface at high forming forces [7].

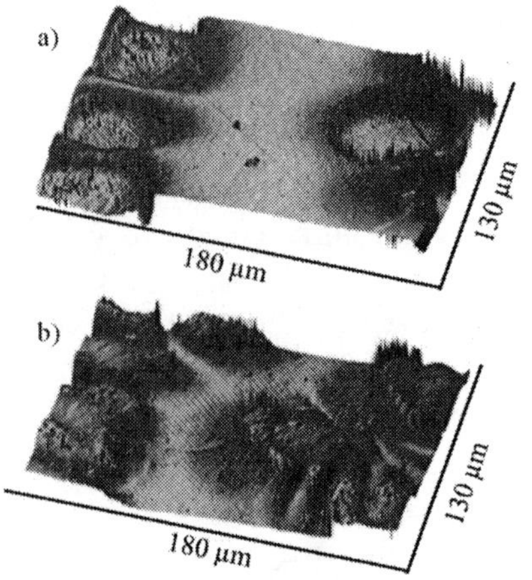

Fig. 9. Wear documentation of the die in new state and after n = 50 forming cycles

4. Outlook

By the use of the presented method, an enabling technology regarding future forgery-proofing was approved. The replication of holographic structures into the surface of a process adapted forming die was established. By the use of this forming die, a manufacturing process for the forming of holographic structures into materials with a certain market relevance was developed and validated. Based on the forming experiments and with a forming force $F_F \leq 150$ kN, a number of cycles $50 \leq n \leq 500$ can be assumed, depending on the material to be embossed.

The objective of the further studies is to produce a reasonable understanding about the microscopic, structural processes during the forming process and the derivation of measures to minimise wear processes.

References

1. Abele, E.; Kuske, P.; Lang, H.: Schutz vor Produktpiraterie: Ein Handbuch für den Maschinen- und Anlagenbau. Springer, (2011)
2. Kugler, H.: Umformtechnik – Umformen metallischer Konstruktionswerkstoffe. Carl Hanser, München, (2009)
3. Ganzherli, N. M.; Maurer, I. A.; Chernykh, D. F.; Gulyaev, S. N.: Forming the surface relief of complex holographic structures on photosensitive material PFG-01. Journal of Optical Technology Vol. 82, (3), (2015)
4. Vollertsen, F.; Hu, Z.; Schulze Niehoff, H.; Theiler, C.: State of the art in micro forming and investigations into micro deep drawing. Journal of Materials Processing Technology Vol. 151 (3), (2004)
5. Wang, Z.; Shuhui, L.; Wang, X.; Cui, R.; Zhang, W.: Modeling of surface layer and strain hardening effects on micro-bending of non oriented silicon steel sheet. Materials Science and engineering Vol. 711, (2018)
6. Li, Y.; Wang, G.; Liu, S.; Yang, J.; Yang, C.; Zhang, K.: Drawability and size effects for micro-arrayed deep drawing of Ni-Co/GO nanocomposite foils. Journal of Materials Processing Technology Vol. 249, (2017)
7. Obikawa, T.; Hayashi, M.: Ultrasonic-Assisted Incremental Microforming of Thin Shell Pyramids of Metallic Foil. Micromachines Vol. 8(5), (2017)

Generation of micro channels in AlMg4.5Mn0.7 sheets using a vibration-assisted micro forming process

*Benedikt Müller** and *Andreas Schubert*

Chemnitz University of Technology, Professorship Micromanufacturing Technology, Reichenhainer Str. 70, 09126 Chemnitz, Germany

Abstract. Micro forming can be used for a highly efficient manufacturing of small parts like screws or caps. Because of the small dimensions of the structures, high stresses and forming forces occur, especially in the case of micro bulk forming. In addition, the influence of friction is significantly higher compared to macro forming processes. A typical example of small structures are micro channels. An approach for the reduction of the necessary forming force to obtain lower loads on the tool consists in a vibration-assisted micro forming process. As vibration source, a piezo power module is placed directly in the force axis of the forming press. For the investigations, micro channels with a depth of more than 300 µm and a width of 300 µm are formed into 1.5 mm thick AlMg4.5Mn0.7 aluminium alloy sheets. The focus of the research lies in the influence of the process parameters like frequency and oscillation amplitude onto the material flow and the achieved channel depths. To investigate a possible influence on the friction conditions due to the vibration assistance, different lubrication conditions are applied. First results show a channel depth increase of 12 % compared to samples formed without vibration assistance.

Keywords: Micro forming, Aluminium, Vibration assistance

1 Introduction

Micro forming is a highly efficient and precise manufacturing method for small parts as well as for small functional structures. Such small parts can be screws, bolts, caps, or other structures like frames, mostly used for electronic devices. In addition, micro forming also allows the generation of small structures into the surfaces for a functional integration. Typical examples are micro grooves or channels that can be used for micro fluidic applications, micro cooling systems, or for the direct integration of active materials like piezo ceramics to generate smart sheets with integrated sensor and actuator functionalities.

Because of the small dimensions of the structures, high stresses and forming forces occur, especially in the case of micro bulk forming of micro channels. In addition, the influence of friction between tool and workpiece is significantly higher compared to macro forming processes.

A reduction of the necessary forming force due to a decrease of the friction leads to a more intense material flow, a higher achievable precision, and a reduction of the tool load. One approach for the reduction of the friction forces is a vibration-assisted micro forming process. Researches show, that ultrasonic vibration assistance can lead to a reduction of the forming force [1], as well as a reduction of surface roughness values [2], or nearly an elimination of the spring back behaviour [3]. These researches are mainly focused on the generation of single micro parts or investigations in micro deep drawing. The applied forming force of those processes is lower, compared to bulk forming.

The application example of this research is the generation of an array of ten micro channels, which is intended to serve as intake cavities for piezo rods. Subsequently, the rods can be placed into the channels. After a joining by forming process, the integrated piezo ceramic rods in combination with the aluminium sheet is supposed to act as a smart sheet with sensor and actuator functionalities [4]. Because of forming forces of 80 kN and above, large high-powered ultrasonic vibration systems would be necessary. Due to this, the paper focuses on the investigation, how a low frequency vibration assistance can improve the forming process.

2 Experimental procedure

The micro channel array consists of ten single channels with a channel width of 300 µm, a channel depth of 300 µm, and a channel length of 10 mm. Within one press stroke also secondary cavities, which are placed perpendicular and in parallel to the micro channels, are generated. The cavities, positioned parallel to the micro channels act as flow barriers to decrease the lateral material flow, whereas the perpendicular to the channels positioned cavities act as space for electrodes of the piezo fibres. The entire area of the micro structure is 7.2×14 mm^2. Fig. 1a) shows the micro channels and the secondary cavities. A cross section of the micro channels is represented in Fig. 1b).

* Corresponding author: benedikt.mueller@mb.tu-chemnitz.de

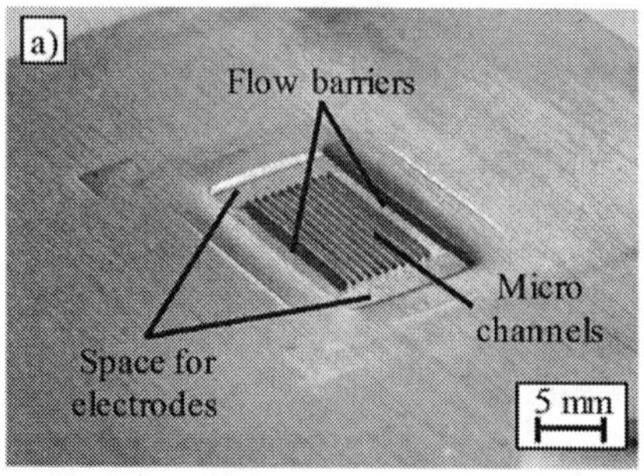

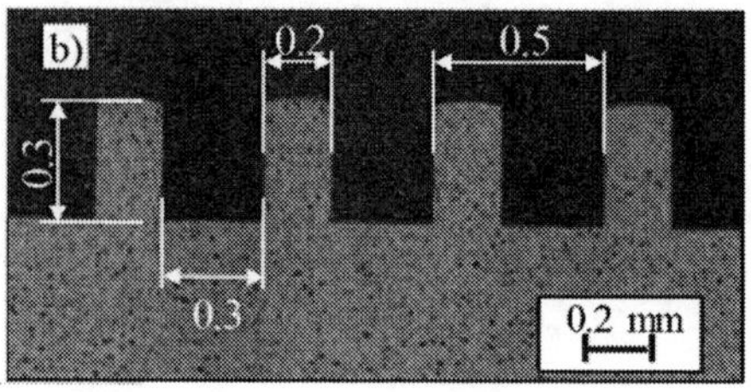

Fig. 1. a) Geometry of the samples; b) Cross section and dimensions of the micro channels.

The micro forming tool, used for this investigation, consists of a upper and a lower die. The tool stamp is made from hardened (64 + 2 HRC) HS6-5-2C (1.3343) high speed steel and is designed as the negative structure of the micro channels. Together with a blank holder, which is set to a blank holder force of 2 kN, the stamp forms the upper die. The test samples are 50 × 50 mm^2 aluminium (EN AW-5083 - AlMg4.5Mn0.7) plates cut out of 1.5 mm thick sheets using water jet cutting. The samples are placed on a flat lower die plate, which is set on top of a piezo power module, acting as vibration source. The power module of the company CeramTec consists of several piezo stack actuators housed in a hermetically sealed steel case. The module is capable of reaching a frequency $f = 50$ Hz and can be used up to a blocking force of $F = 80$ kN. Up to this force, the piezo power module is able to deliver an oscillation amplitude. A sine wave is used as input signal for the oscillation.

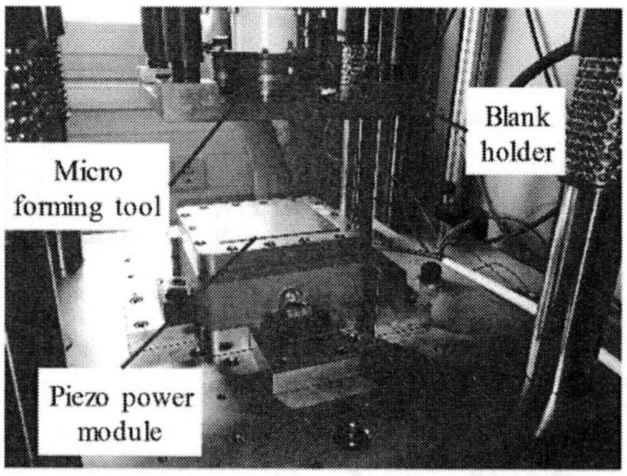

Fig. 2. Micro forming tool and piezo power module (lower die plate not shown).

All the components are integrated into a 200 kN precision servo press P.FÜ.MA from the company Hegewald & Peschke.

A cleaning step with ethanol is performed on the samples before they are lubricated with oil and placed on the lower die plate. After each forming experiment, the tool and the sample are cleaned again. In a next step, the characterisation of the samples is conducted. Therefore the channel depth d_c is chosen as evaluation parameter to obtain information regarding the material flow. For the measurements a Nikon MM400 measuring microscope is used. To visualise the forming results, additional profile scans are made using a 3D laser scanning microscope. The profile data of each measured sample consist of 60 single images, which are stitched together with a special software. To gain comparable results the image processing parameters and scales are held constant for all measurements.

For the experiments, the parameters, shown in Table 1 are varied and their influence on the channel depth and the material flow are investigated. All forming experiments are carried out with a constant stamp velocity v_s up to a maximum forming force of $F = 70$ kN. For each parameter set, two samples are formed.

The forming force is below the blocking force of the power module to ensure that a vibration amplitude is generated during the forming process, even if complete tool filling is not achieved at this point.

Table 1. Investigated parameters.

Parameter	Symbol	Specification
Lubrication condition	-	Plantoform, Platinol
Frequency	f	(10; 20; 50) Hz
Amplitude of sine input signal	y	(1.25; 2) V

For technological reasons, the maximum frequency of the piezo power module is limited to 20 Hz at an amplitude of 2 V and above. To prevent a negative input signal, an offset of 5 V is applied to the input sine signal.

Furthermore, as a reference, samples are formed without vibration assistance. As lubricants, two special forming oils are used, whereas the viscosity is different between both types. With 49 mm^2/s Plantoform MBO 2797 from the company Fuchs exhibits the lower viscosity compared to the viscosity of 163 mm^2/s from Platinol V 1937/11 manufactured by OEST.

3 Results and discussion

During the forming process up to the forming force $F = 70$ kN the stresses in the sample are not sufficient to generate a complete uniform channel array field. This can be seen in Fig. 3, indicated with the dashed line.

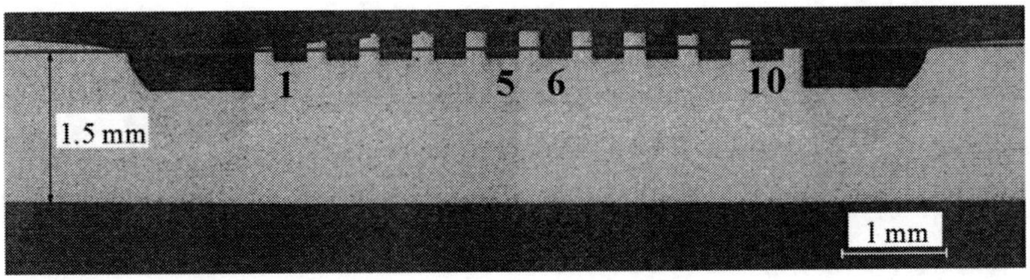

Fig. 3. Section of a formed aluminium sheet with non-homogeneous formed channels.

When the tool is pressed on top of the aluminium sheet, a stress state is generated, which leads to a material flow into the opposite direction of the applied forming force. The solid line in Fig. 3 shows the initial sheet thickness and how the material, forming the webs of the structure, flows into the die. The discussion in this paper will focus on this part of the material flow, which flows into the die to form the channels. All the lateral material flow is not part of this evaluation.

At the beginning of the forming process, the material flow starts in the middle of the structure (channels 5 and 6, see Fig. 3). The highest structure depths are achieved for these channels. Conversely, at the latest, the deformation starts at the edge of the structure (channel 1 and 10), which manifests in the lowest channel depths. In the following evaluation, the maximum channel depth represents the deepest channel of a sample and the minimum channel depth the lowest channels from the edge of the structure. All measurements are conducted at a channel length of 5 mm, which is the lateral centre of the structure.

Fig. 4 and Fig. 5 show the mean minimum channel depths and the mean maximum achieved channel depths for the different vibration parameters, using Plantoform as forming oil. The dashed line in both diagrams stands for the mean minimum or maximum channel depth of the reference samples, formed without vibration assistance. All other parameters like forming force or stamp velocity are the same as for vibration-assisted experiments.

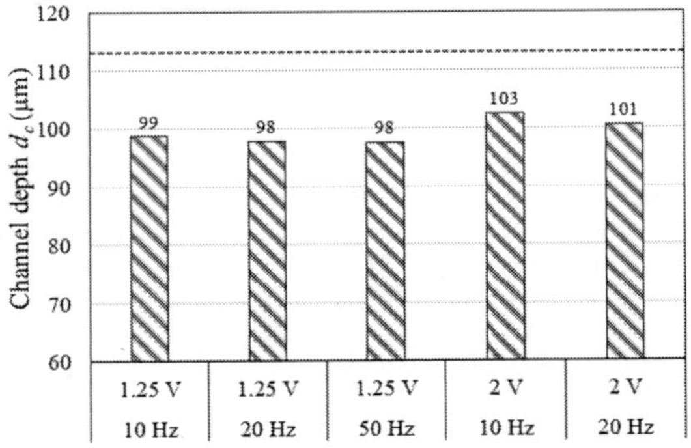

Fig. 4. Minimum channel depth according to amplitude and frequency for samples formed with Plantoform.

The minimum mean channel depth of the reference is $d_c = 112$ µm and the maximum mean channel depth is $d_c = 217$ µm. As can be seen in Fig. 4, the minimum channel depths for samples formed with Plantoform are in a very narrow range from 98 µm (1.25 V and 50 Hz) to 103 µm (2 V and 10 Hz). Because of these nearly identical values, only a slight trend can be observed for this measurement series. By varying the frequency *f*, a slight decrease of the achieved channel depth can be observed for increasing frequencies. This small trend applies both to an amplitude of 1.25 V and to an amplitude of 2 V. Furthermore, it can be seen that higher minimum channel depths are achieved with an increase of the oscillation amplitude y. However, it is very noticeable that the minimum depths of all samples formed with vibration assistance are significantly below the reference (9 % – 14 %).

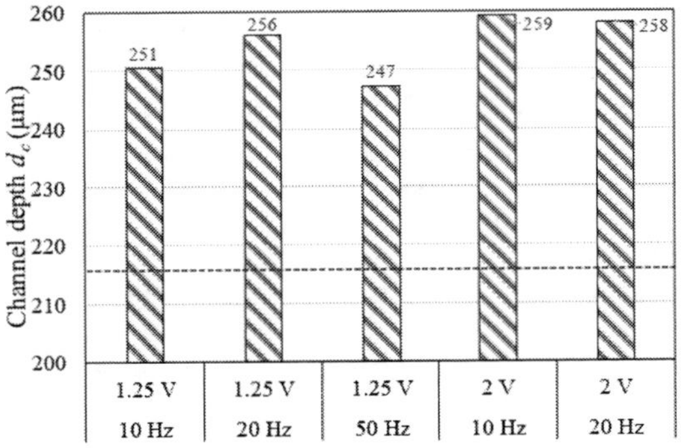

Fig. 5. Maximum channel depths according to amplitude and frequency for samples formed with Plantoform.

When comparing the trends in Fig. 4 with the maximum channel depths in Fig. 5, it is obvious that there is no clear influence of the frequency onto the channel depth. The results of vibration assisted forming experiments using Plantoform show maximum channel depths ranging from $d_c = 247$ µm to $d_c = 259$ µm. Both results achieved with the higher oscillation amplitude indicate slightly higher channels depths, which corresponds to the minimum depths of this test series.

But at the centre of the structure, which is represented by channels 5 and 6 or their depths, all samples formed with vibration assistance samples show a significant increase of the depth compared to the reference samples. An increase in channel depth of about 14 % – 19 % is observable. This nearly correlates to the 9 % – 14 % deeper channels of the reference channels in the border region of the structure. So at this point, it is not clear if the vibration assistance with low frequencies has a positive influence on the forming process. Therefore, Table 2 shows the overall mean channel depth of all channels of an array for each parameter setup. To gain information about the distribution of the achieved channel depths and

 https://doi.org/10.1051/matecconf/201819010003

so the uniformity of the structure, the difference Δd_c from maximum to mean channel depth is calculated. A very small difference between the maximum and the mean channel depth indicates a uniform forming result. In contrast, a large difference indicates an uneven shape of the structure. Furthermore, Table 2 shows the increase in overall mean channel depth compared to the reference.

It can be seen in Table 2 that the overall mean channel depths can be increased about 5 % – 12 % by a vibration assistance.

Table 2. Averaged results for geometry of samples formed with Plantoform.

	Overall mean channel depth (μm)	Δd_c (μm)	Increase
Ref.	170	47	-
1.25 V 10 Hz	182	69	7.1 %
1.25 V 20 Hz	184	72	8.2 %
1.25 V 50 Hz	179	68	5.3 %
2 V 10 Hz	191	68	12.4 %
2 V 20 Hz	186	72	9.4 %

Beneath the increase of the channel depths through the vibration assistance, also the material flow is changed. The reference samples show shallower channels, but a more even channel depth distribution over the entire array. This fact can be seen from the parameter Δd_c, which is shown in Table 2. The difference between maximum and overall mean channel depth for samples formed with vibration assistance is higher, but in a small range of 4 μm (68 μm – 72 μm).

To illustrate the different material flow, profile scans are made, using a 3D laser scanning microscope. Two scans represented in Fig. 6. On the upper profile, a reference sample, formed without vibration assistance is shown, whereas the lower figure represents a sample formed with an oscillation amplitude of $y = 2$ V and a frequency of $f = 10$ Hz. In the figure, all channels and the adjacent flow barriers are shown and measured. Due to this, the indicated maximum channel heights are nearly 400 μm because the height measurement starts in the areas of the deeper flow barriers.

The figures confirm the measurement results that the outer webs of the structure are slightly higher in the reference sample, resulting in deeper channels in this area. Also the benefit of the vibration assistance is clearly visible in the figures. The red and nearly white coloured webs in the lower figure point at significant higher webs and deeper channels than the reference sample. An explanation for this behaviour can be found in the friction. When the material flow starts in the centre region, the material flows into the die and forms the first webs of the structure.

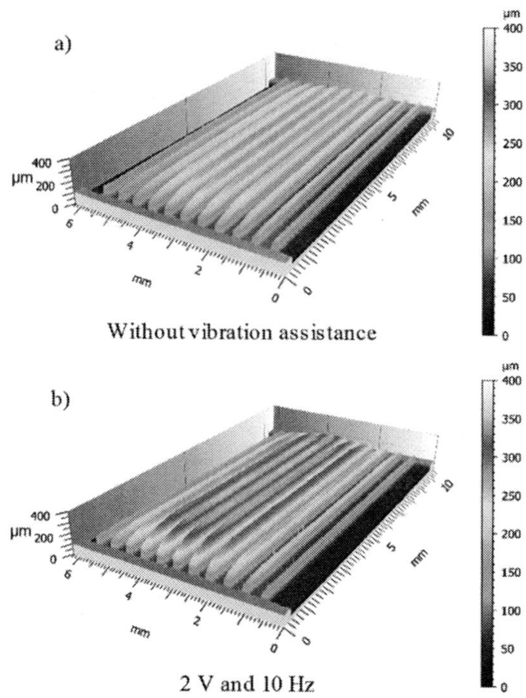

Fig. 6. Profile scan of samples a) Formed without vibration assistance b) Formed with vibration assistance.

The material comes in contact with the sidewalls of the webs in the die, which leads to an increase of friction at this moment. This lowers the material flow, increases the induced stresses and leads to a sideways material flow into other channels in the border region of the structure. With the relative motion during the vibration assisted forming process, the tool and the workpiece are separated again for a little moment, which has a favourable effect on the friction conditions. So the material flow into the centre channels of the die is larger compared to the reference sample and the forming of the outer webs starts later. Due to the fact, that every experiment is done until the maximum force of $F = 70$ kN, in both cases the generated stresses are not high enough to lead to a material flow, which fills the die completely.

The same experiments are also done with Platinol, a forming lubricant with a higher viscosity. It could be possible, that during the relative motion of the tool, more lubricant is able to flow into the contact area to support the forming process. It can be assumed, that with "thicker" oil, this phenomena would not happen in such a significant way. The results regarding the minimum and maximum channel depths of samples formed with Platinol are shown in Fig. 7 and Fig. 8.

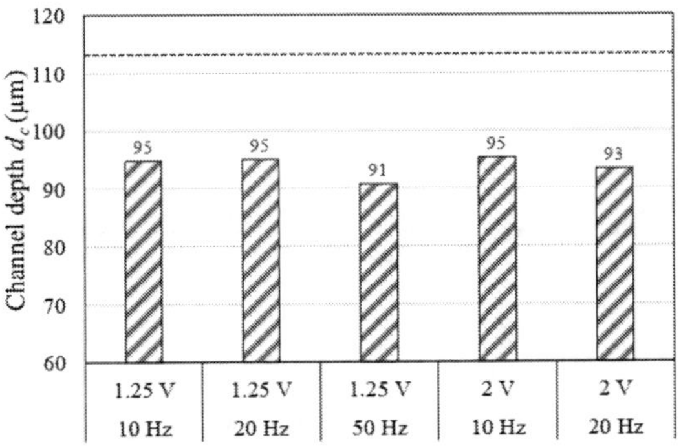

Fig. 7. Minimum channel depth according to amplitude and frequency for samples formed with Platinol.

Similar to the results with Plantoform, the minimum channel depths are also below the reference for the samples formed with Platinol. For no parameter combination, channel depth exceeds d_c = 95 µm, whereas this is the case for all experiments with Plantoform. The same behaviour can also be seen in Fig. 8. Even the highest channel depth of the Platinol test series (d_c = 247 µm) is not larger than the lowest value achieved with Plantoform. Nevertheless, all maximum channel depths of samples formed with vibration assistance using Platinol are above the reference value.

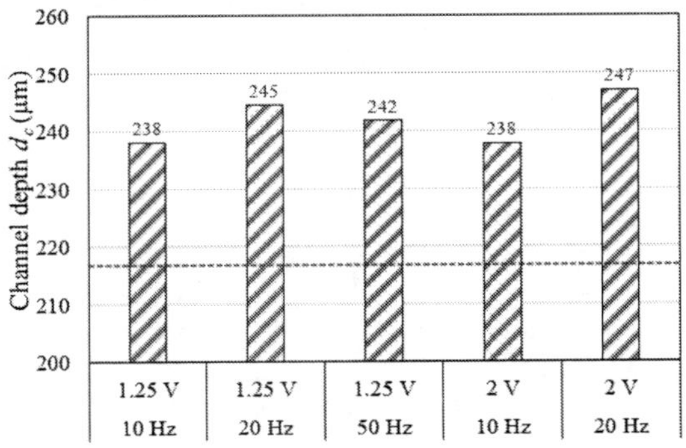

Fig. 8. Maximum channel depth according to amplitude and frequency for samples formed with Platinol.

The overall mean channel depths of all ten channels, shown in Table 3, indicate that the thicker oil (Platinol) lead to almost no improvement due to vibration assistance.

Table 3. Averaged results for geometry of samples formed with Platinol.

	Overall mean channel depth (µm)	Δd_c (µm)	Increase
Ref.	170	47	-
1.25 V 10 Hz	175	63	2.9 %
1.25 V 20 Hz	180	65	5.9 %
1.25 V 50 Hz	172	69	1.2 %
2 V 10 Hz	174	63	2.4 %
2 V 20 Hz	178	69	4.7 %

By investigating the profile scans of samples formed with Platinol (Fig. 9), it can be seen, that the material flow in the lateral direction is not as homogenous compared to the reference or Plantoform samples (Fig. 6). At the channel length of about 10 mm, the achieved channel depths are higher, compared to the centre of the structure.

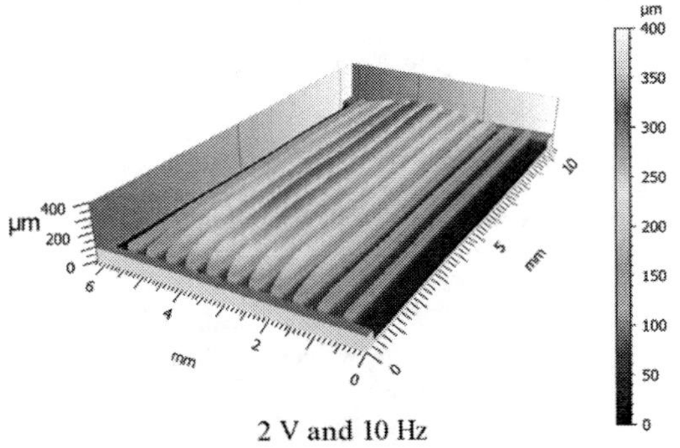

Fig. 9. Profile scan of a sample formed with vibration assistance and Platinol.

4 Summary and conclusions

Experiments concerning vibration assisted micro forming of channels into aluminium sheets are conducted. Therefore, lubrication conditions, the oscillating amplitude and the frequency are varied and their influence on the achievable channel depth and the material flow is studied.

The vibration assistance in combination with the low-viscosity forming lubricant Plantoform shows a significant improvement regarding channel depth compared to reference samples. A slight trend that a higher amplitude can lead to deeper channels was found. Because of the unfinished tool filling, higher deviations Δd_c occur. Due to this, more experiments should be done to improve the statistically reliability of the data.

In summary, it can be stated that with the vibration assistance a larger deformation or material flow can be achieved. Measured by the geometric parameter channel depth d_c, the improvement is slightly more than 12 %.

This research was supported by the Deutsche Forschungsgemeinschaft (DFG) in context of the Collaborative Research Centre/Transregio 39 PT-PIESA, subproject A2.

References

1. Y. M. Huang, Y. S. Wu, J. Y. Huang, Int. J. Adv. Manuf. Technol. **71**, 1455-1461 (2014)
2. Y. Bai, M. Yang, Proc. Eng. **81**, 1475-1480 (2014)
3. S. Amini, A. H. Gollo, H. Paktinat, Int. J. Adv. Manuf. Technol. **90**, 1569-1578 (2017)
4. B. Müller, A. Schubert, Proc. Eng. **207**, 1045-1050 (2017)

https://doi.org/10.1051/matecconf/201819010004

Forming of micro gears by compressing a pure copper sheet through its thickness

Amirreza Asghari Ganji[a], *Mohammad* Bakhshi-Jooybari and *Hamid* Gorji

[a] Mechanical Engineering, Advanced material forming research Center, Babol Noshirvani University of Technology, Babol, Iran.

Abstract. Currently, micro gears are mostly fabricated by LIGA technology and micromachining. These processes have some limitations. Forming processes not only satisfy mass production and appropriate productivity rate but also present superior mechanical properties. A major problem preventing the bulk micro metal forming is the preparation of micro billets and their precise transfer between the forming stages. The purpose of this study is developing a method to form a micro gear without the need to a separate micro billet preparation. In this paper, pure copper sheets were compressed into the predetermined micro gear profiles though their thicknesses, so that there is no need for preparation of micro billets and also its troublesome transforming. The tests were performed at room temperature, in two cases of single extrusion process and extrusion-forging process. Micro gears with 6 teeth and 250μm in module were formed completely with good repeatability in both the cases. A major advantage of the proposed study compared with the blanking process is that, in blanking, the process is merely cutting the edges, while here the material fills the die by deformation. Thus, better mechanical properties will be achieved. Measuring the micro-hardness of the formed parts in comparison with raw material, verified this point. In general, the micro-hardnesses of combined extrusion-forging parts were higher than those of single extrusion ones in the same positions on the micro gears surface.

keywords: Micro forming, Extrusion, Micro gear.

1 Introduction

Due to the wide applications of small gears, there are various methods for the production of micro gears in industries [1]. Among of all, LIGA and micromachining are advanced techniques for producing metallic parts. Nevertheless, these processes challenge with some limitations. On the other hand, microforming techniques are able in mass production with high productivity rate, near net shape manufacturing and better mechanical properties [2, 3].
The first definition of microforming was presented by Geiger et al. [4] as the production of parts with at least two dimensions in the sub-millimeter range [5].
Saotome and Yavazaki [6] produced micro gear shafts from a super plastic of aluminum alloys by means of extrusion process at elevated temperatures. Wang et al. [7] formed a micro gear from bulk metallic glasses by a hot embossing process. After the process, they used a chemical solution to solve the silicon die for ejecting the micro gear. Chen and Qiu [8] studied the forming of micro stepped gears by forging process. Dong et al. [9] used hot extrusion and investigated microstructure and micro-hardness of 7075 aluminum alloy micro gears. They reported that the micro-hardness at the teeth are lower than that of the center of specimens. Chang and Cuo [10] investigated the effects of temperature and grain refinement on the closed-die forging of copper micro gear by ECAPed and anealed material. Kim et al. [11] used ECAPed 6061 aluminum alloy to form micro gears by means of extrusion process. Guo et al. [12] forged micro gears from Zr-based metallic glass. They also measured the nano-hardness of the formed part and reported that the core of micro gear has the maximum of nano-hardness.
In the above researches, micro billets were prepared separately and were positioned in the die before each test. Meanwhile, one of the main problems of industrializing the micro bulk forming is preparing and controlling of billets between the forming stages. This is because at micro scale, the workpiece tends to stick to the grippers [13]. Also, there would be some other questions that must be addressed, for example, in an extrusion of a micro gear shaft how an extruded part would be cut into small ones or how to prepare a micro billet without using separate process to circumvent the problem of positioning and handling.
This experimental study tries to address the mentioned questions based on the concept of progressive microforming of bulk parts using directly sheet metals [14]. In this type of controlling, micro parts are attached to the sheet and in the final stage, the ejection is performed. [15, 16]. On the other hand, it is ideal that a process chain for micro-manufacturing should be short

* Corresponding author: ganjimecheng@gmail.com

as much as possible [17]. Therefore, in this research, the die was designed to perform the forming operation and ejection without transforming the formed part. In fact, at the center of a blanked sheet, extrusion occurs and material will be compressed through thickness to the predefined gear profile. When forming was finished, the ejection was performed by means of a torsion operation of dies which departs micro gear from sheet. In addition, conventional ejection was done too. The torsion method and investigstion of its feasability was performed for the first time in this paper.

2 Experimental procedures

2.1 Material and die setup

Pure copper sheets with the thickness of 3mm were used as received material. The sheets were annealed at 700°C and cut into disc shape to the diameter of 44mm. To eliminate the oxide layers on the sheet surfaces and also for the reduction of thickness to 2.75mm, the discs were grinded. The micro-hardness of the annealed material was measured 55HV.

Either at micro or macro scale, providing a billet is mandatory for the processes like extrusion and forging. But at micro scale, the sheet can be used as a raw material to form a micro gear. The idea is that the material of the sheet which is in the deformation zone, plays the role of micro billet. As illustrated in Fig. 1, the material above the die cavity and beneeth the punch exists in the deformation zone. Thus, a fraction of blank is used instead of a longer micro billet which should be cut from a wire and positioned with too much difficulty in the die. The rest of material plays carrier role for material at the deformation zone like a gripper.

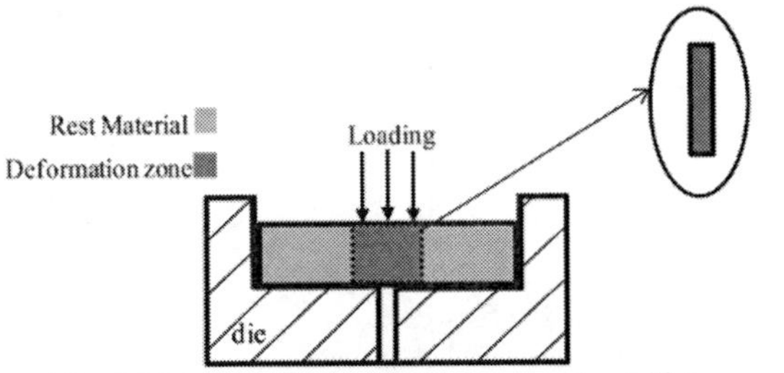

Fig. 1. Material in deformation zone as a micro billet

The schematic of the die set and experimental procedure are shown in Fig. 2. The components of the die set are punches, blank holder, blanking die, inserts and insert holder, and bottom plate. The experimental procedures have two stages for both single extrusion and combined extrusion forging processes. The first stage of each process relates to fine blanking of the sheet with the diameter of 12mm, Fig. 2-a. The extrusion of the material at the deformation zone of the blanked part into the predefined micro gear cavity is done for the second stage, Fig. 2-b or 2-c. Since the two extrusion processes, including single extrusion or combined extrusion-forging, are performed for the second stages, two kinds of inserts were designed. Single extrusion test was designed to determine the maximum capacity of the copper sheet under this test situation and combined extrusion-forging insert was designed to produce a final part from the copper sheet. The thickness of the insert for combined extrusion-forging was considered 2.25mm, Fig. 3. A hard stopper was used to change the extrusion process to the forging one.

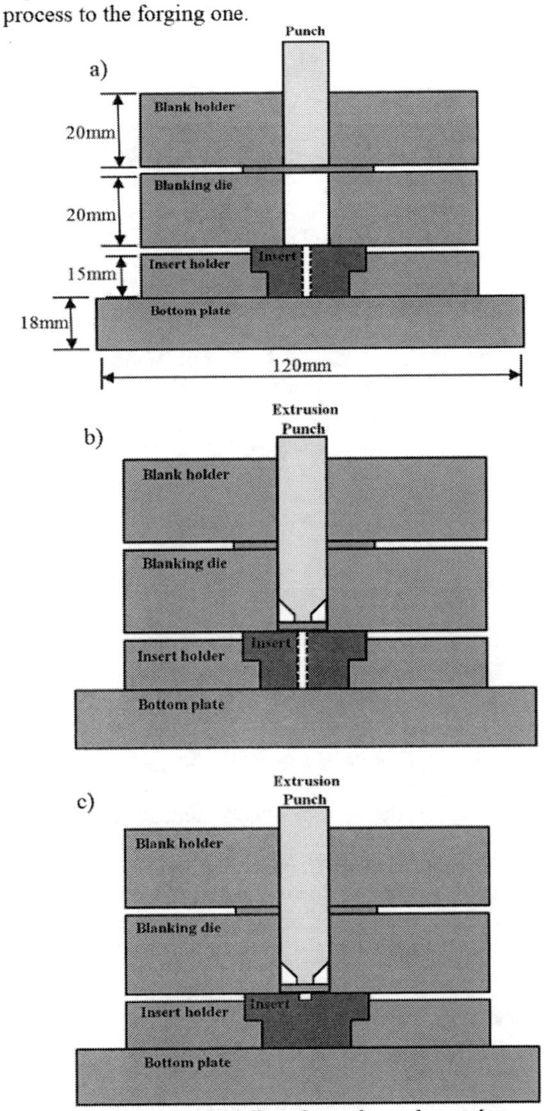

Fig. 2. Schematic of the die and experimental procedure a) blanking operation, b) single extrusion and c) combined extrusion-Forging

Tests were done at room temperature. For dry lubrication condition, the die was cleaned by acetone solution before each test. The ram speed was considered 25mm/min for blanking process, and 3mm/min for extrusion processes. Low speed of extrusion processes was for better controlling and preventing die damage. The size of the blanking clearance was 0.02mm which created a blank with smooth edge. In fact, a fine blanking was occurred. This smooth edge helps to prepare a good lateral constraining and completely blockage of the lateral

https://doi.org/10.1051/matecconf/201819010004

movement of material. Components of die were manufactured from K100 steels by machining.

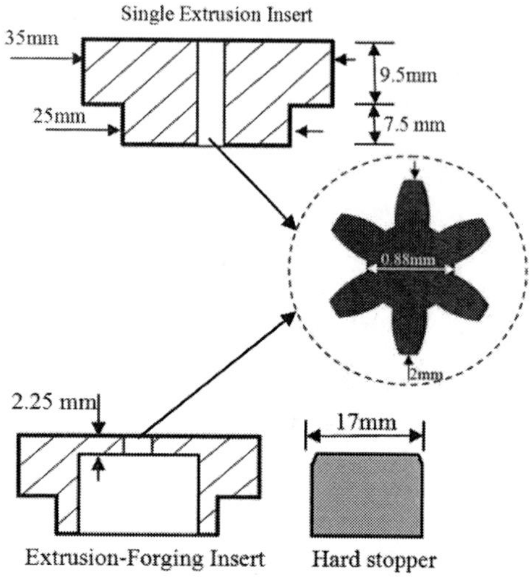

Fig. 3. Inserts a) Single extrusion b) Combined Extrusion-Forging

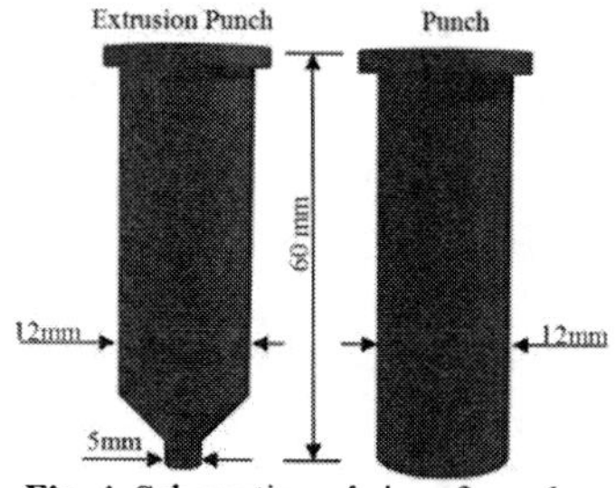

Fig. 4. Schematic and size of punches

The gear female die with 6 teeth and 250μm in module was machined on the inserts by Wire Electro Discharge Machining (WEDM). The tip of extrusion punch is 5mm in diameter which is more than 3 times the micro gear pitch circle diameter. This creates enough material and makes a significant deformation zone for micro extrusion. Schematic of punches are shown in Fig. 4. The die assembly and components of die are shown in Fig. 5.

2.2 Ejection Methods

The ejection of micro parts has been performed by means of torsion operation. The principle of this operation is simple and needs no extra tool which is very significant, because designing and manufacturing of gear shaped punch for blanking needs high precision at micro scale.

Since the micro parts are very small, they could be departed from the sheet at low forces. So, when the insert holder and subsequently inserts keep constant, if the blanking die twists through its axis, the formed part will be departed from the blank, Fig. 6.

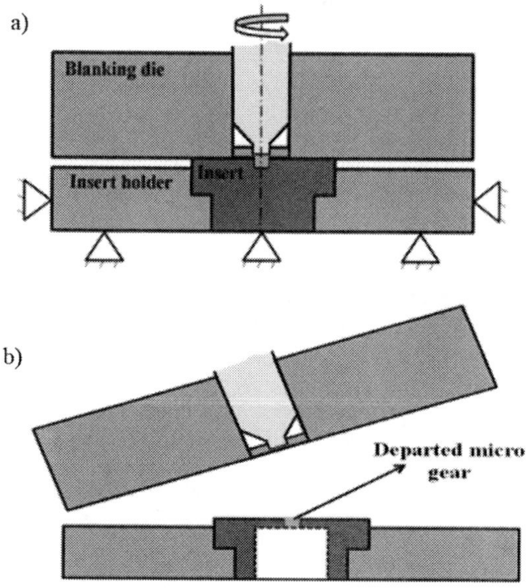

Fig. 6. Torsion operation a) keep constant of lower die and twisting of upper die b) picking up the upper die

Fig. 5. die assembly and components of die

The conventional ejection of micro gears was also done to show the result of tests and to bring out the attached formed part with blank. For this purpose, 3 pins were used to push up the blank, Fig.7. Through this movement, the micro gear that is connected to the blank, exits from the die cavity.

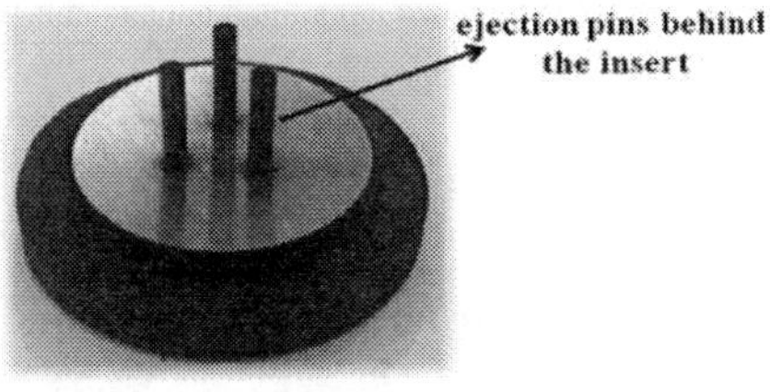

Fig. 7. conventional ejection by three pins

3 Results and discussion

Because of passing the material through a shaped die, extrusion process is one of the manufacturing methods in which parts replicate the geometrical features very well. Therefore, the manufacturing accuracy of the die features intensively affects the quality and accuracy of final parts. Due to the actual difficulty of the WEDM to create sharp corners in the die, the corners were produced by a wire with the radius of 0.06 mm. This affects the accuracy of the die geometry. The tests were done successfully and the micro gears were formed completely.

Images of micro gears produced by single extrusion and combined extrusion-forging are shown in Figs. 8 and 9, respectively. The bulged area around the bottom of micro gears attributes to the lack of vertical constraining of blank during compressing. The punch stroke was 4.15mm which is 150% of the sheet thickness. The reason of the longer stroke was aggregation of material under the punch during the forming operation. This observation was previously reported in reference [16]. As the figures show, there are no defects on the surface of the micro gears and the lines that are observed on the lateral surfaces of the part, parallel to the extrusion direction, were caused by the roughness of the die machining during WEDM. Since the material in the deformation zone must pass a vertical path to enter into the die cavity, a large strain was occurred at the edge of die entrance. Increasing of the stresses leads to increase the strains and furthermore, material flow interference caused a mass of material at the entrance of the die cavity. An appropriate radius at the edge of material flow into the cavity could improve the process and caused better material flow. On the other hand, high ratio of extrusion punch diameter to micro gear pitch circle diameter and dry lubrication condition can be other reasons of material accumulation at the entrance of the cavity. Separation and cutting of these areas from the whole deformation zone, especially the material around the die cavity at the longer stroke, prevents the process to achieve a higher micro gear shaft.

Fig. 8. Micro gear formed by single extrusion process at 2.4mm and 4.15mm stroke

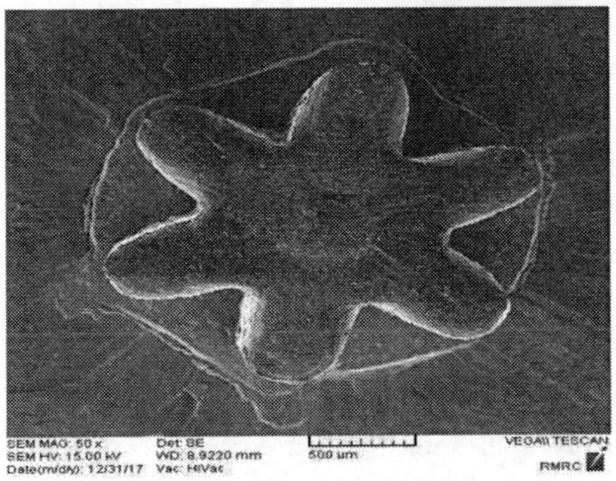

Fig. 9. eliminating the curvature of micro gear top surface by hard stopper

Fig. 10. inscribed Hexagon on the teeth and secondary bulging after 4.15mm stroke

Squeezing the separated area, shaped a secondary bulged area around the micro gear and an inscribed hexagon on the micro gear teeth, Fig. 10. However, maximum capacity of the 2.75mm copper sheet under the situation of this test at the stroke of the 4.15mm created 5mm micro gear shaft height.

The curvature of the micro gear at the top surface was caused by the frictional force of the side wall in the single extrusion process, Fig. 8. This curvature was eliminated in combined extrusion-forging process, Fig.9, by applying the hard stopper on the path of the micro gear.

The measurement of the outer diameter and root diameter size of micro gear by SEM image showed that the features of the cavity were more than the size of the design. Outer diameter was 2.08mm and root diameter was 0.89mm. Also, the walls of micro gear did not conform to the involute curve. The accuracy of the die was directly affected by WEDM accuracy and its setup parameters.

3.1 Load- displacement curve

Based on the experimental observations, the process of sheet compressing through its thickness to a die cavity has two main forming stages. The first stage is upsetting of the material between the surfaces of the punch and die. The second stage relates to material extrusion into the die cavity. In the first stage, some material enters die cavity; however, this amount in comparison to the amount of second stage, is not significant. As a result, the main height of the micro gear forms in the second stage.

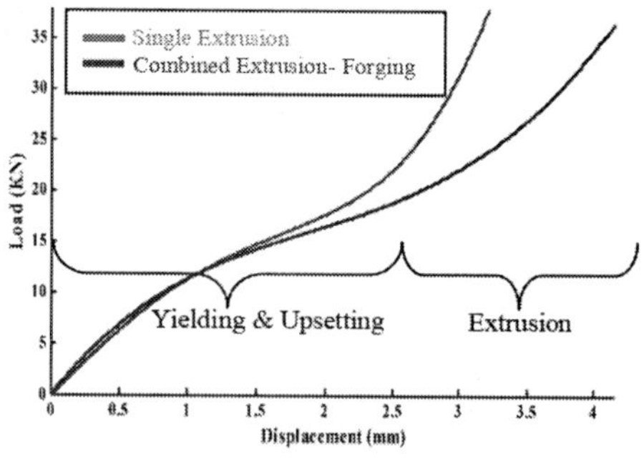

Fig. 9. Load-displacement curve of the micro gears

As illustrated in Fig. 11, both cases of single extrusion and combined extrusion-forging have similar trend. The curves are almost identical till the 1.7 mm stroke. For the single extrusion curve, the first stage of the forming, yielding and upsetting, was done approximately to the 2.5 mm stroke. After this point, the slope of the curve grows gradually which means that the flowing speed of the material into the die cavity is increasing. Since the material entry into the die cavity becomes harder, therefore, the slope of the forming load increases in the second stage.

For combined extrusion-forging, after the stroke of 1.7mm, the load-displacement curve deviates from that of the single extrusion curve gradually. This is because the curvature of micro gear top surface reaches to the hard stopper sooner than the rest of material. Hereafter, the forging process also was performed in the forming zone and the material was forced to fill the die corners and bottom of the teeth. Hence, the load increases sharply. In fact, the material that is forced to come into the die cavity by extrusion process fills the die features through the forging process.

3.2 Micro-hardness tests

In comparison to blanking process, the processes examined in this research do not need a punch like gear teeth and precision clearance. Because of using punch with bigger diameter, the punch has better strength. Also, since the parts are produced by deformation, they would have better mechanical properties than the blanked ones where the process is merely cutting the edges. In order to measure the micro-hardness of the gear tooth, the distance between the center of the tooth to the edge was divided into equal sections and micro-hardness at each position was measured after grinding and polishing of the micro gears. The micro-hardness versus position curves are shown in Fig. 11.

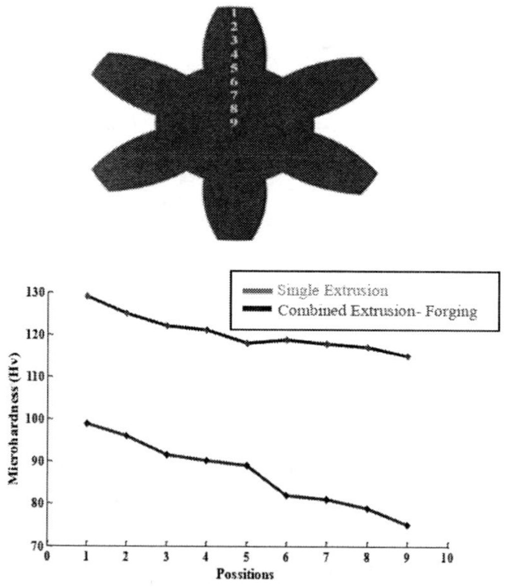

Fig. 10. Micro-hardness – position curve on the cross section of micro gears

For both the extrusion processes, the maximum and minimum hardness was at the tooth edge and at the tooth center. According to the curve, at the tooth edge, the material was under high deformation forces and large strain which makes a large strain hardening, and subsequently, higher micro-hardness. In fact, the grains of material, especially those at the sides, experienced high frictional force caused by contact between the material and the cavity wall. So, the boundary materials were subjected to intense shear stresses over the interior materials. Hard edge teeth could be a promising property for micro gear. By continuing the path and leaving the tooth edge area, a significant decrease was occurred in micro-hardness, especially for single extrusion process, which reveals how much the deformation of grains on this position can be different from the edges area. Micro hardness for combined extrusion-forging was more uniform than the single extrusion. This is because of the influence of the forging process. The center of the formed part has the least micro-hardness that would be predictable due to the less deformation of the center region of the part in the extrusion processes. Totally, micro-hardness of the combined extrusion-forging was higher than the single extrusion at each position. This was due to severe plastic deformation and higher load for the combined extrusion-forging process.

3.3 Ejection method based on torsion operation

According to the concept of progressive microforming, after forming of micro gear it needs a blanking stage to eject the parts from sheet. So, in this case the main question would be how the extra material among the teeth and the bottom of the micro gear can be eliminated. In this study, after forming, the micro gears were ejected from the die cavity with both of the conventional method and the new approach based on the Torsion operation. In this approach, after the forming operation, by keeping constant of insert holder and insert, the blanking die was twisted through its symmetry axis and the micro gear was departed simply from the blank. The Torsion operation was done with hand power. Fig. 12 shows the topography of the tooth surface after the Torsion test.

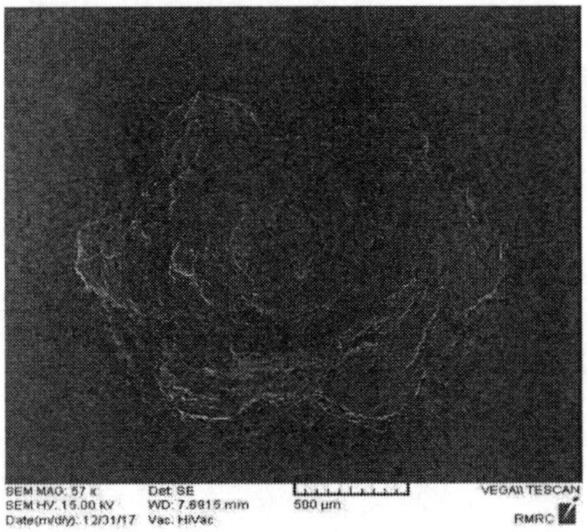

Fig. 11. Topography of blank surface after torsion operation

As it can be seen from Fig. 13, some material remained on the blank which affected the bottom surface geometry of micro gear. The formed micro gear which was still in die cavity can be pushed out by the next extrusion operation.

By means of the presented approach, the mass production of micro gears can be achieved by using sheet metal. Generally, the whole forming and ejecting operations can be done without moving the part. In fact, one stage could be specified for forming, after that, by a torsion operation the formed micro part can be departed, finally by extruding another part to the cavity this formed part pushed out.

It should be pointed out that this study concentrated on the presentation, feasibility and partial analysis of a new method for forming of micro gears. So, the investigation and optimization of parameters such as the relation between tip of extrusion punch diameter, pitch circle diameter and the sheet thickness, friction due to surface roughness of insert and roughness of inner side of cavity, effect of material, constraining condition and blanked part diameter, speed of the test and especially effect of elevated temperature on the test, etc. would be the issues of next researches.

4 Conclusions

- The implementation of using directly sheet metal for producing micro gears was successful. The micro gears were formed completely and filled the die cavity in both the extrusion processes.
- The pure copper sheet with the thickness of 2.75mm can produce a 5mm height micro gear with 6 teeth, module 250μm under the conditions of room temperature, dry condition of lubricant, ram speed of 3mm/min, lateral constraining of blank and 4.15mm stroke.
- The load-displacement curve of the micro gear has two main stages; yielding and upsetting as the first stage and extrusion as the second one. The main height of micro gear was formed in the second stage.
- In both of extrusion processes, the maximum of micro hardness was at the teeth and the minimum was at the center of micro gear. Comparing the hardness of initial material to the micro-hardness of the formed gear shows a promising mechanical property.
- Generally, the whole forming operation and ejecting can be done in one stage. One stage can be specified for forming, after that, by a torsion operation the formed micro part can be departed. Finally, by extruding another part to the cavity this formed part pushed out. However, the topography of cut surface must be studied more.

References

1. C.S. Kumar, J.N. Kumar, State-of-art review of past research on manufacturing of meso and micro cylindrical gears. Precision Engineering, (to be published).
2. F. Vollertsen, H. Schulze Niehoff, Z. H, Int. J. of Machine Tools & Manufacture, **46**, 1172–1179 (2006).
3. Y. Qin, A. Brockett, Y. Ma, A. Razali, J. Zhao, C. Harrison, W. Pan, X. Dai, D. Loziak, Int. J. Adv. Manuf. Tech., **47**, 821–837 (2010) .

4. M. Geiger, M. Kleiner, R. Eckstein, N. Tiesler, U. Engel, Annals of CIRP, **50**, 2–44 (2001).

5. J. Jeswiet, M. Geiger, U. Engel, M. Kleiner, M. Schikorra, J. Duflou, R. Neugebauer, P. Bariani, S. Bruschi, CIRP. J. Manuf. Sci. Tech., **1**, 2–17 (2008).

6. Y. Saotome, H. Iwazaki, J. Micro. Sys. Tech., **6**, 126–129 (2000).

7. D. Wang, T. Shi, J. Pan, G. Liao, Z. Tang, L Liu, J. Mat. Proc. Tech, **210**, 684–688 (2010).

8. C.C. Chen, Y.X. Qiu, Key. Eng. Mat., **626,** 541-547 (2015).

9. X. Dong, F. Chen, S. Chen, Y. Liu, Z. Huang, H. Chen, S. Feng, L. Zhao, Z. Wu, X. ping, Zhang, J. Mat. Proc. Tech, **219**, 199–208 (2015).

10. C. Chang, W. Kuo, *Proceedings of the Institution of Mechanical Engineers, Part B*, J. Eng. Manuf., **224**, 1767–1772 (2010).

11. W.J. Kim, Y.K. Sa, H.K. Kim, U.S. Yoon, Mat. Sci. & Eng, **A 487**, 360–368 (2008).

12. X.L. Guo, D.B. Shan, B. Guo, C.J. Wang, J. Zhou, Z.H. Xu, *Materials Science Forum*, **551-552,** 569-574 (2007).

13. M. W. Fu, W. L. Chan, *Micro-scaled Products Development via Microforming Deformation Behaviours, Processes, Tooling and its Realization*, (2014).

14. M.W. Fu & B. Meng, *MATEC Web of Conferences*, **21**, 09001 (2015).

15. W.L. Chan & M.W. Fu, Materials and Design, **43**, 249–257 (2013).

16. E. Ghassemali, M.J. Tan, A. E. W. Jarfors, S. C. V. Lim, Int. J. Adv. Manuf. Tech, **66**, 611–621 (2013).

17. Y. Qin, *Micromanufacturing Engineering and Technology*, (2015).

Hot embossing of polypropylene micro-tubes into functional tubular components with controlled inner-pore sizes

Jie Zhao[1,*] and *Yi* Qin[1]

[1]Centre for Precision Manufacturing, Department of Design, Manufacture and Engineering Management, The University of Strathclyde, Glasgow G1 1XJ, United Kingdom

Abstract. To meet ever-increasing demands on the micro-components for medical and non-medical applications, a new micro-shaping technology - "hot embossing of micro-tubes", had been developed for the forming of polymeric tubular micro-components. The paper presents the results from the forming of Polypropylene(PP) micro-tubes with outer diameters of 1.3mm and inner diameters 0.6mm, to achieve various reduced inner-features. The study was effected by combining experiment, numerical simulation and SEM analysis. FE simulation was implemented by using the material data obtained from the material characterisation tests. The forming experiment was conducted with a high-precision hot-embossing machine, developed in-house, with automated handling and good reliability and repeatability. The Polypropylene (PP) micro-tubes were successfully formed into the desired features at the temperatures of 60°C and 100°C respectively. The influences of the parameters/factors, such as tool design, temperature, forming pressure and holding time, on the quality of the shaped parts, are discussed in details. Based on this study, it is concluded that PP is an ideal candidate material among the polymeric materials for hot embossing of tubular micro-components, due to its good ductility, low transition temperature and low viscosity.
Keywords: Micro forming, Manufacturing process, Tool geometry

1 Introduction

Technical and market demands of tubular micro-components have led to the development of a hot embossing process and corresponding machine for forming polymeric micro-tubes. Those micro-tubular components are usually with outer diameter of 2mm or below and are designed with various shapes and different features, such as needles, balloons, multi-faceted, closed and open ended at the tips, with a range of plastic materials, and are being employed in various areas, for instance, micro-fluidic devices, drug delivery system, laboratory analysis equipment, and micro-fluidic system [1]. Due to the fact that most of these kinds of components are disposable, cost effectiveness is one important factor for the process development and evaluation. However, the current manufacturing technology is not yet satisfying the demand for economical mass-production of tubular micro-component with reduced inner channel dimension. Injection moulding is one of the most widely used process for polymer tubular component forming, which makes the complex and intricate shaping possible. However, injection moulding includes expensive production line and a high cost for the mould making it only economical for very large production quantities.
An innovative hot embossing process and machine for polymeric micro-tubular component forming have been developed in the recent years and demonstrated a great potential [2-4] in the technical and economical aspects, including the advantages of lower forming temperature, reduced system complexity, shorter production cycle time, quicker die changeover and more cost efficiency compared to other replication processes.

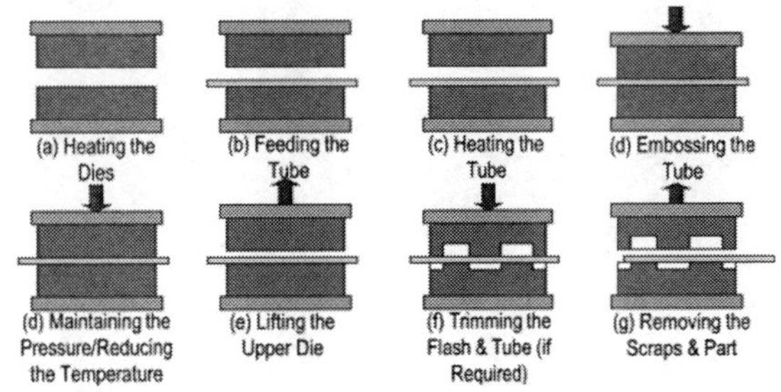

Fig. 1. Polymeric micro-tube hot embossing forming process chain.

With some variations from the conventional hot embossing process, the processing sequence of forming micro-tubes is as shown in Fig. 1: (a) the tool-set would be heated to the required temperature; (b) the tube would be fed into the lower shaping die; (c) the tube would be heated rapidly by contact with the die due to the small volume of the material involved; (d) the lower die would secure the tube properly while it is embossed by the upper tool. A proper pressure would be maintained for a specific time to allow shape-setting and fusion-bonding at the inner folded-interfaces. At this stage, to maintain shape

* Corresponding author: j.zhao@strath.ac.uk

 https://doi.org/10.1051/matecconf/201819010005

stability, a cooling mechanism could be necessary; (e). the upper-tool would then be lifted; (f). the flash formed may also be trimmed off by a following shaving process; and (g). the scrap and shaped part are removed from the lower-die.

The aim of the research reported in this paper is to conduct a comprehensive study on the influences of tool and process factors on the quality of the parts formed with a view to generating a design and manufacture control guideline for this forming technology. The forming principle was investigated by a combination of experimental and numerical approaches. The hot embossing micro-tube process was analysed by focusing on the central-pore forming, the cavity filling and monitoring the transformation of stress, strain and material flow. The in-depth studying of forming result and discussion on its influential factors enhances the understanding of hot embossing micro-tubular component and quality control.

2 PP micro-tube

Polypropylene, acronym PP, is one common polymer with versatile applications. PP has melting temperature of around 150°C, which makes it beneficial for almost all thermal plastic forming process, including hot embossing. Its good ductility and low viscosity also make it suitable for hot embossing process. Besides, another benefit of the PP material is the low material cost compared to other engineering polymers. The original PP tube (Fig. 2) has an outer diameter of 1.3mm and an inner diameter of 0.6mm.

Fig. 2. Original PP tube, a-c) longitudinal tube direction, d) circular cross-section of the tube.

3 FE Simulation

FE simulation of the hot embossing micro-tubes is aiming to examine the shaping capability of the proposed process and in-depth understanding of the forming mechanism, the flow behaviour and its influential parameters. The valid model can also be used to predict the forming result in order to save the time and cost from the experiment.

The simulation was conducted with a commercial FE software DEFORM-3D™ by SFTC. The general setup of the simulation can be seen in Fig. 3. The simulation contains top and bottom dies, with the micro-tube set as the workpiece located between both dies. The die cavities for both top and bottom dies are one variable parameter for this study. The top and bottom dies were set as rigid bodies. The micro-tube is set as an elasto-plastic model. The tube elements are treated as elastic until the yield point is reached. After this point, any portions of the elements that reach the yield point are treated as plastic, while the remainder of the element is treated as elastic.

The plasticity model is defined by the flow stress as shown in equation (1), which represents the flow stress behaviour of polymer micro-tube, where the stress ($\bar{\sigma}$) is dependent on strain ($\bar{\varepsilon}$), strain rate ($\dot{\bar{\varepsilon}}$) and temperature (T).

$$\bar{\sigma} = \bar{\sigma}\,(\bar{\varepsilon}, \dot{\bar{\varepsilon}}, T) \tag{1}$$

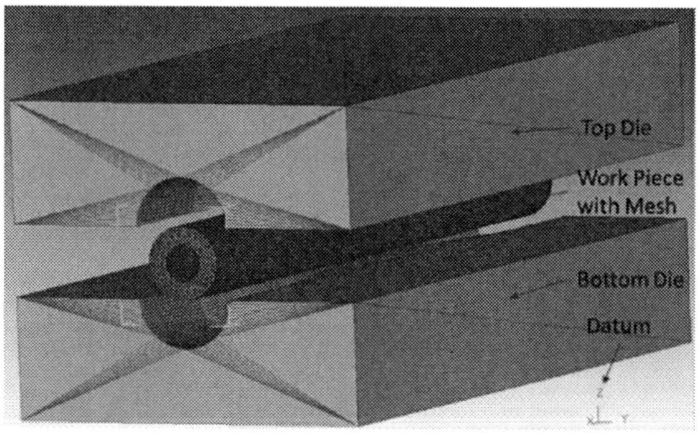

Fig. 3. Hot embossing polypropylene micro-tube simulation setup.

In the DEFORM software, a user defined material model was created with the material properties shown in Table 1. The key mechanical and thermal properties such as the temperature and strain rate dependent flow stress were obtained from the material characterisation tests of the micro-tube used in experiment. The Young's Modulus was calculated according to the different temperatures and applied in the simulation.

Table 1. Material properties for PP micro-tube.

Property	Value
Density [g/cm³]	0.9
Poisson's ratio [-]	0.42
Thermal expansion [μm/m-°C]	100 – 180
Thermal Conductivity [W/m-K]	0.15
Specific Heat Capacity [J/g-°C]	1.8
Heat transfer Coefficient [W/m2K]	24.5
Friction Coefficient [-]	0.1

The boundary condition was set as an interaction between the work piece and the dies and a "self-contact" relationship to simulate the bonding of the tube inner surface caused by temperature and pressure during the forming process. This interaction is dominated by the friction and heat transfer, however the heat transfer to the dies was not considered to reduce calculation time, but the change of heat within the work piece was calculated. The values of friction coefficient and heat transfer can also be found in Table 1. Additionally, heat exchange with the environment was considered to simulate the cooling effect of the area on the work piece which are not in contact with the dies. The analysing temperature of work piece and the dies were set according to the experimental setup.

The vertical movement is set for the top die to the -Y direction with a defined velocity. The velocity is a variable, which has been identified as one of influential forming parameters. The displacement of the top die leads to a moulding pressure of the polymer tube and forces the polymer flows into the die cavity and spread around during the embossing process.

4 Experiment

Forming experiments were conducted by adopting the high precision and fully automatic hot embossing machine developed in-house (Fig. 4), in order to gain a substantial understanding of the hot embossing micro-tube forming process, as well to determine the key forming parameters and their influence to the process result. Each micro-tube was cut to 50mm in length, picked up by a handling device into the machine and formed with pre-set forming parameters, such as temperature, force and holding time.

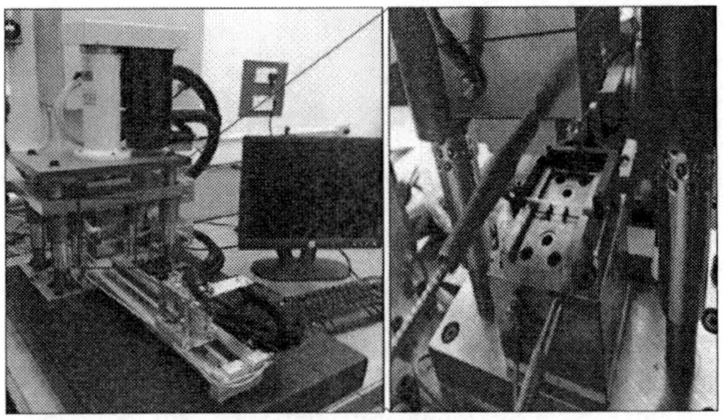

Fig. 4. The micro-tube hot embossing machine system (left) and forming tool (right).

Fig. 5. shows the SEM images of PP tube formed under 60°C, with different magnitudes. The formed tube exhibits smoothly-formed edges on the top right image. The inner pore measurement is 480μm in its width and 164μm in its height. A good fusion bonding result can be observed; where the inner surface line disappeared and the material shows a homogeneous appearance near the bonding area.

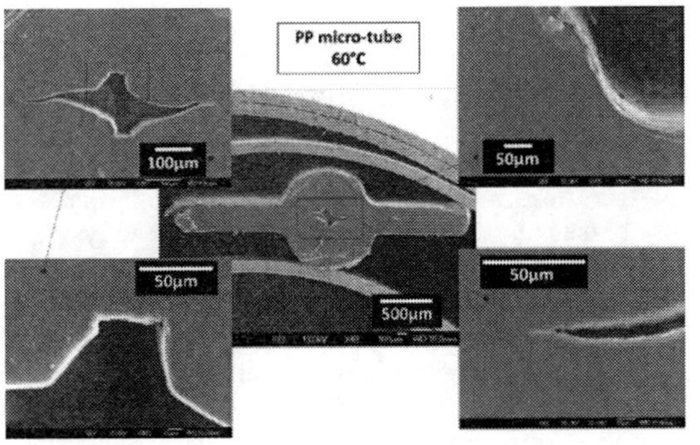

Fig. 5. The SEM images of PP tube formed under 60°C.

5 Result and Discussion

Generally, hot embossing is commonly used in precise and qualitative manufacture, for replication of micro-feature from the forming die to a flat surface [5]. Even though many process variations have been developed over the recent years, the application on direct form 3D features (both outer and inner features) on micro-tubes needs further investigation due to more dedicated tool-design and process control are required. In order to achieve a controllable pore-shape and dimensions at the inner area of the tubes, the forming parameters need to be properly considered.

5.1. The geometries formed

The results obtained on micro-tube forming under same experimental and simulated conditions were compared in order to validate the simulation model. As shown in Fig. 6, six geometrical measurements are taken from two samples, from both experimental and simulation results. From the comparison it can be seen, there is minor deviation of the outer dimension of the tube between the experiment and simulation. However, the rest of the dimensions are very similar, especially the ratio of vertical and horizontal direction for the formed inner pore (H_i/W_i) and the height of the tube material fill into the die cavity (H_f). It shows that the numerical model can predict the trend of the hot embossing process and can be used to assist in the further forming research.

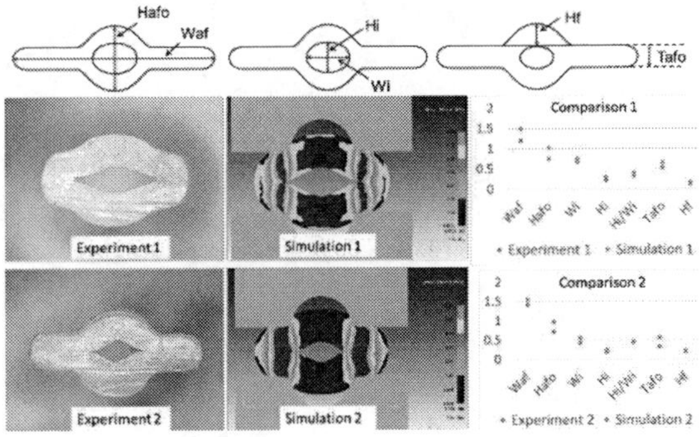

Fig. 6. Geometrical comparison of the experiment and simulation result.

5.2. The die cavity

Unlike the conventional 2D hot embossing on a foil or plate, where the high quality product refers to exact replication of the micro-structure from the die to the polymer, the micro-tube forming is not a process to "print" the concave cavity on the polymer directly. It is a dynamic transformation process, which assists the tube transferring into different geometry from its original size. The geometry of the tube, such as the inner pore size and aspect ratio is changing along with time. It means that, if the top die stroke moving distance and speed are well controlled, different forming results can be achieved with the same micro-tube and die setup.

The die cavity has the most direct impact on the forming result. In order to effectively study the design of die cavity, the micro-forming die has been configured as shown in the Fig.7. The cavity includes three geometrical variations as R_c, R_f and h, which represent the radius of the die cavity, the radius of the die fillet and the offset of the circle centre respectively.

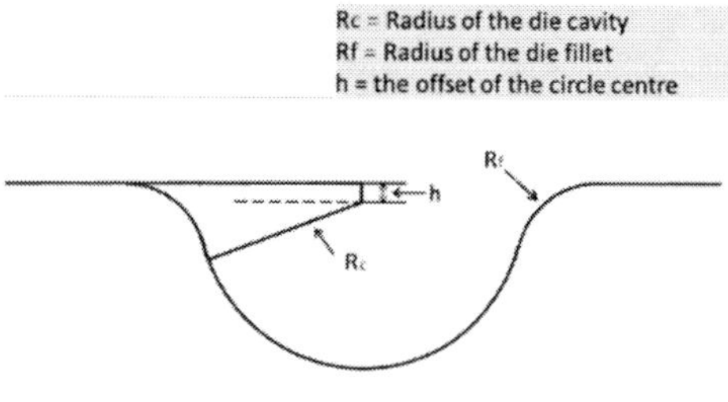

Fig. 7. Hot embossing die cavity configuration with three geometrical variations.

5.2.1 The parameter R_c

A series of simulations with different die cavity configurations are shown in Fig.8. to study the influence of R_c on the forming result.

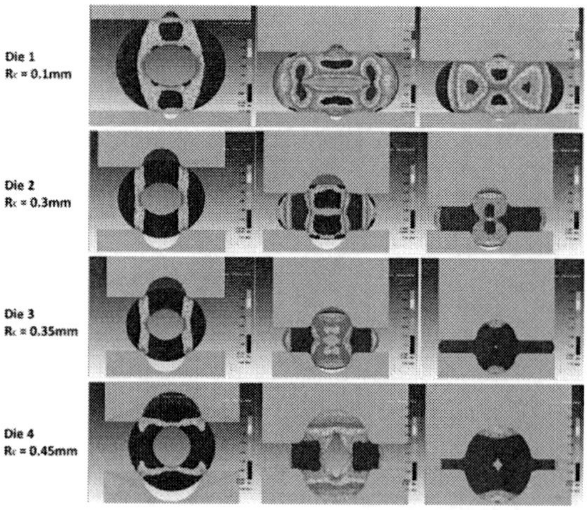

Fig. 8. Illustration of influences on the pore-formation from the die-cavity geometry (PP tube formed under 100°C).

Employing the same PP micro-tube and by changing R_c (radius of die cavity), the formed tubes exhibit large variations in geometry, especially the inner pore size. It can be explained that by controlling the ratio of the material flow in the vertical direction (the direction of the tool-movement) and the transversal direction, it is possible to control the height to the width ratio of the formed central pore as well as its transitional shape of the pore edges. Die 1 for instance, since the die cavity is much smaller than tube diameter, the surface tension generates high resistance force to prevent material flow into the hollow cavity. However, the increase of cavity size from Die 2 to 4 leads to a better material 'filling' into the cavity. It is evident that the geometry of die design has a strong impact to the formed components. In other words, controlling the shape and size of the central pore is possible by controlling the die cavity shape and dimensions properly. In order to further understand the effect of the die cavity geometry R_c on the formation of tube inner pore, the dynamic transformation process of micro-tube inner pore along with the movement of top die has been presented in Fig.9.

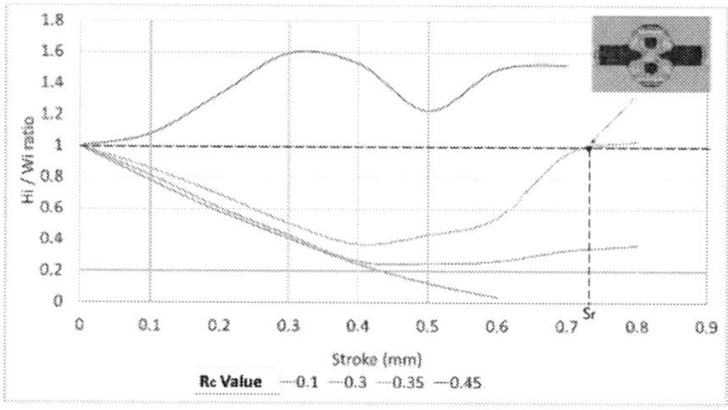

Fig. 9. Simulation result of PP tube under 100°C (change of R_c).

Along with the movement of the top die stroke, the ratio of H_i/W_i reflects the shape of the inner pore at a different stage, in order to find out which transitional direction is dominant to the inner pore forming. Considering the H_i/W_i ratio for original tube is 1, if the ratio is reducing from 1 to the direction of 0, means the pore is becoming flat in horizontal direction; on the other hand, if the ratio is increasing from 1 to the direction of 2, means the pore is stretched in the vertical direction. The S_r in the graph represents the point of top die stroke when the ratio of H_i/W_i inner pore is valued 1. This graph supports the process designer to determine an optimised die cavity design as well as the setting of the top die stroke for achieving the desired forming result.

5.2.2 The parameter R_f

A similar analysis method was used for examining the influence from R_f (the radius of die fillet) to the formation of the tube inner pore. Fig.10. shows three results from the simulation, where all simulation conditions are the same except of variable R_f, which is 0.05mm, 0.1mm and 0.15mm respectively. The other settings for the simulation are: micro-tube PP under 100°C forming

https://doi.org/10.1051/matecconf/201819010005

temperature, 0.1mm velocity, R_c = 0.35mm and h=0. From the results it can be seen that the change of R_f does not have a large impact to the shape of the tube inner pore, especially at the earlier forming stage when stroke is less than 0.4mm. With the increase of stroke, the difference of the three trend lines becomes more visible, which the largest fillet corner leads to the earliest reach of a ratio of 1; and the smallest fillet the last. It demonstrated that, for the hot embossing micro-tube forming, a larger radius of corner fillet produces lower resistance force on the tube material, which allows the material flow into die cavity easier.

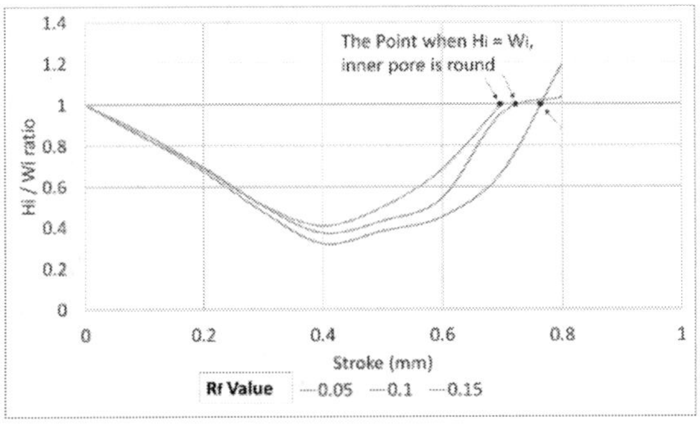

Fig. 10. Simulation result of PP tube under 100°C(change of R_f).

5.2.3 The parameter h

The h represents the offset of the circle centre point for the die cavity, in other words, the depth of die cavity. The variation of h applied in the simulation is -0.1mm, 0 and 0.1mm respectively. The results presented in Fig.11. shows the influence of the h value to the shape of the tube inner pore size. As expected, the depth of the cavity also leads to different forming results of the tube inner pores. The deeper of die cavity leads to a shape transition of tube inner pore tend to vertical direction. When the forming goal is to form a reduced round inner pore, both h = 0 or -0.1mm are achievable. However, when the stroke setting is different, the achieved final inner pore sizes are also different. The inner pore size is around 0.35mm when h = -0.1mm and 0.1mm when h = 0.

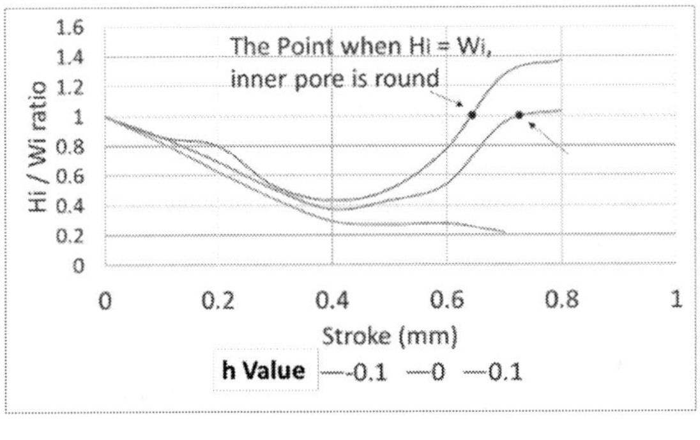

Fig. 11. Simulation result of PP tube under 100°C (change of h).

5.3. The forming temperature

The processing temperature is important for hot embossing as it influences the material flow and fusion bonding significantly. The viscosity of the polymer is directly responds to the applied temperature. Generally speaking, an elevated embossing temperature contributes to a better filling of die cavity, which results in increased forming quality. On the molecular level, when the temperature is raised, secondary bonding forces are diminished, so that the relative movement of adjacent chains is facilitated when a stress is applied. Compared to the hot embossing on a plane area for creating surface micro-structures, shaping a micro-tube requires lower temperature, because it can be characterised by the short flow-path of the polymer to the micro-cavity of the die insert [6]. The heating of the polymer should be maintained at a level to allow for the viscous short flow of material.

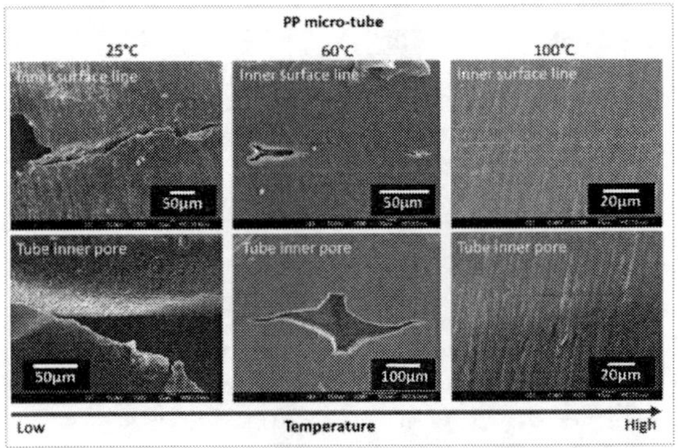

Fig. 12. The SEM images of PP tube formed under different temperature.

The Fig.12. present a group of SEM images for PP micro-tubes (cross-section view) formed under three levels of forming temperatures. Firstly, when the forming was carried out under room temperature, the material micro-structure was rough, especially near the junction between two contacting inner surfaces. By increasing the forming temperature, the material exhibits much smoother appearance. An oval inner pore has been produced and the inner surface boundary has clearly disappeared at 60°C. When the forming temperature was raised to 100°C, there is no sign of the surface line, which means the material on the inner boundary has melted and joined together. The molten material has filled the tube inner pore as well.

5.4. The forming force

Three kinds of forces may be acting on the micro-tube during the hot embossing process [7]. The applied force by movement from top die to bottom helps the polymer material flow into the die cavity. However, the other two forces, interface frictional force between the plastic material and the die cavity, also the surface tension force of the plastic material are resistant forces occurred during the process. Surface tension force becomes larger when the curvature of the die cavity becomes smaller. Increasing the process temperature (which lower the surface tension force), or increasing the applied force may alleviate this problem.

Since the changing of process temperature has been discussed in section 5.3, the discussion here is mainly focusing on the forming force in determining the shape of final product. The higher forming force is corresponding to a longer top die moving distance (the depth of embossing), which also means more vertical compression of the tube. The Fig.13 top three images show the geometry transformation of PP tube when the forming force changed from 2KN to 3KN, by using the die cavity R_c=0.45mm. The thickness of the exceed flash reduced from 0.22mm to 0.17mm, along with the significant dimension change of W_i and H_i. Likewise, the Fig.13. bottom shows the tube dimension change by using the die cavity R_c=0.35mm. In addition, it is worth noting that the inner surface line of the micro-tube is obviously visible when the forming force is 2KN; however, it disappeared when the forming force increased to 3KN. It can be concluded that, the forming force provides energy for micro-tube deformation, as well as the energy for material inner molecular mobilisation and combination.

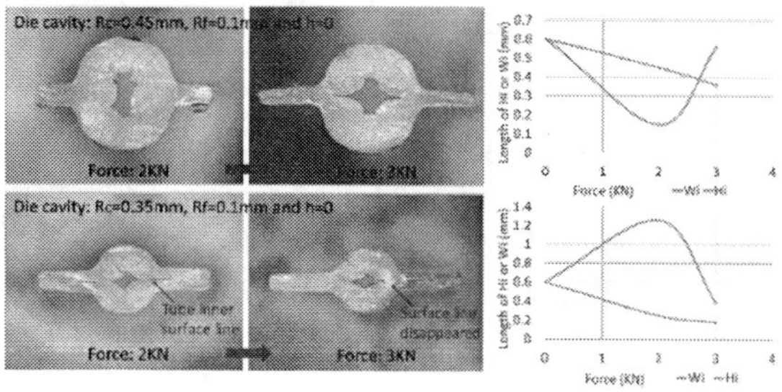

Fig. 13. PP micro-tube formed under room temperature and different forming force.

5.5. The holding time

Holding time is crucial to the formation of fully sealed tube inner interfaces bonding under an elevated temperature. As shown in Fig.14. the PP micro-tube was formed under the hot embossing parameters of 100°C forming temperature, 2KN force, 0.1mm/s forming velocity without hold time. As can be seen, the micro-tube presented a good deformation and formed a uniform shape of the outside profile and central pore. However, due to the viscosity of the polymer, material takes time to flow and fill all the undesired gaps. Hence, several small pores can be observed in the formed micro-tube, which are caused by the time limitation for the material to flow and stabilise the inner surface bonding.

Generally, longer holding time allows for the fusion-bonded interfaces to settle down without separation after cooling. It is obvious a parameter which affects the production rate. To ensure good quality of the part to be produced, a longer holding-time is preferred. On the other hand, a longer hold time may affect the tool life due to the sustained hot-condition of the die/tool. An optimised holding time needs to balance the quality of the product and the tool life in the best interests of productivity.

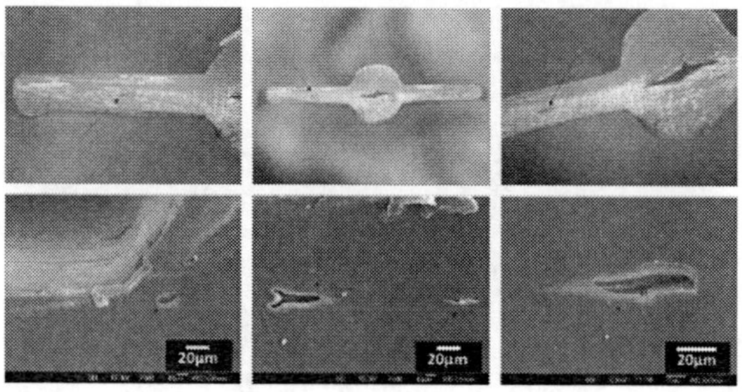

Fig. 14. The PP micro-tube formed under 100°C without holding time shows some small unfilled pores.

6 Conclusion

The hot embossing of PP micro-tube and its processing parameters are discussed in this paper. The results showed that PP is one preferable polymer for the desired hot embossing micro-tube forming. It is also shown that, the inner features (pores) with controllable sizes and shapes can be achieved with a proper die design and processing parameters set. It was further confirmed that the die geometry is a major factor in determining the material flow and, hence, the formation of the central pore. Other process variables, such as temperature, force and holding time, also play vital role for producing high quality end product.

Acknowledgement: Support from the European Commission for conducting research into the "A Process Chain and Equipment for Volume Production of Polymeric Micro-Tubular Components for Medical Device Applications (POLYTUBES)" (NMP2-SE-2009-229266) is acknowledged. Collaboration with all project partners in developing the hot-embossing process and the machine are particularly acknowledged.

References

1. A E. Gubera, M. Heckele, D. Herrmanna, A. Muslija, V. Saile, L. Eichhornb, T. Gietzelt, W. Hoffmann, P.C. Hauser, J. Tanyanyiwa, A. Gerlach, N. Gottschlich, G. Knebel. Chem. Eng. J. **101**, Issues 1–3, 447–453 (2004)
2. Y. Qin, J. Zhao, G. Anyasodor, K.S. Hansen, I. Calderon, K. Konrad, C. Hartl, M. Arentoft, I.S. Chronakis, Micro-Manufacturing Engineering and Technology. 2nd ed. Oxford, 179-200 (2015)
3. J. Zhao, Y. Qin, A. Razali, A.L.K. Yip, Z. Fei, 7th International Conference on Micromanufacturing, 534-541 (2012)
4. J. Zhao, A.L.K. Yip, Y. Qin, A Razali, Z Fei. Trans. Nonferrous Met. Soc. China. **22**, Supplement 2, s214–s221 (2012)
5. M. Worgull. Hot Embossing:Theory and Technology of Microreplication, 1st ed, Elsevier (2009)
6. M. Heckele, W. K. Schomburg. J. Micromech. Microeng. 14, Issue 3, R1–R14 (2004)
7. L. Lin, Y.T. Cheng, C.J. Chiu, Microsyst Technol, 4, 113-116 (1998)

Fabrication of WC-TiC-Co Cemented Carbide at Different Heating Rate by Micro-FAST process

*Yitong Chen, Yi Yang, Gang Yang, Libo Wang, Mingxia Wu**

School of Manufacturing Science and Engineering, Sichuan University, Chengdu, 610065, P.R. China

Abstract. As demands on miniature products increase significantly, a rapid prototyping and production system for highly flexible and cost-efficient production of micro components made from a wide range of materials is needed. In present work, the Micro-forming Fields Activated Sintering Technology (Micro-FAST) with the coupling effects of electrical and pressure fields as the dominant driving force, has been proved to be very successful in the fabrication of the nearly full-density micro-cylinder parts from WC-8Co-xTi- nC powders (x=4, n=2 or x=6, n=0). The influence of heating rate and composition of titanium and carbon has been investigated. Based on the analysis of the relative density, microstructure and mechanical properties, high density (up to 99.7%), fine grain sizes and good mechanical properties could be obtained by Micro-FAST to sinter WC-TiC-Co Cemented Carbide bodies under a low sintering temperature (1200 ℃).

Keywords: Rapid prototyping Sintering Micro structure Micro-FAST

1 Introduction

Cemented carbide is usually an alloy material prepared by powder metallurgy from a carbide powder of high hardness refractory metal such as WC and TiC and the metal such as Co, Fe, Ni or their compound[1]. Due to its high hardness, strength and excellent wear-resistance and corrosion resistance, the traditional tungsten-cobalt-based cemented carbide has been widely used in cutting tools, wear-resistant parts and mining materials and other fields[2-4]. Nevertheless, it also exists problems such as a relatively short life span, working softening and other issues.

The hardness and wear-resistance of WC-TiC-Co alloy are generally better than that of WC-Co cemented carbide due to the hardness and wear-resistance of titanium carbide is higher than that of tungsten carbide. Compared with the WC-Co based cemented carbide, the WC-TiC-Co cemented carbide the flexural strength decreases with increasing temperature to a lesser extent[5]. Therefore, WC-TiC-Co cemented carbide is prepared, which can solve the problems of working softening, short life span and poor red hardness in practical applications[6-9].

The cemented carbide manufactured by the traditional sintering process usually requires a higher sintering temperature and longer sintering period. Prolonged sintering at high temperatures leads to the growth of grain size of the main phase, which seriously affects the further improvement of the properties of the sintered sample. Therefore, development of a new sintering technique with a lower sintering temperature and shorter sintering period, is needed. K. Feng [10,11] have achieved the reaction of combustion synthesis by activating Fe-Ti-C system in electric field.

In this work, a newly developed method, named as Micro-FAST (Micro- forming Fields Activated Sintering Technology), based on the coupling effect of electric and force fields, offers a promising new route for fabricating the cemented carbide. The process (Micro-FAST) uses low voltage and high current, pressure-assisted sintering and synthesis technique, which has been used recently in materials processing[12,13].

In this paper, Micro-FAST was carried out to fabricate the WC-TiC-Co cemented carbide, by adding a certain amount of titanium powder and carbon powder in the raw WC-8Co powder. The effect of the different heating rate during the process of sintering of WC-8Co-xTi-nC(x=4,n=2 or x=6, n=0) in multiple physical fields was investigated, which have been prove to play a critical role in affecting the densification process.

2 Experimental Method

The chemical compositions of cermets in this work were given in Table 1. The characteristics of the starting powders used in the experiment were listed in Table 2. Raw powders were mixed in a planetary ball-mill for 2 hours at a speed of 400 rpm with a ceramic ball-to powder ratio of 5:1.

Table 1. Chemical compositions of WC-Ti-C-Co powders

Cermets	WC	Ti	C	Co

* Corresponding author: wumingxia@scu.edu.cn

https://doi.org/10.1051/matecconf/201819010006

W1	86	4	2	8
W2	86	6	0	8

Table 2. Characteristics of the raw powders

Powder	Particle Size (μm)	Chemical composition (wt.%)
WC	0.6	T.C:5.92, F.C: 0.030, O: 0.15, Cr_3C_2:0.50
Ti	40	C:0.02,O:0.35, others: 0.1
C	30	others: 0.1
Co	2	C: 0.015, O: 0.4, others: 0.002

The sintering process, as shown in Figure 1, was carried out in a Gleeble-1500D thermal simulation instrument. An approximate amount of raw powder was directly placed onto a 4 mm×4mm graphite die. Sintering was started when the vacuum pressure reached 10-3 Pa. A thermocouple device situated on the die wall, was used to measure and adjust the temperature. A summary of the detailed sintering parameters were given in Table 3.

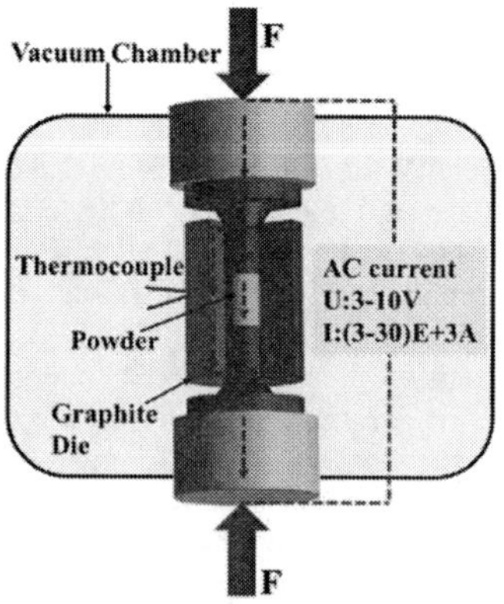

Fig. 1. Schematic diagram of sintering process

Table 3. Processing parameter of the experiment

Powder	Heating Rate (°C/s)	Sintering Temperature (°C)	Pressure (MPa)	Holdin g Time (min)
W1	25	1200	75	12
W1	50	1200	75	12
W1	75	1200	75	12
W1	100	1200	75	12
W2	25	1200	75	12
W2	50	1200	75	12
W2	75	1200	75	12
W2	100	1200	75	12

The density of sintered samples was measured by the Archimedes' method, where each value was checked for at least three times in order to eliminate artificial inaccuracy. Microstructure of polished specimens was observed by SEM (Hitachi S-3000N) in BSE mode coupled with EDS (HORIBA EX-320). Phase identification was carried out by XRD (Rigaku Ultima IV). Hardness was measured using HV-50A Vickers hardness meter. Fracture toughness (K_{IC}) was calculated from Vickers hardness, which was examined by indention method under load of 30 kg, following the equation [14]

$$K_{IC} = 0.15\sqrt{\frac{HV_{30}}{\sum_{i=1}^{4} l_i}} \quad (1)$$

Where HV_{30} is the Vickers hardness (N/mm^2), li is the length of the crack tip from the hardness indent (mm).

3 Results and discussion

3.1 Phase characterization

The XRD patterns of sintered samples with different heating rates are shown in Fig. 2. Only the diffraction peaks of WC and TiC were detected, implying that the TiC was generated by C and Ti reactions in raw powder during the sintering process. As the point of thermodynamic analysis, it shows that the free C in WC can be easier to combine with Ti, which can become TiC phase.

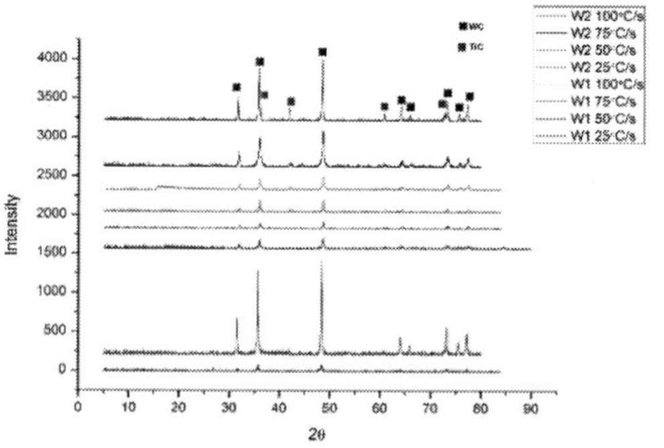

Fig. 2. XRD of samples under different holding times

3.2 Microstructure

The microstructures of sintered samples were observed by SEM in BSE mode. As shown in Fig.3, Although the cermets show similar core-rim structure, it can be seen that there are some gradual changes in the microstructure with different heating rate as shown in Fig. 3. It shows that with the inner rim increases with the increasing of heating rate. The core-rim structures are more obvious with the increasing heating rate, especially, when the heating rates were 75 °C/s and 100°C/s. Because the lower the heating rate is, the more thoroughly the titanium grains and carbon grains of the samples are combined, forming a TiC dark phase. When the heating rate of the sample is accelerated, the titanium powders and the carbon powders are underreaction, and the samples appear core-rim structure.

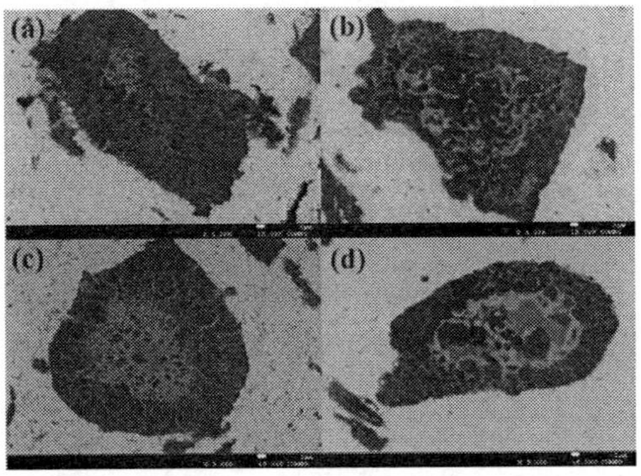

Fig. 3. BSE micrograph of W1 at different heating rate,

(a) 25℃/s, (b) 50℃/s, (c) 75℃/s, (d) 100℃/s

Fig. 4 is an image of W1(WC-8Co-6Ti) sintered with a heating rate of 50°C/s. The EDS results of WC-8Co-6Ti cermets are shown in Table 4. According to the atomic ratio of energy spectrum, the region 1 is a typical WC-Co phase matrix, the region 2 is mainly (W, Ti) C phase and TiC phase. It is showed that Ti atoms replace the W atoms in WC, forming (W, Ti) C phase, at the same time, it is accompanied by the formation of the TiC phase during the WC-8Co-6Ti sintering.

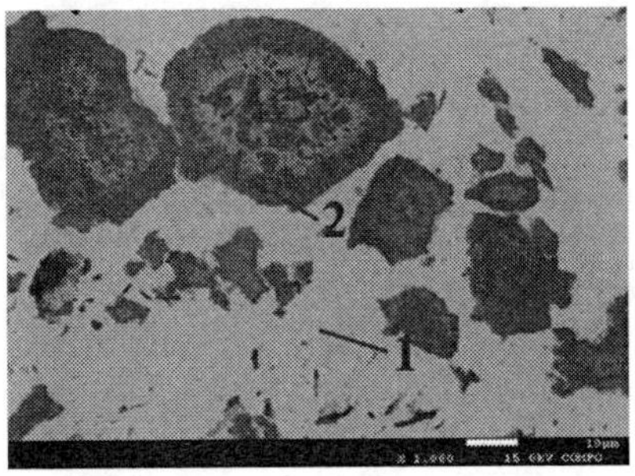

Fig. 4. BSE image of W1 sample sintered with a heating rate of 50℃/s

Table 4. EDS analysis of W1 sample sintered with a heating rate of 50℃/s (at.%)

Element	Region 1	Region 2
C	66.23	58.53
Ti	0	27.67
Co	5.50	0.92
W	28.27	12.88

Fig. 4 is an image of WC-8Co-4Ti-2C sintered with a heating rate of 50°C/s. The EDS results of WC-8Co-4Ti-2C cermets are shown in Table 5. According to the atomic ratio of energy spectrum, it is known that the region 1 is the typical WC-Co phase matrix. Region 2 the atomic ratio of W+Ti to C is roughly 1:1, which is mainly (W, Ti) C phase, and some TiC phase. It is showed that Ti atoms replace a small fraction of W atoms in WC.

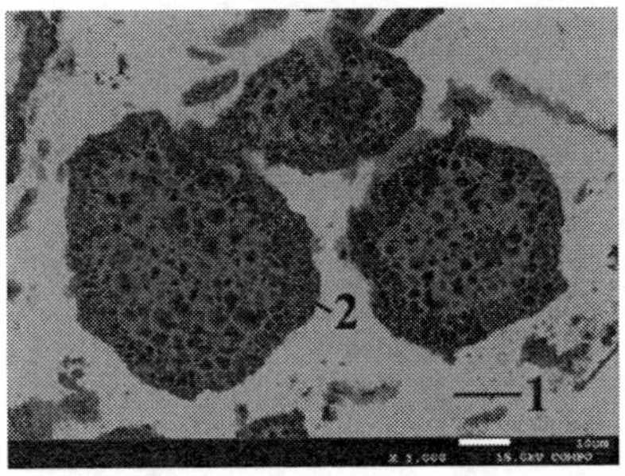

Fig. 5. BSE image of W2 sample sintered with a heating rate of 50℃/s

Table 5. EDS analysis of W1 sample sintered with a heating rate of 50℃/s (at.%)

Element	Region 1	Region 2
C	64.03	35.13
Ti	0	47.36
Co	2.50	0
W	30.27	17.51

3.3. Mechanical properties

Fig. 6, Fig. 7 and Fig. 8 shows hardness, fracture toughness and relative densities of the sintered samples. Compared with the hardness, fracture toughness and relative density, it peaks at 50C°/s. Due to the heating rate increasing, the core-rim structure of TiC phases and (W,Ti)C phases substitute the single TiC phase, which brings better mechanical properties. Furthermore, with the heating rate increasing, the titanium grains and the carbon grains are underreaction, and the core-rim structure has poor connection with WC-Co matrix. Therefore, the mechanical properties of the samples are reduced when the heat rate is high.

The mechanical properties of WC-8Co-6Ti samples are better than that of WC-8Co-4Ti-2C samples both in hardness and fracture toughness. Compared with WC-8Co-4Ti-2C, WC-8Co-6Ti has less TiC phases and more

https://doi.org/10.1051/matecconf/201819010006

(W, Ti) C phases. (W, Ti) C phase has higher modulus of elasticity than TiC phase, which can prevent crack propagation and enhance ceramic toughness.

When sintering is carried out under an electrical field, the rapid heating rate can enhance mass transport by electromigration, which contributes significantly to the remarkable increase in neck growth because the highest density of current passing through the particles could generate larger Joule heating and induce a local high temperature. And it may promote the atomic diffusion. And rapid heating rate can decrease the soaking time and inhibition of coarse grains. However, the heating rate is too high, instant high temperature of the samples may result in coarse grains and reduce performance.

As a result, 50C°/s is the best heating rate to fabricate the WC-TiC-Co samples.

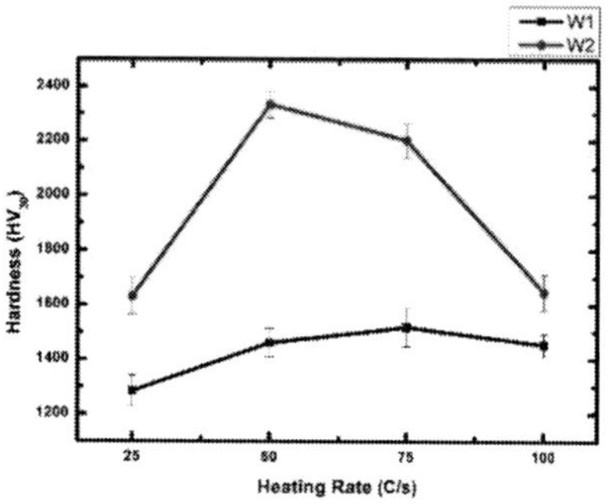

Fig. 6. Hardness of the samples with different heating rate

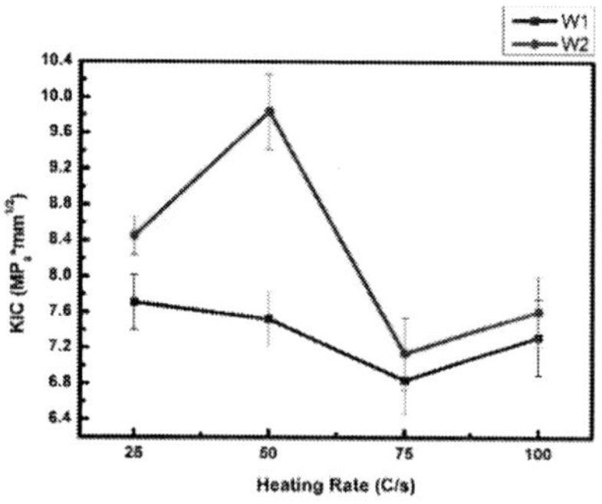

Fig. 7. Fracture toughness of the samples with different heating rate

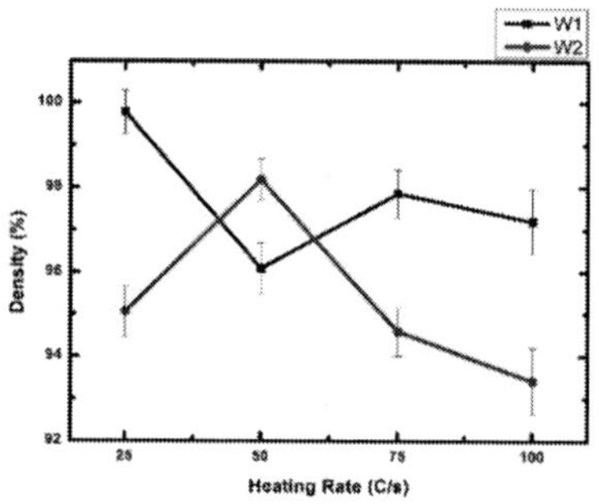

Fig. 8. Density of the samples with different heating rate

4 Conclusion

In summary, we successfully fabricated the nearly full-density (99.70%) micro WC-TiC-Co parts from raw powders using the Micro-FAST technique. With the increasing of heating rate, the hardness, fracture toughness and density almost present a tendency to increase and then decrease. The results show that 50C°/s is the optimal heating rate to fabricate the WC-TiC-Co samples. And the mechanical properties of WC-8Co-6Ti samples are better than that of WC-8Co-4Ti-2C samples, which probably because of the content of TiC and (W, Ti) C phases. In the future work, Micro-FAST will be used to prepare the metal matrix composites and and promote the real engineering applications of this new technique.

Acknowledgements

This work was supported by the Natural Science Foundation of China (Grand numbers 51675357 and 51575369). And the author Mingxia Wu thanks the support of Sichuan University post-doctoral research and development fund (2018SCU12067).

References

1. G. S. Upadhyaya. Mater. Des. **22**, 6 (2001)
2. V. Bounhoure, S. Lay, S. Coindeau, S. Norgren, E. Pauty. Int. J Refract. Met. Hard Mater. **52** (2015)
3. W. Su, Y. Sun, J. Feng, J. Liu, J. Ruan. Int. J Refract. Met. Hard Mater. **48**, 6 (2015)
4. M. R. Rumman, Z. Xie, S. J. Hong, R. Ghomashchi. Mater. Des. **68** (2015)
5. P. Feng, W. Xiong, L. Yu, Y. Zheng, Y. Xia. J. Ruan. Int. J Refract. Met. Hard Mater.**22** (2004)
6. W. M. Merchan, A.V. Saveliev, L.A. Kennedy. Carbon **42**, 3 (2004)
7. S. P. Chen, Q. S. Meng, J. F. Zhao, Z. A. Munir. J. Alloys Compd. **476**,1 (2009)
8. Z. A. Munir. Mater. Sci. Eng. A **287**, 2 (2000)
9. G. Jiang, H. Zhuang, W. Li. Scr. Mater. **50**, 8 (2004)
10. K. Feng, Y. Yang, B. Shen, H. Lin, H. He. Rare Metal Mater. Eng.**34**, 3 (2005)
11. Y. Yang, K. Q. Feng, K. Z. Shen, H. He. Rare Metal Mater. Eng. **34**, 4 (2005)
12. D. Lu, Y, Yang, Y. Qin, G, Yang. J. Microelectromech. **22** (2013)
13. M. X. Wu, G. Yang, J. Liu, D. Q. Yin, Y. Yang. J. Alloys Compd. **692** (2017)
14. W. D. Schubert, H. Neumeister H, G. Kinger, B. Lux. Int. J Refract. Met. Hard Mater.**16** (1998)

Fabrication of Miniature Components from ZrO_2 Powder by Combining Electrical-field Activated Sintering Technique and Micro-forming

Hasan Hijji[1,*], *Yi* Qin[1], *Kunlan* Haung[2], *Muhammad* Bin Zulkipli[3] and *Jie* Zhao[1]

[1]Centre for Precision Manufacturing, Department of Design, Manufacture and Engineering, The University of Strathclyde, 75 Montrose Street, Glasgow, G1 1XJ, UK
[2]School of Manufacturing Science and Engineering, Sichuan University, Chengdu, P.R. China
[3]Department of Mechanical Engineering Technology, Faculty of Engineering Technology, University Tun Hussein Onn Malaysia, 86400 Parit Raja, Batu Pahat, Johor, Malaysia

Abstract. There is an increased demand for miniature/micro products (such as MEMS) and nanotechnology-based products (such as nano-materials). Micro-manufacturing is a link between Macro- and Nano Manufacturing and an effective means for transferring nanotechnology-product designs into volume production. The micro forming has the potential for low-cost, high volume manufacturing applications. In order to meet the high demands on miniaturised products, a rapid production technique and the system, high flexibility, cost-effectiveness and processing a wide range of materials are needed. Recently, a series of studies have been undertaken to investigate forming miniature/micro-components by using a combination of micro-forming and Electrical-field activated sintering (Micro-FAST). The process uses low voltage and high current density, pressure-assisted densification and synthesis technique, which renders several significant merits. The work to be reported in this paper will be focused on the forming of miniature components from Zirconia (ZrO_2) powder, without using binders. Several processing parameters have been investigated, such as pressure, heating rate, heating temperature and holding time, which helped to obtain high-quality parts. Using graphite dies and punches, sample parts (solid cylinders of Ø4.00mm × 4.00 mm) were formed. These were subjected to detailed examinations and analysis, such as analysis of the relative density, hardness at the necks formed among the particles and in the particle bodies, as well as the microstructures. The results showed that directly forming the parts from loose powder is feasible, and by properly designing and control the processing parameters, high-quality parts could be achieved, among which heating temperature and holding time are extremely important. At the same time, due to low conductivity of the powder material, carefully designing the tooling is essentials for ensuring properly heating, pressurisation and cooling.

Keywords: Micro forming, Sintering, Powder

1 Introduction

Zirconia (ZrO_2) is one of the most favourable materials for a new generation of ceramics due to excellent mechanical properties. When compared to other ceramic materials, zirconia has unique strength at room temperature. Moreover, Zirconia has high fracture toughness, high density, high hardness and wear resistance, good frictional behaviour, high-temperature capability, non-magnetic, low thermal conductivity, electrical insulation, the coefficient of thermal expansion similar to iron, and modulus of elasticity similar to steel. Nowadays, zirconia is being used widely in the production of many products, such as such as structural ceramics, dentistry, wear parts, ceramics bearings, etc.... [1-5]. Nowadays, there are high demands for micro- or miniature-components and there is a fast growth in its applications in several areas, such as automotive industry, telecommunications, biomedical industry, information technology and home-use electronics products. According to some market research, microelectromechanical systems (MEMS) market had a double-digit growth after 2010 and the estimation for

* Corresponding author: hasan.hijji@strath.ac.uk

MEMS market in consumer electronics grows at a compound annual growth rate of 19.8% during the forecast period 2009–2015 [6-8]. However, there still some challenges regarding the size-effect related issues, improving the quality of the product, delivering multi-material processing capabilities, and addressing all of these with low-cost. Moreover, the manufacturing of micro- products with traditional manufacture techniques for micro-electronics (e.g., photolithography, deposition and etching processes) is not always suitable since only a limited range of materials can be handled with a particular process, while other manufacturing methods such as conventional sintering and metal-powder injection moulding would need longer process chains. Therefore, the non-traditional manufacturing technologies have been developed to improved productivity and economic effectiveness [6, 9]. The Electric Current Assisted Sintering (ECAS) or Field Assisted Sintering Technique (FAST) is one of the promising technologies that have been used widely these days. The FAST technology is a general term for a class of consolidation methods which combining external electric field/currents with mechanical pressure for powder sintering. Comparing with other sintering processes, such as pressure sintering or hot pressing, the FAST process has many benefits. This methods has faster heating rate, lower sintering temperature, shorter holding time, the consolidation of difficult-to-sinter-powders, the elimination of the need of sintering aids, no need of cold compaction, less sensitivity to the characteristics of the initial powders, and marked comparative improvements in the properties of the consolidated materials [10, 11]. Therefore, a novel process used to produce miniature/micro-parts directly from loose powder by combining (FAST) and Micro-forming, was developed to address the issues raised above and it called and called Micro-FAST. It has been proved that the Micro-FAST process is particularly appropriate for forming micro/miniature components due to using very high heating and cooling rates and is of high flexibility for processing different powder-materials. The process uses low voltage and high current density, pressure-assisted densification and synthesis technique, which render several significant merits. To-date, high-quality parts and high process efficiency for forming with various powder materials have been achieved [12-14]

2 Process Configurations and Tools

The target samples to be formed were designed as size of Φ 4mm x 4mm cylinders. The loose powders were weighted using a precision electric balance according to the calculated value (see Eq. 1) and directly filled into the die cavity.

$$m = \rho \times V \qquad (1)$$

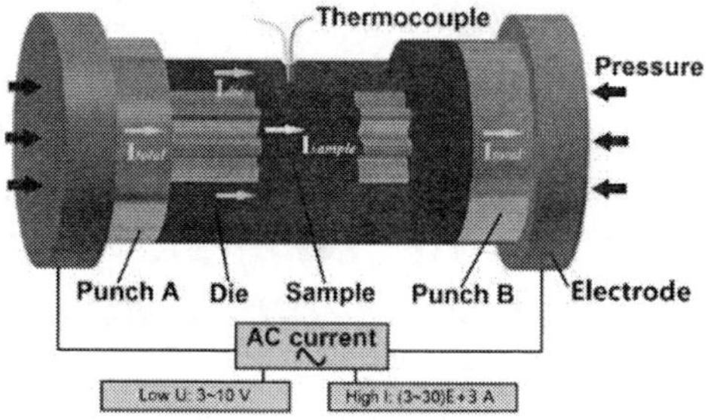

Figure 1 Illustration of the Micro-FAST process for the forming of miniature/micro- components [12]

Fig. 1 shows the principles of the micro-FAST process is being illustrated. The closed die and punch set with filled powder was then placed horizontally between two electrodes on the Gleeble 3800 machine as shown in figure 2. The Gleeble-3800 thermal simulation machine controls the heating process with a computer-controlled system which is able to pre-set a value of the heating rate, and the accuracy of the temperature control is within ±3 °C. The electric field that has been produced by the machine has low voltage (3~10 Volts) and high current (3000~30000A). By applying an external pressure and electrical-field through both punches, the Joule heat induced in a powder material and/or the die by the high electric current, plus possible electric plasticity in the powder material, could enable a combination of powder sintering and forming and hence, produce near-net-shaped micro-products within short time.

Figure 2 Tools set used and experiment setup with Gleeble-3800 [11]

The process parameters such as pressure, heating rate, temperature and time are input through a computer-based interface employing QuikSim Software. The thermocouple was used for measuring the real-time temperature and giving a feedback to the computer system. The temperature and pressure were maintained for a given period of time, in order to consolidate the powder and gain a high-density solid part. This process can produce high-quality components with required structures and functionalities, through appropriate process design and control, with less or no restrictions on the raw-material types and the microstructures, including the use of nano-alloys.

3 Experimental Procedures

Zirconia (ZrO_2) powder with an average particle size of <50 μm was used for experiments. The theoretical bulk density for the used powder is 5.92 g/cm^3. The received powders were sufficient and ready to make up a sample with the size of Φ 4.0mm×4.0mm (solid cylinders). The experiments were conducted with a variation of different key parameters, such as pressure, heating rate, maximum temperature and holding time. The loose powders were weighted prior to being fed into the graphite dies (see figure 3). The graphite cylindrical dies and punches were designed and manufactured for the experiments to form 4.0mm×4.0mm and examine the formability of the material and to improve the operating parameters. A small hole has been set in the middle section of the die, in order to measure the temperature using the thermocouple. As mentioned, the pre-determined process parameters (pressure, heating rate, sintering/forming temperature and time) were set through the machine interfaces. Upon starting a program, a constant pressure was applied onto the powder through applying a force onto a punch, while the powder and die were heated up through a high density AC current passing through the powder and/or the die/tool-component, until the desired temperature was reached and the heating cycle completed. The temperature and pressure may be maintained for a given period of time for forming and densification, which was controlled by closed-loop control facilities. The ZrO_2 material has a very high melting-point and low electric conductivity. Therefore, the sintered temperature that has been set for the experiment was between 1100℃ and 1400 ℃ with a heating rate of 50℃/s.

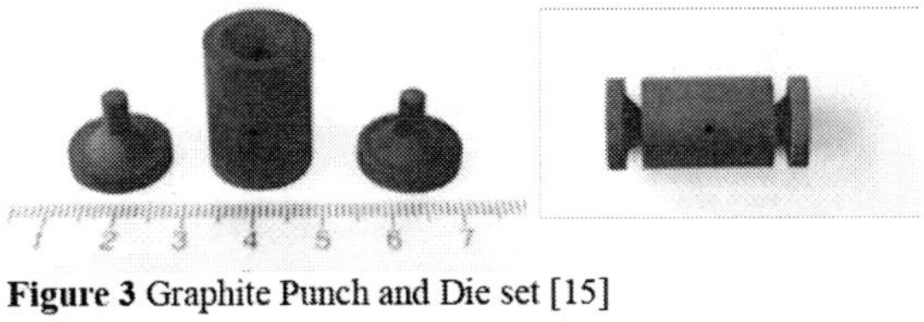

Figure 3 Graphite Punch and Die set [15]

Figure 4 shows the steps and the sequence of the sintering process of ZrO_2 powder. At the beginning of the process, there was a gap between the punches and die. The gap decreased steadily during the heating period as the temperature increased. It can be shown that the gap gone during the temperature holding period which shows an ideal current flow. It is worth mentioning that the temperature distribution indications a high uniformity during the whole sintering process irrespective of whether or not the gap disappeared.

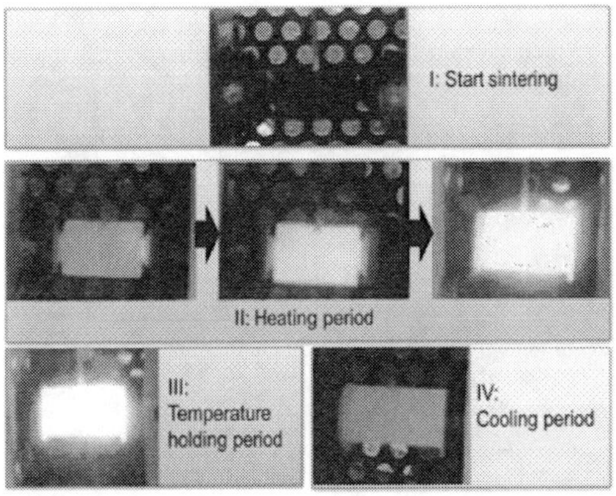

Figure 4 processing Sequence using the Gleeble 3800 [15]

Figure 5 shows the corresponding time-dependent stroke and sintering temperature were achieved with the Gleeble 3800 output setup. The graphs show the setting temperature, the actual temperature and the stroke versus the time during the forming process. In terms of temperature control, it can be seen from the curves that the actual temperature almost matches the setting temperature with only a small delay, from the beginning of the process. Once the stetting temperatures gained, the actual temperatures remain stable without any deviations. Moreover, matching with the applied pressure, the stroke curves decrease constantly: good material flow can be detected as soon as the force is applied. When the maximum temperature is achieved, it can be noticed there is a steep drop in the curve. Then, during the holding time, the stroke has no obvious changes, which means that the compact of the powder can be accomplished rapidly after the powder reaches maximum temperature. Nevertheless, the stroke changes during cooling stage, due to the shrinkage of the part that has been sintered. Comparing with the other sintering methods for ceramic powders [9] , Micro-fast sintering cycle with maximum sintering temperature of 1300°C is considered as a fast process.

https://doi.org/10.1051/matecconf/201819010007

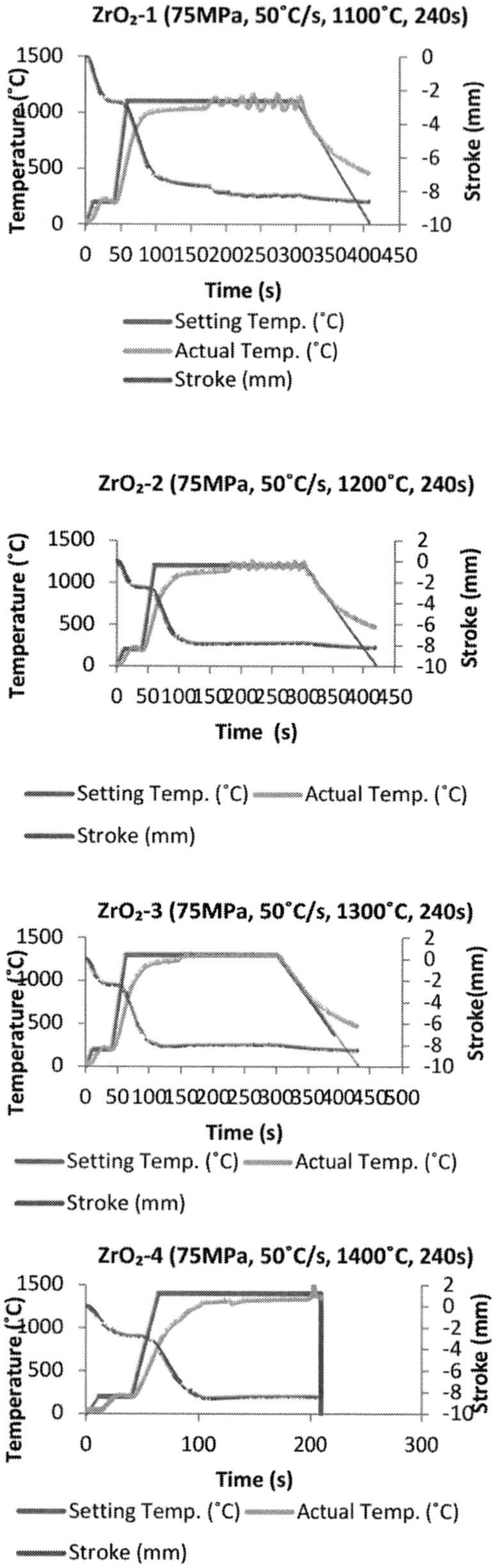

Figure 5 Process graphs of the forming/sintering of four samples parts of ZrO_2

4 Results and Discussion

The formed samples were examined carefully using the following methods: the sample geometry morphological measurement, calculating the relative density by the Archimedean approach, and microstructural observation under SEM. All experiment of ZrO_2 powders have been successfully formed into solid cylindrical components.

Table 1 The results of the formed samples of ZrO_2

Specimen designation	Sintering temperature (°C)	Sintering cycle time (s)	Sintered Part Density (g/cm³)	Sample relative density
ZrO_2-1	1100	405	5.279	94.27%
ZrO_2-2	1200	417	5.255	93.84%
ZrO_2-3	1300	429	5.427	96.91%
ZrO_2-4	1400	441	5.381	96.09%

As shown Table 1, the highest relative density of 96.91%. It can be predicted that the ZrO_2 powders sintering process can be improved by optimising the process parameters, such as by increasing the sintering temperature and holding time. The formed samples are very good and strong and it not easy tobreak it. The formed samples were found to have similar dimensions with the design of the dies (Φ 4.0mm×4.0mm) (see figure 6)

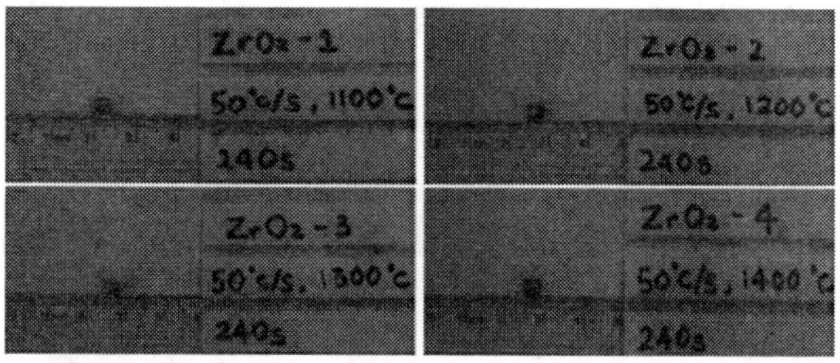

Figure 6 the Formed samples of ZrO_2 (4.0 mm in diameter and 4.0 mm in height).

Figure 7 shows the microstructure and the morphology of ZrO_2-3 sample. The formed samples are very good and strong look very good and shiny. The microstructure of sample shows a really good surface and there are no pores.

The EDS results showed an existence of carbon in the samples. This is might be due to using graphite die and punches where the carbon at the punch and die wall can penetrate into the sample during the sintering process. As a result, it is difficult to achieve a formed sample that is 100% free of any contamination by other elements such as carbon when graphite dies and punches are used.

https://doi.org/10.1051/matecconf/201819010007

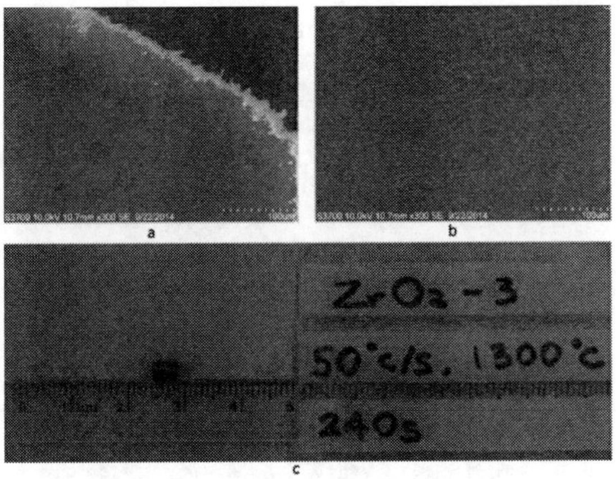

Figure 7 A formed sample (solid cylinder) with a size of Φ 4.0mm×4.0mm and its SEM micrograph

For the hardness test, the nano-indentation is used for the for the formed ZrO_2 -3 sample, 100 indentations were made on the sample: 10 in the sintering neck area and 10 in the particle centre (Fig. 8) It was found that the average Nano-hardness at the sintering for a Zirconia sample was 13.76 GPa. It is shown that the similar hardness values were found at both, the centres of the particles and necks, which indicates that good bonds among the particles had been formed.

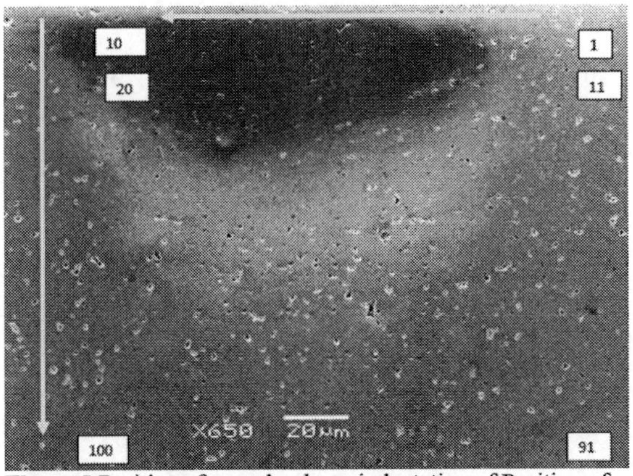

Figure 8 Position of nano-hardness indentation of Position of nano-hardness indentation of ZrO_2 -3 sample by using NanoTest Vantage hardness tester.

Table 2 shows the full datails of the average values for nano-hardness test that has been made using Nano-indintation approach.

Table 2 Average value for nano-hardness test for ZrO_2 -3 by using NanoTest Vantage hardness tester

Results	Units	Means	Errors
Max Depth	nm	373.3352	31.42479
Plastic Depth	nm	301.4467	33.32642
Maximum Load	mN	40.02	0.000001
Hardness	GPa	13.76014	2.329558
Reduced Young Modulus, Er	GPa	209.0603	16.32858
Elastic Recovery Parameter, ERP		0.241675	0.030812
Contact Compliance	nm/mN	2.395085	0.115092
Plastic Work	nJ	2.326256	0.531406
Elastic Work	nJ	301.4467	0.080666

5 Conclusions

The work conducted in this paper demonstrated that the capability of the Micro-FAST as a rapid process for the forming of micro/miniature- components. In general, Zirconia materials can be sintered successfully at a relatively low sintering temperature and sintering time.
The Micro-FAST provides an effective alternative to other technologies. The sintered micro-parts were assessed on morphology, relative density, micro-structures and mechanical properties, which showed good quality having been reached. Comparing the Micro-FAST to other conventional sintering processes and some micro-manufacturing processes, it can be shown that the micro-FAST has several advantages over other processes/technologies. Moreover, it can be suggested as a promising micro-manufacturing technology.

The challenges that face this process and its engineering applications mainly concern tool-design and process awareness in environment of the industry. These are being addressed by the EU Micro-FAST consortium.

Future work will be focussed on the investigation of influential sintering parameters in order to optimise the process, raise its repeatability and increase the quality of the parts to be manufactured.

Acknowledgement

The authors would like to acknowledge the fund from the European Commission through FP7 FOF Micro-FAST Project (GA No. 608720) for supporting the development of Micro-FAST manufacturing facilities. The fund from UK Royal Society through an international exchange programme to support the research in the Micro-FAST process is also acknowledged. The authors would like to especially thank Prof. Jianguo Lin and his colleagues in Imperial College London for providing Gleeble 3800 machine for conducting the experiments, and thank EU Micro-FAST project partners who provided powder materials for conducting the experiments.

References

1. Birkby, I. and R. Stevens. Applications of zirconia ceramics. in Key Engineering Materials. 1996. Trans Tech Publ.
2. Manicone, P.F., P. Rossi Iommetti, and L. Raffaelli, An overview of zirconia ceramics: basic properties and clinical applications. J Dent, 2007. 35(11): p. 819-26.
3. Denry, I. and J.R. Kelly, State of the art of zirconia for dental applications. Dent Mater, 2008. 24(3): p. 299-307.
4. Aytimur, A., et al., Magnesia stabilized zirconia doped with boron, ceria and gadolinia. Ceramics International, 2012. 38(5): p. 3851-3856.
5. Lee, S.-Y., Sintering behavior and mechanical properties of injection-molded zirconia powder. Ceramics International, 2004. 30(4): p. 579-584.
6. Qin, Y., Micromanufacturing engineering and technology 2nd ed. 2015, Oxford: Elsevier.
7. Petzoldt, F., Micro powder injection moulding-challenges and opportunities. Powder Inject Mould Int, 2008. 2(1): p. 37-42.
8. Meng, J., et al., Tribological behavior of 316L stainless steel fabricated by micro powder injection molding. Wear, 2010. 268(7-8): p. 1013-1019.
9. Orru, R., et al., Consolidation/synthesis of materials by electric current activated/assisted sintering. Materials Science and Engineering: R: Reports, 2009. 63(4): p. 127-287.
10. Lange, F., Densification of powder compacts: An unfinished story. Journal of the European Ceramic Society, 2008. 28(7): p. 1509-1516.
11. Hijji, H., et al. Fabrication of micro components with MSZ material using electrical-field activated powder sintering technology. in Advances in Manufacturing Technology XXX: Proceedings of the 14th International Conference on Manufacturing Research, Incorporating the 31st National Conference on Manufacturing Research, September 6–8, 2016, Loughborough University, UK. 2016. IOS Press.
12. Huang, K., et al., A new densification mechanism of copper powder sintered under an electrical field. Scripta Materialia, 2015. 99: p. 85-88.
13. Lu, D., et al., Forming microgears by micro-fast technology. Microelectromechanical Systems, Journal of, 2013. 22(3): p. 708-715.
14. Qin, Y., et al., Forming of Miniature Components from Powders by Combining Field-activated Sintering and Micro-Forming. Procedia Engineering, 2017. 207: p. 1212-1217.
15. Zhao, J., et al. Forming of micro-components by electrical-field activated sintering. in MATEC Web of Conferences. 2015. EDP Sciences.

Effect of annealing on microstructure and hardness of thin aluminium strips fabricated by micro flexible rolling

Mingshuai Huo[1,2], *Jingwei* Zhao[1], *Haibo* Xie[1], and *Zhengyi* Jiang[1,2*]

[1]School of Mechanical, Materials, Mechatronic and Biomedical Engineering, University of Wollongong, NSW 2522, Australia

[2]School of Materials and Metallurgy, University of Science and Technology Liaoning, Liaoning 114051, China

Abstract. Thin strips with varying thicknesses (TSVTs) produced by micro flexible rolling have wide applications in diverse micro manufacturing areas such as micro electromechanical systems (MEMS) and micro system technologies (MST). TSVTs consist of three regions based on different thicknesses, i.e. the thicker, the transition and the thinner zones. Among them, the transition zone plays a key role in evaluating the quality of TSVTs. In the current work, thin 1060 aluminium alloy strips were flexibly rolled under various thickness ratios followed by annealing treatment. The influences of annealing temperature on the microstructural evolution and hardness variation of the transition zone of TSVTs were investigated. The results reveal that the recrystallisation has happened after annealing at 400 °C for 30 min, which induces fine and homogeneously-distributed grains in both of the thinner and transition zones. Only moderate increase in hardness is found from the thicker to the thinner zones after annealing treatment.

Keywords: Aluminium, Microstructure, Hardness

1 Introduction

Miniaturisation has been an ongoing developing trend in the current technical fields [1, 2]. It is not only a slogan for energy saving and emission reduction, but also the reflection of the highly demands of micro products, such as connector pins, micro parts for electronics, chip lead frame, micro screws and lead frame, for applications in micro electromechanical systems (MEMS) and micro system technologies (MST) [3, 4]. Micro flexible rolling is a promising microforming technique for the production of thin strips with varying thicknesses (TSVTs), and has attracted an increasing research interests [5]. Qu et al. [6, 7] conducted numerical investigation on the effects of strip thickness, friction coefficient and rolling velocity on the springback during micro flexible rolling technology with considering the grain properties using the 3D Voronoi tessellation method. Zhao et al. [8] analysed the size effects involved in micro flexible rolling of metals.

Annealing has an important effect on the microstructure, texture, dislocations and stored energy of the plastically deformed materials, and in turn influences the properties of final products. Unfortunately, literature reporting the annealing effect of TSVTs is still unavailable so far. In current study, TSVTs with different thickness ratios (the ratio of thickness at thicker zone to that of thinner zone) have been treated under different annealing conditions in order to delineate the effect of annealing on the characteristics including the microstructure evolution and hardness distribution.

2 Experimental

1060 aluminium strips with a thickness of 464 μm were cut into 150 mm × 15 mm, and then annealed at 500 °C for 2 h in order to obtain fully recrystallised grains. Following this, they were micro flexibly rolled into different thickness ratios (the ratio of thickness at the thicker zone to that of the thinner zone) in terms of the rolling schedule shown in Table 1. In which, $R_1 = T_1/T_0$ and $R_2 = T_2/T_0$. The diameter and roll barrel length of the work roll are 25 and 40 mm, respectively.

Table 1 Rolling schedule of Group 1.

Rolling parameters	S1	S2	S3
Initial thickness T_0 (μm)	464	464	464
Thickness at thicker zone T_1 (μm)	346	346	346
Reduction at thicker zone R_1 (%)	25.4	25.4	25.4
Thickness at thinner zone T_2 (μm)	96	172	245
Reduction at thinner zone R_2 (%)	79.3	62.9	47.2
Thickness ratio	3.60	2.01	1.41
Rolling speed (cm/min)	60	60	60

Annealing heat treatment was applied to all the flexibly-rolled specimens. The detailed procedures are

* Corresponding author: jiang@uow.edu.au

 https://doi.org/10.1051/matecconf/201819011001

shown in Fig. 1. One batch of specimens was isothermally annealed at 400 °C for varying holding times in the range of 10-60 min, and the other batch was isochronally annealed at 200, 300 and 400 °C respectively, for 30 min. All the heat-treated specimens were water-quenched after corresponding annealing processes.

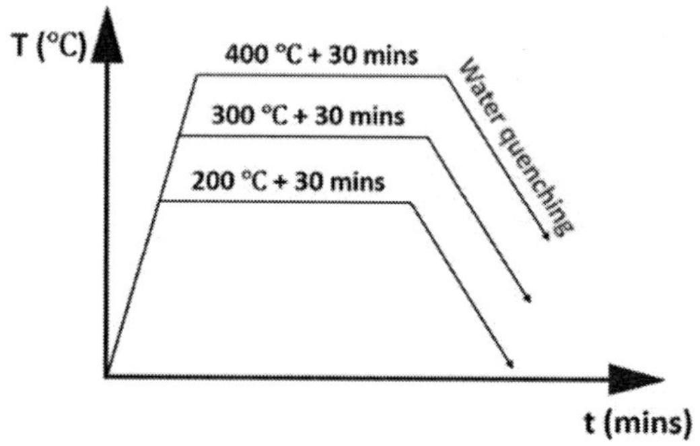

Fig. 1 Flow chart for various target specimens.

All the samples were ground and polished to the colloidal silica stage on a Struers Tegramin and electrolytic etching using Barker's reagent with Struers Lectropol 5. Subsequently, Leica DMRM and Nikon Fluorescence Microscope were used for the identification and acquisition of the polarised light false colour optical micrographs from longitudinal sections as defined by the rolling direction (RD) and the normal direction (ND).

Vickers hardness testing was conducted at five pre-set points on the cross-sectional surfaces of the TSVTs through the thicker, transition and thinner zones along the rolling direction, as shown in Fig. 2. A TIME TH715 Microhardness tester was used with a 4.9 N load and a dwelling time of 10 s for all the specimens with three duplicate tests.

3 Results and discussion

3.1 Microstructural evolution

Fig. 3 shows the initial microstructure of aluminium strip after homogenisation annealing carried out at 500 °C for 2 h. It can be seen that the recrystallised microstructure with equiaxed grains has been obtained.

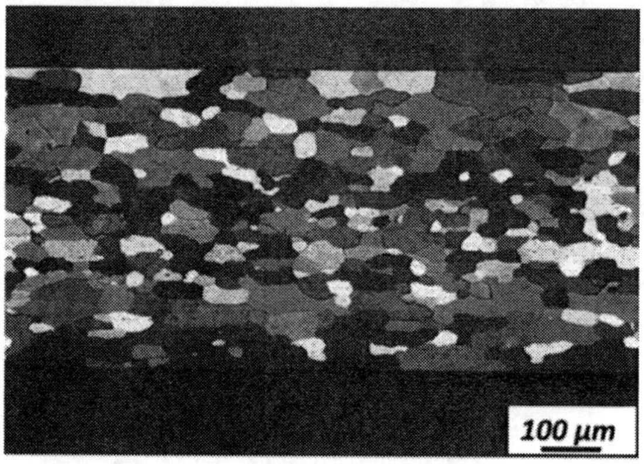

Fig. 3 Initial microstructure after fully recrystallisation at 500 °C annealing for 2 h.

After micro flexible rolling with three thickness ratios, different thickness variations and grain characteristics from the thinner zone to the thicker zone of S1-S3 can be observed, as shown in Fig. 4. Apparently, the grains distribution and the grain size have changed to a large extent in the thinner and transition zones compared with the initial condition (Fig. 3), while only a little variation happened in the thicker zone because of the relatively small rolling reduction. In the thinner zones (Fig. 4a, d and g), nearly entire elongated grains are aligned parallel to the rolling direction have been observed. Largest reduction applied to S1 presents the most severe microstructural deformation with the longest aspect ratio of the elongated bands with the most densely lamellar space. Typical waved characteristics have been found in the middle layer in all the thinner zones of TSVTs, and meanwhile some shear bands have formed at the surface layers, which are caused in the rolling process with a reduction of above 50% [9]. For the transition zones, microstructure near the top and bottom surfaces always deform as the rolling schedules. The lamellar spaces of the grains decrease gradually with the increase of the rolling degrees. Besides, the waved characteristics also arise especially near the thinner zone sides.

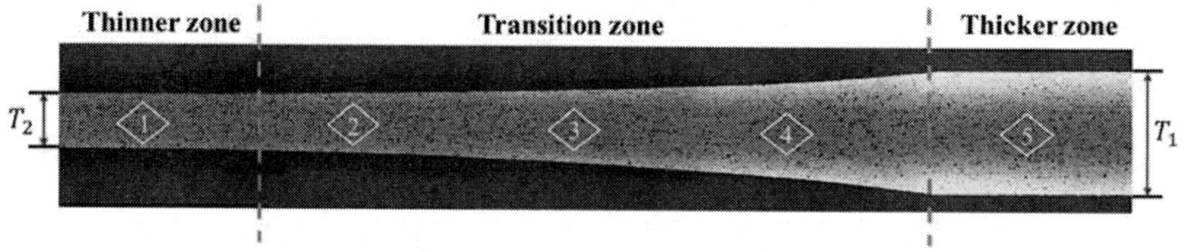

Fig. 2 Measured hardness positions in the thicker zone, transition zone and thinner zone.

 https://doi.org/10.1051/mateconf/201819011001

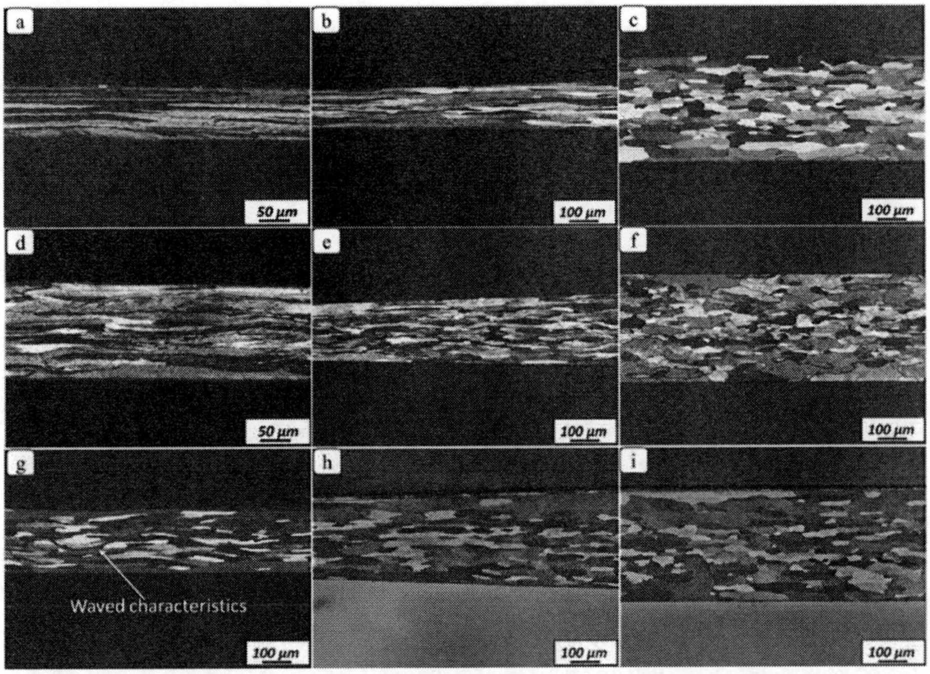

Fig. 4 Microstructure distribution after micro flexible rolling in the thinner zones of (a) S1, (d) S2, (g) S3, the transition zones of (b) S1, (e) S2, (h) S3 and the thicker zones (c) S1, (f) S2, (i) S3.

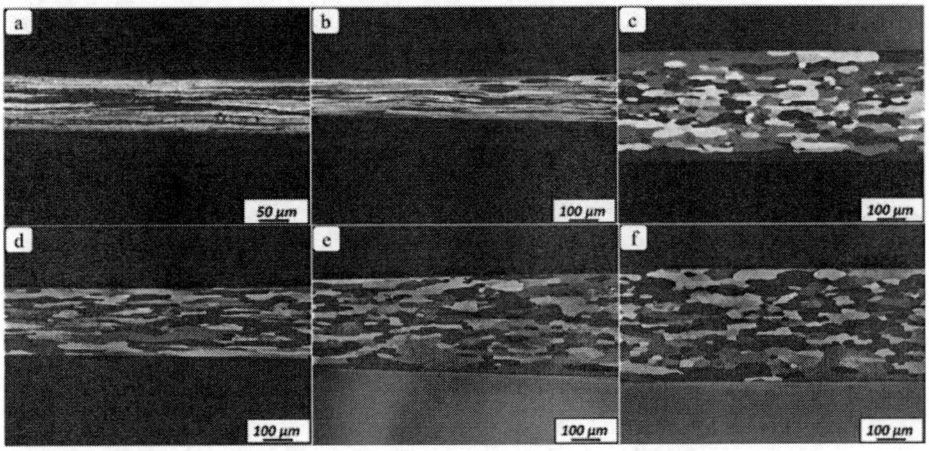

Fig. 5 Microstructure distribution after isochronal annealing at 200 °C in the thinner zones of (a) S1, (d) S3, the transition zones of (b) S1, (e) S3 and the thicker zones (c) S1, (f) S3.

Fig. 5 shows the microstructure of S1 and S3 after annealing at 200 °C for 30 min. Comparing with the micro flexibly rolled results (Fig. 4), no apparent influence on the TSVTs is found, which means annealing at 200 °C is not sufficient for recovery or recrystallisation. Comparatively, a slight change occurs when the annealing temperature arises to 300 °C, as shown in Fig. 6. With regards to the thinner zones (Fig. 4), there is almost no obvious waved characteristics presenting in both thinner and transition zones. This results in the release of the deformation stored energies from strain hardening and residual stress. The grains become more closely aligned with the rolling plane, which means recovery phenomenon has occurred at this annealing temperature. The same situation has appeared in the transition zone (Fig. 6b and e) that the lamellar grains in the top and bottom layers display a similar distribution characteristic along the surface layers apart from the grains in the middle layer parallel to the rolling plane.

When raising the annealing temperature to 400 °C, as shown in Fig. 7, TSVTs in the thinner and transition zones reveal the salient variations although almost no apparent change occurs in the thicker zones. The reason

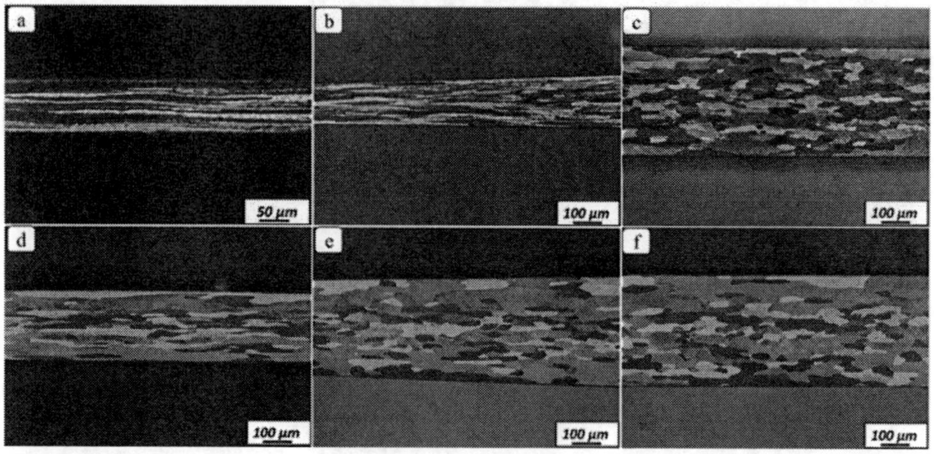

Fig. 6 Microstructure distribution after isochronal annealing at 300°C in the thinner zones of (a) S1, (d) S3, the transition zones of (b) S1, (e) S3 and the thicker zones (c) S1, (f) S3.

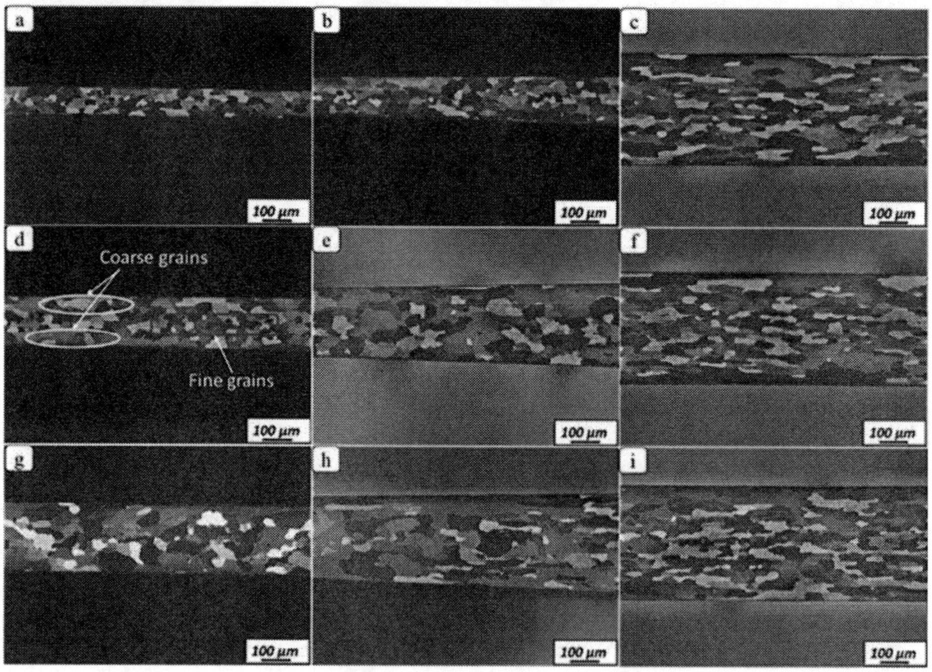

Fig. 7 Microstructure distribution after isochronal annealing at 400°C in the thinner zones of (a) S1, (d) S2, (g) S3, the transition zones of (b) S1, (e) S2, (h) S3 and the thicker zones (c) S1, (f) S2, (i) S3.

is that the higher reductions at the thinner and transition zones lead to more stored energy, which provides more driving force for recrystallisation. This provides large density of nucleating sites for recrystallisation, contributing to the formation of fine grains. Grain refinement is beneficial for the improvement in mechanical properties based on the well-known Hall-Petch relationship [10]. However, fine grains tend to emerge in the middle layer comparing with coarse grains near the surface layers. It is thought that the high dislocation density in the areas resulting from the shear deformation generates more stored energy. This is also the cause of the recrystallisation gradient in the transition zone, as shown in Fig. 7(h).

3.2 Hardness

Hardness results after micro flexible rolling and subsequent annealing processes are given in Figs. 8 and 9, from which a climbing trend from the thicker to the thinner zones has been observed. With the decrease of thickness ratio from S1 to S3, the hardness value reduces gradually at each measured position. This phenomenon can be explained that the increase of reduction leads to the growth of strain hardening. Dislocation tangles with high density are the keys to boost the intensity of this hardening behaviour [11].

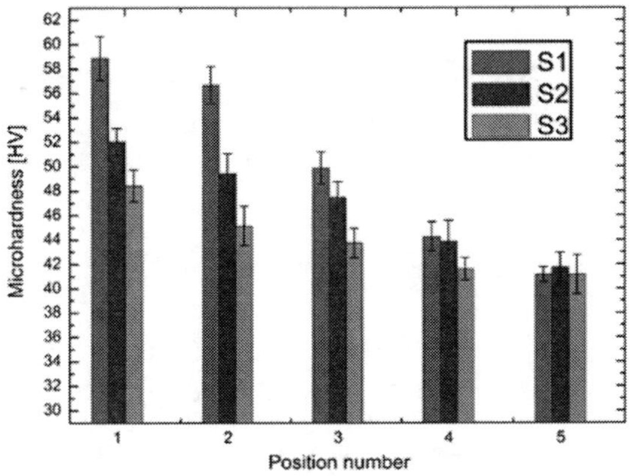

Fig. 8. Variations in Vickers hardness after micro flexible rolling with different thickness ratios.

As the microstructure of TSVTs under different annealing conditions presents a similar trend, S1 specimen with the highest thickness ratio after annealing has been taken for hardness measurements. Fig. 9 shows that all the specimens exhibit a decreasing trend with the increase of temperature from 200 to 400 °C, no matter in the thinner, thicker or transition zones. It is clear that specimens annealed at 400 °C for 30 min exhibit smooth hardness distribution, and only moderate increase in hardness is found throughout the various thickness zones.

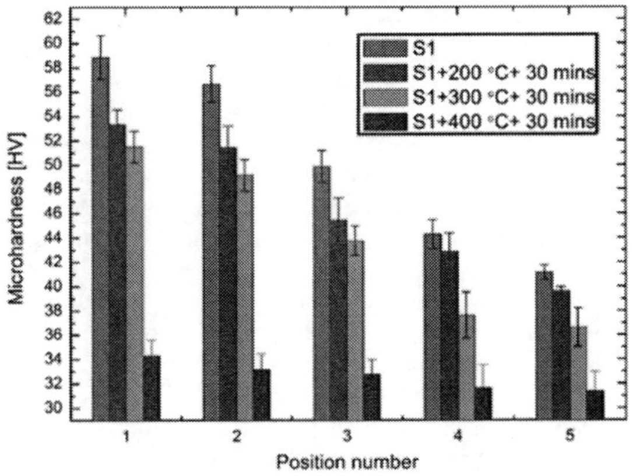

Fig. 9. Variations in Vickers hardness of S1 after annealing at 200, 300 and 400 °C for 30 min.

4 Conclusions

The effect of annealing temperature on the microstructure and hardness of thin 1060 aluminium alloy strips was investigated. The following conclusions can be drawn from the current research:

(1) The annealing temperature significantly affects the grain shapes, size and distribution of TSVTs. Recrystallisation results in fine grains in both of the thinner and transition zones after annealing at 400 °C.

(2) A relatively steady hardness distribution is observed when TSVTs is annealed at 400 °C 30 min, and the hardness increases moderately through the thicker to thinner zones.

Acknowledgements

The first author gratefully appreciates UOW for PhD scholarship. This study is also supported by the grant from Australian Research Council (ARC, Grant No. FT120100432) and National Natural Science Foundation of China under Grant No. 51474127.

References

[1] U. Engel, R. Eckstein, Journal of Materials Processing Technology 125 (2002) 35-44.

[2] M. Geiger, M. Kleiner, R. Eckstein, N. Tiesler, U. Engel, Cirp Annals-Manufacturing Technology 50 (2001) 445-462.

[3] W.L. Chan, M.W. Fu, J. Lu, Materials & Design 32 (2011) 198-206.

[4] Q. Zheng, T. Shimizu, M. Yang, International Journal of Mechanical Sciences 133 (2017) 416-425.

[5] Z. Jiang, J. Zhao, H. Xie, *Microforming Technology: Theory, Simulation and Practice* (2017).

[6] F. Qu, Z. Jiang, H. Lu, International Journal of Mechanical Sciences 105 (2016) 182-190.

[7] F. Qu, Z. Jiang, D. Wei, Q. Chen, H. Lu, International Journal of Mechanical Sciences 123 (2017) 324-339.

[8] J. Zhao, H. Xie, H. Lu, Z. Jiang, Size effects in micro rolling of metals, IOP Conference Series: Materials Science and Engineering, vol 282, 2017.

[9] P.J. Hurley, F.J. Humphreys, Acta Materialia 51 (2003) 1087-1102.

[10] W. Chrominski, S. Wenner, C.D. Marioara, R. Holmestad, M. Lewandowska, Materials Science and Engineering: A 669 (2016) 447-458.

[11] B. Wang, X. Chen, F. Pan, J. Mao, Y. Fang, Transactions of Nonferrous Metals Society of China 25 (2015) 2481-2489.

Internal Flow-Turning – extended manufacturing possibilities in tailored tube production

Eugen Wiens[1,*], *Werner Homberg*[1]

[1]LUF Chair of Forming and Machining Technology, Faculty of Mechanical Engineering, Paderborn University, Warburger Straße 100, 33100 Paderborn, Germany

Abstract. The motivation behind the development of a new approach to the manufacture of tailored tubes is the need for greater flexibility in the wall thickness variation of semi-finished parts compared with present manufacturing processes. In the so-called Internal Flow-Turning (IFT) process, the wall thickness contour of regular tubes with a uniform wall thickness is adjusted by means of a rolling process employing an innovative roller tool on the inside of the tube and a die on the outside of the tube. The gap between the roller element and the die can be adjusted through the axial infeed of the roller tool. The gap defines the current wall thickness of the tube during the manufacturing process. In this way, the wall thickness contour can be set in a precise manner by varying the gap distance as the tube is being pulled through the die. Besides the possibility of locally adjusting the wall thickness, IFT can also be used to form geometrical rib elements in the radial tube direction by using a radially shaped die. As the material flow in the IFT-process is mainly caused by compressive stress, the grooves of the die can be "filled up" by radial material displacement.

Keyword: Manufacturing process, Metal forming, Rolling

1 Introduction

The efficient use of material in present-day manufacturing is driven by the demand for sparing use of resources and the need for lightweight structures at a reasonable cost. One way to improve lightweight design is to adapt the wall thickness of the work pieces to match the expected loads, and hence to manufacture tailored semi-finished parts. By comparison with present manufacturing processes, the Internal Flow-Turning has a greater flexibility regarding the wall thickness variation in the production of semi-finished parts, which holds great potential for the manufacturing of tailored tubes. In the IFT-process regular tubes with an uniform wall thickness are formed to tubes with a tailored wall thickness contour by means of a rolling process using a innovating roller tool on the inside and a die on the outside of the tube. By axially adjustment of the tool the gap between the rolling elements and the die can be set. The gap defines the current wall thickness of the tube during the manufacturing process. By simultaneous movement of the tube relative to the die and the adjustment of the rolling cycle diameter a given wall thickness contour can be formed in a precise manner. As in conventional flow forming, the plastic deformation in the Internal Flow-Turning process is caused mainly by compressive stress, which means that quite a large number of materials can be formed. In experimental investigations, aluminium alloys like EN AW 6060 and several steels (from mild steel (1.0308) through to high-strength steel (1.3548)) were successfully formed using the IFT process [3]. Internal Flow-Turning makes it possible to extend the wall thickness reduction range as well as to increase the freedom of design in terms of the wall thickness distribution or contour when manufacturing wall-thickness-contoured tubes. Local reductions of up to 90% of the initial wall thickness are possible. The transition between the different wall thicknesses can vary from almost "rectangular" transitions to very smooth (conical) transitions of just a few degrees. The varying wall thickness contour in the axial tube direction is only implemented on the inside of the tube wall, and the outer diameter stays constant in the axial direction. In addition, the IFT process can also be used to form geometrical rib elements in the radial tube direction by using a radially shaped die. As the material flow in the IFT process is mainly caused by compressive stress, the grooves of the die can be "filled up" by the radial material flow. In addition the geometrical forming of the tube cross-section, Internal Flow-Turning also improves the mechanical properties and geometrical accuracy of the rolled tube section. The hardness and strength improvement as a function of the wall thickness reduction, generates an increase of up to 70% [1]. Moreover, the surface roughness can be reduced significantly, depending on the related feed rate [mm/rev] of the rolling tool and the wall thickness reduction. IFT could be used to advantage not only for the production of work pieces but also for the production of preforms or semi-finished parts in a highly efficient manner. As a demonstrator for potential applications for

* Corresponding author: ew@luf.uni-paderborn.de

internal flow-turned tubes, a preform for an automotive exhaust manifold in high temperature stainless steel was optimised in terms of weight and material cost. The hydroformed demonstrator has a greatly reduced wall thickness over its area, apart from in the main forming zone, which gives a weight and material reduction of 20% compared to the original part [2, 3].

2 Process principles

The IFT process is based on an internal rolling process with an innovative roller tool, whereby the contour in the axial direction is only implemented on the inner side of the tube. The driven roller tool positioned inside of the tube forms the wall thickness contour in a large number of small local forming steps. Due to the incremental character of the forming process the forces are quite low, thus allowing a far higher wall thickness reduction compared to flexible tube drawing for example. Additionally, the transitions between different wall thickness sections are more flexible, even enabling steep transitions.

The rolling takes place while the tube has an axial movement relative to the rolling elements. The wall thickness is determined by the gap between the roller elements on the inside of the tube and the die on the outer surface of the tube. The die and the roller tool are mounted on two separate sliding supports, which can be moved in an axially synchronized manner at the same speed in relation to the static tube. In addition the tool support can be moved with a superimposed translational displacement. In doing this, the gap can be adjusted while the die is moving over the tube, whereby two different tool concepts can be used. The two process concepts differ in respect of internal flow-turning with integrated tube drawing and a rigid turning circle diameter (Fig.1a)). Here the relative movement between the die and the tube is a necessary process element for the ability to adjust the gap or wall thickness reduction respectively. In the process version without tube drawing the relative movement between the die and the tube is not necessary because the tool has a radially adjustable rolling circle diameter and the drawing process element is not required for the adjustment of the gap (Fig. 1b)).

2.1 IFT with tube drawing

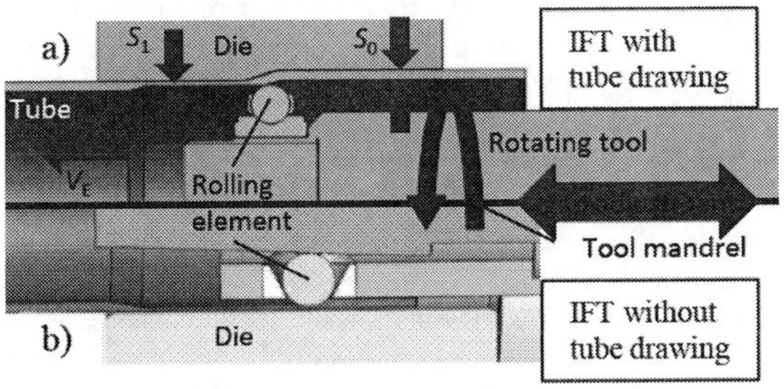

Fig. 1. Process principle of Internal Flow-Turning (IFT) with tube drawing a); and without tube drawing b)

In the version of the process with tube drawing, the roller gap adjustment takes place as a result of the axial movement of the roller tool into the tapered section of the die (see detail a)1 & a)2 in Fig.2). The axial displacement of the roller tool is quite small and depends on the tapering angle of the die and the desired wall thickness reduction Δs. The main component of the tool is a standard groove ball bearing where the external ring is removed. A schematic diagram of the tool is shown as a segment of the cross-section in Fig. 1a). The tube drawing forms part of this process, since it is necessary for the roller gap adjustment. In this process version, therefore, the forming of the wall thickness contour to achieve the desired external diameter has to be performed in a single overrun step.

2.2 IFT without tube drawing

In the version of the process without tube drawing, the tool has a radially adjustable rolling circle diameter, whereby the infeed is achieved through the axial displacement of the conical driving shaft surface. Due to the quite small angle of the conical driving shaft, the axial displacement is higher compared to the process version with tube drawing. The tool design is shown in Fig. 1b and a detail of the gap adjustment in Fig. 2b. Because the processed tube keeps its initial external diameter, the forming process of the wall thickness contour can be divided into several overrun steps. Due to the absence of tube drawing, the axial tension forces on the tube are reduced. This leads to a higher potential wall thickness reduction of almost 90 % of the initial wall thickness. Furthermore, this ITF version allows backward and forward flow-turning by contrast to the IFT version with tube drawing, where only forward flow-turning is possible. This option depends however, on the geometry of the rolling elements employed. Due to the modular setup of the radial adjustable tool, it is possible to use balls or various cone geometries as rolling elements. If cones are used as rolling elements, the conical driving shaft must have the same angle as the attack angle of the rolling element.

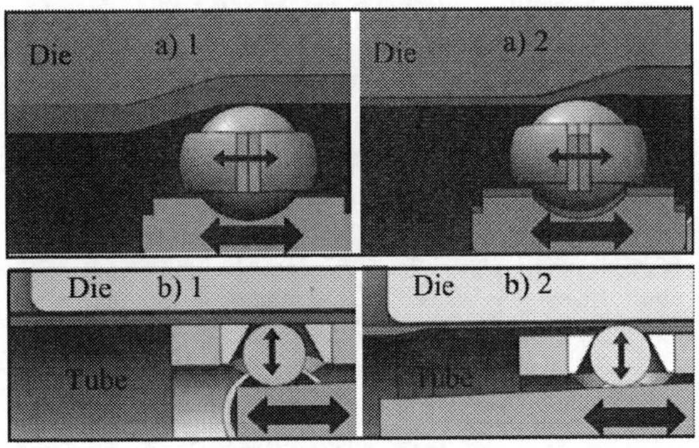

Fig. 2. Details of the gap adjustment principle for the IFT process with tube drawing a) 1&2 and the IFT process without tube drawing and a radially adjustable roller tool with balls as rolling elements b) 1&2

https://doi.org/10.1051/matecconf/201819011002

2.3 Effects on the processed tube properties

The IFT process has various impacts on the geometrical and mechanical properties of the processed tube. With the mentioned increase of the hardness and strength of the rolled tube section the remaining formability decreases as a function of the wall thickness reduction Δs due to work hardening. As the strength increase only takes place within the rolled tube section for the IFT process without tube drawing, the cold work hardening impacts the whole tube for the process with tube drawing and has a greater impact in rolled sections in relation to the wall thickness reduction Δs. The tube drawing itself also leads to a reduction of the remaining formability of the tube. The choice of IFT version (with/without tube drawing) thus depends on the downstream manufacturing process in addition to other factors. If further forming processes are needed, the IFT version without tube drawing is suitable, due to the greater remaining formability properties. If a high strength is needed, the IFT process with tube drawing is advantageous [2].

The rolling process also leads to an improved surface, whereby the processed surface depends mainly on the related feed rate f (millimetres per revolution) and the wall thickness reduction Δs. In addition, the geometry of the rolling element affects the surface roughness. The IFT process also improves the roundness of the tube, irrespective of the wall thickness reduction Δs [1].

3 Manufacturing of contoured outer surfaces with the IFT process

Internal Flow-Turning is in various ways comparable and related to flow forming. In both processes, the material flow is based mainly on compressive stress, and very high wall thickness reductions of up to 90 % of the initial state are possible [4,5]. The concept of using flow forming to create tubular parts with internal gearing elements was developed decades ago [6,7] and is applied today in the series production of plate carriers with internal gears for gearboxes in the automotive industry for example [8,9]. In the flow forming of tubes with inner ribs it is also common to use balls as rolling elements [10-14].

Experimental investigations relating to the manufacture of contoured outer surfaces using the IFT process with tube drawing revealed that the rib element shaping results mostly from a bending process during tube drawing. With an increasing wall thickness reduction Δs, the level of groove filling even decreases. Moreover, the cross section on the inner side of the tube is not circular, but has a local deviation from circularity in the rib area (see Fig.3) which increases with the size of the groove. The deviation decreases with an increasing wall thickness reduction Δs. The height of the groove has to be quite small to achieve a high enough wall thickness reduction Δs to create a cylindrical inner cross section before any tearing of the tube occurs. Fig. 3 shows an example with a circular inner cross section (top right). Here the outer contour is an oval with wall thickness differences of 0.4 mm in 90-degree shifts. The remaining pictures in Fig 3 show the forming behaviour of higher rib elements with a non-circular cross-section in the rib area [15].

Fig. 3. Specimens with rib elements on the outer contour processed by IFT with tube drawing [15]

The risk of non-circular inner cross-sections developing could be avoided using the IFT process without tube drawing. For the experimental investigation of this variant, a die with three different-shaped grooves was designed. The grooves differ in width from 3 mm to 20 mm and have various edge geometries. The maximum height of each groove, at 1.5 mm, is the same for all geometries. The different-shaped grooves ought to show the influence of the edge geometry and the length or volume, respectively, of the groove on the material flow and the filling ability by IFT. Fig. 4 shows the inner contour of the die ring. The tubes used in the experimental investigations were seamless tubes with a 48 mm external diameter an initial wall thickness of 4 mm. The material of the tubes was ether mild steel (E235 (1.0308), seamless drawn precision tube) or extruded aluminium (EN AW 6060 T4).

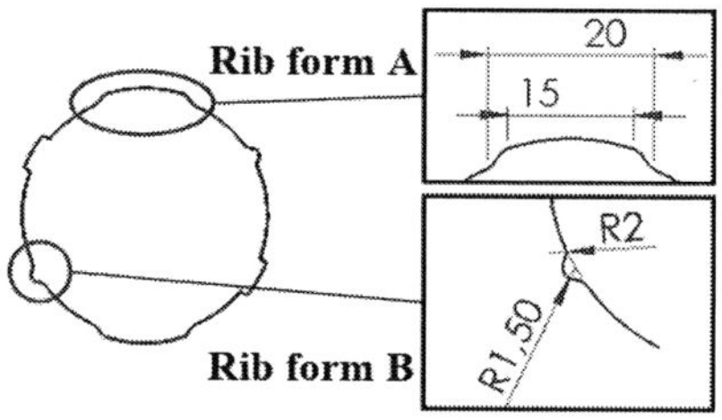

Fig. 4. Inner contour of the die ring; measurements in [mm]

Initially the experiments were carried out with balls as rolling elements that had a diameter of 11.9 mm. The results showed that the increase in the wall thickness reduction Δs and the increase in the related feed f has a positive impact on the filling of the grooves. An even bigger impact than the overall wall thickness reduction comes from the wall thickness reduction in each individual overrun step. Here a maximum wall thickness reduction of $\Delta s_{max} = 0.8$ mm was achieved in each overrun step.

Fig. 5 shows, by way of example, the result of rib form A with the maximum wall thickness reduction (overall and in each overrun) with balls. Complete filling of the groove could not be achieved for all the rib geometries examined, only rib form B displayed almost full filling.

https://doi.org/10.1051/matecconf/201819011002

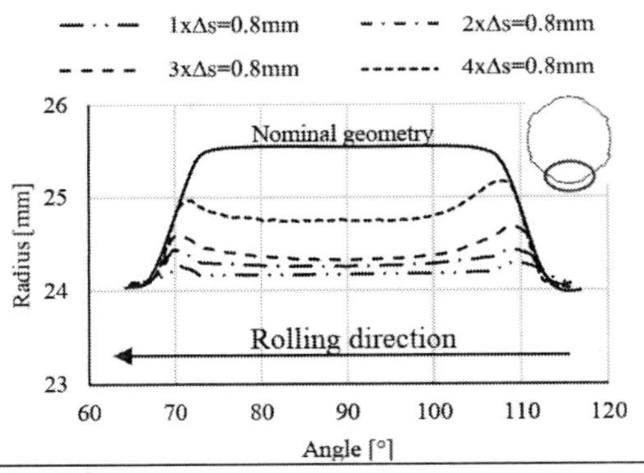

Material: Steel (1.0308)	Tube type: seamless precision
External Ø:48 [mm]	Initial wall thickn.: $s_0 = 4$ [mm]
Roller: Ball (Ø11.9 [mm])	Rib form: A

Fig. 5. Influence of the overrun number on the contour accuracy

Due to the fact that conical roller elements enable a bigger wall thickness reduction Δ*s* in a single overrun step, further investigations were carried out using cone elements with an attack angle of α = 8° and a transition radius r_{trans} of 12 mm. The attack angle α of a ball can be approximated as the connecting line of the two endpoints of the contact surface between the ball and the tube [7] (see Fig. 6).

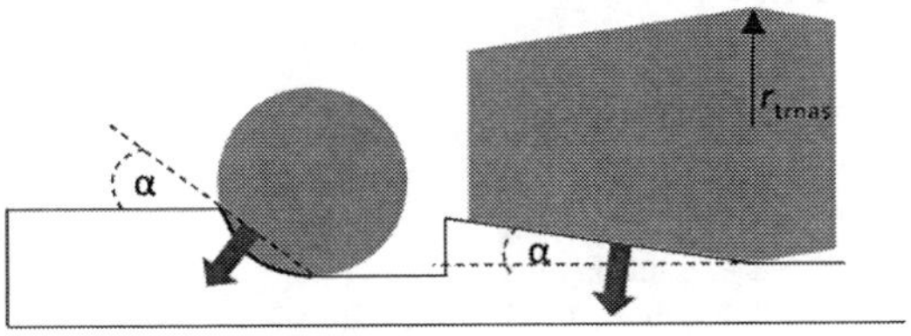

Fig. 6. Entry / Attack angle α of ball and cone roller element and resulting compressive force

According to this definition, the attack angle α of the ball increases with an increasing wall thickness reduction and leads to a resulting compression force with a similarly increasing axial component. This is counterproductive for the radial material flow, which normally allows good filling of the radial grooves. For the wall thickness reduction of Δ*s* = 0.8 mm, the attack angle α for the balls used is α = 15°. A comparison of groove filling for an equal wall thickness reduction of Δ*s* = 0.8 mm after a single overrun with balls and cones confirms the assumption of the resulting compression force. Irrespective of the groove geometry the filling using cone-rolling elements is significantly higher than that using balls. Fig. 7 shows, by way of example, a comparison for filling in the rib form A.

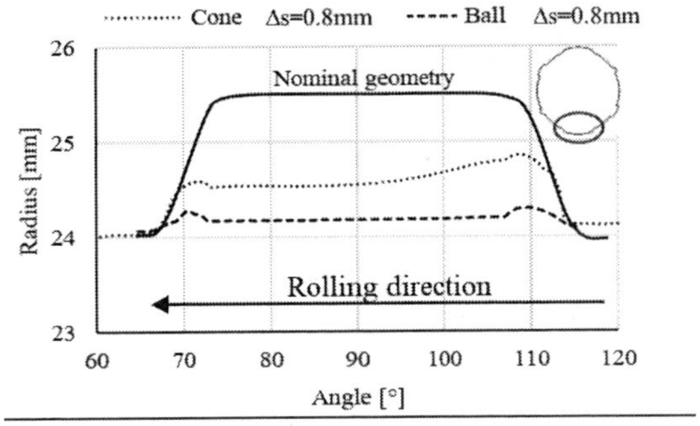

Material: Steel (1.0308)	Tube type: seamless precision
External Ø:48 [mm]	Initial wall thickn.: $s_0 = 4$ [mm]
Roller: Ball (Ø11.9 [mm])	Rib form: A
Roller: Cone (8 [°])	Wall thickn. red.: $\Delta s_1 = 0{,}8$ [mm]

Fig. 7. Influence of the rolling element geometry

These results could be confirmed by a look at the cross section of the specimens shown in the picture in Fig. 8. It was observed that a wall thickness reduction of Δ*s* = 1.5 mm through conical rolling elements in a single overrun leads to almost complete filling of the groove geometries. A wall thickness reduction that was twice as high, of Δ*s* = 3.0 mm, achieved with ball rolling elements in four overrun steps, leads to clearly inferior groove filling. Both specimens are made of the same seamless steel tube with an initial geometry of 48 mm external diameter and a wall thickness of 4 mm.

Fig. 8. Comparison of rib forming by IFT with balls (left) and cones (right) as rolling elements

Beside the tactile measurement of the cross section with a coordinate measuring machine (Model Carl Zeiss Contura G2), the complete filling of the groove was also determined by visual testing.

Figs. 9 and 10 show, by way of example, the influence of the wall thickness reduction Δ*s* on the contour accuracy and groove filling for rib form A and B. The results reveal that a bigger volume / length of the rib cross section at the same maximum height requires a greater wall thickness reduction Δ*s* for complete groove filling. For small rib elements (form B) the wall thickness reduction has to be only slightly higher than the rib height itself.

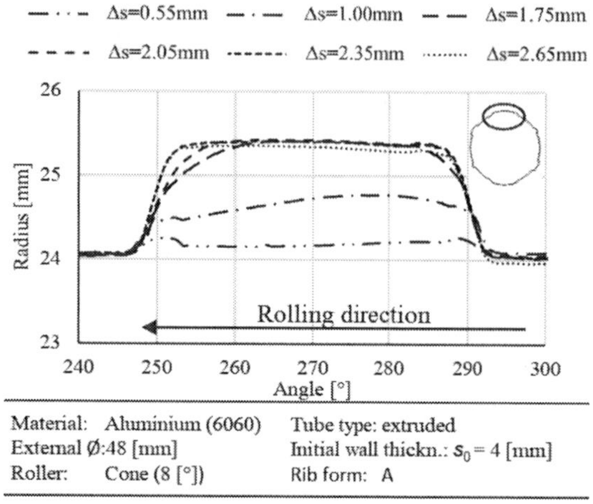

Fig. 9. Influence of the wall thickness reduction Δs on the contour accuracy; The curve with a wall thickness reduction Δs = 2.65 [mm] is the nominal rib contour

With small wall thickness reduction stages in all groove geometries the corner region starts to build up first. With a further wall thickness reduction, the centre of the rib starts to rise faster than the corners. By increasing the overall wall thickness reduction in a number of overrun steps, the shape with the bulging edges continues to exist even with very high overall wall thickness reductions (compare Fig. 5). With an increasing wall thickness reduction in a single overrun step, the chronologically first over-rolled corner of the groove "fills up" first for longer ribs (see e.g. form A in Fig. 9). For small grooves (form B in Fig. 11) the chronologically second over- rolled edge "fills up" first. This corresponds to investigations conducted for internal gear forming by flow forming [6].

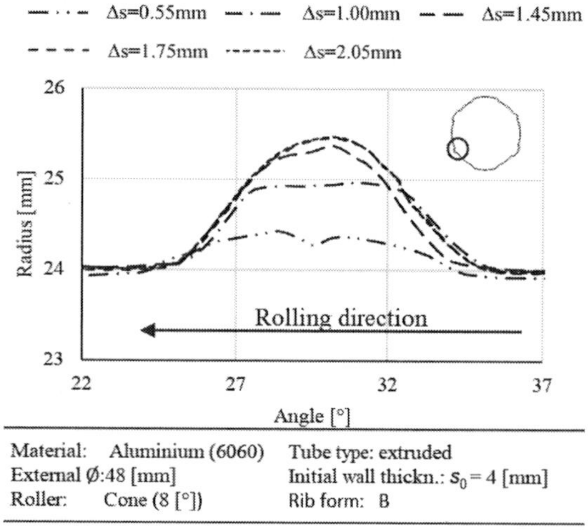

Fig. 10. Influence of the wall thickness reduction Δs on the contour accuracy; The curve with a wall thickness reduction Δs = 2.05 [mm] equals the nominal rib

Due to the rolling process, there is a varying wall thickness distribution inside the groove elements in the circumferential direction. The different groove filling is demonstrated in Fig. 11 on the basis of a tactile measured cross-section of a tube section with semi-filled grooves. For a better understanding, the tool and the rotation of the tool elements is illustrated as well as the compression direction and the direction of the tangential material flow.

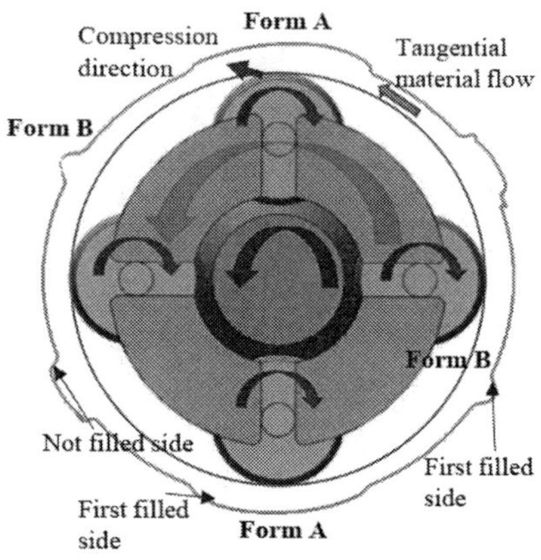

Fig. 11. Effect of the tangential material flow on the contour

Tests that were conducted with an increase in the wall thickness reduction up to the tearing of the aluminium tube (EN AW6060 T4) at Δs = 3.7 mm (92.5 % of the initial wall thickness) revealed that there is on optimum wall thickness reduction Δs beyond which any further wall thickness reduction causes a deterioration in the filling quality of the groove. This is most noticeable with the widest rib (rib form A). By increasing the wall thickness reduction to above a certain point, the corners and flanks of the rib will no longer be completely filled. The reason for this phenomenon is presumed to be the increasing tangential material flow.

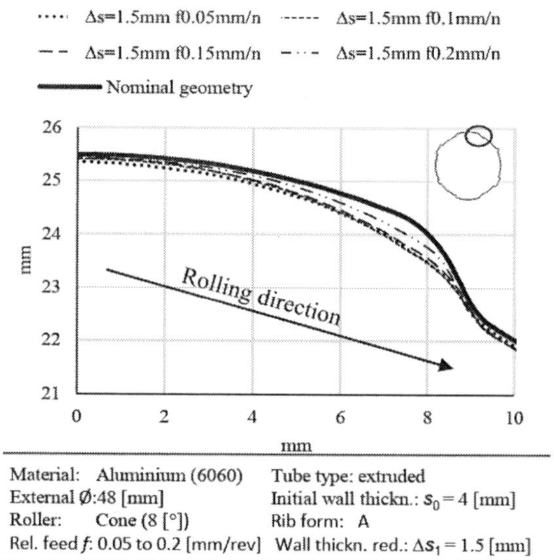

Fig. 12. Improvement of corner filling by increasing related feed f with a constant wall thickness reduction Δs = 1.5 [mm]

With an increasing related feed *f* the "filling level" of the grooves improves to the extent previously mentioned for the tool version with balls. Fig. 12 shows the improvement of the corner filling for rib form A with a constant wall thickness reduction of $\Delta s = 1.5$ mm. The wall thickness reduction in this investigation is deliberately set to a not complete "filling level", to show the improvement achieved by increasing the related feed *f* from 0.05 mm/rev to 0.2 mm/rev.

To identify the influence of the IFT process on the grain texture, microsections of the three different rib elements were made. The enlargements shown have an optical magnification of 600 times and are taken from each chronologically second-over rolled corner of the rib elements (Fig. 13). For a better comparison, the initial state of the steel tube, which has a ferritic pearlitic microstructure, is shown too. Because the initial tube is a drawn precision tube, the mostly isometric grain structure already shows a slight grain refinement towards the outer tube surface in the initial state.

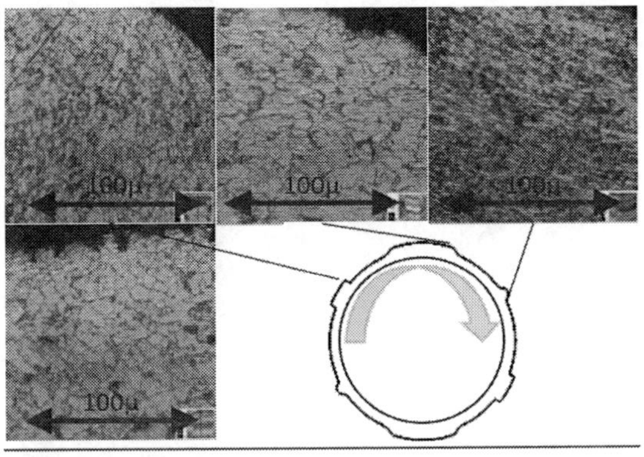

Material: Steel (1.0308)	Tube type: seamless precision
External Ø:48 [mm]	Initial wall thickn.: $s_0 = 4$ [mm]
Roller: Cone (8 [°])	Rib form: A, B, Square
Magnification: 600x	Wall thickn. red.: $\Delta s_1 = 2.0$ [mm]

Fig. 13. Grain structure in the rib element corners at $\Delta s = 2.0$ mm of the steel tube (top) and the outer edge of the initial steel tube (bottom)

The grain deformation shows an increase with a decreasing rib volume at the wall thickness reduction stage of Δs = 2.0 mm. This is the consequence of greater compression at the smaller grooves after these grooves achieve complete filling first. The appearance of the grain structure also confirms the contrary filling order of grooves of different lengths by showing a higher deformation at each first-filled corner. In addition, the "transition sharpness" at the edges of the grooves also has an influence on the grain deformation. The wide rib element (form A) has a very smooth splined transition that merges tangentially between the outer tube surface and the groove surface. Therefore, the material flows into the groove without any significant deformation of the grain structure. The sharp edges (square rib) show severe elongation and bending of the grains in the transition area. Here, the grain deformation on the chronologically second-rolled edge is far more distinct than the first. This results due to the compression direction of the material flow against the groove edge.

The edge hardness of the formed element rib is measured with a minimum distance of 0.1 mm from the outer surface. The test specimen is taken from an steel tube with a wall thickness reduction of $\Delta s = 2.55$ mm ~~which showed the best filling for all the groove geometries~~. The findings of the filling pattern and the studies of the grain structure reflect the hardness distribution on the edge of the formed rib elements. The smallest rib shows the highest hardness with 270 HV0.1 at the first filled edge, which is the chronologically second-over rolled corner of the rib element. The hardness distribution also shows that the first-filled edge of the rib is subject to the highest compression and therefore has the highest hardness. This also applies to the diagram of the wide rib element form A, shown by way of example, where the first-rolled side is filled first and has a slightly higher hardness (see Fig. 14). For all rib elements, the lowest hardness is seen in the centre of the rib outer surface.

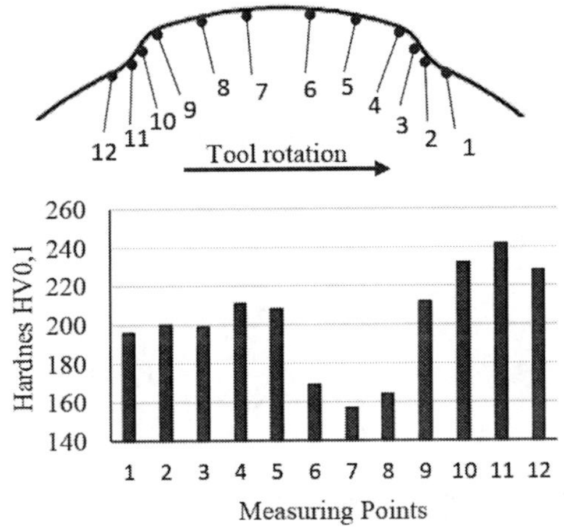

Material: Steel (1.0308)	Tube type: seamless precision
External Ø:48 [mm]	Initial wall thickn.: $s_0 = 4$ [mm]
Roller: Cone (8 [°])	Wall thickn. red.: $\Delta s_1 = 2.55$ [mm]
Edge distance: 0.1 [mm]	Rib form: A

Fig. 14. Edge hardness for rib element form A of an IFT formed steel tube (1.0308) at a wall thickness reduction Δs = 2.55 mm

4 Conclusions

The internal flow-turning process is not only suitable for manufacturing tailored tubes that have a contoured wall thickness in the longitudinal direction and improved mechanical and geometrical properties but is also suitable for producing tubes with a contoured outer surface e.g. with longitudinal ribs.

Experimental results showed that the production of ribs by the IFT process is greatly influenced by the relative feed *f*, the roller geometry and the wall thickness reduction Δ*s*. Here, the wall thickness reduction Δ*s* in a single overrun step gives better groove filling than a big overall wall thickness reduction Δ*s* in several overrun steps. A low attack angle increases the radial material flow which, in addition to an increasing relative feed *f*, ensures a better filling level.

The influence of the element geometry can be divided up into the edge form and the length of the groove effect. A smooth transition facilitates good formability and precise shaping of the groove corners. As the length of the element cross-section increases, a bigger wall thickness reduction Δs is required to achieve complete groove filling, even though the maximum groove height is the same.

Research project P948 / IGF 17753 N was carried out at the Chair of Forming and Machining Technology (LUF), Paderborn University, with technical and scientific steering by the Research Association for Steel Application (FOSTA), Düsseldorf, and was set up by the Federal Ministry for Economic Affairs and Energy following a resolution of the German parliament.

References

1. W. Homberg, T. Rostek, E. Wiens, Internal Flow-turning – an Innovative Technology for the Manufacture of Tailored Tubes, Key Engineering Materials Vol. **639** (2015) pp 65-70
2. W. Homberg, E. Wiens, Internal Flow-Turning – a new approach for the manufacture of tailored tubes with a constant external diameter, Procedia Engineering of the 12th ICTP, Vol. **207** (2017) pp 1755-1760
3. W. Homberg, E. Wiens, Internal Flow-Turning – efficient manufacture of load-adapted tubes with a constant external diameter, conference proceedings of 5th International Conference on Steels in Cars and Trucks, Amsterdam, The Netherlands (2017)
4. M. Sivanandini, S.S. Dhami, B.S. Pabla, Flow Forming Of Tubes – A Review, International Journal of Scientific & Engineering Resurch, Vol. **3**, Issue 5 (2012)
5. C.C. Wong, T.A. Dean, J. Lin, A review of spinning, shear forming and flow forming processes, Int. Journal of Machine Tools & Manufacture **43** (2003) pp 1419-1435
6. C. Dahmen, Fertigungstechnologie und Bauteilverhalten drückgewalzter Innenverzahnungen, PhD Thesis, WZL RWTH Aachen, Shaker Verlag (2005)
7. P. Groche, D. Fritsche, Application and modelling of flow forming manufacturing process for internally geared wheels, Int. Journal of Machine Tools & Manufacture **46** (2006) pp 1261-1265
8. U. Wirtz, DuoBond® is ready for production – New composite material provides stable drives , Compact, pp 28-29, **1** (2013)
9. C. Brinkmann, Drückwalzen Innovativer Werkstoff für ein effizientes Verfahren , Ingenieur forum, pp 8-10, **1** (2013)
10. K.I.E. Ahmed, A new ball set for tube spinning oft hin-walled tubular part with longitudinal inner ribs, Journal of Engineering Sciences, Vol. **39** pp 15-32, (2011)
11. K. H. Heidel, S. Kühmel, Verfahrensoptimierung des Abstreckdrückens mit Kugeln für tiefgezogene Näpfe, PhD Thesis, Karl-Marx-Stadt, (1979)
12. D. Zhao, J. Deng, H. Mao, D. Qian, H. Ou, FE simulation and experiment study on flow forming of inner-splined flange, Procedia Engineering of the 12th ICTP, Vol. **207** (2017) pp 621-626
13. S. Jiang, Y. Zheng, Z. Ren, C. Li, Multi-pass spinning oft hin-walled tubular part with longitudinal inner ribs, Trans Nonferous Met. Soc. China **19** (2009) pp 215-221
14. S. Jiang, Z. Ren, C. Li, K. Xue, Role of ball size in backward ball spinning of thin-walled tubular part with longitudinal inner ribs, Journal of Materials Processing Technology, Vo. **209** (2009) pp 2167-2174
15. E. Wiens, Herstellung innovativer Stahlhalbzeuge mit wanddicke- und festigkeitsveränderlichen Eigenschaften für den Leichtbau durch Innendrückwalzen; final report oft he resurch project P948/IGF 17753 N; 2017

The effect of radial-shear rolling on microstructure and mechanical properties of stainless austenitic steel AISI-321

Abdrakhman Naizabekov[1,]*, *Sergey* Lezhnev[1], *Alexandr* Arbuz[2], *Evgeniy* Panin[2]

[1] Rudny industrial institute, 50 let Oktyabrya, 111500, Rudny, Kostanay region, Kazakhstan
[2] Karaganda state industrial university, Republic avenue 30, 101400, Temirtau, Karaganda region, Kazakhstan

Abstract. Improving the quality of metal products by crushing of the microstructure of material is one of the promising areas of modern metallurgy. The basic idea consists in refinement the grain structure of the material to sizes less than a micron, i.e. the obtaining of ultrafine-grained (UFG) materials, offering higher strength properties of the material under preservation or a small loss of ductility. Stainless austenitic steel AISI 321 is widely used in all the above areas as well as in chemical, vacuum and nuclear technology. For the obtaining of UFG structure in this material the method of radial-shear rolling is used. For the purpose of identifying the influence of radial-shear rolling on microstructure and mechanical properties of stainless austenitic steel AISI-321, the experiment was conducted where at the radial-shear rolling mill SVP-08 at 800 °C in several passes of the workpieces with a diameter of 30 mm rolled till a diameter of 13 mm with following cooling in water. The analysis of the microstructure of deformed samples showed the presence of equiaxed ultrafine-grained structure in the peripheral areas of the workpiece and the presence of elongated fibrous texture in the axial zone. The strength characteristics of the workpiece has increased more than 2 times, with a slight decrease of plasticity.

Keywords: metal forming, rolling, micro structure.

1 Introduction

Most of industrial processes of metal production are oriented to technological effectiveness of deformation process, and necessary level of mechanical properties is obtained by using appropriate steel grades. Generally, after traditional plastic processing metal products have coarse-grained structure. In the meantime, it is known that ultra-fine grain stage of metals and alloys with grain size less than 1 μm and special condition of edges can significantly (2-3 times) increase strength of pure metals and 1,5-2 times increase strength of alloys along with quite high plasticity [1-3].

From all kinds of severe plastic deformation which are used to receive long products with significant changes in microstructure and mechanical properties there is one that should be noted – cross rolling, particularly one of its kinds which is defined by its authors as a separate way called radial-shear rolling (RSR) [4-5]. Its difference from cross rolling (also known as a skew or helical rolling) used, for example, in pipe production [6-7] is that there is rolling of solid bar using three-high mill arrangement with large feed angles.

2 Peculiarities of cross rolling and used equipment

In process of cross rolling, stress state close to all-around pressure with big shear deformations appears in the deformation zone. The main peculiarity of cross rolling is non-monotony and turbulence of deformation; there are also differences in plastic flow and structure elaboration of different bar zones due to trajectory speed features of the process [8]. Because of this features of metal flow the most intensive shear deformations are concentrated in the metal flow lines crossing zone – the cross-section circle common for three-roll scheme. In the outer layer every small trajectory-oriented element is exposed to compression in direction of bar radius, compression in direction of metal flow (along cross rolling trajectory) and, accordingly, tensile strain across the cross rolling trajectory. It is important that there is constant radial gradient of velocity and flow direction which adds more shearing elements into overall complex strain-stress state. Metal structure composition elements exposed to dilatable flow with double-sided sinking strain (along the trajectory and along radius) obtain the form of isotropic insulated high dispersion particles [4].

* Corresponding author: naizabekov57@mail.ru

Speed of particles in axial grain and its length increases proportionately with elongation ratio in the same way as in longitudinal rolling. Cross section of central flow tubes decreases. Metal structure elaboration works in a way similar to longitudinal rolling in multisided grooves or compression. Structural composition elements become longer and thinner, obtaining distinctive structural streaky [9]. These peculiarities are described and illustrated in details in the works of S.P. Galkin [4-5, 8-9]. Based on that works, family of radial-shear rolling mills using intensive plastic deformation for solid round bar rolling were created in Moscow Institute of Steel and Alloys. These devices were used for obtaining fine grain structure of pure magnesium in South Korea [10] and aluminium in Poland [11]. These family of mills include SVP-08 mill, which was purchased to Rudny Industrial Institute. The exterior view and stand scheme of this mill is shown at the Figure 1.

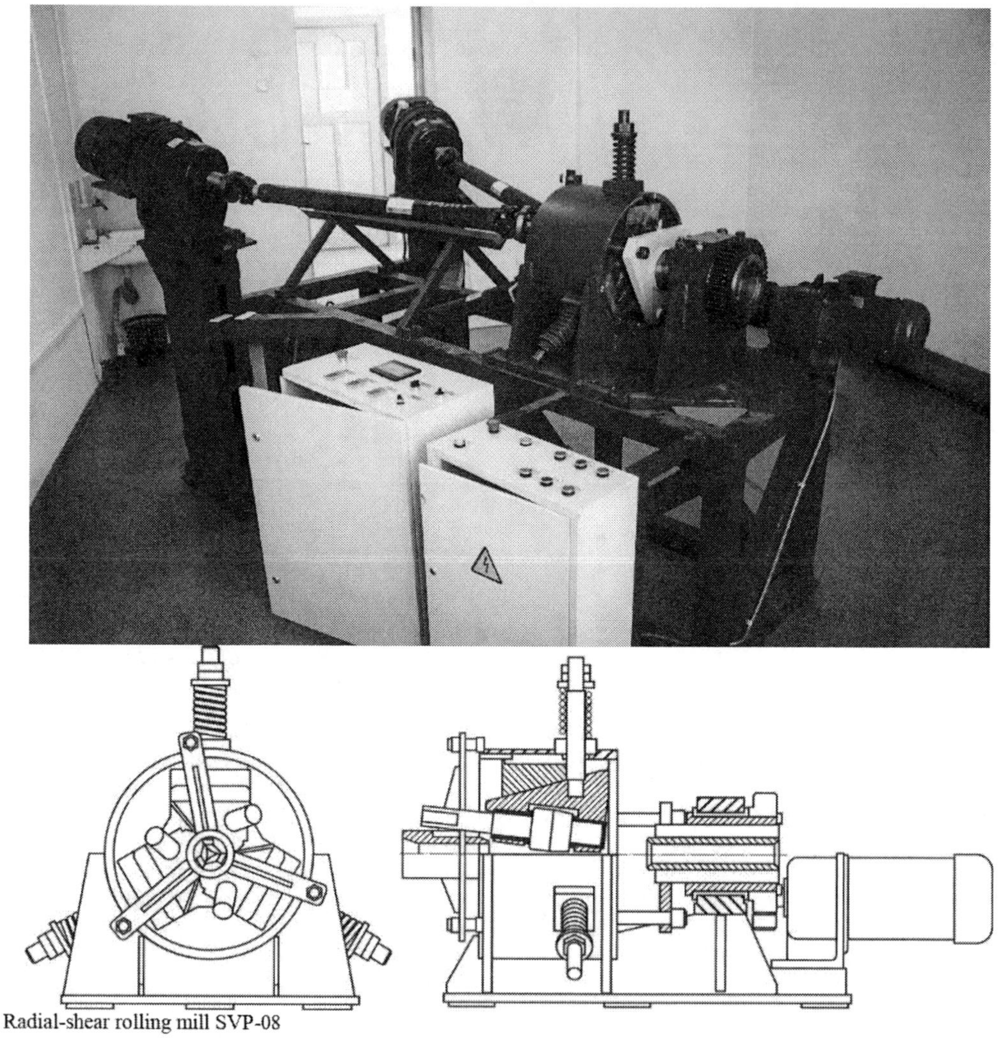

Fig. 1 Radial-shear rolling mill SVP-08

This mill is designed for hot deformation of solid round bars of practically any materials, including low-ductile, continuously cast and powder-metallurgical. Rolling of bars with 10-25 mm diameter is done in three-high mill of special rigid structure from 15-30 mm billets by means of their diametrical pressing in one or several passes using special calibrated rolls and, if necessary, with intermediate heating. Rolls diameter is 56 mm, elongation ratio reaches up to 1,1-5,0; mill capacity is 0,1-0,3 tons per hour; main drives power is 3×7,5 kW [5]. The main feature of the mill of this design, as can be seen from fig 1 scheme, is extremely increased constructions stiffness. This is a favorable factor for experiments to obtain UFG materials, because it requires forming operations with reduced temperature and therefore increased deformation resistance.

This mill was selected for running experiments on looking into impact of cross rolling on steel microstructure because it provides wide range of sizes, rigid structure of the stand and is convenient to use.

3 Experimental part

For the experiment a bar of AISI-321 steel with 30 mm initial diameter was used. Chemical composition of AISI-321 steel - 0,08 % C; 17-19 % Cr; 9-11 % Ni; 2 % Mn; 0,8 % Si; 0,5-0,7 % Ti. It is used for making equipment working in extremely aggressive environment (heat-exchanging units, pipes, parts of furnace and reactor carcass, electrodes of spark ignition plugs).

Rolling temperature was chosen to be constant and equal to 800 ^{0}C. In several passes the billet was rolled from 30 mm to 15 mm with intensive water cooling of the bar. Similar temperature setting was used in works [12-13] for receiving ultra-fine grain structure of stainless steel. After the rolling some slices were cut off from the bar in longitudinal direction which were used to make samples for investigating the structure using transmission electron microscope.

After rolling, cylinders 30 mm in length were cut from the rod. On a high-precision cutting machine Struers AccuTom-5 along the rod samples for mechanical testing were cut, in the form of strips 30x3x0.3 mm (Figure 2 (a), the bottom strip on the right). Dimensions of tensile sample were chosen according to internal standard of Institute of nuclear physics (Almaty, Kazakhstan) because tensile test was provided there and then it is planned to use such samples for neutron irradiation. The central (largest) section was used to prepare TEM samples for studying the fine structure. The places of sampling for microstructure research are shown in Figure 2 (B). The remaining massive half (Figure 2 (a) on the left) was used to measure the microhardness along the cross-section of the rod.

a)

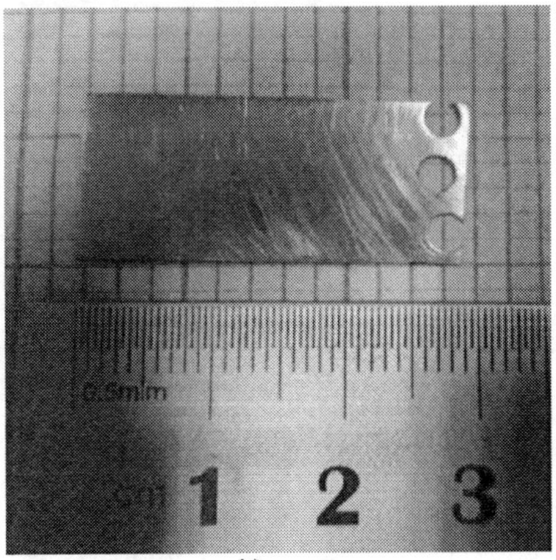

b)

Fig. 2 Sample preparation for tensile testing (a) and for TEM-study (b)

4 Results and discussion

The samples were prepared by electrolytical method (solution composition 600 ml Methanol, 360 ml Butyl cellosolve, 60 ml Perchloric Acid) and examined on a transmission electron microscope JEM-2100 (JEOL, Japan) at an accelerating voltage of 200 kV. Photos of typical microstructure species in the center and on the periphery of the rod are shown in Fig. 3, 4.

The initial structure of AISI-321 steel in the delivery state has a coarse-grained structure with a grain size of about 40-60 μm. After deformation with total strain 11-13 (according to known FEM-models [9, 14]) under favorable stress state, an ultrafine-grained microstructure was obtained in the peripheral part of the bar with a grain size (300-600 nm) comparable with the results of known studies [9, 12-13, 15-17]. In this case, it should be noted that the grains of the peripheral part are equiaxed. The structure of the central zone of the rod is the long and narrow grains that are elongated in the direction of rolling, resembling a texture of conventional rolling.

 https://doi.org/10.1051/matecconf/201819011003

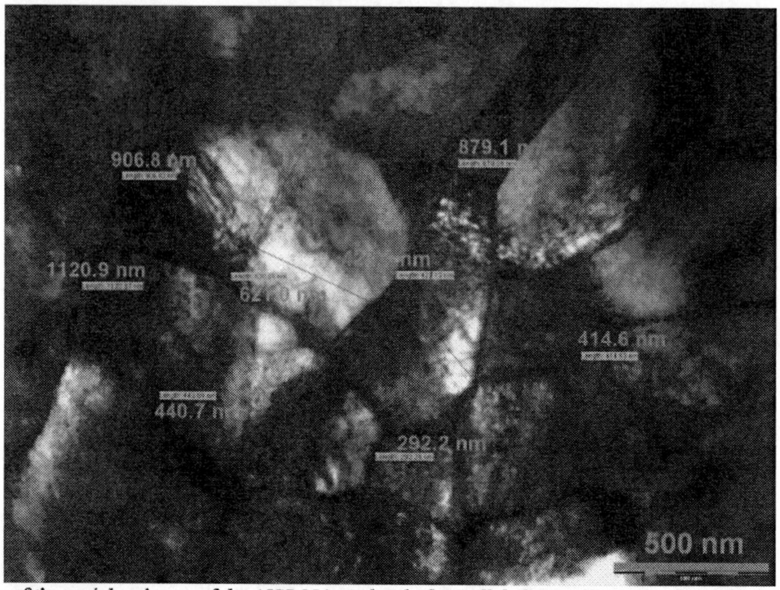

Fig. 3 Microstructure of the peripheral zone of the AISI-321 steel rod after radial-shear rolling

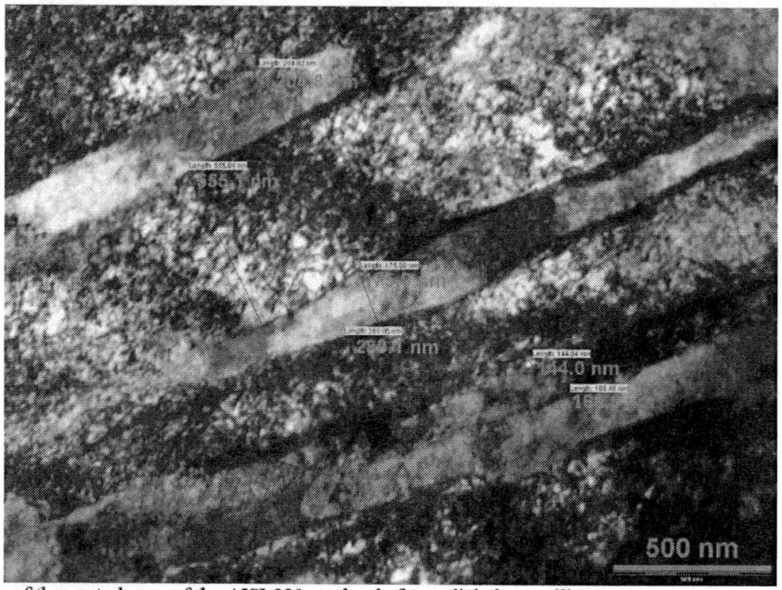

Fig. 4 Microstructure of the central zone of the AISI-321 steel rod after radial-shear rolling

The mechanical properties were determined by tensile testing flat samples on an Instron-1195 test machine. For the tests, strips were selected, by the next rule. The central line of sample located at a distance of 0.5 diameter of the rod, which will allow obtaining the most objective results, taking the detected heterogeneity of the microstructure along the cross-section of the rod. Based on the results of testing 5 samples, the average value of the tensile strength was 1073 MPa, which is approximately 2 times greater than the original value of 580 MPa. The elongation was 21%, which is approximately 2 times less than the initial value of 40%. The reduction in plasticity in this case is within the limits of the norm for materials that have undergone similar processing. One of the stress-strain curve is shown in Figure 5.

The microhardness was measured on a HVS-1000B testing machine. Microhardness testing was providing by Vickers method with a force of 9.87 N for 15 seconds exposition. Taking the heterogeneity of the microstructure, it was decided to investigate the change in microhardness along the cross-section of the rod. For this purpose, measurements were made through each millimeter of the sample cross section, 3 times at each point of the graph. The results for each point were averaged. The microhardness graph in the rod cross section is shown in Figure 6. The initial microhardness is 160 HV. After radial displacement rolling, the microhardness level rose to 288-321 HV and as a whole increased 2 times compared to the initial one. In this case, due to the structural inhomogeneity along the cross-section of the rod, the microhardness level of the central zone of the rod is smoothly reduced by 10.2% (33 HV). This fact according to theoretical expectations and correspond to the received data on microstructure in different zones of bar. Receiving graph suggest that changing of microstructure type across the sample section from equiaxed UFG grains on periphery to elongated fibrous structure of central zone take place also without sharp borders. Small symmetrical decreasing of microhardness level (less than 1 %) on the bar edge probably caused by thermal decarbunization.

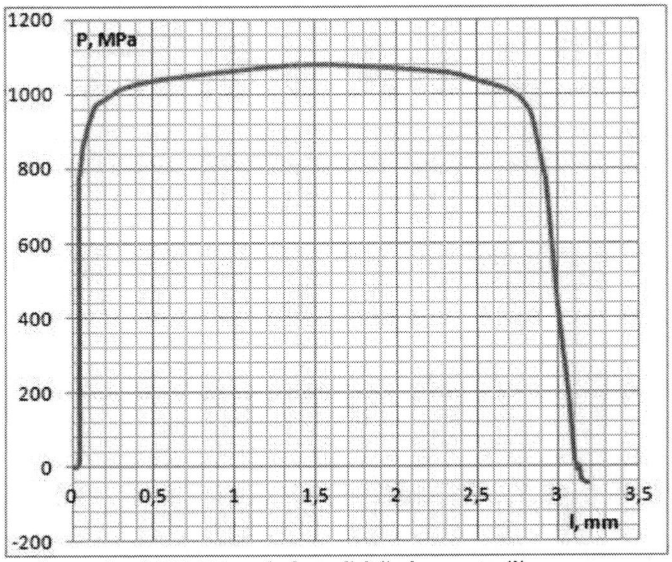

Fig. 5 The stress-strain curve of a sample of AISI-321 steel after radial displacement rolling

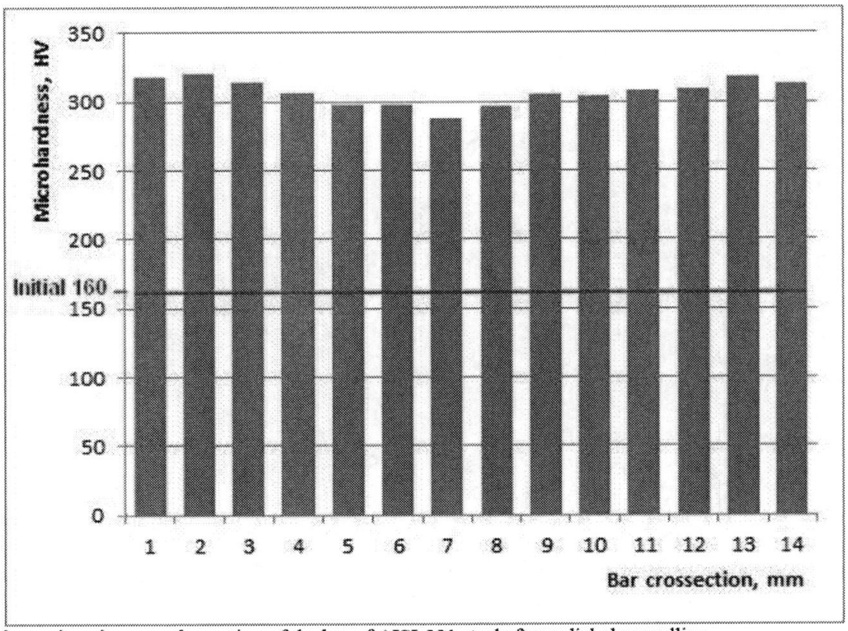

Fig. 6 Microhardness changing over the section of the bar of AISI-321 steel after radial-shear rolling

5 Conclusions

A radial-shear rolling with a total stretching of 4, in a single rod with initial grain size of 60 μm, a microstructure of two different types was obtained. On the periphery of the rod was a more and less equiaxed ultrafine-grained structure with a grain size of 300-600 nm, while in the central zone of the rod there was an oriented texture, similar with conventional rolling.

The obtained microstructure was made first time for this material on the radial-shear mill of SVP-08 design. Results correlate with the data of the other studies on radial-shear rolling for other materials [4-5, 7, 9] and with the data on this material for other severe plastic deformation (SPD) processes conducted at same temperature and deformation conditions [10]. Obtaining such a structure by one of the most common SPD methods – equal channel angular pressing (ECAP), requires at least 6-8 pressing cycles [1, 3, 12, 13] and is possible only for short workpiece, while at the mill of radial displacement rolling it is obtained for 3-4 passes for rods of unlimited length. The problem is in the heterogeneity of the structure in the central and peripheral zones of the rod.

Further improvement of cross rolling ways with purpose of receiving more homogeneous structure in bar cross section will provide an opportunity to get large amounts of ultrafine-grain materials with the least time and energy consumption, which will make commercial efficiency and cheapening of UFG materials production available.

ACKNOWLEDGMENT

Work is performed under the state budget-funded theme № AP05131382 "Research and development of the technology for obtaining ultrafine-grained materials with improved mechanical properties and increased radiation resistance for their use as materials of the first wall of thermonuclear reactors and in nuclear power engineering" with the program "Grant financing of scientific researches on 2018-2020" (Customer - the Ministry of education and science of the Republic of Kazakhstan).

References

1. R. Valiev, R. Islamgaliev, I. Alexandrov. Prog Mater Sci **45**, **2**, 103-189 (2000)
2. C.C. Koch, I.A. Ovid'ko, S. Seal, S. Veprek. *Structural nanocrystalline materials: Fundamentals and applications* (Cambridge University Press, 2007)
3. M. Furukawa, Z. Horita M. Nemoto, T.G. Langdon, Journal of Materials Science, **36 (12)**, 2835-2843
4. S.P. Galkin, Patent RF № 2293619, (2007)
5. S.P. Galkin, Steel in Translation, **44-1**, 61-64 (2014).
6. S. Li, W. Meng, L.Hu, B. Ding, Advanced Materials Research, **145**, 238-242 (2011)
7. H. Dyja, A. Kawalek, K. Ozhmegov, S. Sawicki, K. Laber, *METAL 2017*, **2017-January**, 1701-1707 (2017)
8. S.P. Galkin, Steel in Translation **34-7**, 57-60 (2004).
9. N.V. Lopatin, G.A. Salishchev, S.P. Galkin, Russ J Non Ferr Met, **52-5**, 442-447 (2011)
10. M. Diez, H.-E. Kim, V. Serebryany, S. Dobatkin, Y. Estrin, Materials Science and Engineering A, **612**, 287-292 (2014)
11. A. Stefanik, P. Szota, S. Mróz, H. Dyja, Sol. State Phenomena **220-221**, 892-897 (2015)
12. O.V. Rybalchenko, *Influence of severe plastic deformation on structure, mechanical and service properties of steel 08X18H10T* (Dissertation, Moscow, 2014)
13. C.X Huang, G.Yang, B.Deng, S.D. Wu, S.X. Li, Z.F. Zhang, Philosophical Magazine **87-31**, 4949-4971 (2007)
14. C.-K. Shin, C. Hung, R.-Q.Hsu, Journal of Materials Processing Technology **113-1-3**, 115-123 (2001)
15. A.B. Naizabekov, S.N. Lezhnev, H. Dyja, T. Bajor, K. Tsay, A. Arbuz, N. Gusseynov, R. Nemkaeva, Metalurgija **56-1-2**, 199-202 (2017)
16. K. Tsay, A. Arbuz, N. Gusseynov, R. Nemkaeva, N. Ospanov, I. Krupenkin, J. Chem. Technol. Metall., **51-4**, 385-392 (2016)
17. B. Radiguet, A. Etienne, P. Pareige, X. Sauvage, R. Valiev, Journal of Materials Science **43-23-24**, 7338-7343 (2008)

Analysis of deformation characteristics and microstructure variation of thick plate during asymmetric shear rolling

Tao Zhang[1,*], *Lei* Li[1], and *Shihong* Lu[1]

[1]College of Mechanic and Electrical Engineering, Nanjing University of Aeronautics and Astronautics, 210016 Nanjing, China

Abstract. Aluminium alloy thick plate is widely used in field of aerospace and automobile industry for its light weight, high strength, good corrosion resistance and formability. Hot rolling is the key process of preparation for thick plates, however, the conventional symmetrical rolling process causes large deformation on the surface and small deformation in the central portion of the thick plate, resulting in inhomogeneous deformation, microstructure and property distributions through the thickness direction of the plate. Asynchronous rolling is beneficial for deformation permeation into the central part of thick plate with introduction of strong shear strain due to velocity asymmetry of two work rolls. However, the plate will bend towards the slower roll due to deformation asymmetry at upper and lower surface of the plate and serious bending is bad for multi-pass rolling process. Asymmetrical shear rolling with velocity asymmetry and geometry asymmetry is adopted in this study. Compared to asynchronous rolling, there is a horizontal offset distance of upper roll in outlet direction. A moment will be applied to the plate in the opposite direction of bending towards the slower roll by an offset distance of the slower roll. The comparisons of distributions of equivalent strain, temperature, DRX fraction and average grain size between symmetrical rolling and asymmetrical shear rolling are studied. The effects of velocity asymmetry and offset distance on microstructure variation and microstructure distribution at different positions of the plate are studied. Due to larger velocity of lower work roll, larger equivalent strain and higher temperature appear at lower layer of the plate, which results in higher DRX fraction and finer grain size. The speed ratio contributes greatly to sufficient DRX process and finer grain size; while the ascending offset distance decreases the DRX fraction. Therefore, speed ratio plays significant role in deformation permeation and grain refinement in asymmetrical shear rolling and the offset distance is mainly used to decrease the bending behaviour of the plate. This study can provide guidance to thick plate preparation with good properties in asymmetrical rolling process.

Keywords: Rolling、Deformation、Microstructure

1 Introduction

With the development of aerospace industry, aluminum alloy thick plates with large thickness and high performance (even to 250 mm and 700 MPa) are in great need. Hot rolling is the main process for thick plate preparation. However, large deformation is mainly generated on the surface of the plate in symmetrical rolling and the properties at the centre are poorer than that on the surface, which cause severe property inhomogeneities. Differential speed rolling is useful for deformation permeation into the centre of the thick plate by inducing strong shear strain. However, there is severe bending behaviour of the plate as the deformation between upper and lower surface of the plate is uneven, which may cause accidents in production process. Therefore, a new method of asymmetrical shear rolling is adopted to solve this problem [1]. The bending behaviour can be effectively controlled with the basis of obtainment of strong shear deformation.

Finite element model (FEM) was widely used to study the deformation distribution [2] and temperature variation [3-4] during asymmetrical hot rolling. Serajzadeh [5] studied the strain inhomogeneity in asymmetrical hot strip rolling through FEM simulation. Zuo [6] analyzed the shear strain quantificationally of pure Al in asymmetric rolling by experiments and observed the grain refinement after strong shear deformation. Ding [7] studied the temperature distribution of Mg alloy and the temperature variation at different positions of the plate in the hot rolling by experiment and FEM. In hot rolling process, there are different deformation mechanisms, such as work hardening (WH), dynamic recovery (DR) and dynamic recrystallization (DRX) [8]. Yin [9] studied the mechanism of continuous DRX during hot compressive deformation by experiments. Lin [10] establishend a

* Corresponding author: 297zhangtao@nuaa.edu.cn

numerical model to simulate the microstructure evolution in a three-dimensional hot upsetting process and conducted experiments to verigy the simulation results. Cellular automata (CA) method is a new method to simulate the microstructure variation due to DRX and it can be used to obtain the microstructure topology [11]. Zheng [12] used CA model to simulate the austenite recrystallization during multi-pass steel strip hot rolling and established a model to predict the microstructure variation. Li [13] studied the relationship between mechanical properties and microstructure of AA1060 aluminum plate during snake rolling.

However, the comparisons of strain and temperature between symmetrical rolling and asymmetrical shear rolling were rarely published as there are velocity asymmetry and geometry asymmetry in asymmetrical shear rolling. Most studies focused on the effect of speed ratio on deformation distribution of the plate. Furthermore, the effects of speed ratio and offset distance on microstructure evolution in asymmetrical shear rolling need to be further studied. In this study, coupled numerical models are established to compare the distributions of equivalent strain, temperature, DRX fraction and average grain size between symmetrical rolling and asymmetrical shear rolling. CA models are established to simulate the microstructure topology and grain variation at different parameters due to DRX process.

2 Asymmetrical shear rolling

Fig.1 shows the diagrams of symmetrical rolling and asymmetrical shear rolling. In symmetrical rolling, the deformation zone is divided into backward slip zone (BSZ) and forward slip zone (FSZ) as the velocities of two work rolls are the same. However, in asymmetrical shear rolling, the velocity of lower work roll is larger than that of upper lower roll and there is an offset distance (*S*) in outlet direction of two rolls. Therefore, a new zone of cross shear, the so called cross shear zone (CSZ) appears as the neutral points of the upper and lower surface are not in the same verticle position. In CSZ, the frictional direction is the same at the upper and lower surface, which is different from that in BSZ and FSZ. As a result, the plate is subjected to compressive deformation and strong shear deformation in CSZ, which is beneficial for deformation permeation of thick plate. The plate will bend up for the smaller velocity of upper roll. In asymmetrical shear rolling, the offset distance of the upper roll will bring a moment in the oppsite direction of bending up, which is useful to decrease the bending behaviour of the plate.

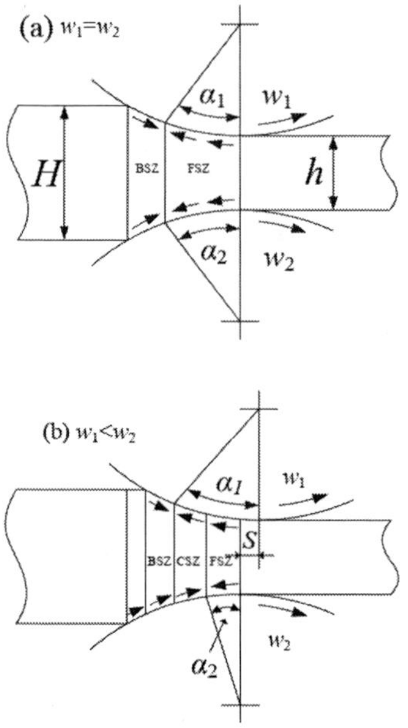

Fig. 1. Schematic of different rolling methods: (a) symmetrical rolling (b) asymmetrical shear rolling

3 Model descriptions

3.1. Coupled numerical model

In order to simulate the deformation distribution of the plate in symmetrical rolling and asymmetrical shear rolling, single pass rolling FEM models are established. The work rolls are defined as rigid bodies and the plate is 7055 aluminum alloy and its material constants and constitutive equation are shown in author's previous article [14]. The rolling parameters of asymmetrical shear rolling are shown in Table 1; while the speed ratio is set 1 and offset distance is set 0 in symmetrical rolling.

In hot rolling, DRX occurs when the true strain reaches the critical value and DRX plays a vital role in grain refinement as new dynamic recrystallized grains are formed at grain boundary of the matrix. DRX fraction, recrystallized grain size and average grain size were widely used to describe the DRX behaviour by Sellars and Yada [15-16], as shown in Eqs. (1)-(6).

$$\varepsilon_p = 1.697\times10^{-5}\dot{\varepsilon}^{0.15975}\exp(\frac{42448}{RT}) \quad (1)$$

$$\varepsilon_c = 0.8\varepsilon_p \quad (2)$$

$$X_d = 1-\exp\left[-3.4\times10^{-2}(\frac{\varepsilon-\varepsilon_c}{\varepsilon_{0.5}})^{1.065}\right] \quad (3)$$

$$\varepsilon_{0.5} = 7.57\times10^{-4}\dot{\varepsilon}^{0.09349}\exp\left(\frac{25707}{RT}\right) \tag{4}$$

$$D_d = 1.0473\times10^{3}\dot{\varepsilon}^{0.129}\exp(\frac{-21542}{RT}) \tag{5}$$

$$\overline{D} = X_d D_d + (1-X_d)D_0 \tag{6}$$

Where ε_p is peak strain, ε_c is critical strain, X_d is DRX fraction, $\varepsilon_{0.5}$ is the strain corresponding to 50 % DRX fraction, D_d is recrystallized grain size, D_0 is initial grain size, $\overline{D}$ is average grain size after deformation.

Table 1. Single rolling parameters

Parameter	Value
Initial rolling temperature, T /℃	410
Work roll radius, R /mm	500
Shear friction coefficient, m	0.4
Velocity of lower roll, ω_2/(rad/s)	2~2.6
Velocity of upper roll, ω_1/(rad/s)	2
Initial thickness, H /mm	250
Pass reduction, Δh /mm	30，50
Speed ratio, $i=\omega_2/\omega_1$	1.00~1.3
Offset distance, S /mm	0~70
Heat transfer of convection and radiation, h_1 / (W·m^{2}·K^{-1})	5
Heat transfer of contact and cooling, h_2 /(W·m^{2}·K^{-1})	30000

3.2 Cellular automata model

CA model coupled with mathematical random algorithm and physical metallurgy theory is used to simulate the microstructure evolution. The plate is divided into cells discreted in time and space and rules can be added to simulate the cell change in each CA step. There are several parameters in each cell, such as crystal orientation, dislocation density, temperature, strain rate and recrystallized flag.

The dislocation density increases with strain by WH and decreases at high temperature by DR and DRX, which can be described by Kocks-Meching (KM) model [17] in Eq. (7). When the dislocation density reaches a critical value, new dynamic recrystallized nuclei appear at grain boundary. The nucleation rate can be described by Ding and Guo model [18], as shown in Eq. (8). After the nucleation, the nuclei will grow into recrystallized grains with certain sizes and the dislocation density difference between recrystallized grains and matrix provides the driving force for the growth. The grain growth velocity is calculated by Eqs. (9)-(10).

$$\frac{d\rho}{d\varepsilon} = k_1\sqrt{\rho} - k_2\rho \tag{7}$$

$$\dot{n} = C\dot{\varepsilon}^{m}\exp(\frac{-Q_{act}}{RT}) \tag{8}$$

$$v = Mp \tag{9}$$

$$M = \frac{b}{kT}\lambda D_b \exp\left(\frac{-Q_b}{RT}\right) \tag{10}$$

Where ρ is the dislocation density, k_1 and k_2 are work hardening and dynamic softening coefficient, respectively; $\dot{n}$ is nucleation rate, m is the sensitivity coefficient of strain rate, Q_{act} is active energy,T is temperature, v is grain growth velocity, M is grain boundary (GB) mobility, p is pressure, b is the modulus of Burger's vector, k is Boltzmann constant, λ is the thickness of characteristic GB, D_b is the boundary self-diffusion coefficient.

4 Results and discussion

4.1. Comparisons of two rolling technologies

In asymmetrical shear rolling, the speed ratio and offset distance will change the deformation zones, and then change the deformation and temperature distribution of the plate. Fig.2 shows the comparison of equivalent strain and temperature of two rolling ways. The equivalent strain distribution is symmetrical and it is near to "M" shape in symmetrical rolling. Due to the contact friction force, the strain at the surface is smaller than that at sub-surface (nearly 1/8 from surface). However, the equivalent strain at lower layer of the plate is much larger than that at upper layer and the strain distribution is obvious asymmetrical in asymmetrical shear rolling. The metal flow velocity at lower layer is larger as the lower roll has a faster velocity. In asymmetrical shear rolling, the maximum strain at lower layer is 15.1 % larger than that in symmetrical rolling, while the strain at upper layer is 8.9 % smaller. Due to strong shear strain in CSR, the equivalent strain in asymmetrical shear rolling is larger than in symmetrical rolling, which illustrates that speed ratio is beneficial for deformation permeation. The temperature distribution is also asymmetrical and the temperature at lower layer is higher than the upper layer. The temperatures at both surfaces decrease due to heat exchange with environment and work rolls. The temperature variation is related to plastic deformation heat and heat exchange, especially for the contact heat exchange. In asymmetrical shear rolling, the heat generated by plastic deformation at lower layer is larger than in symmetrical rolling; meanwhile, the contact time is shorter for the larger velocity of lower roll. More plastic deformation heat and less heat exchange results in a higher temperature at lower layer.

Large strain and high temperature are beneficial to DRX process and grain refinement in hot rolling. Fig.3 shows the comparison of DRX fraction and average grain size of two rolling ways. The DRX fraction distribution is similar to that of equivalent strain. The DRX fraction at lower sub-surface is 32 % and it is much larger than upper sub-surface of 25 % due to its large deformation and high temperature. As a result, the average grain size reduces from 100 μm (initial value) to 80.4 μm at lower sub-surface. As for the centre point of the plate, the DRX fraction increase and the average grain size decreases slightly in asymmetrical shear rolling.

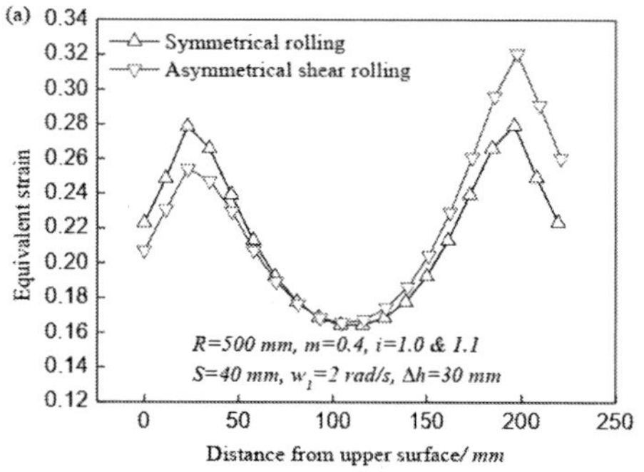

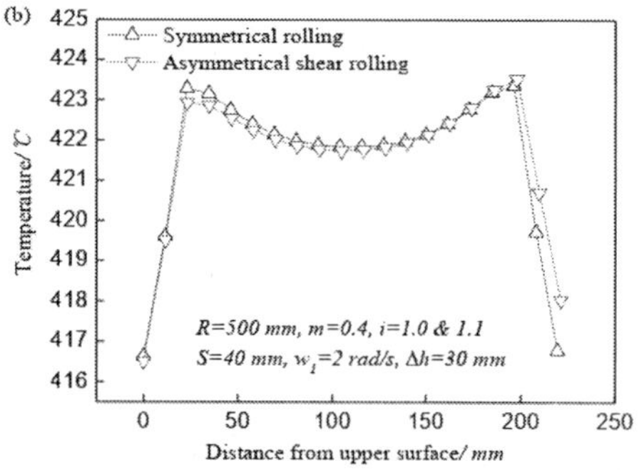

Fig. 2. Comparisons of (a) equivalent strain (b) temperature between two rolling ways

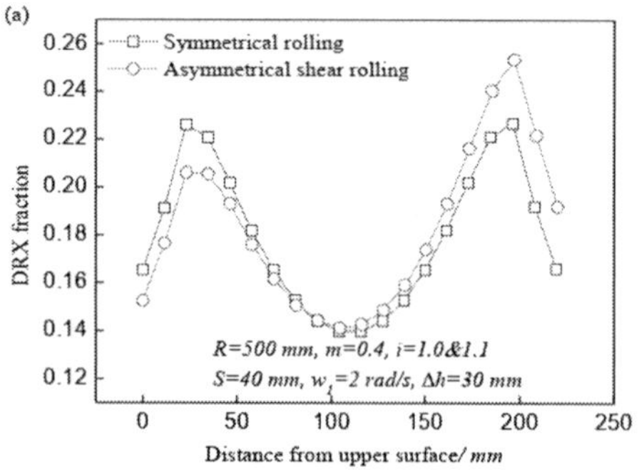

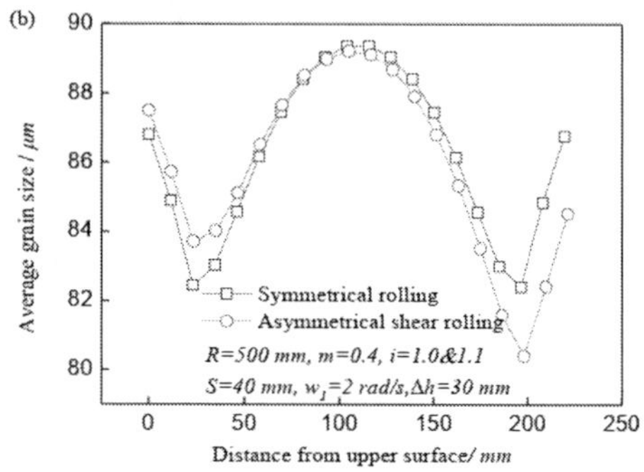

Fig. 3. Comparisons of (a) DRX fraction (b) average grain size between two rolling ways

4.2 Microstructure visualization

The microstructure visualization can be achieved in CA models and Fig. 4 shows CA simulated microstructure evolution at different strains under temperature of 450 °C and strain rate of 0.01 s^{-1}. The white region represents matrix and the colored region is recrystallized grain, different colors represent grains with different orientations. With increase of strain, the nuclei start to appear at grain boundary for its high energy and then grow into recrystallized grains with a certain size. Finally, the recrystallized grains grow into a stable size and will not grow with further increase of strain. The deformation energy inside the material increases sharply with ascending strain, which provides sufficient driving force for the growth. The nuclei grow into recrystallized grains by absorbing grains of the matrix. The dislocation densities of the new recrystallized grains and the matrix both increase with ascending strain. When they reach the saturation values, the dislocation density difference between the matrix and recrystallized grains will decrease to zero, in which the new grains will stop growing up. It is obvious that the DRX fraction increases with ascending strain, as a result, the grains are significantly refined.

Fig.5 shows microstructure variation at different deformation zones at center point of plate under i=1.1 and S=40 mm in asymmetrical shear rolling. It is obvious that the strain increases from BSZ to FSZ, which can be seen from the shape of the matrix grains. A few nuclei start to appear under small deformation in BSZ because the plate is just bited into the deformation zone. The grains are almost equiaxed in BSZ. The DRX fraction and the recrystallized grain size increase distinctly in CSR. The strong shear deformation contributes greatly to strain enlargement; meanwhile, the temperature rises quickly due to large plastic deformation heat, which obviously decrease the critical strain for DRX and enhances thermal motion of molecules and atoms inside the material. The DRX fraction is 1.8 % and the average grain size is 100 μm in Fig.5(a), while they change to 10 % and 93.2 μm in Fig.5(b). In FSZ, there is a slight increase of DRX fraction and then the grain can be further refined.

 https://doi.org/10.1051/matecconf/201819011004

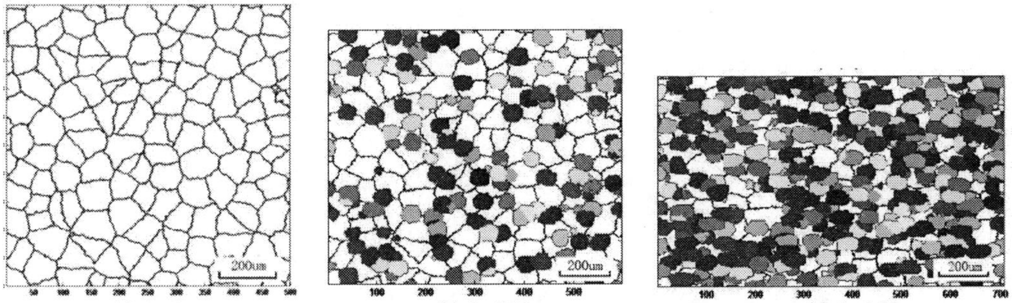

(a) ε =0 (b) ε =0.19 (c) ε =0.35

Fig. 4. Microstructure evolution at different strains under temperature of 450 °C and strain rate of 0.01 s^{-1}

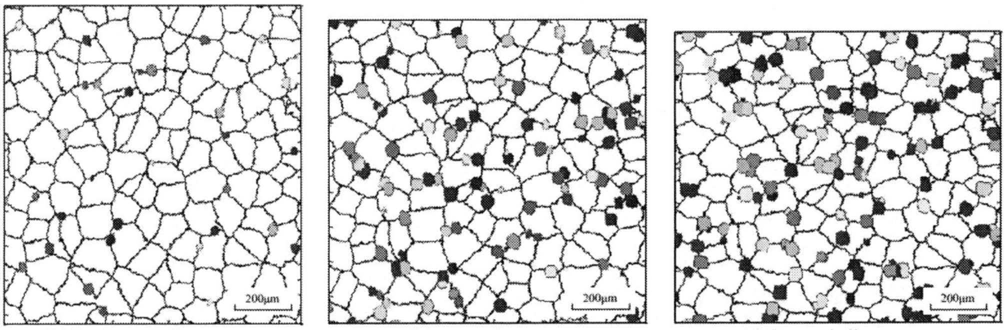

(a) backward slip zone (b) cross shear zone (c) forward slip zone

Fig. 5. Microstructure variation at different deformation zones at center point of plate under i=1.1 and S=40 mm

The speed ratio and offset distance are important parameters in asymmetrical rolling. Fig. 6 shows the effect of speed ratio on DRX fraction and average grain size. The length of CSR and shear strain increase dramatically with ascending speed ratio [19]. The DRX fraction increases sharply and the average grain size decreases when the speed ratio changes from 1.0 to 1.2. Due to increase of velocity difference, the equivalent strain and temperature both increase obviously, which provide more deformation energy for DRX process and reduce the critical strain for occurence of DRX. As recrystallized grain size is much smaller than initial grain size, the increasing DRX fraction results in severe grain refinement. However, the equivalent strain almost keeps steady by further increase of speed ratio from 1.2 to 1.3 [20]. Therefore, there are slight variations of DRX fraction and average grain size.

The effect of offset distance on DRX fraction and average grain size is shown in Fig.7. The DRX fraction decreases and average grain size increases with ascending offset distance. For a given pass reduction, the existence of offset distance will decrease the actual rolling reduction, which reduces the equivalent strain of the plate. Therefore, the offset distance is mainly used to decrease the bending behavior of the plate during asymmetrical shear rolling and it should be properly adjusted according to the value of speed ratio.

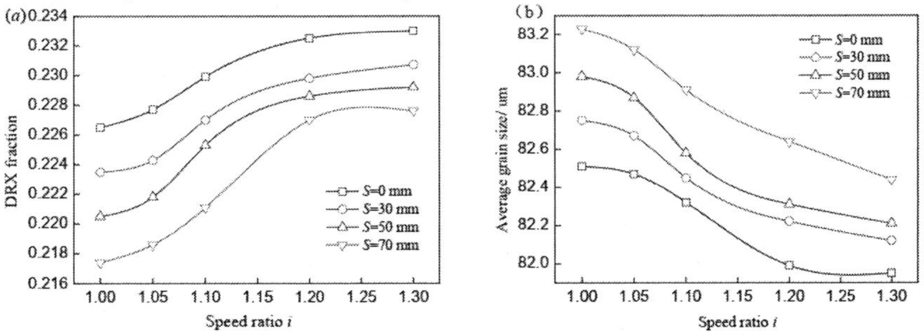

Fig. 6. Effect of speed ratio on DRX fraction and average grain size at center point of plate

* Corresponding author: 297zhangtao@nuaa.edu.cn

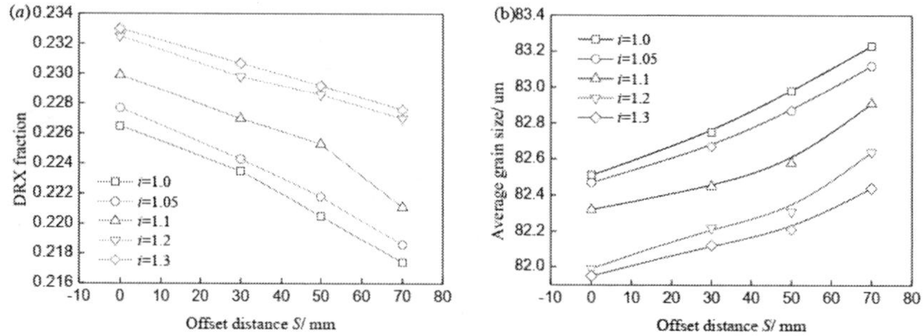

Fig. 7. Effect of offset distance on DRX fraction and average grain size at center point of plate

Conclusions

Coupled numerical models are established to compare the distributions of equivalent strain, temperature, DRX fraction and average grain size between symmetrical rolling and asymmetrical shear rolling. CA models are established to simulate the microstructure topology variation of the plate due to DRX process in hot rolling.

1. The distributions of equivalent strain, temperature are asymmetric in asymmetrical shear rolling. The strain and temperature at lower layer of the plate is higher due to faster velocity of lower work roll.

2. More sufficient DRX and finer grain appear at lower layer of the plate; meanwhile, the grain size at the centre is smaller in single asymmetrical shear rolling, which illustrates that asymmetrical shear rolling is potential for grain refinement.

3. Increasing speed ratio is beneficial to enlarge DRX fraction and decrease average grain size; while the ascending offset distance reduce the DRX extent. Therefore, speed ratio is the vital factor for grain refinement and offset distance is mainly used for bending control of the plate.

Acknowledgments: This project is supported by National Natural Science Foundation of China (Grant No.51705248); Natural Science Foundation of Jiangsu pvovince, China (Grant No.BK20170785) and Scientific Research Staring Foundation for Talent Introduction of Nanjing University of Aeronautics and Astronautics (Grant No. 90YAH17038)

References

1. Y. Fu, S.S. Xie, B.Q. Xiong, G.J. Huang, L. Cheng, J. Wuhan. Univ. Technol. **27**(02): 247-251 (2012).
2. S. Serajzadeh, A. K. Taheri, M. Nejati, J. Izadi, M. Fattahi, J. Mater. Process. Tech. **128**(1): 88-99 (2002).
3. M. Bagheripoor, H. Bisadi, Appl. Therm. Eng. **31**(10): 1556-1565 (2011).
4. R. B. Mei, C. S. Li, X. H. Liu, B. Han, J. Iron. Steel. Res. Int. **17**(2): 17-21 (2010).
5. H. Yang, M. Wang, L.G. Guo, Z.C. Sun, Comp. Mater. Sci. **44**(2): 611-621(2008).
6. F.Q. Zuo, J.H. Jiang, A.D. Shan, J.M. Fang, X.Y. Zhang, T. Nonferr. Metal. Soc. **18**(4): 774–777 (2008).
7. Y. Ding, Q. Zhu, Q. Le, Z. Zhang, L. Bao, J. Cui, J. Mater. Process. Tech. **225**: 286-294 (2015).
8. C. Shi, J. Lai, X.G. Chen, Materials. **7**(1), 244 (2014).
9. X. Q. Yin, C. H. Park, Y. F. Li, J. Alloy. Compd. **693**:426-431(2017).
10. Y. C. Lin; M. S. Chen, J. Mater. Process. Tech. **209**(9):4578-4583 (2009).
11. M. Sitko, M. Pietrzyk, L. Madej, J. Comput. Sci. **16**: 98-113 (2016).
12. C. Zheng, N. Xiao, D. Li, Comp. Mater. Sci. **44**(2), 507-514 (2008).
13. S. Li, N. Qin, J. Liu, Mater. Design. **90**: 1010-1017 (2016).
14. T. Zhang, Y. X. Wu, H. Gong, Mechanika. **22**(5):359-365 (2016).
15. C. M. Sellars, J. A. Whiteman, Met. Sci. **13**(3-4):187-194 (1979).
16. T. Senuma, H. Yada, H. Yoshimura, Adv. Mater. Res. **626**(6):430-435 (1986).
17. Y. Liu, Y. Lin, H. Li, D. Wen, X. Chen, M. Chen, Mat. Sci. Eng. A-Struct. **626**: 432-440 (2015).
18. R. Ding, Z.X. Guo, Acta. Mater. **49**(16): 3163-3175 (2001).
19. T. Zhang, Y. X. Wu, H. Gong, J. Cent. South Univ. **24**(2):296-302 (2017).
20. Zhang T, Wu Y X, Gong H, J. Cent. South Univ. **22**: 2463-2469 (2015).

Investigation of a plain ball burnishing process on differently machined Aluminium EN AW 2007 surfaces

Marco Posdzich[1], *Rico* Stöckmann[1], *Florian* Morczinek[1], and *Matthias* Putz[2]

[1]Chemnitz University of Technology, Institute for Machine Tools and Production Processes (IWP), Reichenhainer Str. 70, 09126 Chemnitz, Germany
[2]Fraunhofer Institute for Machine Tools and Forming Technology IWU, Reichenhainer Str. 88, 09126 Chemnitz, Germany

Abstract. Burnishing is an effective chipless finishing process for improving workpiece properties: hardness, vibration resistance and surface quality. The application of this technology is limited to rotationally symmetrical structures of deformable metals. Because of the multiaxial characteristics, the transfer of this force controlled technology on to prismatic shapes requires a comprehensive process development. The main purpose of this paper is the characterization of a plain burnishing process on aluminium EN AW 2007 with a linear moved, spherical diamond tool. The method of design of experiments was used to investigate the influence of different machined surfaces in conjunction with process parameters: burnishing force, burnishing direction, path distance and burnishing speed. FEM simulation was utilized for strain and stress analysis. The experiments show, that unlike the process parameters the initial surface roughness as 3rd order shape deviation does not have a significant influence on the finished surface. Furthermore a completely new surface is created by the process, with properties independent from the initial surface roughness.

Keywords: Rolling, Micro forming, Free forming

1 Introduction

Burnishing is a force controlled forming process, which is widely applied on finishing of rotationally symmetrical parts e.g. found in the automotive industry. A burnishing tool is pressed under a force onto the workpiece surface for this finishing process. As seen in Fig. 1 a tangential sliding or rolling of the tool deforms the material at the surface, which results in a reduction of the roughness [1]. At the same time an increase in compressive residual stress at the boundary layer is achieved, which leads to an increase of surface hardness and strength [2]. With this procedure a surface quality similar to the results after grinding for ductile materials can be reached [3].

In comparison to other finishing processes burnishing has the advantage that it can be applied on conventional turning and milling machines without great investments. Furthermore a variety of different surface structures, which are only limited by the geometry of the burnishing tool and the machine kinematics, can be processed.

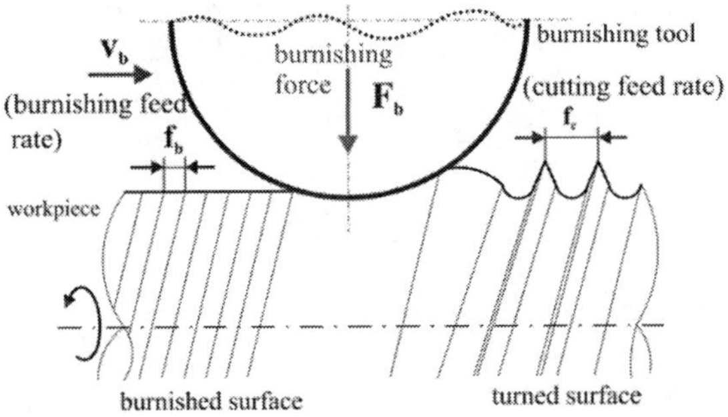

Fig. 1. Schematics of a burnishing process on a turned surface

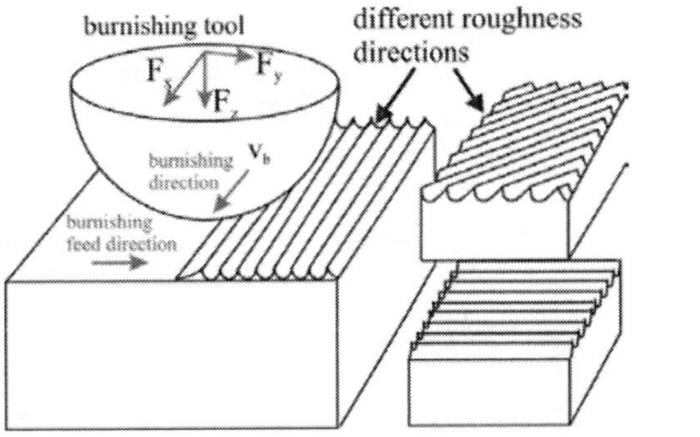

Fig. 2. Schematics of a burnishing process on a plain surface

* Corresponding author: marco.posdzich@mb.tu-chemnitz.de

Additionally it is a chipless process in which only a negligibly small amount of heat is introduced into the workpiece. As seen in Fig. 2 the process can be adapted from rotationally symmetrical to prismatic parts, because of its potentially high flexibility and the produced positive material properties [4].

Because of the different kinematics of turning in comparison to milling the direction of the roughness-grooves is not constant [5].Therefore the influence of the different parameters of the burnishing and the structure of the pre-machined surface will be studied.

The aim of presented investigations is to show and interpret the dependencies between the surface structure before and after the burnishing process and the influence of the process forces. Because residual stresses are not easily measurable the finite element method (FEM) is often utilized for stress and strain analysis in burnishing [6]. FEM studies of the burnishing process can be done in 2D [7] or 3D [8], but do not necessarily lead to the same results [9]. In this paper a 2D-FEM was applied to analyse stresses during and after the burnishing process.

2 Experimental Investigations

A tool concept for a 3-axis milling machine was developed for the investigation of the dependency between the surface structure before and after the burnishing process. The main components of the burnishing tool are a spherical dome made out of diamond with a diameter of 10 mm and a spring. The spring is loaded during the process to control the necessary burnishing force. The axis of the burnishing tool was equal to the rotation axis of the main spindle. The workpiece was set on top of a 4 component dynamometer of the type 9272 to measure the forces during the burnishing process. In addition a hydraulic overload protection was used under the force measurement system. As workpiece material the aluminium alloy 3.1645 (EN AW 2007) with a yield strength of $R_{p0.2} = 220$ MPa and a tensile strength of $R_m = 340$ MPa was chosen. The experiments were done utilizing design of experiments consisting of a fractional factorial screening experiment and a t-test. Therefore various factors were investigated. The feed of the cutting process f_c was varied to create specific surfaces of the specimen as input for the burnishing process. The kinematic roughness Rt can be calculated and adjusted with the cutting edge corner radius r_ε of the cutting tool by using the Eq. 1.

$$Rt = f_c^2/(8r_\varepsilon) \qquad (1)$$

During the experiments the kinematic pre-roughness only depends on f_c, because r_ε was kept constant with a value of 0.4 mm. The angle γ_b describes the angle between the direction of the pre-machining process (cutting) and the finishing process (burnishing). This factor has been chosen to check whether the direction of the pre-machining has an influence on the burnished surface. In addition the effects of the used lubricant strategy on the burnishing process forces F_z were investigated. An increasing forming speed leads to an increasing yield stress and therefore the burnishing speed v_b was varied to investigate its effects. The burnishing feed f_b describes the distance of the parallel burnishing paths. As seen in Eq. 1 the feed has an influence on the kinematic roughness and therefore is used as a factor. The burnishing infeed a_{pb} represents the amount whereby the tool is set in z-direction. This causes the spring in the burnishing tool to be loaded and ultimately controls the burnishing forces. The forces in the main directions of the Cartesian coordinate system were measured. Those three measured variables are the force in machining direction F_x, the force in burnishing feed direction F_y and the force F_z which works perpendicular to the surface. The burnishing infeed a_{pb} was to 0.1, 1.5 and 2 mm resulting in burnishing forces F_z of 170, 320 and 400 N. Further measurands are the surface-parameters Ra and Rz, which are specified in DIN EN ISO 4287. All process parameters are listed in Table 1. The pre-tests were done with a full combination the 1st and 2nd factor for all process parameters. On the basis of the pre-test, additional investigations with focus on burnishing force F_z, burnishing feed f_b and angle between cutting and burnishing γ_b were extended by an additional 3rd factor. Those investigations were done with the 1st factor for the process parameters f_c, lubricant and v_b. The other parameters were varied according to Table 1.

Table 1. Factor combinations for the pre- and the main tests

factor	1st	2nd	3rd
f_c	0.36 mm	0.18 mm	-
angle γ_b	0°	90°	45°
lubricant	dry	emulsion	-
v_b	2 m/min	20 m/min	-
f_b	0.5 mm	0.125 mm	0.25 mm
F_z	170 N	320 N	400 N

First the surface of the specimen was machined via a cutting process using a loop motion, seen in Fig. 3. A defined and constant surface structure was created and a surface parallel to the burnishing process was ensured by using a constant clamping position. This machined surface is comparable to a turned structure with an infinite radius. The direction of the cutting process is rotated around the z-axis to create the angle between cutting and burnishing process.

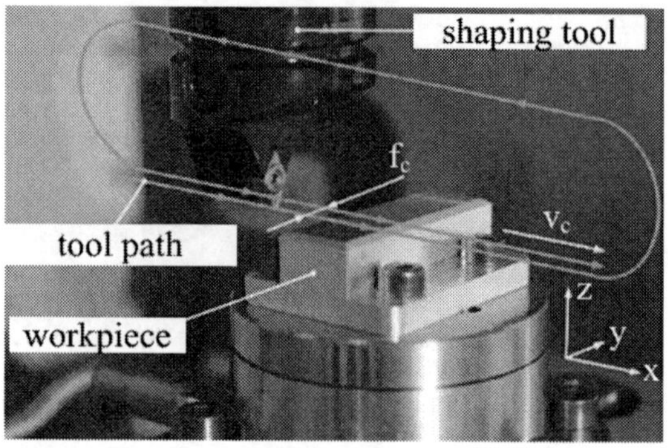

Fig. 3. Experimental setup and kinematics for the cutting

The burnishing process was done in a similar way to the cutting process as shown in Fig. 4. The only difference was that the direction of the burnishing process was unchanged to ensure the comparability of the force recordings.

Fig. 4. Experimental setup and kinematics burnishing process

3 Numerical analysis

The numerical analysis was done by the finite element method (FEM) using a 2D plain-strain model, which is shown with mesh and boundary conditions in Fig. 6. The burnishing tool was modelled as a ridged sphere with a diameter of 10 mm. The workpiece was a 5 by 30 mm rectangle modelled as elastic-plastic. EN AW 2007 was used as material, which was present in the material database of the simulation software. A nonlinear plasticity material law and work hardening was used.

As seen in Fig. 5 the ratio of the measured forces $F_x = 41$ N to $F_z = 410$ N was 0.1. In the simulation the contact between tool and workpiece used a Coulomb friction model with this ratio as friction coefficient. For the creation of a 2D simulation model a methodology similar to Röttger [7] was used. The kinematics of the burnishing process was split in 2 different components, which are the burnishing speed v_b and the burnishing feed f_b. Together with the burnishing force F_z those three components are perpendicular to each other. The component f_b and v_b were observed in separate simulation models, f_b in model A and v_b in model B. For both models the movement is fixed in z-direction and depending on the model a different movement in x-direction is enforced. The movement of model A takes place in three steps, first loading the tool with the force F_z, second unloading the tool and third moving the workpiece by the burnishing feed f_b.

This procedure is repeated 10 times. The burnishing feed was set to 0.125 mm and 0.5 mm. The movement in model B takes also place in three steps. The first step is loading the tool with the force F_z, second moving the workpiece with a constant speed v_b of 20 m/min while keeping a constant force F_z and third unloading the tool. An ideal plain and a rough surface, which was modelled according to the pre-processed surface used during the experiment with a surface roughness of Rz = 40 µm, was used for each model. The burnishing force F_z was set to 200 N and 400 N, as seen in Fig. 5. The mesh was divided in to 3 zones with a total of 4000 elements. The first zone is the area of the roughness with an element size of 0.04 mm (relative element size 0.02). The second zone is the contact area between tool and workpiece with an element size of 0.18 mm (relative element size 0.1). The remaining area is modelled with an element size of 1.86 mm (relative element size 1).

4 Results

The aim of the pre-test was to detect the influencing factors on the surface properties and forces. A change of the burnishing infeed a_{pb} from 0.1 mm to 1.5 mm increases the burnishing force perpendicular to the surface from 170 N to 400 N, taking the spring constant of the tool and the stiffness of the setup into account. The movement sequence of the machine results in a force diagram, which is shown in Fig. 5. The contact between tool and workpiece surface, where the preload of the spring is overcome is shown in the contact section (1). In the loading section (2) the burnishing force perpendicular to the surface is set by reaching the burnishing infeed. Afterwards the tool is moved in x-direction in the burnishing section (3) and generates the force in processing x- and feed y-direction. In the unloading section (4) a fast lifting movement occurs, where the process forces are lifted.

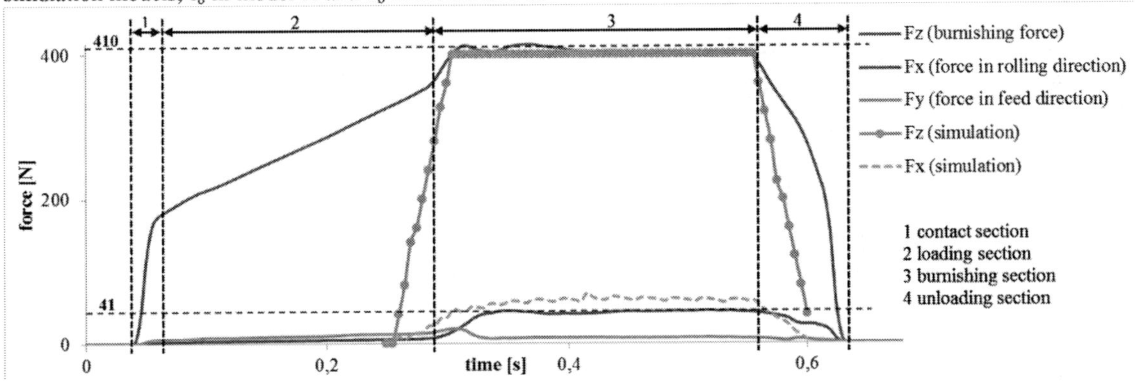

Fig. 5. Different force diagrams during a single burnishing path ($f_c = 0.36$ mm, $\gamma_b = 90°$, $v_b = 2$ m/min, dry, $f_b = 0.125$ mm)

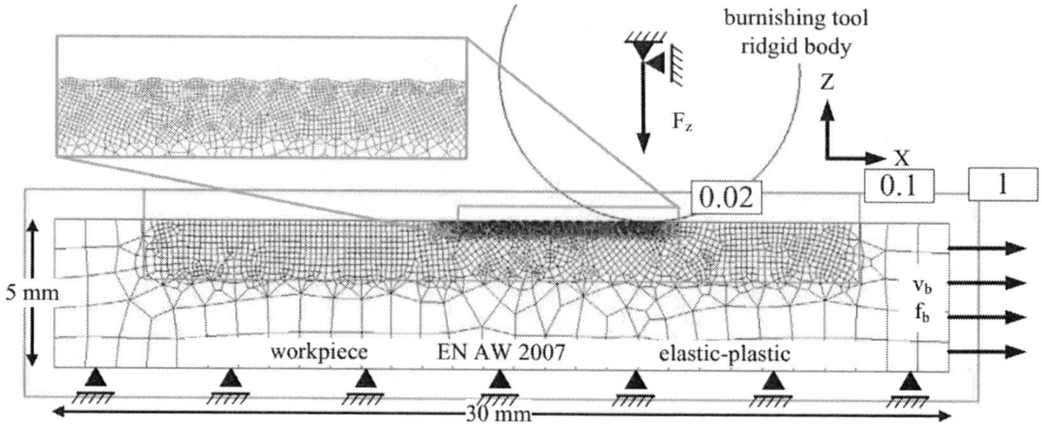

Fig. 6. 2D FEM model for the burnishing simulation with boundary conditions and mesh with three different sizes

In general it can be observed, that the oscillation of the force F_z in section 3 is damped very well and therefore the process is stable. Furthermore the burnishing forces do not rise during the process, which indicates that the bead stays at constant height. Also cracks and damage to the burnished surface by too large forces is prevented. It was possible to reduce the roughness value Rz by 0.81 µm with an increase in burnishing infeed. On the other hand the roughness value Ra was only reduced by 0.06 µm. The burnishing feed f_b has a great influence on the mean value of the roughness profile Ra. A reduction of the factor f_b from 0.5 mm to 0.125 mm reduces the Ra value by an average of 1.44 µm as seen in Fig. 7. A similar relationship can be observed for the roughness value Rz. The factors burnishing infeed a_{pb} (and the resulting burnishing force F_z), the angle between cutting and burnishing γ_b and the roughness of the pre-machined surface only have a small influence on the measured roughness. For example a change of the angle γ_b from 0° to 90° only reduces the roughness values Ra by 0.04 µm and Rz by 0.2 µm, which are within the measurement uncertainty. The factors cutting feed fs, lubricant strategy and burnishing speed v_b show no significant effect on the roughness.

The Pearson correlation coefficient r was calculated to show dependencies between the investigated factors and the surface properties. It is a value between -1 and 1, which show the linear correlation between two investigated variables. In this context r with a value of 0 means no correlation between the variables, whereas -1 or 1 means a negative or positive linear correlation. The error probability p is a value between 0 and 1 that shows how randomly distributed the investigated variables are. A higher value means a more random distribution and therefore p has to be lower than 0.05 for the correlation to be valid. The angle between cutting and burnishing γ_b does not correlate with the roughness values Ra and Rz with a $r_{x,y}$ (Ra) = 0.008 and $r_{x,y}$ (Rz) = 0.012 ($p < 0.05$). In the observed range up to F_z = 400 N the burnishing infeed a_{pb} does not have a significant effect on the roughness of the machined surface, but it does have a big impact on all process forces. The burnishing feed f_b correlates with most surface values measured according to DIN EN ISO 4287 and DIN EN ISO 13565-2, with values of $r_{x,y}$ (Ra) = 0.956 and $r_{x,y}$ (Rz) = 0.94 ($p < 0.05$). Conclusively the evaluation of the statistical experiments shows that the condition of the pre-processed surface has no significant influence on the roughness after the burnishing process within the frame of the established factors. A correlation with the target values Ra and Rz could only be proven for the burnishing feed f_b.

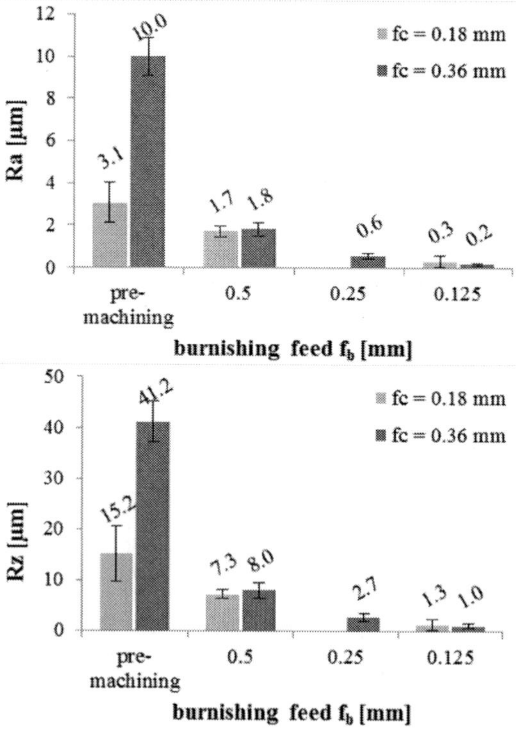

Fig. 7. Results of the target values Ra and Rz under different burnishing feeds (γ_b = 90°, dry, v_b = 2 m/min, F_z = 320 N)

https://doi.org/10.1051/matecconf/201819011005

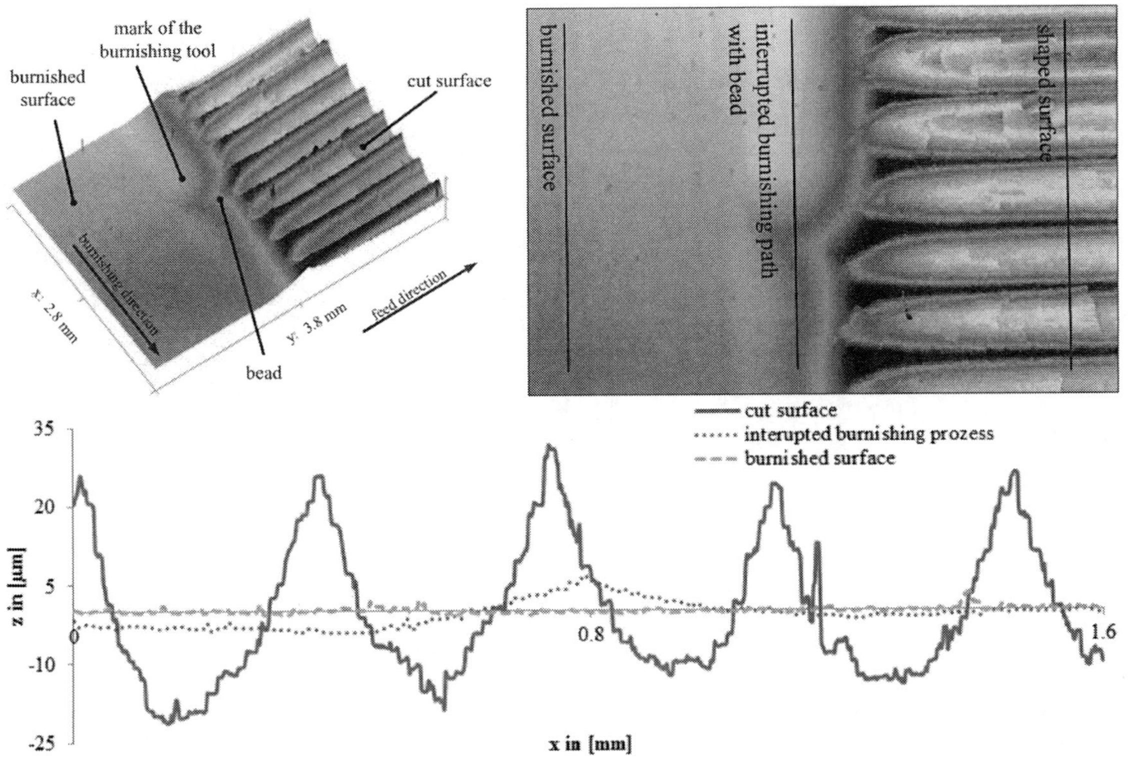

Fig. 8. 3D-measurement of a burnishing process and roughness profiles for three different zones ($F_z = 320$ N, $f_b = 0.125$ mm, $\gamma_b = 90°$, $f_c = 0.36$ mm)

In the following the burnishing process was interrupted and their surfaces were optically measured in three dimensions for a detailed process analysis. This graphical representation of the measured surface is seen in Fig. 8 and show the remaining imprint of the burnishing tool after the elastic recovery of the workpiece material.

Due to the difference in height between the interrupted burnishing path and burnished surface it can be stated, that the material "flows" around the tool, but at the same time forms a bead in front of the tool. Therefore the profile tips are not smoothed and the valleys lifted like it is the case with harder materials. Instead a completely new surface structure is created. This supports the statement that the parameters of the pre-process have low influence on the roughness of the surface after the burnishing, especially for such ductile materials like aluminium. The described relationship of the resulting roughness with the burnishing feed f_b is also visible by the adjacent burnishing paths.

The aim of the FEM-simulation was to determine the stress and strain distribution during the burnishing process and the residual stresses after the burnishing. An example of the simulation results of model B with the rough surface is seen in Fig. 9. Step I, II and III show the deformation of a single asperity. During all three shown steps the von Mises stress for the asperity is higher than the yield strength of $R_{p0.2} = 220$ MPa represented by the area C. Therefore the whole asperity as well as a large area under it is in a state of plastic deformation. The maximum tensile strength of $R_m = 340$ MPa is not reached, which means that the workpiece is not damaged. In all done simulations the average maximum von Mises stress under load is 263 MPa with a standard deviation of 14 MPa. Higher values of the von Mises stress are reached when the rough surface is present, with an average value of 273 MPa. The burnishing of the flat surface reaches an average value of 252 MPa. Furthermore the results of the simulation show that the burnishing force does only have a miniscule influence on the maximum von Mises stress. In all simulation cases $R_{p0.2}$ is reached and therefore plastic deformation occurs. It is also possible to apply higher loads in further investigations, because the maximum von Mises stress reached is at around 80% of R_m. The diagrams in Fig. 9 show different stresses during and after the process (marked in Step II) in comparison to the depth of the workpiece. After the burnishing process the maximum absolute stress in x-direction is 7.7 times higher than in z-direction. Therefore the stress in x-direction has the most contribution to residual stress after the burnishing process. The stress in x-direction also shows a prominent point where it switches from compressive (negative stress values) to tensile stress (positive stress values). This point is used to determine the depth of the residual stresses.

https://doi.org/10.1051/matecconf/201819011005

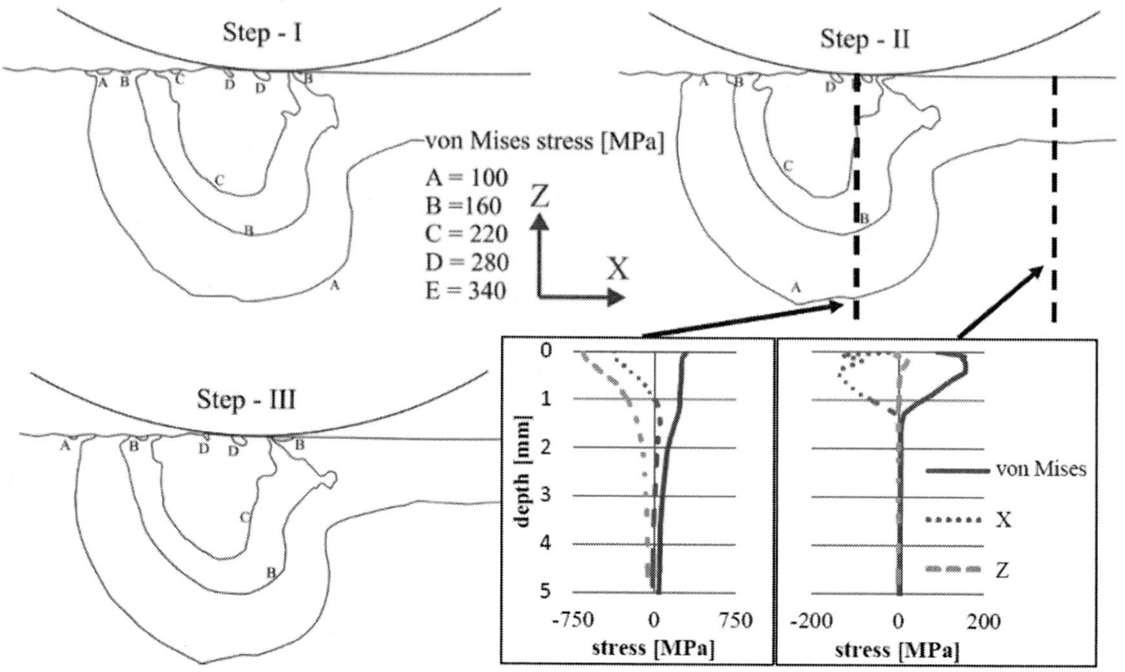

Fig. 9. Step I, II and III: von Mises stress during burnishing, diagrams: von Mises, x- and z-stress during and after burnishing (residual stress) in comparison to the depth of the workpiece (F_z = 400 N, model B, v_b = 20 m/min, pre-machined surface)

The average depth of residual stress is identical between the models A and B, as well as between a burnishing feed f_b of 0.125 mm and 0.5 mm, which means that the type of simulation model and the burnishing feed have no influence on the average depth of the residual stress. However the burnishing force has influence on depth of residual stress. A burnishing force of 200 N leads to a depth of 0.5 mm while a force of 400 N leads to a depth of 1.3 mm. As seen in Fig. 5 the resulting force F_x in burnishing direction was at around 50 N. This value is higher than the measured value of the experiments and its cause has to be investigated in further studies.

5 Conclusion

The experimental investigation and the numerical analysis of burnishing of EN AW 2007 show that the burnishing direction and surface structure before burnishing have no significant influence on the burnishing process and the investigated surface parameters, especially for such ductile materials like aluminium. Therefore the process can be viewed as direction-independent. The investigation at the surface border area of the process shows that the material forms a constant bead in front of the tool. This behaviour leads to the forming of a new surface, with roughness values manly depending on the burnishing feed.

The stresses during and after the burnishing process were analysed with FEM-simulation. The results show that the maximum von Mises stress during burnishing is nearly independent from the burnishing force, but slightly depends on the surface roughness. Furthermore it was shown, that at the investigated burnishing forces the von Mises stress only reaches the tensile strength of the investigated aluminium by 80%, which means that even higher loads can be applied without damaging the surface. The depth of the residual stress is largely influenced by the applied burnishing force, with higher forces leading to deeper residual stresses. Therefore it is possible to increase the depth of the residual stresses by applying higher forces. Furthermore the experimental investigation also shows, that a two dimensional simulation is not sufficient to reproduce the deformation effects which are present during the burnishing of aluminium.

Further investigations should utilize a 3D FEM-simulation. Also the range of the experimental investigation should be increased by additional factors regarding the burnishing force and structure of the pre-machined surface. Materials with other properties should be taken into account as well.

The research was funded by the European Union (European Social Fund) and the Free State of Saxony.

Europe funds Saxony.
European Union
ESF
European Social Fund

References

1. W. Degner, *Handbuch Feinbearbeitung* (1979)
2. F. Grünwald, *Fertigungsverfahren in der Gerätetechnik* (1985)

3. U. Virkus, *Oberflächengestalltung durch die Verfahrensfolge Spanen - Glattwalzen unter stofflichen, geometrischen und kinematischen Aspekten, Technische* Universität Dresden (2001)

4. L. N. López de Lacalle, A. Lamikiz, J. Muñoa, J. A. Sánchez, Int. J. Mach. Tools Manuf. **45**, 1659 (2005)

5. L. N. López de Lacalle, A. Rodríguez, A. Lamikiz, A. Celaya, R. Alberdi, Mater. Manuf. Process. **26**, 997 (2011)

6. P. Sartkulvanich, T. Altan, F. Jasso, C. Rodriguez, J. Manuf. Sci. Eng. **129**, 705 (2007)

7. K. Röttger, „Walzen hartgedrehter Oberflächen", Technische Hochschule Aachen, 2003.

8. P. Balland, L. Tabourot, F. Degre, V. Moreau, Precis. Eng. **37**, 129 (2013)

9. Y.-C. Yen, P. Sartkulvanich, und T. Altan, CIRP Ann. - Manuf. Technol. **54**, 237 (2005)

A New Cross Wedge Rolling Process for Producing Rail Axles

Zbigniew Pater[1,*], and *Janusz* Tomczak[1]

[1]Lublin University of Technology, Faculty of Mechanical Engineering, Nadbystrzycka 36, 20-618 Lublin, Poland

Abstract. Rail axles are large-size parts produced in large batches. Currently, these parts are produced by metal forming techniques such as rotary forging, open die forging with hydraulic presses and open die hammer forging (minimum ram weight: 3 Mg). Nevertheless, not only are the above methods far from being efficient, they also lack accuracy (open die forging). As a result, new techniques for producing rail axles are constantly developed. One of such alternative techniques is based on the use of cross wedge rolling (CWR), which is the subject of the present study. An innovative roll design for producing rail axles by CWR is proposed. The rolls are provided with three pairs of wedge tools that act simultaneously on the workpiece and form the part in one revolution of the rolls, i.e., during 20 s. The numerical modelling of a CWR process with the proposed roll design reveals that the solution can be used to produce railway axles with the desired geometry. This technique, however, requires relatively high loads and torques. To decrease the force parameters, the forming process was modified and ran in two operations. The first operation consists in forming the central step of the workpiece while the other one involves the formation of steps on the ends of the workpiece. The results of the new simulation show a significant decrease in the loads and torques, which is caused, among others, by reducing the nominal diameter of the rolls from 1600 mm to 1200 mm. The numerical findings can be used to design a rolling mill for producing rail axles.

Keywords: Rolling, Finite element method (FEM), Tool geometry

1 Introduction

Rail axles, which come as solid or hollow pieces, are large-size parts manufactured in large batches. For instance, the annual production of rail axles in China alone amounts to approx. 300,000 pieces [1], which – taking into account their weight (approx. 400 kg) – gives an annual production of 120,000 tons.

Currently, rail axles are produced by open die forging on hammers or hydraulic presses or by rotary forging. Nonetheless, not only are these methods far from being efficient, they can lack accuracy (open die forging processes). For this reason, new methods for producing rail axles are sought, with emphasis on techniques based on skew rolling [2] and cross wedge rolling (CWR) processes [1, 3-5].

In this paper, the authors propose novel methods for producing rail axles described by the parameters shown in Fig. 1. The proposed solutions are based on the CWR process and allow for practically waste-free manufacturing of these parts. The proposed solutions are verified by FEM analysis.

2 New trends in the CWR technique

Cross wedge rolling (CWR) is widely used for producing parts such as stepped axles and shafts and for manufacturing preforms [6]. In recent years, however, numerous studies have been performed to develop the technological potential of this effective manufacturing technique. New trends in cross wedge rolling include:

- CWR for parts with steps with cross sections different than circular (e.g. square, oval, hexagonal), formed by tools with specially profiled forming surfaces [7-9]. The rolling of such parts is characterized by oscillatory variations in the forces resulting from cyclic changes in reduction ratio;
- CWR for toothed or worm shafts using tailor-made inserts fixed behind the sizing zone of the tools [10-14];
- CWR for hollow parts performed with an elongated sizing zone of the tools [3, 15-21];
- CWR for products made of non-ferrous materials such as titanium alloys [22-24], aluminum alloys [12], magnesium alloys [25], superalloys based on nickel [26] and copper [27];
- multi-wedge rolling in which the workpiece is

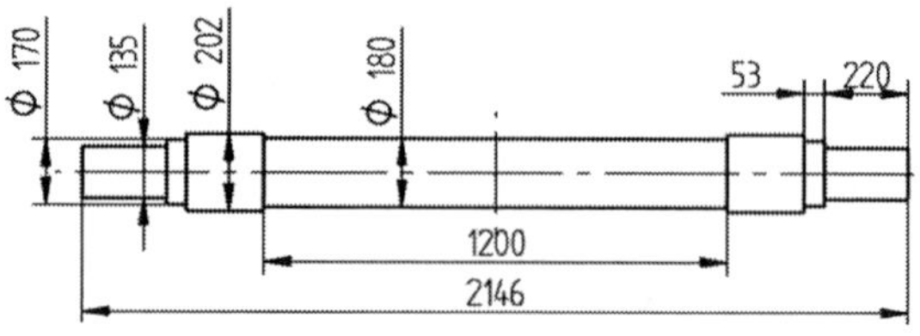

Fig. 1. Rail axle analyzed in the study

* Corresponding author: z.pater@pollub.pl

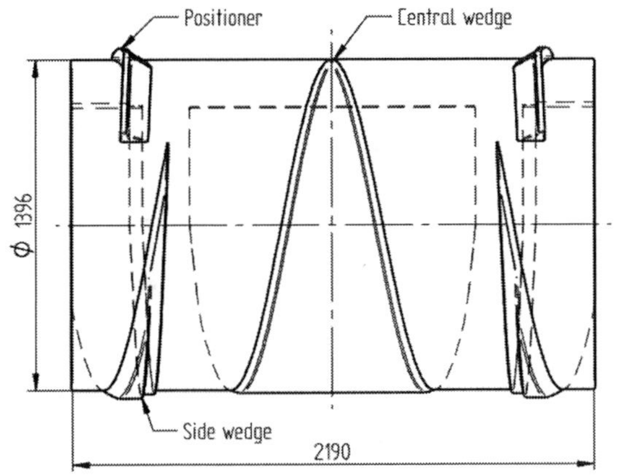

Fig. 2. Roll with 1600 mm nominal diameter used in one-stage CWR for rail axles

formed simultaneously by more than one pair of tools. This rolling technique is used in the production of very long axles and shafts (e.g. for the automotive or railway industry) [1, 3-5, 28] and in the processes wherein several short parts are formed simultaneously (e.g. balls for ball mills) [29-31].

3 CWR for rail axles

3.1 One-stage CWR process

First of all, the study investigated whether it was possible to produce a rail axle in one CWR operation, i.e., using a pair of identical tools, one of them being

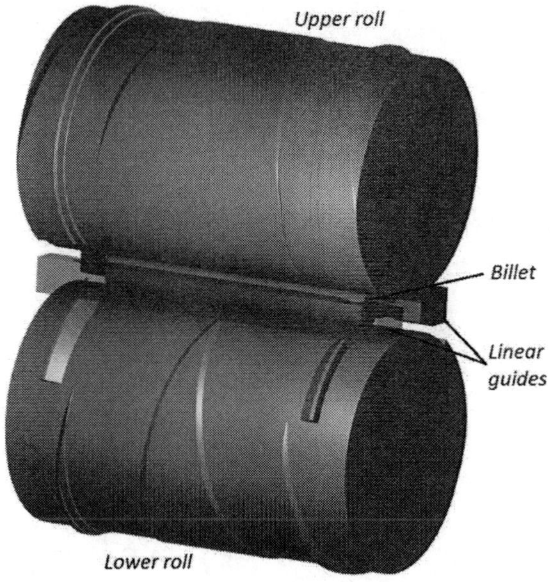

Fig. 3. Geometric model of CWR for rail axles formed with 1600 mm diameter rolls

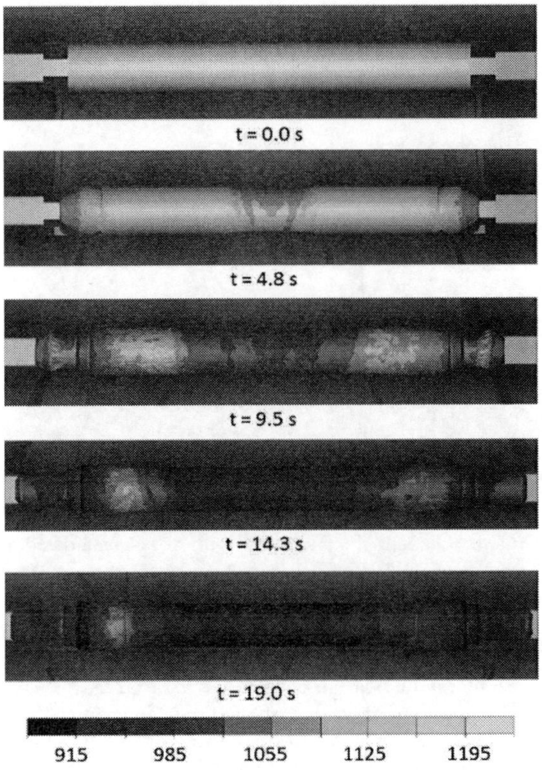

Fig. 4. Workpiece shape changes and temperature distribution (in °C) in CWR for producing rail axles

shown in Fig. 2. Given the overall dimensions of the product, a multi-wedge rolling technique was employed in which the axle is deformed simultaneously by three pairs of tools that form the central step and end steps at the same time. In addition, due to product elongation caused by the action of the central wedge tool, the side wedge tools are deflected from the tool motion direction.

The proposed solution was verified by a numerical simulation in Forge NxT 1.1. This software had previously been used for the modeling of skew and cross rolling processes [32-34].

Fig. 3 shows a geometric model of the analyzed rolling process. The model consists of two rolls (shown in Fig. 2), two guides (for maintaining the billet in the working space between the rolls) and a billet with a diameter of 202 mm and a length of 1660 mm. The billet was assigned the properties of 42CrMo4 steel, the model of which was obtained from the material database of the employed software. The billet was preheated to a temperature of 1180 °C, and the temperature of the tools during the rolling process was maintained constant at 250 °C. The rotational speed of the rolls was fixed at 3 revolutions per minute, the heat transfer coefficient between the workpiece and the tools was set to 10 kW/m^2K, and the friction factor on contact surface was set equal to 0.99.

The numerical simulation allowed for close examination of the proposed rolling process for producing a rail axle shown in Fig. 4. Examining the

https://doi.org/10.1051/matecconf/201819011006

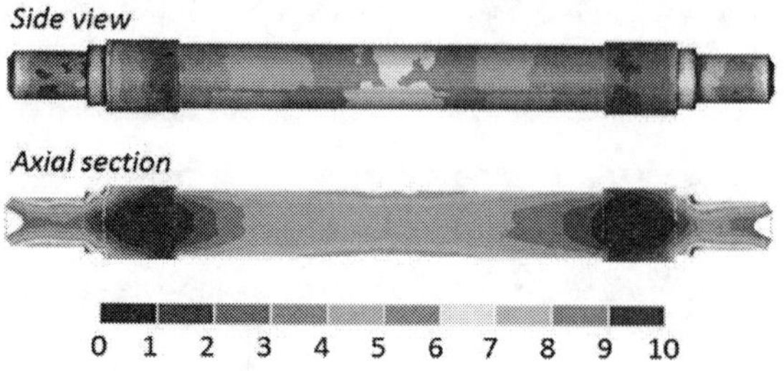

Fig. 5. Distribution of effective strains in a rail axle produced by CWR

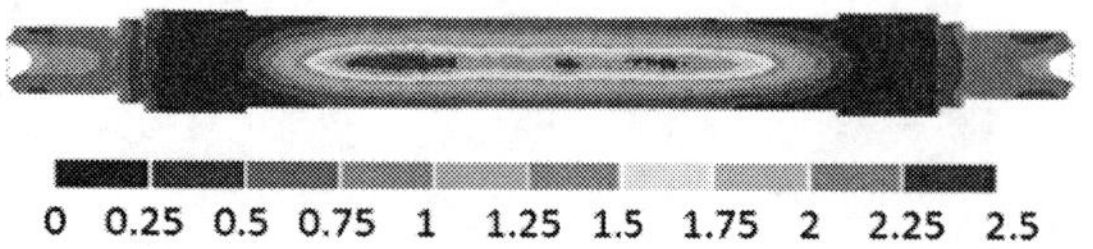

Fig. 6. Damage function distribution, as calculated in accordance with the Cockcroft-Latham criterion

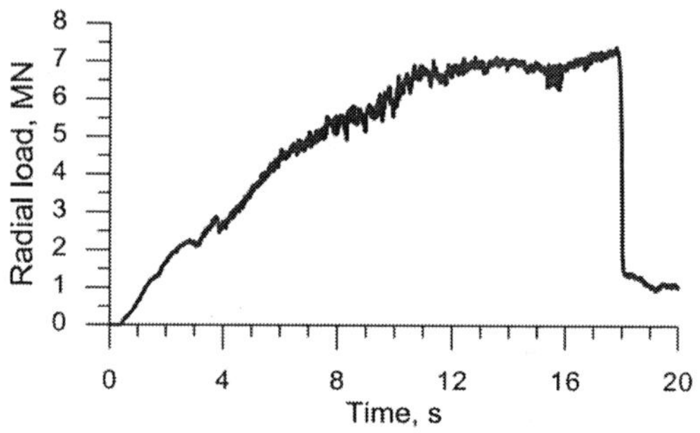

Fig. 7. Radial load distribution on the roll

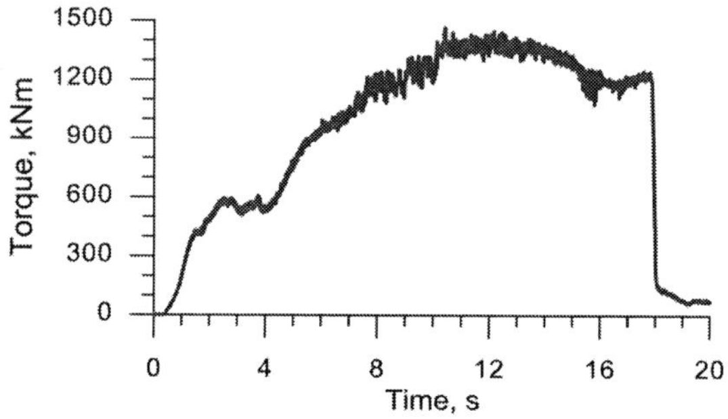

Fig. 8. Torque distribution on the roll in one-stage CWR

data in the figure the following can be observed: first of all, tapered ends are formed on the billet to reduce the size of cavities on the ends of the workpiece. Next, all steps are formed at the same time, and, finally, the workpiece is formed into the desired shape. It is worth stressing that the temperature of the workpiece during the rolling process remains in the hot working range. Significant temperature drops only occur in the surface layers of the workpiece.

Fig. 5 shows the distribution of strains in the produced rail axle. It can be observed that the highest strains are located in the surface layers of the steps on the workpiece ends and at the center of the central step, hence where the tools cut into the billet. This distribution of strains results from rapid metal flow in a circumferential direction and is typical of CWR processes [6].

The employed software allowed for investigating the probability of crack occurrence at the center of the workpiece. To this end, we examined the distribution of the damage function calculated in accordance with the Cockcroft-Latham criterion, as shown in Fig. 6. It can be observed that the function reaches very high values in the axial zone, which indicates the probability of material cracking in this region. This stems from the difficulty with removing ovalization of the cross section of the central step, which leads to cyclic compression as is the case in the Mannesmann process.

Figs. 7 and 8 show the distributions of the force parameters, i.e., the radial load (acting on the rolls) and the torque. Analysis of these figures reveals that the realization of the proposed rolling process for rail axles will require the application of high loads and torques. What may be particularly difficult to obtain in practice is a torque amounting to 1.4 MNm, as this requires a power of 900 kW.

3.2 Two-stage CWR process

To decrease force parameters in the rolling of rail axles, we also investigated a CWR process performed in two

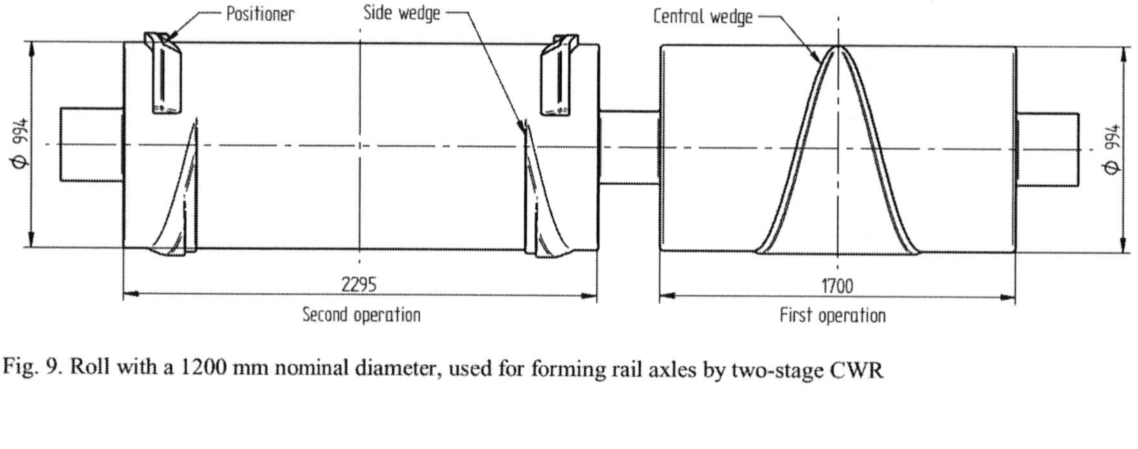

Fig. 9. Roll with a 1200 mm nominal diameter, used for forming rail axles by two-stage CWR

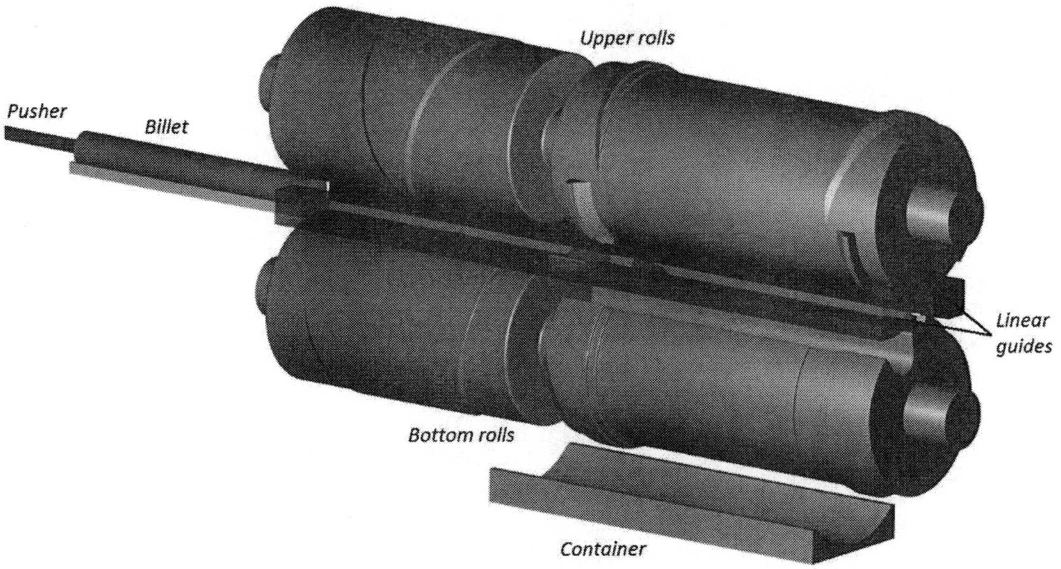

Fig. 10. Model of two-stage CWR for producing rail axles by rolls with 1200 mm nominal diameter

operations with two pairs of rolls shown in Fig. 9. The first stage involves forming the central step of the part. Second, following the positioning of the billet between the guides, steps are formed on the ends of the workpiece. A side effect of dividing the forming process into two operations is that the nominal diameter of the rolls (equal to the axial spacing between the rolls) must be decreased from 1600 to 1200 mm.

Similarly to the one-stage rolling process, the proposed design solution was first verified by a numerical simulation. The geometric model of the two-stage CWR process is shown in Fig. 10. This time, the entire production cycle was modelled, from the feeding of the billet and passing the workpiece to the other pair of the rolls, to the removal of the finished part from the working space of the tools. The parameters of the forming process were identical to those applied in one-stage CWR. The speed of the pusher for billet displacement was set to 100 mm/s.

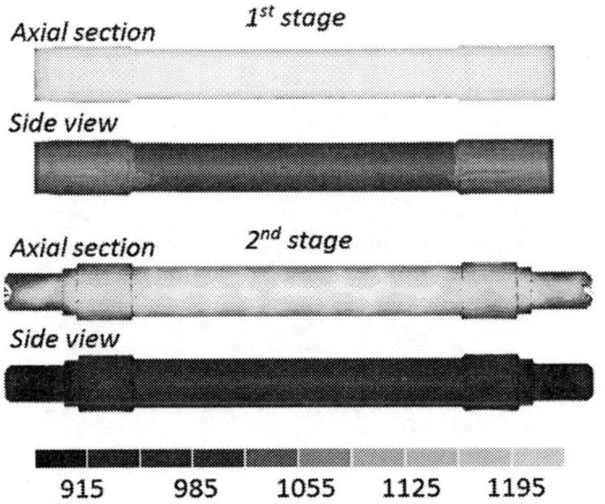

Fig. 11. Temperature distribution (in °C) in a rail axle produced by two-stage CWR

Analysis of the numerical results of the proposed two-stage CWR process for producing a rail axle demonstrates that despite extending the forming time to 90 s, the temperature of the billet is in the hot working range (Fig. 11). This may results from a large accumulation of heat. The highest drops in temperature can be observed in the end steps of the product, however the drops only occur in the surface layers of the material.

The proposed change to the CWR technique also led to reducing strains in the product (Fig. 12). This results from decreased shear stresses (considered to be redundant) due to reduced diameter of the roll, as a result of which individual steps on the rail axle are formed by shorter wedge tools. In addition to this, one must also take account of the fact that cross-sectional ovalization in individual steps of the axle is reduced.

As a result of dividing the rolling process into two operations, the probability of crack occurrence in the central step of the product is greatly reduced (Fig. 13). This is reflected in a two-and-a-half decrease in the damage function calculated according to the Cockcroft-

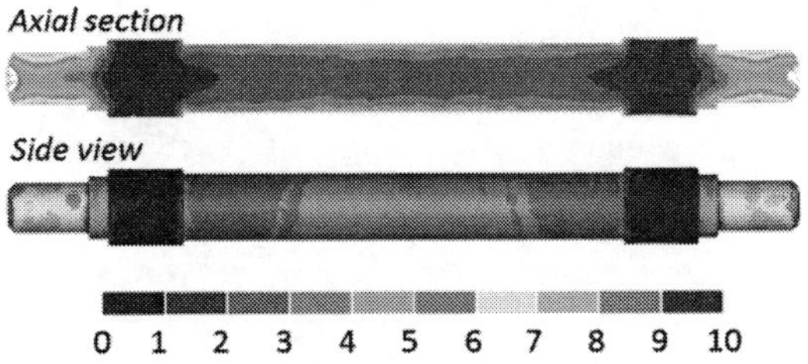

Fig. 12. Distribution of effective strains in a rail axle produced by two-stage CWR

https://doi.org/10.1051/matecconf/201819011006

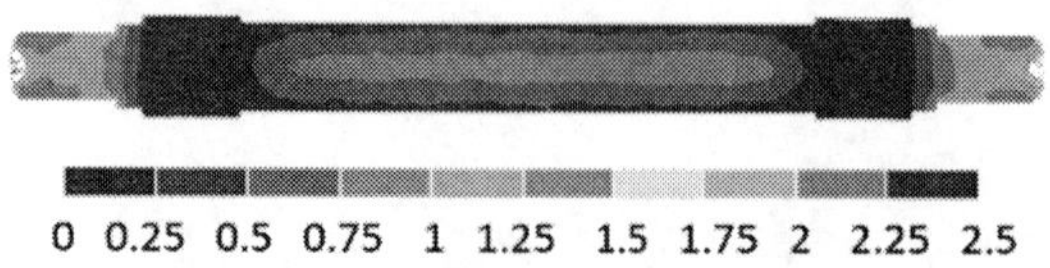

Fig. 13. Damage function distribution in a rail axle produced by two-stage CWR, as calculated in accordance with the Cockcroft-Latham criterion

Latham criterion. Therefore, it can be assumed that the two-stage CWR helps prevent material cracking.

In addition, the two-stage CWR technique for producing rail axles has a positive effect on the force parameters of the process. The maximum radial load (which ensures stiffness of the rolling mill) is reduced by 25% while the torque is decreased by as much as 60%. In this case, the estimated power of the rolling mill is only 350 kW. In connection with this, it should be noted that rolling mills currently used in industrial practice have similar parameters [35]. For example, the H1400 rolling mill manufactured in China is equipped with 1400 mm diameter rolls, and its driving power is 220 kW. Another Chinese rolling mill, D46-165x1200, is provided with the rolls described by a diameter of 1200 mm and the engine with a power of 315 kW. Given the above, the construction of a rolling mill for two-stage rolling of rail axles is feasible.

4 Conclusions

The results of the study lead to the following conclusions:

- rail axles can be produced by the CWR method performed in either one or two stages;
- the two-stage CWR technique is more effective for two reasons: (i) it prevents material cracking in the axial zone of the workpiece, and (ii) it requires lower loads and torques;
- among rolling mills available on the market, one can find machines with parameters similar to those of rolling mills for producing rail axles. Therefore, the construction of a tailor-made rolling mill for producing these parts is feasible;
- the research on the CWR method for producing rail axles should be extended to include experimental tests on real material.

References

1. X. Shu, X. Wei, Ch. Li, Y. Hu, Appl. Mech. Mater., **37-38**, 1482-1488 (2010)
2. Z. Pater, J. Tomczak, T. Bulzak, Arch. Metall. Mater., **60**, 415-418 (2015)
3. W. Peng, S. Zheng, Y. Chiu, X. Shu, L. Zhan, Rare Metal Mat. Eng, **45**, 836-842 (2016)
4. B. Sun, X. Zeng, X. Shu, W. Peng, P. Sun, Appl. Mech. Mater., **201-202**, 673-677 (2012)
5. B. Hu, X. Shu, P. Yu, W. Peng, Appl. Mech. Mater., **494-495**, 457-460 (2014)
6. Z. Pater, in *Comprehensive Materials Processing*, Ed. Elsevier Ltd., **3**, 211-279 (2014)
7. Z. Zheng, B. Wang, Z. Hu, Appl. Mech. Mater., **217-219**, 1713-1718 (2012)

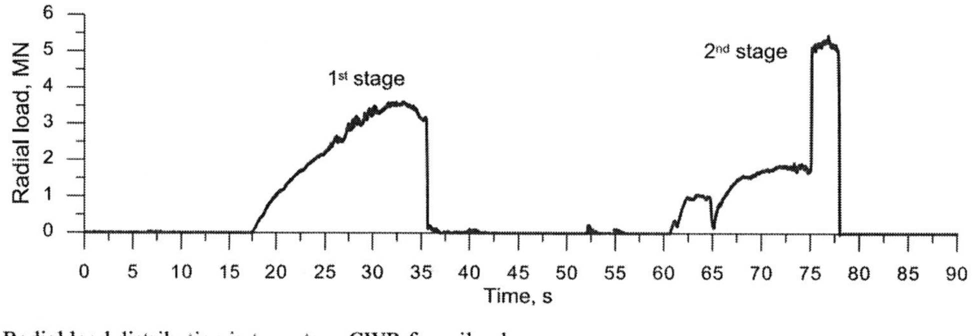

Fig. 14. Radial load distribution in two-stage CWR for rail axles

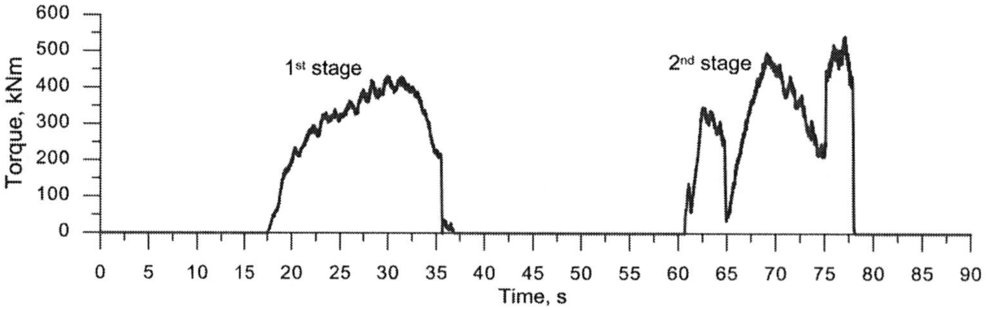

Fig. 15. Torque distribution on the roll in two-stage CWR for rail axles

8. T. He, B. Wang, Z. Hu, Adv. Mat. Res., **154-155**, 1796-1799 (2011)
9. W. Baoyu, Z. Zheng, H. Zhenghuan, L. Jianguo, Steel Res. Int., **81**, 222-225 (2010)
10. F. Ying, J. Shen, L. Wu, Adv. Mat. Res., **497**, 55-60 (2012)
11. Z. Pater, Acta Mechanica Slovaca, **15**, 50-58 (2011)
12. Z. Pater, A. Gontarz, A. Tofil, Journal of Shanghai Jiaotong University (Science), **16**, 162-166 (2011)
13. H. Yan, L. Wang, Y. Liu, G. Li, J. Liu, Z. Hu, Appl. Mech. Mater., **440**, 177-181 (2014)
14. H. Yan, H. Mao, G. Li, X. He, J. Liu, Z. Hu, Adv. Mat. Res., **853**, 605-609 (2014)
15. S. Urankar, M. Lovell, C. Morrow, Q. Li, K. Kawada, J. Mater. Process. Tech., **177**, 545-549 (2006)
16. H. Ji, J. Liu, B. Wang, Z. Zheng, J. Huang, Z. Hu, The International Journal of Advanced Manufacturing Technology **77**, 15-26 (2015)
17. J. Ma, C. Yang, Z. Zheng, K. Zhang, W. Ma, Int. J. Min. Met. Mater., **23**, 1302-1314 (2016)
18. C. Yang, J. Ma, Z. Hu, J. Mater. Process. Tech. **239**, 346-358 (2017)
19. H. Ji, J. Liu, B. Wang, A. Fu, W. Xiao, Z. Hu, J. Mater. Process. Tech. **240**, 1-11 (2017)
20. J. Bartnicki, Z. Pater, J. Mater. Process. Tech. **164-165**, 1154-1159 (2005)
21. J. Bartnicki, Z. Pater, J. Mater. Process. Tech. **155-156**, 1867-1873 (2004)
22. Z. Pater, A. Gontarz, G. Samołyk, J. Bartnicki, Arch. Metall. Mater., **54**, 617-626 (2009)
23. Z. Pater, T. Bulzak, J. Tomczak, Key Eng Mater., **687**, 125-132 (2016)
24. R. Neugebauer, B. Lorenz, J. Steger, D. Holstein, Steel Res. Int., **79**, 375-381 (2008)
25. Z. Pater, J. Tomczak, Arch. Metall. Mater., **57**, 919-928 (2012)
26. S.J. Mirhamadi, M. Hamedi, S. Ajami, The International Journal of Advanced Manufacturing Technology, **74**, 995-1004 (2014)
27. A. Tofil, Z. Pater, Advances in Science and Technology Research Journal, **11**, 72-86 (2017)
28. J. Zhao, L. Lu, Appl. Mech. Mater., **101-102**, 1002-1005 (2012)
29. Z. Pater, J. Iron Steel Res. Int., **20**, 46-50 (2013)
30. Z. Pater, A. Tofil, J. Tomczak, Metalurgija, **52**, 103-106 (2013)
31. Z. Pater, Int. J. Mater. Prod. Tec., **47**, 12-22 (2013)
32. M. Meyer, M. Stonis, B.A. Behrnes, Key Eng Mater., **504-506**, 205-210 (2012)
33. M. Meyer, M. Stonis, B.A. Behrnes, Production Engineering. Research and Development, **9**, 61-71 (2015)
34. M.F. Novella, A. Ghiotti, S. Bruschi, P.F. Bariani, J. Mater. Process. Tech. **222**, 259-267 (2015)
35. G. Kozhevnikova, *Cross-Wedge Rolling* (Belorusskaya naukam Minsk, 2012)

EBSD study of microstructure evolution of ferritic stainless steel during cold rolling and annealing

Chi Zhang[*], *Yijie*Xu, *Liwen Zhang*,and *YinlinGu*

School of Materials Science and Engineering, Dalian University of Technology, Dalian 116024, China

Abstract. The microstructure and texture of ferritic stainless steels (FSSs), formed during cold rolling and annealing processes, determine the mechanical properties of final sheet, especially the deep drawing formability. In this work, aNb, Ti stabilized17%Cr FSS was cold rolled with the reductions of 20%~70% and annealed for periods at 700℃. EBSD technique was used to characterize the microstructure evolution and inhomogeneous deformation strain distribution of the sheet during cold rolling. Partially annealed sheets were also analyzed to observe the nucleation and growth of recrystallized grains. Special attentions were paid on the crystal orientation of the deformed grains and recrystallzed grains. The results infer that in-grain shear band was formed in the cold rolled sample with the reduction higher than 30%, associated with the formation of high deformation strains. And the recrystallized grains prefer to form at some unique grain boundaries and in-grain shear bands. The orientations of recrystallized grains relates to the deformed grains.

Key words: stainless steel, rolling, texture

1. Introduction

Ferritic stainless steels (FSSs), which have lower cost as compared to austenitic stainless steels due to the absent addition of Ni, can be widely used for elevator panel, kitchen equipments, home appliance, automotive exhaust manifolds, and so on [1]. They possess excellent properties, such as high stress corrosion cracking (SCC) resistance, high thermal conductivity, low thermal expansion, and preserving magnetic. However, the formabilityof FSSs, especially deep drawing formability, is lower than the austenitic stainless steels. Researches confirmed that the formability of FSSs is closed related to the texture. The component ofγ-fibre texture (<111>//ND) is supposed to be benefit for the improvement of formability [2, 3, 4, 5]. Therefore, many efforts have been devoted to develop the texture of FSSs through processing [6, 7]. The microstructure and texture undergo intensive changes during cold rolling and annealing processes [8, 9]. Recently, the EBSD technique is increasingly used to clarify the micro-texture and grain boundary character since it's powerful in analyzing crystal orientation. The aim of this paper is to characterize the microstructure and texture evolution during cold rolling and annealing processes using EBSD technique.

2. Experiments

Table.1 shows the chemical compositions of the tested 17%Cr FSS, in which Nb and Ti were added to stabilize C and N atoms. The hot rolled and annealed sheets with thickness of 5mm were obtained from Bao steel company. The sheets were then cold rolled with the reduction 0f 20%, 30%, 40%, 50%, 60% and 70% by a 6-high pilot cold rolling mill with the work roll diameter of 110mm in lab. Lubricant oil was used during cold rolling. The cold rolled sheets were annealed at 700℃ for a period. Specimens were cut from the sheets for microstructure and texture observations. The recrystallized fractions of partially annealed samples were calculated from the microstructure observations. EBSD technique, which equipped in a field emission SEM, was used for the orientation detection. The data were typically obtained with a step size of 0.15-0.25μm depending on the grain size. Then the data were processed using HKL Channel 5 software.

Table.1 Chemical compositions of tested material

C	Si	Mn	P/S	Cr	N	Ti	Nb
≤ 0.015	≤ 0.6	≤ 1.00	≤ 0.05	16.0~ 18.0	≤ 0.015	≤ 0.30	≤ 0.30

3. Results and Discussions

3.1 Microstructure characters during cold rolling

Fig.1 shows the orientation maps of cold rolled sheets with different reductions. After cold rolling, the microstructures were composed of elongated ferrite

* Corresponding author: zhangchi@dlut.edu.cn

grains. The width of the deformed grains decreases dramatically with the increasing reduction. In-grain shear bands, which have inclination to the rolling direction by 20-35°, were formed at the cold rolling reduction higher than 30%.

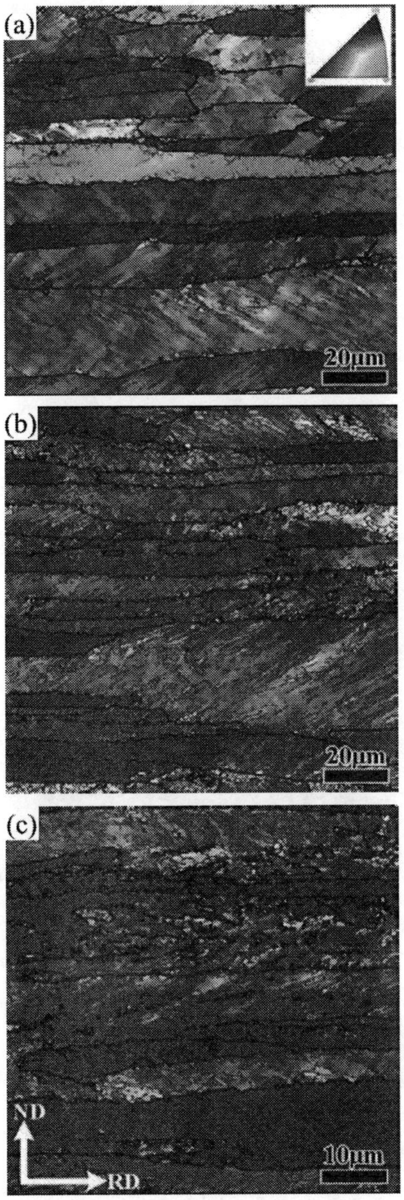

Fig.1 Orientation imaging maps of cold rolled sheets with different reductions. a 30%, b 50%, c 70%.

Based on the quantitative data of crystal orientations, the nature of grain boundary and misorientation between adjacent detected points can be calculated using HKL channel 5 software. Fig.2 shows the grain boundary characters of the cold rolled sheet corresponding to Fig.1. The high angle grain boundary (HAGB) is defined by a misorientation between adjacent grains $\theta > 15°$, and low angle grain boundary (LAGB) is defined by $\theta < 15°$ .It is found that with the increasing of cold rolling reduction, the numbers of LAGB increase intensively, especially at the grains with in-grain shear bands. This infers that more fragmented grains and sub-grain structure were obtained by cold rolling. And these fragmented grains preserve more stored energy for recrystallization in the subsequent annealing process.

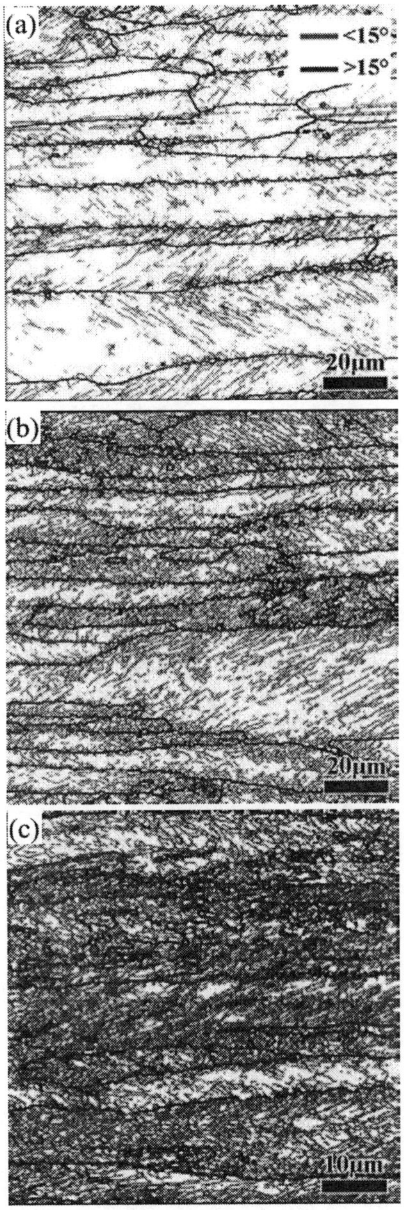

Fig.2 Grain boundary maps of cold rolled sheets with different reductions. a 30%, b 50%, c 70%.

Fig.3 shows the $\varphi_2 = 45°$ sections of the ODFs calculated from the EBSD data. Strong rolling texture components near $\{112\}<110>$ were displayed in all sheets. The fraction of γ-fibre texture increases with the increasing reduction, indicating that compression

 https://doi.org/10.1051/matecconf/201819011007

deformation promotes the grains rotating to the <111>//ND direction. It should be mentioned that the ODFs just indicate the texture of the EBSD detected areas which is supposed to be micro-texture. Nevertheless, the evolution of texture with cold rolling reduction is obvious.

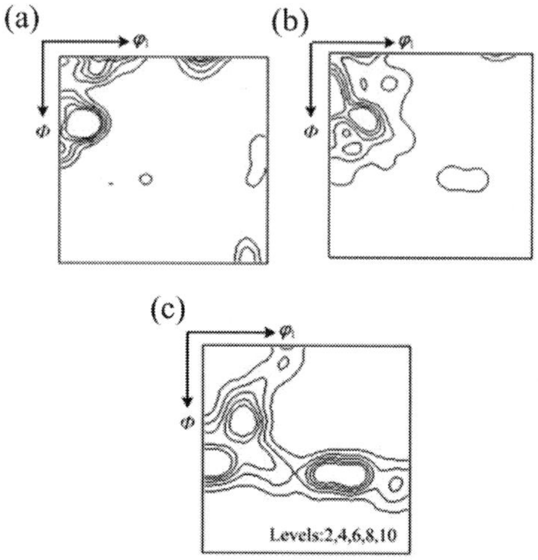

Fig.3 $\varphi_2 = 45°$ sections of the ODFs for cold rolled sheets with different reductions.a 30%, b 50%, c 70%.

3.2 Microstructure after annealing

The cold rolled sheets with reductions of 50%, 60% and 70% were annealed at 700℃ for different times. The microstructures of the specimens were observed and the recrystallization fraction was calculated as the recrystallized fraction ratio. Then the recrystallization kinetics of the cold rolled tested steel was obtained, as shown in Fig.4.

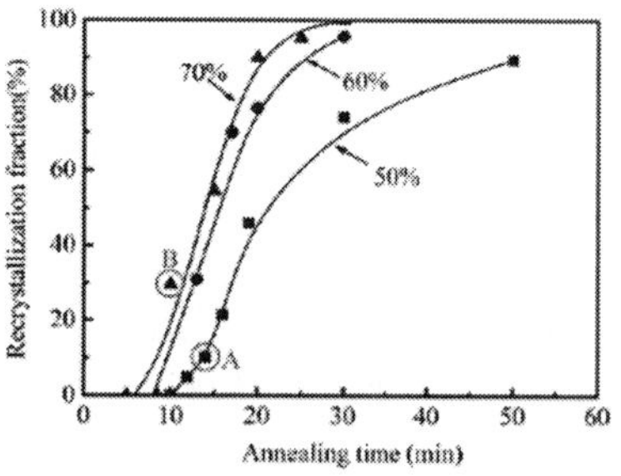

Fig.4 The recrystallization kinetics of cold rolled sheets during annealing.

After an incubation period for nucleation of recrystallized grains, the recrystallization fraction increases rapidly with the increasing time. At the later period of annealing process, the increasing rates of recrystallization fraction decreases. The curves indicat that the kinetics follow the Johnson-Mehl-Avrami-Kormogorov (JMAK)-type kinetics. It founds that the recrystallization kinetics incrases dramatically with the increasing cold rolling reduction. For example, the recrystallization fraction of 50% cold rolled sheets is about 45% when annealed for 20min while the recrystallization fractions increase to 75% for the sheet with 60% reduction and 90% for the sheet with 70% reduction. This is because that more energy is stroed in the sheet with the increasing reduction, stimulating the recrystallization kinetics of FSS sheet.

The partially annealed sheets, noted as A and B in Fig.4, were examined using EBSD technique to comprehensively characterize the recrystallization evolution during annealing, as shown in Fig.5. For the sheet with 50% reduction and annealed for 14min, only a few recrystallized grains were formed. And most of the recrystallized grains were formed at grain boundaries. Orientation analysis indicates that the recrystallized grain prefer to nucleate at the grain boundary of γ/α and γ/γ boundaries. This result is similar to the observation by Sinclair et al [10, 11]. While for the sheet with 70% reduction and annealed for 10min, more recrystallized grains formed. The newly formed recrystallized grains not only nucleate at grain boundary but also grain interiors with in-grain shear bands as shown in the center area in Fig.5b. This infers that the grains with in-grain shear bands can also be served as nucleation sites for recrystallization.

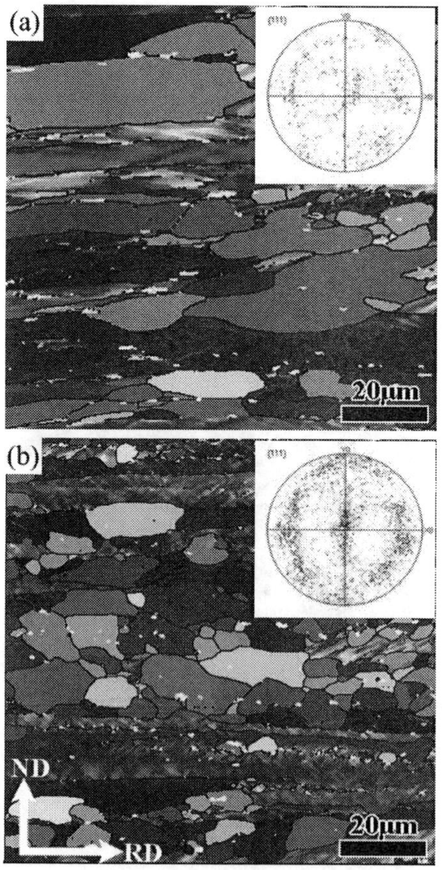

Fig.5 Orientation imaging maps ofpartially annealed cold rolled sheets. a sheet with 50% and annealed for 14min, b sheet with 70% reduction and annealed for 10min. ({111} pole figures indicate the orientations of the recrystallized grains)

Orientation analysis indicates that the orientations of newly formed recrystallized grains in Fig5a and b have realtions with the deformed grains. The deformed grains with blue colors, which have orientations near to <111>//ND, have prioriety to recrystallization. The newly formed grains inherit the orientations of deformed grainswith {111}<11-2> and {111}<1-10>. This character is same to the IF steel [12]. It is suggested that these grains seem to grow faster than other grains through the mobile of low energy CSL boundaries [13, 14]. The resutls in this work indicate that the orientations of recrystallized grain with <111>//ND at nucleation period already have priorieties. And this is supposed to contribute the formation of<111>//ND texture in the final sheets, which is benefit for the impovemetn of formability of FSSs. Furthermore, the intensity of <111>//ND texture for the recrystallized grains increases with the increasing reduction. This suggests the <111>//ND texture can be intensified by increasing the cold rolling reduction.

4. Conclusions

The microstructure and texture evolution during cold rolling and annealing were investigated in this work. The main conclusions are as follows.

1. With the cold rolling reduction higher than 30%, in-grain shear bands were formed in the cold rolled FSS sheets. With the increasing cold rolling reduction, more LAGBs were formed in the deformed grain interiors.

2. Microstructure observations of the annealed sheets indicate that the recrystallization kinetics of FSS sheet during annealing follow JMAK kinetics and increase with the increasing cold rolling reduction.

3. The recrystallized grains prefer to nucleate at grain boundaries of γ/α and γ/γboundaries. With increasing reduction, the recrystallized grains also nucleate at grain interior with in-grain shear bands. And the orientation of recrystallized grains has closed related to the deformed grains. <111>//ND texture can be intensified by increasing the cold rolling reduction

This work was supported by a grant from the National Natural Science Foundation of China (Nos. 51604058), the Fundamental Research Funds for the Central Universities of China, the Open Research Fund from the State Key Laboratory of Rolling and Automation, Northeastern University.

References

1. K.H. Lo, C.H. Shek, J.K.L. Lai. Materials Science and Engineering: R. 65, 39 (2009).
2. I. Tikhovskiy, D. Raabe, F. Roters. ScriptaMaterialia. 54, 1537 (2006).
3. P.D. Wu, S.R. MacEwen, D.J. Lloyd, K.W. Neale. Materials Science and Engineering: A. 364, 182 (2004).
4. F. Gao, Z.-Y. Liu, G.-D. Wang. Journal of Materials Science. 48, 2404 (2013).
5. H. Liu, Z. Liu, G. Wang. ISIJ International. 49, 890 (2009).
6. Y. Yazawa, Y. Ozaki, Y. Kato, O. Furukimi. JASE Review. 24, 483 (2003).
7. C. Zhang, Z.Y. Liu, G.D. Wang. Journal of Materials Processing Technology. 211, 1051 (2011).
8. H.T. Yan, H.Y. Bi, X. Li, Z. Xu. Materials Characterization. 60, 65 (2009).
9. H.T. Yan, H.Y. Bi, X. Li, Z. Xu. Journal of Materials Processing Technology. 209, 2627 (2009).
10. C.W. Sinclair, J.D. Mithieux, J.H. Schmitt, Y. Brechet. Metallurgical and Materials Transactions A. 36A, 3205 (2005).
11. C.W. Sinclair, F. Robaut, L. Maniguet, J.-D. Mithieux, J.-H. Schmitt, Y. Brechet. Advanced Engineering Materials. 5, 570 (2003).
12. S.-H. Hong, D.N. Lee. ISIJ International. 42, 1278 (2002).
13. D. Raabe, K. Lücke. Materials science and Technology. 9, 302 (1993).
14. D. Raabe. Modelling and simulation in materials science and engineering. 15, 39 (2007).

Electrohydraulic sheet metal forming with flexible tools

Holger Pegel[1,3*], *Lasse* Langstädtler[1,3], *Marius* Herrmann[1,3], *Christian* Schenck[1,2,3], and *Bernd* Kuhfuss[1,2,3]

[1]bime, Bremen Institute for Mechanical Engineering, Badgasteiner Str. 1, 28359 Bremen
[2]MAPEX Center for Materials and Processing
[3]University of Bremen

Abstract. In cold metal forming usually a die and a punch are required to form sheet metal in the desired shape. These two parts of the tool need to be produced of hardened materials in high precision. In this paper it is shown, that by using electrohydraulic forming the punch is replaced by a hydraulic punch and the die can be made easily out of softer materials that seem absolutely not suited at first sight. This allows for rapid and cheap manufacturing of sheet metal parts.

Keywords: Metal Forming, Die, Rapid tooling

1 Introduction

Cold forming of sheet metals is a convenient and high efficient way for bulk production of a broad variety of parts. A conventional forming process is the mechanical deep drawing. In this process, a blank is formed using two corresponding tool parts. These tools directly reproduce their geometry and surface into the parts. Hence, both die parts must provide a sufficient quality concerning these features as well as a high durability to endure high process loads during forming. In consequence, the production of those tools themselves are cost and time intensive, since they are usually made of hard working steel and require narrow tolerances. To reduce the costs and the manufacturing time one tool side is replaced by a shapeless and contactless working shock wave like it is provided by the electrohydraulic forming (EHF) process. Doing so also the costly tolerances between the die parts become obsolete.

Electrohydraulic forming is a high speed forming process first researched by Yutkin in 1955 [1]. The forming energy is produced by shorting an electric circuit. When the circuit is shorted an electric arc is discharged or an aluminum wire is vaporized within a liquid resulting in a shock wave. The effect of the electric wire explosion is known since 1774 [2]. The shock wave is unlike the most shock waves not spherical but cylindrical. This is a result of the initiation of the shock wave that is not located in a single point but along the whole wire or along the travel of the electric arc with a velocity of up to 3.5 km/s [1, 3-6]. In this process the shock wave within the fluid replaces a mechanic punch. Due to this only a one sided die is needed. At conventional hydraulic and electrohydraulic forming the die is produced from metal with money and time consuming methods like hard milling. This is only profitable for mass production of the formed parts by this tools. In recent times, customizability and smaller production series as well as rapid production become more and more important. Since in electrohydraulic forming the force in not applied quasi static but with a very short impulse the die material needs to fulfill the requirement of a high resistivity against short-time and high frequency excitation and not necessarily of high rigidity against static loads. Hence, the paradigm of designing dies can change with the changing requirements. These thoughts inspired an experimental study in which different materials were tested concerning their ability to be applied for fast and cheap die manufacturing.

2 Experimental setup

For all experiments a modular forming setup was used which is illustrated in Figure 1. The modular design of the setup allowed changing of all components to test a diversity of different parameters and die geometries. The capacitor bank with a capacity of 104 µF and a maximal loading voltage of 2500 V was connected to one connector on each side of the setup. The connectors were equipped with a spring mechanism to hold the wire in place. The wire featured a diameter of d_w = 0.3 mm and was made out of 99.5 % aluminum (aluminum alloy 1050). The connectors were insulated from the steel pressure chamber with PLA (polylactic acid) plastic bushings. The chamber was basically a cavity in a steel cylinder. On top of the chamber was an opening which

* Corresponding author: pegel@bime.de

could be sealed and opened with a screw for filling up the chamber with distilled water.

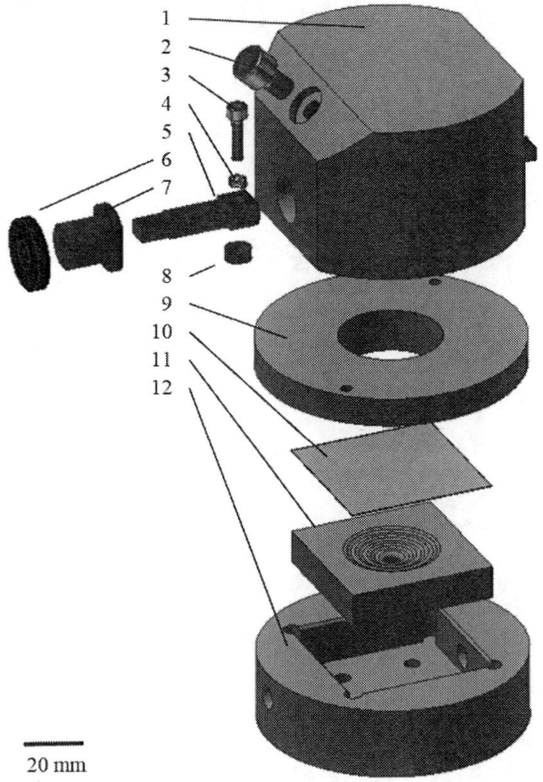

Fig. 1 Experimental setup: 1 pressure chamber; 2 closing screw; 3,4,8 wire clamp; 5 connector; 6,7 insulation; 9 aperture; 10 sample; 11 die; 12 base plate.

Underneath the chamber was an aperture made of Al7075 restricting the shock wave and as a result the forming area to a 34 mm circle. The sample was directly located under the aperture. The sample was a square piece of sheet aluminum with a side length of L = 50 mm and a thickness of $s_0 = 0.5$ mm. Under the sheet metal was the forming die. The die was fixed in a base plate. This base plate distributed the closing forces of $F_{cl} = 50$ kN that were provided by a hydraulic press across the die.

The dies were basically shaped as a prismatic square (50x50x12 mm³) with a circular forming geometry in the center. The forming geometry which is displayed in Figure 2 was a triple-staged bore with a stage size of 1 mm each and a transition radius of 1.7 mm. Located in the center is a 5.1 mm bore to handle with the compressed air during the process which is important to receive a good result [7]. The cavity in stages was chosen to enable a fast and easy evaluation of the forming results.

For some tested materials a 3D printed frame structure was applied which is shown in Figure 3. This die holder provided a cylindrical cavity that could be filled with liquid or pasty materials to be shaped off with finished inlets. The plug is necessary to be able to remove the inlets from the forming frame easily.

Dies made from five different materials were tested which were produced in three different ways. PLA (polylactide) was 3D printed with an infill of 90 %, silicone and gypsum were casted in molds and sand and modeling clay were embossed with a 3D printed forming stamp into the right couture. The die made out of PLA is shown in Figure 2 and the other four are shown in Figure 4. All dies were tested with loading voltages of 1000, 1500, 2000 and 2500 V.

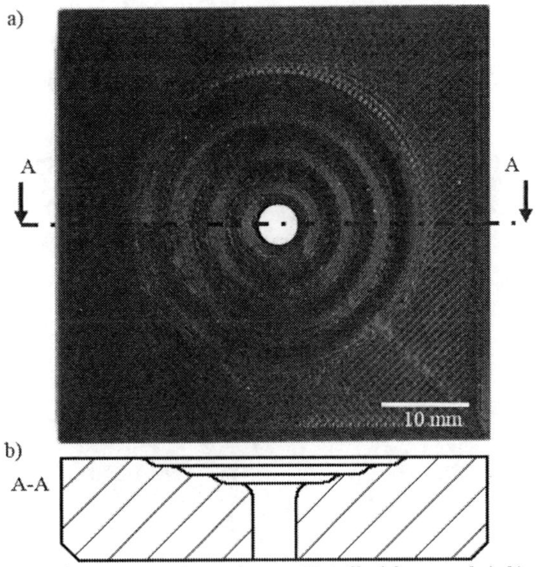

Fig. 2 Die geometry: a) top view on PLA die (photography), b) cross section.

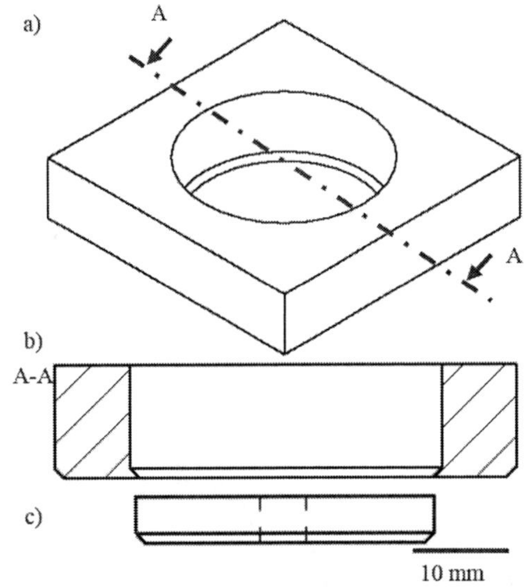

Fig. 3 Sketch: a) the forming frame, b) cross section and c) the plug.

 https://doi.org/10.1051/matecconf/201819012001

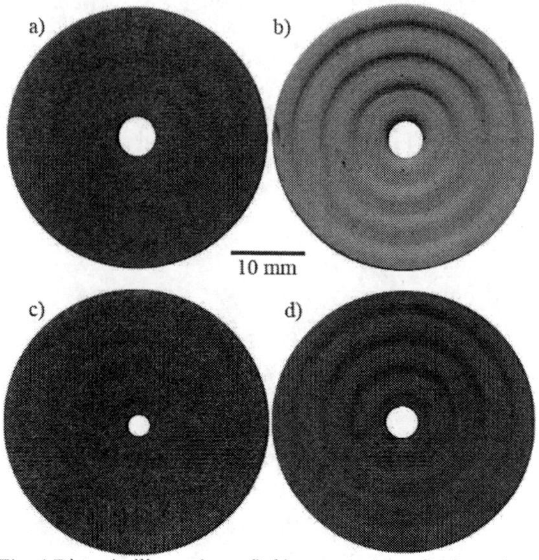

Fig. 4 Dies: a) silicone (casted), b) gypsum (casted), c) sand (embossed), d) modeling clay (embossed).

3 Execution and observation

3.1 Tool Handling

During the conduction of the experiments it became obvious that some of the die materials were easier to handle than others especially taking into consideration the involvement of water in the process. The handling of the 3D printed die was the easiest. It was reusable, ready as soon as it was printed and it was waterproof. The gypsum and silicone dies were also easy to handle. After filling the casting molds, they just had to cure before they were ready to use. The gypsum die was single-use whereas the silicone die was reusable. Modeling clay proved to be easy to shape and easy to reshape for reuse. The most complicated handling was observed with the sand die. The shape turned out to be very fragile and the surface structure varied with every new die embossing. Additionally, the used kind of sand became instable and not shapeable when it became wet.

3.2 Measurement procedure

All samples had been scanned with a 3D scanner from the side that was touching the die during the process, which is the top side in Figure 5 a). The measurements were made with a confocal microscope, see Figure 5 b) which has the same perspective like Figure 5 a). As this research was conducted to explore the feasibility, every experiment was repeated only three times and only one sample was measured exemplarily for each set of parameters.

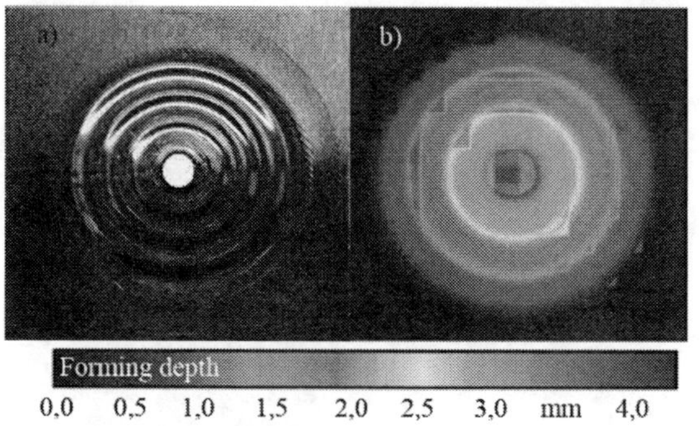

Fig. 5 Sample formed with 1500 V in a 3D printed PLA die: a) Picture, b) 3D scan.

4 Result and discussion

Best results were achieved with the 3D printed dies, Figure 5 a). All stages of the die were properly formed and the circularity was quite well. The PLA die surpassed expectations. Besides the form also the surface was imprinted. The structure of the printed die is clearly visible in the sample (compare Figure 2 a) and Figure 5 a)). In Figure 6 the height profiles of the samples are displayed. These are deducted from the 3D scans. Further, the PLA die was capable to be applied as a cutting tool. The center bore of the die worked as a cutting edge despite the smooth transition radius. During the conducted experiments no tear or wear was visible at the PLA die after the twelve conducted experiments.

With an increase of the loading voltage the shock wave got stronger and quicker as a result the forming force increased which usually results in a more detailed forming result [8]. In case of the PLA dies the forming result stayed more or less the same except for the test with 1000 V. In this test the sheet metal was not cut in the center due to the lower forming energy.

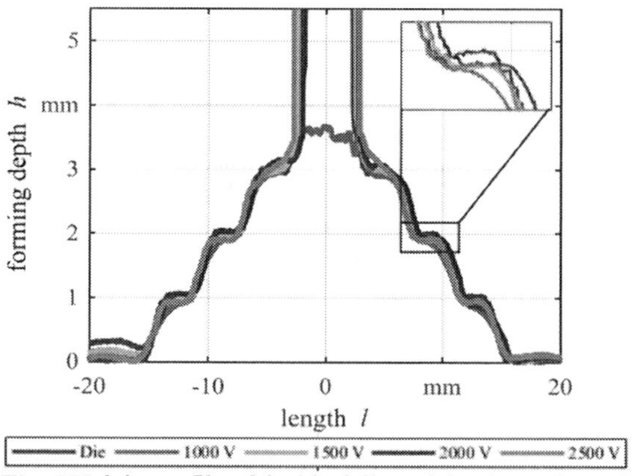

Fig. 6 Height profile of the ideal die and the sheets formed in the PLA die with different loading voltages.

The 2D height profiles of the samples made with gypsum dies show that the individual stages of the die were smoothened and not very defined see Figure 7. Also the contour is always above the designed die outline. This means that the die was compressed during the process.

https://doi.org/10.1051/matecconf/201819012001

Most likely this is a result of the porose structure of gypsum. Due to deformation of the die for every test a new die was used. By the higher average forming depth with the same base diameter, the whole sample appeared pointier and the steps were formed only rudimentarily. Similar to the PLA die cutting with the gypsum die was possible at U ≥ 1500 V.

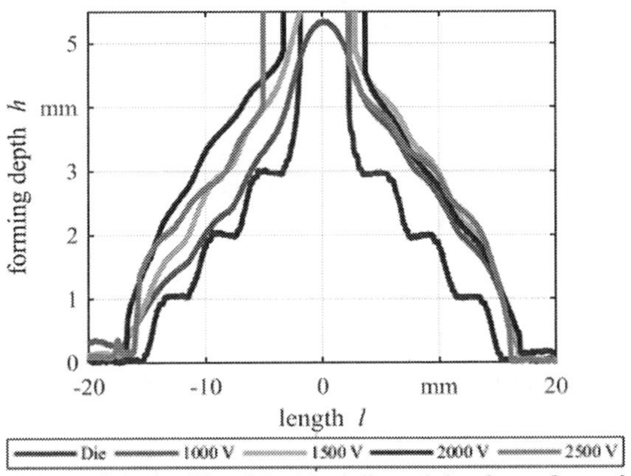

Fig. 7 Height profile of the ideal die and the sheets formed in the gypsum die with different loading voltages.

The samples made with a die out of modeling clay showed a very different result. The shape of the die was reduced to a cone like shape with concave sides and the stages were not recognizable, see Figure 8. Unlike the previous results none of the samples were cut in the center. This shows that with the modeling clay die the very rough shape of the cone could be formed but no precise stages. The die was deformed during every experiment and the center hole closed, but the die could easily be reshaped with the forming stamp and then used again. The sample made with 2500 V showed a hutch in the center against the forming direction. This may be a result of trapped and compressed air between the die and the sample due to the closing of the center bore. Summarized, the tested modeling clay is not a suitable die material for detailed tool geometries.

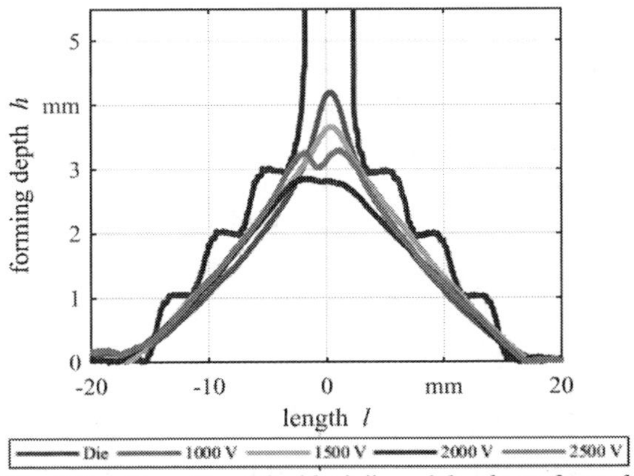

Fig. 8 Height profile of the ideal die and the sheets formed in the modeling clay die with different loading voltages.

The dies made out of sand produced a forming result worse than the one made with the modeling clay die. The stages are again not visible instead a cone like shape was reached with a major stage at a forming depth of about 1 mm, see Figure 9. This over all stages outside perimeter seems to be restricted by the forming frame and not the die itself. Only the test with 2000 V loading voltage has a smoother stage at an even higher forming depth. It seems like the whole sand die was compressed down about 1 mm and then a small, pointy cone formed in the center. Cutting with sand dies is also possible at U ≥ 2000 V. The surface of the sample is very rough due to the structure of the sand that was transferred onto the sheet metal. The die looked completely destroyed after every experiment and needed to be remodeled. In combination with the difficulties in the die handling, sand or at least this kind of sand does not seem to be suitable as a die material for electrohydraulic forming without further preparation like preconsolidation or the use of binder.

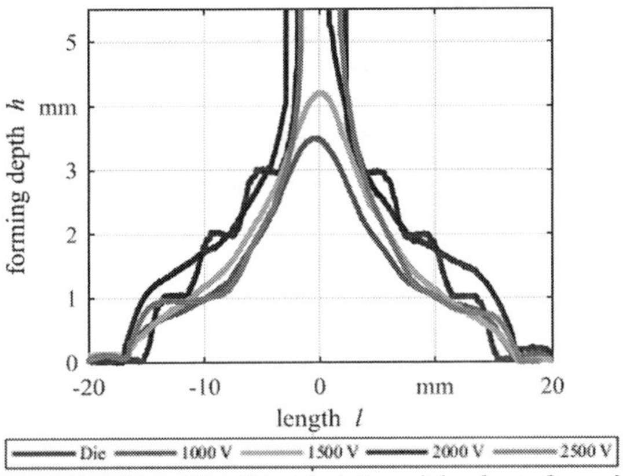

Fig. 9 Height profile of the ideal die and the sheets formed in the sand die with different loading voltages.

The fifth tested die material was silicone. Figure 10 shows that the shape of the samples was also smoothened and the stages were not recognizable. Tree of the samples showed a very pointy tip in the center. When the sample was made with U ≥ 2000, the silicone die had a darker color in the center and the smell of burned plastic was noticeable. Besides that, no wear and tear was detected

For a better comparison the contour accuracy was evaluated by defining the quality measure RIQ (relative integral quotient) by equation (1). The formed cross section was related to the cross section of the ideal die. The values in the interval $l \in [-5, 5]$ were neglected to consider the partial cuttings in the center.

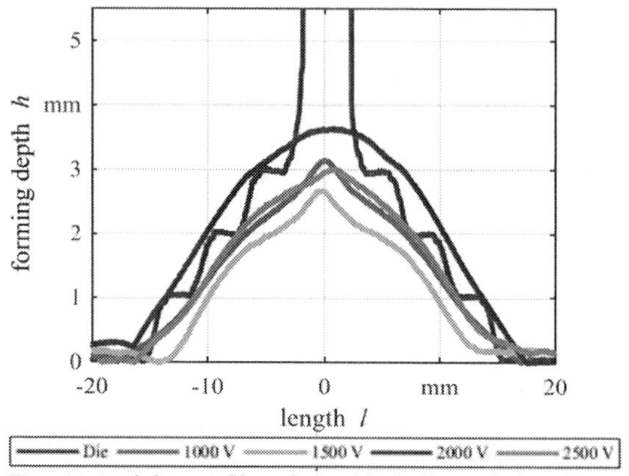

Fig. 10 Height profile of the ideal die and the sheets formed in the silicone die with different loading voltages.

$$RIQ = \frac{\int_{15}^{-5} h_{sample}\,dl + \int_{5}^{15} h_{sample}\,dl}{\int_{-15}^{-5} h_{die}\,dl + \int_{5}^{15} h_{die}\,dl} * 100\ \% \qquad (1)$$

The relative integral quotients in Figure 11 show for the samples made with the PLA die a nearly perfect RIQ between 95 % and 102 %. Interestingly the highest RIQ was achieved with 2000 V and not with the highest loading voltage.

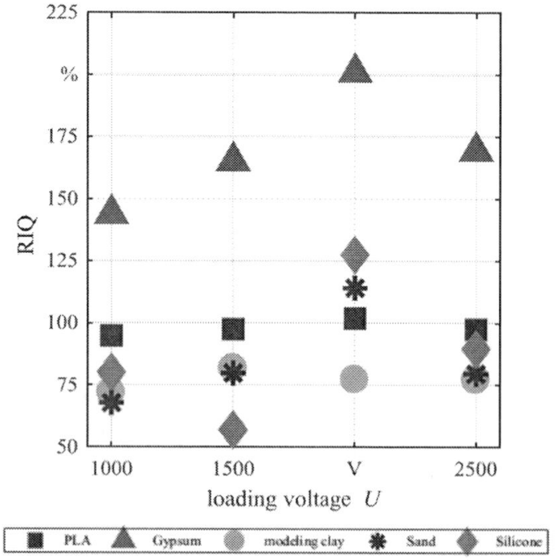

Fig. 11 RIQ for different loading voltage and different dies.

The RIQ of the samples made with a gypsum die are always considerably above 100 % with values between 144 % and 201 %. The RIQ rises with an increase in loading voltage up to 2000 V and then drops back down at 2500 V. When modeling clay was used as a die material the RIQ is very stable with values between 72 % and 82 %. Especially the values for 1000 V, 1500 V and 2500 V vary only about 4%. The samples made with a sand die show a larger variation with the RIQs between 68 % and 114 % with the highest value at U = 2000 V. If silicone is used as material for the die the RIQs are the most inconsistent of all tested die materials. With a variation between 57 % and 127 % and no recognizable trend.

It is an unexpected result, that except for modeling clay which produced a very stable RIQ all other die materials reached the highest RIQ at U = 2000 V and not at the highest loading voltage. It is assumed that, with U = 2500 V the sample reached a very high speed when it hit the die and bounced back of the die.

5 Excursion

To illustrate a farfetched possibility of the new approach of flexible tools a jelly bear was attempted to be used as a forming tool. Instead of a die a jelly bear was directly placed in the die holder. The attempt was realized with 1500 V and with an aluminum sheet with s_0 = 200 µm instead of the previously used 500 µm. The result is a nearly perfect copy of the jelly bear shown in Figure 12). Even the small details like the eyes and the dents in the jelly bears belly are visible. This shows that even with very flexible die materials a good forming result can be achieved and using very thin sheet metal the level of details can be increased.

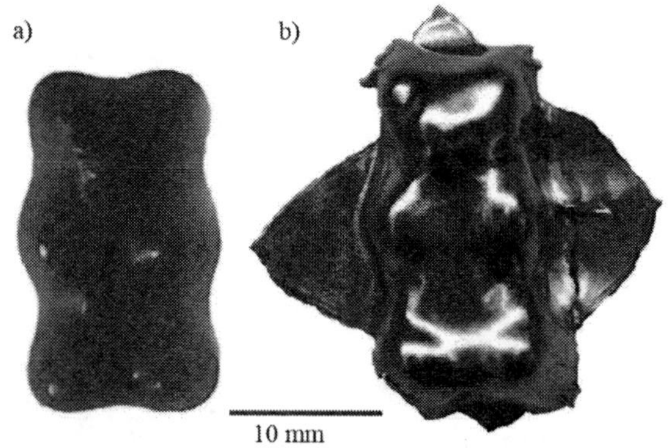

Fig. 12 Molding of a jelly bear: a) tool (common candy); b) shaped sheet metal (s_0 = 200 µm).

6 Conclusion and outlooks

To discover the possibility of forming sheet metal in dies made out of a variety of much softer materials compared to hardened steel experiments were executed. The chosen forming technique was a high speed forming process, the electrohydraulic forming. Five chosen materials PLA plastic, gypsum, modeling clay, sand, and silicone provided different flexibilities and structures. Dies from all materials were tested with a circular staged shape with variation of loading voltages that resulted in different forming energies in the process. The main results of the conducted experiments are:

- Electrohydraulic forming with the presented die materials is generally possible.
- 3D printed PLA dies provided the best forming results.

- Even cutting was possible with PLA, gypsum and sand dies.
- Elastic flexibility and even plastic deformation of the die are not criterions of exclusion but must be considered in the design process of the die.
- Softer die materials provide usable results if no small details or sharp transitions are desired.
- With new die materials, rapid sheet metal forming operations are possible when electrohydraulic forming is applied.

In further experiments more complex die geometries should be investigated, as well as different sample sizes. Also other high speed forming processes like electromagnetic forming should be considered for sheet metal forming with flexible tools. In addition, other operations like cutting with those tools need to be investigated. In the design process, the dynamic behavior of the dies in the range of high frequency excitation must be taken into account to find optimized solutions.

7 References

1. K. Siegert, *Blechumformung Verfahren, Werkzeuge und Maschinen*, (2015)

2. Y.A. Kotov,. Nanotechnologies in Russia. **4**, Nos. 7-8, (2009).

3. M. Woetzel, M.J. Löffler, E.Spahn, H. Ritter *Preliminary examination of high-velocity metal-shaping with electrical wire explosion* (2006)

4. G.S. Sarkisov, S.E. Rosenthal, K.R. Cochrane, K.W. Struve, C. Deeney, D.H. McDaniel PHYSICAL REVIEW. **71**, (2005)

5. Y.E. Krasik, A. Grinenko, A. Sayapin, S. Efimov, A. Fodotov, V. Gurovich, I. Oreshkin, IEEE Transactions on plasma Science Vol.**36** No.2 (2008)

6. W. Müller, Zeitschrift für Physik, Bd. **149**, p. 397-411. (1957)

7. W. Homberg, C. Beerwald, A. Pröbsting, 4th International Conference of High Speed Forming. (2010)

8. A. Abdelkefi, P. Malécot, N. Boudeau, N. Guermazi, N., The International Journal of Advanced Manufacturing Technology. (2016)

 https://doi.org/10.1051/matecconf/201819012002

Influence of material damage during the forming process on the vibration fatigue behaviour of a clinched connection.

Bernd Maier[1,*], *Markus* Klingler[1], *Sabine* Böhm[1], and *Birgit* Awiszus[2].

[1]Robert Bosch GmbH, Automotive Electronics, 72770 Reutlingen, Germany
[2]Chemnitz University of Technology, Department of Virtual Production Engineering, 09126 Chemnitz, Germany

Abstract. This paper presents the results of an analysation of forming damage during the joining process on the fatigue failure behaviour of a clinched connection. The two joining partners out of aluminium, AlMg0.7Si on the punch side and AlMg3 on the die side, are clinched with a flat clinching system and the use of a preformed hole at the die side. By the use of two different geometry variants it could be shown, in fatigue Wöhler tests, that the fatigue failure behaviour is influenced by the material damage in the deformed material of the clinch. The material damage during the clinching process could be calculated with the help of FE-Simulations. So it was possible to show that more forming damage during the joining process causes less lifetime of the connection.
Keywords: Joining, Damage, Fatigue

1 Introduction

The mechanical joining process clinching is getting more and more important for the automotive industry. Especially for light weight designs, clinching is a very interesting joining method to connect sheet metal components. The main advantages of clinching are that there is no need of additional joining parts like screws, there is no heat injection in the joining partners, there is no emission of welding gases during the clinching process and there are no problems with coated surfaces of the joining partners. [1, 2]

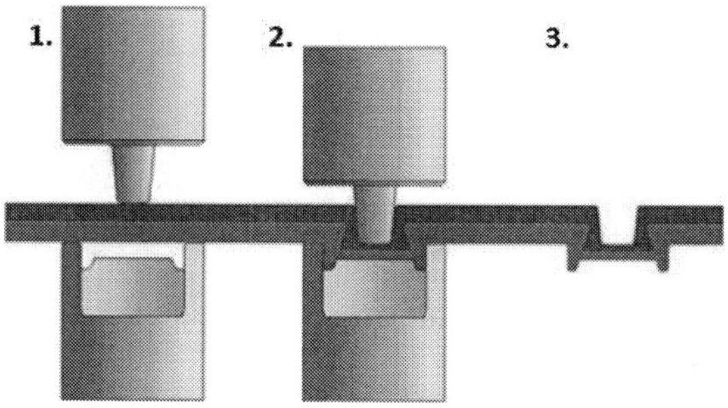

Fig. 1. Sequences of a no-cutting, single part die clinching process [3].

During the clinching process, the material of the joining partners is deformed by a punch and a die to get an interlocking connection. The joining process is illustrated in picture 1. The die is placed under the bottom metal sheet, the punch is placed above the upper metal sheet. When the punch moves downwards it presses the material into the die. Because of the geometry of punch and die the material is deformed in a way that an interlocked connection is formed. During this process the material is highly deformed and this deformation could possibly affect the material properties. During the forming process the accumulated plastic strain of the material reaches 400%. This is only possible with ductile metals and under pressure dominated load. [1, 2, 4]

In this work a special clinching variant was analysed which will later be used to connect the housing parts of an electronic control unit (ECU). The main reasons why clinching was chosen as a joining method for the product were, the low costs per part, that it is possible to join partners with coated surfaces and that no finishing of the connection area is necessary.

1.1 Special Clinch Method

In this work, a special clinching variant was examined. Due to the lack of space, it was not possible to use a standard clinching method with a single or a multiple part die to connect the two parts of the ECU housing. The available space was sufficient only to place in a nearly flat anvil. The used clinching method is shown in picture 2. The bottom metal sheet was provided with a preformed hole including a chamfer at the lower side of the hole. Instead of a die, a nearly flat anvil was used. During the clinching process, material of the upper metal sheet is formed through the hole and behind the chamfer. So it is possible to build up an interlocking connection.

1.2 Fatigue Loads

Automotive components are subjected to high vibration loads. Consequently, it is very important to have a closer look at the fatigue failure behaviour of the connection under these high vibration loads. It is necessary that these vibration loads cause no failure of the clinch during the lifetime of the vehicle. The main focus of this work lies in understanding how the change in material properties during clinching influences fatigue failure behaviour. The question is, to what extent changes in the material properties can be allowed to ensure that there will be no fatigue failure of the connection over lifetime.

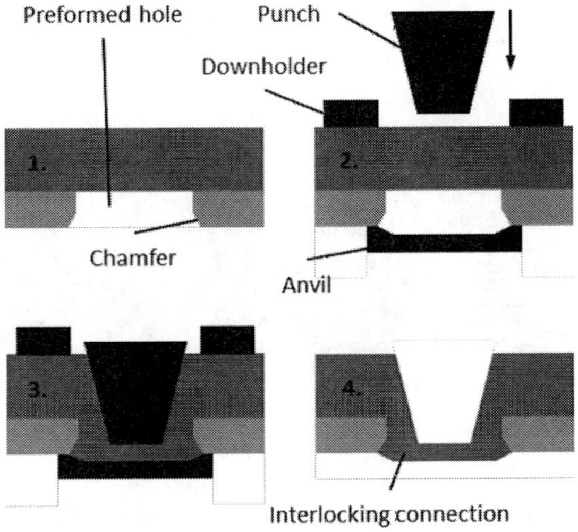

Fig. 2. Special clinch geometry used for electronic control unit housing.

2 Research method

To reach the goal there were different research methods used. The mainly used method was the Finite Element Method (FEM) to calculate the changes in the material properties during the forming process. With this method it is possible to have a look inside of the material during the clinching process. It is possible to predict the effects of cold hardening and also to calculate the damage during the forming process. To use FEM to calculate the hardening effects, it is necessary to have the hardening curves of the used material which have to be measured in experiments. The material hardening can be analysed with the help of micro Vickers hardening tests on cross sections, these cross sections will also enable validation of the forming simulations.

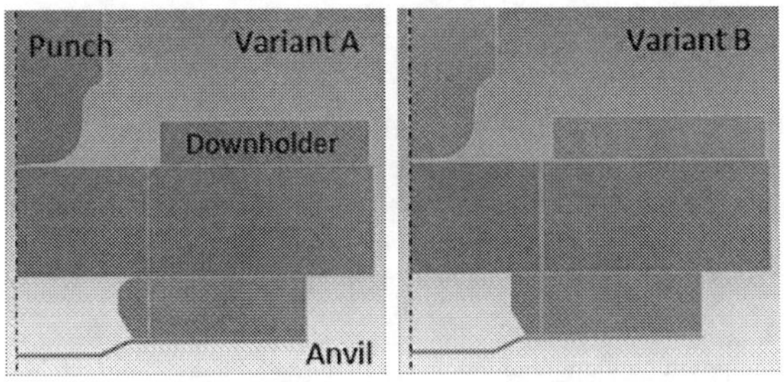

Fig. 3. Geometric difference between the two analysed samples variant A and variant B.

The fatigue failure behaviour can be analysed with fatigue Wöhler tests. To show the influence of the material damage during the clinching process two different geometries of the preformed hole were compared. Picture 3 shows the geometric difference between the two samples. On the left side variant A is printed with a rounded edge on the upper side of the bottom metal sheet. The right side shows variant B with a sharp edge on the upper side of the preformed hole.

2.1 Material and Data

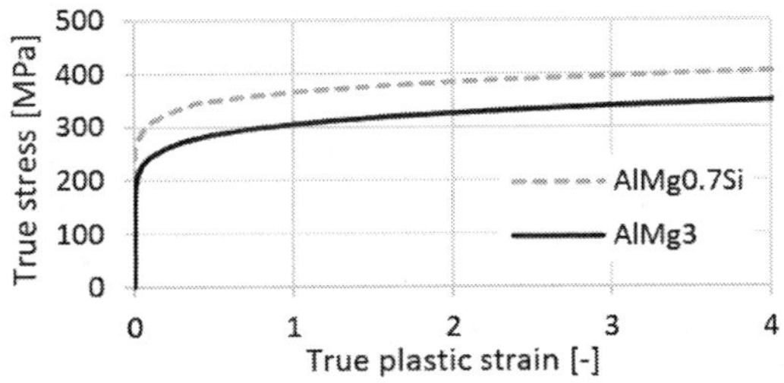

Fig. 4. Hardening curves used for the forming simulations.

The used material in this work was aluminium. More specifically, AlMg3 for the bottom metal sheet with the preformed hole and AlMg0.7Si for the upper metal sheet with the high deformations during the forming process. To simulate the forming process it is necessary to have hardening curves of the used material until high deformations. These hardening curves cannot only be calculated out of tensile stress test, because with tensile stress tests the specimens fail at a deformations value of 10-15%. But for the simulations the hardening curves are needed until a deforming value of 400%. Because of this a test program was done to measure the hardening behaviour until an accumulated plastic strain of 200%. In this test program tensile tests have been done of different metal sheets. The metal sheets have been preformed in a rolling process to get specimens with a different value of accumulated plastic strain. So it is possible to measure the stress strain curves for different values of accumulated plastic strain and put them finally together to one curve. Out of this stress and strain curve hardening cure can be calculated. The sector between 200% and 400% strain has been calculated with an exponential law. In picture 4 the used hardening curves are shown.

2.2 Forming simulation

To simulate the forming process, the FE-Program Ansys was used. With Ansys mechanical it is possible to simulate the forming process in an axis-symmetric 2D-Model. The remeshing tool of Ansys (since Ansys 17.0) allows to simulate forming processes with high deformations in the implicit Ansys code. With this method it is possible to get the results of the simulation very fast. The model was build up with the hardening curves shown before for the two different aluminium metal sheets. The tools (steel) were modelled as a linear elastic material model with an E-module of 210 GPa. The friction between the parts was modelled with the coulomb law of friction. Between the two aluminium metals a friction factor of μ=0.2 was set. Between the aluminium metals sheets and the tools, the friction factor was set to μ=0.15, except between the punch and the aluminium metal sheet. This friction factor was set to μ=0.1 because the punch is oiled in the clinch process.

2.3 Damage Model

To calculate the damage in the material while the forming process the Oyane criterion (1) was used. Behrens et al. have shown that this damage model works for clinching processes [5].

$$C = \int_0^{\bar{\varepsilon}_{pl}} \left(\frac{\sigma_m}{\bar{\sigma}} + B\right) d\bar{\varepsilon}_{pl} \geq C_{crit} \quad (1)$$

The Oyane criterion uses stress triaxiality $\frac{\sigma_m}{\bar{\sigma}}$, equivalent plastic strain $\bar{\varepsilon}_{pl}$ and two material parameters B and C_{crit} to calculate the damage during the forming process. The two material parameters have to be defined. Behrens used 2.36 for the parameter B out of the work from Chen [6] and set C_{crit} to 2.13 for his studies [5].
In his studies he used an aluminium material similar to the material which is used in this work.

2.4 Micro structure analyse

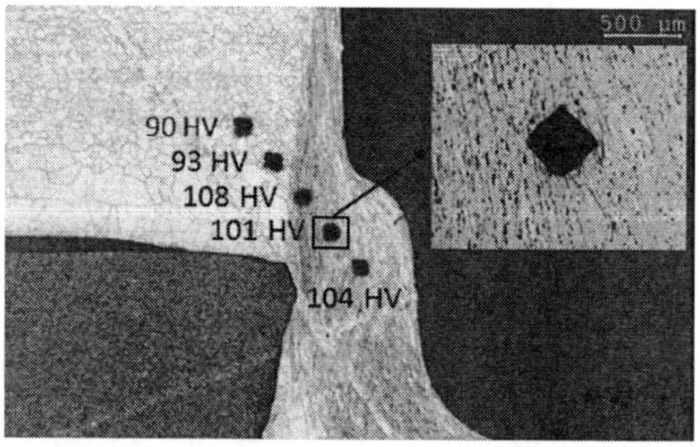

Fig. 5. Micro Vickers hardness test of deformed material after clinching process (exemplified by variant B).

To analyse the micro structure of the clinch and to validate the simulated results clinch specimens were built up. With the help of cross sections it was possible to observe the microstructure inside the clinch. A micro Vickers hardness test was done to show the effects of cold hardening during the forming process. The micro hardness measurement was carried out according to DIN EN ISO 6507-1. The impressions were generated with a load of 10 kg. The micro Vickers hardening test on a cross section is shown in figure 5 exemplified by variant B. Unfortunately, the method is not exact enough to represent differences between variants A and B. Picture 5 also shows the microstructure of the clinch. It is shown which area of the two joining partners is deformed and how strong. The form of the grains changes form nearly square to rectangular grains which could be 5 times longer than it is wide. The comparison between the micro Vickers hardening measurement and the deformed grains shows that the hardening increases with more deformations in microstructure. So it is possible to evaluate the hardening effects on the deformations in microstructure.

2.5 Fatigue Damage Failure

To analyse the fatigue behaviour of the clinched connection Wöhler fatigue tests have been done. In this study the main attention was on the fatigue behaviour under peeling tensile load of the clinched connection.

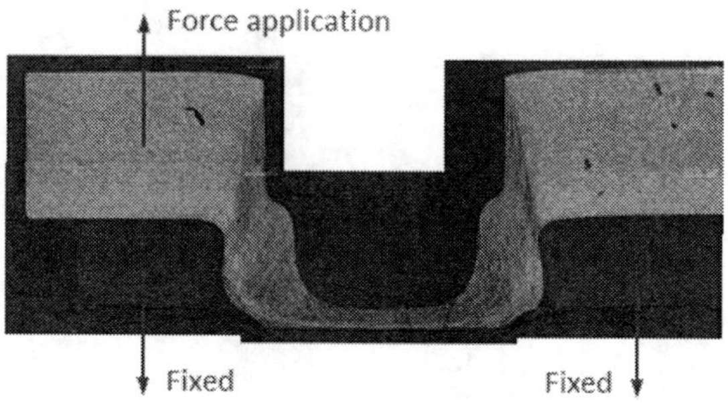

Fig. 6. Force injection of the peeling tensile loaded fatigue Wöhler tests (exemplified by variant A).

Therefore, specimens of clinched connections of variant A and B have been tested under peeling tensile swelling load as shown in picture 6. In the Wöhler test the bottom metal sheet is fixed. The upper metal sheet is loaded on one side with the tensile force. As a result, the clinch is under a peeling tensile load.

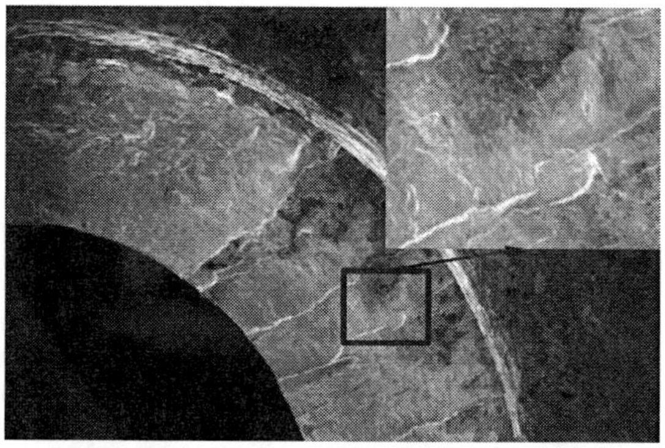

Fig. 7. SEM picture of fatigue crack fracture surface of the clinch (exemplified by variant A).

After the Wöhler tests, the fracture surfaces were analysed under a scanning electron microscope (SEM) which clearly shows that the fracture is a fatigue crack and not a static crack. Figure 7 shows a typical image of a fatigue crack, with fatigue striations. The picture shows that the whole fracture surface failed under fatigue.

3 Results and Discussion

The results of this work start with the results from the fatigue Wöhler experiments. Subsequently, the results of the FE-Simulations are presented.

3.1 Fatigue failure behaviour

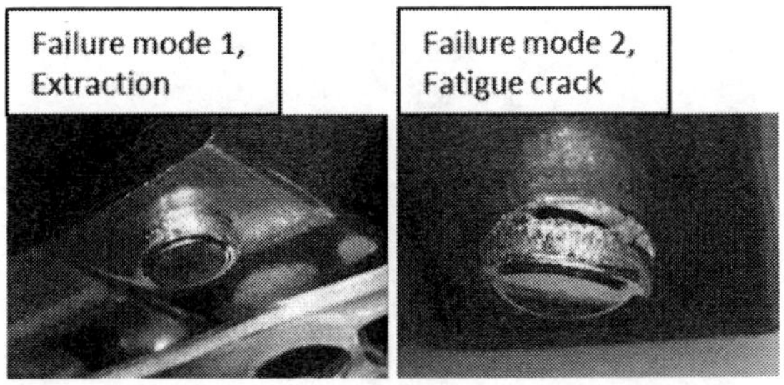

Fig. 8. Failure modes of fatigue tests, extraction on the left and fatigue crack on the right side (exemplified by variant A).

The fatigue Wöhler test has shown two different failure modes for the clinched connection. The first failure mode is the extraction of the upper metal sheet form the bottom metal sheet. This failure is shown on the left side of figure 8. The second failure mode is a fatigue crack through the deformed material of the upper metal sheet, shown on the right side of figure 8. The occurring failure mode depends on the testing force. For higher forces the clinch failed under mode 1 and for lower forces the clinch failed under mode 2. In the used product only the lower force is expected. So that failure mode 2 is the interesting one.

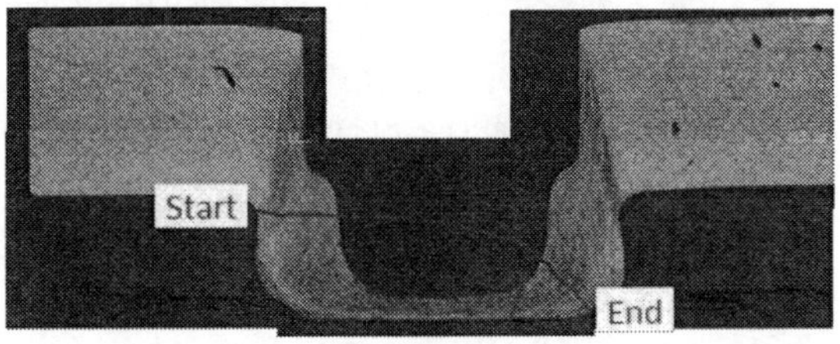

Fig. 9. Fatigue failure crack path of failure mode 2 (exemplified by variant A).

The path of the fatigue crack of failure mode 2 was similar in every clinch specimen, as shown in Figure 9. The picture shows where the fatigue crack starts and where it ends. Important for the clinch as connection is the area where the fatigue crack initiates. The Wöhler test has been done for both geometry variants A and B. The compression between the two variants shows that the specimens of variant B failed earlier than the specimens of variant A. Variant B has nearly 50% less lifetime versus variant A. A possible reason is that because of the sharp edge of variant B it is a more cutting forming process than in variant A. Even though there are no forming process cracks it is likely that the sharp edge causes a higher forming process damage in the area were the fatigue crack initiates. So that the fatigue crack can initiate earlier or grow faster in the area with higher forming damage. This gives the idea that the loss of lifetime might depend on a higher predamage of the connection during the forming process. If it is possible to calculate the predamage during the forming process and correlate it to a loss of lifetime, it would be possible to predict the lifetime of the clinch only based on the forming simulation. This would help to improve the punch and anvil geometry keeping the fatigue behaviour in mind. But for this first a working forming simulation model is needed.

3.2 Comparison between simulation and experiment

To ensure that the simulation model gives the right results it has been compared with the experiments. Figure 10 shows the experimentally measured and simulated force displacement curves of the clinching process. In the figure it can be seen that the simulation model is able to calculate the force displacement curve to a very good degree of accuracy. The difference between the two curves can be explained by the fact that the forming process is simulated in a 2D-Model and the real process is not completely axial-symmetric.

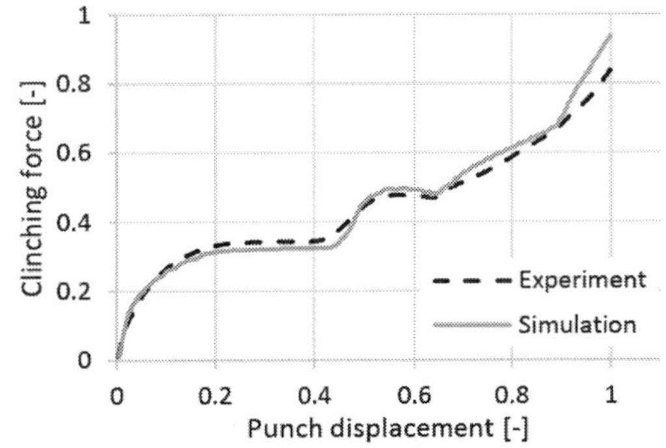

Fig. 10. Force displacement curves of the clinching process in experiment and simulation (exemplified by variant A)

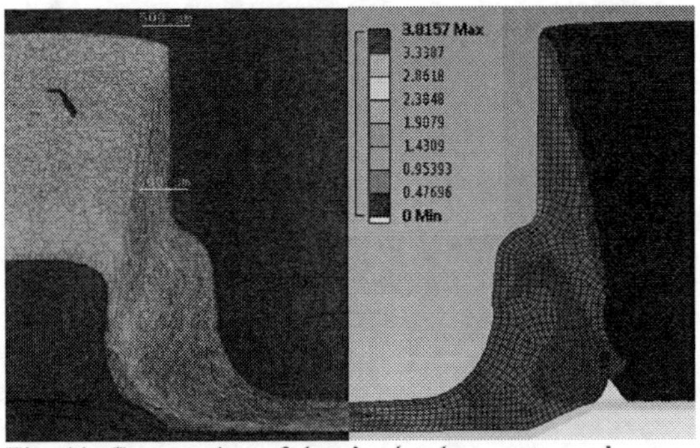

Fig. 11. Compression of the simulated geometry and a cross section of the clinch (exemplified by variant A).

Figure 11 compares the geometry of the clinch. On the left side the cross section of the real clinch is shown while on the right side the simulated geometry is shown. The colours show the value of equivalent plastic strain of the different area. The compression shows that the simulation model is able to describe the forming process. The left side of the figure also shows the deformation of the grains after the forming process. The higher deformed grains indicate a higher degree of hardening. The simulation model is also able to calculate this effect. The areas with the highly deformed microstructure in the real clinch match with the areas with high plastic strain in the simulation. The used simulation model is able to calculate the right force displacement curve of the forming process

 https://doi.org/10.1051/matecconf/201819012002

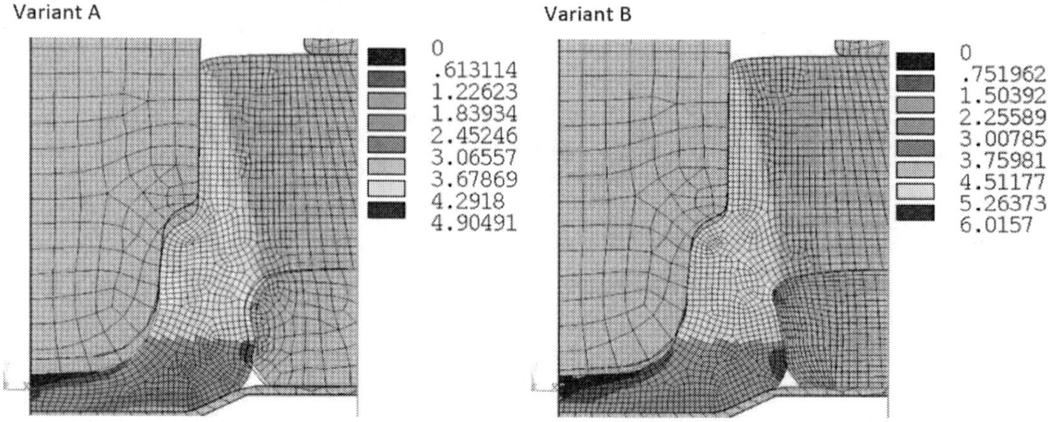

Fig. 12. Simulated damage values after the clinch process for variant A (left) and variant B (right).

and also the right material flow during the forming process including the effects of cold hardening. Now it has to be analysed if it is also possible to calculate a fitting damage value including the presented damage law.

3.3 Calculation of the damage value

In chapter 3.1 we presented that clinches of variant B fail more quickly than the clinches of variant A. A possible reason for this effect is that because of the sharp edge, the material in the area where the fatigue crack starts is more damaged during the clinching process. So the fatigue crack process starts earlier or grows faster. To demonstrate this effect, the forming process of two variants are simulated and the damage was calculated with the help of the Oyane damage model. Figure 12 shows the calculated damage values for the two variants A and B. The figure shows that the damage value of variant B (6.0) is higher than the damage value of variant A (4.9). This fits to the idea that the process forming damage of variant B is higher than in variant A. But both damage values exceeded the value of C_{crit} from Behrens. This would mean that with the used damage model including the used parameters both variants should have process forming cracks. As the cross sections shows no cracks in the areas with the high damage values, the damage model or the parameters from Behrens cannot be used to calculate cracks occurring during the forming process in this use case. Because of this other parameters for the value of B were tested, like B=1 or B=0.3. For all tested parameters of B the value of C_{crit} was higher for variant B than for variant A. But without tests where the value of C_{crit} is really reached it is not possible to adjust the parameter setup of the damage model for the used material.

4 Conclusion

This paper has shown that the vibration fatigue failure behaviour of a clinch connection can be influenced by damage effects during the clinching process. It could also be shown that more forming damage leads to an earlier failure in the fatigue Wöhler tests. The work also shows that with FE-simulations it is possible to calculate the forming process including effects of hardening. It was possible to simulate the right force displacement curve of the clinch process and the resulting geometry. This shows that the simulation model used works properly and is able to simulate the process sequence of clinching in a realistic way. A higher damage value corresponding to Oyane law is an indication towards the reason for an earlier fatigue failure of variant B. It was not possible to use the value of C_{crit} from literature (Behrens) to calculate forming cracks. Here the used parameter setup needs correction. As a result of such a correction, it could be possible to calculate the critical damage value encountered during the forming process leading to forming process cracks.

References

1. H. Fahrenwaldt, V. Schuler, J. Twrdek, *Praxiswissen Schweißtechnik*, Edition **5**, Chapter 6 *Fügen durch Umformen*, (2014)
2. E. Doege, B-A. Behrens, *Handbuch Umformtechnik*, Edition **1**, Chapter 3.12 *Fügen*, (2007)
3. TOX Pressotechnik, *TOX-Joining-Systems*, Product Catalogue, (2014)
4. B. Awiszus, U. Beyer, M. Todtermuschke, F. Riedel, *Flach-Clinchen – Simulationsbasierte Optimierung und Weiterentwicklung einer einseitig ebenen, einstufig gefügten Clinch-Verbindung*, UTFscience, **2** (2009)
5. B. Behrens, A. Bouguecha, M. Vucetic, S. Hübner, D. Yilkiran, Y. Jin, I. Peshekhodov, *FEA-based optimisation of a clinching process with an open multiple-part die aimed at damage minimization in CR240BH-AlSi10MnMg joints*, MATEC Web of Conferences **21**, EDP Sciences (2015)
6. D. Chen, C. You, F. Gao, *Analysis and experiment of 7075 aluminum alloy tensile test*, **11th** International Conference on Technology of Plasticity, Sciences Direct (2014)
7. Y. Bao, T. Wierzbicki, *On the cut-off value of negative triaxiality for fracture*, Engineering Fracture Mechanics **72**, Sciences Direct (2005)

Improving the formability of magnesium by cushion-ram-pulsation

Anja Rautenstrauch[1*], *Johannes Luft*[2], *Peter Birnbaum*[1], *Markus Baumann*[1], *Madlen Ullmann*[2], *Verena Kräusel*[1], *Dirk Landgrebe*[1], *Rudolf Kawalla*[2] *and Ulrich Prahl*[2]

[1] Professorship for Forming and Joining, Technische Universität Chemnitz, Germany
[2] Institute of Metal Forming, Technische Universität Bergakademie Freiberg, Germany

Abstract. The application of forming processes using lightweight magnesium alloys is known to be difficult with regard to minor formability caused by a small number of active slip systems, especially at low temperatures. However, a new approach for deep drawing at elevated temperatures considers flexible motion profiles of a servo-screw press. The so-called cushion-ram-pulsation (CRP) is a newly developed method of position-controlled motion. As a result, the process limitations for deep drawing are significantly extended. This study investigates the deep drawing process of AZ31 cups by variable motion profiles of cushion and ram, focusing on enhanced drawability. Therefore the initial AZ31 sheet was fabricated by twin-roll casting. Furthermore, the adjustment of the forming temperature for an improved forming process and optimized component properties are described. In addition, the results were evaluated in comparison to conventional deep drawing, and the superior technological potential will be outlined.

Keywords: Deep drawing, Magnesium, Sheet metal

1 Introduction

Due to the continuing trend towards lightweight design, magnesium as a construction material is predicted to become increasingly widespread [1]. Despite major advances in manufacturing technology, alloy development and corrosion protection, a broad range of applications is still limited in terms of formability [1–4]. Thanks to its crystal structure, the hexagonal solidifying magnesium has only a small number of independent slip systems at low temperatures: two instead of the required five [5,6]. Additionally, the flow performance of AZ31 sheet metal depends very much on the strain rate. . Higher strain rates lead to a lower logarithmic strain at fracture [2,7]. Therefore, in the past magnesium has only been successfully deep-drawn at low forming speeds and high process temperatures as described in [3,8–10].

Accordingly, the objective of the present study is to extend the process window of this process, to increase productivity and open up new areas for magnesium application. For this purpose, the investigation deals with the effect of cushion-ram pulsation (CRP) on deep drawing of magnesium blanks. Cushion-ram pulsation is a new method to extend the process limitations, despite the fact that it takes more time to conduct the position-controlled process by using a servo-screw press [11].

2 Experimental procedure

2.1 Production of the blanks

The initial sheets of magnesium alloy AZ31 were produced by a twin-roll casting process (TRC) [16]. After the TRCs the material (t = 5.35 mm) were hot rolled in three passes with a heat treatment between the first two passes (430° C, 8 h) and after the last pass (330° C, 0.5 h).

A sheet with a thickness t of 1.5 mm and a blank diameter of 220 mm also of 240 mm were used. The chemical composition was obtained of 3.09 Al, 1.02 Zn and 0.456 Mn (all wt.%) and cut in the samples.

2.2 Cushion-ram-pulsation and utilized process parameters

A DUNKES servo-screw press with 800 kN maximum ram force and 300 kN maximum cushion force was used (Fig. 1) to investigate the formability of magnesium by cushion-ram-pulsation (CRP), The servo-screw is characterized by a special axial powertrain layout.

Fig. 1. Servo-screw press [12].

The servo-driven spindle press technology allows high variability in motion of the ram and the cushion. Using

* Corresponding author: anja.rautenstrauch@mb.tu-chemnitz.de

the press technology, special movements are possible such as CRP and conventional deep drawing. The motion and speed of cushion and ram are individually programmable, which enables the adjustment of variable flange gaps.

The process of CRP is a variable kinematic process and consists mainly of two steps. The first step starts with a ram dwell time. Therein, the flange gap increases as the cushion moves away from the ram. In the next stage, the ram moves to draw the part while the gap between the die and the blankholder is opened by a defined value. In the following second step, the ram is stopped during the holding time and the blankholder flattens the flange by closing the flange gap. The resulting wrinkles are formed back by the movement of the blankholder. For the investigation in the present study, a modified CRP technique was used (Fig. 2) to close the flange gap to initial sheet thickness. The process cycle was adapted from former investigation of press hardening using CRP technology [12]. The final holding time is not implemented in this present study of forming magnesium sheet.

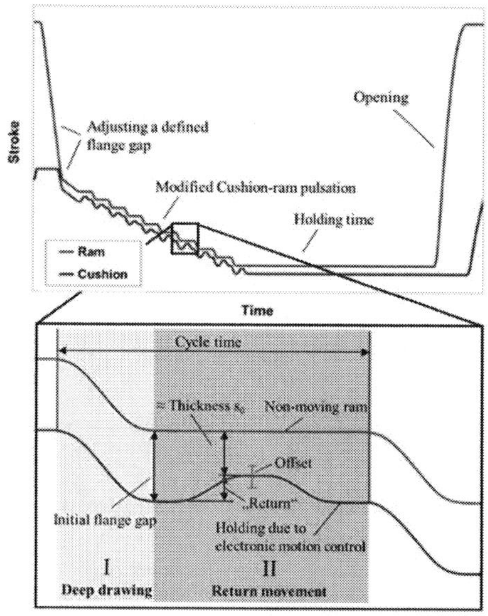

Fig. 2. Stroke-time-curve of the cushion ram pulsation [12].

A ram-speed of 85 mm/s and a flange gap (gap between die and blank holder) of 4 mm was defined for CRP as well as the conventional deep drawing process. The forming tool for the processes includes active elements such as die, blankholder and punch which are guided by pillars and bushings. Furthermore, die and the blank holder were heated up with integrated heating elements. The forming temperature was varied with a constant value of 150 °C or 250 °C, which are controlled by temperature regulation. The punch diameter is 100 mm and for the die 104 mm which results in a drawing gap of 2 mm. The radius at the punch and the die is 10 mm. The specimens were heated up for 5 minutes in a test furnace up to 350 °C for a forming temperature of 250 °C and 200 °C for 150 °C respectively. This is necessary due to the cooling during the transfer between furnace and forming tool. After the subsequent transfer to the tool the forming operation starts. The drawing experiments were realized with lubricant MoS2. (Fig. 3). After completion of the operations, the sheet thickness distribution was investigated using the GOM ATOS 3D scanning and inspection system.

The influence of forming profiles on the variation of the thickness can be stated clearly by optical coordinate measuring. After forming, the specimen were cut in half and clamped in a measuring frame.

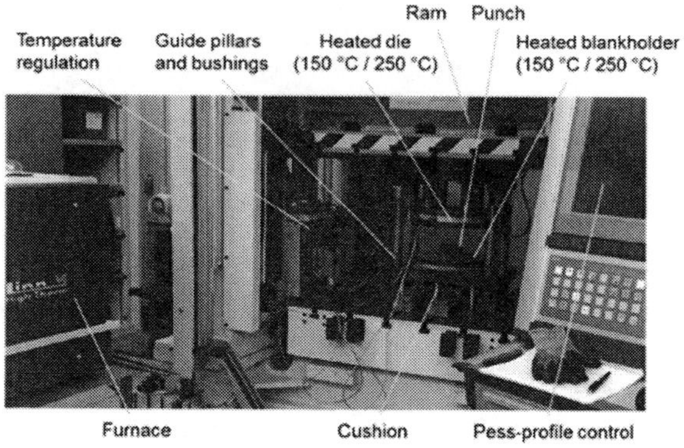

Fig. 3. Experimental setup of the deep drawing investigations.

The experimental process parameters are summarized in table 1. According to Dröder [2], the Limiting Drawing Ratio of the AZ31 at 225 C is 2.5. For this reason, a diameter of 220 and 240 mm was chosen for the investigations.

Table 1. Process parameters.

Process parameter	
material	AZ31
material thickness	1.5 mm
blank diameter	220 mm, 240 mm
lubricant	MoS_2
ram speed	85 mm/s
forming temperature	150 °C, 250 °C
tool temperature	150 °C, 250 °C
motion profile	conventional, CRP
flange gap	4 mm

https://doi.org/10.1051/matecconf/201819012003

2.3 Characterisation of the drawn cups

After the drawing experiments, the formed cups were analyzed in different positions (1 - 6) with regard to the local and microstructure and hardness (Fig.4). Since it turned out in preliminary trials that the material behaves symmetrically, only one side of the cups is discussed below. For analyzing the microstructure, the samples were grinded in several steps using SiC sandpaper, polished with OPS suspension, etched using picric acid and finally studied with an optical microscope. To measure the grain size the linear intercept method was used by analysing approx. 500 grains per sample in the centre of the sheet. For mechanical characterization, HV1 Vickers hardness measurements were carried out with the testing device Zwick ZHU 250.

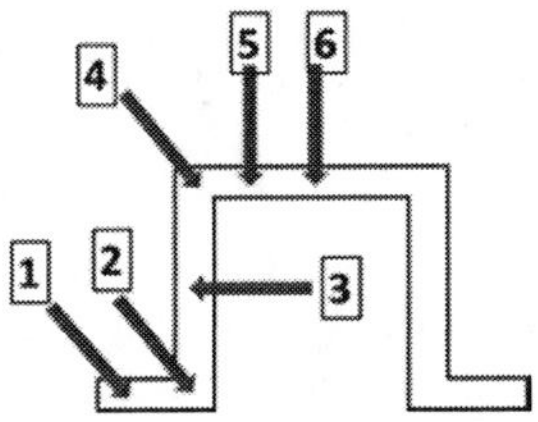

Fig. 4. Schematic illustration of the removal positions.

3 Results and discussion

3.1 Material characterization

The sheets of the initial material have a uniform microstructure with an average grain size in the middle of the sample of 8.2 ± 3.0 μm (Fig 5.).

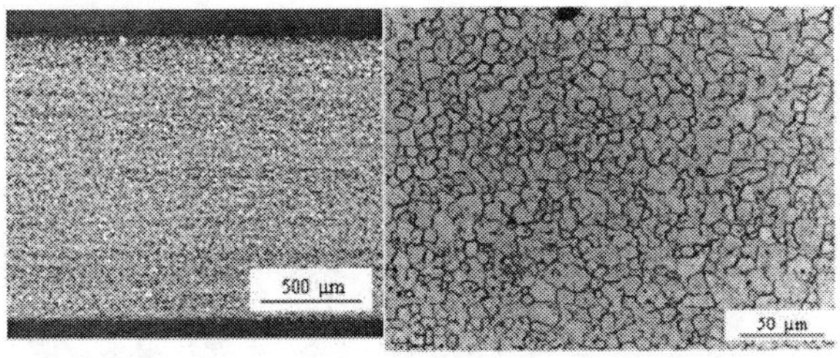

Fig. 5. Microstructure of the source material. Left: overview; right: detailed view.

3.2 Comparison between conventional process and CRP

Two different forming technologies are investigated. The first is a conventional deep drawing process with a ram speed of 85 mm/s and a constant flange gap of 4 mm (Fig. 6).

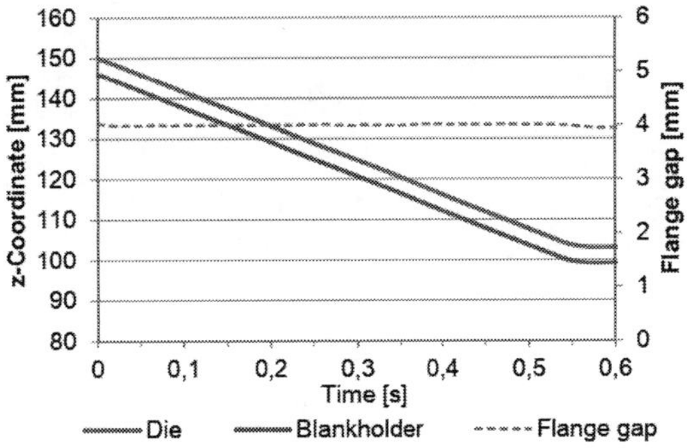

Fig. 6. Displacement - time dependent motion for Conventional deep drawing.

The second is a CRP-profile with a flange gap of 4 mm in the drawing process and a closing gap to 1.5 mm during the holding time of the ram. The CRP motion profiles are divided into forming steps of 5 mm depth and steps to form back the wrinkles which result in a cycle time of 0.4 s. The ram speed in the drawing step reaches 85 mm/s maximum. The process starts with an adjustment of a starting gap between the die and the blank holder. Afterwards the forming process begins (Fig. 7).

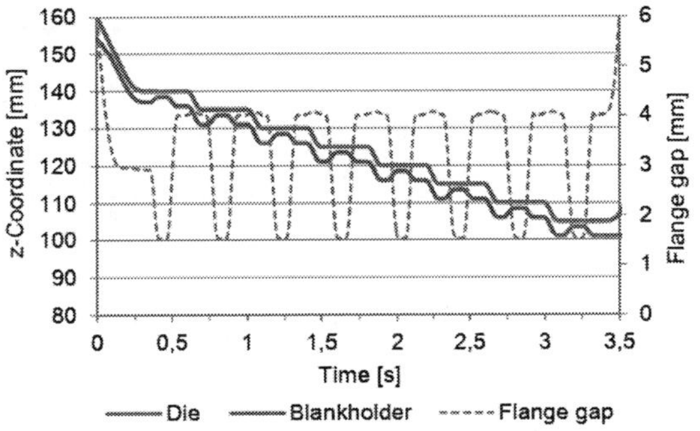

Fig. 7. Displacement - time dependent motion for Cushion Ram Pulsation (CRP).

Table 2 shows cups with the maximum reached drawing depths depending on the motion profile and the forming temperature. The different test parameters show a clear influence on formability. The maximum drawing depth achieved with the conventional process and CRP by a forming temperature of 250 °C was 45 mm. Because the limitations of the experimental set-up, this was the maximum possible depth. At the forming temperature of 150 °C, the maximum drawing depth achieved with the CRP and a blank diameter of 240 mm was 32 mm (Fig. 8). The samples have wrinkles. This is due to the opening of the die and blank holder between the drawing stages.

Fig. 8. Appearance of the deep draw cups for CRP (left) and conventional process (right) at 150 °C.

Whereas the influence of the blank diameter is of secondary importance, a clear impact of the process temperature is evident. CRP at both temperatures results in well-shaped cups in contrast to conventional deep drawing where cracking occur at 150° C tool temperature. However at 250 °C the maximum drawing depth is reached. It is assumed that pyramidal and prismatic sliding systems can be activated at this temperature [13], leading to an improved formability.

Table 2. Maximum drawing depth.

blank diameter [mm]	forming temp. [° C]	tool temp. [° C]	motion profil	drawing depth [mm]
220	250	250	conv.	maximum possible depth: 45
220	250	250	CRP	maximum possible depth: 45
240	250	250	conv.	maximum possible depth: 45
240	250	250	CRP	maximum possible depth: 45
220	150	150	conv.	crack at 25
220	150	150	CRP	27
240	150	150	conv.	crack at 30 (Fig. right)
240	150	150	CRP	32 (Fig. left)

3.3 Sheet thickness profiles after deep drawing

According to the applied press motion profile, a significant influence on the thickness distribution can be stated. Forming of cups at a tool temperature of 250 °C by conventional forming leads to commencing necking at the cup wall, whereas the bottom keeps its initial thickness (see figure 9 a). The amount of thinning can be estimated with up to 0.2 mm. In contrast, a CRP motion profile at the same forming conditions exhibits a homogeneous thickness reduction of about 0.05 mm at the bottom and the wall of the cup (see figure 9 b). The deep drawing process at a lowered tool temperature of 150 °C causes fracturing of the specimen at conventional forming. Cracking starts in the corner of the most heavily loaded zone, where a material flow in thickness direction is required [8].

As a result of the limited forming depth of 30 mm, an increased thinning as observed at a forming depth of 45 mm cannot be stated (see figure 9 d). In contrast, deep forming with CRP at 150 °C tool temperature causes again homogeneous thinning at the bottom and the wall with about 0.05 mm thickness reduction. Due to the reduced forming depth of 32 mm, enhanced thickening at the flange is not observed. All specimens exhibit an increasing thickness at the flange where wrinkles occur.

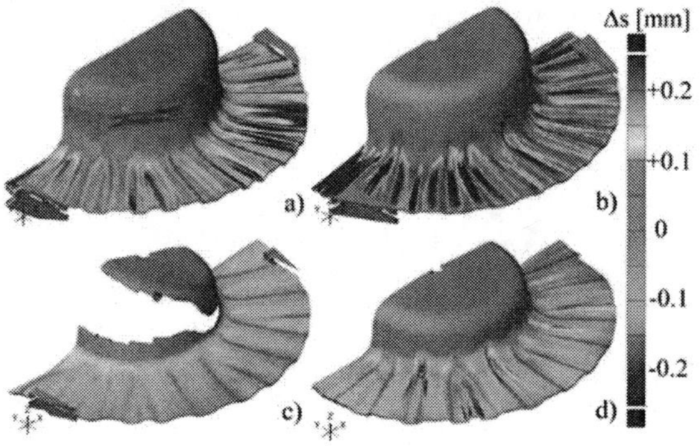

Fig. 9. Sheet thickness distribution of cups by varying press motion profiles. a) T = 250 °C, conv. b) T = 250 °C, CRP c) T = 150 °C, conv. d) T = 150 °C, CRP

An explanation for that is the material flow in the feed section. The same phenomenon was observed for the deep drawing of steel [12].

3.4 Characterization of the formed parts

All examined states of the formed sheet showed the same microstructure characteristic: a fine structure in the middle of the sheet as well as large grains with diameters of approx. 50 micrometers in the edge areas. These coarse grains already originated from the initial material.

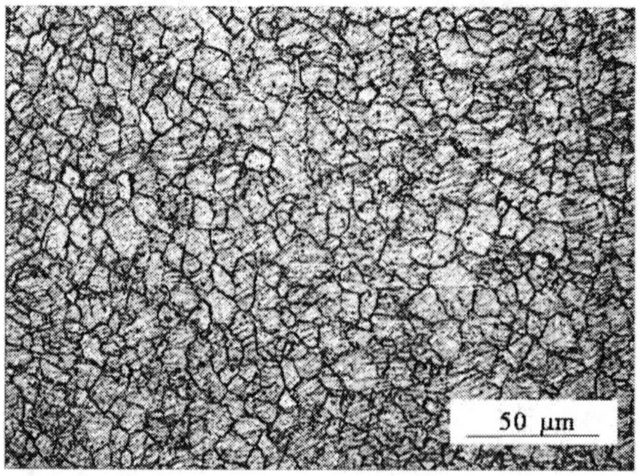

Fig. 10. Microstructure of the at 250 °C conventional-drawn cup at position 4.

The grain size of positions 2 - 6 does not show any significant differences with regard to the process parameters temperature and pulsation (Fig. 12). In figures 10 and 11 this fact is exemplarily demonstrated for position 4 at a forming temperature of 150 °C. The measured deviations were all within the scattering range.

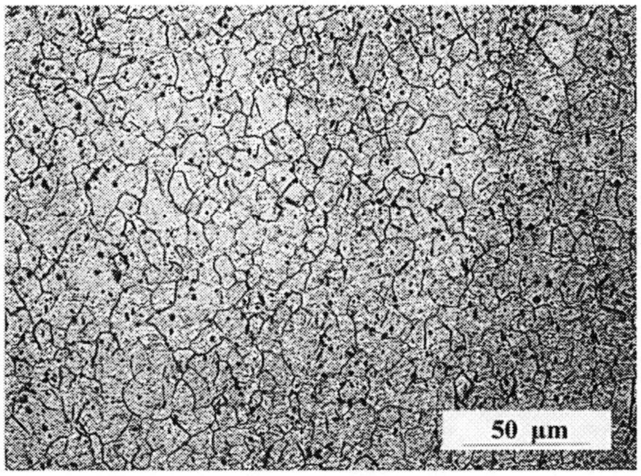

Fig. 11. Microstructure of the at 250 °C CRP-drawn cup at position 4.

However, there are major differences for position 1 (and to a lesser extent in position 2). The authors assume that this effect is not due to the process parameters but to the random local wrinkling at the point of sample extraction.

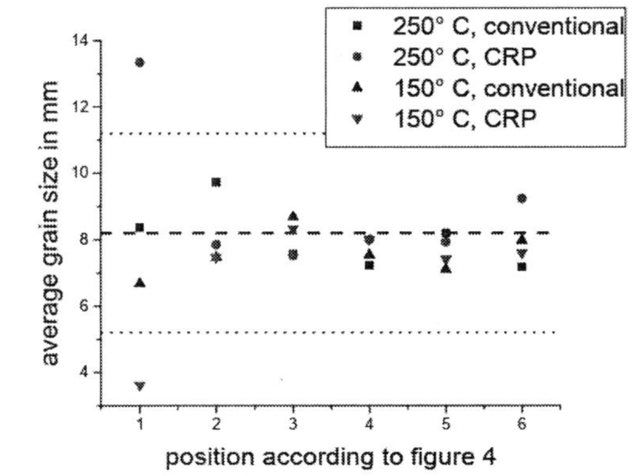

Fig. 12. Dependence of the average grain size from the position according to figure 4. The dashed lines symbolize the average grain size of the initial material and the associated scattering.

The cups drawn at 150 °C show a strong twinning formation over the entire sheet thickness (both conventional and CRP). These are either evenly distributed over the entire cross-section of the sheet or arranged into twin bands (see Fig.13).

At 250 °C deformation temperature, twins are observed only sporadically in the edge grains, especially on the exposed sheet metal side. The local temperature gradient could cause this to occur.

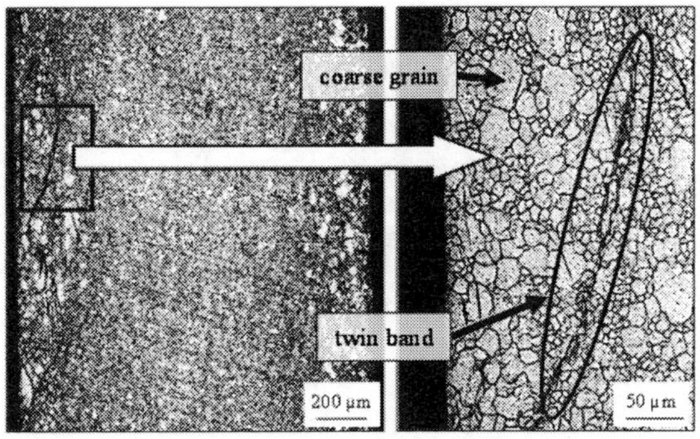

Fig. 13. Cross section through the conventional deep draw cups at 150° C with the occurrence of twin bands.

No indications for recrystallization are visible. Consequently, the increase in hardness (Fig. 14) occurring in the process could be caused by hardening effects. The strong scattering of the hardness values at position 1 are probably similarly, caused by random local wrinkling. The highest degree of deformation is at the cup walls (see Fig. 9) and therefore the biggest hardness increase is observed at position 3. This explanation is also supported by the fact that no pronounced temperature dependence was spotted. The process temperature of 250 °C lies above the recrystallization temperature, 150 °C below [7].

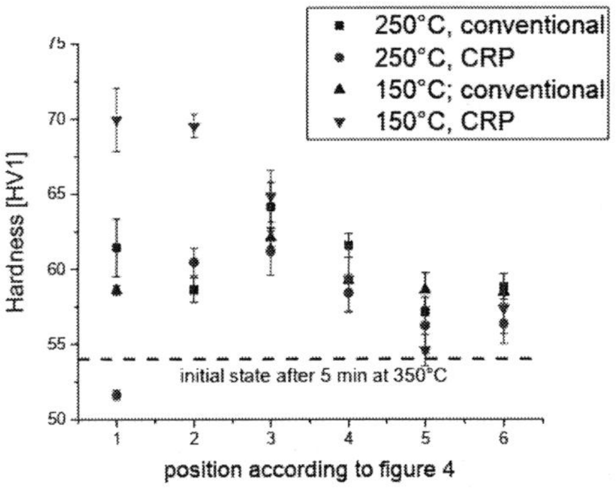

Fig. 14. Dependence of the hardness from the position according to figure 4.

4 Conclusion and outlook

The potential of pulsation superimposed deep drawing for magnesium is demonstrated. While it was possible to achieve a more uniform constriction of the cup walls compared to conventional deep-drawing, the potential of a process temperature reduction was also illustrated. As shown in [13], the critical shear stress of both the basal slip and the twin formation has only a very low temperature dependence for magnesium. As a result, a further reduction of the process temperature seems possible without significant losses in formability. In addition, the transfer of the method to other forming operations such as forging seems conceivable.

However, the underlying mechanism for the substantially improved forming ability were not clarified with the investigation methods used. On the one hand, the changed friction conditions between sheet and punch certainly play a decisive role. On the other hand, there may also be softening effects between the impacts. This mechanism was introduced for CRP of 22MnB5 by Birnbaum et al. [14]. With help of the inclusion of additional investigation methods such as EBSD measurements, a better process understanding on the microscale could be achieved.

As a part of future research, further investigations on the depth drawabilty of this magnesium alloy should be carried out to identify the process window. In this context it is important to determine the Limiting Drawing Ratio (LDR) as a function of tool and forming temperature as well as tribological system (tool material, lubricant e.g.).

References

[1] J.F. King, Magnesium, Materials Science and Technology, **23** (2013)
[2] Doege, E., K. Dröder, Journal of Materials Processing Technology, **115** (2001).
[3] K.F. Zhang, D.L. Yin, D.Z. Wu, International Journal of Machine Tools and Manufacture, **46** (2006)
[4] S. You, Y. Huang, K.U. Kainer, N. Hort, Journal of Magnesium and Alloys, **5** (2017)
[5] R. v. Mieses, Zeitschrift für angewandte Mathematik und Mechanik, **8** (1928)
[6] G. Gottstein, *Physical foundations of materials science* (Springer, Berlin, 2004)
[7] F. Berge, L. Krüger, H. Ouaziz, C. Ullrich, Transactions of Nonferrous Metals Society of China, **25** (2015)
[8] S. Yi, J. Bohlen, F. Heinemann, D. Letzig, Acta Materialia, **58** (2010)
[9] J.S. Lee, H.T. Son, Y.K. Kim, I.H. Oh, C.S. Kang, J.C. Bae, Material Science Forum (2007)
[10] S.H. Zhang, K. Zhang, Y.C. Xu, Z.T. Wang, Y. Xu, Z.G. Wang, Journal of Materials Processing Technology, **185** (2007)
[11] K. Osakada, K. Mori, T. Altan, P. Groche, CIRP Annals, **60** (2011)
[12] D. Landgrebe, A. Rautenstrauch, A. Kunke, S. Polster, S. Kriechenbauer, R. Mauermann, *Proceedings of the 19th International ESAFORM Conference on Material Forming*, 2016
[13] M.R. Barnett, Metallurgical and Materials Transactions A, **34** (2003)
[14] P. Birnbaum, M. Baumann, A. Kunke, V. Kraeusel, D. Landgrebe, MRS Advances, 2 (2017)
[15] H. Kirchner, P. Birnbaum, J. Quellmalz, H. Schlegel, V. Kräusel. Proceedings of the 10th ICIT&APT, 2017
[16] Ullmann, M., R. Kawalla, and H. P. Vogt, Metallurgia Italiana ,**106** (2014)

Numerical modelling approach for the temperature dependent forming behaviour of Ti-6Al-4V

Thomas Papke[1], *Matthias* Graser[1], and *Marion* Merklein[1]

[1]Institute of Manufacturing Technology, Friedrich-Alexander-Universität Erlangen-Nürnberg, Egerlandstr. 13, 91058 Erlangen, Germany

Abstract. Titanium alloys offer several beneficial characteristics, such as high specific strength, metallurgical stability at elevated temperature, biocompatibility and corrosion resistance. With regard to these superior properties, Ti-6Al-4V is a commonly used titanium alloy for aerospace components and medical products. The production of parts made of Ti-6Al-4V can be done in various ways. One approach is forming at elevated temperature, which requires a focused design of parts, processes and numerical modelling of the forming process. Essential input parameters for the numerical models are temperature dependent material parameters. Since, the yield stress and Young's modulus of the material decrease significantly at elevated temperature, the forming limits are enhanced. For the characterization of the forming behaviour, uniaxial tensile tests at temperatures from 250 °C to 400 °C have been conducted. The samples are heated by conduction in a thermal-mechanical simulator for the tensile test. However, the resulting inhomogeneous temperature distribution along the longitudinal axis of the specimen is a challenge in order to measure proper material properties. Inhomogeneous temperature distribution leads to varying mechanical properties and temperature dependent forming behaviour. To overcome this issue, simple numerical models based on experimental data are necessary, which allow the estimation of the influence of the inhomogeneous temperature distribution. In this paper, therefore, the temperature distribution and the subsequent tensile test are investigated using electrical-thermal and mechanical numerical simulations of the tensile test at elevated temperature. With the combined approach of experimental tests and numerical simulations, the forming behaviour of Ti-6Al-4V can be modelled.
Keywords: Modelling; Thermal effects; Titanium

1 Introduction

The titanium alloy Ti-6Al-4V has become a popular material for medicine or aerospace applications. Beneficial material properties are biocompatibility, corrosion resistance and high specific strength. Despite of these advantages, narrow forming limits at room temperature of titanium are challenges for metal forming processes. Ti-6Al-4V consists of densely packed hexagonal α- and body-centred cubic β-phase. Due to the crystal structure of the α-phase, the ductility is reduced, which leads to narrow forming limits at room temperature [1]. However, by forming at elevated temperatures, the yield stress decreases and the forming limits are enhanced [2]. In [3] experiments on Ti-6Al-4V sheet metal parts are conducted at temperatures below 600 °C, which results in a significantly lower yield stress compared to forming at 20 °C. This indicates that forming of Ti-6Al-4V is also possible at elevated temperatures below the β-transus temperature [1].

A commonly used device for conducting hot tensile tests is the thermal-mechanical simulator Gleeble 3500 from Dynamic Systems Inc. (DSI). For testing at elevated temperatures, the specimen is clamped between copper grips and heated by conduction, which means the specimen is part of an electrical circuit and represents the resistance. However, the temperature distribution from the centre of the specimen to the clamping is not homogenous. The temperature distribution can be described by analytical equations [4]. One challenge resulting from the temperature gradient along the axis of the sample is the displacement of the forming zone relative to the zone of the target temperature. This influences the forming behaviour and thus leads to differing results regarding mechanical properties [5]. Since the test conditions according to DIN EN ISO 6892-2 [6] require a temperature deviation of less than ±3 °C for test temperature lower than 600 °C, the inhomogeneous temperature distribution is critical in order to gain valid testing results. An approach to ensure the failure of the specimen under tensile load at the area of target temperature is to reduce the gradient of temperature along the axis of the sample. Therefore, Ganapathy et al. [8] modified the shape of the specimen by adding extra copper wires at the edges of the specimen which leads to a significantly lower temperature gradient. However, the influences of the modified specimen geometry remain unclear.

It can be seen, that hot tensile tests in accordance with [6] are challenging. Hence, combined experimental

* Corresponding author: thomas.papke@fau.de

and numerical methods are applied to determine material properties influenced by temperature gradients. Kardoulaki et al. [9] used the commercial software Abaqus to implement separate models for the electro-thermal and mechanical simulation of a hot tensile test in a Gleeble testing system for free cutting steel (FCS). Furthermore, the implementation of a complex electro-thermal-mechanical coupled simulation of the hot tensile test for 22MnB5 using Abaqus is shown by Gerhardt [10]. As a result, the forming behaviour from the experimental tensile test can be simulated by the coupled model. The electro-thermal modelling approach based on the experimental setup should be investigated for LS-Dyna in this work.

The aim of this paper is to analyse the influence of a temperature gradient on the forming behaviour of Ti-6Al-4V sheet metal parts by using a combined experimental and numerical approach. Since the tensile test is a simple forming test, the results could be used for the investigation of the influence of temperature gradients on sheet metal parts in more complex forming processes as deep drawing [11] or bending [12].

Within this work, the tensile test setup in a Gleeble 3500 for Ti-6Al-4V is modelled by using the electromagnetism module of LS-Dyna. The numerical model is divided into two subsequent models, an electro-thermal for simulating the temperature distribution and a mechanical model for the tensile test. The models are not coupled, because the modelling approach should be considered as a separated analysis of both process steps. In the first step, an electro-thermal model is set up to simulate the temperature distribution along the sample's longitudinal axis. Based on the resulting temperatures, flow curves from experimental tensile tests are implemented into the mechanical model, which takes the temperature dependent flow behaviour into account. Prior to the simulations, experimental data of electrical current and flow curves are derived from tensile tests at elevated temperature with Ti-6Al-4V sheet metal parts.

2 Experimental Setup

2.1. Tensile test

For the tensile test at elevated temperature, a Gleeble 3500 from Dynamic Systems Inc. (DSI) is used. The machine is equipped with a 100 kN load cell for the measurement of the forming force. Furthermore, for the optical strain measurement, an ARAMIS System from GOM is applied. For the measurement of the strain on the specimens' surface, the samples are covered with a stochastic pattern [3]. The change of the stochastic pattern is recorded by a CCD-Camera with 40 Hz and distortion of the pattern is used to calculate the true stress-true strain curve. The specimen geometry has a width of 12.5 mm, a parallel length of 62.5 mm and a sheet thickness of 1.0 mm. Since the test is focused on the forming behaviour at elevated temperature; the rolling direction (RD) for all specimens is 0 °. The strain rate is 0.1 s^{-1}. The specimens are clamped between a pair of copper grips without cooling, which is sufficient for the maximum temperature of 400 °C investigated within this work.

The temperature during the test is measured by up to four thermocouples type-K Ni/Cr-Ni, to investigate the temperature distribution along the longitudinal axis of the specimen. The first thermocouple is located in the centre of the specimen, whereas the others are placed along the longitudinal axis with 10 mm, 20 mm and 32.5 mm distance from the centre.

Before the tensile test, the samples are heated from room temperature to the testing temperature within 30 s. This temperature is held for 15 s, in order to ensure a consistent test condition. The tensile test is conducted subsequently as shown in Fig. 1.

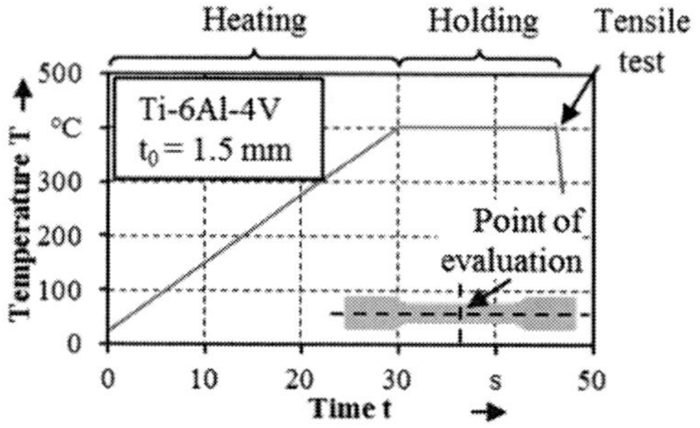

Fig. 1. Temperature-time curve for the tensile test

In addition to the tests at elevated temperature, tensile tests at room temperature (20 °C) are conducted in order to compare the forming behaviour. The tests are performed with a universal testing machine Zwick Z 100 Allround Line, which is equipped with a 100 kN load cell. Like in the tensile test at elevated temperature, the strain on the surface is measured by an ARAMIS System. To avoid influences of the sample geometry, the same geometry as for the tensile test at elevated temperature is used.

2.2 Electrical properties

Besides the temperature distribution electrical current during the heating process is measured. The experimental setup is shown in Fig. 2.

Fig. 2. Experimental setup for tensile test in Gleeble 3500

For measuring the current through the specimen, a current clamp VC605 from Voltcraft with a measurement range of 1000 A for AC is used, since the sample is heated by alternating current [10]. The result is the root mean square (RMS) value of the AC, which is equal to the value of a direct current that would produce the same average power in the resistor.

3 Numerical Model

The modelling approach is based on two separate numerical models. The first model is used to simulate the temperature distribution in the specimen during the resistive heating process, whereas the second model is a mechanical model for the simulation of the tensile test. Based on the temperature distribution, which results from the electro-thermal model, the specimen is subdivided into several sections with different flow curves to simulate the tensile test with the mechanical model. For the numerical models, the specimen geometry from the experiments is used with a sheet thickness of 1.5 mm.

3.1. Electro-thermal Model

For the electro-thermal model, the electromagnetism module [14] including resistive heating [15] from LS-Dyna is applied. The simulation model consists of two pairs of grips and the tensile specimen as shown in Fig. 3. The mesh of the specimen consists of constant stress solid hexahedral elements with a length of 0.5 mm in z-direction and up to 1 mm in x- and y-direction.

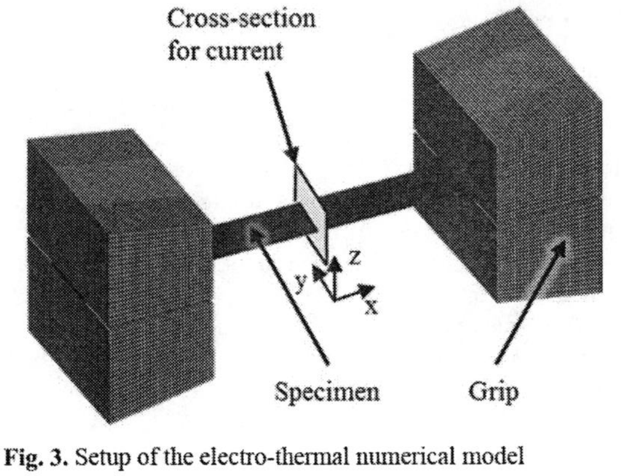

Fig. 3. Setup of the electro-thermal numerical model

The grips and the specimen are connected with nodes in the contact area. The maximum element size at the grips is 1.5 mm, which is bigger than for the specimen to reduce the total number of elements and the temperature increase of the grips is neglected in this work. In total, the model consists of 157920 elements. As boundary conditions, the initial temperature for the grips and the specimen is set to 32 °C, which is selected due to the temperature in the Gleeble process chamber. The material keyword MAT_T01 is used for all parts in the electro-thermal model. The material card allows thermal isotropic properties and includes constant values for specific heat (hc) and thermal conductivity (tc) [16]. The electrical current is imposed through the cross-section at the symmetry plane of the specimen and implemented by a current-time-curve. In table 1 mechanical, thermal and electrical properties are listed. The material properties for copper and Ti-6Al-4V are taken from literature. Parameters at room temperature are chosen in order to keep the numerical models simple.

Table 1. Properties of Copper (Cu) and Ti-6Al-4V (Ti)

Element	Style	Ref.
Young's modulus	Cu: 117 GPa Ti: 110 GPa	[17] [1]
Mass density	Cu: 8950 kg/m³ Ti: 4430 kg/m³	[17] [1]
Poisson ratio	Cu: 0.35 Ti: 0.342	[17] [1]
Specific heat capacity	Cu: 400 J/(kgK) Ti: 546 J/(kgK)	[17] [18]
Thermal conductivity	Cu: 400 W/(mK) Ti: 6.6 W/(mK)	[17] [18]
Electrical conductivity	Cu: 58e6 1/(Ωm) Ti: 5.8e5 1/(Ωm)	[17] [18]

3.2. Mechanical Model

The second part of the modelling approach uses a mechanical model for simulating the tensile test at elevated temperature. The aim is to investigate the forming behaviour in dependence of the temperature gradient of the experiment compared to ideal isothermal conditions at 400 °C. Based on the mesh of the electro-thermal simulation, the maximum temperatures at the elements are evaluated and the specimen is subdivided into four major sections with different flow curves depending on the temperature. The assignment of flow curves from the experimental tensile tests is based on a range of temperatures, which is shown in table 2.

Table 2. Temperature range for flow curves

Flow curve for section	Temperature range for section
$FC_{20\,°C}$	20 °C – 225 °C
$FC_{250\,°C}$	226 °C – 300 °C
$FC_{360\,°C}$	300 °C – 380 °C
$FC_{400\,°C}$	381 °C – 500 °C

Since for the forming behaviour the target temperature 400 °C is investigated, a small temperature range for the subdivision is selected at high temperatures and wider at lower temperatures. Furthermore, an equidistant interval between the temperatures would include a linear temperature distribution along the x-axis, which is not the case in the experiment as shown in Fig. 6. The resulting subdivision of the mesh and the setup for the tensile test is shown in Fig. 4. In accordance with the tensile test experiments, one side of the specimen is clamped to avoid movement, whereas the other side is moved with 5 mm/s, where the tensile force is applied. The boundary conditions are implemented with an SPC

keyword node set for the nodes at the outer side in the y-z-plane.
Since the stress state is uniaxial tension; the material model is based on von Mises. The corresponding keyword in LS-Dyna is MAT_024, which is indicated as piecewise linear plasticity model with elasto-plastic material behaviour in LS-Dyna [16]. The material model includes the flow curves for 20 °C, 250 °C, 360 °C and 400 °C, which result from the tensile tests in section 4.1. Besides flow curves, mass density, Young's modulus and Poisson's ration of table 2 are considered.

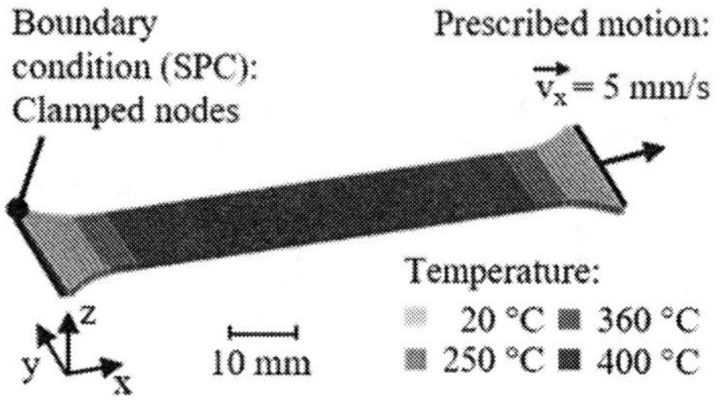

Fig. 4. Setup for the numerical model of the tensile test

4 Results

4.1. Tensile test

The tensile tests are conducted at 20 °C, 250 °C, 360 °C and 400 °C, where the maximum true strain reach from 0.10 to 0.15 for all temperatures. However, for the numerical model in LS-Dyna flow curves with a true strain of 1.0 are necessary. Hence, the experimental data is extrapolated to higher true strains. Within this work, the model of Hockett-Sherby [21] is used. In prior investigations by Merklein and Schaub [3], the extrapolation models presented by Swift [19] and Nemat-Nasser [20] are applied. In comparison to Swift and Nemat-Nasser, the model by Hockett-Sherby comprises a lower hardening effect for higher true strain values. Essentially, the model is based on four parameters [21].

$$\sigma(\varphi) = \sigma_s - (\sigma_s - \sigma_y) exp(-m\varphi^p) \qquad (1)$$

To calculate the flow curve, equation (1) includes σ_s, representing the steady-state flow stress, the yield stress σ_y, which results from the tensile test and the free parameters m and p. The parameters for calculating the flow curves are shown in table 3.

Table 3. Used parameters for Hockett-Sherby extrapolation

Flow curve	σ_s in MPa	σ_y in MPa	m	p
$FC_{20\,°C}$	1300	900	1	0.2
$FC_{250\,°C}$	1000	713	1	0.2
$FC_{360\,°C}$	1200	600	3	0.7
$FC_{400\,°C}$	1000	600	3	0.7

4.2 Electrical measurement

Besides the temperature, the electrical current is measured during the tensile test at elevated temperature to link thermal and electrical values. In Fig. 5 are shown the measured values for current and temperature in the centre of the specimen. The current increases during the heating process and drop significantly after 30 s, when the holding phase starts. The increase in current during the heating phase could be explained by energy losses due to convection and radiation as well as temperature dependent physical properties. Hence, for an increase in temperature during the heating phase more current is needed.

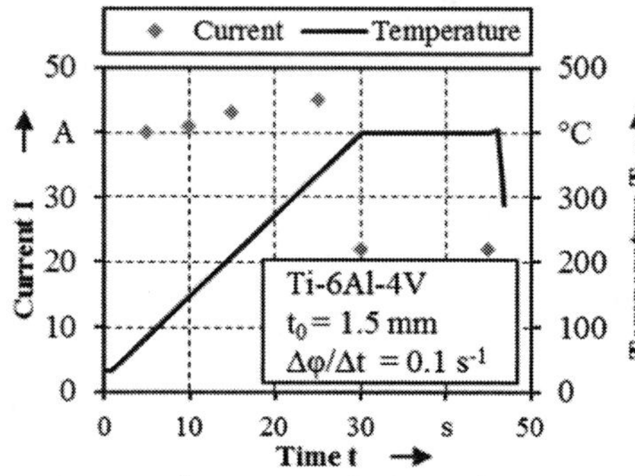

Fig. 5. Current and temperature for the tensile test at 400 °C

4.3 Numerical model

The experimental data from the electrical measurement and the tensile tests are implemented in the numerical models. In the first step, the electro-thermal model is used for numerical modelling of the temperature distribution and the result compared to experimental data. Secondly, the mechanical model is set up to investigate the influence of inhomogeneous temperatures in a forming process demonstrated on a tensile test at elevated temperature. For the evaluation, the temperature distribution on the specimen after the heating process, the force-displacement curve and the strain distribution are analyzed. Finally, the mechanical numerical model is validated by experimental results.

4.3.1 Electro-thermal model

To evaluate the electro-thermal model, the temperature distribution along the x-axis of the specimen is analysed. In accordance with the location of the thermocouples in the experiment, the elements at the surface of the specimen are investigated by a section plane. Since the specimen is symmetric, only one half of the sample is presented. In Fig. 6 the temperature distributions for experiment and simulation are shown for t = 30 s, which represents the end of the heating process. In the simulation, the maximum temperature is reached after the heating process and a continuous temperature distribution is received, which is indispensable for implementing temperature dependent flow curves into the mechanical model. The experimentally measured

https://doi.org/10.1051/matecconf/201819012004

temperature at the centre (x = 0 mm) is 400.5 °C, which is almost reached in the simulation with 399.2 °C. The deviation between the experimental values and calculated temperatures increases with higher distance from the centre. This indicates that the energy, which is introduced by the current is too small. Hence, the temperature in the simulation decreases faster for lower temperatures than in the experiment. Higher energy supply in the simulation could reduce the difference from the experimental data. The maximum deviation is about 17 % at x = 32.5 mm. One explanation could be that the material parameters implemented in the simulation are kept constant. The result could be improved by temperature dependent parameters. Therefore to improve the model it is mandatory to analyse the physical properties of the sheet material used in this work and compare them to literature values.

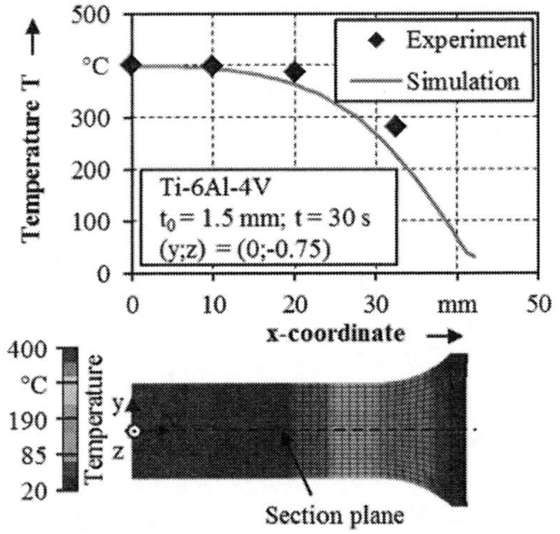

Fig. 6. Temperature distribution along the x-axis

4.3.2 Mechanical model

Based on the electro-thermal simulation, the specimen is subdivided into four sections with different temperature fields. For each section different flow curves are assigned in the mechanical model thus emulating the thermal gradient of the real forming process. To show the influence of the thermal gradient the results are compared to an isothermal simulation with only one flow curve at 400 °C. The mechanical model is validated by two criteria. On the one hand the results of the simulation are compared to the force-displacement curve from the experiment and on the other hand, the strain distribution along the longitudinal axis is taken into account. To show the influence of isothermal conditions the comparison is done for the specimen with temperature gradient subdivision and the isothermal sample. Fig. 7 shows the force-displacement curve for experiment and simulations including the maximum force. The elastic part of the curve, which is represented by a linear rise of the force is not fitted well by both simulations. One explanation could be that the elastic deformation of tool and machine of the experimental setup is neglected in the simulation. After the elastic part, both simulations show lower forces compared to the experiment. However, for the simulation with temperature dependent flow curves, the maximum force is reached for shorter displacement values, which fits the experimental results in a better way. For the simulation with isothermal conditions, the displacement at maximum force is much higher and therefore the deviations from experimental data as well.

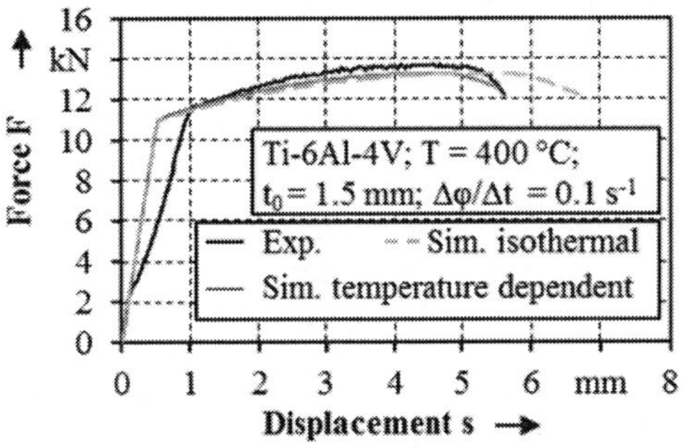

Fig. 7. Force-displacement for experiment and simulation

Based on these results, the force-displacement curve can be fitted better by using the gradient specimen, when the deviations in the elastic region are neglected. For a comprehensive investigation, the strain distribution along the longitudinal axis is evaluated at the point of maximum force, which is shown for the experiment and both simulations in Fig. 8.

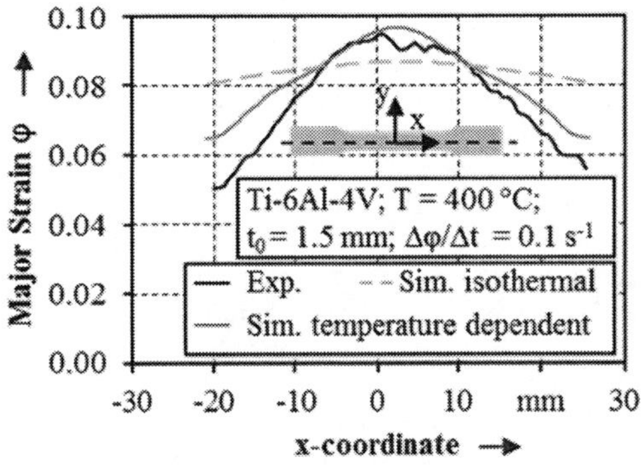

Fig. 8. Strain distribution along the x-axis

The deviation from the experimental data of maximum major strain 0,094 is 2.1 % for the simulation with gradient properties 0,096. Whereas, the deviation is 7.5 % in case of isothermal conditions 0,087. However, under isothermal conditions, the major strain is underestimated, which is crucial, because the real strains are higher. Besides the underestimation of the maximum strains, the simulation with temperature dependent flow curves calculates the strain distribution along the x-coordinate of the experiment more precisely. Again the simulation with temperature gradient dependent flow curves shows a much better representation of the experimental data. This shows that the forming behaviour is strongly influenced by the temperature gradient. Therefore, the temperature gradient should be considered in simulations of hot forming processes. To handle non-uniform temperatures, which may occur during a hot forming process, the temperature deviation from the target temperature should be small.

5 Conclusion

The major findings of this work are that the temperature distribution and the strain behaviour can be simulated by simple electro-thermal and mechanical numerical models in LS-Dyna. With this methodology, the determination of mechanical properties in a resistance heating process can be supported. It can be clearly seen that the forming behaviour depends strongly on the temperature distribution. When using a homogeneous temperature in the forming process, the maximum major strain is underestimated, whereas the experimental results are fitted in good approximation by the temperature dependent mechanical model. This means more real forming operations at elevated temperature as warm bending or deep drawing of Ti-6Al-4V should consider the temperature distribution on the specimen during the forming process.

6 Summary and Outlook

In this work, the influence of a temperature distribution on the forming behaviour of Ti-6Al-4V for forming at elevated temperature is demonstrated with a tensile test at 400 °C by electro-thermal and mechanical numerical models using LS-Dyna and experimental results. As an experimental input, tensile tests at elevated temperature are conducted with a Gleeble 3500 with resistance heating. During the test, the current through the specimen is measured as an input parameter for the simulation. Afterwards, the mesh with the temperature distribution is subdivided into several sections with different flow curves and implemented into a mechanical model to investigate the temperature-dependent flow behaviour. By using the models, the temperature distribution after the heating process, the force-displacement-curve and the strain behaviour approximately represent the experimental data. To evaluate the influence of the temperature gradient, the results are compared to a sample with homogeneous material properties. With the temperature gradient specimen, the force-displacement-curve and the strain distribution is calculated more realistic. For the homogeneous material, the maximum major strain is underestimated. Hence, the forming behaviour is strongly influenced by the temperature gradient during the tensile test at elevated temperature.
Within further investigations, physical effects as convection and radiation should be implemented into the numerical models, which could reduce the deviations in temperature distribution. Furthermore, temperature dependent physical properties as thermal and electrical conductivity should be considered. Since the electro-thermal and the mechanical model for the use of the temperature dependent material properties represent the experimental result, a coupled numerical model including heating and forming should be set up.

The authors acknowledge the support by German Research Foundation within CRC 814 – "Additive Manufacturing" subproject B5.

References

1. M. Peters, C. Leyens, *Titanium and Titanium Alloys* (2003)
2. J. Beal, R. Boyer, D. Sanders, *Metalworking: Sheet Forming* 656-669 (2006)
3. M. Merklein, H. Hagenah, M. Kaupper, A. Schaub, *Key Engineering Materials* **549** 311-316 (2013)
4. S. Novotny, M. Celeghini, M. Geiger, *Proc. of SheMet Int. Conf.* 363-370 (2000)
5. K. Müller, *Werkstoffkundliche Qualifizierung des Randschichthärtens mit Laserstrahlung* (1999)
6. DIN EN ISO 6892-2, *Metallic materials - Tensile testing* part 2 (2017)
7. R. S. Lee, Y. K. Lin, T. W. Chien, *Procedia Engineering* **81** 1682-1688 (2014)
8. M. Ganapathy, N. Li, J. Lin, M. Abspoel, H. Guido, D. Bhattacharjee, *Matec Web of Conferences* **21** (2015)
9. E. Kardoulaki, J. Lin, D. Balint, D. Farrugia, *Journal of Strain Analysis* **49** 521-532 (2014)
10. K. Gerhardt, *Untersuchung zur konduktiven Erwärmung für Warmzugversuche an Blechen* (2015)
11. A. Schaub, J. Degner, A. Bhrigu, M. Schmidt, M. Merklein, *Int. Conf. on Competitive Manufacturing*, (2016)
12. M. Merklein, P. Dubjella, A. Schaub, L. Butzhammer, M. Schmidt *6th Int. Conf. on Additive Technologies* 309-316 (2016)
13. A. Krawczynska, M. Lewandowska, R. Kuziak, K. Kurzydlowski, *Phys. Status Solidi* **5** 1380-1383 (2010)
14. P. Eplattenier, G. Cook, C. Ashcraft, M. Burger, J. Imbert, M. Worswick *Steel Research Int.* **5** 351-358 (2009)
15. P. Eplattenier, I. Caldichoury, *Int. LS-Dyna Users Conf.* 1-10 (2012)
16. LSTC, *LS-Dyna Keywords Users's Manual R9.0 – Material Models* (2016)
17. M. Li, S. Zinkle, *Comprehensive Nuclear Materials* **4** 667-690 (2012)
18. J. Valencia, P. Quested, *ASM Handbook* **15** 468-481 (2008)
19. H. Swift, *Journal of the Mechanics and Physics of Solids* **1** 1-18 (1952)
20. S. Nemat-Nasser, W. Guo, V. Nesterenko, S. Indrakanti *Mechanics and Materials* **33** 425-439 (2001)
21. J. Hockett, O. Sherby, *Journal of Phys. Solids* **23** 87-98 (1975)

Stress-level-dependent microstructure evolution and deformation behaviors during creep age forming of AA2219 alloy

Ziyao Ma[1,2], *Lihua* Zhan[1,2,3,*], *Chunhui* Liu[1,2,3*], *Lingzhi* Xu[1,2]

[1] School of Mechanical and Electrical Engineering, Central South University, Changsha 410083, China
[2] State Key Laboratory of High-Performance Complex Manufacturing, Central South University, Changsha 410083, China
[3] Light Alloy Research Institute, Central South University, Changsha 410083, China

Abstract. Creep age forming (CAF) can shape and strengthen the sheet metals loaded externally at elevated temperature by corresponding creep/stress-relaxation and age hardening. Understanding the deformation and strengthening behavior at different stress levels plays a pivotal role in accurate prediction of CAF. This work experimentally investigates the effect of stress level on the evolution of both properties and microstructures, including precipitates and dislocations, during CAF. The microstructural evolution is characterized in detail by scanning/transmission electron microscopy (S/TEM) and X-ray diffraction (XRD). Stress orienting effect of θ" occurs at low-stress loaded samples and results in a decline in the hardening potential of creep-aging. A bimodal distribution of precipitates consisted of θ" homogeneously formed in the matrix and θ' heterogeneously formed on dislocations is found in the high-stress loaded samples, giving rise to a significant enhancement of hardening response. In addition, the creep strain increases drastically when the loading stress is higher than the initial yield strength. The relevant mechanisms for the transition of the obtained strength and creep behaviors with the applied stress are discussed based on microstructure observations.

Keywords: **Creep age forming; Microstructure; Aluminum alloy**

1 Introduction

Creep age forming (CAF), originally developed to manufacture large integral components of aluminum alloys in the aerospace industry(Holman, 1989; Li et al., 2017; Zhan et al., 2011a), has aroused increased attentions as a promising technique for high performance and reduced cost. The process can simultaneously realize shaping and strengthening of the sheet metals loaded externally at elevated temperature by corresponding creep/stress-relaxation and age hardening. Elastic deformation transforms into plastic deformation through creep/stress-relaxation mechanism(Ho et al., 2004; Xu et al., 2017b), while the alloy strength improves mainly by solid state precipitation and marginally by dislocation multiplication during CAF (Zhan et al., 2011b; Zhang et al., 2013). The above processes determine the springback and mechanical properties of the CAF formed alloys (Lei et al., 2017; Zhan et al., 2011b; Zhang et al., 2013). When the rib-stiffened sheets are pressed onto the die surface, they will be subject to a complex and non-uniform stress distribution. For instance, the stress at the ribs is much higher than the other areas and can even exceed the yield strength of the Al alloy during the initial loading stage of CAF. The aim of this study is to experimentally investigate the effect of stress level on the evolution of both properties and microstructures, including precipitates and dislocations, during CAF. These results will serve as input for accurate modeling and simulation of CAF process, especially that of complex structured component.

Aluminum-Copper AA2219 alloy is currently widely used to manufacture light weight structural components for aerospace applications. The precipitation sequence in an Al-Cu alloy during artificial ageing treatment is generally accepted as (Biswas et al., 2011; Bourgeois et al., 2011; Guinier, 1938; Liu et al., 2014; Ma et al., 2016): Supersaturated solid solution→ G.P. Zones→ θ"→ θ' (Al_2Cu). Different from simple artificial ageing process, the applied stress during CAF creates a thermal-mechanical coupled situation and thus adds to the complexity of the precipitation involved. Stress generally could promote the solutes precipitation in many aluminum alloy systems (Hu et al., 2017; Xu et al., 2017a; Yang et al., 2016; Zhang et al., 2013), leading to the acceleration of hardening kinetics. The stress orienting effect, which means precipitates preferentially grow on specific habit planes (only one or two variants are present), of θ' in single-crystal Al-Cu alloy was detected by Zhu and explained by the stress-induced change in lattice parameters and thus interfacial energy anisotropy (Eto et al., 1978; Zhu and Starke, 2001a; Zhu

* Corresponding author: yjs-cast@csu.edu.cn (L. Zhan), chunhuiliu@csu.edu.cn (C.H. Liu)

 https://doi.org/10.1051/matecconf/201819012005

and Starke, 2001b). Creep behavior is dictated by the evolution of the related microstructural variables in CAF. At the primary creep stage, the decrease in creep strain rate is associated with the strain hardening caused by a fast increase in dislocation density and the dislocations reconfiguration at later time.

This study systematically investigates the effect of stress, ranged from within the elastic regime to those higher than the initial yield strength, on the creep and ageing process in CAF of an AA2219 alloy. The microstructural evolution is characterized in detail by scanning/transmission electron microscopy (S/TEM), differential scanning calorimetry (DSC) analysis and X-ray diffraction (XRD).

2 Experimental procedures

The experiments were conducted on a commercial AA2219 alloy whose composition is listed in Table 1. The samples were firstly solution treated at 535°C for 35min and then water quenched at room temperature. The uniaxial tensile stress-ageing treatments (creep-ageing) were performed on a RMT-D10 electronic creep testing machine, at 165 °C under various stress levels (60MPa, 90MPa, 120MPa, 150MPa, 180MPa, 210MPa). The stress-free-aged samples were placed in the creep testing machine and experienced the same thermal procedures with the stress-aged samples. According to the applied stresses, the stress-ageing treatments are labeled as 60SA, 90SA, 120SA, 150SA and 180SA, respectively. The stress-free-ageing treatment is labeled as SFA. A special group of experiments were also carried out: the samples were loaded (pre-deformed) at the beginning of heating process to 180MPa, which lasts for only 2 min, and then stress-free-aged for the rest of the isothermal ageing at 165 °C (labeled as 180SFA).

Table 1. Chemical Composition of the AA2219 alloy (wt. %)

Cu	Mg	Mn	Si	Zr	Ti	Al
6.09	0.0032	0.29	0.090	0.14	0.070	Bal.

The aged samples were subjected to the uniaxial tensile tests on a CMT-5105 test machine and the yield strengths were obtained as 0.2% offset yield stress. For microstructural observations, the S/TEM samples were prepared by standard method. STEM images were acquired on a Titan G^2 60-300 Transmission Electron Microscope. The X-ray diffraction (XRD) tests were carried out on an Advance D8 X-Ray Diffractometer with a scan range from 20° to 100° and a scan speed of 0.02° each step. Differential Scanning Calorimeter (DSC) analysis was performed, using a PerkinElmer DSC8500 instrument, on samples subjected to the same treatment as the S/TEM ones.

3 Results and discussion

3.1 Mechanical property

Figure 1 depicts the evolution of the room-temperature yield strengths of AA2219 stress-free-aged and stress-aged at 165°C under various stresses. The yield strengths of different stress-aged samples all increase rapidly upon ageing and arrive at the peak values after about 24h of ageing. The effect of stress varies with the stress levels: the lower stresses (60MPa, 90MPa, 120MPa) weaken the hardening effect while the higher stresses (150MPa, 180MPa) exhibit different levels of enhancements, as compared with the SFA samples. The weakening degree of different low stresses is almost the same, however the enhancements varies with different high stresses. To clarify this interesting phenomenon, the yield strength at the water-quenched state, which is 141MPa at room temperature and 138MPa at 165°C after a 35min non-isothermal heating process, needs to be noticed. Since the higher stresses (150MPa, 180MPa) are larger than the yield stress at 165°C, plastic deformation and dislocations should occur during the loading process. It can be confirmed that stresses of different levels all pose a negative effect on the age-hardening response of the alloy, while the positive effect of the high stresses (150MPa, 180MPa) mainly results from dislocation multiplication. This phenomenon is related to the effect of high-stress-induced dislocations on the microstructure evolution and will be discussed later.

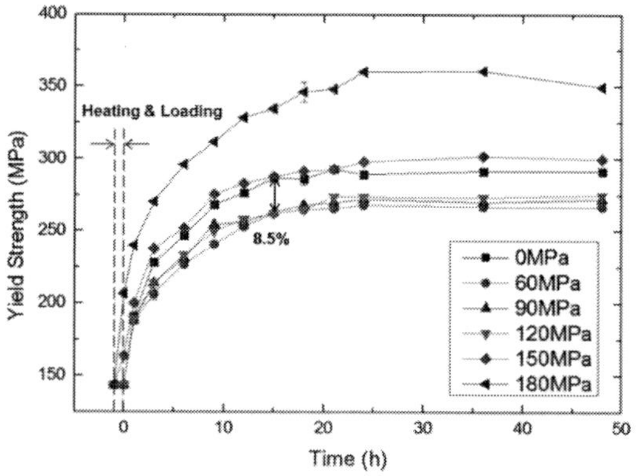

Fig. 1. Hardening curves with time for the creep-aged samples: stress-free-aged at 0MPa, stress-aged at 60MPa, 90MPa, 120MPa, 150MPa and 180MPa, respectively.

3.2 Creep deformation

The creep strain curves and the creep strain rate curves under different stresses are shown in Fig. 2 and Fig. 3, respectively. When the applied stress is lower than the yield strength (138MPa) at 165°C, the total creep strain after 24h is very small and does not exceed 0.1%. When the applied stress is higher than the yield strength, the creep strain increases rapidly upon heating. The duration of the primary creep stage increases significantly.

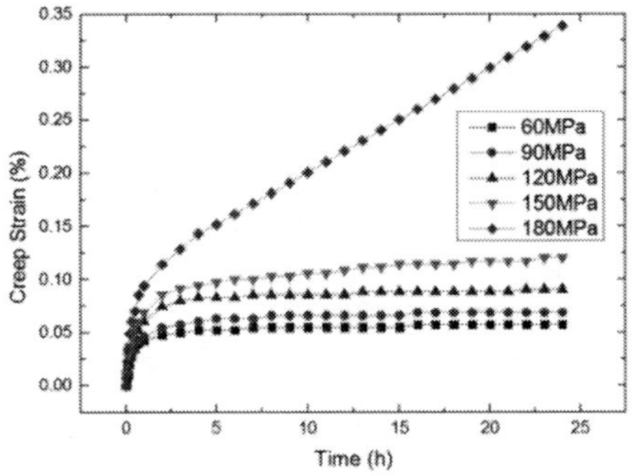

Fig. 2. Creep strain curves of the stress-aged samples.

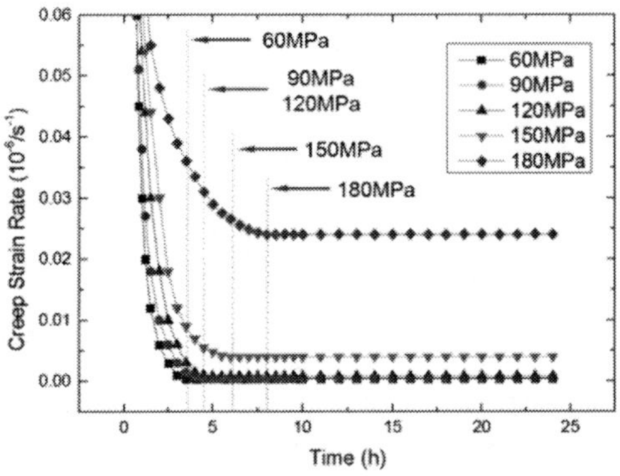

Fig. 3. Creep strain rate curves of the stress-aged samples.

The stress exponent, calculated by $n_s = \frac{\ln \dot{\varepsilon}_s}{\ln \sigma_0}$, where $\dot{\varepsilon}_s$ is the steady creep strain rate and σ_0 is the applied stress, is introduced to describe the creep mechanisms (Frost and Ashby, 1982). As displayed in Fig. 4, in the elastic region, the stress exponent is 0.94, indicating that diffusion creep mechanism controls the creep deformation. In the plastic region, the stress exponent is 9.5, suggesting that dislocation climbing mechanism is controlling. The substantial difference between low-stress and high-stress creep behaviors is caused by the stress-induced dislocations.

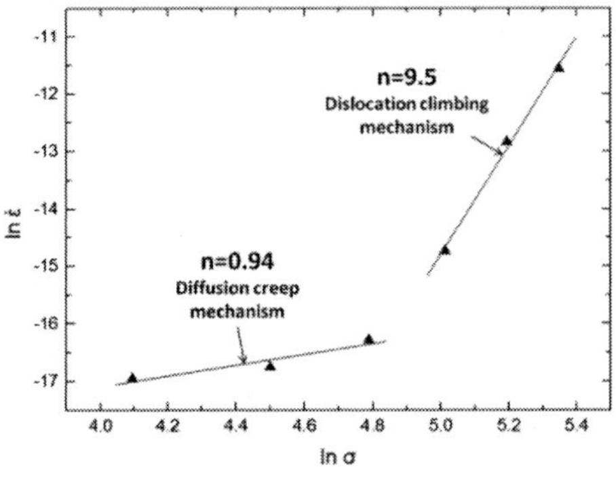

Fig. 4. Log-log plot of creep strain rate versus applied stress: the line slope represents the stress exponents of the creep-ageing process.

3.3 Microstructure evolution

3.3.1 Dislocation evolution

The XRD spectrums were analyzed with commercial JADE software to obtain the peak full width at half maximum (FWHM) which serves as input to the modified Williamson-Hall equation (Ungár, 2001) to calculate the dislocation density. The evolutions of dislocation density of different samples are shown in Fig. 5. The dislocation density decrease in a declining rate. The difference results from the dislocation multiplication during creep-ageing. Moreover, the 180MPa-stress-induced dislocation density is 0.78E14 m^{-2}, much higher than that of the quenched-in dislocation which is too small to be measured accurately using present method.

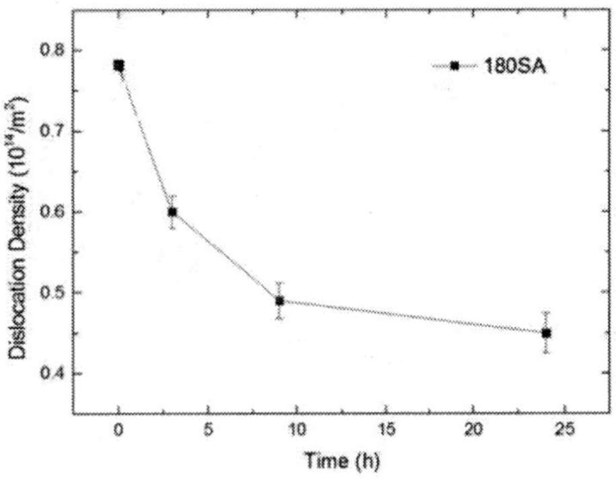

Fig. 5. Dislocation density evolution with creep ageing time in high-stress loaded samples

3.3.2 Precipitation behaviors

3.3.2.1 Bi-modal precipitation microstructure

High-angular dark-field STEM (HAADF-STEM) imaging provides the atomic number (*Z*) dependent contrast which can be easily interpreted (Howie, 1979; Pennycook and Jesson, 1991). In our case, the region enriched in Cu would appear brighter than the Al matrix. HAADF-STEM images of the samples under different conditions are shown in Fig. 6 for comparison. By comparing with HAADF images reported in the literature (Bourgeois et al., 2011; Liu et al., 2014), the precipitates in SFA and 60SA samples are identified as G.P. Zones and θ" phase, while those in 180SFA and 180SA samples are G.P. Zones, θ" phase and θ' phase. According to the literatures (Biswas et al., 2011; Chen et al., 2013), the ageing temperature in this work (165°C) is too low to activate θ' nucleation in the Al matrix. In Fig.

https://doi.org/10.1051/matecconf/201819012005

7, it can be seen that the unexpected θ', boxed in the yellow frames, nucleate and grow at the dark regions, whose contrast are usually caused by stress-field around dislocations. Dislocation has a relatively high energy and could facilitate the nucleation of stable θ' (Kumar et al., 1996). As the 180MPa stress induces a plastic-deformation of about 2% at the initial loading stage, a certain amount of dislocations is generated in 180SA samples. While in SFA and 60SA samples, the dislocations have a much lower density and can't provide preferential nucleation sites for θ', leading to the absence of θ'. Therefore, the homogeneous precipitation from the Al matrix and the heterogeneous precipitation at dislocations results in the formation of bi-modal sized precipitates comprised of fine G.P. Zones and θ" as well as large θ' in the high-stress-aged samples (Fig. 6e and f).

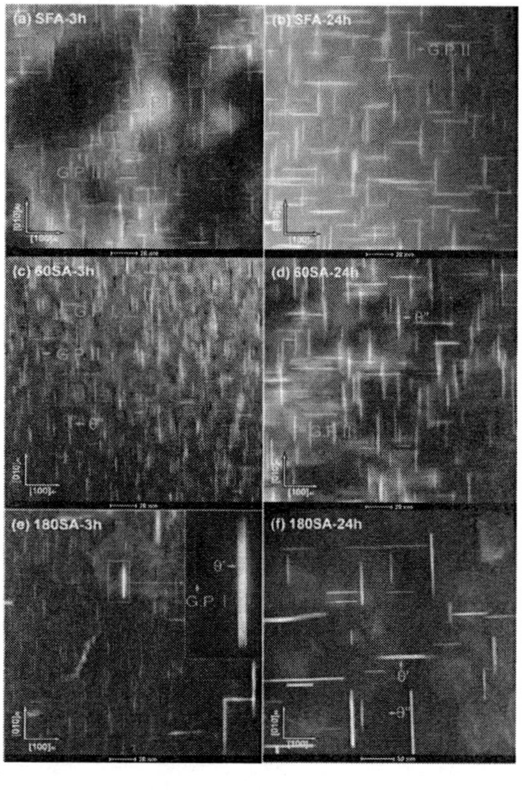

Fig. 6. Low-magnification STEM images of the aged Al-Cu alloy samples:(a) stress-free-aged for 3h; (c) 60MPa stress-aged for 3h; (e) 180MPa stress-aged for 3h; (b) stress-free-aged for 24h; (d) 60MPa stress-aged for 24h; (f) 180MPa stress-aged for 24h.

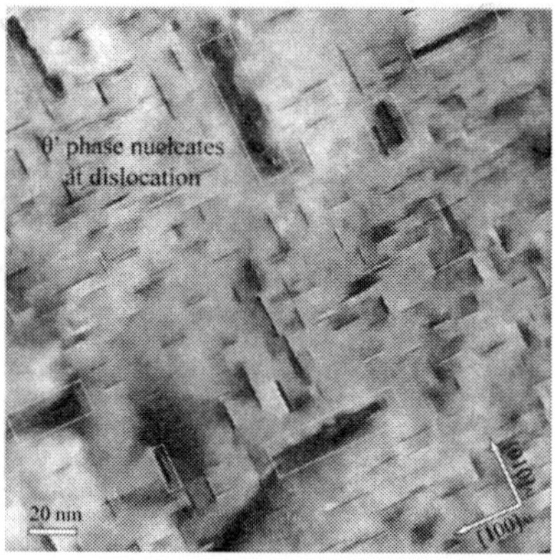

Fig. 7. A bright-field diffraction contrast TEM image of the sample stress-aged for 3h under 180MPa. The boxed precipitates represent θ' precipitates.

3.3.2.2 Stress orienting effect

When observed along $[001]_{Al}$ direction, two edge-on variants of the plate-like precipitates are visible in the STEM images (see Fig. 6). Generally, the proportion of these two variants is equal to each other, as shown in Fig. 6a and b for the SFA samples. In contrast, the proportion of one particular variant is much higher than the other one in the stress-aged samples (Fig. 6c-f). Another interesting point is that this stress orienting effect (SOE) mainly occurs on the *G.P.* Zones and *θ"* phase, while that of the large θ' precipitates in the high-stress-aged sample is not appreciable. As the major microstructural difference, the SOE of precipitates should be responsible for the negative effect of stress on the age hardening. The stress orienting effect was examined in detail in this study. As shown in Fig. 8, the extent of SOE for the stress-aged sample is varied in different grains and even in different regions of the same grain. There are more $(100)_{Al}$ variants in region A while more $(010)_{Al}$ variants in region C. In region B, the numbers of two variants are almost the same. This phenomenon is caused by the varying micro-stress field in different grains, even in the same grain. SOE is caused by stress-induced anisotropy of interfacial energy of precipitates on various $\{001\}_{Al}$ planes (Zhu and Starke, 1999, 2001a). A changing orientation relationship between micro-stress and the habit planes is thus expected in the investigated Al-Cu alloy, resulting in a variation of the observed SOE.

The absence of SOE for θ' precipitates suggests a negligible stress effect on the nucleation and growth of precipitates at the randomly distributed dislocations. The stress-field around dislocations should have a much larger stress value than that imposed by the externally applied stress. Therefore, the pre-existed dislocations should play a decisive role on the orientations of the precipitates growing around them. The main strengthening phase in 60SA samples is θ" while 180SA samples are mostly strengthened by θ', which has no evident SOE. The reduction of strength is thus lower when the applied stress changes from 60MPa to 180MPa, relative to their respective stress-free-aged samples.

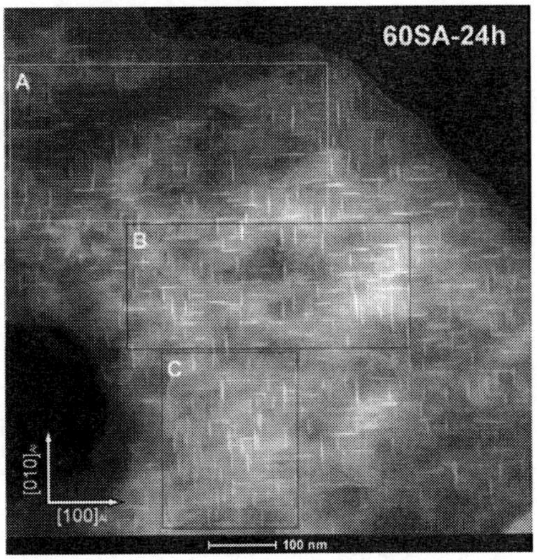

Fig. 8. A low-magnification STEM image of the sample stress-aged for 24h under 60MPa.

3.3.2.3 Relative volume fraction

For purpose of simplicity, as G.P. Zones share similar SOE, crystal structure and strengthening mechanism with θ" phase, G.P. Zones were categorized as θ" phase hereafter. There are several ways to measure the volume fraction of the precipitates in alloys (Chen et al., 2013; Donnadieu et al., 2011; Rodgers and Prangnell, 2016). In our work, differential scanning calorimeter (DSC) was used to quantify the relative volume fractions (RVF) of the different type precipitates (Rodgers and Prangnell, 2016). The volume proportion of θ" to θ' can be determined with the statistical analysis in S/TEM images, i.e. size and number proportion. In this way, the RVF of θ" and θ' with different ageing treatments are obtained.

In this bimodal precipitation, the evolution of the RVF is very complex due to the θ"/θ' competition for Cu atoms, which is essential to the nucleation and growth of precipitate. As shown in Fig. 9, the RVF of θ' increases monotonically and differs with that of θ". The RVF of θ' is smaller than that of θ" during the first several hours. After that, the RVF of θ' continues to increase while that of θ" begins to decrease. At the initial creep ageing stage, both θ" and θ' consume Cu atoms from the Al matrix, so the RVFs increase together. Since it is harder for semi-coherent θ' to nucleate and grow than coherent θ", the volume fraction of θ" increases faster than that of θ'. The θ"/θ' competition intensifies as ageing proceeds. The metastable θ" phases dissolve while the stable θ' can survive. Thus, the RVF of θ' continues to increase while that of θ" begins to decrease.

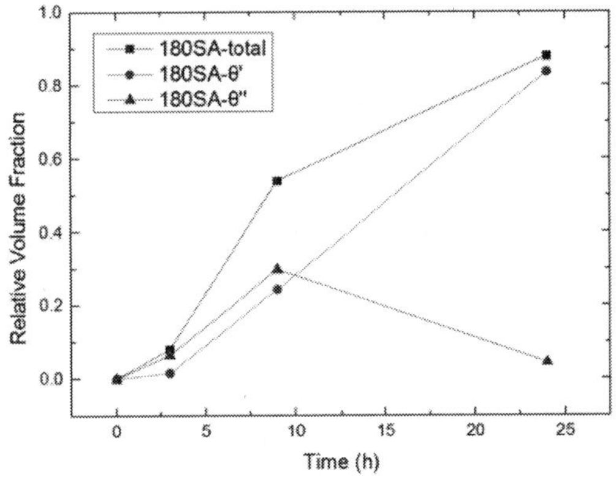

Fig. 9. The evolution of relative volume fraction of precipitates in 180SA samples.

3.3.2.4 Precipitates size distribution

Precipitate sizes were measured with Nano Measurer software for statistical analysis. At least 5 HAADF-STEM images acquired at different regions for each sample were chosen for measuring precipitate sizes. Fig. 10 presents the precipitates size distribution (PSD) of the bi-modal precipitation in the high-stress-aged samples. θ" is dominant at the early ageing stage in 180SA. The proportion of θ" shrinks as the ageing proceeds.

The θ"/θ' competition and the applied stress perplex the evolution of the average precipitate diameters (APD) during creep ageing. During the first several hours of ageing, the growth of both θ" and θ' consumes Cu atoms from the Al matrix which still contains abundant Cu solute atoms. During the later several hours (Oswald ripening regime), when the concentration of Cu atoms in the Al matrix is too low, θ" will dissolve rather than grow further if more energetically stable θ' form (Nie and Muddle, 2008; Polmear, 1987). In 180SA sample, the APD of θ" particles increases firstly and then decreases. Fig. 11 shows the similar PSD in the low-stress-aged samples with the stress-free-aged ones. The APD of θ" particles are only slightly larger in 60SA samples (Fig. 12), verifying that the applied low stress has a marginal effect on the growth of θ" particles.

 https://doi.org/10.1051/matecconf/201819012005

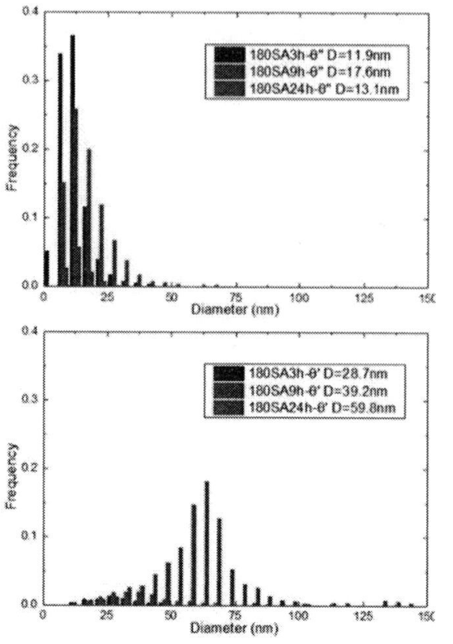

Fig. 10. Size distribution evolution of precipitates (top: θ"; bottom: θ') during the ageing process (D is the average diameter) in 180SA samples.

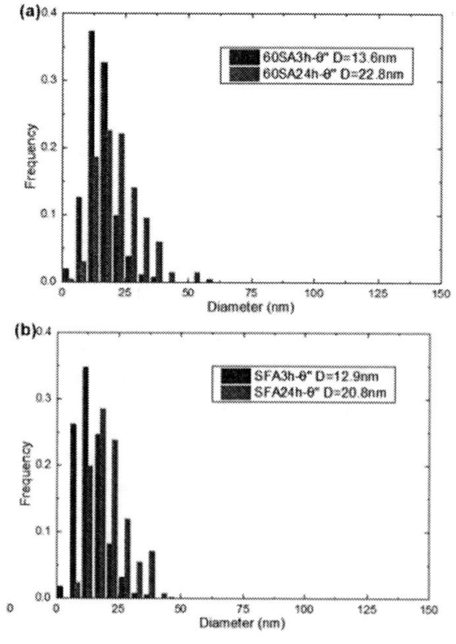

Fig. 11. Size distribution of precipitates in the samples with and without low-stress ageing (D is the average diameter). (a) θ" in 60SA samples; (b) θ" in SFA samples.

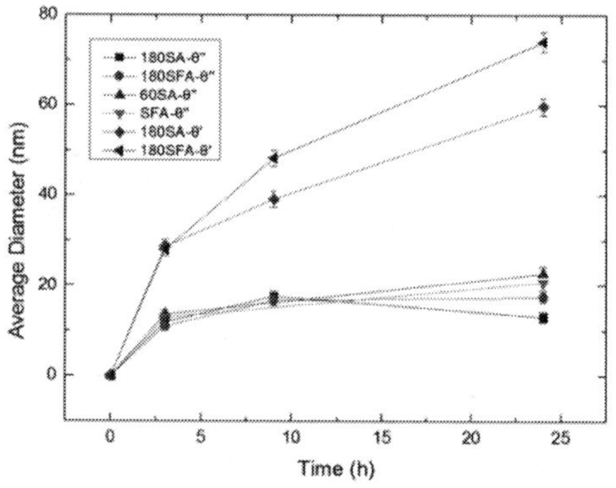

Fig. 12. Average diameters of precipitates under different creep ageing treatments.

3.4 Microstructure-property relationship

After careful examination of the microstructure evolution, including the dislocation evolution and precipitation behaviors, it is possible to explain the creep-ageing response observed in Fig. 1. The average strength loss due to the negative stress effect before the peak-aged time decreases from 8.5% to 3.7% when the stress increases from elastic-stress-regime (60MPa-120MPa) to plastic-stress-regime (180MPa). The weakening of the strength contribution from age hardening mainly stems from the apparent SOE of θ" precipitates. The high-stress-induced dislocations give rise to the heterogeneous growth of θ' precipitates, which have no detectable SOE. Therefore, the negative stress effect is reduced when the applied stress exceeds the yield strength. The θ' is more effective in strengthening than θ" (Nie and Muddle, 2008). θ' precipitates are the major strengthening contributor in the high-stress-aged samples, while θ" plays a primary role in low-stress-aged samples. The significant strength enhancement in the high-stress-aged samples results from mainly by θ' strengthening and partly by dislocation hardening.

The sketch in Fig. 13 depicts the microstructure evolution in stress ageing process. The precipitates occurring during low-stress ageing are dominated by θ" particles with obvious SOE. Under high-stress ageing, besides preferentially oriented θ" particles, randomly distributed θ' particles at dislocations form continuously. The percentage of θ" is very high in the first several hours but decreases after then due to the gradual growth of θ'. Microstructure evolution, including dislocation evolution and precipitation behavior, and their relationships with the creep deformation and the mechanical properties are discussed qualitatively in this section. The quantitative relation needs to be investigated to guide the practical application in forming process.

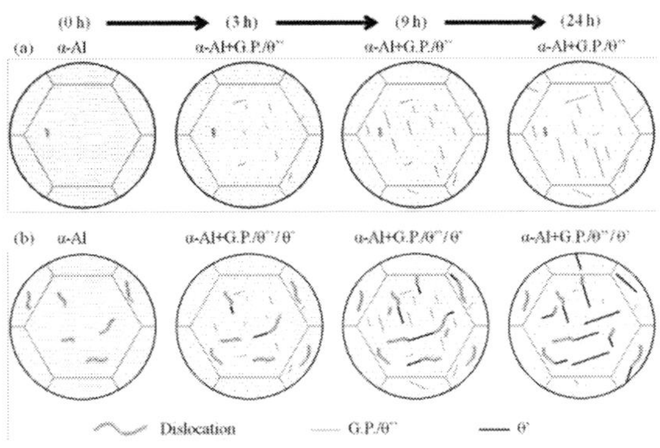

Fig. 13. Evolution of microstructures during stress ageing of AA2219 at: (a) low stress; (b) high stress.

4 Conclusions

This study systematically investigated the creep ageing behaviors of AA2219 Al alloy under varied applied stresses. Microstructural evolution including dislocations and precipitates observed in detail by X-ray diffractometer, Transmission electron microscopy and Differential scanning calorimetry provides new insights into the creep deformation and strengthening mechanisms. The main conclusions can be summarized as follows:

(1) Applied stress of different levels generally exerts a negative effect on the age hardening response of the alloy. The considerable increase of hardening potential for the high stress applied samples mainly results from dislocations generated at the loading stage.

(2) The creep mechanism changes from diffusion creep to dislocation climbing when the stress exceeds the initial yield strength. Creep strain is significantly increased and the primary creep stage is prolonged notably due to the high-stress-induced dislocations.

(3) The strength loss in the stress-aged samples is ascribed to the apparent preferential alignment of θ" precipitates (stress-orienting effect, SOE). The dislocations induced at high-stress loading stage give rise to the heterogeneous growth of relatively large θ' precipitates without detectable SOE, leading to the formation of a characteristic bimodal precipitate morphology. As a result, the negative stress effect is reduced evidently when the applied stress exceeds the yield strength. The θ" dominates in the first several hours but dissolves gradually due to the continuous growth of more stable θ'.

This work is supported by National key R&D Program of China (2017YFB0306300), National Natural Science Foundation of China (No. 51235010, 51601060), National Key Basic Research Development Plan Funded Project of China (No. 2014cb046602), National Defence Program of China (JCKY2014203A001) and the Fundamental Research Funds for the Central Universities of Central South University (1053320170568).

References

Biswas, A., Siegel, D.J., Wolverton, C., Seidman, D.N., 2011. Precipitates in Al–Cu alloys revisited: Atom-probe tomographic experiments and first-principles calculations of compositional evolution and interfacial segregation. Acta Materialia 59, 6187-6204.
Bourgeois, L., Dwyer, C., Weyland, M., Nie, J.-F., Muddle, B.C., 2011. Structure and energetics of the coherent interface between the θ" precipitate phase and aluminium in Al - Cu. Acta Materialia 59, 7043-7050.
Chen, Y., Weyland, M., Hutchinson, C.R., 2013. The effect of interrupted aging on the yield strength and uniform elongation of precipitation-hardened Al alloys. Acta Materialia 61, 5877-5894.
Donnadieu, P., Shao, Y., De Geuser, F., Botton, G.A., Lazar, S., Cheynet, M., de Boissieu, M., Deschamps, A., 2011. Atomic structure of T1 precipitates in Al–Li–Cu alloys revisited with HAADF-STEM imaging and small-angle X-ray scattering. Acta Materialia 59, 462-472.
Eto, T., Sato, A., Mori, T., 1978. Stress-oriented precipitation of G.P. Zones and θ' in an Al-Cu alloy. Acta Metallurgica 26, 499-508.
Frost, H.J., Ashby, M.F., 1982. Deformation mechanism maps: the plasticity and creep of metals and ceramics. Pergamon press.
Guinier, A., 1938. Structure of age-hardened aluminium-copper alloys. Nature 142, 569-570.
Ho, K.C., Lin, J., Dean, T.A., 2004. Modelling of springback in creep forming thick aluminum sheets. International Journal of Plasticity 20, 733-751.
Holman, M.C., 1989. Autoclave age forming large aluminum aircraft panels. Journal of Mechanical Working Technology 20, 477-488.
Howie, A., 1979. Image Contrast And Localized Signal Selection Techniques. Journal of Microscopy 117, 11-23.
Hu, L., Zhan, L., Shen, R., Liu, Z., Ma, Z., Liu, J., Yang, Y., 2017. Effects of uniaxial creep ageing on the mechanical properties and micro precipitates of Al-Li-S4 alloy. Materials Science and Engineering: A 688, 272-279.
Kumar, K.S., Brown, S.A., Pickens, J.R., 1996. Microstructural evolution during aging of an AlCuLiAgMgZr alloy. Acta Materialia 44, 1899-1915.
Lei, C., Yang, H., Li, H., Shi, N., Zhan, L.H., 2017. Dependences of microstructures and properties on initial tempers of creep aged 7050 aluminum alloy. Journal of Materials Processing Technology 239, 125-132.
Li, Y., Shi, Z., Lin, J., Yang, Y.-L., Rong, Q., Huang, B.-M., Chung, T.-F., Tsao, C.-S., Yang, J.-R., Balint, D.S., 2017. A unified constitutive model for asymmetric tension and compression creep-ageing behaviour of naturally aged Al-Cu-Li alloy. International Journal of Plasticity 89, 130-149.
Liu, L., Chen, J.H., Wang, S.B., Liu, C.H., Yang, S.S., Wu, C.L., 2014. The effect of Si on precipitation in Al–Cu–Mg alloy with a high Cu/Mg ratio. Materials Science and Engineering: A 606, 187-195.

Ma, P.P., Liu, C.H., Wu, C.L., Liu, L.M., Chen, J.H., 2016. Mechanical properties enhanced by deformation-modified precipitation of θ′-phase approximants in an Al-Cu alloy. Materials Science and Engineering: A 676, 138-145.
Nie, J.F., Muddle, B.C., 2008. Strengthening of an Al–Cu–Sn alloy by deformation-resistant precipitate plates. Acta Materialia 56, 3490-3501.
Pennycook, S.J., Jesson, D.E., 1991. High-resolution Z-contrast imaging of crystals. Ultramicroscopy 37, 14-38.
Polmear, I.J., 1987. Role of Trace Elements in Aged Aluminium-Alloys. Materials Science Forum 13-14, 195-214.
Rodgers, B.I., Prangnell, P.B., 2016. Quantification of the influence of increased pre-stretching on microstructure-strength relationships in the Al–Cu–Li alloy AA2195. Acta Materialia 108, 55-67.
Ungár, T., 2001. Dislocation densities, arrangements and character from X-ray diffraction experiments. Materials Science and Engineering: A 309, 14-22.
Xu, Y., Zhan, L., Li, W., 2017a. Effect of pre-strain on creep aging behavior of 2524 aluminum alloy. Journal of Alloys and Compounds 691, 564-571.
Xu, Y., Zhan, L., Xu, L., Huang, M., 2017b. Experimental research on creep aging behavior of Al-Cu-Mg alloy with tensile and compressive stresses. Materials Science and Engineering: A 682, 54-62.
Yang, Y., Zhan, L., Ma, Q., Feng, J., Li, X., 2016. Effect of pre-deformation on creep age forming of AA2219 plate: Springback, microstructures and mechanical properties. Journal of Materials Processing Technology 229, 697-702.
Zhan, L., Lin, J., Dean, T.A., 2011a. A review of the development of creep age forming: Experimentation, modelling and applications. International Journal of Machine Tools and Manufacture 51, 1-17.
Zhan, L., Lin, J., Dean, T.A., Huang, M., 2011b. Experimental studies and constitutive modelling of the hardening of aluminium alloy 7055 under creep age forming conditions. International Journal of Mechanical Sciences 53, 595-605.
Zhang, J., Deng, Y., Zhang, X., 2013. Constitutive modeling for creep age forming of heat-treatable strengthening aluminum alloys containing plate or rod shaped precipitates. Materials Science and Engineering: A 563, 8-15.
Zhu, A.W., Starke, E.A., 1999. Strengthening effect of unshearable particles of finite size: a computer experimental study. Acta Materialia 47, 3263-3269.
Zhu, A.W., Starke, E.A., 2001a. Stress aging of Al–xCu alloys: experiments. Acta Materialia 49, 2285-2295.
Zhu, A.W., Starke, J.E.A., 2001b. Stress aging of Al–Cu alloys: computer modeling. Acta Materialia 49, 3063-3069.

MATEC Web of Conferences **190**, 12006 (2018)
ICNFT 2018
https://doi.org/10.1051/matecconf/201819012006

Effect of initial temper on mechanical properties of creep-aged Al-Cu-Li alloy AA2050

Yong Li[1,*], *Yo-Lun* Yang[1], *Qi* Rong[1], *Zhusheng* Shi[1], *Jianguo* Lin[1] *and Rajab* Said[2]

[1]Department of Mechanical Engineering, Imperial College London, London SW7 2AZ, UK

[2]ESI Group, 100-102, Avenue de Suffren, 75015, Paris, France

Abstract. The evolution of mechanical properties of a third generation Al-Cu-Li alloy, AA2050, with different initial tempers (T34 and as-quenched (WQ)) during creep-ageing has been investigated and analysed in this study. A set of creep-ageing tests under 150 MPa at 155 °C for up to 24 h was carried out for both initial temper conditions and tensile tests were performed subsequently to acquire the main mechanical properties of the creep-aged alloys, including the yield strength, ultimate tensile strength (UTS) and uniform elongation. Precipitation behaviour of the T34 and WQ AA2050 alloys has been summarised and successfully explains the detailed evolutions of the obtained mechanical properties of the alloy with these two initial tempers during creep-ageing. The results indicate that the T34 alloy can be a better choice for creep age forming (CAF) process compared with WQ alloy, as it provides better yield strength and uniform elongation properties concurrently after creep-ageing. In addition, a work hardening rate analysis has been carried out for all the creep-aged alloys, helping to reveal the detailed dislocation/precipitates interaction mechanisms during plastic deformation in the creep-aged T34 and WQ AA2050 alloys.

Keywords: Sheet metal; Simulation; Forming.

1 Introduction

Creep age forming (CAF) is an advanced forming technology originally proposed to manufacture large/extra-large panels in the aerospace industry, in which the plastic deformation and age hardening occur simultaneously in the material during the process [1]. The creep-ageing behaviour of the recently developed third generation Al-Cu-Li alloy AA2050 has been investigated previously [2, 3], which indicated that the alloy is promising to be used for the aerospace components through CAF. For potential industrial applications, the detailed mechanical properties of the alloy after CAF are important, especially for the aerospace components, and therefore need be investigated and evaluated.

It has been widely reported that initial tempers could highly affect the precipitation progress of Al-Cu-Li alloys during artificial ageing and result in different yield strength properties of the aged alloys [4, 5]. Kumar et al. [4] investigated the effect of pre-stretch on ageing behaviour of a naturally-aged Al-Cu-Li alloy and indicated that pre-stretch can not only accelerate the age hardening speed of the alloy, but also assist the nucleation of T_1 (Al_2CuLi) precipitates, enhancing the yield strength of the alloy. Similar phenomenon has also been observed in an as-quenched Al-Cu-Li alloy by Ringer et al. [5]. While for CAF process, the creep-ageing behaviour of an Al-Cu-Li alloy (AA2050) with different initial tempers, namely the as-quenched (WQ), T34 and T84, has also been investigated and compared, showing different creep strain and age hardening properties under the same stress level at 155 °C [2]. However, only the yield strength of the alloys after creep-ageing was considered in previous studies, other mechanical properties, such as ultimate tensile strength (UTS) and uniform elongation, are also important for the CAFed alloys and should be considered as well, so as to provide a comprehensive way to select alloys with a proper initial temper condition for CAF applications.

Meanwhile, the plastic deformation behaviour of Al-Cu-Li alloys also attracts a lot of interest, as it can explain the strength and elongation properties of the alloys and also indicate the detailed dislocation/precipitates interaction mechanism during plastic deformation. Some earlier studies reported that Al-Cu-Li alloys with different precipitates after artificial ageing show different work hardening behaviour during plastic deformation, depending on the main precipitates in them (such as T_1 and/or δ' (Al_3Li)) [6-8]. As different precipitates exist in the alloys with different initial tempers after creep-ageing, their effects on the plastic deformation behaviour of particular alloys are also of interest.

Hence, in this paper, an Al-Cu-Li alloy, AA2050, with two commonly reported initial temper conditions, as-quenched and naturally-aged, was used to investigate the

* Corresponding author: yong.li14@imperial.ac.uk

initial temper effect on evolutions of mechanical properties, including yield strength, UTS and elongation, during creep-ageing in CAF. The plastic deformation behaviour of the creep-aged alloys is then discussed based on a work hardening rate analysis, so as to reveal the detailed dislocation/precipitates interaction mechanisms related to the mechanical properties of the creep-aged AA2050.

2 Material and experiments

2.1. Material and heat treatment

AA2050, a third generation Al-Cu-Li alloy provided by Embraer (Brazil), was used in this study. The main chemical composition of the alloy is listed in Table 1 below.

Table 1. Main chemical composition of AA 2050 (wt%).

Al	Cu	Li	Mg	Ag	Mn	Zr
Balance	3.6	0.9	0.34	0.35	0.34	0.08

The material was received with a naturally-aged T34 initial temper, which has been solution heat treated (SHT), water quenched, pre-stretched and finally naturally-aged. The material with another initial temper condition, as-quenched, was prepared from the as-received T34 material by solution heat treating (SHT) at 500 °C for 1 h, followed by water quenching. Fig. 1 shows the heat treatment procedures for material preparation and subsequent creep-ageing test in this study.

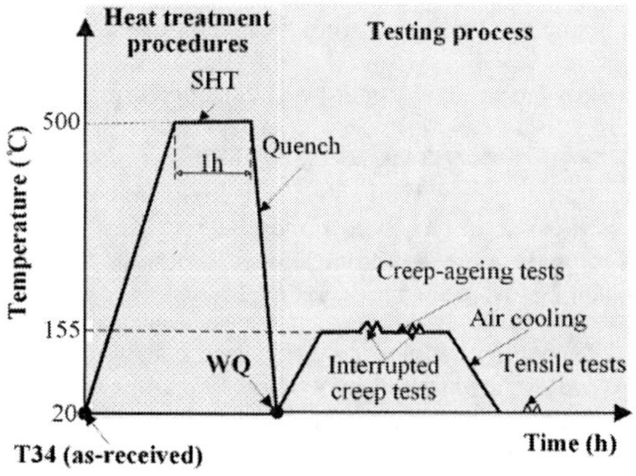

Fig. 1. Heat treatment procedures and test conditions

2.2 Creep-ageing and tensile tests

AA2050 alloy with T34 and WQ initial tempers was creep-aged under a constant stress level of 150 MPa at 155°C for different durations, as listed in Table 2. For the alloy with WQ initial temper, creep-ageing tests were performed immediately after water-quenching. Subsequent tensile tests were carried out at room temperature to obtain the main mechanical properties of the creep-aged alloys. The heat treatment condition for these tests is shown in Fig. 1.

Table 2. Experimental plan for creep-ageing and tensile test.

Initial temper	Creep-ageing time (h)
T34	0, 2, 5, 8, 12, 18, 22
WQ	0, 5, 12, 18, 24

Both creep-ageing and tensile tests were performed on an Instron 5584 machine. Fig. 2 shows the sample geometry used for both tests. A high temperature extensometer Instron 2620-601 was attached onto the sample to capture the strain during both tests.

For creep-ageing tests, an assisting furnace was used to provide the high temperature environment and the constant load was applied after the sample reached 155 °C. For tensile tests, a strain rate of 10^{-4}/s was used.

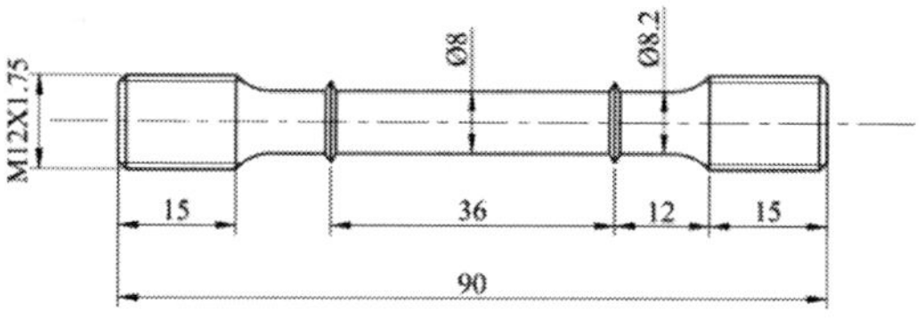

Fig. 2. Specimen geometry and dimension for creep-ageing and tensile tests (units: mm).

3 Evolution of precipitates

The details of precipitation progress of AA2050 and other similar Al-Cu-Li alloys with T34 and WQ initial tempers during artificial ageing/creep-ageing have been investigated and discussed before [2, 5, 6]. A brief summary of the precipitation is given in this section to assist the further analysis of the current study, as listed in Table 3. Different microstructures have been observed in the initial state of T34 and WQ tempered AA2050 alloys. Cu-rich clusters exist in T34 alloy and the WQ alloy shows a supersaturated solid solution (SSSS) condition. During artificial ageing, T34 alloys experience an initial dissolution of Cu-rich clusters and accelerated nucleation of T_1 (Al_2CuLi) precipitates (with some minor θ' (Al_2Cu) / S' (Al_2CuMg)) occurs subsequently because of the dislocations in the alloys from pre-stretching. While for the WQ alloys, nucleation and growth of GP zones and δ' (Al_3Li) occur first and T_1 precipitates (with some θ'/S') start to nucleate and grow afterwards. Hence, in the peak-aged state, different precipitates are observed in the T34 and WQ alloys. The peak-aged T34 alloys contain mainly T_1 precipitates with some θ'/S', while the peak-aged WQ alloys have both T_1 and δ' precipitates, along with some θ'/S'.

Table 3. Summary of main precipitates during artificial ageing/creep-ageing of AA2050 with T34 and WQ initial tempers at 155 °C (summarised from [2, 5, 6]).

Initial temper	Initial	Under-aged	Peak to Over-aged
T34	Cu clusters	Dissolution of Cu clusters → Nucleation and growth of precipitates (mainly T_1 with some θ'/S')	Majority of T_1 with some θ'/S'
WQ	SSSS	Nucleation and growth of GP zones and δ' → Nucleation and growth of T_1 and some θ'/S'	T_1 and δ' with some θ'/S'

4 Results and discussion

4.1 Mechanical properties

Test results for the yield strength, UTS and uniform elongation of the creep-aged T34 and WQ AA2050 alloy are presented and discussed in this section.

4.1.1 T34 initial temper

Fig. 3 shows the tensile test data of the T34 AA2050 alloy creep-aged under 150 MPa at 155 °C for different time. According to the stress-strain curves shown in Fig. 3(a), the as-received T34 alloy shows a relatively high yield strength of about 269 MPa with a significant work hardening behaviour. In addition, apparent serrations, known as the Portevin-Le Chatelier effect, can be observed in the stress-strain curve of the as-received T34 alloy. This phenomenon may indicate that only minor stable clusters exist in the as-received alloy and hence, sufficient solutes in the alloy matrix can diffuse around dislocations to restrain its movement, leading to the observed serrations during plastic deformation [9, 10].

After 2 h of creep-ageing, the yield strength of the alloy drops to about 245 MPa, which has been explained as the dissolution of the naturally-aged Cu clusters in the T34 alloy. After that, yield strength increases with the creep-ageing time and nearly recovered to the as-received level (269 MPa) at about 5 h creep-ageing. Significant serrations are also observed in the stress-strain curve of the 2 h creep-aged T34 alloy, reflecting the dissolution of clusters in the first 2 h of creep-ageing. For the 5 h creep-aged T34 alloy, only slight serrations can be found, indicating that nucleation and growth of precipitates begin and less free diffused solutes exist in the matrix [9].

After 5 h creep-ageing, accelerated nucleation and growth of T_1 precipitates occurs, resulting in the dramatical increase of yield strength until peak-ageing occurs after 18 h creep-ageing, with a maximum yield strength of about 498 MPa. In addition, serrations in stress-strain curves disappear with the present of T_1 precipitates after 5 h creep-ageing. Subsequently, yield strength starts to slightly decrease along the creep-ageing time, which corresponds well with some previous studies [8].

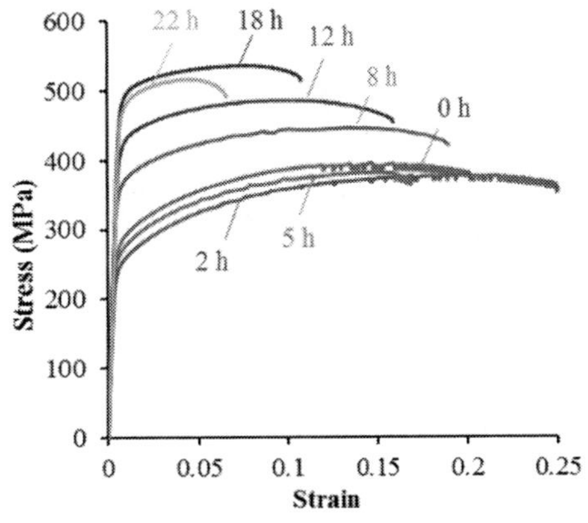

(a)

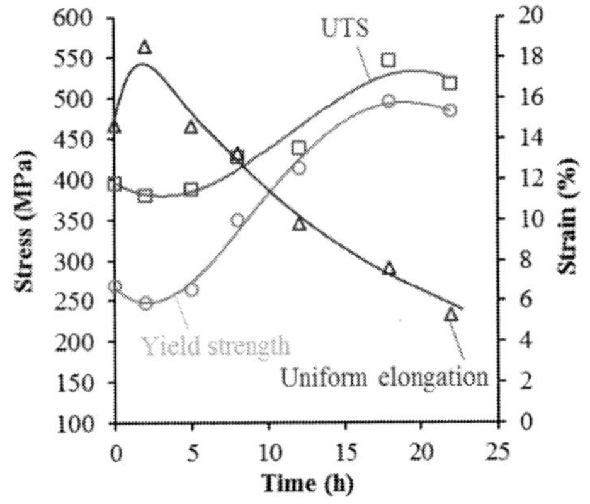

(b)

Fig. 3. (a) Stress-strain curves for T34 AA2050 creep-aged under 155 MPa at 155 °C with indicated durations and (b) resultant yield strength, UTS and uniform elongation of the creep-aged alloys.

Fig. 3(b) illustrates the evolutions of yield strength, UTS and uniform elongation data during creep-ageing of T34 alloys. UTS shares the similar pattern with yield strength along the creep-ageing time, dropping down in the first 2 h creep-ageing and increasing substantially up to 18 h. However, the difference between UTS and yield strength becomes smaller with increasing creep-ageing time, indicating a weakened work hardening behaviour because of the well-developed precipitates [8].

* Corresponding author: yong.li14@imperial.ac.uk

Meanwhile, the uniform elongation evolves oppositely with yield strength and UTS, increasing at the first 2 h and decreasing afterwards.

4.1.2 WQ initial temper

Fig. 4 shows the same tensile data for various creep-aged AA2050 with WQ initial temper. The initial WQ alloy is with SSSS state and exhibits a comparative low yield strength of 117 MPa. However, substantial work hardening behaviour occurs in the alloy and its stress-strain curve demonstrates significant serrations, reflecting its precipitates free condition [8, 10].

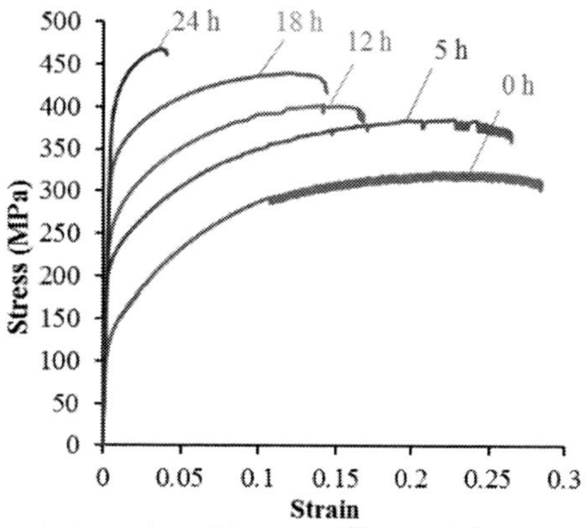

(a)

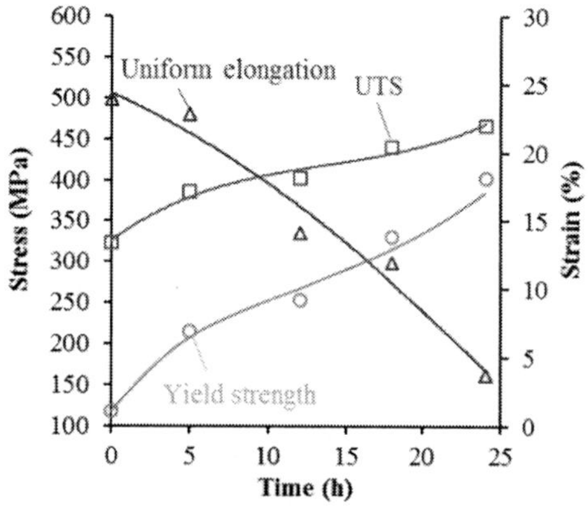

(b)

Fig. 4. (a) Tensile curves of AA2050 with WQ initial temper creep-aged under 155 MPa at 155 °C for different time durations and (b) resulted yield strength, UTS and uniform elongation of the creep-aged alloys.

The yield strength increases continuously along the creep-ageing time for WQ AA2050 alloys, because of the nucleation and growth of precipitates. However, in the early stage of creep-ageing, within 12 h at 155 °C in this study (Fig. 4(a)), serrations still exist in the stress-strain curves of the creep-aged WQ alloys. Deschamps et al. [8] observed the same phenomenon in artificial ageing of AA2196 at the same stage where δ' precipitates control yield strength and indicated that it may relate to the properties of precipitates. After 12 h creep-ageing, yield strength increases with a higher speed and serrations disappear in corresponding stress-strain curves because of the nucleation and growth of T_1 precipitates, as explained in [8]. The alloy is still in under-aged condition after 24 h creep-ageing and the yield strength reaches 402 MPa.

The UTS data shares the similar trend with yield strength, as demonstrated in Fig. 4(b). Similar to T34 alloy results in Fig. 3(b), the level of work hardening strength (difference between UTS and yield strength) decreases with increasing creep-ageing time and a much higher decreasing speed can be observed after 12 h, when T_1 precipitates start to nucleate. The uniform elongation of the creep-aged WQ alloys, on the other hand, decreases continuously with the creep-ageing time.

4.1.3 Comparison

As CAF is mainly used for manufacturing structural panel products, high properties of both strength and ductility of the creep-aged material are preferred. According to the tensile data in Figs. 3 and 4, it seems that after the same creep-ageing time, T34 AA2050 alloy has a higher yield strength, but with a lower uniform elongation compared to the WQ alloy.

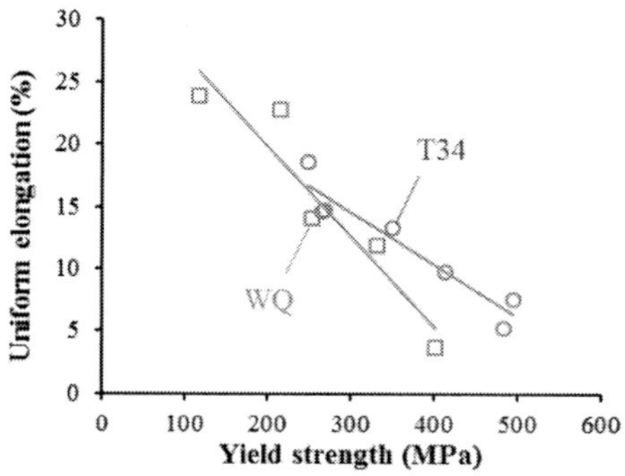

Fig. 5. Relationship between uniform elongation and yield strength for the creep-aged AA2050 with T34 and WQ initial tempers.

To evaluate the actual performance and relationship, the uniform elongation is plotted against the yield strength in Fig. 5 for the creep-aged T34 and WQ alloys. An inverse relationship can be seen between the yield strength and uniform elongation for both alloys. However, an interesting result is observed that with the same yield strength level, the creep-aged T34 alloy has a comparatively higher uniform elongation than the creep-aged WQ alloy, indicating that AA2050 alloy with T34 initial temper could be a better choice for CAF process, as both comparatively high yield strength and uniform elongation can be achieved concurrently after creep-ageing. This phenomenon is believed to be related to the different precipitates in the creep-aged T34 and WQ alloys. As T_1 precipitates contribute to a higher hardening effect and nucleated more homogenously than δ' precipitates [11], better mechanical properties can be

https://doi.org/10.1051/matecconf/201819012006

expected for the creep-aged T34 alloys that contain mainly T_1 precipitates, rather than WQ alloys, in which both T_1 and δ' exist.

4.2 Work hardening analysis

During tensile tests of the creep-aged alloys, the interaction between dislocations and precipitates determines the plastic deformation behaviour of the alloys. Work hardening analysis is carried out in this section to reveal the possible dislocation/precipitates interaction mechanism during plastic deformation of the creep-aged T34 and WQ AA2050 alloys. Kocks-Mecking plots [12] are used in this study for the work hardening analysis, as shown in Fig. 6, illustrating the evolution of work hardening rate ($d\sigma/d\varepsilon$, where σ is the applied stress and ε is the strain) against the work hardened strength (σ-σ_y, where σ_y is yield strength) for both T34 and WQ AA2050 alloys.

Generally, the creep-aged T34 and WQ alloys share similar work hardening behaviour, where the work hardening rate decreases with increasing hardened strength levels. For T34 alloys, when only solute clusters or minor precipitates (such as GP zones) exist during the early stage of creep-ageing (within 5 h in this study), similar work hardening rate can be observed in the alloy in Fig. 6(a). A very high initial work hardening rate (6000 to 8000 MPa) is observed at this stage because of the high free solute contents in the alloy matrix [8]. Similar work hardening rate is also observed in the as-quenched alloy with a SSSS condition, as shown in Fig. 6(b).

The work hardening rate behaviour starts to change when the main strengthening phases (T_1 and δ') start to substantially nucleate in the alloys. For T34 alloys, a lower initial work hardening rate with a higher dropping speed at the end part of the curves in Fig. 6(a) is observed after 5 h creep-ageing. The same trend can be observed in the creep-aged WQ alloy after 12 h creep-ageing according to Fig. 6(b). Moreover, the rate dropping speed increases in the material with longer creep-ageing time. Similar behaviour with ageing time has been reported in some Al-Cu-Li [8] and Al-Zn-Mg [13] alloys with shearable precipitates (such as δ' and η') and has been explained as the occurrence of shearing of precipitates during plastic deformation in tensile tests. Deschamps et al. [8] has reported that T_1 precipitate is a semi-shearable precipitate, which exhibits a single-time shearing property. Hence, shearing can be observed in creep-aged T34 alloys with dominant T_1 precipitates during tensile tests, leading to the lower initial work hardening rate with a higher dropping speed at the end of the Kocks-Mecking curves after 5 h creep-ageing [8]. Meanwhile, for the creep-aged WQ alloys, in which both T_1 and δ' exist, a much higher dropping speed of the work hardening rate is expected as δ' has a higher shearable property than T_1, as shown in Fig. 6(b) for the 24 h creep-aged WQ alloy.

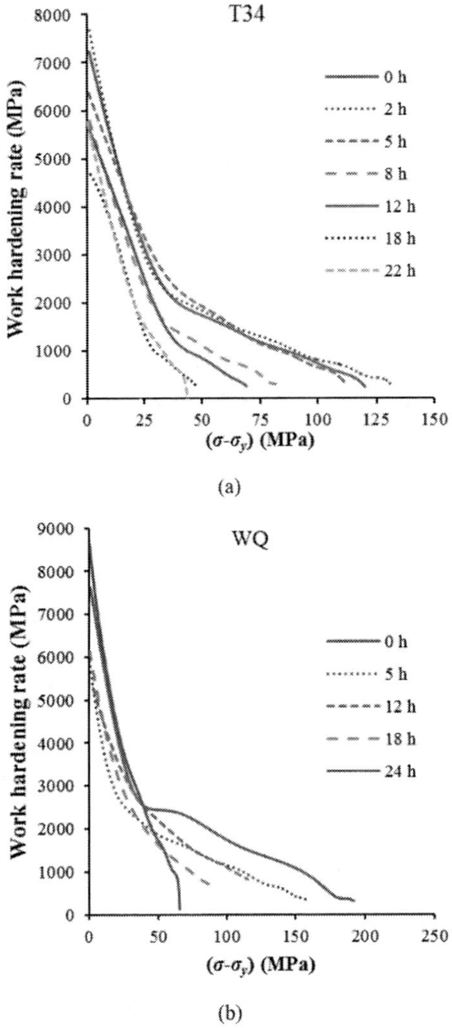

Fig. 6. Evolution of work hardening rates with the hardening strength (σ-σ_y) for AA2050 with (a) T34 and (b) WQ initial tempers creep-aged under 155 MPa at 155 °C for indicated durations.

5 Conclusions

In this study, main mechanical properties of AA2050 with T34 and WQ initial tempers, creep-aged under 150 MPa at 155 °C for various time, have been investigated and a work hardening analysis has been carried out to reveal the dislocation/precipitates interaction mechanism during tensile tests. The following conclusions can be drawn:

1) The yield strength of T34 AA2050 alloys drops from 269 to 245 MPa within the first 2 h of creep-ageing because of dissolution of Cu-clusters. It subsequently increases and reaches peak-ageing state with a yield strength of 498 MPa after 18 h. UTS shares similar behaviour with yield strength while the uniform elongation demonstrates an exactly opposite behaviour. For WQ AA2050 alloys, the yield strength and UTS

increase with creep-ageing time in the tested range up to 24 h, while uniform elongation decreases conversely.

2) T34 alloy is shown to be a better choice for CAF process than WQ alloy, as the creep-aged T34 alloy achieves a higher uniform elongation than the creep-aged WQ alloy with the same yield strength.

3) The different mechanical properties and work hardening behaviour of creep-aged alloys with T34 and WQ initial tempers can be explained by their precipitation behaviour. Creep-aged T34 alloys contain mainly T_1 precipitate, resulting in a higher strength and slight lower work hardening rate dropping speed than the creep-aged WQ alloys, in which both T_1 and δ' precipitates are present concurrently.

Acknowledgements

The authors would like to thank the financial support from ESI Group (France) and the provision of material by Embraer (Brazil).

References

1. L. Zhan, J. Lin and T. A. Dean, Int. J. of Mach. Tool. Manu. **51**, 1 (2011): 1-17.
2. Y. Li., Z. Shi, J. Lin, Y-L. Yang, B-M. Huang, T-F. Chung and J-R. Yang, Mater. Sci. Eng.: A. **657** (2016): 299-308.
3. Y. Li, Z. Shi, J. Lin, Y-L. Yang, Q. Rong, B-M. Huang, T-F. Chung, Ch-S. Tsao, J-R. Yang and D. Balint, Int. J. Plasticity. **89** (2017): 130-149.
4. K. S. Kumar, S. A. Brown and J. Pickens, Acta Mater. **44**, 5 (1996): 1899-1915.
5. S. P. Ringer, B. C. Muddle and I. J. Polmear, Metall. Mater. Trans. A. 26, 7 (1995): 1659-1671.
6. B. Decreus, A. Deschamps, F. De Geuser, P. Donnadieu, C. Sigli and M. Weyland, Acta Mater. 61, 6 (2013): 2207-2218.
7. A. Csontos and E. A. Starke, Int. J. Plasticity. **21**, 6 (2005): 1097-1118.
8. A. Deschamps, B. Decreus, F. D. Geuser, T. Dorin and M. Weyland, Acta Mater. **61** (2013): 4010-4021.
9. P. G. McCormick, Acta Metall. **19**, 5 (1971): 463-471.
10. D. Thevenet, M. Mliha-Touati and A. Zeghloul, Mater. Sci. Eng.: A. 266, 1 (1999): 175-182.
11. W. A. Cassada, G. J. Shiflet and E. A. Starke, Metall. Trans. A. 22, 2 (1991): 299-306.
12. U.F. Kocks and H. Mecking, Prog. Mater. Sci. **48** (2003): 171-273.
13. G. Fribourg, Y. Bréchet, A. Deschamps and A. Simar, Acta Mater. **59** (2011): 3621-3635.

Experimental Investigation on Creep Forming of A New Al-Li-Cu-Mg Alloy Plate

BAI Xuepiao[1,2,3,*], LI Hong[1,2,3], ZENG Yuansong[1,2,3], WANG Mingtao[1,2,3], LYU Fenggong[1,2,3], MENG Qiang[1]

[1] AVIC Manufacturing Technology Institute

[2] Aeronautical Key Laboratory for Plastic Forming Technology

[3] Beijing Key Laboratory of Digital Plasticity Forming Technology and Equipment

*Corresponding author: baixp@163.com

Abstract: Fundamental creep forming experiments on a new Al-Li-Cu-Mg alloy plate were carried out to investigate the forming parameters of integral reinforced panel with thin skin. The effects of process parameters on springback ratio and material properties are investigated by orthogonal design method. The result data show that the springback ratio decreases with the temperature and time period of creep age forming, increases with radius of pre-bend. The tension strength and yield strength increase and then decrease with temperature of creep forming. The elongation decreases with the temperature of creep forming. A mathematical equation model for springback ratio is proposed and resolved thorough regression analysis method. A typical structure of fuselage panel is creep formed, which verified the effectiveness and accuracy of the mathematical equation model.

Key Words: Metal forming; Alloy; Thermal effects;

1 Introduction

Aluminium-lithium(AL-Li) alloys has comprehensive advantages including relatively low density, high elastic modulus, high specific strength and stiffness, excellent fatigue strength, superior corrosion resistance and good weld ability, and lower cost than composite materials. After three generations of development, the new Al-Li alloys are applied in modern aerial and space vehicles to meet their requirement of weight reduction [1-4]. For example, Aluminium alloy AA2124 is replaced by Al-Li alloy 2197 in the rear fuselage and other parts of F-16 Fighting Falcon to reduce weight for 5%, improve fracture toughness for 7% [5-6]. Almost all parts for fuselage of Bombardier's C Series airplanes are manufactured by Al-Li alloys, and its Al-Li alloy usage rate accounts for 23% of aircraft materials. Al-Li alloy 2060 is applied to fuselage of C919 airplane of China for reducing weight.

Creep forming is a kind of new manufacturing technology, which restrain a component to a specific shape during heat treatment allowing the component to relieve stresses and creep to contour. There are many advantages for creep forming, including lower residual stress, good shape stability, high stress corrosion resistance, good surface quality and so on. Thereby, it can be applied to manufacture skin components and integral stiffened panels of airplane [7-10].

In this study, fundamental creep forming experiments are carried out on a new Al-Li-Cu-Mg alloy. After just one heat treatment process, the material is age-hardened, and the part is also shaped. The effect of creep aging parameters on springback rate is studied, and a function relationship between springback rate and the process parameters is obtained. Experiment on typical part of welding fuselage panel is used to validate the model, which establish the technical basis for the manufacturing of large Al-Li alloy structures especially of welded fuselage panels.

2 Fundamental creep forming experiment

https://doi.org/10.1051/matecconf/201819012007

The experiment material is a new kind of Al-Li-Cu-Mg alloy which is aged to T8 status, with t=1.8 mm in thickness. Its chemical components are listed in table 1, and its mechanical properties in longitudinal direction (LD) and transversal direction (TD) of rolling fibre at room temperature are listed in table 2 which shows a little anisotropy properties.

Table 1. Chemical components of the new Al-Li alloy

Al	Si %	Fe %	Cu %	Mn %	Mg %	Zn %	Zr %	Ti %	Li %
Bal.	≤0.05	≤0.07	3.4~4.2	0.10~0.50	0.6~1.1	0.30~0.45	0.10~0.50	0.6~0.9	0.04~0.18

Table 2. Mechanical Properties of the new Al-Li alloy

Materials	**Tension Direction**	**Tensile strength MPa**	**Yield strength MPa**	**Elongation %**	**Modulus of elasticity GPa**
Al-Li-Cu-Mg	LD	520.2	491.2	16.1	74.1
	TD	514.2	424.8	19.8	74.72

Key parameters influencing the creep forming result are target material thickness, temperature of creep forming, time period of creep forming, radius of pre-bend. Orthogonal design method is used to setup experiments due to the quantity of influence factors. In this paper, as the target material thickness is fixed, time period of creep forming (Factor A), temperature of creep forming (Factor B) and radius of pre-bend (Factor C) are selected as the designing variables with 4 response levels which are shown in Table 3. Experiments of variable creep forming parameters are carried out by the orthogonal array of $L_{16}(4^5)$ [11] which are shown in Table 4. Pre-bend is imposed with special tool as shown in figure 1, then creep formed in electric thermostat-box. The dimensions of the test Al-Li alloy plates are 260×70 mm.

Table 3. Orthogonal array of experimental parameters

Trials	Time Period (hour)	Temperature (°C)	Radius of Pre-bend (mm)
	A	B	C
1	12	140	170
2	16	160	190
3	20	180	210
4	24	200	230

Fig. 1 Specimen in pre-bent

3 Result and discussion

 https://doi.org/10.1051/matecconf/201819012007

3.1 Springback ratio

The contour precision of part component is one of the most important objective for sheet metal forming and creep forming is no exception. The general contour error is 0.5 mm or less for sheet panel component. Experiments are carried out to obtain the relationship between the process parameters and springback ratio (SR), which can established the fundamental technical base for precise creep forming. The effects of key parameters on creep forming in the dominant deforming direction are investigated using range method. The springback is defined as shown in figure 2.

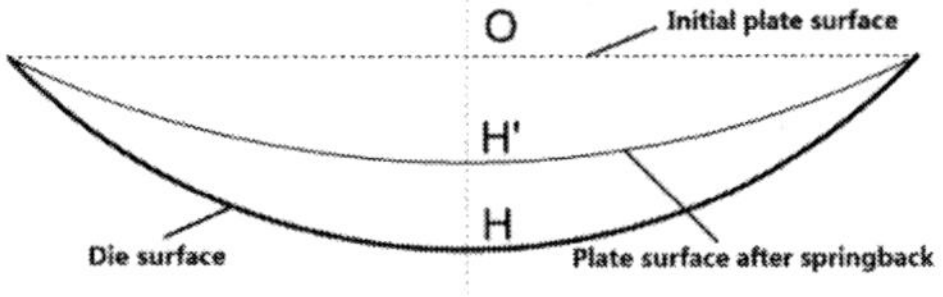

Fig. 2 Definition of springback

So the calculation of springback rate is shown as follows,

$$SR = \frac{\overline{HH'}}{\overline{OH}} \times 100\% \quad (1)$$

The experimental range analysis results are shown in table 4. In this table, K_{ij} under column A is the sum of SR with $A_i(i=1,2,3,4)$, K_{ij} under column B is the sum of SR with $B_i(i=1,2,3,4)$, K_{ij} under column C is the sum of SR with $C_i(i=1,2,3,4)$. $K_{ij}/4$ is the average of K_{ij}. T means the sum of SR with the whole column. R_j under column A is the difference between the maximum and the minimum of $K_{ij}/4$ under column A. Comparing the effect of each parameter on forming result, the most influential factor is the temperature of creep forming, the next is time period of creep forming, and the last is radius of pre-bent.

Table 4. Experimental data and range analysis results of creep forming

Range analysis results	Time Period (h)	Temperature (°C)	Radius of Pre-bend (mm)	**Springback ratio(%)**
	A	B	C	SR
1	12(A1)	140(B1)	170(C1)	97.12
2	12(A1)	160(B2)	190(C2)	93.46
3	12(A1)	180(B3)	210(C3)	84.89
4	12(A1)	200(B4)	230(C4)	73.36
5	16(A2)	140(B1)	210(C3)	97.26
6	16(A2)	160(B2)	230(C4)	94.63
7	16(A2)	180(B3)	170(C1)	77.95
8	16(A2)	200(B4)	190(C2)	66.49
9	20(A3)	140(B1)	230(C4)	94.63
10	20(A3)	160(B2)	210(C3)	89.33
11	20(A3)	180(B3)	190(C2)	78.84
12	20(A3)	200(B4)	170(C1)	61.61
13	24(A4)	140(B1)	190(C2)	93.98
14	24(A4)	160(B2)	170(C1)	88.90
15	24(A4)	180(B3)	230(C4)	80.51

https://doi.org/10.1051/matecconf/201819012007

16		24(A4)	200(B4)	210(C3)	60.28
Sum	**K_{1j}**	348.83	382.99	325.58	
	K_{2j}	336.33	366.32	332.77	
	K_{3j}	324.41	322.19	331.76	
	K_{4j}	323.67	261.740	343.13	T=1333.24
Average	**$K_{1j}/4$**	87.2075	95.7475	81.395	
	$K_{2j}/4$	84.0825	91.58	83.1925	u=T/16=83.3275
	$K_{3j}/4$	81.1025	80.5475	82.94	
	$K_{4j}/4$	80.9175	65.435	85.7825	
Range	***R_j***	6.29	30.3125	4.3875	

The influential ability of each parameter on creep forming can be obtained by the experimental data analysis. As shown is figure 3, the horizontal component is parameter level of each factor and the vertical component is the springback ratio which shows as $K_{ij}/4$ in figure 3. The springback ratio decreases with the time period of creep forming and the temperature, increased with the radius of pre-bend, which shows the viscoplastic deformation characteristic from T8 status.

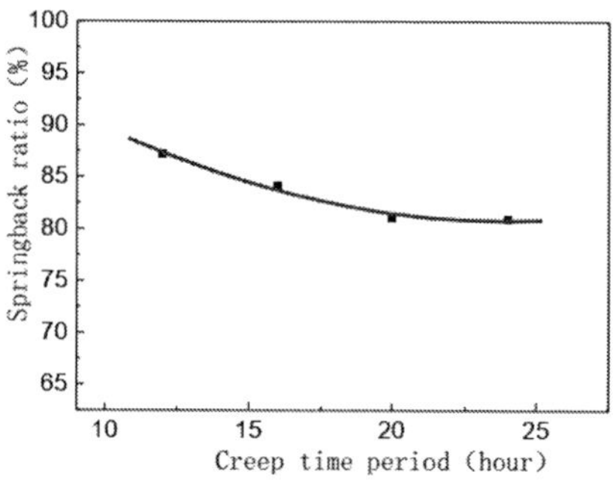

(A) Time period vs. springback ratio

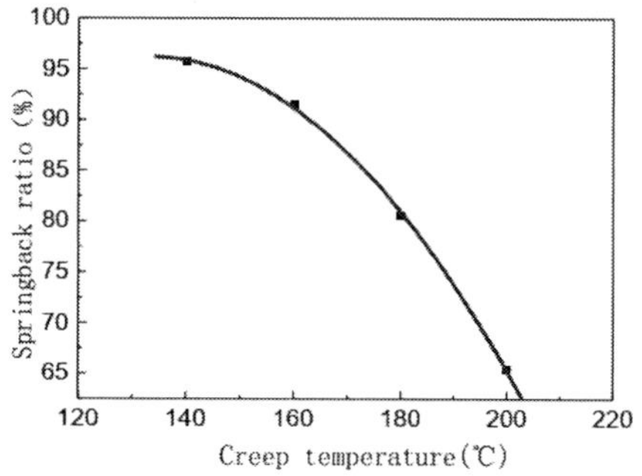

(B) Temperature vs. springback ratio

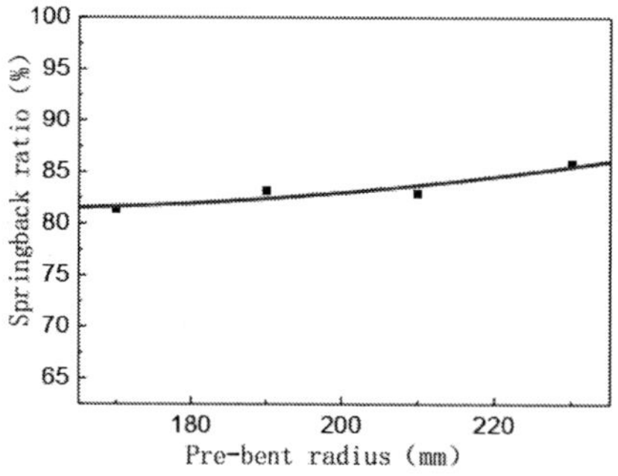

(C) Radius of pre-bend vs. springback ratio

Fig. 3 Relationships between parameters and springback ratio

The regressive analysis method is used to analyse the exact relationship between each parameters and springback ratio for optimization. The experiments are designed with 4 variables and 4 response levels, and each variable has an effect function as,

$$\eta = \bar{\eta} + \varphi(A) + \varphi(B) + \varphi(C) \quad (2)$$

where

$$\bar{\eta} = \frac{1}{p} \bullet \frac{1}{n}\left(\sum_i y_i\right) \quad (3)$$

and p is times of repetition, n is number of level.

Thorough the formula deduction, the equation with values for the relationship between springback ratio and process parameters is,

$$\begin{aligned}\eta &= \bar{\eta} + \varphi(A) + \varphi(B) + \varphi(C) \\ &= 83.3275 - 0.54625(A-18)\text{-}0.50985(B-170) - 0.0023(3B^2 - 1020B + 85200) + 0.06455(C-200)\end{aligned}$$

where A is time period of creep forming (hour), B is temperature of creep forming ($^{\circ}$ C), C is radius of pre-bend (mm).

The equation model shows that the springback ratio decreases with the time period of creep forming and the temperature, increases with the radius of pre-bend, which are in agreement with previous analysis. Substituting the process values of experiments into the equation model, the regressive springback ratio are calculated and compared with the experimental data. As shown in Fig. 4, the regressive springback ratio from the equation model and the measured experimental data are very close. So this equation model derived from experimental data is accurate and can be referenced in engineering manufacturing.

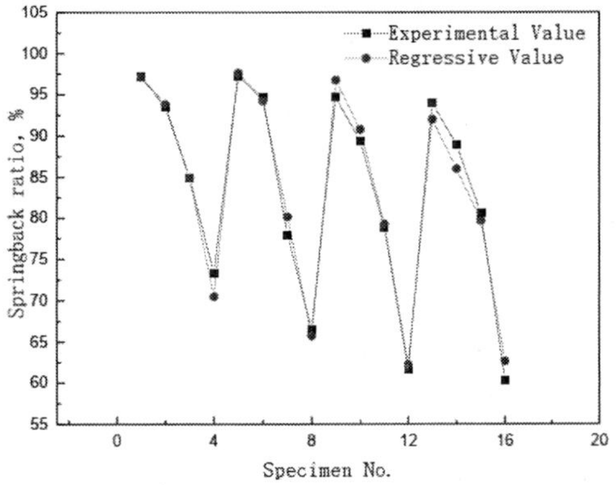

Fig. 4 Comparison of the regressive and experimental springback ratio

3.2 Material property test

Specimens of uniaxial tension test are heat treated with the fundamental experiment as well. The influence rule of the process parameters on material properties are shown in Fig. 5 and Fig. 6. The tension strength, yield strength and elongations rate of TD direction decrease with time period of creep forming. The tension strength and yield strength of LD direction decrease and then increase with time period of creep forming. Elongation of LD direction increases and then decreases with the time period of creep forming. The tension strength and yield strength of both directions increase and then decrease with temperature of creep forming. Elongation decreases with the temperature of creep forming.

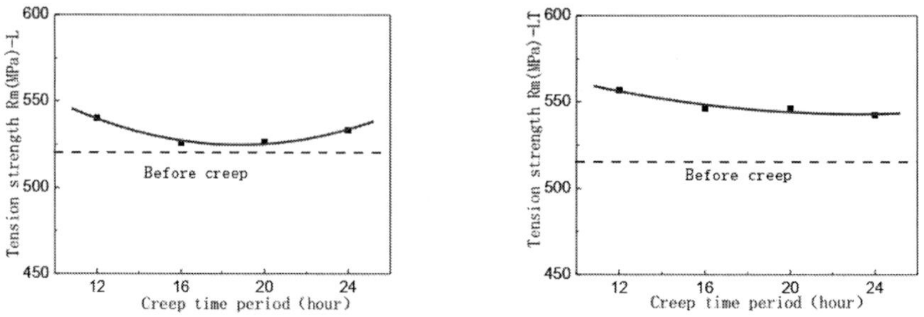

(a) Creep time period vs. Tension strength

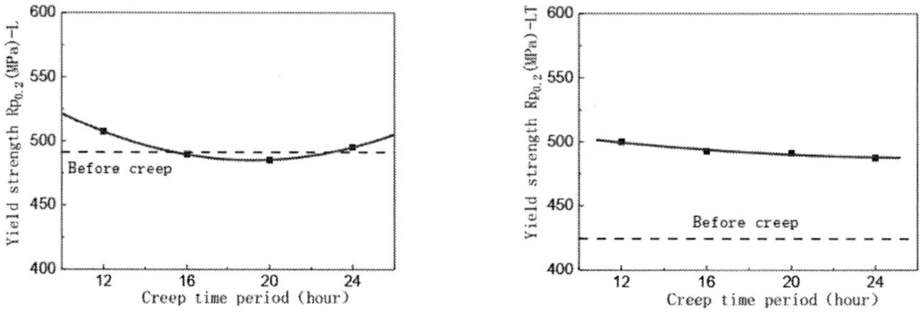

(b) Time period vs. Yield strength

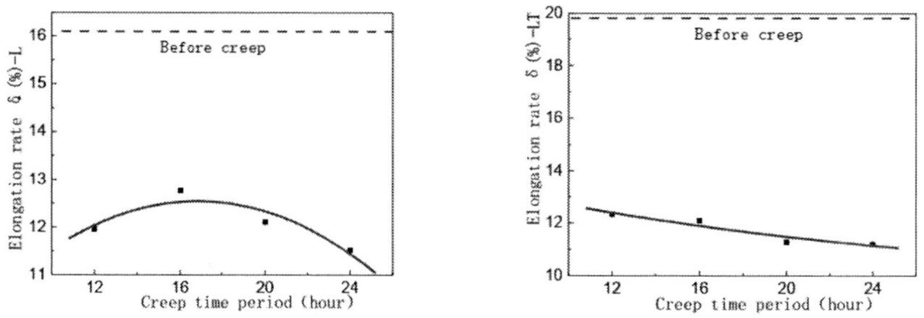

(c) Time period vs. Elongation rate

Fig. 5 Relationships between creep time period and material properties

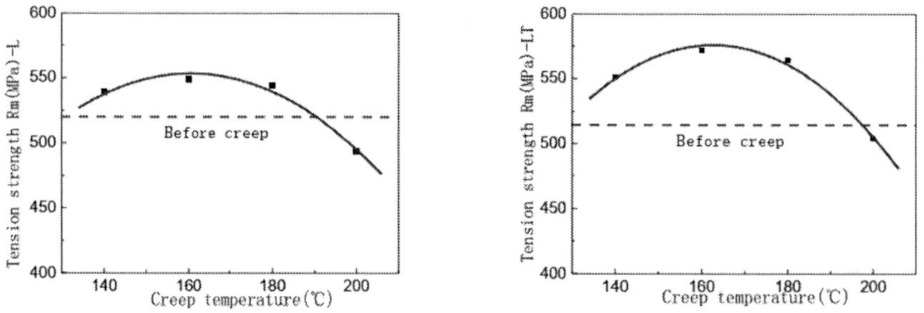

(a) Creep temperature vs. Tension strength

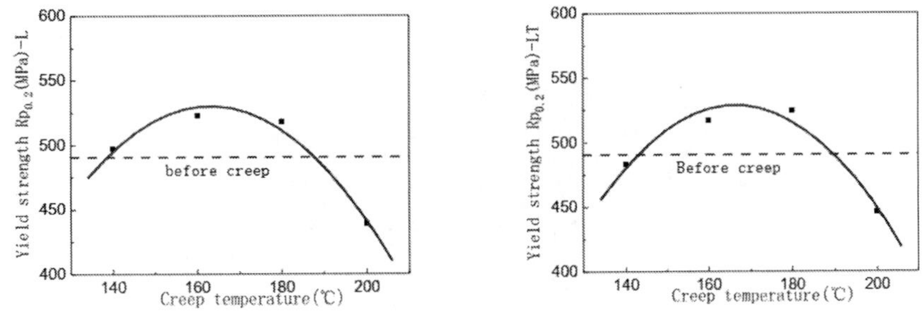

(b) Creep temperature vs. Yield strength

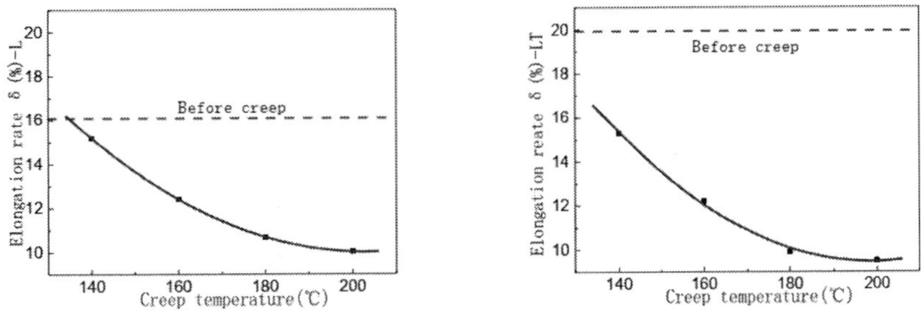

(C) Creep temperature vs. Elongation rate

Fig. 6 Relationships between creep temperature and material properties

Based on the influencing laws of the process parameters on material properties, the heat treatment parameters of 172 °C/24h is selected. There are no significant changes in grain size and the microstructures before and after creep forming, which is shown is Fig. 7. Because the temperature of creep forming is low and the time period of creep forming is short. The mechanical properties test also can demonstrate that this Al-Li alloy is no much overaged from T8 status.

Fig. 7 Microstructure of the material before and after creep forming

4 Creep forming experiment of a typical structural component

Fig. 8 shows a partial welded component of aircraft fuselage with single curvature radius of R=1000 mm. The dimension size is 420×400 mm and it has four "L"-section welded ribs spacing 120 mm. The thickness of skin part is 1.5 mm. The ribs are welded by friction stir welding.

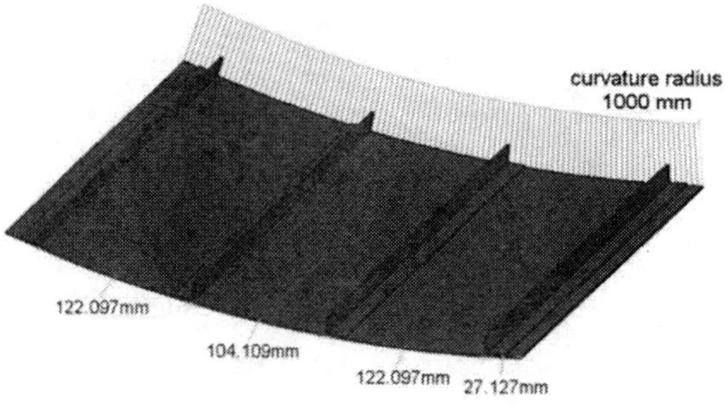

Fig. 8 Typical structure part component

Based on the relationship between the process parameters and formability obtained from fundamental experiments, the parameters of typical structural part component are designed as table 5. Pre-bend is imposed with special die as shown in figure 9 before creep forming.

Table 5. Creep forming parameters

Parameter	**Value**
Time period, hour	24
Temperature, °C	172
Radius of pre-bend, mm	185

Fig. 9 Pre-bend part with tool

Fig. 10 and 11 show the part component before and after experiment. There are gaps of 15 mm and 18.5 mm between the bottom surface of part and the plane before creep forming. After forming process, the deformation caused by welding is reduced obviously due to the stress relief during creep aging. The gap between gauge and part surface is less than 0.5mm in the acceptable error range, which confirmed the effective of equation model derived from fundamental experimental data. The main creep deformation occurs perpendicular to the friction stir welding direction, so the effect of welded region on creep forming can be ignored.

The uniaxial tension samples were cut from the creep formed structural panel and table 6 shows the mechanical properties. The tension strength and yield strength of both directions are increased, which demonstrate the effect of the process.

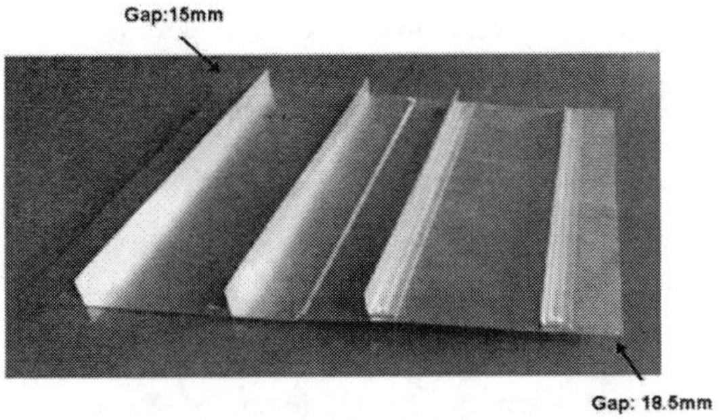

Fig. 10 Part before creep forming

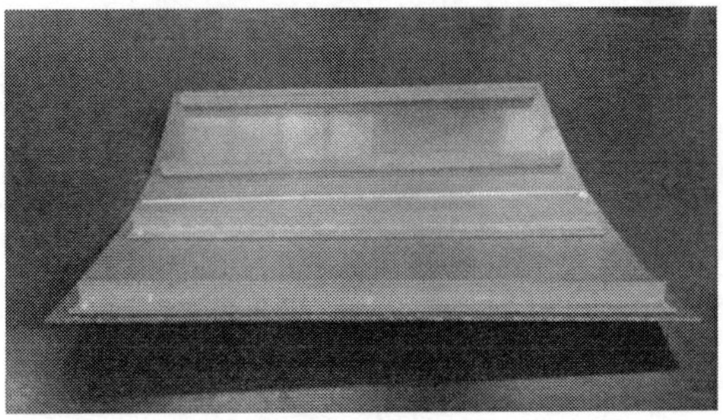

Fig. 11 Part after creep forming

Table 6. Mechanical properties of the alloy after creep forming

Materials	Tension Direction	Tensile strength MPa	Yield strength MPa	Elongation %
Al-Li-Cu-Mg	LD	536.4	510.6	13.3
	TD	569.2	520.6	10.6

5 Conclusion

For creep forming on the new Al-Li-Cu-Mg alloy, the most influential factor is the temperature, the next is time period, and the last is radius of pre-bend. The springback ratio decreases with the temperature and time period, increases with radius of pre-bend.

The equation model derived from experimental data by regressive analysis method is accurate and can describe the relationship between the springback ratio and creep parameters, which is validated by creep forming experiment of a typical welded reinforced panel component.

References

1. WU Xiuliang, LIU Ming, ZANG Jinxin, et al. Research Progress and Aerospace Application of Aluminium Lithium Alloys [J]. Materials Review, 2016, 30(28): 571-585.
2. ZHENG Zi-qiao, LI Jin-feng, CHEN Zhi-guo, et al. Alloying and microstructural evolution of Al-Li alloys [J]. The Chinese Journal of Nonferrous Metals, 2011, 21(10): 2337-2351.
3. YANG Shoujie, LU Jian, FENG Zhaohui, et al. The History of Al-Li Alloys and the Research Development in China [J]. Materials Review, 2014, 28(24): 430-435.

4. XU Jinjun, KANG Wei, DOU Changbing. Research status and development trends of Al-Li alloys for aeronautic and astronautic industry [J]. Ordnance Material Science and Engineering, 2017, 40(3): 132-136.
5. Wei Xiuyu, Zheng Ziqiao, Pan Zhengrong, et al. The Role of Plastic Deformation on Microstructure and Mechanical Properties of 2197 Al-Li Alloy [J]. Rare Metal Materials and Engineering, 2008, 37(11): 1996-1999.
6. GAN Zhong XIONG Wei ZHANG Zhi-guo. Springback prediction of age-forming for 2124 Aluminium alloy [J]. Journal of Plasticity Engineering, 2009, 16(3): 140-144.
7. LI Hong, BAI Xue-piao, ZENG Yuan-song, MENG Qiang. Effect of processing parameter on springback during creep forming of Al-Li alloy 2198 butt-joined by FSW [J]. Journal of Plasticity Engineering, 2012, 20(6): 108-111.
8. Holman M C. Autoclave age forming large Aluminium aircraft panels [J]. Journal of Mechanical Working Technology, 1989, 20(9): 477-488.
9. Foroudastan S, Peddision J, Holman M C. Application of a unified viscoplastic model of simulation of autoclave age forming [J]. Journal of Engineering Materials and Technology, 1992, 114(1): 71-76.
10. Lin J, Ho K C, Dean T A. An integrated process for modelling of precipitation hardening and springback in creep forming [J]. International Journal of Machine Tools and Manufacture, 2006, 46(11): 1266-1270.
11. Department of mathematics and mechanics of Peking University. Orthogonal design methods [M]. Beijing: Petroleum Chemical Industry Press, 1976:17-43

Investigation of a partial, inductive short-time heat treatment of thin metal sheets integrated into the forming process

Benjamin Clausius*, and *Petra* Maier

University of Applied Sciences Stralsund, D-18435 Stralsund, Germany

Abstract. Flanging is a widespread method in the sheet metal working industry to connect same or different materials by forming. Especially the sealing technology makes high demands on the flanging process: a low sheet thickness of the inner eyelet is necessary for proper sealing. The outer edges of the neck rings are mostly manufactured by shear cutting. The quality of the cut surface and the level of the local strain hardening influence decisively the limit of the flanging process by possible cracking. This paper is focused on the dependencies of these factors regarding thin metal sheets of different materials with a thickness down to 100 µm. It could be shown that strain hardening has a stronger effect on the process limits compared to the notch effect of the sheet edges when using standard values for the clearance of the shear cutting tool. Furthermore, a process is investigated with a partial inductive short-time heat treatment of the most deformed edge area. Due to the low thickness of the material and low heat capacities related thereto, it is possible to integrate a recrystallization annealing as single step into the forming process. As a result, the strain hardening can be removed from the affected zone directly between two forming steps to increase the process limits.

Keywords: Sheet metal, Cutting, Heat treatment

1 Introduction

For the manufacturing of permanent, form-locked connections for rotationally symmetric parts made of metal sheets, flanging is often applied because no heat input to the parts or additional joining material is required and high quantities can be produced because of the short joining times [1]. Quality of pre-punched holes which are expanded during the stretch-flanging process, sheet thickness, material and degree of deformation are influencing factors for flanging [1-3]. Regarding the sealing technology in the field of chemical and food industries, corrosion-resistant materials in combination with low sheet thicknesses are required [4]. In this work, the quality of pre-punched holes which can be divided into strain hardening and notch effect, achievable degrees of deformation and an increase of the process limits by inductive short-time heat treatments are investigated [5, 6].

2 Experimental

The material used for the investigation is an austenitic stainless steel 1.4571 with a thickness of 0.1 mm. To ensure a maximum ductility, a bright annealing at 1110 °C and quenching under inert gas was applied after the cold rolling process. The 2D surface specification according to DIN EN 10088-2 allows a cold re-rolling after the heat treatment. Table 1 shows the chemical content and table 2 the mechanical properties and grain size of the coil.

Table 1. Analysis of 1.4571 – 0.10x105.00 mm coil

Analysis					
%	%C	%Mn	%Si	%P	%S
max	0.080	2.000	1.000	0.045	0.015
	0.022	1.160	0.440	0.033	0.003
%	%Cr	%Ni	%Mo	%Ti	-
min	16.500	10.500	2.000	-	
max	18.500	13.500	2.500	0.700	
	16.600	10.560	2.010	0.283	

Table 2. Mechanical properties and grain size of 1.4571 – 0.10x105.00 mm coil

Test results			
	$TYS_{0.2}$ N/mm²	TYS_{l} N/mm²	UTS N/mm²
	277	289	622
	Elong. A 80 mm	Hardness HV	Grain size ASTM E112
min	58	-	6.5
max	-	-	8.0
	59.0	125	8.0

* Corresponding author: benjamin.clausius@hochschule-stralsund.de

Among others the 0.2% tensile yield stress ($TYS_{0.2}$), 1.0% tensile yield stress ($TYS_{1.0}$), ultimate tensile stress (UTS) are given. By using different methods of cutting, round holes were punched into the metal sheet. The manufactured edges of these pre-punched holes were further investigated. In addition to shear cutting, the manufacturing of thin metal sheets provides other types of cutting. First, a method of punching with a combination of cutting ring and cutting pad and second, a steel rule die cutting process was investigated for comparison.

a.) The shear cutting tool consists of a flat punch, die and blank holder made of 1.2842 with a hardness of 60±2 HRC. The clearance is set to 10 µm which is 10 % of the sheet thickness (Fig. 1 a) [7, 8].

b.) The cutting ring tool consists of a 5 mm thick steel ring, made of 1.2842 with a hardness of 58±2 HRC, which represents the die and a cutting pad, made of 30 mm polyurethane with a Shore hardness of 90 A, which works as inner and outer cutting punch (Fig. 1 b).

c.) The steel rule die tool consists of a pre-hardened and pre-sharpened steel rule, which is fitted into a multiplex wood base. Rubber blocks, positioned on each side of the cutting rule work as ejectors. The cutting base is made of EN AW-5754 (Fig 1 c).

A Hans Schoen UTE/B160 hydraulic press with a maximum press force of 1600 kN was used for the cutting processes.

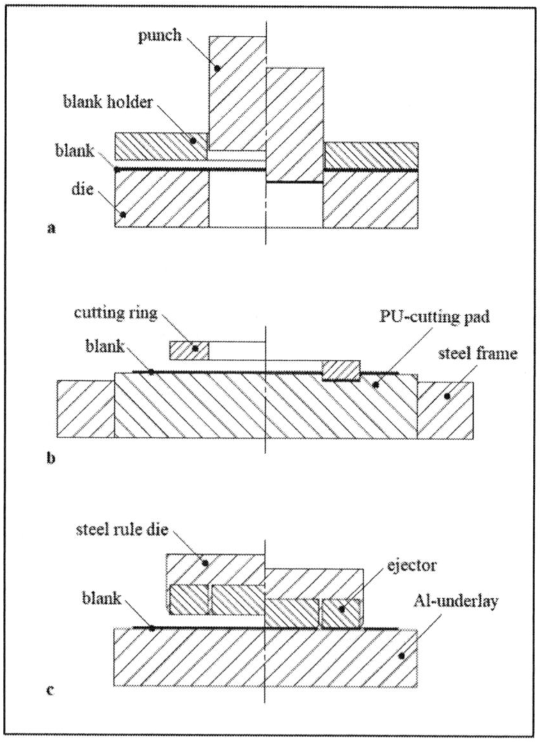

Fig. 1. Schematic drawings of cutting methods; a – shear cutting tool, b – cutting ring tool, c – steel rule die tool

Regarding the notch effect, the cutting edges were investigated with a Keyence VW-9000 digital microscope. The notch depth was measured over a distance of 1 mm at four points each turned 90 °. By measuring six single sections on each point, an averaged notch depth for each measuring distance and sample and a total averaged notch depth for each cutting method could be determined (Fig. 2).

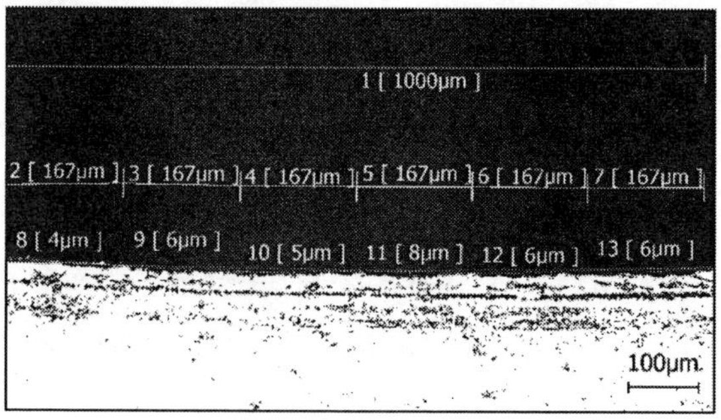

Fig. 2. Microscopic image of notch depth measurements at six sections over 1 mm distance of shear cut sample, top view

The effect of strain hardening was investigated by hardness measurements with a Zwick/Roell BZ2.5/TS1S hardness testing machine in the area of the cutting edge. As testing process, the Vickers HV0.2 was used with a test force of 1.96 N. The indents were made on the cross-sections of the macrosections (Fig. 3).

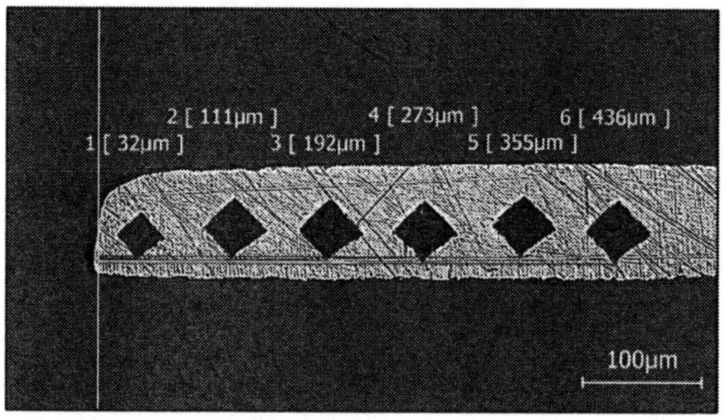

Fig. 3. Microscopic image of cross-section with distances of hardness indents to cutting edge

Due to the sample geometry and the purpose of measuring the strain hardened area, the positioning differs from the standard DIN EN ISO 6507. The first indent was set with a distance of 30 µm to the outer edge of the ring and the following five indents are set with a distance of 80 µm to each other.

Samples of the different cutting processes were then treated with a solution annealing at 1020 °C for 10 min with a Nabertherm N11H annealing furnace and quenched in air at room temperature. The hardness measurements were repeated as mentioned before.

To evaluate the risk of cracking, a hole expanding test following the standard ISO 16630 was used (Fig. 4). A flat drawing punch with a diameter of 60 mm and a radius of 5 mm centered the samples by the pre-punched hole with a diameter of 23.6 mm, which was produced by the shear cutting tool. To minimize the friction during the forming process, an underlay of 0.25 mm PTFE was used. The process was conducted with a mechanical testing machine TIRAtest 28100 at a strain rate of 5 mm/min.

After a reduction in force of 20 N, which indicates cracking, the process was stopped. The expansion ratio of cracking diameter and pre-punched hole diameter provides information on cracking resistance [9, 10]. To investigate the influence of strain hardening and the notch effect, samples with deburred cutting edges, solution annealed samples and solution annealed samples with deburred cutting edges were compared with untreated samples.

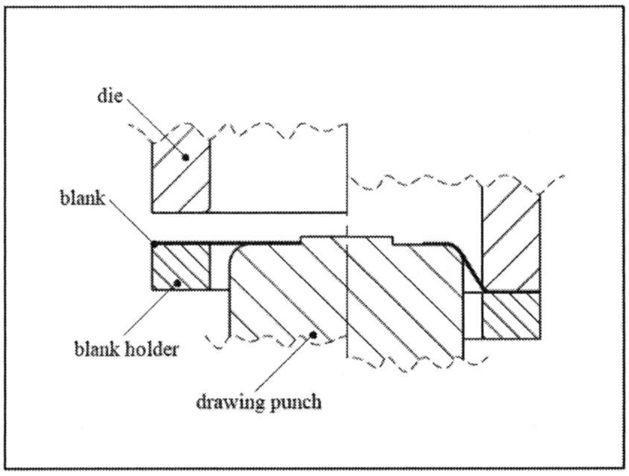

Fig. 4. Schematic drawing of hole expanding test

In the context of the short-time heat treatment, the inductive heating system iew TT2Ht was used with a flat, 50 mm spiral inductor to apply a partial annealing limited to the area of the cutting edge. The temperature was regulated by a SensorTherm Regulus RP10 14 closed-loop control with pyrometer. To compare the results of the inductive short-time heat treatment with the solution annealing, hardness measurements and hole expanding tests were conducted as mentioned.

3 Results and discussion

The measurements of the notch depth give a total averaged notch depth of about 6 µm for the shear cutting tool, 10 µm for the cutting ring tool and 12.5 µm for the steel rule die tool (Fig. 5).

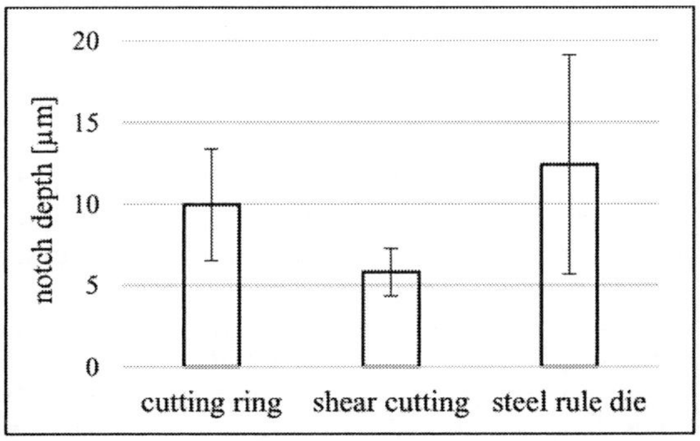

Fig. 5. Total averaged notch depth of cutting methods

Besides the averaged notch depth, the maximum to minimum depth varies considerably from 51 µm to 7 µm regarding the steel rule die tool. Except the notch depth, the form also influences the stress condition. The cutting ring and the steel rule die tool cause deep and sharp notches whereas the shear cutting tool causes flat and rounded notches, which effect significant lower stress peaks (Fig. 6).

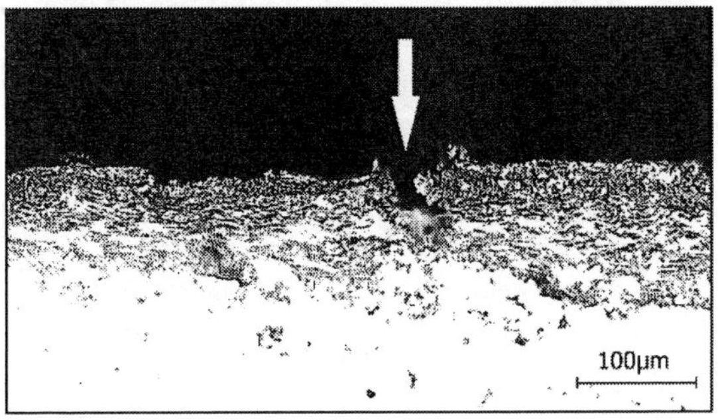

Fig. 6. Microscopic image of deep notch of steel rule die cut sample, top view

Consequently, the shear cutting method with a clearance of 10 % of the sheet thickness causes the lowest notch effect for the investigated material.

The hardness measurement results show a small strain hardened area within a distance of about 50 µm to the cutting edge whereas the area behind remains unaffected (Fig. 7).

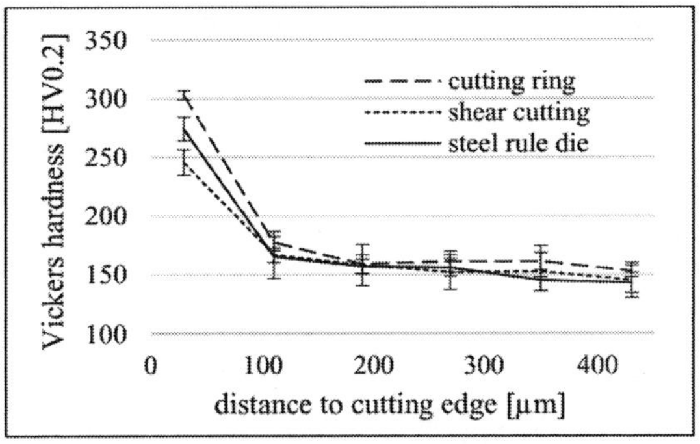

Fig. 7. Hardness profile of investigated cutting methods on distance to cutting edge

The minor scattering at higher distances and the marginal deviation from the specified value of 125 HV0.2 of hardness result from the positioning of the indents on the cross-section with a maximum possible distance of 50 µm to each surface. Regarding the punching methods, the cutting ring tool causes the highest strain hardening which can be traced to the highest hardness of about 300 HV0.2. The steel rule die cut edges show a hardness of about 275 HV0.2 and with a hardness of about 245 HV0.2 the shear cutting tool causes the lowest strain hardening for the considered methods.

To reduce the strain hardening which is caused by the cutting process, it is necessary to achieve a recovery, even better a recrystallization. Recrystallization requires a critical degree of deformation in the area of the cutting edge. The measurement results of the hardness after solution annealing show a complete degradation of the strain hardening (Fig. 8).

 https://doi.org/10.1051/matecconf/201819012008

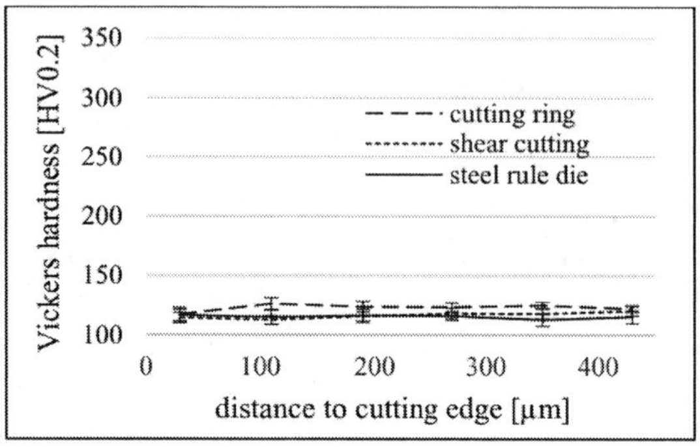

Fig. 8. Hardness profile of investigated cutting methods on distance to cutting edge after solution annealing

With an averaged hardness of about 120 HV0.2 no significant differences between the cutting methods or measuring points exist. The lower hardness values compared to the values of the untreated samples could result from a cold re-rolling, which is allowed due to the specified surface finish. The solution annealed condition is the reference for the inductive short-time heat treatment. Due to the previous findings, the shear cutting method was applied for further investigations.

The hole expanding test results for the solution annealed and deburred condition show the process limits with a maximum expansion ratio of 1.52 (Fig. 9)

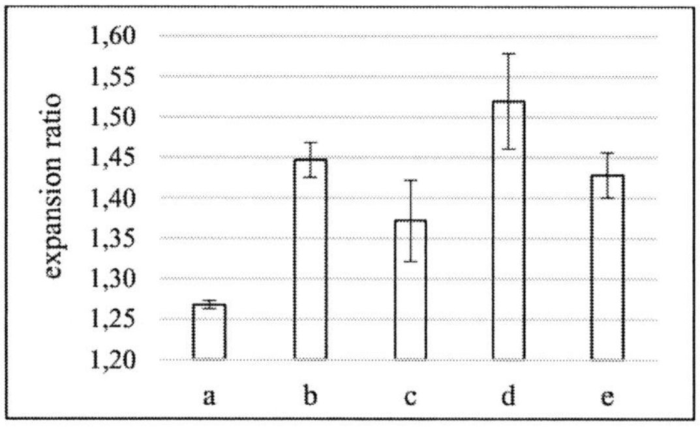

Fig. 9. Expansion ratio of different conditions – a.) untreated b.) solution annealed c.) deburred d.) solution annealed, deburred e.) inductive annealed

The expansion ratio of the deburred condition with 1.37 is lower than the solution annealed condition with 1.45. To ensure that the strain hardening is unaffected by the deburring, hardness measurements are conducted which show values inside the standard deviation of the untreated shear cut samples (Fig. 10). Consequently, the strain hardening has a higher influence on the risk of cracking than the notch effect. The untreated samples as reference measurements show the lowest expansion ratio with a value of 1.27.

To remove the strain hardening by an inductive short-time heat treatment, an optimized heating process for the investigated material is determined. The samples were placed with a distance of 5 mm under the inductor. With a high frequency generator power of 1.94 kW, the cutting edge reached a temperature of 1020 °C in 1.8 s (Fig. 11).

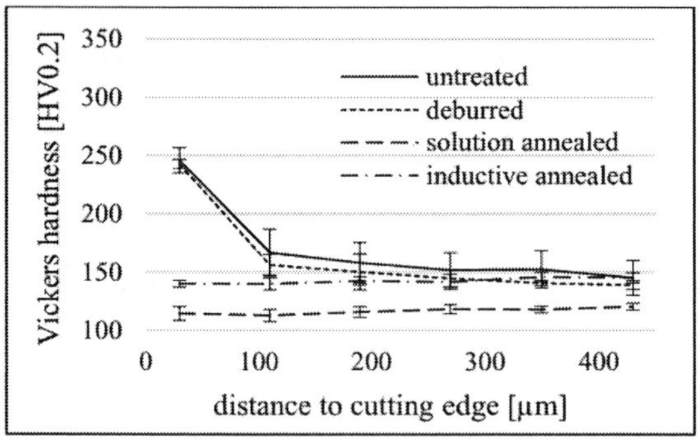

Fig. 10. Hardness profile of different conditions on distance to cutting edge

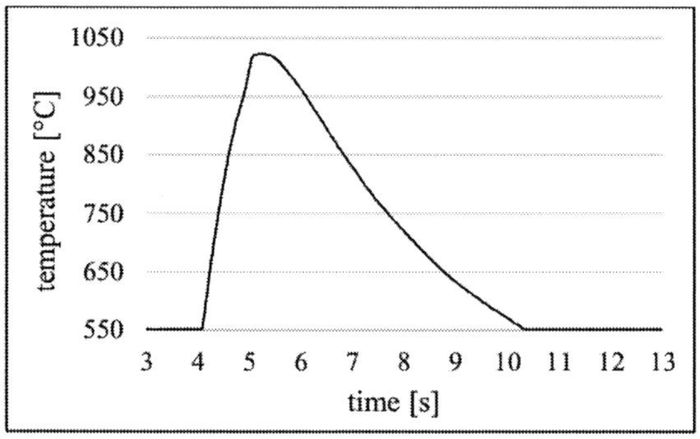

Fig. 11. Temperature profile of an inductive annealed sample with a hold time of 0.5 s

After testing different heating parameters, a hold time of 0.5 s was considered to be suitable to remove the strain hardening as shown by the reduced hardness values (see Fig. 10). The differences of the hardness values between the solution annealed and inductive short-time annealed samples must be caused by the duration of the heat treatment. The expansion ratio of 1.43 for the inductive annealed condition almost achieves the value of the solution annealed condition which is 1.45 (see Fig. 9). The marginal higher expansion ratio of the solution annealed samples is caused by the higher reduction of strain hardening.

4 Conclusion

In this work, different processes which are usually applied for flanging were investigated: First, the manufacturing of the pre-punched hole which is an important part for the further forming process and second, the risk of cracking during hole expansion and opportunities to increase the process limits. It could be shown, that for thin 1.4571 stainless steel sheets with a thickness of 100 μm it is possible to apply unconventional cutting methods like cutting ring or steel rule die tools. The quality of the different cutting methods is based on the factors strain hardening and quality of the cutting surface which causes the notch effect. The shear cutting tool provides with about 6 μm the lowest averaged notch depth including a significant lower statistic variance. The hardness in the area of the cutting edge increased from about 155 HV0.2

of the basic material to about 245 HV0.2 which is a lower strain hardening than the next-best alternative cutting method with 275 HV0.2. To separate the influence of notch effect and strain hardening, the hole expanding test was conducted which showed that the solution annealed samples have a higher expansion ratio of 1.45 compared to the deburred samples with a ratio of 1.37. In order to integrate an annealing into the forming process, the inductive short-time heat treatment was adapted to reduce the strain hardened area with an annealing for 0.5 s at 1020 °C. The inductive annealed samples reached an expansion ratio of 1.43 which is marginal lower than the ratio of the solution annealed condition. As a result, the shear cutting method should be used for flanging or comparable forming processes when higher expansion ratios of pre-punched holes are required. The strain hardening due to the cutting process has a greater influence regarding the limits of hole expansion than the notch effect. The integration of the inductive annealing into forming processes without impermissible heat input to the forming tools seems possible due to the low heat capacity of the thin raw material which has to be considered in further investigations.

References

1. E. Doege, B.A. Behrens, *Handbuch Umformtechnik*, 386–388 (2006)
2. K. Lange, *Umformtechnik*, 522–526 (1990)
3. C.H. Phan, S. Thuillier, P.Y. Manach, Metall and Mat Trans A **46**, 3502-15 (2015)
4. P. Gümpel, *Rostfreie Stähle*, 3 (2016)
5. W. Weißbach, *Werkstoffkunde*, 48–49 (2015)
6. H.J. Bargel, G. Schulze, *Werkstoffkunde*, 106–108 (2000)
7. S.M. Yi, B.Y. Joo, M.S. Park et al., Microsyst Technol **12**, 877-82 (2006)
8. J. Xu, B. Guo, D. Shan, Int J Adv Manuf Technol **56**, 515-22 (2011)
9. J.I. Yoon, J. Jung, H.H. Lee et al., Met. Mater. Int. **22**, 1009-14 (2016)
10. X. Chen, H. Jiang, Z. Cui, C. Lian, C. Lu, Procedia Engineering **81**, 718-23 (2014)

https://doi.org/10.1051/matecconf/201819013001

Geometrical design of micro-textured DLC coatings toward lubricant-free metal forming

Tetsuhide Shimizu[1,*], *Hironori* Kan[1], *Hamza* Messaoudi[2], *Frank* Vollertsen[2,3] *and Ming* Yang[1]

[1]Department of Intelligent Mechanical Systems, Tokyo Metropolitan University, Tokyo, Japan
[2]BIAS – Bremer Institut für angewandte Strahltechnik, Bremen, Germany
[3]Faculty of Production Engineering-Mechanical Engineering & Process Engineering, University of Bremen, Bremen, Germany

Abstract. To achieve a stable and high wear resistive functional surface under dry sliding friction, the present study proposes micro-textured diamond like carbon coatings fabricated by ionized physical vapour deposition (I-PVD) using metallic masks. To clarify the optimized geometrical dimensions under dry sliding friction, geometrical conditions of textured array patterns are varied by using metallic masks with different circular hole array patterns fabricated by picosecond pulsed laser processing. The effect of texturing on friction and wear properties is evaluated by ball-on-disk type friction tests for the condition of a constant DLC-coverage per unit area. Thereby, textured DLC pattern with diameters of 50 μm, 100 μm and 150 μm are applied. Within the first 10,000 laps stable and lower coefficient of friction is obtained with the smallest diameters of 50 μm. However, at further rotation of more than 40,000 laps, the wear of the smaller diameters becomes more pronounced due to the increase of stress concentration at the edge of the structure. Based on these findings, geometrical design of micro-textured DLC coating is discussed with regard to the suppression of three-body ploughing and the prevention of stress concentration.
Keywords: Surface modification, Tribology, Wear

1 Introduction

With the growing awareness of environmental issues, the efforts towards a lubricant-free metal forming are intensifying [1]. Since the predominant tribological factors in the case of the dry sliding friction are adhesion and deformation resulting from the direct interaction of surface asperity peaks in contact with each other, the mechanical properties and the geometry of the surface asperities directly affect the friction and wear behaviour [2]. In particular under the plastic deformation of the work materials, as it is the case within a metal forming operation, wear debris of various sizes is generated and the trapped particles at the interface induce a three-body ploughing behaviour, which significantly increases the friction and its large deviations [3].

As a basic concept to suppress the influence of the three-body ploughing caused by generated wear particles, the authors focused on a theoretical model of scale effect of the three body deformation [4, 5]. It is supposed that wear particles at the edge of an apparent contact area can detach from the surface if these have diameters twice as high than the width of the contact area. In contrast, particles at the centre of the contact region are likely to be trapped and plough the contact surface. Thus, a down scaling of the apparent contact area can help in reducing the three-body ploughing effect. For this purpose, functionalized surfaces are required.

To realize stable dry friction conditions, the authors proposed an application of surface texturing with micro structured diamond like carbon (DLC) coatings to decrease the apparent contact area during sliding [6]. In previous works, micro-textured array patterns fabricated by using metallic wire meshes have demonstrated the low friction with less deviation under dry friction and either in application to sheet bending process [6,7]. Thereby, it was noticed that it is essential to consider the ejection of wear particles from the contact interface as one key factor in the design of suitable surface texturing due to the crucial effect of wear debris generation and its entrapment at the interface. However, the influence of the texture dimensions under dry sliding friction on the tribological properties has not been discussed in details for the representative contact state in metal forming.

The present study performed a systematic approach to vary the geometrical conditions of textured array patterns of DLC coatings. Therefore, metallic masks consisting of different circular hole arrays were fabricated using picosecond laser processing, as flexible and high precise machining method, which ensures a minimized thermal impact as well as distortion-free material ablation [8]. Based on the comparison of tribological properties between different texture dimensions, geometrical design of micro-textured DLC coatings towards lubricant-free metal forming is discussed.

* Corresponding author: simizu-tetuhide@tmu.ac.jp

2 Experimental

2.1. Picosecond laser fabrication of metallic masks

To ensure a precise and multi-design DLC-deposition procedure, the manufacture of the metallic masks was provided using ultrashort pulsed laser machining. Therefore, a picosecond laser system (TruMicro 5050 from Trumpf GmbH) was used. It operates at a wavelength of 1030 nm in its fundamental Gaussian mode and delivers an average maximum power of 50 W at 200 kHz. It produces pulse widths of < 10 ps and maximum pulse energy of 250 µJ. The collimated laser beam is first expanded and then coupled into a frequency conversion system (Xiton box from Xiton GmbH), which allows the second harmonic generation (wavelength of 515 nm) in a separated optical path. Thereby, the maximum average power is about 33 W and the maximum pulse energy is 150 µJ. Depending on the used wavelength the collimated beam is coupled into one of two programmable galvanometer scanners (HurriScan 14 from ScanLab), which enable process speeds up to 12 m/s. The orthogonal arrangement of the scanner mirrors directs the beam down towards the workpiece using an F-Theta lens (160 mm focal length). This enables both a large scan fields (100 mm x 100 mm) and small focused laser spot sizes (48 µm and 24 µm, for 1030 nm and 515 nm, respectively) at perpendicular incidence of the laser beam. In addition, the workpiece is mounted on a z-axis to ensure the adjustment of the focal plane.

In this work the manufacture of the metallic masks was applied on 10 µm thin stainless steel foils over an area of 8 mm x 8 mm. Thereby, the texturing was based on the drilling of circular micro holes having diameters D between 10 µm and 150 µm. Thereby, it has to be mentioned that the targeted diameters were realized at the laser exit side. Moreover, the interspace distance between the holes was varied between 15 µm and 75 µm in order to realize different DLC-coverage during the deposition procedure. The different mask configurations realized are illustrated in Fig.1.

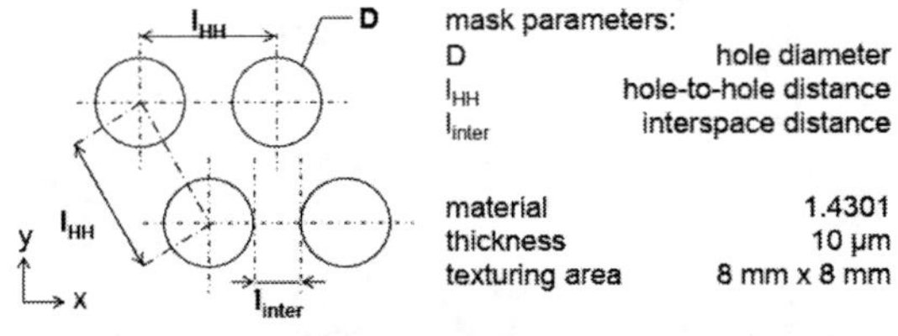

	D [µm]	l_{HH} [µm]	l_{inter} [µm]
mask type 1	10	40	30
mask type 2	50	75, 80, 100	25, 30, 50
mask type 3	100	130, 150	30, 50
Mask type 4	150	225	75

Messaoudi 2018 BIAS ID 180068

Fig. 1. Schematic illustration of the texturing parameters as well the applied mask configurations

A detailed investigation was performed to identify the influence of different process parameters on the resulting hole properties. With respect to the targeted configurations suitable laser parameters (wavelength λ, pulse energy E_p, pulse repetition rate f_{rep}, number of pulses n_p, drilling time per hole t_{hole} and focal position Δz) were selected, as illustrated in Table 1.

Table 1. Applied laser parameters for the texturing of the different mask configurations

D [µm]	λ [nm]	E_p [µJ]	f_{rep} [kHz]	n_p [-]	t_{hole} [s]	Δz [mm]
10	515	7.5	10	800	0.08	0
50	1030	37.5	10	600	0.06	0
100	1030	112.5	2	700	0.35	+ 2
150	1030	137.5	2	900	0.45	+6

The characterization of the realized metallic masks at the laser exit side was performed using optical microscopy records, from which the hole diameters D, the hole-to-hole distance l_{HH} and the interspace distance l_{inter} were measured.

2.2 Ionized physical vapour deposition of DLC films

Depositions were performed in an ionized physical vapoured deposition (I-PVD) system (NPS-330S from Nanotec Corp.) with a usable coating volume with a size of φ150 mm × 150 mm. In this system, the direct current ion source, composed of a tantalum (Ta) filament and anode electrodes, was installed to ionize the benzene (C_6H_6) gas. Bulk specimens of sintered tungsten carbide-cobalt (WC-Co) hard alloy, JIS: V20, with a size of 14 mm × 14 mm and a thickness of 5 mm were used as the substrate material. Prior to the deposition, the WC-Co substrates were mirror polished, cleaned sequentially in acetone and ethanol, and mounted on a table with fixing jigs. Starting with a base pressure of < 5 × 10^{-3} Pa, argon (Ar) gas of 10 sccm was introduced to initiate an Ar plasma and remove the adsorption layers on the substrates by plasma etching process.

Table 2. Conditions of DLC deposition by I-PVD process.

Pre-treatment of substrate		Ar^+ bombard DC -1kV for 30min
Gas species		C_6H_6
Gas flow rate	[sccm]	0.75
Gas pressure	[Pa]	2.5×10^{-2}
Bias voltage	[kV]	-1
Filament current	[A]	30
Temperature	[°C]	250
Deposition time	[h]	2

The pre-treatment was performed using a DC-voltage of -1 kV for 30 min. Subsequently, the DLC deposition was carried out in a C_6H_6 atmosphere at 250 °C. The

C_6H_6 flow was set to 0.75 sccm to maintain a total pressure of 2.5×10^{-2} Pa. Throughout the deposition, the substrate bias voltage was maintained at -1 kV. The conditions of the complete process are summarized in Table 2.

2.3 Tribological evaluations

To characterize the basic tribological properties of the fabricated non-textured and micro-textured DLC films under dry sliding conditions, ball-on-disk type tribology tests were carried out with a commercial tribometer (TRIBOMETER, CSM Instruments). A constant normal force of 5 N, which corresponds to an initial Hertzian mean contact pressure of about 1.2 GPa and elastic contact diameter of approx. 110 μm, was applied for the all testing samples. This contact pressure is equivalent to that generally assumed in the severe contact state in metal forming, such as ironing or forging process. As a counterpart, an alumina (Al_2O_3) ball with a dimeter of 6mm was used. The tests were performed under unlubricated condition and with a linear speed of 42 mm/s at room temperature in ambient air. Wear tracks of the substrates and sliding balls were analyzed by optical microscopy and scanning electron microscopy (SEM). In addition, the chemical composition at the contact surface was also analyzed by energy dispersive X-ray spectroscopy (EDS).

3 Results and discussion

3.1. Laser manufactured metallic masks

The first step in this work was the manufacture of the metallic masks following the design required for the subsequent DLC-deposition. Figure 2 shows selected microscopic images of different mask configurations. The images show that the laser machining enables a high reproducibility as well as high dimensional accuracy.

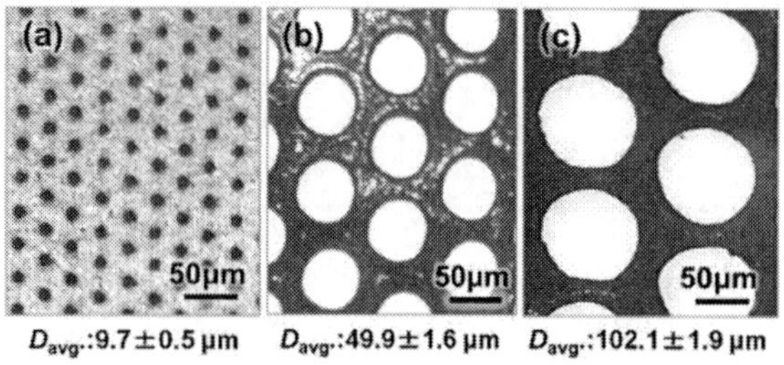

Fig. 2. Optical microscopic images of picosecond laser processed metallic masks with diameters D of 10 μm (a), 50 μm (b) and 100 μm (c) and a constant interspace distance l_{inter} of 30 μm. *D_{avg}: Averaged diameter.

3.2. Fabrication of micro-textured DLC films

By using the micro-perforated metallic masks, a micro-texturing of thin DLC films was realized during the deposition process. Figure 3 shows examples of micro-textured DLC films and their related surface profile obtained using a 3D surface profiler. The circular island structure patterns with diameters of approx.10 μm, 50 μm and 100 μm and a constant interspace distance of 30 μm are characterized by high dimension accuracy as well as sharp edges. Thereby, the thickness of the DLC coating was ≈ 300 nm for all geometrical conditions.

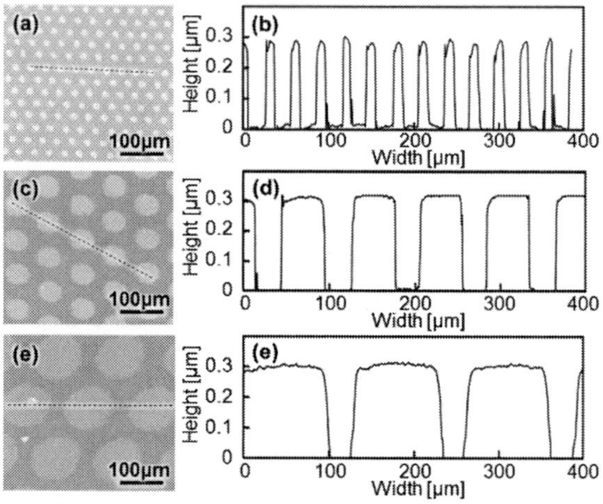

Fig. 3. Microscopic images of micro-textured DLC coatings and their surface profile with a texture diameter of 10 μm (a),(b), 50 μm (c),(d) and 100 μm (e),(f).

3.2. Comparison of tribological properties between different textured dimensions

Ball-on-disk tests were carried out to characterize the tribological properties in dependence of the applied texturing. In order to evaluate the effect of texture diameter under the same apparent contact pressure, following three geometrical conditions were chosen to keep a constant ratio of the DLC-coverage in unit area (40%); D= 50 μm with l_{inter}= 25 μm, D= 100 μm with l_{inter}= 50 μm and D= 150 μm with l_{inter}= 75 μm.

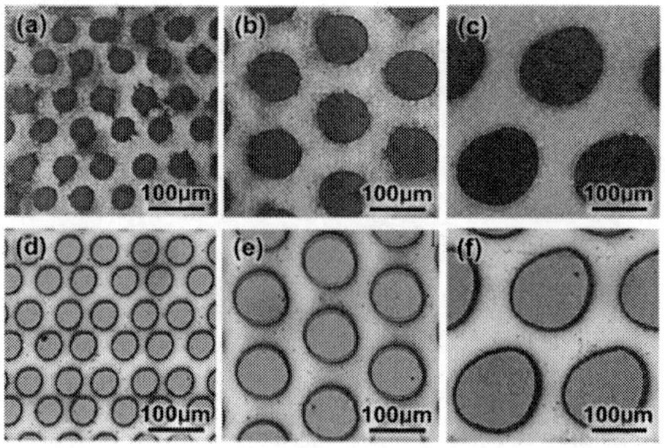

Fig. 4. Optical microscopic images of the laser processed metallic masks with different diameter D of 50 μm (a), 100 μm (b) and 150 μm (c) with a constant DLC coverage and that of the corresponding micro-textured DLC (d), (e), (f).

Fig. 4 shows the optical images of the laser machined metallic masks and the corresponding micro-textured DLC. Thereby, it is visible that the precise micro-textured array patterns were successfully obtained as they were originally designed for metallic masks.

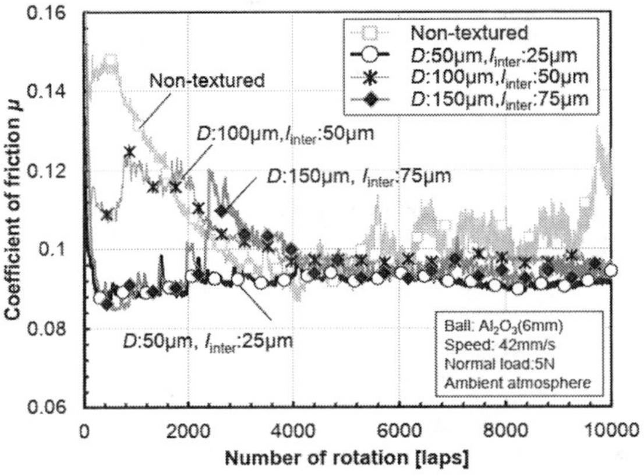

Fig. 5. Evolution of COF during initial 10,000 laps of ball-on-disk friction test of non-textured and micro-textured DLC films with different texture dimensions.

Fig. 5 shows the evolution of the coefficient of friction (COF) during 10,000 laps of rotations in ball-on-disk friction tests for both, non-textured and micro-textured DLC films. As expected, a high COF with relatively large deviation is shown for the non-textured films. In contrast, the use of textured DLC shows a significant influence of the texture diameter on the resulting COF in the first 5,000 laps of rotation. However, the variation of COF reduces and becomes more stable above the second 5,000 laps. In particular, the texture with the smallest diameter of 50 μm shows a relatively low COF value of $\mu = 0.09$ with almost no deviations from the initial rotation. This stability in smaller diameter texture is assumed to be due to the suppression of ploughing by the wear debris at contact interface [3]. Taking into account the scale effect of the three-body deformation, the real area of contact of three-body ploughing is approximately 1/6 at the original scale dimensions by scaling down the apparent contact area to the half [9]. Thus, the entrapment probability of wear debris decreases and thereby contributes to the stable and reduced friction under dry sliding conditions.

Above 1000 laps of rotation in the ball-on-disk friction tests, however, the micro-textured DLC with smallest texture diameter of 50 μm shows a limited durability compared to that with the largest texture diameter of 150 μm, see Fig. 6. After approx. 20,000 laps of rotation, the COF of the texture diameter of 100 μm increases significantly, while that of 50 μm increases after 40,000 laps. In comparison, the texture patterns with diameter of 150 μm reveal more than twice longer durability up to > 80,000 laps. Thereby, the COF level is maintained constant around $\mu = 0.08$.

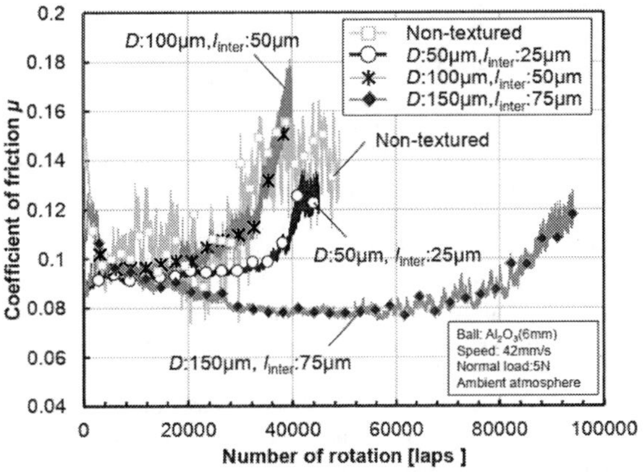

Fig. 6. Evolution of COF during ball-on-disk friction test of non-textured and micro-textured DLC films with different texture dimensions.

As can be seen in Fig. 7(a) to (c), the DLC coating on the patterns with the largest diameters ($D = 150$ μm) is remained after the 90,000 laps of rotation, while on the pattern with smaller diameters of 50 μm and 100 μm the DLC layer is almost worn after 40,000 laps of rotation. In fact, the wear debris including large amount of carbon contents was accumulated beside the sliding tracks for the textured diameter of 50 μm and 100 μm, which shows the wear of the textured DLC structure even under the same coverage of DLC films at unit area. According to the classical empirical equation of wear volume proposed by Archard [10], those higher wear rates in smaller textured structure seem to be attributed to the higher local surface pressure for the smaller diameters during the sliding of the spherical surface of the ball. In particular, the stress concentration at the edge of the texture during the sliding promotes the wear of the texture structure. As seen in the Fig. 7(b), the right edge of the texture structure is preferentially worn, where the stress concentration of the ball is more pronounced due to the sliding direction.

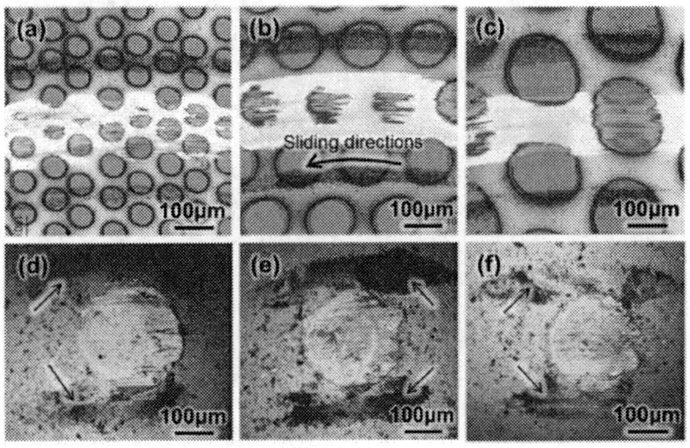

Fig. 7. Optical surface images of wear tracks of micro-textured DLC films with different texture diameters of 50 μm (a), 100 μm (b), 150 μm (c) as well as the corresponding counterparts of alumina balls (d), (e), (f), respectively, after the ball-on-disk friction tests.

In addition, the comparison between Fig. 7(d) and (e) indicates that the accumulation of carbon wear debris beside the worn spots at the alumina balls is less for textured diameter of 50 µm than that of 100 µm. Thus, it is suggested that the generated wear debris did not accumulate owing to the effective ejection of wear debris by the scale effects of three-body ploughing and it might contribute to the slightly higher durability for the texture diameter of 50 µm than 100 µm. In view of this accumulation of the carbon particles, it is clearly seen the less amount of accumulation for the texture diameter of 150 µm, which has the longest durability, as shown in Fig, 7 (f).

For the design of micro-texture that fulfil the requirements of a durable dry sliding, the balance between the efficient ejection of wear debris from contact interface, which can be realized by scaling down of the apparent contact area, and the suppression of the stress concentration by decreasing the interspace to distribute the stress with surrounding textures represent the key parameters to determine the geometrical dimensions. As an approach to ensure the suppression of stress concentration at the structure edges, the structural analysis can be used to modify both, the structure geometry and dimensions.

4 Conclusions

Through the tribological evaluation of ball-on-disk type friction tests under constant DLC-coverage and varied structure diameter, following conclusions can be drawn:

- Within the 10,000 laps of rotation, smaller texture diameter ($D = 50$ µm), stable and relatively low COF of $\mu = 0.09$ was measured, which is probably the results of an effective ejection of wear debris from the contact interface.
- Regarding the wear resistance, larger DLC structures ($D = 150$ µm) showed more than twice longer durability than that of smaller diameter, which was characterized by low COF of $\mu = 0.08$. The lower wear resistance of the textures with smaller diameters is attributed to the higher stress concentration at the edge of the structure, which can be modified by structural design of the micro-texture patterns.

In summary, stable and reduced friction under dry sliding condition can be achieved by a miniaturization of apparent contact area of textured surface, while durable wear resistivity by taking into account the ejection of wear debris from contact interface and the stress concentration. Future works will focus on quantitative structural analysis in order to design appropriate geometrical structures.

The authors gratefully thank the financial support from Tokyo Metropolitan University in the framework of the Co-Tutorial Program in 2016 as well as from the minister for science, health and consumer protection of the free hanseatic city Bremen.

References

1. N. Bay, A. Azushima, P. Groche, I. Ishibashi, M. Merklein, M. Morishita, T. Nakamura, S. Schmid & M. Yoshida CIRP Ann. **59**, 760-780 (2010)
2. B. Bhushan, *Introduction to Tribology 2nd Edition* (Wiley, New York), (2013)
3. N.P. Suh & H.C. Sin, Wear **69**, 91-114 (1981)
4. B. Bhushan & M. Nosonovsky, Acta Mater. **52**, 2461-2474 (2004)
5. M. Nosonovsky, B. Bhushan, Scale effect in dry friction during multiple-asperity contact, ASME J. Tribol. **127**, 37–46 (2005)
6. T.Shimizu, T. Kakegawa & M. Yang, Procedia Eng. **81**, 1884-1889(2014)
7. T. Kakegawa, T. Shimizu & M. Yang, J. Jpn Soc. Technol. Plast. **56**-657, 891-896 (2015) (in Japanese)
8. J.Meijer, K. Du, A.Gillner, D. Hoffmann, V.S. Kovalenko, T.Masuzawa, A.Ostendorf, R.Poprawe, W.Schulz, CIRP Annals **51**(2), 531-550 (2002)
9. T. Shimizu, M. Yang & K. Manabe, Wear **330-331**, 49-58 (2015)
10. E. Rabinowicz: *Friction and Wear of Materials*, (Wiley, New York), (1995)

Experimental studies of the efficient use of flexible tool in creep age forming

Qi Rong[1,*], *Yong* Li[1], *Xiaoguang* Sun[2], *Zhusheng* Shi[1], *Lichun* Meng[2] *and Jianguo* Lin[1]

[1]Department of Mechanical Engineering, Imperial College London, London SW7 2AZ, UK

[2]National Engineering Research Center for High-Speed EMU, CRRC Qingdao Sifang Co., Ltd, Qingdao 266111, China

Abstract. Application of a newly developed flexible forming tool to creep age forming (CAF) process has been investigated in this study. The flexible tool mainly consists of sparsely distributed forming pins, splines and elastomeric sheet. The effect of key factors related to the forming tool on the shape of the formed parts has been studied through various CAF experiments. The key factors investigated in this study include: the interval between forming pins, the arrangement of pins, the accuracy requirement of pin height and the material of splines. It has been found that reducing the interval between pins can efficiently smooth the shape of CAFed plates. The feasibility of asymmetric arrangement of pins has been proven, which can decrease the number of used pins, reduce tool weight, and increase efficiency. The forming results are very sensitive to the pin height, thus the experimental set-up error should be carefully controlled. Additionally, compared with mild steel, spring steel is more suitable as the spline material.

Keyword. Aluminium, Sheet metal, Flexible tool

1 Introduction

Creep age forming (CAF) is a forming method combing forming and age hardening in one process, and has become one of the most popular manufacturing methods for the production of extra-large panel components [1], such as upper wing skins of Airbus A380 [2], and Gulfstream G-IV Aircraft [3].

An important issue in the CAF process is the forming tool. One of the characteristic tools for various forming technologies is solid dies. However, the cost of solid dies for CAF is significant, as extra-large tools are generally required and tool modification is also frequently needed due to springback compensation. Levers [4-6] developed a semi-flexible forming tool for CAF of extra-large skin panels, replacing the solid dies into a set of rib boards and leading to the lighter weight and better flexibility than the solid dies, and its feasibility has already been proven to apply to the production of large aircraft panel components [7-9]. However, since the forming surface is dependent on the rib boards, when the shape of wing skin is changed, a new set of rib boards needs to be manufactured, which is costly and time-consuming. Thus, changing tool surface is a big challenge, which leads to the need of flexible tools for the CAF investigation, especially in a laboratory environment.

Lam et al. [10] have developed a multi-point flexible forming tool composed of sparsely distributed forming pins, splines, and elastomeric sheet. The forming pins whose heights can be adjusted, are used to control the tool shape, splines are attached to the pins to maintain the continuity of the shape and an elastomeric sheet is used to more uniformly distribute the stress on the workpiece in the forming process. Compared to the tools mentioned above, this tool is much more flexible and with lower cost, making it a suitable choice for CAF study. However, it has been found that some kinks may occur on the deformed plates with this multi-point flexible tool when the target curvature of the workpiece is large and complex. Furthermore, how to use this tool in a more efficient way, i.e., using less pins and simpler arrangement to attain the required forming shape, is also of interest. The forming results will depend on each part of the tool, including forming pins, splines and elastomeric sheet, however no studies have been reported to investigate their detailed effects on the forming results currently.

For the forming pins, three factors are important to control the forming results: the interval between pins, increasing which can reduce the number of pins required and lower the tool weight; asymmetric arrangement of pins in top and bottom dies, which may largely decrease the number of pins required; and the accuracy requirement of pin height, which may affect the final forming shape. For the splines, the material used is also worth investigating, so as to obtain a smoother curved shape of the formed plate.

* Corresponding author: q.rong15@imperial.ac.uk

Thus, in this study the effects of four different factors in the flexible forming tool on CAF forming results with aluminium alloy 6082 (AA6082) have been experimentally investigated, including the interval between neighbouring pins, asymmetric arrangement of pins, the error of pin height, and the material used for splines.

2 Experimental programme

2.1. Experimental Material

The material used in this study was AA6082-T6, whose chemical composition is listed in Table 1. The as-received material had undergone solution heat treatment at 525 °C for 22 min, water-quenching, then artificial ageing at 160 °C for 10 h and air-cooling. The workpiece with a dimension of 550 x 180 x 4 mm was used in this study, which was machined from large 4 mm thick plates, and the width of the machined workpiece (180 mm) is parallel with the rolling direction.

Table 1. Main chemical composition of AA6082 (wt%).

Al	Si	Mg	Mn	Fe	Cr	Zn	Cu
Balance	0.82	0.82	0.64	0.24	0.09	0.024	0.021

2.2 Flexible tool

The investigated flexible tool in this study was designed by Lam [10], as shown in Fig. 1(a) [11]. This flexible tool mainly consists of the top and bottom die holders, sparsely distributed forming pins, splines, and an elastometic sheet. More details about the tool design can be found in [11] and [12].

(a)

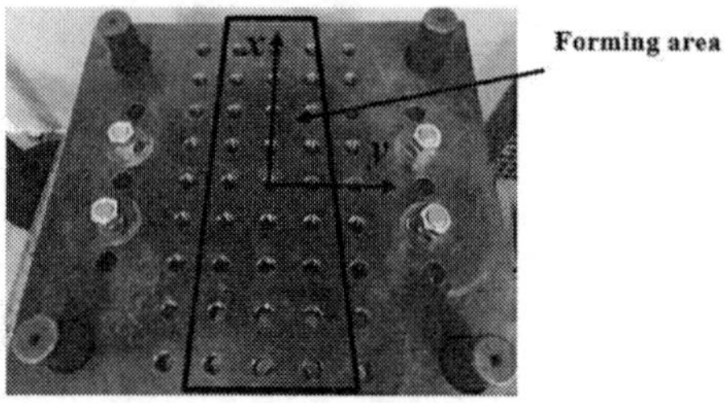

(b)

Fig. 1: (a) The flexible tool used for CAF and test setup [11]; (b) The top view of the tool without pins.

There are five rows of nine uniformly spaced holes on both the top and bottom die holders, where forming pins are located as shown in Fig. 1(b). The interval between each row and column of pins is 60 mm. The number of pins can be changed according to the actual need, and the heights of pins can be adjusted, based on the target forming shape. In this study, three rows of pins in the centre of die holders were used as the forming area, as marked in Fig. 1(b).

A spline with a dimension of 500 × 30 × 3 mm is attached onto each row of pins in both upper and bottom die holders, to maintain the continuity of the surface generated by sparse pins. An elastomeric sheet (silicone rubber) with a dimension of 500 × 300 × 12 mm is placed between the workpiece and bottom splines to decrease the stress concentration on the plate, as shown in Fig. 1(a). Two materials of splines were used in this study, mild steel with a yield strength of 450 MPa and 75Cr1 spring steel with a yield strength of 1000 MPa.

2.3 Creep-age forming tests

Four groups of CAF experiments were carried out to investigate the exact function of each part of the tool and the key factors in the multi-point flexible tool that affect the forming results, as shown in Table 2. The same tool shape, a single curved surface with a radius of 700 mm along the x axis, was used for all CAF tests in this study.

Table 2: Details of four groups of CAF experiments.

Test Group	Key factor	Test information
1	Interval between pins	(a) Five uniformly distributed pins in each row (5 ×3), uniform pin interval: 120 mm; (b) Seven pins in each row (7×3): pin interval: 60 and 120 mm; (c) Nine uniformly distributed pins in each row (9×3); pin interval: 60 mm. (Fig. 2)
2	Arrangement of pins	(a) Asymmetric: nine pins at the top, three pins at the bottom (9/3); (b) Asymmetric: nine pins at the top, five pins at the bottom (9/5); (c) Symmetric: nine pins at both the top and bottom (9/9). (Fig. 4)
3	Accuracy of the pin height	Same symmetric pins arrangement (7×3). ± 2 mm error of height were added to the pins in the x+ area (x = +120, +180 and +240 mm). (Fig. 6)

4	Material of the splines	Same symmetric pins arrangement (5 × 3). (a) Mild steel; (b) Spring steel.

The main procedures of the CAF tests include: (a) adjust the pin height to achieve the target tool surface, and prepare the whole set of flexible tool as shown in Fig. 1(a); (b) load the tool by a 250 ton Instron servo-hydraulic testing machine until the plate is deformed to the target shape (the same maximum load of 100 KN was applied for all CAF tests); (c) use bolts and nuts to lock the tool, and release the pressure from Instron machine; (d) move the tool to a fan-assisted Aerotherm box oven and run the creep-ageing thermal cycle at 160 °C for 3 hours; (e) cool down, unload the tool, and let springback occur; (d) measure the dimensions of the deformed plate along its centre line (x = 0) with a Nikon LK G90C coordinate measuring machine (CMM).

3 Results and discussion

3.1 Effects of the interval between pins

Fig. 2 shows the pin number and interval details of the first group of test in Table 2. From five pins to nine pins in each row, the interval between pins decrease from 120 mm to 60 mm.

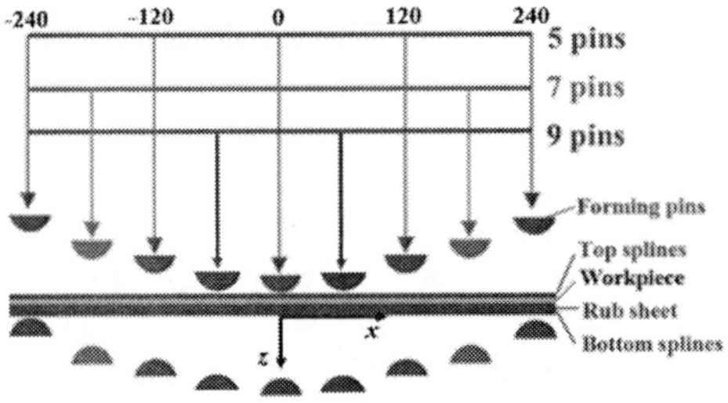

Fig. 2: Different pin number and interval conditions used in the first group of CAF tests.

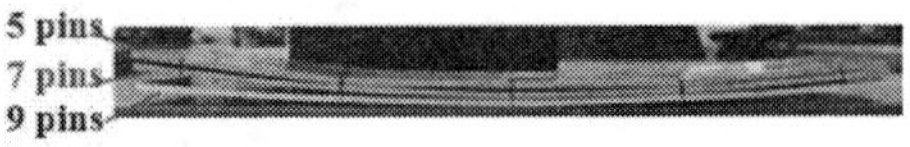

(a)

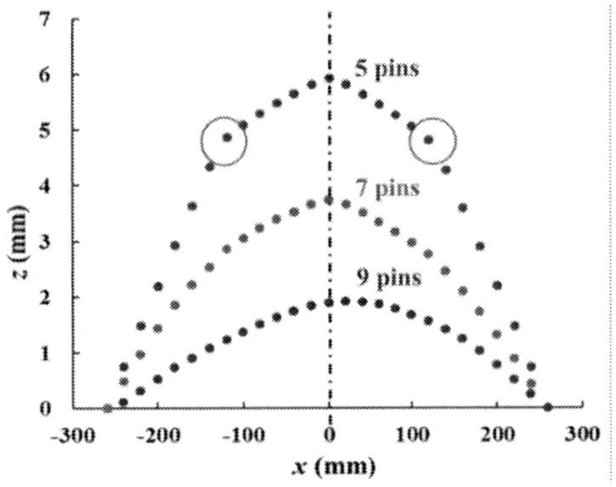

(b)

Fig. 3: (a) Side view of the CAFed plates under different pin interval conditions; (b) deflection results of the centre line (x = 0) of CAFed plates obtained by CMM.

Fig. 3 shows the shape of the CAFed plates with different pins number and interval conditions. It can be observed from Fig. 3 that significant kinks (red circles in Fig. 3(b)) were present in the CAFed plate when using five forming pins in each row of the tool, gentle kinks occurred in the formed plate with seven forming pins in each row, while almost smooth curve was achieved in the formed plate when using nine pins. It is also found that with less forming pins, i.e., larger pin interval along the x-axis, larger deflections were achieved. From Fig. 3(b), it can be seen that kinks are mainly present around $x = \pm$ 120 mm, where pins were located, indicating that the stress concentration may have occurred in the pin areas. As higher stresses were generated in these regions, more significant deformation (kinks) occurred. This can explain why smaller but less smooth curvature was generated by using more pins. It can be concluded that the interval between neighbouring pins can significantly affect the shape of the plates after CAF and smaller interval between pins generate a shape with smoother but smaller curvature; more pins can benefit the forming results, but make it more towards solid tools, reducing the benefit of flexible tool at the same time. The suitable interval between pins should be dependent on the target forming curvature, and further investigations are needed to quantify the exact relationship between the proper pin interval and the target forming curvature.

3.2 Effects of the arrangement of pins

Fig. 4 shows the three different arrangements of pins used for the CAF tests in this study: 9/3, 9/5, and 9/9, as indicated in Table 2.

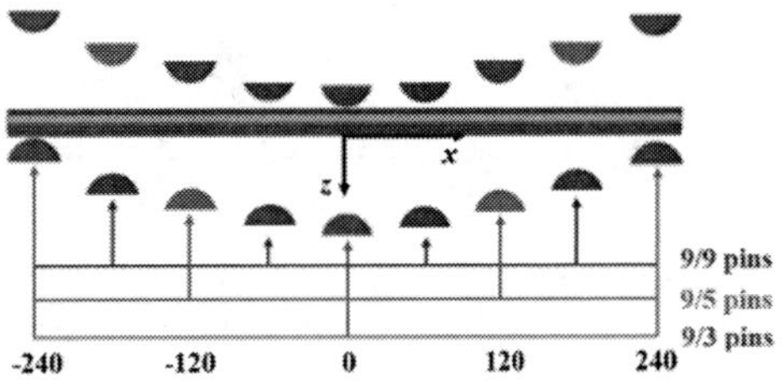

Fig. 4: Arrangement of pins: 9/3, 9/5 and 9/9.

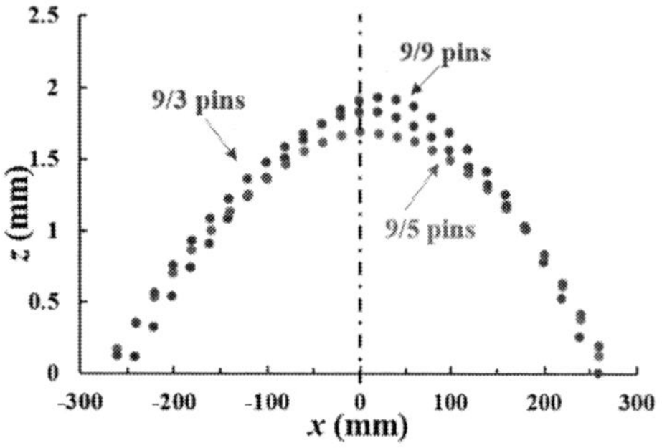

Fig. 5: Shape of the CAFed plates under three different arrangement of pins: 9/9, 9/5 and 9/3.

Fig. 5 shows the surface profile along centre line (*x*=0) of the plates after CAF under three different pins arrangements. It can be observed that there is slight difference among the final shape of the CAFed plates under these three different pins arrangements. It indicates that reducing the number of pins on one side will not largely affect the forming results. This proves the feasibility of using asymmetric number of pins in the multi-point flexible tool to deform the plates in the CAF process. Thus, for the future CAF tests, it is viable to decrease the number of pins on one side, which can reduce more weight and increase the efficiency.

3.3 Effects of accuracy of pin height

Errors of ±2 mm errors were intentionally introduced to some target pin height in the *x*+ region, as shown in Fig. 6, so as to investigate the sensitivity of the pin height accuracy on the formed shape.

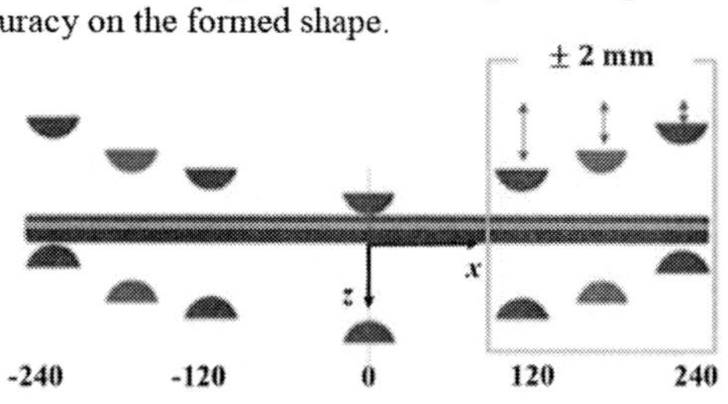

Fig. 6: Arrangement of pins: ± 2 mm errors added to the target pin height for the pins on the *x*+ direction.

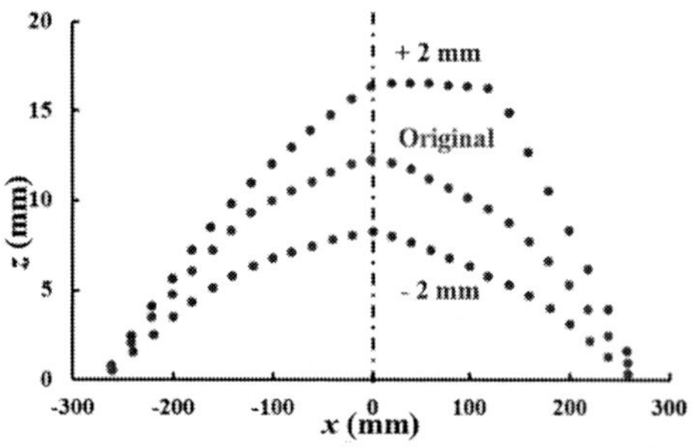

Fig. 7: Comparison of the shapes of CAFed plates between forming tools with original pin height and intentionally adjusted pin height with 2 and -2 mm error added to the pins on *x*+ direction.

Fig. 7 shows the shape of the formed plates with different conditions, as configured in Fig. 6. A significant difference of final shape can be observed when ± 2 mm errors were introduced. It can also be seen that errors in *x*+ side affect the shape of the whole plate. Thus, it can be concluded that the final plastic deformation is quite sensitive to the pin height and small error in pin height can lead to significant difference of the final formed shape. As a result, during the set-up stage of the experiments, the pin height should be adjusted as accurate as possible to achieve best results.

3.4 Effects of the material of splines

Another component of the flexible tool for the CAF process is splines. The initial intention of adding splines between workpiece and pins is to keep the continuity of the surface of the tool and help to avoid indentations in the plate by reducing stress concentration. Thus, the material chosen for the spline needs to have sufficiently high strength to prevent it from deforming plastically. In this study, two materials were used for splines in the CAF tests: mild steel and 75Cr1 spring steel, which are much stronger than the Al alloy workpiece. The experimental results are shown in Fig. 8. Two cases were investigated: one is measured after loading, the other is after complete CAF process. It can be observed that larger deflections were produced when using mild steel splines. This is because the mild steel with lower strength is easier to deform rather than the other one. Smoother curves were obtained by using spring steel splines than mild steel splines. Additionally, visual examination of the splines of these two materials after CAF, revealed that plastic deformation occurred on the mild steel splines, but no plastic deformation was present on the spring steel splines. The latter means that the spring steel splines can be repeatedly used which is beneficial for the running cost. As a result, to generate smoother shape, and take into consideration of reusability of the splines, spring steel is a better choice as the material of the splines.

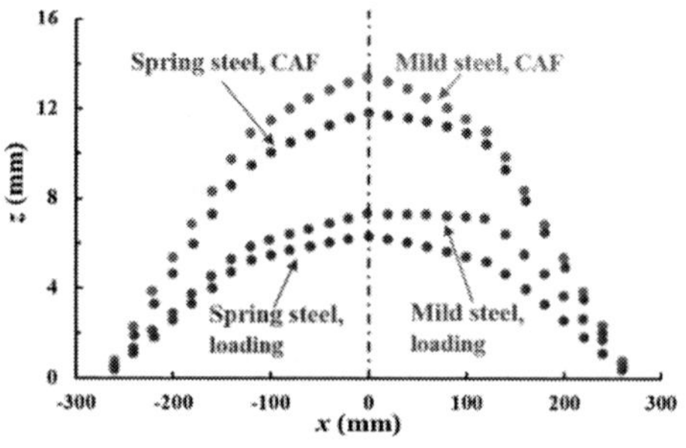

Fig. 8: Shape of the deformed plates after different process (CAF/ loading) with different material of the splines (mild steel/ spring steel).

5 Conclusions

In this study, four different groups of CAF tests related to the multi-point flexible tool were carried out to investigate the effects of the interval between pins, the asymmetric arrangement of pins, the error requirement of the pin height, and the material of splines on the forming results. The experimental study in this work helps to reveal how these key factors affect the forming results with flexible tool during the CAF process, providing a practical guidance on the setup of forming tools, and facilitating further applications of this kind of flexible tool in industry. The following conclusions can be drawn:

1. Reducing the interval between pins can efficiently smooth the shape of the plates after CAF, but decrease the achieved deflections. When the target curvature is large and complex, small interval between pins is preferred to avoid kinks.
2. The feasibility of asymmetric arrangement of pins for CAF has been proven, which means one-to-one correspondence between pins at top and bottom is not necessary. Asymmetric arrangement of pins can largely decrease the number of used pins, reduce the weight of tool, and increase the efficiency.
3. The forming results are very sensitive to the pin height, and the error of pin height should be carefully controlled. Otherwise a large difference between the final forming shape and the target shape will occur.
4. Compared with mild steel, spring steel, which has higher strength, is more suitable as the spline material. It can help to generate smoother shape of the deformed plate and can be used after CAF process.

Acknowledgements

Much appreciated is the strong support received from CRRC Qingdao Sifang Co., Ltd. The research was performed at the CRRC Sifang-Imperial Centre for Rail Transportation Manufacturing Technologies at Imperial College London.

References

1. L. Zhan, J. Lin, T. Dean. Int. J. of Mech. Tool. Manu. **51**, 1 (2011) 1-17.
2. K. Watcham. Mate. World. **12** (2004) 10-11.
3. M. Holman. J. of Manu. Working Tech. **20** (1989) 477-488.
4. A. Levers, J. of Manu. Eng. **82** (2003) 42-45.
5. A. Levers, PAT US 7,322,223 B2, 29 Jan 2008.
6. A. Levers, Int. J. for History of Eng. & Tech. **80** (2010) 55-79.
7. H. Bornschlegl, PAT US 6,264,771. 24 Jul 2001.
8. TL. Newkirk, MC Holman, PAT US5, 729,462, 17 Mar 1998.
9. HM. Brewer, PAT US 5, 168, 169, 1 Dec 1992.
10. A. Lam, Z. Shi, J. Lin, X. Huang, Y. Zeng, T. Dean. Int. J. of Adv. Manu. Tech. **80** (2015) 361-372.
11. A. Lam, Z. Shi, J. Lin, X. Huang. Int. J. of Mech. Sci. **103** (2015) 115-126.
12. A. Lam, PhD thesis: A flexible tool design and integrated modelling techniques for springback compensation in creep age forming. (2016)

Material flow control in sheet-bulk metal forming processes using blasted tool surfaces

Marion Merklein[1], *Maria* Löffler[1], *Daniel* Gröbel[1], and *Johannes* Henneberg[1,*]

[1] Institute of Manufacturing Technology, Friedrich-Alexander-Universität Erlangen-Nürnberg (FAU), 91058 Erlangen, Germany

Abstract. Highly-integrated and closely-tolerated functional components can be produced by sheet-bulk metal forming which is the application of bulk forming operations on sheet metals. These processes are characterized by a successive and/or simultaneous occurrence of different load conditions such as stress and strain states which reduce the geometrical accuracy of the functional elements. Thus, one main challenge within sheet-bulk metal forming is the identification of methods to control the material flow and thus to improve the product quality. One suitable approach is to control the material flow by local modifications of the tribological conditions. Within this study requirements regarding the needed adaption of the tribological conditions for a specific sheet-bulk metal forming process were defined by numerical investigations. The results reveal that a local increase of the friction leads to an improved die filling of the functional elements. Based on these results abrasive blasting as a method to modify the tool surface and thus influencing the tribological behaviour was investigated. For the determination of the tribological mechanism of blasted tool surfaces, the influence of different blasting media as well as blasting pressures on the surface integrity and the friction were determined. The correlations between surface properties and friction conditions were used to derive the mechanisms of blasted tool surfaces.

Keywords: Tribology, Cold forming, Surface modification

1 Introduction

Nowadays the manufacturing industry is enforced to upgrade their products due to rising customer and legal requirements. Lightweight construction and minimization of material usage is a promising approach to deal with these new requirements. Weight reduction is achievable by the usage of highly integrated and closely tolerated parts. Due to rising economic competition, parts must be produced by time and cost efficient manufacturing processes. Metal forming is suitable to fulfil these economic requirements, because it enables an efficient production due to high material usage and low production time per part [1]. Conventional sheet and bulk forming operations are limited while producing highly integrated and closely tolerated parts. The existing limitations motivate the development of the new process class sheet-bulk metal forming (SBMF). SBMF processes are defined as the application of bulk forming operations on sheet metal [2]. By this process class sheet metal parts with integrated functional elements can be produced in an efficient way. In SBMF processes high tribological loads which are characteristic for bulk forming operations occur simultaneously and/or successive with lower tribological loads like in sheet metal forming processes. As a result the plastic material flow is negatively influenced into areas with lower tribological loads. Hence, the part quality in terms of geometrical accuracy is insufficient. As shown in figure 1 the die filling of the functional elements is incomplete. Consequently the in service behaviour of the produced parts is impaired. Stress concentration due to cavities in the die and high yield stresses due to forming at room temperature are already causing high tool loads. Thus, the enhancement of the die filling by higher forming forces is limited by high tool loads. Hence, other ways to control the material flow during forming need to be developed. The local adaption of the friction to control the material flow seems to be a promising approach [3]. On tool side, the occurring friction can be influenced by the material and the surface properties [4].

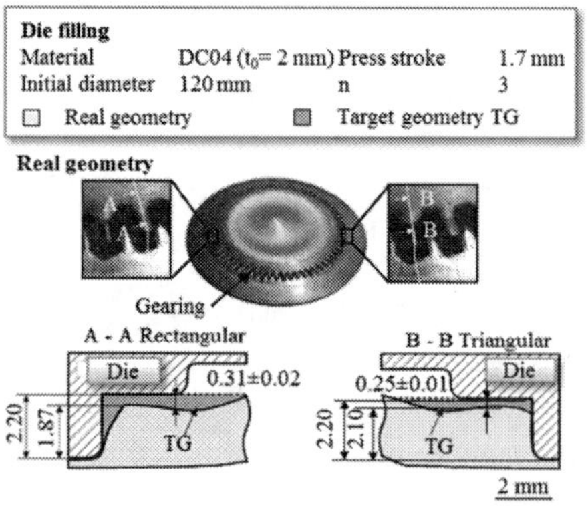

Fig. 1. Die filling.

* Corresponding author: Johannes.Henneberg@fau.de

 https://doi.org/10.1051/matecconf/201819013003

Locally modified surfaces, also named tailored surfaces, which adapt the friction have the potential to improve the part quality by increasing the die filling of functional elements [3]. The material flow is led into the cavities by a local reduction of friction and/or an inhibition of the material flow in other areas of the part by a local increase of friction. Numerical investigations can be utilized to analyse the requirements for tailored surfaces by examining the tribological loads during SBMF. In addition, the influence of locally varying friction on the forming process can be evaluated by numerical investigations.

The adjustment of tool- and workpiece-surfaces is one possibility to locally influence the friction. Workpiece-sided modifications are disadvantageous because every workpiece has to be adapted. Hence, the SBMF-process chain is extended and economic competitiveness declines. By contrast, tool-sided surface modifications do not lengthen the process chain. In this paper the requirements for tool-sided tailored surfaces in a specific SBMF process are analysed by numerical investigations and the influence of tool-sided abrasive blasting as a method to influence the surfaces integrity and friction is analysed in a laboratory test. Tool sided abrasive blasting is normally applied as a method to prepare the tool surfaces for plasma coating. The aim of these investigations is to identify fundamentally functional relations between the tool surfaces integrity and the friction.

2 Methodology and experimental setup

The aim of this study is to evaluate the potential of tailored surfaces to improve a specific SBMF process. General requirements for tailored surfaces should be derived. As shown in figure 2, the tribological loads of a specific SBMF process are analysed by numerical investigations. Based on these results, a numerical model with modified friction is designed. The potential of locally adapted friction for controlling the material flow and increasing geometrical accuracy of parts is investigated. The generated knowledge is used to derive requirements for tailored surfaces. Subsequently, the potential of abrasive blasting of tools to tailor the surfaces is evaluated in a laboratory test. Functional relations between tool surface integrity and friction are fundamentally analysed.

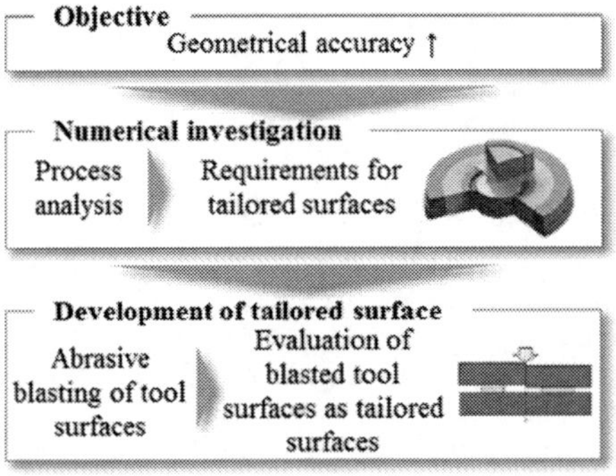

Fig. 2. Methodology.

As an example for SBMF a full forward extrusion process, which produces parts with rectangular and triangular functional elements, is numerically examined. The process is illustrated in figure 3.

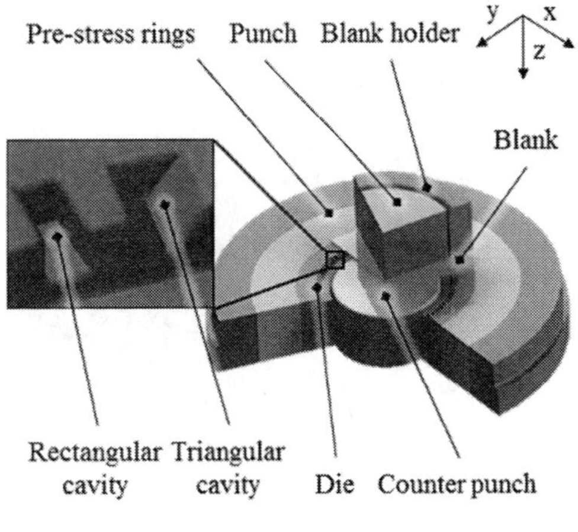

Fig. 3. Tool setup of the forward extrusion process [5].

The blank which is made out of the deep drawing steel DC04 (1.00338) with a diameter of 120 mm and a thickness of 2 mm is formed by a punch with a stroke of 1.7 mm [5]. During the forming process, the blank is supported by a counter punch with a force of 40 kN to reduce the material flow in the inner area of the part. In addition, it is supported by a blank holder with a force of 360 kN. To reduce the simulation time, a 10°-segment of the part was processed. The validation of the model was done in [5].

For the design of tailored surfaces, the tribological loads, more specifically the contact normal stresses, the sliding paths and the surface enlargement, of the SBMF process were analysed. A model with locally adapted tribological conditions was simulated for different friction factors. For the evaluation of the effectiveness of a local adaption, the results were compared with simulations with globally homogeneous friction. The material flow in the cavities of the functional elements was analysed by calculating the ratio between the volume of the formed functional element and the volume of the die cavities. In addition, the volumes of the outer and inner areas as shown in figure 4 were compared to the volume of these areas before forming. Based on the material flow analysis, requirements for tailored surfaces were derived.

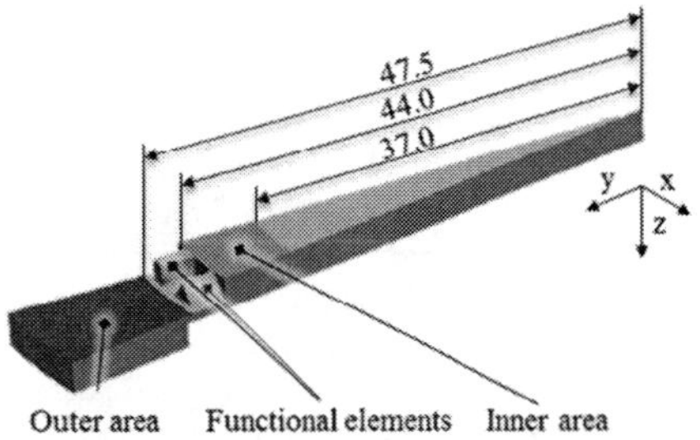

Fig. 4. Investigated areas for the material flow analysis.

Abrasive blasting was chosen as a method to modify the tool surfaces. The potential of abrasive blasting to fulfil the requirements for tool-sided tailored surfaces and the influence on the friction was investigated. Six differently blasted surfaces were analysed in the ring compression test (RCT). In the RCT, a ring made out of DC04 is compressed from 2 mm to 1 mm height as shown in figure 5.

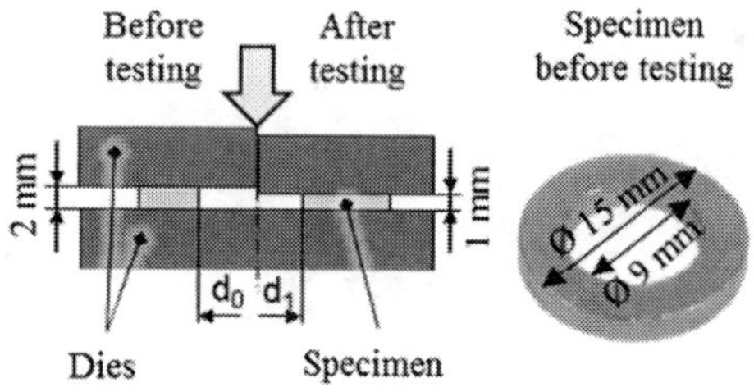

Fig. 5. Setup of the ring compression test.

The inner diameter after forming is sensitive to the friction [6]. In case of high friction a smaller inner diameter occurs. The geometry of the specimen before forming was selected based on a numerical study guaranteeing a high sensitivity for the specific forming conditions of SBMF processes [7]. An outer diameter of 15 mm and an inner diameter of 9 mm were selected. The friction factor m is calculated based on equation 1, which was fitted for the selected specimen geometry and material in [3].

$$m = (-0.07722 * d_1 + 0.9696) / (d_1 - 5.834) \quad (1)$$

The influence of the lubricant Beruforge 150 DL on the RCT was investigated in preliminary tests. An amount of 10 g/m² was chosen, since saturation is reached for this amount of lubricant. For the RCTs five specimens were formed per tool surface to guarantee statistic relevance. The RCT was conducted on the hydraulic deep-drawing press Lasco 100SO with a forming velocity of the upper die of 2 mm/s. The forming tools are made out of powder metallurgical tool steel ASP2023 (1.3344) with a hardness of 61 ± 2 HRC. The tool surfaces were ground before blasting. For the blasting process, two different blasting media and three different blasting pressures were used with the Peenmatic 620S. The topography of the tools before and after the RCT were analysed by confocal microscopy with a Keyence VK X200. To characterize the tool surfaces, the reduced peak height R_{pk} as a parameter describing the average height of the roughness peaks which are higher as the core roughness are analysed [8]. This parameter was chosen because it describes the tendency for mechanical interlocking of roughness peaks. The influence of the abrasive blasted surfaces on the friction is analysed by comparing the resulting friction factors. The relation between tool topography and friction enables the evaluation of abrasive blasting for tailored surfaces. The analysis of tool surfaces after the forming process enables the estimation of the wear resistance of blasted tool surfaces. Based on the result the potential of abrasive blasted tool surfaces for processes of SBMF is rated.

3 Numerical investigations

For the design of tailored surfaces, the tribological loads of the specific SBMF process have to be analysed. The occurring contact normal pressure, the sliding path and the surface enlargement are analysed for a friction factor of 0.1. Three areas with different loads are identified and presented in figure 6 and table 1.

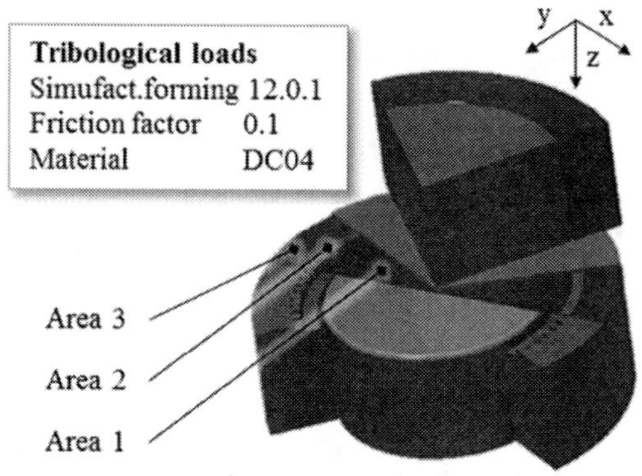

Fig. 6. Areas with different tribological loads.

The maximum tribological loads in area 1 as well as in area 3 are low, due to the great contact area between tool and workpiece and thus lower resulting strains. In addition, the surfaces enlargements and the sliding paths are low. In both areas, maximum contact normal pressures between 183 and 282 MPa occur. These low contact normal pressures are characteristic for sheet metal forming processes [9]. The forming of functional elements causes high contact normal pressures up to 2803 MPa as they occur in bulk metal forming operations. As a result of the long sliding paths and high surface enlargements, the risk of a disconnection of the lubrication film rises. A disconnected lubrication film would increase the tribological loads even further.

Table 1. Maximum tribological loads of the SBMF process.

Maximum values	Area 1	Area 2	Area 3
Contact normal pressure in MPa	282	2803	183
Surface enlargement in %	0.00	70.00	5.38
Sliding path in mm	-0.61	2.09	1.60

The varying tribological loads influence the material flow. The material flow is directed into areas with low tribological loads. Thus, the rectangular functional elements reach a die filling of only 83.63% because the material flows into the lower loaded areas 1 and 3 next to the functional element. Hence, the volume of the workpiece in the inner and outer area rises up to 106.36 % and 106.04 % compared to the volume before the forming. To reduce the material flow in these areas and consequently increase the volume of the functional elements the friction has to be adapted. The die filling of the triangular functional element is 97.57 % since the triangular cavity is more easily filled than the rectangular cavity due to its geometry. The influence of adapted friction is examined with two simulation models illustrated in figure 7 and table 2.

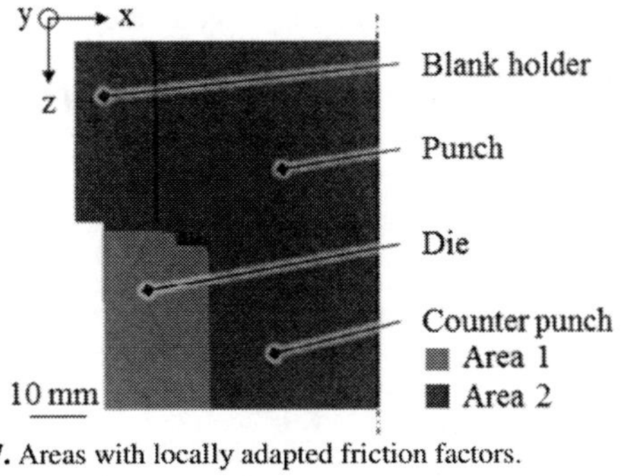

Fig 7. Areas with locally adapted friction factors.

By means of a local increase of friction in all areas except for the die, the material flow into the functional elements should be supported. For the examination of the potential of a material flow control by tailored surfaces the friction factor of all tool surfaces except of the die are increased from 0.1 up to 0.6 in six steps (locally adapted model 1, No. 1 to 6).

Table 2. Experimental design of the numerical material flow study.

Locally Adapted Model 1			Global Model 2	
No.	m_{Area1}	m_{Area2}	No.	m_{global}
1	0.1	0.1		
2	0.1	0.2	7	0.2
3	0.1	0.3	8	0.3
4	0.1	0.4	9	0.4
5	0.1	0.5	10	0.5
6	0.1	0.6	11	0.6

There is no knowledge about the maximum friction factor after blasting of tool surfaces. For the numerical investigations 0.6 is chosen as highest friction factor. The simulations with locally adapted friction are evaluated by a comparison to simulations with globally homogeneous friction factors between 0.1 and 0.6 (global model 2, No. 7 to 11). The friction factor 0.1 is chosen as the lowest friction factor, since lapped tool surfaces cause approximately this friction factor in the RCT.

The areas which are examined for the material flow analyses are defined in figure 4. These areas are analysed since they are critical in terms of geometrical accuracy. The local increase of the friction in model 1 reduces the volume of the outer area from 106.36 % at a friction factor of 0.1 to 104.18 % at a friction factor of 0.6 as illustrated in figure 8. Hence, the die filling of the rectangular functional element increases consistently from 83.63 % up to 92.39 %. Since the die filling of the triangular functional element is already 97.93 % at a low friction factor and variation of the friction only causes minor changes of the die filling up to 98.32 %, further investigations are focused on the rectangular element. The volume of the inner area increases for the locally adapted model from 106.04 % up to 110.34 %. In the inner area lower friction occurs, due to the one-sided contact between the workpiece and the tool in this area. Hence, the volume of the outer area is not only transferred into the functional elements but also into the inner area. A global increase of the friction from 0.1 to 0.6 causes a higher decrease of the volume of the outer area. The volume at a friction factor of 0.6 is 103.51 %. The volume is lower compared to the locally adapted model, since the friction of the die is also increased. Hence, the friction force in the outer area increases on both sides of the workpiece. In comparison to the locally adapted model, more material is transferred to other parts of the workpiece. Due to this fact, the die filling of the rectangular functional element and the volume of the inner area is in the global model for all determined simulations slightly higher than in the locally adapted model.

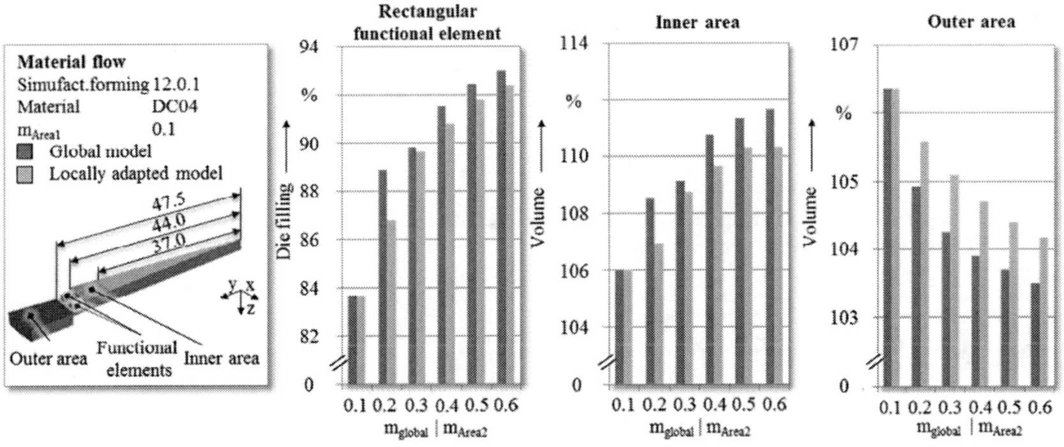

Fig. 8. Material flow for the globally and locally adapted models for different friction factors.

 https://doi.org/10.1051/matecconf/201819013003

As shown in figure 9 the friction also influences the required forming forces. Due to a higher friction, the material flow is inhibited. The increase of the friction in in both investigated models causes higher forming forces. A global variation of the friction factor from 0.1 to 0.6 increases the necessary forming forces from 1320 kN up to 2317 kN. A local adaption of the friction causes lower forming forces. The required forming force in the simulations with locally adapted friction is for all variants about 200 kN lower compared to the variants with globally homogenous friction. Since the maximum forming force is an indicator for the external load of a forming tool [10], a lower forming force due to a local adaption of the friction may reduce the tool load. For instance, a global increase of the friction factor from 0.1 to 0.3 raises the reduced stress in the highest loaded radius of the triangular cavity from 2713 ± 205 MPa ($n_{Elements} = 5$) up to 3060 ± 130 MPa, whereas a local adaption of the friction factor only increases the reduced stress to 2744 ± 66 MPa.

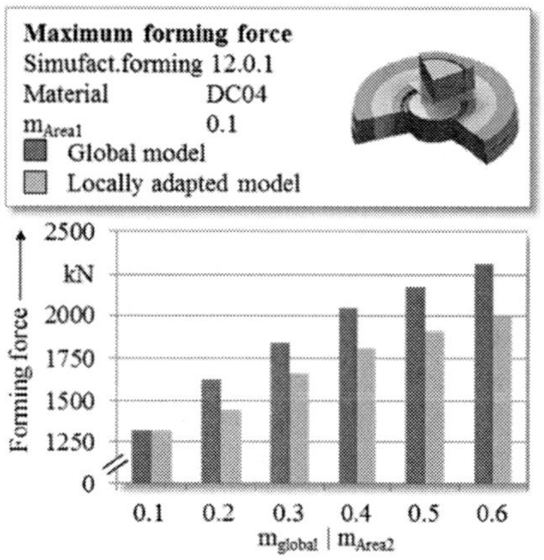

Fig. 9. Influence of friction on maximum forming forces for the globally and locally adapted models.

The results of the numerical investigations are concluded in figure 10. The numerical study has shown that an increase in friction results in reduced material flow in the outer area. Hence, the die filling of the rectangular functional elements is improved.

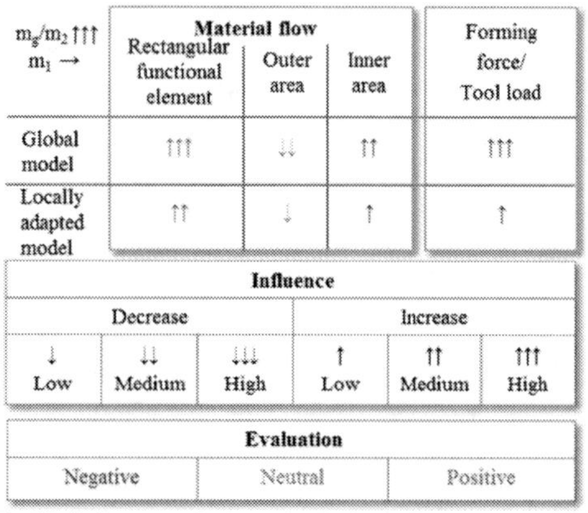

Fig 10. Influence of adapted tribological systems on SBMF process.

Nevertheless, the volume of the inner area as well as the required forming forces increase in both analysed models. By a local adaption of the friction, the necessary forming forces could be reduced with slightly lower die filling of the rectangular functional element compared to a model with globally homogenous friction. For instance, a local adaption of the friction reduces the necessary forming forces for a friction factor of 0.3 by 9.55 %, whereas the die filling of the rectangular functional element is only reduced by 0.20 %. Additionally, an increase of the friction of the die as simulated in model 2 for example by abrasive blasting of the tool surface would probably reduce tool life by producing surfaces defects. These surfaces defects can initiate cracks and therefore reduce fatigue strength [11]. Hence, the friction has to be adapted locally on all tool surfaces except for the die.

4 Tailored Surfaces

The numerical study has shown that tailored surfaces can be used to improve SBMF processes. The surface modifications should increase friction to reduce the material flow in areas with low tribological loads. Further requirements are a local applicability as well as an efficient production of the tailored surfaces. Consequently, abrasive blasting has potential as a surfaces modification because it can be applied locally and is a cost efficient technique to influence the surfaces integrity of tools [12]. Abrasive blasting roughens the surfaces. Hence, the aim of the experiments is to investigate the potential of tool-sided abrasive blasting as a surfaces modification to increase friction.

For the blasting of the tool surfaces two different blasting media with a particle size between 15 and 35 µm and a hardness of 3000 HV as well as 50 to 70 µm and 2400 HV were used. The material of both blasting media is Al_2O_3. The particles are angular to roughen the surfaces. The pressure of the blasting process is varied from 0.1 to 0.3 MPa in three steps. Lower pressures are impeded by the blasting process. Higher pressures would destroy the blasting particles. The ground tool surfaces are blasted for 47 s with a distance of 40 mm to create uniform surface properties. The blasting time was investigated in preliminary tests to create a surface with homogenous properties. A 90° angle between nozzle and surfaces was chosen, since this angle causes the highest roughening of the surfaces [13].

https://doi.org/10.1051/matecconf/201819013003

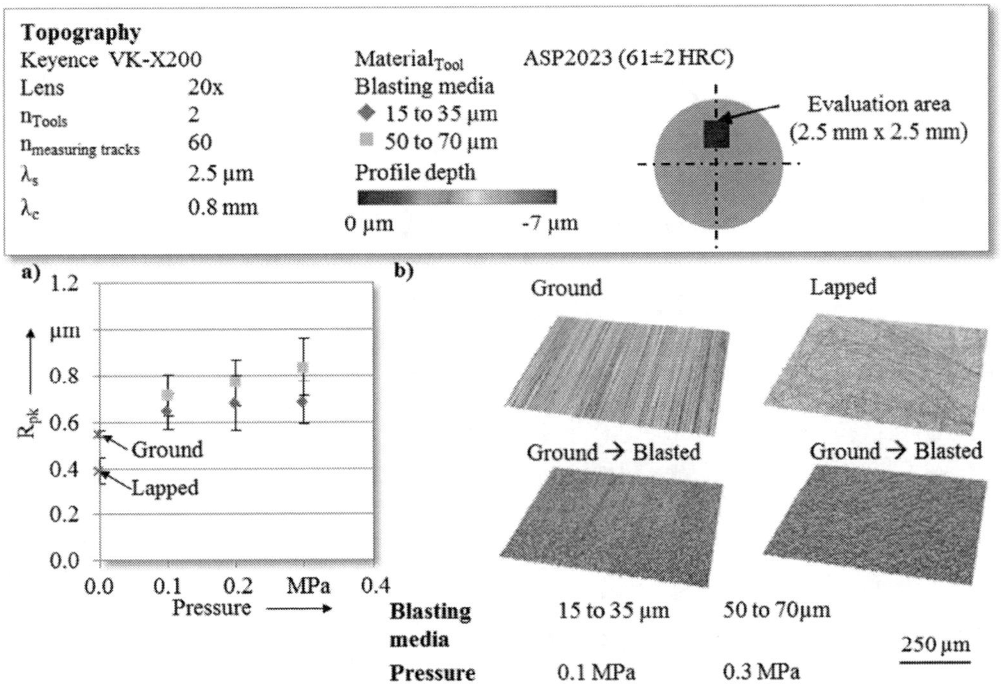

Fig 11. a) Roughness and b) topography images of tool surfaces.

4.1 Analysis of blasted tool surfaces before RCT

The ground tool surfaces have a reduced peak height of 0.55 ± 0.05 µm before blasting. The abrasive blasted tool surfaces are compared to lapped surfaces in figure 11 a) as a reference since lapping is a widespread surface finishing for forming tools.

The lapped surfaces have a reduced peak height of 0.39 ± 0.08 µm. The roughness of all blasted surfaces is higher than the roughness of lapped or ground surfaces. The lowest reduced peak height of 0.65 ± 0.07 µm compared to other blasted surfaces is achieved with the fine blasting media and the lowest pressure of 0.1 MPa. Blasting media with a greater particle size and a blasting pressure of 0.3 MPa causes the roughest surface with a reduced peak height of 0.84 ± 0.12 µm. Both, the particle size and the blasting pressure positively influence the roughness of the surface, since they increase the kinetic energy of the particles ablating the surface. These effects can also be seen in [14]. The high standard variation of the roughness of the blasted surfaces can be explained by the particle sizes varying between 15 and 35 µm as well as between 50 and 70 µm. The results demonstrate that abrasive blasting can be used to increase the roughness of tool surfaces.

As shown in figure 11 b), the blasting process changes the topography of the tools. The topography before the blasting is characterised by orientated grinding marks. Lapped surfaces are smooth and have no preferential direction. The blasting process creates a topography with stochastically dispersed peaks. This topography is created by a combination of platelet formation and micro-machining by the blasting media [15]. After blasting with low pressure and fine particles the surfaces is characterised by small peaks. Grinding marks are partially still visible. The highest blasting pressure in combination with the greatest particle size causes a topography with greater peaks. The grinding marks are no longer apparent, since they were removed by a blasting process with a higher ablation rate due to higher blasting pressure and particle size. A homogeneous surface is created. The same effect of the removal of orientated grinding marks by shot peening of tool surfaces can be seen in [16].

4.2 Functional relation between tool surface topography and friction

The RCT is used to analyse the influence of the tool surface on the friction. As a reference, lapped tools are utilized. Lapped tool surfaces cause a friction factor of 0.1332 ± 0.0071. As shown in figure 12 a) the friction factors of all abrasive blasted surfaces are higher. Hence, abrasive blasting of tool surfaces increases the friction. The highest friction factor of 0.1794 ± 0.0179 occurs for tools which are blasted with 0.3 MPa pressure and a particle size between 50 and 70 µm. As shown in figure 12 a) an increase in blasting pressure and particle size both cause higher friction.

Since both varied blasting parameter positively influence the tool roughness, the functional relation between tool roughness and friction is analysed in figure 12 b). A higher reduced peak height of the tool surfaces causes higher friction. The positive functional relation between the height of the roughness peaks and the friction is caused by the mechanical interlocking of tool- and workpiece-sided roughness peaks.

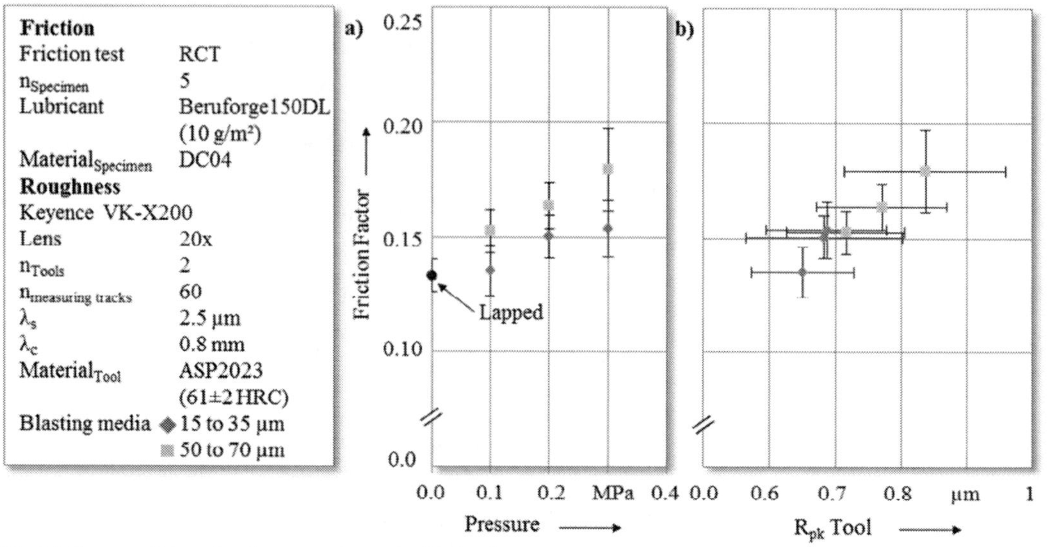

Fig. 12. a) Influence of blasted tool topography on the friction factor and b) relation between tool roughness and friction factor.

4.3 Analysis of blasted tool surfaces after RCT

After the forming of five specimens, the tool surfaces were analysed to estimate the wear resistance of the tool surface modification by abrasive blasting. As illustrated in figure 13 a) the reduced peak height of all tool surfaces is reduced by forming of five RCT specimens. The reduced peak height of the tool surface which is blasted with 0.1 MPa pressure and fine blasting media declines by 20 %. The surfaces with the initially highest roughness have a reduced peak height of 0.71 ± 0.09 µm after the usage. For all surfaces the roughness declines between 7 and 20 %. There is no interrelation identifiable between the initial surfaces roughness and the amount of the decline. By the comparison of the topography images before and after RTC in figures 13 b) it can be seen that the tool surfaces smoothens. The decline of the roughness is caused by the tribological loads during the forming. High contact normal pressure smoothens the roughness peaks. In addition, peaks are sheared off by the sliding of the workpiece on the tool surfaces.

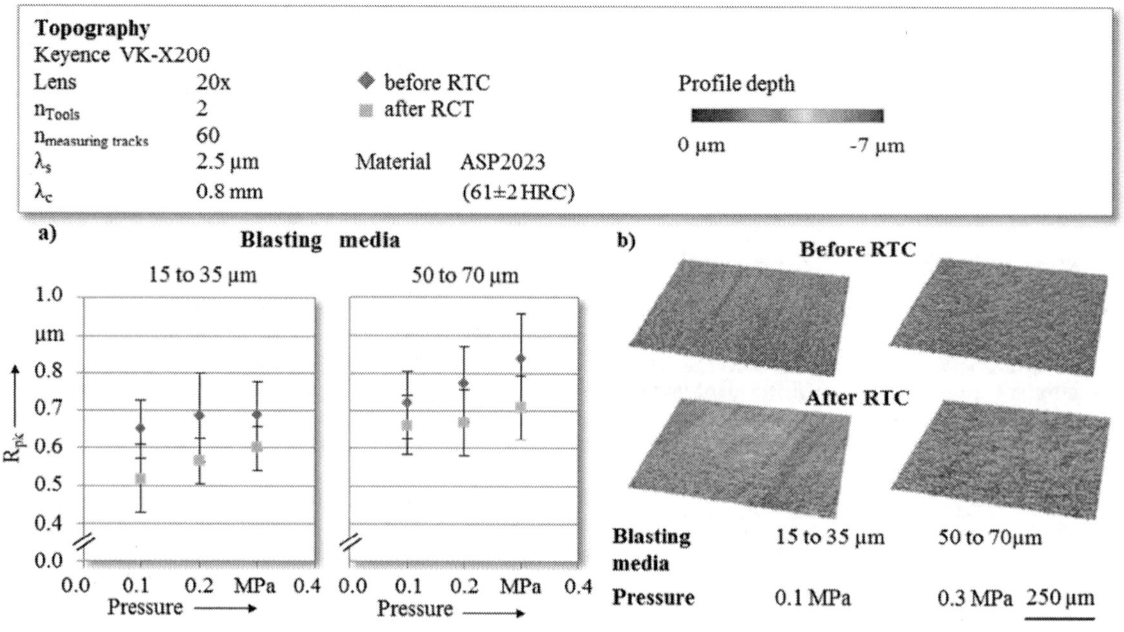

Fig. 13. a) Roughness and b) topography images of tool surfaces before and after RCT.

5 Conclusions and Outlook

The numerical investigations have shown that SBMF processes are characterised by areas with widely varying tribological loads. High tribological loads which are characteristic for bulk forming operations are caused by the forming of functional elements. The material flows in adjacently areas with lower tribological loads. Hence, the die filling and as a consequence the quality of the produced part in terms of geometrical accuracy are reduced.

It is shown that a local increase of the friction in areas with lower tribological loads controls the material flow and increases the quality of the produced parts by inhibiting the material flow away from the functional elements and enhancing the material flow into the die cavities. In addition, a locally adapted tribological system reduces the increase of forming forces due to higher friction compared to a globally adapted tribological system. Hence, surface modifications for a local increase of friction need to be developed.

The potential of abrasive blasting of tool surfaces was analysed in RCT. Abrasive blasting of ground tool surfaces increases the roughness. Higher surfaces roughness due to abrasive blasting increase the friction by mechanical interlocking of tool- and workpiece-sided roughness peaks. Both, the blasting pressure and the particle size of the blasting media positively influence the friction since both increases the roughness and the size of the roughness peaks. Abrasive blasting of tool surfaces is suitable to locally increase the friction in SBMF processes.

On the other hand the roughness of abrasive blasted surfaces is reduced up to 20 % by forming of five specimens. Hence, the wear behaviour has to be analysed and options to reduce wear such as coatings need to be developed. In addition, it has to be shown that the results of the RCT can be used to control the material flow in SBMF processes, before abrasive blasting of tool surfaces can be utilized for batch production.

This work was supported by the German Research Foundation within the scope of the Transregional Collaborative Research Centre on sheet-bulk metal forming (CRC/TR 73, Subproject C1 and A2).

References

1. K. Lange, Handbook of Metal Forming (McGraw-Hill, New York, 1985)
2. M. Merklein, J. Allwood, B.-A. Behrens, A. Brosius, H. Hagenah, K. Kuzman, K. Mori, A. E. Tekkaya, A. Weckenmann, CIRP ANN.-Manuf. Techn. **61**, pp. 725-745 (2012)
3. M. Löffler, R. Schulte, D. Freiburg, D. Biermann, D. Stangier, W. Tillmann, M. Merklein, Journal of Production Engineering (to be published)
4. L. Kirkhorn, V. Bushlya, M. Andersson, J.E. Stahl, Wear **302**, pp. 1268-1278 (2013)
5. D. Gröbel, J. Koch, H.-U. Vierzigmann, U. Engel, M. Merklein, Procedia Engineering **81**, pp. 401-406 (2014)
6. F. Klocke, *Manufacturing processes: Forming* (Springer-Verlag, Berlin, 2013)
7. U. Vierzigmann, "Beitrag zur Untersuchung der tribologischen Bedingungen in der Blechmassivumformung - Bereitstellung von tribologischen Modellversuchen und Realisierung von Tailored Surfaces." Ph.D. thesis, Friedrich-Alexander-University Erlangen-Nürnberg (2016)
8. H. Bodschwinna, "Oberflächenmesstechnik zur Beurteilung und Optimierung technischer Funktionsflächen." Habilitation thesis, University Hannover (2000)
9. N. Bay, A. Azushima, P. Groche, I. Ishibashi, M. Merklein, M. Morishita, T. Nakamura, S. Schmid, M. Yoshida, CIRP ANN.-Manuf. Tech. **59**, pp. 760-780 (2010)
10. U. Engel, "Beanspruchung und Beanspruchbarkeit von Werkzeugen der Massivumformung." Habilitation thesis, Friedrich-Alexander-University Erlangen-Nürnberg (1996)
11. Y. Murakami, Int. J. Fatigue **41**, pp. 2-10 (2012)
12. K. Chander, Mater. Des. **30**, pp. 2895-2902 (2009)
13. J. Wigren, Surf. Coat. Technol. **34**, pp. 101-108 (1988)
14. M. Mellali, A. Grimaud, A. C. Leger, P. Fauchais, J. Lu, J. Therm. Spray Technol. **6**, pp. 217-227 (1997)
15. A. W. Momber, Y. C. Wong, R. Ij, E. Budidharma, Tribol. Int. **34**, pp. 271-281 (2002)
16. M. Babic, S. Mitrovic, D. Adamovic, D. Dzunic, F. Zivic, "The effect of shot peening on tribological behavior of alloyed steel," in *11th International Conference on Shot Peening,* edited by J. Champaigne, pp. 12-15 (2011)

Observation of the a-C:H run-in behaviour for dry forming applications of aluminium

Tim Abraham[1,*], *Günter* Bräuer[1], *Felix* Kretz[2], and *Peter* Groche[2]

[1]Fraunhofer-Institute for Surface Engineering and Thin Films, 38108 Braunschweig, Germany
[2]Institute for Production Engineering and Forming Machines, Darmstadt University of Technology, 64287 Darmstadt, Germany

Abstract. Amorphous hydrogenated carbon coatings (a-C:H) are well known for their exceptional tribological properties and are established as tool coatings for numerous forming applications. However, utilized in dry forming processes of aluminium a premature failure of an a-C:H coated tool often occurs due to strong adhesive wear. In this paper the run-in behaviour of a-C:H is investigated and as a possible reason for the premature tool failure evaluated. Therefore, oscillating ball-on-disc tribometer tests and strip drawing tests, for a more realistic emulation of real forming processes, will be conducted. According to these tests, the run-in period of a-C:H coatings is characterized by a high friction value and adhesion tendency and thus is decisive for the tool performance. Based on a subsequent analysis of the coating wear, the predominating wear mechanisms during the run-in period are discussed. The intrinsic nanomater-scale a-C:H roughness is identified as a crucial factor determining the tribological properties of the run-in behaviour. By reducing the coating roughness prior to the forming process, the adhesion tendency and friction value can be reduced significantly. The results demonstrate the tribological performance of pre-treated a-C:H coatings for dry sheet metal forming of aluminium EN AW-5083.

Keyword: Metal forming, Coating, Tribology

1 Introduction

Due to its weight advantage and excellent energy absorption capacity, aluminium is an excellent material for a wide range of applications, for example in the automotive or aerospace industry. However, the design of forming technologies for aluminium alloys presents some challenges. Especially in dry forming processes, the strong adhesion tendency of aluminium alloys leads to high adhesive wear on forming tools [1] and affects the surface quality of components, the process stability and the targeted tolerances. As a consequence of the adhesive wear, the production of aluminium alloy components is currently limited to lubricated forming processes. In order to develop effective strategies to reduce adhesive wear and enable dry sheet metal forming processes, the relevant wear mechanisms and influencing parameters must be considered.

One solution to optimize the tribological properties of forming tools and to enable dry sheet metal forming of aluminium alloys is the deposition of hard thin film coatings. Amorphous hydrogenated carbon coatings (a-C:H) are well known for their exceptional tribological properties [2]. Combining a low friction value and a high wear resistance, a-C:H coatings are already in use for several forming applications. Numerous investigations reported a low adhesion tendency in contact with aluminium [3, 4, 5]. The low adhesion tendency of a-C:H coatings was attributed to the intrinsic capability of self-lubrication and high inertness of the passivated surface [6,7]. Nevertheless, forming tools coated with state of the art a-C:H coatings still fail due to a rapid formation of aluminium adhesions comparable with uncoated forming tools.

A commonly known characteristic of hard thin film coatings is the run-in behaviour. In general, this behaviour describes a short period at the beginning of a sliding contact with differing tribological properties. Luo et. al. [8] and Goldsmith et. al. [9] reported an increased friction value and wear during the run-in period of a-C:H sliding against sapphire ($\mu \sim 0.20$) and aluminium oxide ($\mu \sim 0.63$) without lubrication. This differing tribological behaviour could explain the rapid formation of aluminium adhesions in dry forming processes of aluminium and may be the reason for a premature tool failure.

In this paper an a-C:H coating system is systematically investigated to define the tribological properties and wear mechanisms during the run-in behavior. Furthermore, the impact of these results on dry forming processes of aluminium is evaluated in strip drawing tests, which enable a more realistic emulation of real forming processes.

2 Experimental setup

2.1. Coating preparation and characterisation

The a-C:H coating system was deposited by a plasma assisted chemical vapour deposition process (PACVD) with acetylene (C2H2) as the precursors gas combined with argon as process gas. To enhance the coating adhesion, the substrates were sputtered clean for 30 min at the beginning of the coating process and an interlayer

* Corresponding author: tim.abraham@ist.fraunhofer.de

based on titanium with a thickness of 0.2 µm was applied by magnetron sputtering. The thickness of the complete a-C:H coating system measured 2.2 µm. A more detailed description of the deposition process and the coating characteristics is stated in [10].

Additional to the substrates for the tribological tests, polished flat samples made of bearing steel (1.3505) were coated to allow a subsequent analysis of the coating properties. Coating hardness HUpl was determined with a commercial instrument (Fischerscope H 100) recording load versus depth curves up to 30 mN measuring 34.1 GPa. The abrasive wear rate of 1.0 $10^{-15}m^3/Nm$ was determined with the ball cratering test [11] operating with an alumina (Al2O3) suspension (mean alumina grain size 1 µm).

2.2 Tribological tests

The run-in behaviour of the a-C:H coating was investigated with an oscillating ball-on-disc tribometer test using a Universal Mechanical Tester (UMT-3, Bruker Corporation). Therefore, the a-C:H coating was deposited on steel balls and tested in a dry sliding contact against aluminium sheets. The coated steel balls measured 10 mm in diameter and were made of bearing steel (1.3505). The aluminium sheets were made of EN AW-5083 in a mill finished H111 condition with a thickness of 1.5 mm. Roughness data of the test substrates were collected using a profilometer (Taylor-Hubson) and are summarized in Table 1. The ball-on-disc tests were conducted three times to improve the statistical quality of the friction and wear values. The amplitude of the oscillation measured 20 mm.

Table 1. Roughness of the test substrates.

	Ra [µm]	Rz [µm]
Coated steel ball	0.023	0.240
Strip drawing tool	0.028	0.180
Aluminium sheet	0.308	1.612

The a-C:H coating system was tested afterwards by a strip drawing test to verify the observations of the ball-on-disc tribometer tests under an application nearer test condition. The strip drawing test enables emulating the typical tribological load spectrum of deep drawing and stretch forming processes [12]. In this test, a pressure is applied to a strip of sheet material by an upper and lower tool while the strip is drawn with the defined sliding speed and sliding distance summarized in Table 2. Figure 1 shows the test principle schematically. In order to apply a characteristic load spectrum for sheet metal forming the typical linear load and corresponding high contact pressure at the die radius of a deep drawing tool is reproduced through the cylinder-plane geometry in the strip drawing. The upper tool has a cylindrical surface with a radius r = 258 mm while the lower tool is flat. The basic measurements of both tools are equal to 40 mm x 40 mm. Tool material (1.2379) and tested coatings are identical in both tribometer tests and it is equally tested against EN AW 5083 aluminium.

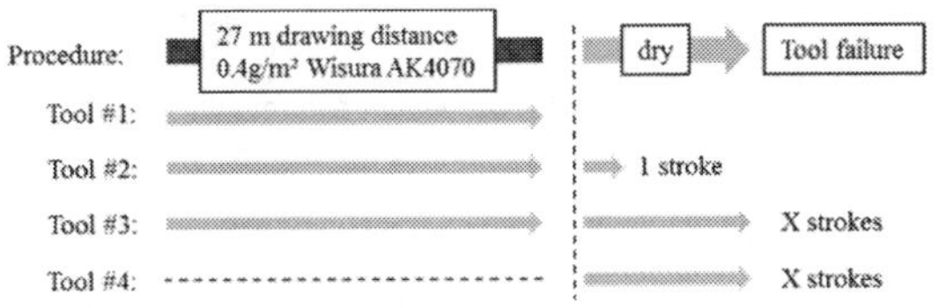

Fig. 1. Test procedure for the strip drawing tests.

The test procedure was chosen according to Fig. 1. After drawing 27 m under lubricated conditions, tool #1 was cleaned from lubricant and underwent further investigations regarding the conditions of the coating. Tool #2 and Tool #3 were also tested with a lubricated drawing distance of 27 m and cleaned. While Tool #2 was tested in one stroke under dry conditions, Tool #3 was tested until severe wear or strip failure occurred. Tool #4 were directly tested without lubrication and similar to Tool #3 continued until severe wear were observable. All tools were coated with the standard a-C:H coating. Every strip-drawing test setup were performed once.

The test parameters were deduced from an industrial forming process of EN AW-5083 aluminium and are summarized in Table 2. The initial contact pressure in the ball-on-disc contact is higher than in the strip drawing test but decreases drastically due to wear of the coating and aluminium sheet. All test substrates were cleaned with ethanol and acetone to ensure a technical pure contact during the tribometer tests.

Table 2. Test parameters of tribological tests.

	Ball-on-disc Tribometer	Strip drawing test
Load [N]	4	9000
Initial contact stress [MPa]	480	100
Average velocity [mms^{-1}]	50	50
Humidity [%]	45 ± 5	-
Temperature [°C]	23 ± 1	-

3 Results and discussion

3.1 Run-in behaviour in tribometer test

Oscillating tribometer tests revealed high friction values during the run-in period of the a-C:H coating in an unlubricated sliding contact against aluminium, see Fig. 2. Regarding the friction value, the run-in period can be divided in two phases: a high friction phase (up to 0.9)

 https://doi.org/10.1051/matecconf/201819014001

within the first 1.5 m and a medium friction phase (~ 0.4) until the end of the run-in period (6 to 7.5 m). After the run-in period, the friction value decreased to a steady state level of 0.1 which is a well-known friction value of a-C:H coatings in a dry sliding contact against aluminium [5]. The high friction value during the first period equals the high friction values of uncoated forming tools sliding against aluminium [3]. In case of uncoated forming tools, the high friction values are correlated with the formation of aluminium adhesions on the tool surface, which is the main reason for an early tool failure. However, no aluminium adhesions were detectable in the main contact zone on the a-C:H coated steel ball. Only around the main contact zone did aluminium adhere to the a-C:H coating.

For an investigation of the wear progression during the run-in period further oscillating tribometer tests with incrementally increased sliding distances were conducted, see Fig. 3. Similar to the high friction values already indicated, aluminium adhesions were formed on the a-C:H coated steel ball within a sliding distance of 0.5 m - covering the complete contact zone, see Fig. 4. During the medium friction phase (> 1.5 m) the amount of aluminium adhesions on the coated steel ball decreased and no further wear of the aluminium sheet was observable with an increasing sliding distance, see Fig. 5. Instead, aluminium accumulated randomly along the aluminium wear track, indicating a retransfer of the aluminium adhesions from the coated steel ball to the wear track. Furthermore, a black transfer layer formed on top of the aluminium accumulations. Reaching the steady state friction level (> 6.5 m), nearly no aluminium adhesions were detectable in the main contact zone comparable to the wear characteristic after 15 m in Fig. 2.

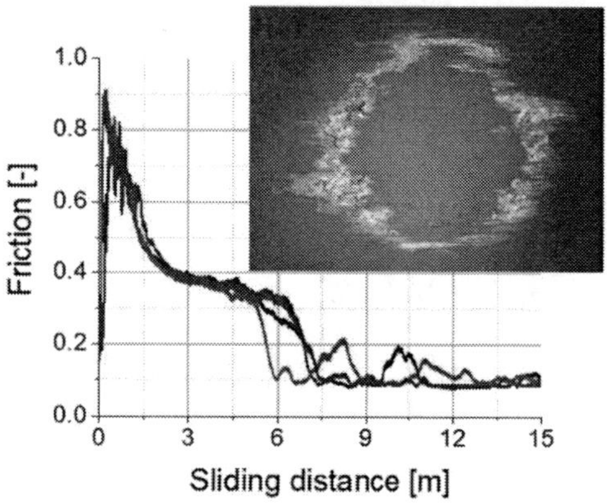

Fig. 2. Oscillating ball-on-disc tribometer tests: Friction development and wear of the a-C:H coated steel ball.

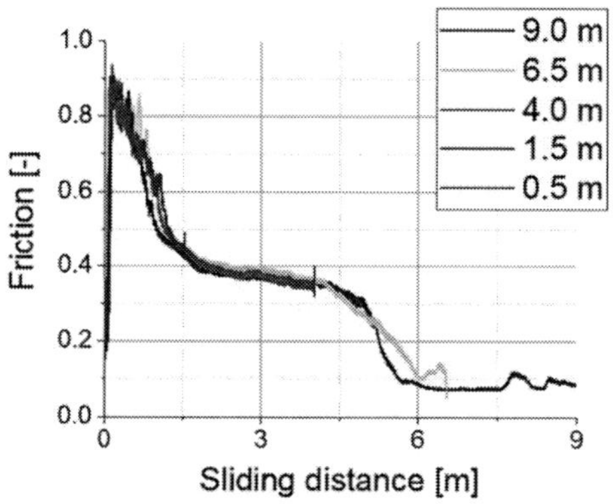

Fig. 3. Friction development of a-C:H during run-in period - incrementally increased sliding distances.

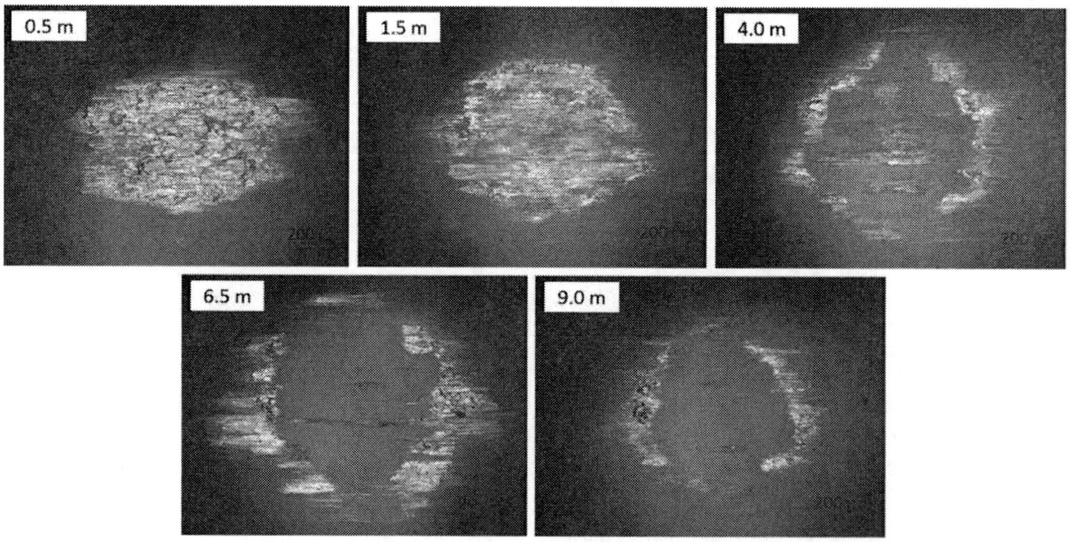

Fig. 4. Microscope pictures of the a-C:H coated steel ball after tribometer tests with incrementally increased sliding distances.

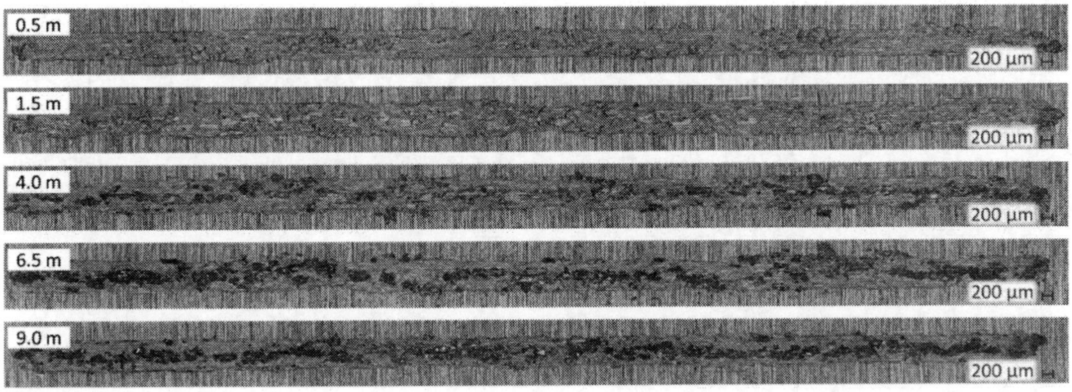

Fig. 5. Microscope pictures of the wear tracks on the aluminium sheets after tribometer tests with incrementally increased sliding distances.

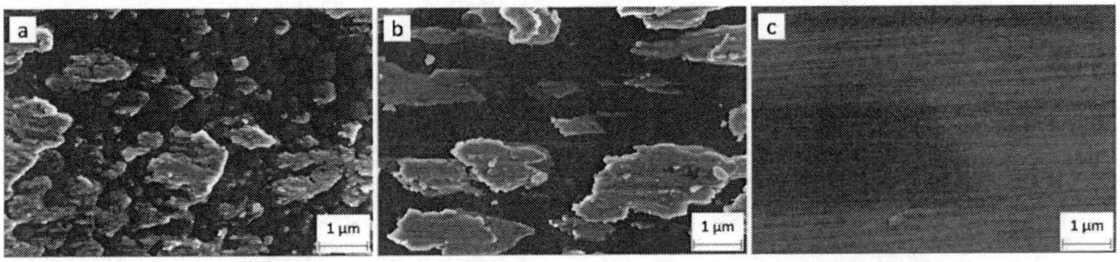

Fig. 6. Scanning electron microscopy (SEM) of the aluminium adhesions formed on the a-C:H coated steel ball after a sliding distance of 0.5 m (a), 4 m (b) and 9 m (c).

At the beginning of the run-in period, small aluminium adhesions adhered particularly on microscopic dome-shaped asperities, see Fig. 6a, which are characteristic for the intrinsic surface structure of the tested a-C:H coating system [13]. These small adhesions may be the initial state for the large laminar structured aluminium adhesions. Meanwhile as the amount of adhesions was reduced, a smoothening of the underlying a-C:H surface structure was observable with an increasing sliding distance, see Fig. 6b and Fig. 6c. At the end of the run-in period the main contact area was completely covered with grooves indicating an abrasion of the a-C:H coating. According to this observations, the a-C:H surface structure may promote the formation of aluminium adhesions and thus, may be an important factor for the run-in behaviour of the a-C:H coating in a dry sliding contact against aluminium.

For a more detailed investigation of the observed smoothening progression, the nanometer-scale roughness of all tested steel balls were determined. Therefore, the aluminium adhesions were removed by etching with a sodium hydroxide solution at room temperature. The surface roughness was measured afterwards with an atomic force microscope (AFM) (Veeco Dimension 3100) in tapping mode with a scan size of 10 x 10 µm. Focusing on the nanometer-scale a-C:H roughness, the collected data were subsequently processed with a Gaussian filter to subtract the roughness of the steel ball and macroscopic grooves in the a-C:H coating. The measurements revealed a significant reduction of the a-C:H coating roughness during the run-in period, see Fig. 7. The a-C:H asperities were mainly abraded during the medium friction phase dropping from an initially mean height of 15.3 nm to 1.9 nm after a sliding distance of 6.5 m.

According to Harris et. al. [14] the intrinsic nanometer-scale roughness of a-C:H coatings is an important tribological property influencing the abrasion of the counter material. In particular, the a-C:H asperities penetrate and consequently abrade the surface of the counter material. In a dry sliding contact against aluminium the a-C:H surface contacts initially the passivated aluminium surface. The passivated surface consists of an inert aluminium oxide layer (Al2O3) with a thickness ranging between 2 to 12 nm [15]. In comparison the reduced peak height of an untested a-C:H coating measured 21.1 nm. Thus, the asperities may penetrate the aluminium oxide layer and contact the pure aluminium matrix. The aluminium matrix is highly reactive and as a consequence may adhere to the a-C:H asperities. However, the a-C:H surface is highly inert due to a passivation by hydrogen and hydroxide [7]. Furthermore, the soft aluminium matrix may be pressed into the surface valleys of the a-C:H coating. Providing a mechanical attachment, the surface valleys may promote the formation of aluminium adhesions. This thesis would explain the observed small aluminium adhesions particularly formed on the a-C:H asperities (Fig. 6a).

As already stated, the aluminium accumulations along the wear track (Fig. 5) indicate a retransfer of the aluminium adhesions from the coated steel ball to the aluminium sheet. Analysis of these accumulations revealed a higher hardness HUpl of 3.3 GPa compared to the untested aluminium surface with a HUpl of 1.2 GPa. During the transfer processes, these accumulations were

https://doi.org/10.1051/matecconf/201819014001

stressed and frequently exposed to air leading to work hardening and a stronger oxidation. With an increasing amount of aluminium accumulations, the contacting area between the coated steel ball and the aluminium sheet were gradually reduced. In regard to the postulated theory about the promoted adhesion formation by the a-C:H asperities, the adhesion tendency decreased in present of the stronger oxidised aluminium accumulations. Instead, the accumulations abraded the a-C:H coating and the carbonic wear particles formed the observed transfer layer. These wear mechanisms would explain the smoothening effect and lowered friction value during the midsection of the run-in period.

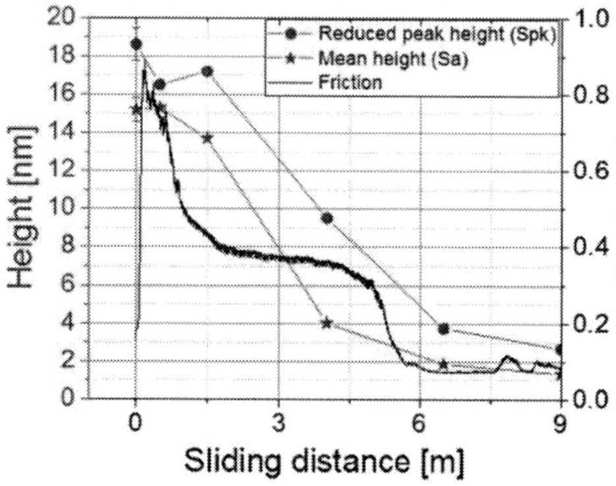

Fig. 7. Smoothening progression of the a-C:H coated steel ball during the run-in period.

Raman spectroscopies were conducted to investigate the atomic structure of the carbonic transfer layer on the aluminium wear track. The measurements were conducted with a laser wavelength of 532 nm and a laser power of 0.9 mW revealing a ID/IG-ratio shift from 0.24 for the untested a-C:H coating to 0.81 for the transfer layer [16]. The ratio shift indicates a higher content of the graphitic sp^2-configuration in the transfer layer. According to Liu et al. [17], the graphitization leads to a self-lubricating effect lowering the friction value and wear. The delayed graphitization of the a-C:H coating and subsequent formation of a carbonic transfer layer (Fig. 5) are further possible explanations for the high adhesion tendency and high friction values during the run-in period.

In case of real forming processes there is no stationary tribological contact between the coated tool and the aluminium sheet. Thus, the beneficial tribological effect of the carbonic transfer layer should be lowered. To evaluate the impact of the smoothening effect and the graphitization for dry forming processes, further ball-on-disc tribometer tests were conducted with an additional transversal offset added to the oscillating motion to ensure a steady contact with an untested aluminium surface. The elongation of the aluminium sheet in real forming processes enhances significantly the formation of aluminium adhesions [18]. However, there is no elongation of the aluminium sheet in ball-on-disc tribometer tests. To enhance the adhesion tendency, the aluminium sheets were preconditioned by a grinding pad and afterwards cleaned with ethanol and acetone. Consequently, the roughness of the aluminium surface increased (R_a = 0.463 µm and R_z = 4.717 µm) and the thickness of the native aluminium oxide layer was minimal.

Figure 8a shows the friction development of the a-C:H coating tested under the modified tribometer test conditions. The friction value and adhesion tendency remained for the complete sliding length of 3 m on a high level similar to the oscillating test with a sliding length of 0.5 m (Fig. 4). Neither a smoothening of the intrinsic a-C:H roughness nor the formation of a carbonic transfer layer were observable. Aluminium adhesions were steadily formed inhibiting a direct sliding contact at the a-C:H surface. These mechanism may have mitigated a run-in of the a-C:H coating. In an additional test, the a-C:H coating was pre-treated by a prior oscillating test with a sliding length of 9 m, see fig 8b. The adhesive wear was similar to the a-C:H coating tested for a sliding distance of 9 m in the stationary test setup (Fig. 5) and no carbonic transfer layer was observable. In addition, the friction coefficient of the pre-treated a-C:H coating continued nearly on the same low level like at the end of the prior oscillating test. Only one sample showed a higher friction value at the beginning which can be explained by a high friction value at the end of the prior oscillating test. As it can be seen in Fig. 2, the friction value sometimes increases for a short distance after the run-in period. This may be caused by a short-term adhesion formation and continued in the following modified tribometer test.

Although the aluminium sheet was changed and the test conditions were non-stationary, the pre-treated a-C:H coating still possessed the optimized tribological properties. According to these results, the run-in behaviour of a-C:H is a unique and temporary effect at the beginning of a dry sliding contact against aluminium. Furthermore, the run-in behaviour and the adhesion tendency of a-C:H are not affected by the formation of a carbonic transfer layer. Still, the graphitization of the a-C:H boundary surface may influence the tribological properties given that the graphitized state would remain after changing the tribometer test setup. In regard of the fundamental mechanism of the self-lubricating effect, the coating wear would be enhanced due to the graphitization. However, the coating wear did not increase during the modified tribometer tests. Thus, the influence of the graphitization on the run-in behaviour and adhesion tendency of a-C:H coatings is low leaving the smoothening of the intrinsic nanometer-scale a-C:H coating roughness as the most important factor.

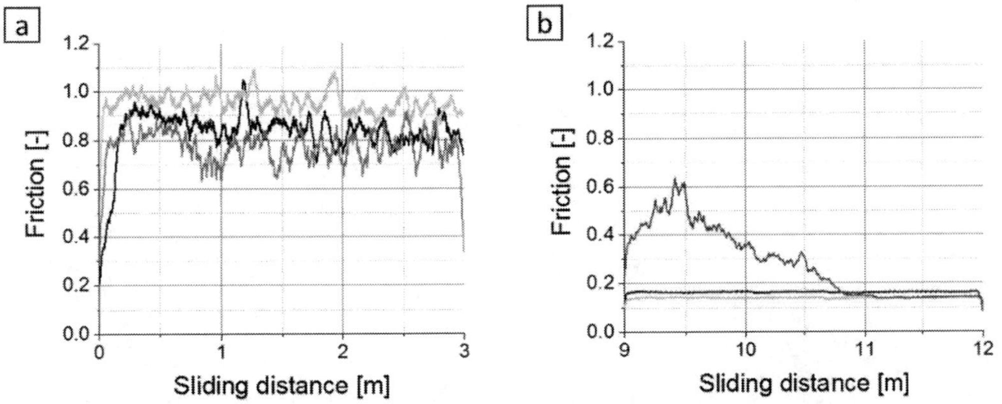

Fig. 8. Friction development of a-C:H (a) and pre-treated a-C:H (b) coated steel balls in modified tribometer tests with forming process nearer test conditions.

As already mentioned, the rapid formation of aluminium adhesions often lead to a premature tool failure in real forming processes. Even the deposition of a state of the art a-C:H coating system shows no significant improvements of the adhesion tendency. According to the tribometer tests, the rapid formation of aluminium adhesions can be explained by a missing smoothening of the intrinsic a-C:H roughness. Consequently, the forming tool fails during the high friction phase of the run-in period when a strong adhesion tendency were observable. To evaluate the impact of the tribological properties during the run-in period on dry forming processes of aluminium further tests were conducted with a strip drawing test, while emulating sheet metal forming conditions more realistically.

3.2 Run-in behaviour in strip drawing test

In order to assure lubricant free test conditions for the dry forming processes in Fig. 1, X-ray photoelectron spectroscopies (XPS) of the tool surfaces were performed similar to [19]. Therefore, the tested Tool #1 were analysed at three different points along the wear track (A, B and C) and compared with an untested a-C:H coating. According to these tests, no significant residues of lubricant additives were detectable after cleaning.

The pre-treated Tools #1, #2 and #3 (before stroke 14) showed significantly less wear than the corresponding Tool #4 without a prior run-in. This can be observed in the microscopic images taken of the worn tool surface shown in Fig. 9. The darker colour on the Tool #1 marks the contacting area with the aluminium sheets. Nearly no aluminium adhesions were formed and some isolated deep grooves in sliding direction were observable. Partly the a-C:H coating showed cracks and signs of delamination along the groove edges. These effects may be caused by a contamination of the aluminium surface with hard particles leading to a coating overload. Tool #2 additionally was drawn one stroke under dry conditions. In comparison to Tool #1, Tool #2 slightly more aluminium adhesions adhered to the surface but the amount of adhesions was significantly lower compared with an untreated a-C:H coated tool. Here, the dry sliding contact against aluminium leads to a severe adhesive wear already after one stroke (100 mm sliding length), see [3]. After 75 strokes, the wear appearance did not differ, presented by Tool #4. In order to find the maximum of possible sliding length without severe adhesive wear the dry sliding length with pre-treated a-C:H coating was extended. After 89 strokes under dry conditions the strip drawing test was interrupted due to a severe adhesive wear which already started to form after 14 strokes. In comparison to the untreated tools, small areas between the adhesions still showed no adhesions. Furthermore, no transfer layer was observable in all tests. Thus, the transfer layer did not affect the tribological performance.

For an investigation of the wear mechanisms during the dry strip drawing tests, the friction values were calculated based on the normal load and the pulling force. Figure 10 shows the average friction development for Tool #3 during the first 24 strokes under dry conditions. The dashed line represents the friction value of the untreated Tool #4 ($\mu \sim 0.37$). The dash-dotted line represents friction value of Tool #1 tested under lubrication. The squares show the development of the average maximum peak roughness RZ of the sheet metal for each stroke.

For the first 14 strokes under dry conditions, the friction value laid steadily around 0.1 - slightly above the lubricated case. Beside of small grooves on the sheet metal caused by minor adhesions on the tool surface, there was almost no wear observable, especially compared to the results of Tool #4. After the first groove in the right side had developed, this groove dominated the R_Z value of the sheet metal. Thus, the R_z value did not change much during stroke 4 to 16 corresponding with the observed wear and friction values. After stroke 14 the friction value and R_z values raised indicating a formation of aluminium adhesions on the tool surface. After 21 strokes, the friction value and the wear of the aluminium sheet was similar to the untreated Tool #4. In respect to real forming processes, these tribological properties may lead to a tool failure.

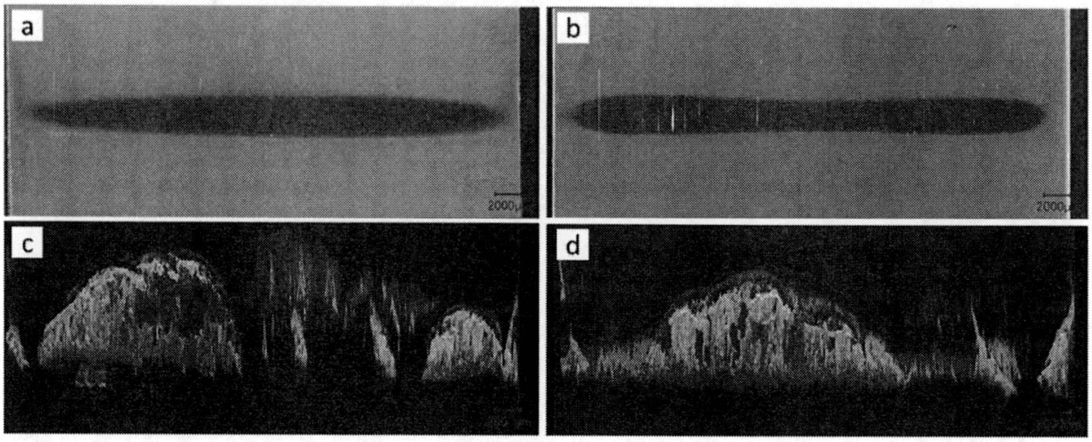

Fig. 9. Microscopic images of the worn tool surfaces. a: Tool #1, b: Tool #2, c: Tool #3, d: Tool #4.

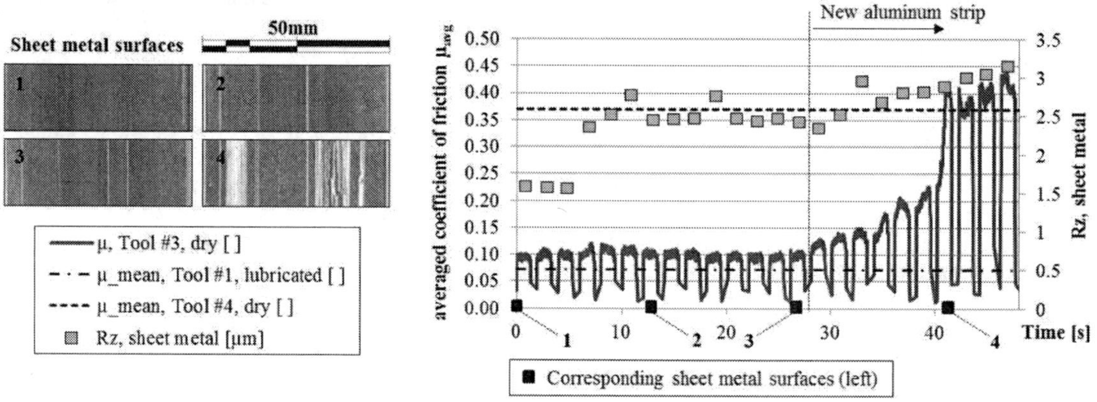

Fig. 10. Development of the coefficient of friction for Tool #3 including the lubricated reference case of Tool #1 and #5.

Analog to the oscillating tribometer tests, the nanometer-scale roughness of the tested strip drawing tools were determined by AFM measurements, see Fig. 11. All tools showed a smoothened surface roughness. The mean height of Tool #1, #2 and #3 was equal to the surface roughness of the tested steel balls at the beginning of the medium friction phase (Fig. 7) and most of the intrinsic nanometer-scale a-C:H surface structure was intact, see Fig. 12. The eased smoothening effect may be caused by the lower contact pressure of the strip drawing test and the lubricant. Furthermore, numerous line-shaped areas with unworn a-C:H surfaces were observable in the contact zone. Each line started at coating defects, which were caused by contaminations during the deposition process. These defects possibly deformed the aluminium sheet and lowered the contact stress in the following contact zone. Thus, the smoothening process was partially inhibited. The surface roughness of Tool #4 was also reduced. Like in the modified tribometer tests, the aluminium adhesions may have mitigated the abrasive wear of the a-C:H coating.

The initial tribological properties of the preconditioned Tool #2 and #3 differed from the untreated Tool #4 and [3]. The reduced initial surface roughness, represented by Tool #1, lowered the adhesion tendency and friction value significantly for a sliding distance of 1.4 m matching the performance of the pre-treated a-C:H coating in the modified tribometer test. The abrupt aggravation of the tribological performance of Tool #3 can possibly be explained by the setup of the strip drawing test. After 14 strokes, the aluminium test strip had to be changed for a new strip (Fig.10). Therefore, the press was opened and the next strip was fed to the gripping device. By changing the strip, the line-shaped unworn a-C:H surface areas on Tool #3 contacted the aluminium sheet. These unworn a-C:H areas still possessed the high adhesion tendency and may have promoted the adhesion formation. Regarding Tool #2, this effect may already have caused the formation of small adhesions during the first aluminium strip but the amount of adhesions were maybe too little to enhance the adhesion formation. Continuing with the second strip, a certain amount of adhesions was reached promoting the adhesion formation progressively.

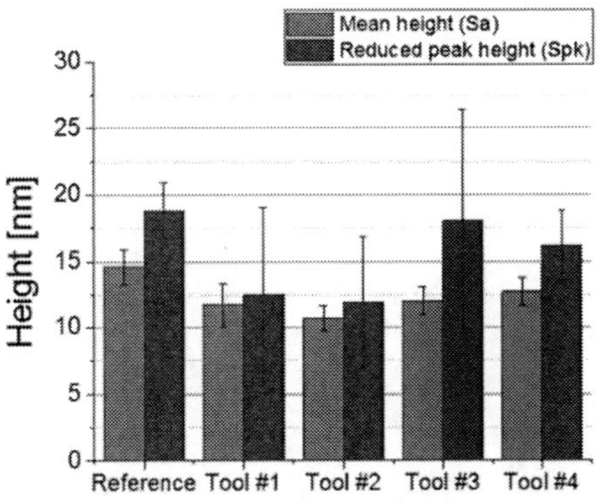

Fig. 11. Nanometer-scale roughness of the tested strip drawing tools.

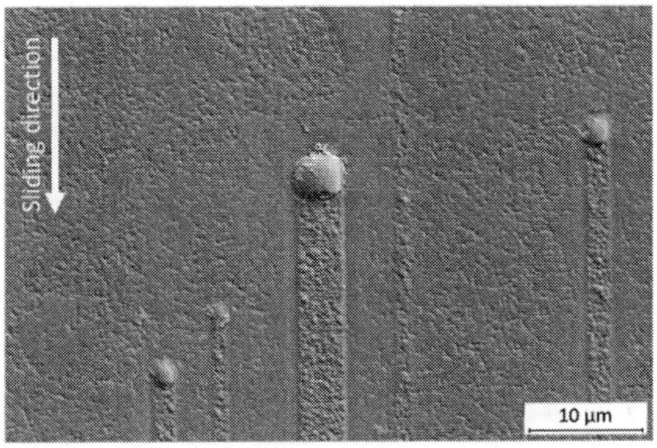

Fig. 12. SEM image of Tool #1 – coating defects inhibited the smoothening process during lubricated strip drawing tests (sliding direction: top to down).

4 Conclusion

In this paper the run-in behaviour of a-C:H used as tool coating for dry forming processes of aluminium sheets was investigated and as a possible reason for a premature tool failure evaluated. Oscillating ball-on-disc tribometer tests revealed a high friction value and strong adhesion tendency during the run-in period. The aluminium adhesions adhered particularly on the dome-shaped asperities of the intrinsic nanometer-scale a-C:H surface roughness. During the run-in period, the intrinsic coating roughness was completely abraded leading to a significantly smoother surface and a graphitic transfer layer connected with a very low adhesion tendency and a low friction value after the run-in period. In subsequent non-stationary tribometer tests, these properties remained on the same level and no transfer layer was observable. Thus, the run-in period of a-C:H is a unique and temporary effect at the beginning of a dry sliding contact against aluminium. Furthermore, the intrinsic nanometer-scale a-C:H surface roughness may be a crucial factor determining the adhesion tendency and friction value during the run-in period. These interdependencies can be explained by a grooving of the aluminium oxide layer caused by the coating asperities. The subsequent contact with pure aluminium may promote a chemical attachment whereas the surface valleys may promote a mechanical attachment for the deformed aluminium. Although no transfer layer was observable, a tribological induced graphitization of the a-C:H boundary surface connected with self-lubricating properties cannot be excluded as an additional factor. Though, no wear of the coating was observable after the run-in period, the influence of a tribological induced graphitization may be little.

These wear mechanisms were also verified by strip drawing tests, which emulate sheet metal forming conditions more realistically. Based on a tool pre-treatment, the intrinsic nanometer-scale a-C:H roughness were smoothened leading to a significantly reduced friction value and adhesion tendency. However, the pre-treatment was insufficient to enable dry forming of aluminium. Due to coating defects, small areas of the intrinsic a-C:H roughness were still intact promoting a delayed adhesion formation.

According to these results, the intrinsic nanometer-scale a-C:H surface roughness promotes the rapid adhesion formation in dry sliding contacts against aluminium and thus determines the tribological performance of a-C:H tool coatings in dry sheet metal forming of aluminium. The rapid formation of aluminium adhesion mitigates the smoothening due to abrasion. In consequence, the premature tool failure occurs before falling under a certain critical a-C:H coating roughness. Thus, dry metal sheet forming of aluminium may be enabled by the usage of very smooth a-C:H tool coatings.

In order to validate this hypothesis, further tests with smoother a-C:H coatings are necessary. By extending the tool pre-treatment method, the obtainable a-C:H surface roughness can be improved but is limited to the amount of surface defects. A more appropriate approach is a direct deposition of smooth a-C:H coatings based on an optimization of the deposition parameters. Alternatively, the intrinsic a-C:H roughness can possibly be lowered by plasma etching in a post-treatment process.

Acknowledgements

The authors would like to thank the German Research Foundation (DFG) for funding of the project "Functionalising of a-C:H tool coatings and homogenization of the aluminium passive layer for the dry forming of aluminum" received within the priority program SPP 1676 "Dry metal forming - sustainable production through dry processing in metal forming".

References

1. P. Groche, F. Resch, Mat.-wiss. u. Werkstofftech **46**, 813-828 (2015)
2. K. Bewilogua, D. Hofmann, Surface & Coatings Technology **242**, 214-225 (2014)
3. T. Abraham, G. Bräuer, F. Kretz, P. Groche, Dry Met. Forming OAJ FMT **2**, 18-24 (2016)
4. A. Wank, G. Reisel, B. Wielage, Surface & Coatings Technology **201**, 822-827 (2006)

5. H. Ronkainen, S. Varjus, K. Holmberg, Wear **222**, 120-128 (1998)
6. H. Ronkainen, J. Likonen, J. Koskinen, S. Varjus, Surface & Coatings Technolgy **79**, 87-94 (1996)
7. A. Erdemir, Surface & Coatings Technology **146-147**, 292-297 (2001)
8. S.Y. Luo, J.K. Kuo, T.J. Tsai, C.W. Dai, Wear **249**, 800-807 (2001)
9. J. Goldsmith, E. Sutter, J.J. Moore, B. Mishra, M. Crowder, Surface & Coatings Technology **200**, 2386-2390 (2005)
10. K. Taube, Surface & Coatings Technology **98**, 976-984 (1998)
11. T. Michler, C. Siebert, Surface & Coatings Technology **163-164**, 546-551 (2003)
12. P. Groche, J. Filzek, G. Nitzsche, Annals of the German Academic Society for Production Engineering XI/1 **11**, 55-60 (2004)
13. M. Weber, K. Bewilogua, H. Thomsen, R. Wittorf, Surface & Coatings Technology **201**, 1576-1582 (2006)
14. S.J. Harris, A.M. Weiner, C.H. Olk, M. Grischke, Wear **219**, 94-104 (1998)
15. K. Wefers, Aluminium **57**, 722 (1981)
16. A.C. Ferrari, J. Robertson, Physical Review B **61**, 20 (2000)
17. Y. Liu, A. Erdemir, E.I. Meletis, Surface & Coatings Technology **86-87**, 564-568 (1996)
18. N. Bay, AWS Annual Meeting **63**, 137-142 (1982)
19. A. Almohallami, M. Arghavani, F. Böhmermann, H. Freiße, M. Herrmann, S.A. Mousavi, S. Schöler, P. Scholz, J. Tenner, M. Teller, G. Umlauf, D. Wulff, D. Yilkiran, H.J. Maier, Dry Met. Forming OAJ FMT **3**, 90-94 (2017)

Adaption of tribological behavior of a-C:H coatings for application in dry deep drawing

Tom Häfner[1, 2], *Benedict* Rothammer[3*], *Jennifer* Tenner[4], *Kim* Krachenfels[4], *Marion* Merklein[4], *Stephan* Tremmel[3], *Michael* Schmidt[1, 2]

[1]Institute of Photonic Technologies, Friedrich-Alexander-Universität Erlangen-Nürnberg (FAU), Konrad-Zuse-Str. 3/5, 91052 Erlangen, Germany
[2]Erlangen Graduate School in Advanced Optical Technologies (SAOT), Friedrich-Alexander-Universität Erlangen-Nürnberg (FAU), Paul-Gordan-Str. 6, 91052 Erlangen, Germany
[3]Institute of Engineering Design, Friedrich-Alexander-Universität Erlangen-Nürnberg (FAU), Martensstr. 9, 91058 Erlangen, Germany
[4]Institute of Manufacturing Technology, Friedrich-Alexander-Universität Erlangen-Nürnberg (FAU), Egerlandstr. 13, 91058 Erlangen, Germany

Abstract. Nowadays the sheet metal forming industry faces challenges regarding efficient usage of resources and sustainability. One strategy to increase the environmental friendliness is to abandon the application of lubricants. The direct contact between tool and workpiece leads to an intensive interaction which increases friction. Especially for deep drawing processes with long sliding distances, this causes distinctive wear. The tool sided application of carbon based coatings is a well-known approach to reduce friction and wear. Former studies have shown a beneficial behavior of hydrogenated amorphous carbon based coatings (a-C:H) to improve the tribological conditions in contact with steel sheets and aluminium alloys under dry conditions. Within this study the coating process and the resulting coating properties will be analyzed. Afterwards mechanical and laser based surface treatment processes prior and after the deposition process will be investigated to reduce the coating roughness. Different roughness values were achieved by varying the surface treatment processes. The laser based finishing enables a reduction of the S_{pk} values by removing single roughness asperities. In order to identify the necessary process parameters for the laser treatment, an analytical model of the material removal was applied. The laser surface treatment achieved similar roughness characteristics compared to mechanical treatment. In this study the tribological behavior of a-C:H coated tools was analyzed under dry conditions within strip drawing tests. The tribological investigations revealed that for dry deep drawing of zinc coated DC04 a broader range of S_{pk} values leads to acceptable tribological conditions whereas for AA5182 a smoother tool surface has to be ensured to prevent adhesion and utilize the full potential of a-C:H coatings.

Keywords: Tribology, Surface Modification, Dry Deep Drawing

1 Introduction

The deep drawing process is characterized by low energy consumption due to short production cycles and high material efficiency. Facing future economic and environmental challenges, such as the scarcity of fossil fuels, further improvements in today's manufacturing processes are essential. One approach to optimizing sustainability and environmental protection is the elimination of lubricants in deep drawing processes. In this way, environmentally harmful substances in the lubricant can be dispensed. Furthermore, the process chain can be shortened by eliminating time-consuming cleaning steps of the workpiece after forming. The lubricant-free contact leads to solid friction between tool and workpiece, which results in a drastic increase in friction and wear. As a result, forming forces increase, tool life is shortened and workpiece quality is reduced [1]. The role of lubricant can be replaced by a promising approach of tool sided coatings. Diamond-like carbon (DLC) coatings in particular showed a high potential for reducing friction under dry conditions in contact with stainless steel [2] and aluminium alloys [3]. Different tribological conditions are desirable for deep drawing in order to locally impede or facilitate the flow of material and to achieve increased wear resistance in critical tool areas. For this purpose, the development of locally adapted tribological conditions is necessary. In prior studies, tetrahedral amorphous carbon (ta-C) coatings showed the most promising results in minimizing friction and wear in contact with the tested sheet materials: DC04, a mild deep drawing steel and the aluminium alloy AA5182 [4]. Furthermore, a-C:H coated tool surfaces contribute to a certain reduction of friction by reducing adhesion in comparison to blank tool surfaces. In general, a-C:H coatings have a higher adhesive strength than ta-C coatings [5]. DLC modifications in different scales are possible to adapt the tribological properties. The friction behavior can be modified by changing the coating's chemical structure

*Corresponding author: rothammer@mfk.fau.de

https://doi.org/10.1051/matecconf/201819014002

with regard to its hybridization states or by reducing the roughness asperities. In order to investigate this influence, a laser based smoothing of coated surfaces is analyzed. On a microscopic scale, the roughness of the coating influences the resulting friction and wear [6]. For this reason, the deposited coatings are subsequently treated by means of surface finishing mechanical brushes or finished by laser ablation. In this study, the tribological behavior of surface modified DLC coatings is evaluated by determining the friction coefficient in strip drawing tests and characterization of the surfaces. Two different initial topographies and two different surface finishes will be investigated.

2 Experimental setup and methodology

Within this study the cold working steel 1.2379 (X155CrVMo12) with a hardness of about HRC 60 ± 1 was used as tool steel for discs and friction jaws. The substrate surfaces were polished, representing the surface qualities of conventional deep drawing tools without preferential direction. Subsequently, the polished surfaces were cleaned in an ultrasonic bath with isopropanol and acetone each for 10 min and 2 min, respectively. As sheet materials the deep drawing steel DC04 with a zinc coating and the aluminium alloy AA5182 was analyzed. Both sheets have an initial thickness of 1 mm and an electrical discharge texture (EDT). While DC04 is commonly used for challenging geometries [7], AA5182 is applied for non-visible car body parts due to its tendency for strain markings [8].

As already mentioned, DLC coatings exhibit a beneficial tribological behavior in dry sheet metal forming [9]. These coatings are mainly characterized by their chemical inertness and high hardness [10]. According to the ternary phase diagram in [11], DLC coatings are distinguished by their graphite (sp^2), diamond (sp^3) and hydrogen content. Previous investigations have shown that in lubricant-free contact conditions, the adhesion tendency between metal surfaces can be successfully reduced by undoped a-C:H coatings [12]. Therefore, this study focuses on the surface modification of this coating type.

2.1 Deposition and properties of a-C:H coatings

The a-C:H coatings were deposited using a hybrid physical vapor deposition (PVD)/ plasma enhanced chemical vapour deposition (PECVD) coating machine (H-O-T, TT-300) with a twofold rotating charging rack. The basic coating system consists of a chromium (Cr) adhesive layer, a tungsten-carbide (WC) interlayer, an a-C:H:W interlayer and the a-C:H functional layer, as shown in Fig. 1. For the Cr adhesive layer and the WC/a-C:H:W interlayers arc evaporation and (reactive) unbalanced magnetron sputtering were used as coating technologies, respectively. The a-C:H coating functional layer was deposited by a pure PECVD process using C_2H_2 as precursor gas. The substrate bias voltage U_{bias} and the deposition temperature were -550 V and 100 °C, respectively. The deposition time for a-C:H layers were kept constant at 8580 s. The C_2H_2 gas flow was 220 sccm and the Ar gas flow was 40 sccm, resulting in a C_2H_2/Ar ratio of 5.5:1.

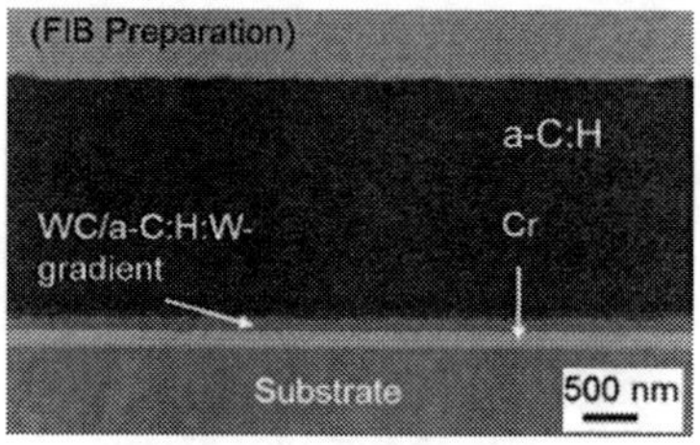

Fig. 1 Setup of the a-C:H coating system using FIB preparation

After deposition, the coating samples were characterized regarding coating thickness and adhesion to substrate using calotte grinding method [13] and Rockwell C adhesion test [14], respectively. Hardness and indentation modulus were measured by micro hardness indentation according to [15]. The a-C:H functional layer was deposited between 80 °C and 100 °C. The mechanical properties of the a-C:H coating system and the substrate are summarized in Table 1.

Table 1 Mechanical properties of coated and uncoated discs

	a-C:H	Substrate
Thickness t of the coating system in µm (n = 5)	2.34 ± 0.07	–
Thickness t of a-C:H functional layer in µm (n = 5)	1.78 ± 0.04	–
Adhesion HF (n = 5)	< HF4	–
Hardness HV 0.001 (n = 12)	1936.8 ± 109.5	1007.0 ± 113.0
Indentation modulus E_{IT} in GPa (n = 12)	502.7 ± 28.9	507.41 ± 48.1

Table 1 shows that the thickness of the total coating system is 2.34 ± 0.07 µm whereby the a-C:H functional layer is 1.78 ± 0.04 µm. All coating samples have an adhesion better than HF4, which represents a sufficient adhesion with small cracks and delamination around the Rockwell indentation according to [14]. The micro hardness and indentation modulus of coating samples are 1936.8 ± 109.5 HV 0.001 and E_{IT} = 502.7 ± 28.9 GPa (n = 12), respectively.

2.2 Surface treatment and characterization of a-C:H coated surfaces

Former studies revealed that the roughness of the tool surface has an essential influence on the tribological conditions especially under dry conditions [16]. Therefore, the aim of this study is to further analyze how the tribological behavior of coated surfaces can be adapted by modifying the topography of the tools. An overview of the investigated surface treatments is given in Table 2.

Table 2 Variants of sequences of mechanical surface treatment steps prior and after coating deposition

Variant	Treatment before deposition	Treatment after coating deposition
(A)	Rough polished	Mechanical
(B)	Smooth polished	Mechanical
(C)	Smooth polished	Laser based

One approach was to change the tool roughness by means of a varying mechanical surface treatment (variant A/B). In a first step, the tool roughness before the coating deposition was varied. All substrate surfaces were polished with a minimum grain size of 9 µm to a roughness of $R_z = 0.9 \pm 0.1$ µm and $R_{pk} = 0.1 \pm 0.01$ µm ($n = 5$), which represents the surface qualities of conventional deep drawing tools without preferential direction. Some of the tool surfaces were additionally polished with a minimum grain size of 3 µm to achieve a smoother topography before coating deposition (variant B). The rough and smooth polished surfaces were afterwards coated according to the process described in 2.1. As the coating process leads to an increase of the surface roughness a post treatment of the coated surfaces was necessary. Therefore, the rough and the smooth polished a-C:H coated surfaces were mechanically post treated for variants A and B. The post treatment was performed manually with 3 µm and 1 µm diamond suspensions each for 15 min. Variant C was post treated by the laser based process which is described in the next section. In order to analyse the influence of the varying surface treatment on the tool roughness a characterization of the tool surfaces was performed prior and after the tribological experiments. The tactile stylus and optical surface measurements of the tool surfaces are conducted with the *Mar Surf GD 120* and the nanofocus *µSurf* microscope, respectively.

2.3 Setup and method for laser based finishing

For ultrashort pulsed laser finishing a mode-locked Nd:YVO_4 laser system (Fuego, Time-Bandwidth Products) with a pulse duration $\tau_P = 10$ ps (FHWM) and wavelength $\lambda = 1064$ nm was used. The laser beam was focused by a telecentric f-Theta lens with a focal length of 160 mm resulting in a beam waist radius ($1/e^2$) $w_0 = 30$ µm. The focused laser beam was deflected by the galvanometer scanner hurryScan 14 II (Scanlab AG). Laser based finishing was performed at room temperature and in normal atmosphere. After laser finishing the laser treated surfaces were cleaned for 3 min in an ultrasonic bath with isopropanol to remove any residues.

2.4 Method for tribological characterization

The tribological behavior of a-C:H coated tool surfaces with varied topographies was analyzed in the flat strip drawing test. This setup models the tribological conditions in the flange area of a deep drawing process and thus enables more process like conditions than tribometer tests with a closed tribological system. The tribological behaviour modelled in the strip drawing test is comparable to the behaviour in the strip bending rotation test [17]. Thus the general findings regarding the tribological relations can be transferred also to the die radius. The friction jaws have a contact area of 55 mm x 100 mm. The strips are 65 mm wide. The basic setup is schematically shown in Fig. 2. The strips are located between an upper and a lower friction jaw and are clamped on one side. The lower friction jaw applies the defined normal force F_N by moving upwards. Then the strips are drawn through the jaws with the drawing force F_D which equals the sum of upper and lower friction forces on both friction jaws F_F. By applying the Coulomb friction law the friction coefficient μ is determined. As evaluation area the measured forces between 100 mm to 170 mm drawing length are considered because here constant normal forces F_N and constant drawing velocities v_{rel} are ensured.

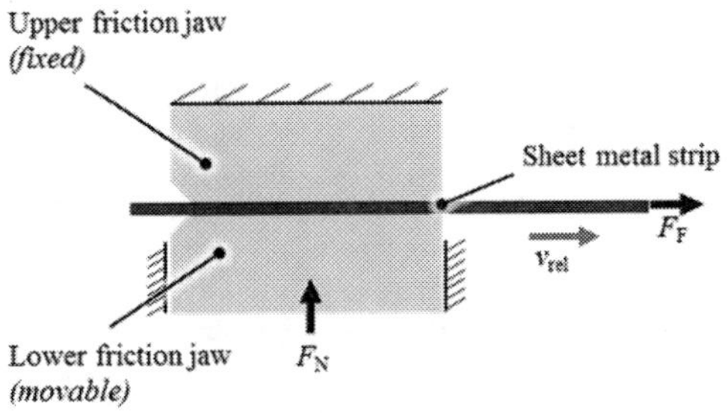

Fig. 2. Experimental setup of flat strip drawing tests (schematically)

Dry strips which were cleaned in an acetone bath were used in all strip drawing tests. The proper removal of anti-corrosive oil and dirt particles was checked using an infrared sensor. A constant drawing velocity v_{rel} of 100 mm/s was applied which is a typical velocity for deep drawing processes. According to characteristic values of the binder pressure, the contact pressure p_N for DC04 was set to 4.5 MPa while for AA5182 a lower pressure of 1.5 MPa was chosen. The lower pressure was choosen due to the lower tensile strength and higher adhesion tendency of AA5182 which leads to a hampered material flow in the flange area. Thus a lower binder pressure is beneficial to improve the material flow.

2.5 Characterization of post treated a-C:H coated surfaces

The micro hardness of a-C:H coatings was determined by Vickers indentation (Fischerscope HM2000, Helmut Fischer GmbH) with an indentation force of 10 mN in 10 s according to DIN EN ISO 14577-1 [15] at 7 points for each modified region. The analyzation of structural properties of the (modified) a-C:H coatings was done by micro Raman spectroscopy (alpha 300RA, WITec). Inelastic scattering of the laser excitation wavelength $\lambda_{Raman} = 532$ nm was recorded to investigate chemical bond configuration. The Raman spectrum was evaluated in the range of 800 to 2700 cm^{-1} where a-C:H coatings have their characteristic peaks [18]. Undesired modifications of the coating structure were avoided by focusing the applied laser power of 10 mW by a 10x objective on the coating surface [19]. The intensity of the G and D peaks was calculated after decomposition of the recorded spectra by the combination of a linear fit for the background as well as a Breit-Wigner-Fano (BWF) and a Lorentzian model [18].

3 Surface quality and tribological behavior of a-C:H coated tool surfaces

3.1 Ultrashort pulsed laser finishing

Finishing by ultrashort pulsed laser based material removal was investigated as an alternative process to mechanical finishing technologies of a-C:H coatings. In general, laser finishing offers the opportunity of contactless and fast machining due to high-speed beam deflection. Thus, the laser applied as finishing tool is no subject to wear. Therefore, reproducibility of reduced roughness parameters should be increased by automated contactless laser machining compared to mainly manually performed mechanical finishing processes. Particularly, ablation by picosecond laser pulses enables modification of the surface topography with almost no heat input. Therefore, structural and mechanical properties of the carbon-based coatings are expected to remain unaffected by this finishing technology. Each of the laser finished surfaces was characterized by areal roughness parameters within the following investigations. Areal roughness parameters were used to prevent any unwanted influence of existing preferential direction of laser based generated surface topographies on roughness measurements. According to ISO 25178 [20] roughness parameters were optically ascertained by confocal laser scanning microscopy (LSM) with 20x magnification and lateral dimensions of the evaluated region of interest about 540x640 µm ($\lambda_C = 250$ µm; $\lambda_S = 2.5$ µm).

3.1.1 Material parameters to describe picosecond laser ablation of a-C:H coatings

The selection of appropriate peak fluences to reduce the surface roughness as well as to achieve minimal dimensions of possible heat affected zones requires the knowledge of the material dependent ablation threshold. Furthermore, this threshold fluence F_{th} and the energy penetration depth κ were used to calculate the surface topography and the microscopic shape of the treated region (Fig. 3a) which is generated by the superposition of subsequent crater geometries (Fig. 3b) resulting from the material removal with a Gaussian intensity profile by a single pass. The achieved depth t at position (x,y) is calculated according to eq. (1) which represents an advancement of a published model describing pulsed micromachining of metals [21]:

$$t(x,y) = \sum_{i=1}^{N_x}\sum_{j=1}^{N_y}\left[\kappa(N_{xy})\cdot ln\left(\frac{F(x+i\cdot p_x,y+j\cdot p_y)}{F_{th}(N_{xy})}\right)\right] \quad (1)$$

where $\kappa(N_{xy})$, $F_{th}(N_{xy})$ and N_{xy} describe the pulse number dependent energy penetration depth, the pulse number dependent threshold fluence and the effective pulse number per area. The effective pulse number per area N_{xy} along the beam paths for a single pass of the laser treated region is calculated according to eq. (2) [22]:

$$N_{xy}(w,v_S,f_P,p_y) = \frac{(1.25\cdot w)^2}{p_x\cdot p_y} = \frac{(1.25\cdot w)^2\cdot f_P}{v_S\cdot p_y} \quad (2)$$

where w, v_S, f_P and p_y describe the beam radius, the scanning speed of the laser beam, the pulse repetition rate and the hatch distance applied for subsequent parallel beam paths, respectively. N_x and N_y are the ratios of the two-fold beam radius w and the number of laser pulses which are positioned per $2\cdot w$ depending on p_x and p_y: $N_x = 2\cdot w/p_x$ and $N_y = 2\cdot w/p_y$. The fluence distribution $F(x+i\cdot p_x, y+j\cdot p_y)$ of the Gaussian beam depending on the positions of the laser spot can be calculated according to eq. (3):

$$F(x+i\cdot p_x, y+j\cdot p_y) = F_0 e^{-2\frac{(x+i\cdot p_x)^2+(y+jp_y)^2}{w^2}} \quad (3)$$

where F_0 is the applied peak fluence. The logarithmic dependence of the ablation depth on the applied peak fluence according to eq. (1) is based on the simplified correlation between the optical penetration depth and the exceeded threshold intensity for material ablation under the constraint of no heat conduction by the lattice as it is reasonable for ultrashort laser pulses [23].

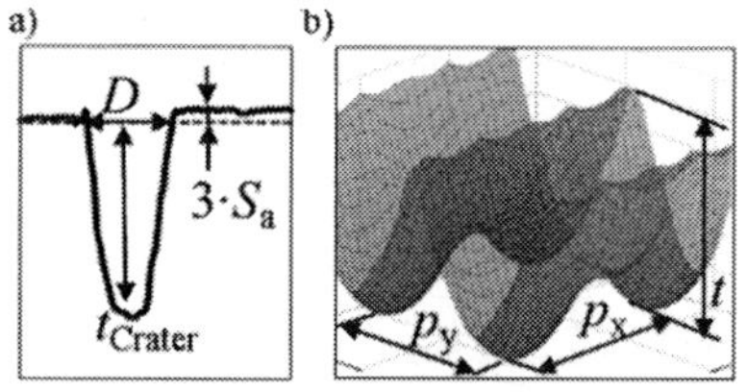

Fig. 3. a) Cross-section of an ablated crater and b) typical ablated topography generated by several laser pulses superpositioned with pitch p_x and hatch p_y

The threshold fluence F_{th} and the energy penetration depth κ were determined by the evaluation of the crater diameter and depth according to the zero-damage-method [24] and logarithmic correlation between depth and peak fluence based on Beer's law [25]. The craters were generated on a-C:H coated samples which were mechanically polished after coating deposition to achieve low roughness $S_a = 0.02 \pm 0.01$ µm and $S_{pk} = 0.06 \pm 0.01$ µm ($n = 5$) to enable automatized detection and evaluation of the crater dimensions with high precision. Both parameters F_{th} and κ were investigated for different pulse numbers N which were used to generate craters at $f_P = 1$ kHz. The determined pulse-number dependent threshold fluences F_{th} are fitted according to eq. (4) [26] (Fig. 4).

$$F_{th}(N) = F_{th}(1) * N^{S-1} \quad (4)$$

The fit is used to consider the incubation behavior which describes the change of the material properties with material-dependent parameter S [26]. S represents a fit parameter which covers several phenomena affecting material incubation which leads to the pulse number dependence of the threshold fluence. Exemplary phenomena are the change of reflectivity and structural modifications [27] due to the mechanical damage by subsequent laser pulses. The fit function published by Jee et al. is applied because this model shows good coincidence for the relevant low pulse numbers $N_{xy} \leq 30$ [25]. The results of the determined penetration depths reveal no significant pulse number dependence as the determined values fluctuate about $\kappa = 29.5 \pm 3.8$ nm. Therefore, constant penetration depth $\kappa = 29.5$ nm is chosen for the calculation of the structure topography.

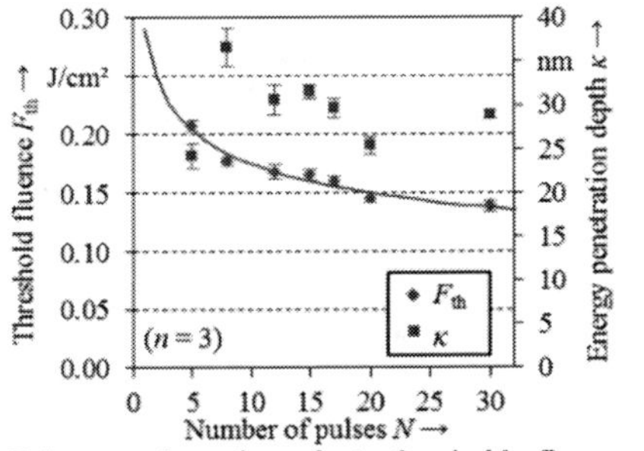

Fig. 4. Pulse number dependent threshold fluence F_{th}, corresponding fitting curve according to Jee et al. and energy penetration depth κ

3.1.2 Influence of Gaussian intensity profile on surface quality

In this chapter, investigations aiming at the tailored energy input and spatial positioning of subsequent laser pulses to minimize the reduced peak height (S_{pk}) of a-C:H coated tools are described. Accordingly, in the order of prioritization low S_{pk}, minimum heat input into the a-C:H coating and minimum reduction of the layer thickness are criteria to assess the laser process and its result. The thickness should be minimally reduced to have a sufficient remaining layer thickness. Low heat input is indicated by low hardness reduction and constant structural properties analyzed by Raman spectrometry. An ablation process with low heat input is characterized by high ablation efficiency which describes the ratio of removed volume divided by the supplied laser pulse energy. Hence, a large part of the laser energy impinging the sample surface causes material ablation.

The samples used for investigations on laser finishing of a-C:H coated tools were smooth polished before the coating process but no mechanical post treatment was applied after layer deposition. The a-C:H coated samples have an arithmetic mean height $S_a = 0.04 \pm 0.01$ µm and reduced peak height $S_{pk} = 0.15 \pm 0.02$ µm ($n = 5$) which are significantly higher compared to additionally mechanically polished surfaces (chapter 2.2). First experimental investigations aim at the selection of an appropriate peak fluence to reduce the surface roughness. For this purpose, the peak fluence was varied at a comparatively low pulse repetition rate $f_P = 1$ kHz to avoid an accumulation of residual heat (Fig. 5a). At peak fluences below 1.0 J/cm² the measured S_{pk} is in the range of the initial roughness of the sample. Additionally, the increased deviation indicates an inhomogeneous material removal. Higher peak fluences lead to significant roughness decrease. The roughness of non-post treated surfaces is mainly reduced due to the removal of asperities by laser ablation. Therefore, the topographies before and after laser finishing differ significantly from each other (Fig. 5c).

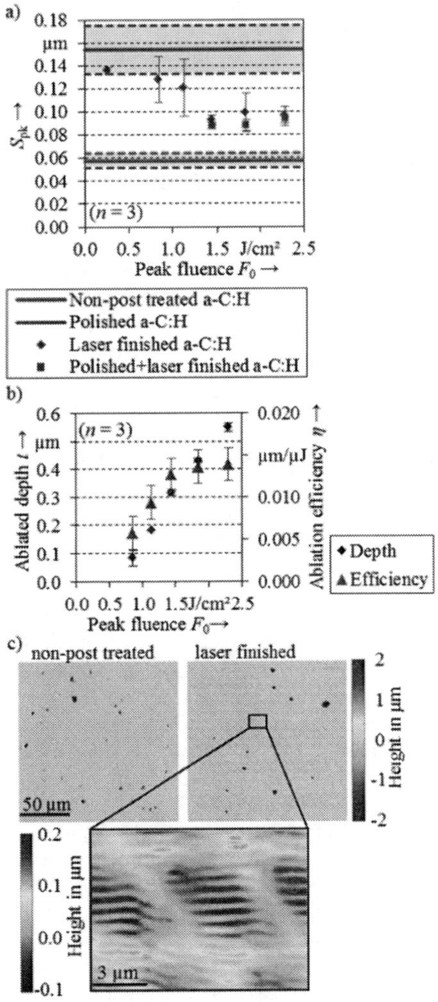

Fig. 5. Influence of peak fluence on a) S_{pk} and b) ablation efficiency; c) topographies of initial non-post treated and laser finished a-C:H surfaces

Acquisitions of the surface topography reveal that the non-post treated coating surface shows more roughness peaks. In contrast to that, these roughness peaks do not appear in regions which were laser finished at higher peak fluences (Fig. 5c). The roughness peaks result from unavoidable droplet deposition during arc evaporation process of the Cr adhesion layer. The removal of droplets is also detectable by a changing reduced valley depth (S_{vk}) due to laser finishing. Initial non-post treated surfaces reveal $S_{vk} = 0.06 \pm 0.01$ µm ($n = 5$) while the amount and depth of the valleys on areas which were laser finished at 1.5 J/cm² are increased, so that $S_{vk} = 0.14 \pm 0.01$ µm ($n = 3$) is ascertained.

The analysis of the ablated depth reveals the logarithmic increase with increasing fluence (Fig. 5b). Maximum ablation efficiency is achieved at F0 = 1.8 J/cm². At this peak fluence the largest portion of the incident laser pulse energy is removed by ionized and vaporized ablated material and therefore does not remain as residual heat around the generated crater geometry. Any additional increase of the pulse energy would mainly lead to higher distinct heating and a larger heat affected zone. For investigation purposes on the influence of the initial surface roughness of the coating, the peak fluence was varied to process samples which were also mechanically polished after coating deposition. For these samples, S_{pk} values of laser finished samples are higher compared to mechanically post treated samples (Fig. 5a). The measured roughness parameters are similar to the parameters of the non-post treated samples. The minimum fluence dependent S_{pk} seems to be independent from the initial roughness. Thus, the roughness is mainly affected by the topography resulting from laser ablation of superpositioned craters and additional ripples on the processed surfaces (Fig. 5c). Ripple formation is caused by the interference of laser light scattered by the surface roughness and the incident laser beam [28]. On the finished a-C:H surfaces the arising periodic ripples have a width of about 800 nm and a depth of about 150 nm. They are oriented perpendicularly towards the linear polarized laser light. These nanometer scaled structures can limit the achievable minimum S_{pk}. As the minimum S_{pk} and almost maximum ablation efficiency are achieved by using $F_0 = 1.5$ J/cm², this fluence is applied in the following investigations on the tailored spatial separation of the laser pulses to achieve minimum roughness resulting from microscopic topography. For this purpose, hatch distance p_y was varied while scanning speed and pulse repetition rate were kept constant at $v_s = 8$ mm/s and $f_P = 1$ kHz resulting in a pulse-to-pulse pitch of 8 µm. Roughness measurements do not reveal any significant influence on S_{pk} because these parameters are in a range of 0.08 to 0.10 µm for hatch distances 5 µm $\leq p_y \leq$ 11 µm (Fig. 6a). A larger hatch of 14 µm leads to slightly higher S_{pk} and larger standard deviation. Higher fluctuations can indicate locally resisting peaks within regions which are only irradiated with lower fluence, e.g., in the range of flanks of the Gaussian profile. In order to select the proper hatch distance, the surface topographies resulting from ablation by subsequent laser pulses with a Gaussian intensity profile were additionally simulated (Fig. 3b). Based on the fitted and determined parameters $F_{th}(N_{xy})$ and κ described in chapter 3.1.1 the depth of the induced material removal and consequently the arithmetic mean height are calculated. Theoretically calculated depths correspond well with the experimentally evaluated depths (Fig. 6a), so that the applied model is suited to simulate the resulting microscopic topography. However, the calculated average mean heights are about one magnitude lower than the measured ones (Fig. 6b) although the lateral resolutions in the model and the measurements are similar with 0.5 µm and 0.625 µm, respectively. The deviation can be explained by the ripple pattern which occurs at the ground of each superpositioned crater (Fig. 5b). The model does not include the formation mechanisms of ripples and thus, this pattern on the nanometer scale is not illustrated. Nonetheless, the model substantiates the choice of tailored hatch distance, as minimal measured as well as calculated S_a due to minimal roughness on the micrometer scale is achieved by $p_y = 8$ µm.

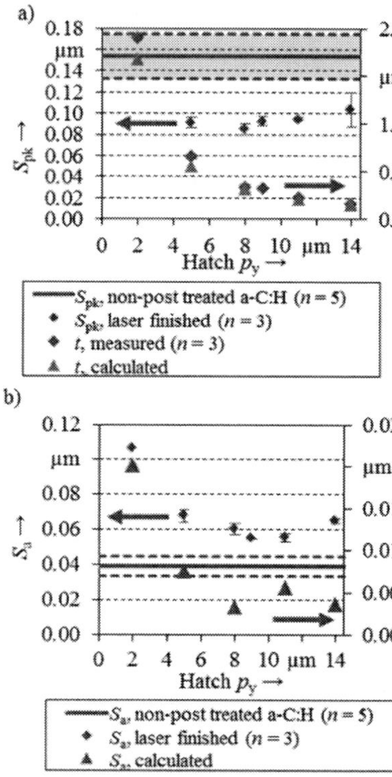

Fig. 6. Influence of hatch distance on a) measured S_{pk} and ablated depth t as well as on b) arithmetic mean height S_a received from experiments and simulations

Based on these results, further investigations on process acceleration to allow fast finishing of tool surfaces are carried out with a hatch $p_y = 8$ µm. For this purpose, the scanning speed is scaled proportionally to the increased pulse repetition rate to keep the pulse pitch p_x constant. These experiments aim at constant roughness as well as structural and mechanical properties. For the examined pulse repetition rates, S_{pk} remains constant within the standard deviation (Fig. 7a). The structural properties of the initial and the laser finished a-C:H coatings are evaluated by the I_D/I_G-ratio derived from Raman spectroscopy according to the methodology described in chapter 2.5. As the I_D/I_G-ratio is drastically higher for laser finished a-C:H coatings than for non-post treated coatings (Fig. 7b), picosecond laser based ablation with $\lambda = 1064$ nm seems to induce formation of heat affected zones which are characterized by an increased degree of sp² hybridization. The

 https://doi.org/10.1051/matecconf/201819014002

transformation of sp^3 in sp^2 hybridized phases due to laser based ablation and heating is independent from the applied pulse repetition rate because I_D/I_G-ratio remains constant for each laser parameter set. Therefore, accumulation of residual heat after laser ablation seems to have no influence on the structural properties although the time interval between two consecutive laser pulses is reduced up to factor 100. The increase of the degree of sp^2 hybridization due to laser finishing is also detectable by an hardness decrease of laser treated areas compared to initial a-C:H coatings (Fig. 7c). This hardness drop is also independent of the pulse repetition rate between 1 kHz and 100 kHz.

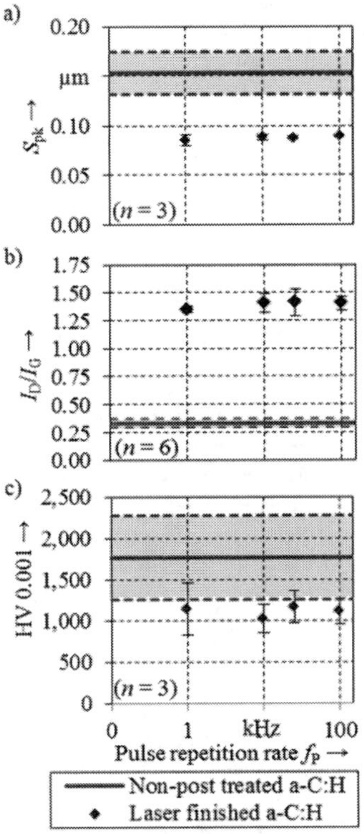

Fig. 7. Dependence of a) S_{pk}, b) I_D/I_G-ratio and c) Vickers hardness on the pulse repetition rate applied for laser finishing

Thus, a high pulse repetition rate can be used for laser finishing enabling fast processing. Exemplary application of the parameter set with $f_P = 100$ kHz for finishing of an a-C:H coated friction jaw would need less than 15 minutes. As the applied pulse energy of 5.1 µJ is much lower than the pulse energy in the range of several hundred µJ of commercially available ultrashort pulsed laser sources, there is further potential to significantly enhance the productivity by process parallelization. For manufacturing purposes of friction jaws for tribological investigations in chapter 3.3 the following parameter set is used: $F_0 = 1.5$ J/cm², $f_P = 25$ k Hz, $v_s = 200$ mm/s, $p_y = 8$ µm. LSM based measurements of regions laser finished with this parameter set reveal $S_{pk} \approx 0.07$ µm ($n = 6$).

3.2 Surface properties of post treated and a-C:H coated tools

The optical measured surface properties after the mechanical and laser based post treatment and before the strip drawing tests are shown in Fig. 8.

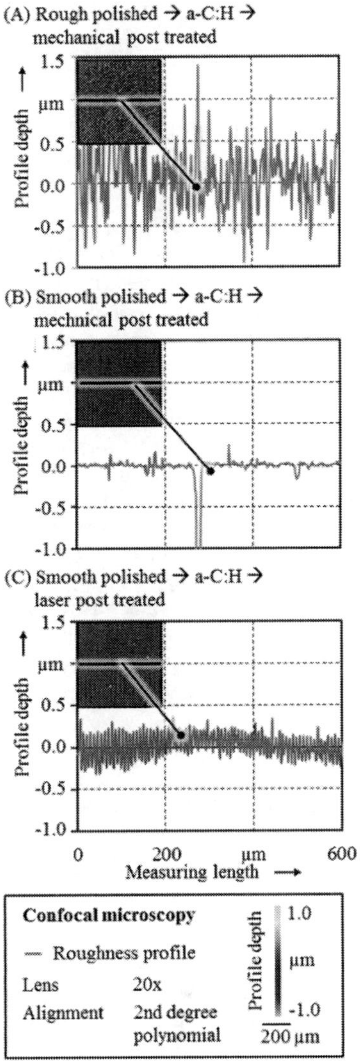

Fig. 8. Surfaces of friction jaws after applied variants of post treatment

Both topographies with mechanical post treatment reveal a surface without preferential direction. A smooth polishing before the coating deposition leads to a surface with few and small roughness asperities after the mechanical post treatment. In contrast, a significantly rougher surface occurs after post treatment of the rough polished substrates. Single roughness peaks achieve heights of over 1 µm whereas the smooth surfaces (variant B) achieve maximum heights of 0.25 µm. The laser based post treatment revealed a preferential direction in form of ripples in drawing direction according to the direction of the laser beam movement.

The height and depth of these ripples which occur due to the laser treatment have an amplitude of 0.25 µm. The height of single roughness asperities is in the same range as for variant B. Additionally some deep valleys result from the coating process which can clearly be seen in the smoother variants B and C. Reason for these pores on the coated surfaces are determinations in the coating chamber or the formation of droplets during arc evaporation of the chromium layer.

3.3 Influences on lubricant-free sliding in strip-drawing tests

The above described methods of mechanical surface treatment and laser based finishing were used to manufacture friction jaws with a-C:H coatings and varying surface roughness. Hence, the influence of the post treatment strategy and the tool roughness on the tribological behavior could be evaluated with the help of strip drawing tests. For the laser based finishing the parameter set which was mentioned in the last section of 3.1.2 was selected to achieve a reduced peak height. The resulting friction coefficients for the different tools surfaces are shown in Fig. 9. In order to describe the tool roughness the reduced peak height of each tool surface is related to the resulting friction. This value is especially important under dry conditions because mainly the roughness asperities of tool and sheet surface are in contact. Thus more or higher asperities have a direct influence on the mechanical interlocking and the resulting resistance towards relative movement.

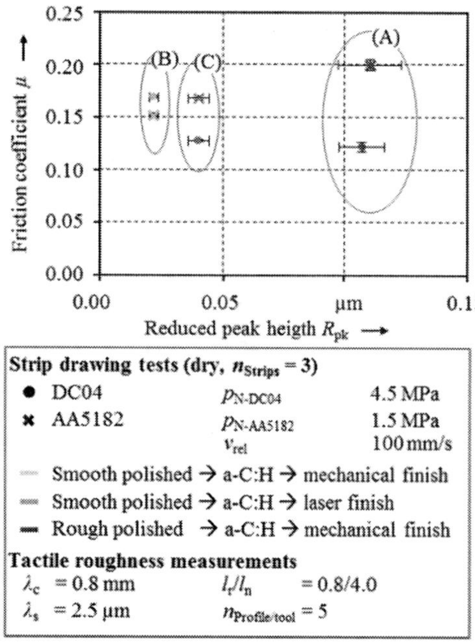

Fig. 9. Influence of tool roughness of a-C:H coated jaws on friction coefficients

Compared to the laser finishing with a reduced peak height of around 0.039 µm a slightly lower roughness is achieved with the mechanical finishing with R_{pk} values around 0.022 µm. A much higher tool roughness with a reduced peak height above 0.10 µm reveals for the friction jaws of variant (A). With smooth polished mechanical finished friction jaws similar friction coefficients between 0.15 and 0.17 are determined for DC04 and AA5182 although a higher contact pressure was applied for DC04. A comparison of the influence of the tool surface treatment on the friction depending on the sheet material is possible even for different levels of contact pressure because the impact of the pressure is low in dry tests [4]. For an increasing tool roughness both sheet materials reveal a different behavior. For DC04 the friction coefficient decreases slightly to a level of 0.13 when a laser based finishing is applied. In contrast, for strips out of AA5182 the friction coefficient increases from 0.15 to 0.17. When the roughness of the friction jaws exceeds 0.1 µm for the R_{pk} value a further increase of friction to 0.2 is measured for AA5182. In contrast, for DC04 the friction coefficient remains at the same level as for the smoother tool surfaces. Thus, there seems to be dependence between the influence of the tool roughness and the sheet material. The different behavior can be explained by the varying surface and mechanical properties between DC04 and AA5182. DC04 has a topography with many plateaus whereas the surface of AA5182 consists of many roughness asperities which are more likely to interact with the asperities of rougher tool surfaces e.g. at the ripples of variant (C). Furthermore, the applied DC04 has a soft zinc coating which smoothens more easily in contact with the tool surfaces. Thus, the risk of interlocking asperities is reduced by a smoothening of the sheet surface. Due to the above mentioned properties of DC04 rougher tools can be applied without risking an increase of friction and wear which can be seen in Fig. 7 for variant (A). In order to ensure that the coated tools lead to a generally beneficial tribological behavior independent from the sheet surface a topography with lower R_{pk} values should be preferred.

The tool surfaces were characterized after the strip drawing tests in contact to AA5182. On a macroscopic scale no distinctive wear was visible. Therefore, a microscopic analysis was performed. With the normally used lens with 20x magnification no distinct wear marks could be detected. However, a small amount of adhesive material was identified using 50x magnification. In a previous study [29] such metallic adhesions were investigated by EDX analysis. It was observed that the adhesion on the a-C:H coating surface against AA5182 consists mainly of oxygen, magnesium and aluminium and against DC04 mainly of oxygen and zinc [29]. Representative tool topographies for the different surface treatments are compared in Fig. 10 regarding the profile depth and their reflectivity. For variant (A) a locally increased profile depth was measured in the range of 1 µm. For variant (B) and (C) it was not possible to clearly identify wear in the topography pictures. Looking at the reflection pictures shown in the second row in Fig. 10, the brighter regions indicate metallic adhesion which has higher reflectivity than the coating itself. Adhesive wear occurred also for variant (C) especially on pores resulting from the coating deposition process.

For the smoother tool surfaces which were achieved with variant (B) no adhesion occurred. Overall, the highest amount of adhesive wear was measured for the rough polished surface of variant (A). This is accordance with the strip drawing tests where these friction jaws revealed the highest friction coefficient for AA5182.

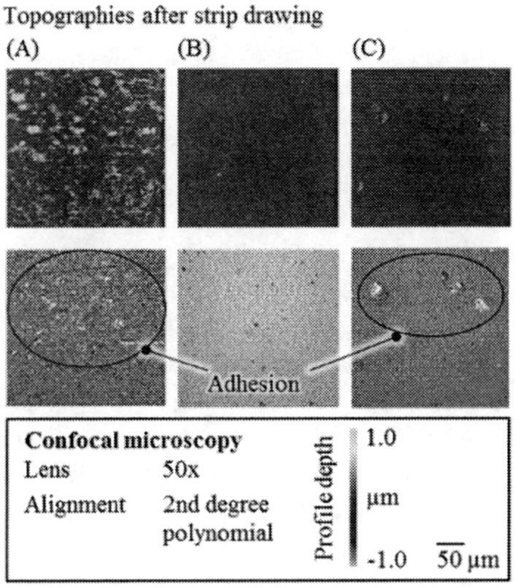

Fig. 10. Topographies and reflectivity of friction jaw surfaces after strip drawing tests in contact to AA5182

Besides the roughness the Raman spectra and hardness of the tools was measured before the strip drawing. The results are shown in Fig. 11 a) and b). An increasing ratio of I_D/I_G indicates a growing amount of sp^2 (graphite-like) content which means that the amount of sp^3 hybridization (diamond-like) phases is reduced. The mechanical finished surfaces reveal nearly the same I_D/I_G ratio as the not finished friction jaws before the laser based treatment. After the laser based finishing process the ratio increased from around 0.3 to 1.4. The changed hybridization is the reason for the changes of the coating hardness which is depicted in Fig. 11 b). For laser finished friction jaws the hardness was measured before and after the laser based finishing. The rough and the smooth polished a-C:H coated surfaces revealed a hardness between 1560 and 1860 HV0.001 after the mechanical post treatment. The smooth polished variant (B) revealed a lower standard deviation due to the improved surface quality which influences the results of hardness measurements. For the friction jaws of variant (C) a hardness of 1725 HV0.001 was measured before the laser based finishing. After the post treatment, the hardness decreased to a significantly lower level of around 1000 HV0.001 due to the heat impact of the laser treatment. It is known from former studies that with decreasing hardness the friction coefficient increases for hydrogenous carbon based coatings deposited in PVD processes [30]. Thus, besides the slightly higher R_{pk} values the lower hardness might be a further reason for the increase of friction from 0.151 to 0.169 for AA5182. This hardness drop caused by laser treatment should be reduced by appropriate measures concerning ultrashort pulsed laser finishing in future investigations.

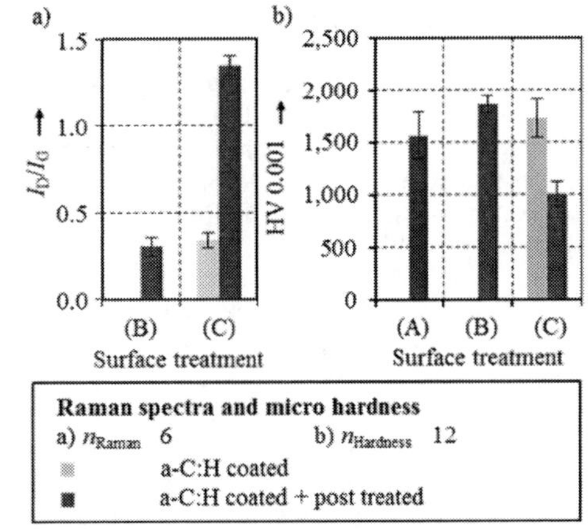

Fig. 11. a) Raman spectra and b) Vickers hardness of tool surfaces in dependence of surface treatment

4 Conclusion and outlook

In this study the tribological behavior of a-C:H coated tool surfaces was analyzed under dry conditions with the help of strip drawing tests. Especially, the influence of a varying roughness of the coated surfaces was investigated. Different roughness values were achieved by varying the pre and post treatment processes. In addition to commonly used mechanical surface treatments a laser based finishing process was investigated. The coating process leads to an increased amount and height of roughness asperities. Reasons for the increasing roughness are droplets occurring during the arc evaporation process of the chromium adhesive layer and possible contaminations within in the coating chamber. The laser based finishing enables a reduction of the S_{pk} values by removing single roughness asperities. In order to identify the necessary process parameters for the laser treatment an analytical model of the material removal was applied. Supported by the model a suitable spatial separation of the pulses was determined to achieve a minimum roughness. However, the model is not able to predict specific roughness values but trends for the roughness development depending on the spatial separation. The final roughness values are not only influenced by removal of asperities but by the formation of ripples. These ripples are caused by interferences of the linear polarized laser beam on the surface and cannot be avoided with the applied system technology. Besides the formation of ripples which influence the tool topography, the hardness of the a-C:H coatings is reduced due to the heat impact of the laser pulses. The heat input leads to a structural change of the coating composition by increasing the sp^2 content. Thus, the coating consists of more graphite and less diamond phases which explains the reduction of the hardness. The tribological behavior of laser treated a-C:H coatings in direct contact with DC04 and AA5182 was compared to mechanical treated friction jaws. Smooth polished surfaces revealed only slightly higher S_{pk} values for laser based finishing compared to mechanical finishing after coating deposition. Rough polished substrates revealed much higher S_{pk} values after the mechanical finishing. For DC04 an increasing tool roughness did not lead to higher friction coefficients. In contrast, for AA5182 the friction coefficient increases with increasing S_{pk} values. Additionally, the adhesion tendency increases which results in adhering of the sheet material on the coated surfaces at a microscopic scale. Reason for the 13 % higher friction coefficient for laser finished compared to mechanical finished tools in contact with AA5182 might be the increased sp^2 content evoking a lower hardness in combination with the slightly higher S_{pk} values. The tribological investigations revealed that for dry deep drawing of zinc coated DC04 a broader range of S_{pk} values leads to acceptable tribological conditions whereas for AA5182 a smooth tool surface has to be ensured to utilize the full potential of a-C:H coatings. In future investigations additional developments regarding the coating deposition process and the surface finishing will be analyzed to further improve the tribological behavior of a-C:H coatings under dry conditions. The influence of a structural multilayer design on coating properties and tribological behavior will be analyzed. The wear resistance will be increased by targeted adjustment of the hardness to indentation modulus ratio through variation of coating parameters such as bias voltage and gas flow. Regarding the laser based finishing further tests with a homogenous intensity profile [31] and femtosecond pulse duration will be performed to reduce the heat-affected zone and restrict the hardness reduction and structural changes. Future investigations have to show to which degree a hardness reduction and structural changes can be limited [17].

The authors thank the German Research Foundation (DFG) for supporting the presented investigations by funding the priority program SPP 1676 project ME 2043/50-1, SCHM 2115/49-1 and TR 1043/5-1 with the project title "Lubricant free forming with tailored tribological conditions".

References

1. F. Vollertsen, F. Schmidt, IJPEM-GT, **1**, 1 (2014)
2. A. Ghiotti, S. Bruschi, Wear, **271**, 9–10 (2011)
3. V. Weihnacht, A. Brückner, S. Bräunling, VIP, **20**, 3 (2008)
4. M. Merklein, M. Schmidt, S. Tremmel, K. Andreas, T. Häfner, R. Zhao, J. Steiner, OAJ FMT, **3**, pp. 42–56 (2015)
5. J. Vetter, Surf. Coat. Technol., **257**, pp. 213–240, (2014)
6. J. Steiner, K. Andreas, M. Merklein, ASHMET (Eds.) Proc. Int. Deep Drawing Research Group Conf., pp. 679–688 (2016)
7. A. Birkert, S. Haage, M. Staub, *Umformtechnische Herstellung komplexer Karosserieteile*, Springer (2013)
8. E. Romhanji, M. Popović, D. Glišić, M. Stefanović, M. Milovanović, Metallurgija, **10**, pp. 205–216 (2004)
9. J. Steiner, K. Andreas, M. Merklein, IOP Conf. Ser. Mater. Sci. Eng., **159**, 1 (2016)
10. Association of German Engineers (VDI), *VDI Guideline 2840*, Beuth (2012)
11. Y. Lifshitz, G. D. Lempert, E. Grossman, Phys. Rev. Lett., **72**, 17 (1994)
12. J. C. Sánchez-López, A. Fernández, Tribology of Diamond-Like Carbon Films, C. Donnet, A. Erdemir (Eds.), Springer (2008)
13. German Institute for Standardization (DIN): DIN EN ISO 26423. Beuth (2016)
14. Association of German Engineers (VDI), *VDI Guideline 3198*, Beuth (1992)
15. German Institute for Standardization (DIN), DIN EN ISO 14577, Beuth (2003)
16. B. Podgornik, J. Jerina, Surf. Coatings Technol., **206**, 11–12, pp. 2792–2800 (2012)
17. J. Tenner, T, Häfner, B. Rothammer, K. Krachenfels, R. Zhao, M. Schmidt, S. Tremmel, M. Merklein, OAJ FMT, **4**, pp. 35–46 (2018)
18. A.C. Ferrari, J. Robertson, Phil. Trans. R. Soc. Lond. A, **362**, pp. 2477–2512 (2004)
19. S. Prawer, K.W. Nugent, Y. Lifshitz, G.D. Lempert, E. Grossman, J. Kulik, I. Avigal, R. Kalish, Diamond Relat. Mater., **5**, pp. 433–438 (1996)
20. German Institute for Standardization (DIN): DIN EN ISO 25178, Beuth (2011)
21. D.A. del Cerro, Dissertation, Mechanical Automation, University of Twente, Enschede (2014)
22. L. M. Machado, R. E. Samad, W. de Rossi, and N. D. Vieira Junior, Opt. Expr., **20**, pp.4114–4123 (2012)
23. S. Nolte, C. Momma, A. Jacobs, A. Tünnermann, B.N. Chichkov, B. Welleghausen, N. Welling, J. Opt. Soc. Am. *B*, **14**, 10, pp. 2716–2722 (1997)
24. J.M. Liu, Opt. Lett., **7**, pp. 196–198 (1982)
25. T. Häfner, J. Heberle, M. Dobler, M. Schmidt, J. Laser Appl., **28**, pp. 0226051–0226056 (2016)
26. Y. Jee, M.F. Becker, R.M. Walser, J. Opt. Soc. Am. B, **5**, pp. 648–659 (1988)
27. A. Naghilou, O. Armbruster, W. Kautek, Appl. Surf. Sci., 418, pp. 487–490 (2017)
28. J. Bonse, S. Höhm, S.V. Kirner, A. Rosenfeld, J. Krüger, IEEE J. Sel. Top. Quantum Electron., **23**, pp. 1–15 (2017)
29. R. Zhao, S. Tremmel, TEWISS (Eds.) Proceedings of THE «A» Coatings 2016, **IFW 3**, pp. 193–200 (2016)
30. C. Zhang, M. Fujii, J. Surf. Eng. Mater. Tech., **5**, pp. 110–123 (2015)
31. T. Häfner, J. Strauß, C. Roider, J. Heberle, M. Schmidt, Appl. Phys. A, 124:111 (2018)

Selective oxidation of tool steel surfaces under a protective gas atmosphere using inductive heat treatment

S. Schöler[1], D. Yilkiran[2], D. Wulff[1], F. Özkaya[2], K. Möhwald[1], B.-A. Behrens[2], H.J. Maier[1]*

[1] Institut für Werkstoffkunde (Materials Science), Leibniz Universität Hannover, 30823 Garbsen, Germany
[2] Institut für Umformtechnik und Umformmaschinen (Forming Technology), Leibniz Universität Hannover, 30823 Garbsen, Germany

Abstract. For the realization of liquid lubricant free forming processes different approaches are conceivable. The priority program 1676 "Dry forming - Sustainable production through dry machining in metal forming" addresses this issue in the context of metal forming processes. The present study reports results from one subproject of the priority program that employs selective oxidization of tool steel surfaces for the implementation of a dry sheet metal deep drawing process. Within the present study, specimen surfaces of the tool steel (1.2379) were heat-treated to optimize their tribological properties with respect to sliding wear behaviour in contact with drawn sheet metal (DP600+Z). The heat treatment was designed to result in the formation of selective oxide layers that can act as friction reducing separation layers. The heating setup employed an inductive heating under protective gas atmosphere. Selective oxidation was realized by controlling the residual oxygen content. Specifically, the specimens were heated in the near-surface region just above the annealing temperature, thus avoiding the degradation of mechanical properties in the bulk. Evaluation of hardness along cross-sections of each specimen revealed suitable initial temperatures for the inductive heat treatment. Oxide layer systems were analyzed regarding their tribological sliding wear behaviour after selective oxidation, as well as their morphology and chemical composition before and after the sliding wear tests.

Keywords: Dry metal forming, sheet metal forming, tribology

1 Introduction

For modern industrial manufacturing, economic and ecological objectives become more and more significant. In this respect, lubricant free dry metal forming is a promising approach and is examined within the scope of the priority program 1676 founded by the German Research Foundation (DFG). However, the sheet metal forming and the bulk metal forming industries are two of the most affected sectors in the context of related manufacturing failures [1-3]. In sheet metal forming, deep drawing processes feature large contact areas between tool surface and sheet metal, and thus, lubricant oils are used to reduce friction and increase tool wear resistance. Lubricants are ecological polluting products which are inconsistent with requirements of sustainable production. Moreover, process chains are extended by additional cleaning processes. Selective oxidation of tool steel surfaces promises sustainability in modern manufacturing. The specific loading case of deep drawing processes is very complex, but depends in general on friction. Yet, the direct contact of the tool surface and the drawn sheet metal should be avoided. Therefore thin oxide layers were generated on the tool steel surfaces using a heat treatment under defined atmospheres.In earlier studies, positive effects of α-Fe_2O_3 oxide layers on friction and wear behaviour have already been reported. Lubricant free pin on disc tests at temperatures between 20 °C and 600 °C showed that oxidized tool steel had friction-reducing properties. However, α-Fe_2O_3 was also generated from metallic debris, so that homogeneous oxide layers could not be realized [4-6].

The present study focuses on the wear behaviour of selectively oxidized tool steel specimens (1.2379), which were heat treated inductively under an argon process atmosphere with an oxygen content of 0.03 vol.-%. These process conditions were selected to favour the formation of α-Fe_2O_3 oxide layer systems, which can act as a friction reducing separation layers. The friction coefficients and wear resistance were determined to compare the systems' behaviour with results obtained in previous studies using convective tempering methods.

* Corresponding author: schoeler@iw.uni-hannover.de

2 Experimental

2.1 Materials and preparation

For investigations of sliding wear, cylindrical specimen of the tool steel X153CrMoV12 (EU alloy grade 1.2379) with an elemental composition of 12% Cr, 1.55% C, 0.9% V and 0.8% Mo (in wt.-%) were hardened to 56 ± 2 HRC ((600 ± 30) HV30). The specimens were circular ground to generate a surface roughness with an arithmetical mean height value of S_A = 1.04 µm ± 0.09 µm, which was measured using a 3D scanning laser microscope (Keyence VK-9710). Prior to the oxidizing heat treatments, contaminations were removed from the specimen surfaces by different cleaning steps in an ultrasonic bath. After the first cleaning step with ethanol (> 96 %) for 10 minutes, lipophilic contaminations were removed in a further cleaning step with acetone (> 99 %) before cleaning with pure ethanol (> 99.8 %) to remove hydrophilic contaminations.

Afterwards the specimens were marked at the front to localize positions for subsequent surface analysis of selected areas. The design of the specimen is shown schematically in Fig. 1 and is described in detail in [7].

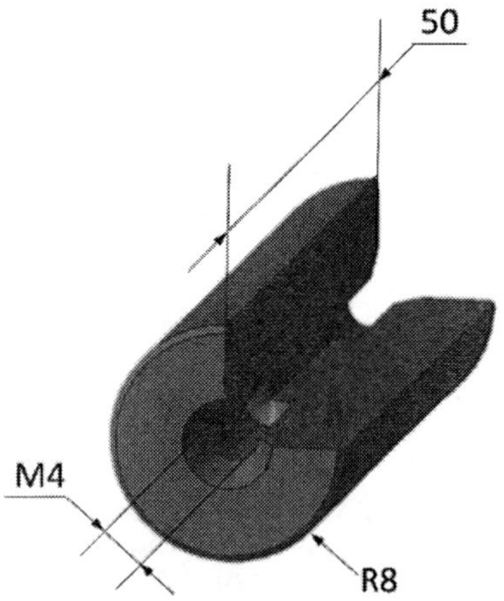

Fig. 1. Planar cross section of the specimen geometry used

The dual-phase-steel DP600+Z (EU alloy grade 1.0936 with an elemental composition of 2.0 % Mn, 1.5 % Si, 1.0 % Cr + Mo + Ni, 0.14 % C, 0.07 %P, min. 0.015 % Al, 0.015 % S, 0.005 % B; in wt.-%) was used as sheet metal for the wear investigations. The metal strips had a nominal thickness of 0.96 mm and a width of 35 mm and were factory-provided hot-dip galvanised with a Zn coating of 9.5 µm ± 0.5 µm thickness.

2.2 Selective oxidation

Selective oxidation of the specimen surfaces was performed in a tubular furnace. Stationary conditions were achieved by controlling the gas flow, the composition of the atmosphere and the process temperature. Argon (Ar) with a purity of 99.996% was used as inert shielding gas. The heat treatments were conducted inductively with a coupled electrical power of 3000 W at a frequency of 500 kHz. The inductive heat treatment was performed at a constant oxygen content of 0.03 vol.-% measured by a lambda probe applying different peak temperatures above the annealing temperature (530 °C) of the tool steel at a constant heating rate of 10 °C/s. After holding the specimen at the selected target temperature for one minute, the temperature was decreased to 500 °C. Subsequently the specimens were hold isothermally for 15 minutes prior to cooling down to ambient temperature at a rate of 5 °C/s, cf. Fig. 2.

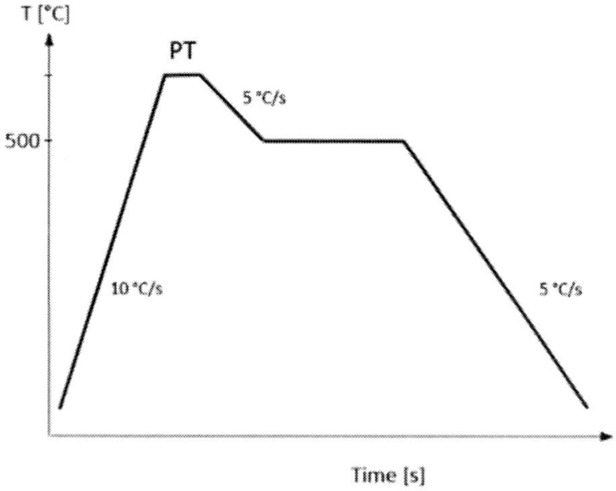

Fig. 2. Schematic showing the temperature-time-regime used for the heat treatments, PT: peak temperature

The actual peak temperatures of the inductive heat treatment are shown in Table 1 along with the parameters used for the convectively tempered specimen. For sake of completeness, the native specimen is included as well.

Table 1. Process parameters of the actual heat treatments used

specimen/ identifier	duration of isothermal hold/ peak temperature (PT)	heating rate	heat treatment type
R1 (native)	/	/	/
A1	60 min/ 510 °C	5 °C/s	convective
B1	15 min/ 570 °C	10 °C/s	inductive
B2	15 min/ 620 °C	10 °C/s	inductive
B3	15 min/ 650 °C	10 °C/s	inductive
B4	15 min/ 700 °C	10 °C/s	inductive

2.3 Measurement of hardness

After the heat treatment, the hardness of the specimens was measured along cross sectional areas from the edge to the centre, in order to investigate the influence of different peak temperatures on the mechanical properties. The measurements were conducted according to the ISO 6507-1 standard (Metallic materials - Vickers hardness test - Part 1).

2.4 Strip drawing tests

Strip drawing tests were carried out in order to determine the friction coefficient μ of the selectively oxidized specimen surfaces. The tests were performed with 90° re-direction, to reproduce the loading case of deep drawing and related wear investigations. For this purpose, steel strips of DP 600+Z were slid over the corresponding specimen surfaces at a drawing speed of 20 mm/s with a contact pressure of 18 MPa. The strip drawing setup is illustrated schematically in Fig. 3. The friction coefficient μ was calculated from the measured forces using the equation

$$\mu = \frac{2}{\pi} \cdot ln \frac{F_p - F_b}{F_c} \quad (1)$$

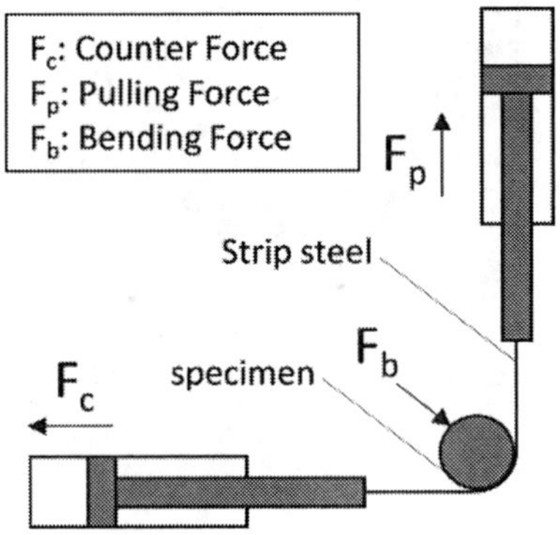

Fig. 3. Schematic of the strip drawing tests

2.5 Wear investigations

The wear investigations were carried out at a hydraulic wear test bench to analyze the sliding wear behaviour of the tribological system formed by the wear specimen and the sheet metal. By pulling the sheet metal from a coil over the oxidized specimen with 90° re-direction, a loading case is generated that resembles the one of modern deep drawing processes. For the investigations, the sheet metal used was provided from a coil, which was cleaned inline to remove the prelube-oil from the surface. A 10% solution of a cleaning fluid (Tickopur R33) was used. To avoid residues of the cleaning solution – especially phosphates - on the surface of the sheet metal, the sheet metal was then cleaned with an ethanol (> 96%) soaked sponge prior to getting in contact with the tool steel specimen. Detailed information of the wear test bench procedure can be found in [8]. The actual setup of the wear test bench is shown in Fig. 4.

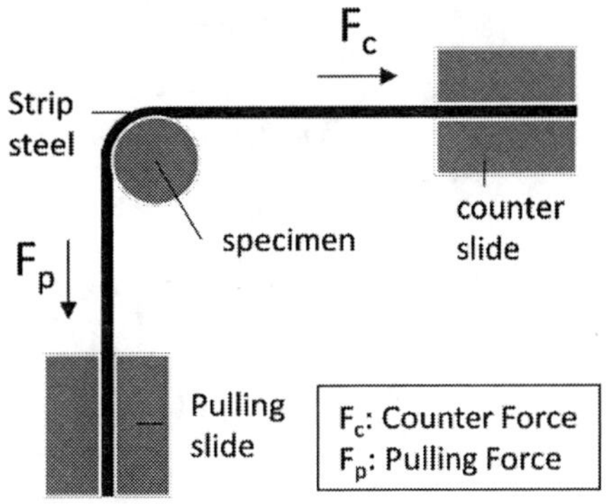

Fig. 4. Wear test bench with 90° re-direction (top), principle of wear investigations (bottom)

2.6 Surface analysis

Different methods were employed to analyze the specimen surface. The specimen were initially characterized regarding their topography using 3D microscopy (Keyence VK-9710). Further investigations were performed using a scanning electron microscope (SEM: Zeiss Supra 55 VP), which was equipped with different detectors. Inlens and secondary electron detectors were employed prior to energy dispersive X-ray spectroscopy (EDX) to determine the chemical composition of selected areas. EDX data were collected at an acceleration voltage of 3 kV in areas of 90 μm x 60 μm. For representative cases, high resolution cross sections were analyzed using a focussed ion beam SEM (Zeiss Auriga). Low acceleration voltages were used in order to minimize contribution to the signal from the underlying substrate when analyzing the very thin oxide layer systems.

 https://doi.org/10.1051/matecconf/201819014003

3 Results

3.1 Oxide layer characterisation

For comparison with results from previous investigations that had employed convective tempering under defined argon atmospheres (O_2 content = 0.03 vol.-%), the surface morphology of the inductively tempered specimen surfaces was characterized in detail. Reference samples (diameter: 12 mm, height: 2 mm) of the hardened (56 ± 2 HRC) tool steel (1.2378) used were polished with 3 µm diamonds dispersion and inductively tempered under using the conditions of the selective oxidation process of specimen B1, cf. Table 1.

Secondary electron images of the surface are shown in Fig. 5. The oxide layer exhibit the typical structure of α-Fe_2O_3 layers, which have been characterized in detail in [8]. The structure of the oxide layer generated seems homogeneous in most regions. However, it is not uniform throughout, and several areas have not been oxidized. The inset in Figure 5 shows a cross section prepared from the marked area using focused ion-beam cutting. The selected area was coated with a platinum layer before sectioning in order to protect the oxide during the process. The oxide layer has an average thickness of about 450 nm, while precipitations can also be detected at the surface that vary in size and are not oxidized. Beneath the oxide layer, a thin reaction zone occurs (dark appearing areas) suggesting a near-surface enrichment of light elements.

The EDX mappings (Fig. 6.) revealed a superposition of iron (Fe) and oxygen (O) in the areas corresponding to the generated oxide layer. The precipitations seen in uncovered regions consisted mainly of chromium (Cr), carbon (C) other alloying components like vanadium (V), which is typical for carbides in this material.

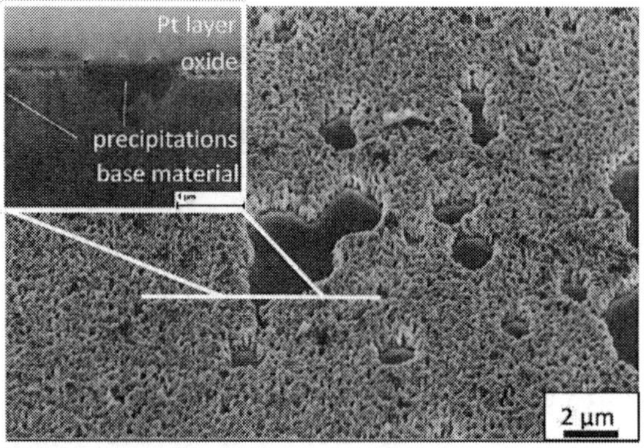

Fig. 5. SEM micrograph of the oxide layer formed after inductive heat treatment with FIB-cut cross sectional secondary electron image shown in the inset

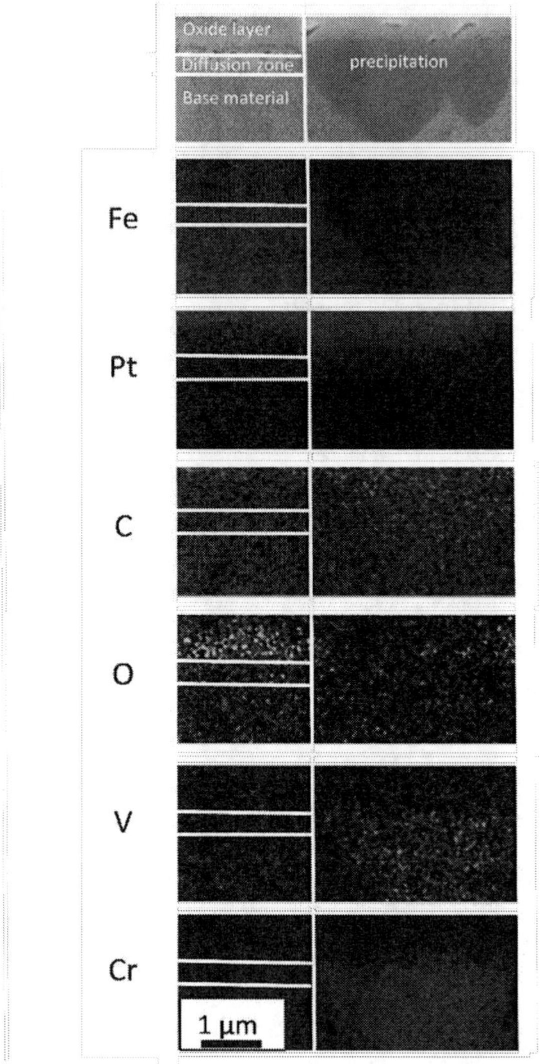

Fig. 6. Secondary electron image of a cross section (top) prepared by focused-ion beam cutting of a reference sample; the corresponding elemental mappings were recorded at 15 keV

For comparison, a convectively heat-treated reference sample (diameter: 12 mm, height: 2 mm) was characterized. The selective oxidation process was performed using the conditions (510 °C for 60 minutes) described in detail in [8]. In this case, the oxide formed was also an α-Fe_2O_3 layer system. Morphologically, the layer was similar to the layer on the inductively tempered specimen. However, the chain like bondings of the oxide to the substrate had coarsened. This can be attributed to a thin oxide layer, which was detected with an average thickness of about 150 nm in a cross section using focused ion-beam cutting (Fig. 7, inset). Compared to inductive tempering chromium precipitations were also detected, but in a higher extend preventing the oxide layer to cover the surface at these areas. So, large-scaled areas, which are not covered by the oxide layer can be detected after convective tempering.

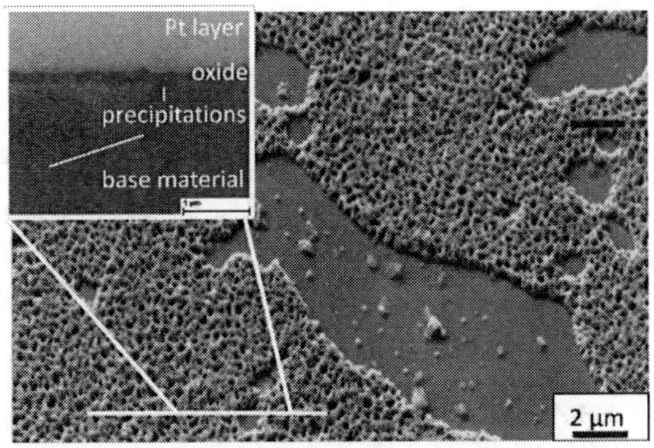

Fig. 7. SEM micrograph of convectively heat treated oxide layer formed along with a cross sectional secondary electron image in the inset

3.2 Measurement of hardness

Hardness measurements were performed along cross sectional areas of the tempered specimen in order to determine the effect of different peak temperatures above the annealing temperature on the mechanical properties of the tool steel. Figure 8 present hardness data measured on the specimen in direction from the specimen surface to the centre. Clearly, tempering procedures with short initial temperatures between 570 °C (specimen B1) and 620 °C (specimen B2) have negligible influences on the material, while a heat treatment up to 700 °C decreases the HV30 value by nearly 40% (Specimen B4) .

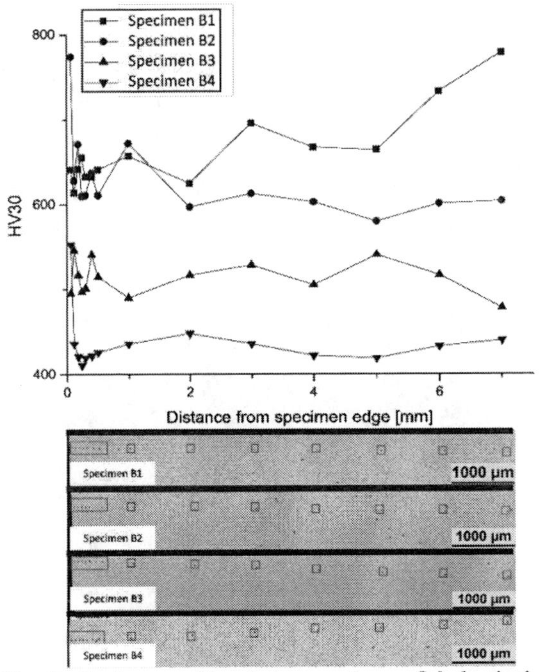

Fig. 8. Hardness profile of cross sections of inductively tempered specimens

3.3 Surface analysis after wear

The wear investigations were performed at constant loading conditions at 80 °C specimen temperature and a normal stress of 80 MPa. Selected specimens that have kept a HV30 value within the scope of tolerance after the hardening procedure were tested with 500 strokes. Therefore, specimen B1 (initial temperature: 570 °C) and specimen B2 (620°C) , which seem to have no noticeable losses of hardness after the tempering process were characterized regarding their sliding wear resistance in comparison to conventional heat treated (A1) and the reference specimen (R1).

The surface topography of the specimen was measured before and after the wear investigations in order to characterize the morphological changes of the surfaces. For this purpose, the arithmetical mean height S_A and the dale void volume V_{VV} measurements were determined in the area at the central 45° position of the specimen prior and after the wear investigations. This position represents a constant normal stress loading case, while the contact area of the tool steel and the drawn sheet metal amounts to $\pi/2$ of the specimen surface. In Fig. 9 the arithmetical mean height S_A for different specimens before and after the wear investigation are compared. In general, selectively oxidized surfaces seem to feature smaller initial S_A-values than untreated reference samples before wear testing. Furthermore, the convectively tempered specimen A1 shows the lowest initial arithmetical mean height. Clearly, the magnitude of reduction in the S_A-values after wear testing is nearly constant (at about 10 %) for selectively oxidized specimen, while the reference sample R1 exhibits a loss of nearly 30 %. Thus, the arithmetical mean height of the oxidized specimens is not significantly changing during wear, but even appears to decrease slightly. The clear decrease of the S_A-value of the reference specimen R1 after the wear tests could be related to two mechanisms that seem to govern the wear process. As a major effect, smoothening of the surface appears. Secondly, transferred zinc coating from the sheet metal can result in adhesive zinc pick-up, which was also observed in previous studies [8]. This is confirmed by analyzing the dale void volume V_{VV}, which is a parameter that provides information about the empty volume of surface valleys. In Fig. 10 the dale void volumes determined prior to and after the wear tests for each specimen are given. The dale void volume V_{VV} shows a trend that can be related with the arithmetical mean height. Following wear, V_{VV} decreases substantially for the reference surfaces, whereas only small changes were obtained for the oxidized surfaces. The friction coefficient decreased slightly for the selectively oxidized specimen compared to the virgin reference. Furthermore, the inductively tempered specimen showed lower friction coefficients than convectively tempered specimen. As a result an increasing effect of thicker oxide systems as friction reducing separation layers occurs.

https://doi.org/10.1051/matecconf/201819014003

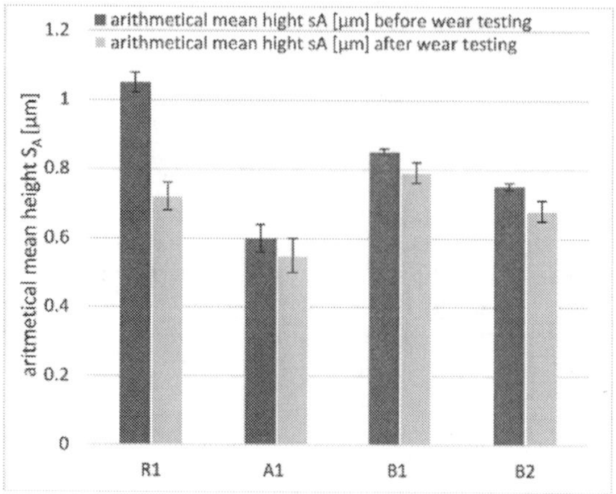

Fig. 9. Arithmetical mean height S_A prior to (dark grey) and after (grey) wear tests for each type of specimen

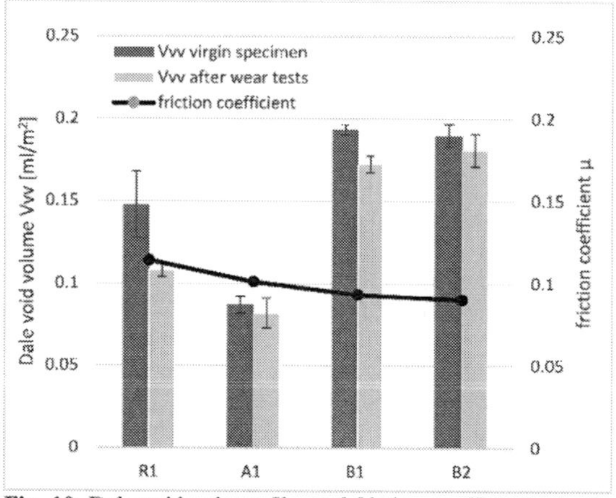

Fig. 10. Dale void volume V_{VV} and friction coefficient prior to (dark grey) and after (grey) wear tests for each type of specimen

To gain further insight into the wear mechanisms, SEM micrographs of the specimen B1 and specimen B2 were taken prior to and after the wear tests. For condition B1, an untested specimen is shown in Fig. 11 (top) along with a SE image of the critical 45° area in a tested one. The surface is completely covered, with a chain-like bonding oxide structure, which is characteristic for α-Fe_2O_3 layer systems. After the wear tests, the layer appears smoothed (Fig. 11 bottom). The SE micrograph shown in Fig. 11 was recorded at the central position of the worn area. The layer structure is still recognizable and the chain-like oxide bondings are even more concentrated, and oxide particles seem to have been picked up during the wear tests at uncovered substrate areas.

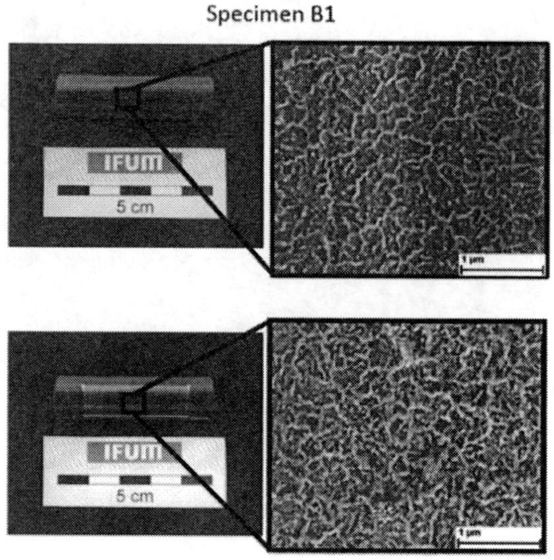

Fig. 11. Specimen B1 prior to (top) and after (bottom) the wear test (500 strokes); SE images were taken a low acceleration voltage of 3 kV to minimise charging effects

This effect was also confirmed for specimen B2. The micrograph (Fig. 12 top) also showed a layer system with a similar structure of chain-like oxide bondings prior to the wear tests, but the occurrence is more intense. Again small particles have accumulated in between the chain structure of the oxide compounds after the wear test (Fig. 12 bottom).

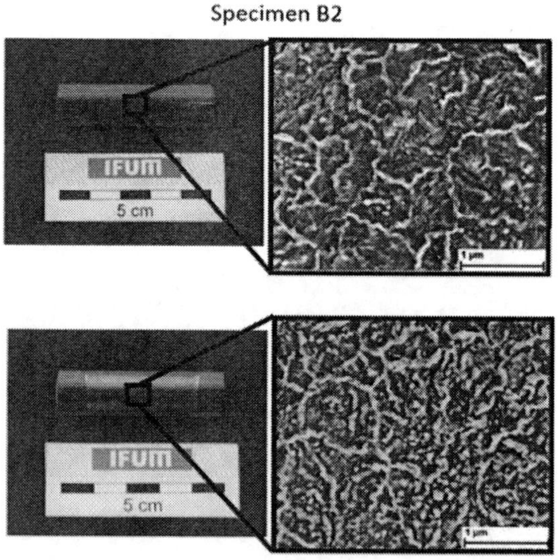

Fig. 12. Wear specimen B2 prior to (top) and after (bottom) the wear tests (500 strokes); SE images were taken at 3 kV

4 Discussion and outlook

Depending on the actual heat treatment parameters, the oxide layer systems generated inductively at higher peak temperatures showed substantially different behaviour. On the one hand, peak temperatures above the annealing temperature of the tool steel can favour surface activation, and thus, selective oxidation processes that are caused by thermal acceleration of diffusion and higher reactivity of the tool steel elements with oxygen. On the other hand, even short dwell periods at peak temperatures above 620 °C reduced the hardness of the tool steel by nearly 40% in the bulk. In this context, it is important to note that the hardened tool steel features precipitations of chromium carbides, which also occur in the near surface areas. The focussed ion-beam prepared cross sections of the inductively tempered reference specimen showed that the carbides correspond to the areas that were not covered by the α-Fe_2O_3 oxide layer. As this may negatively affect the long-term wear behaviour, powder metallurgically manufactured tool steel with fewer or smaller chromium carbides might provide for even better performance in this respect. This approach is based on the assumption that layer defects increase constantly as a function of wear.

The measured arithmetical mean height and dale void volume of inductively tempered specimen indicated promising wear behaviour. Decreasing arithmetical mean height and dale void volume were determined after the wear tests and SEM micrographs also showed a characteristic oxide structure with several particle pick-ups in between the oxide compounds. It is assumed that these result from tribooxidation effects. Increased surface temperatures at the contact areas between tool steel and drawn sheet metal should favour oxide growth. Thus, it appears feasible that an erosion of the oxide layer might be counterbalanced by a growth of oxide particles.

The present study demonstrates that inductive heat treatment could be a viable approach generating reproducible oxide layer systems by selective oxidation processes. Compared to conventional heat treatments, the higher heating rates in combination with short isothermal dwell periods would reduce overall processing times. More importantly, however, the hardness of the bulk is less affected by this approach. Albeit the current study indicated promising wear properties, it should be noted that the contact pressures in industrial deep drawing processes could be substantially higher than those used in the present set-up for measuring sliding wear resistance, and work is underway to address this effect.

Acknowledgements

Financial support of this study by the German Research Foundation (DFG) (reference numbers BE1690/170-2 and MA1175/41-2) within the framework of the priority programme Sustainable Production through Dry Processing in Metal Forming (SPP 1676) is gratefully acknowledged.

References

[1] B.-A. Behrens, A. Bouguecha, B. Homan, Investigations of surface topography evolution of sheet aluminum under pressure and tension, Comput. Plast. XII (2013) 1230-1237.

[2] G. Braeuer, H. Paschke, M. Weber, B.-A. Behrens, T. Yilkiran, Surface modifications for optimized forming operations, Key Eng. Mater. 611-612 (2014) 231-239, http://dx.doi.org/10.4028/www.scientific.net/KEM.611-612.231.

[3] B.-A. Behrens, A. Bouguecha, L. Lüken, J. Mielke, M. Biston, Tribology in hot forging, Compr. Mater. Process. 5 (2014) 211-234, http://dx.doi.org/10.1016/B978-0-08-096532-1.00538-0.

[4] J. Glascott, F. H. Stott, G. C. Wood, The effectiveness of oxides in reducing sliding wear of alloys, Oxid. Met. 24 (1985) 99-113, http://dx.doi.org/10.1007/BF00664227.

[5] J. Glascott, G. C. Wood, F. H. Stott, The influence of experimental variables on the development and maintenance of wear-protective oxides during sliding of high-temperature iron-base alloys. Proc. Inst. Mech. 199 (1985) 43-41, http://dx.doi.org/10.1243/PIME_PROC_1985_199_088_02.

[6] F.H. Stott, M. P. Jordan, The effects of load and substrate hardness on the development and maintenance of wear-protective layers during sliding at elevated temperatures, Wear 250 (2001) 391-400, http://dx.doi.org/10.1016/S0043-1648(01)00601-9.

[7] D. Yilkiran, A. Almohallami, D. Wulff, S. Hübner, M. Vucetic, H. J. Maier, B.-A. Behrens: New Specimen Design for Wear In-vestigations in Dry Sheet Metal Forming, Dry Met. Forming OAJ FMT 2 (2016) 62-66.

[8] D. Yilkiran, D. Wulff, A. Almohallami, F. Özkaya, A. Bouguecha, S. Hübner, K. Möhwald, H. J. Maier, B.-A. Behrens, Wear behaviour of thermally oxidized tool surfaces as low-friction separation layers for dry sheet metal forming , Wear Vol. 376 – 377 (2017)

The influence of process parameters on the temperature development in the forming zone

Philipp Tröber[1,*], *Markus* Welm[1], *Hannes Alois* Weiss[1], *Peter* Demmel[2], *Roland* Golle[1], and *Wolfram* Volk[1]

[1]Technical University of Munich, Chair of Metal Forming and Casting, 85748 Garching, Germany
[2]MAN Truck & Bus AG, 80995 Munich, Germany

Abstract. Cold metal forming is a fast and economical way of producing a wide range of precise components. Its profitability mainly depends on part quality, process stability and service intervals of tools. As these factors are all determined by tool wear, detailed process knowledge is indispensable to maximize profitability by minimizing wear. One of the most crucial factors in this context is temperature. During every forming process, a temperature rise occurs between tool and workpiece due to frictional heating and a large part of plastic work dissipating into heat. This temperature affects the whole forming process but especially tool wear. Currently, there is little solid information about temperatures occurring during forming operations. Therefore, the temperature was measured based on varying process parameters in several embossing and blanking examinations. The use of a tool-workpiece-thermocouple enabled accurate and instantaneous measurement during the process. The results presented show the strong influence of process parameters on temperatures in the forming zone.

Keywords: Temperature, Cold forming, Tool-workpiece-thermocouple

1 Introduction

Cold metal forming and blanking are among the most frequently used mass production processes for sheet metal. To meet increasing requirements concerning part quality, cycle times and service life, precise process knowledge is essential. During every forming operation, a temperature rise occurs in the workpiece's forming zone. This effect has three causes: First, up to 95 % of the plastic work dissipates into heat [1]. Only a small part is stored as dislocations and other lattice defects in the sheet metal [2]. Second, macroscopic friction between contact surfaces and finally, microscopic friction on the atomic level also generates heat [3]. Especially during forming operations with high surface pressures and deformation degrees, such as embossing or blanking, temperatures of several hundred degrees can occur, as shown below. The temperature rise strongly influences the process. Material behaviour [4] and tool wear are just two examples. Furthermore, temperature is a decisive factor in the choice of tool coatings and lubricants [5]. Therefore, precise knowledge of the height and profile of the temperature that occurs is indispensable with respect to process parameters.

Several reports have dealt with this topic across various forming and blanking investigations but the determined temperatures differ significantly, even during similar manufacturing processes. Groche et al. calculated maximum temperatures of 36 °C for deep drawing, which is characterized by smaller surface pressures than embossing [6]. Furthermore, when processing aluminium, lower temperatures occur due to the low mechanical properties compared to steel. During blanking of steels, authors indicate mostly higher values from 50 °C [7] to 600 °C [8] and even 1000 °C [9]. This discrepancy is mainly due to undefined geometric and temporal resolution of the measurement signal. Demmel et al. were the first to use a tool-workpiece-thermocouple during blanking of S355MC. In combination with a punch having undercuts, this method enables an instantaneous measurement of the temperature in-situ at the punch edge. They observed a temperature of up to 250 °C with a 4 mm thickness and almost 300 °C with 6 mm [10]. The maximum temperature always occurs with the end of clean cut formation, just before the final separation of the sheet metal. Tröber et al. measured temperatures of up to 264 °C for embossing 4 mm carbon steel with a similar tool [11]. They confirmed the significant influence of process parameters on emerging temperatures [12].

Especially the wide range of measured temperatures shows a high research demand concerning this topic. For that reason, a tool-workpiece-thermocouple was implemented in a new tool and several embossing examinations under varying process parameters were conducted. Investigations of smaller clearances and faster velocities compared to former examinations have enriched knowledge of temperature and improved understanding of the process.

* Corresponding author: philipp.troeber@utg.de

2 Experimental Setup

2.1. Measuring principle

There are several measurement techniques for determining temperatures but the cramped conditions in a forming tool hamper precise temperature measurement. High surface pressures prevent the use of foil or fine wire thermocouples on punch or sheet metal because of insufficient robustness. Another possibility is the integration of thermocouples in small holes in a punch which weakens the tool. Other methods cannot be used because of limited space. In sum, the challenge is to ensure high-quality measurement that is instantaneous and at a defined location without influencing the process.

Thermoelectric phenomena offer the possibility of transforming tools to temperature sensors. Based on different thermoelectric properties of sheet metal and tool, these materials can form a thermocouple. Thermoelectric voltages always arise if two different electrical conductors are put together at one end that has a different temperature than the other ends. In this case, the thermoelectric voltage is proportional to the temperature gradient along the conductors. The proportional factor is the Seebeck coefficient, which represents thermoelectric material properties [10].

During cold metal forming or blanking, an electrical contact between punch and sheet metal emerges. The dissipation of conducted inelastic forming work and friction between sheet metal and punch result in local heating of the contact area. Subsequently, charge carriers in the conductors reach different energy levels. A thermodiffusion starts. The resulting potential difference changes with the temperature difference between the junctions and can be measured with a voltmeter [13]. As the measurement area is identical to that of heat generation, an instantaneous measurement of the temperature occurring during every kind of forming process can be guaranteed.

2.2. Calibration of the thermocouple

To deduce a temperature from the measured voltage, calibration of the thermocouple is necessary. As the thermoelectric material properties can only be determined experimentally [14], a special measurement device was used. Its basis is the integral measuring method, which requires a defined temperature gradient along the sample. Figure 1a illustrates the electric circuit used for calibration. One end of the sample is heated to the defined temperature T_{hot} with a maximum of 600 °C. The other end is kept at the reference temperature T_{ref} of 0 °C in an ice bath. Both ends are attached to a pure platinum wire, which serves as reference material. Depending on the Seebeck coefficients of sample (S_B) and reference material (S_A), a defined thermoelectric voltage according to Equation 1 arises [15]:

$$V_{AB}(T,S) = \int_{T_{ref}}^{T_{hot}} (S_B(T) - S_A(T))\, dT = $$
$$= (S_B(T) - S_A(T)) * (T_{hot} - T_{ref})\ [V] \quad (1)$$

This voltage lies within the range of a few millivolts for metals. Therefore, the signal is first amplified and then recorded with a precision voltmeter. Figure 1b shows the implementation of the setup used. The sample is pushed against a block of pure copper by a pneumatic cylinder. To ensure a well-defined and reproducible electric contact, the adjustable force is observed with a piezoelectric sensor. Another key factor is temperature. As the calibration quality directly depends on temperature measurement accuracy, two high-precision thermocouples are placed directly at both junctions. They are small to minimize time delay due to heating. An argon atmosphere prevents measurement errors because of chemical effects at the sample surface and platinum wires. Characterizing both standardized type K thermocouple legs showed a maximum measurement deviation of the measured thermovoltage of 1.5 %.

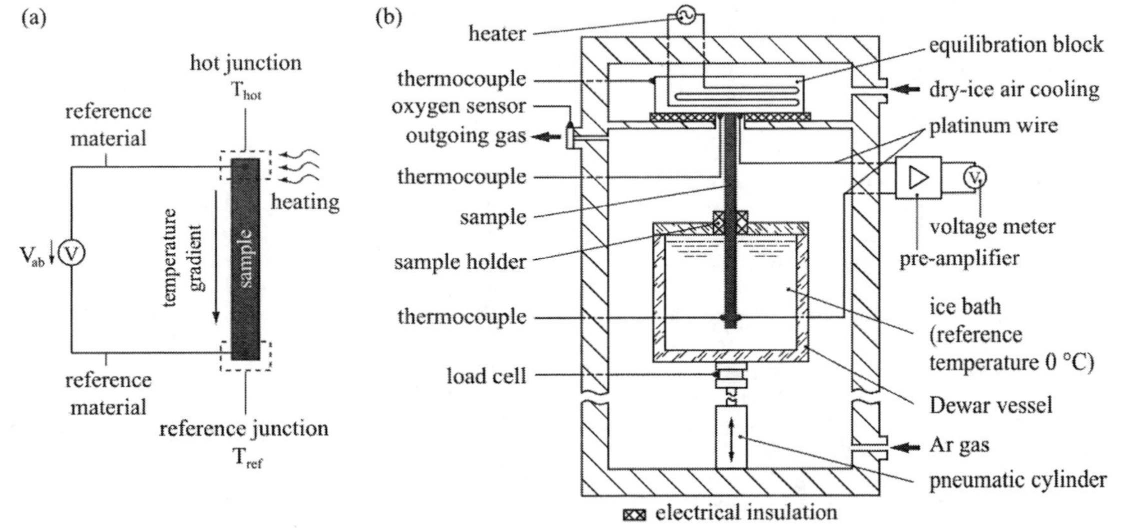

Fig. 1. (a) Schematic diagram of an integral measuring setup for thermocouple calibration and (b) Apparatus for measuring relative Seebeck coefficients [13]

https://doi.org/10.1051/matecconf/201819014004

2.3. Tool and press

The experimental tool has a four-pillar structure providing high stiffness and the application of smallest cutting clearances down to 25 µm. Due to its modular design, the tool is very variable and suitable for parameter variation. A configurable blankholder force allows analysis of different materials. The punch and sheet metal are electrically insulated from the remaining tool to prevent interfering signals. This is done with zirconium oxide applications on the die, blankholder and punch (Fig. 2).

The whole tool consists of four stations where both forming and blanking operations can be implemented. A fifth station can be used for cutting sheet metal strips. This enables instantaneous and simultaneous temperature measurement during continuous stroke experiments. The forming forces are measured by piezoelectric load cells in every station and the punch travel by a contact-free eddy current sensor.

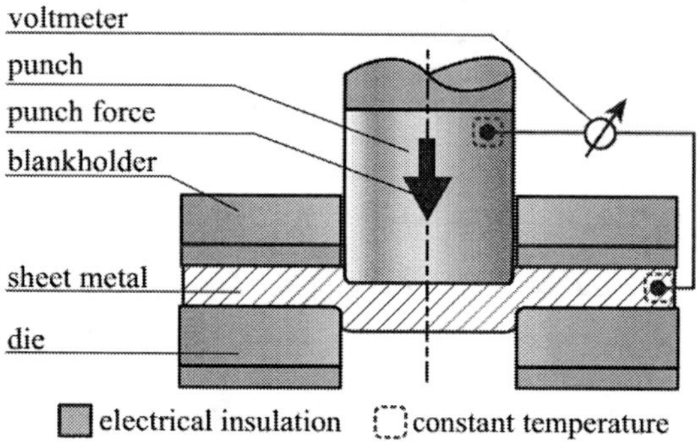

Fig. 2. Schematic drawing of a tool-workpiece-thermocouple for temperature measurement during embossing and blanking operations

For the investigations in this paper a circular punch with a diameter of 15 mm and a punch edge radius of 50 µm was used. The immersion depth was 4 mm. No lubricants were applied during any tests.

The investigation was made on a BSTA 1250-181 built by Bruderer AG (Frasnacht, Switzerland). This high-performance stamping press has a maximum press force of 1250 kN and an infinite variable stroke rate between 60 1/min and 600 1/min.

2.4. Investigated materials

Austenitic stainless steel 1.4301 (X5CrNi18-10) with a thickness of 2.5 mm has been chosen as sheet material. A high tensile strength of up to 720 MPa and good corrosion resistance make this material suitable for many applications.

The punch is made of CF-H40s, a powder metallurgical cemented carbide. Its fine grain structure and high homogeneity are especially important because constant thermoelectric properties across the whole punch are required for high-quality temperature measurement. Hardness of 1400 HV10 allows use for forming and blanking processes.

Table 1 shows the chemical composition of CF-H40s and 1.4301 determined by an optical emission spectrometer.

Table 1. Chemical composition (elements over 0.1 %) of the materials used in weight percent.

1.4301	Si	Mn	Cr	Mo	Ni	Cu	Fe
	0.7	1.3	19.0	0.4	10.8	0.3	balance
CF-H40s	WC						Co
	88.0						balance

3 Results and discussion

In this investigation, the focus is on forming velocity and die clearance. To find out their influence on emerging temperatures in the forming zone, various experiments were done with the aforementioned experimental setup. As the highest temperatures occur during processes with high surface pressures and deformation degrees, embossing was chosen as manufacturing process. As maximum temperatures occur with the end of clean cut formation and thus the maximum embossing depth, a complete sheet metal separation is done. A die clearance of 1 % (0.025 mm), 5 % (0.125 mm) and 10 % (0.25 mm) is chosen for the experiments. The stroke rate is varied between 60 1/min, 150 1/min and 300 1/min which corresponds to an approximate punch impact velocity of 50 mm/s, 140 mm/s and 270 mm/s. Every experiment was repeated at least three times.

3.1. Temperature profile

Figure 3 shows representative curves of the temperature occurring in the forming zone as related to the punch travel. Negative values stand for a punch travel downwards until bottom dead center at 0 mm. In figure 3a, a 1 % die clearance was used. As long as the contact between sheet metal and punch remains, slight adjustments in process parameters only lead to changed slopes and higher or lower maximum temperatures. With the contact of punch and sheet metal at -6.34 mm, a thermoelectric voltage emerges and a temperature of 22 °C occurs which corresponds to room temperature. During elastic deformation at the beginning, a levelling of roughness asperities takes place, which results in a slight temperature rise until -5.90 mm. With the plastic deformation of the sheet metal, the temperature increases faster due to dissipating inelastic work. The maximum temperature of 230 °C is reached shortly before the complete separation of the sheet metal at the maximum embossing depth at -3.93 mm. Due to the high embossing depth of the sheet metal thickness (Tab. 2) and a large contact area between punch and sheet metal, temperature briefly remains at the same level. With the end of plastic deformation, temperature decreases until bottom dead center. During the return stroke, temperature increases again until a maximum of 171 °C at 5.92 mm. With the disconnection of the electric circuit, no more temperature

https://doi.org/10.1051/matecconf/201819014004

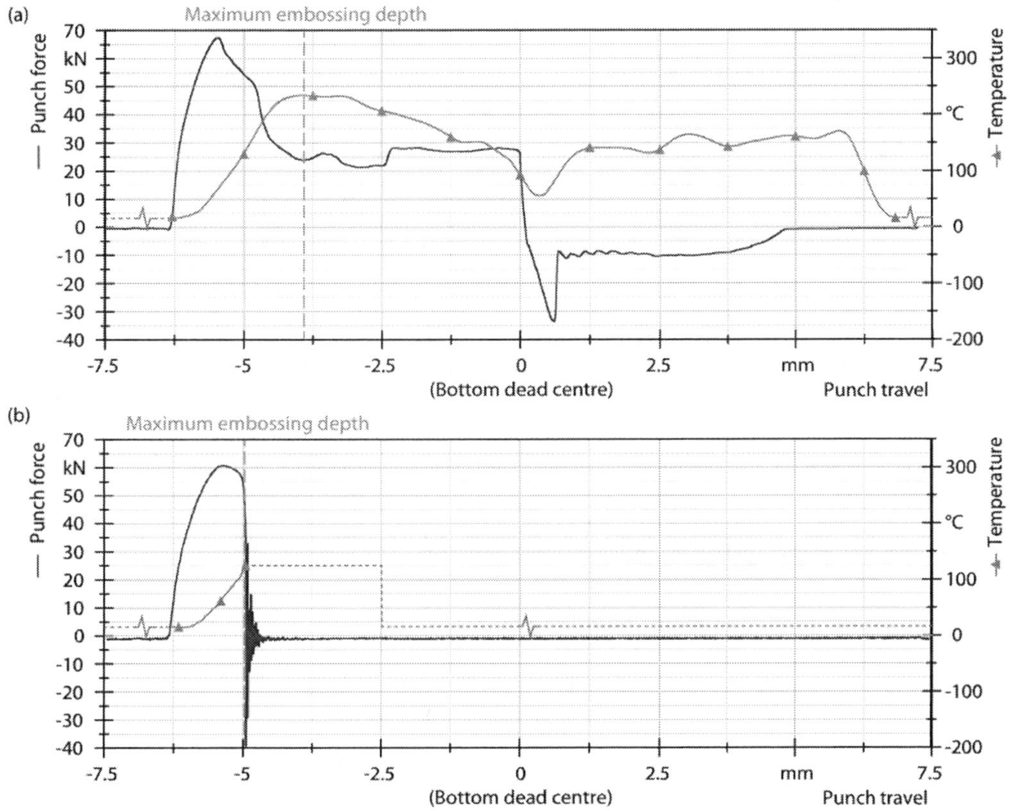

Fig. 3. One representative temperature profile and punch force over the punch travel for a die clearance of (a) 1 % and (b) 10 % blanked with a punch stroke rate of 300 1/min

is measurable. A die clearance of 5 % showed almost the same temperature curve as 1 % because the achieved cutting surface and thus, the embossing depth is equivalent to the sheet thickness, as can be seen in table 2.

Figure 3b shows the representative temperature curve for 10 % die clearance, which is used for very small embossing depths. The temperature trend at the beginning shows a comparable increase to smaller die clearances. Its maximum of 126 °C is reached at -5 mm right before the complete separation of the sheet metal. Contrary to smaller die clearances, the maximum embossing depth is not equivalent to the sheet metal thickness. The sudden rupture and the loss of contact results in strong vibrations due to the release of elastic energy. Due to undefined contact conditions, no temperature can be measured until the end of the progress.

Table 2. Cutting surfaces and corresponding maximum embossing depth for 1 %, 5 % and 10 % die clearance

Die clearance	Maximum embossing depth (MED)	Cutting surface
1 %	MED is equal to sheet metal thickness	
5 %	MED is equal to sheet metal thicknes	
10 %	MED reached at the dashed line	

3.2. Maximum temperatures as a function of chosen process parameters

Figure 4 illustrates the maximum temperatures as a function of the stroke rate for all die clearances investigated. As already mentioned, the temperature trend of 1 % and 5 % are very similar. Regarding the maximum temperatures, minor differences are ascertainable. For 60 1/min, 1 % die clearance reaches 119 °C (SD = 2 °C) which is 14 % higher than 5 % (104 °C (SD = 6 °C)). In case of 150 1/min the difference amounts about 7 %. At the fastest stroke rate (300 1/min), the smallest die clearance results in a temperature of 230 °C (SD = 3 °C) which corresponds to a 11 % increase compared to 5 % clearance (208 °C (SD = 6 °C)). This shows the influence of process parameters. A good choice allows similar embossing depths with a simultaneous reduction of the occurring temperatures, which for example, can positively affect tool wear. A 10 % die clearance results in a significantly lower temperature of 84 °C (SD = 6 °C) increasing to 101 °C (SD = 1 °C) in the case of 150 1/min. The maximum temperature of 126 °C (SD = 1 °C) could be measured at 300 1/min.

There are two different effects explaining the development of the maximum temperatures occurring during embossing. On the one hand, punch velocity has a great influence on the thermal conduction. The faster the forming velocity is, the less time exists for heat equalization. This results in higher temperatures in the forming zone. Varying die clearances have different consequences. First of all, smaller clearances enable higher deformation degrees of the sheet metal for which more inelastic work must be conducted. This is in accordance with increasing blanking work for decreasing clearances [16]. As a result, more work dissipates into heat and higher temperatures occur. This effect is strengthened by the changing size of the forming zone. A higher clearance entails a larger forming zone in the sheet metal. Consequently, inelastic work dissipates in a smaller area which results in higher temperatures.

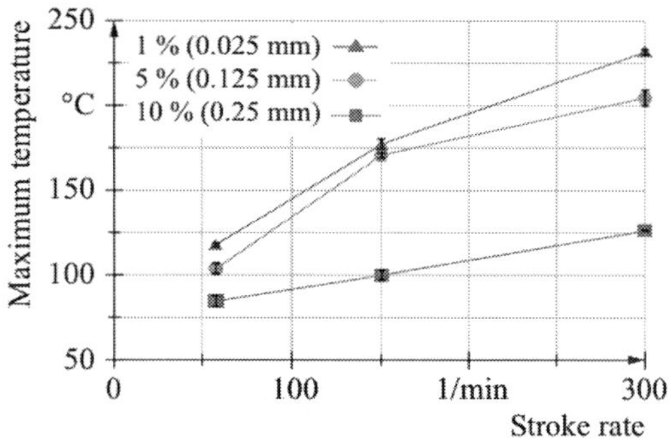

Fig. 4. Maximum temperatures occurring during embossing with clearances of 1 %, 5 % and 10 % over stroke rate

4 Conclusion and outlook

In this paper, a method was presented for measuring temperatures occurring in the forming zone during embossing based on thermoelectricity. The temperature trend shows a characteristic curve that is almost independent of the chosen process parameters. Only slope and maximum temperatures change during all experiments. Due to the dissipation of plastic work, the temperature rises steadily until the maximum embossing depth where the temperature reaches its maximum. The peak value of these temperatures strongly depends on both punch velocity and die clearance.

In the case of punch velocity, no changes in the maximum embossing depth could be observed. In contrast, the occurring temperature at the highest stroke rate and therefore punch velocity increases by 95 % compared to the lowest one. As the maximum embossing depth does not change, higher temperatures emerge because less time is available for temperature equalization in the sheet metal.

The die clearance also shows a significant influence on the peak temperatures. With a stroke rate of 300 1/min, temperature increases by 83 % when embossing with the smallest die clearance. This increase appears for two reasons. First, the maximum embossing depth decreases at larger die clearances, which means that less plastic work is conducted. Consequently, less work dissipates into heat. Secondly, the forming zone grows at bigger clearances and the dissipation is distributed over a larger volume.

Using the smallest die clearance of 1 % and the highest stroke rate of 300 1/min, a temperature of 230 °C was measured. Due to the geometry of the punch, this temperature is a mean value over the whole contact area of punch and sheet metal. Therefore, local temperatures can be even higher, which confirms the significance of the temperature in the forming zone. Besides wear behavior and part quality, the functionality of lubricants can especially be affected, which proves the necessity of temperature consideration during forming operations.

Acknowledgement

The authors would like to express their gratitude to the German Research Foundation (DFG) for the financial support under the grant number VO14876.

References

1. G.I. Taylor, H. Quinney, Proc. Roy. Soc. Lon., **413**:307-325 (1933)
2. A.T. Zehnder, Mech. Res. Comm., **18-1**:23-28 (1991)
3. J. Schulz, W. Holweger, *Wechselwirkung von Additiven mit Metalloberflächen (2010)*
4. P. Demmel, H. Hoffmann, R. Golle, C. Intra, W. Volk, CIRP Ann. – Man. Tech., **64**:249-252 (2015)
5. P. Demmel, T. Kopp, R. Golle, W. Volk, H. Hoffmann, Proc. 14th Int. Con. Met. For. (2012)
6. P. Groche, G. Nietsche, A. Elsen, CIRP Ann. – Man. Tech., **57**:295-298 (2008)
7. S.-H. Kim, J.J. Kang, D.-J. Lee, K.-Y. Lee, H.-Y. Kim, AIP Con. Proc., **908**:877-882 (2007)
8. Z.H. Chen, L.C. Chan, T.C. Lee, C.Y. Tang, J. Mat. Proc. Tech., **138**:610-614 (2003)
9. P. Huml, Draht Fachz., **27**:273-276 (1976)
10. P. Demmel, *In-situ Temperaturmessung beim Scherschneiden*, Dissertation, TU München (2014)
11. P. Tröber, H.A. Weiss, P. Demmel, R. Golle, W. Volk, Dry Met. For. O.A.J., **3**:13-16 (2017)
12. P. Tröber, M. Welm, H.A. Weiss, P. Demmel, R. Golle, W. Volk, Dry Met. For. O.A.J., (to be pulished)
13. P. Demmel, P. Tröber, T. Kopp, R. Golle, W. Volk, H. Hoffmann, Mat. Sci. For., **755**:1-7 (2013)
14. P. Tröber, P. Demmel, H. Hoffmann, R. Golle, W. Volk, CIRP Ann. – Man. Tech., **66**:293-296 (2017)
15. P. Tröber, H.A. Weiss, T. Kopp, R. Golle, W. Volk, Int. J. Mach. T. & Man., **118-119**:91-97 (2017)
16. H.A. Weiss, P. Tröber, R. Golle, S. Steentjes, N. Leuning, K. Hameyer, W. Volk, IEMDC Conf. Proc., 1-7 (2017)

Influence of alternating bending on springback behavior of parts in deep drawing process with macro-structured tools

Ali Mousavi[1,*] *Kevin Sven* Ridolfi[1], and *Alexander* Brosius[1]

[1]Institute of Manufacturing Technology, Chair of Forming and Machining Processes, Technische Universität Dresden, George-Bähr-Str. 3c, 01069 Dresden, Germany

Abstract. In today's industry, waste prevention and an efficient use of resources are becoming more important due to economic and environmental requirements. Especially in forming processes such as deep drawing, a reduced use of lubricants is highly promising for saving resources and reducing production costs. Deep drawing with macro-structured tools is a novel approach to reduce the amount of friction forces and realize a lubricant free process. The induced alternating bending in sheet metal during the deep drawing process with macro-structured tools leads to an increase of its geometrical moment of inertia and consequently stabilizes the sheet against wrinkling. The induced alternating bending increases the back force in the sheet metal during the process and as a result affects the springback behaviour of the formed part. Furthermore, the resulted cyclic loading in flange area can leads to a kinematic hardening of some material. Within the scope of this paper, the influence of alternating bending on springback behaviour of parts in deep drawing process with macro-structured tools is studied. For this purpose, numerical simulations as well as experimental tests to form a U-channel with macro-structured tools are carried out.

Keywords: Springback, Deep drawing, Hardening

1 Introduction

Lubricant is essential in the conventional metal forming to reduce the friction between tool and workpiece and forming energy, increase the forming limit, and to prolong the tool life by prevention of galling and seizure. High amount of wasted lubricants has become a great nuisance of environmental issues [1]. Therefore, lubricant free deep drawing is attractive toward zero emission of lubricants in sheet metal forming [2]. As one of the most promising methods, use of macro-structured deep drawing tools [3] is highlighted because of its performance to reduce the friction forces and simplicity in application. In [4] it is shown that in order to decrease the amount of friction force for a given friction coefficient, the integral of the contact pressure over the contact area can be reduced by means of macro-structured tools. In this way, the contact area between the workpiece and tools will be reduced to a small number of lines.

By reducing the contact area in flange area, the risk of wrinkling in the unsupported sheet metal parts will be increased, because the usually utilized blankholder force is not applicable. By increasing the geometrical moment of inertia of the sheet, this effect can be avoided. In this new developed process, it is achieved by immersing the macro-structured blankholder slightly into the drawing die and inducing an alternating bending mechanism in flange area. The induced alternating bending leads to a cyclic loading in the sheet metal, which can cause a kinematic hardening for some materials. Figure 1 compares the conventional and lubricant free deep drawing process with macro-structured tools.

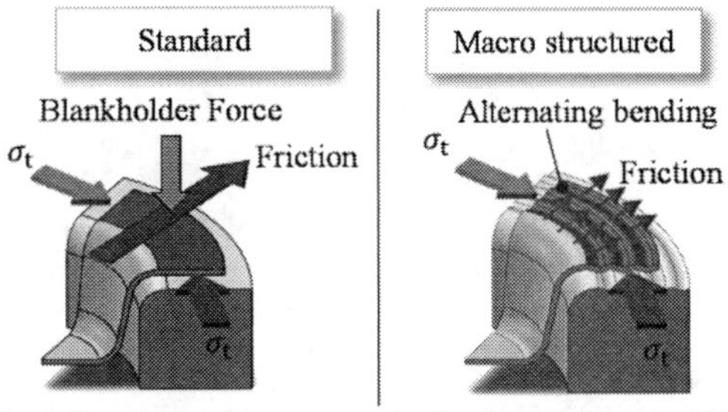

Fig. 1. Comparison between standard and macro-structured deep drawing process.

Depending on the geometry of macro-structures, wave length λ and immersion depth δ (as shown in Figure 2) are two parameters in macro-structured deep drawing process which determine the bending mechanism and can be used as setting parameters to ensure a stable process.

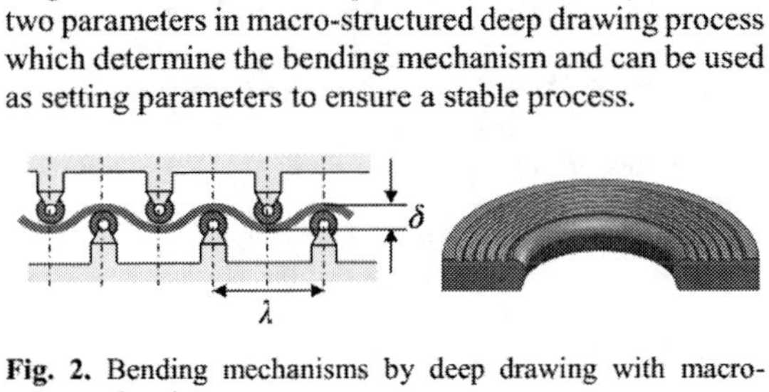

Fig. 2. Bending mechanisms by deep drawing with macro-structured tools.

* Corresponding author: seyed_ali.mousavi@tu-dresden.de

https://doi.org/10.1051/matecconf/201819014005

By increasing the immersion depth or decreasing the wavelength the bending curvature will be increased. Therefore, the restraining force in flange area which can be used to compensate the springback will be increased. From the other side, a kinematic hardening as well as Bauschinger effect due to cyclic loading can tack place. This should be also considered as a secondary effect of process on springback behaviour of products. Figure 3 illustrates schematically the difference between isotropic and kinematic hardening behaviour of materials.
Generally, the Bauschinger effect is very important parameter in sheet metal forming because it has been associated to the influencing parameters on final part geometry such as springback. This phenomenon, which refers to the elastic distortion of a part when forming loads are removed, is a major issue in sheet metal forming and its reduction is of importance [5].

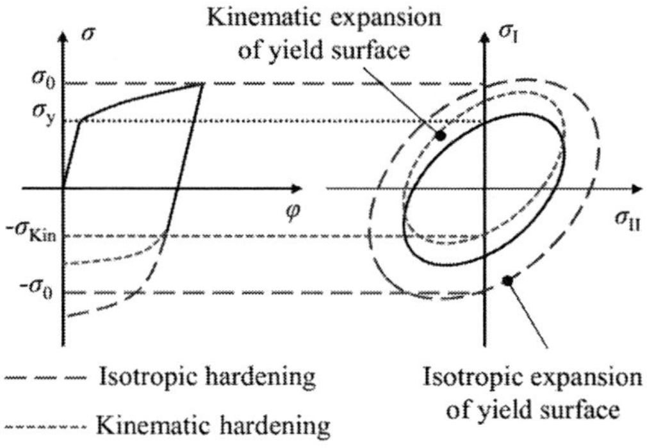

Fig. 3. Comparison between pure isotropic and kinematic hardening.

In order to control the accuracy of parts made by deep drawing with macro-structured tools, it is necessary to have an appropriate investigation about the phenomena that occurs during cyclic loading, such as the Bauschinger effect, the isotropic and kinematic hardening. Within the scope of this paper, the influence of restraining force caused by alternating bending and also the Bauschinger effect as a result of cyclic loading on springback behaviour of parts in lubricant free deep drawing process with macro-structured tools is studied.

2 Methods for reduction of springback in sheet metal forming operations

In sheet metal forming, the shape of the blank obtained at the end of the forming step closely conforms to the geometry of tools. However, as soon as the loads are removed, elastically-driven change in the blank shape and so called springback takes place. In the automotive industry, engineering guidelines and finite element software are used in the design process for new sheet metal parts. Very often during the design process the level of springback is numerically predicted. Based on this prediction, the tools' geometry and process parameters are modified to obtain the desired product shape [6]. However, the current accuracy of springback prediction is not sufficient [7]. Therefore, there is a need to start an extensive experimental trial and error process to determine the necessary tools' geometry and other variables, which will enable production of the required product shape. As a result, the product cost and the time from design to production are increased considerably. There are several commonly used methods to reduce the springback in conventional deep drawing process:

- Blankholder force control: The method is based on increasing the friction force during forming, and thus increasing the tension in the sheet materials flange [8]. However, this can increase the risk of bottom crack in the specimen due to the high friction force.
- Through thickness deformation: In this method the springback is decreased by deforming locally certain corners of the parts [9]. This method can also leads to a locally overloading of the specimen and as a result causes surface defects or even bottom crack.
- Forming in multiple steps: In this method a product is produced by means of several tool sets or by means of one tool set with some additional mechanisms. This way increases the process complexity and results in a time-consuming process.

As it mentioned above, conventional methods for reduction of springback are either difficult to apply or even cause other geometrical problems in the part. However, lubricant free deep drawing process with macro-structured tools can besides the reduction of friction also provide further advantage regarding springback reduction by increasing the restraining force through alternating bending. In order to determine the influence of alternating bending on springback reduction of products, some numerical and experimental investigation are carried out in this study.

3 Objective and approach

In order to determine the effect of restraining force as a result of induced alternating bending on springback behaviour of deep drawn parts with macro-structured tools, draw bending of a U-Chanel is considered. This method is attractive because the level of springback due to the 90° bending is relative large and it can easily be measured. Sensitivity of springback to basic parameters, such as tool radius, sheet thickness, geometric parameters of the tools, mechanical properties of sheet material and friction parameters is usually studied by means of this technique [10]. Furthermore, the U-Channel parts are common elements in industrial sheet metal forming. It appears in many auto-body cover panels like side members and beams.

To get an information about the effect of material on springback in deep drawing with macro-structured tools, three different types of industry-relevant materials like cold rolled DC04 (1.0338) as a low-carbon steel, DP600 (1.0936) as a multiphase high-strength steel and also the aluminium alloy AlMg4.5Mn0.4 (EN AW-5182) are subjected for numerical and experimental tests. In order to study about the effect of restraining force caused by alternating bending on springback behaviour of workpiece, the process without alternating bending (δ =

0.0 mm) is compared to the process with alternating bending (δ = 0.2 mm). Furthermore, in order to examine the effect of kinematic hardening on springback of the workpiece during the process, the materials are considered with pure isotropic, pure kinematic and also combined hardening behaviour in the numerical simulation. The Bauschinger coefficient of each material for the combined hardening rule are taken from literature [11], see Table 1. The process is simulated with a 2D FEM-Model (Marc Mentat v. 2017) with plane strain solid elements (7 elements across the sheet thickness of 1 mm) and a constant friction coefficient of μ = 0.1 based on Coulomb friction model.

Table 1. Bauschinger coefficient of three testing materials [11].

Material	Bauschinger coefficient
DC04	0.70
Al 5182	0.75
DP600	0.40

4 Experimental setup

The springback behaviour of the specimens is investigated by means of draw bending of a U-Channel with macro-structured tool from tool steel 1.2162 (21MnCr5) and die edge radius of 10 mm. The contact area of the tool is polished to reach a roughness of R_a = 0.3 µm and R_z = 0.5 µm. Here a rectangular punch (82 mm × 82 mm) with an edge radius of 10 mm is used. The drawing gap is then 1.75 mm from each side of tool. The gap between blankholder and drawing die can be adjusted by using the spacer strips on the unstructured area of the tools. Figure 4 shows an overview of macro-structured tool with a constant wave length of λ = 9 mm.

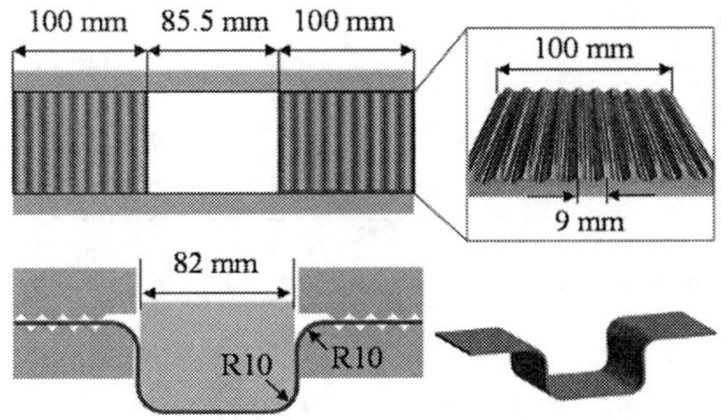

Fig. 4. Geometry of macro-structured draw-bending tool.

All tests are carried out with hydraulic press machine BUP 600 from Zwick / Roell at room temperature with a constant drawing velocity of 5 mm/s and in dry condition. Workpiece strips are cut from DC04, DP600 steel and Al 5182 aluminium alloy with 1 mm thickness, 60 mm width and 280 mm length. Each strip is cold-rolled, cleaned using a citrus based cleaner and finally treated with acetone to remove all traces of pre-lubricants and to assure a comparable test condition. All tests are repeated five times in order to get an information about the statistical repeatability of the results.

5 Results and discussion

In order to simulate the draw bending process with FE-method, the flow curve of all three specimens is determined experimentally through tensile test. The curves are extrapolated up to the equivalent plastic strain of 1.0 regarding the isotropic hardening model of voce based on following relation [12]:

$$\sigma_y = \sigma_\infty - (\sigma_\infty - \sigma_{y0}) \cdot e^{-m\varepsilon} \quad (1)$$

where σ_∞ is the asymptotic value for saturation strength (ultimate strength), σ_{y0} is the initial value of isotropic hardening, m and ε are the characteristic strain constant and equivalent plastic strain respectively. The corresponding parameters for the voce model are listed in Table 2.

Table 2. Parameters for the Voce model for three testing materials.

Material	σ_∞	σ_{y0}	m
DC04	405	175	9
Al 5182	375	150	9.5
DP600	750	350	22

The resulting flow curves for all three materials are shown in Figure 5.

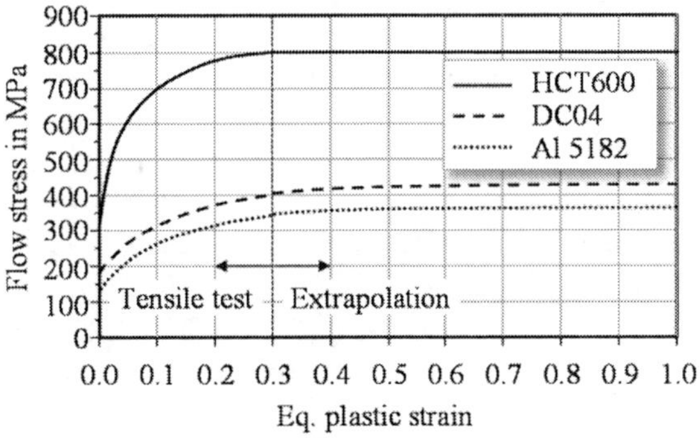

Fig. 5. Flow curve of three testing materials.

The springback behaviour of U-Channels can be characterized regarding the final part geometry with three parameters: the angle between the bottom and the wall θ_1, the angle between the wall and the flange θ_2 and the radius of curvature of the sidewall ρ. As θ_1 and θ_2 increases and ρ decrease, springback increases. Fig. 6 shows the springback geometry of a U-Channel cross section. In this study the effect of process parameters on deflection angel θ_1 is studied.

https://doi.org/10.1051/matecconf/201819014005

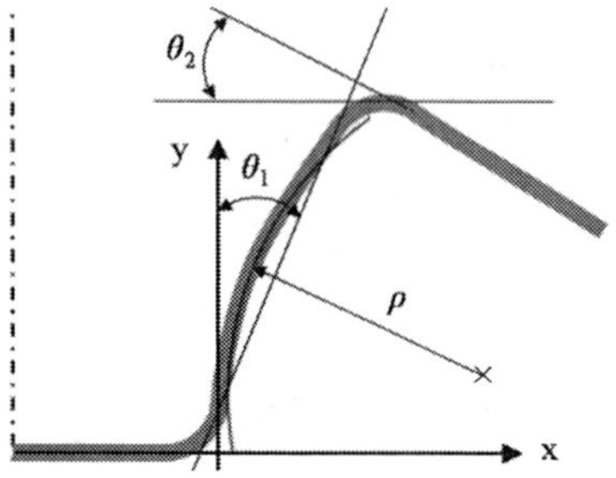

Fig. 6. Schematic overview for springback geometry of a U-Channel.

5.1 Investigation about the effect of back force caused by alternating bending on springback

In order to compare the materials regarding their springback behaviour, specimens from different materials are subjected to experimental test with macro-structured tools under immersion depth of $\delta = 0.0$ and $\delta = 0.2$ mm. As the Figure 7 shows, the springback of workpieces will be reduced through generating an alternating bending mechanism. This is due to the induced tensile force in flange area. The experimental results reveal that the elastic springback of aluminium alloy is higher than DC04 and DP600 because of its smaller value of Young's modulus. Furthermore, the elastic springback of DP600 is higher than DC04 because of its higher yield strength. Therefore, it can be concluded that springback is a severe problem in high-strength aluminium alloys as it is in high-strength steels because of the high strength-to-modulus ratio.
In order to investigate the effect of kinematic hardening as a result of alternating bending on the springback of workpieces numerical simulations are performed.

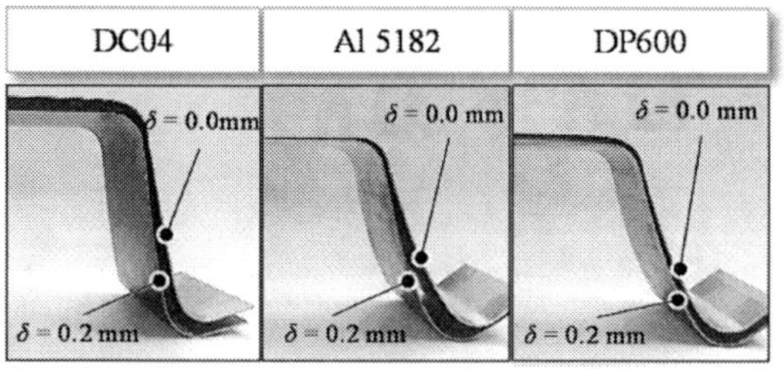

Fig. 7. Springback behaviour of workpieces with and without alternating bending.

5.2 Investigation about the effect of kinematic hardening on springback

The numerical results based on FEM-simulation for draw bending of U-Channel with macro-structured tools reveals the fact that the restraining force caused by alternating bending in flange area compensate the springback of workpiece regardless of hardening type. As the table 3 shows, the springback will be reduced by generating alternating bending for all hardening types in all testing materials. The results show that the alternating bending can reduce the springback effect of materials with pure isotropic hardening more than materials with pure kinematic hardening behaviour. Generally, when the pure kinematic hardening is used for numerical simulations, the springback amount is considerably underestimated, suggesting that the pure kinematic hardening is not adequate to predict the material response of the specimens [13].
Since springback is a function of yield strength, material softening during the alternating bending can slightly affect the amount of springback of materials with predominate kinematic hardening behaviour [14]. However, the experimental as well as numerical results show that considering these opposite effects, the restraining force has a major influence on the springback behaviour of materials.

Table 3. Springback angel of U-Channels under different hardening type.

Material	Numerical						Experiment	
	Pure isotropic		Pure kinematic		Combined		-	
	δ in mm		δ in mm		δ in mm		δ in mm	
	0.0	0.2	0.0	0.2	0.0	0.2	0.0	0.2
DC04	8.8°	4.1°	9.8	7.5°	9.2°	5.0°	9.7°	7.4°
Al 5182	14.2°	9.6°	15.6°	11.3°	14.5°	10.2°	17.8°	12.1°
DP600	14.8°	11.2°	18.0°	13.6°	16.9°	12.9°	18.4°	15.2°

6 Summary and conclusion

Today, more than 15% of the total cost of sheet-metal forming process is spent on lubricating liquids (buying the lubricants, lubrication process, cleaning the parts for following processes, waste disposal) [15]. Moreover, lubrication has a great impacts on the environment and the health of the workers. Thus, it is of importance to develop a process which permit a significant reduction of the amount of lubricating agents or even dry metal operation. Deep drawing with macro-structured tools is a new developed process which can reduce the friction force through minimization of contact area. The induced alternating bending can besides the stabilization of material against wrinkling leads to compensation of springback. A part of total punch force in deep drawing process with macro-structured tools is to bend and unbend of sheet metal between the tools structures. This leads to a tension in sheet metal in drawing direction which can compensate the springback behaviour of workpiece. Experimental results reveals that the springback can be reduced between 17 to 32% depending on the material by $\delta = 0.2$ mm. The amount of springback reduction can be slightly affected through Bauschinger effect by materials with predominate kinematic hardening behaviour.

Acknowledgement

The German Research Foundation (DFG) is greatly acknowledged for financial support (Schwerpunktprogramm "Trockenumformen – Nachhaltige Produktion durch Trockenbearbeitung in der Umformtechnik" (SPP 1676)).

References

[1] S. Kataokaa, M. Murakawa, T. Aizawa, H. Ike, J. Surf. and Coat. Techn, **177-178**, pp 582-590 (2004)
[2] F. Vollertsen, F. Schmidt, I. J. Prec. Eng. And Manu. Techn, **1**, pp 59-62 (2014)
[3] A. Mousavi, T. Kunze, T. Roch, A. Lasagni, A. Brosius, ICTP, **207**, pp 48-53 (2017)
[4] A. Brosius, A. Mousavi, CIRP, **65**, pp 253-256 (2016)
[5] F. Barlat, J. J. Gracio, M.-G. Lee, E. F. Rauch, G. Vincze, I. J. Plast., **27**, pp 1309-1327 (2011)
[6] I. Burchitz, *Springback: improvement of its predictability Literature study report* (NIMR project MC1. 02121, 2005)
[7] S. Sumikawa, A. Ishiwatari, J. Hiramoto, T. Urabe, JMPT **230**, pp 1-7 (2016)
[8] G. Liu, Z. Lin, W. Xu, Y. Bao, JMPT, **120(1-3)**, pp 259-264 (2002)
[9] I. N. Chou, C. Hung, I. J. Mach. T. and Manu. **39(3)**, pp 517-536 (1999)
[10] T. Meinders, A. W. A. Konter, S. E. Meijers, E. H. Atzema, H. Kappert, I. J. F. Process, **9(3)**, 365-402 (2006)
[11] Q. Yin, *Verfestigungs- und Schädigungsverhalten von Blechwerkstoffen im ebenen Torsionsversuch,* (Technischen Universität Dortmund 2014)
[12] E. Voce, J. Inst. Met. **74**, pp 537-562 (1948)
[13] M.-G. Lee, D. Kim, C. Kim, M. L. Wenner, K. Chung, I. J. Plast. **21**, pp 915-953 (2005)
[14] K. Ahn, D. Yoo, M. H. Seo, S.-H. Park, K. Chung, Met. Mater. Int. **15(4)**, pp637-647 (2009)
[15] K. Taube, J. Surf. and Coat. Techn. **98**, pp 976-984 (1998)

Tool surface with a supporting plateau of hard particles for deep drawing of high alloy steel

Hannes Freiße[1,*] and *Thomas* Seefeld[1]

[1]BIAS – Bremer Institut für angewandte Strahltechnik GmbH, 28359 Bremen, Germany

Abstract. Sheet metal forming normally requires the application of lubricants to protect the tool and the sheet against wear. The parts must be cleaned to remove the lubricants before joining and coating. This process step wastes energy and water resources. In the case of non-lubricated sheet metal forming, cleaning processes would not be necessary anymore and the process chain could be optimized regarding ecological and economical aspects. However, forming without lubrication leads to an intensive contact between the tool and the sheet. Thus, higher wear occurs and process reliability cannot be ensured for industrial mass production. High alloy steels are applied for mass-market products e.g. for appliances. Because of the higher strength, strain hardening and galling effects the austenitic steels are comparatively difficult to form. For dry metal forming of high alloy steels new tool concept must be developed to withstand the higher loads. In this work, a laser generated tool surface with a supporting plateau of hard particles (metal matrix composite (MMC-surface)) is presented. Spherical fused tungsten carbides were injected into the surface by laser melt injection. The metallic matrix of the composite was rejected by applying laser ablation. In consequence, the hard particles stood out of the matrix and were in direct contact with the sheet material. The surface of hard particles had a high hardness about 3000 HV and less metallic character. Cold working steel and aluminium bronze were tested as reference tool materials. Dry and lubricated forming experiments were carried out by strip drawing with bending and deep drawing of cups. Dry deep drawing of cups was not possible by using cold work tool steel. This can be traced back to the occurrence of wrinkles and cup base fracture at the same time. Applying aluminium bronze as tool material for dry metal forming resulted in high adhesive wear. Within this work the feasibility of dry metal forming of high alloy steel could be demonstrated by applying the MMC-surface whereby adhesive wear could be reduced.

Keywords: deep drawing, laser beam machining (LBM), metal matrix composite

1 Introduction

The high alloy steel (1.4301, X5CrNi18-10) is applied in industrial mass production e.g. in automotive industry or home appliance manufacturing because of the corrosion resistance and high fracture strain. This steel has metastable austenitic structure at room temperature. During forming the austenite can transform into martensite (TRIP-effect (Transformation Induced Plasticity)). An approach of calculating the martensite content depending on the strain is given in [1]. Because of the martensitic strain hardening, flow stress increase and higher forming forces are required [2]. The formation of martensite depends, among other, on the temperature [3]. In industrial forming processes the temperature may rise to 100 °C within the first 100 strokes [4]. Calculated heat transfer coefficient (HTC) of aluminium bronze and cold working tool steel against high alloy steel depending on the surface pressure are given in [5]. Aluminium bronze is a proofed tool material to form high alloy steel in lubricated deep drawing because of the high HTC [6].

However, to realize dry metal forming new tool concepts must be developed to withstand the higher loads [7]. In scientific investigations the surface condition before forming is rarely described, e.g. how the tools and sheets were cleaned which amount of residual contaminations were present on the surface [8].

Dry sliding of high alloy steel against tool steel 1.2379 resulted in high friction coefficients and adhesive wear [9]. In contrast, good results in dry deep drawing were achieved by applying ceramic tools [10]. Ceramics are expensive and add complexity to set-up. Furthermore, ceramics are brittle and particularly susceptible to cracking. Another approach to achieve a higher wear resistance is to reinforce the metallic surface with hard particles (metal matrix composite (MMC)) [11]. MMC combine the hardness and wear resistance of the particles with the ductility of the metal and can be generated e.g. by laser melt injection [12]. Investigations in strip

* Corresponding author: freisse@bias.de

drawing tests with bending revealed that MMC tool surfaces are suitable for lubricated sheet metal forming [13]. Literature presents only a few investigations on lubricated as well as in dry sheet metal forming by using MMC tool surfaces.

2 Experimental Details

2.1 General

Forming experiments were carried out by strip drawing with bending and deep drawing of cups. The description of the forming apparatus is given in [14].

In this work, three tool materials were applied. The MMC tool surface was made of aluminium bronze (CuAl10Ni5Fe4) reinforced with spherical fused tungsten carbides (SFTC)) with a particle content of MMC of about 50%. Experimental details of reinforcing the surface with hard particles by laser melt injection and of producing a supporting plateau out of hard particles by ablation of the matrix using ultra short pulse laser are described in [15]. Two reference tool materials were tested: aluminium bronze (CuAl10Ni5Fe4) and cold working tool steel (1.2379). The workpiece sheet material was cold-rolled stainless steel 1.4301 with a sheet thickness of 0.5 mm. The sheet surface was not textured and the roughness Sa was 0.35 ± 0.01 µm. The hardness was 173 ± 4 HV0.5. Mechanical properties were determined in tensile tests according to DIN EN 10002-1 (Table 1 and Figure 1). The average of the normal anisotropy coefficient R was calculated by equation (1) and the planar anisotropy factor R_p was calculated by equation (2). The average of the normal anisotropy coefficient was 1.21 and the planar anisotropy coefficient was -0.57.

$$R = \frac{1}{4} \cdot (R_0 + 2 \cdot R_{45} + R_{90}) \quad (1)$$

$$R_p = \frac{1}{2} \cdot (R_0 - 2 \cdot R_{45} + R_{90}) \quad (2)$$

Table 1. Mechanical properties of the sheet.

property	load direction	value
tensile strength	rolling direction (0°)	729 ± 1.1 MPa
	diagonal direction (45°)	663 ± 1.5 MPa
	transverse direction (90°)	671 ± 10.2 MPa
yield strength	rolling direction (0°)	281 ± 0.9 MPa
	diagonal direction (45°)	262 ± 0.8 MPa
	transverse direction (90°)	264 ± 5.9 MPa
fracture strain	rolling direction (0°)	52 ± 0.3%
	diagonal direction (45°)	60 ± 0.8%
	transverse direction (90°)	60 ± 1.9%
normal anisotropy coefficient	rolling direction (0°)	0.81 ± 0.009
	diagonal direction (45°)	1.49 ± 0.05
	transverse direction (90°)	1.04 ± 0.36

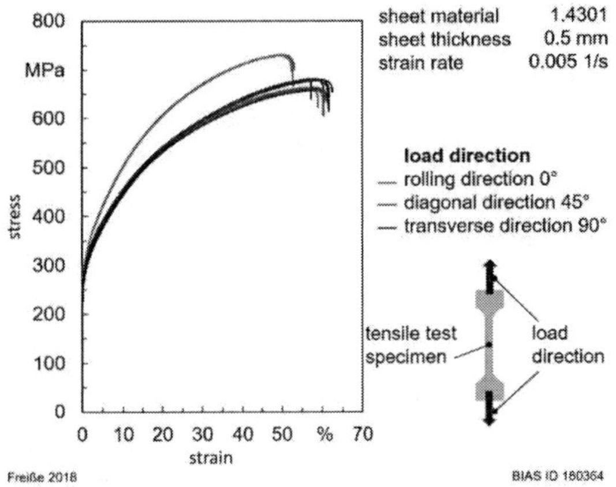

Fig. 1. Stress-strain diagram of the sheet metal

To investigate the behaviour for dry metal forming the sheets and the tools were cleaned in an ultrasonic bath with 10% solution of Tickopur R33 cleaning agent for 10 minutes. Afterwards, the parts were rinsed with deionized water, submerged in pure ethanol for 5 seconds and dried with a fan.

At first all dry forming experiments and thereafter all lubricated experiments were carried out with the same tool. The tools were not removed and mounted again between the dry and lubricated experiments to avoid assembly-related influences. In each case three parts were formed. The maximum forming forces were measured, the average and the standard deviation were investigated.

The surface roughness Sa was measured according to EN ISO 25178 with a measurement area of 200x200 µm² using 3D laser scanning confocal microscope Keyence VK-9700 with a magnification of 50x. The low-pass filter (S-filter) was adjusted to 0.5 µm (lowest value) and the high-pass filter (L-filter) was set to 0.2 mm (edge length of the measurement field).

2.2 Strip drawing with bending

A sketch of strip drawing with bending is depicted in Figure 2. The tools were machined by wire cut electrical discharge machining (EDM). The punch diameter was Ø 30 mm and the punch radius was 4 mm. The sheets were of 52x20 mm² and were cut in rolling direction (0°).

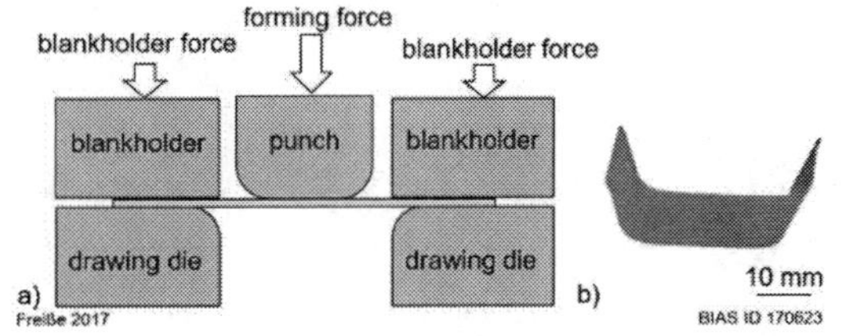

Fig. 2. a) Sketch of strip drawing with bending, b) bended strip as a U-shaped component.

A strip drawing die with MMC surface is shown in Figure 3. After EDM machining, the surface roughness Sa amounted to 0.75 ± 0.04 µm. Pores were visible in the

MMC surface which resulted from the process of laser melt injection. At first forming experiments were performed without depression. Afterwards the surfaces were laser ablated and a depression of 5 µm was set.

Forming experiments were carried out by variation of the drawing radius (from 5 mm to 7 mm) and two different initial blankholder pressures were applied (1.6 MPa and 3.2 MPa).

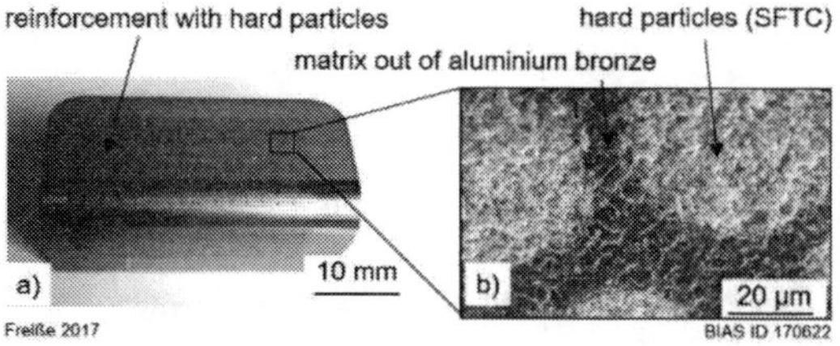

Fig. 3. a) drawing die reinforced with hard particles machined by wire electrical discharge machining (EDM), b) scanning electron microscope (SEM) image of the surface.

The tools of aluminium bronze and cold working tool steel were hardened. The roughness Sa of the bronze surface amounted to 0.36 ± 0.05 µm and of the steel surface was 0.45 ± 0.03 µm. Forming experiments were carried out by variation of the drawing radius (3 mm to 7 mm), by variation of the drawing clearance (0.7 mm to 1.5 mm) and two different initial blankholder pressure were applied (1.6 MPa and 3.2 MPa).

2.3 Deep drawing of cups

Details of the experimental set-up are given in [14]. The die radius was 4 mm. The initial blankholder pressure was 3.2 MPa. Drawing dies were tested with a MMC surface, aluminium bronze and cold working steel (Table 2). Eight tools were applied: three tools with a drawing clearance of 0.7 mm and five tools with a drawing clearance of 1 mm. The tools were machined by milling and die-sinking electrical discharge machining (EDM).

Table 2. Material, manufacturing process and drawing clearance of the drawing dies.

tool no.	material	machined	drawing clearance
1	bronze	milled	0.7 mm
2			
3			1 mm
4			
5	steel		0.7 mm
6	bronze	die-sinking EDM	1 mm
7	MMC		
8			

Dry and lubricated forming was performed (3 sheets each) and afterwards the tools were removed, cleaned and the set of experiments (dry and lubricated) was repeated one more time. The maximum drawing ratio in dry metal forming was investigated by varying the blank diameter and using bronze tool 3.

The milled drawing diess out of steel had a surface roughness Sa of 0.3 ± 0.03 µm and the milled drawing dies out of bronze had a surface roughness Sa of 0.6 ± 0.02 µm.

The micro geometry of the MMC surface produced by die sinking electrical discharge machining (EDM) differed significantly from the wire cut EDM machined surface of the tools for strip drawing with bending. No plane surface was produced by die sinking EDM (see Figure 4). The aluminium bronze matrix was removed stronger. The depression varied between 20 µm and 100 µm. The wire cutting resulted in uneven removal of the surface, where removal of the matrix material was more pronounced than of the hard particles, see Figure 4. The height difference was 20-100 µm.

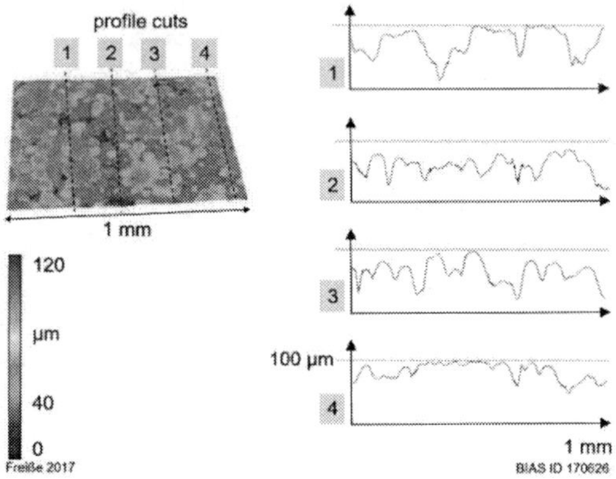

Fig. 4. 3D laser scanning confocal microscope image and profile cuts of the MMC surface machined by die-sinking electrical discharge machining (EDM).

2.4 Calculated true blankholder pressure

The initial blankholder force was 3.2 MPa for this particular calculation. The blankholder force was kept constant and the parts were completely drawn in. So, the area under the blankholder decreased and the true blankholder pressure tended towards infinity. Exemplarily calculated blankholder pressures up to a sheet draw-in of 11 mm are depicted in Figure 5. The larger the drawing radius and the wider the drawing clearance, the earlier was the increase of the true blankholder pressure.

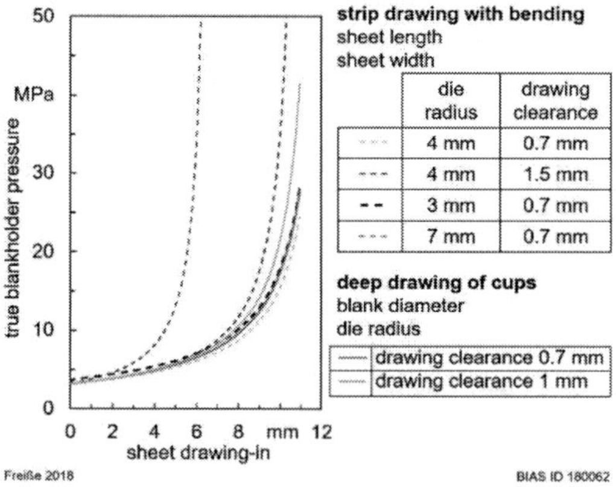

Fig. 5. Example on calculated blankholder pressures up to a sheet draw-in of 11 mm.

Furthermore, the amount of sheet under the blankholder depends on the die geometry. The smallest sheet drawing-in of 6.6 mm occured in strip drawing when bending with a die radius of 7 mm and whereas the largest drawing-in of 12.6 mm occurred when bending with a die radius of 3 mm.

3 Results

3.1 Strip drawing with bending

3.1.1. Reference tool materials

The influence of the drawing clearance on the forming force by using bronze tool is shown in Figure 6 and by using steel tool is shown in Figure 7.

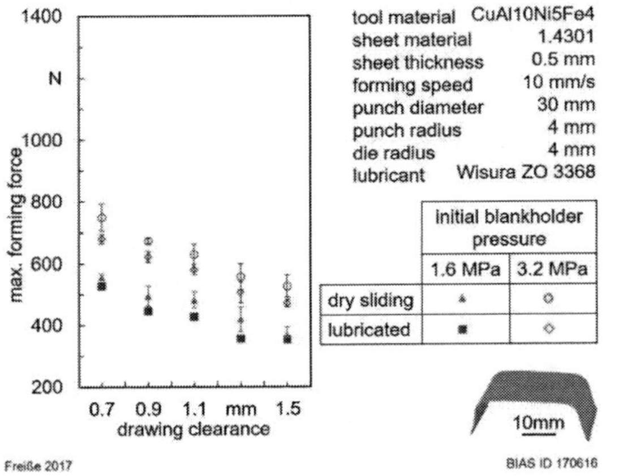

Fig. 6. Max. forming forces using bronze tool under variation of the drawing clearance.

No significant influence of the tool material was investigated even in dry forming. The forming forces fluctuated between 353 N and 800 N. Applying either lubrication or smaller drawing clearance or lower blankholder pressure resulted in lower forming forces.

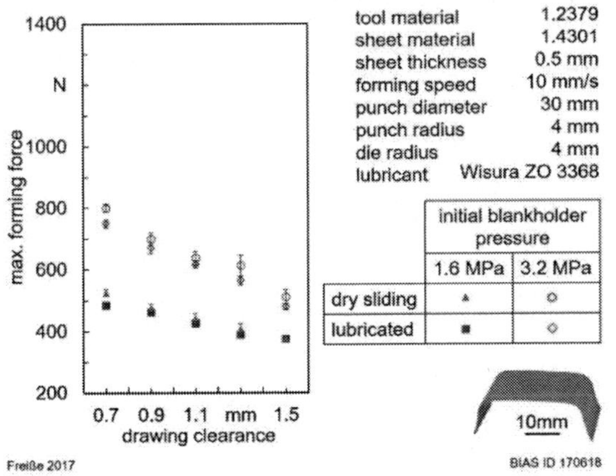

Fig. 7. Max. forming forces using steel tool under variation of the drawing clearance.

A significance analysis was carried out for each experimental variation: increasing the initial blankholder pressure, forming without lubrication and reducing the drawing clearance.

For calculating the impact factor on the increased forming force by higher initial blankholder pressure and forming without lubrication, at first the ratio between the high and low max. forming force for each drawing clearance was determined. E.g. the influence factor for increasing the initial blankholder pressure for a drawing clearance of 0.7 mm by using steel tool (Fig. 7), the ratio in dry sliding was calculated by: 800 N/525 N = 1.52 and the ratio for lubricated sliding was calculated by: 749 N/485 N = 1.54 N. Thereafter, the averages and standard deviations of these ratios for all drawing clearances were calculated (1.45 ± 0.07).

E.g. the influence factor for forming without lubrication for a drawing clearance of 0.7 mm by using steel tool (Fig. 7), the ratio for forming by using a low initial blankholder pressure was calculated by: 525 N/485 N = 1.08 and the ratio for forming by using a high initial blankholder pressure was calculated by: 800 N/749 N = 1.07. Thereafter, the averages and standard deviations of these ratios for all drawing clearances were calculated (1.05 ± 0.02).

Increasing the initial blankholder pressure had a higher influence compared to forming without lubrication (Table 3).

Table 3. Factors for increasing the forming forces in the case of using bronze tool and steel tool.

variation	bronze tool	steel tool
increasing the initial blankholder pressure	1.36 ± 0.05	1.45 ± 0.07
without lubrication	1.10 ± 0.04	1.05 ± 0.02

The significance analysis of variation the drawing clearance was carried out by a linear regression analysis. The results of the linear regression analyse in the case of using bronze tool is given in Table 4 and in the case of using steel tool is given in Table 5. The influence of the drawing clearance was higher in the case of higher initial blankholder pressure. The slopes of the regression were slightly higher in the case of forming without lubrication.

Table 4. Regression analysis of variation of the drawing clearance in the case of using bronze tool

lubrication	initial blankholder pressure	slope of the regression	determination coefficient R^2
with	1.6 MPa	y = -44.2x + 554.9	0.923
without		y = -45.7x + 601.3	0.973
with	3.2 MPa	y = -53.2x + 731.9	0.991
without		y = -56.4x + 796.8	0.984

Table 5. Regression analysis of variation of the drawing clearance in the case of using steel tool

lubrication	initial blankholder pressure	slope of the regression	determination coefficient R^2
with	1.6 MPa	y = -26.6x + 515.7	0.986
without		y = -36.7x + 555.9	0.993
with	3.2 MPa	y = -64.6x + 809.5	0.992
without		y = -66.6x + 851.47	0.964

The maximum forming forces under variation of the drawing radius using bronze tool are shown in Figure 8 and for the steel tool in Fig. 9.

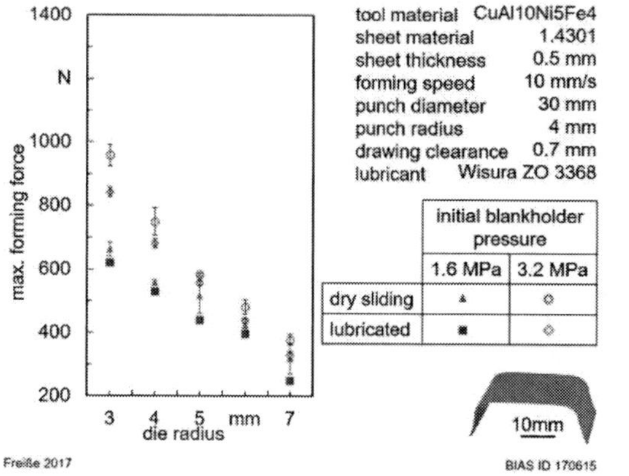

Fig. 8. Max. forming forces using bronze tool under variation of the drawing radius.

The forces were slightly higher in the case of using steel tool with smaller drawing radius compared to bronze tool. Forming forces from 247 N to 1058 N were measured. The forces were lower in the case of lubricated forming compared to dry metal forming. By applying higher blankholder pressure the forming forces were increased. Forming with smaller drawing radius resulted in higher forming force.

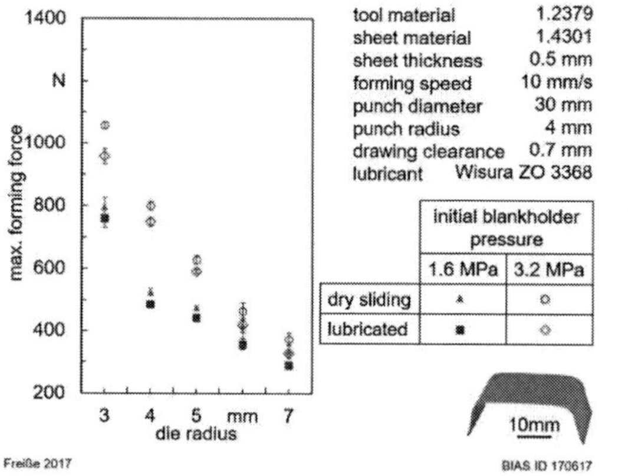

Fig. 9. Max. forming forces using steel tool under variation of the drawing radius.

The main influence factor increasing the forming force was increasing the initial blankholder pressure (Table 6). The lowest impact on the forming forces was determined by dry forming.

Table 6. Factors for increasing the forming forces in the case of using bronze tool and steel tool.

variation	bronze tool	steel tool
increasing the initial blankholder pressure	1.25 ± 0.1	1.28 ± 0.15
without lubrication	1.19 ± 0.06	1.10 ± 0.04

The linear regression analysis of variation the drawing radius in the case of using bronze tool is given in Table 7 and in the case of using steel tool is given in Table 8. The influence of the drawing radius was higher in the case of higher initial blankholder pressure. It is remarkable that the slope of the regression was lower by forming without lubrication in the case of lower initial blankholder pressure for both tool materials. Furthermore, the influence of the drawing radius was higher when steel was used as tool material.

Table 7. Regression analysis of variation of the drawing radius in the case of using bronze tool.

lubrication	initial blankholder pressure	slope of the regression	determination coefficient R^2
with	1.6 MPa	y = -87.9x + 710.2	0.974
without		y = -82.1x + 742.43	0.983
with	3.2 MPa	y = - 126.7x + 949.4	0.994
without		y = - 142.8x + 1058.9	0.973

Table 8. Regression analysis of variation of the drawing radius in the case of using steel tool.

lubrication	initial blankholder pressure	slope of the regression	determination coefficient R^2
with	1.6 MPa	y = - 107.1x + 787.9	0.878
without		y = - 103.6x + 819.9	0.867
with	3.2 MPa	y = - 159.1x + 1086	0.984
without		y = - 170.6x + 1176	0.97

3.1.3 MMC tool

In Figure 10 the forming forces are shown in the case of using MMC tool without depression. Forming forces from 365 N to 1130 N were measured. The results of the forming tests by applying the MMC tool surface with a depression of 5 µm are depicted in Figure 11.

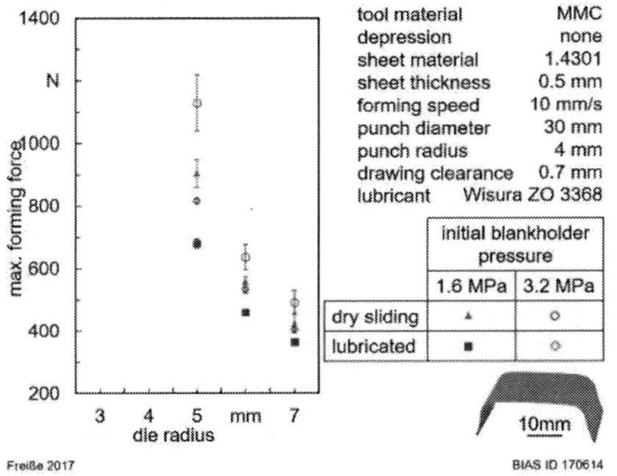

Fig. 10. Max. forming forces using MMC tool without depression.

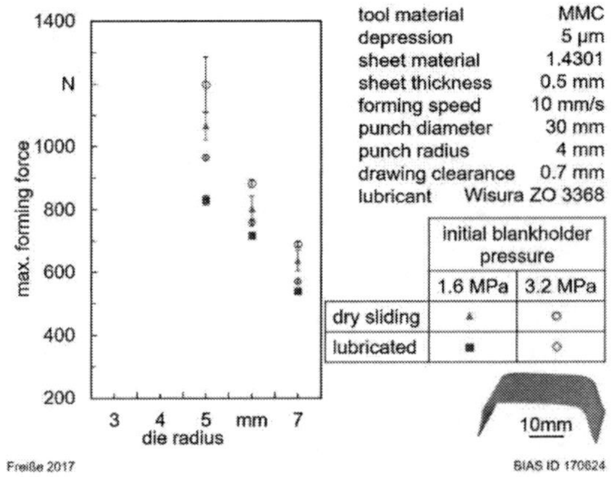

Fig. 11. Max. forming forces using MMC tool with 5 µm depression.

The values of the forming forces varied between 536 N and 1197 N. The forces increased when using smaller drawing radius, higher blankholder pressure and no lubrication. These effects can be observed by applying the MMC tool surface without as well as with a supporting plateau of hard particles.

Factors for increasing the forming forces by variation in the case of using MMC tool surface with and without lubrication are given in Table 9. The lowest impact on the forming force resulted by applying higher blankholder pressure. The influence of forming without lubrication had higher influence on the forming force compared to the increased initial blankholder pressure. Particularly, using a MMC surface with depression led to lower impact of the increased initial blankholder pressure and avoidance of lubrication on the forming force.

Table 9. Factors for increasing the forming forces in the case of using MMC tool surface.

variation	without depression	5 µm depression
increasing initial blankholder pressure	1.17 ± 0.04	1.09 ± 0.03
without lubrication	1.25 ± 0.08	1.20 ± 0.05

The results of the linear regression analysis of variation the drawing radius in the case of using MMC tool without depression is given in Table 10 and in the case of using MMC tool with depression is given in Table 11.

Table 10. Regression analysis of variation of the drawing radius in the case of using MMC tool without depression.

lubrication	initial blankholder pressure	slope of the regression	determination coefficient R^2
with	1.6 MPa	y = -157.4x + 1131.2	0.951
without		y = -238.8x + 1585.5	0.941
with	3.2 MPa	y = -205.5x + 1406.8	0.955
without		y = -319.4x + 2030.3	0.909

Table 11. Regression analysis of variation of the drawing radius in the case of using MMC tool with depression.

lubrication	initial blankholder pressure	slope of the regression	determination coefficient R^2
with	1.6 MPa	y = -145x + 1275.7	0.983
without		y = -215x + 1695.1	0.984
with	3.2 MPa	y = -197.2x + 1553.2	0.999
without		y = -254.4x + 1940.5	0.982

The slope of the regression was increased either by applying higher initial blankholder pressure or by forming without lubrication. It is remarkable that the influence of the drawing radius on the forming force was lower by using MMC with depression in the case of higher initial blankholder pressure.

3.2. Deep drawing of cups

3.2.1. Milled tools

In Figure 12 the deep drawing forces are shown by using two tools in each case with a drawing clearance of

https://doi.org/10.1051/matecconf/201819014006

0.7 mm and 1 mm. The deep drawing forces were lowered about 15% by applying higher drawing clearance in the case of lubricated deep drawing. A reproducibility of the lubricated forming results could be achieved by applying different tools. In contrast, in dry metal forming there were high deviations between the results of applying different tools with the same drawing clearance. On the average, the forces were lower in the case of applying higher drawing clearance. Significant higher standard deviations of the forming forces were found in the case of dry metal forming.

The experiments were repeated with the tool 3 and 4 (Figure 13). The tools were removed, cleaned and mounted again. The deep drawing forces were reproducible in the case of lubricated forming.

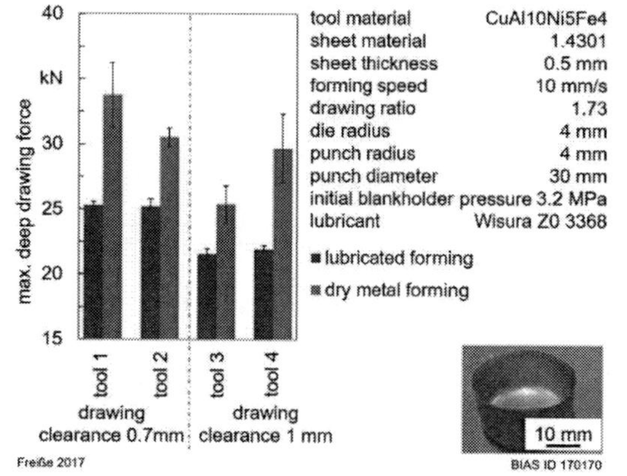

Fig. 12. Max. deep drawing forces using milled bronze tool under variation of the drawing clearance.

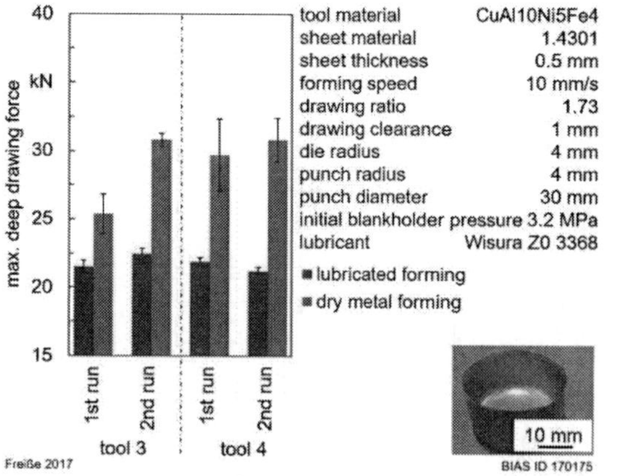

Fig. 13. Max. deep drawing forces using milled bronze tool by repeated forming tests.

The highest difference was about 4% by using tool 3. However, the results differ strongly in the case of dry metal forming. The highest difference was about 22% by using tool 3.

By forming different blank diameters from 52 mm to 62 mm and using bronze tool 3, the drawing ratio was varied from 1.73 to 2.07. In dry metal forming, a maximum drawing ratio of 1.93 was investigated (Figure 14). Using larger blank diameter than 58 mm in dry metal forming resulted in cup base fractures. By reducing the initial blankholder pressure, the simultaneous occurrence of cup base fracture and wrinkles formation of first order was observed which was an indication for the process limit in dry metal forming.

Blank diameter of 62 mm could be formed in lubricated deep drawing. Larger diameters could not be applied because of geometric circumstances of the deep drawing apparatus.

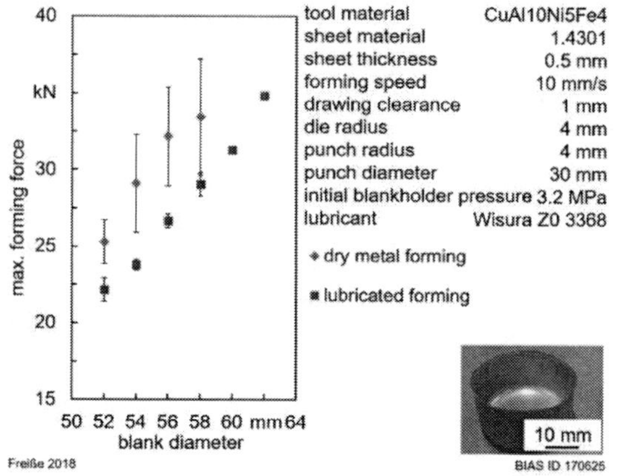

Fig. 14. Max. deep drawing forces using milled bronze tool under variation of the blank diameter using bronze tool 3.

The max. forming forces in lubricated forming by using steel tool were comparable to those in the case of using bronze tool. Dry metal forming by using steel tool was not possible. Applying a blankholder pressure of 3.2 MPa resulted in cup base fracture. When the blankholder pressure was reduced to 2.8 MPa, wrinkle formation of the first order and cup base fracture occurred at the same time (Figure 15).

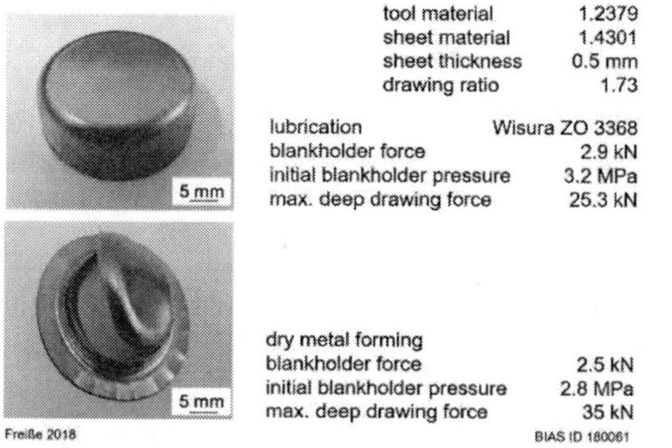

Fig. 15. Deep drawing by using tool out of tool steel and a drawing clearance of 0.7 mm.

3.2.2 EDM tools

In Figure 16 the max. forming forces are shown by using tools machined by die-sinking EDM out of bronze and MMC. The forming forces by using the EDM-machined bronze tool were about 20% lower compared to the forming forces investigated by using the milled bronze

https://doi.org/10.1051/matecconf/201819014006

tool. Dry metal forming led to higher forming forces compared to lubricated forming. The forces were higher by applying the MMC tools. Two MMC tools were tested and a reproducible result could be achieved.

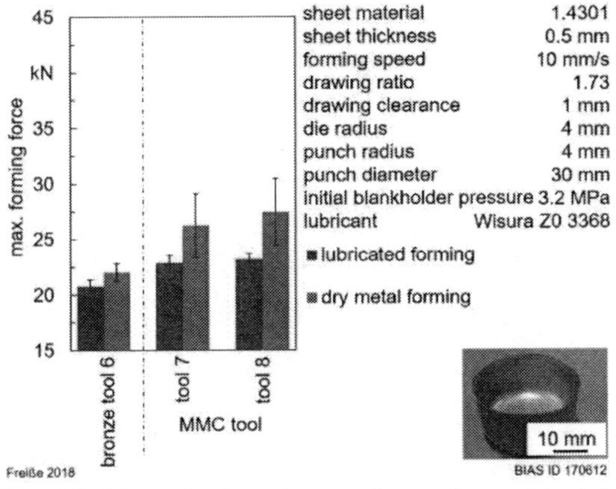

Fig. 16. Max. forming forces using tools machined by die-sinking EDM.

The formed cups were analysed by EDX (Electron Diffraction X-Ray) analysis (Figure 17). Adhesive wear was determined. This can be traced back to the detected copper (Cu) from the tool on the steel sheet (Fe). The adhesive wear could be significantly reduced by applying the MMC tool. However, more scoring occurred when MMC acted as tool material. It is assumed that the particles penetrated the sheet material and caused the scoring. The anisotropic properties of the cold rolled sheet led to earing at the cups. Exemplarily hardness measurement revealed that the hardness in the sheet was increased up to 360 HV0.5. The cups became magnetic which can be traced back to the formation of martensite.

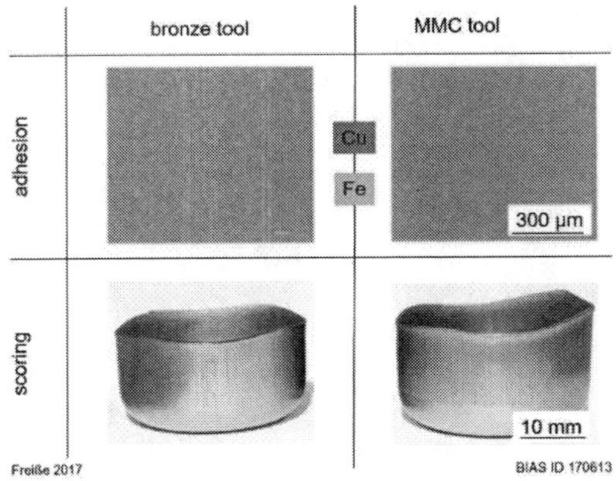

Fig. 17. Adhesive and abrasive wear in the deep drawn cups using bronze tool with and without hard particles.

4 Discussion

Sheet forming experiments were carried out by variation of the drawing clearance and drawing radius to investigate the influence on the forming force. Lower drawing clearances [16] and smaller drawing radius [17] lead to higher forming forces. This could be confirmed in this work by strip drawing with bending and deep drawing of cups. The deep drawing forces were about 15% lower in lubricated deep drawing of cups by increasing the drawing clearance from 0.7 mm to 1 mm. This is comparable to the results of strip drawing with bending. By using a regression analyse of the results of strip drawing with bending, it could be shown that the influence of the drawing radius was significantly higher compared to the influence of the drawing clearance on the forming force. Furthermore, the influence of the drawing radius as well as the influence of the drawing clearance was higher in all experiments in strip drawing with bending when the initial blankholder pressure was increased.

Dry deep drawing of cups could be carried out by using the reference tool material bronze. There was a high difference of the forming forces in the case of dry metal forming when two different tools with the same geometry were applied. It was assumed that this was caused by production-related tolerances of the drawing dies. However, by using the same tool after removing, cleaning and mounting again the forming forces differed in dry metal forming. This may be traced back to assembly-related tolerances. However, there was no impact of production-related and assembly-related influences in lubricated forming.

In this work, lubricated forming by using tool steel led to comparable forming force to those by using bronze tool. However, dry forming with steel tool resulted in cup base fracture and wrinkle formation. So, there was no influence of the tool material in lubricated forming but a high influence of the tool material in dry deep drawing of cups. This agrees with the results of strip drawing without bending in previous work [18]. Bronze showed low friction coefficient against high alloy steel even at higher blankholder pressure. Dry sliding using tool steel led to high friction coefficient and adhesive wear even at low contact pressure. In contrast, in this work the forming forces were in some cases higher in strip drawing test with bending when bronze was used as tool material. However, slightly higher forming forces resulted when cold working tool steel was applied as tool material. The factor by which the forces were higher in the case of steel tool amounted to 1.04 ± 0.05. This was not influenced by variation of the drawing radius nor of the drawing clearance. It can be concluded that the results of the strip drawing with bending cannot be compared with results of strip drawing without bending or with the results of forming cups.

Avoiding lubrication increased the forming forces by 5% to 19% in strip drawing with bending using the reference tool materials. By using the MMC surface in strip drawing with bending the forming forces were increased from 22% to 23% by avoiding lubrication even in the case the surface had no depression. It is assumed that this was caused by the different tool properties in the composite material. The Young's modulus of the spherical tungsten carbide [19] is about five times higher compared to the Young's modulus of the aluminium bronze [20]. It is assumed that the matrix of the MMC was slightly impressed when high contact pressure occurred during forming particularly under the blankholder. In consequence the particles stood out of the surface and penetrated in the sheet material. This acted as a mechanical hindrance to the movement and higher forming forces were investigated. By the regression analysis it could be shown that ablation the matrix by 5 µm and using a plateau of hard particles had no significant influence on the forming forces. Furthermore, in deep drawing of cups with higher penetrations of 20 µm to 100 µm in the die-sinking EDM-machined MMC surface showed the same difference of dry forming to lubricated of around 22%. So, higher penetrations do not significantly influence the forming forces. It was shown that adhesive wear could be reduced by using MMC surface with higher penetration for deep drawing of cups. It is assumed that the high penetration reduced the contact area between the sheet and the bronze matrix. However, using penetration in MMC surface lead to scoring in the cup surface even in lubricated sliding. This agrees with the results of previous work by applying strip drawing without bending. The scoring effect occurred particularly at high contact pressure [18].

5 Conclusion

Dry metal forming of high alloy steel was investigated by strip drawing with bending and deep drawing of cups. The feasibility of dry metal forming by using a MMC tool surface could be demonstrated. Higher forming forces were required compared to a conventional bronze tool. However, adhesive wear of the tool could be reduced by applying MMC tool surface.

It can be concluded that it is appropriate to ablate the matrix of the MMC surface because there is no significant influence on the forming force. The contact area with the matrix can be reduced and thus adhesive wear in dry metal forming can be reduced by a plateau out of hard particles with high penetration.

Acknowledgement

This work was supported by Deutsche For-schungsgemeinschaft (DFG) within priority program SPP 1676 and the project Se1435/2-2.

References

1. P. Schmid, M. Liewald, AIP Conference Proceedings, **1383**, 446-452 (2011)
2. B.-A. Behrens, S. Hübner, C. Sunderkötter, J. Knigge, K. Weilandt, K. Voges-Schwieger, Adv. Mat. Res., **22**, 5-15 (2007)
3. K. Voges-Schwieger, S. Hübner, B.-A. Behrens, AIP Conf. Proc., **1353**, 224-228 (2011)
4. N. Bay, E. Ceron, Advanced Materials Research, **966-967**, 3-20 (2014)
5. P. Schmid, Advanced Materials Research, **769**, 221-228 (2013)
6. H. Freiße, A. Langebeck, H. Köhler, T. Seefeld, F. Vollertsen, Manufacturing Rev., **3**, 13, 1-10 (2016)
7. N. Bay, A. Azushima, P. Groche, I. Ishibashi, M. Merklein, M. Morishita, T. Nakamura, S. Schmid, M. Yoshid, CIRP Ann. Manuf. Technol., **2**, 760-780 (2010)
8. A. Almohallami, M. Arghavani, F. Böhmermann, H. Freiße, M. Herrmann, S.A. Mousavi, S. Schöler, P. Scholz, J. Tenner, M. Teller, G. Umlauf, D. Wulff, D. Yilkiran, H.J. Maier, Dry Met. Forming OAJ FMT **3**, 90-94, (2017)
9. H. Hetzner, J. Koch, S. Tremmel, S. Wartzack, J. Manuf. Sci. Eng., **133**, 1-11 (2011)
10. S. Kataoka, M. Murakawa, T. Aizawa, H. Ike, Surface Coating Techn., **177–178** 582–590 (2004)
11. A. Pramanik, Trans. Nonferrous Met. Soc. China, **26**, 348–358 (2016)
12. M. Cabeza, G. Castro, P. Merino, G. Pena, M. Roman, Surf. Interface Anal., **46**, 861–864 (2014)
13. A. Grueninger, N. Weidlich, O. Meier, M. Deutschmann, Int. J. Microstructure and Materials Properties, **5**, 2/3, 178-192 (2010)
14. H. Freiße, V. Hohenäcker, T. Seefeld, F. Vollertsen, Dry Met. Forming OAJ FMT, **1**, 5-10 (2015)
15. H. Freiße, J. Vorholt, T. Seefeld, F. Vollertsen, Dry Met. Forming OAJ FMT, **3**, 41-44 (2017)
16. E. Doege, B.-A. Behrens, *Handbuch der Umformtechnik, Grundlagen, Technologien, Maschinen*, **2**, 324 (2010)
17. M.T. Farr, *Zieh- und Stempelkantenradien beim Tiefziehen*, Ph.D. thesis, University Stuttgart, 67 (2002)
18. H. Freiße, S. Schmidt, T. Seefeld, F. Vollertsen, Dry Met. Forming OAJ FMT, **4**, 052-058 (2018)
19. H. Berns, *Hartlegierungen und Hartverbund-werkstoffe* (1998)
20. H.J. Meigh, *Cast and Wrought Aluminium Bronzes, properties, processes and structure*, **697** (2000)

https://doi.org/10.1051/matecconf/201819014007

Lubricant-free deep drawing using CO_2 and N_2 as volatile media injected through laser-drilled microholes

Ehsan Zahedi[1,*], *Christoph Woerz*[2], *Gerd Reichardt*[2], *Georg Umlauf*[3], *Mathias Liewald*[2], *Jakob Barz*[3], *Rudolf Weber*[1], *Thomas Graf*[1]

[1]Institut für Strahlwerkzeuge, Universität Stuttgart, Pfaffenwaldring 43, 70569 Stuttgart, Germany
[2]Institut für Umformtechnik, Holzgartenstraße 17, 70174 Stuttgart, Germany
[3]Fraunhofer Institute for Interfacial Engineering and Biotechnology, Fraunhofer IGB, Nobelstraße 12, 70569 Stuttgart, Germany

Abstract. Most metal forming processes use lubricants based on mineral oils as an intermediate medium to reduce friction and wear. To avoid the well-known drawbacks of oil lubrication, a novel and environment friendly lubrication system for deep-drawing processes was demonstrated at the University of Stuttgart. Liquid carbon dioxide and gaseous nitrogen are being used as volatile lubrication during the deep-drawing process, locally injected at high pressure through laser-drilled microholes. This new tribological system provides a significantly enlarged working range and at least 15% larger drawing depths compared to conventional oil lubrication.

Keywords: Deep drawing, Laser micro machining, Drilling.

1. INTRODUCTION

In metal deep-drawing, usually mineral oil- or wax-based lubricants are used. These lubricants may contain toxic additives, and the formed components have to be cleaned in subsequent process steps in a cost and time-consuming manner [1]. To avoid these issues, a novel tribological system for sheet metal deep drawing that eliminates the use of conventional lubricants was developed at the University of Stuttgart. A sketch of the process is shown in Figure 1.

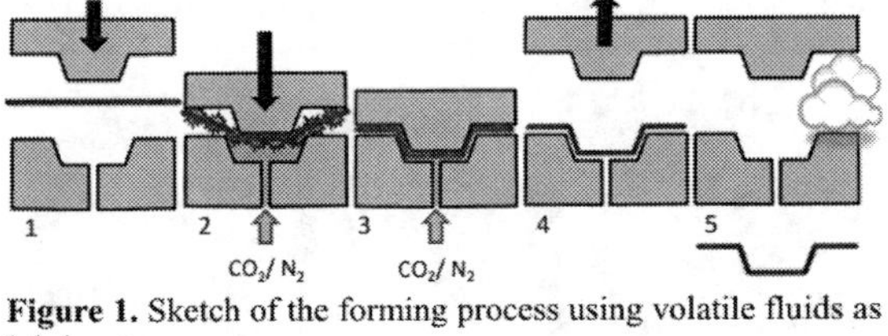

Figure 1. Sketch of the forming process using volatile fluids as lubricant.

During the forming of the material, liquid CO_2 or N_2 is supplied at high pressure through diffusor-shaped, laser-drilled microholes. The liquid gas that evaporates close before or in the contact zone between sheet metal and tool surface, serves as lubricant, and subsequently evaporates without residue.

This paper summarizes the results achieved in the three basic contributions from the institutes involved. This includes (1) the flow behaviour of N_2 and CO_2 for different hole geometries, (2) the basic issues for laser drilling of shaped deep microholes, and (3) fundamental deep drawing tests with rectangular cups to demonstrate the capability of volatile lubricants. A CAD-drawing of a cross section through the segmented die of the deep drawing tool is shown in Figure 2.

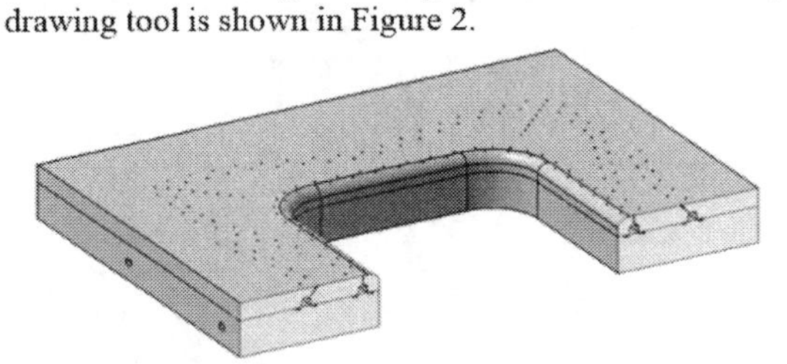

Figure 2. Cross section through the segmented die of the deep drawing tool.

2. PROPERTIES OF CO_2 AND N_2

Since the beginning of the 90s, the use of carbon dioxide as a working medium in refrigeration, air conditioning and heat pump processes has become the focus of technical interest. In the chemical industry CO_2, e.g. as industrial gas, long used as raw material - also as protective gas for packaged food, as fertilizer or for the production of medicines it is used. Finally, in the late 1990s, the discussion on atmospheric greenhouse gas emissions in general and the capture, transport and storage of Carbon Capture and Storage (CCS) in particular became more important. This puts carbon dioxide at the center of technical and political interest.

* Corresponding author: ehsan.zahedi@ifsw.uni-stuttgart.de

 https://doi.org/10.1051/matecconf/201819014007

First of all, however, another aspect will be briefly discussed: the danger potential associated with the large-scale handling of carbon dioxide. It is well known that carbon dioxide can be toxic and heavier than air. In the past, this has repeatedly resulted in fatal accidents in cellars or feed silos. In fact, the standard density (density at $T_n = 273.15$ K and $p_n = 0.10$ MPa) of carbon dioxide is $\rho_n = 1.97$ kg /m³. CO_2 is 1.529 times heavier than dry air. However, a dangerous accumulation of escaping carbon dioxide is usually limited to poorly ventilated rooms or cellars. Outdoors, mixing with the moving ambient air takes place quickly. CO_2 is not toxic only by the displacement of atmospheric oxygen, as a metabolite of the body, it also has an immediate effect on vital functions. It should be noted that although the maximum workplace concentration for carbon dioxide is already reached at 0.5% (based on the partial pressure), only concentrations above 20% are lethal in the short-term. While gaseous CO_2 dissolves only a few substances, the solution behavior improves in the liquid and especially in the supercritical state.

Figure 3, shows the p-T diagram of CO_2 and N_2. On the sublimation curve, solid and gas exist simultaneously. Solid and liquid phases exist simultaneously on the melting curve. The sublimation and melting curves coincide with the vapor pressure curve. At the triple point, all three states of matter are gaseous, liquid and firmly in equilibrium [2, 3]. At the upper end of the vapor pressure curve is the critical point. This point is characterized by the critical temperature T_k and the critical pressure p_k. If the temperature and the pressure are increased beyond the critical point, the pure substance is in the supercritical state.

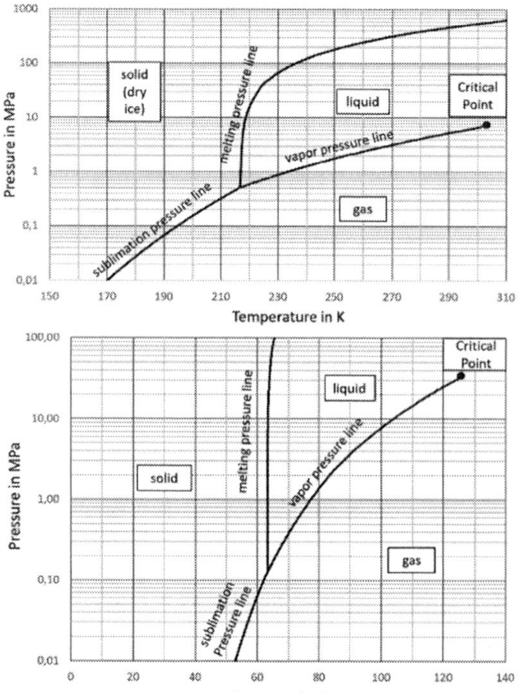

Figure 3. p-T-diagram for CO_2 (upper) and N_2 (bottom).

In contrast, for nitrogen, the critical point for the temperature is 127.15 K and a pressure of 3.39 MPa. This requires that only a strong decrease in temperature at atmospheric pressure can liquefy N_2. A large increase in pressure above 3.39 MPa causes a transition to the supercritical region. In this area, the gas and liquid phases can no longer be separated. In addition, both properties of the states of aggregation are present in parallel. The supercritical state is as dense as a liquid but has the same viscosity as a gas.

Due to the numerous phase transitions in the working window from 1 bar to 60 bar, CO_2 is examined more closely. Nitrogen is in the gaseous state at any point in the processing chain.

Table 1. Comparison of physical properties of CO_2 and N_2.

Property	Value for CO_2	Value for N_2
Crit. Temperature (K)	304.13	126.19
Crit. Pressure (MPa)	7.37	3.39
Crit. Density (kg/m³)	467.6	313.3
Triple point temperature (K)	216.59	63.15
Triple point pressure (MPa)	0.52	0.01

For these experiments, a CO_2 pressure bottle filled with 6 MPa and a riser is used. This pressure bottle is filled with 60 bar and has a riser.

This pressure bottle is filled with 60 bar and has a riser. Due to the phase differences, the lighter gas phase is above the heavy liquid phase. The bottle is equipped with a riser to insure the state of CO_2 stays liquid in the whole period of time that the bottle has been consumed. When removing the liquid phase flows through the riser upwards, the leads and the microholes directly into the forming tool. The microholes are considered to be open ends and there is atmospheric pressure. It comes to an expansion of the liquid phase. It forms the so-called dry ice in solid form, this can be seen very well by the whitish color as shown in Figure 4. Then sublimates the solid CO_2 in the gas phase.

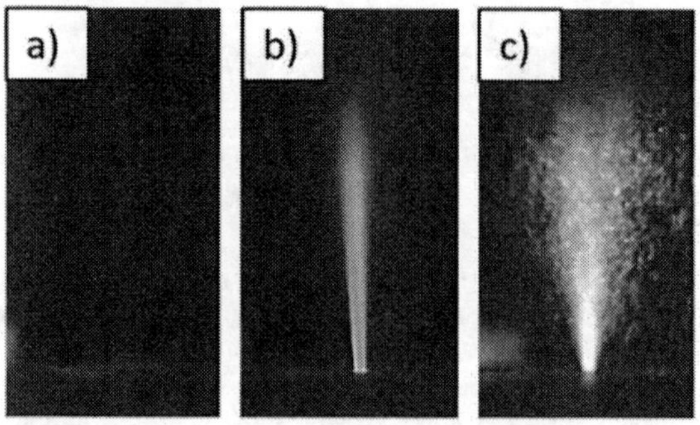

Figure 4. Comparison between the aggregate states gaseous (a), solid (b) and liquid (c) during the flow of CO_2 through a laser drilled mircohole.

The expansion at the mircohole takes place adiabatically, i.e without heat transfer to the surrounding material inside the mircohole. There is just a temperature decrease in the flow medium. This phenomenon is called the Joule-Thomson effect. This is largely based on the binding energies between atoms and molecules - the so-called van der Waals forces. For CO_2, the substance-specific Joule-Thomson coefficient is 1.1K / bar [4]. This means that during an expansion the medium cools down by 1.1 K per bar. When the CO_2 at the exit of the microholes expands freely to atmospheric pressure, it cools to 194.65 K and forms the described dry ice.

Nozzle shapes are generally divided into two types of geometry, a confusor and a diffuser. The diffusor gradually widens its cross section in the flow direction, the confusor becomes narrower (see Figure 5).

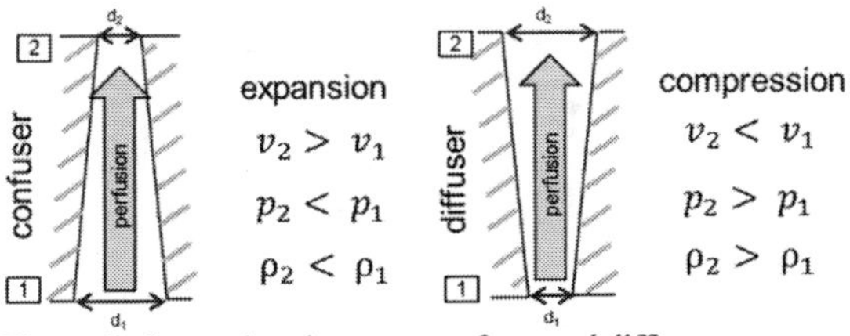

Figure 5. Comparison between confuser and diffusor .

The laws of mass conservation apply:

$$\dot{m} = \rho * \dot{V} = \rho * A * \vartheta = \rho * \frac{\pi}{4} d^2 * \vartheta \quad (1)$$

$$\dot{V} = \frac{\pi}{4} d^2 * \bar{\vartheta} = constant \quad (2)$$

From this, it follows that:

$$d_1 * \bar{\vartheta}_1 = d_2 * \bar{\vartheta}_2 \quad (3)$$

Here, the term for the diffusor as a widening nozzle shape in flow direction is retained regardless of the flow rate. The same applies to the confusor as tapering nozzle geometry in flow direction (Figure 4).

The microholes were drilled with the laser in order to allow arbitrary shaping of the holes according to the needs described before. However, drilling of very deep microholes holes with the laser is not yet established and requires special attention due to basic physical limits which determine the resulting quality.

3. BASIC LASER DRILLING LIMITS

Laser-drilling of deep holes into tool steel is challenging: On the one hand, tool steel suffers from local phase transformation when the temperature exceeds 600 K. On the other hand, drilling of micro holes with high aspect ratio requires specific considerations regarding energy distribution in the capillary. Decreasing quality with increasing depth of the holes was observed and reported in numerous publications, e.g. in [5]. One of the major reasons for reduced quality during laser processing with ultra-short laser pulses is the so-called heat accumulation [6]. In addition, the limited pulse energy limits the maximum drilling depth: With increasing hole depth the total surface area of the wall of the microhole increases. This means that the average laser fluence (i.e. the laser pulse energy divided by the wall surface area) decreases. If the average fluence reaches the ablation threshold fluence, only very slow and localized drilling can occur. This limit is called "quality depth limit" in the following.

Heat accumulation and the quality depth limit were investigated for the required geometry of the microholes. For the experiments, the prototype kW-class ps-laser of the IFSW was used [7]. The passive disk-amplifier of this laser allows very high pulse energies which are necessary for laser drilling of deep holes. The diameter of the holes was adjusted with a helical drilling optics. This optics moves the beam on a circle with adjustable diameter. Furthermore, the optics allows adjusting the inclination angle of the beam relative to the surface which is necessary for longitudinal shaping of the holes. The properties of the laser, the helical drilling optics, and the focusing optics are summarized in Table 2.

Table 2. Properties of the IFSW kW-class-ps laser system and the helical drilling optics.

IFSW kW-ps laser	
Pulse duration	8 ps
Wavelength	1030 nm
Average power	≤ 650 W
Max. Pulse energy	≤ 2.2 mJ
Repetition rate	≤ 300 kHz
M^2	< 1.3
Polarization	Linear
Collimated beam diameter	5.2 mm
Helical drilling optics (GL-Trepan, GFH GmbH)	
Rotation speed	0-30'000 rpm
Hole diameters	0-1500 µm
Inclination angle	0°-4°
Focusing and ablation threshold	
Focal length	400 mm
Focus diameter	140 µm
Polarization	Circular
Focus position	On the surface
Max. Fluence	15 J/cm²
Ablation threshold (Steel, 8 ps, 1030 nm)	0.1 J/cm²

As it is described in Chapter 2, the inlet and outlet diameters of the microholes have to be 600 μm and 200 μm. Drilling microholes with a specific geometry needs precise moving of the laser beam. A laser drilling optic assists to fulfil this task. Here, we have used GL Trepan from GFH GmbH, Germany. The principle of a laser-drilling optic is described in [8]. As the collimated beam hits the focusing lens with a specific angle (relative to the optical axes of the lens), the focal point will be consequently at a certain distance of tens to hundreds of micrometer from the optical axis. If the collimated beam revolve in a circular route on the surface of a lens, the focal point will revolve on the surface of work piece. Therefore depends on the inclination angle of collimated beam and focal distance of the lens, the revolving diameter can be precisely adjusted.

3.1 Heat accumulation

Heat accumulation was identified as one of the major quality-reducing effects for deep-hole drilling [5].

The laser system that was described above was used to determine the limits where heat accumulation effects begin to reduce the quality. The optically measurable occurrence of phase transitions in the material was defined as an indicator for a reduced quality. These phase transformations include the formation of melt (T ≈ 1500 K) as well as a solid phase transformation (T ≈ 600 K). While the formation of melt is visible on the sample surface, the solid state phase transformations were analyzed by polishing and etching of the cross sections with Nital 5% for 40 seconds.

Figure 6 shows the effect of a decreasing repetition rate on heat affected zone and the drilling progress for a constant pulse energy of 2.2 mJ. At 300 kHz and an average power of 660 W the depth of the microhole was about 8 mm after $5 \cdot 10^5$ pulses (2 s of drilling time) (Figure 6a). However, a large heat-affected zone of about 1 mm can be seen. If the repetition rate is reduced to 100 kHz (which leads to an average power of 220 W), the hole depth after $1 \cdot 10^6$ pulses (10 s of drilling time) was reduced to about 5 mm. A heat affected zone is still visible as a bright area at the walls and in particular in the bottom of the hole (Figure 6b). Further reducing the repetition rate to 60 kHz yielded the same hole depth of about 5 mm after $1 \cdot 10^6$ pulses (17 s of drilling time) but without a visible heat affected zone and good hole quality.

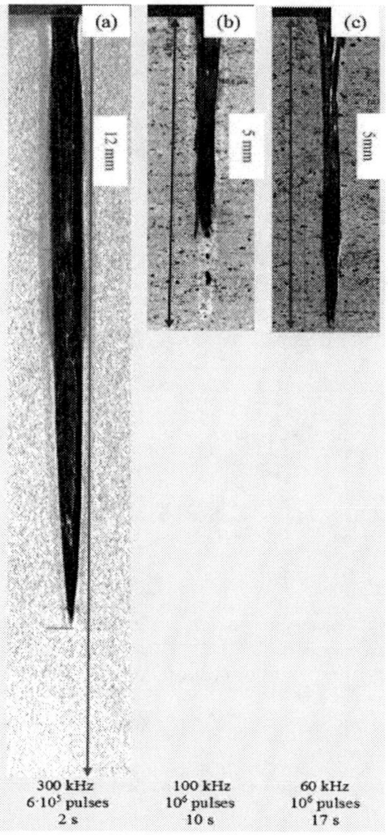

Figure 6. Cross-sections of laser-drilled microholes with a pulse energy of 2.2 mJ. All samples were etched with Nital 5% for 40 seconds.

These results lead to the conclusion that 60 kHz is the limiting frequency at the pulse energy of 2.2 mJ to maintain high quality of the drilled holes.

It is worth to mention that when drilling with a repetition rate of 300 kHz, melt ejection contributes to the drilling process. However, melt ejections lead to a reduced quality of the hole. Figure 7 shows ejected melt on the inlet of the microholes.

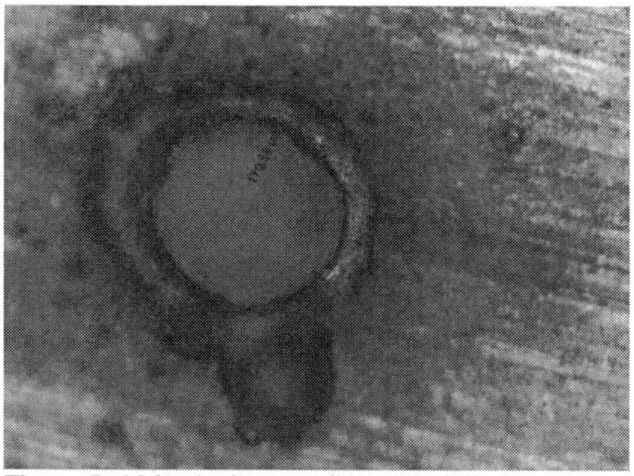

Figure 7. Melt ejection contributes to drill deeper using 300 kHz, however, a resolidified ring around the inlet is visible.

https://doi.org/10.1051/matecconf/201819014007

3.2 Depth limit

The hole depth as a function of the number of laser pulses for different repetition rates is shown in Figure 8. The pulse energy was 2.2 mJ.

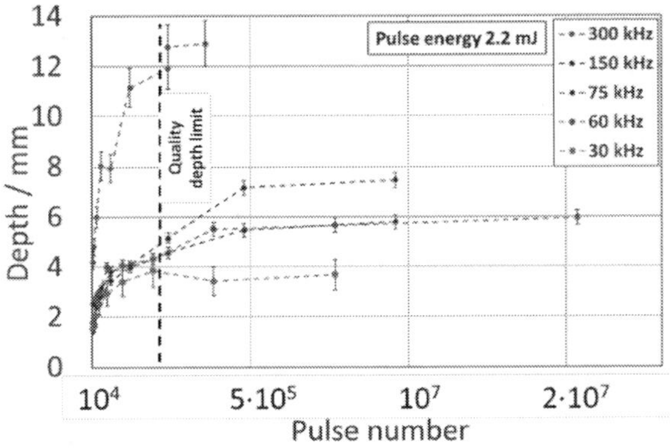

Figure 8. Hole depth as a function of the number of pulses for the pulse energy of 2.2 mJ for different repetition rates.

The drilling progress is very similar for repetition rates of 150 kHz and below. In the beginning of the drilling process, up to about $1{\cdot}10^5$ pulses, the drilling progress is very fast down to a depth of about 4 mm. As the depth increases, the process becomes slow and irregular. The faster drilling progress for $>2{\cdot}10^5$ pulses for the frequencies >30 kHz suggests, that heat accumulation gains influence after a very large number of pulses. It can be concluded that the above-mentioned "quality depth limit" is reached at about 4 mm for pulse energies of 2.2 mJ (marked by the dashed line in Figure 8).

As the required depth exceeded the above-discussed quality depth of 4 mm, we have to apply much more number of pulses to drill through. Inevitably, the extra amount of energy will be deposited into the capillary. To minimize the risk of damage, an additional break time should be apply to the drilling strategy to optimize the hole quality: After drilling for 0.5 s a break was introduced. The required duration of the break was determined experimentally. Figure 10 shows that decreasing the break between the drilling intervals from 2.0 s to 0.5 s results in the appearance of dross around the hole on the surface of the sample, marked with a red arrow.

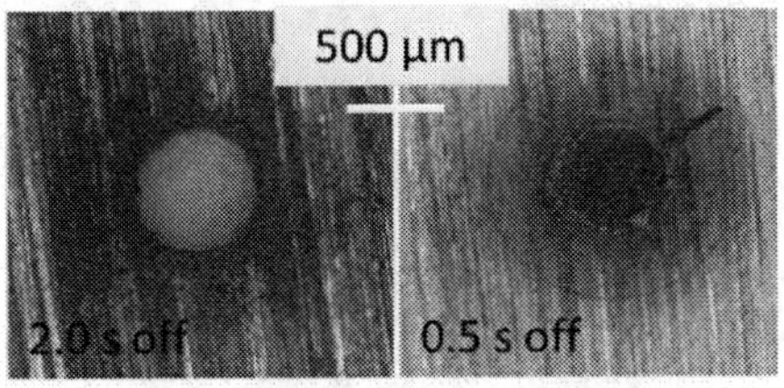

Figure 10, Decreasing the break between the drilling intervals of 0.5 s from 2.0 s to 0.5 s results in the appearance of dross on the surface of the sample, marked by the red arrow.

The influence of the longitudinal shape of the microholes feeding the liquid gases was discussed in Chapter 2. For the deep drawing tool the microholes shoud be diffusor-shaped with an output diameter of 600 µm and an input diameter of 200 µm. The longitudinal shaping of the hole was done with a helical drilling optics.

4. LASER DRILLING OF MICROHOLES IN THE DEEP-DRAWING TOOL

4.1 Requirements

A sketch of the required geometries of the microholes are shown in Figure 9. The required hole depths and angles of the holes with respect to the surface of the deep drawing tool can be seen. The depth of the holes was either 5 mm or 6.7 mm, depending on the position of the hole in the tool. The angle of the hole relative to the tool surface was either perpendicular or 35°. In total, 254 microholes were drilled.

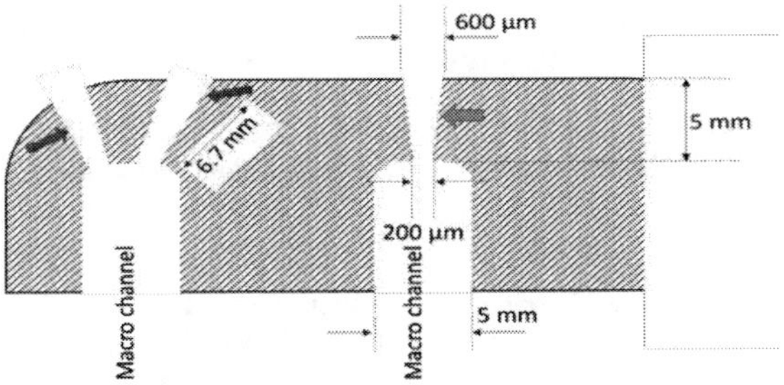

Figure 9. Geometry and position of the microholes: depth of 5 mm for perpendicular (assigned with blue arrow) and depth of 6.7 mm for angled microholes (assigned with red arrows). The microholes are connected to the macro channel which were manufactured by conventional mechanical tool.

4.2 Laser-drilled microholes

Based on the results of the basic investigations described above, the microholes in the deep drawing tool were drilled with 60 kHz, the maximum available pulse energy of 2.2 mJ, and the above-mentioned drilling strategy of a 0.5 s drilling interval follow by a 2.0 s break (for the depth of 6.7 mm). In Figure 11, typical cross-sections of the laser-drilled, deep microholes suitable for the deep drawing application are shown.

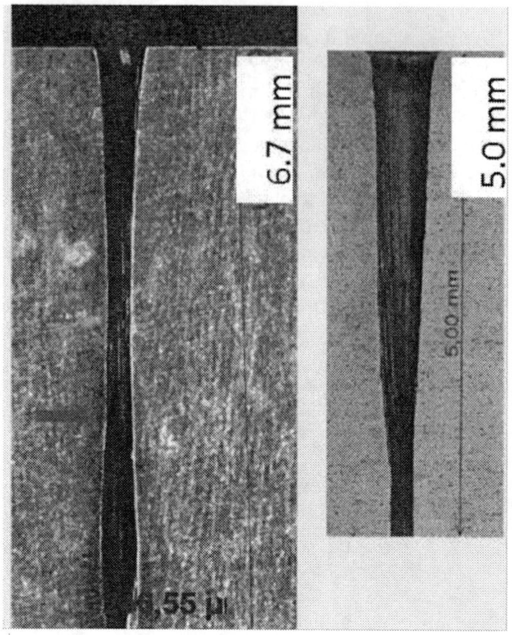

Figure 11. Typical cross-sections of the laser-drilled, 6.7 mm (left) and 5 mm (right) deep microholes for the deep drawing application. The green arrow indicates the "quality depth limit", the red arrow the widening of the hole due to the very long drilling time.

The net drilling time was 60 s and 320 s for the 5 mm and the 6.7 mm holes, respectively. Although the depth of 6.7 mm was reached, the respective drilling time was very long. That the quality depth limit was exceeded manifests in the longitudinal shape of the hole as seen in Figure 10. Down to a depth of about 4 mm (marked with a green arrow), the shape is conical and very regular. This depth corresponds to the "quality depth limit" as described above. For depths larger than 4 mm, the very long drilling time leads to an irregular widening of the hole cross-section. However, the holes were free of heat affected zones and free of burr on the surface. Overall, the quality was acceptable for the deep drawing experiments.

4.3 Manufacturing of the deep drawing tool

The completely drilled deep drawing tool is shown in Figure 12.

Figure 12. Processed deep drawing tool.

Due to constraints of the micro-processing station, positioning of the blank with respect to the absolute coordinates of the holes and the angle relative to the surface was done manually using auxiliary tools. This caused in particular inaccuracy of the focus position relative to the surface of the blank, which resulted in scattering of the hole diameters of about ±15%.

5. STRIP DRAWING INVESTIGATIONS

5.1 Experimental Setup

In order to evaluate the lubricating effect of gases during deep drawing, a strip drawing testing rig was modified to determine the coefficient of friction under different conditions in such way that the fluid could be induced into tool contact zone with work piece via laser-drilled micro holes inserted in strip drawing jaws, as it is shown in Figure 13.

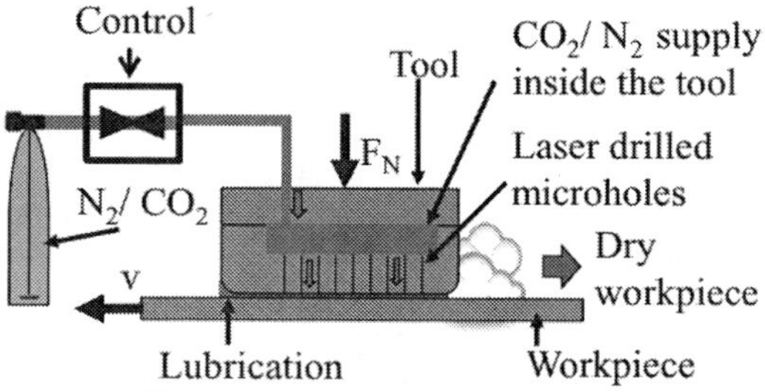

Figure 13. Experimental setup of strip drawing test using volatile media.

With the help of this test rig, various parameter combinations, such as nozzle type, borehole arrangement, borehole diameter, contact normal pressure, drawing speed and fluid pressure could be investigated for their influences. For all executed strip drawing investigations electrolytic galvanized sheet material DC05 was used. Because carbon dioxide in the conventional gas cylinder is present as a 2-phase area, where gaseous and liquid CO_2 are present at the same time. CO_2 is taken out of a standard gas bottle in liquid state, so that the pressure cannot be adjusted and remains constant depending on level of exit temperature. The pressure of nitrogen is set to the same value as for carbon dioxide at the supply valve (T=20°C, p≈ 60 bar) to maintain comparability.

The strip drawing test is an experimental model to investigate the influence of different parameters on the friction coefficient. However, the elastic and plastic deformations of the sheet metal material that occur in the actual deep-drawing process (e.g. thickness reduction) cannot be reproduced. Nevertheless, this experiment is very suitable for determining the qualitative influences of the individual parameters.

5.2 Results of strip drawing experiments

In order to gain a better understanding of the interdependencies of the tribological system using volatile media as lubricant substitutes, a large number of parameters were investigated [9, 10]. The following gives a compact overview of the results obtained from the strip drawing experiments.

The effects of number of conical holes (3, 7, 9, 21) and the diameters of (200 µm, 600 µm) and nozzle types (diffuser, confuser) were investigated. The main effects of these parameters on the coefficient of friction are depicted in Figure 14 and Figure 15.

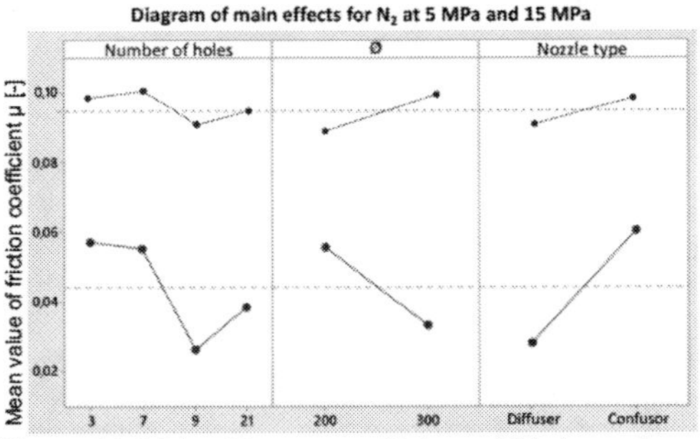

Figure 14. Main effects on coefficient of friction µ determined with strip drawing investigations at 5 MPa (lower, blue values) and 15 MPa (higher, green values) surface pressure, drawing speed 100 mm/s, lubrication medium N_2, All values represent averaged values for the respective configurations.

Using nitrogen (N_2) or carbon dioxide (CO_2) as lubrication a significant reduction of the friction coefficient is achieved at low contact pressure (5 MPa) with an increasing number of bores. In the tests with 21 holes using N_2, friction increases slightly again, which can be justified by the fact that the pressure in the supply channels decreases sharply due to the increased outflow of the gas with a large number of holes and therefore a lubricating effect is only achieved in a weakened manner. CO_2 is not that sensitive on the diameters of the supply channels, because the specific volume increase at 20°C, due to the phase change from liquid to gaseous, is nearly four fold. That's why decreasing pressures in the supply channels will occur at markedly higher output rates of CO_2 compared to N_2.

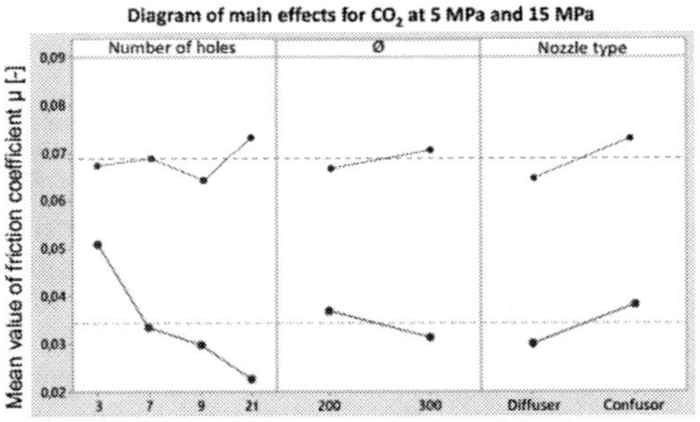

Figure 15. Main effects on coefficient of friction µ determined with strip drawing investigations at 5 MPa (lower, blue values) and 15 MPa (higher, green values) surface pressure, drawing speed 100 mm/s, lubrication medium CO_2, All values represent averaged values for the respective configuration.

For both fluids, diffusors obtain lower friction coefficients at 5 MPa as well as at 15 MPa. This result was used, among all the experiments, for the design of the forming tool of the rectangular cup and so there all laser-drilled microholes were drilled as diffusers [11]. Dashed lines in both diagrams represent the averaged value of all friction coefficients obtained by strip drawing experiments at a contact pressure of 5 MPa and 15 MPa. As can be seen in the diagrams, CO_2 achieves lower averaged friction values than N_2 at both high and low surface pressures. A possible explanation of this effect could be the formation of dry ice. To get a better understanding of these effects, further investigations on the effects of dry ice formation when using CO_2 as a lubricant substitute [9] and the pressure conditions in the tool contact zone [11] were carried out in the course of this research work.

6. DRY DEEP DRAWING OF A RECTANGULAR CUP

Tool design

In order to reduce the drilling depth for the laser process without reducing the tool strength a segmented tool design was chosen (see Figure 2). The base plate contains the holes for the media supply and a sealing to avoid an uncontrolled flow out between the plates while the upper plate contains different supply channels and laser-drilled microholes. The position of the supply channels determines the location of microholes. In order to avoid a free flow out of the media through the microholes at the end of the forming process the position of the supply channels was optimized using a sheet metal forming simulation of the blank draw-in. As a result, one ring channel was integrated next to the die radius and four additional channels are arranged with regard to the blank draw-in. Each supply channel can be controlled separately by a valve to stop free flow out of liquid during deep drawing.

One row of microholes was drilled perpendicular into the supply channel having a depth of 5 mm and one row was incorporated by an angle of 35° and a length of around 7 mm. Additional microholes were placed in the die radius area. In total, 154 microholes were distributed over the tool surface of the die. For the blankholder, the same design was chosen though without radius. Here about 100 microholes were drilled into the tool.

Figure 17. Free flow out of the CO_2 through laser-drilled microholes in the die of the deep drawing tool.

Due to some problems in the tool manufacturing, it was necessary to re-mill the die radius after die laser drilling. For this purpose, the tool was connected to a compressed air system applying a pressure level of 2 bar during the

https://doi.org/10.1051/matecconf/201819014007

milling process. By doing so, surprisingly none microhole was blocked due to the machining process. Also after the final hardening process of the tool, no blockage of the microholes could be observed. The final assembly of the die tool is shown in Figure 17 while a free flow out of the CO_2 through the laser-drilled microholes is activated.

6.1 Experimental setup

In order to understand the limits of this new tribological system, several deep drawing tests have carried out and compared with conventional deep drawing process using mineral oil- or wax-based lubricants. An established method is the determination of the process window. Usually, the blank holder force and the drawing ratio is varied in the deep drawing process in order to find maximum drawing depth without wrinkles or cracks. By doing so, the valid process window can be determined where no cracks and no wrinkles in the flange area appear. For non-rotational parts the drawing depth is used instead of the drawing ratio.

All drawing experiments were carried out using electrolytic galvanized sheet material DC05. The sheets were cleaned manually and then degreased in an acetone bath in order to remove remaining oil on the sheet. Liquid CO_2 having an initial pressure level $p_{initial}$ of 60 bar and gaseous N_2 at the same pressure level were used as temporarily acting lubricant for dry metal forming. The supply of the media was controlled by the ram movement. The supply was switched on in the moment of the first contact between the tool and the sheet and shut down at the bottom dead center of the press. Additionally, drawings with the lubricant Wisura ZO3368 (1.5 g/m²) were investigated using the same tool. All tests were carried out on an AIDA servo press at a ram speed of 6 strokes per minute which corresponds approximately to 100 mm/s at the beginning of the forming process.

6.2 Experimental results und discussion

Presented approach for dry metal forming was tested successfully and so, a rectangular cup was deep drawn using N_2 and CO_2 as temporarily acting lubrication for the very first time. In accordance with former tests results investigating the coefficient of friction [10] and the deep drawing of a U-shaped profile geometry [9], the new lubrication system performed better than the conventional one using a mineral oil-based lubricant (see Figure 18).

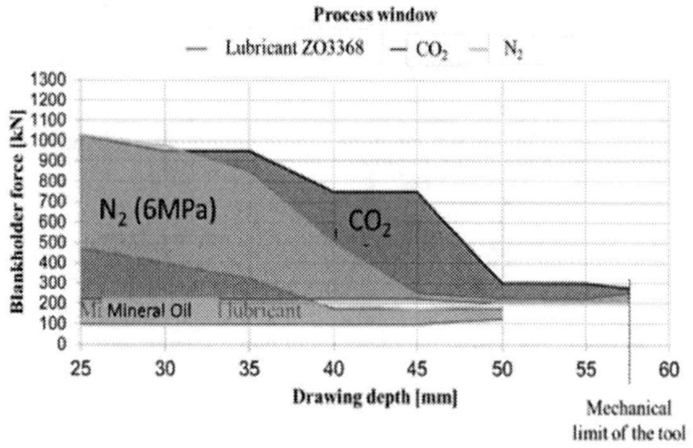

Figure 18. Process windows for deep drawing of a rectangular cup using different lubrication systems.

The maximum drawing depth could be increased to 57.5 mm by using CO_2 and N_2. This depth corresponds to the mechanical limit of the tool. By using lubricant ZO3368 only 50 mm as a maximum drawing depth was achieved. Also the fracture limit was raised up to 50% depending on the drawing depth by using CO_2 as well as N_2. Thereby, CO_2 performs better for deeper cups than N_2. It is assumed that this is caused by different pressure levels acting in the gap between the sheet and the tool resulting by the different media [11]. In general this pressure in the gap p_{gap} is influenced by microholes (position, numbers, nozzle type), the initial pressure level of the injected media $p_{initial}$ and the sealing effect between tool surface and the sheet. Also the sealing effect is influenced by many factors such as the blankholder force $F_{Blankholder}$, the tool and sheet surface roughness and the thinning and thickening of part flange during deep drawing. Finally, there are interactions between the sealing effect, the pressure level p_{gap} and the height of the mean gap h_{gap} between the sheet asperities and the forming tool. A schematic illustration of the conditions in the gap and the applied designations are shown in Figure 19.

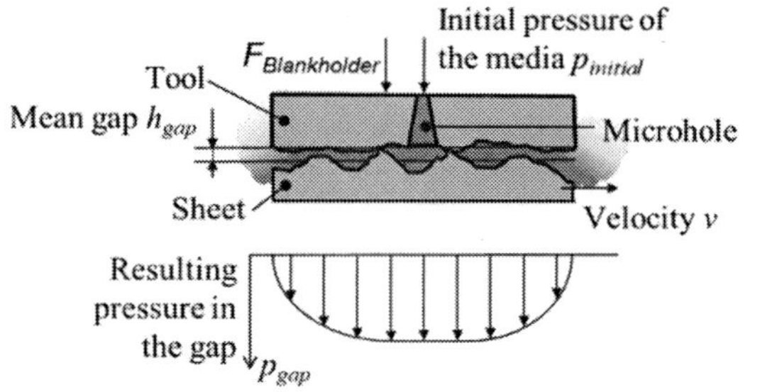

Figure 19. Schematic illustration of the resulting pressure between sheet and tool.

By deep drawing with volatile media acting as lubrication not only the fracture limit, but also the wrinkle limit was increased noticeably (see Figure 18). Also this effect can be explained by the assumption that the mean gap h_{gap} and the pressure p_{gap} mainly do influence the sheet metal forming behaviour. For lower blank holder forces a higher gap h_{gap} occur without or with minimum contact areas between the tool and the sheet asperities. While forming, wrinkles of 1st order develop due to increasing tension stress in the blank without prevention by the blank holder. Thereby, the pressure in the gap, which is a scalar quantity acting in all directions, cannot avoid the local development of wrinkles. The outflowing gas through the wrinkles support such development additionally. Other tests indicate that the wrinkle limit is raised further by using a pressure $p_{initial}$ of N_2 higher than 60 bar. However, also the fracture limit can be increased extremely. This confirms the assumption that the resulting pressure level p_{gap} of the volatile media between the sheet and tool mainly influences the friction behaviour and therefore the forming limits in the deep drawing process of this new approach.

In addition to those investigations on process limits in deep drawing, the cooling effect on the sheet surface caused by the Joule-Thomson effect during the relaxation of CO_2 and N_2 was investigated. The temperature was measured visually after the deep drawing process using a thermal camera.

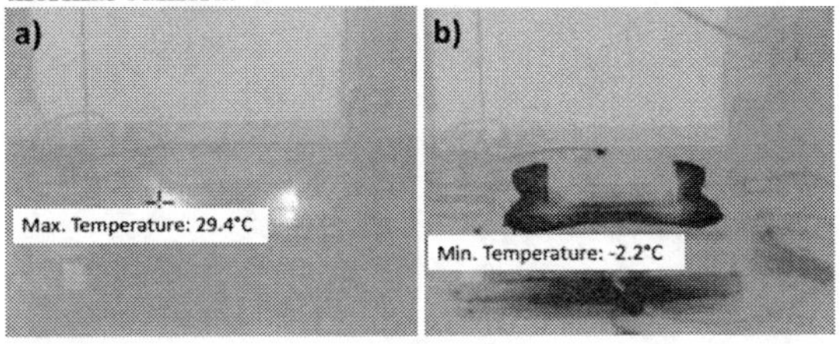

Figure 20. Measured temperature on the sheet surface after deep drawing using (a) nitrogen and (b) liquid carbon dioxide as lubricant.

In Figure 20, the visualized temperature is shown using nitrogen (a) and liquid carbon dioxide (b) as lubricant for the same process having a drawing depth of 35 mm and a blank holder force of 275 kN. The maximum measured temperature (29.4°C) after deep drawing using nitrogen is almost similar to the sheet temperature when using mineral oil-based lubricants (30.1°C). The heating in the corner of the cup is caused by dissipating forming energy and friction heat. By contrast, using liquid CO_2 as temporally acting lubricant reduces the sheet temperature in the flange area down to -2.2°C (see Figure 20(b)). Compared to the first deep drawing experiments using a U-profile geometry [8], a cooling effect of only a few degrees Celsius were measured. The cooling effect merges much more significantly when deep drawing the rectangular cup. Additional to that the control of valves also plays an important role. Finding the right timing to switch on and off the flow of media appear more complex for the drawing process of a rectangular cup having curved arrangements of microhole positions onto the tool surface and more complex draw-in compared to previous drawn part as described. Another reason might be given by effect of changing flange thickness during drawing counteracting to local blank holder force. Small gaps between the tool and the sheet induced by small wrinkles of 1st order can occur supporting the flow out of the CO_2 and therefore the cooling of the sheet due to high velocity of media. According to these results, the flow control and also the position of microholes have to be optimized in further research work in order to reduce observed extreme cooling of the sheet. A different approach is given to use the cooling effect actively to reduce the heating in drawing process, while heating of parts due to friction heat and dissipating forming energy is undesirable.
However, overall presented results show that a dry forming by means of temporally acting volatile media is possible, which not only allows an equivalent substitution of mineral oil-based lubricants, but also an enhancement of the process limits in sheet metal forming. Especially using nitrogen with a pressured level adjusted to the respective process conditions are extremely promising.

Conclusions

The progress in the use of volatile media as lubricant substitutes in the deep drawing process in this paper was demonstrated based on achieved results. Thus, numerous technological challenges in the manufacturing of precise micro-drilling by laser drilling were mastered, but also deepened knowledge could be gained experimentally and simulatively by examining emerging friction conditions in the tool contact zone. Furthermore, after deep drawing of approximately 300 cups under dry as well as lubricated conditions no microholes were blocked by zinc abrasion or other effects. So it can be assumed that blockage of the microholes is not a limitation factor for this new approach. In a next step, these results have to be confirmed by prospective deep drawing endurance tests. Further performed measures also will be implemented into the test stand in the coming funding period. The tool radii of a new testing rig, which are subjected to high friction and wear loads, will be provided with feedholes for volatile media flow too. Complex friction conditions in corner areas and interaction with deep-drawing process will be investigated in this new special testing rig. A further goal of future research is to expand the range of applications for zinc-coated steel sheets to include selected aluminium sheet materials, which poses enormous challenges with regard to a more complex tribological system and its susceptibility to failure. Furthermore, the theoretical knowledge of the friction behaviour of volatile media in the tribological system is to be further deepened in order to understand and specifically influence the occurring effects and thus to ensure robust and stable deep drawing processes.

Acknowledgement

The scientific investigations of this paper are funded by the German Research Foundation (DFG) within the priority program SPP 1676 Dry Metal Forming - Sustainable Production by Dry Processing in Metal Forming.

References

1. F. Vollertsen, F. Schmidt, "Dry Metal Forming: Definition, Chances and Challenges,, IJPEM-GT, **1**, 59-62 (2014).
2. Kapler, P.: Partikelbildung und -morphologie bei der Hochdruckmikronisierung gashaltiger Lösungen, Dissertation, Ruhr-Universität Bochum, 2002.
3. Fachlabor UTRM SoSe 2010, Hochdruckextraktion mit CO2.
4. Atkins, P. W. and de Paula J., "Physical Chemistry" Oxford University Press, 10th Revised Edition, Oxford, 2014.
5. S. Döring, S. Richter, A. Tünnermann, S. Nolte, "Evolution of hole depth and shape in ultrashort pulse deep drilling in silicon", Appl. Phys. A **105**, 69–74, (2011).
6. R. Weber, T. Graf, P. Berger, V. Onuseit, C. Freitag, M. Wiedenmann, A. Feuer, "Heat accumulation during pulsed laser materials processing", Optics Express **22**

(9), 11312–11324 (2014), and Erratum thereto in Optics Express 22 (23), 28232, (2014).

7. J. P. Negel, A. Voss, M. Abdou ahmed, D. Bauer, D. Sutter, A. Killi, T.Graf. "1.1 kW average output power from a thin-disk multipass amplifier for ultrashort laser pulses", Optics Letters **38**, 5442-5445, (2013).
8. S. Nolte, F. Schrempel, F. Dausinger, *Ultrashort Pulse Laser Technology,* 201-205, (2016).
9. C. Woerz, E. Zahedi, G. Umlauf, M. Liewald, R. Weber, T. Graf, G. E. M. Tovar, "Tiefziehen eines U-Profils mit flüchtigen Medien als Schmierstoffer- satz", Dry Met. Forming OAJ FMT, **3**, 50-61, (2017).
10. C. Woerz, M. Liewald, M. Singer, "Investigation of Tribological Conditions in the Strip Drawing Test Using Liquid CO_2 and N_2 as a Volatile Lubricant", *Proceed. of 7th ICTMP*, 140-148, (2016).
11. C. Woerz, G. Reichardt, M. Liewald, E. Zahedi, R. Weber, T. Graf, "Dry deep drawing of a rectangular cup assisted by volatile media injected from laser-drilled microholes" Dry Met. Forming OAJ FMT, **4,** (2018).

Influence of the surface microstructure on the adhesion of a CVD-diamond coating on steel with a CrN interlayer

Markus Prieske[1,*], *Richard* Börner[2] and *Andreas* Schubert[2]

[1]BIAS – Bremer Institut für angewandte Strahltechnik GmbH, Klagenfurter Str. 5, 28359 Bremen, Germany
[2]Chemnitz University of Technology, Reichenhainer Straße 70, 09126 Chemnitz, Germany

forming tool coating. Most of the forming tools are made of steel, so that especially the coatability of steel by a polycrystalline diamond coating would rise the range of fields of application. The polycrystalline CVD-diamond coatings are deposited by a laser induced plasma CVD process, without a vacuum chamber. Various surface microstructures were investigated regarding their influence on the residual stresses to prevent a flaking of the coating: on the one hand, deterministic structures generated by ultrasonic vibration assisted milling (UVAM) and on the other hand, stochastic structures manufactured by blasting and polishing processes. For the UVAM, a surface prediction tool was used to design the surface microstructure beforehand. All steel substrates (material no. 1.2379) were coated in one batch by high-power impulse magnetron sputtering with a chromium nitride coating with a thickness of 2.4 µm. The specimens were analysed by laser microscopy, scanning electron microscopy, energy-dispersive X-ray spectroscopy and Raman spectroscopy. None of the microstructures investigated in this study was able to prevent delamination of the coating entirely. It could be shown that a roughness higher than Sa 0.1µm supports the interlocking between coating and surface as well as that sharp peaks inhibit a homogenous diamond coating deposition.

Keywords: Micro structure, Diamond coating, Ultrasonic vibration assisted milling

Nomenclature

A_{C1}	first temperature of phase transformation
A_{US}	Ultrasonic amplitude: peak-to-peak
AR	Aspect ratio of the microstructure (A_{US} / λ)
D_{tool}	Tool diameter
f_{US}	Ultrasonic frequency
f_z	Feed per tooth
n_{tool}	Spindle speed
Sa	Arithmetic mean height
Sdr	Developed interfacial area ratio
v_c	Cutting speed
v_f	Feed
v_{US}	Maximum speed of the oscillation
α	Clearance angle
λ	Wavelength of the oscillation

1 Introduction

The high hardness and chemical inertness of diamond coatings lead to a high interest in use as forming tool coating. Most of the forming tools are made of steel, so that especially the coatability of steel by a polycrystalline diamond coating would rise the range of fields of application a lot.

The deposition of polycrystalline diamond coatings onto steel has already been researched by various scientists [1], but is still challenging due to the large mismatch in thermal expansion coefficient. Direct deposition of polycrystalline diamond onto steel substrates was investigated by Buijnsters et al. [2], with the conclusion that diffusion barrier layers are essential in the deposition of high-quality and adherent polycrystalline diamond coatings on steel. Several different materials that could serve as a diffusion barrier have been investigated, such as among others CrN [3], TiC/TiN [4], and SiC [5]. The thickness of the diffusion barrier varies in the different researches. Buijnsters et al. [3] deposited successfully a 1.5 µm thick polycrystalline chemical vapour deposition (CVD)-diamond coating on high-speed steel of the quality no. 1.3207 with a 2.5 µm thick chromium nitride interlayer. They used a CVD-diamond deposition temperature of 650 °C. By the use of the same process conditions and coating system an AISI-type 316 stainless steel substrate could not be diamond coated without delamination. This shows that the type of steel has a strong impact on the coatability.

Using a 20 µm thick CrN intermediate layer Glozman et al. deposited a 2 µm thick diamond coating on steel specimens by HF-CVD. In subsequent tests on the adhesive strength, delamination took place between the steel substrate and the intermediate layer. [6]

Kellermann et al. [1] showed on 41Cr4 steel specimen with a chromium carbide intermediate layer, that austenite

* Corresponding author: prieske@bias.de

https://doi.org/10.1051/matecconf/201819014008

to ferrite transformation has a high impact on the residual stress and thereby on the delamination of the coating. Diamond coatings deposited at temperatures underneath the first temperature of phase transformation A_{C1} (738 °C) caused cracks and flaking. Higher deposition temperatures between 760 °C and 920 °C reduced the residual stress in the diamond coating due to the austenite to ferrite transformation in the steel substrate which results in a volume expansion during cooling down.

All processes applied for the successful deposition of polycrystalline diamond coatings onto steel substrates use a vacuum chamber. According to the dimensions of, for example, forming tools and their complex geometries, a chamber-free deposition process is required for the application of local wear protection.

Multiple studies examine the impact of the substrate surface microstructure on the adhesion strength of CVD-diamond coatings. The term *surface microstructure* is used in this publication with the meaning of a defined surface topography. The effect of the substrate roughness on the CVD-diamond coating of Si_3N_4 specimens was reported by Amaral et al. The combination of different pre-treatments and seeding processes led to the conclusion that surfaces roughened by plasma etching increase the layer adhesion of the CVD-diamond coating due to "anchoring effects", whereby no closer information on the substrate roughness was given. [7] Singh et al. observed that the generation of a three-dimensional (3D) thermally and compositionally graded interface between diamond and the substrate material leads to improved adhesion of the diamond film [8]. Under Wako et al. investigations were carried out regarding the adhesion or the layer adhesion of CVD-diamond coating on differently pre-treated molybdenum specimens (99.9% Mo) [9]. The molybdenum substrates were pre-treated by polishing and grinding to achieve both different roughness and anisotropic as well as isotropic surface structure. An ultrasonic cavitation test under water exposed the CVD-diamond coating (thickness 2.5 μm) to stress. The substrates with the higher surface roughness (Ra 0.32 μm and 0.23 μm) showed a better layer adhesion basis. In addition, it was concluded from the crack growth parallel to the surface that anisotropic structures pass it in one direction only, while isotropic structures tend to exhibit circular layer delamination in this type of stress. [9] Investigations on the influence of the surface roughness on the layer adhesion were also done under Xu et al. In this case, CVD-diamond coatings were deposited on cemented carbide substrates with smooth (Rq 0.04 μm) as well as blasted (Rq 0.27 μm) surfaces using a Cr-CrN interlayer system. While the diamond coating of the specimens without any interlayer system flaked off directly on cooling, adherent CVD-diamond with a thickness of approx. 9 μm was deposited on both with diffusion barriers. In a Rockwell C indentation test the diamond coating without surface pre-treatment showed significantly poorer attachment (delamination at the interface to the interlayer), suggesting that increased surface roughness prior to CVD-diamond coating enhances layer adhesion. [10] Gomez et al. found in cutting experiments that a preferred direction of the substrate roughness, resulting from the delivery state or the finishing of the cemented carbide inserts, has a negative impact on the subsequent adhesion of the CVD-diamond layer (> 20 μm) under mechanical stress [11]. By means of a more intensive mechanical interlocking and a mechanically graded transition from the substrate material into the CVD-diamond coating, the compressive stresses in the transitional area can be significantly reduced compared to a smoother substrate surface. This was found by Lee et al. in their investigations in the CVD-diamond coating of laser induced micro-rough surfaces on cemented carbide substrates. [12]

In recent studies it was shown that diamond crystals can be deposited on steel with a silicon carbide interlayer by a laser induced plasma CVD process without a process chamber [13]. But as soon as a closed polycrystalline CVD-diamond coating is formed at the process temperature of 1100 °C the diamond coating delaminates during the cooling down phase after the process. To lower the impact of thermal expansion of the steel substrate, the deposition temperature of the process could be reduced to 700 °C. The investigations examine the impact of the surface microstructure, especially regarding the surface roughness and isotropy on preventing a delamination of the diamond coating.

2 Methods

The specimens were made of the tool steel 1.2379 (X135CrMoV12) in the soft annealed state (255 HB / approx. 268 HV) which is for example used for forming tools. Seven different microstructures on the surfaces of the steel specimens were produced. One is the as-delivered condition with a ground surface. A polished surface was prepared by grinding with 1200-grit abrasive paper and subsequent polishing by a diamond suspension of 3 μm grain size. Two different surface microstructures were produced by a blasting process of polished substrates by the company Oerlikon Balzers. A blasting by glass beads with a pressure of 4 bar as well as a corundum blasting with a pressure of 3 bar.

To generate reproducible surface microstructures in the range of the surface roughness, ultrasonic vibration assisted milling (UVAM) is a suitable process. This process was used to produce three different surface microstructures. Due to its kinematics, the microstructure is adequately predictable. Also it is possible, to vary the geometric properties regarding the isotropy as well as the aspect ratio or the mean height of the surface structure by adjusting various process parameters.

Face milling with vibration superimposition in the direction of the tool axis could be determined by a defined trajectory of the tool cutting edge. This results from the (ultrasonic) oscillations per revolution as well as the cycloidal trajectory, which reflects the rotational speed of the tool via a translatory feed (schematically shown in Fig. 1). On the basis of this trajectory, the simulation of the resulting surface microstructure takes place.

 https://doi.org/10.1051/matecconf/201819014008

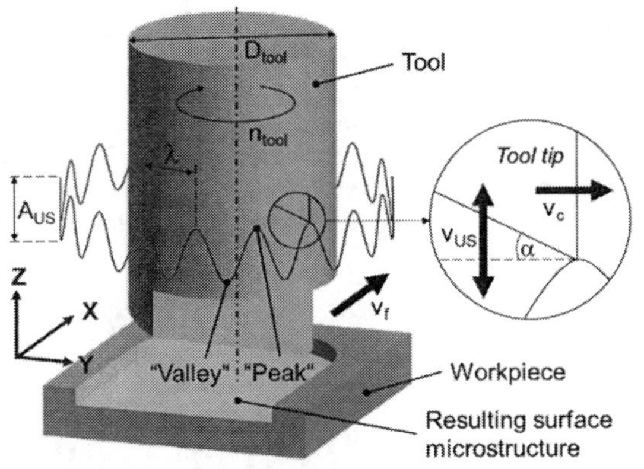

Fig. 1. Kinematics of ultrasonic vibration assisted face milling.

A simulation model was developed in MATLAB to calculate the surface microstructure of ultrasonic vibration assisted face milling. Therefore, a virtual tool was intersected along a defined trajectory with the virtual workpiece. This so called "kinematic model" does not consider dynamic effects nor physical effects such as burr formation, flaking or temperatures. Hence, a dexel model is sufficient for the volumetric representation of both the tool and the workpiece with regard to the spatial accuracy. Each dexel describes a discrete point with one height value aligned over a regular grid in the X-Y-plane. The advantage of this form of spatial discretisation is that due to the two-dimensional, equidistant dot distribution, the storage requirement increases not during the simulation and only quadratically with the increasing number of dexel points. Additionally, the surface simulation tool allows the integration of various tool geometries and kinematic paths. For the calculation, only the engaged part of the tool is relevant, which is initially constructed by means of a CAD software. The description of the tool movement along the trajectory was temporally discretized, too. In [14] the approach for the stepwise calculation of the tool position and orientation as a path vector was presented. Considering the requirement for a time efficient simulation, the computational effort is reduced by the simulation of one single rotation, which was afterwards patterned in the distance and the direction of the feed. This method allows an implementation and a variation of UVAM kinematics. In this way, the effects of displacement as well as re-cuts can be simulated and evaluated.

The experimental investigations were carried out at a high precision machining centre (KERN Pyramid Nano). A common transducer (Hielscher UIP 2000 hdT) was used to generate the ultrasonic vibration. According to Fig. 2 the set-up was clamped on the machine table. Inside the transducer are inert masses, a bolt for preload, piezoelectric discs and electrodes. The electrodes transfer the voltage to the piezoelectric discs and the discs transform the applied electric field into a linear motion using the inverse piezoelectric effect.

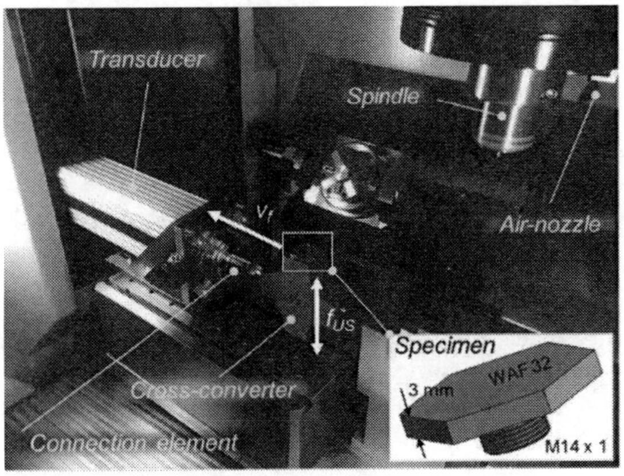

Fig. 2. Experimental set-up for UVAM.

To reduce the mass of the specimens and ensure the resonant vibration mode the specimens were designed only with a very small height (3 mm, Fig. 2). The direction of the vibration should coincide with the rotational tool axis to generate the desired microstructure on the specimen's surface. Thus, a cross-converter was required. On the one hand, the direction of the oscillation was turned 90°. This is based on the eigenfrequency of the converter: an elongation in the horizontal direction will result in a contraction in the vertical direction. On the other hand, the converter carries the specimen by a screw connection (thread M14 x 1, Fig. 2).

The minimum amplitude (A_{US}) which could be set by the transducer was about 4.5 µm. To achieve an aspect ratio of approx. ¼ of the microstructure, this results in a structural distance of about 18 µm: in one direction, it represents the feed, in the direction of the cutting speed it corresponds to the wavelength λ. For this reason it was necessary to use tools with a special geometry to prevent a collision between the tool's minor flank face and the generated surface microstructure, which is explained in detail in [14]. With this geometric specifications, the tool have a clearance angle of 40° and therefore, to ensure the stability of the cutting edge, a negative rake angle (-10°). The angle of the minor cutting edge is 15°. To protect the sharp cutting corner they have a very small chamfer (14 µm x 60°). Those geometric specifies of the tool tip are shown in Fig. 3.

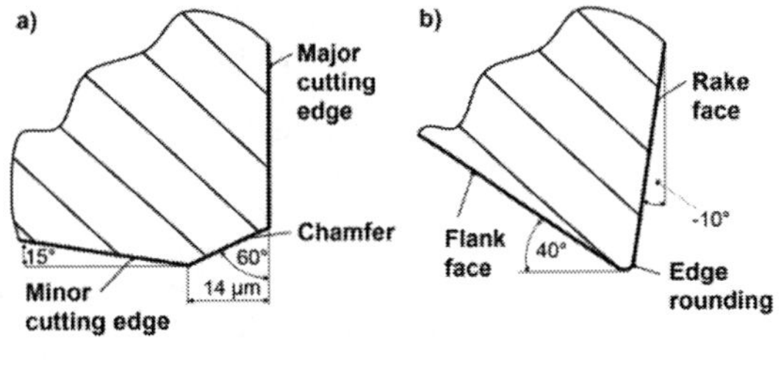

Fig. 3. Schematical tool tip shown in tool reference plane (a) and tool orthogonal plane (b)

Only single-edged tools were used in order to exclude a possible height offset between cutting edges due to the manufacturing process. The cemented carbide end milling cutters with a diameter of $D_{tool} = 6$ mm were manufactured by voha-tosec Werkzeuge GmbH and have a TiAlN / TiSiN coating system ("Hardlox"). The "wavelength" (λ) of the microstructure in the circumferential direction is defined by the ratio of the rotational speed and the oscillation frequency. Due to a nearly constant eigenfrequency, this ratio could be determined by the cutting speed. According to the maximum ultrasonic speed, the amplitude and the clearance angle, the cutting speed ($v_{c,\, min} = 21$ m/min) was calculated by (1):

$$v_{c,\, min} = v_{US} / \tan 40° = (2 \cdot \pi \cdot f_{US} \cdot A_{US}) / \tan 40°. \quad (1)$$

Slight deviations of the measured values of the amplitude are attributed to the transmission characteristics of the oscillation system. The vibration amplitude was transformed in the orthogonal direction (parallel to the rotational tool axis) with negligible vibration in the parallel direction (in feed direction, but only in the sub µm range). During the process, one additional piezoelectric disc is used as a sensor element to ensure operation in resonance mode. The resonance frequency f_{US} was about 19.25 kHz. Additionally, two more process parameters were investigated. These are presented in Table 1. The specimens include the two parameters A_{US} (ultrasonic amplitude: peak-to-peak) and the wavelength λ.

Table 1 Experimental plan and machining parameters

#	A_{US} [µm]	f_{US} [kHz]	f_z [µm]	v_c [m/min]	AR
UVAM 4.5 18	4.5	19.25	18	21	0.25
UVAM 4.5 26	4.5		26	30	0.17
UVAM 6.5 26	6.5		26	30	0.25

All steel specimen are coated by the company Oerlikon Balzers with a 2.4 µm thick chromium nitride (BALIQ® CRONOS) layer by high-power impulse magnetron sputtering (HIPIMS) in one batch. The thickness and the type of coating were chosen regarding to the publication of Buijnsters et al. [3]. A high CrN interlayer thickness of 20 µm as used by Glotzman et al. [6] was avoided to be able to evaluate the impact of the microstructure on the adhesion of the diamond coating. Diamond nucleation was carried out with a dispersion consisting of 200 ml of isopropanol and 210 mg of diamond powder with an average crystal size of 0.25 µm to 0.50 µm from the company Microdiamant AG. The substrates were put into the dispersion for ten minutes in an ultrasonic bath and subsequently into isopropanol for three minutes.

A CO_2 laser-based plasma CVD process, shown in Fig. 4 was used at atmospheric pressure without a vacuum chamber for the deposition of polycrystalline CVD-diamond coatings. A 6 kW high-power CO_2 laser with a wavelength of 10.6 µm was used for energy supply of the plasma jet. The ignition chamber was flooded by argon gas through a gas nozzle. The ignition of the plasma flame was achieved by thermionic emission through inserting a tungsten lanthanum ignition rod into the laser focus. This generates an optical breakdown. The laser radiation that was not absorbed by the plasma flame was directed to a water-cooled beam trap. After the ignition of the argon plasma flame, the process gases 0.02 standard litres per minute (slm) methane and 1.98 slm hydrogen, which are required for CVD synthesis of polycrystalline diamond, are supplied.

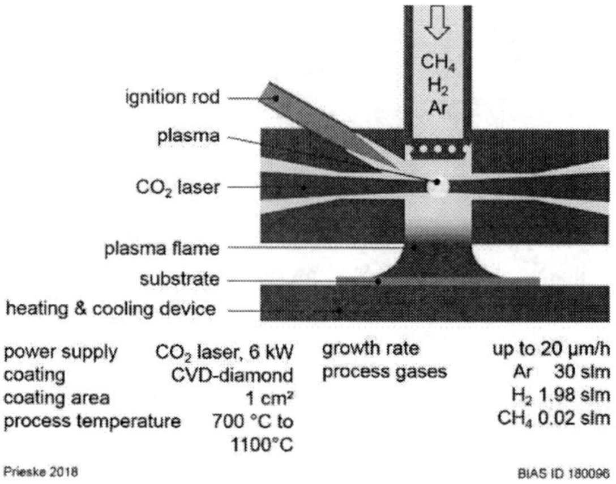

Fig. 4. Schematic layout of the laser induced plasma CVD process at atmospheric pressure.

To ensure a constant deposition temperature over the whole deposition time, a feedback control was implemented, which regulates the laser power according to the measured substrate temperature of the IMPAC pyrometer IGAR 12-LO. Detailed information of the implementation of the feedback control are published in [15]. For temperature monitoring, three pyrometer were used, IMPAC pyrometer IGAR 12-LO, IP 140 and IGA 10. To determine the emission coefficient for each pyrometer, one of the CrN coated steel substrates was heated up to 800 °C by a ceramic hot plate, which was measured by a thermocouple. The emission coefficients were set to 0.34 for pyrometer IGAR 12-LO, to 0.37 for pyrometer IP 140 and to 0.25 for pyrometer IGA 10, so that each of them measures 800 °C.

Scanning electron microscopy (SEM) (Carl Zeiss Microscopy EVO MA-10) and 3D laser microscopy (Keyence VK-9710 and Keyence VK-9700) were used to take images of the pre-treated as well as coated surfaces and to determine the roughness parameters according to ISO 25178 part 2 and 3 on a measuring area of 0.5 mm x 0.5 mm. This ensures that a minimum of five form elements in both the X-direction and the Y-direction are included in the calculation of the surface parameters. For the roughness analysis an S-Filter of 2 µm and an L-Filter of 0.5 mm was used

The characterisation of both the milled and the simulated surface microstructure of the UVAM specimens was carried out using the surface analysis software VK Analyzer. The selected measuring field was located centric along the milling path. In order to remove inclination deviations, the measured surfaces of the specimens were aligned by means of a subtraction method and a form filter (2nd degree polynomial) was used to eliminate shape deviations.

Element analysis was done by an energy-dispersive X-ray spectroscope (EDX) (Bruker Nano GmbH XFlash Detector 610M). Raman spectroscopy (Renishaw system 1000) with an excitation wavelength of 514 nm was used to determine diamond and graphite. The existence of diamond was evidenced by the measurement of the first order Raman line of diamond at 1332 cm^{-1} [16]. Cryofractures were produced by eroding a predetermined breaking point into the steel substrate up to 0.5 mm underneath the coating area. Afterwards the specimen was cooled down by liquid nitrogen and broken into two parts. The hardness of the steel specimen was measured by a Vickers hardness test (Durascan 50G5) according to DIN EN ISO 6507-1.

Dilatometer measurements were performed with a quenching dilatometer DIL 805A according to ASTM A1033-10. The steel specimens for the dilatometer tests were used in form of pipes and had a length of 10 mm, an outer diameter of 4 mm and an inner diameter of 3 mm.

One batch of steel specimens was coated at a process temperature of 840 °C with a coating thickness of 8 µm and another one at 700 °C with a thickness of 1 µm. 700 °C was the lowest achievable temperature by the use of a water cooled copper specimen holder. The cooling down rate after deposition was 33 K/s from 840 °C to 350 °C measured by the pyrometer IGAR 12-LO. Cooling down was done under a protective argon atmosphere to prohibit oxidation.

3 Results

In the visual comparison of the UVAM microstructured surfaces with the simulated, it was found that the milled surfaces correspond to those simulated without re-cutting, but rotated by 180°. It seems likely that the microstructure results entirely from the re-cut, visible on individual form elements. The valleys as well as the peaks show a similar preferred direction. This could be attributed to an elastic material spring back, whereby the pre-cut compresses the material thus the re-cut enters full material. In addition, a slight displacement of the microstructure occurs (Fig. 5, down). This results from a non-controllable, non-integer ratio of the rotational speed and the oscillation frequency. The two variants of the simulation in Fig. 5 show the case without (SIM_1) or with a maximum structural displacement (SIM_2).

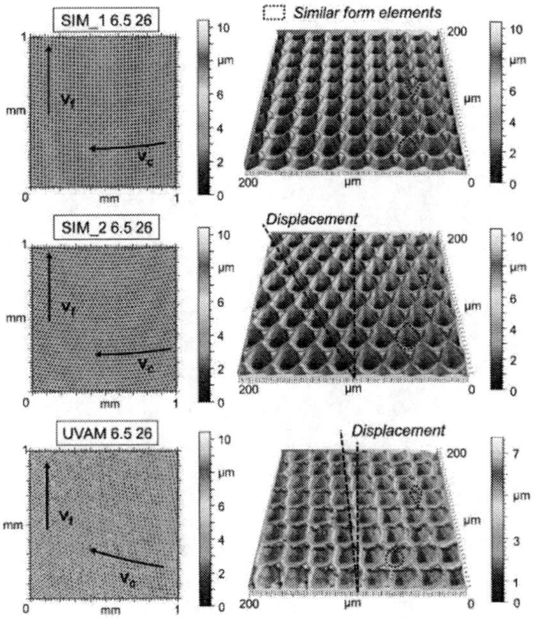

Fig. 5. Visual surface comparison of SIM_1 6.5 26 (top), SIM_2 6.5 26 (middle) and *UVAM 6.5 26* (down).

Furthermore, it is shown, that the profile height of the generated surface is smaller than the simulated height. This could be due to the damping of the amplitude by the tool intervention. The quantitative comparison of the surface parameters Sa and Sdr confirms the considerations of the surface design (Fig. 6). It could be noted that the smallest value for the arithmetic mean height corresponds with the smallest developed interfacial ratio (for the three with UVAM generated surfaces). The highest value for the developed interfacial area ratio of the specimens (UVAM 6.5 26) results form a higher amplitude.

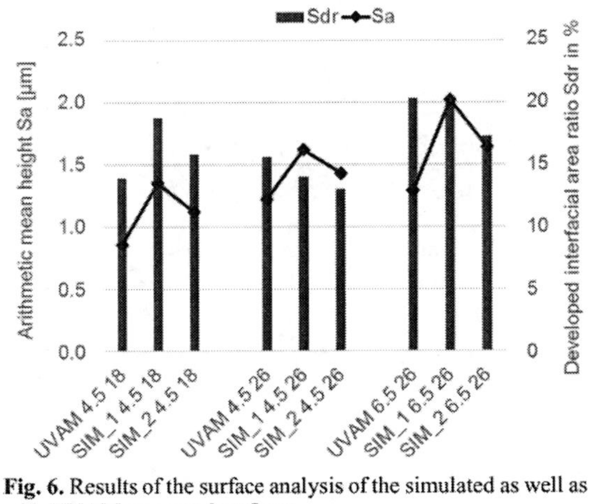

Fig. 6. Results of the surface analysis of the simulated as well as with UVAM generated surfaces.

Although the qualitative (as well as the quantitative) accordance between the designed, simulated and experimentally manufactured surface microstructure is sufficient, it has to be noted that this only applies for the middle area of the milled path. In the overlapping area, the microstructure is distorted and this leads to strongly anisotropic areas (Fig. 7). In addition to the microstructure, further elements on the surface, such as burr formation, occur.

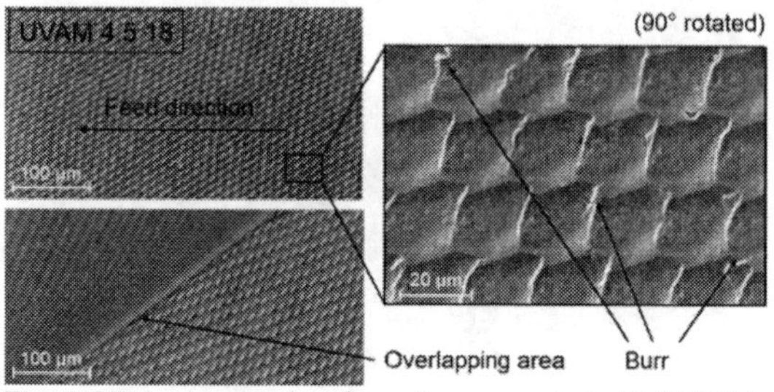

Fig. 7. SEM micrographs of a surface generated with UVAM.

In Fig. 8 differently manufactured microstructures coated by a chromium nitride layer are shown as SEM micrographs as well as the measured Sa and Sdr values. Both parameters vary over a wide range from the very smooth polished surface with Sa = 0.1 µm and Sdr = 2% to the rough corundum blasted surface with Sa = 3.0 µm and Sdr = 297%. In comparison to Fig. 6 the developed interfacial area ratio (Sdr) values increase and the Sa values decrease due to the chromium nitride coating. EDX measurements of the chromium nitride coating detected 47 at.% nitrogen and 53 at.% chromium, which verifies a chromium nitride coating.

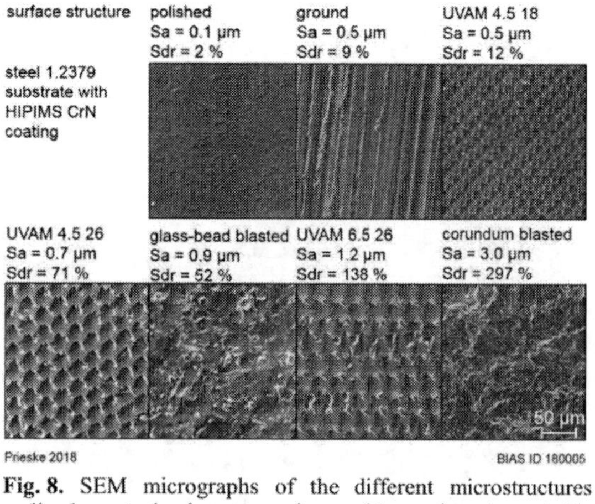

Fig. 8. SEM micrographs of the different microstructures realized on steel substrates and coated by a chromium nitride layer.

The result of the dilatometer measurement shown in Fig. 9 a) with a slower heating rate of 2 K/s shows an austenite to ferrite transformation and thereby a nonlinear change in length. The temperatures of phase transformation A_{C1} and A_{C3} are measured as 840 °C and 871 °C. In Fig. 9 b) the test parameters are adapted to the process conditions, with a heating from room temperature to the deposition temperature in 2.5 minutes and a subsequent holding time of 40 minutes at the deposition temperature. The tests show that a holding temperature of 840 °C leads to the smallest length change of 0.86%. With the LaPlas CVD process 700 °C is the lowest achievable temperature, where a diamond coating can be deposited. During the heating process of the dilatometer test the length change at 700 °C shows a length change of 0.93%.

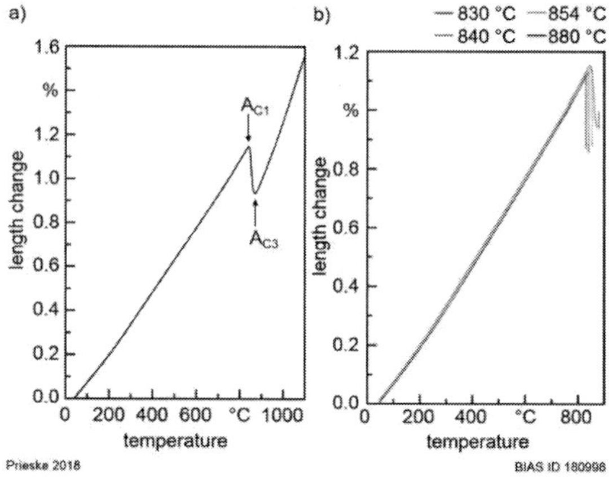

Fig. 9. Dilatometer measurements of 1.2379 steel specimen a) with a heating rate of 2 K/s up to 1100 °C and b) with a heating rate of 7 K/s and different holding temperatures for 40 minutes.

Fig. 10 shows the result of the local CVD-diamond deposition at 840 °C with a thickness of 8 µm. In the photography of the coating, the bright grey area is the CrN layer, where the diamond coating delaminated. The dark grey area is the remaining diamond coating. The black box shows the position of the measured height profile. The scale of the height profile in Fig. 8 and Fig. 12 is individually chosen to show the optimum contrast, with the low structures in blue colour and the high structures (usually the diamond coating) in red. The four microstructured specimens with the lowest roughness (Fig. 10 left side) led to a delamination of the diamond coating in form of a diamond foil, so that nearly no diamond is left on the substrate. The glass-bead blasted substrate led to a flake like delamination. The *UVAM 6.5 26* specimen shows that especially in the overlap area with a distorted microstructure a complete delamination occurs. The diamond coating on the corundum blasted specimen looks as a closed diamond coating in the macroscopic image.

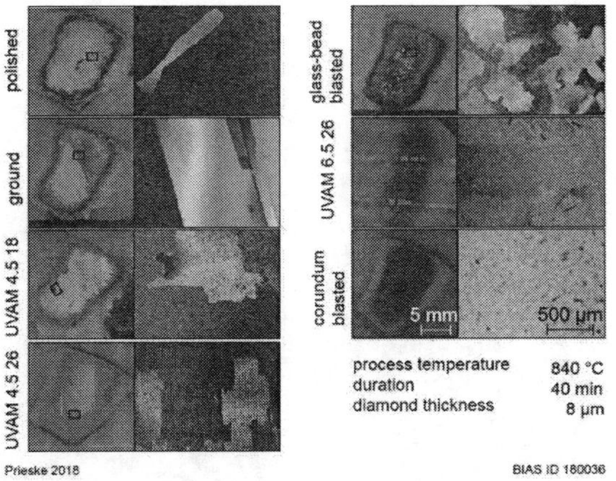

Fig. 10. Photography and height profile of the diamond coating deposited at 840 °C.

A higher magnification of the remaining diamond coating on the substrate is shown in Fig. 11. On each image the polycrystalline character of the coating can be seen.

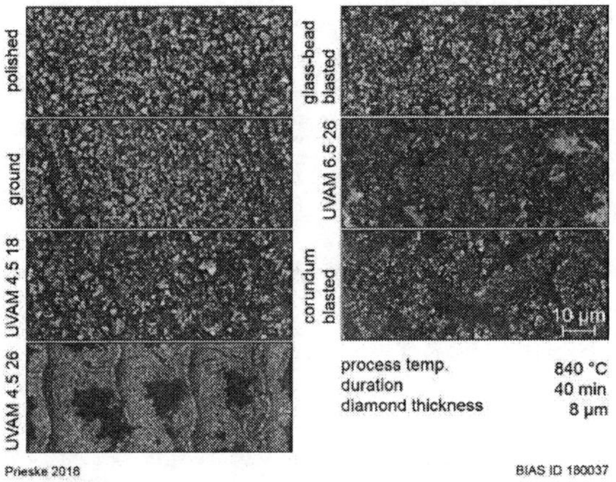

Fig. 11. Microscopic images at high magnification of the diamond coatings deposited at 840 °C, which show the crystalline structure.

The two structured specimens *UVAM 4.5 26* and *UVAM 6.5 26* show that in the valleys of the surface structure the diamond coating is not a closed layer. Especially the higher peak to valley height of specimen *UVAM 6.5 26* leads to the growth of large diamond crystals on the peaks and just small diamond crystals in the valleys, as it is marked in Fig. 12 on the right image. The diamond coating on the corundum blasted substrate has small uncoated or delaminated areas as can be seen in Fig. 13. The white dots in Fig. 13 a) represent the CrN coating, which is not coated by diamond as can be seen in the detailed image in Fig. 13 b) and was verified by an EDX measurement.

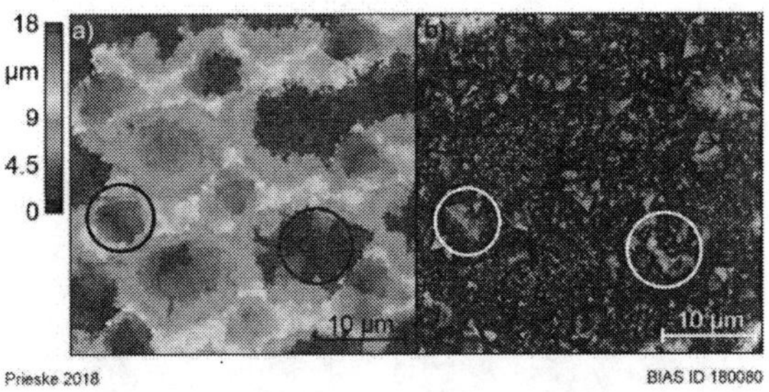

Fig. 12. Diamond coating deposited at 840 °C on an *UVAM 6.5 26* substrate a) as a height profile and b) microscope image on the same position.

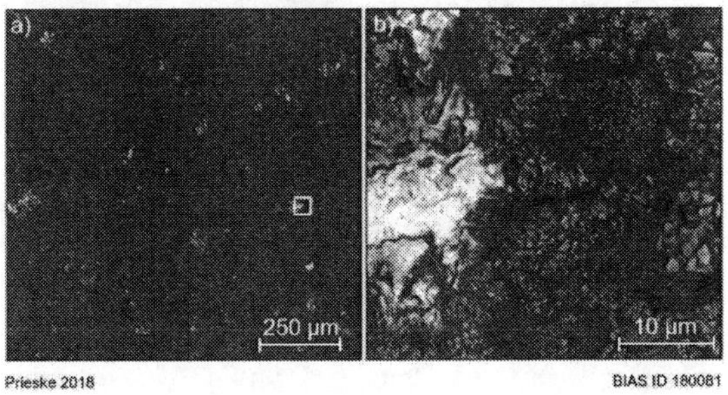

Fig. 13. a) Overview image and b) detailed image (box shown in a) of the diamond coating deposited at 840 °C on corundum blasted substrate.

Fig. 14 shows the result of the local CVD-diamond deposition at 700 °C with a thickness of 2 µm. Compared to the diamond coatings deposited at 840 °C (Fig. 10) the area of delamination decreases. In the height profiles of all specimen clear delamination can be seen except on the ground and corundum blasted specimen. On the ground substrate a closed diamond layer is deposited with lots of small delaminated areas (black dots in the height profile). The overlap area of the UVAM specimens promotes the delamination of the diamond coating, as can especially be seen in the photography of specimen *UVAM 4.5 18*.

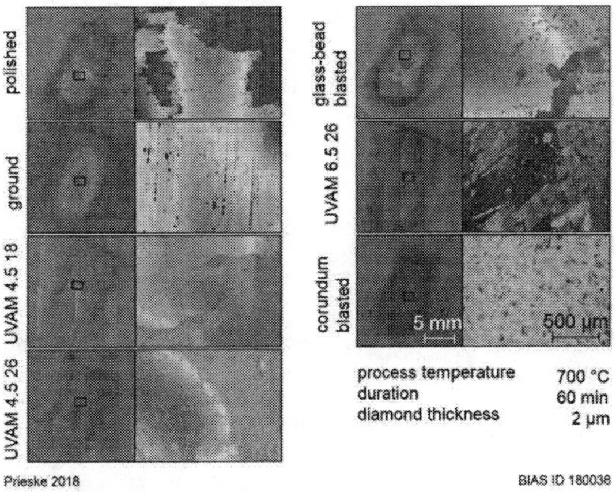

Fig. 14. Photography and height profile of the diamond coating deposited at 700 °C.

The highly magnified images of the diamond coatings deposited at 700 °C in Fig. 15 show a nanocrystalline structure. In all images, except of the diamond coating on the polished substrate, small delaminated or uncoated areas can be seen, which have a length smaller than 10 µm.

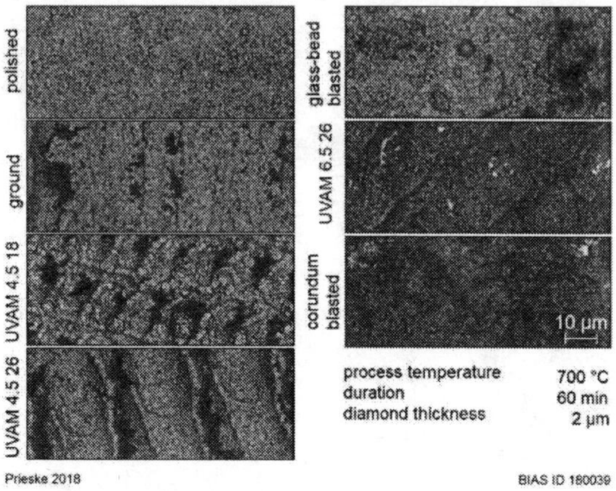

Fig. 15. Laser scanning microscope images at high magnification of the diamond coatings deposited at 700 °C, which show the crystalline structure.

A representative Raman spectrum of a diamond coating deposited at 840 °C and 700 °C is shown in Fig. 16. The spectrum of the coating deposited at 840 °C shows the peak at 1332 cm^{-1}, which proofs residual stress free microcrystalline diamond [16]. No residual stress is measured, because the measurement is done on a diamond coating which is partly delaminated. The peak at 1560 cm^{-1} is the so called G-peak [17] which shows that the coating also has a graphitic content. The peak at 1190 cm^{-1} can be related to incorporated nitrogen in the diamond coating, which is introduced by the CrN interlayer. Zhang et al. [18] detected an increase of two peaks at 1190 cm^{-1} and 1550 cm^{-1} with increasing nitrogen content in the diamond coating.

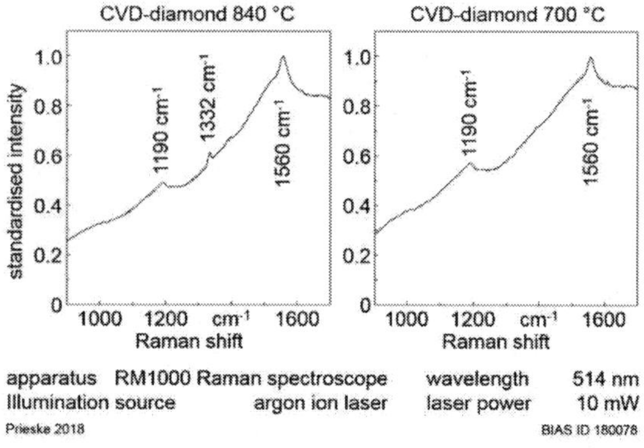

Fig. 16. Raman spectra of the diamond coating deposited at 840 °C (left) and at 700 °C (right).

The cryofractures shown in Fig. 17 are produced after the diamond deposition. It can be seen, that the CrN interlayer has a coating thickness of 2.4 µm. Even on specimen *UVAM 6.5 26* the CrN coating is very homogenous. The diamond coating delaminates by producing the cryofracture. Just at the boarder of the coated area on the polished substrate Fig. 17 b) the diamond coating remained.

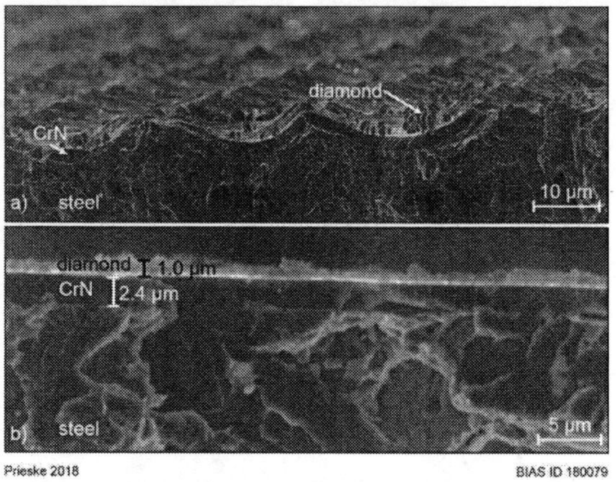

Fig. 17. a) Cryofracture of *UVAM 6.5 26* specimen diamond coated at 840 °C and b) cryofracture of the polished specimen after diamond coating at 700 °C.

Hardness measurements in Fig. 18 show that the steel is in a soft annealed state after the deposition process, with a hardness of approximately 220 HV0.1.

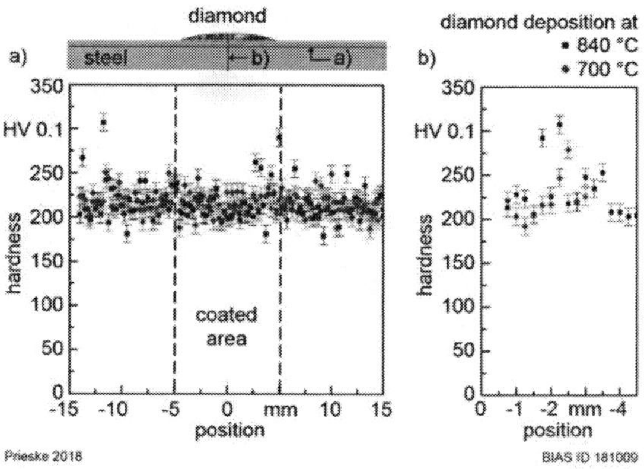

Fig. 18. Hardness measurements of the steel substrate a) parallel and b) perpendicular to the surface after the deposition process at 700 °C and 840 °C.

4 Discussion

CVD-diamond deposition of steel provides several challenges. To enable the coatability of steel by a diamond layer a diffusion barrier is necessary, which prevents the diffusion of carbon into the steel specimen and the diffusion of iron into the diamond coating. The investigations show, that the used HIPIMS chromium nitride coating with a thickness of 2.4 µm successfully worked as a diffusion barrier and has a high adhesion to the steel substrate. In all cases the delamination took place at the interface between CrN interlayer and diamond coating.

The used tool steel 1.2379 has a thermal linear expansion coefficient of $13.0 \cdot 10^{-6}$ m/(m·K) in the temperature range from 20 °C to 400 °C [19]. Diamond has a thermal expansion coefficient between

$1.0 \cdot 10^{-6}$ m/(m·K) and $4.5 \cdot 10^{-6}$ m/(m·K) [1]. This leads to a big mismatch of thermal expansion coefficients and induces large residual stresses.

The delamination can be influenced by the surface microstructure. The diamond coating deposited at 840 °C with a coating thickness of 8 µm shows larger delamination than the coating deposited at 700 °C with a thickness of 2 µm. Fig. 9 shows that the prolongation of the steel specimen is less for a deposition at 840 °C with 0.86% than for 700 °C with 0.93%, which is due to the austenite to ferrite transformation. Uncoated areas in the 2 µm thick diamond coatings prevent the generation of high stresses. A higher coating thickness of the diamond coating as well as a closed coated area leads to an increase of the residual stresses. Hardness measurements of the steel after the deposition processes showed that the steel is in annealed condition. The cooling down rate from deposition temperature to room temperature in two minutes would be fast enough for a hardening of the steel sample. That leads to the assumption that the total length change could be lower with a higher deposition temperature than 840 °C due to the austenite to ferrite transformation in the steel substrate which results in a volume expansion during cooling down in case of a hardening process.

Fig. 10 shows, that the coating adhesion is increased with increasing arithmetic mean height (Sa) to a certain extent. This agrees with the literature cited in the introduction [7-10]. Regarding the coating thickness of 8 µm higher roughness values result in a flake-like delamination, with remaining areas of the diamond coating. Whereas low roughness values lead to a delamination of the diamond coating as a foil. The overlap area of *UVAM 6.5 26* reduces the surface roughness (Fig. 7). It can be seen that at the positions of the overlap areas a complete delamination takes place, whereas inside the milling path just local delamination is detected (Fig. 10). Furthermore, the *UVAM 6.5 26* substrate shows that sharp peaks lead to the formation of large diamond crystals (Fig. 12). At the same time the valleys are not coated by a closed diamond layer. This can be explained by a local build-up temperature on the sharp peaks, which results in a local higher diamond growth rate. The diamond coating growth in microwave plasma CVD processes also strongly interacts with mountains, which leads to a reduced diamond growth rate in valleys. To achieve a homogenous coating the chamber pressure is increased, which is not possible in the atmospheric LaPlas CVD process. It is assumed that the higher the peak to valley height is, the higher is the probability that the carbon atoms bind to the larger diamond on the peak rather than to the small diamond crystals in the valley.

In Fig. 13 it can be seen that very high Sa and Sdr values, as achieved by corundum blasting, results in an inhomogeneous diamond coating which varies between nano- and microcrystalline diamond. The size of the delaminated areas on the corundum blasted substrates decrease from around 20 µm at 840 °C to 5 µm at 700 °C (compare Fig. 13 and Fig. 15).

Interestingly, the ground specimen showed quite good results at a deposition temperature of 700 °C. There are small delaminated areas of maximum 10 µm diameter of the closed diamond coating, but it did not lead to a complete delamination of the coating, even though it has the second smallest roughness values of the investigated microstructures.

It can be seen that the delamination always takes place in accordance to the structure. The delaminated area of the diamond coating has the form of stripes on ground substrates, of a foil on a polished substrates and is random shaped on the stochastic structures of blasted substrates.

5 Conclusion

UVAM can be used to generate defined and reproducible microstructures that can be simulated by a surface prediction tool and varied - in technological limits - in their geometrical properties.

It could be shown that the surface microstructure has a high impact on the interlocking between coating and surface. Low surface roughness values up to 0.9 µm lead in case of 8 µm thick diamond coatings to a delamination of the diamond coating in form of a diamond foil. Higher roughness values result in a flake-like delamination, with remaining areas of the diamond coating. None of the microstructures investigated in this study was able to overcome the residual stress induced by thermal expansion of the steel substrate in such a way, that it prevents delamination of the coating entirely.

The following conclusions of expedient parameters can be drawn to achieve a closed diamond coating on steel:

- Before diamond deposition of a steel specimen, the thermal expansion of the specific type of steel needs to be investigated. By taking into account that the austenite to ferrite transformation has a strong influence on the prolongation of the steel, the most advantageous deposition temperature and cooling down rate needs to be estimated.
- For the deposition of the steel 1.2379 the deposition temperature needs to be either lower than 650 °C where the length change of the steel specimen is 0.85% or higher than 840 °C with a with hardening of the steel and reducing the thermal expansion.
- Sharp peaks at the microstructured specimen surface should be avoided due to local build-up temperature, which leads to an inhomogeneous diamond coating thickness.
- A certain roughness higher than Sa 0.1 µm supports the adhesion.

Acknowledgements

This work was supported by the Deutsche Forschungsgemeinschaft (DFG) under contract no. VO 530/75-2 which the authors gratefully acknowledge. We would like to thank Oerlikon Balzers for the support regarding the surface preparation and CrN coating as well as the Department of Surface Engineering of the IWT – Leibniz Institut für Werkstofforientierte Technologien, especially Mr. Henning

Hasselbruch who perfomed the dilatometer tests in cooperation within the priority program SPP 1676.

References

1. K. Kellermann, C. Bareiß, S.M. Rosiwal, R.F. Singer, Advanced Engineering Materials **10** 7 (2008) 657–660
2. J. G. Buijnsters, P. Shankar, J. J. Ter Meulen, Surface and Coatings Technology **201** 22–23 (2007) 8955–8960
3. J. G. Buijnsters, P. Shankar, W. Fleischer, W. van Enckevort, J. Schermer, J.J. ter Meulen, Diamond and Related Materials **11** 3–6 (2002) 536–544
4. R. Polini, F. P. Mantini, M. Braic, M. Amar, W. Ahmed, H. Taylor, M.J. Jackson, 2006, Proceedings of the 32nd International Conference on Metallurgical Coatings and Thin Films, ICMCTF 2005, Thin Solid Films **494** 1–2 (2006) 116–122
5. H. Hei, Y. Shen, J. Ma, X. Li, S. Yu, B. Tang, W. Tang, Vacuum **109** (2014) 15–20
6. O. Glozman, A. Hoffman, Diamond and Related Materials **6** (1997) 796–801
7. M. Amaral, F. Almeida, A. J. S. Fernandes, F. M. Costa, F. J. Oliveira R. F. Silva, Surf. Coat. Technol. **204** 21–22 (2010) 3585–3591
8. R. K. Singh, D. R. Gilbert, J. Fitz-Gerald, S. Harkness, D.G. Lee, Science **272** (1996) 396–398
9. H. Wako, T. Abe, T. Takagi, und T. Ikohagi, Appl. Surf. Sci. **256** 5 (2009) 1466–1471
10. Z. Xu, L. Lev, M. Lukitsch, A. Kumar, Diamond and Related Materials **16** 3 (2007) 461–466
11. H. Gomez, D. Durham, X. Xiao, M. Lukitsch, P. Lu, K. Chou, A. Sachdev, A. Kumar, J. Mater. Process. Technol. **212** 2 (2012) 523–533
12. D. G. Lee, D. R. Gilbert, S. M. Lee, R. K. Singh, Compos. Part B Eng. **30** 7 (1999) 667–674
13. M. Prieske, Proc. of 12th Intern. Conf. The "A" coatings 2016 TEWISS-Technik und Wissen GmbH (PZH Verlag) Hannover **3** (2016) 99–106
14. R. Börner, A. Schubert, S. Winkler, T. Junge, C. Titsch, A. Schubert, W.-G. Drossel, Journal of Materials Processing Tech. 255C (2018) 749–759
15. M. Prieske, C. Wiegmann, M. Schwander, F. Vollertsen, Dry Met. Form. OAJ FMT **1** (2015) 1-4
16. A.C. Ferrari, J. Robertson, Mathematical, Physical and Engineering Sciences **362** (2004) 2477–2512
17. A.C. Ferrari, Solid State Communications **143** (2007) 47–57
18. Q. Zhang, S.F. Yoon, J. Ahn, Rusli, Y.-P. Guo, Microelectronics Journal **29** (1998) 875–879
19. http://www.dorrenberg.es/download/aceros/DOE/1.2379_deu.pdf, opened 19th Jan. 2018

https://doi.org/10.1051/matecconf/201819014009

Influence of surface integrity on the plastic flow in analogous testing of 42CrMo4

Rafael Hild[1,*], *Andreas Feuerhack*[1], *Daniel Trauth*[1], *Dennis C. Hoffmann*[2], *Nathan C. Kruppe*[2], *Tobias Brögelmann*[2], *Kirsten Bobzin*[2] *and Fritz Klocke*[1]

[1]Laboratory for Machine Tools and Production Engineering (WZL), RWTH Aachen University
[2]Surface Engineering Institute (IOT), RWTH Aachen University

Abstract. Surface modification of workpieces for cold forming processes is common in industrial practice. Specific surface topographies manufactured by means of shot peening form lubricant pockets, which provide a continuous lubricating film throughout the process. Furthermore, hydrostatic as well as hydrodynamic lubrication effects emerge, which further reduce the friction during the process and protect the tool from wear effects. These effects are well researched and applied in order to enable cold forming processes. In dry forming processes, no lubricants are used, and the positive lubrication effects do not appear. Surface modifications by shot peening or knurling change surface integrity in terms of hardness, residual stresses and geometry. The effects of the changed surface integrity on the forming process are not researched. In tribometer tests, a hard and rough surface texture lead to a lower frictional shear stress, whereas a rather smooth and hard surface texture leads to a lower punch force in an extrusion process. The mechanisms between surface textures and the smoothing and forming process are unknown. Furthermore, findings of the tribometer tests cannot be fully transferred to the extrusion process. As in tribometer tests a pin is used as tool and in extrusion a whole die, a discrepancy between the most suitable surface texture for a friction reduction results out of different flow conditions. Still, the tribometer tests give a first reference on a surface pretreament. The paper deals with an investigation of surface textures in order to distinguish characteristics of surface textures respectively a friction reduction. By means of indentation tests of the quenched and tempered steel 42CrMo4 the plastic flow in dependency of different surface integrities are researched. The results show which characteristics help to lower the punch force the most, so for further investigations an optimized surface modification is set up in order to enable a dry metal forming process.

Keywords: Cold forming, Tribology, Surface integrity

1 Introduction

Cold forming processes are highly effective due to a high ratio of material utilization, high product quality and short production cycles. At the same time, cold forming processes are highly reproducible and suitable for the production of high quantities [1]. Nowadays, lubricants in combination with surface texturing of workpieces are used to enable cold forming processes [2]. Lubricants reduce process loads and prevent wear [3]. By means of surface texturing like shot peening a surface enlargement is achieved [4]. The surface enlargement supports a higher lubricant capacity and at the same time provides a lubricant film throughout the forming process by means of lubricant pockets [4]. In order to raise the process productivity of cold forming even further and at the same time reduce ecological problems of lubricants, cold forging without lubricants is investigated [5]. Cold forging without lubricants would reduce process steps as applying and removing of lubricants as well as recycling steps [5]. Due to the absence of lubricants the process loads may increase substantially and could cause damage to the tools or deteriorate the process [6].

There is some research on dry metal forming regarding sheet metal forming. By means of DLC-coatings on deep drawing tools, a dry deep drawing process was realized by Murakawa [6]. Murakawa et al. extended this research by a review of different tribological conditions and found non-silicon films most suitable [7]. Osakada et al. investigated the impact of tool surface roughness on the coefficient of friction in dry sheet metal forming and found that a higher roughness resulted in a higher coefficient of friction [8]. Kataoka et al. made researches regarding ceramic tools and achieved comparable results to the lubricated sheet metal forming of low-alloyed steels [9]. Tamaoki et al. developed electroconductive ceramic tools for deep drawing [10] and was able to reach drawn quantities as high as with

* Corresponding author: r.hild@wzl.rwth-aachen.de

lubricants [11]. Due to higher process loads and surface extension a transfer of these findings to cold metal forming is not directly possible [12]. Furthermore, the investigation of the influence of surface textures on cold forging processes are rather rudimental. Wang et al. investigated surface texturing of billets in upsetting and ironing processes and found a higher lubricant capacity for textured surfaces. An investigation on the dependency of lubricant capacity or forming force reduction in favor or the surface roughness or hardness was not done [13]. Groche et al. investigated the influence of surface textures on forming tools in order to reduce friction and wear. It was possible to reduce the friction ratio up to 40 %, but an investigation of a surface textured workpiece was not done [14].

The aim of the authors' work is to investigate the surface textures of workpieces and their influence on the forming capability and friction reduction. The understanding of the mechanisms between a textured workpiece and the forming tool helps to enable a dry metal forming process, see Fig. 1 [13]. In former investigation the influence of surface textured workpieces on the use of self-lubricating PVD-coated tools is done [19]. The boundary conditions of the tribosystem are either changed by the use of a self-lubricating coating (Cr,Al)N+X:S (X = Mo,W) on tools or surface textures by means of shot peening. In this work, former experimental and numerical researches are extended with an analysis of surface textures and their forming behavior. The paper demonstrates how a dry metal forming process using surface textures on workpieces and self-lubricating tool coatings can be realized, see Fig. 1.

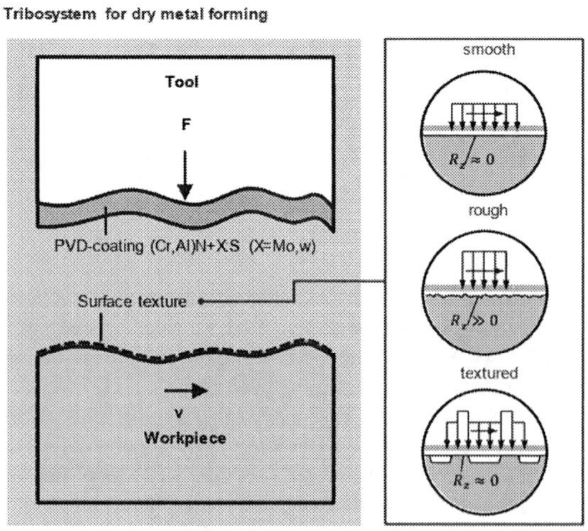

Fig 1. Research objectives for a novel tribosystem for dry metal forming in view of the tool from the surface engineering view (IOT) and surface texturing view (WZL) [13]

2 Previous Work

The current research is realized in the second phase of the priority program SPP 1676 of the German Research Foundation. In phase I, new approaches to enable dry metal forming processes were investigated initially using stationary loads. In this section an explanation of a novel pin-on-cylinder tribometer, which was used to analyze the influence of surface textures on the frictional shear stress is given. To do so, a long frictional path has to be analyzed and ensure that an untreated surface texture is in contact with the pin along the frictional path. The experimental results are further complemented by the development of a numerical FE-model to analyze the influence of different surface textures on the friction shear stress. These results a correlated with shot peening parameters of the peening process to distinguish a parameter set to reduce the frictional shear stress the most for each shot peening medium. After the first step of tribometer tests, the surface textures are applied on specimens for an extrusion process in order to distinguish the most suitable surface texture to reduce friction and by that the punch force.

2.1 Deposition process of the (Cr,Al)N+Mo:S coating

The (Cr,Al)N+Mo:S-coatings were deposited by means of the hybrid direct current magnetron sputtering/high power pulsed magnetron sputtering (dcMS/HPPMS) technology in an industrial scale coating unit CC800/9 Custom, CemeCon AG, Wuerselen, Germany. The unit is constructed with a chamber volume of $V = 1\ m^3$. For the deposition five CrAl20 (Cr-base plate with 20 Al plugs) with a purity of $w_{Cr} = 99.9$ wt.-%, $w_{Al} = 99.5$ wt.-% and one MoS_2 target with a purity of $w_{MoS2} = 99.5$ wt.-% were used. The MoS_2 target was mounted on one of the four dcMS cathode. In order to ensure a sufficiently high adhesion strength between the top layer and the substrate, a (Cr,Al) bond coat and a (Cr,Al)N interlayer were deposited first. The process parameters for the deposited coatings are listed in Table *1*. The process parameters of the (Cr,Al)N+Mo:S top layer are explained in more detail in Bobzin et al. [19].

Table 1. Process parameters for deposition of the (Cr,Al)N+Mo:S-coatings

Process parameter	Unit	
Depostion time, t	Min	107
Argon flux, j(Ar)	sccm	200
Nitrogen flux, $j(N_2)$	sccm	Pressure controlled
Pressure , p	mPa	710
dcMs cathode power, $P_{dcMS-MoS2}$	kW	2
dcMs cathode power, $P_{dcMS-CrAl20}$	kW	3
HPPMS mean cathode power, $P_{HPPMS-CrAl20}$	kW	5
Pulse length, t_{on}	µs	40
Frequency f_{pulse}	Hz	500
Bias voltage, U_{Bias}	V	100

2.2 Analysis of the influence of shot peening parameters on evolving surface textures

Shot peening by steel casks is commonly used as workpiece preparation in industrial cold forging processes [13]. The surface modification supports a friction reduction in combination with the use of a lubricant. In dry metal forming, as a result of the missing lubricant a higher friction is expected. In order to get the most suitable surface texture for a friction reduction in cold forging, different peening media were investigated [13]. Besides steel casks, ceramic beads and corundum particles were used with different peening parameters, Table 2.

Table 2. Shot peening parameters in dependency of shot media, (Steel = St, Ceramic = Ce, Corundum = Co)

Medium	Particle form	Particle size [µm]	Density [g/cm³]	Hardness [HV]
St	round	700-1,000	7.8	390 - 535
Ce	round	125 – 250	3.8	~ 1,200
Co	sharp	425-600	3.9 – 4.1	~ 2,600

The best peening parameters for a reduction of the fricitonal shear stress in tribometer tests are the ones shown in Table 3 [14]. Furthermore, a rather sharp and hard surface texture as the one shot by corundum particles, reduced the frictional shear stress the most.

Table 3. Resulting surface integrity after shot peening (Steel = St, Ceramic = Ce, Corundum = Co, Ref = Reference)

Surface	Peening parameters		Surface integrity		
	p [bar]	Ο [%]	HV10 [HV]	Sa [µm]	Sz [µm]
St	3	100	280	2.23	25.7
Ce	2	100	410	1.38	26.8
Co	4	100	260	4.09	59.7
Ref			200	1.63	29.5

2.2 Development and Use of a Pin-on-Cylinder Tribometer

For an explicit investigation of the influence of the surface textures on the frictional shear stress conventional tribometers as a pin-on-disc-tribometer cannot be used since textures are flattened after the first revolution of the disc. To analyze the surface textures intensively, a new frictional path per revolution has to be used. Thus, a pin-on-cylinder (POC) tribometer concept was realized on a lathe, see Fig. 2.

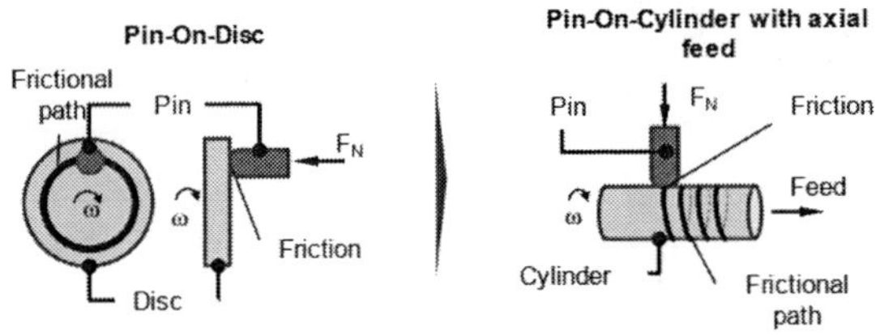

Fig. 2. Schematic overview of Pin-On-Cylinder Tribometer

The normal load is applied by means of a hydraulic actuator [15]. Due to these investigations, results regarding the influence of the contact between self-lubricating coatings and surface textures on the friction shear stress were obtained [18]. Shot peening by corundum particles as the best treatment to reach lower frictional shear stresses in this interaction was confirmed, see Fig. 3.

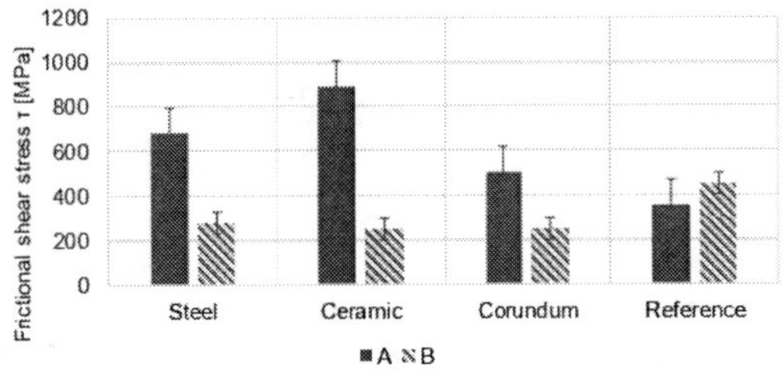

Fig. 3. Frictional shear stress during tribometer tests with uncoated (A) and coated (B) tools

2.3 Effect of surface textures on the punch force in an extrusion process

After investigating surface textures using tribometer tests, the surface textures were applied in an extrusion process [16]. As in the tribometer experiments, the named surface textures were used. As coating a slighty adapted one compared to the tribometer tests were used. The punch velocity was a constant $v_P = 5$ mm/s and the punch force was measured throughout the whole process. The punch force in extrusion processes is greatly influenced by the friction [17]. A lower punch force corresponds to a lower friction during the process. The results showed, that the lowest punch force was achieved from the combination of ceramic shot peened surface texture with self-lubricating tool coating, Fig. 4.

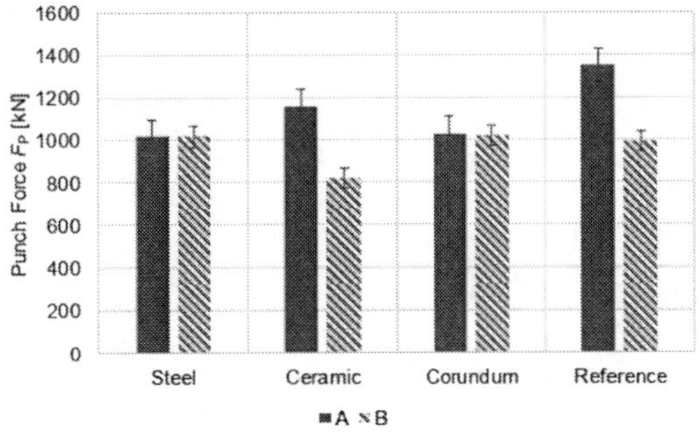

Fig. 4. Maximum punch forces of workpieces with different surface structures and uncoated (A) and coated (B) tools

The comparison between the results of the tribometer tests and the extrusion process show a discrepancy. A hard and rough surface structure results in low frictional shear stresses in the tribometer tests, whereas a softer and slightly rough surface texture leads to lower punch forces in the extrusion processes. This lack of knowledge needs to be overcome. To do so, smoothing effects for each surface texture need to be investigated. In dependency of the contact normal force, smoothing of the different surface textures manufactured by shot peening with different media is researched.

3 Experimental set-up

This section describes both the testing of the specimens and the evaluation of the surface textures using

https://doi.org/10.1051/matecconf/201819014009

roughness and hardness measurements. As specimen material the quenched and tempered steel 42CrMo4 (DIN: 1.7225, AISI: 4140) was used. This steel is used for highly loaded parts in automotive or aircraft industries, e.g. gears, spindles, rods, and rams. Therefore, the material is very important for the forming industry. As a tool material the cold working tool steel X155CrMoV12 (DIN: 1.2379, AISI D2) was used for the pin. It is a typical tool material in cold forging [17]. X155CrMoV12 was hardened to 62 HRC and the tool surfaces were polished (Rz = 1) according to the industrial standard. The experiments were conducted using the Pin-on-Cylinder Tribometer, without any revolution, see Fig. 5.

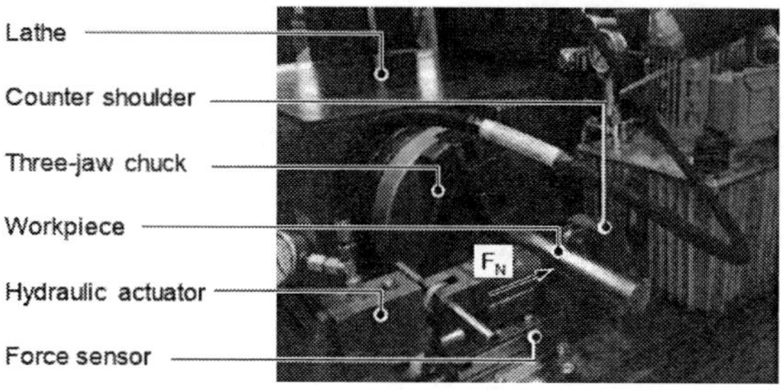

Fig 5. Pin-On-Cylinder Tribometer with different active parts

In order to analyze the impact of different surface textures during the smoothing process, different normal contact forces were applied. Before and afterwards the experiments, the surface roughness and hardness of the specimens were analyzed. For the experiments a turning lathe was used and the contact pressure was applied by means of a hydraulic actuator. Five different normal contact forces were applied: F_N = [460;640; 1180; 1850; 2600 N]. Each experiment was repeated three times to account for statistical effects.

Surface properties

The surface hardness was quantified by means of Vickers hardness according to DIN EN ISO 6507 and ASTM E384 using a Wilson Universal Hardness Tester UH250. 16 measurements using a testing load of 30 N were performed on polished cross-sections of each specimen. The 16 measurements were then averaged.

The surface roughness was measured using a combined roughness and contour measurement system Hommel Etamic nanoscan 855 made by Jenoptik AG, Germany. According to preliminary work, the surface parameters Sa and Sz were evaluated based on standard to ISO 25178 and EUR 15178N in a measurement grid of 5 x 5 mm. A contact tip with a tip angle of 60 degrees and a tip radius of 2 µm was used. Sa describes the averaged roughness of a surface, Sz its maximum height.

Furthermore, laser scanning microscopy analysis was conducted in order to measure the size of each indentation. Laser scanning microscope Keyence VK-X100 was used. Each indentation has a specific size. In order to get the maximum resolution, for each indentation a specific measuring area was selected.

4 Results and Discussion

The reference specimen (Ref) represents the initial state of the workpiece after the turning process and before any shot peening or other machining process, see Fig. 6.

Fig 6. Surface textures before machining (Steel = St, Ceramic = Ce, Corundum = Co, Ref = Reference)

The other specimens are shot peened by steel casks (St), ceramic casks (Ce), and by corundum particles (Co). Scanning electron microscopy images (SEM) show, that the reference structure is even. The steel and ceramic shot peened specimens have some craters, but remain in a flat surface. The corundum shot peened surface texture has many craters and is rough at the same time.

After indentation tests were done, laser scanning microscopy analysis were made. The results for a normal contact force of F_N = 2,600 N show the indentation clearly, see Fig. 7. The indentations were measured according to every specimen and contact normal force. This indicates, which surface texture has the biggest resistance against the plastic deformation. The elliptical extent was measured and the area of the ellipsoid was calculated via:

$$A = \pi \cdot w \cdot h \tag{1}$$

A is the area, w is the width and h the height of the contact ellipse. In this example, The indentation on the ceramic shot peened surface texture resulted in an area A_{Ce} = 4.7 mm², the steel shot peened one in A_{St} = 6.6 mm², the corundum shot peened one A_{Co} = 7.1 mm², and the reference surface texture in A_{Ref} = 8.3 mm². Each experiment was repeated three times to account for statistical effects.

https://doi.org/10.1051/matecconf/201819014009

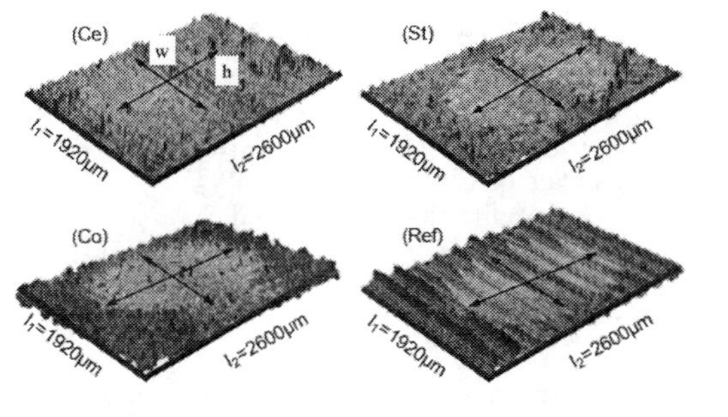

Fig 7. Laser Scanning Microscopy pictures with indentation by a normal force of F_N = 2,600 N and their resulting area (Steel = St, Ceramic = Ce, Corundum = Co, Ref = Reference)

Table 4. Extent of indentation area generated by a forming force of FN = 2,600 N (Steel = St, Ceramic = Ce, Corundum = Co, Ref = Reference)

Medium	w = width [μm]	h = *height* [μm]	A = Area [mm²]
Ce	809	1,344	4.7
St	1,108	1,730	6.6
Co	961	1,772	7.1
Ref	1,299	1,827	8.3

The evaluation of all indentations shows that due to a higher contact normal force greater indentation areas evolve, see Table 4. The rise of the area is not linear. The reason for these irregularities is the varying surface roughness on each specimen. The smoothing mechanism starts with a contact of the tool with the roughness peaks. In the beginning, there is a plastic deformation on a microscopic level. The more the roughness peaks are smoothed in, the more material gets into contact with the tool and a plastic deformation on a macroscopic level proceeds. Depending on the section, some higher roughness peaks are smoothed in faster than others and the plastic deformation on a macroscopic level proceeds earlier. Overall, the indentations on the reference surface texture evolve into the greatest indentations. The reference surface structure was not pre-treated, so no strain hardening took place due to surface modifications. The roughness of the surface was low, so from the beginning on, nearly the whole surface was in contact with the tool. The surface texture shot peened by ceramic casks resulted in the smallest indentation areas. The surface texture is a bit rougher than the reference structure in the beginning, but not as rough as the steel or corundum shot peened surface texture. At the same time it is the hardest surface texture. Due to a higher roughness and a higher hardness than the reference surface texture, the smallest indentation area evolve. The ceramic shot peened specimen has a lower roughness than the other peened specimens have. Due to the lower roughness, more material gets into contact with the pin from the beginning on. Therefore, the contact normal force results in a lower contact normal pressure, as in a rougher surface texture, which results in the lower indentation area. The surface roughness is the dominant criteria for the evolution of the indentation area in the beginning of the process. When the forces rise the strain hardening due to the pretreatment becomes the dominant criterion and supports a lower plastic deformation. In combination with the results of the extrusion process, a surface texture, which results in a smaller indentation area is the most suitable one to lower punch forces and friction in extrusion at the same time.

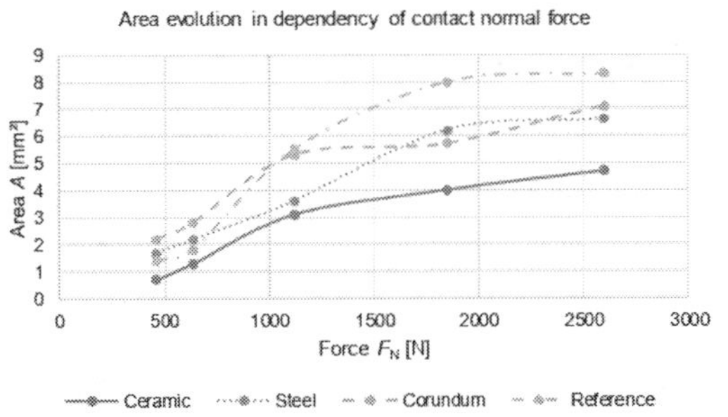

Fig. 8. Indentation area in dependency of shot peening medium and contact normal force

5 Summary and Outlook

The resulting discrepancy between the frictional shear stress during tribometer testing and the punch force in an extrusion process led to an investigation of the plastic deformation of different surface textures. A reference surface texture machined by turning as well as several shot peened specimens, peened by steel, and ceramic casks and by corundum particles were investigated. Due to different pretreatments, different surface integrities evolved. These surface textures were tested with increasing normal contact forces. Rising contact normal forces resulted in larger indentation areas. Furthermore, a harder and rather flat surface texture resulted in the smallest indentation area, whereas a flat but soft surface texture resulted in the greatest indentation area. Respective to the lowest punch force, an ideal workpiece has a hard surface texture with a low roughness. In order to separate the effects of roughness and hardness of the surface texture, further researches have to be done. Hardened and unhardened with an identic roughness as well as rougher or flatter surface textures with an identic hardness need to be investigated.

Acknowledgments

The research was funded by the German Research Foundation (Deutsche Forschungsgemeinschaft DFG) within the priority program "Dry metal forming – sustainable production through dry processing in metal forming" (SPP 1676). The authors also thank Presswerk Krefeld GmbH for their support supplying tools and workpieces as well as conducting the full forward extrusion process.

References

1. N. Bay, A. Azushima, P. Groche, I. Ishibashi, M. Merklein, M. Morishita, T. Nakamura, S. Schmid, M. Yoshida: Environmentally benign tribo-systems for metal forming. In: *Annals of the CIRP* 59/2 (2010) 760-780

2. F. Vollertsen, F. Schmidt: Dry Metal Forming: Definition, Chances and Challenges. In: *Int. J. Precision Engineering and Manufacturing –Green Technology* 1/1 (2014) 59-62
3. Vollertsen, F. and Schmidt, F. (2014). Dry metal forming: Definition, Chances and Challenges. *Int J Precis Eng Man* - Green Technology 1(1):59-62
4. K. Müller (2003) Praktische Oberflächentechnik – Vorbehandeln – beschichten – Beschichtungsfehler – Umweltschutz; JOT Fachbuch
5. M. Teller, M. Bambach, G. Hirt: A compression-torsion-wear-test achieving contact pressures of up to eight timest he initial flow stress of soft aluminium. In: *CIRP Annals – Manufacturing Technology 64* (2015) 280-292
6. M. Teller, M. Bambach, G. Hirt, I. Ross, A. Temmler, R. Poprawe, H. Bolvardi, S. Prünte, J. M. Schneider: Investigation of the suitability of surface treatments for dry cold extrusion process-oriented tribological testing. In: *Key Engineering Materials Vols. 651-653* (2015) 473-479
7. M. Murakawa, N. Koga, T. Kumagai: Deep-drawing of aluminum sheets without lubricant by use of diamond-like carbon coated dies. *Surf Coat Tech 76* (1995) 553–558
8. M. Murakawa, S. Takeuchi: Evaluation of tribological properties of DLC films used in sheet forming of aluminum sheet. *Surf Coat Tech 163* (2003) 561–565.
9. K. Osakada R. Matsumoto: Fundamental Study of Dry Metal Forming with Coated Tools. *CIRP Ann. Manuf. Technol 49/1* (2000) 161–164.
10. S. Kataoka, A. Motoi. Improvement in DLC thin film adhesion and its application to dry deep drawing. *J. Jpn. Soc. Technol. Plast 46/532* (2005) 412–416.
11. K. Tamaoki, S. Kataoka: Study of deep drawing using diamond coated tools. *Journal of Material Testing Research Assoc. of Japan 53/4* (2008) 247–253.
12. K. Tamaoki, S. Kataoka, K. Minamoto: Dry deep-drawing with use of electroconductive ceramic tools. *Proc. Int. Conf. Trib. Manuf. Pro* (2007) 175–179.
13. Z. G. Wang, S. Komiyama, Y. Yoshikawa, T. Suzuki, K. Osakada (2015) Evauluation of lubricants without zinc phosphate precoat in multi-stage cold forging, In: CIRP Annals – Manufacturing Technology 64
14. P. Groche, J. Stahlmann, J. Hartel, M. Köhler (2009) Hydrodynamic effects of macroscopic deterministic surface structures in cold forging processes, In: Tribology International 42
15. Klocke, F., Trauth, D., Schongen F. and Shirobokov, A. (2014). "Analysis of friction between stainless steel sheets and machined hammer peened structured tool surfaces." *Prod. Eng. Res.* Devel. 8: 263-272.
16. Bobzin, K., Brögelmann, T., Bastürk, S., Klocke, F., Mattfeld, P., Trauth, D. (2015). "Development of an insitu Plasma Treatment of X155CrMoV12 for a (Cr,Al)N PVD Tool Coating for Dry Metal Forming in Cold Forging." *Dry Metal Forming Open Access Journal* 1(1): 57-62.
17. Bobzin, K., Brögelmann, N.C., Bastürk, S., Klocke, F., Mattfeld, P., Trauth, D. (2015). "Tribological Behavior of $(Cr1\text{-}_xAl_x)N/WS_y$ PVD Tool Coatings for the Application in Dry Cold Forging of Steel." *Dry Metal Forming Open Access Journal* (1)1:152-158.
18. Trauth, D.; Hild, R.; Mattfeld, P.; Bastürk, S.; Brögelmann, T.; Bobzin, K.; Klocke, F.: Advances in dry metal forming of low alloyed steels for cold forging using a (Cr,Al)N tool coating and surface structures on workpieces. In: Proceedings of the 12th International Conference THe „A“ coatings in Manufacturing Engineering, Hannover, Germany, March 31st - April 1st (2016)
19. Bobzin, K., Kruppe, N.C., Arghavani, M, Hoffmann, D.C., Klocke, F., Mattfeld, P., Trauth, D., Hild R. (2017). "Mechanical and tribological characterization of self-lubricating $(Cr1\text{-}_xAl_x)N$ coatings for deposition on complex-shaped forging toolsMechanical and tribological characterization of self-lubricating $(Cr1\text{-}_xAl_x)N$ coatings for deposition on complex-shaped forging tools." *Dry Metal Forming Open Access Journal* (3) 81-89.

Probing the potential of structured and surface functionalized tools for dry cold forging of aluminium

Marco Teller[1,*], *Ingo* Ross[2], *Stephan* Prünte[3], *André* Temmler[4], *Moritz* Küpper[4], *Reinhart* Poprawe[2,4], *Jochen M.* Schneider[3], and *Gerhard* Hirt[1]

[1]Institute of Metal Forming (IBF), RWTH Aachen University, Intzestr. 10, 52072 Aachen, Germany
[2]Fraunhofer Institute for Laser Technology (ILT), Steinbachstr. 15, 52074 Aachen, Germany
[3]Materials Chemistry (MCh), RWTH Aachen University, Kopernikusstr. 10, 52074 Aachen, Germany
[4]Chair for Laser Technology (LLT), RWTH Aachen University, Steinbachstr. 15, 52074 Aachen, Germany

Abstract. In order to reduce friction and wear in cold forging of aluminium and its alloys a tremendous amount of lubricants is used. The main task of the liquid lubricants is the separation of tool and workpiece surfaces to avoid adhesion and abrasion. From an ecological and economical point of view, a reduction of the consumption of lubricants is meaningful. Novel surface modification techniques have been investigated to enable dry cold forming. In previous work, a reduced wear by adhesion of aluminium to an AISI H11 ESR tool steel surface was detected for a combination of laser polishing and a subsequent surface functionalization. Here, structured surfaces combined with surface functionalization are investigated regarding their wear behaviour. Therefore, different structures are realized on AISI H11 ESR tool steel surfaces by laser structuring. These structures are aligned parallel and perpendicular to the material flow direction. Furthermore, different structure depths are generated. It is evident that the laser induced surface melting causes both, a reduced micro roughness as well as a homogenized elementary distribution. Moreover, the surfaces are functionalized with an octadecylphosphonic acid selfassembled monolayer (C18PA-SAM). The SAM's distal alkyl termini functionalize the tool steel surface und thus only weak van-der-Waals interactions are present in the aluminium-tool steel interface rendering an interaction, which is expected to be similar to the interaction in conventionally lubricated aluminium forming. Finally, the surfaces are appraised regarding their wear behaviour in a compression-torsion-wear tribometer. The different experimental results document the potential of the here investigated combined tool surface modification. Although, the performance of a lubricated process is not completely reached a reduction in friction and wear can be achieved compared to unmodified tools.

Keywords: Tribology, cold forming, aluminium

1 Introduction

Tribological phenomena between tool and workpiece can significantly influence metal forming processes. For example, friction effects the material flow in the manufacturing process and due to wear of the tool surfaces, the boundary conditions of the process can change. In order to control friction and wear and to guarantee a stable production lubricants are widely used. Especially in cold forging, a high effort is spent to several process steps for lubricant application, subsequent removal and recycling or disposal, respectively [1].
Current research activities aim for a significant reduction of the lubricant usage culminating in dry metal forming [1]. Ecological and economic reasons are driving forces for the development of novel surface treatments and modifications to realize this purpose.
Prior to the application of new tribological solutions to the manufacturing process an extensive testing and characterization is done on lab-scale. Tribometers reproduce specific features of the real process and are widely used for a preliminary evaluation of tribologically optimized surfaces [2].
In previous research works, the authors documented the positive effect of laser polishing [3], a surface functionalization with self-assembled monolayers (SAM) [4,5] and the combination of both strategies [6] for the reduction of aluminium adhesion and friction in a compression-torsion-wear tribometer simulating the load case in cold forging. Especially dry cold forging of aluminium is challenging due to the severe load case including high relative normal stresses and a significant surface expansion [7]. Both leading to a break-up of the passivating oxide layer and a direct contact between virgin aluminium and the tool surface [8]. The highly reactive behaviour of aluminium is documented in several papers, e.g. by Bolvardi et al. via ab initio calculations [9,10] and experimentally by Fuentes et al. [11].
In this paper, another surface modification concept in order to reduce friction and wear is presented. Instead of a laser polishing treatment, which can decrease the micro and macro roughness, surface structuring by laser

* Corresponding author: teller@ibf.rwth-aachen.de

remelting is used on specimens from AISI H11 ESR in order to adjust the surface structure and to prepare it for a functionalization with self-assembled monolayers.
The paper is structured as follows: First, the specimen geometry, the laser structuring process for several surface structures and the subsequent functionalization step are described. Afterwards, results from the tribometer experiments are presented and discussed. The results show a positive effect of a circular laser structure with 1 µm depth combined with SAM. The experiments show a reduced interaction between tool and workpiece.

2 Specimen preparation

2.1 Tribometer specimens

The wear behaviour of the different surface modifications is tested in a compression-torsion-wear tribometer. The specimen geometry is shown in Fig. 1.

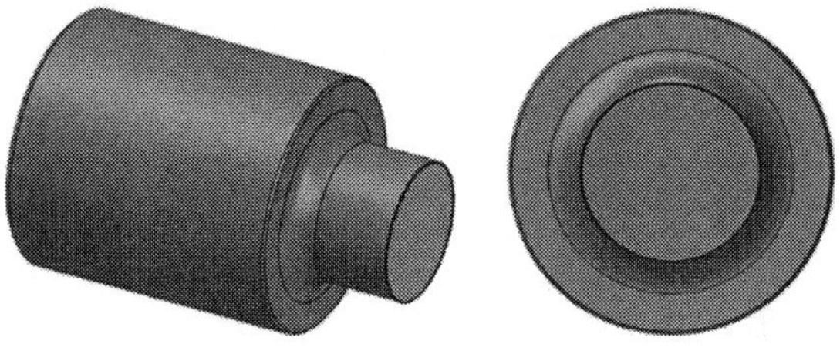

Fig. 1. Tool specimen for the compression-torsion-wear test with marked contact surface (blue) for the structuring process in side and top view.

The compression-torsion-wear tribometer reproduces main features of the cold forging process like a high relative contact pressure. The presented tool specimen from steel will be pressed on a workpiece specimen from pure aluminium. The workpiece specimen enables a surface expansion and as a result the break-up of the oxide layer. The work presented in [12] depicts further details regarding the compression-torsion-wear tribometer.

2.2 Laser structuring

Surface structuring by laser remelting is based upon the physical interrelationship between a modulation of melt pool volume and the dependent movement of the three-phase line that determines the resulting surface topography while the molten material solidifies. In laser polishing a change of the melt pool volume is undesirable [13,14]. Surface structuring by remelting however, aims for a precise modulation of the melt pool volume, e.g. by a modulation of the laser power. As a consequence of the induced modulation of melt pool volume, the resulting surface topography is generated by the same mechanism as shown for ripple formation during laser polishing [15]. Figure 2 shows the process principle of surface structuring by laser remelting. By moving the laser beam relatively to the workpiece surface, a thin surface layer (<100 µm) is molten. After the laser beam has passed, the material solidifies. The direction of the solidifaction is defined by the melt pool surface. At constant laser power, as used for

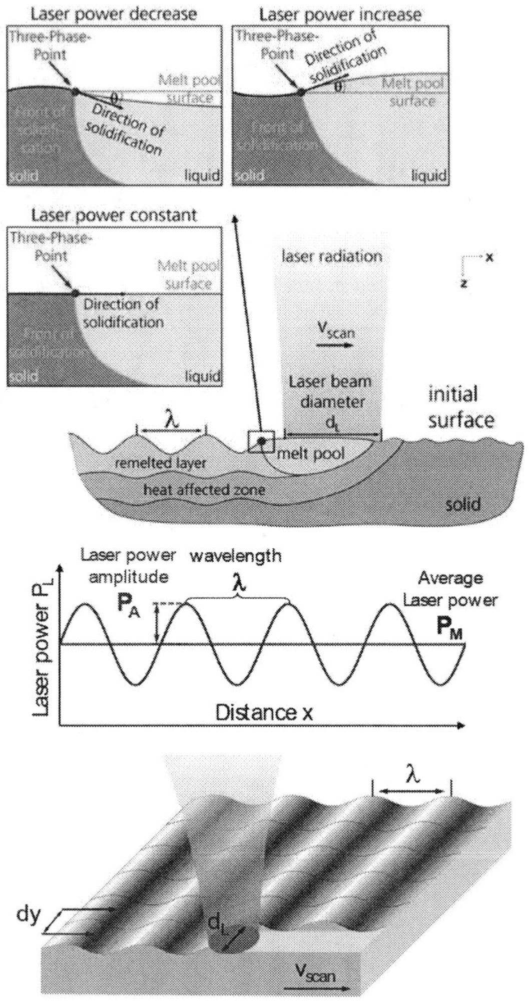

Fig. 2. Top: Schematics of the active principle for surface structuring by laser remelting; Bottom: Schematics of the process principle. [17]

laser polishing [14], the melt pool volume of the steady state process does not change, resulting in an approximately flat melt pool surface. Thus, no structuring occurs. Increasing the laser power will increase the amount of molten material and, due to density changes and thermal expansion, will lead to a bigger melt pool volume. The melt pool surface bulges outwards (see Fig. 2). Following the bulged surface, the material solidifies in an upwards-slope at the three-phase-point. Decreasing the laser power will lead to the opposite effect. Thus, structuring can be achieved by modulation of the melt pool volume. The precise control of the spatial and temporal deposition of the laser energy is essential for structuring by remelting. Height, size and form of the produced surface topography depend directly on the average melt pool volume and on the absolute change and time-dependent alteration rate of the melt pool volume. In order to achieve periodic structures, the laser power is modulated sinusoidal at an average laser power P_M with an amplitude P_A and a wavelength λ (Fig. 2 middle). In

https://doi.org/10.1051/matecconf/201819014010

contrast to the Surfi-Sculpt process, which takes advantage of the keyhole welding effect and uses both vapour pressure and surface tension variation for creating singular structures [16], surface structuring by laser remelting stays in the regime of heat conduction welding and allows areal modification of surfaces with smooth transitions between the periodical features and without evaporation of material.

The tribometer specimens are processed on the "POLAR" 5-axis laser polishing machine (Fig. 3). The integrated 3D galvanometric laser scanner deflects the laser beam on the workpiece and moves the laser spot at a defined scanning velocity v_{scan} and track offset dy. The wavelength of the created structures is a function of the wavelength λ of the laser power modulation and the scanning velocity v_{scan}. Areal structures can be obtained by overlapping the remelted tracks (Fig. 2 bottom) [17].

Tribometer specimens investigated in this work are made from AISI H11 ESR and are structured with circular and star-shaped patterns by surface structuring by laser remelting (wave shape). During the tribometer test, the circular structures are parallel to the material flow and the star-shaped structures are perpendicular to the material flow. For structuring, a focus diameter $d_L = 125$ µm, scan velocity $v_{scan} = 206$ mm/s, average laser power $P_M = 50$ W and amplitude $P_A = 5$ W is used. For the circular patterns ($\lambda_{circular} = 328$ µm) an oscillating process strategy is used where each remelting scan vector is turned by 1°. The star-shaped patterns consist of 100 modulation cycles, resulting in a radius-dependent wavelength between $\lambda_{star,r=1mm} = 63$ µm and $\lambda_{star,r=4.5mm} = 283$ µm. This pattern is applied with a circular strategy with a track offset dy = 20.5 µm. By changing the number of processing steps, specimens with average structure heights of $h_{structure} = 1$ µm and 10 µm are realized.

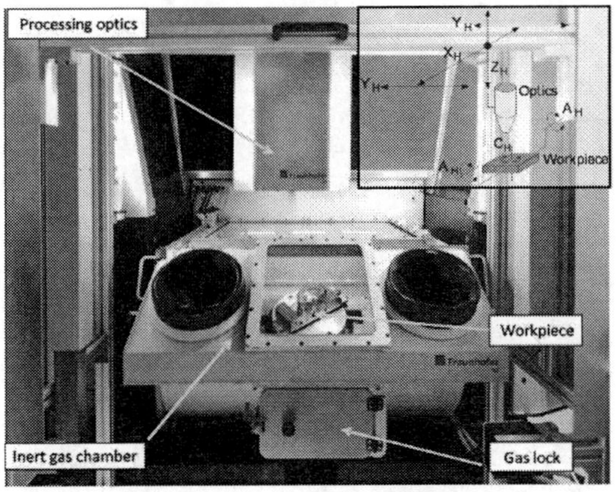

Fig. 3. Fraunhofer ILT's "POLAR" 5-axis laser polishing machine suitable for structuring by laser remelting. [18]

2.3 Surface functionalization

Octadecylphoshonic acid (C18PA) are deposited on laser structured surfaces of AISI H11 ESR by vacuum thermal evaporation at residual pressures of $5 \cdot 10^{-5}$ Pa, followed by an annealing of the processed surface at 160 °C. Reactant molecules are anchored to the tool surface forming P-O-M (M=Metal) bonds between the phosphonic acid head-group and the oxide layer of the surface [19]. Henceforth, the surface is functionalized with a dense self-assembled monolayer (SAM) consisting of C18PA molecules having their alkyl moieties oriented distally to the tool surface. Previously, we showed that functionalization in such fashion lubricates the tool-workpiece interface and reduces Al-adhesion on the tool surface due to acting van-der-Waals interactions between alkyl moieties and the aluminium workpiece [4]. Contrarily, atomic scale onset of Al-adhesion was observed in interfaces of aluminium and tool coatings in theory [9] and experiment [11]. Ab-initio DFT calculations suggest the adsorption energies of Al-atoms to be two orders of magnitude higher in the latter interface than for interfaces with only van-der-Waals interaction present [10].

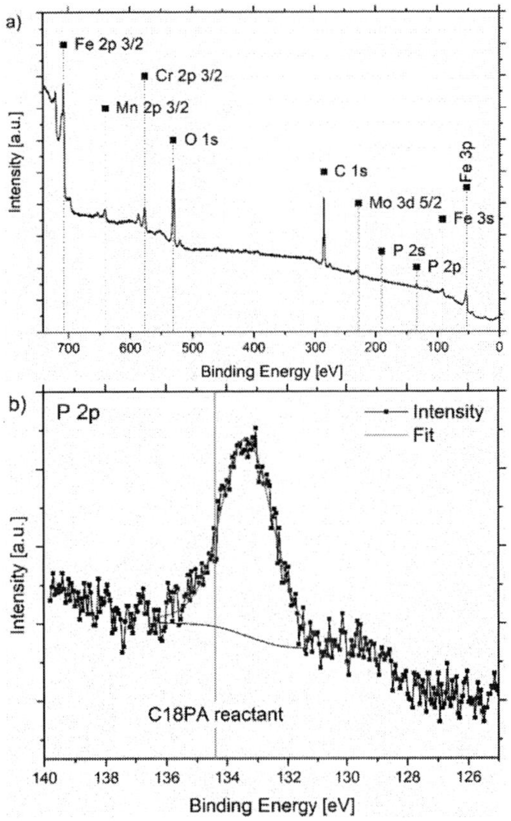

Fig. 4: (a) overview and (b) P 2p spectra obtained by XPS of 1 µm circular laser structured surface functionalized with a C18PA-SAM.

Functionalized surfaces were investigated by X-ray photoelectron spectroscopy using a JEOL JAMP 9500F equipped with a non-monochromatic Al Kα X-ray source. Acquired spectra were recorded with a step-size of 0.25 eV and 5 measuring cycles for overview and 0.05 eV and 20 cycles for detailed spectra, respectively while the pass-energy of the detector was set to 20 eV. Spectra analysis was carried out using the CASA XPS software. XPS-obtained surface chemistry of the 1 µm circular structured tool discloses the presence of P beside M-oxides on the surface originating from functionalization

(Fig. 4a). The detected P 2p signal located at binding energies of 133.3 eV (Fig. 4b) reveals an off-set of 1.1 eV to the P 2p signal detected for C18PA reactant indicative for formation of P-O-M bonds [4]. Thus, the surface of laser structured AISI H11 ESR tool steel is functionalized with C18PA-SAM anchored by P-O-M bonds, while van-der-Waals interactions mediate the tool-workpiece due to distal alkyl moieties of the C18PA-SAM.

3 Results

The different modified tool surfaces were tested in the compression-torsion-wear tribometer at a normal pressure of three times the initial yield stress of pure aluminium for two rotations at a rotation velocity of one rotation per second. The normal pressure is less than in a typical cold forging process in order to allow for a more significant validation of the influence of the different surface modifications on tribological parameters.

The following sections present the transferred torque between tool and workpiece specimen, the surface characteristics measured via white light interferometry (WLI) before and after the wear test and the adhered aluminium volume on the tool specimen. These parameters allow for an assessment of the different surface modifications.

3.1 Torque curves

Figure 5 shows torque curves for the different structures perpendicular to the material flow direction. The star-shaped structure leads to an oscillating shape with the same frequency as the rotational velocity of the tribometer. The deeper structure ($h_{structure}$= 10 µm) without functionalization generates the highest maximum torque M of approximately 30 Nm. The functionalization on the same structure reduces the maximum torque to circa 24 Nm. Moreover, the peak-to-valley value is less. The smooth structure with $h_{structure}$= 1 µm depth reaches approximately 26 Nm. The minimum value (17 Nm) is found for the smooth structure combined with functionalization. Nevertheless, the functionalized surfaces show an increase for each peak value. This could be an indicator for an ongoing failure of the SAM layer.

Figure 6 presents the torque curves for the different structures circular to the material flow direction. The several surfaces are determined in the same order as the star-shaped structures. The maximum torque values are found for the non-functionalized deep structure (22 Nm) and the non-functionalized smooth structure (22 Nm) followed by the functionalized deep structure (15 Nm) and the functionalized smooth structure (9 Nm). Again, an increase of the transferred torque for each peak can be detected for the SAM specimens. Moreover, the functionalized specimens show a more pronounced oscillation compared to the non-functionalized. Nevertheless, the circular structuring reveals a lower torque than for the star-shaped structure. Hence, the alignment of the structure to the material flow direction has a significant influence.

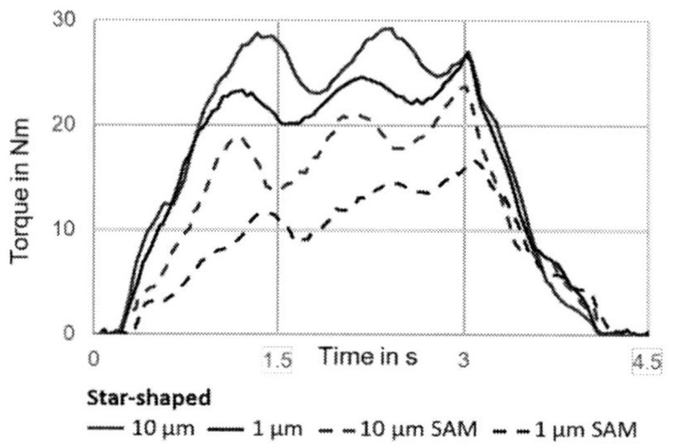

Fig. 5. Torque curves for star-shaped structures with different structure depths and functionalization.

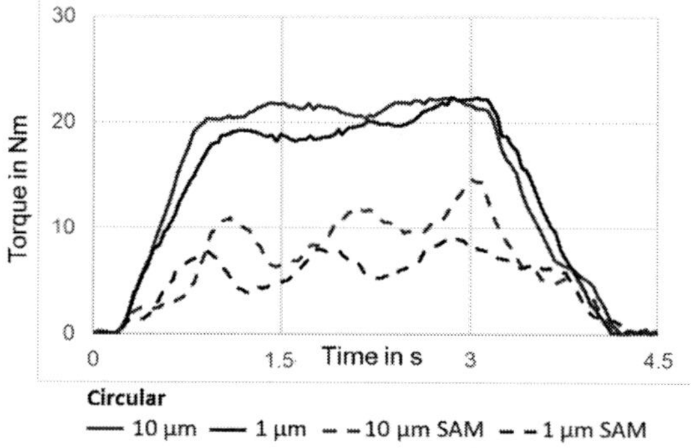

Fig. 6. Torque curves for circular structures with different structure depths and functionalization.

3.2 WLI surfaces

The surfaces of the tribometer specimens are measured with a Zygo NewView 7300 white light interferometer. Due to the known boundary effects at the specimens' outer radius [12], only the surface within an inner radius of 5 mm is evaluated. Figure 7 shows the resulting contact surfaces of the different surfaces with and without SAM coating after the test. The laser structured specimens with non-functionalized deep structures ($h_{structure}$ = 10 µm), circular and star-shaped, show a significant amount of adhered aluminium on the surface. However, the distribution of the adhered material differs. For the circular structure, the material agglomerates to several larger lumps, which do not seem to have a preferred position on the underlying structure (Fig. 7 A2). For the star-shaped structures however, the areas with adhered material are distributed more evenly over the whole specimen and can be found predominately at the top of the structures and smearing down to the bottom (Fig. 7 A1). Taking the measured torque into account, these wear characteristics imply an increased tendency to shearing off aluminium for the star-shaped structure.

The specimens with functionalized deep structures show a significant reduction of adhered material. This is reflected in the reduced torque measured during the test for the coated specimens (Fig. 5 and 6). The amount of adhered aluminium is reduced for both structure types and the characteristics of the distribution is analogue to the non-functionalized specimens: on the one hand, scattered, albeit larger agglomerations for the circular structures, on the other hand, adhesions beginning the top and smearing to the bottom of the star-shaped structures.

This general behaviour can also be found on the specimens with smooth structures ($h_{structure} = 1$ µm). While the amount of adhered material is, compared to the deeper structures, already reduced for the non-functionalized specimens, the SAM functionalization further decreases the volume of the adhered material. However, this decrease is less pronounced for the star-shaped structure.

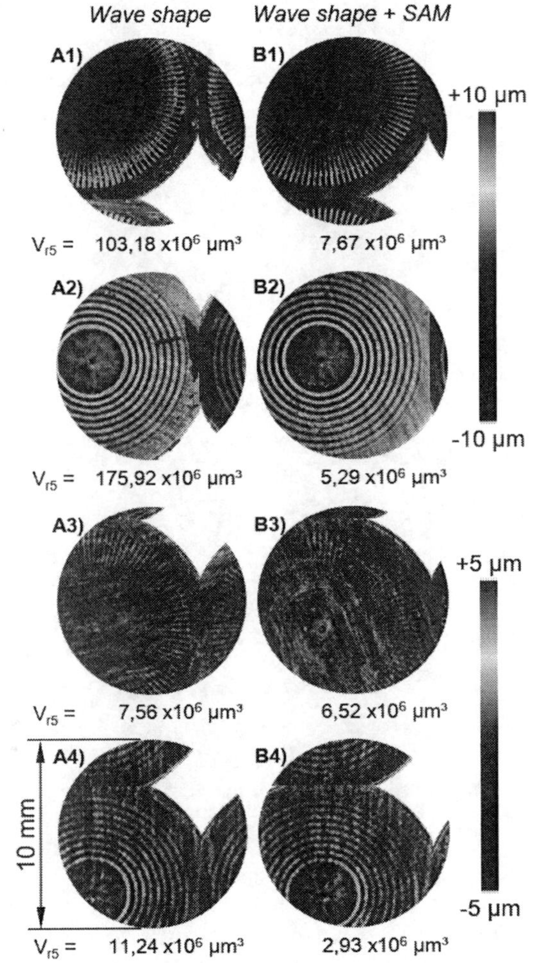

Fig. 7. WLI measurements of the overall adhered Al volume of different surfaces after testing in the tribometer. *Columns:* A) structured (Wave shape); B) structured and functionalized; *Rows:* 1) deep star-shaped structure (10 µm); 2) deep circular structure (10 µm); 3) smooth star-shaped structure (1 µm); 4) smooth circular structure (1 µm)

On the surface of the functionalized smooth star-shaped specimen, the above described adhesion pattern as well as several small agglomerations can be found (Fig. 7 B3). Similar agglomerations can also be found on the functionalized deep circular structure and may hint at a local failure of the SAM functionalization. This kind of wear characteristics cannot be detected on the functionalized smooth circular specimen (Fig. 7 B4). In addition, the circular specimen produces the smallest torque during the test and shows the smallest amount of adhered material. This leads to the hypothesis that the SAM functionalization did not fail under these specific load conditions.

3.3 Adhered volumes

By subtracting the surface topography of the unused specimen from the topography after the tribometer test, the volume of the adhered aluminium can be measured. This procedure was described in detail in [3].

A summary of the torque and adhered volume is given in Table 1. The analysis of the adhered volume shows that the deep structure lead to the largest amount of adhered material. A possible explanation may be the increased shear load in the aluminium workpiece induced by the deep structures and a hypothetical increase of contact with non-oxidized aluminium due to the penetration of the structures into the natural oxide layer. Reducing the structure height as well as choosing a structure, which is parallel to the material flow results in a reduced amount of adhered aluminium. For all tested surfaces, the SAM functionalization is effectively reducing the adhered volume and measured torque. However, the reduction of the measured torque is less significant for the deep than for the smooth structures.

The smallest volume was measured for the functionalized smooth circular structure: $V_{circular,1µm,SAM} = 2.93 \times 10^6$ µm³, which is, an almost 75 % decrease in adhesive volume compared to the same surface without the C18PA-SAM. $V_{circular,1µm,SAM}$ is slightly higher than the best result on a functionalized manually polished surface with $V_{polished,SAM} = 2.11 \times 10^6$ µm³ [5]. However, there is a greater than 2-fold reduction in $M_{circular,1µm,SAM}$ compared to both, the 1 µm, circular structured specimen without SAM and the manually polished surface with SAM, supporting the notion that combined functionalization with SAM and surface structuring decreases the overall process load in a real extrusion process.

Tab. 1. Maximum torque M and adhered aluminium volume V for different tested specimens

Specimen	Torque M [Nm]	Volume V [x10⁶ µm³]
Manually polished*	38*	5.12*
Manually polished + SAM*	20*	2.11*
Circular 1 µm	22	11.24
Circular 1 µm + SAM	9	2.93
Circular 10 µm	22	175.92
Circular 10 µm + SAM	15	5.29
Star-shaped 1 µm	26	7.56
Star-shaped 1 µm + SAM	17	6.52
Star-shaped 10 µm	30	103.18
Star-shaped 10 µm + SAM	24	7.67

*published by the authors in [5]

3.4 EDX measurements

Additionally, the two SAM-functionalized specimens with 1 µm structures, star-shaped and circular, respectively, were investigated by scanning electron microscopy (SEM) and energy-dispersive X-ray spectroscopy (EDX) after tribological experiments to characterize adhesion using a Zeiss Σigma SEM equipped with an Oxford Instruments X-MaxN 50 EDX-detector.
Fig. 8 depicts SE-micrographs of aluminium adhesion on star-shaped and circular structures, respectively, identified by EDX (Tab. 2). All investigated areas are localized 3 mm off the centre of each specimen, in the middle of the structured zone. Obviously, Al is attached to a larger degree on the star-shaped structures than on the circular surface (please note the factor 2-fold larger magnification necessary to show an overview of Al-adhesion on the circular surface). Beside the macroscopic star-shaped or circular structure of the surface, a substructure is visible which is circular for the star-shaped structured surface and star-shaped for the circular structured surface (Fig. 8). These features represent the sides of the remelting tracks of the laser process as a result of different process strategies for star-shaped and circular structures, which include the boundary condition of aligning the laser-tracks non-parallel to the desired structure. Using white light interferometry, depths between 100 nm and 300 nm were measured for these undercuts.
In addition to the observed preference for adhesion at the top of the star-shaped and circular structures (Fig. 7), high resolution SE- and BSE-micrographs (Fig. 9 and Fig. 10) also suggest a preferred adhesion parallel to the undercuts as indicated by the low intensities of Al in the BSE-micrographs. However, the Al-adhesion is more severe for the star-shaped structure than for the circular one. Both, Fig. 8a and 9 indicate continuously build-up of Al in sliding direction of the Al-workpiece during tribological experiments leading to the formation of facetted Al-adhesion. Contrarily, only thin Al-adhesion spots are located on features of the circular structure (Fig. 8b and 10). Moreover, the EDX-signal of the Al-adhesion consists of information of both, aluminium and tool steel beneath, which is not the case for the EDX-measurement of the star-shaped structure (Tab. 2). Thus, SEM- and EDX-investigation of the SAM-functionalized surfaces reflect the result obtained by WLI of a 2-fold larger Al-volume adhered on the star-shaped structure compared to the circular one.

Tab. 2. Chemical composition in at.% derived by EDX measured at spots indicated in Fig. 8.

Pos.	Al	Si	V	Cr	Mn	Fe	Mo
1	99.0	-	-	-	-	1.0	-
2	4.8	1.6	0.3	5.3	0.8	86.4	0.8
3	51.1	0.7	0.3	3.3	-	44.7	-
4	-	1.8	0.7	6.3	-	90.1	1.1

Fig. 8. SE-micrographs of laser structured surfaces after tribological tests (a) star-shaped, 250x mag. (b) circular, 500x mag. Arrow indicates the sliding direction of the Al-workpiece.

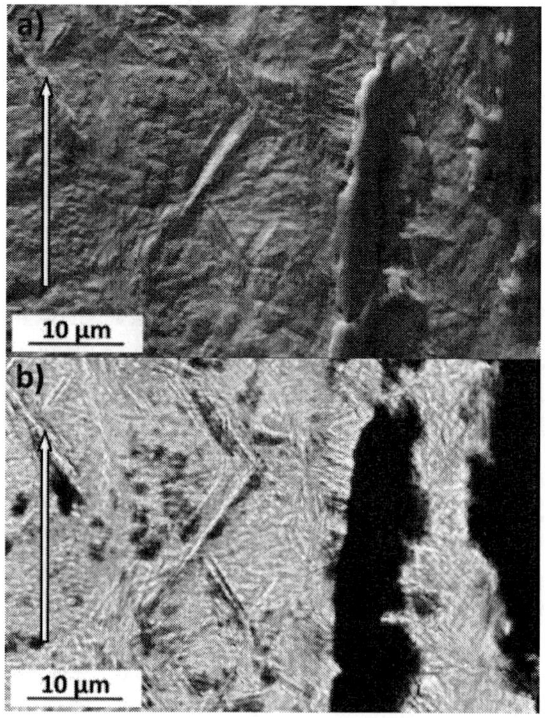

Fig. 9. High resolution (a) SE- and (b) BSE-micrograph of Al attached to a feature on the star-shaped structured surface. Arrow indicates the sliding direction of the Al-workpiece.

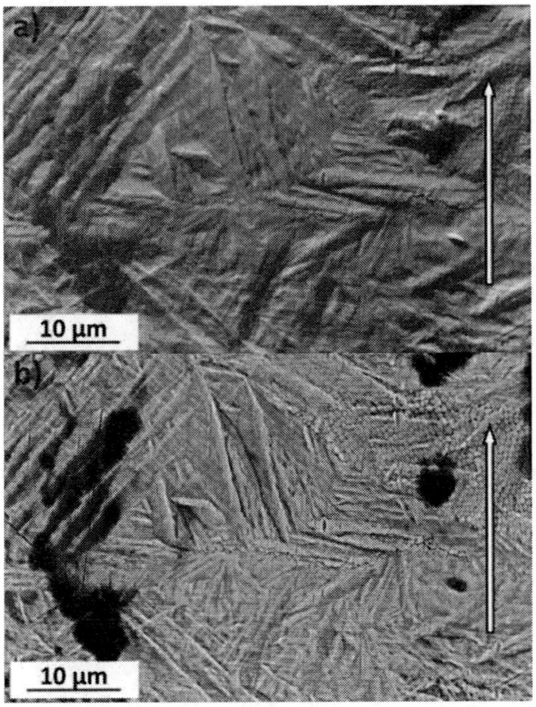

Fig. 10. High resolution (a) SE- and (b) BSE-micrograph of Al attached to a feature on the circular structured surface. Arrow indicates the sliding direction of the Al-workpiece.

4 Conclusion and future work

In this paper, star-shaped and circular structures with different depths were generated by laser surface processing of compression-torsion-tribometer specimens. The star-shaped structure is oriented perpendicular to the aluminium flow direction, while the circular structure is oriented parallel to the material flow direction. Besides evaluating the effect of laser structuring on the tribological behaviour, also structured tools surfaces, which were functionalized with C18PA-SAM, were tested against pure aluminium in dry conditions. The results show that both, laser structuring and surface functionalization influence the transferred torque and the amount of adhered aluminium. A structure parallel to the material flow direction is favourable, while a lower structuring depth transfers less torque. Moreover, functionalized, 1 µm circular structured surfaces render a 2-fold reduction in transferred torque and almost 75 % decrease in aluminium adhesion compared to the same surface structure without the C18PA-SAM. In future work, the different mechanisms responsible for aluminium adhesion on modified surfaces will be analysed and other structural patterns will be investigated with the goals to utilize the above reported finding on geometrically complex cold forging dies.

Acknowledgement

The authors gratefully acknowledge the financial support of the Deutsche Forschungsgemeinschaft (DFG) within the priority programme SPP 1676 "Dry Metal Forming". Moreover, the authors would like to thank the Department of Ferrous Metallurgy (IEHK) of RWTH Aachen University for their support in the EDX measurements.

References

[1] F. Vollertsen, F. Schmidt, International Journal of Precision Engineering and Manufacturing-Green Technology **1** (2014) 59-62.
[2] P. Groche, C. Müller, J. Stahlmann, S. Zang, Tribology International **62** (2013) 223-231.
[3] I. Ross, A. Temmler, E. Willenborg, R. Poprawe, M. Teller, in: T. Graf, L. Overmeyer, F. Vollertsen, Lasers in Manufacturing Conference 2015, 2015.
[4] S. Prünte, D. Music, M. Teller, G. Hirt. P.H. Mutin, G. Ramanath, J.M. Schneider, J. Vac. Sci. Technol. A **35** (2017) 020605.
[5] M. Teller, S. Prünte, I. Ross, A. Temmler, J.M. Schneider, G. Hirt, AIP Conference Proceedings **1896** (2017) 140001.
[6] I. Ross, A. Temmler, M. Küpper, S. Prünte, M. Teller, J. M. Schneider, R. Poprawe, KEM **767** (2018) 69-76.
[7] N. Bay, "*International Cold Forging Group – 1967-1992 – Objectives, History, Published Documents*", 1992.
[8] W. Zhang, N. Bay, CIRP Annals - Manufacturing Technology **41** (1992) 293-297.
[9] H. Bolvardi, D. Music, and J. M. Schneider, Thin Solid Films **589** (2015) 707-711.
[10]H. Bolvardi, D. Music, and J. M. Schneider, Appl. Surf. Sci. **332** (2015) 699-703.
[11]G. G. Fuentes, M. J. Dıaz de Cerio, R. Rodriguez, J. C. Avelar-Batista, E. Spain, J. Housden, and Y. Qin, J. Mater. Process. Technol. **177** (2006) 644-648.
[12]M. Teller, M. Bambach, G. Hirt, CIRP Annals - Manufacturing Technology **64** (2015) 289-292.
[13]A. Temmler, E. Willenborg, and K. Wissenbach, "*Laser Polishing*" Proceedings of the SPIE, Volume **8243**, DOI: 10.1117/12.906001, 2012a.
[14]E. Willenborg, "*Polieren von Werkzeugstählen mit Laserstrahlung*" Ph.D. Thesis, RWTH Aachen, Aachen, 3-8322-4896-x, 2006.
[15]N. Pirch, S. Höges, and K. Wissenbach, "*Mechanisms of surface rippling during laser polishing*" Proceedings of the 8. International Seminar on Numerical Analysis of Weldability, Graz-Seggau, Austria, 2006.
[16] J. Blackburn, P. Hilton, Physics Procedia **12** (2011) 529-536.
[17]A. Temmler, "*Laserumschmelzstrukturierung*" PhD. Thesis, RWTH Aachen University, Aachen, ISBN 978-3-8440-1837-0, 2013.
[18]R. Ostholt, "*Laserpolieren metallischer Freiformen*", PhD. Thesis, RWTH Aachen University, Aachen, ISBN 978-3-8440-0696-4, 2012.
[19]G. Guerrero, J. G. Alauzun, M. Granier, D. Laurencin, and P. H. Mutin, Dalton Trans. **42** (2013) 12569-12585.

 https://doi.org/10.1051/matecconf/201819014011

Reducing Abrasive Particle Generation in Dry Rotary Swaging by Utilizing DLC Hard Coated Dies

Florian Böhmermann[1,*], *Marius* Herrmann[2], *Oltmann* Riemer[1], and *Bernd* Kuhfuss[2,3]

[1]IWT Leibniz Institute for Materials Engineering, Laboratory for Precision Machining, Badgasteiner Straße 3, 28359 Bremen, Germany
[2]bime Bremen Institute for Mechanical Engineering, University of Bremen, Badgasteiner Straße 1, 28359 Bremen, Germany
[3]MAPEX Center for Materials and Processing, University of Bremen, Am Fallturm 1, 28359, Bremen, Germany

Abstract. The emphasis of this paper is the investigation of the impact of the diamond like carbon (DLC) hard coating system on the amount of abrasive particles being generated during dry rotary swaging. Rotary swaging experiments applying coated and uncoated macro structured forming dies were carried out against aluminum and steel work pieces varying the process parameter feed velocity. It was found that DLC coatings effectively reduce the generation of abrasive particles from the work piece. For dry machining of aluminum the amount was reduced to a tenth of the original quantity achieved with uncoated dies. The results are discussed with regard to the mechanics of interfacing surfaces. Additionally, forming dies exhibiting macro structures surfaces of improved design were introduced and applied in dry rotary swaging experiments, which allowed minimizing the abrasive particle generation.
Keyword: Sustainable machining, Cold forming, Die

1 Introduction

Rotary swaging is an incremental bulk metal forming process for the manufacture of rotational symmetric lightweight components, e.g. axles and steering spindles. Compared to those machined by cutting, work pieces machined by rotary swaging exhibit advantageous material properties. These properties are a result of strain hardening and an undisturbed material fiber flow. Besides, rotary swaging allows for an optimal use of material resources through the flexible adaption of wall thicknesses of work pieces. Common for bulk metal forming processes, rotary swaging is associated with high pressures per unit area and high process forces. A robust process layout mandatorily requires for the adequate lubrication of forming dies and work pieces [1]. Functions of the lubricant are mainly the avoidance of abrasive die wear when machining steel, the avoidance of adhesive die wear when machining aluminum, the provision of work piece's surface finish and dimensional accuracy, and at last the purging of particles abraded from the work pieces out of the swaging unit. Recent scientific research dealt with the development of a dry rotary swaging process layout. Functionalized rotary swaging dies were introduced, exhibiting both hard coated and structured surfaces to encounter complex and opposing tribological requirements [2]. Tungsten doped amorphous diamond like carbon coatings (a-C:H:W) with additional a-C:H top layer characterized by excellent hardness and fracture toughness were applied to forming dies for friction reduction and wear protection [3, 4]. Furthermore, macro structured reduction zones of forming dies were used to control the axial reaction force that counteracts the feed force. The features, here, were sine wave structures with propagation parallel to the feed direction, amplitudes of several tenths of micrometers, and wave lengths of about one millimeter [5, 2]. It should be noticed, that structured reduction zones are necessary, as obligatory friction increasing rough layers from tungsten carbide as applied to conventional rotary swaging dies, cannot be used in dry rotary swaging due to immediate clogging. The principles of this new approach are sketched in Figure 1.

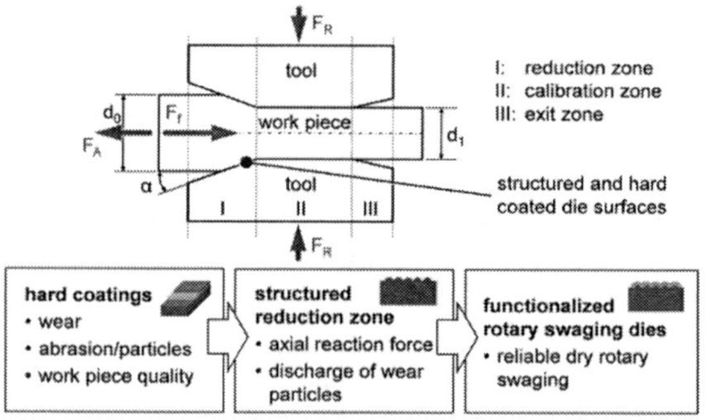

Fig. 1. Principles of infeed rotary swaging displayed in cross-sectional view and approaches for a dry process layout (in accordance with [6]).

The conduction of first dry rotary swaging experiments was shown in multiple studies. Solely DLC coated rotary swaging dies allowed for the successful

* Corresponding author: boehmermann@lfm.uni-bremen.de

 https://doi.org/10.1051/matecconf/201819014011

machining of aluminum 3.3206 tubes. The coating suppressed adhesion of aluminum on the dies' surfaces while providing good surface finish of the work pieces, similar to conventional, lubricated rotary swaging [7]. Forming dies exhibiting sine wave structures (amplitude $A = 0.05$ mm, wavelength $\lambda = 1.3$ mm) were applied to the machining of tubes from 1.0038 construction steel. In comparison to unstructured rotary swaging dies, the dies' design reduced the tracking error of the feed system associated with the axial reaction force F_A by 50 % for dry machining and up to 400 % when machining under lubricated conditions [8]. Furthermore, such structured forming dies with additional tungsten doped amorphous diamond like carbon coating (a-C:H:W) plus a-C:H top layer were tested against steel tubes. Successful dry rotary swaging was achieved with maintainable defects of the hard coating under given harsh conditions [7]. However, dry rotary swaging with functionalized forming dies is subject to an increased abrasive particle generation from the work piece material and accumulation of such particles in the forming zone, associated with a loss in work piece quality and a clogging of the swaging unit. The generation of abrasive particles was explained with the particular contact conditions of sine wave structured dies and the work piece. Induced local strain in the work pieces lead to a local embrittlement of the surface near material and ultimately to the detachment of tinsels from the work pieces' surfaces [6].

The emphasis of this work is the investigation of the impact of the diamond like carbon (DLC) hard coating system on the amount of abrasive particles generated in dry rotary swaging with forming dies. Dry rotary swaging experiments are conducted with uncoated and coated surfaces varying the material machined and the feed velocity. Abrasive particles are collected and investigated by means of optical microscopy. Furthermore, functionalized forming dies of improved macros structure design are presented and applied to dry rotary swaging experiments. Results of the presented work are of major concern for the development of robust rotary swaging processes under dry conditions.

2 Experiments

All dry rotary swaging experiments were carried out on a Felss HE-32 machine tool with a four die swaging unit setup. The swaging frequency is depending on the rotational speed of the swaging unit and was set to 37.5 Hz. The stroke height h_t of the dies was 1 mm. The machine tool is equipped with a linear direct driven feed unit for axial work piece feed, able to generate a maximum feed force F_f of 20 kN.

Two sets of functionalized rotary swaging dies were applied in the experiments. These die sets were both of the same geometry, but one set additionally exhibited hard coated surfaces. The bulk material of the dies was 1.2379 hardened tool steel with a hardness of 62 ± 0.6 HRC. The dies exhibited a die angle of $\alpha = 10$ ° and allow for the diameter reduction of rods and hollow shafts from $d_0 = 20$ mm down to $d_1 = 15$ mm. The reduction zone of the dies exhibited a sine wave structure with an amplitude of $A = 0.05$ mm and a wavelength of $\lambda = 1.3$ mm for the control of the axial reaction force F_A. The uncoated set of forming dies hereafter is notated "S" for structured. Physical vapor deposition (PVD) was used for coating the second set of dies (hereafter notated: "SC" for structured and coated). The hard coating system was a Cr/CrN/Cr+WC/a-C:H:W multilayer system with additional a-C:H top layer.

Two different types of work pieces were applied to the machining experiments: Aluminum 3.3206 (AlMgSi0.5) tubes and 1.0038 construction steel tubes. All work pieces were of the same geometry: initial length $l_0 = 300$ mm, initial diameter $d_0 = 20$ mm, and initial wall thickness $s_0 = 2$ mm.

The general experimental procedure comprising of the actual dry rotary swaging as well as the methodology of abrasive particle collection and analysis was introduced before in [8]. This procedure is adopted for the presented work, but for the sake of completeness will hereinafter again be explained in detail. The experimental procedure comprises of the dry rotary swaging of two work pieces in a row and subsequently the disassembly of the four rotary swaging dies from the swaging unit, investigation of the dies, and collection of abrasive particles. For each set of dies and work piece material experiments were conducted at two feed velocities: $v_f = 500$ mm/min and $v_f = 2000$ mm/min; the first is considered a moderate feed velocity for infeed rotary swaging, while the latter is considered an advanced feed velocity. This summed up to an overall number of experiments of eight and the machining of 16 work pieces. In preparation for every new experiment the forming dies were thoroughly cleaned by rinsing and wiping using ethanol. After the experiments the dies were visually inspected. Particles were first collected from the reduction zone and the calibration zone from each of the four forming dies. Here, actual contact of dies' and work piece's surfaces occur when rotary swaging. The particles were rinsed into Petri dishes for later investigation using acetone. Particles adhering to the tapered rotary swaging dies' flanks were collected by carefully removing those using a brush, compare Figure 2.

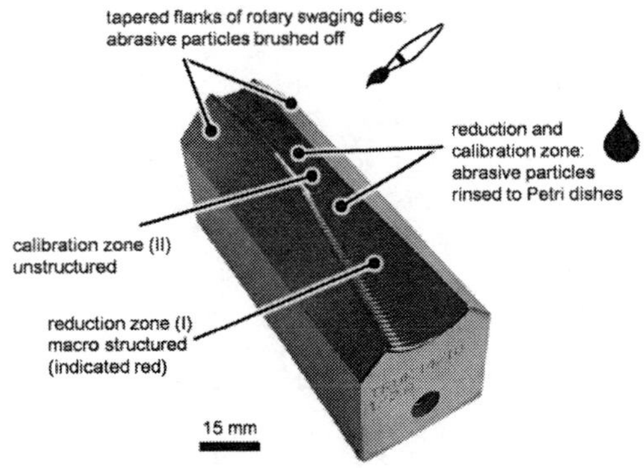

Fig. 2. Functionalized rotary swaging die exhibiting a sine wave structured reduction zone and indication of abrasive particle collection procedure.

The particle analysis was carried out using a Sensofar Plu 2300 optical profilometer in microscope operation mode. The maximum particle areal expanses A_{max} in square millimeter were derived for each the samples taken from the reduction and calibration zone, and from the tapered flanks, respectively. Furthermore, a comparative analysis of the amount of abrasive particles generated in the various experiments was conducted. For this purpose, the categories A to J for the assessment of the amount of abrasive particles are introduced, where A is attributed to the minimum amount of particles over all 12 experiments and J attributed to the maximum amount of particles. This analysis is solely qualitative. Within limits, it allows for the derivation of trends regarding the abrasive particle generation when rotary swaging in dependence of the work piece material and the feed velocity, but is accompanied with a comparably high uncertainty.

3 Results

Distinct abrasive particle generation was found for the experiments applying the uncoated forming dies “S”. The amount and the characteristics of particles strongly differed with the machined material. After the machining of aluminum tubes excessive amounts of aluminum flakes were found. In the context of this work particles referred to as flakes are of planar shape and comparably high rigidity, where the rigidity is determined by the particles’ thickness. Their overall amount strongly depends on the feed velocity. In contrast, when machining steel the feed velocity of the experiment had barely any impact on the amount and the size of abrasive particles generated. Besides steel abrasive particles were of finer structure and should be referred to as tinsels. Tinsels are as well of planar shape but less rigid, what indicates lower particle thicknesses. The results of the optical abrasive particle analyses for all experiments are given in Table 1.

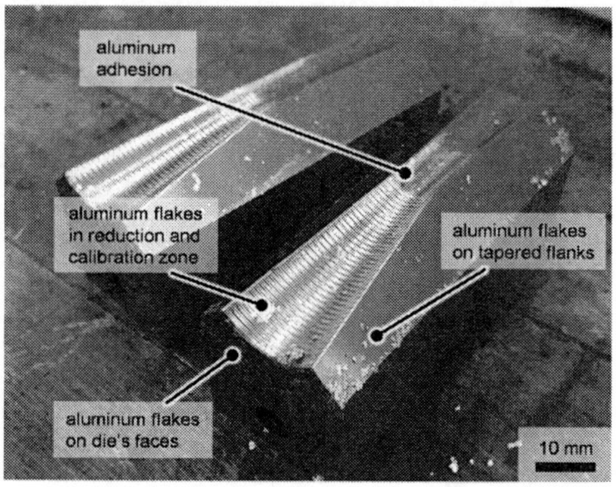

Fig. 3. Uncoated rotary swaging dies “S” after die machining of aluminum 3.3206 work pieces (v_f = 500 mm/min) exhibiting aluminum flakes and aluminum adhesion in the reduction and calibration zone.

Figure 3 shows an uncoated rotary swaging die after the machining of aluminum 3.3206 work pieces. Representative for results of all experiments applying this particular work piece and forming die combination, the die is distinctively covered with flakes from aluminum abraded form the work pieces in the reduction and calibration zone, the tapered flanks, and additionallyon the faces of the forming dies. Those flakes can be characterized as of medium expanse – about five to 20 mm² – and as relatively rigid. Besides distinctive aluminum adhesion was found in the reduction and the calibration zone of the dies. Whilst the particle expanse remained nearly unchanged, the overall amount of particles strongly increased towards the higher feed velocity of 2000 mm/min. The amount of

Table 1. Results of particle analyses in dependence on the applied forming dies, the work piece material, and the feed velocity.

forming die type	work piece material	feed velocity v_f mm/min	particles collected from reduction and calibration zone: areal expanse A_{max} mm²	particles collected from reduction and calibration zone: amount category - -	particles collected from the tapered flanks of the dies: areal expanse A_{max} mm²	particles collected from the tapered flanks of the dies: amount category - -
uncoated forming dies “S”	aluminum 3.3206	500	7.8*	H	10.0*	I
		2000	14.7*	J	16.5*	J
	steel 1.0038	500	4.9	C	5.0	G
		2000	9.2	C	11.0	G
hard coated forming dies “SC”	aluminum 3.3206	500	1.2	A	27.7	A
		2000	7.0	E	234.3	C
	steel 1.0038	500	3.5	C	3.0	C
		2000	1.3	C	4.4	B
novel design “nSC”	aluminum 3.3206	500	0.6	A	0.2	A
		2000	0.5	A	7.2	A
	steel 1.0038	500	0.4	B	0.1	A
		2000	0.6	B	0.2	A

*Abrasive particles characterized as rigid flakes.

particles found here was the largest throughout all experiments carried out in this work and consequently assigned to the amount category J. The number of particles generated was more than double than those generated when machining with the moderate feed velocity of 500 mm/min (amount category H), compare Figure 4. Strong aluminum adhesion was found in the reduction and calibration zones of the forming dies.

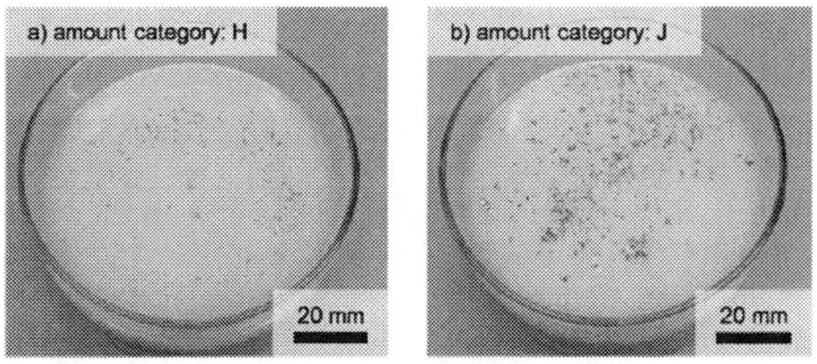

Fig. 4. Petri dish with abrasive particles from aluminum 3.3206 rinsed from the reduction and calibration zone of the uncoated forming dies "S" after the experiments with a) vf = 500 mm/min and b) vf = 2000 mm/min.

Abrasive particles of relatively small expanse were found on both the reduction and calibration zone and the tapered flanks of the uncoated forming dies after the machining of work pieces from 1.0038 construction steel. The maximum expanse of a single particle was $A_{max} = 11.0\ mm^2$ generated during the machining with a feed velocity of $v_f = 2000$ mm/min. However, a significant influence of the selected feed velocity on the amount or the expanse of abrasive particles cannot be determined form the experiments, compare Figure 5. Generally, particles found for the particular combination of uncoated forming dies and steel work pieces are less rigid compared to those generated during the machining of aluminum and, here, were characterized as tinsels.

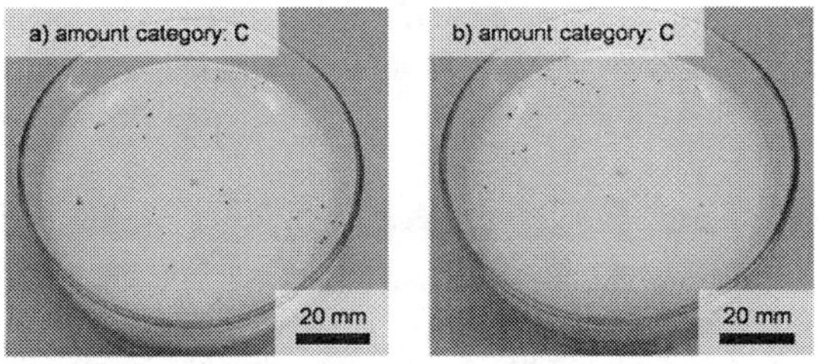

Fig. 5. Petri dishes with abrasive particles from steel 1.0038 rinsed from the reduction and calibration zone of the uncoated forming dies "S" after the experiments with a) $v_f = 500$ mm/min and b) $v_f = 2000$ mm/min.

The experiments applying the structured and coated forming dies "SC" were also subject to distinct abrasive particle generation from the work pieces. The expanse, the amount, and characteristics of such particles were primarily determined by the work piece material, while the influence of the feed velocity on the particle generation is of second order. Only when machining aluminum work pieces a more distinct dependency of the particle size on the feed velocity was found, reflecting those results gained from the machining with uncoated dies. Most important to notice is the capability of the hard coating system on the forming dies to effectively reduce the amount of abrasive particles generated when machining 3.3206 aluminum work pieces to about a tenth of the original value indicated by the amount categories. The results of the experiments regarding the abrasive particle analyses are given in Table 1.

Fig. 6. Coated rotary swaging die "SC" after die machining of aluminum 3.3206 work pieces ($v_f = 500$ mm/min) exhibiting aluminum tinsels of comparably great areal expanse on its tapered flanks.

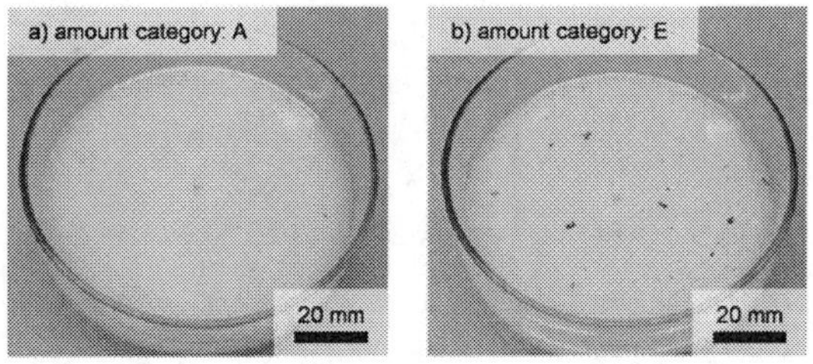

Fig. 7. Petri dishes with abrasive particles from aluminum 3.3206 rinsed from the reduction and calibration zone of the coated forming dies "SC" after the experiments with a) $v_f = 500$ mm/min and b) $v_f = 2000$ mm/min.

In Figure 6 one of the four rotary swaging dies after the machining of aluminum 3.3206 tubes is displayed. Representative for all four experiments machining aluminum and varying the feed velocity, the forming dies exhibit abrasive particles of greater areal expanse on the tapered flanks. These particles, other than those generated during the machining of aluminum work pieces with uncoated tools, were of relative fine structure and should be characterized as tinsels. The increase of the feed velocity to $v_f = 2000$ mm/min resulted in a strong increase of the amount and the areal expanse of those particles. The largest tinsel had an areal expanse of $A_{max} = 234.3\ mm^2$. In comparison, the reduction zone and the calibration zone were free of larger tinsels and exhibited comparably low numbers of abrasive particles. This is reflected in the small amount of particles rinsed into the Petri dishes after the experiments; compare Figure 7. It is expected that the majority of the abrasive particles generated while forming is pushed in the gaps

* Corresponding author: boehmermann@lfm.uni-bremen.de

between the forming dies and compressed to tinsels of greater areal expanse at the tapered flanks when the dies repetitively come into contact at every incremental forming operation. The primary particle areal expanse generated from the work piece may not exceed $A_{max} = 7.0\ mm^2$ for the machining of aluminum 3.3206 tubes.

For all rotary swaging experiments machining 1.0038 construction steel work pieces the rotary swaging dies exhibited greater amounts of tinsels with smaller areal expanse on their tapered flanks after machining; compare Figure 8).

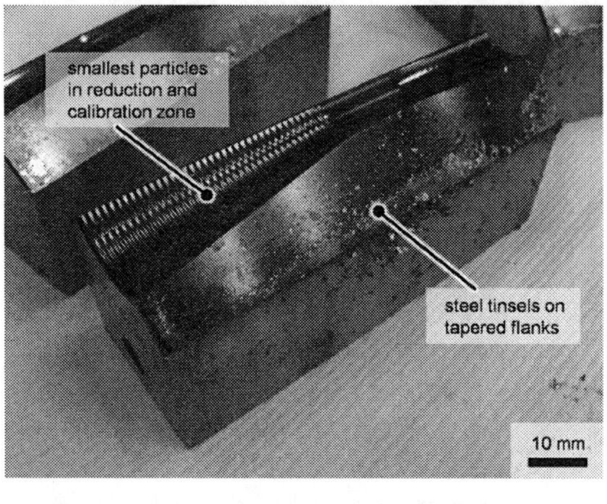

Fig. 8. Coated rotary swaging die "SC" after die machining of 1.00038 work pieces ($v_f = 500$ mm/min) exhibiting distinct amounts of tinsels of smaller areal expanse on its tapered flanks.

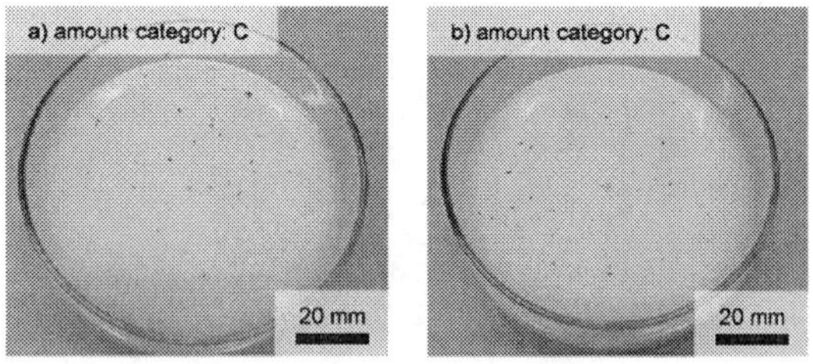

Fig. 9. Petri dishes with abrasive particles from steel 1.0038 rinsed from the reduction and calibration zone of the coated forming dies "SC" after the experiments with a) $v_f = 500$ mm/min and b) $v_f = 2000$ mm/min.

Independent from the chosen feed velocity, the abrasive particles found on both, the tapered flanks and in the reduction and calibration zone, exhibited almost the same maximum areal expanse A_{max} in a range from $1.3\ mm^2$ to $4.4\ mm^2$. The tinsels at the tapered flanks, as in the other experiments, are expected to be formed by the incremental forming operation, after being pushed in the gaps between the rotary swaging dies. However, far smaller tinsels were found compared to the experiments machining aluminum work pieces with the coated forming dies. Small tinsels were rinsed from the reduction and the calibration zone into the Petri dishes for optical inspection after the experiments machining 1.0038 steel tubes, compare Figure 9. These were assigned to the amount category C for both the feed velocities 500 mm/min and 2000 mm/min, respectively. The maximum areal expanse of these particles did not exceed $3.5\ mm^2$ for any of the applied feed velocities

4 Discussion

Dry rotary swaging is generally associated with an increased generation of abrasive particles form the work piece material. The phenomenon was observed for all conducted experiments, independent from the machined work piece material. The most distinct particle generation was observed for the machining of aluminum tubes with uncoated forming dies at increased feed velocities. The amount of particles found after the experiment was about ten times of those found for the machining with hard coated rotary swaging dies.

The abrasive particle generation in dry rotary swaging is traced back to the particular contact conditions of sine wave structured reduction zone and the work piece during the forming process. Here, alternating local strain and compression of the surface near work piece material layer occurs. Local strain occurs once the material is pressed into a valley of the sine wave structure, local compression when the material is pressed onto a peak of the sine wave structure. Local strain and compression result in a distinct strain hardening of the surface near work piece material layer accompanied with ongoing local embrittlement. This finally leads to the failure of the surface near material layer and with that the detachment of abrasive particles from the work piece's surface. The explained mechanism of particle generation is supported by the findings from finite element method (FEM) simulations, investigating stress states in the work piece material emerging while rotary swaging with sine wave structured rotary swaging dies [9]. The abrasive particles generated in the experiments can be characterized as fine tinsels, except the case of machining of aluminum work pieces with uncoated forming dies. Besides different characteristics of generated abrasive particles – these were flakes of relatively rigid structure – the amount of generated particles increased tenfold at a feed velocity of $v_f = 2000$ mm/min. Here, an additional mechanism of particle generation must be taken into account. Besides the local stain occurring while the work piece material is pressed into the sine wave structure, the adhesion between the work piece and the forming dies plays a major role. During the deformation of the work piece new surface layers on the aluminum work piece are generated, which are free of protective oxide layers. These new surfaces are able to chemically bond, or adhere, to the steel surfaces of the forming dies. The breakup of these to cold welds requires for considerably increased shearing of the work piece material during the forming process, intensifying its local embrittlement and such the detachment of abrasive particles. The findings are underlined by the presence of aluminum adhesive layers in the reduction and the calibration zone of the uncoated rotary swaging dies after the experiments. It can be concluded that dry rotary swaging of aluminum work pieces with uncoated forming dies is not feasible. Besides, the experimental results impressively underline the capability of the a-C:H:W/a-C:H hard coated dies to effectively suppress adhesion. In direct comparison to the experiments applying uncoated dies, the number of particles generated was cut by ten when machining aluminum 3.3206 work pieces.

For all experiments the largest amount of abrasive particles from the work piece material was found on the tapered flanks of the rotary swaging dies. Particles predominantly generated in the reduction zone are subsequently pressed into the gaps between the forming dies and are formed to tinsels of greater areal expanse by the incremental forming strokes. Exceptions were the result of the experiments forming aluminum work pieces with uncoated forming dies. The here generated particles generated were more rigid. Contact pressures between dies were not enough to deform particles sufficiently in order to generate tinsels; the particles maintained their original areal expanse of about $A_{max} = 15\ mm^2$. The revolving motion of the swaging unit is expected to facilitate the phenomenon of pressing particles into the gaps between forming dies. For ongoing machining the buildup of greater amounts of work piece material in the gaps is expected. This ultimately hinders the complete closing of the dies whilst the die stroke and leads to geometrical inaccuracy of the machined work pieces and a significant earlier termination of the forming process. A robust dry rotary swaging cannot be achieved for the given conditions and requires for a further development of the rotary swaging dies' design.

An enhanced design of the rotary swaging dies must encounter both, the reduction of abrasive particle generation and measures preventing particles to be pressed into the gaps between the dies and their agglomeration on the tapered flanks; compare Figure 10. A reduced particle generation can be achieved by a re-design of the sine wave structures in the dies' reduction

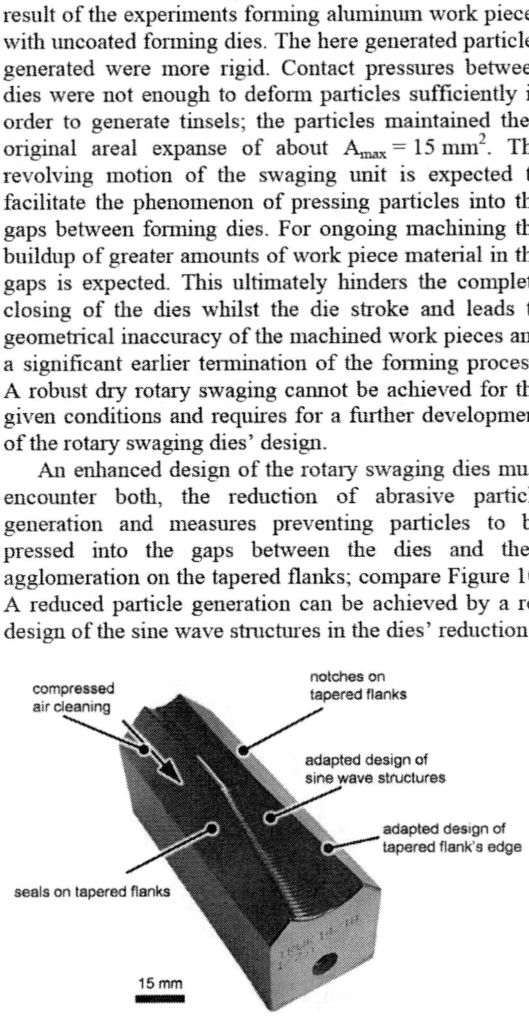

Fig. 10. Improvements of dies' design reducing abrasive particle generation and abrasive particle agglomeration on the tapered flanks in dry rotary swaging.

zone. Sine wave structures with fading, amplitude towards the calibration zone are a possible compromise between control of process forces and minimizing the generation of abrasive particles. Besides, this particular structure design can help improving the surface finish of the machined work pieces. An adaption of the edge between reduction and calibration zone and the tapered flanks could additionally help reducing the amount of particles pressed in the gaps between the rotary swaging dies. Particles still able to enter the gaps between the dies should be entrapped in notches on the tapered flanks. Thus, a buildup of greater amounts of abrasive particles on the tapered flanks can effectively be avoided. A cleaning system utilizing compressed air could be applied to blow particles out from the notches and out of the swaging unit. Additional seals alongside the tapered flanks shall avoid abrasive particles to penetrate deeper into the mechanics of the swaging unit. The presented changes in the forming dies should consequently allow for a robust dry rotary swaging process layout.

5 Improved dies' design

Based on the findings from the experiments presented in the sections above, functionalized forming dies of improved design were developed, encountering a redesign of the macro structures applied to their reduction zone. These novel forming dies hereafter should be notated "nSC" for novel, structured, and coated. Other than the primary forming dies with a sine wave structure of constant amplitude ($A = 0.05$ mm) applied to the reduction zone, newly developed dies exhibit a sine wave structure, with fading amplitude towards the calibration zone. Starting from a maximum amplitude of again $A = 0.05$ mm, the structure's height decreased to $A = 0$ mm over a distance of $s = 10.4$ mm until reaching the transition between deduction and calibration zone (unified amplitude decrease about 5 µm/mm), compare Figure 11. The wavelength λ of the sine wave structure and the forming die angle α were kept unchanged at 1.3 mm and 10 °, respectively. The set of improved forming dies again was made from 1.2379 hardened tool steel with a hardness of 62 ± 0.6 HRC. Besides, the novel rotary swaging dies were hard coated with the well-established Cr/CrN/Cr+WC/a-C:H:W multilayer system with a-C:H top layer. Such functionalized rotary swaging dies of improved design were first introduced by Hermann et al. [10].

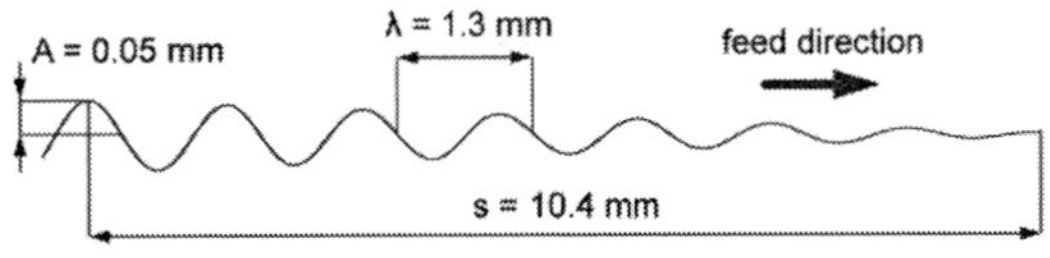

Fig. 11. Cross-sectional sketch of the fading sine wave macro structure applied to the reduction zone of the novel functionalized forming dies nSC (illustration not to scale).

The novel dies were applied to the same experimental procedure as explained above for abrasive particle generation assessment in dry rotary swaging. Again, work pieces from 3.3206 aluminum alloy and 1.0038 construction steel were applied. The feed velocities in the dry rotary swaging experiments were 500 mm/min and 2000 mm/min, respectively.

https://doi.org/10.1051/matecconf/201819014011

6 Results: Improved dies

Only lowest amounts of abrasive particles were found for the experiments applying the novel forming dies with redesigned reduction zone and DLC hard coated surfaces "nSC". The results of the abrasive particle analyses after experiments are given in Table 1. Furthermore, the Petri dishes containing the abrasive particles rinsed from the reduction and the calibration zones are displayed in Figure 12 and Figure 13.

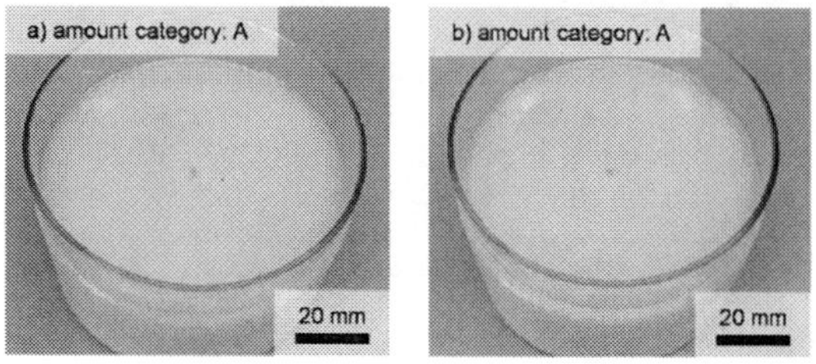

Fig. 12. Petri dishes with abrasive particles from aluminum 3.3206 rinsed from the reduction and calibration zone of the novel functionalized forming "nSC" dies after the experiments with a) $v_f = 500$ mm/min and b) $v_f = 2000$ mm/min.

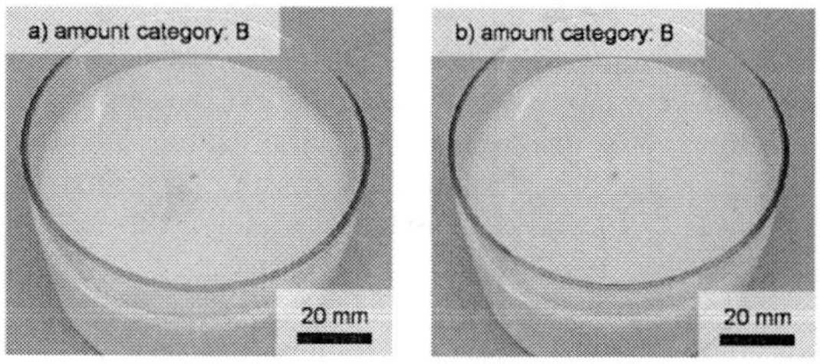

Fig. 13. Petri dishes with abrasive particles from steel 1.0038 rinsed from the reduction and calibration zone of the novel functionalized forming dies "nSC" after the experiments with a) $v_f = 500$ mm/min and b) $v_f = 2000$ mm/min.

The expanse of particles found usually did not exceeded $A_{max} = 0.6$ mm^2, the amount of particles found did not exceeded category B, neither for the machining of aluminum nor for 1.0038 steel work pieces. Only after the machining of aluminum tubes at the advanced feed velocity of $v_f = 2000$ mm/min a slight tendency towards the formation of tinsels of greater expanse on the tapered flanks of the rotary swaging dies was detected. The particle of greatest expanse found here had a projected area of $A_{max} = 7.2$ mm^2. Generally, the tapered flanks of the dies exhibited barely any abrasive particles after the machining procedures. Compared to the experiments conducted first, the obtained results indicate a distinct reduction of abrasive particle generation and impressively underline the capability of the novel forming dies with improved design to contribute to the development of a robust dry rotary swaging process layout. Furthermore, the findings from the experiments using the novel rotary swaging dies clearly underpin the before formulated theory of particle generation when dry rotary swaging with macro structured dies. This theory states that particle generation can be traced back to the local strain and compression induced in the surface near material layers of the work piece, occurring when the sine wave structured reduction zone and the work pieces come into contact. These processes are associated with ongoing material embrittlement and ultimately the detaching of abrasive particles from the work pieces' surfaces. The less pronounced geometry of the sine wave structures in the reduction zone of the novel functionalized rotary swaging dies allowed to distinctively cut the number of abrasive particle generated, which confirms the explained mechanism of particle generation.

7 Conclusion

In this work the generation of abrasive particles for dry rotary swaging with functionalized forming dies was investigated. Experiments were conducted applying uncoated and hard coated rotary swaging dies exhibiting macro structured reduction zones to the machining of 3.3206 aluminum and 1.0038 steel tubes. Due to their macro structured reduction zone such forming dies are suitable to effectively control the axial reaction force during the process. However, with regard to abrasive particle generation, this forming dies design is not sufficient for the machining of larger quantities of work pieces. Distinct amounts of abrasive particles were generated predominantly in the structured reduction zone. Particles subsequently penetrate the gaps between the rotary swaging dies, agglomerate to tinsels of greater expanse, and ultimately hinder the dies from complete closing. Alternating strain and compression induced in the surface near material layers of the work pieces while machining was identified as the abrasive particle generation mechanism. The local deformation results in a distinct strain hardening of the work piece material accompanied with ongoing local embrittlement and at last the detachment of abrasive particles from the work piece's surface. This mechanism is distinctively amplified for the machining of aluminum work pieces with uncoated rotary swaging dies due to strong adhesion accompanied with increased shearing of the surface near material layer of the work pieces. Based on the findings, forming dies of improved design were introduced, exhibiting sine wave structures of fading amplitude and DLC hard coated surfaces. This less pronounced macro structure allowed to distinctively reducing the number of abrasive particles generated in dry rotary swaging experiments. The application of such functionalized forming dies is one key step towards the successful development of robust dry rotary swaging processes accompanied with major economic and ecological benefits.

Acknowledgements: The authors would like to thank the German Research Foundation (DFG Deutsche Forschungsgemeinschaft) for funding this work within the sub-project "Potentials of Dry Rotary Swaging" of the priority program SPP 1676 "Dry metal forming - sustainable production through dry processing in metal forming." The experimental assistance by Mr. Christian Rösner is greatly appreciated.

References

1. P. Groche, F. Heislitz, Kraftbedarf beim Kaltrundkneten – Abschlussbericht zum FKM-Vorhaben Nr. 224, FKM-Heft **224** (2000)
2. F. Böhmermann, H. Hasselbruch, M. Herrmann, O. Riemer, A. Mehner, H.-W. Zoch, B. Kuhfuss, Trockenrundkneten – Funktionalisierte Werkzeug-oberflächen für eine schmierstofffreie Prozess-auslegung, wt Werkstattstechnik online **105**, 11/12 (2015) 830-835
3. H. Hetzner, Systematische Entwicklung amorpher Kohlenstoffschichten unter Berücksichtigung der Anforderungen der Blechmassivumformung, Dissertation, Universität Erlangen (2014)
4. H. Hasselbruch, M. Herrmann, A. Mehner, B. Kuhfuss, Development, characterization and testing of tungsten doped DLC coatings for dry rotary swaging, MATEC Web of Conferences **21** (2015)
5. M. Herrmann, C. Schenck, B. Kuhfuss, FEM simulation of infeed rotary swaging with structured tools, MATEC Web of Conferences **21** (2015)
6. F. Böhmermann, M. Herrmann, O. Riemer, B. Kuhfuss, Abrasive Particle Generation in Dry Rotary Swaging, Dry Metal Forming OAJ **3** (2017) 001-006
7. M. Herrmann, F. Böhmermann, H. Hasselbruch, B. Kuhfuss, O. Riemer, A. Mehner, H.-W. Zoch, Forming without Lubricant – Functionalized Tool Surfaces for Dry Forming Applications, Procedia Manufacturing **8** (2017) 533-540
8. M. Herrmann, C. Schenck, B. Kuhfuss, Dry rotary swaging with structured tools, Procedia CIRP **40** (2016) 654-659
9. Y. Liu, M. Herrmann, C. Schenck, B. Kuhfuss, Plastic Deformation History in Infeed Rotary Swaging Process, AIP Conference Proceedings **1896** (2017) 080013
10. M. Herrmann, C. Schenck, B. Kuhfuss, Graded Structured Tools for Dry Rotary Swaging, Dry Metal Forming OAJ **4** (2018) 000-000

A new approach for dry metal forming: CO_2 as volatile lubrication in combination with hard and low friction coatings

Georg Umlauf[1,*], *Henning* Hasselbruch[2], *Jan Hinnerk* Henze[1], *Jakob* Barz[1], and *Andreas* Mehner[2]

[1]Fraunhofer Institute for Interfacial Engineering and Biotechnology, Fraunhofer IGB, Nobelstraße 12, 70569 Stuttgart, Germany
[2]Leibniz Institute for Materials Engineering IWT, Badgasteiner Straße 3, 28359 Bremen, Germany

Abstract. Nowadays, it is more important than ever to meet the increasing technical and statutory requirements and to develop new process strategies. In sheet metal forming the low consumption of oil lubrication gets an increasingly important role. The aim in sheet metal forming is to reduce the amounts of lubricants. In the long term, the future sheet metal processing should be able to completely dispense without mineral oil-containing lubricants. Two promising approaches for a dry process design are combined in this paper for the first time and the fundamental feasibility of this new hybrid technology is shown. On the one hand, a novel approach for temporary lubrication of deep-drawing processes with CO_2 as a volatile medium is used. On the other hand, it is supported by the application of an additional hard coating system such as a silicon nitride (Si_3N_4) or tungsten doped a-C:H multilayered (Cr/CrN$_x$/a-C:H:W/a-C:H) coating system to further reduce tool wear and wear debris of the formed sheets made of DC04 (mat. no.: 1.0338). The results show a low coefficient of friction and reduced wear. Especially for the carbon coating system, there is minor tool wear at a higher surface pressure. By means of the graphite constituent, even a smoothening of the roughness peaks can be recorded. The next step would be the implementation of this hybrid technology on a tool for deep drawing a rectangular cup.

Keywords: Tribology, Lubrication, Dry metal forming

1 Introduction and State of the Art

Tribology is defined as the science of interacting surfaces in relative motion, friction and wear [1]. The system is composed of the three subcomponents (base-) body, counter-body, interfacial element and surrounding medium (environment), see figure 1.

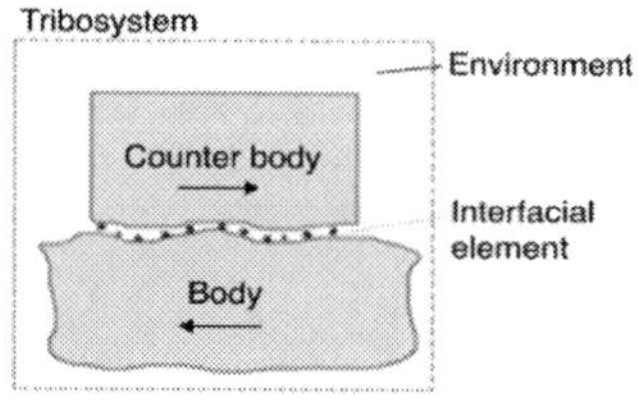

Fig. 1. The Tribosystem provides the basis for the description and analysis of tribological problems [2].

In most forming processes, mineral oil-based lubricants are used as the interfacial medium. The lubricant is able to reduce friction and heat formation in the contact zone, to reduce overall deformation force and energy, and to extend tool life and prevent seizure [3, 4]. Sometimes toxic additives are contained in the lubricant and the formed components have to be cleaned cost intensely and time-consuming for further process steps [5]. Today, there is a great awareness of the environment and its conservation. This inevitably leads to increased requirements and the need for solutions to develop new environmentally friendly tribological systems. Already since the beginning of the 2000s, the legislation in Europe has been changed. Companies are particularly limited in the use of harmful lubricants. In 2007, the EU introduced a new REACH (Registration Evaluation Authorisation and Restriction of Chemicals) law aimed at achieving a high level of protection of human health and the environment from the risk of chemical pollution [6, 7, 8]. Forming of modern high-strength or stainless steels results in a complex tribological system. Reducing the amount of lubricant increases the risk of fretting. In the worst case, it causes a premature tool failure. Nevertheless, attempts are being made to reduce lubricant quantities in order to simultaneously reduce the health and environmental impacts in the manufacturing process [9, 10]. In addition, the use of lubricants subsequent or between two forming operations requires for a complex cleaning process to obtain a clean and oil-free surface. This aspect plays a special role when the components are painted and bonded as a final production [11]. In the past ten years, semi-dry processes and minimum quantity lubrication (MQL) have been implemented in the forming process [12]. The specification of a completely lubricant-free metal forming design could not implement with these strategies. This shows the fact that today still large

* Corresponding author: georg.umlauf@igb.fraunhofer.de

amounts of mineral oil are used for lubrication in the forming process.

Metal forming processes, that completely avoid lubricants, could offer economic and environmental benefits. The definition of dry metal forming does not exclude the use of additives in the process, but focuses on the aspect that no additional purification or drying is required in the subsequent processing step [5]. It is clear that dry metal forming processes present a challenge, since the absence of lubricants typically results in a reduced processing window with respect to the forming limit of the workpiece and its quality. The negative effects of lubricant removal must be compensated with a novel tool concept to ensure a stable and complete forming process.

For this reason, the institutes Fraunhofer IGB and Leibniz-IWT have joined forces for a sub-research project in order to further develop an effective tribosystem for dry forming. In addition to the use of liquid CO_2 as a volatile lubrication a hard material layer is applied on the tool surfaces. The union of these two technologies should result in a new hybrid system with improved wear resistance. The results presented in the following serve first of all as proof in principle. The hybrid system presented here is presented for the first time in this form.

2 Two-way hybrid system

2.1 Lubrication with liquid CO_2

In the priority program (SPP1676) funded by the German Research Foundation (DFG), the institutes IGVP, IFU and IFSW at the University of Stuttgart are working together to develop a manufacturing approach for a mineral oil free forming process. The work explores an environmentally friendly sheet metal forming lubrication system using liquid CO_2 as a lubricant.

The phase diagram of the CO_2 shows, according to figure 2, that CO_2 is in a liquid state at room temperature and a pressure of 57.3 bar. This represents the initial state (red X in figure 2).

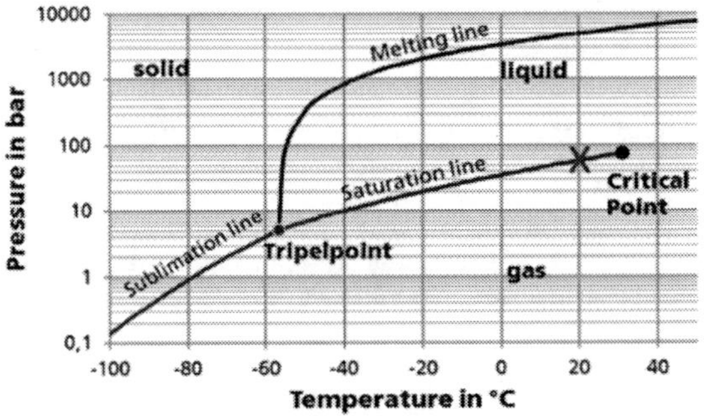

Fig. 2. CO_2 phase diagram, include solid, gas und liqiud phase [13].

Figure 3 shows the overall process sequence of the new CO_2 lubricating system presented here in a simplified form.

First, the liquid lubricant is passed directly into the contact zone and expands there to atmospheric pressure. This causes a cooling and it produces dry ice snow, which acts as a lubricant, achieves a separation of the tool and workpiece, and further reduces the restraining forces.

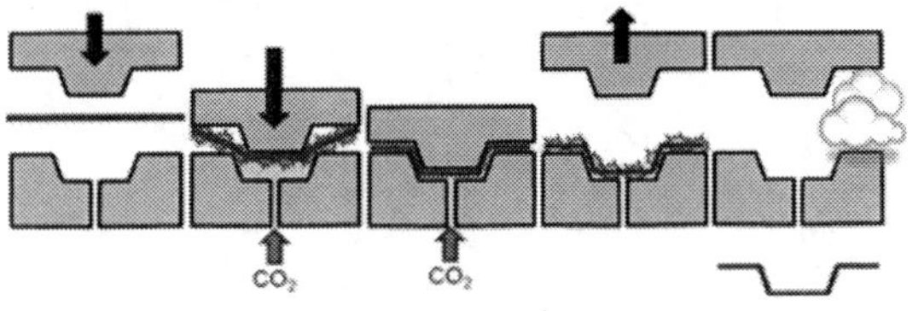

Fig. 3. Process sequence during forming with liquid CO_2 as lubricant replacement [11].

After the forming process, the dry ice is heated by the ambient temperature of the production hall and by the workpiece itself due the deformation induced cohesive frictional heat. So it reverts to the gaseous state of aggregation, as can be seen from the marking in the phase diagram at approximately 1 bar and 23 °C. Thus, there is no oiled component after the forming and further process steps, such as cathodic dip coating or painting, can be carried out without additional purification steps.

2.2 Depostion of a hard coating system

In addition to lubrication with liquid CO_2, a coating of hard material is applied on the active tool surfaces. This subsequent coating should improve the tool life. It may be possible to achieve an additional reduction in friction coefficients when combining a coating with CO_2 compared to a single uncoated CO_2 lubrication. For this purpose, the performance of two different types of hard coatings are tested. The combination of CO_2 lubrication and hard coatings results in the described 2-way hybrid system.

2.2.1 Si_3N_4 coating

Silicon nitride as non-oxide ceramic material has several positive material properties. In addition to the low density (3.44 g/cm^3), it has a very high fracture toughness (6 MPa*m$^{0.5}$) and at the same time good flexural strength (> 800 MPa at 20 °C). Due to the inherent microstructure a high thermal shock resistance is given. These properties make silicon nitride insensitive against thermal shocks and impacts. Si_3N_4 is already used in ball or roller bearings. Also in forming technology Si_3N_4 is used in high stressed tools. The approach for producing this coating will be explained in chapter 3.2.

2.2.2 a-C:H:W/a-C:H PVD coating

The hydrogenated amorphous carbon (a-C:H:W/a-C:H) coating consists of a multi-layered design with a total coatings thickness of about 2.2 µm. As shown in figure 4 b the precise coating design is illustrated. First a thin Cr layer of a few nanometers is deposited acting as soft glue. Then a CrN_x layer is applied with increased hardness of about 8-10 GPa $HIT_{0.01}$ acting as a bonding agent for the upper layers. To further increase the adhesion strength between substrate and functional layers a graded (Cr,W)C_y intermediate layer is introduced. This step also realizes a smooth transition from the metallic into the non-metallic character of the coating. The reason for doping with tungsten of both the graded intermediate and the overlaying a-C:H:W layer (thickness of about 1 µm) is the reducing of the residual coating stresses during the deposition process [14]. This objective is designed to increase the fracture toughness of the whole coating system and to increase the hardness stepwise [15]. Finally, a hard functional a-C:H layer of about 22 GPa $HIT_{0.01}$ is applied for ensuring low dry coefficients of friction of $\mu < 0.2$ and wear against steel and aluminum counter bodies of $k < 3\cdot10^{-7}mm^3/Nm$ [16]. In general, a-C:H coatings have a low tendency to adhesion due to the saturation of carbon bonds by hydrogen of the OH groups on the surface [17, 18]. Further details were previously published [19].

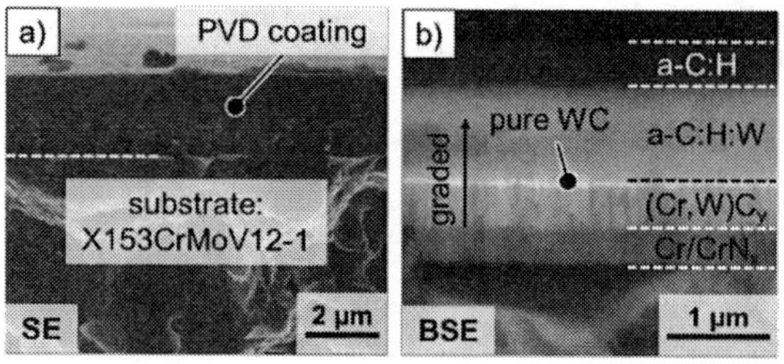

Fig. 4. SEM fracture patterns: a) SE image and b) BSE image at higher magnification of the tested Cr/CrN_x/(Cr,W)C_y/a-C:H:W/a-C:H coating system denoted as a-C:H:W/a-C:H [16].

3 Experimental setup

In order to prove the technical fundamental feasibility in a forming process, the exchangeable drawing jaws with laser drilled micro-holes are tested in a strip drawing system. In each experiment CO_2 as volatile lubrication was used. For the new hybrid system, the contact surfaces of the exchangeable tool jaws are additionally applied with different hard coatings. The coating made of Si_3N_4 was carried out at the Fraunhofer-Institute IGB in an ICP plasma reactor. The a-C:H(:W) coating were produced in Bremen at the Leibniz-IWT. For this purpose, an PVD reactive plasma system was used. After the strip drawing tests, a comparison of the determined friction values takes place. Likewise, the roughness values are measured and analyzed.

3.1 Strip drawing test bench

On a strip drawing test system with flat track, it is possible to determine coefficients of friction depending surface pressure per unit area respectively CO_2 flow rate. At the Institute for Forming Technology (IFU, University of Stuttgart), the test bench was modified so that the restraining forces of the CO_2 lubrication system could be measured. For the approach, an experimental tool for supplying the liquid CO_2 directly into the contact zone between tool and sheet was designed and constructed, see figure 5. The strip drawing test bench is described in detail in [20] and [21]. All tests were carried out at room temperature using sheet material DC04 (mat. no.: 1.0338) in rolling direction and repeated three times for statistical reasons.

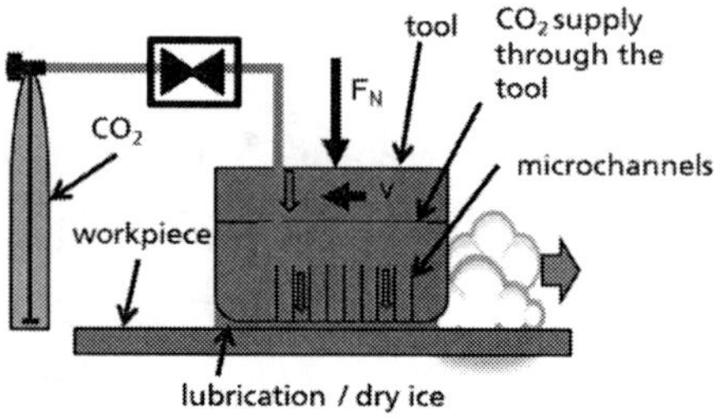

Fig. 5. Schematic illustration of the experimental setup for the operation in the strip drawing test with CO_2 lubrication [11].

In order to apply the liquid CO_2 directly in the effective area between the tool and the workpiece, it must first be passed through the tool. A two-part design of the experimental tool was selected, composed of a hollow body and at 5 mm thin exchangeable jaws made of tool steel X153CrMoV12 (mat. no.: 1.2379). Furthermore, good accessibility for the production of micro-holes is ensured. The construction of the experimental tool can be seen in figure 6. The production of micro-holes is carried out at IFSW, University of Stuttgart and will be explained in more detail in [21].

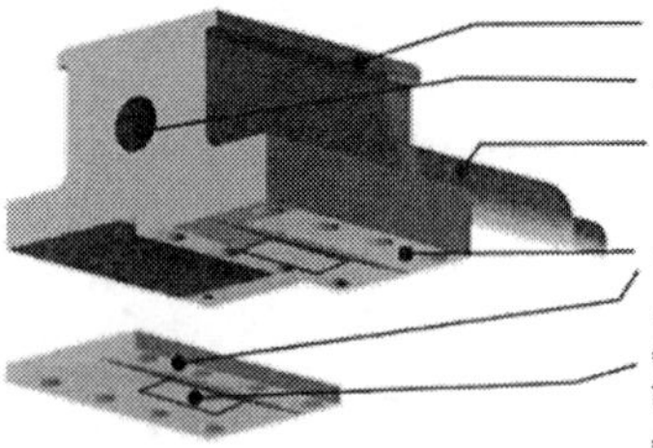

Fig. 6. Modified drawing tool for CO_2 lubrication in strip drawing test [21].

The exchangeable jaws have a contact area of 400 mm^2 in total. Before the exchangeable jaws are used, they are coated with a hard material layer.

3.2 ICP plasma reactor for Si_3N_4 coating

The term ICP-CVD denotes a chemical vapour deposition, system in which an inductively coupled plasma (ICP) source is used. The deposition plasma is ignited via an alternating electromagnetic field within the recipient, which is induced by an inductance coil created by radio frequency (RF). In an ICP reactor, the electrons

https://doi.org/10.1051/matecconf/201819014012

collide numerous times by the electric field, causing a Joule heating. This is the main mechanism for energy absorption. [22]

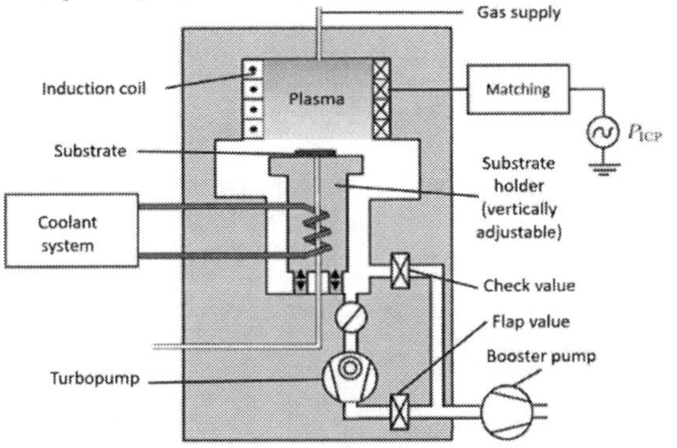

Fig. 7. Simplified illustration of an ICP-PECVD plasma system with the main components, based on [23].

In this work, silicon nitride layers were deposited on the exchangeable jaws by means of the ICP-CVD process and the influence on the frictional behavior in combination with volatile lubricating media was investigated.
The chemical reaction equation for the vapor deposition by ICP-CVD is said to be:

$$3\,SiH_4 + 4\,NH_3 \rightarrow Si_3N_4 + 12\,H_2 \qquad (1)$$

The process duration was 10 minutes with a power input of 2000 W. During the process, the pressure was 0.2 μbar. For optimum oxide-free layer formation, the gas flow was set to 190 cm³/min for SiH_4. The gas mixture ratio was 1:10 (NH_3:SiH_4).

3.3. Reactive magnetron sputtering device for a-C:H:W/a-C:H coating

An industrial reactive magnetron sputtering device from the company CemeCon of the type CC 800/9 SinOx was used for the deposition of the a-C:H:W/a-C:H multilayer coating system. The general principle of the process is illustrated in figure 8. When reactive magnetron sputtering the target material is vaporized due to a momentum exchange caused by ionized process gases. As process gases the noble gases argon and krypton (Ar, Kr) are used. The starting pressure of the device is set to 1 mPa. By applying a negative voltage at the cathodes behind the target materials, released respectively free electrons cause an impact ionization cascade of the noble gases (a). The positive noble gas ions (Ar^+, Kr^+) are accelerated on the target material with high kinetic energy, which initializes the sputtering process of the target atoms (b). Finally, the deposition of the target atoms on the substrate takes place (c). The deposition of the CrN_x layer is realized by the incorporation of pure nitrogen (N_2) as reactive gas. The precursor gas acetylene (C_2H_2) is used for deposition of the a-C:H:W and a-C:H layers. Further details of the deposition process are published in previous publications [16, 19].

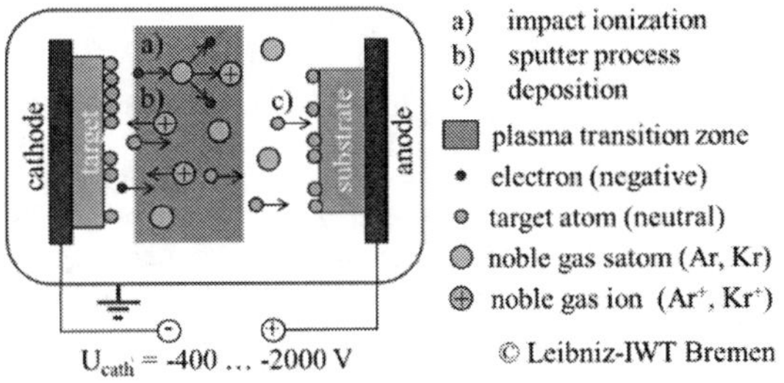

Fig. 8. Simplified illustration of the reactive magnetron sputtering process (source: Leibniz-IWT).

In figure 9 a) the used target configuration is shown in order to apply the mentioned a-C:H:W/a-C:H coating system on a substrate material. The substrate adhesion was determined on a 1.2379 cold working steel, hardened to an overall hardness of about 60±2 HRC (≈800 $HIT_{0.01}$), by Rockwell indentation and scratch tests according to VDI 3198 and DIN EN ISO 1073-3. As illustrated in figure 9 b), the adhesion class HF was 1-2 and the critical load Lc2 was in the range of 30 N [24]. Due to its good adherence no coating failures should be expected due to the modified drawing tool tests.

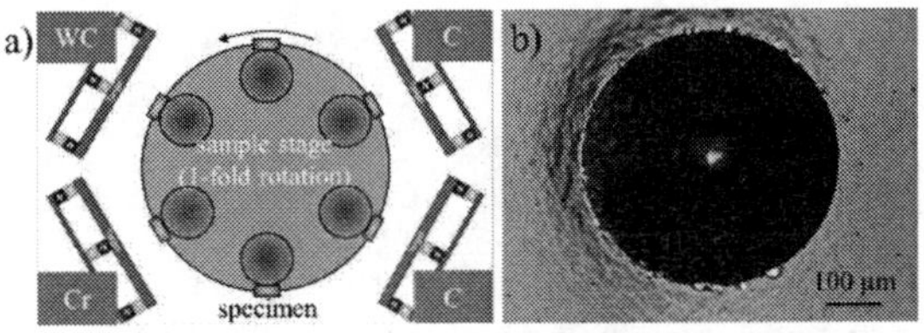

Fig. 9: a) Used target configuration in the CC 800/9 Si_nO_x and b) adhesion strength class (~ HF 1-2) according to VDI 3198.

Figure 10 shows the direct comparison between an exchangeable jaw with an a-C:H:W/a-C:H and a Si_3N_4 coating.

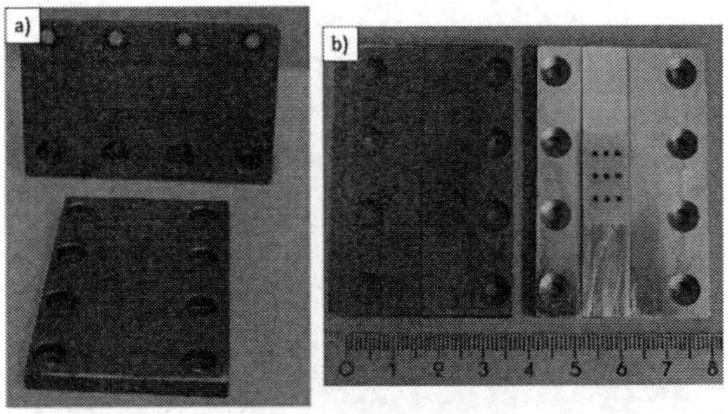

Fig. 10. a) A pair of exchangeable jaws coated with a-C:H:W/a-C:H, b) comparison between a jaw coated with a-C:H:W/a-C:H (left) and Si_3N_4 (right). Both exchangeable jaws are unused.

The next chapter presents the results of the strip drawing tests. Likewise, the roughness was measured after the load and the contact angle with water was determined.

4 Results and Discussion

In this chapter the results of the strip drawing tests are shown and discussed. Figure 11 shows the detailed propagation of the coefficients of friction depending on the applied surface pressure during a volatile CO_2 lubrication in addition with the two used hard coatings. In general, the coefficient of friction increases at higher surface pressures for both applied coatings.

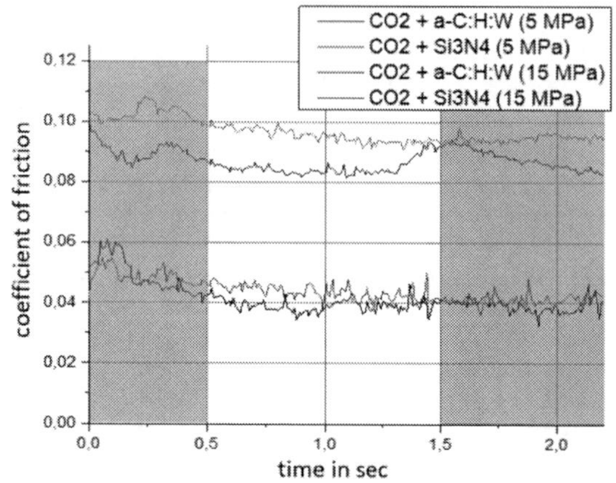

Fig. 11. Detailed coeffiencent of friction for a volatile CO_2 lubrication in addition with two hard coatings at two different surface pressures.

For an averaged coefficient of friction, not the entire drawing distance is considered, otherwise the running-in behavior would change this value too much. Only the recorded values of 0.5 to 1.5 sec are included in the determination. This results in a friction coefficient of 0.039 for the Si_3N_4 coating at a surface pressure of 5 MPa. For comparison, the friction coefficients for exchangeable jaws without hard material coating are plotted in figure 12.

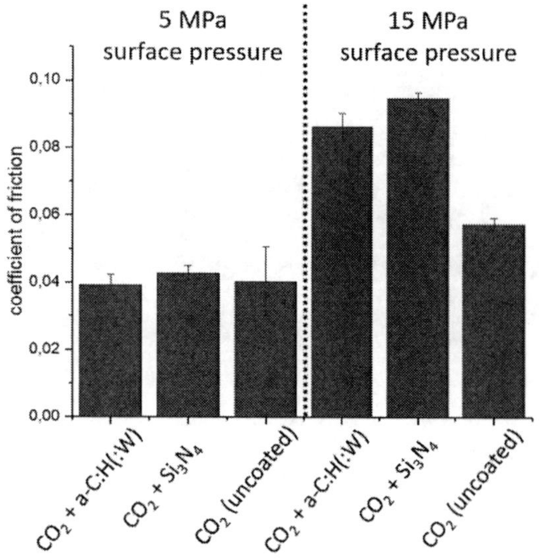

Fig. 12. Representation of the coefficients of friction for coated and uncoated exchangeable jaws in combination with a volatile CO_2 lubrication, each with 5 and 15 MPa surface pressure.

For a surface pressure of 5 MPa, all three exchangeable jaws are at the same low level. Only at a surface pressure of 15 MPa, the coated exchangeable jaws have a higher coefficient of friction compared to the uncoated (just CO2 lubrication). This can be attributed to excessive wear of the surface around the micro-hole (see figure 13). So there was no smooth flat surface on the entire contact area.

However, these results are only partially comparable, since the uncoated replaceable plates have been previously hardened and thus more resistant.

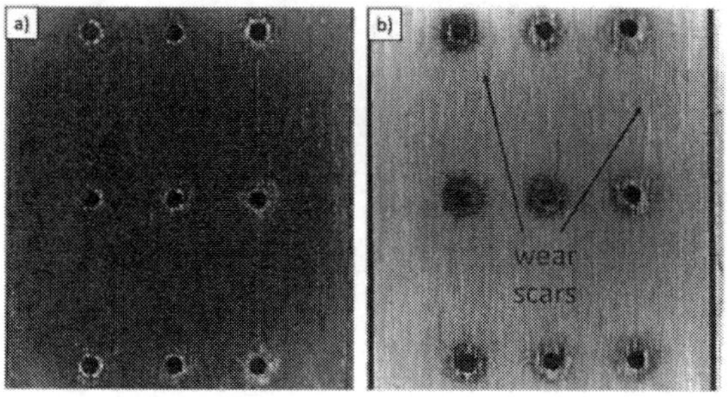

Fig. 13. Wear scars and wear around the mirco-holes at the a-C:H:W/a-C:H coating system (a) and Si_3N_4 coating (b) after strip drawing test.

Figure 13 shows the wear patterns of the applied hard coatings. It can be observed that coating failure took place especially around the mirco-holes. Previous quality investigations of uncoated exchangeable jaws around the mirco-holes at the outlet showed a bulge (see figure 14). This is up to 0.025 mm. This increase significantly influences the coefficient of friction and explains the abrasion of the hard material layers around the mirco-holes. The current object of the investigations at IFSW is to produce the mirco-holes without a material bulge.

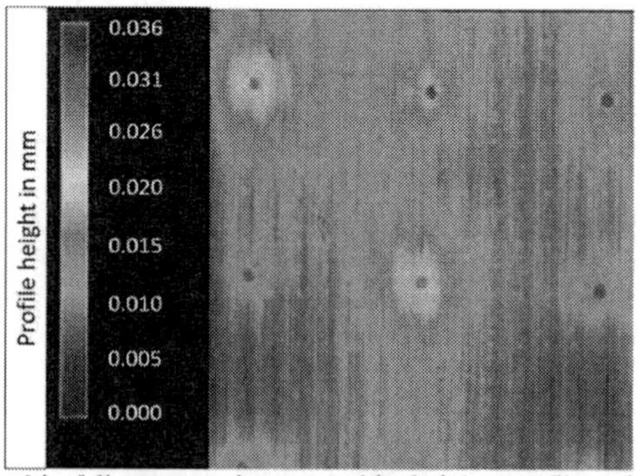

Fig. 14. Microscope images with height profile of the exchangeable drawing jaws show an significant increase around the mirco-holes of up to 0.025 mm (source : IFU; University Stuttgart).

However, additional line-shaped wear scars are found mainly located between the drill holes in drawing direction of the tools. Further verification measurements were carried out on a profilometer determining the roughness profiles and the resulting surface roughness R_a. In addition, measurements of the contact angle were performed to identify the wetting behavior.

The results of the roughness measurements were determined by four single lines whereas each line has a length of about 5 mm at the contact area (perpendicular to the drawing direction). Figure 16 shows representative roughness profile lines measured at the a-C:H:W/a-C:H coating system (a) respectively at the Si_3N_4 coating (b).

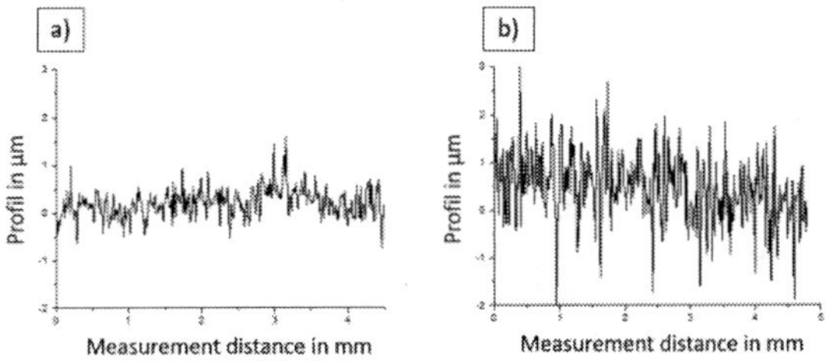

Fig. 15. Roughness R_a after a strip drawing test for a) a-C:H:W/a-C:H coating and b) Si_3N_4 coating.

The mean roughness value for a-C:H:W/a-C:H coating system R_a is between 0.19 and 0.21 µm (Si_3N_4 coating: R_a is between 0.52 to 0.57 µm; 3 measurements per line) after the strip drawing test. The water contact angles before and after the strip drawing test were measured in order to get conclusions about the wear and roughening. At the a-C:H:W/a-C:H coating system the water contact angle drops from 74.98 ± 1.23° to 53.13 ± 2.03° degrees. Whereas with a Si_3N_4 coating, the contact angle of water increases from 52.68 ± 0.96° to 84.47 ± 3,13°. The decrease in an a-C:H:W/a-C:H coating indicates a leveling and shifting of the applied graphite coatings. In contrast, the Si_3N_4 coating must have been roughened, as the surface becomes more hydrophobic. This effect can be attributed to the so-called Cassie-Baxter case. [25] This means that the surface is finely structured and it increases the water contact angle. In this special case, a surface micro-structuring (uniform or irregular) is not desired because it also increases the coefficient of friction.

5 Conclusion

Previous researches has shown that a deep drawing process of a U-profile [21] as well as for a rectangular cup [26] with pure CO_2 as a volatile lubrication without mineral-oil are possible. With an additional coating, no further reduction in the coefficient of friction compared to pure CO_2 lubrication with uncoated tools in a strip drawing test bench was achieved. Especially at 15 MPa surface pressure, the coefficient of friction increases with regard to the uncoated reference. However, a roughening effect takes place especially at the Si_3N_4 coating. This leads to increased wear and subsequently to higher coefficients of friction. However, another important observation is that at both applied coatings the layers are completely or partially worn around the drill holes (compare to figure 13). Despite the additional CO_2 lubrication, this could be an indication that there is no complete surface separation of the DC04 sheet metal and the coated exchangeable tool jaws. Solid state contact around the drill holes would be the consequence which explains the observed wear patterns. Formed wear debris, e. g. removed coating or tool material, at the mirco-holes is ploughed through the coating in strip drawing direction which could be an adequate explanation of the observed wear patterns in figure 13. For this reason, further wear analyzes have to be carried out at the mirco-holes and between by scanning electron microscopy (SEM) and energy dispersive X-ray spectroscopy (EDS) in order to finally clarify the reasons of this wear pattern. Next to that, the quality of flatness of the jaws surfaces in the area of the drill holes has to be quantified by laser confocal microscopy to ensure solid body separation during strip drawing. Also the fine structure of the wear scars have to be measured to derive a link of the dimension of ploughed wear debris. Related to the wear analyzes it should be possible to derive sufficient strategies with regard to the coating development. It needs to be clarified, whether an insufficient coating adhesion strength to the substrate material, a too low coating fatigue resistance or failures of the tool material due to the laser drilling process itself are responsible for such premature wear patterns.

The hybrid approach for dry lubrication with a primary CO_2 lubrication and additional hard material coating, which is presented for the first time in this paper, is initially used exclusively for basic investigations. Further investigations must be carried out for a final evaluation. It can be concluded:
The coatings presented in this purpose could a promising approach. The layer durability must be improved. In addition, the laser drilling process must be material compatibility so that there is no material bulge around the borehole. With regard to initial endurance tests at the IFU, an additional plasma technological layer should be applied to the hardened deep drawing tool.

The financial support of this study by the German Research Foundation (DFG) within the framework of the priority programme *Sustainable Production through Dry Processing in Metal Forming* (SPP 1676) is gratefully acknowledged. The authors thank Christoph Wörz from the Institute for Metal Forming Technology for provision of strip drawing test bench. Thanks also to Andreas Schulz from the Institute of Interfacial Process Engineering and Plasma Technology for assistance in measuring with the profilometer.

References

[1] Gesellschaft für Tribologie (Hrsg.): Tribologie – Verschleiß, Reibung, Definitionen, Begriffe, Prüfung. GfT-Arbeitsblatt 7, Aachen, (2002)

[2] Zum Gahr K.-H. : Microstructure and wear of materials. Tribology series, 10. New York: Elsevier, (1987)

[3] P. Skakun, I. Kacmarcik, T. Pepelnjak, O. Luzanin, A. Inavisevic, M. Milutinovic: Compersion of conventional and new lubricants for cold forming. 34th International conference on production engineering, (2011)

[4] M. Arentoft, N. Bay, P.T. Tang, J.D. Jensen: A new lubricant carrier for metal forming. CIRP Annals - Manufacturing Technology 58,243–246 (2009)

[5] F. Vollertsen, F. Schmidt: Dry Metal Forming: Definition, Chances and Challenges. Int. J. Precision

Engineering and Manufacturing – Green Technology 1/1, 59-62 (2014)
[6] N. Bay, A. Azushima, P. Groche, I. Ishibashi, M. Merklein, M. Morishita, T. Nakamura, S. Schmid, M. Yoshida: Environmentally benign tribo-systems for metal forming. CIRP Annals - Manufacturing Technology 59, 760–780 (2010)
[7] European Parliament, Council REACH, EC Regulation No. 1907/2006 of the European Parliament and of the Council, (2006)
[8] ICFG Doc. 10/95: Lubrication Aspects in Cold Forging of Aluminium and Aluminium Alloys. The International Cold Forging Group. WIRE 1, 1–12 (1996)
[9] Institute of Health and Consumer Protection (IHCP): European Union Risk Assessment Report, alkanes, C10-13, chloro (CAS No. 85535-84-8), (2008)
[10] A. Mousavi, A. Brosius: Track Tool concepts for dry forming. Dry Met. Forming OAJ FMT 1, 79–82 (2015)
[11] M. Liewald, T. Graf, T. Hirth, M. Singer, A. Feuer, E. Zahedi, G. Umlauf: Tribosysteme für die Kaltumformung auf der Basis von flüchtigen Schmiermedien und laserstrukturierten Oberflächen. Dry Met. Forming OAJ FMT 1, 22-33 (2015)
[12] P. C. Priarone, M. Robiglio, L. Settineri, V. Tabeldo: Milling and turning of titanium aluminides by using minimum quantity lubrication. New Production Technologies in Aerospace Industry - 5th Machining Innovations Conference, Procedia CIRP 24, 62 – 67 (2014)
[13] ChemicaLogic Corporation, http://www.chemicalogic.com/Pages/default.aspx (2015)
[14] X. Chen, Z. Peng, Z. Fu, S. Wu, W. Yue, C. Wang: Microstructural, mechanical and tribological properties of tungsten-gradually doped diamond-like carbon films with functionally graded interlayers. Surface and Coatings Technology, Vol. 205, 3631-3638 (2011)
[15] A. Czyzniewski: Optimising deposition parameters of W-DLC coatings for tool materials of high speed steel and cemented carbide. Vacuum, Vol. 66, 2140-2147 (2012)
[16] H. Hasselbruch, M. Herrmann, A. Mehner, H.-W. Zoch, B. Kuhfuss: Development, characterization and testing of tungsten doped DLC coatings for dry rotary swaging. MATEC Web of Conference, Vol. 21, 1-7 (2015)
[17] M. Murakawa, M. Jin, M. Hyashi: Study on semidry/dry wire drawing using DLC coated dies. Surface and Coatings Technology, Vol. 177-178, 631-637 (2014)
[18] H. Hetzner: Systematische Entwicklung amorpher Kohlenstoffschichten unter Berücksichtigung der Anfoderungen der Blechmassivumformung. Dissertation, University of Erlangen, (2014)
[19] H. Hasselbruch, M. Herrmann, A. Mehner, B. Kuhfuss, H.-W. Zoch: Incremental dry forging – Interaction of W-DLC coatings and surface structures for rotary swaging tools. Proceedia Manufacturing, Vol. 8, 541-548 (2017)
[20] M. Liewald, S. Wagner, D. Becker: Ausstattung und Kompetenzen des Tribologie-Labors am Institut für Umformtechnik der Universität Stuttgart. IDDRG-Sitzung Düsseldorf, (2005)
[21] C. Wörz, E. Zahedi, G. Umlauf, M. Liewald, R. Weber, T. Graf, G. Tovar: Tiefziehen eines U-Profils mit flüchtigen Medien als Schmierstoffersatz. Dry Met. Forming OAJ FMT 3, 50-61 (2017)
[22] T. Okumura: Inductively Coupled Plasma Sources and Applications, Hindawi Publishing Corporation Physics Research International, Vol. 2010, (2010)
[23] S. M. Jatta, Plasmaunterstütze Abscheidung von dielektischen Schichten zur Entwicklung von mikromechanisch abstimmbaren optischen Komponenten. Dissertation, TU Darmstadt, (2010)
[24] H. Hasselbruch, A. Mehner, H.-W. Zoch: Influence of plasma nitriding parameters and subsequent surface finishing on the adhesion strength of tungsten doped PVD a-C:H coatings. Proceedings of the 13th International Conference of THE"A" Coatings, Greece, 169-179 (2017)
[25] W. M. Sigmund, SH. Hsu: Cassie-Baxter Model. in Encyclopedia of Membranes, Springer, Berlin, (2016)
[26] C. Wörz, G. Reichardt, M. Liewald, E. Zahedi, R. Weber, T. Graf: Dry deep drawing of a rectangular cup assisted by volatile media injected from laser-drilles mircoholes. Dry Met. Forming OAJ FMT 5, (2018)

Tensile properties and drawability of thin bimetallic aluminum-scandium-zirconium / stainless steel foils and monometallic Al-Sc-Zr fabricated by magnetron sputtering

Julien Kovac[1,*], *Lukas* Heinrich[2], *Bernd* Koehler[1], *Andreas* Mehner[1], *Brigitte* Clausen[1] and *Hans-Werner* Zoch[1, 3]

[1]Leibniz Institute for Materials Engineering IWT, Bremen, Germany
[2]Bremer Institut für angewandte Strahltechnik (BIAS), Bremen, Germany
[3]University of Bremen, Faculty of Production Engineering, Bremen, Germany

Abstract. Al-Sc-Zr alloys are interesting for the production of high strength micro components by micro deep drawing. These alloys show a good hardenability due to the formation of nanometer-scale spheroidal $Al_3(Sc, Zr)$ precipitates, which are highly coherent with the aluminum matrix. However, the formation of these precipitates in Al-Sc-Zr foils fabricated by conventional metallurgical methods dramatically reduces their ductility and drawability. In this work, magnetron sputtering was used to produce Al-Sc-Zr foils and Al-Sc-Zr / stainless steel bimetallic foils which are nearly free of these precipitates. Tensile tests were carried out to measure and compare the mechanical properties of monometallic Al-Sc-Zr foils and bimetallic Al-Sc-Zr / stainless steel foils deposited with varying plasma target powers and containing different volume fractions (layer thickness) of Al-Sc-Zr. Micro deep drawing was used to determine the drawability of selected monometallic and bimetallic foils. The results show that the density of monometallic Al-Sc-Zr foils can be improved significantly by increasing the DC target power and by using the high power impulse magnetron sputtering (HiPIMS) technology, resulting in foils with higher ductility. Bimetallic foils achieved higher strength and ductility than monometallic Al-Sc-Zr foils. Their mechanical properties vary with the target power and the volume fraction (thickness) of Al-Sc-Zr. The limit drawing ratio of HiPIMS deposited monometallic foil was 1.7 or 1.8 depending on the side of the foil facing the die, whereas a limit drawing ratio of 1.9 was observed for bimetallic foils.

Keywords: Micro forming, Sheet metal, PVD-coating

1 Introduction

Micro deep drawing is an efficient cold sheet forming process for the mass production of hollow microscopic parts with complex shape, which are widely used in micro system technologies (MST) and micro electro-mechanical systems (MEMS) such as connector pins, resistor caps, lead chip frames or parts of pressure sensors [1]. The principle of micro deep drawing is essentially similar to the conventional (macroscopic) deep drawing process: a metallic sheet is maintained by a blank holder and put in shape through the action of a punch into a die. However, the microscopic dimensions of the finished parts require the use of thin foils with a thickness generally below 50 μm instead of sheets of a few millimeters thickness. This thickness reduction leads to the occurrence of size effects which greatly influence the mechanical and tribological behavior of the drawn foil [2]. Therefore, a considerable research effort is made to improve the efficiency and reliability of micro deep drawing and of other micro forming processes [3]. Foil materials employed in micro deep drawing process are quite restricted to a few types of highly available cold rolled foil strips such as pure copper, pure aluminum or austenitic stainless steel [4] although studies were conducted to investigate the drawability of other special materials [5, 6].

Among all the typical alloying elements in aluminum alloys, scandium is one of the most effective elements to improve the mechanical properties. Indeed, the addition of scandium into the aluminum matrix results in the formation of nanometer-scale Al_3Sc precipitates, which lattice parameters and structure are close to those of the aluminum matrix [7]. This low mismatch ensures a high coherency of the precipitates with the aluminum matrix. Because of their high coherency and dispersivity, the Al_3Sc precipitates efficiently oppose the mobility of dislocations and grain boundaries, which considerably increases the hardness and the yield strength as well as the recrystallization temperature of the alloy [8]. Beside

* Corresponding author: kovac@iwt-bremen.de

these effects on the mechanical properties, it has also been reported that scandium also improves the corrosion resistance [9] and the weldability [10] of aluminum. Nevertheless, the hardness of Al-Sc alloys also depends on the size of the Al_3Sc precipitates. During heat treatment, the precipitates may grow up to a critical radius at which, the interface between the precipitates and the matrix becomes incoherent [11]. Consequently, dislocations and grain boundaries are less efficiently pinned, and the hardness of the alloy starts to decrease.

This issue can be addressed by the addition of zirconium to the Al-Sc alloys. As a result, small amounts of zirconium are incorporated into Al_3Sc precipitates. $Al_3(Sc_{1-x}Zr_x)$ precipitates are more stable and coarsen slower due to the lesser diffusivity of Zr in aluminum [12]. Moreover, substantial augmentations of strength have been reported in Al-Sc-Zr alloys compared to zirconium free Al-Sc alloys [13].

Due to their excellent mechanical properties, Al-Sc-Zr alloys can be employed for deep drawing high strength micro components for mechanically demanding applications. For instance, body parts of micro valves may be submitted to large fluid or contact pressures from other parts of the valve and therefore, the body has to sustain relatively large elastic deformations and avoid permanent plastic deformation. Due to the combination of high yield strength and low Young's modulus, Al-Sc-Zr, can achieve larger elastic deformations than austenitic stainless steel. Moreover Al-Sc-Zr alloys show a higher resistance to fatigue then other conventional aluminum alloys [14] and may thus be used to improve the reliability of micro parts, since a considerable demand exists in the industry [15].

Drawing high strength Al-Sc-Zr alloys is however, not an easy task since precipitates can dramatically reduce the ductility and thus, the formability of these alloys. Furthermore, the formation of precipitates is difficult to avoid in cast alloys, due to the low maximum solubility of Sc and Zr in aluminum. Therefore, alternative methods were developed to prepare thin precipitates free Al-Sc-Zr foils for micro deep drawing. In previous investigations magnetron sputtering was successfully used to prepare freestanding Al-Zr foils [16]. Their microstructure consisted of a nearly precipitate free oversaturated aluminum solid solution which could be hardened by heat treatment after deep drawing [17]. The limit drawing ratio of these foils was 1.7, which is comparable to conventional pure aluminum foils [18]. This was less than other materials like copper or stainless steel, which drawability is higher.

One possible way to increase the formability of aluminum or aluminum alloy foils would be to bond these foils to materials with better formability like stainless steel and thus making bimetallic foils instead of monometallic foils. Bimetallic sheets are composed of a stacking of layers of two different materials. Usually one of the materials is hard and brittle, as the other is soft and ductile and the mechanical properties of bimetallic sheets are a compromise of those of the components. More precisely, it was shown that some properties of a bimetallic sheet can be deduced from the rule of mixtures, which means that the property of the bimetallic sheet is an average value of the properties of the respective components weighted by their volume fractions. Hence, in aluminum / stainless steel bimetallic sheets, the apparent elastic modulus, the tensile strength and the plastic strength coefficient (according to Hollomon's law) follow the rule of mixtures, whereas the uniform elongation and the strain hardening exponents could be deduced by a force weighted law of mixtures [19]. In every case, a significant improvement of strength and ductility compared to pure aluminum was observed. The deep drawing behavior of aluminum / stainless steel bimetallic sheets was investigated by various authors. Parsa et al. investigated the evolution of the limit drawing ratio in single drawing and redrawing for bi-layered sheets with various proportions of aluminum and stainless steel through simulation and experiments [20]. They found that with increasing steel layer thickness (i. e. decreasing aluminum layer thickness) the limit drawing ratio increased. However, the limit drawing ratio depended on the experimental set-up and was lower when the aluminum layer faced toward the punch. A similar behavior was obtained by simulation by Takuda et al. [21] and Afshin et al. in warm deep drawing [22].

To the best of our knowledge, up to now only thick (around 1 mm) bimetallic sheets were obtained by the conventional methods like roll bonding [23]. In this study, direct current magnetron sputtering (DCMS) and high-power impulse magnetron sputtering (HiPIMS) are used to produce 30 µm thick monometallic Al-Sc-Zr foils and two layer bimetallic Al-Sc-Zr / stainless steel foils under various conditions. Tensile testing was carried out to compare their mechanical properties and micro deep drawing experiments were made to compare their drawability.

2 Experimental methods

2.1. Preparation of Al-Sc-Zr monometallic foils and Al-Sc-Zr / stainless steel bimetallic foils by magnetron sputtering

Different monometallic Al-Sc-Zr foils and bilayer Al-Sc-Zr / steel foils were produced by DCMS and HiPIMS. For the production of thin ($\approx$ 30 µm) monometallic Al-Sc-Zr foils, different Al-Sc-Zr alloys with $\approx$ 1.8 mass-% Sc and $\approx$ 0.25 or 0.6 mass-% Zr were deposited onto 100 µm polished sheets (1300 × 150 mm²) of low carbon steel. The steel substrates were removed by etching in concentrated nitric acid after the deposition process. The nitric acid did not attack the remaining freestanding Al-Sc-Zr foils. Nevertheless hydrogen absorption and diffusion might affect the mechanical properties of the foils. For the production of bimetallic foils, Al-Sc-Zr coatings with thicknesses between 5 and 15 µm were deposited onto commercial austenitic stainless steel foils (X5CrNi18-10: 8 mass-% Cr, 10 mass-% Ni and 0.05 mass-% C). The length of the stainless steel strips was 1.3 m and the width was 50 mm. The thickness of the substrate strips was chosen between 15 µm, 10 µm and 25 µm according to the Al-Sc-Zr coating thickness, so that the thickness of the resulting bimetallic foils was always close to 30 µm.

The deposition process was carried out in a commercial CemeCon CC800/9 magnetron sputtering unit fitted with four DC cathodes and two HIPIMS cathodes. For direct current magnetron sputtering (DCMS) processes, three of the DC cathodes were fitted with aluminum targets alloyed with 2 mass-% Sc and the remaining DC cathode was fitted with an aluminum target alloyed with 2 mass-% zirconium as depicted in Figure 1a. For HiPIMS processes the two HiPIMS cathodes were fitted with two copper-bonded aluminum targets alloyed with 1.5 mass-% Sc 0.5 mass- % Zr. (Figure 1b). The dimensions of the targets were 488 × 88 × 9 mm³. A homogeneous alloying element distribution is ensured by setting the different substrate foils around a rotating substrate table (Figure 2), which rotation speed was set to 1 rotation per minute. The distance between the different targets and the substrate was in every case about 50 mm.

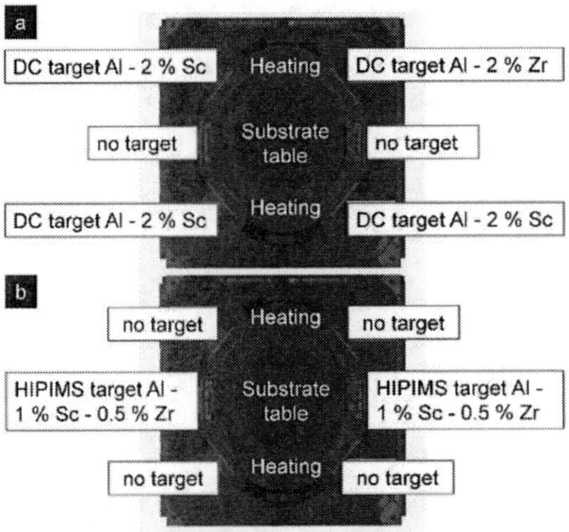

Fig. 1. Target setup in the magnetron sputtering unit for the production of Al-Sc-Zr monometallic foils and Al-Sc-Zr / stainless steel bimetallic foils: a) setup for DCMS, b) setup for HIPIMS.

All deposition processes were carried out as follows: First, the different substrates were cleaned with ethanol and fixed to the substrate table. Then, the deposition chamber was evacuated up to a residual air pressure of about 5 10^{-7} mbar. Then, the substrate was heated for 2 hours to evaporate traces of humidity. Next, a flow of 250 sccm of argon and 25 sccm of krypton has been let into the chamber and the substrate table was biased with a middle frequency (MF) voltage of 160 V at a frequency of 650 kHz. Consequently, plasma is created, and the substrate is bombarded with accelerating argon and krypton ions, which remove impurities from the substrate surface. This ion etching phase was carried out for 1 hour. Afterwards, the substrate was cooled down for 3 hours with the same constant flow of argon. Finally, for the deposition process, the argon pressure was set to 3 10^{-4} mbar. The different cathodes were biased with negative voltage, whereas the substrate and the rest of the unit parts were earthed. Plasma of positive argon ions and electrons was created, and argon ions were accelerated to the target which results in the sputtering (vaporization) of the target atoms which diffuse through the chamber and condense on the substrate. The coating thickness was varied by adapting the deposition time.

Fig. 2. Picture of the rotating substrate table in the magnetron sputtering unit fitted with a polished steel sheet substrate.

With this setup three different monometallic Al-Sc-Zr foils and 5 different Al-Sc-Zr / stainless steel bimetallic foils were produced. Among the three monometallic foils, two foils were deposited with DC power with target powers at respectively 2 kW for 18000 seconds and 4 kW for 9000 seconds. In this case a constant voltage is continuously applied. The third variant was deposited with HiPIMS. The average target power was set up to 2 kW and the deposition time was 50 000 seconds. In this case the voltage varies periodically. During off-phase the target is not biased, and the current is loaded in a capacitor and released during the duty phase. The time of the duty cycle was 200 µs (20 % of the period) and the frequency was 1000 Hz. Three of the bimetallic foils consist of 5 µm of Al-Sc-Zr and 25 µm of stainless steel. The Al-Sc-Zr layer deposition was done with similar target powers (respectively 2 kW DC, 4 kW DC and 2 kW HiPIMS) as for the monometallic foils. Two supplemental bimetallic foils were deposited at 4 kW DC with different layer thicknesses, respectively 10 µm Al-Sc-Zr / 20 µm stainless steel and 15 µm Al-Sc-Zr / 15 µm stainless steel. The characteristics of the different monometallic and bimetallic foils are summarized in Table 1.

The chemical composition of the different Al-Sc-Zr foils was measured by glow discharge optical emission spectroscopy GDOES (LECO GDS-750A). The results are summarized in table 2.

Table 1. Data of the different monometallic Al-Sc-Zr foils and of the different Al-Sc-Zr / stainless steel bimetallic foils.

Specimen	Al-Sc-Zr layer thickness [µm]	Stainless steel layer thickness [µm]	Target power and deposition time
MDC2	28.8	0	DC 2 kW 18000 s
MDC4	26.8	0	DC 4 kW 9000 s
MHP	28.7	0	HIPIMS 2 kW 50000 s
BDC2	5	25	DC 2 kW 3168 s
BDC4a	5	25	DC 4 kW 1677 s
BDC4b	10	20	DC 4 kW 3378 s
BDC4c	15	15	DC 4 kW 5000 s
BHP	5	25	HIPIMS 2 kW 8881 s

Table 2. Chemical compositions of the different monometallic Al-Sc-Zr foils.

Foil	mass-% Al	mass-% Sc	mass-% Zr
MDC2	Balance	1.82	0.26
MDC4	Balance	1.87	0.25
MHP	Balance	1.78	0.57

2.2 Determination of the tensile properties and other characteristics

The tensile properties of the different monometallic and bimetallic foils and of a 30 µm thick stainless steel foil were determined using special tensile test specimens which were produced by cutting each foil into 10 rectangular pieces (50 × 10 mm²). The ends of the foil pieces were reinforced by sticking four pieces of 200 µm of steel as illustrated in figure 3. The testing unit was an INSTRON E1000 tensile tester which is adapted for the measurement of low thickness specimens. During testing, load and strain were measured by a calibrated 1000 N load cell meeting the requirements of ISO 7500-1 and ASTM E4 and a video extensometer, respectively. Additional details about the unit can be found in [24]. The average values of elastic modulus, yield strength, tensile strength and fracture strain of each foil variant were determined from 10 specimens.

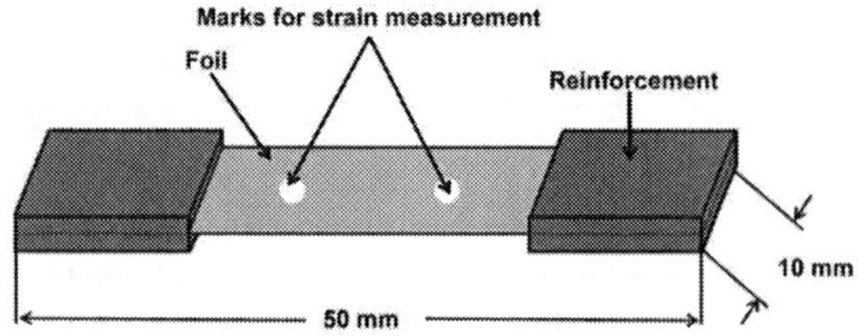

Fig. 3. Sketch of a tensile specimen.

The morphology of the foils was investigated by scanning electron microscopy (SEM) (Camscan MV2300, EO Scan). Transmission electron microscopy was used to determine the presence of precipitates (Phillips CM12).

2.3 Determination of the formability in micro deep drawing

The geometric data of the tools are summarized in table 3. A sketch of the used setup can be seen in Figure 4. The deep drawing process was carried out on a single axis micro forming press with a maximum punch force of 500 N and a constant punch velocity of 10 mm/s using HBO 947/11 as lubricant. The punch force was measured with a Kistler 9217A piezo load cell with an accuracy of 0.01 N. Punch displacement was measured with a Heidenhain LS477 linear scale with an accuracy of 1 µm. The press is driven by a servo motor controlled by a NI 9514 servo drive interface. The blank holder acts passively. It uses its own weight and is supported by two springs. Blank holder pressure can be adjusted by changing the spring tension and was set to P_{NHD} = 5 MPa. The blank positioning was realized by an automated positioning system. The blank was positioned with a pneumatic gripper that is driven by a linear direct drive cross table. The position of the blank was then measured using an Allied Vision G 917 B monochrome CCD camera with a resolution of 9 Mpx equipped with a telecentric lens with built-in coaxial illumination and a magnification of 0.75. With this setup the blanks can be positioned within a radius of 10 µm from the center of the drawing die.

Table 3. Specifications of tool geometry used in experiment.

Punch diameter (mm)	1.005
Punch radius (mm)	0.110
Drawing die diameter (mm)	1.061
Average drawing die radius (mm)	0.077
Drawing gap (mm)	0.028

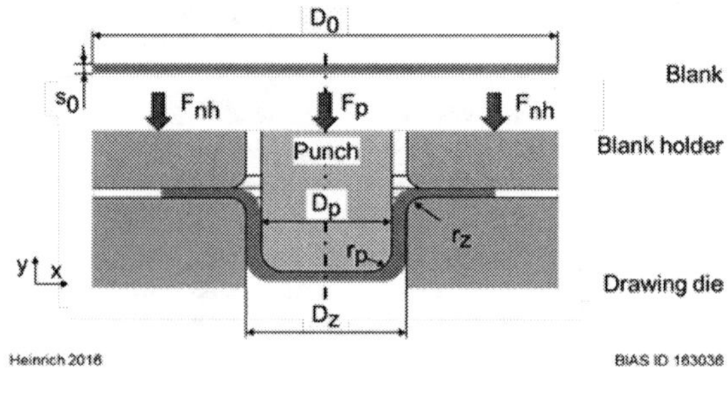

Fig. 4. Sketch of the deep drawing setup.

3 Results and discussion

3.1 Properties of monometallic Al-Sc-Zr foils and Al-Sc-Zr / stainless steel bimetallic foils

3.1.1 Morphological and microstructural characteristics of monometallic and bimetallic foils.

 https://doi.org/10.1051/matecconf/201819015001

The morphology of the different monometallic foils after deposition and separation from substrate has been investigated by scanning electron microscopy. Pictures of both the cross section and free surface of the different monometallic foils are shown in Figure 5.

The foils show the classical columnar morphology of sputter deposited coatings [25]. This implicates that monometallic sputtered Al-Sc-Zr foils present two different surfaces. The bottom surface, which was in contact with the substrate, usually consists of tiny close packed amorphous grains, and this surface reflects the topology of the substrate. At the top surface (in Figure 5) or free surface, the columnar crystallites grow freely and the tapered shape of these crystallites results in a rough surface. The foil sputtered with 2 kW DC power (MDC2) shows a relatively dense morphology in the bottom region near the substrate contact surface. About 10 µm beneath the free surface however, the boundaries between the columnar crystallites are more open voided and the foil is more porous. In comparison, the foil produced at 4 kW (MDC4) and the HiPIMS deposited foil (MHP) are overall denser since porosities are less frequent and confined to the near free surface region. This observation was confirmed by optical microscopy observations of the cross sections of the foils which are featured on Figure 6.

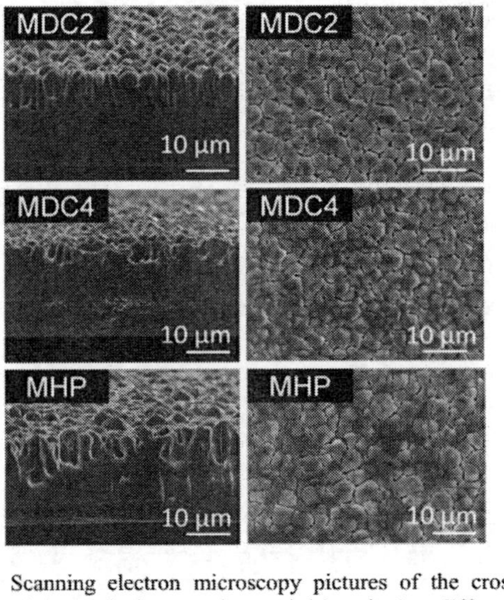

Fig. 5. Scanning electron microscopy pictures of the cross section (left) and free surface (right) of the different monometallic Al-Sc-Zr foils.

This increase of density is to be related to the higher kinetic energy of the sputtered atoms when the target power is at 4 kW. Indeed, when the target power increases, the voltage at the target increases as well and the argon ions striking the targets have a higher kinetic energy. This higher kinetic energy is also transmitted to the sputtered atoms. The mobility of the atoms impinging on the coating surface, is higher and the voids in the coating are more likely to be filled [26]. The same mechanism can be applied to the HiPIMS deposited foil (MHP). Indeed, in high power impulse magnetron sputtering, the average value of the target power is only 2 kW, while the energy of the pulse may reach more than 10 kW [26].

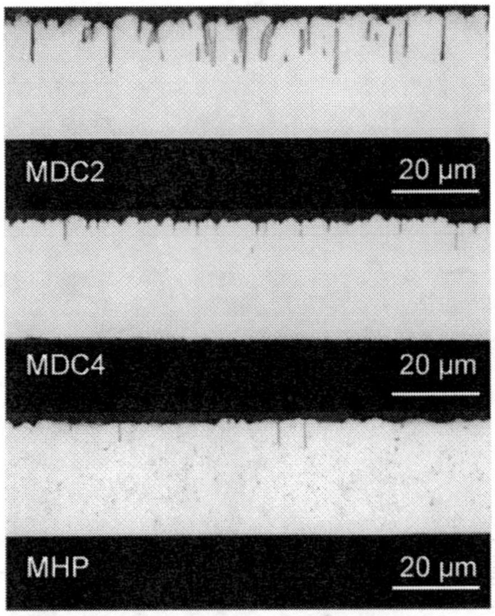

Fig. 6. Optical microscopy pictures of the cross sections of the different Al-Sc-Zr monometallic foils after embedding and polishing.

Compared to monometallic foils, scanning electron microscopy observations carried out on bimetallic foils (Figure 7) show a finer and denser columnar morphology. This morphology can be explained by the thinner Al-Sc-Zr coating thickness in the bimetallic foils. The diameter of the columnar crystallites increases when the coating grows. Porosities are the result of shadowing effect from already formed crystallites, which prevent impinging atoms to cover homogeneously the whole surface. Since the top diameter of crystallites is smaller in thin coatings, it can be assumed that shadowing effects are less severe for thin coatings, which results in les surface porosities than in thick coatings.

Transmission electron microscopy was used to determine the phase content of the Al-Sc-Zr foils after deposition and after heat treatment at 500 °C for 1 hour for comparison purpose. Pictures and the corresponding electron diffraction patterns taken according to {1 0 0} zone axis are featured in Figure 8. After deposition, no Al_3(Sc, Zr) precipitates are visible on the picture and only reflections corresponding to the face centered cubic lattice of the α-aluminum phase are to be found on the diffraction pattern. However, after heat treatment, precipitates with diameters between 30 and 40 nm are clearly visible on the picture and besides the α-aluminum; further reflections can be identified on the pattern. These have a structure in agreement with the simple cubic $L1_2$ structure of Al_3(Sc, Zr) precipitates. These results confirm that Al-Sc-Zr foils elaborated by magnetron sputtering process consist of an oversaturated α-aluminum solution as for Al-Sc and Al-Zr alloys in a previous work [16].

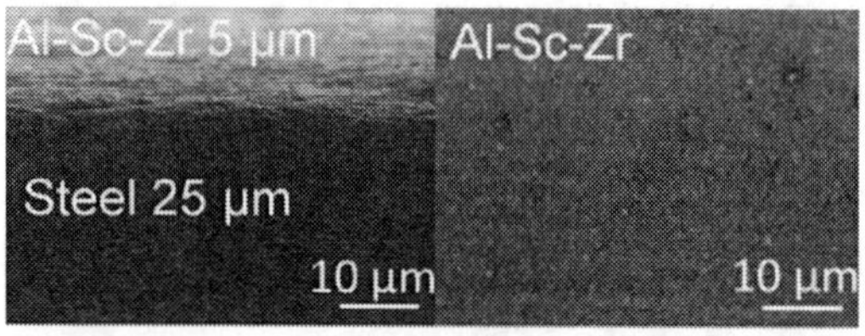

Fig. 7. Scanning electron microscopy pictures of the cross section (left) and free surface (right) of a bimetallic Al-Sc-Zr (5 µm) / Steel (25 µm) foil.

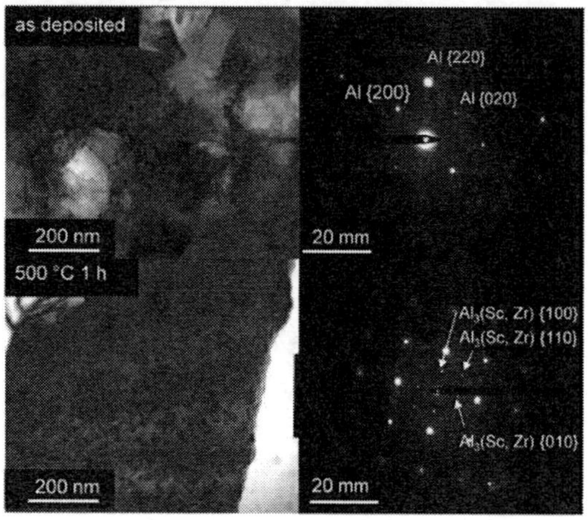

Fig. 8. Transmission electron microscopy pictures (left) and electron diffraction patterns (right) of a sputtered Al-Sc-Zr foil after deposition and after heat treatment for 1 h at 500 °C. Camera constant: 4,233 nm×mm, zone axis {1 0 0}.

3.1.2 Tensile properties of monometallic Al-Sc-Zr foils.

The mechanical properties of the monometallic Al-Sc-Zr foils produced with different DC and HIPIMS powers were determined by tensile tests. Various mechanical properties, namely the Young's modulus (E), the 0.01 % offset yield strength ($R_{p0.01}$), the ultimate tensile strength (R_m) as well as the deformation at fracture (A) were extracted from the data. The average values for 10 specimens for each foil are summarized in Table 4. Selected tensile tests curves of each foil are featured in Figure 9.

A foil sputtered at 2 kW DC power (MDC2) shows a relatively moderate tensile strength of 292 MPa and fails at only 0.58 % strain. Furthermore, its elastic modulus is lower than the expected value of 70 GPa for aluminum. These poor mechanical properties are explained by the high porosity of the foil as shown in Figures 4 and 5. When the DC power is increased up to 4 kW (MDC4), the kinetic energy of the impinging species increases as well as the mobility of the atoms on the coating surface and the porosities are more efficiently filled [26]. The denser structure of this foil improves its ductility, tensile strength and Young's modulus, which values reach respectively 1.45 %, 314 MPa, 400 MPa and 69 GPa. The strain at fracture remains however under 2 %, which can be related to the high yield and tensile strength of the foil. This higher strength is also the consequence of the higher energy of the vapor atoms. Indeed, the higher energy of the vapor atoms may also results in impulse driven displacements in the coating and therefore increases the density of dislocations and points defects of the coating [28]. The foil prepared with HiPIMS process shows by contrast much more interesting characteristics. Its average ductility is much higher than in the DC foils. Young's modulus is higher than in the foil MDC2. The yield and tensile strengths are nevertheless weaker than in the other foils. In the used HiPIMS process the peak power may reach 10 kW which is far more than in the DCMS process. According to the above mentioned mechanism a higher density is expected too.

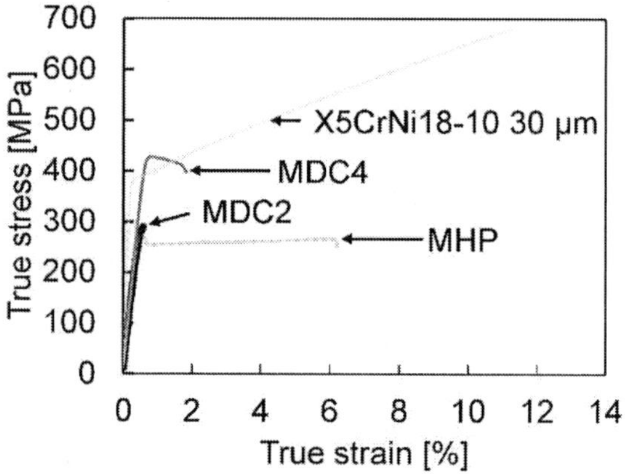

Fig. 9. Selected tensile test curves of each of the produced Al-Sc-Zr foil and of a commercial 30 µm stainless steel foil.

Table 4. Average values and standard deviations of the mechanical properties of Al-Sc-Zr foils and of a 30 µm stainless steel foil.

Foil	E [GPa]	$R_{p0.01}$ [MPa]	R_m [MPa]	A [%]
MDC2	58 ± 4	269 ± 42	292 ± 63	0.6 ± 0.2
MDC4	69 ± 2	314 ± 29	400 ±38	1.5 ± 1.3
MHP	65 ± 1	230 ± 9	260 ±12	5.7 ± 3.9
Steel	197 ± 25	355 ± 13	684 ± 41	11.5 ± 1.8

3.1.3 Tensile properties of bimetallic Al-Sc-Zr / stainless steel foils.

In total, five different bimetallic foils were produced. Three of them (BDC2, BDC4a and BHP) consist of a 25 µm thick stainless steel substrate foil coated with 5 µm of sputtered Al-Sc-Zr and were produced in the same conditions as the three monometallic Al-Sc-Zr foils in order to properly compare the mechanical properties of bimetallic foils to those of monometallic foils and to check if the changing mechanical properties of the differently deposited Al-Sc-Zr layers also affect the properties of the Al-Sc-Zr / stainless steel bimetallic foils. In two other bimetallic foils (BDC4b and BDC4c) the stainless steel and Al-Sc-Zr layer thicknesses were varied at respectively 10 µm of Al-Sc-Zr for 20 µm stainless steel and 15 µm of Al-Sc-Zr for 15 µm of stainless steel. These bimetallic foils were produced in the same conditions as the foil BDC4a so that the validity of the law of mixture can be tested.

The mechanical properties of the different bimetallic foils were evaluated by tensile testing in the same way as for monometallic Al-Sc-Zr foils. Exemplary tensile test curves of each of the considered bimetallic foils are featured on Figure 10. For comparison purpose, exemplary tensile test curves of the monometallic Al-Sc-Zr sputtered at 4 kW DC and of a 30 µm thick stainless steel foil were also featured on the same figure. The average values and standard deviations of the mechanical properties of the bimetallic foils over 10 tensile specimens are summarized in Table 5.

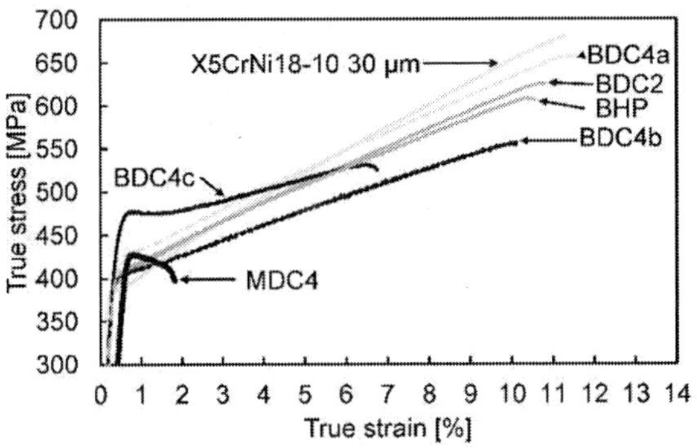

Fig. 10. Selected tensile test curves of each of the produced Al-Sc-Zr / stainless steel bimetallic foils, of a commercial 30 µm stainless steel foil and of a monometallic Al-Sc-Zr foil sputtered at 4 kW DC (MDC4)

The bimetallic foils BDC2, BDC4a and BHP sputtered at respectively 2 kW DC, 4 kW DC and 2 kW HiPIMS show much higher apparentYoung's modulus, deformations at fracture, yield and tensile strengths than the three monometallic Al-Sc-Zr foils produced with similar parameters (MDC2, MDC4, MHP). Furthermore, by comparing the properties of these three bimetallic foils, one can notice that apparent Young's modulus and the tensile strength of these bimetallic foils vary similarly as Young's modulus and tensile strengths of their corresponding monometallic foils. Hence. by increasing the target power from 2 to 4 kW DC, both the apparent elastic modulus and tensile strength increase. The bimetallic foil BHP shows an intermediary apparent Young's modulus of 180 GPa and a lower tensile strength than the bimetallic foils BDC2 and BDC4a, which is also the case for the monometallic foil MHP compared to MDC2 and MDC4. This similar trend between monometallic and bimetallic foils seems less evident for the yield strength, whereas for the deformation at fracture, the trend observed for monometallic foils is not reproduced in bimetallic foils. In each bimetallic foil, the deformation at fracture approximates the deformation at fracture of the monometallic stainless steel foil (Table 4).

The comparison of the mechanical properties of the bimetallic foils BDC4a, BDC4b and BDC4c sputtered at 4 kW DC and containing Al-Sc-Zr layers of respectively 5 µm, 10 µm and 15 µm, shows as expected, that the apparent Young's modulus, the yield and tensile strength of the bimetallic foils increase when the volume fraction (ratio of the Al-Sc-Zr layer thickness to the total thickness of the bimetallic foil) of Al-Sc-Zr in the bimetallic foil decrease or reciprocally when the volume fraction of stainless steel increases. The evolution of the mechanical properties in bimetallic foils with the volume fraction of components can be predicted by the rule of mixtures [19]:

$$P_{AB} = V_A P_A + V_B P_B \quad (1)$$

With V_A and V_B designating the volume fractions of component A and B in the bimetallic foil and P_A and P_B representing a particular mechanical characteristic of components A and B like Young's modulus, yield strength, strain at fracture or the ultimate tensile strength. The validity of the law of mixture for the prediction of respectively apparent Young's modulus, the deformation at fracture, the yield and the ultimate tensile strength of the produced Al-Sc-Zr / stainless steel bimetallic foils has been tested by using the data of Tables 3 and 4. Therefore, the theoretical evolution of the different characteristics of the bimetallic foils were predicted from the measured characteristics of monometallic Al-Sc-Zr and stainless steel foils (Table 3) by taking their standard deviations in account, which results in the different plots featured on Figure 10. The experimental characteristics measured in bimetallic foils BDC4a, BDC4b and BDC4c containing respectively 16.7, 33.3 and 50 % of Al-Sc-Zr were also featured on the different diagrams on Figure 11.

As shown, the rule of mixtures is statistically in quite good agreement with the experimentally extracted yield and tensile strengths of the produced bimetallic foils. Apparent Young's modulus shows a slightly positive deviation from the rule of mixtures, which can be attributed to the higher density of the Al-Sc-Zr layer (Fig. 7) in bimetallic sheet due to the lower thickness of the Al-Sc-Zr coating in bimetallic foils, as previously explained. Finally, the experimental values of the deformation at fracture are not in good agreement with the rule of mixtures.

https://doi.org/10.1051/matecconf/201819015001

Table 5. Average values and standard deviations of the mechanical properties of Al-Sc-Zr / stainless steel bimetallic foils.

Foil	E [GPa]	$R_{p0.01}$ [MPA]	R_m [MPa]	A [%]
BDC2	175 ± 17	323 ± 17	626 ± 35	12 ± 1.3
BDC4a	192 ± 8	339 ± 12	662 ± 22	12.6 ± 1.1
BDC4b	177 ± 14	327 ± 9	548 ± 10	12.1 ± 3
BDC4c	148 ± 25	329 ± 43	526 ± 21	11.1 ± 8
BHP	180 ± 19	347 ± 26	572 ± 21	10.8 ± 1.2

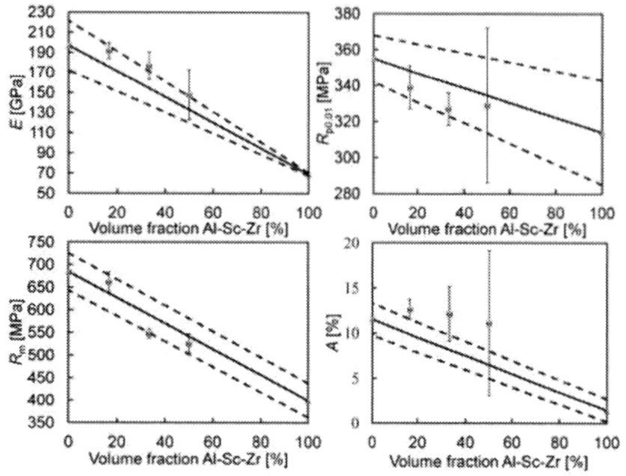

Fig. 11. Comparison of the experimental values of apparent Young's modulus, the yield and tensile strengths and deformation at fractures measured in bimetallic foils BDC4a, BDC4b and BDC4c with those predicted by the law of mixtures. Full line: law of mixture calculated from the average values for pure Al-Sc-Zr and stainless steel in Table 3. Dotted lines: standard deviations of the law of mixtures. Square marks: experimental points.

Broken tensile test specimens of the different bimetallic foils were examined with scanning electron microscopy after tensile testing. As depicted in Figure 12, the Al-Sc-Zr coating presents no crack indicating a premature coating failure during the tensile testing of the bimetallic foils, despite being submitted to deformations well above the deformation at fracture of monometallic Al-Sc-Zr foils. In other terms, this proves that lamination dramatically improve the ductility of Al-Sc-Zr foils. According to Choi et al., this improvement of mechanical properties is due to warping of the tensile specimen during tensile testing as a consequence of lateral stress resulting from the different contraction behaviors of aluminum and steel [29]. By making the assumption that the aluminum and steel layers are strongly bonded together and that consequently, the axial strain in both layers is equal, these authors could show the validity of the rule of mixture for the prediction of the mechanical properties of bimetallic sheets.

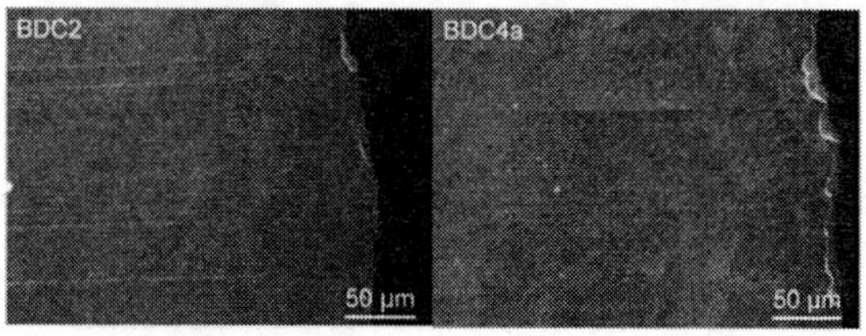

Fig. 12. Scanning electron microscopy pictures of the Al-Sc-Zr free surface of the bimetallic foils BDC2 and BDC4a after tensile testing.

3.2 Micro deep drawing

Deep drawing experiments were carried out on monometallic Al-Sc-Zr foils and a bimetallic Al-Sc-Zr / stainless steel foil produced by HiPIMS in order to determine the drawability of the foils. As shown in the exemplary process window in Figure 13 a limiting drawing ratio of 1.9 was reached by the bimetallic Al-Sc-Zr / stainless steel foil (BHP). In this case the stainless steel side was facing the punch.

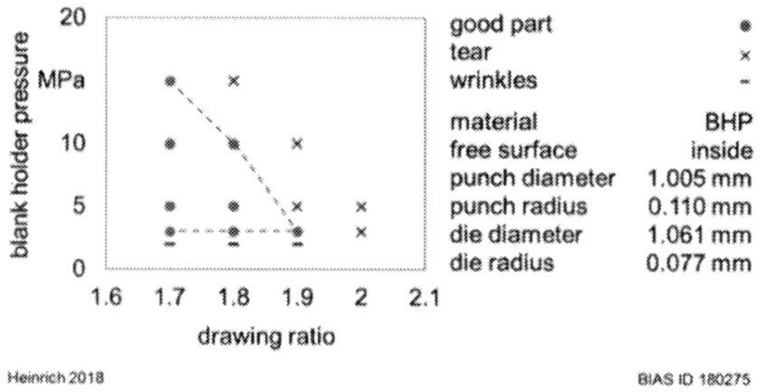

Fig. 13. Process window of a bimetallic Al-Sc-Zr / stainless steel foil determined in micro deep drawing experiments

A comparison of all deep drawn materials is shown in Figure 14. In case of the bimetallic Al-Sc-Zr / stainless steel foil (BHP) a limiting drawing ratio of 1.9 is achieved independently of the side facing the punch. However, in the monometallic Al-Sc-Zr (MHP) the limiting drawing ratio decreases from 1.8 to 1.7 if the free surface is facing away from the punch.

The maximum punch force achieved during deep drawing experiment is shown in Figure 15. The drawing ratio was set to 1.7 and a blank holder pressure of 5 MPa was used. If the the monometallic Al-Sc-Zr (MHP) is used, no significant difference can be found depending on the side facing the punch. In the case of the bimetallic Al-Sc-Zr / stainless steel foil (BHP) foil, the punch force decreases from 52 N to 45 N if the steel side is facing the die.

https://doi.org/10.1051/matecconf/201819015001

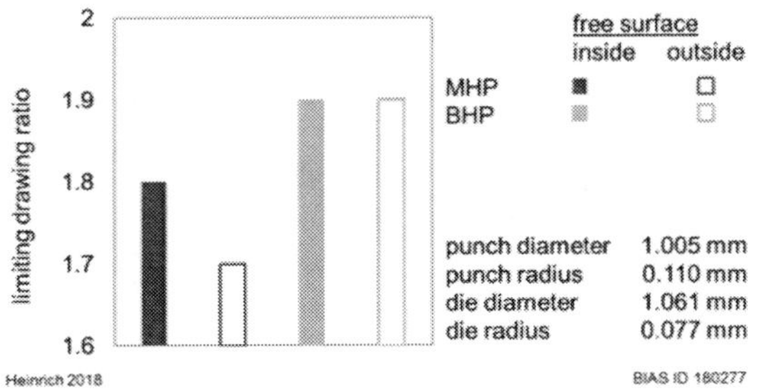

Fig. 14 Limiting drawing ratios determined in micro deep drawing experiments monometallic Al-Sc-Zr foils and a bimetallic Al-Sc-Zr / stainless steel foil produced by HiPIMS

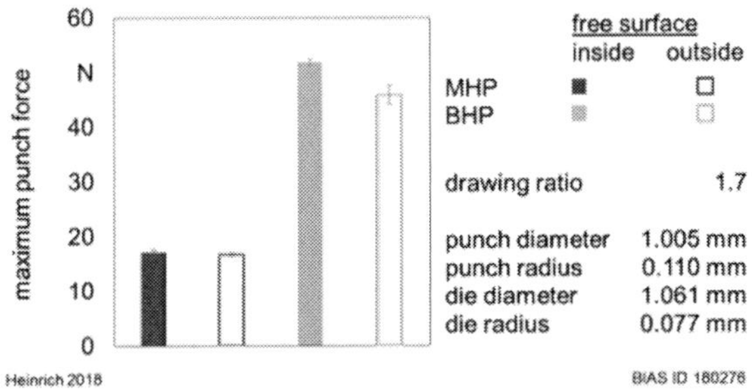

Fig. 15 Maximum punch forces determined in micro deep drawing experiments of monometallic Al-Sc-Zr foils and a bimetallic Al-Sc-Zr / stainless steel foil

Figure 16 shows a comparison between the inner and outer surface of two cups made of monometallic Al-Sc-Zr (MHP). In the first case (a and c) the free surface was facing the punch. Both inner and outer surface are free of cracks. If the free surface is facing the die (b and d) small cracks can be found on the outer surface. As the maximum punch force does not change in case of the monometallic Al-Sc-Zr (MHP) foil, the difference in limiting drawing ratio depended on the side that is facing towards the die cannot be explained by varying friction. It can however be seen that the porosity has significant influence on the process. If the more porous free surface side is facing towards the die, tears along the initial porosity appear at the area of the punch radius causing the part to fail at a lower limiting drawing ratio. If the bimetallic Al-Sc-Zr / stainless steel foil (BHP) is used, no negative influence on the limiting drawing ratio can be found and the deep drawn cups achieve the same limiting drawing ratio as the substrate material tested in [4].

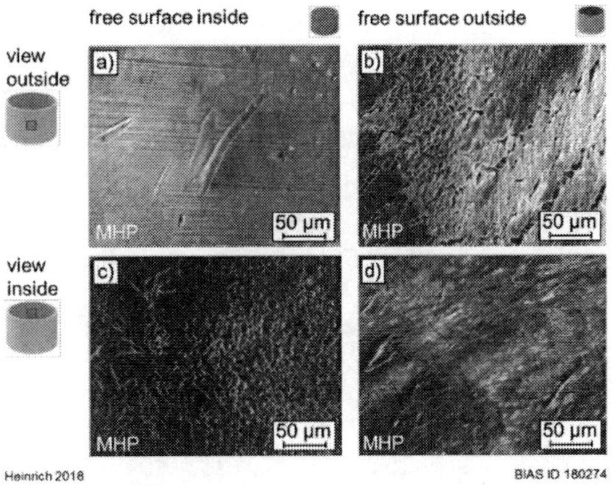

Fig. 16. Surface textures of cups made of MHP monometallic foils with a drawing ratio of 1.7 near the bottom edge.

4 Summary

In this study, different Al-Sc-Zr monometallic foils and Al-Sc-Zr / stainless steel bimetallic foils were fabricated by using direct current (DC) and high power impulse magnetron sputtering (HiPIMS). Their mechanical properties were investigated by tensile testing. The results of this investigation have shown that the foils sputtered by DCMS show surface porosities, reducing their ductility. Denser foils with better ductility were obtained by increasing the DC target power or by using HiPIMS. Bimetallic Al-Sc-Zr / stainless steel foils show in comparison to monometallic Al-Sc-Zr foils higher yield and tensile strengths and a much higher ductility. By varying the target power and the sputtering mode (DC or HiPIMS) in a same manner as for the monometallic sheet, it was shown that the mechanical properties of the bimetallic sheets vary accordingly to the mechanical properties of the components layers. Variations of the layer thickness of both Al-Sc-Zr and stainless steel have shown that at least the yield and tensile strengths of the bimetallic sheets can be predicted by the rule of mixture as it was already shown by other authors for thick Al / steel bimetallic sheets. The drawability of both a monometallic Al-Sc-Zr foil and a bimetallic Al-Sc-Zr / stainless steel foil produced by HiPIMS were compared in micro deep drawing experiments. Both foils have shown a higher limit drawing ratios than DCMS Al-Zr foils which were previously produced [18]. However, in case of the monometallic foil, different limiting drawing ratios were determined according to the side of the foil (free surface or substrate) facing toward the deep drawing die, which was attributed to surface porosity. If bimetallic Al-Sc-Zr / stainless steel foil is used, the limiting drawing ratio of the substrate material is achieved independently of the orientation of foil.

The authors gratefully acknowledge financial support by DFG (German Research Foundation) for projects A1 B7 and B4 within the Collaborative Research Center SFB 747 Micro Cold Forming — Processes, Characterization and Optimization.

References

1. Frank Vollertsen, *Micro Metal Forming* (Springer, Berlin, Heidelberg, 2013)
2. F. Vollertsen, Z. Hu, H. Schulze Niehoff, C. Theiler, J. Mat. Proc. Tech. **151**, 70 (2004)
3. http://www.sfb747.uni-bremen.de
4. G. Behrens, F. O. Trier, H. Tetzel, F. Vollertsen, Int. J. Mater. Form. **9**, 253 (2016)
5. G. Brabie, M. Costache, N. Nanu, B. Chirita, Int. J. Mech. Sci. **68**, 277 (2013)
6. S. Hadi, A. Kiet Tieu, C. Lu, H. Zhu, Int. J. Mat. Prod. Tech., **47**, 175 (2013)
7. J. Royset, N. Ryum, Int. Mat. Rev., **50**, 19 (2005)
8. K. Yu, W. Li, S. Li, J. Zhao, Mat. Sci. Eng.A, **368**, 88 (2004)
9. V. Neubert, B. Smola, I. Stulikova, A. Bakkar, J. Reuter, Mat. Sci. Eng. A, **464**, 358 (2007)
10. A. Cabello Munoz, G. Rückert, B. Huneau, X. Sauvage, S. Marya, J. Mat. Proc. Tech., **197**, 337 (2008)
11. S. Iwamura, Y. Miura, Acta Mat., **52**, 591 (2004)
12. C. B. Fuller, J. L. Murray, D. N. Seidman, Acta Mater., **53**, 5401 (2005)
13. M. Song, Y. He, S. Fang, J. Mater. Eng. Perform., **20**, 377 (2011)
14. O. Roder, T. Wirtz, G. Gysler., G. Lutjering, Mat. Sci. Eng. A, **234-236**, 181 (1997)
15. http://www.sme.org/MEMagazine/Article.aspx?id=31841&taxid=1419
16. J. Kovac, H.-R. Stock, B. Koehler, H. Bomas, H.-W. Zoch, Surf. Coat. Tech., **215**, 369 (2013)
17. R. von Bargen, Z. Hu, H.-W. Zoch, F. Vollertsen, Materialwissenschaft und Werkstofftechnik, **42**, 1035 (2011)
18. G. Behrens, J. Kovac, B. Koehler, H. Bomas, H.-W. Zoch, Trans. Nonferrous Met. Soc. China, **22**, 268 (2012)
19. D. N. Lee, Y. K. Kim, J. Mat. Sci., **23**, 1436 (1988)
20. M. H. Parsa, K. Yamaguchi, N. Takakura, Int. J. Mech. Sci., **43**, 268 (2001)
21. H. Takuda, K. Mori, H. Fujimoto, N. Hatta., J. Mat. Proc. Tech., **60**, 291 (1996)
22. E. Afshin, M. Kadkhodayan, Mat. Des., **87**, 25 (2015)
23. D. Pan, K. Gao, J. Yu, Mat. Sci. Tech., **5**, 934 (1989)
24. B. Köhler, H. Bomas, J. Stalkopf, H.-W. Zoch, J. Eng. Integrity Soc., **31**, 18 (2011)
25. J. A. Thornton, J. Vac. Sci. Technol. A, **4**, 3059 (1986)
26. K. H. Mueller, J. Appl. Phys., **62**, 1796 (1987)
27. K. Sarakinos, J. Alami, S. Konstantidinis, Surf. Coat. Tech., **204**, 1661 (2010)
28. M. Ohring, *Material Science of Thin Films (second edition)* (Academic Press, San Diego, San Francisco, New York, Boston, London, Sydney, Tokyo, 2002)
29. S.-H. Choi, K.-H. Kim, K. H. Oh, D. N. Lee, Mat. Sci. Eng A, **222**, 158 (1997)

High productivity micro rotary swaging

Eric Moumi[1,3*], *Marius* Herrmann[1,3], *Christian* Schenck[1,2,3], *and Bernd* Kuhfuss[1,2,3]

[1]bime, Bremen Institute for Mechanical Engineering, Badgasteiner Str. 1, 28359 Bremen, Germany
[2]MAPEX Center for Materials and Processing
[3]University of Bremen

Abstract. Rotary swaging is an incremental forming process with two main process variations plunge and infeed rotary swaging. With plunge rotary swaging, the diameter is reduced within a limited section whereas the infeed rotary swaging enables a diameter reduction over the entire workpiece length. The process is now subject to intensive investigation for manufacturing of micro parts. By increasing the process speed, failures occur particularly due to inappropriate material flow. In plunge rotary swaging, the workpiece material can flow radially into the gap between the dies and thus the workpiece quality suffers. In infeed rotary swaging the workpiece material flows against the feeding direction and can provoke bending or braking of the workpiece. Therefore, additional measures to control both the radial and the axial material flow to enable high productivity micro rotary swaging are investigated. The radial material flow during plunge rotary swaging can be controlled by elastic intermediate elements that enable an increase of productivity by factor three. A spring-loaded clamping device that enables an increase of the productivity by factor four can temporarily buffer the axial material flow in infeed rotary swaging against the feeding direction.

Keywords: Cold forming, Micro forming, Feed rate

1 Introduction

Rotary swaging is an incremental open die forging process for rods and tubes [1]. The process is subject to intensive investigation in the micro range with focus on machine and workpiece behavior, achievable dimensions und workpiece properties [2, 3]. According to CIRP micro parts are those with at least two dimensions smaller than 1 mm [4], which means for cylindrical workpieces that their diameter is smaller than 1 mm. However, many challenges related to the stiffness of the workpiece and the tolerances of the forming machine have to be overcome. In rotary swaging, two or more synchronously oscillating dies act on the workpiece. Material flow takes place mainly in radial (reduction of the diameter) and in axial direction (elongation of the part). Figure 1 shows the two most important process variants, the infeed and the plunge rotary swaging. An outer ring encloses the dies and other mechanical components that are necessary for the die motion and the forming. The two process variants differ particularly in the direction of the feed. In plunge rotary swaging, the workpiece is positioned between the dies prior to the forming, then a radial displacement with a feed rate v_r in the direction of the workpiece and an oscillating movement with an amplitude h_T (stroke) of the dies take place simultaneously.

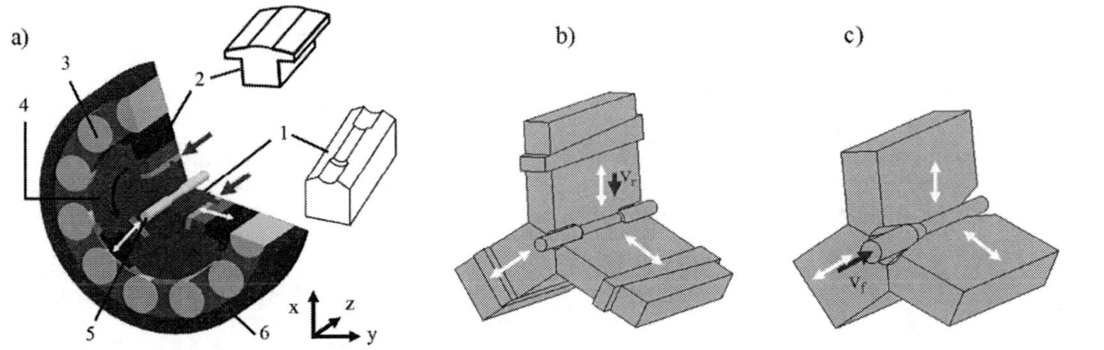

Fig. 1. Principle rotary swaging; a) swaging head 1 die, 2 base jaw,3 cylinder roller, 4 drive shaft, 5 workpiece, 6 outer ring; b) plunge rotary swaging; c) infeed rotary swaging.

* Corresponding author: moumi@bime.de

In infeed rotary swaging however, the workpiece is fed with an axial feed rate v_f in the swaging head and the dies perform only the oscillating motion. In both variants, a relative rotation between the workpiece and the dies is necessary to produce cylindrical workpieces.

It was found in previous investigations that the utmost limitations in micro rotary swaging are due to bending, break or the emergence of flashes on the surface of the workpiece [5]. These failures occur particularly due to inappropriate material flow in the forming zone. When the forming volume per stroke is increased by increasing the radial feed rate v_r in plunge rotary swaging the workpiece material can flow radially into the gaps between two adjacent dies and thus generates flashes on the workpiece surface [6]. A high surface quality is not guaranteed any more. In axial direction, the dies do not enclose the workpiece and by this the material flow is completely unguided in this direction. When the forming volume per stroke is increased by increasing the axial feed rate v_f in infeed rotary swaging the workpiece material flows in and against the feeding direction. The material flowing backwards leads to high loads in the workpiece and the feeding system, which provokes bending of the workpiece.

To increase the feed rate and thus the productivity while avoiding process failures and enabling even higher workpiece quality, additional measures to control both the radial and the axial material flow in micro rotary swaging are presented.

2 Plunge rotary swaging

2.1 Initial situation

In classical plunge rotary swaging the cross section of a region of a bar or tube with an initial diameter of d_0 is reduced to a final diameter of d_1. The length changes from l_0 to l_1. The forming die with the geometry to be reached is generally made out of hard metal or of tool steel. The die with its characteristic regions as well as the workpiece before (Figure 2a)) and after forming (Figure 2b)) are given below.

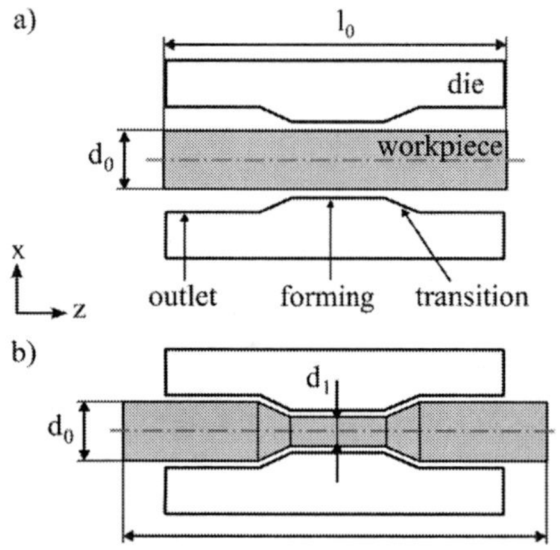

Fig. 2. Geometry of the die and the workpiece in plunge rotary swaging; a) before forming, b) after forming.

This process variant is ideal for generating local geometrical denting and allows a broad diversity of possible part designs. However, a limitation of the process is the discrepancy between the geometry of the closed set of dies and the formed workpiece. In micro range the forming accuracy could be improved by controlling the axial material flow [7]. Furthermore, it was found during manufacturing of micro parts that the process was highly constrained by the radial feed rate and the formation of flashes. Generally, the flashes are parallel to the longitudinal axis of the workpiece and their number is equal to the number of dies that are used [6]. For example, three flashes will occur in the case of a set three dies as presented in Figure 3. A cross section in the forming zone of the workpiece in which the contours of the dies are schematically drawn is presented. The geometry can be approximated to a polygon (triangle).

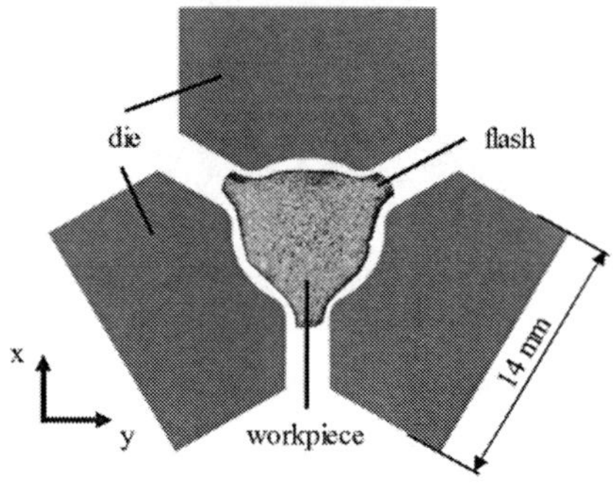

Fig. 3. Flash formation in plunge rotary swaging with three dies

To achieve the goal of a high radial feed rate, one measure is to control the flow of material in the radial direction. From this consideration, a new concept has emerged, based on the use of elastic intermediate elements (EIE) in rotary swaging. Comparable approaches for other forming processes are mentioned in [8, 9]. The use of elastomers in metal forming was summarized in [8] and in [9] micro patterning of thin metallic plates using rubber in a forming container was investigated. For micro plunge rotary swaging, the basic idea is to position an elastic intermediate element in the set of dies to prevent any workpiece material to flow into the gaps between the single dies. Conceivable intermediate elements could be made of super-elastic metallic or polymeric substances with a high tear resistance.

2.2 Experimental procedures and results

As workpiece material austenitic steel 304 alloy and pure copper CW004A were used. The initial diameter was $d_0 = 1.5$ mm for both materials, and was reduced using dies with a nominal diameter of $d_{nom} = 985$ µm. The nominal diameter is defined as the inside diameter of the forming zone of a closed die set. The forming zone was 20 mm long. A stroke heights of $h_T = 0.3$ mm for each die was applied. The workpieces were lubricated during the process. During the forming, the elastic deformation of the outer ring of the swaging machine was monitored by strain gauges. The response of the strain gauges is

* Corresponding author: moumi@bime.de

 https://doi.org/10.1051/matecconf/201819015002

proportional to the deformation of the ring and thus can be also taken as a measure of the radial forming force. Prior to the forming the workpiece is clamped into a chuck and the elastic intermediate elements (EIEs), which are available as hose, are cut into the desired length and fitted on the workpiece area, that has to be reduced. Thermoplastic polyurethane (Elastollan C78A1000, C98A10000 and 1174D11000) intermediate elements were tested. They were available with three different hardness values (low hardness = ILH; middle hardness = IMH and high hardness = IHH). The outer diameter of the thermoplastic hoses was 2.5 mm and the wall thickness was 0.5 mm. Some characteristics of the EIEs are summarized in Table 1.

Table 1. Characteristics of the elastic intermediate elements (according to manufacturer).

	units	ILH	IMH	IHH
hardness	Shore A	80		
	Shore D	(~30)	52	75
tensile elongation	%	650	550	380
E-modulus	MPa	-	160	560

Forming experiments were conducted with different radial feed rates v_r. Through a gradual increase of v_r, possible processing windows in which workpieces could be produced without failure were framed. The stroke frequency of the swaging machine was kept constant at 102 Hz and the same bottom dead center of the dies was applied for all experiments. After forming, the final diameter (d_1) of the deformed area of the workpiece was measured by means of an optical micrometer.

Forming results for 304 alloy

With the stroke height $h_T = 0.3$ mm and without intermediate element the maximum feed rate of $v_r = 0.26$ mm/s was reached. However, this boundary was caused by strong vibrations of the machine. For the mentioned reason higher feed rates as well as IMH and IHH were not tested for this alloy. An assessment of the radial extension of the outer ring of the rotary swaging machine during forming with and without EIE is given in Figure 4. The measured signals were graphically synchronized with respect to the bottom dead center of the dies, which also corresponded to the end of the closing motion of the dies (point C in Figure 4). The starting points of the forming are different for both cases, earlier with EIE (point A) and later without EIE (point B). The die closing takes place between the points A and C respectively B and C. From the point C the dies reopen. During the forming with EIE a small extension of the ring occurs first, which then dropped again. This can be attributed to a compacting of the EIE (extension) followed by a flow into the gap (release of the extension). The forming of the workpiece started almost equally with or without EIE. Apart from the points A and B the course and the amplitude of the ring extensions are comparable.

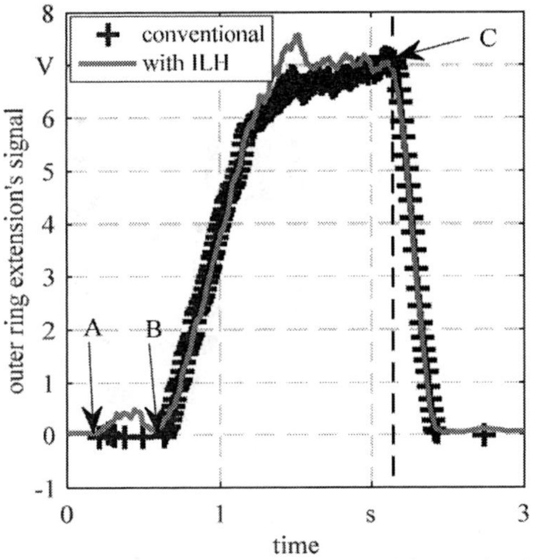

Fig. 4. Elastic extension of the outer ring of the swaging head during forming of 304 alloy with $v_r = 0.18$ mm/s without and with elastic intermediate element.

The mean final diameters of five samples each, measured in the middle of the deformed area of the workpiece (see Figure 2) were 1123 ± 6 µm for conventional plunge rotary swaging and 1108 ± 9 µm for plunge rotary swaging with ILH. The difference between the two values is in the range of the standard deviations whereas nominal diameter of $d_{nom} = 985$ µm was not reached. The time for forming without EIE is about 1.48 s compared to an expected time of 1.42 s. The expected time for the forming with ILH cannot be properly estimated. However, the actual time from the measurement was 1.92 s. The greater diameters after forming with and without EIEs can be explained by the resilience of the swaging head, which leads to an extension of the outer ring, thus the die closure is limited. A measure to reduce these two effects could be by using a calibration time at the end of radial feed of the dies towards the workpiece. The ILHs were not reusable as presented in Figure 5.

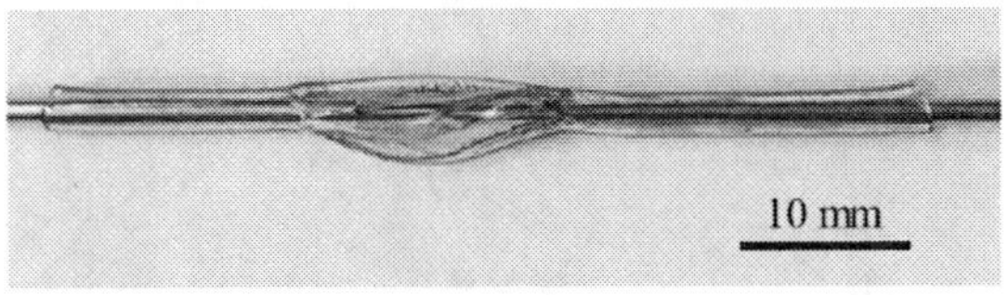

Fig. 5. ILH and workpiece after forming with $v_r = 0.07$ mm/s.

Forming results for copper CW004A

Whit $h_T = 0.3$ mm a feed rate of $v_r = 0.34$ mm/s was reached before flash or break occurred. The feed rate was increased when EIE was applied, see in Figure 6.

https://doi.org/10.1051/matecconf/201819015002

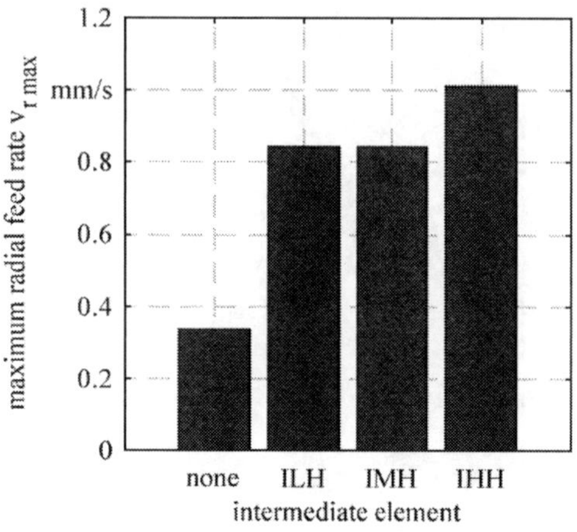

Fig. 6. Maximum radial feed rate in plunge rotary swaging of copper CW004Awith different elastic intermediate elements.

The EIEs allowed to increase the feed rate up to about 200 % for h_T = 0.3 mm (IHH). It was noticed during forming with higher stroke that the EIEs were easily rotating around the workpiece. It can be assumed that the EIEs will flow first in the gap of the dies, but if they are hindered in their motion early in the process flashes and breaks can still occur, for example at high feed rates. As for the 304 alloy, the EIEs were not reusable. After forming, they presented a structure like on Figure 5 in some cases they were torn apart. The type of EIE influenced not only the feed rate but also the elastic radial extension of the outer ring of the swaging head. For a feed rate of 0.44 mm /s, the extensions of the outer ring given in Figure 7 were measured. Again, the starting point of the forming is earlier for the process with EIE.

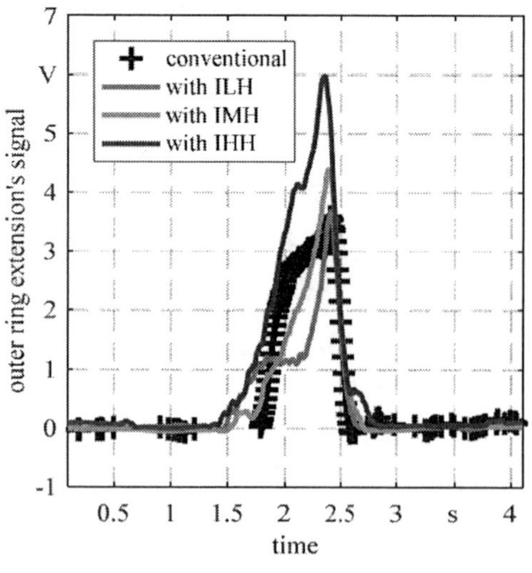

Fig. 7. Elastic extension of the outer ring g of the swaging head during forming of copper with v_r = 0.44 mm/s with different elastic intermediate elements.

A relationship between the hardness of the EIE and the reaction of the outer ring can be seen. Harder EIE induced higher extensions of the ring. Nevertheless, the ring extension during forming with the weakest EIE is comparable to the one of the conventional forming. For this same feed rate the smallest diameter of d_1 = 991 ± 16 µm was achieved with ILH and the highest d_1 = 1041 µm with IHH. The increase of the diameter with the hardness of the EIEs correlates with extension of the outer ring. The harder EIE flowing into the gaps between the dies contributed to increase the ring extension that led to a reduced die closure and as result, the workpiece diameter increased.

3 Infeed rotary swaging

3.1 Initial situation

In infeed rotary swaging the cross section of a bar with an initial diameter of d_0 was reduced over the complete feeding length to a final diameter of d_1. In this procedure, the length changed from l_0 to l_1. The die with its characteristic regions as well as the workpiece before (Figure 8a)) and after forming (Figure 8b)) are given below.

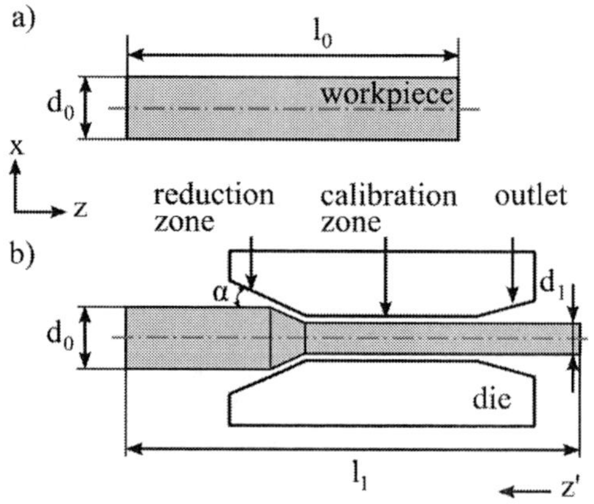

Fig.8. Geometry of the die and the workpiece in infeed rotary swaging; a) before forming, b) after forming.

Previous investigations had shown that the process is suitable for micro manufacturing. Workpieces with final diameter down to 300 µm were achieved [3]. However, the process is constrained by bending or torsion of the workpiece between the collet and the reduction area of the dies. An example of workpiece with bending is given in Figure 9.

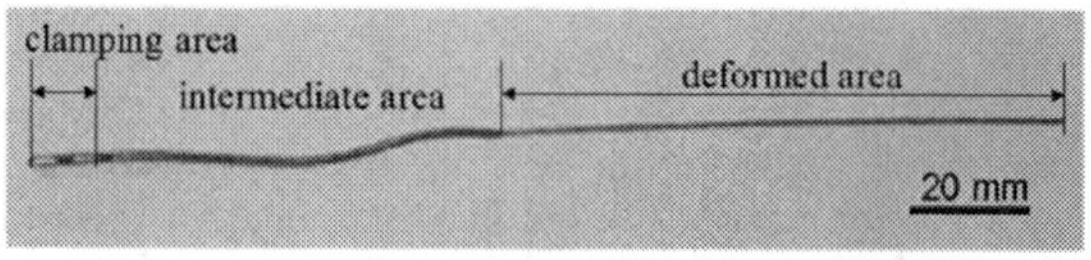

Fig. 9. Workpiece with bending

Concerning the different phases of rotary swaging, bending can be related to the closing phase of the dies. The clamping device in which the workpiece is fixed is fed with a constant feed rate in all process phases and the workpiece material flows axial in (forwards) and against (backwards) the direction of the feed motion. This material flow increases not only with diameter reduction but also with the feed rate. Bending will occur when the critical bending load is reached. This load depends on the Young's modulus (E-modulus), the initial diameter and the length of the workpiece. Especially for the micro range, the critical load is particularly low due to the small initial diameters. The theoretically critical bending load for different initial diameters with respect to the Young's modulus is given in Figure 10. In this case, one end is fixed (clamping), and the other end is guided. A buckling/bending length of 60 mm was considered.

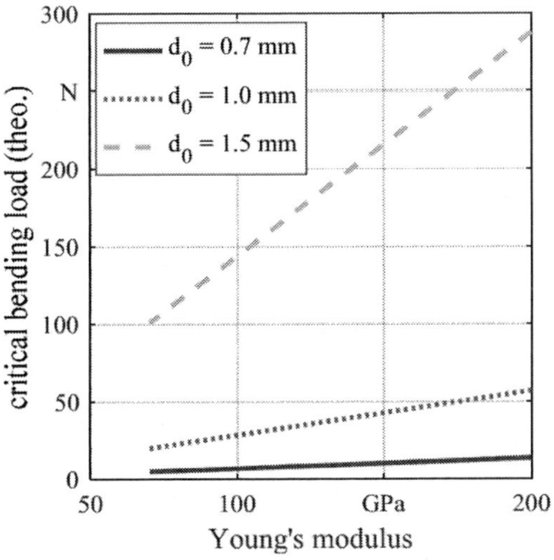

Fig. 10. Critical bending load with respect to the Young's modulus and the workpiece diameter.

A measure for reducing the bending and hence increasing the productivity is by controlling the backward material flow. This can be done in a small range by increasing the friction in the reduction zone [10]. This contribution focuses on another approach: the spring-loaded clamping, which allows an axial shift of the workpiece during forming and reduces the compressive force. The concept of the spring-loaded clamping device is presented in Figure 11.

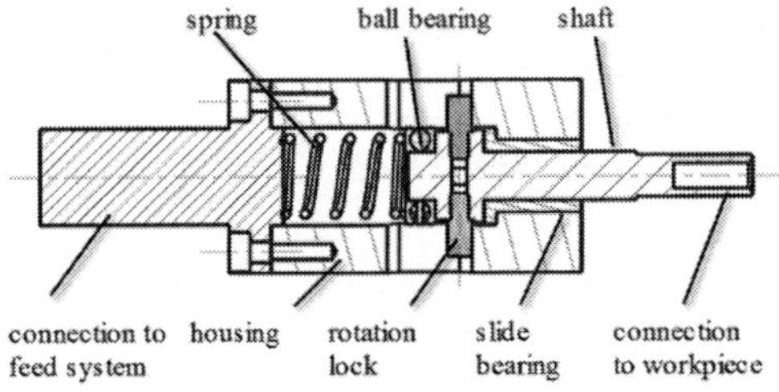

Fig. 11. Spring-loaded clamping device.

By using the spring directly between the feed system and the workpiece, the rebound that is caused by the forming in the reduction zone is absorbed and stored in the spring. When the dies reopen, the spring is decompressed and the buffered back pushing is added to the feed into the swaging head.

3.2 Experimental procedures and results

The workpiece material was the austenitic steel 304 alloy with an initial diameter of d_0 = 1.0 mm. The set of dies had a nominal diameter of d_{nom} = 750 µm. Forming experiments were proceeded with a stroke frequency of 102 Hz and a stroke height of h_T = 0.2 mm. The workpieces were lubricated during the process. The feed rate was gradually increased from test to test until a process failure like bending occurred. Beside of the elastic radial extension of the outer ring, the final diameter was characterized contactless by an optical micrometer. The diameter of the workpiece was measured starting from the tip every 2 mm along the workpiece axis at three different rotation angles.

Feed rate and final diameter

It was found that the spring-loaded clamping device could substantially increase the feed rate. Workpieces were formed with a feed rate up to 100 mm/s compared to a maximum feed rate of v_f = 25 mm/s that was possible with a fixed clamping. At even higher feed rates, bending and flashes occurred again.

Figure 13 shows the diameter evolution along the deformed workpiece for different axial feed rates. The tip of the workpiece was located at Z' = 0 mm. It can be noticed that the final diameters increased not only according to the feed rate but also according to the position in the workpiece, especially at high feed rates. The diameter values rose monotonically with the distance to the tip.

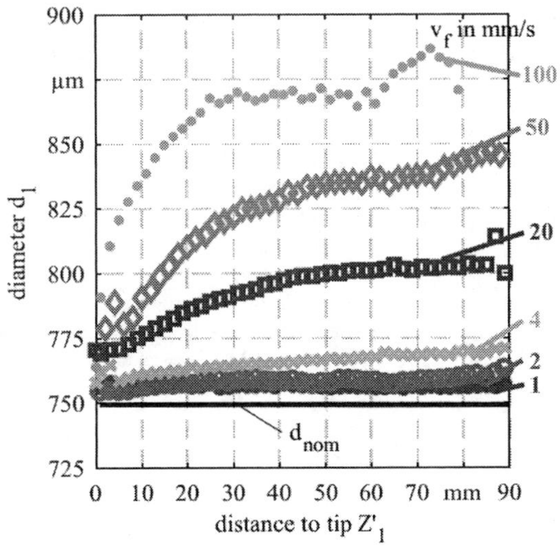

Fig. 13. Diameter evolution with regard to the axial feed rate.

Workpieces with final diameters close to d_{nom} and constant within the processed length were those manufactured with feed rates up to 2 mm/s. For these settings, the stiffness of the outer ring of the swaging head was high enough to prevent the diameter evolution. In Figure 14 the envelope function of the strain gauges measurements clearly indicates the correlation between the diameter evolution and the equivalent radial forming force.

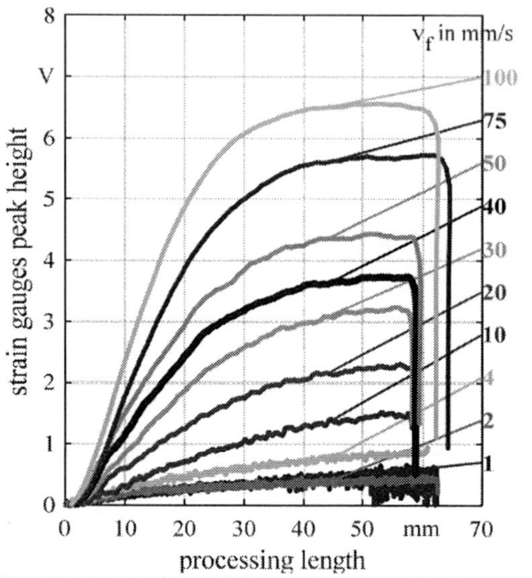

Fig. 14. Peak height of the strain gauge signals over the processed length of the workpiece.

Due to the increase of the deformation of the outer ring, the closing of the dies was not ensured any more with higher axial feed rates. This could be confirmed by high-speed video recording.

From this result, it can be stated that the size respectively the stiffness of the used swaging head without further measures was not suitable for forming with high feed rates with high quality. Therefore, solutions for compensating the diameter increase are necessary to reach high geometry accuracy during high feed rates. Among others, a closed-loop control of the feed motion of the workpiece considering the actual load of the outer ring could be a solution. The possibility of influencing the final diameter by an in-process adjustment of the wedges was proved by experiments. A combined feeding in radial and axial direction with constant feed rates of $v_r = 8$ µm/s and $v_f = 20$ mm/s resulted in a more homogeneous diameter chart (Figure 15). Better results are to be expected if the motion of the wedges is closed-loop controlled.

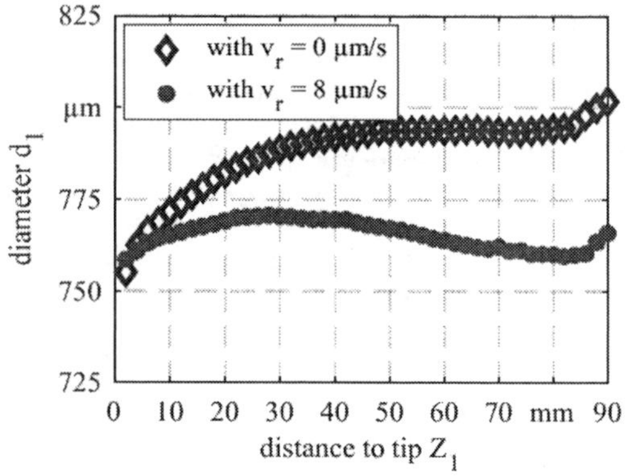

Fig. 15. Forming with $v_f = 20$ mm/s without and with simultaneous radial feeding.

4 Conclusions

In this study, measures to control the material flow in plunge and infeed rotary swaging to increase the productivity are presented. Elastic intermediate elements were used in plunge rotary swaging to control the radial material flow and a spring-loaded clamping device was deployed to control the axial material flow in infeed rotary swaging. Following main results could be drawn:

For plunge rotary swaging:

- Elastic intermediate elements enabled dependent on their hardness higher radial feed rates v_r up to three times compared to forming without these elements.
- The final diameter depends on the stiffness of the elastic intermediate element with bigger diameter for harder elements.

For infeed rotary swaging:

- A spring-loaded clamping device enabled four times higher axial feed rates v_f, however the final diameter of the workpiece increased and was not constant along the length.
- The rising value of the final diameter over the length of the workpiece could be positively manipulated by a continuous radial feed during the process.

The presented results facilitate high productive micro rotary swaging for plunge as well as for infeed rotary swaging. Further studies need to focus on the adjustment of the final diameter after plunge rotary swaging by including a calibration time in the process. In addition, an implementation of a reusable elastic intermediate element is of interest. In the case of infeed rotary swaging, the regulation of the diameter evolution by the process control will be examined.

Acknowledgments

The authors gratefully acknowledge the financial support by German Research Foundation (DFG) for the Subproject A4 "Material displacement" within the Collaborative Research Center SFB 747 "Micro Cold Forming – Processes, Characterization, Optimization".

References

1. A. Uhlig, Ph.D. thesis, Technical University of Hannover (1964)
2. B. Kuhfuss, E. Moumi, V. Piwek, Microsyst Technol, **14**, (2008)
3. B. Kuhfuss, E. Moumi, Incremental forming in F. Vollertsen (edt.) *Micro Metal Forming* (Springer, Berlin, 2013)
4. M. Geiger, M. Kleiner, R. Eckstein, N. Tiesler, U. Engel, CIRP Ann., **50**, 2, (2001)
5. B. Kuhfuss, E. Moumi, V. Piwek, *Procedings ICOMM2008*, 86-91, (2008)
6. R. W.Bolz, *Production Processes: The Productivity Handbook* (5th ed, Industrial Press Inc.1977)
7. E. Moumi, C. Schenck, P. Wilhelmi, M. Herrmann, B. Kuhfuss, *Proceedings of ICNFT* 2018 (to be published)
8. S. Thiruvarudchelvan, J. Mater. Process. Technol., **39** (1993)
9. C. K. Jin, M. G. Jeong, C. G. Kang, Proc. Eng. **81** (2014)
10. M. Herrmann, C. Schenck, B. Kuhfuss, Dry Met. Forming OAJ FMT 1 (2018)

Eccentric rotary swaging variants

Anastasiya Toenjes[1*], *Svetlana* Ishkina[2], *Christian* Schenck[2,3], *Axel* von Hehl[1,3], *Hans-Werner* Zoch[1,3] and *Bernd* Kuhfuss[2,3]

[1]Leibniz Institute for Materials Engineering IWT
[2]bime Bremen Institute for Mechanical Engineering
[3]MAPEX Center for Materials and Processing
[1,2,3]University of Bremen, Faculty of Production Engineering, 28359 Bremen, Germany

Abstract. Rotary swaging is an incremental cold forming process that changes beneath the geometry also the microstructure and mechanical properties of workpiece. Especially a new process design with Eccentric Flat Shaped Dies (EFSD) influences both the kind and amount of stress and plastic strain and consequently the material structure and hence the material and workpiece properties. Eccentric rotary swaging typically provides a helical material flow. According to the process parameters the microstructure features a typical eddy pattern with a spiral shaped grain orientation. The forming process can be carried out in one or more process steps. In a multi-stage process, it is possible to change the feed direction and, hence, the material flow helix direction. This approach can be used as a possibility to improve the homogeneity of the workpiece and material properties. In addition, for this aims an intermediate heat treatment in multi-stage forming operations could be realised. Following the goal of optimising the final properties, the question arises how these mechanical and thermal treatments affect the material microstructure and the forming properties of the workpiece and how they interact. Experiments were conducted with austenitic stainless steel rods of grade AISI304. The effects of the varied feed direction, feed velocity and heat treatment between the forming operations are discussed.
Keywords: Cold forming, Micro structure, Metal forming.

1 Introduction

Metastable austenitic steels are characterized by an elevated phase hardening due to strain-induced transformation of austenite into martensite especially by multi stage processing. An improvement of material characteristics and a better formability of parts are typical for Ultra Fine Grained (UFG) materials that are characterized by good strength in combination with sufficient ductility [1, 2]. Changing of material properties, i.e. decrease of the strain and increase of the ultimate tensile strength more than twice compared to the initial values, could be achieved by cold forming processes like rotary swaging [3]. In this process the workpiece is fed axially into the swaging head with a feed velocity v_f. The tools revolve around the workpiece and oscillate radially with a small amplitude h_T, see Fig. 1. By means of the movable wedges the closing pressure and the final diameter can be adjusted. Swaged parts are well used in automobile industry due to the dynamics and stress capacity, for example, steering shaft. The variation of the process parameter during rotary swaging leads to changing of the mechanical properties and of the swaged parts [4]. It can be used at preparation of semi-finished products with both a calibrated diameter as well as predefines material properties of the parts. To modify the microstructure of swaged rods with small diameter about 1 mm an approach of eccentric rotary swaging was applied [5]. The eccentric machined parts featured very fine grains. For these components, besides the phase hardening, a work softening could also be detected. Such material characteristics provide advantages for downstream forming processes like for example extrusion [6]. Nevertheless the eccentric swaged parts had an inhomogeneous structure with well distinguishable grains in a core area and a non-defined structure with spiral shaped orientation in a boundary area [5]. A disadvantage of eccentric rotary swaging was found as deep cracks on the surface [7]. In order to minimize these cracks, the newly developed tools named Eccentric Flat Shaped Dies (EFSD) were integrated into the swaging process.

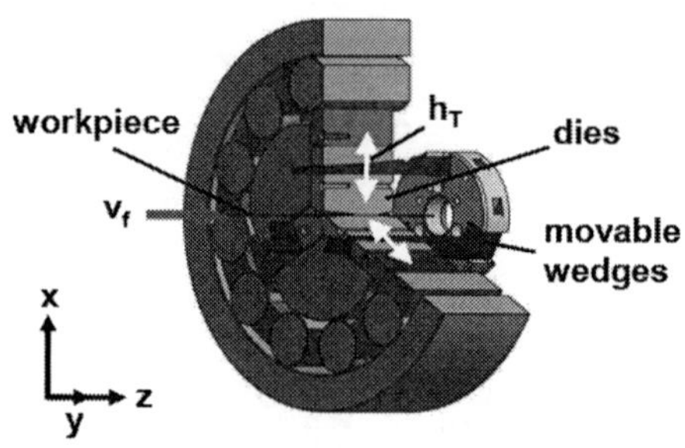

Fig. 1. Rotary swaging head.

* Corresponding author: toenjes@iwt-bremen.de

https://doi.org/10.1051/matecconf/201819015003

Also, subsequent heat treatment influences the microstructure. During cold forming like rotary swaging, the formability of the metal decreases with increasing forming ratio. To regain formability of the material for further forming steps, a heat treatment process like recrystallization annealing is often necessary in industrial applications. The energy that was consumed by the material during swaging is the driving force for the grain growth and dislocation movement during recrystallization [8]. The aim of this study is to analyse the evolution of the microstructure and the mechanical properties after modification of the eccentric rotary swaging by means of incremental two steps process with change of feed direction and intermediate heat treatment. In the present work, the microstructure of a variety of eccentrically swaged components is investigated microscopically. The micro hardness was obtained for the cross-section of samples as well as for the longitudinal section. The influence of the intermediate heat treatment between two swaging steps on the microstructure and material properties was analysed. The influence of eccentric rotary swaging variants on the homogeneity, material properties and the surface of the components was studied.

2 Experiments

For the experimental tests, sections from the austenitic stainless steel 1.4301 (AISI304) with an initial diameter of $d_0 = 2.7$ mm and a length of $l_0 = 110$ mm were used. The chemical composition of the material is shown in Table 1 and the microstructure before forming is shown in Fig. 2.

Table 1. Chemical composition of stainless steel 1.4301 (AISI304) acc. to melt analysis.

	C	Si	Mn	P	Cr	Ni
Ma.-%	0.018	0.50	1.00	0.027	18.19	10.44

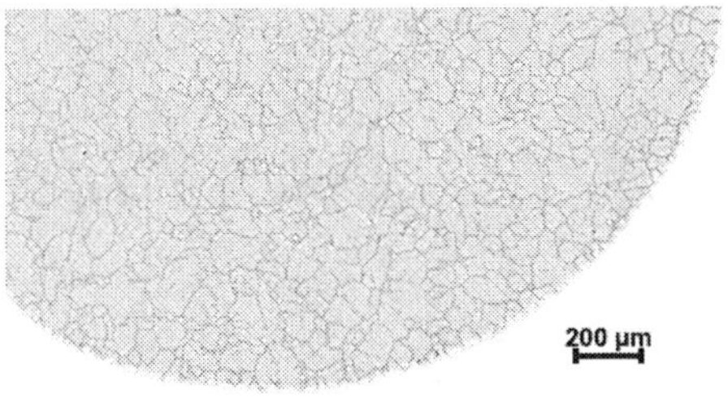

Fig. 2. Transverse section of the microstructure of stainless steel 1.4301 before forming.

Experiments were conducted in a rotary swaging machine using three eccentric flat shaped dies (Fig. 3). The workpieces were fed into the swaging head with a constant velocity of $v_f = 1.0$ mm/s or $v_f = 2.0$ mm/s. The strokes of all swaging dies were equal to $h_T = 0.4$ mm and the stroke frequency was kept constant at $f_{st} = 102$ Hz for each experiment.

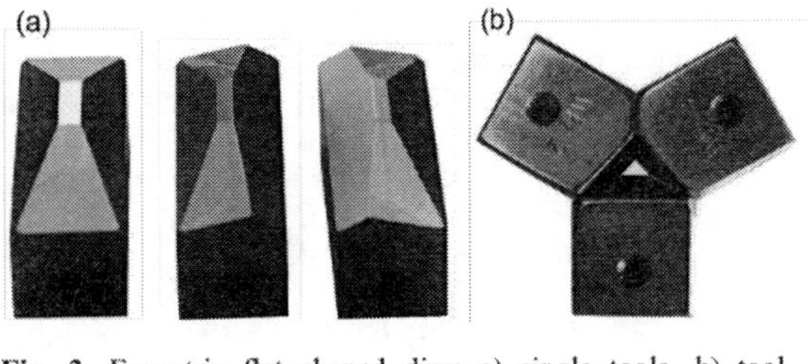

Fig. 3. Eccentric flat shaped dies: a) single tools, b) tool assembly.

All samples were formed in two steps. After the first step, the feed direction was converted and the second forming step was performed. In one part of the experiments the feed velocities were also varied between two steps. In addition, a second series of experiments with the same forming parameters was investigated in which a heat treatment at 600°C for 1 h using a vacuum furnace (Producer ALD, Type VKNQ) was carried out between step 1 and step 2. The forming parameters of all experiment series are summarized in Table 2.

Table 2. Forming parameters.

	Step 1		Intermediate…			Step 2	
Parameter	Initial diameter d_0 [mm]	Feed velocity v_f [mm]	diameter d_b [mm]	heat treatment	change of the feed direction	Feed velocity v_f [mm]	Final diameter d_1 [mm]
Setting 1-1	2.70	1.0	1.80	no	no	1.0	1.25
Setting 2-2		2.0				2.0	
Setting 1-1 a		1.0		no	yes	1.0	
Setting 1-2 a		1.0				2.0	
Setting 2-1 a		2.0				1.0	
Setting 2-2 a		2.0				2.0	
Setting 1-1 b		1.0		600°C 1h	yes	1.0	
Setting 1-2 b		1.0				2.0	
Setting 2-1 b		2.0				1.0	
Setting 2-2 b		2.0				2.0	

For comparison, an additional experiment series with a constant velocity of $v_f = 1.0$ mm/s (setting 1-1) or $v_f = 2.0$ mm/s (setting 2-2) were carried out without changing the feed direction and without heat treatment.

A characterization of the swaged microstructure is done by metallographic analysis and micro hardness measurements after the second forming step. For this purpose, small pieces of the workpieces were embedded in a polymer matrix and corroded electrolytically in a 60 % aqueous nitric acid solution for 25 s with a voltage of 1.5 V. Finally, hardness measurements were performed on the mechanically polished transverse section as well as longitudinal section of each embedded sample using a micro hardness tester (FISCHERSCOPE H100C) with a Vickers indenter. For the transverse section the determined lines with measured points lay perpendicular to each other along the whole diameter of the samples, Fig. 4 (a). The longitudinal section contains a rectangular area of 2.0x1.2 mm^2 with ten lines along x-coordinate direction and ten measured points along z-direction, Fig 4 (b).

The indentation force was 10 mN and the duration of the penetration, holding and indenter withdrawal phases was 10 seconds for each measurement.

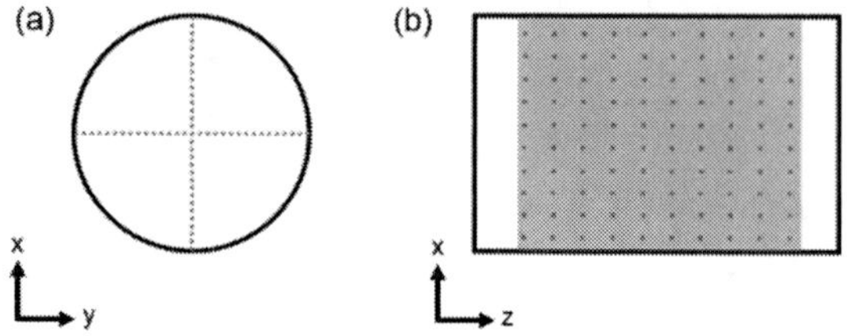

Fig. 4. Points of the hardness measurement on the a) transverse section and b) longitudinal section of the specimen.

3 Results and discussion

The application of the eccentric flat swaging dies generated a significant reduction of the number and the depth of cracks on the surface of the specimens in comparison with eccentric process without EFSD [7]. In a case of two-stage formed rods without any change of direction (setting 1-1) and (setting 2-2), the crack direction was found to be identical to the rotation direction of the tools, (Fig. 5). In addition, the microstructure in the transverse micro section could be divided into three zones: I) uniformly shaped center of the workpiece; II) severely deformed and swirled middle section of the workpiece; and III) outer section with cracks.

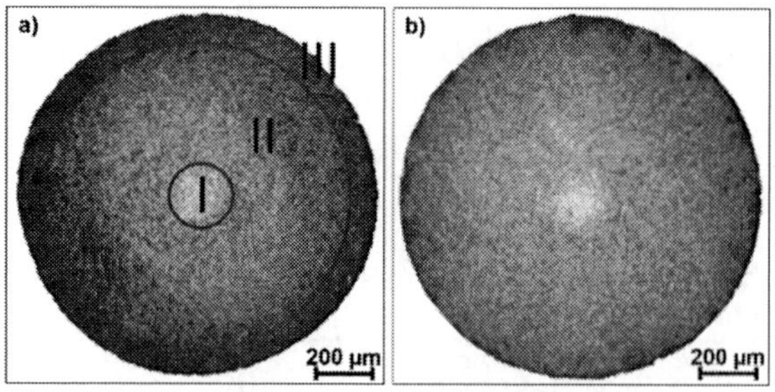

Fig. 5. Microstructure of stainless steel after forming with setting a) 1-1 and b) 2-2, both with the same feed direction in step 1 and step 2.

The increase of the feed velocity leads to more supplied material into the swaging head during each opening of the tools and thus, to a rise of the formed volume during one stroke. The process caused a changing of some properties of the workpieces after swaging, such as diameter, hardness or martensite content [5, 9]. In the case of two-steps swaging without direction change the microstructure of the formed parts without intermediate heat treatment was less affected by the feed velocity. However, the slow process (setting 1-1, Fig. 6a) resulted in a more pronounced inhomogeneous structure with more whirl-like pattern in Zones II and III as with higher feed velocity (setting 2-2, Fig, 6d), which corresponds to the number of strokes per processing length that is higher with lower feed velocity.

With changing the direction between the two steps, the workpieces exhibited cracks after the second forming step, which were either oriented in the direction of rotation or against the direction of rotation of the tools in the second step. This is a clear evidence that some of the cracks already occurred during the first forming step, (Fig. 6).

The microstructure was developed more homogeneously with direction change (Fig. 6) than without direction change (Fig. 5). It could be seen both visually by the micro structure and by the micro hardness measurements.

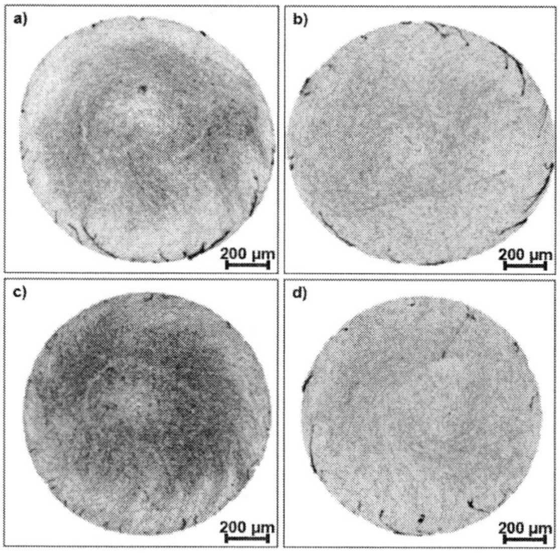

Fig. 6. Microstructure of stainless steel after forming with setting a) 1-1 a, b) 1-2 a, c) 2-1 a and d) 2-2 a with change of feed direction after step 1.

Figure 7 shows the results of the hardness measurement of the two-stage formed workpieces with setting 2-2 without changing the feed direction between the two forming steps. The hardness average along the transverse micro section is about 431±28 HV 0.001 in zones II and III. In zone I, the hardness is in the range of 400±26 HV 0.001. Thus, the specimen exhibited an inhomogeneous distribution of hardness along the transverse micro section.

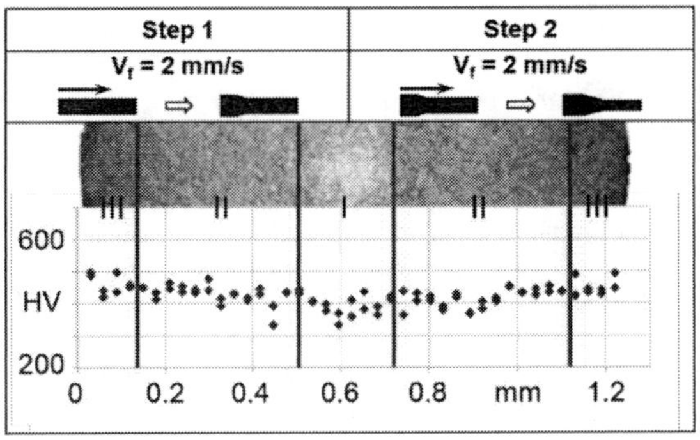

Fig. 7. Hardness tests of the transverse section after forming with setting 2-2.

The measurements of the longitudinal section at setting 2 -2 are shown in Fig. 8. The average value of the hardness is 416±32 HV 0.001. This value lies within the hardness distribution of the transverse section. The shapes in x- and z- directions show only slight differences.

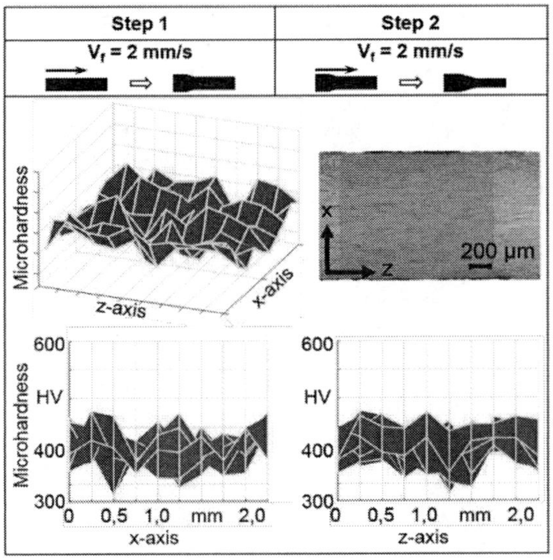

Fig. 8. Hardness tests of the longitudinal section after forming with setting 2-2.

By changing the feed direction between the two forming steps, a higher hardness was achieved in both transverse and longitudinal sections. The hardness of the workpiece with setting 2-2 a is approximately 423±21 HV 0.001 in zone I and 472±23 HV 0.001 in zone II and III (Fig. 9). However, hardness values for direction change (Fig. 9) seem comparable homogenous to no direction change (Fig. 7).

Just as for the transverse section results of the specimen swaged using setting 2-2 a the hardness values for the longitudinal section for the same setting were also higher as for the workpieces formed without changing the feed direction between the two forming steps, and obtained an average value of 435±25 HV 0.001, see Fig. 10. The deviations of the measurements were smaller in comparison to this at setting 2-2 and they became more homogenous at large. The shape of these measurements in x-direction is similar to the hardness distribution of the transverse section.

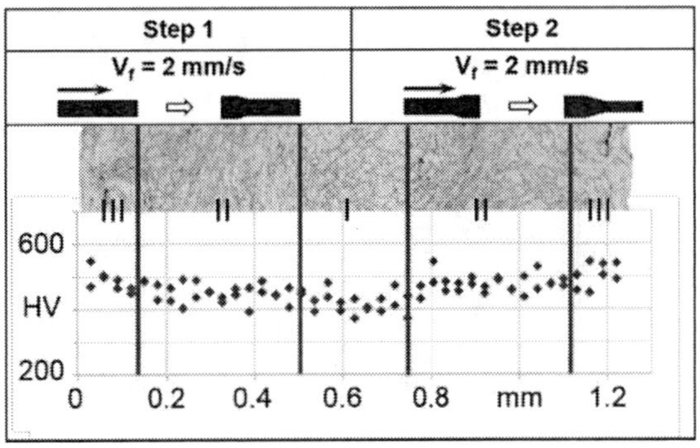

Fig. 9. Hardness tests of the transverse section after forming with setting 2-2 a with change of feed direction after step 1.

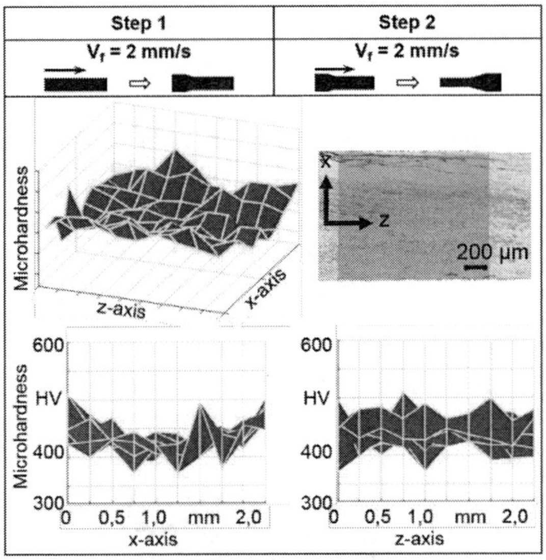

Fig. 10. Hardness tests of the longitudinal section after forming with setting 2-2 a with change of feed direction after step 1.

In the case of the change of the feed direction and additional heat treatment between two forming steps, the workpiece with setting 2-2 b showed a higher hardness of about 457±19 HV 0.001 in zone I and 488±38 HV 0.001 in zone II and III as for settings 2-2 and 2-2 a.

However, the middle zone I was not still centric and shifted to the side. Moreover, the area of the zone III became greater due to deeper cracks on the surface. The structure for setting 2-2 b was more homogeneous in the middle area and had less scattering of the measured values, Fig. 11. The change of the feed direction caused additional work hardening. That could be explained by three different mechanisms. The first mechanism is that the material flow changed with the different forming procedures. In consequence, the real deformation degree differed from the nominal deformation degree, which is defined only by the resulting overall change of the geometry and was the same for all experiments. The second mechanism is that the kind and direction of the stress and strain differed significantly. Hence, the hardness generation would differ especially if it was caused by martensite formation [10]. The third mechanism is that the crystallographic orientation of the constituent grains of the microstructure changed during the variation of the feed direction.

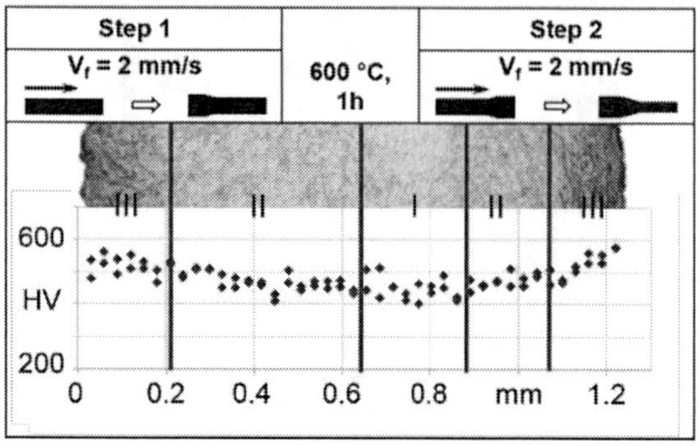

Fig. 11. Hardness tests of the transverse section after forming with setting 2-2 b with change of feed direction after step 1 and heat treatment between step 1 and step 2.

The analysis of the samples based on setting 2-2 b with intermediate heat treatment provided the highest values of the hardness in the longitudinal section with about 457±23 HV 0.001. The shape of the measurements in x-direction showed distinct patterns which as well shifted from the middle low hardness values, see Fig. 12. It could be one more evidence of the non-centric zone I, see Fig. 12.

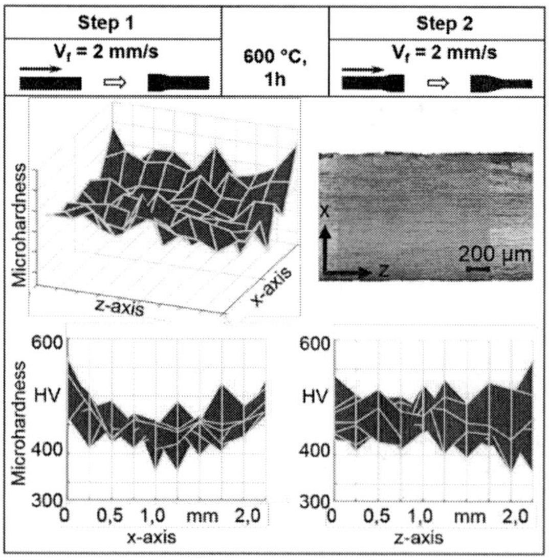

Fig. 12. Hardness tests of the longitudinal section after forming with setting 2-2 b with change of feed direction after step 1 and heat treatment between step 1 and step 2.

The heat treatment reduced the stress and restored the material's reshaping capability. Indeed, the heat treatment after the first forming step resulted in significant degradation of the workpiece surface after the second forming step. The heat treated workpieces (setting 1-1 b, 1-2 b, 2-1 b and 2-2 b) also featured more cracks that also penetrated significantly deeper into the structure than for the parts formed without heat treatment. Some of the heat treated samples showed parts on the surface that were torn off. Wherein, the deterioration of the surface was much stronger with the workpieces 1-1 and 2-1, than with the workpieces 1-2 and 2-2. This is due to the fact that with lower feed velocity also less volume was formed during a single stroke and the twisting effect by eccentric rotary swaging was increased. Hence, the probability increased that the cracks opened by the heat treatment and that these opened cracks could be trapped and torn between the swaging tools within the second forming step. This effect was also increased by a higher ductility of the heat-treated samples.

With regard to the homogeneity of the samples, the workpieces with the change of feed direction showed the best results. The combination 2-1 a resulted in significantly less cracks than the other combinations. In the transverse section as well as in the longitudinal micro section it could be observed that the cracks had formed only at the surface of the workpieces with the forming parameters 2-1 a.

In this way, the process kinematic and the intermediate heat treatment influence the microstructure and the material characteristics of the swaged parts. Individual setups give the opportunity to improve the swaging process and to produce the parts with adjustable characteristics depended on the function of the components.

4 Conclusions

A two-stage process of eccentric rotary swaging using eccentric flat shaped dies (EFSD) was analyzed in this study. Experiments were conducted with austenitic stainless steel rods AISI304 with an initial diameter d_0 = 2.7 mm. The influence of changing the feed direction and a heat treatment between the two consecutive forming steps was examined. It could be stated that:

- A change of the feed direction influences the material characteristics of the swaged workpieces. It conducted to a higher hardness and a more homogeneous structure and, hence, changed the material properties of the swaged workpieces.
- In this way multistage forming allows new variations at conditioning of components through the process control.
- The heat treatment between the two forming steps in combination with the change of the feed direction led to a further increase of the hardness and a lower variation of the measured values. This behavior was explained by three mechanisms: the material flow during the different forming procedures; the kind and direction of stress and strain creation; and the change of the crystallographic grains orientation.
- The heat treatment that is commonly used in consecutive cold forming procedures was found to be harmful in multistep eccentric rotary

swaging. The intermediate heat treated workpieces featured deep cracks. The cracks which occur during the first forming step become "open" by the heat treatment and negatively interacted with the tools in the following forming step.

- The multistage eccentric rotary swaging gives an opportunity to produce rods with a fine grain structure homogeneously over the transverse cross section.

For future investigations the influence of the stage number will be analyzed. The investigation of the microstructure and material properties variation by one-stage forming compared to two-stage forming with the same deformation degree should give a better understanding of the material flow by micro rotary swaging. It should be also discussed which adjustments of the process realization lead to highest homogeneity as well as to finest grained microstructure.

5 Acknowledgments

The authors would like to thank the German Research Foundation (DFG) for financial support of the project T2 and A2 within the collaborative research centre SFB 747 "Micro Cold Forming".

References

1. R.Z. Valiev, R.K. Islamgaev, I.V. Aleksandrov, Prog. In Mat. Sc., **45**, 103-189 (2000)
2. R.J. Hellmig, Matwiss. und Werkstofftechnik, **3/04** (2008)
3. E. Moumi, B. Kuhfuss, *Proc. Of 14th Int. Conf on Nanoman* (2014)
4. B. Kuhfuss, E. Moumi, B. Clausen, J.Epp, B. Koehler, Key Eng. Mat., 651-653 (2015)
5. S. Ishkina, C. Schenck, B. Kuhfuss, E. Moumi, K. Tobeck, Int. Journal of Pres. Ing. and Manuf., **18- 7**, 1035-1041 (2017)
6. S. Ishkina, C. Schenck, B. Kuhfuss, *Proc. Of 18th Int. Conf on Mat. Form ESAFORM (2018)*
7. A. Toenjes, S. Ishkina, A. von Hehl, C. Schenck, H- W Zoch, B. Kuhfuss, *8. Kolloquium Mikroproduktion* (2017)
8. R. von Bargen, A. von Hehl, H.-W. Zoch, *Micro Metal Forming*, 56-68 (2013)
9. B. Kuhfuss, E. Moumi, V. Piwek, *Proc. Of 3th Int. Conf on Manuf. ICOMM (2008)*
10. A. A. Lebedev, V. V. Kosarchuk, Int. Journal of Plasticity, **16**, 7-8, 749-767 (2000)

Interactions between feed system and process in production of preforms as linked micro parts

Philipp Wilhelmi[1,3,*] , *Christine* Schattmann[4], *Christian* Schenck[1,2,3] and *Bernd* Kuhfuss[1,2,3]

[1]University of Bremen
[2]MAPEX Center for Materials and Processing
[3]bime, Bremen Institute for Mechanical Engineering, Badgasteiner Str. 1, 28359 Bremen, Germany
[4]Bremer Institut für angewandte Strahltechnik GmbH, Klagenfurter Str. 5, 28359 Bremen, Germany

Abstract. This contribution deals with interactions between feed system and process in the production of preforms as linked parts, which is the first step of a multi-stage process chain for cold forming of micro parts. Due to the interconnection of the parts, the feed system is not only used for part transport, but also for the positioning during the generation of the preforms by laser rod melting. Thereby, the influence of the feed system on the production is more significant. Absorbed laser energy melts a wire, so that a melt pool is formed. While the wire is fixed on one side, the other side is fed into the melt pool whose volume increases. The production can be divided in the steps of preheating, active melting, solidification and transportation. The positioning takes place in parallel to the melting. Until now, the increase of the output rate was based especially on the consideration of the melting process and higher feed velocities. In this contribution, the interactions between feed and process are analyzed with the goal of further increasing the output rate. For that reason, the positioning behavior and its influence on the geometry of the produced preforms are analyzed. Finally, a method is presented, which unites the steps of transportation, preheating and melting. It is shown, that by a favorable coordination of the individual process steps, a further increase of the output rate is achievable without significantly worsening part quality.

Keywords: Handling, Positioning, Laser micro machining

1 Introduction

The technical trend towards increasing functional integration and at the same time more compact products leads to an increasing demand for micro parts. However, technical or economic reasons may complicate the adaption of processes to micro range. To meet the micro specific requirements, not only alternative processes, but also new production methods are necessary. Like stated by Alting in [1] a great variety of possible manufacturing processes for micro production is available. But as the product design may completely change depending on the manufacturing process, it is very important to take these processes into account at an early stage of the product development. The appearing of size effects [2] and the often high sensitivity to mechanical damages complicate handling and thereby make a production at high output rates very difficult [3]. Precise damage-free gripping of micro components as well as overcoming the dominant adhesion forces when releasing is technically challenging [4, 5]. The single manufacturing processes can only be successfully applied, if they go along with a production concept enabling an economically rentable production. Arentoft introduces a multistage micro bulk forming system based on individual part handling in [6]. The system is specified at a rate of 50 parts per minute. As an alternative approach in order to simplify the handling, the production as linked parts is investigated for micro cold forming [7]. In [8] investigations for the production of bulk formed micro parts as linked parts based on sheet metal with a thickness of 1, 2 and 3 mm are presented. In this case, several forming steps are required to produce billets within the strip material. These billets, which remain interconnected by the strip, serve as basis for the actual forming process. In this work, a process chain is considered, which could be used for example to produce micro valve tappets or camshafts. In a first step preforms are generated within a wire by laser rod melting [9, 10], see Fig. 1.

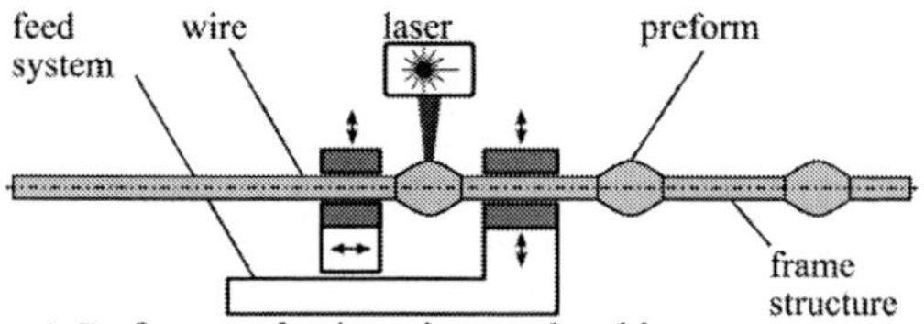

Fig. 1. Preform production – laser rod melting

In a second step, the linked parts can be positioned in a forming tool [11] and the preforms can be cold formed for example by rotary swaging [12, 13], see Fig. 2.

* Corresponding author: wilhelmi@bime.de

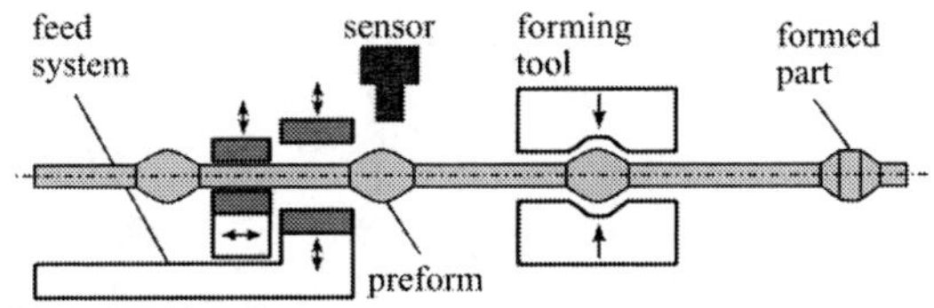

Fig. 2. Forming process

This paper deals with the laser rod melting. The basic principle of the process is analyzed in [9]. In [10] the currently considered process is presented and measures for influencing the part geometry are investigated at a feed velocity of $v_{feed} = 5.9$ mm/s. In [14] the feed velocity could be raised to $v_{feed} = 50$ mm/s and a cycle time according to Table 1 could be reached, by using the new developed feed system, which is also applied for the investigations in the current work.
This contribution addresses the interactions between feed system and process in the production of the linked micro parts. The parts remain interconnected and are fed as a string through all production stages. The feed system is not only used for transportation and positioning between single stages, but also for feeding during processes. It is analyzed, how the feed system behaves, when further reducing cycle times and how this affects the process.

Table 1. Process times for $v_{feed,max} = 50$ mm/s, $x_{feed} = 1.7$ mm and $x_{transport} = 10$ mm [14]

Process step	time
preheating	7 ms
melting	34 ms
solidification	130 ms
transport	100 ms
cycle time	271 ms

2 Methods

The process of laser rod melting is based on the effect that in micro range surface tension dominates against the gravitational force [9]. Thereby, when melting a wire with absorbed laser energy, the melt pool forms a spherical drop and the resulting preform solidifies connected to the wire. For the rod melting of linked parts the feed system feeds the wire during the melting and consequently has a significant influence on the part geometry. The process is illustrated in Fig. 3 and additionally the single actions are listed in Table 2. The feed system is based on two grippers, where one is mounted on a feed axis (Fig. 3, gripper 1) and the other one is fixed in place (Fig. 3, gripper 2). In the initial position, both grippers are closed. Then the laser is switched on with a constant set power P_{laser}. The laser beam is focused on the wire between both grippers at an invariant position relative to the feed system. When the wire starts melting, gripper 1 is moved towards gripper 2. Thereby, further wire material is molten and forms an increasing spherical melt pool. Reaching a certain position x_{feed}, the feed is stopped, the laser is switched off and the melt solidifies. Afterwards, a return stroke follows: Gripper 1 is opened, the axis moves back and gripper 1 is closed again. Then the forward stroke is performed: Gripper 2 is opened and the generated intermediate form is transported by moving gripper 1 to the initial position again. Finally, gripper 2 is closed and a new cycle can start. Considering the motion of the linked parts, two positioning processes are performed. The positioning during the melting step is named feed in the following. The transport is composed of the return stroke and the forward stroke and also involves the gripping. For the scheme in Fig. 3 the part distance $x_{transport}$ is defined by the forward stroke distance $x_{forward}$.

$$x_{transport} = x_{forward} \quad (1)$$

The return stroke is defined as the sum of the forward stroke $x_{forward}$ and the feed x_{feed}.

$$x_{return} = x_{forwrward} + x_{feed} \quad (2)$$

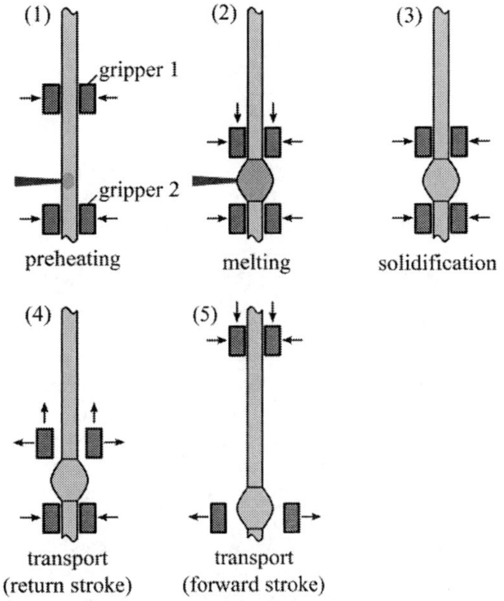

Fig. 3. Laser rod melting as linked parts, sequential process

Table 2. Cycle time in detail

process step	single actions	detailed time	step time
preheating	laser on	-	$t_{preheat}$
melting	laser on, feed	-	t_{feed}
solidification	laser off, wait	-	t_{solid}
transport	open gripper 1	$t_{gripper1,o}$	$t_{transport}$
	return stroke	t_{return}	
	close gripper 1	$t_{gripper1,c}$	
	open gripper 2	$t_{gripper2,o}$	
	forward stroke	$t_{forward}$	
	close gripper 2	$t_{gripper2,c}$	
total cycle			t_{cycle}

In order to reduce the cycle time, the scheme from Fig. 3 can be modified by parallelizing the steps of preheating (1) and transport (5), as illustrated in Fig. 4. In this case, a cycle starts with step (5). The preheating and the closing of gripper 2 are initiated during the forward stroke and a direct transition to the feed is performed without stopping the motion. After the time necessary for closing gripper 2 $t_{gripper2,c}$, the gripper stops the lower part of the wire, which contemporaneously starts to melt. At that time the velocity is reduced from $v_{transport}$ to v_{feed}. When x_{feed} is reached, the feed is stopped and the laser is switched off. The solidification and return stroke are equal to Fig. 3. The parallelization leads to a reduction of the times for closing gripper 2 $t_{gripper2,c}$ and the preheat time $t_{preheat}$.

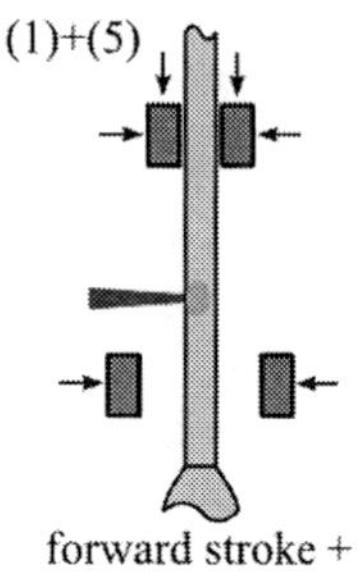

Fig. 4 Parallelization of preheating and transport

For this purpose, an appropriate feed system, which provides adequate acceleration and feed velocity, has been developed. According to the aforementioned principle it is based on two grippers. The feed axis of the moving gripper is driven by a linear direct drive with a direct position sensor. The grippers are actuated pneumatically and their alignment is realized by constructive measures. Segmented prismatic jaws with a center bar enable a self-centering and deformation free gripping. This is especially important to maintain the collinear orientation of the wire ends, when the wire is molten by the laser. To demonstrate the challenges, in Fig. 5 possible process failures that may appear in consequence of an incorrect process coordination are illustrated.

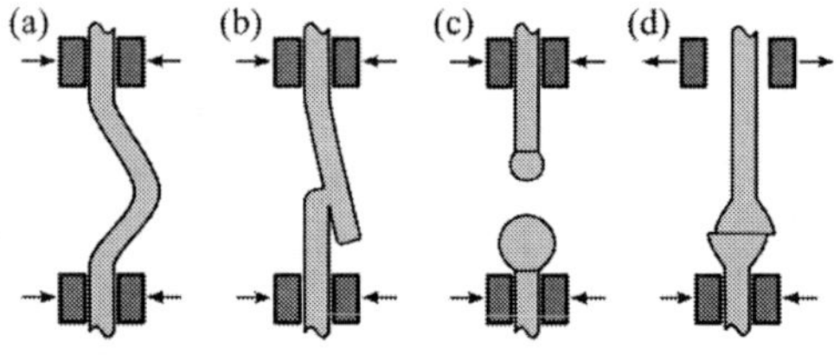

Fig. 5. Process failures

(a) The preheating energy is too low. - The material does not melt and the wire buckles when the feed starts.
(b) The laser power is too low respectively the feed velocity is too high. - The wire is molten, but when the feed starts, the wire ends slide off and no melt drop is formed.
(c) The laser power is too high respectively the feed velocity is too low. - The melt drop moves along the wire in both directions and the wire is separated into two parts.
(d) The material is not solidified completely when the gripper opens for transport. – The wire ends misalign or even separate.

The failures (a), (b) and (c) do not allow producing proper preforms at all. Failure (d) however destroys the preform even though it has been generated properly. Considering this, it becomes clear that the feed system has a high impact on the process results as well as the cycle time and that the coordination of the single process steps is very important. Consequently, in this work, the feed system and different kinds of interactions are considered under the aspect of reducing the cycle time, while keeping the part quality.
The experiments are evaluated by reading out the direct position sensor of the feed system and by measuring the produced linked parts with an area scan camera in an external measurement setup, which is shown in Fig. 6. The probes are clamped in a probe holder, which is moved with a feed axis while the camera is fixed in place. This way, a measurement of the entire linked parts is performed.

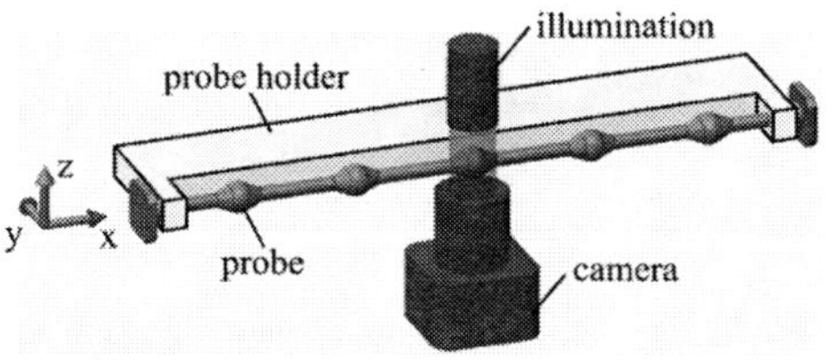

Fig. 6. Probe measurement

As the transmitted light method is applied, the measurement result is a black and white image of the contour. These measurements are two-dimensional, but the samples are not necessarily ideally rotationally symmetric, consequently each sample is measured twice. The first measurement is taken from the perspective, from which the laser beam is aimed on the wire during the experiments. In the second measurement, the sample is rotated around its longitudinal axis (x-axis) by 90°. Fig. 7 shows the measurement of a produced part. Within the image the geometrical parameters, which are determined on basis of an image evaluation, are marked. Those are the part diameter d_{part}, the part length l_{part}, the part volume V_{part} (Fig. 7a)and the net part volume $V_{part,net}$ (Fig. 7b).

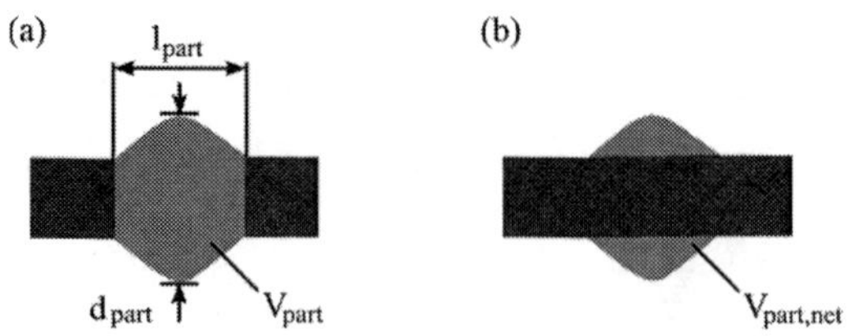

Fig. 7 Measurement of a part - characteristic values

All the parameters are calculated on basis of the diameter curve of the linked parts. The diameter is determined in discrete steps according the effective resolution of the camera. The part diameters are detected by a peak search. The part length is determined by detecting the diameter transitions between wire and part with the help of a diameter limit. Due to the smooth transition, the determination of the part length is more difficult, which leads to higher standard deviations. As a single characteristic value, the shape number G introduced by Schattmann et al. in [13] is used to compare the parts.

$$G = d_{part} / l_{part} \quad (3)$$

The part volume V_{part} is calculated by numerical integration. At each position x, for the measured diameter d_x a cylindrical volume is calculated. These volumes are summated over the part length l_{part}.

$$V_{cyl} = (d_x/2)^2 \cdot \pi \cdot res_x \quad (4)$$

$$V_{part} = \Sigma \ V_{cyl} \quad (5)$$

By substracting a cylindrical volume with l_{part} and wire diameter d_{wire} from V_{part} the net part volume $V_{part,net}$ is calculated.

$$V_{part,net} = V_{part} - (d_{wire}/2)^2 \cdot \pi \cdot l_{part} \quad (6)$$

The advantage of using $V_{part,net}$ is, that it reduces the influence of l_{part} and thereby is more precise than the part volume. It should ideally be equal to the volume $V_{wire,feed}$ of the wire fed during the melting process, which can be calculated on base of the feed $x_{feed} = 2$mm and the wire diameter $d_{wire} = 360$ µm.

$$V_{wire,feed} = (d_{wire}/2)^2 \cdot \pi \cdot x_{feed} = 0.20 \text{ mm}^3 \quad (7)$$

3 Experiments

Table 3 summarizes the parameters, which are used in all experiments. The linear motor provides a maximum force of $F_{max} = 137$ N, which results in a maximum acceleration of $a_{max} \approx 300$ m/s² of the feed system. The assumption is made, that a constant feed velocity and thereby a constant fed volume per time increment is favorable, if a constant laser power is applied. For the rod melting experiments presented in this paper a trapezoidal velocity profile is used, the acceleration is set to $a_{feed} = 100$ m/s² and the feed velocity is varied up to a maximum of $v_{feed,max} = 90$ mm/s. These parameters result in a maximum acceleration time of $t_{acc,max} = 0.9$ ms. The distance in which the acceleration takes place is $x_{acc} = 81$ µm. Between a phase of acceleration and deceleration there is a phase of constant feed velocity. Under the assumption of melting $x_{feed} = 2$ mm of wire as done during the following presented experiments, the time of constant feed is about 20 ms. For lower feed velocities t_{acc} is even shorter both absolute as well as in relation to the feed time t_{feed}, which is needed to cover the feed distance x_{feed}.

Table 3. Common experimental parameters

parameter	value
a_{feed}	100 m/s
v_{feed}	30, 40, ... 90 mm/s
P_{Laser}	100 ... 300 W
$E_{preheat}$	~ 2.1 J
E_{feed}	~ 6.7 J
x_{feed}	2 mm
$a_{transport}$	300 m/s
$v_{transport}$	600 mm/s
$x_{transport}$	10 mm
d_{wire}	360 µm
$V_{part,net,theretical}$	0.2 mm³
material	stainless steel - AISI304
laser	Trumpf TruFiber 300
wave length	1085± 5 nm (cw)
scanner system	Raylase Turboscan 30

The time composition of the current state of the optimized transport cycle is listed in Table 4. The times given in the table indicate when the next step is initiated by the control unit. These times are based on measurements of the single times, like for example the time for closing gripper 1, but they are not necessarily equal to the measured values. It is already considered that some actions can overlap. For example, gripper 1 does not need to be opened completely to start the return stroke (10 ms instead of 21 ms) or the closing can be initiated before the return stroke is completely finished (14 ms instead of 18 ms). The acceleration is set to $a_{transport} = 300$ m/s² and the maximum velocity is set to $v_{transport} = 0.6$ m/s.

Table 4. Transport time in detail

open gripper 1	10 ms
return stroke	14 ms
close gripper 1	21 ms
open gripper 2	17 ms
forward stroke	18 ms
close gripper 2	17 ms
transport time	97 ms

In the first experiments the behavior of the feed unit is analyzed without involving the process. Like described in the previous section a constant velocity is aspired and consequently a trapezoidal velocity profile with high acceleration is applied. The velocity profile serves as base of the reference input for the controller of the feed unit. The actual position profile respectively velocity profile deviates from this depending on the control parameters and the mechanical properties of the system. This is analyzed by varying the velocity between

v_{feed} = 30 mm/s and v_{feed} = 90 mm/s and comparing the reference values to the measured values from the position sensor. The evaluation is based on the velocity profiles, the position profiles as well as a consideration of specific deviations and the positioning time in dependency of the velocity. For each velocity six repeated measurements are performed. The optical position sensor of the feed unit consists of a Renishaw T1000 encoder and a RGSZ scale with a scale pitch of 20 µm. The digital interface has a resolution of 0.2 µm. The feed velocity is gained by numerical differentiation of the position measurement.

For the laser rod melting experiments, a Trumpf TruFiber 300 laser is used. The laser beam is deflected to the machining position with a Raylase Turboscan 30 scanner system. The laser focus position remains fixed at the same point during the process for all experiments. The theoretically induced energy E_{ind} under the assumption of no losses is kept constant throughout all experiments regardless of v_{feed}. This is done by adapting the laser power P_{Laser} and the preheat time $t_{preheat}$ according to the applied feed velocity and the resulting feed time.

$$E_{preheat} = t_{preheat} \cdot P_{Laser} \approx 2.1 \text{ J} \quad (8)$$

$$E_{feed} = t_{feed} \cdot P_{Laser} \approx 6.7 \text{ J} \quad (9)$$

$$E_{ind} = E_{preheat} + E_{feed} \approx 2.1 \text{ J} + 6.7 \text{ J} = 8{,}8 \text{ J} \quad (10)$$

To see the influence of a position deviation Δx from the set value x_{feed}, experiments with v_{feed} = 60 mm/s are performed where an offset in a range of +/- 100 µm is added to the feed distance x_{feed} and the produced linked parts probes are measured. In this case, twenty-two parts for each parameter set are measured. For the measurements a Dalsa Genie TS-1920M area-scan camera is used together with a telecentric camera lens (magnification: 3x). The effective resolution is res_x = 3.67 µm/pixel in x-direction and res_y = 1.83 µm/pixel in y-direction, see Fig. 6. The probes are clamped in the probe holder with a preload force of $F_{pl} \approx 15$ N. As for the performed experiments no significant difference between the measurement of the two perspectives (0° and rotated by 90°) could be noticed, in the illustrated diagrams, the average values of both measurements is used to illustrate the results.

Finally, the modified process scheme with the parallelization of preheating and transport according to Fig. 4 is tested. Parameters for a stable process are determined. These are in this case especially the correct time for closing gripper 2 and switching on the laser. The evaluation is based on the position curve and by measuring twenty-two produced parts.

4 Results and discussion

4.1 Trajectory deviation / positioning behaviour

In Fig. 8 for reasons of clear visibility only three exemplary velocity measurements of the totally seven investigated feed velocities are illustrated. The not illustrated measurements show an analog behavior. The aspired trapezoidal velocity profile with short acceleration and deceleration phases is generally achieved, but a decaying oscillation around the target value of the feed velocity v_{feed} appears. The amplitude of the oscillation increases with increasing v_{feed}. Thereby, for higher feed velocities, the aspired value is strongly exceeded and the oscillation does not end before reaching the target position. Finally, the velocity becomes negative for a short time, which indicates an overshoot, the axis overruns the target position and moves back. Aside from the oscillation, the average velocity and thereby the average fed volume per time increment is constant.

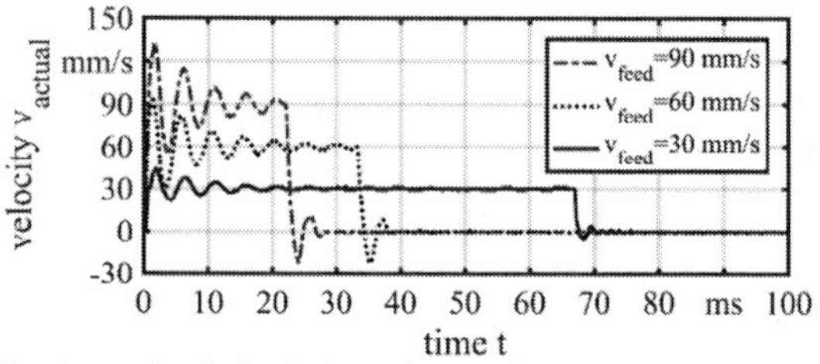

Fig. 8 Actual velocity in dependency of v_{feed}

For a more detailed analysis, the position measurements are considered in the following. Fig. 9 illustrates a curve of the set input position x_{input} and the measured actual position x_{act} for v_{feed} = 60 mm/s as well as the according position deviation in the diagram beneath.

$$\Delta x = x_{input} - x_{act} \quad (11)$$

The highest deviations from the input curve given to the controller appear especially in the first section and when the final position is reached. A controller needs some deviation to build up a force. This happens, when the feed starts. In consequence of the fast increasing input velocity, the deviation increases fast and an oscillation results, as already observed in the velocity curve. The oscillation decays and the mean deviation does not significantly change until the final position is reached at t ≈ 34 ms. In the end of the positioning an overshoot Δx_o is observed, the target position is run over before the position decreases again and the deviation goes back to zero. The positioning behavior is strongly influenced by the control parameters, which in this case are optimized under the two aspects of a fast reaching of the target position and a high stiffness against disturbing forces. Hence, the oscillation was not in focus, when setting up the controller. A further possible explanation for the strong oscillation is the cycle time of the controller, which is 0.2 ms for the position controller and 0.1 ms for the velocity controller. This is close to the acceleration time, which is smaller or equal $t_{acc,max}$ = 0.9 ms depending on the velocity. Consequently, the input value of the velocity controller is almost a step function.

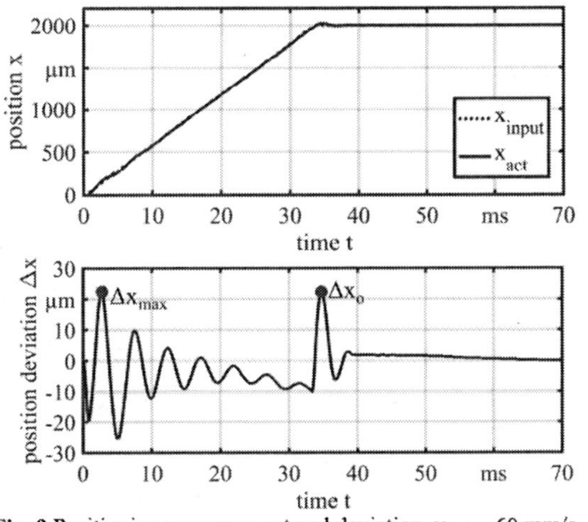

Fig. 9 Positioning measurement and deviation, $v_{feed} = 60$ mm/s

Apart from the oscillation, the maximum deviation Δx_{max} of the position of gripper 1, which appears at the beginning of the feed, and the overshoot Δx_o when reaching the target position are considered as critical values. They are illustrated in Fig. 10. The maximum deviation Δx_{max} increases with v_{feed} and reaches a mean value of more than 30 µm for the maximum tested velocities $v_{feed} = 80$ mm/s and $v_{feed} = 90$ mm/s. However, the overshoot Δx_o only increases until $v_{feed} = 60$ mm/s, where it reaches a maximum of 22 µm. Then decreases slowly and stays on a relative constant level of about 15 µm up to $v_{feed} = 90$ mm/s.

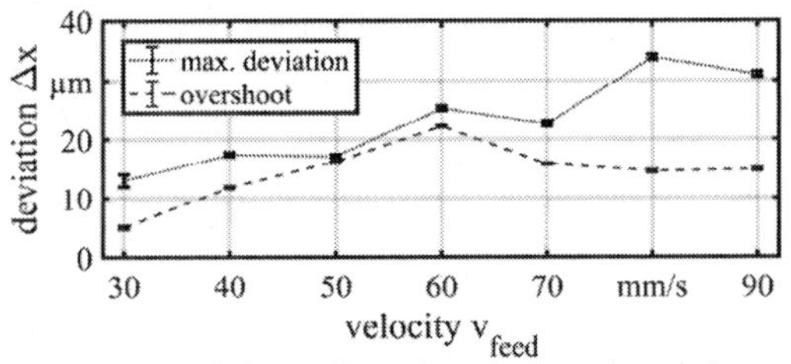

Fig. 10. Max. deviation and overshoot in dependency of v_{feed}

The maximum position deviation Δx_{max} directly influences the melting, where Δx_o may have an effect on the solidification. In the current setup, the laser is switched off after the theoretical ideal positioning time independently from the actual position. Thereby, the overshoot and the moving back happens during the solidification. Taking this into account an analysis of the exact times is done on basis of three characteristic values, which are illustrated in the exemplary curve in Fig. 11. t_{pos} indicates the time when the final value is reached the first time. $t_{5.0}$ indicates the time after the overshoot at which the deviation from the target position stays below $\Delta x = 5$ µm. $t_{0.2}$ indicates the time after which the deviation from the target position stays below $\Delta x = 0.2$ µm, which is equal to the position sensor resolution. The difference between t_{pos} and t_{feed}, after which the laser is switched off, is below 1 ms. It decreases with increasing feed velocity v_{feed}, which is not illustrated here.

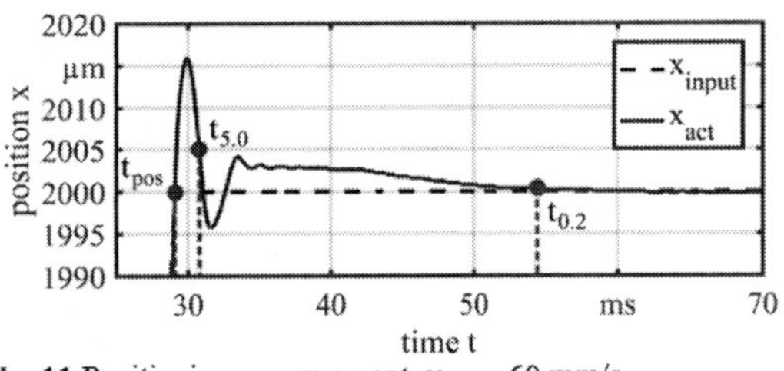

Fig. 11 Positioning measurement, $v_{feed} = 60$ mm/s

Regarding the process, what happens after switching off the laser is more important. For that reason, the following time differences are considered.

$$\Delta t_{5.0} = t_{5.0} - t_{pos} \tag{12}$$

$$\Delta t_{0.2} = t_{0.2} - t_{pos} \tag{13}$$

The results are illustrated in Fig. 12. $\Delta t_{5.0}$ is increasing with increasing feed velocities, from 0.3 ms at $v_{feed} = 30$ mm/s to 5.3 ms at $v_{feed} = 90$ mm/s. $\Delta t_{0.2}$ however firstly increases and reaches a maximum of 37.8 ms for $v_{feed} = 40$ mm/s. Afterwards, it decreases and reaches a value of 23.7 ms for $v_{feed} = 90$ mm/s.

The maximum value of $\Delta t_{0.2}$ is smaller than the solidification time, which in Table 1 is specified to 130 ms. Consequently, there is still a motion during the solidification, which may influence the part geometry. $\Delta t_{5.0}$ is more significant. The maximum time $\Delta t_{5.0}$ of 5.3 ms, within the deviation falls below 5 µm, is relative small compared to the solidification time. Nevertheless, after that there still is a deviation in the range of 5 µm.

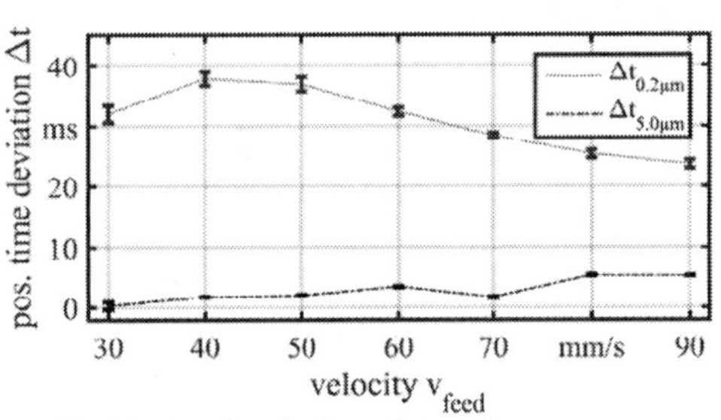

Fig. 12. Positioning time in dependency of v_{feed}

Finally, the laser rod melting has been tested for all given velocities and preforms could be produced without appearing of process failures up to $v_{feed} = 90$ mm/s. Nevertheless, the oscillation could cause problems in case that v_{feed} is further increased. Thus, induced vibrations can have an effect on the geometry.

In Table 5 the sum of preheat time $t_{preheat}$ and melting time t_{melt} is compared between the previous and the actual state. The first row shows the reached result from the previous work [14], which is summarized in Table 1. The second row shows the value reached during the currently presented experiments. As the feed x_{feed} was increased, for reasons of comparability to further investigations in the third row the current result is converted to the feed from the former experiments. This is done by keeping the preheat time $t_{preheat}$, which is independent of the feed and calculating the feed time t_{feed}

for x_{feed} = 1.7 mm. In this case, a cycle time reduction of 15 ms respectively 5 % is achieved.

Table 5. Transport time in detail

source	v_{feed}	x_{feed}	$t_{preheat}+t_{melt}$
[14]	50 mm/s	1.7 mm	7ms+34ms=41ms
actual work	90 mm/s	2.0 mm	7ms+22ms=29ms
act. work converted	90 mm/s	1.7 mm	7ms+19ms=26ms

4.2 Influence of position deviations on the parts

In the previous section, positioning deviations in context of feed velocity have been discussed. With respect to the final geometry especially the solidification and thereby the final positon is expected to be important. To analyze the influence of a position deviation Δx_{feed} from the set feed x_{feed} the net part volume $V_{part,net}$ (see Fig. 13) is considered based on experiments with v_{feed}= 60 mm/s. The data shows that in a range of Δx_{feed} = +/-20 µm the preforms change only slightly. It can be stated that for Δx_{feed} = 0 the net part volume is equal to the volume of the wire fed during the melting $V_{wire,feed}$ = 0.20 mm³. Thereby, the material must have been molten properly despite the oscillations observed from the measurements in the previous section. In the mentioned range of Δx_{feed} = +/-20 µm the net part volume deviation stays within $\Delta V_{part,net} \approx$ +/-1 %. A negative Δx_{feed} clearly reduces the part volume. Considering the maximum negative position deviation Δx_{feed} = -100 µm, a theoretical volume deviation of ΔV_{theo} = 0.01 mm³ results.

$$\Delta V_{theo} = (d_{wire}/2)^2 \cdot \pi \cdot \Delta x_{feed} = 0.01\ mm^3 \quad (14)$$

This is a theoretical deviation of 5% and again conforms to the measurement in the diagram. In contrast, a positive position deviation does not change the volume significantly. A saturation, where no further material is molten appears, is also described in [14].
Relating these results to the results from the previous section, it can be stated that the deviations that appear in consequence of an increased feed velocity do not have a significant effect on the geometry of the preforms. The process seems to be quite robust against small position deviation in the range of Δx_{feed} = +/-20 µm. Especially the overshoot seems to be uncritical.

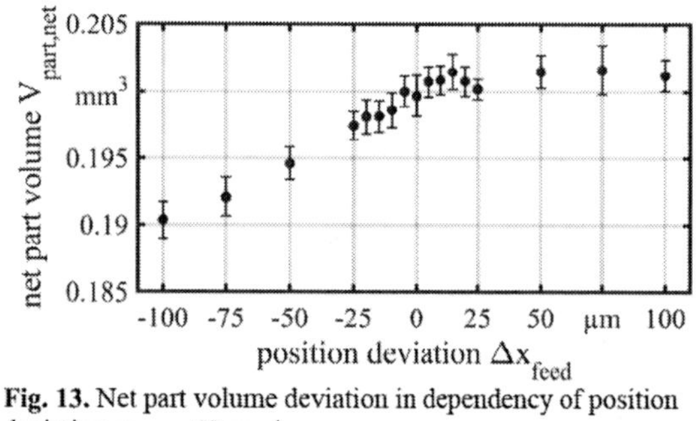

Fig. 13. Net part volume deviation in dependency of position deviation, v_{feed} = 60 mm/s

4.3 Parallelized process

The objective of the investigations in this section is a further cycle time reduction by parallelization. Concretely, the time for closing gripper 2 (17 ms) can be saved. The effect of the parallelization on the feed is considered and probes are produced. Fig. 14 illustrates an exemplary position curve for the parallelized process. In contrast to Fig. 3 it begins with the return stroke (4) and then continues with the parallelized forward stroke (5) and preheat (1) as mentioned in section 2. The transport distance is $x_{transport}$ = 10 mm and the feed distance is again x_{feed} = 2 mm. The return stroke is the sum of both values $x_{return} = x_{transport} + x_{feed}$ = 12 mm. The interesting zone in the curve is the direct transition between transport and feed at t ≈ 81 ms and x = 0 mm, which saves the time for closing gripper 2.

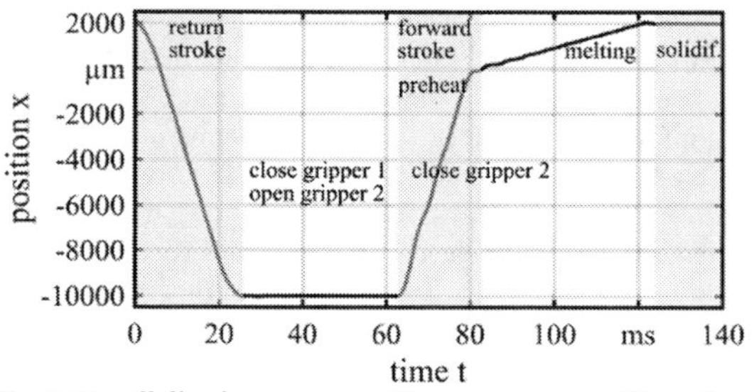

Fig. 14 Parallelized process - position curve, v_{feed} = 50 mm/s and $v_{transport}$ = 600 mm/s

The solidification time and the time for opening gripper 1 are not illustrated completely in the curve, because the feed axis stands still during this time. Values are given in Table 1 and Table 2. Fig. 15 shows the correspondent velocity curve. The feed velocity is v_{feed} = 50 mm/s in this case. The curve corresponds to the curves in Fig. 8 where the sequential process is considered.The same oscillations appear and for the return stroke as well as the transport they are even stronger, which has no significant influence on the feed during the melting. Nevertheless, a stable process could be realized. Table 3 compares the measured parts for the sequential (section 4.3) and the parallelized process. As the coordination of the parallelized process is much more difficult, it is more complex to adapt the part geometry. The differences for the measured characteristic values can mainly be explained by a different feed x_{feed}, which can be recognized by the different net part volume. Nevertheless, the standard deviation is on a similar level for both process variants. A significant difference can only be seen in the part length. However, as explained before, the determination of the part length from the measurements is more complicated, which might be a reason for the difference here. This leads to the assumption, that the new process layout can be applied to reduce cycle time further by 24 ms respectively 8.8 % compared to Table 1 (the sum of preheat time and the time for closing gripper 2). A comparison of the standard deviations shows, that the reproducibility does not suffer by this measure. The deviation in the part geometry is explained by a difference in the effective feed x_{feed}, which for the new approach depends on the time when closing gripper 2.

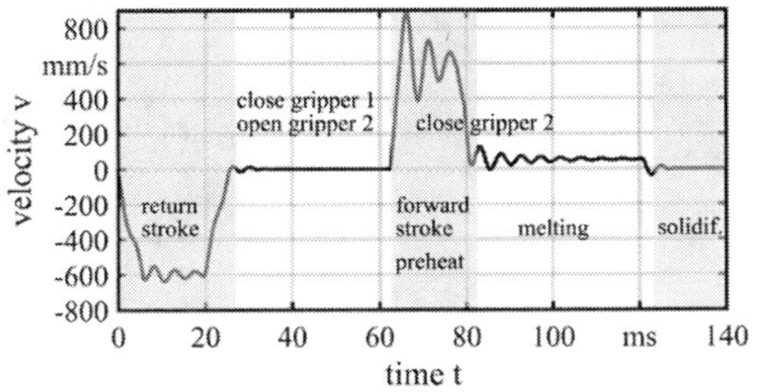

Fig. 15 Parallelized process – velocity curve

Table 3. Comparison sequential and parallelized process

	Sequential process	**Parallelized process**
v_{feed} [mm/s]	60	50
L_{part} [µm]	1135 +/- 32	1281 +/- 37
d_{part} [µm]	743 +/- 5	702 +/- 5
G	0.66 +/- 0.02	0.55 +/- 0.02
V_{part} [mm³]	0.317 +/- 0.004	0.312 +/- 0.005
$V_{part,net}$ [mm³]	0.20 +/- 0.001	0.181 +/- 0.002

5 Conclusion and outlook

The behavior of the feed unit has been investigated in a velocity range between $v_{feed} = 30$ mm/s and $v_{feed} = 90$ mm/s. The applied control parameters were optimized under the two aspects of a fast reaching of the target position and a high stiffness against disturbing forces. Under these conditions a significant oscillation of the actual feed velocity v_{feed} could be observed, which resulted in a maximum position deviation Δx_{max} of more than 30 µm for the maximum tested velocities $v_{feed} = 80$ mm/s and $v_{feed} = 90$ mm/s. The maximum measured overshoot was $\Delta x_o = 22$ µm. Nevertheless, linked parts could be produced for all tested velocities without appearing of process failures. Regarding the influence of a position deviation, it was found, that in a range of $\Delta x_{feed} = +/-20$ µm the net part volume deviation stays within $\Delta V_{part,net} \approx +/-1$ %. Especially a positive deviation seems to be uncritical, due to the appearing of a saturation. Thereby, the deviation in consequence of the overshoot, which falls below 5 µm in less than 5.3 ms is assumed to be uncritical. Finally, a new parallelized process approach could be tested successfully, which allows a further cycle time reduction without worsening the reproducibility of the geometry of the produced parts. By combining all the described measures and under the assumption of a feed of $x_{feed} = 1.7$ mm, the cycle time could be reduced by 39 ms respectively 14.4% compared to [14]. Like this, cycle times are below $t_{cylce} < 232$ ms respectively output rates of more than 258 parts/minute can be reached. The process resulted to be quite robust against the appearing oscillations and the resulting deviations. Nevertheless, for the tested velocities the feed unit is operated at its limits and process failures could appear, when further increasing the feed velocity. Measures that could be tested for reducing those oscillations are mainly seen in the controller. This could be a controller with shorter cycle times or an adaptive control.

The authors gratefully acknowledge the financial support by Deutsche Forschungsgemeinschaft (DFG, German Research Foundation) for the subprojects C5 "Linked Parts" and A3 "Material Accumulation" within the SFB 747 (Collaborative Research Center) "Micro Cold Forming—Processes, Characterization, Optimization".

References

1. L. Alting, F. Kimura, H.N. Hansen, G. Bissacco, CIRP Annals Manuf. Tech. **52**, 635 (2003)
2. F. Vollertsen, D. Biermann, H.N. Hansen, I. S. Jawahir, K. Kuzman, CIRP Annals Manuf. Tech. **58**, 566 (2009)
3. H. van Brussel, J. Peirs, D. Reynaerts, A. Delchambre, G. Reinhart, N. Roth, M. Weck, E. Zussman, CIRP Annals Manuf. Tech. **49**, 451 (2000)
4. G. Fantoni, M. Porta, *IFIP Micro Assembly Technologies and Applications*, 223 (Springer, New York, Berlin, Heidelberg, 2008)
5. M. Tichem, D. Lang, B. Karpuschewski, Ass. Aut. **24**, 88 (2004)
6. M. Arentoft, R.S. Eriksen, H.N. Hansen, N.A. Paldan, CIRP Annals Manuf. Tech. **60**, 335 (2011)
7. B. Kuhfuss, E. Moumi, K. Tracht, F. Weikert, F. Vollertsen, A. Stephen, *Proceedings of 14th international ESAFORM conference on material forming*, (AIP, 2011)
8. M. Merklein, T. Stellin, U. Engel, Key Eng. Mat. **504-506**, 587 (2012)
9. F. Vollertsen, R. Walther, CIRP Annals Manuf. Tech. **57**, 291 (2008)
10. C. Schattmann, C. Hennick, H. Bruening, F. Pfefferkorn, F. Vollertsen, Journal for Technology of Plasticity **42(1)**, (2017)
11. P. Wilhelmi, C. Schenck, B. Kuhfuss, Journal of Mech. Eng. and Autom. 7, **44–49**, (2017)
12. H.Lin, C. Lee, Y. Lin, C.-H. Li, Key Engineering Materials, **623**, 725 (2014)
13. P. Wilhelmi, E. Moumi, C. Schenck, B. Kuhfuss, *Proceedings of 5. Kolloquium Mikroproduktion (german)*, (2015)
14. C. Schattmann, P. Wilhelmi, C. Schenck, B. Kuhfuss, F. Vollertsen, *Proceedings of 8. Kolloquium Mikroproduktion (german)*, (2017)

Application of Cause-Effect-Networks for the process planning in laser rod end melting

Daniel Rippel[1,*], *Christine* Schattmann[2], *Mischa* Jahn[3], *Michael* Lütjen[1] and *Alfred* Schmidt[3]

[1]BIBA – Bremer Institut für Produktion und Logistik GmbH at the University of Bremen, 28359 Bremen, Germany
[2]BIAS – Bremer Institut für angewandte Strahltechnik GmbH, 28359 Bremen, Germany
[3]University of Bremen, Center for industrial mathematics (ZeTeM), 28359 Bremen, Germany

Abstract. In micro manufacturing, a precise configuration of manufacturing processes constitutes an essential factor for success. The continuing miniaturization of work pieces results in ever decreasing tolerances, whereas machines and processes become more and more specialized. As a result, a precise determination of each process result is important to guarantee the final product quality. Unfortunately, so called size effects often prevent the direct transfer of knowledge from the area of macro manufacturing. To cope with these effects, finite element simulations provide a suitable tool to simulate forming processes and their results in advance and to perform parameter studies in order to analyze the process and effect interdependencies. Unfortunately, these simulations usually require a rather long computing time, so that only simulations for a small subset of the available parameter range can be performed in a reasonable planning interval. In this context, this article presents an application of the method "Micro – Process Planning and Analysis" (µ-ProPlAn) for the configuration of laser rod end melting, which is used to create preforms for further forming processes. This method uses cause-effect networks, to combine expert knowledge with methods from artificial intelligence to estimate the result of laser melting processes quickly. For this purpose, the cause-effect networks are trained using a finite element simulation of the laser process using different process parameters and varying rod diameters. Results show a high accuracy for the prediction of the finite element simulation results. This article focusses on the validation of these cause-effect networks in comparison to the real laser rod end melting process and demonstrates how these models can be used to predict the resulting volume, eccentricity and the largest diameter of the solidified preform for different process configurations.
Keyword: Laser Micro Machining, Simulation, Predictive Model

1 Introduction

During the last decades, the demand for metallic micro parts continuously increased. On the one hand, these components become increasingly smaller while on the other hand, their shape's complexity and their functionality constantly increases [1], [2], [3]. Major factors contributing to this development are a rising number and complexity of applications for micro components as well as an increasing demand for such components in the markets of medical- and consumer-electronics [3]. Besides the growing demand for Micro-Electro-Mechanical-Systems (MEMS), generally produced using methods from the semi-conductor industry, the demand for metallic micromechanical components grows similarly. Often used as connectors for MEMS, casings, or contacts, these components cannot be manufactured using semi-conductor technology but are manufactured by applying processes from micro forming, micro injection, micro milling etc. [2], [4]. In this context, cold forming processes constitute an option for an economic mass production of metallic micromechanical components, as these processes generally provide high throughput rates at comparably low energy and waste costs [5].

An efficient industrial production of such components usually requires high throughput rates up to several hundred parts per minute [6], whereby very small tolerances have to be achieved. These tolerances result from the components' small geometrical dimensions, which, by definition, are below one millimeter in at least two dimensions [7]. Additionally, so called size-effects, defined as "deviations from intensive or proportional extrapolated extensive values of a process, which occur when scaling the geometrical dimensions" [8], can result in increasing uncertainties and unexpected process behaviors when processes, work pieces, and tools originating from the macro domain are miniaturized.

As a result, the planning and configuration of process chains is seen as a major factor of success for an industrial production of metallic micromechanical components [9]. To cope with the occurrence of size-

* Corresponding author: rip@biba.uni-bremen.de

effects, companies require tools and methods for a highly precise planning, not only of the single processes and their configurations but also spanning the complete process chain. Thereby, they have to consider interrelationships between processes, materials, tools and devices. In micro manufacturing, small variations in single parameters or characteristics can have a significant influence along the process chain and can finally impede the compliance with the respective tolerances [10].

Consequently, the planning and configuration of processes requires highly detailed information on the effects and behaviors of the involved processes. As companies often face a lack of long-standing expertise with particular processes, physical or mathematical simulation models are typically used to provide this information so that different process configurations or process chains can be analyzed and compared thoroughly. Unfortunately, these simulations often take a long time to compute and have to be performed individually for varying process parameters and different process configurations. This is especially disadvantageous since the continuous development of new processes and machines leads to a broad range of available process chains for the same product. Each of these alternative process chains may provide different advantages and disadvantages regarding the product's quality as well as the performance of the production system. Hence, it is necessary to design, configure and evaluate several process chains in order to determine the best solution. In order to facilitate the planning and configuration process, this article presents the application of the "Micro – Processes Planning and Analysis" (μ-ProPlAn) methodology as an additional layer of abstraction between the process planning and the configuration of process chains via Finite-Element Models (FEM). The methodology itself provides tools and methods to model, configure and evaluate process chains in micro forming [11]. Whereby the process design relies on classical approaches, the configuration is performed by the application of so-called cause-effect networks, detailing the interrelationships between relevant process-, machine-, tool- and work piece parameters. These cause-effect networks can act as surrogates for numerical FEM simulations, thus increasing the planning speed. The aim is to provide approximate results with less computational efforts during the planning stages.

2 Process Planning and Configuration in Micro Manufacturing

There exist only few approaches that support a joint planning of process chains and its technological and logistic configuration. In contrast, during the last years several articles focused on the configuration of specific processes (compare e.g. [12]). Most of these approaches rely on detailed models of the corresponding processes and are usually supported by very detailed physical models in form of finite element simulations (e.g. [9], [13]). A different kind of approach found in the literature focusses on the use of historical data as templates for the configuration (e.g. [14]). Although both of these approaches allow a precise configuration of single processes, they are not directly applicable to process chains without conducting comprehensive studies of all involved processes. Even then, the interrelationships between these processes cannot be addressed directly. In addition, the construction of finite element simulations as well as the direct application of historical information requires a comprehensive understanding of the processes, their behaviors as well as of the physical backgrounds, which, in many cases, is unavailable due to size-effects or the novelty of processes.

The planning of process chains is generally conducted using methods like event-driven process chains, Unified Modelling Language, or simple flow charts. As these methods do not allow a configuration of processes, Denkena et.al. proposed an approach that indirectly addresses this topic [15]. This approach uses the modelling concept for process chains [16]. Thereby, a process chain consists of different process elements, again consisting of operations. These operations connect to each other by so-called technological interfaces, describing sets of pre- or post-conditions for each operation. While these interfaces can be configured manually, Denkena et.al. propose the use of physical, numerical or empirical models to estimate an operation's post-conditions based on its pre-conditions [15], [17]. Although this approach bridges the gap between process chain planning and the configuration, the creation of these models requires very detailed insight into the processes as stated before. In addition, a single model can be unsuitable to capture all relevant interrelations between process-, machine-, tool- and work piece-parameters, particularly due to the influence of size-effects.

Based on the literature review, it can be concluded, that there exists no method enabling a joint planning and configuration of process chains for the micro domain, which provides the necessary level of detail and generality to cope with size-effects as well as with the continuous development of new technologies in this domain. While classic methods from the area of process planning lack the required level of detail to enable process configurations, methods for the configuration of (forming) processes usually come with comparably high demands in computational power and time, rendering them unsuitable for an application in process chain planning tasks.

One approach to achieve the required computational speed as well as the level of detail is the use of surrogate functions. Surrogates usually substitute computationally expensive simulations with an easier to compute function or model to increase the speed [18]. In general, two different approaches to construct a surrogate model can be observed in the literature. On the one hand, surrogates can be acquired by model reduction. On the other hand, surrogates can be created by learning an approximate model based on a small set of evaluations. In the context of finite-element simulations, both approaches can be found in the literature [19].

For complex FEM simulations, so called low-fidelity models or physical surrogates can be constructed in various ways. Examples are the focusing on relevant parts of the model, by reducing the models degree of detail (selecting coarser meshes) or by replacing certain structures with adequate, faster to compute structures [19]. In this context, recent work focused on the combination (fusion) (e.g. [20]) or automatic deduction (e.g. [21]) of high and low-fidelity models to achieve higher computational speeds while maintaining the required precision. While the use of low-fidelity models can achieve high decreases in the required computational time, the according simulations are still time consuming. For the task of designing and configuring process chains, several of such simulations are required to first determine the most promising sequence of operations or machines and second, to evaluate different configurations for those. Consequently, the use of approximate models, so called response models is a more suitable approach for this task.

Response models usually try to approximate the simulation results based on selected properties. Therefore, a set of FEM simulations is conducted and the response model is then modelled or learned according to the simulation results. In this context, statistic methods from the Design of Experiments are usually applied to plan and execute the simulations. In the literature, there exist several studies on the relationship between problems and suitable methods to learn the response model (e.g. [22], [23], [24]). According to Goel et al. [25], methods commonly proposed in the literature come from the area of statistics (e.g. Kriging, polynomial Response Surfaces) or use methods from the field of Artificial Intelligence (e.g. Artificial Neural Networks or Radial Basis Functions) to learn the relation between the simulations input settings and the simulation results (e.g. [26] [27]). While these models lack the descriptive properties of FEM simulations and thus can only be validated using key performance indicators common to regression techniques (see [24] for a summary of common measures), they can be calculated very quickly.

In the context of process chain planning and configuration, response models can provide a substantial decrease in computational times, which is particularly important to evaluate different process chains and configurations. In order to structure the process of deriving surrogate models and to integrate them into a consistent framework for the planning, configuration and evaluation of process chains in micro manufacturing, the µ-ProPlAn methodology can be used.

3 Alternative upsetting process: Two-stage cold forming

In order to upset a certain length l_0 of a rod with diameter d_0 in macro scale, conventionally a multi-stage cold forming process is used. While the achievable upset ratio $s = \frac{l_0}{d_0}$ is already limited to approximately 2.1 in macro scale [28], this upset ratio decreases significantly in micro range due to size effects.

Therefore, an alternative upsetting process based on laser rod end melting and a subsequent forming step has been developed. Here, the shape-balance effect enables the production of a material accumulation. As shown in Fig. 1 and described in [8], a laser beam is deflected laterally to a thin rod and the material melts because of the absorbed laser energy. Due to the shape-balance effect, i.e. surface tension predominates gravitational force, the melt pool forms a spherical and moves upwards simultaneously to the laser beam while staying connected to the rod [29]. The thereby accumulated material retains its spherical shape even after the laser is switched off. After solidification, it is called preform. Depending on the material mass as well as the accumulated volume, comparable upset ratios of $s > 200$ can be reached [30]. The completely solidified preform is then formed into the requested final geometry during a subsequent cold forming step.

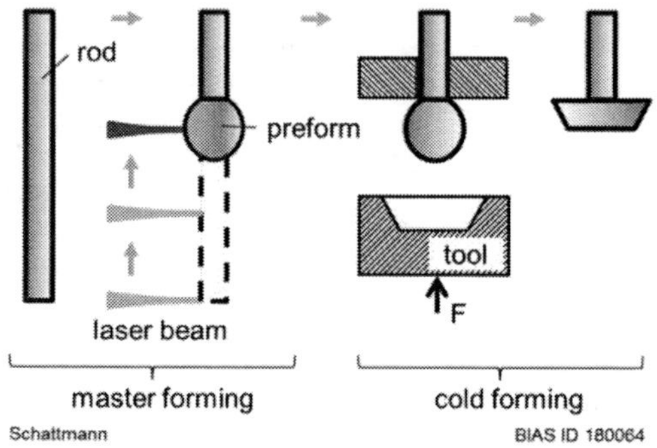

Fig. 1. Laser rod end melting as alternative to upsetting

Since the master forming step defines the shape, eccentricity, and volume of the preform, a thorough understanding of the process is mandatory to match small tolerances requested by the industry. Therefore, a partial differential equation based model of the process coupling the Stefan problem [31] with the Navier-Stokes equations including a free capillary surface, cf. [32] has been established and a corresponding finite element method has been developed. This simulation allows for a numerical computation of suitable parameters and process results and is used among other things to analyze the energy balance in the work-piece and the dynamics in the melt (e.g [29], [33], [34]).

4 µ-ProPlAn

In order to create surrogate models for the FEM simulation, the modelling methodology µ-ProPlAn is applied. In general, it covers all phases from the process and material flow planning to the configuration and evaluation of the processes and process chain models [10]. It focusses on an integrated planning of manufacturing, handling and quality inspection activities at different levels of detail, from process chains, to the level of cause-effect relations between single parameters. At this, µ-ProPlAn applies so-called cause-effect networks. In contrast to holistic approaches, like FEM simulations described in the state of the art section or artificial neural networks, each network consists of a set of parameters and a set of cause-effect relationships, forming a directed graph of sub-models. The set of parameters consists of all technical and logistic characteristics that are relevant to describe the influence on the production process. In case of work pieces, these are e.g. costs per unit, material properties or geometrical characteristics. As for machines and devices, these parameters include velocities, forces or other characteristics that can be selected, calculated or measured. The cause-effect networks are constructed hierarchically. Each material flow object (work pieces, machines, tools, workers, etc.) holds its own cause-effect network, or at least a set of parameters. When combining these single elements to operations, process elements or process chains, higher-level cause-effect networks are created by introducing additional relationships between parameters of the networks or by connecting them through previously specified process interfaces.

The creation of cause-effect networks is divided into two steps: the qualitative modelling and the quantification. The qualitative model is designed by collecting all relevant parameters and denoting their influences among each other. The second step concerns the quantification of the cause-effect networks. The objective is to enable the propagation of different configurations throughout all connected networks. For instance, to estimate the characteristics of a product for different materials or machining strategies. Through quantifying the cause-effect relationships, it is possible to estimate the results of parameter changes to all connected parameters along complete process chains. In contrast to holistic approaches like artificial neural networks, cause-effect networks use a different model for each parameter instead of a single, large model for the overall process. As a result, cause-effect networks are highly customizable as each parameter can be described using different formulas or model types.

In case of well-known relations, μ-ProPlAn allows the direct input of mathematical equations for cause-effect relations. Thereby, the network calculates the value of a parameter directly, based on its input parameters' values. For example, a continuous manufacturing process generally has a duration according to the total length of the work piece divided by its feed velocity. More complex but well-established cause-effect relations can be included from literature, e.g. the calculation of the static friction between tools and work pieces based on their surface roughnesses. Nevertheless, in micro manufacturing some parameters can have a significant impact on the process chain, which can be neglected in macro manufacturing. In addition, size effects may induce a different behaviour than expected in macro forming. As a result, it is often impossible to describe all parameters and cause-effects relations directly. In such cases, μ-ProPlAn provides several methods to derive or learn prediction models from experimental or production data for the respective parameters. Therefore, the software prototype offers a variety of regression techniques from the area of Artificial Intelligence (e.g. Support-Vector Machines or Artificial Neural Networks) as well as statistical methods (e.g. least-square linear or polynomial regressions as well as localized weighted linear regressions). In practice, the application of locally weighted linear regression models has shown a high precision in estimating unknown or hard-to-describe cause-effect relations (see e.g. [10], [35]). After quantification, each parameter holds a model (mathematical or empiric) that allows to estimate is value based on the remaining values on the cause-effect network.

As a result, these networks enable a fast and, depending on the selected models, precise evaluation of different process configurations (e.g. the use of different materials or different production velocities). They enable an assessment of the impacts of different choices on follow-up processes or the production system in general. As cause-effect networks and material flow elements are closely connected, μ-ProPlAn can directly reflect changes to the configuration within the material flow simulation and evaluate these configurations e.g. regarding work-in-progress-levels, lead times or the products' estimated qualities.

5 Cause-Effect Networks as Surrogate for Laser Rod End Melting

While cause-effect networks constitute an option to achieve a suitable process configuration as part of the overall process chain design, their quantification requires a sufficiently extensive set of production or experimental data. Within a running production system, this dataset can be acquired relatively easy by relying on production and measurement data for already existing processes. For new machines, technologies or products, this information may not be available directly. As a result, it often becomes necessary to resort to physical or numerical models (e.g. finite element models) during the process chain design. These models can be used for the construction and during the quantification of cause-effect networks. In such cases, the cause-effect networks can act as surrogates during the process chain design, to achieve a fast configuration of alternative process chains. This section demonstrates the design and quantification of a cause-effect network for a laser based melting process, specifically designed for micro manufacturing.

5.1 Characteristics of generated preforms

As described before, the laser rod end melting process is the essential process step of the two-stage forming process. The geometry of the preforms depends primarly on the laser power, the deflection velocity as well as the rod diameter. Brüning and Jahn have investigated the influence of the aforementioned parameters on different geometrical features in a series of articles:

In [29] the specific molten volume in dependency of the thermal upset ratio is determined for different rod diameters. It is shown that a decreasing laser intensity with an increasing thermal upset ratio leads to a constant specific melting volume for all rod diameters with appropriate laser powers.

In [33] the influence of the cold forming stage on the eccentricity is shown. The diameter of preforms varied between 0.38 mm and 0.43 mm by constant rod diameter $d0 = 0.2$ mm, processed with a laser power of 102 W, a deflection velocity of 72 mm/s and a variation in thermal upset length of 0.9 mm to 1.2 mm. The results show that the eccentricity varies up to 98 μm. After the forming stage, it is still in the range of 30 μm.

In [34] the influence of the keyhole formation on the eccentricity of the preform is investigated. It is shown, that the preforms generated with a laser power of 51 W exhibit a higher eccentricity than the preforms generated with a laser power of 102 W, which leads to a keyhole formation. It is assumed that the heat-affected zone is smaller when using the keyhole configuration instead of the Fresnel absorption. Furthermore, the melt front is oriented almost parallel to the keyhole. During Fresnel absorption, the heat is absorbed and reflected at the surface, which leads to an inclination of the melt front and therefore to a high eccentricity.

5.2 Experimental setup and characterization

In order to design and quantify a cause-effect network for the laser rod end melting process, a parameter study is performed experimentally and numerically. Using the parameters listed in Table 1, preforms for each set of parameters are generated and simulated. Afterwards, their diameter, volume, and eccentricity are measured.

Table 1. Process parameters

	Simulation	Experiment
material	AISI304 (X5CrNi18-10)	
shielding gas	Argon	
gas flow	20 Nl/min	
beam radius	25.5 μm	
upset length	3 mm	
rod diameter	0.2 … 0.4 mm	
laser power	40.3 … 223 W	
deflection velocity	30 … 75 mm/s	

Maximum radius: The diameter of the preform is determined experimentally by the largest diameter of the preform orthogonal to the rod as shown in Fig. 2a. The simulation determines the deviations between the surface area and the center of the preform and determines for each quadrant the maximum deviation l_S as shown in Fig. 2b. To account for possibly skewed preformes, the mean value of all maximum radii is determined and used as maximum estimate for the overall radius r_p resp. the preform's diameter.

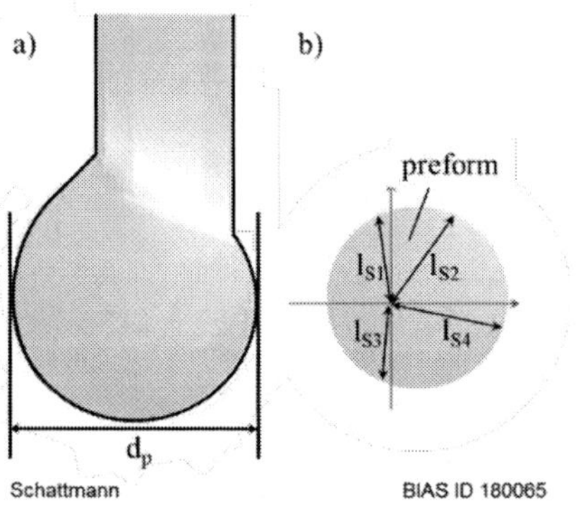

Fig. 2. Maximum diameter

Volume: While the volume of the preform can be identified easily within the finite element simulation by adding together the volume of all elements in the corresponding subdomain, measuring the volume of the experimentally generated preforms is more complicated. For this purpose, an additional line is marked at the rod as shown in Fig. 3. To calculate the final accumulation length l_F, equation (1) is used

$$l_F = (l_L + l_P) - \frac{\sum l_n}{n}, \tag{1}$$

with l_L as theoretical accumulation length, l_P as distance between line and preform, l_n as measurement $1..n$ of the real distance between the additional line and the preform as it is shown in Fig. 3. Using this averaged accumulation length, the molten volume constituting the preform can be calculated by the equation for a cylinder volume as shown in equation (2)

$$V = \pi \cdot \left(\frac{d_0}{2}\right)^2 \cdot l_F, \tag{2}$$

with the diameter of the rod d_0.

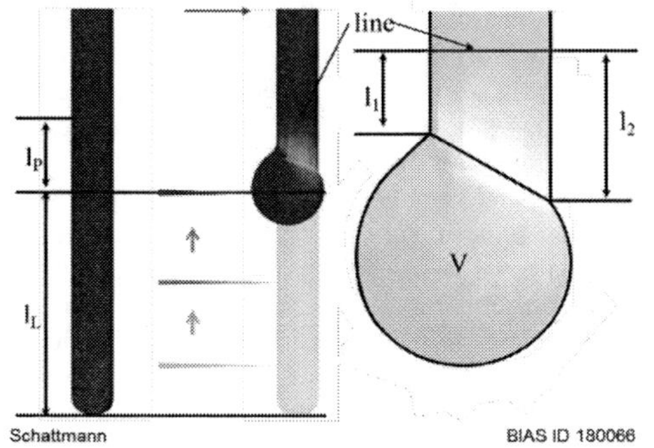

Fig. 3. Volume

Eccentricity: In regards to the eccentricity, the distance between the center of the rod and the center of the preform Δx is determined for all generated and simulated preforms as shown in Fig. 4. Therefore, the center of the preform is determined experimentally by fitting a circle around the spherical shape. Since the mount of the rod is installed permanently, the initial rod position is constant and thus the distance can be measured with a microscopy.

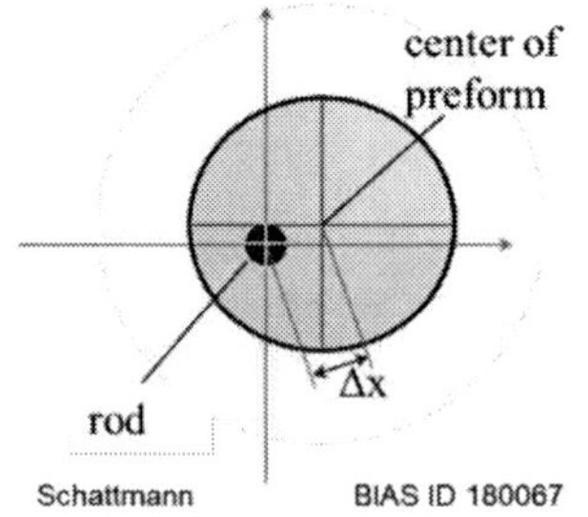

Fig. 4. Eccentricity

6 Results

Based on the parameters of the process and the FEM simulation, the cause-effect network is constructed using the laser power, the deflection velocity as well as the rod diameter (radius) as inputs. The cause-effect network is evaluated in terms of its accuracy with respect to the simulation results (verification) as well as to the experimental results (validation).

6.1 Quantification

Using a training set of 56 simulation runs, the cause-effect network learns a prediction model for each output parameter using a linear regression. Thereby, the cause-effect network tries to estimate the resulting radius r_p of the preform, the eccentricity e, the volume of the preform V, as well as the accumulation length l_F of rod material that is molten during the process. For the estimation of the length, the energy per unit length (see equation 3) as well as the rod's diameter were selected as input parameters.

$$energy\ per\ unit\ length = \frac{laser\ power}{deflection\ velocity}, \quad (3)$$

For the other dependent parameters r_p, e, V, the rod's diameter, the laser power and the deflection velocity were selected as input parameters (see Fig. 5). The linear regression was chosen to construct the prediction models for all the parameters, as the resulting models already showed a high accuracy concerning the training and evaluation datasets. Although more complex models, e.g. locally weighted linear regressions, achieve an even higher accuracy, linear regression models provide several advantages. For example, they can be used to weight the importance of the input parameters, or can be reformed to calculate specific inputs for a desired result.

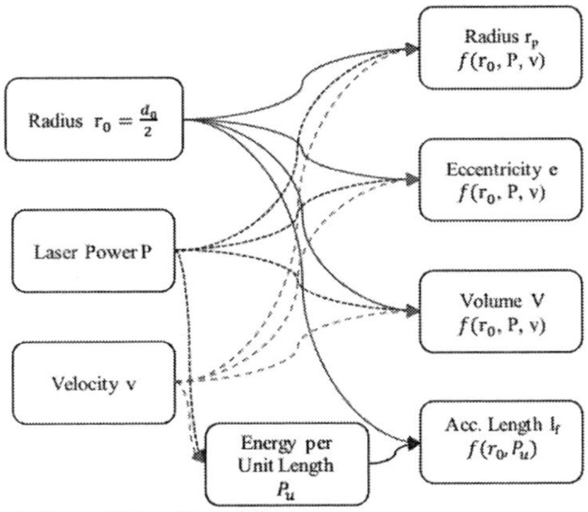

Fig. 5. Cause-Effect Network

Table 2 summarizes the evaluation of the prediction models. The first two columns present the Pearson product-moment correlation coefficient (PPMCC) as well as the Spearman rank correlation coefficient (SRCC) to assess linear and non-linear correlations between the prediction model and the training set. Thereby, values close to 1 represent a high correlation, i.e. the regression model closely represents trends in the presented dataset. Values close to 0 mean that there is no correlation between the model and the dataset. The next column depicts the mean absolute error (MAE), i.e. the sum of residuals between the models predictions and the provided dataset. At this, a low value represents a good approximation of the model in relation the single values within the dataset. Nevertheless, as prediction models try to generalize and smooth the input data, a certain error is to be expected, depending on the training data's variance. The last column provides each parameters value range to allow a better assessment of the error values. For example, an error of 1.0 can be high if the value range only ranges between 0 and 1, whereby it is small if the value range lies between 0 and 100.

Table 2. Quantification results

	PPMCC	SRCC	MAE	Range
r_p	0.949	0.908	14.499	[284; 325]
e	0.824	0.839	17.082	[5.9; 133.2]
V	0.997	0.957	0.0066	[0.08; 0.12]
l_F*	0.983	0.946	60.166	[2959; 4111]
* The length was only trained using data for rods with a diameter of 0.2mm.				

The results of the quantification show very high correlation coefficients and comparably low errors. Consequently, the cause-effect network provides a good approximation concerning the FEM-simulation results.

6.2 Validation

To evaluate the prediction model, the cause-effect network is presented a dataset containing production and measurement data of 281 preforms generated using the settings described in Table 1.

For each dataset, the cause-effect values input parameters are set according to the dataset, i.e. the rod diameter, the laser power and the deflection velocity. Afterwards, the cause-effect network predicts the corresponding results. These predictions are then compared to the measurements, noted within the evaluation dataset. Overall, the evaluation of the complete dataset took less than a second on a standard desktop computer (i7-3770k, 8GB RAM, JAVA8). The results of this evaluation in terms of correlation and error are presented in Table 3. Further details are presented in the respective subsections. Thereby, each of these subsections provides a figure of the predictions and the expected results. In order to generate these figures, the prediction models were reduced in dimensionality. Originally, each model uses two or three parameters as input. To generate these figures, the models response, as well as the measured values are provided only for a single input (energy per unit length for the radius, volume and length or laser power for eccentricity). As a result, the variance of measurements in these figures is usually higher, as each mark represents several configurations. The same accounts for the prediction model. While these usually represent a plane in 3D space or a hyper-plane in 4D space, they are depicted only as a single line in these figures, neglecting the models' additional dimensionality. As a result, the figures are only meant to provide a general impression of how the prediction models conform the measurements. Their overall performance can be seen in Table 3.

Table 3. Evaluation results

	PPMCC	SRCC	MAE	Range
r_p	0.989	0.828	27.69	[277; 527]
e^*	0.08	0.0	19.38	[47; 111]
V	0.937	0.879	0.0544	[0.08; 0.55]
l_F**	0.731	0.703	1422.15	[2747; 3970]

* The eccentricity was only evaluated for some rods with a diameter of 0.3mm. See text for more details.
** The length was evaluated for all rod diameters. Nevertheless, the model was only trained using data for diameters of 0.2mm. See Table 2 for more details.

Maximum radius

With regard to the maximum radius of the preform r_p, Table 3 shows a high correlation between the cause-effect network and the evaluation dataset. Nevertheless, Fig. 6 shows, that the prediction model increasingly underestimates the measured radius with an increasing energy per unit length. This effect was also observed in regards to the FEM simulation [36] and is caused by assuming material parameters to be constant for each physical condition. The same is true for all parameters controlling heat dissipation mechanisms.

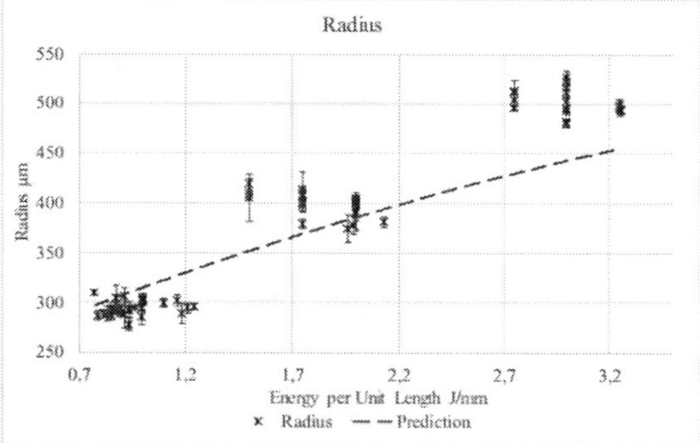

Fig. 6. Maximum radius - Predictions and Measurements for all rod diameters

Eccentricity

In regards to the eccentricity, it turned out that, firstly, the eccentricity varies a lot due to the high dynamics of the melt during the process and, secondly, that finding a reliable and precise measuring method is rather complex, especially for rods with small diameters because of handling and buckling effects. Therefore, experiments were only conducted for a rod diameter of d_0 = 0.3mm. Fig. 7 presents the comparison of the prediction with 13 measurements. As can be seen in the figure, the standard deviation of the measurements is still comparably large. Despite the low correlation coefficients given in Table 3, a general trend for a decreasing eccentricity with increasing laser power can be observed in the experimental data and likewise in the cause-effect networks' predictions.

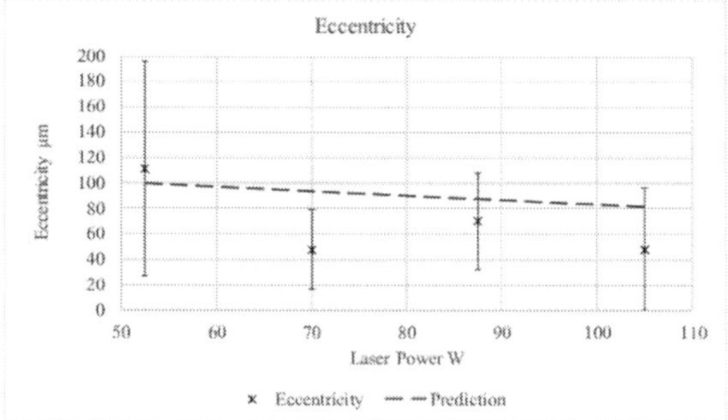

Fig. 7. Eccentricity - Predictions and Measurements for rods with a diameter of 0.3mm

Volume

In general, the results for the volume are comparable to the results concerning the maximum preform radius. Table 3 shows a high correlation between the cause-effect network and the evaluation dataset, whereby Fig. 8 depicts the same behavior of underestimation as with the radius. As the volume of the melt pool forms approximately spherical relating to the surface tension, it is expected that the radius of the spherical melt pool is comparable to the maximum radius of the preform. The underestimation is explainable again by the constant parameters for the physical condition of the simulations.

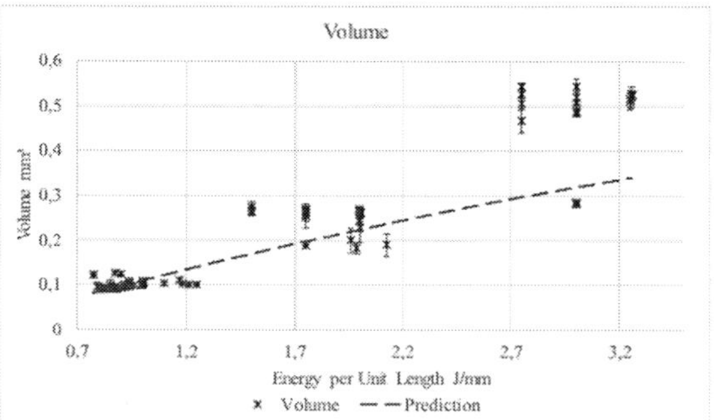

Fig. 8. Volume - Predictions and Measurements for all rod diameters

Accumulation length

In general, Table 3 shows a good but comparably low correlation between the prediction model and the measurements. The cause effect network was only trained using data for rods with a diameter of d_0=0.2mm, whereby the evaluation included rods of all diameters. At this, the difference between the predictions and the measurements increases for higher diameters. In contrast, Fig. 9 depicts only the predictions and measurements for rods with a diameter with d_0=0.2mm. The figure shows, that the trend between the measurements and the predictions is comparable, whereby the prediction has a slight offset of ~200µm compared to the majority of measurements. Consequently, it can be assumed that the estimation quality would increase if training data for additional rod diameters would be used.

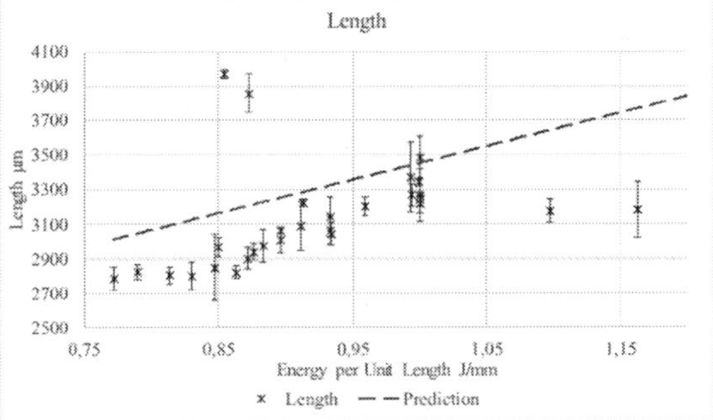

Fig. 9. Accumulation length - Predictions and Measurements for rods with a diameter of 0.2mm

7 Conclusion

As a result it can be stated, that the application of cause-effect networks as surrogates is a suitable way to decrease computational time for predicting process results. The results demonstrate, that a cause-effect network can be efficiently trained using only a small set of simulation data and achieve a very high prediction quality and generalization, in order to sufficiently reflect the simulation. Depending on the complexity of the network and the chosen prediction models, an estimation can be acquired in way under a second, allowing for a broad range of parametrizations to be evaluated. Thereby, the results show that the composition of training data constitutes a major factor for the generalizability of the cause-effect network. In particular, the estimation of the accumulation length demonstrates that the prediction model can be applied comparably well for rods conforming to the training set, but generates insufficient predictions for rods with a larger diameter.

In regards to predicting real-world process data, it is known that the algorithms highly depend on the training data and small deviations of the experimental results. This explains the effect of an increasing difference between prediction and experimental data for the preforms' final radius and volume when using higher energies. Despite this effect, it has been shown that cause-effect network are an interesting alternative to provide an approximation of the final preforms characteristics.

8 Acknowledgment

The authors gratefully acknowledge the financial support by Deutsche Forschungsgemeinschaft (DFG, German Research Foundation) for the Subprojects C4 'Simultaneous Engineering' and A3 'Material Accumulation' within the CRC 747 (Collaborative Research Center) "Mikrokaltumformen - Prozesse, Charakterisierung, Optimierung".

9 References

[1] J. P. Wulfsberg, T. Redlich, and P. Kohrs, "Square Foot Manufacturing. a new production concept for micro manufacturing," Production Engineering - Research and Development **4** ,1 (2010), 75–83.

[2] H. N. Hansen, K. Carneiro, H. Haitjema, and L. de Chiffre, "Dimensional Micro and Nano Technology," Annals of the CIRP **55** 2 (2006), 721–734.

[3] E. Mounier and A. Bonnabel, *Press Release Emerging MEMS, 29. August 2013*, 2013.

[4] M. W. Fu and W. L. Chan, "A review on the state-of-the-art microforming technologies," The International Journal of Advanced Manufacturing Technology **67** 9 (2012), 2411–2437.

[5] E. P. DeGarmo, J. T. Black, and R. A. Kohser, *Materials and Processes in Manufacturing*, 9th (Wiley, 2003).

[6] H. Flosky and F. Vollertsen, "Wear behaviour in a combined micro blanking and deep drawing process," CIRP Annals - Manufacturing Technology **63** ,1 (2014), 281–284.

[7] M. Geiger, M. Kleine, R. Eckstein, N. Tieslerl, and U. Engel, "Microforming," CIRP Annals - Manufacturing Technology **50** 2 (2001), 445–462.

[8] F. Vollertsen, "Categories of size effects," Production Engineering **2** 4 (2008), 377–383.

[9] S. M. Afazov, A. A. Becker, and T. H. Hyde, "Development of a Finite Element Data Exchange System for chain simulation of manufacturing processes," Advances in Engineering Software **47** (2012), 104–113.

[10]D. Rippel, M. Lütjen, and B. Scholz-Reiter, "A Framework for the Quality-Oriented Design of Micro Manufacturing Process Chains," Journal of Manufacturing Technology Management **25** 7 (2014), 1028–1049.

[11]D. Rippel, M. Lütjen, and M. Freitag, "Local characterisation of variances for the planning and configuration of process chains in micro manufacturing," Journal of Manufacturing Systems **43, Part 1** (2017), 79–87.

[12]S. M. Afazov, "Modelling and simulation of manufacturing process chains," CIRP Journal of Manufacturing Science and Technology **6** ,1 (2013), 70–77.

[13]M. Pietrzyk, L. Madej, and S. Weglarczyk, "Tool for optimal design of manufacturing chain based on metal forming," CIRP Annals - Manufacturing Technology **57** ,1 (2008), 309–312.

[14]I. Sabotin, J. Valentincic, M. Junkar, and A. Sluga, "Process planning system for micro-products," in *Proceedings of the 10th International Conference on Management of Innovative Technologies* (2009).

[15]B. Denkena, H. Rudzio, and A. Brandes, "Methodology for Dimensioning Technological Interfaces of Manufacturing Process Chains," CIRP Annals - Manufacturing Technology **55** ,1 (2006), 497–500.

[16]B. Denkena and H. K. Tönshoff, "Prozessauslegung und -integration in die Prozesskette," in *Spanen - Grundlagen*, B. Denkena and H. K. Tönshoff, eds. (Springer Verlag, 2011), pp. 339–362.

[17]B. Denkena, J. Schmidt, and M. Krüger, "Data Mining Approach for Knowledge-based Process Planning," Procedia Technology **15** (2014), 406–415.

[18]Kim, Y.-H.: Moraglio, A., A. Kattan, and Y. Yoon, "Geometric Generalisation of Surrogate Model-Based Optimisation to Combinatorial and Program Spaces," Mathematical Problems in Engineering **2014** ,1, 10.

[19]F. Duddeck and E. Wehrle, "Recent Advances on Surrogate Modelling for Robustness Assessment of Structures with respect to Crashworthiness Requirements," in *Proceedings of the 10th European LS-DYNA Conference 2015* (DYNAmore GmbH, 2015), p. 11.

[20]B. M. Colosimo, L. Pagani, and M. Strano, "Reduction of calibration effort in FEM-based optimization via numerical and experimental data fusion," Structural and Multidisciplinary Optimization **51** 2 (2015), 463–478.

[21]E. S. Hung and S. D. Senturia, "Generating efficient dynamical models for microelectromechanical systems from a few finite-element simulation runs," Journal of Microelectromechanical Systems **8** 3 (1999), 280–289.

[22]T. W. Simpson, J. D. Poplinski, N. P. Koch, and J. K. Allen, "Metamodels for Computer-based Engineering Design. Survey and recommendations," Engineering with Computers **17** 2 (2001), 129–150.

[23]R. Jin, W. Chen, and T. W. Simpson, "Comparative studies of metamodelling techniques under multiple modelling criteria," Structural and Multidisciplinary Optimization **23** ,1 (2001), 1–13.

[24]S. Shan and G. G. Wang, "Survey of modeling and optimization strategies to solve high-dimensional design problems with computationally-expensive black-box functions," Structural and Multidisciplinary Optimization **41** 2 (2010), 219–241.

[25]T. Goel, R. T. Haftka, W. Shyy, and N. V. Queipo, "Ensemble of surrogates," Structural and Multidisciplinary Optimization **33** 3 (2007), 199–216.

[26]S. Giurgea, D. Fodorean, G. Cirrincione, A. Miraoui, and M. Cirrincione, "Multimodel Optimization Based on the Response Surface of the Reduced FEM Simulation Model With Application to a PMSM," IEEE Transactions on Magnetics **44** 9 (2008), 2153–2157.

[27]C. Huang, B. Radi, and A. E. Hami, "Uncertainty analysis of deep drawing using surrogate model based probabilistic method," The International Journal of Advanced Manufacturing Technology **86** 9 (2016), 3229–3240.

[28]A. Messner, *Kaltmassivumformung metallischer Kleinstteile: Werkstoffverhalten,Wirkflächenreibung, Prozeßauslegung. Fertigungstechnik* (Meisenbach Verlag, 1998).

[29]H. Brüning and F. Vollertsen, "Energy efficiency in laser rod end melting," in *Proceedings of the LiM - Lasers in Manufacturing 2015*, Vol. 138 (2015).

[30]A. Stephen and F. Vollertsen, "Influence of the Rod Diameter on the Upset Ratio in Laser-based Free Form Heading," in *Proceedings of the 10th Int. Conference on Technology of Plasticity 2011*, G. Hirt and E. A. Tekkaya, eds. (Wiley, 2011), pp. 220–223.

[31]C. M. Elliot, "On the finite element approximation of an elliptic variational inequality arising from an implicit time discretization of the Stefan problem," IMA Journal of Numerical analysis **1981** ,1, 115–125.

[32]E. Bänsch, "Finite element discretization of the Navier-Stokes equations with a free capillary surface," Numerische Mathematik **88** 2 (2001), 203–235.

[33]H. Brüning and F. Vollertsen, "Form filling behavior of preforms generated by laser rod end melting," CIRP Annals - Manufacturing Technology 64 (2015), 293–296.

[34]H. Brüning, M. Jahn, F. Vollertsen, and A. Schmidt, "Influence of laser beam absorption mechanism on eccentricity of preforms in laser rod end melting," in *Proceedings of the 11th Int. Conference on Micro Manufacturing* (2016), p. 77.

[35]D. Rippel, E. Moumi, M. Lütjen, B. Scholz-Reiter, and B. Kuhfuss, "Application of Stochastic Regression for the Configuration of a Micro Rotary Swaging Processes," Mathematical Problems in Engineering (2014), 12.

[36]M. Jahn and A. Schmidt, *Finite element simulation of a material accumulation process including phase transitions and a capillary surface - Berichte aus der Technomathematik 12-03* (Universität Bremen, 2012).

Increased output in micro production by tolerance field widening and synchronisation

Ann-Kathrin Onken[1,2*], *Philipp* Wilhelmi [1,2], *Kirsten* Tracht[1,2,3], and *Bernd* Kuhfuss[1,2,3]

[1]University of Bremen, Bibliothekstr. 1, 28359 Bremen, Germany
[2]Bremen Institute for Mechanical Engineering, Badgasteiner Str. 1, 28359 Bremen, Germany
[3]MAPEX Center for Materials and Processing, Am Fallturm 1, 28359 Bremen, Germany

Abstract. The manufacturing of linked micro parts simplifies the handling and facilitates a consideration of trends due to the maintaining of the manufacturing order. These known changes of geometrical characteristics of parts are the basis for softening tolerances. The basic idea of the tolerance field widening is familiar with selective assembly. Instead of single parts, trend-based sections are used for increasing the number of assemblies under consideration of the clearance. The outcome of the assembly of the identified trend-based sections is influenced by the occurring trends and the changes of trends. Therefore, based on simulated data the effects of occurring trends are investigated. For sorting and matching the identified trend-based sections in high quantities, a case specific pre-assembly in the synchronisation point is required. Possible synchronisation scenarios as well as a synchronisation station for the synchronisation of two types of linked parts are presented. The results for the influence of trends show that the tolerance field widening increases the outcome especially for intersecting and opposing trends.

Keywords: Assembly, Handling, Tolerances

1 Introduction

Mass production of micro parts at high output rates is still challenging in industrial applications. On the one hand, the occurrence of size effects [1] and sensitive parts [2] rises the complexity of handling. On the other hand, in most cases, small tolerances [3] are required for assembly. The high demand for micro parts, as well as the resulting demand for economic production and handling processes, [4] shows the necessity for new methodologies in micro production. From the technological point of view, the output rates can be increased by simplifying and speeding up handling operations. The method of handling micro parts as linked parts [5] avoids micro-specific challenges directly by concept. The design of linked micro parts depends on the linkage of the micro parts and is case specific, with a possible differentiation between the three basic types ladder type, line type and comb [2].

The production with linked parts is investigated at the University of Bremen for two exemplary production processes within the collaborative research centre 747. The manufacturing process of ladder linked parts is shown in Figure 1. The parts are produced on base of sheet metal by stretch drawing. The manufacturing process of the line linked parts is based on wire material. In the first step, preforms are manufactured by laser rod melting and in the second step, the preforms are processed by rotary swaging, see Figure 2.

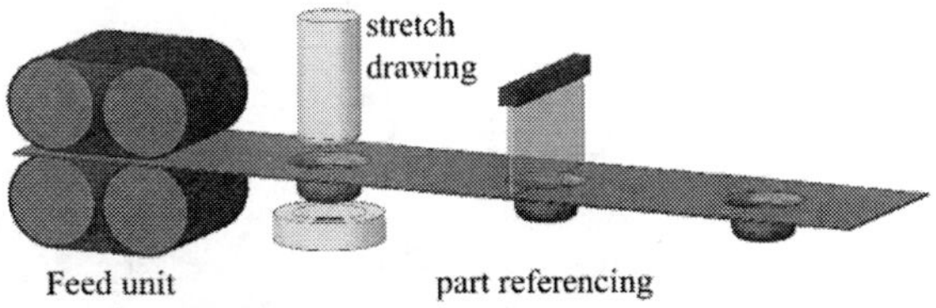

Fig. 1. Manufacturing process of ladder linked parts

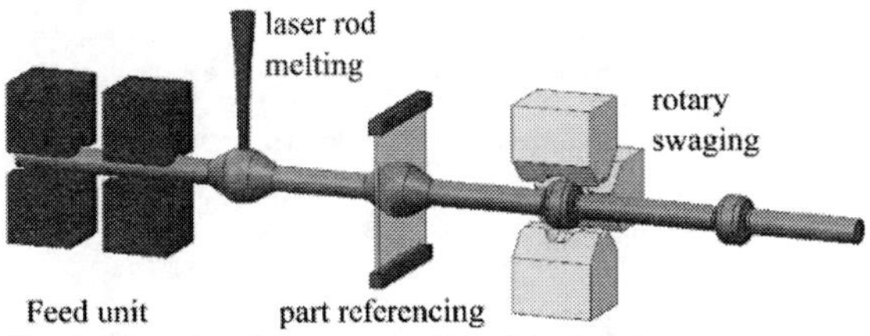

Fig. 2. Manufacturing process of line linked parts

The interconnection of the parts decreases the complexity of handling [5] and maintains the manufacturing order. The sustainment of the manufacturing order enables an identification and consideration of trends. The basic idea behind this, is familiar with selective assembly. Until now, methodologies of selective assembly are mainly applied for processes that cannot guarantee the production of parts within required tolerances. Kumar and Kannan investigated genetic algorithms for obtaining optimal

* Corresponding author: onken@bime.de

manufacturing tolerances [6]. Studies for applying selective assembly and adaptive production are described by Lanza [7] or Colledani [8]. Within these approaches, the tolerances are used for direct consideration in further manufacturing processes. Raj et al. [9] investigated the selective assembly of batches, by using a non-dominated sorting genetic algorithm. This approach is only applicable for small or medium sized batches.

The fact that linked parts are arranged according to the production time enables an extension of the application spectrum of selective assembly by tolerance field widening to established processes like stretch drawing. These processes are generally able to produce parts within required variances. In this case, an increased output can be achieved by softening tolerances. In case of defects in micro parts like valves, the whole product is interchanged. For this reason, the tolerances could be defined for every single assembly. The basic idea of the tolerance field widening is illustrated in Figure 3. It is assumed that the geometrical features of the produced parts show a mean variance, which is small compared to the tolerances, but that the mean value may show trends. These trends can be caused for example by tool wear. In [10], the concept for tolerance field widening for the production of linked parts in three steps is presented. Firstly, trends of the first batches of the process chains are considered and the linked parts are divided into clusters. Then clusters from linked parts of both types are matched. Finally, the process parameters are adapted, in order to maximize the number of successfully matched parts, for the following produced parts. Trends are occurring, especially in forming processes, due to e.g. wear. Therefore, this approach is specifically useful in this case.

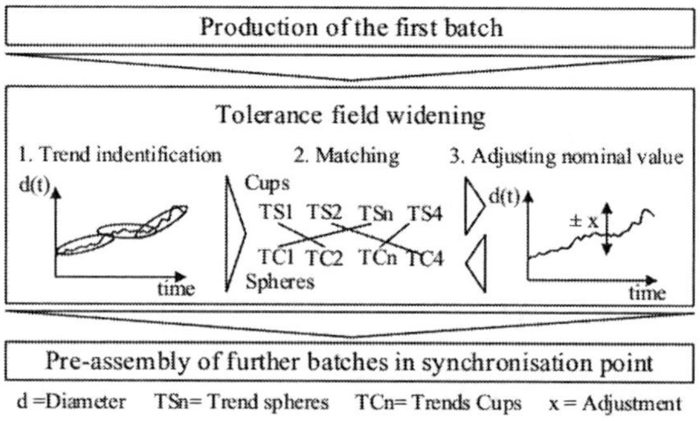

Fig. 3. Concept of tolerance field widening in linked parts production

For assuring a cost effective micro mass production, technological aspects, as well as the softening of tolerances, has to be considered. These two areas of research interdepend, especially in the synchronisation station, where the produced parts of the different process chains are synchronised. Possible scenarios for synchronisation as well as a design of a synchronisation station based on the example of cups and spheres are introduced. Trends that occur during production processes, influence the geometrical characteristics of micro parts. Different distribution will lead to varying numbers and sizes of sections. The number of sections is limited due to the objective of short cycle times in the synchronisation station. For identification of a trends-specific number of production sections, the influence of trends on the number of pre-assemblies in interdependence with the number of sections, is investigated based on simulation studies.

2 Appearance of trends

Changing states of components over time, are already addressed in research. The first field of research is within through life engineering services and the second one is within tools. In both fields, occurring wear and timing of tool or part changes are predicted. Up to sixteen wear mechanisms are considered in literature, that are caused by mechanical action, chemical action, or thermal influences [11]. Hsu et al. identified 32 different parameters that have been used for describing and investigating wear [12]. Looking at through life engineering services, the prediction of wear is based on the consideration of various parameters and environmental influences. The Weibull curve [13] is often used within prediction models and approaches. In the research field of manufacturing processes, the emphasis is the endurance of tools. Areas of interest are for example, the prediction of wear and end of lifetime. Belfiore et al. introduces an approach for wear and fatigue prediction in metal forming, by multilayer neural networks [14]. Furthermore, the influence of surface properties on wear is investigated. Yilkiran et al. present the influence of oxidised tool surfaces [15]. The mechanisms are of further interest. Liu et al differentiate wear patterns and mechanisms of cutting tools in milling [16]. The delineation of wear is also described in literature. Ber and Kaldor describe the course of wear in cutting processes [17]. Boher et al. investigated the influence of wear behaviour on the radius portion and identified that the surface degradation is caused by adhesion and ploughing [18]. The necessity of lubricant on wear in combined blanking and deep- drawing process of micro parts, due to the occurrence of flakes, was investigated by Flosky and Vollertsen [19]. In general, the influence of tool wear on the geometrical changes of parts, over the manufacturing time, is not described. Weikert et al. identified a trend in the measured diameter of produced micro parts that were investigated for statistical studies [20]. Changes within the trend were not described since only a minor number of parts were considered. Due to the missing information, data with trend changes were simulated. For identifying in which way trends and their changes influence the number of pre-assemblies and the number of sections, especially if they are limited, the scenarios for identifying boundaries are simulated.

3 Synchronisation of micro parts

Synchronisation in the context of different process chains is usually understood as an adaption of the cycle rates, in order to have the same output. For the following considerations the term of synchronisation is defined different and respectively extended to the production of

 https://doi.org/10.1051/matecconf/201819015006

linked parts, in combination with tolerance field widening. It is aspired to maintain the interconnection, not only during the single process chains, but also for the part supply in assembly. For that purpose, at least two parts, for example type A sphere and type B cup, need to be provided at a certain time to a certain point (synchronisation point) and ideally with a defined orientation. When applying the tolerance field widening to the linked parts production, an additional condition counts for the synchronisation. In order to produce a functional assembly, it is not longer sufficient that a random part from type A is synchronised with a random part from type B. A part from a specific cluster of the linked parts of type A must be synchronised with a part from a specific cluster of the linked parts of type B. The operation performed in the synchronisation point must not necessarily be an assembly, but can also be the creation of a new type of linked parts, containing pairs of synchronised parts. In case of creating a new type of linked parts there are theoretically the three following options.

- The linked parts of both types are joined directly without separating them before
- The linked parts of both types are separated and a new linkage is created
- The linked parts of one type are separated and joined to the linked parts of the other type

For a direct joining as proposed, in case one the distances between the parts in both linked parts must be exactly the same. Furthermore, the separation of such linked parts would be more difficult compared to the other cases. For the creation of a completely new type of linked parts, the number of necessary operations is much higher. Firstly, both linked parts must be separated and then joined again. By these considerations it results, that in most cases the third option is the most convenient. For the exemplary process chains described before there are the two options illustrated in Figure 4.

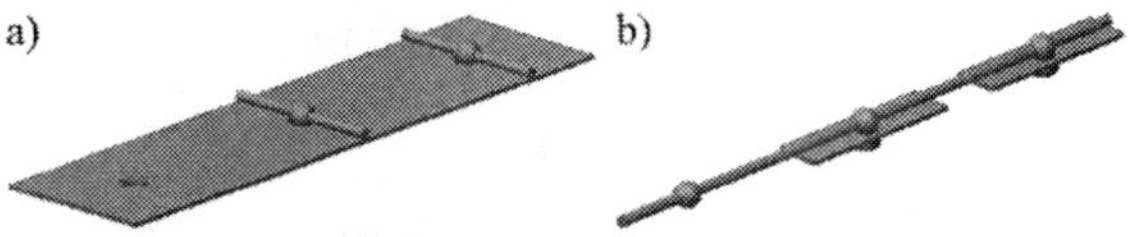

Fig.4. Synchronised linked parts

As they are produced of thin metal foil, the cups in the ladder linked parts are much more sensitive against mechanical damage. Consequently, for the further handling, it is easier to maintain the ladder linked parts and separate the line linked parts (Figure 4a). To investigate the synchronisation, a synchronisation station has been designed for that case. It applies all the developed technologies from the single process chains, such as feed units [21] and high speed referencing system, which enable a fast and accurate positioning within the synchronisation point [22]. The central module of the synchronisation station is illustrated in Figure 5. The separation of the line linked parts is realised by mechanical shearing and the joining of the linked parts is performed via capacitor discharge welding. Pneumatic cylinders realise the mechanical actuation.

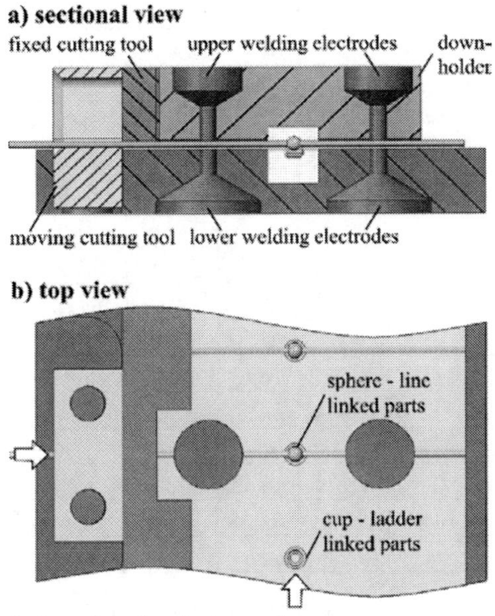

Fig.5. Central synchronisation module

The synchronisation module is designed for assembly of more than 200 parts per minute. While implementing the results of the tolerance field widening on the synchronisation station, every change of linked parts and therefore the total number of sections that results in the identified clusters causes an interruption of the assembly process. For assuring high quantities, the number of sections should be as low as possible. For this reason, the influence of trends on the number of sections is investigated in the following.

4 Simulation setup

Depending on the process chains for the produced linked parts, different trends may be identified in the part dimensions. Based on the example of cups and spheres, four different trend scenarios and three different intensities of trends are assumed, which result in twelve data sets. For each of the twelve data sets, two thousand data-points are simulated. Variances are considered by adding a random value between -0,0005 µm and +0,0005 µm. The aspired fitting tolerance is assumed to be between 0 µm and 10 µm. This small clearance assures that not all parts could be matched, for showing the influence of the trends on the number of sections and on the number of assemblies.

In Figure 6 the four different trend scenarios are shown. In all twelve data sets the nominal value of the diameter of the spheres starts at 908.2 µm and the diameter of the cups at 890.8 µm. Within the scenarios 1 and 2, changing linear trends and in the scenarios 3 and 4 the combination of linear and quadratic trends are simulated.

The three datasets of a scenario have the same direction of trends with varying intensities as listed in Table 1. The

Table 1. Intensity of trends in µm.

T	D 1.1	D 1.2	D 1.3	D 2.1	D 2.2	D 2.3	D 3.1	D 3.2	D 3.3	D 4.1	D 4.2	D 4.3
C 1	0.001	0.0001	0.0005	-0.001	-0.0001	-0.0005	0.001	0.0001	0.0005	-0.001	-0.0001	-0.0005
C 2	0.0011	0.00011	0.00055	-0.0011	-0.00011	-0.00055	0.0011	0.00011	0.00055	-0.0011	-0.00011	-0.00055
C 3	0.0013	0.00013	0.00066	-0.0013	-0.00013	-0.00066	0.0013	0.00013	0.00066	-0.0013	-0.00013	-0.00066
C 4	0.00135	0.000135	0.00074	-0.00135	-0.000135	-0.00074	0.1	0.2	0.2	-0.1	-0.2	-0.2
C 5	0.00145	0.000145	0.0008	-0.00145	-0.000145	-0.0008	0.00145	0.000145	0.0008	-0.00145	-0.000145	-0.0008
C 6	0.0015	0.00015	0.001	-0.0015	-0.00015	-0.001	0.0015	0.00015	0.001	-0.0015	-0.00015	-0.001
C 7	0.0016	0.00016	0.0015	-0.0016	-0.00016	-0.0015	0.12	0.3	0.4	-0.12	-0.3	-0.4
C 8	0.0017	0.00017	0.0017	0.0017	-0.00017	0.0017	0.15	0.4	0.7	-0.15	-0.4	-0.7
S 1	0.0009	0.0009	0.0003	0.0009	0.0009	0.0003	0.0009	0.0009	0.0003	0.0009	0.0009	0.0003
S 2	0.001	0.001	0.00035	0.001	0.001	0.00035	0.001	0.001	0.00035	0.001	0.001	0.00035
S 3	0.0012	0.0012	0.0005	0.0012	0.0012	0.0005	0.0012	0.0012	0.0005	0.0012	0.0012	0.0005
S 4	0.00129	0.00129	0.00129	0.00129	0.00129	0.00129	0.00129	0.00129	0.00129	0.00129	0.00129	0.00129
S 5	0.00133	0.00133	0.00133	0.00133	0.00133	0.00133	0.00133	0.00133	0.00133	0.00133	0.00133	0.00133
S 6	0.0015	0.0015	0.0015	0.0015	0.0015	0.0015	0.0015	0.0015	0.0015	0.0015	0.0015	0.0015
S 7	0.00155	0.00155	0.00155	0.00155	0.00155	0.00155	0.00155	0.00155	0.00155	0.00155	0.00155	0.00155
S 8	0.0016	0.0016	0.0016	0.0016	0.0016	0.0016	0.0016	0.0016	0.0016	0.0016	0.0016	0.0016

first dataset of every scenario shows similar trends with lower trend variations.

Within the second dataset, the trend of one kind of parts is distinguished by higher trend intensities and higher changes. For the third data set of the scenarios, the diameters of both kind of parts has intensive trends and intensive changes. The trends (T) in µm of the data sets (D) are changing every 250 parts of cups (C) and spheres (S). The table elements with grey background do have a quadratic trend. These trends were considered using the following equation:

$$d(t) = d(t-1) + T * x^2. \quad (1)$$

X starts at 0.01 and increases in 0.01 steps over the 250 iterations to 0.250. Due to the fact, that in accordance to minimal cycle times only a limited number of sections is contemplated a limited number of sections is considered. For the investigation of trends the data sets were divided into 2, 4, 8, 10, and 16 sections. The sections include 1000, 500, 250, 200, and 125 parts. For the matching of these classes the Munkres algorithm is used [23]. The algorithm for bipartite graphs calculates all possible combinations and the assemble able parts and returns the best matching of the used sections. Due to the fact that the algorithm uses a N × N matrix, only scenarios where the parts have the same number of sections could be used.

5 Results

The results show that the length of sections has a high impact on the number of the matched parts. An increasing number of sections leads to an increasing number of assemblies in all 4 scenarios, as shown in Figure 7. The influence of the number of sections is much higher in the scenarios 2 and 4, where opposing trends are simulated. By the intersection of these two trends in total a higher number of parts could be matched successfully.

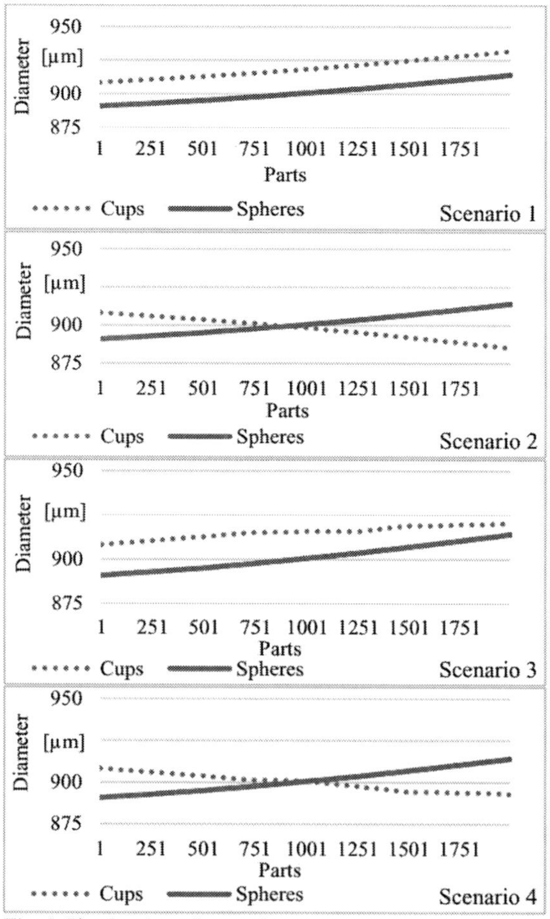

Fig. 6. Simulated trend scenarios.

In the first and third scenario, where the trends show the same direction, the number of sections has a lower influence on the number of successfully matched parts. The scenarios 1.2, 2.2, 3.2, and 4.2, where one trend has a much higher intensity than the other one, the influence of the number of section has also a lower impact on the successful matches.

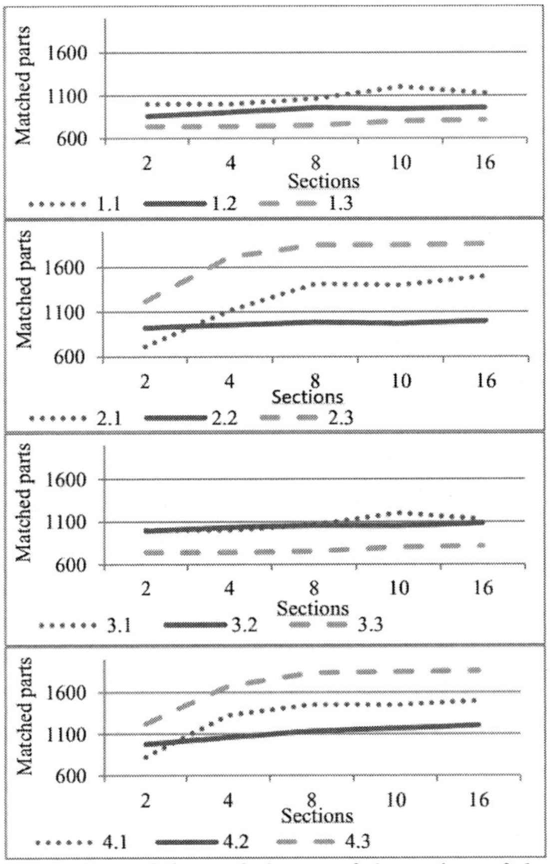

Fig. 7. Successfully matched parts of the sections of the scenarios.

Furthermore, the results indicate that the consideration of the trends influences the output while matching sections. The increases of the output are more significant in the first sections. The number of eight sections is equal with the number of the trends. Therefore, these sections are trend-specific.

Looking at the successfully matched parts at a number of eight sections, it is significantly recognizable that due to the consideration of trends in six scenarios local optima occurs. As listed in Table 2 in the scenarios 1.2, 2.1, 2.2, 3.2, 3.3, and 4.1 the output decreases by the increase of eight to ten sections.

Especially in the scenario 1.2, where both parts show an increasing trend and one of them has a higher intensity, the trends have an influence on the results. Although the number of sections increases further, the found number of matched parts using 8 sections could not be enhanced by the consideration of 10, or 16 sections.

Table 2. Rise/fall of the output from 8 to 10 sections.

Scenario	Rise/fall of the output from 8 to 10 sections
1.1	12,89%
1.2	-1,26%
1.3	6,67%
2.1	-0,99%
2.2	-1,32%
2.3	-0,11%
3.1	12,78%
3.2	-0,57%
3.3	6,67%
4.1	-0,28%
4.2	3,01%
4.3	0,71%

Due to the similar intensities of quadratic and linear trends in the third and fourth scenarios, these changes do not have a high impact on the ability of matching these sections.

6 Summary and Outlook

The paper introduces an approach for micro mass production that considers conveyance and synchronisation technologies as well as an adapted process design. The design of the parts as well as the synchronisation of these parts interdepends on each other. Therefore, different synchronisation scenarios as well as the synchronisation station for the example of cups and spheres are introduced.

For an additional improvement of the output rates, the tolerances can be softened by building trend-specific sections. For this reason, the influence of trends and trend changes on the number of successfully match able parts is investigated. The results show that especially occurring opposite trends with similar intensities according to the gradient show an increasing number of joinable parts by matching sections. This is caused by the intersection of the two parts. If only one type of linked parts shows a high intensity, appearing intersections are low and for this reason, the effect of building sections is less intensive. In further research it has to be investigated how the number of successfully matched parts of these scenarios could be increased by the adjustment of the nominal value.

The results of the tolerance field widening are implemented in the synchronisation station. Looking at the limited numbers of interruptions for enhancing mass production, the local optima confirm that the trend changes also have an impact in half of the simulated scenarios and that the consideration of the trend changes for the building of sections is an appropriate reference point for the building of sections.

For an assessment of a number of allowed interruptions and synchronisation scenarios, the cycle times of the synchronisation station are implemented in a simulation scenario in future research. Furthermore, the influence of real trends on the process times of the synchronisation station and the diameter changes of parts has to be considered. Therefore, parts have to be produced and measured. Due to the fact that varying sizes of trend-specific sections are possible, the matching of flexible section sizes is required.

Acknowledgements

The authors gratefully acknowledge the financial support by Deutsche Forschungsgemeinschaft (DFG, German Research Foundation) for Subproject C5 "Teileverbunde" within the SFB 747 (Collaborative Research Center) "Mikrokaltumformen - Prozesse, Charakterisierung, Optimierung".

References

1. F. Vollertsen, D. Biermann, H.N. Hansen, I.S. Jawahir, K. Kuzman, Size effects in manufacturing of metallic components, CIRP Annals – Manufacturing Technology, **58**/2, 566-587 (2009)
2. K. Tracht, F. Weikert, Handling of Micro Parts, *Micro Metal Forming*, Springer-Verlag Berlin Heidelberg 331-342 (2013)
3. H.N. Hansen, K. Carneiro, H. Haitjema, L. De Chiffre, Dimensional Micro and Nano Metrology, Annals of the CIRP, **55**/2, 721-743 (2006)
4. P. Groche, R. Schneider, Method for Optimization of Forming Presses for the Manufacturing of Micro Parts, Annals of the CIRP, **53/1**, 281-284 (2004)
5. K. Tracht, C. Schenck, F. Weikert, B. Kuhfuss, Förderung mikrokaltumgeformter Teile im Verbund (Conveyance of micro-cold-formed parts in a linkage – Overcoming of size effects by connecting micro-cold-formed parts), wt werkstattstechnik online, **100/11/12**:864-868 (2010) – in German
6. M.S. Kumar, S.M. Kannan, Optimum manufacturing tolerance to selective assembly technique for different assembly specifications by using genetic algorithm. Int J Adv Manuf Technol, **32**, 591-598 (2007)
7. G. Lanza, B. Haefner, A. Kraemer, Optimization of selective assembly and adaptive manufacturing by means of cyber-physical system based matching, CIRP Annals – Manufacturing Technology, **64/1**, 399-402 (2015)
8. M. Colledani, D. Ebrahimi, T. Tolio, Integrated Quality and Production Logistics Modelling for the Design of Selective and Adaptive Assembly Systems, CIRP Annals – Manufacturing Technology, **62/1**, 5–8 (2014)
9. M.V. Raj, S.S. Sankar, S.G. Ponnambalam, Minimizing Clearance Variations and Surplus Parts in Multiple Characteristic Radial Assembly Through Batch Selective Assembly. Int J Adv Manuf Technol, **57**, 1199–1222 (2011)
10. A.-K. Onken, P. Rückert, C. Perl, K. Tracht, Joining linked micro parts by tolerance field widening and synchronization, *Tagungsband des 2. Kongresses Montage Handhabung Industrieroboter*, Springer-Verlag Berlin Heidelberg, 1-8 (2017)
11. H.C. Meng, K.C. Ludema, Wear models and predictive equations : their form and content, Wear, **181-183**, 443-457 (1995)
12. S.M. Hsu, M.C. Shen, A.W. Ruff, Wear prediction for metals, Tribology International, **30/5**, 377-383 (1997)
13. R.B. Abernethy, *The New Weibull Handbook*, **5**, (2010)
14. N.P. Belfiore, F. Ianniello, D. Stocchi, F. Casadei, D. Bazzoni, A. Finzi, S. Carrara, J.R. González, J.M. Llanos, I. Heikkila, F. Peñalba, X. Gómez, A hybrid approach to the development of a multilayer neural network for wear and fatigue prediction in metal forming, Tribology International, **40** 1705–1717 (2007)
15. D. Yilkiran, D. Wulff, A. Almohallami, F. Özkaya, A. Bouguecha, S.Hübner, K. Möhwald, H.J. Maier, B.-A. Behrens, Wear behaviour of thermally oxidised tool surfaces as low-friction separation layers for dry sheet metal forming, Wear, **376-377**, 1789–1803 (2017)
16. Z.Q. Liu, X. Ai, H. Zhang, Z.T. Wang, Y. Wan, Wear patterns and mechanisms of cutting tools in high-speed face milling, Journal of Materials Processing Technology, **129**, 222-226 (2002)
17. A. Ber, S. Kaldor, The First Seconds of Cutting, Wear Behaviour, Annals of the CIRP, **31/1**, 13-17 (1982)
18. C. Boher, D. Attaf, L. Penazzi, C. Levaillant, Wear behaviour on the radius portion of a die in deep-drawing: Identification, localisation and evolution of the surface damage, Wear, **259**, 1097-1108 (2005)
19. H. Flosky, F. Vollertsen, Wear behaviour on a combinded micro blanking and deep drawing process, CIRP Annals- Manufacturing Technology, **63**, 281-284, 2014
20. F. Weikert, J. Weyhausen, K. Tracht, Statistical Study for Micro Forming Technologies Used for Linked Parts Production, *5th International Conference on Changeable, Agile, Reconfigurable and Virtual Production*, Springer International Publishing Switzerland,41-46 (2016)
21. P. Wilhelmi, C. Schenck, B. Kuhfuss, Diameter adaptive guides for wire based linked micro parts. Proceedings of the 4M/IWMF Conference, Lyngby, Denmark, 13.09.-15.09.2016
22. P. Wilhelmi, C. Schenck, B. Kuhfuss, High Speed Referencing System for Linked Micro Parts, Proceedings of the ICOMM – International Conference on Micro Manufacturing 2016. Orange County, California (USA), March 29.03.-31.03.2016.
23. J. Munkres, Algorithms for the Assignment and Transportation Problems, Journal of the Society for Industrial and Applied Mathematics, **5/1**, 32-38 (1957)

Inverting Prediction Models in Micro Production for Process Design

Phil Gralla[1,*], *Iwona* Piotrowska-Kurczewski[1], *Daniel* Rippel[2], *Michael* Lütjen[2], and *Peter* Maaß[1,3]

[1]University of Bremen, Bremen, Center for Industrial Mathematics (ZeTeM), Porstfach 330440, 28334 Bremen, Germany
[2]BIBA – Bremer Institut für Prduktion und Logistik GmbH at the University of Bremen, Hochschulring 20, 28359 Bremen, Germany
[3]University of Bremen, Bremen, MAPEX Center for Material and Processes, Bremen, Germany

Abstract. Databased prediction models are used to estimate a possible outcome for previously unknown production parameters. These forward models enable to test new production designs and parameters virtually before applying them in the real world. Cause-effect networks are one way to generate such a prediction model. Multiple inputs and stages are being connected to one large prediction model. The functional behaviour and correlation of inputs as well as outputs is obtained through data based learning. In general, these models are non-linear and not invertible, especially for micro cold forming processes. While already being useful in process design, such models have their highest impact if inverted to find process parameters for a given output. Combining methods from the mathematical field of inverse problems as well as machine learning, a generalized inverse can be approximated. This allows finding process parameters for a given output without inverting the model directly but still using inherit information of the forward model. In this work, Tikhonov functionals are used to perform a parameter identification. The classical approach is altered by changing the discrepancy term to incorporate tolerances. Thereby, small deviations of a certain pattern are being neglected and the parameter finding process is being stabilized. In addition, different types of regularization are taken into consideration. Besides theoretical aspects of this method, examples are provided to demonstrate advantages and boundaries of an application for the process design in micro cold forming processes.
Keywords. Predictive Model, Optimization, Process Control

1 Introduction

Modern production processes for micro components depend on many process parameters. Each of them is affecting process quality and cost in its own way. Estimating the outcome for a set of process parameters needs a large amount of experiments, as former knowledge from macro production can only partly be applied due to size effects (see [1]). Prediction models can help reducing necessary experiments by using knowledge from other parameter sets as well as former experiments to estimate the influence of a new parameter set on the process outcome.

In this work, cause-effect networks are used to generate a process model and allow the exhibition of new process strategies. These networks consist of a set of interconnected technological and logistic parameters, e.g. representing forces, times or material properties, which are relevant for the process. Thereby, each parameter contains a prediction model, allowing to calculate its value using values from connected parameters.

Cause-effect networks allow estimating the outcome of a complex process for a new parameter set, however they are generally not invertible. To find a suitable set of parameters for a desired output new strategies need to be developed. In mathematics this is considered an inverse problem where the forward operator and an output are known but the causing parameter is to be identified. In application these are called ill-posed problems, which are either non-invertible or amplify small perturbations of the output in the reconstructed parameter. Since prediction models for complex processes, such as production processes for micro components, tend to lead to ill-posed problems, this issue needs to be solved via regularization methods. Different approaches to deal with ill-posed problems have been developed over time (see [2,3,4]). In this work Tikhonov type functionals will be used to tackle the ill-posed nature of the operators at hand.

Besides estimating the expected value of a production process, its variance in the output is of great interest as well. To ensure a low rate of non-usable components the variance should stay within a predefined area. Thus, the classic approach for parameter identification from inverse problems needs to be adjusted. In this work Tikhonov functionals are being altered in order in to include variances and to further stabilize the parameter identification.

The structure of this work follows the procedure of the presented method as shown in Fig. 1. First prediction models and how these are obtained is described. In this section the state of the art is summarizes focusing on the models which are later used in this work. In the following

* Corresponding author: phil.gralla@uni-bremen.de

 https://doi.org/10.1051/matecconf/201819015007

section the planning tool μ-ProPlAn, which generates process models through cause-effect networks, is inaugurated. Cause-effect relations are models by the means of prediction models as introduced in section 2. In addition mathematical notations are introduced and mathematical properties stated. It forms a basis for section 4 where the introduced notations and stated properties are needed.

In the next section, section 4, the forward model is used for parameter identification. A short overview about parameter identification and its challenges is provided before the applied method is introduced. This method uses classic approaches and alters them to further stabilize the parameter identification as well as incorporate variations in the process outcome. Section 5 provides two examples to demonstrate its functionality as well as difference to existing methods.

The overall objective of this work is to derive a reliable method for process parameter identification for micro components while keeping the necessary amount of experiments to a minimum.

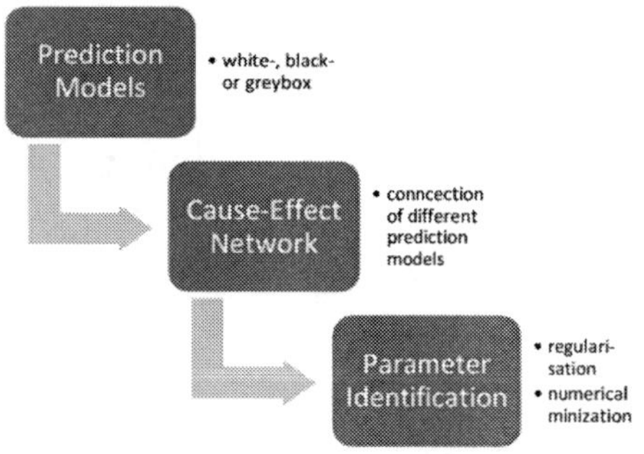

Fig. 1. Workflow of building prediction models and cause-effect networks for performing parameter identification.

2 Prediction Models

Prediction models take an input parameter and map this parameter on a related output. They are the connection of input to output. In this work, prediction models are restricted to production steps and chains. The input parameters are used material and process parameters and the output is the state of manufactured good.

Different approaches to form a suitable model exist and can be classified into two main groups, data based models, also referred to as black box models, and models based on physical laws, also known as white box models. A combination of both approaches is called a grey box model. While each approach has its own advantages and disadvantages, the following section will focus on black box models, however the results and methods of the following sections can be applied on white and grey box models as well, as long as the stated necessary mathematical properties are fulfilled.

Black box models are based on experimental data and are thus often statistical regression models. Other approaches include but are not restricted to linear- or spline interpolation as well as convolution. For this work local regression models are used to obtain a prediction model, also known as LOWESS (see [10]). Instead of fitting one regression model for the whole data, for each point a separate model is fitted that takes only the nearest neighbours into account.

Rather than giving all points in the neighbourhood equal weight, weights that die off smoothly with distance from the target point can be used. One way to achieve this is by using Epancehnikov quadratic kernels as defined in (1) and (2).

$$K_\lambda(x_0, x) = D\left(\frac{|x-x_0|}{\lambda}\right) \quad (1)$$

$$D(t) = \begin{cases} \frac{3}{4}(1-t^2) & if\ |t| \leq 1 \\ 0 & otherwise \end{cases} \quad (2)$$

Locally weighted polynomial regression solves a separate weighted least square problem at each target point x_0 by minimizing function (3). The maximum degree of the polynomial is denoted d.

$$\min_{\alpha(x_0),\beta_j(x_0),j=1,\dots,d} \sum_{i=1}^{N} K_\lambda(x_0, x_i) \cdot J$$

$$J = \left[y_i - \alpha(x_0) - \sum_{j=1}^{d} \beta_j(x_0) x_i^j\right]^2 \quad (3)$$

An example of a regression using the above method is given in figure 2. The polynomial degree is set to 2 and $\lambda = 0.2$. Data was generated using a sinus function with added normal noise. The applied method will produce a differentiable function as a regression. For graphical representation of the regression, it is evaluated at a set number of points and linear interpolation is used to draw the missing lines in the graph to achieve the desired resolution, which results in a graph that seems to be only continuous instead of differentiable. For the following parameter identification in section 5, each function evaluation will be done by fitting the desired local quadratic regression. It is therefore necessary to safe all data points instead of only one regression function.

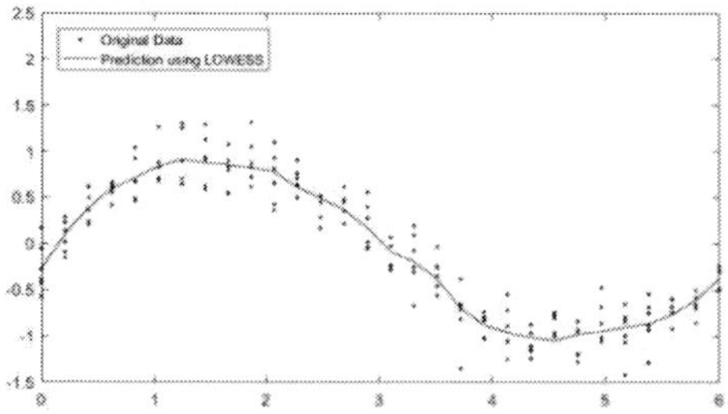

Fig. 2. Example of locally weighted regression to estimate the expected value of a sinus function with additive normal random noise.

This type of regression is very flexible as it does not require one function that fits all data but instead uses a type of smoothing operator. It is therefore suited for complex processes with none or insufficient theoretical models. It is however computational intensive and does not provide an easyly comparable regression function, which depends on few parameters, as do classic least square methods (for further detail refer to [11]). Due to the complexity of micro production processes, LOWESS with Epancehinkov quadratic kernels and local polynomial degree of 2 will be used as a default option in cause-effect networks generated by the μ-ProPlaAn software, which will be described in more detail later.

So far the prediction model is estimating the expected value of the process outcome. For its application in process design additional information of variance is needed as well. For data based prediction models this can be achieved by learning a second model on the sample variance of the given data, i.e. the output of trainings data is $var_j = VAR(y_{j,i})$, where $y_{j,i}$is the i-th measurements of the input x_i. In this work, LOWESS will be used to find a suitable approximation for the variance over the whole function domain and to keep consistency in the applied methods.

3 Cause-Effect Models

The introduced prediction models estimate the expected value and variance of one process by combing multiple processes into a process chain to form a new process. Cause-effect networks can be used to combine prediction models of single process into a larger process chain and estimate the expected value of the whole process chain. This section introduces one methodology to generate such cause-effect networks and outlines some mathematical basics of such networks.

3.1 Micro – Process Planning and Analysis (µ-ProPlAn)

This article focuses on cause-effect networks generated as part of the methodology “Micro – Process Planning and Analysis” (µ-ProPlAn) (c.f. [5] for further details on the methodology). µ-ProPlAn covers all design phases from the process and material flow planning to the configuration and evaluation of the processes and process chain models. The methodology itself consists of a modeling notation, a procedure model as well as a set of methods and tools for the evaluation of the corresponding models.

The modeling notation consists of three views, representing different levels of detail. The first view focuses on top level process chains. This view’s notation closely follows the classic notation of process chains, as described in [6]. In contrast to the classical approach, process elements as well as operations are connected using process interfaces. These interfaces, which additionally include logistic parameters. Extending the classical approach, operations act as interfaces to the second view: the material flow view.

The material flow view further details operations by assigning those material flow objects that are used to conduct the operation (e.g. machines/devices, work pieces, tools, operating supplies or workers). This enables the modelling of specific production scenarios with specified resources and therefore allows an evaluation of the models regarding logistic aspects. Therefore, µ-ProPlAn offers the option to conduct material flow simulations based on the specified production system and the modelled process chains.

The third view focuses on the configuration of the processes and process chains using cause-effect networks. Each network consists of a set of parameters and a set of cause-effect relationships, forming a directed graph. The set of parameters consists of all technical and logistic characteristics that are relevant to describe the object’s influence on the production process. In case of work pieces, these are e.g. material properties, costs per piece or geometrical characteristics. As for production processes, these parameters include production speeds, forces or other characteristics that can be set, calculated or measured. From a modelling perspective, the cause-effect networks are modelled hierarchically. Each material flow object (work pieces, machines, tools, workers, etc.) holds its own cause-effect network, or at least a set of describing parameters. When combining these single elements to operations, process elements or process chains, higher-level cause-effect networks are created by describing additional relationships between the parameters of the networks or by connecting them to previously specified process interfaces (c.f. figure 3).

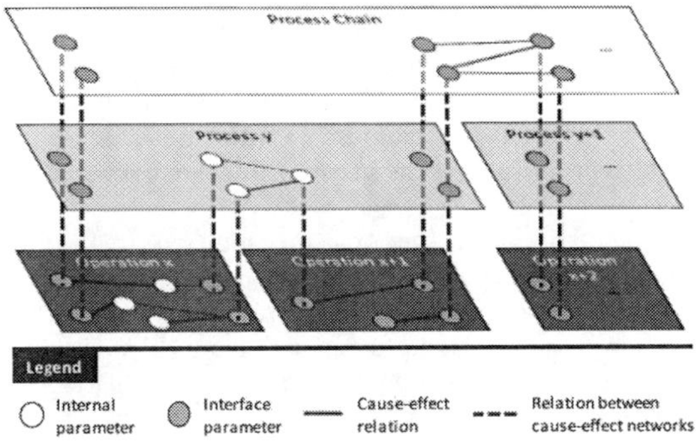

Fig. 3. Structure of Cause-Effect Networks.

The design of cause-effect networks is divided into two steps: qualitative modeling and quantification. The qualitative model of the corresponding network is created by collecting all relevant parameters and denoting their influences among each other. At this point the knowledge of a process expert can facilitate the creation of networks by point out the most relevant parameters and indicating the relationships.

The second step concerns the quantification of the cause-effect networks. The objective is to enable the propagation of different parametrizations throughout the network. In case of simple or well-known relations, µ-ProPlAn allows to input mathematical formulas directly (white-box models). However, in the area of micro manufacturing different parameters can have a more significant impact than in the macro domain, resulting in the inclusion of parameters that can be neglected in macro manufacturing. In addition, size-effects may induce a different behavior then usually observed. Therefore, it is often impossible to describe all parameters and cause-effect relations comprehensively directly.

As a result, μ-ProPlAn offers the capabilities to quantify cause-effect relations from experiment- or production data by applying methods form the areas of data mining and statistics. Therefore, the qualitative cause-effect network is subdivided into a set of sub-problems. Each of these consists of one single dependent parameter and its independent parameters. Thus objective of the quantification is the determination of a function or model to estimate the value of the dependent variable based on the values of the independent variables. To achieve this objective, μ-ProPlAn offers a set of statistical regression methods (e.g. linear or polynomial least-square regressions) or learning methods (e.g. artificial neural networks, support vector machines, regression trees or local regression methods). While statistical regression methods usually calculate a mathematical function to describe the expected value of the referring dependent variable, learning methods usually provide a model for a mean estimator. While these estimators can predict the value of the dependent variable based on the values of the independent variables, their models usually provide less insight to mechanics behind the cause-effect relation and are thus treated as black-box models. Their advantage shows in their ability to learn arbitrary relations without prior knowledge of the relation's shape (no prototype of the function has to be provided) and regardless of the problems dimensionality. In practice, the application of locally weighted linear regression (LOWESS) models has yielded promising results in estimating complex relations (e.g. in [7]).

In addition to the mean estimator, μ-ProPlAn characterizes the (local) variance for each set of training data as described in [8]. Using these variances, μ-ProPlAn applies an interpolation to estimate the local variance in addition to the expected mean value for each parameter, provided a given parametrization for the cause-effect network.

3.2 Mathematical Properties

In order to perform a parameter identification on a full prediction model, including cause-effect networks, certain mathematical properties need to be fulfilled. To understand how properties of single process steps are carried over into the full model, a proper mathematical formulation is needed.

In section 2, different ways of generating data based prediction models are introduced and the LOWESS method is explained in more detail. Each prediction model is a mathematical operator, denoted by F, that maps an input, denoted by x, on an output, denoted by y. In most cases, x, y will be vectors of scalars, but are not restricted to it, e.g. one input could be a time-dependent heat source.

For a full cause-effect network, multiple prediction models, denoted by F_j, for different production steps, are linked together. To apply a second process F_i on a process F_j the output space of F_j must match with the domain of F_i. This may not be the case if, for example, new process parameters are introduced for process F_i. To analyse the network, the first operator F_j is extended by mapping the new process parameter of F_i onto itself, i.e. an identity operator for missing inputs is applied. Thus the new model F for the linked process is as defined in (4).

$$F = F_i(F_j(x)) \tag{4}$$

Parallel processes that do not influence each other can be linked the same way. The order does not matter as long as the input and output space are adjusted accordingly via applying identity operators.

As a result, cause-effect networks can be expressed as one operator, which inherits its mathematical properties from each prediction model within. If, for example, the network consists of a continuous operator applied on a differentiable operator, the whole network will be differentiable.

4 Parameter Identification

Prediction models are useful to find outputs for new parameters. Thus reducing the need of experiments to estimate outputs of new parameters. Finding parameters for a given output is called parameter identification. In an ideal case, this can be done by inverting the model to directly calculate parameters to a given output. In practice, this does not always work due to several reasons. The most common ones are that the model is not invertible, and that, due to model approximations or variations in measurements, the given or observed output has small deviations from the predicted output if both are based on the same parameters. These small deviations can lead to large or even unbounded errors in parameter identification. These two cases are examples of ill-posed problems (see definition and examples in [2, 3]). One class of ill-posed problems is the inversion of integral operators. As shown in earlier sections, black-box models can be represented as integral operators and thus lead to ill-posed problems for the parameter identification. To overcome the ill-posed nature of the problem at hand regularization is used. Various methods of regularization exist (for examples see [3, 4, 9]) and depend on a priori knowledge of the parameter and data. In this work Tikhonov functionals are used to stabilize the parameter identification progress.

Tikhonov functionals consist of two terms, a discrepancy and a regularization term. By minimizing both terms, weighted with a coefficient α on the regularization term, a suitable approximation of the true parameter, i.e. the parameter that causes the given or observed data, is found. Let X be the space in which the prediction model F maps onto and all data resides. The discrepancy term then describes the distance of given or observed data y^δ to the predicted data $F(x)$ to the parameter x in the space X. The regularization term $R(x)$ uses some a priori information and maps the parameter u onto the positive real numbers. A common choice for $R(x)$ is (5), where x_0 is often set to 0. For prediction models $x_0 \neq 0$ but instead a density point of known data is applicable as well to avoid extrapolation in the prediction.

$$R(x) = \|x - x_0\|_q^q, \ 1 \leq q \leq 2 \tag{5}$$

The prediction models in this work do not only give a point estimation of the outcome but a variance as well. To fully utilize this additional information, the given data should not be a single point but a closed set of feasible data points. This set will be denoted as tolerance area of the point v^δ. To compare two sets while maintaining good numerical properties, the distance measurement given in (6) is used. This distance is differentiable and measures the overlapping area, i.e. it is only equal to zero if S_1 is a subset of S_2. This is important to ensure that the determined parameter is within the tolerance area if such a parameter exists.

$$d(S_1, S_2) = \int_{S_1} \min_{y \in S_2} \|x - y\|_2^2 \, dx \qquad (6)$$

In the following, $S_{F(y)}$ denotes the tolerance set of $F(y)$ obtained in the same way as the tolerance set of y^δ is obtained, which will be denoted by S_δ. For quality management the three sigma area is a common choice to define a process quality if an additive normal error is assumed. This area can be used as a tolerance set as it is closed and generated using expected value and variance, which are both estimated by the cause-effect network.

To perform the parameter identification the functional (7) is minimized. In this functional S_δ is the given target and F the cause-effect network operator. The minimiser of (7) is then the identified parameter.

$$T(x) = d\left(S_{F(x)}, S_\delta\right) + \alpha R(x) \qquad (7)$$

The minimisation of (7) is done numerically using a standard non-linear solver, such as gradient decent with adaptive step size, inner point algorithm or sequentially quadratic problem solver. Depending on the properties of F the solver may need to be adapted to be able to minimize the functional. In all cases only a local minimum is identified which may not be global minimum, as all algorithms are of iterative nature and depend on the starting value. Different strategies, such as hill climbing and trust region, can be used to soften the dependency on the starting value but yield higher numerical cost.

5 Numerical Results

In this section two numerical examples are outlined. The first one is a theoretical example to showcase the influence of varying variances on parameter identification. The second example is a toy example of a production process with two process parameters and one output quality parameter. A toy example is used to illustrate the whole process of the presented method and still be useful to provide a quick understanding of the workflow of the process. In addition, the true functional dependency of input and output are known which allows to evaluate the results.

In the example a logarithmic grow of the output $y \in \mathbb{R}$ is observed as the input $x \in \mathbb{R}$ is increasing. The variance in y is increasing as the distance between x and 1 is increasing. The prediction function output including estimated variance is shown in figure 4.

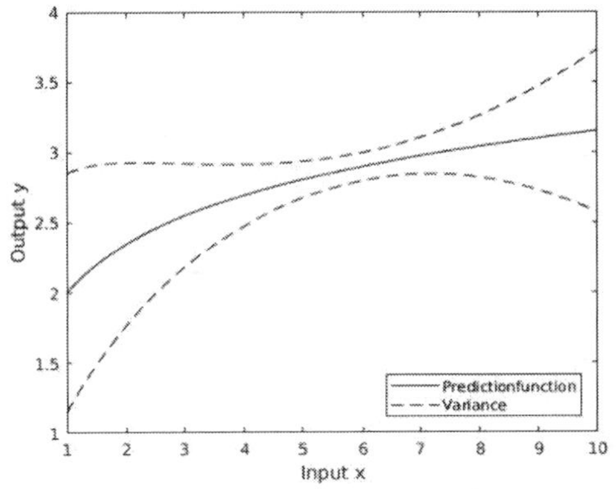

Fig. 4. Example of influence of varying variance in the output on the parameter.

The prediction function can be inverted directly to find a parameter for a desired output. In this example the desired output is $y = 3.07$ with a tolerance area of $[2.83, 0.24]$. This derived from the expected value plus minus allowed variance of 0.24. An input of $x = 8.5$ produces the desired expected value of $y = 3.07$, the variance is 0.2875 and does exceed the given boundary. To find a better parameter set considering the tolerance area a parameter identification using the method introduced in this work is performed. The identified parameter is $x = 6.74$ which produces the output $y = 2.954$ with a variance of 0.1164 and therefore a tolerance area of $[2.8376, 3.0705]$ which lies within the desired boundaries. This example shows how performing a parameter identification on the expected value does not provided the best parameter for a desired outcome with a set acceptable variance.

For the second example a two-step process is considered. The first process F_1 maps $x_1 \in \mathbb{R}$ onto $y_1 \in \mathbb{R}$. The second process F_2 maps $(x_2, x_3) \in \mathbb{R}^2$ onto $y_2 \in \mathbb{R}$. For the cause-effect network process F_2 is applied on the outcome of process F_1 where $y_1 = x_2$ is set. The expanded operator as well as the whole process model is given in (8).

$$\begin{aligned} F_1(x_1, x_3) &= (y_1, x_3)^T \\ F_2(x_2, x_3) &= y_2 \\ F(x_1, x_3) &= F_2(F_1(x_1, x_3)) \end{aligned} \qquad (8)$$

The true model without noise is given in (9). This will be used to compare the numerical results based on the prediction models with the true solution. The noise is an additive normal noise on each prediction model with mean 0 and a standard deviation of 0.2. The domain of x_1 is restricted to $[0,4]$, the domain of x_2 is restricted to $[0,6.25]$, and the domain of x_3 is restricted to $[0,1]$.

$$\begin{aligned} \widetilde{F_1}(x_1) &= (x_1 - 1.5)^2 \\ \widetilde{F_2}(x_2, x_3) &= 5 \cdot \sin(3 \cdot x_3) + \log(x_2^3 + 2) \end{aligned} \qquad (9)$$

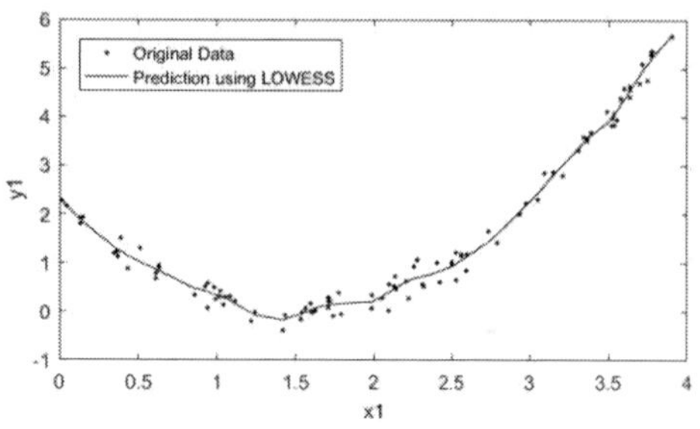

Fig. 5. Data and regression for the first process.

The first model is shown in figure 5, the second model is shown in figure 6. The output of the whole network matches figure 6 as the transformation on x_1 is not visible in such a graph.

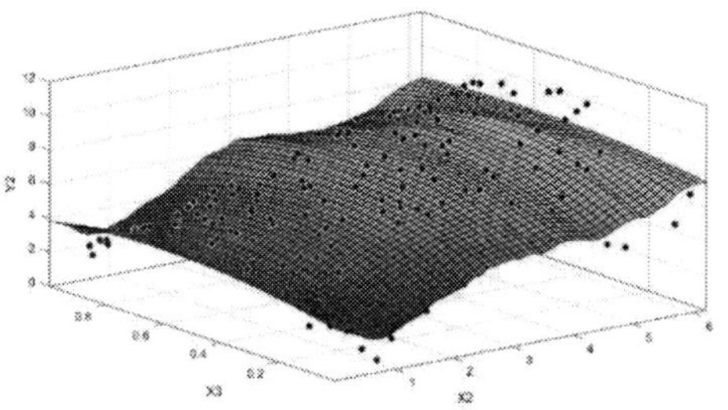

Fig. 6. Data and regression for the second process.

The variance in the data was analysed for the whole cause-effect network at once. The variance is stable for the input x_3 and decreases as the distance of $x_1$to 1.5 decreases. This matches the expected behaviour of a sensitivity analysis of (9) connected as in (8). As a true parameter $[1, 0.2]$ is set which will result in $y = 3.5241$. The variance is set to be 0.2. The calculated parameter is $[0.993, 0.182]$ which has an estimated expected value of $y = 3.5661$ and a variance of 0.18. It is therefore within the set limit. The identified parameter does match the true parameters within a reasonable span. Without any regularization the identified parameter will be $[0.89, 1.62]$ which has an ideal match in the expected value but does not reconstruct the true parameter so well. It is therefore more influenced by the ill-posed nature of the cause-effect network.

6 Summary and Conclusion

This work presents a new method to invert prediction models in micro production for process design. It alters classic approaches from the mathematical field of inverse problems to account for variations in production and enables a successful parameter identification for a given desired output.

The forward model is generated by cause-effect networks which are generated using the μ-ProPlAn methodology and GUI. As the generated models are usually non-linear and not invertable, special numerical method have to be applied for an parameter identification. To choose a suitable method further knowledge about mathematical properties of the model are needed. Therefore, prediction models and how they are derived is discussed in this work in detail. LOWESS as a method for generating data based prediction models is presented in further detail, as it is used as a default option for the μ-ProPlAn method. To demonstrate the challenges and advantages of the presented method two numerical examples are provided.

The presented method allows comprehending complex structures and dependencies in micro production and uses mathematical methods to estimate suitable parameters for a complex process with desired output. It is restricted by the amount of information, theoretical or data based, on each production step available. However, through novel approaches to parameter identification and forward modelling the amount of necessary experiments can be greatly reduced. While its application is not only limited to micro production it was developed and tailored to overcome the challenges faced in micro production.

Due to the possible ill-posed nature of prediction models alternative methods, such as Design of Experiments or Artificial Neural Networks yield a large amount of required experiments and data if applicable. In addition Neural Networks have limitations in solving ill-posed inverse problems as the inverse operator may not exist or is non-differentiable as well as non-injective (see [12]). They can be used in combination with classic approaches but are not suitable to solve the inverse problem on their own.

Overall, this work provides a first proof of a novel concept for parameter identification based on prediction models in micro production for process design. Current work focuses on optimizing numerical efficiency as well as generating new data for further validation.

Acknowledgement

The authors acknowledge the financial support by Deutsche Forschungsgemeinschaft (DFG, German Research Foundation) for the SFB 747 (Collaborative Research Center) "Mikrokaltumformen – Prozesse, Charakterisierung, Optimierung".

P. Gralla is supported by the Deutsche Forschungsgemeinschaft (DFG, German Research Foundation) within in the framework of GRK 2224/1.

References

1. F. Vollertsen, Product Engineering, 2, 2008, 4, S.377-383
2. A. K. Louis, Inverse und Schlecht Gestellte Probleme, Teubner Studienbücher, 2012
3. A. Rieder, Keine Probleme mit Inversen Problemen – Eine Einführung in ihre stabile Lösungen, Vieweg, 2003
4. J. Bangti and P. Maaß, Inverse Problems 28, 2012, 123001
5. D. Rippel, M. Lütjen and B. Scholz-Reiter, Journal of Manufacturing Technology Management 25, 2014

6. B. Denkena and H.K. Tönshoff, Prozessauslegung und –integration in die Prozesskette, in : B. Denkena and H.K. Tönshoff (Eds.), Spanen – Grundlagen, Springer Verlag, Berlin Heidelberg, 2011, pp. 339-362
7. D. Rippel, M. Lütjen and B. Scholz-Reiter, Journal of Manufacturing Technology Management, 25, 2014
8. D. Rippel, M. Lütjen and M. Freoitag, Journal of Manufacturing Systems, 43, 2017, pp. 79-87
9. M. Grasmair, M. Haltmeier and O. Scherzer, Inverse Problems 24, 2008, 055020
10. J. Fox, Nonparametric Regression in R : An Appendix to An R Companion to Applied Regression, 2010, 2nd Edition
11. T. Hastie, The Elements of Statistical Learning – Data Mining, Inference, and Prediction, 2016, Springer Series in Statistics, p. 194-207
12. J. Adler, O. Öktem, Solving ill-posed inverse problems using iterative deep neural networks, arXiv:1704.04058v2, 2017.

Fast Quality Inspection of Micro Cold Formed Parts using Telecentric Digital Holographic Microscopy

Mostafa Agour[1,2,*], *Claas* Falldorf[1], *Benjamin* Staar[3], *Axel* von Freyberg[4], *Andreas* Fischer[4], *Michael* Lütjen[3] and *Ralf* B. Bergmann[1,5]

[1]BIAS – Bremen Institute of Applied Beam Technology, Klagenfurter Straße 5, 28359 Bremen, Germany
[2]Aswan University, Faculty of Science, Department of physics, 81528 Aswan, Egypt
[3]BIBA - Bremer Institut für Produktion und Logistik GmbH at the University of Bremen, Hochschulring 20, 28359 Bremen, Germany
[4]University of Bremen, Bremen Institute for Metrology, Automation and Quality Science (BIMAQ), Linzer Str. 13, 28359, Bremen, Germany
[5]University of Bremen, Faculty 01: Physics and Electrical Engineering and MAPEX Center for Materials and Processes, 28359 Bremen, Germany

Abstract. Quality inspection is an integral part of the production process and often part of the quality management agreements between manufacturer and customer. Especially when it comes to safety-relevant parts, i.e. in the automobile or medical industry, often a 100% quality inspection is mandatory. Here, we present a solution comprised of a digital holographic measurement system, as well as fast algorithms for geometric evaluation and surface defect detection that paves the way for the inspection of metallic micro cups in less than a second. By use of a telecentric lens instead of standard microscope objective, we compensate scaling effects and wave field curvature, which distort reconstruction in digital holographic microscopy. Due to limited depth of focus of the microscope objective, depth information from different object layers are then stitched together to yield 3D data of its geometry. The resulting point cloud data is automatically decomposed into simple geometric shapes in order to analyse geometric deviations. Amplitude as well as phase distribution images are then analysed for surface defects. Our approach is demonstrated by inspecting cold formed micro cups. Defects larger than 2 µm lateral resolution and 5 µm depth can be detected.

Keywords: *Inspection, Interferometry, Metrology*

1 Introduction

Quality inspection is an integral part of the production process and often part of the quality management agreements between manufacturer and customer. Especially when it comes to safety-relevant parts, i.e. in the automobile or medical industry, often a 100% quality inspection is mandatory [1].

Achieving this goal is especially challenging in the micro-domain, i.e. when measurement uncertainty has to be less than a few µm as there is usually a trade-off between precision and speed [2]. Processes like e.g. micro cold forming, however, allow for production rates of multiple parts per second. Presently, there exist no solutions that are fast, precise and suitable for in-process measurements at the same time [3].

Due to these indusial requirements, methods such as tactile measurements, confocal microscopy and phase retrieval [4, 5] are not suited for the inspection task since they are too slow. In contrast to these methods, interferometric methods are established for a full field measurement. Examples of interferometric techniques include Digital Holography (DH) [6], White-Light Interferometry (WLI) [7], Optical Coherence Tomography (OCT) [8] and computational shear interferometry (CoSI) [9]. These methods are based on the determination of the optical path difference (OPD) of light scattered by or transmitted through a test object. Since available cameras can only measure intensities of light, the OPD must be encoded consequently. Based on the utilized encoding technique and the geometrical complexity of the test object, measurement uncertainties down to a fraction of the illumination's wavelength can be achieved [10].

Compared to other interferometric techniques, DH is considered fast since the OPD can be retrieved from only a single digital hologram. DH is based on the evaluation of the lateral phase distribution (within the object plane) of monochromatic light. For this purpose, light coming from a test object and a reference wave having known characteristics are coherently superimposed to convert the phase information into detectable intensity modulations. From the resulting intensity pattern, information on the phase can be extracted using phase

* Corresponding authors: agour@bias.de

shifting techniques. Thereafter, the optical path difference is identified and thus the shape of the test object is obtained. However, the measurement of the OPD is ambiguous with respect to multiples of the wavelength. Hence, for optically rough surfaces, the result of the evaluation cannot be used to reconstruct the shape of the test object if the height variations exceed half of the wavelength. One way to solve this problem is measuring two phase distributions associated with two different wavelengths. This approach is referred to as two-wavelengths contouring technique.

Here, we present a solution comprised of a digital holographic measurement system, as well as fast algorithms for geometric evaluation and surface defect detection that paves the way for the inspection of metallic micro cups in less than a second. By use of a telecentric objective instead of standard microscope objective, we compensate scaling effects and wave field curvature, which distort reconstruction in digital holographic microscopy. Depth information from the measured holograms is then stitched together to yield 3D data of its geometry. The depth information is converted to a 3D point cloud, which is automatically decomposed into simple geometric shapes in order to analyse geometric deviations. Amplitude as well as phase distribution images are then analysed for surface defects.

In section 2, hologram recording using the proposed DHM, digital hologram reconstruction and reconstructing the 3D surface of micro cups will be discussed. The geometry and surface inspection of micro cups under test will be discussed in section 3. Finally, a brief conclusion will be given in section 4.

2 Optical metrology of micro parts using DHM

2.1 Holographic contouring

Commonly engineering objects' surfaces exhibit peaks and steps larger than half of the wavelength. Since the measured phase values are located within the interval [-π, π[, such surfaces are considered to be optically rough. Thus, the evaluation process becomes ambiguous and the measurement is not unique. The solution to this dilemma is referred to as the synthetic wavelength. In this approach, two phase distributions are retrieved from two measurements associated with two different wavelengths λ_1 and λ_2. Calculating the phase difference $\Delta\varphi$ between the reconstructed phase distributions corresponding to the two measurements, the 3D height map z_p of the test object can be directly calculated by utilizing

$$\Delta\varphi = -\frac{2\pi}{\Lambda} z_p(1+\cos\alpha) \qquad (1)$$

Here, α is the angle between the observation and illumination direction and Λ is referred to as the synthetic wavelength which is given by

$$\Lambda = \frac{\lambda_1\lambda_2}{|\lambda_2-\lambda_1|} \qquad (2)$$

The resulting phase difference distribution map given by Eq. (1) equivalents a single measurement with Λ. This map contains fringes and it is referred to as the phase contouring map. Adapting the difference between the two wavelengths λ_1 and λ_2 is required to enable the investigation of objects with step heights of several millimeters.

2.2 Digital Holographic Microscopy

The digital holographic microscope given in Fig. 1 is used to capture the required two holograms with two different wavelengths. The setup contains a long distance microscope objective (LDM) with a magnification of 10×, a numerical aperture (NA) of 0.21 and a working distance (WD) of 51 mm. Optical fibers were used to illuminate the test object and to provide the required reference wave. A beam splitter (BS) is used to combine object wave, light diffracted from the surface under investigation, and the reference wave, resulting in a hologram, which is captured across the camera plane. It is noted that there exists an angle α between the observation and illumination direction. Numerically the phase distribution corresponding to each measurement, is reconstructed utilizing the spatial phase shifting method, where the carrier frequency is controlled by shifting the position of the reference wave with respect to the optical axis [3]. Thus, the setup was used to register the phase information in the plane, which is close to the surface of the object under test.

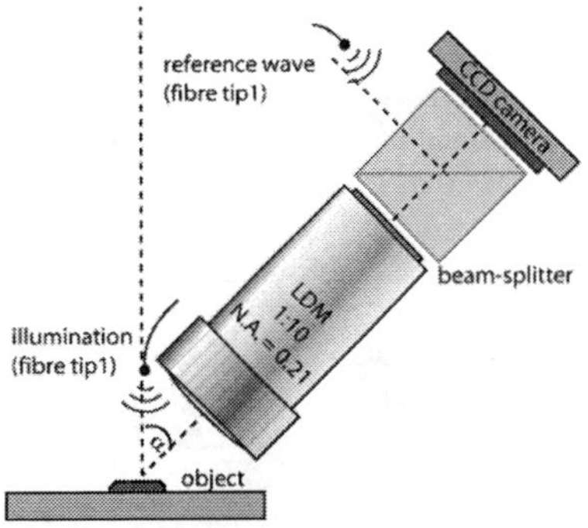

Fig. 1. Digital holographic microscope setup (DHM): To image the surface of the test object onto the utilized camera sensor, a long distance microscope objective (LDM) having 10× magnification, a numerical aperture (N.A.) of 0.21 and a working distance of 51 mm was used. Fiber technology is used to illuminate the object under test and provide a spherical reference wave. For simplicity, only one illumination and one reference is shown. A beam splitter was used in order to combine the object and the reference waves producing a hologram across the camera plane.

2.3 Surface measurements of micro components

The micro deep drawn cup shown in Fig. 2 insert is as an example of a micro component. Geometry inspection of such micro parts is needed in order to improve and optimize the micro cold forming process. In the following, the 3D surface measurements of the micro cup based on digital holographic microscopy, which is considered to be the backbone of geometric inspection, will be discussed.

An experimental setup is realized based on the sketch shown in Fig. 1. An image of the setup is shown in Fig. 2. Four laser diodes supplied by the company Thorlabs two with $\lambda = 638.13$ nm and the other two with $\lambda = 644.08$ nm were used. According to Eq. 2, a synthetic wavelength of 69.07 µm is used for the evaluation process. Utilizing a fiber switch and a 1×2 fiber splitter, object and reference waves are realized.

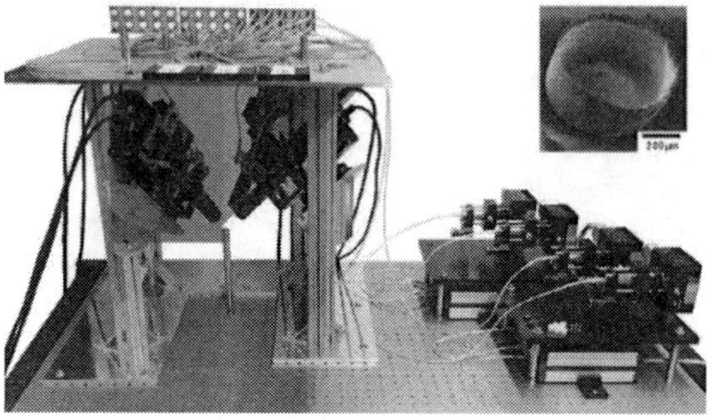

Fig. 2. Digital holographic microscope setup based on the scheme shown by Fig. 1. The setup consists of four DHM units distributed around the test object. Thus each unit delivers part of the test object. These four parts can then be used to reconstruct the whole 3D shape of the test object. The insert is an SEM image of a micro deep drawn cup with a diameter of ca. 1 mm and a depth of 0.5 mm, which is shown as an example of a test object.

The LDM is an object side telecentric objective with a numerical aperture of 0.21, a magnification of 10× and a working distance of 51 mm. From Rayleigh formulae, the latera resolution (0.61λ/N.A) of each DHM unit is 1.87 µm and the axial resolution ($1.67\lambda/N.A.^2$) is 24.4 µm at $\lambda = 644$ nm. This limits the lateral size of the defects measured by the DHM. It is noted that the limited axial resolution is enhanced by utilizing numerical focusing properties of digital holography. The camera sensor is an AVT prosilica GT 2750 having 2750×2200 pixels, where the pixel pitch is 4.54 µm. It should be noted that the test object is illuminated from four different directions and four holograms are recorded on single-shot using four reference waves by applying the digital holography multiplexing principle [3]. These holograms are used to reduce speckle noise in two wavelength contouring. Accordingly, each holographic unit from the four units will capture two successive multiplexed holograms. The two successive multiplexed holograms are captured, one for each wavelength, using only one DHM observation unit. This means for each observation direction, eight holograms are captured in two shots for the two wavelengths. In the following, results captured using only one observation direction will be presented and discussed. The time required for the capturing process and for the switching between the two wavelengths is 70 ms. Using the spatial carrier frequency method [11, 12] one can numerically reconstruct the phase distributions $\phi_{\lambda 1}$ and $\phi_{\lambda 2}$ which correspond to the two measurements. Errors with lateral extensions from 2 µm and minimal depth of at least 5 µm can be detected [3]. An example of the resulting real amplitude reconstructed from the hologram captured for $\lambda = 638.13$ nm is shown in Fig. 3a. The phase difference $\Delta = \phi_{\lambda 2}-\phi_{\lambda 1}$ between the two reconstructed phase distributions across the capturing plane which represents the countering map across that plane is shown in Fig. 3b. As it can be seen in Fig. 3b, fringes are only sharp across the in focus area of the object which can be clearly recognized by the real amplitude shown in Fig. 3a. This is expected since the microscope objective has a limited depth of focus.

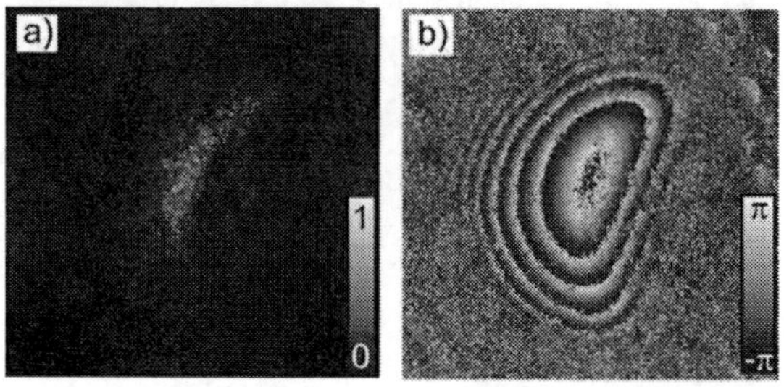

Fig. 3. a) Shows the real amplitude of the reconstructed complex amplitude across the capturing plane of the recorded hologram for $\lambda = 638.13$ nm. b) Shows the phase difference between the two reconstructed $\Delta = \phi_{\lambda 2}-\phi_{\lambda 1}$ phase distributions across the capturing plane. The image size is 2200 × 2200 pixels with a pixel pitch of 4.54 µm.

In contrast to microscopy, digital holography offers the extension of the objective depth of focus, since holograms give access to the complex amplitude, digital refocusing across the whole object by means of numerical propagation is possible. Thus, in order to completely reconstruct a sharp contouring propagation, autofocus algorithms are used. For fast evaluation, an automated process was proposed and implemented within a graphics processing unit (GPU). The autofocus algorithm is implemented by scanning within small windows throughout all propagated planes to define where the object is in-focus by estimating the standard deviations of Δ which is relatively high within those windows, where the object is out of focus because of the speckle decorrelation. The result of this process is shown in Fig. 4.

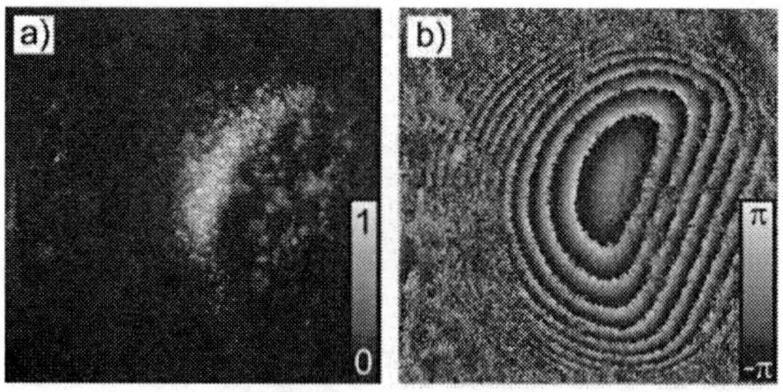

Fig. 4. a) Shows the real amplitude, which represents a sharp image of the micro cup under test with respect to the observation direction. b) Shows the phase difference distribution which represents a sharp contouring phase map across the whole object.

The contouring phase map shown in Fig. 4b is then unwrapped. Form the quantitative comparison which is based on the standard deviation across different areas of the constructed contouring map shown in Fig. 4b, the height error ranges between 2.5 and 4.6 µm. The unwrapped phase values are substituted into Eq. 1 to determine the 3D height map. The result is shown in Fig. 5. Such a height map is converted to a 3D point cloud which is used as the input for both the dimensional and the surface inspection process which will be discussed in the next section.

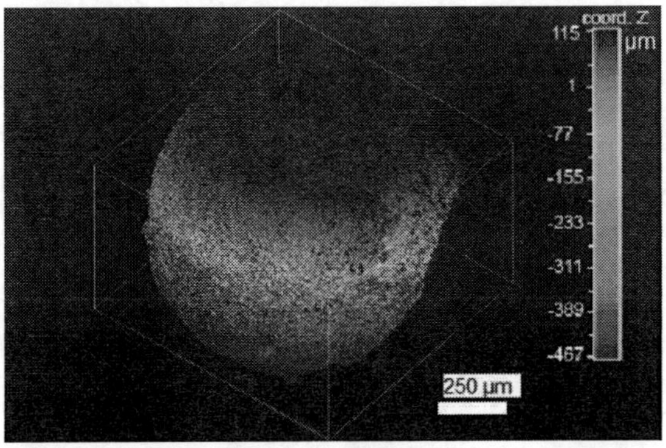

Fig. 5. The 3D height map calculated after unwrapping the countering phase map shown in Fig 4b.

3 Dimensional and surface inspection of micro parts

This section describes the evaluation of optically acquired data with respect to geometric deviations (subsection 3.1) as well as regarding surface defects (subsection 3.2).

3.1 Dimensional inspection

Optical measurement data contains a high number of surface coordinates of one or multiple observation directions and represents either a free-form surface or a combination of several geometric elements. The evaluation of free-form surfaces, on the one hand, consists in aligning the measurement data to the nominal CAD-data [13] and to calculate and visualize the deviations of each measurement point. For this kind of quality inspection, several commercial solutions exist. On the other hand, in order to analyse the optical measurement data regarding geometric parameters like dimensional or shape deviations, respectively, the measured points have to be segmented. This means assigning the individual measurement points to the approximating geometric elements. However, a manual segmentation is time consuming and not suitable for an automated analysis within a mass production. Furthermore, due to the following context, it is to aim for an automated segmentation.

For reducing the measurement uncertainty, the geometric objects to be analysed should be acquired thoroughly. For instance, a decreased central angle of a circle directly leads to an increased uncertainty of the approximated geometric parameters [14]. In case of micro features or measurement data with a low point density, a further reduction of measurement points additionally raises the uncertainty [15]. This can happen during the manual segmentation due to a doubtful assignment, e.g. by leaving out points which cannot be reliably correlated to a certain geometric element. Only by an automated segmentation of the measurement data, the individual measuring points can be assigned in a reproducible and optimal way to the corresponding geometric elements. Two approaches are known for such an automated segmentation:

1. Neighbouring measurement points are rated based on their curvature and assigned to corresponding geometric elements [16]. This method can provide accurate solutions, but it is sensitive to noisy data and not able to distinguish between spheres and cylinders with certain radii.

2. A holistic approximation can evaluate a composed set of data under the present boundary conditions in a single approximation task [17]. By the definition of separating functions, an optimal assignment of the measurement points to the corresponding geometric elements (segmentation) can be carried out simultaneously. The method is presented for different applications in [18, 19].

It was proved that the holistic approximation with automated segmentation (2nd approach) is only little sensitive regarding the initial values of the approximation and at the same time converges reliably within wide ranges [18]. Furthermore, this method was successfully tested for the evaluation of micro-measurements [20], and it allows the automatic detection of outliers by a combination with statistic methods [21]. Thus, the 2nd approach is particularly suited for noisy optical measurement data. However, the algorithms were not yet implemented for the evaluation of optical data acquired with DH.

The presented method is described in detail in [19] for the evaluation of tactile acquired geometries of microforming tools. As the geometry of micro-cups is composed in the same way by a combination of a plane, a torus and a cylinder, the holistic approximation can be applied with the same geometric model for the dimensional evaluation of optically acquired micro-cup surfaces.

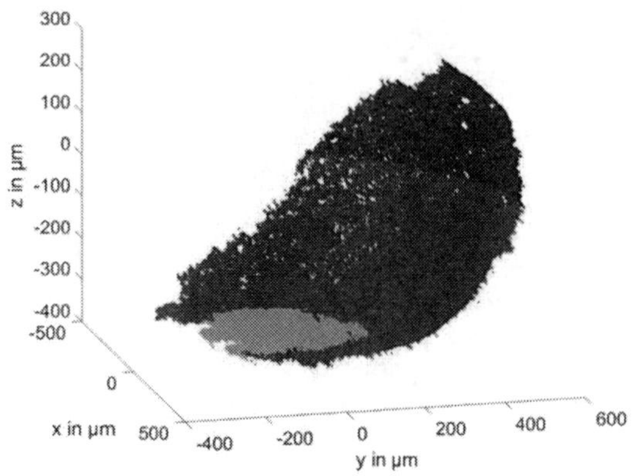

Fig. 6. Result of holistic approximation: The measured points are decomposed into a plane (green), a torus (red) and a cylinder (blue).

The approximately 3 million measured points of the micro-cup's surface were evaluated with the holistic approximation. The decomposition result is presented in Fig. 6. The torus wall radius was calculated to $r_t = 317.7$ µm and the cylinder radius was determined to $r_c = 542.2$ µm. A calculation time of 0.9 s could be achieved by using the MATLAB Parallel Computing Toolbox calculating on a graphics processing unit (GPU) and reducing the density of the acquired point cloud. Tests with simulated data showed that a reduction of the measured points by the factor 100 results in deviations of the approximated radii less than 0.09 µm.

3.2 Surface inspection

Surface defects like e.g. scratches or dirt might be too small to cause a detectable change in the measured phase distribution. Hence, reliable detection necessitates additional methods, which incorporate the measured amplitude image.

In order to realize fast and precise detection of surface defects, we opted for convolutional neural networks (CNN), as they have drastically increased the abilities of machine vision in many areas, as e.g. object classification [23, 24, 25], semantic segmentation [26, 27] or the generation of text descriptions for images [28]. In contrast to methods that rely on laborious feature engineering, CNNs are able to automatically extract useful. CNNs have hence already been successfully applied for automatic surface inspection [29, 30, 31].

3.2.1 Data and pre-processing

We created a data set comprising of 69 amplitude measurements taken from top view, which we hand-labelled for surface defects.

All measurements were downscaled to a size of 512×512. Backgrounds were removed by applying a strong low-pass-filter (i.e. convolution with a 55×55 matrix of ones) to the measurement and subsequent thresholding with the mean of the resulting matrix. In the resulting binary matrix the largest contour was detected using methods, provided by the OpenCV software library [32]. An example for the pre-processing is shown in figure 7.

3.2.1 Network architecture

We opted for a modified „U-Net" architecture [33], as it achieves pixel-level detection while being computationally efficient.

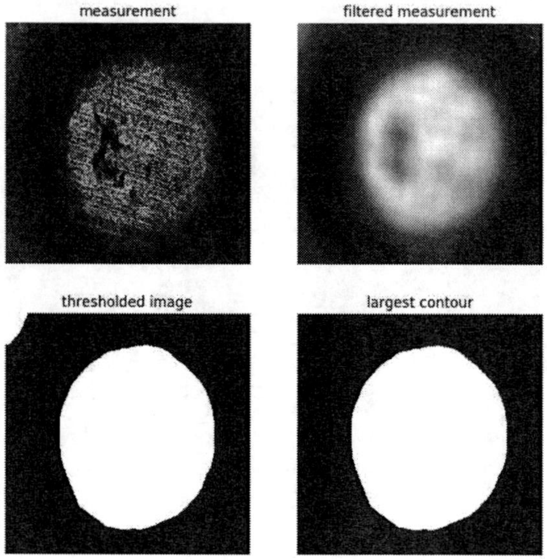

Fig. 7. Top left: measurement. Top right: measurement after application of a low-pass filter. Bottom left: result of applying a binary threshold. Bottom right: largest contour.

This architecture augments the classical CNN architecture with additional layers for up-sampling the results. Fine-grained spatial information is thereby fed into the up-sampling layers via skip connections from the earlier layers. Figure 8 shows a schematic graph of the network used in this work. In the down-sampling part of the network we used blocks of one separable convolutional layer [34] followed by Max-pooling [31] and batch normalization [35].

In the up-sampling we used blocks of one up-sampling layer, one concatenation layer for combining the up-sampled result and the corresponding feature layer, and one separable convolutional layer to further process the results. The parameter values at each layer are provided in table 1.

As a modification we not only predicted local defect maps but also used the features of the middle layer (block 4 in tale 1) for global defect detection, i.e. deciding if a measured cup is overall defective or not. To achieve this, we used global max pooling, i.e. used the maximal activation of each feature over all positions and fed this into a single sigmoid unit.

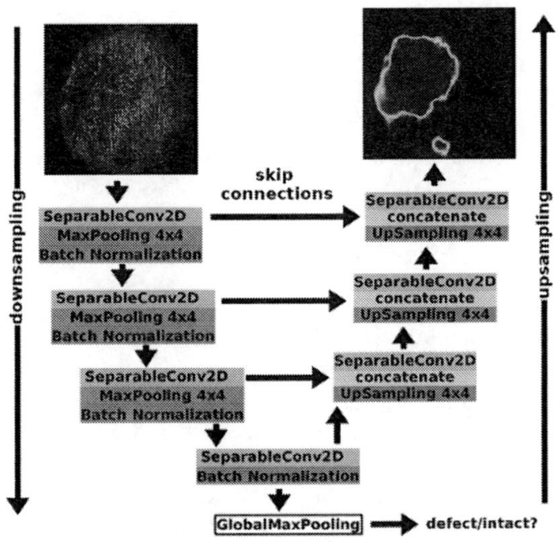

Fig. 8. Schematic of model architecture. The pre-processed measurement is fed into the network (top left) and downscaled while extracting features for defect detection (left part). These features are then used for global, i.e. measurement-wise defect classification (bottom). To also enable defect localization the features are upsampled again (right part), yielding a map with defect probabilities (blue = 0, red = 1).

Table 1. Network architecture and parameters

Name	Units	Input resolution	Output resolution
Block 1	8×elu	512×512×1	128×128×8
Block 2	16×elu	128×128×8	32×32×16
Block 3	32×elu	32×32×16	8×8×32
Block 4	64×elu	8×8×32	8×8×64
Up block 1	64×elu	8×8×64	32×32×64
Up block 2	32×elu	32×32×64	128×128×32
Up block 3	16×elu	128×128×32	512×512×16
Output (local)	1×sigmoid	512×512×16	512×512×1
Output (global)	1×sigmoid	64 (output block 4)	1

3.2.2 Model parameters

As activation function, we used exponential linear units (elu) [36] in all but the output layers were we used sigmoid units to constrain the output to 0-1. Filter sizes where chosen to be 5×5, except for the output layer with 1×1. For all but the last convolutional layer we used depthwise separable convolutions [34].

Max pooling was conducted over 4×4 windows, effectively decreasing the result by a factor of 0.25.

The network was trained to minimize binary crossentropy using the Adam optimizer [40] with minibatch size of 5.

All experiments were conducted using the keras library for Python.

3.2.4 Evaluation

In order to utilize as much of the data as possible for training, we opted for a modified leave-one-out cross validation scheme, i.e. we used 68 samples for training and the remaining sample for validation and repeated this process 10 times. Each model was trained until perfect discrimination between intact and defective parts was reached for the training set. We then computed the lowest threshold for achieving perfect discrimination and used it to determine if the validation sample would be correctly classified.

3.2.5 Results

All 10 validation samples were correctly classified. Examples for defect localization are shown in figure 9. It can be seen that all defects are labelled correctly, although some labels are still rather coarse. We hypothesize that this is due to the relatively small amount of training data and will hence improve as larger amount of data are being acquired.

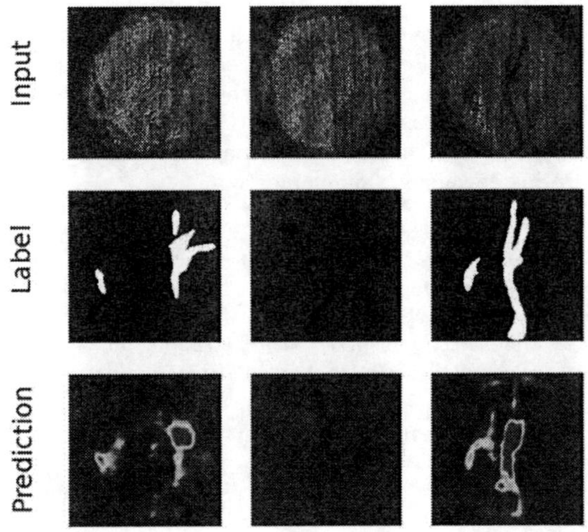

Fig. 9. Top row: input measurements. Middle row: hand-drawn defect labels. Bottom row: the networks predictions. Red indicates high defect probability and blue indicates low defect probability

The algorithm is also fast: On a Desktop PC (CPU: AMD 6100 Six-Core, memory: 8 GB RAM, GPU: GeForce GTX 1060 6GB) preprocessing takes <5ms and defect detection takes ~60ms. The total runtime of the defect detection routine is hence ~65ms per measurement.

4 Conclusions

We demonstrated here a complete method for the automatic measurement, geometric decomposition and surface inspection of micro parts. The experimental configuration is based on two-wavelength holographic contouring technique. Amplitude as well as phase distribution images recovered from the captured holograms are utilized for dimensional and surface defects. The setup can detect defects from 2 μm lateral resolution and 5 μm depth. Algorithms for evaluation are parallelized in order to speed up computing time. The proposed method is demonstrated by inspecting cold formed micro cups. Complete evaluation process for holographic capturing and acquisition is 12 s, for dimensional evaluation is 0.9 s and for surface inspection is 0.065 s. This means in total the evaluation process could be achieved in less than 13 seconds. The next steps would be the characterization and calibration of data acquisition. Stitching of multiple views will likely further decrease uncertainty for incomplete acquired features but will also bring an extra individual uncertainty contribution, dependent on the chosen method. This has to be analysed / characterized. This optimisation and characterization is part of ongoing research.

Acknowledgements

The authors gratefully acknowledge the financial support by Deutsche Forschungsgemeinschaft (DFG, German Research Foundation) for Subproject B5 "Sichere Prozesse" within the SFB 747 (Collaborative Research Center) "Mikrokaltumformen - Prozesse, Charakterisierung, Optimierung".

References

1. C. von Kopylow and R. B. Bergmann, "Optical Metrology" in *Micro metal forming*, pp. 392–404, (Springer, Berlin Heidelberg 2013).
2. M. Agour, R. Klattenhoff, C. Falldorf, R. B. Bergmann, Proc. SPIE **10233**, 102330R (2017).
3. M. Agour, R. Klattenhoff, C. Falldorf, R. B. Bergmann, Opt. Eng. **56**, 124101 (2017)
4. C Falldorf, M Agour, C Von Kopylow, RB Bergmann, J. Opt. **14**, 065701(2012).
5. M. Agour, P. Huke, C. Kopylow, C. Falldorf, AIP Conference Proceedings 1236, 265-270 (2010).
6. U. Schnars and W. Jüptner, Appl. Opt. **33**, 179–181(1994).
7. J. C. Wyant, Proc. SPIE **4737**, 98 (2002).
8. D. Huang, E. A. Swanson, C. P. Lin, J. S. Schuman, W. G. Stinson, W. Chang, M. R. Hee, T. Flotte, K. Gregory; C. A. Puliafito, J. G. Fujimoto, Science, 254, 1178–1181 (1991).
9. C. Falldorf, M. Agour, and R. B. Bergmann, Opt. Eng. **54**, 024110 (2015).
10. P. Marquet, B. Rappaz, P. J. Magistretti, E. Cuche, Y. Emery, T. Colomb, and C. Depeursinge, Opt. Lett. **30**, 468–470(2005).
11. M. Takeda and H. Yamamoto, Appl. Opt. **33**, 7829–7837 (1994).
12. M Agour, K El-Farahaty, E Seisa, E Omar, T Sokkar, Appl. Opt. 54, E188–E195 (2015).
13. Savio, E.; De Chiffre, L.; Schmitt, R.: Metrology of freeform shaped parts. CIRP Annals-Manufacturing Technology 56 (2007), Nr. 2, S. 810-835.
14. Hernla, M.: Messunsicherheit bei Koordinatenmessungen: Abschätzung der aufgabenspezifischen Messunsicherheit mit Hilfe von Berechnungstabellen. expert verlag, 2007. - ISBN 3816926762.
15. Hernla, M.: Abschätzung der Messunsicherheit bei Koordinatenmessungen unter den Bedingungen der industriellen Fertigung. VDI-Verlag, 1992. - ISBN 3181474029.
16. Westkämper, E.; Stotz, M.; Effenberger, I.: Automatische Segmentierung von Messpunktwolken in regelgeometrische Elemente (Automatic Segmentation of Measurement Point Clouds to Geometric Primitives). tm-Technisches Messen 73 (2006), Nr. 1/2006, S. 60-66.
17. Goch, G.: Algorithm for the combined approximation of continuously differentiable profiles composed of straight lines and circle segments. Annals of the CIRP 40/I (1991), S. 499-502.
18. Lübke, K.; Sun, Z.; Goch, G.: Ganzheitliche Approximation eines Gerade-Kreis-Gerade-Profils mit automatischer Trennung in Einzelprofile. In: Scholl, G. (Hrsg.): XXIV. Messtechnisches Symposium des Arbeitskreises der Hochschullehrer für Messtechnik e.V. (AHMT), Hamburg. Shaker Verlag, Aachen, 2010. - ISBN 978-3-8322-9453-3, S. 77-90.
19. Lübke, K.; Sun, Z.; Goch, G.: Three-dimensional holistic approximation of measured points combined with an automatic separation algorithm. CIRP Annals - Manufacturing Technology 61/I (2012), S. 499-502.
20. Zhang, P.; Mehrafsun, S.; Lübke, K.; Goch, G.; Vollertsen, F.: Laserchemische Feinbearbeitung und Qualitätsprüfung von Mikrokaltumformwerkzeugen. In: Kraft, O.; Haug, A.; Vollertsen, F.; Büttgenbach, S. (Hrsg.): 5. Kolloquium Mikroproduktion und Abschlusskolloquium SFB 499, Karlsruhe. 2011. - ISBN 978-3-86644-747-9 S. 169-176.

21. Grubbs, F. E.: Procedures for detecting outlying observations in samples. Technometrics 11 (1969), Nr. 1, S. 1-21.
22. A. Mecke, I. Lee, J.R. Baker jr., M.M. Banaszak Holl, B.G. Orr, Eur. Phys. J. E **14**, 7 (2004)
23. K Krizhevsky, A., Sutskever, I., & Hinton, G. E. (2012). Imagenet classification with deep convolutional neural networks. In Advances in neural information processing systems (pp. 1097-1105).
24. Simonyan, Karen, and Andrew Zisserman. "Very deep convolutional networks for large-scale image recognition." arXiv preprint arXiv:1409.1556 (2014).
25. He, K., Zhang, X., Ren, S., & Sun, J. (2016). Deep residual learning for image recognition. In Proceedings of the IEEE conference on computer vision and pattern recognition (pp. 770-778).
26. Long, J., Shelhamer, E., & Darrell, T. (2015). Fully convolutional networks for semantic segmentation. In Proceedings of the IEEE conference on computer vision and pattern recognition (pp. 3431-3440).
27. Yu, Fisher, and Vladlen Koltun. "Multi-scale context aggregation by dilated convolutions." arXiv preprint arXiv:1511.07122 (2015).
28. Vinyals, O., Toshev, A., Bengio, S., & Erhan, D. (2015, June). Show and tell: A neural image caption generator. In Computer Vision and Pattern Recognition (CVPR), 2015 IEEE Conference on (pp. 3156-3164). IEEE.
29. Weimer, D., Scholz-Reiter B., Shpitalni M.. "Design of deep convolutional neural network architectures for automated feature extraction in industrial inspection." CIRP Annals-Manufacturing Technology 65.1 (2016): 417-420.
30. Staar, B.; Lütjen, M.; Freitag, M.: Präzise Oberflächendefekterkennung in Mikrobauteilen mit neuronalen Netzen. In: Vollertsen F.; Hopmann C.; Schulze V.; Wulfsberg J. (Hrsg.): 8. Kolloquium Mikroproduktion. BIAS Verlag, Bremen, 2017, S. 11-16
31. Masci, J., Meier, U., Ciresan, D., Schmidhuber, J., & Fricout, G. (2012, June). Steel defect classification with max-pooling convolutional neural networks. In Neural Networks (IJCNN), The 2012 International Joint Conference on (pp. 1-6). IEEE.
32. Bradski, Gary, and Adrian Kaehler. "OpenCV." Dr. Dobb's journal of software tools 3 (2000).
33. Ronneberger, O., Fischer, P., & Brox, T. (2015, October). U-net: Convolutional networks for biomedical image segmentation. In International Conference on Medical image computing and computer-assisted intervention (pp. 234-241). Springer, Cham
34. Chollet, François. "Xception: Deep learning with depthwise separable convolutions." arXiv preprint (2016).
35. Ioffe, S., & Szegedy, C. (2015, June). Batch normalization: Accelerating deep network training by reducing internal covariate shift. In International conference on machine learning (pp. 448-456).
36. Clevert, Djork-Arné, Thomas Unterthiner, and Sepp Hochreiter. "Fast and accurate deep network learning by exponential linear units (elus)." arXiv preprint arXiv:1511.07289 (2015).
37. Kingma, Diederik P., and Jimmy Ba. "Adam: A method for stochastic optimization." arXiv preprint arXiv:1412.6980 (2014).

Inline tool wear measurement in lateral micro upsetting

Joseph Seven[1,*], *Lukas* Heinrich[1], *Hendrik* Flosky[2]

[1]BIAS GmbH, Klagenfurter Strasse 5, Bremen, Germany
[2]Former: BIAS GmbH, Klagenfurter Strasse 5, Bremen, Germany

Abstract. Micro metal parts are usually produced in large lot sizes at high production rates. In order to achieve sufficient product quality, excessive tool wear has to be avoided. Conventional methodologies for tool wear measurement usually require tools to be disassembled and measured separately. Since accuracy of handling and assembly is limited in micro range, a reproducible tool positioning between measurements cannot be guaranteed. Therefore, analysis of tool wear development is hardly possible using conventional methods. In this study, an inline tool wear measurement methodology is investigated. Instead of measuring the tool itself, it is based on the measurement of the forming product. Therefore, lateral upsetting experiments on 1.4301 stainless steel wires with a diameter of 0.3 mm are carried out on a micro forming press. The indentation geometry on the wire is measured optically and compared to the tool geometry. It is shown that both are in good agreement for different states of tool wear and that features with geometrical characteristics above 2 μm are transferred onto the indentation surface. Thus, the investigated methodology presents a feasible way for inline tool wear measurement.

Keywords: Micro forming, Cold forming, Wear

1 Introduction

The demand of micro parts in the areas consumer electronics, communication and in medicine technology increases continuously [1]. Products like smart phones become lighter and thinner while the quantities become higher. The sales of smart phones increased from 2013 with 969 million pieces [2] to 1.537 billion pieces in 2017 [3]. These products contain micro parts which have to be manufactured in a fast and cost efficient production [4]. A preferably high tool life is essential to increase the quantity and reduce the production costs. One approach to increase the tool life is the improvement of the tool wear. So a characterisation of the tool wear is required to identify the wear mechanisms and to improve in a following step the wear behaviour.

Especially in micro range the accuracy of handling and assembly is more difficult than in macro range. A dismounting and mounting of the micro tool can result in positioning errors. Therefore, to assure a reliable tool measurement, the tool has to be measured without removing it from the press.

Micro tools can be used for the production of micro parts in a combined blanking and deep drawing process [5]. In this case the outer diameter of a micro blanking and deep drawing die was decreased through a wrong positioning. The contact between the tool and the tool guide became bigger resulting in a stronger abrasion. A method which can be used for the tool wear measurement is the measurement of negative indentations [6]. Here the geometry of the tool is transferred in silicon with an accuracy of less than 1 μm but the production has to be stopped for that indentation. Another technique offers the comparative digital holography with an inline technique due to the accessibility of that measurement setup [7]. The holographic measurement setup is integrated into the press so that the micro tool can be measured inline.

In contrast of measuring the tool itself the forming product can be measured as well to analyse the tool wear. In this paper lateral upsetting experiments in micro cold forming are used to form a 0.3 mm thin wire in a micro forming press. The advantages for this method are no dismounting and mounting of the tool, lower downtimes and better accessibility. Both the geometry of the indentation and the tool geometry are measured and compared to each other by a laser scanning microscope. The goal is to show the feasibility of this method for an inline tool measurement.

2 Method

2.1 Lateral micro upsetting

Lateral upsetting experiments on 1.4301 stainless steel wires with a diameter of 0.3 mm were carried out on a highly dynamic micro forming press [8]. The machine axes are driven by electrical linear motors with a

* Corresponding author: seven@bias.de

maximum acceleration of 10 g and a maximum velocity of 3.2 m/s. The positioning error is below 3 µm at maximum acceleration and the repeat positioning error is below 1 µm up to a stroke of 8 mm. Fig.1 shows a near overview of the forming process. The forming punches are made of HSS and mounted to the tool holder. The tool holder is linked axial to a piezo force sensor, type Kistler 9311B, which has an accuracy of 0.01 N. The forming force F_U is monitored for each sample with a frequency of 5 kHz. The wire is feed through the guide and positioned by a feed system (Zehnder und Sommer EDV 040 – 220 DG). To find the contact point between wire and punch the punch was driven manually until a pre-load force of ~10 N was monitored. This point was used as the contact point in the automatically experiments. The counter plate is made of case hardened steel (C15) and mounted beneath the guide.

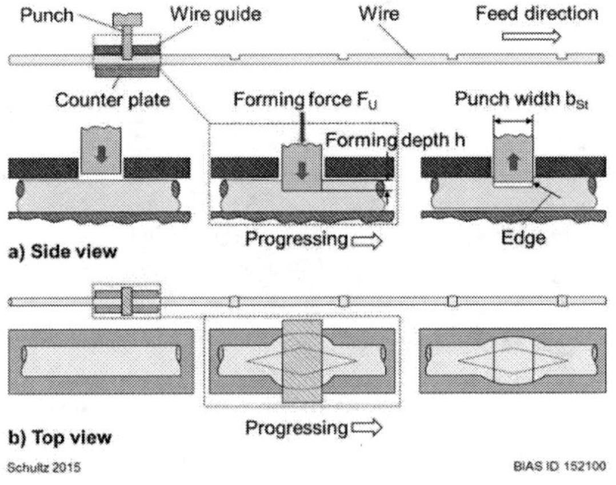

Fig. 1. Schematic view of the lateral upsetting process

Fig. 1 shows one stroke of the forming process in three steps schematically, a.) side view and b.) top view. In the first state (left) the wire is positioned by the feed system below the punch, without any contact between wire and punch. In the second step the punch moves downwards and penetrates the wire to a defined depth of *h*. Before the punch contacts the wire the force measuring system starts to monitor the force. The wire is deformed elastically and plastically and material freely flows to the sides as well as in wire direction. The deformed wire has typically the form of a rhombus, as shown in the middle and on the left side of the top view. When the wire is stripped of from the punch the elastic deformation reverts. At the end (step 3) the punch moves upwards to the starting position and the process starts again with a production rate of 350 strokes per minute. Table 1 shows the process and geometric parameters.

Table 1. Geometries, materials and process parameters for lateral upsetting

punch width	0.50 mm
wire thickness	0.30 mm
forming velocity	18 mm/s
penetration depth	0.15 mm
tool material	HS 6-5-2
wire material	X5CrNi18-10
counter plate	C15 (case hardened)
lubricant	none

2.2 Analysis

The surface measurements of the tool and the wire are done by an optical and contact-free 3D laser scanning microscope (Keyence VK-9710). Depending on the number of strokes surface measurements were done on both parts.

3 Results

Fig. 2 shows a comparison of the tool and wire surface. The punch penetrated the wire with 300000 strokes, a penetration depth of 150 µm and a forming velocity of 18 mm/s.

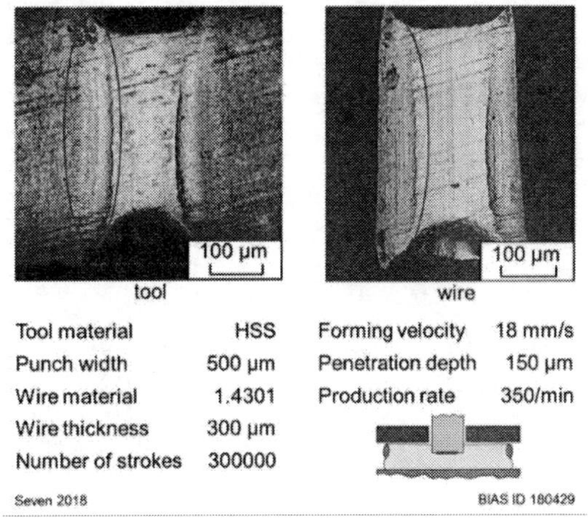

Fig. 2. Comparison of surfaces between tool (left) and wire (right) after 300000 strokes

On the punch surface (Fig.2, left) there are two parallel traces visible (red circles). The two traces of the punch are impressed in the wire which can be seen on the microscopic picture (Fig. 2, right). Furthermore on the dashed red lines in the pictures tool wear can be seen on the punch edges

To measure the tool wear the wire geometry is optically analysed and compared to the tool geometry. For that on both microscopic pictures (tool and wire) similar profile lines were chosen which were orientated through the sanding marks of the surface. Along these profile lines height profiles were created for both components. Both

 https://doi.org/10.1051/matecconf/201819015009

height profiles were fitted and compared related to height differences (see in Fig. 3). The pictures were created with a 20x magnification, a z-pitch of 0.1 µm and a xy-resolution of 346 nm.

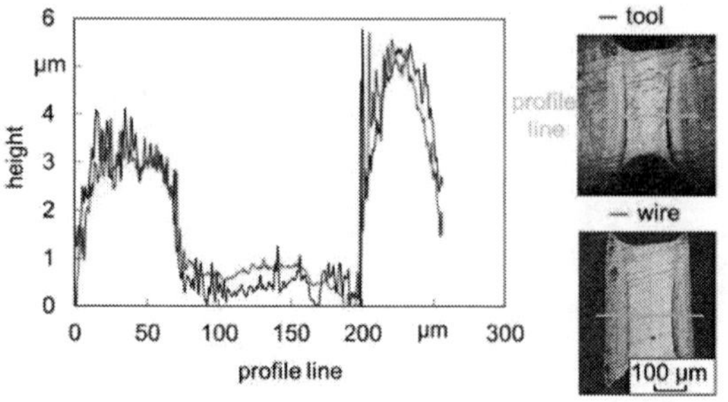

Tool material	HSS (S 6-5-2)	Forming velocity	18 mm/s
Punch width	500 µm	Penetration depth	150 µm
Wire material	1.4301	Production rate	350/min
Wire thickness	300 µm	Number of strokes	300000

Seven 2018 BIAS ID 180430

Fig. 3. Fit of height profiles of tool and wire along similar profile lines

The difference in the height profiles of the tool and wire is determined and averaged over the profile line. In Fig. 4 height difference for an exemplary profile line is shown. With the exception of three peaks around 200 µm the height difference is below 1.5 µm. The average over the profile line is 0.45 µm with a standard deviation of 0.43 µm.

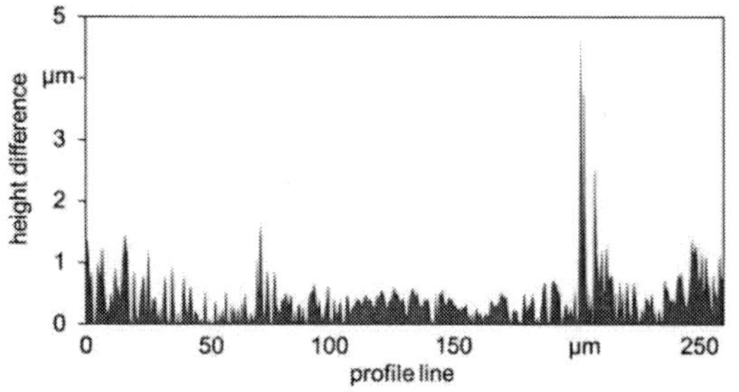

Tool material	HSS (S 6-5-2)	Forming velocity	18 mm/s
Punch width	500 µm	Penetration depth	150 µm
Wire material	1.4301	Production rate	350/min
Wire thickness	300 µm	Number of strokes	300000

Seven 2018 BIAS ID 180432

Fig. 4. Difference between the height profiles of tool and wire

The determination of the averaged height difference is done for three different profile lines. That three averaged height differences are also arithmetically averaged. The result is defined as the forming accuracy. Fig. 5 shows the forming accuracies of two equal samples for several numbers of strokes. In both cases a punch manufactured with HSS is used on a wire of 1.4301.

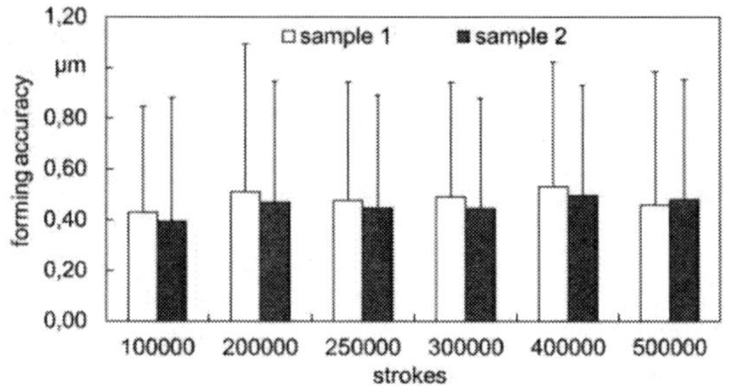

Tool material	HSS (S 6-5-2)	Forming velocity	18 mm/s
Punch width	500 µm	Penetration depth	150 µm
Wire material	1.4301	Production rate	350/min
Wire thickness	300 µm	Number of strokes	500000

Seven 2018 BIAS ID 180431

Fig. 5. Forming accuracy of the geometry of two equal punches HSS on the wire until 500000 strokes

The forming accuracy of each sample is similar for the number of strokes. The averages of the forming accuracies over the strokes are 0.48 µm (sample 1) and 0.45 µm (sample 2). The standard deviations are located between 0.42 µm and 0.58 µm. The forming accuracy of both samples is for each number of strokes below 1.5 µm by addition of average and standard deviation.

4 Discussion

The measurement surfaces show two parallel traces on the punch and wire. By penetration of the punch the wire material flows to the sides where both traces occur through abrasion. The punch edges show a stronger wear. This may be affected by a higher normal force on the punch edges than on the front surface.

The forming accuracies depending on the number of strokes are similar. That means that different high tool wear can be measured. Moreover the results between sample 1 and sample 2 are similar. These similar results are indications for a reproducible measurement system.

The standard deviations of the forming accuracies range between 0.42 µm and 0.58 µm at mean forming accuracies around 0.5 µm. This may be influenced by the inaccuracies during the fitting of the height profiles (see in Fig. 6). Both height profiles are shown with the corresponding height difference. Especially around 200 µm there are three peaks (height difference). Both tool and wire show high height values in that area. One curve is displaced so that tool and wire are not exactly fitted which leads to extremely high standard deviations.

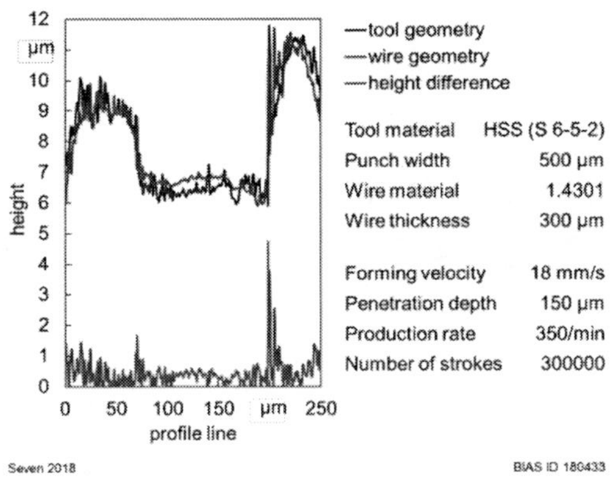

Fig. 6. Forming accuracy of the geometry of two equal punches HSS on the wire until 500000 strokes

The reason could be the elastic rear deformation of the wire. By removing the punch from the wire the elastic deformation of the wire reverts. Through the elastic rear deformation the wire geometry could be contracted and thus the peaks of tool and wire displaced.

In this example (Fig. 6) the averaged difference is $0.45\ \mu m \pm 0.43\ \mu m$. Without the three peaks around $200\ \mu m$ the value is changed to $0.41 \mu m \pm 0.29\ \mu m$. Furthermore artefacts, which are contained in the surface measurements, lead to peaks in the height profiles and disturb the result.

Finally the tool wear can be analysed on the wire indentation with an accuracy of less than 1.5 µm. The advantage is no dismounting and mounting of the tool compared to other measurement methods like the negative indentation. The negative indentation in silicon can also be used for tool wear measurement with accuracies below 1 µm [6], but the production has to be stopped for the silicone indentation. For the inline wear recording with the digital holography the production needs to be stopped as well [7]. However the inline tool wear measurement on the wire indentation offers an accurate measurement without stopping the production.

5 Conclusion

In lateral micro upsetting features of the punch are formed onto the forming product with an accuracy below 1.5 µm. Hence, assessment of tool wear can be done by analysing the forming product rather than the punch itself without stopping the production.

The authors gratefully acknowledge the financial support by Deutsche Forschungsgemeinschaft (DFG,German Research Foundation) for the Subprojects B3 Werkzeuglebensdauer and B7 Prozessstabilität within the SFB 747 (Collaborative Research Center) "Mikrokaltumformen - Prozesse, Charakterisierung, Optimierung".

References

1. M. Geiger, M. Kleiner, R. Eckstein, N. Tiesler and U. Engel, "Microforming," *CIRP Annals - Manufacturing Technology 50,* pp. 445-462 (2001)
2. N.N., "Gartner Says Smartphone Sales Surpassed One Billion Units in 2014," (2015) [Online]. Available: https://www.gartner.com/newsroom/id/2996817. [Accessed 18 04 2018].
3. N.N., "Gartner Says Worldwide Sales of Smartphones Recorded First Ever Decline During the Fourth Quarter of 2017," (2018) [Online]. Available: https://www.gartner.com/newsroom/id/3859963. [Accessed 18 04 2018].
4. H. Hoffmann, "Tensile Test of very thin Sheet Metal and Determination of Flow Stress Considering the Scaling Effect," *CIRP Annuals - Manufacturing Technology 55,* pp. 263-266 (2006)
5. H. Flosky and F. Vollertsen, "Wear behaviour in a combined micro blanking and deep drawing process," *CIRP Annals Manufacturing Technology 63,* pp. 281-284 (2014)
6. H. Flosky, M. Krüger and F. Vollertsen, "Verschleißmessung in der Mikroumformung mittels Silikonabformverfahren an einem kombinierten Schneid-Ziehring," in *Tagungsband 7. Kolloquium Mikroproduktion*, Aachen (2015)
7. S. Huferath-von Lüpke and S. Bergmann, "Digital holography for inline wear recording on deep drawing tools," *Proceedings of the 4th International Conference on Nanomanufacturing (nano-Man 2014)* (2014)
8. H. Schulze Niehoff, Entwicklung einer hochdynamischen, zweifachwirkenden Mikroumformpresse, Dissertation, BIAS Verlag (2008)

Chances and Limitations in the Application of Laser Chemical Machining for the Manufacture of Micro Forming Dies

Hamza Messaoudi[1,*], *Florian* Böhmermann[2], *Merlin* Mikulewitsch[3], *Axel* von Freyberg[3], *Andreas* Fischer[3], *Oltmann* Riemer[2], and *Frank* Vollertsen[1,4]

[1]BIAS – Bremer Institut für angewandte Strahltechnik, Bremen, Germany
[2]Leibniz Institute for Materials Engineering, Laboratory for Precision Machining LFM, Bremen, Germany
[3]University of Bremen, BIMAQ – Bremen Institute for Metrology, Automation and Quality Science, Bremen, Germany
[4]University of Bremen, Faculty of Production Engineering-Mechanical Engineering & Process Engineering, Bremen, Germany

Abstract. Laser chemical machining, a non-conventional processing method based on thermally activated electrochemical material dissolution, represents a promising technology for manufacturing metallic dies for micro forming applications. Prior to widespread industrial acceptance the machining quality of laser chemical machining should be characterized. For this purpose, laser chemical machining is compared with micro milling regarding both the dimensional accuracy and the surface quality. Therefore, square micro cavities exhibiting side walls between 100 µm and 400 µm in length and 60 µm in depth are machined with both manufacturing processes into the cobalt-chrome alloy Stellite 21. The geometrical features are investigated using laser-scanning confocal microscopy and scanning electron microscopy. On the one hand, laser chemical machining is more suitable for manufacturing cavities with dimensions < 200 µm due to higher shape accuracy with stable mean edge radii of (11.2 ± 1.3) µm as a result of roughing and finishing steps. On the other hand, the finish quality of micro milling with mean surface roughness *Sa* of 0.2 µm could not be achieved with laser chemical machining due to in-process induced waviness. Finally, the metallographic analysis of the surface-near layers reveals that both manufacturing processes ensure gentle machining without any noticeable micro structural impact.
Keywords: Micro machining, Machine tool, Quality

1 Introduction

Manufacturing of metallic parts by micro forming is a key approach to meet the ever-increasing demand for micro mechanical and micro electrical components, required for, e.g., consumer products or biomedical devices. Forming dies applied to the manufacture of those components usually exhibit geometrical features of several tenths of micrometers with tolerances in the sub-micrometer range [1]. Therefore, the achievable dimensional accuracy as well as the resulting surface roughness represent crucial figures of merit for micro forming processes [2].

Moreover, tool materials for forming processes require appropriate ductility combined with high hardness and wear resistance [3]. These include mono crystals (e.g. sapphire) as well as ceramic materials (e.g. alumina, zirconia etc.) and cemented carbides (e.g. tungsten or titanium carbide), which are characterized by complex, time- and cost-intensive production procedures. Besides, metallic alloys such as iron-, nickel- and cobalt-based alloys represent widely used alternative materials. However, the manufacture of micro forming tools from the mentioned alloys is still challenging due to additionally occurring size effects in the microscale [4].

Using conventional machining like milling and turning, which are based on mechanical material removal, the hardness of applied tools should significantly exceed that of the machined material. Thereby, the induced thermomechanical loads can induce additional residual stresses and change the micro structure in the surface-near layers. Thus, the chemical inertia as well the stability could be limited [5]. For forming dies exhibiting geometries or features smaller than 1 mm micro milling is widely used, especially in mold manufacturing, due to its high removal rates (some mm^3/min) and its excellent surface finish [6]. To ensure high surface quality fine grained tool steels with dispersed carbides are used. Thus, a surface roughness *Ra* of some 10 nm can be realized. Nevertheless, micro milling is limited regarding the dimensional accuracy and the surface finish; especially when tools of ≤ 0.2 mm in diameter are applied. This can be explained by the incidence of ploughing effects and feed marks [6].

* Corresponding author: messaoudi@bias.de

Alternatively, hard materials can be manufactured using non-conventional machining. Among others this includes laser machining and electrochemical machining (ECM) [7]. Laser machining, based on material ablation, is characterized by an increased machining precision and quality with shorter laser pulses. However, as the removal rate and the processing quality are reciprocal, the economic use of ultrashort pulsed lasers is still inefficient due time- and the high procurement costs and the low removal rates (typically 10^{-3} mm^3/min) [8]. Besides, ECM is characterized by its smooth and low-cost machining. Due to the purely chemical removal the micro structure properties at the workpiece surface remain unaffected. This makes ECM very suitable for the manufacture of temperature sensitive materials such as NiTi-alloys. Nevertheless, in the microscale the size of achievable structures is limited by the size of the electrode, which itself has to be manufactured in advance [9].

One process that unifies the advantages of laser and electrochemical machining is the laser chemical machining (LCM). Within this process the laser beam heats up the workpiece, which is immersed in a continuously pumped electrolyte [10]. Depending on the laser parameters such as the laser power and spot diameter, the induced thermal heating of the workpiece surface can shift the electrochemical potential to the transpassivation region, in which the anodic material dissolution is enhanced. This laser-activated chemical dissolution was observed for different self-passivating metals e.g. stainless steel and titanium alloys [11]. Within suitable process windows material removal occurs without micro structural changes, significant heat affected zones or recast formation. However, LCM is also labelled through its high sensitivity against small parameter variation due to the limited process window and by the low removal rates (up to 10^{-3} mm^3/min) [12]. Further, previous works have shown that the surface roughness depends strongly on the applied scan velocity. The areal roughness *Sa* at the ground of micro cavities machined with velocities < 20 µm/s into the chrome-cobalt alloy Stellite 21 was found to be > 2 µm [13,14], whereby at velocities > 1 mm/s a polishing effect was demonstrated in titanium resulting in *Sa*-values down to 0.1 µm [15].

Hence, the combination of low and high scan velocities could increase the surface quality of LCM-machined micro parts. Pursuing a strategy that comprises roughing and finishing steps, the quality of laser chemical machining is characterized and compared with micro milling taking the dimensional accuracy, the surface roughness and the impact on the material's micro structure into account. For this reason, square micro cavities exhibiting lengths between 100 µm and 400 µm and a depth of 60 µm, representing forming cavities of micro cold forging dies, are machined with both manufacturing technologies into Stellite 21. Subsequently, the geometrical features, the surface roughness as well as the micro structural impact on the surface-near layers are investigated. Furthermore, the machining results are discussed and the process-related limitations are identified.

2 Methodology

2.1. Micro cavity manufacturing

2.1.1. Material

As sample material the cast chrome-cobalt alloy Stellite 21 from "Deloro Wear Solutions GmbH" with the chemical composition (> 60% Co, 27% Cr, 5.5% Mo, 2.5% Ni and 1.5% Fe) was used. On the surface of eroded samples having a diameter of 33 mm and a thickness of 5 mm, square micro cavities with side lengths between 100 µm and 400 µm and depths of 60 µm were machined using micro milling and laser chemical machining. Fig. 1 shows the targeted dimensions as well the quality parameters to be investigated.

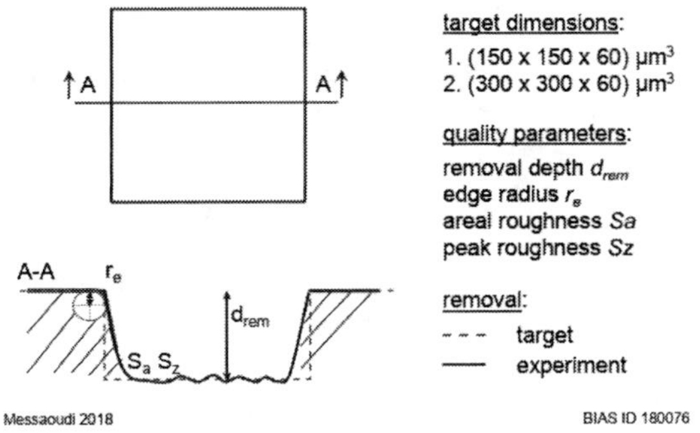

Fig. 1. Schematic illustration of the cavity geometry as well as the investigated quality parameters

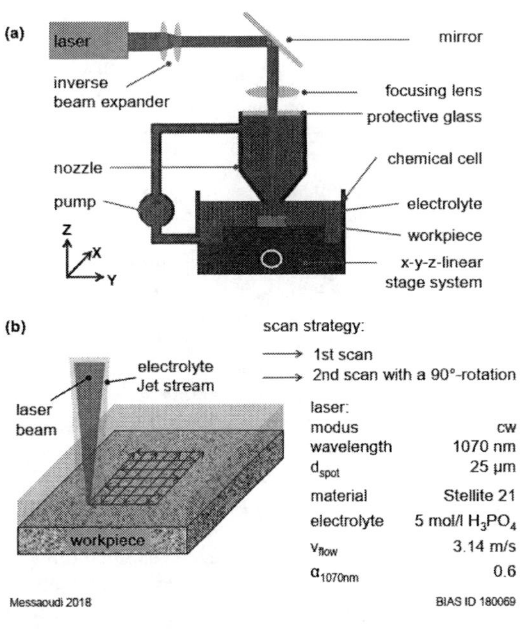

Fig. 2. Schematic illustration of the main components for the experimental set up (a) as well as the relevant process parameters for LCM (b).

2.1.2. Laser chemical machining

A cw-fiber laser TruFiber 300 (from Trumpf) was applied as laser source. Its gaussian laser radiation of 1080 nm is first collimated to a beam diameter of 8 mm, reduced in size up to a value of 2 mm using an inverse telescope and then focussed using a high aperture lens system (focal length of 50 mm). With this system the focal spot size d_{spot} could be varied between 25 µm and 75 µm, whereby a laser spot diameter of 25 µm was applied within this work. As can be seen in Fig. 2, the electrolyte, a 5 molar (28.7% vol.) phosphoric acid solution (H_3PO_4), is pumped through a 2 mm wide nozzle that is arranged coaxially to the laser beam with a velocity v_{flow} of 3.14 m/s. The propagation of the laser beam through the electrolyte affects the power density due to an identified absorption coefficient of 0.6.

A detailed investigation was performed in order to identify the influence of the relevant process parameters on the removal quality. With respect to the targeted dimensions suitable laser and scan parameters were selected (see Table 1). Thereby, the cavity manufacture consists of two steps. As first step a roughing is applied to achieve the aimed removal depth. In a following step, laser chemical finishing is applied to improve the surface quality. Using the parameters illustrated in Table 1 three cavities of dimensions (150 x 150 x 60) µm^3 and (300 x 300 x 60) µm^3 were manufactured.

Table 1. Summary of the selected parameters for the laser chemical manufacture of the micro cavities

	P_L [W]	v_{scan} [µm/s]	$d_{overlap}$ [µm]	n_{scan} [-]	$t_{process}$ [min]
LCM Roughing	0.6	10	6	2	12:30, 50:00
LCM finishing	0.3	50	6	30	37:30, 150:00

*P_L: laser power (after the propagation through the electrolyte), v_{scan}: scan velocity, $d_{overlap}$: lateral overlapping, n_{scan}: scan repetitions, $t_{process}$: processing time

2.1.3. Micro milling

Machining of the cavities by micro milling was carried out on a DMG Sauer Ultrasonic 20 linear five-axes machine tool. Equipped with a high speed spindle with spindle speeds up to 41,500 min^{-1} and direct driven feed systems this machine is well suited for micro milling operations. CAD/CAM-programming was used for tool path generation. For micro machining the procedure of tool path generation is essential to achieve desired shape accuracy and surface finish [16]. To meet the required machining quality, in particular the cavity radii, the tool path generation was carried out with the line pitch a_e arranged tangential to the targeted cavity contour.

For all machining operations hard coated tungsten carbide ball-end mills of 0.2 mm and 0.1 mm in diameter were used. The spindle speed was kept constant at 40,000 min^{-1}. The machining procedure of a single cavity comprises three roughing and three finishing steps, respectively. The roughing steps were the clearing of the cavity, roughing of the bottom area, and the roughing of the contour of walls and edge radii, compare Fig. 3. The finishing operation was comprises first: the finishing of the edge radii to a target value of 50 µm, second: the finishing of the contour of wall and finally the finishing of the bottom area, compare Fig. 3. Material allowance for the finishing cuts was 0.004 mm. Within one experiment five cavities of the same dimension were machined in a row. This allows investigating e.g. the impact of tool wear on the machining results. A detailed overview of all machining operations is given in Table 2.

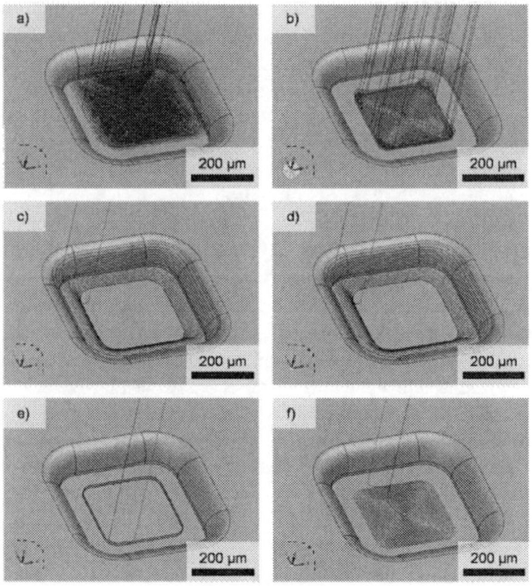

Fig. 3. Tool paths of six consecutive steps for the manufacture of a (300 x 300 x 60) µm^3 micro cavity with an edge radius of 50 µm (Screen-shots form CAD/CAD-software); a) clearing of the cavity, b) roughing of the bottom area, c) roughing of the contour of walls and edge radii, d) finishing of the contour of edge radii, e) finishing of the contour of walls, and f) finishing of the bottom area.

2.2. Determination of manufacturing quality

To examine the manufacturing quality geometrical properties such as removal depth d_{rem}, edge radius r_e and shape accuracy were recorded and characterized using a laser scanning confocal microscope (VHX970-Keyence) and a scanning electron microscope (SEM, EVO M10-Zeiss).

For the determination of the edge radius a 2D holistic approximation was used [17]. Based on an algorithm that uses continuous transitions and an automatic separation of the composite geometries measured points of a 2D profile line (cross section of the cavity) are assigned dynamically to the individual control geometries (line-circle-line). The approximation minimizes the orthogonal point distances to these standard geometries according to the L2-norm and iteratively improves the parameters of the composite profiles to fit the measured

Table 2: Machining steps and associated machining time for the generation of five micro cavities of the dimension (300 x 300 x 60) μm³and a radius edge a_e of 50 μm

machining operation	tool	tool diameter	feed velocity	depth of cut	width of cut	material allowance	machining time
		d [mm]	v_f [mm/min]	a_p [mm]	a_e [mm]	- [mm]	t [min]
clearing of the cavity	ball-end mill	0.2	150	0.005	0.005	0.006	4:30
roughing bottom area	ball-end mill	0.1	25	0.004	0.004	0.004	13:00
roughing walls / radii	ball-end mill	0.1	150	0.003	0.003	0.004	1:00
finishing radii	ball-end mill	0.1	150	0.004	0.004	-	1:00
finishing walls	ball-end mill	0.1	25	0.004	0.004	-	1:15
finishing bottom area	ball-end mill	0.1	25	0.004	0.004	-	2:45

profile [18]. In this work, the mean upper edge radius r_e of each cavity was determined by approximating 20 profile lines spread over the cavity, half horizontally and half vertically aligned. The profile lines, for which the holistic approximation did not converge, due to e.g. excessive burrs along the edges, were removed from consideration for determining the mean edge radius.

In addition, to identify the influence of the manufacturing method on the micro structure, metallographic investigations were carried out. Therefore, the SEM-records were complemented by an energy dispersive spectroscopic (EDS) analysis to show the chemical composition on selected regions within the cavities. Moreover, metallographic cross-sections of the cavities were prepared to visualize possible thermal effects as well as micro structural changes in the surface-near layers.

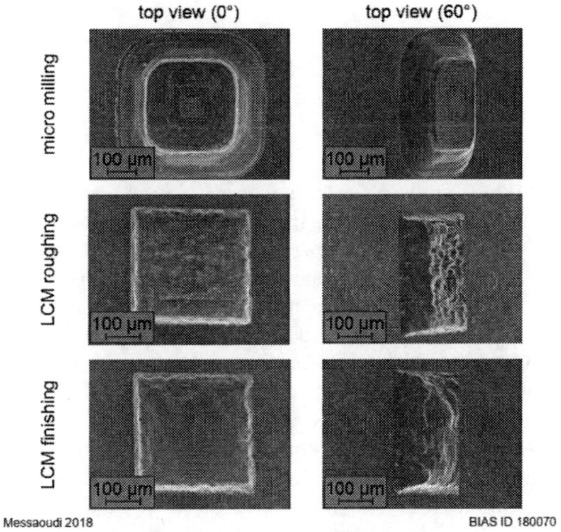

Fig. 4. SEM images of the cavities manufactured using micro milling and laser chemical machining, including the two steps roughing and finishing (top views under 0 ° and 60 °)

3 Results

3.1 Geometrical properties

Figure 4 shows exemplarily the captured SEM images of micro milled, laser chemically roughened and finished micro cavities with the targeted dimensions of (300 x 300 x 60) μm³. Optically first process-related differences can be identified. The micro milling stands out through its very high finish quality at the cavity ground. In comparison, the LCM roughing results in grossly achieving the cavity dimensions while the LCM finishing provides the final contouring and smoothing of the cavity surface. Moreover, laser chemical machining shows sharper and more accurate contours than micro milling.

The visual impression from Fig. 4 was investigated quantitatively by determining the removal depth d_{rem} as well as the edge radius r_e, as depicted in Fig. 5. Both, LCM and micro milling ensured the required depth of 60 μm (compare Fig. 5.a). It has to be noticed that the removal depth within the LCM finishing step was increased additionally by about 10 μm. Thus, the depth of the resulting cavities was ranging between 75 μm and 78 μm. Moreover, the depth standard deviation was similar for all methods applied and less than 10 μm (see Fig. 4.a).

In addition, results of the 2D holistic approximation (see Sect. 2.2) of the differently manufactured cavities are illustrated in Fig 5.b. The determined mean edge radii confirm the visual impression from Fig. 4. For both cavity dimensions, the mean edge radii of the cavities manufactured by micro milling are larger than those manufactured by laser chemical machining. This can be explained by the fact that a certain edge radius, whose minimum is determined by the diameter of the used tool, is an essential factor in micro milling. Thereby, the results show that the mean edge radius of the larger micro milled cavities (color-coded in hatched red) with r_e = 44.94 μm is close to the target value of 50 μm, which could not achieved for the smaller cavities (r_e = 23.75 μm).

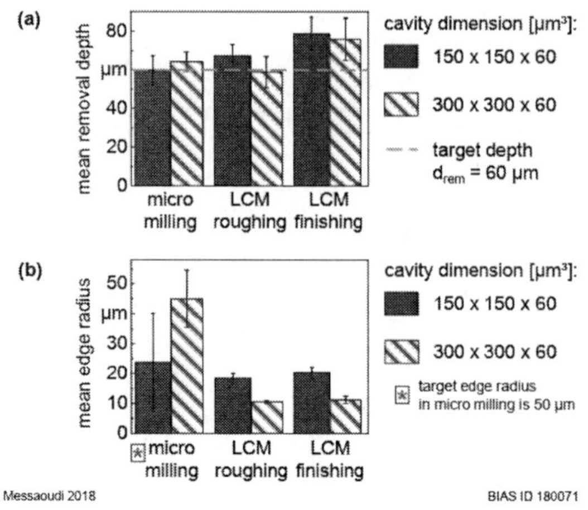

Fig. 5. Achieved (a) mean cavities depth and (b) mean edge radii of both cavity dimensions in dependence of the applied manufacture method.

In comparison, the edge radii of LCM manufactured cavities are significantly lower. They amount 10.49 µm and 11.21 µm for roughing and finishing, respectively. Note that the LCM finishing process does not increase significantly the edge radius for both cavity sizes. Comparing the uncertainties of the mean cavity edge radii manufactured by micro milling (16.4 µm) with those manufactured with LCM (1.6 µm), the former is found to be about an order of magnitude higher in general. This is caused by the wide spread of edge radii for each single manufactured cavity and can be traced back to tool wear and burr formation during the micro milling process at such small dimensions. The increased deviation from the assumed simple edge model of line-circle-line can thus result in different radii regarding the different profile lines within one cavity.

3.2 Surface quality

In order to compare the resulting surface quality, the areal surface roughness *Sa* on the cavity ground was determined following the ISO 25178. As depicted in Fig. 6.a, the micro milling shows the higher surface quality with *Sa* values ranging between 0.07 µm and 0.3 µm. In contrast LCM roughing results in *Sa* > 2 µm, which is clearly reduced during the finishing step down to 0.7 µm. Thereby, the waviness (Fig. 6.a) has not been totally removed.

To better distinguish between micro and macro roughness, surface features were segmented regarding spatial wavelengths between 0.25 µm and 100 µm. Fig. 6.b shows the characteristic roughness spectra with respect to the machining process. For micro milled cavities, the average roughness is *Sa* < 0.2 µm over the complete spatial spectrum, while laser chemical machining is labelled by micro roughness *Sa* of 0.2 µm and 0.4 µm. Further, spatial features between 10 µm and

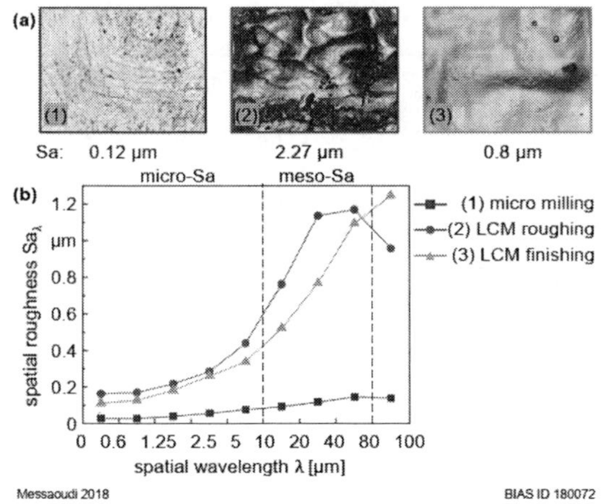

Fig. 6. Microscopic images of the ground surface of micro milled, LCM roughened as well as a finished cavity (a), and their averaged roughness spectra in dependence of the spatial wavelength (b)

100 µm determine the resulting roughness and explain the higher obtained values for laser chemical machining.

3.3 Metallographic properties

To examine possible thermal impacts of micro milling and laser chemical machining on the surface-near layers, cavities cross-sections were prepared. As can be seen in Fig. 7, neither process has a noticeable influence on the micro structure.

Fig. 7. Microscopic images of transversal cross sections of micro milled and laser chemically roughened cavities

In addition, the chemical composition at different regions from the ground, the walls and the upper borders of selected cavities were analysed by energy dispersive spectroscopy (EDS) and were compared with the base material. The mean values obtained from all measurements are illustrated in Table. 3.

Table 3. Mean values of the atomic mass % of the element distribution determined from the EDS-analysis

	base material	after micro milling	after LCM roughing	after LCM finishing
Co	61.2	52.4	48.7	59.1
Cr	28	28.7	28	28.7
Mo	3.6	4.9	10.7	3.2
Si	1.3	2.1	1.4	1
O	2.2	8.5	5	3.2
P	0	0	2.1	0.1

 https://doi.org/10.1051/matecconf/201819015010

Figure 8 shows the atomic mass of the elements Co, Mo, O and P, which were found to be dependent on the machining process. Thereby, an inverse behaviour of Co and Mo has been observed. This is clearly visible in particular during the laser chemical roughing, in which on the one hand Co is reduced from 61.2 % down to 49 %. On the other hand, the amount of Mo is tripled to 10.7 %. However, the laser chemical finishing results again in values similar to the base material. Moreover, it removes also the residues of phosphor detected after the roughing step. In contrast, micro milling is characterized by a noticeable increase in oxygen, especially in the bottom radius between ground and wall.

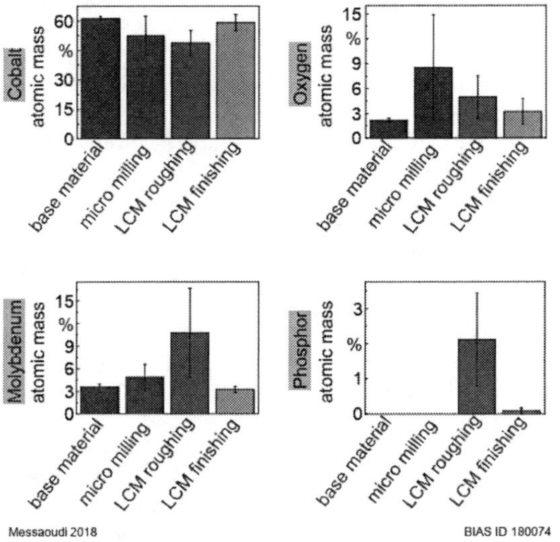

Fig. 8. Determined mean atomic mass distribution of Co, Mo, O and P in dependence of the used processing methods.

4 Discussion

The comparison of the machining quality of laser chemical machining with micro milling for the manufacture of micro cavities has shown the pros and cons of both processes.

Main advantages of micro milling are the structuring flexibility, especially with five-axes milling, and the wide range of machinable materials. Furthermore, compared to other micro machining processes, micro milling provides relatively high material removal rates, associated with short machining times and economic benefits. In addition, micro milling stands out for its excellent surface finish. Laser scanning confocal measurements applied to the micro cavities allowed to determine mean areal roughness of $Sa < 0.2$ µm throughout the complete spatial wavelength spectrum.

However, micro milling faces some major disadvantages. Other than for laser chemical machining, the manufacture of micro cavities by micro milling required for the adaption of the cavities geometries, i.e. the additional realization of corner edge radii of 50 µm as part of the cavity's geometry. This is to avoid critical engagement states of the cutting tool at the moment of machining the cavity's corner edge radii, ultimately leading to tool breakage. As displayed in Fig. 9, the required sharp contours of the square geometry could not be machined. Similar limitations are also relevant for the larger cavities (compare Fig. 4).

Generally the achievable edge radius in a cavity is directly dependent on the diameter of the applied tool. The lower limit of diameter of commercially available tools is 50 µm, what limits the achievable radius to 25 µm. However, due to decreasing tool stiffness and associated risk of tool breakage, machining with such tools enables only the realization of smallest depth of cut and width of cut. This ultimately leads to a distinct increase of machining time.

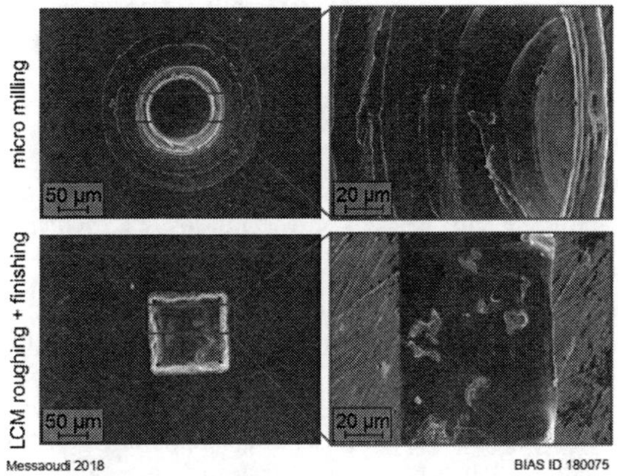

Fig. 9. Top view SEM images of a micro milled (above) and laser chemically machined (below) cavity with targeted dimensions of (150 x 150 x 60) μm^3 as well as under 60 ° magnified sections showing the cavity wall.

Furthermore, burr was found on the machined surfaces. This is expected to be a result of local undershooting of the minimum uncut chip thickness accompanied with ploughing of the cutting edge in the workpiece material inadequate material removal. The minimum uncut chip thickness is determined by the properties, i.e. the hardness, of the machined material and the cutting edge radius of the milling tool. For the machining of metallic workpieces of high hardness (> 50 HRC) the uncut chip thickness should be at least of the same dimension as the cutting edge radius to ensure adequate chip formation [19]. A new commercially available titanium aluminum nitride (TiAlN) hard coated cemented carbide cutting tool usually exhibits cutting edge radii of about 4 µm, which equals the values for the width of cut and the depth of cut applied for the finishing procedure. This leads to the conclusion that the minimum uncut chip thickness was not reached at every state of the finishing procedure resulting in the formation of burr. A further increase of width and depth of cut for the applied cutting tools for the presented machining task was not possible due to the limited stiffness of cutting tools of smallest diameter.

Surfaces of the cavities machined by micro milling were subject to an increased presence of oxygen. This indicates thermal loads applied to the workpiece during the machining as a result to friction between the flank faces of the cutting edges and the workpiece material. These effects possibly intensify with ongoing wear of the flank faces [20]. However, a negative impact on the surface near work piece material layer, such as crack induction, was not determined after the machining experiments.

In comparison, laser chemical machining has proven its potential and flexibility towards the manufacture of micro metallic parts. It offers the opportunity to use one single tool, the laser beam, that can be varied in its properties (spot size, intensity profile and cw/pulsed mode) much easier compared to micro milling, which necessitates a tool change within the process. This represents an advantage especially in manufacturing micro parts with dimensions < 100 µm. This can be clearly seen in Fig. 8. The square geometry of the cavity is much sharper than compared with the micro milled one.

Depending on the machining task an enhanced material removal, which is characterized by a poor surface quality, can be assured with slow scan velocities. On this basis the roughing step was realized. Besides, controllable and low removal rates can be realized with increased scan velocities and reduced laser powers. This combination was applied during the ensuing finishing step and led to an improved surface quality. The mean roughness *Sa* was reduced from > 2 µm down to 0.7 µm, while the peak roughness *Sz* from 45 µm down to 8 µm. Further, it has to be noted that surface features with spatial wavelengths > 10 µm could not be efficiently removed (compare Fig. 6). Considering the micro roughness (spatial wavelengths < 5 µm), the results show that laser chemical machining is able to achieve values down to 0.2 µm. The LCM-induced surface irregularities are related to two different effects. The first one is due to the used Gaussian laser beam profile, which results in an inhomogeneous removal even with closed hatching distance [21]. The second effect is related to the unequal chemical dissolution rates of the different alloy elements with respect to the used electrolyte solution [22] as well as the dependence of the achievable surface quality on the micro structure (grain size and element distribution) [15]. Nevertheless, as Fig. 4 shows, the improvement of the quality of the machined cavities as result of LCM finishing step is clearly visible. In accordance with [15], the minimum roughness level is first obtained after a removal depth exceeding the *Sz*-values. Knowing that the additional removal depth due to the finishing step was 10 µm, it is still about 30 µm to remove in order to obtain the best surface quality.

In addition, the comparison of the edge radii as well as the cavity walls with micro milling indicates the good dimensional and shape accuracy that LCM can provide. The square geometry with sharp edges and straight sides can be achieved even for dimensions < 100 µm. Thereby, the upper edge become sharper with an increased cavity dimensions (18 µm for the smaller and 10 µm for the bigger cavity size) due to the better conditions for the transport of reaction products and the provision of fresh reactants. Moreover, the cavity walls are much smoother with LCM compared to micro milling, in which the single cutting paths are visible (compare Fig. 9). Thus, a zigzag wall profile as well as a burr formation was obtained. The effect of the burr formation especially on the upper cavity edge was reflected significantly in the higher deviations of the edge radius (see Fig. 5.b).

Furthermore, in view of the machining time it can be seen that micro milling appears much more efficient than laser chemical machining, e.g., 5 min and 200 min are needed for the machining of one cavity of (300 x 300 x 60) μm^3, using micro milling and LCM, respectively. However, when considering the tool interaction area ($31.5 \cdot 10^3$ μm^2 for the 100 µm ball-end mill and $2 \cdot 10^3$ μm^2 for the laser beam), which is about 16 times larger for micro milling, the LCM removal rates are close to those of micro milling. With regard to the tool area the average removal rates within the applied machining conditions amounts to $2.7 \cdot 10^{-5}$ mm^3/min for LCM and $6.75 \cdot 10^{-5}$ mm^3/min for micro milling.

In summary, laser chemical machining offers higher accessibility and accuracy in the dimension range of < 200 µm. Nevertheless, it is still challenging to minimize the waviness effect during the roughing procedure. In contrast, micro milling is predestined for the machining of structures with dimensions > 500 µm. The main quality characteristics of LCM in comparison with micro milling are summarized in Table. 4. Further, it is still providing the better surface finish as well as the more homogeneous material removal. Therefore, a combination of both methods to avoid burr formation and to realize better wall curvature is an approach to follow towards the manufacturing of high-accurate micro metallic parts that can be applied during micro forming applications.

Table 4. Summary of the quality characteristics of laser chemical machining in comparison with micro milling

	Laser chemical machining	**micro milling**
tool diameter [µm]	25	100, 200
cavity geometry [-]	 	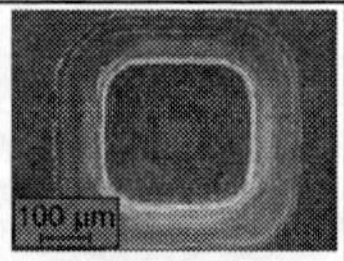 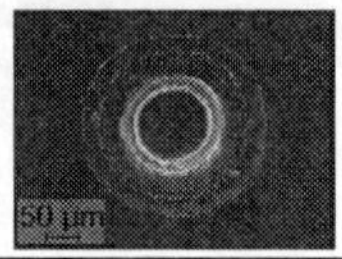
mean edge radius [µm]	25 … 45	11 … 18
mean roughness [µm]	> 0.12	> 0.7
machining time* [min]	200	5
mean removal rate [mm^3/min]	$1.1 \cdot 10^{-3}$ $2.7 \cdot 10^{-5}$**	$6.75 \cdot 10^{-5}$
microstructural impact [-]	negligible	negligible

*for cavity dimensions of (300 x 300 x 60) μm^3
**considering equal interaction areas (similar to those of LCM)

5 Conclusions

Laser chemical machining (LCM) consisting of roughing and finishing steps shows higher shape and dimensional accuracy compared with micro milling due to sharp edge contours with mean edge radii of (11.2 ± 1.3) µm. Based on this, LCM is particularly suited for the manufacture of micro cavities with dimensions < 200 µm. However, the induced surface waviness features (at spatial wavelengths between 20 µm and 100 µm) affects the surface quality and limits the process competitiveness with micro milling although small micro roughness of 0.2 µm.

The authors would like to thank the German Research Foundation (Deutsche Forschungsgemeinschaft "DFG") for funding this work within the sub-projects A5 "Laser Contour", B5 "Reliable Processes" and C2 "Surface Optimization" of the Collaborative Research Centre 747 "Micro Cold Forming – Processes, Characterisation, Optimisation" at the University of Bremen. Furthermore, the authors would like to thank Mr. Daniel Brand and Mr. Christian Rösner for their support during the experimental work. Last but not least greatest thanks, too, to Mrs. Denise Loske for her support during the metallographic investigations.

References

1. M. Geiger, F. Vollertsen, R. Kals, CIRP Annals **45**(1), 277-282 (1996)
2. M. Geiger, M. Kleiner, R. Eckstein, N. Tiesler, U. Engel, CIRP Annals **50**(2), 445-462 (2001)
3. E. Brinksmeier, A. Schulz, *micro metal forming* (Springer Verlag), 204-223 (2013)
4. F. Vollertsen, D. Biermann, H.N. Hansen, I.S. Jawahir, K. Kuzman, CIRP Annals **58**(2), 566-587 (2009)
5. D. Dornfeld, S. Min, Y. Takeuchi, CIRP Annals **55**(2), 745-768 (2006)
6. M.A. Câmara, J.C. Campos Rubio, A.M. Abrão, J.P. Davim, J. Mat. Sci. & Tech. **28**(8), 673-685 (2012)
7. M. Manjaiah, S. Narendranath, S. Basavarajappa, Trans. Nonferrous Met. Soc. China **24**(1), 12-21(2014)
8. M. Weikert, C. Fohl, F. Dausinger, T. Abeln, *SPIE* 5063, 208-213 (2003)
9. R.J Leese, A. Ivanov, Adv. Mech. Eng. **8**(1), 1-13 (2016)
10. A. Stephen, T. Lilienkamp, S. Metev, G. Sepold, *Proc. 1st Int. Conf. EUSPEN* (Shaker Verlag), 20-23 (1999)
11. D. Bäuerle, *Laser Processing and Chemistry*. (Springer Verlag) (2011)
12. A. Stephen, F. Vollertsen, CIRP Annals **59**(1), 251-254 (2010)
13. H. Messaoudi, O. Hauser A. Matson, S. Mehrafsun, F. Vollertsen, *Proc. 8th. Kolloquium Mikroproduktion,* 171-178 (2017)
14. P. Zhang, A. von Freyberg, A. Fischer, Int. J. Adv. Manuf. Tech. **93**, 3693-3703 (2017)
15. S. Eckert, A. Köhnsen, F. Vollertsen, *Proc. Laser in Manufacturing Conference (LIM)*, 1-8 (2017)
16. E. Brinksmeier, F. Vollertsen, O. Riemer, H. Flosky, G. Behrens, F. Böhmermann, *Proc. 6th. Kolloquium Mikroproduktion,* 1-8 (2013)
17. K. Lübke, Z. Sun, G. Goch, *XXIIII. Messtechnisches Symposium des Arbeitskreises der Hochschullehrer für Messtechnik eV (AHMT)*, 70-90 (2010).
18. K. Lübke, Z. Sun, G. Goch, CIRP Annals **61**(1), 499-502 (2012)
19. A. Aramcharoen, P.T. Mativenga, Precision Engineering **33,** 402-407(2009)
20. H.K. Tönshoff, H.-G. Wobker, D. Brandt, Härterei-Technische Mitteilungen-HTM **50**(3), 176-181 (1995)
21. O. Hauser, S. Mehrafsun, F. Vollertsen, *7th Kolloquium Mikroproduktion*, 8-14 (2015)
22. F. Klocke, S. Harst, F. Karges, M. Zeis, A. Klink, *CIRP Conference on Modelling of Machining Operations* **58**, 169-174 (2017)
23. S. Eckert, A. Köhnsen, F. Vollertsen, *Proceedings of Laser in Manufacturing Conference (LIM)*, 1-8 (2017)

Methodology for Reliable Tribological Investigations Applying a Micro Tribometer in Ball-on-Plate Configuration

Florian Böhmermann* and *Oltmann* Riemer

IWT Leibniz Institute for Materials Engineering, Laboratory for Precision Machining, Badgasteiner Straße 3, 28359 Bremen, Germany

Abstract. Aim of this work is the development of a methodology for reliable tribological investigations when applying a micro tribometer. Experiments were conducted in ball-on-plate configuration with linear reciprocating motion. Two different sphere materials were applied: Al99.9 aluminum alloy and 1.4301 austenitic stainless steel. A textured surface from 1.2379 hardened tool steel machined by micro milling exhibiting an areal arithmetic mean height of Sa = 295 nm was used as counterpart. The experiments comprised of the investigation of the coefficient of friction and the evolution of the facet area on the spheres in contact with the textured surface depending on the normal load applied and the number of reciprocating cycles. For the early stage of the experiments an ongoing increase of the area of the facet on the spheres was found; occurring friction was manly governed by abrasion of the sphere's material and three-body deformation. This was considered as unstable state of the tribological investigation process, not producing meaningful results. For the later stage equilibrium facet areas on the spheres were found and an ongoing conduction of the tribological experiments did not provoke any increase of their sizes. Here, occurring friction was mainly governed by adhesion. The final facet area on a sphere was directly dependent on the sphere's materials and the applied normal loads. For all subsequent investigations applying a micro tribometer, only spheres exhibiting an equilibrium facet area should be used to gain robust results of tribological investigations, required for the development of e.g. dry deep drawing processes.

Keyword: Surface, Texture, Friction

1 Introduction

Innovative micro replication techniques such as micro deep drawing allow for the economic mass manufacture of smallest parts fulfilling mechanical functions, and, thus, to cover new market demands. However, miniaturized processes are devoted to size effects. Hu et al. found a reduction of the achievable limit drawing ratio from β = 2.0 to β = 1.6 for the lubricated deep drawing of micro cups from aluminum Al99.5 when reducing the punch diameter from 5 mm to 1 mm. The reduction of the limit drawing ratio was traced back to changes in tribological conditions on microscopic level [1]. Engel related this tribological size effect to the invariance of surface roughness to scaling [2]. This leads to a change in lubrication conditions together with an increase of friction between the die and the work piece during forming. Recent research work focused on the development of dry micro deep drawing processes. A dry conduction of forming processes e.g. allows reducing process steps in production and is associated with major economic and ecological benefits [3]. Hu et al. successfully performed dry micro deep drawing applying diamond like carbon coated (DLC) forming dies. The work piece material was austenitic stainless steel 1.4301 (X5CrNi18-10); the punch diameter was 1 mm [4]. The friction reducing properties of DLC coatings was explained by their graphitization on nanometer scale in dependence on velocity of relative motion and applied loads provoking a self-lubricating effect [5]. Shimizu et al investigated the development of the work piece material adhesion to the die surface in successive dry deep drawing experiments against 1.4301 stainless steel foils [6]. Two different punch diameters were applied: 0.85 mm and 5.1 mm. Distinctively reduced adhesive wear was found for the experiments applying the smaller diameter punch, what was explained with the relative short sliding length in geometrically downscaled deep drawing processes. The researchers also introduced an analytical friction model to generally describe the size effect in friction for dry micro deep drawing [7]. The model was based on that of Bhushan and Nosonovski but was extended about the conditions of downscaled micro strip drawing tests against phosphor bronze foil [8]. The model showed good agreement with the results from the actual strip drawing experiments. Furthermore, Shimizu et al. in their work confirmed the scale dependence of the plowing component contributing to the overall friction force. A decrease of friction was found for geometrically downscaled strip drawing experiments. This was be explained with the lower probability of particles abraded from the work piece dwelling at the die-work piece interface. Micro textured forming die surfaces can

* Corresponding author: boehmermann@lfm.uni-bremen.de

support the forming process by controlling the frictional behavior, especially under dry conditions. Brinksmeier et al. demonstrated the tribological function of micro textured surfaces generated by raster micro milling with ball-endmills in strip drawing tests under dry conditions [9]. Compared to polished reference (Sa = 30 nm), a 20% reduced coefficient of friction was found for selected micro textured samples exhibiting an arithmetical mean height Sa of 200 nm to 400 nm. The design of such deterministic surface textures generated by micro milling is primarily determined by the properties of the machined material, the cutting strategy (up- or down-milling), and the machining parameters width of cut a_e and feed per tooth f_z [10]. The transfer of micro textured surfaces to tribologically adapted forming dies is highly desirable, especially for dry forming tasks. The major advantage when applying micro milling is the potential to manufacture the forming die's geometry and the tribological active micro textured surfaces in one process step. A first attempt towards micro textured forming dies was presented by Böhmermann and Riemer [11]. However, the identification of most suitable micro textured surfaces characterized by areal roughness parameters according to ISO 25178 standard still remains a research topic. Even though strip drawing tests are suitable for necessary tribological investigations, they suffer from time consuming preparation for each single test. The application of a micro tribometer instead can significantly reduce the testing time. However, friction experiments conducted with these delicate instruments under the use of comparably low normal forces (10 mN to several hundreds of mN) are prone to noise and also misinterpretation by the operator.

Aim of this work is the development of a methodology for reliable tribological investigation when applying a micro tribometer. Experiments are carried out in ball-on-plate configuration with reciprocating motion observing the evolution of the coefficient of friction and the facet area on the spheres due to wear in dependence on the number of cycles. Results of the investigations are one key step in design process for textured micro deep drawing dies, suitable for dry forming.

2 Experimental section

2.1 Micro tribometer

The instrument applied in the tribological investigations was a Tetra BASALT MUST micro tribometer in ball-on-plate configuration. The relative motion between the sphere and the sample surface is provided by the reciprocating movement of a motorized stage that carries the sample. The instrument allows for the simultaneous recording of the normal force F_N applied and the friction force F_R as a result of the relative motion of sphere and surface under test, at which F_N and F_R are indirectly determined by deflection measurement of a cantilever (double leaf spring design, fixed position and rotation) using fiber optical sensors. The cantilever is characterized by spring rates of 1200 N/m in normal direction (F_N) and 1200 N/m in tangential direction (F_R), respectively. The normal load F_N applied can be set in a range of 1 mN to 600 mN, the measurement range of the friction force F_R is ± 300 mN. The resolution for the measurement of the normal force and the friction force is given by the manufacturer with 0.36 μN. The maximum stroke length for tribological testing in ball-on-plate configuration is l_{max} = 10 mm; the maximum speed of the reciprocating motion is $v_{lin,max}$ = 10 mm/s. An overview of the general setup is given in Figure 1.

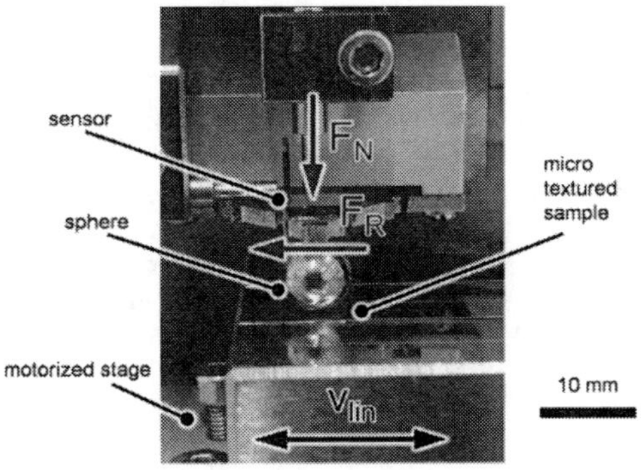

Fig. 1. Ball-on-plate setup for frictional testing on a Tetra BASALT MUST micro tribometer (in accordance with Ref. [11]).

2.2 Textured sample

The textured sample for tribological investigations was made from 1.2379 (X153CrMoV12) hardened tool steel with a hardness of 60 HRC ± 0.6 HRC. The texture was generated by micro milling using a 1.5 mm diameter ball-endmill; the cutting strategy was down milling. The feed velocity v_f, the rotational speed n, the cutting depth a_p and the width of cut a_e were set at v_f = 600 mm/min, n = 25,500 min^{-1}, a_p = 0.06 mm and a_e = 0.03 mm, respectively. Subsequently, the machined surface was measured by a Sensofar Plu 2300 optical profilometer. The raw data were processed by an image processor (Scanning Probe Image Processor, SPIP™) to derive areal roughness parameters according to ISO 25178 standard (cut-off λ_s = 0.25 μm, cut-off λ_c = 0.08 mm).

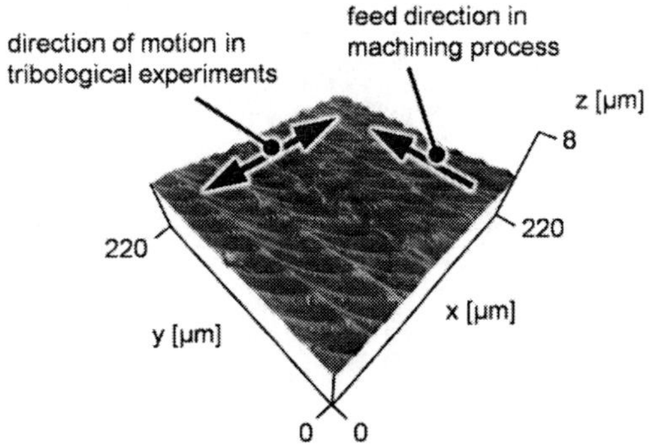

Fig. 2. Surface plot of the textured sample applied in tribological testing.

The arithmetical mean height of the surface was found to be Sa = 295 nm. This textured surface applied in the tribological investigations is representative for a class of micro structured or textured surfaces, known for their friction reducing properties in dry frictional contact under moderate loads [9]. A plot of the textured surface is given in Figure 2.

2.2 Experimental procedure

The experiments were conducted in ball-on-plate configuration with linear reciprocating motion. The selected stroke length l for all experiments was 4 mm; the speed v_{lin} of the reciprocating motion was set at 4 mm/s. All experiments were conducted under dry conditions; both, the spheres and the textured sample surface were thoroughly cleaned using isopropyl alcohol before every single experiment. The testing direction with respect to the micro textured surface was orthogonal to the feed direction of the micro milling process, compare Figure 2. The normal force F_N and the resulting friction force F_R were continuously recorded by the instrument at a sampling rate of 100 data points for each forward and backward motion. The average coefficient of friction μ as well as the maximum and minimum deviation was automatically calculated from the collected data for each cycle.

A new sphere of 5 mm in diameter was used for every single experiment. The spheres were attached to the cantilever by bonding using cyanoacrylate adhesive. Two different sphere materials were investigated: Al99.9 aluminum alloy (hereafter notated: Al) and 1.4301 austenitic stainless steel (X5CrNi18-10, hereafter notated: St). Furthermore, the applied normal force F_N was varied at 15 mN, 40 mN, 70 mN, and 100 mN, respectively.

The wear induced evolution of the facet area A_f on the spheres was determined in dependence on the number of cycles n by graphic evaluation from microscopy images. For the microscopic investigation the whole cantilever including the sphere was dismounted from the micro tribometer in order to maintain the orientation of the facet area on the sphere relatively to the sample surface. The actual facet area A_f was calculated from the mean of the smallest diameter d_1 and the diameter d_2 perpendicular to d_1 detected from the microscopy images in accordance with DIN EN 1071-13 standard, compare Figure 3.

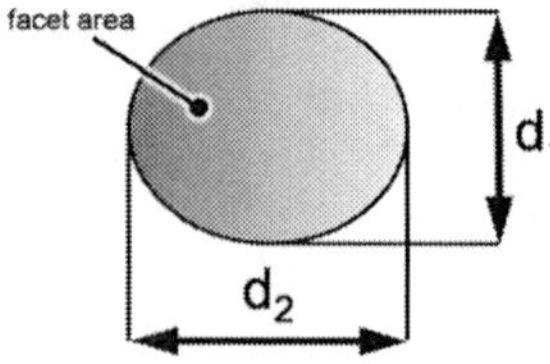

$$\bar{d} = \frac{d_1 + d_2}{2}$$

$$A_f = \frac{1}{4}\,\pi\,\bar{d}^2$$

Fig. 3. Sketch of facet area and procedure for facet area determination.

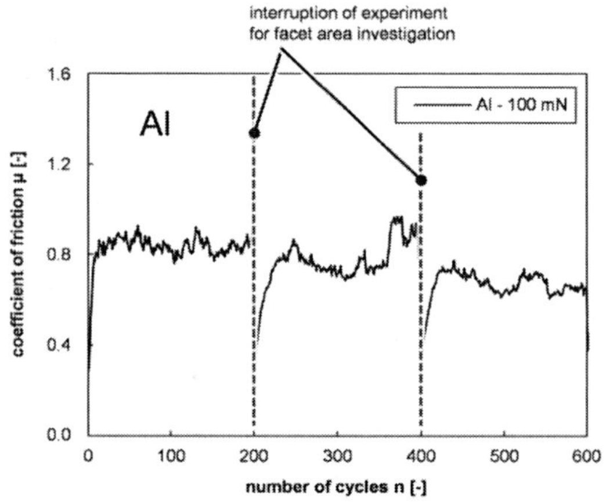

Fig. 4. Coefficient of friction in dependence on the number of cycles n (backward motion only) for the exemplary experiment applying an aluminum Al99.9 ball and a normal force of F_N = 100 mN.

The actual pressure per unit area p was calculated from the facet area A_f and the normal load F_N applied in the experiments. Usually, this procedure was carried out in steps of n = 200 cycles, compare Figure 4. The step width was broadened for experiments at advanced stages. Furthermore, the step width was increased to n = 600 cycles for the testing of the steel spheres, based on the experience gained from the testing of aluminum spheres. For the continuation of an experiment the cantilever was mounted again to the micro tribometer. Once a steady state of the facet area was detected, the experiment was terminated.

3 Results

3.1 Aluminum spheres

Figure 5 shows the development of the coefficient of friction μ for the experiment applying an Al99.9 aluminum alloy sphere and a normal force F_N of 100 mN, starting from cycle 400 after the interruption for facet area investigation. For advancing numbers of cycles, the coefficient of friction increased until reaching equilibrium at about μ = 0.8, usually after ten to 20 cycles. All experiments using the aluminum spheres were subjected to stick slip, what is pictured by the distinct error bars indicating the minimum and maximum values of the coefficient of friction μ for each cycle. Figure 6 shows exemplary the evolution of the facet area A_f on an Al99.9 sphere applied in the tribological experiment with a normal load of 100 mN after n = 600, 1200, and 1500 cycles. The investigation of the facet areas on the spheres throughout the experiments generally revealed an increase of the areas towards advancing numbers of cycles. Equilibrium of the facet area A_f was reached after about 1000 to 1500 cycles for normal forces F_N of 15 mN, 40 mN, and 100 mN, compare Figure 7 a).

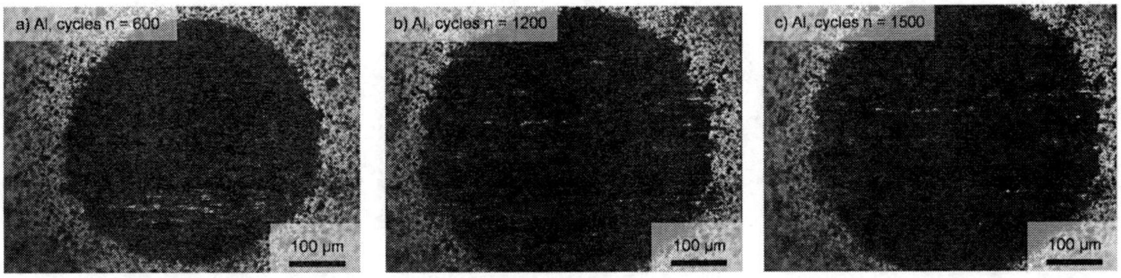

Fig. 6. Microscopic images of the facet areas on an Al99.9 aluminum alloy sphere for the experiment with an applied normal force F_N of 100 mN; a) after 600 cycles, b) after 1200 cycles, and c) after 1500 cycles.

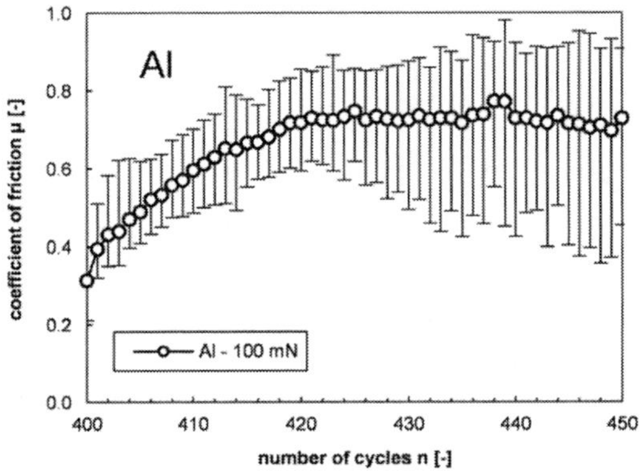

Fig. 5. Coefficient of friction µ and maximum and minimum deviation recorded in the experiment applying an Al99.9 aluminum alloy sphere and a normal force of $F_N = 100$ mN in dependence on the number of cycles n (backward motion only).

Ongoing conduction of the tribological experiments did not provoke any further increase of the size of the facet area. The equilibrium state for the experiment with a normal load of 70 mN was reached after about 2000 cycles. The results of the facet area investigation in dependence on the number of cycles for all experiments applying Al99.9 aluminum alloy spheres are summarized in Figure 7 a). After the determination of the facet areas the actual pressures per unit area were calculated taking the normal force F_N into account. With exception of the experiment conducted with a normal load of $F_N = 40$ mN, the equilibrium pressure for all experiments was found to be about 0.0005 N/mm^2 to 0.0007 N/mm^2. The results of the calculation are summed up in Figure 7 b).

3.2 Austenitic stainless steel spheres

The exemplary development of the coefficient of friction µ for the experiment applying a 1.4301 austenitic stainless steel sphere and a normal force F_N of 100 mN is shown in Figure 8, beginning from cycle 1200 after the interruption for facet area investigation. Similar to those experiments conducted with aluminum spheres, the coefficient of friction increases within the first couple of cycles and reaches equilibrium after five to ten cycles. The experiments were also subjected to stick slip, but less distinct. Those characteristics were found for all

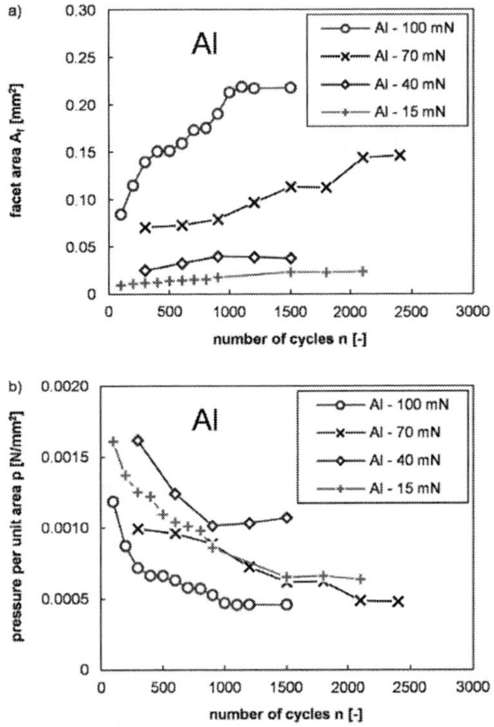

Fig. 7. a) Evolution of the facet area A_f, and b) the pressure per unit area p in dependence on the number of cycles n for the experiments applying Al99.9 spheres.

experiments, regardless of the predetermined normal force F_N. Furthermore, it was found that the absolute values of the coefficient of friction recorded during the experiments applying steel spheres – usually not exceeding $\mu = 0.6$ – are generally lower than those recorded for the experiments with the Al99.9 spheres. The exemplary evolution of the facet area on a steel sphere (normal force $F_N = 100$ mN) in dependence on the number of cycles is given in Figure 9. Again, the establishment of equilibrium in facet area was found after an increase of the facet area with increasing number of cycles. Other than for the aluminum spheres, this state was reached after about n = 1800 cycles. Furthermore, the absolute sizes of the facets' areas on the steel spheres are generally smaller than those on the aluminum spheres. The results of the facet area investigation in

* Corresponding author: boehmermann@lfm.uni-bremen.de

https://doi.org/10.1051/matecconf/201819015011

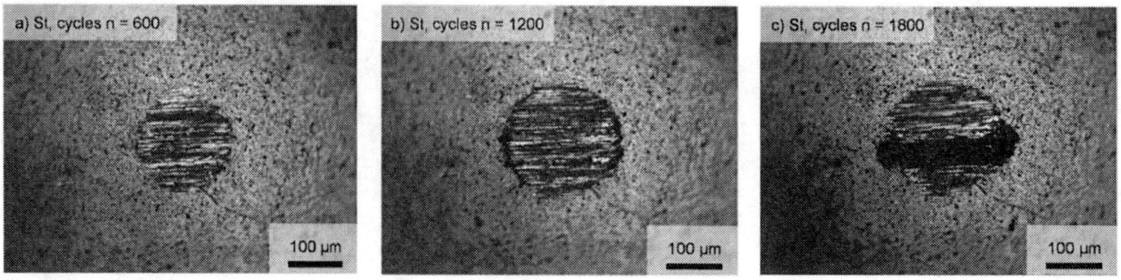

Fig. 9. Microscopic images of the facet areas on an 1.4301 austenitic stainless steel sphere for the experiment with an applied normal force F_N of 100 mN; a) after 600 cycles, b) after 1200 cycles, and c) after 1800 cycles.

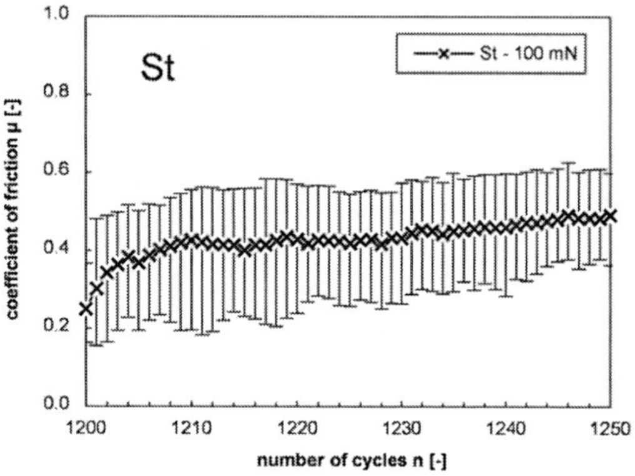

Fig. 8. Coefficient of friction µ and maximum and minimum deviation recorded in the experiment applying a 1.4301 stainless steel sphere and a normal force of F_N = 100 mN in dependence on the number of cycles n (backward motion only).

dependence on the number of cycles for all experiments applying 1.4301 stainless steel spheres are summarized in Figure 10 a). The actual pressures per unit area were calculated from the determined facet area A_f and the normal force F_N. The results of the calculation in dependence on the number of cycles are displayed in Figure 10 b). Here, similar characteristics as in the experiments with the Al99.9 aluminum alloy spheres were found, but generally showing greater numbers of pressure per unit area, coinciding with the smaller facet area.

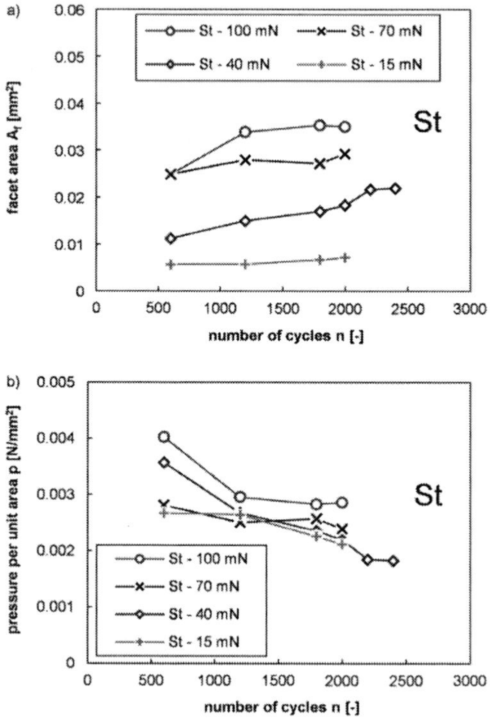

Fig. 10. a) Evolution of the facet area A_f, and b) the pressure per unit area p in dependence on the number of cycles n for the experiments applying 1.4301 stainless steel spheres.

4 Discussion

Tribological experiments using a micro tribometer in ball-on-plate configuration were conducted and the coefficients of friction as well as the evolution of the facet areas on the spheres were investigated throughout the experiments. For both types of spheres – Al99.9 aluminum alloy and 1.4301 austenitic stainless steel – similar characteristics were found for both, the development of coefficient of friction and the evolution of the facet area on the spheres. It is obvious that at the start of every new experiment there was only point contact between the sphere and the textured surface. This is associated with distinct pressures per unit area enforcing the immediate abrasion of the spheres material at the onset of relative motion and leading to the formation of a facet area. Subsequently, the contact conditions between the sphere and the textured surface changed from point to areal contact (formation of an interface). For later stages of the experiments, respectively after every investigation of the facet area, the pressure per unit area was still adequate to provoke sufficient elastic and plastic penetration of the textured surfaces' asperities into the softer sphere material. Relative motion between both the tribological partners in contact lead to plowing and cutting type wear on the spheres, compare work of Kayaba et al. [12]. As a result, wear debris was generated, trapped at the interface between sphere and textured surface. Wear particles are generally subjected to strain hardening thus contributing

* Corresponding author: boehmermann@lfm.uni-bremen.de

to plowing and cutting into the soft sphere material during the experiment. This process is referred as three-body plowing, compare work of Bhushan and Nosonovski [8]. The occurrence of three-body deformation is reflected in the distinct increase of the coefficient of friction for the first couple of cycles after continuation of the conducted experiments. At a stage, where friction is predominantly governed by abrasion of the sphere's material and three-body deformation, the tribological investigation process is considered unstable. No meaningful results can be obtained when tribological testing is conducted under such conditions. For later stages of the experiments – after about 1200 cycles for aluminum spheres and 1800 cycles for steel spheres – equilibrium facet areas were found. A further conduction of the experiments did not provoke any further increase of the facet area's sizes. It is assumed that friction now is predominantly governed by adhesion; a stable state of the tribological investigation process was reached. This allows for reliable tribological testing when applying a micro tribometer, i.e. the determination of the impact of adhesion to friction, which is of major concern for the development of dry micro deep drawing processes.

5 Conclusion

Fresh spheres applied in tribological investigations using a micro tribometer in ball-on-plate configuration are generally subject to wear, forming a facet area. At this initial stage the setup is not suitable for reliable tribological investigation, due to the strong dependence of the coefficient of friction on changing wear and friction mechanisms, e.g. three-body plowing. For later stages, after the execution of greater numbers of cycles, equilibrium of facet areas establishes at which the final size of the facet area is directly dependent on the sphere's material and the applied normal force. It is assumed that friction now is predominantly governed by adhesion. Reliable tribological investigations can be carried out under these preconditions as no further changes in the contact conditions between sphere and the sample surface will occur. It can be concluded, that reliable tribological investigations applying a micro tribometer require for the preparation of the spheres in order to provide equilibrium facet areas. This should be achieved by a run-in procedure. The actual pressures per unit area for equilibrium facet areas derived from the experiments shown in this work can be applied as control values. These were $p = 0.0005\ N/mm^2$ for Al99.9 aluminum alloy spheres and $p = 0.002\ N/mm^2$ for 1.4301 austenitic stainless steel spheres. The presented methodology can now be applied to the comprehensive tribological investigation of micro textured surfaces by using a micro tribometer in ball-on-plate configuration.

Acknowledgements: The authors would like to thank the German Research Foundation (Deutsche Forschungsgemeinschaft "DFG") for funding this work within sub-project C2 "Surface Optimization" of the Collaborative Research Centre 747 "Micro Cold Forming - Processes, Characterisation, Optimisation" at the University of Bremen. Furthermore, the authors would like to express their greatest thanks to Ms. Juelie Fritsch for the support in conducting the tribological experiments.

References

1. Z. Hu, R. Walther, F. Vollertsen, Influence of Size Effects on the Process Window for Deep Drawing, Proceedings of the International Deep Drawing Research Group IDDRG International Conference **28** (2009) 785-796
2. U. Engel, Tribology in microforming, Wear **260** 3 (2006) 265-273
3. F. Vollertsen, F. Schmidt, Dry Metal Forming: Definition, Chances and Challenges, International Journal of Precision Engineering and Manufacturing-Green Technology **1** 1 (2014) 59-62
4. Z. Hu, A. Schubnov, F. Vollertsen, Tribological behaviour of DLC-films and their application in micro deep drawing, Journal of Materials Processing Technology **212** (2012) 647-652
5. Y. Liu, A. Erdemir, E. Meletis, An Investigation of the Relationship between Graphitization and Frictional Behavior of DLC Coatings, Surface and Coatings Technology **86-87** (1996) 564-568
6. T. Shimizu, Y. Murashige, S. Iwaoka, M. Yang, K.I. Manabe, Scale Dependence of Adhesion Behavior under Dry Friction in Progressive Micro-Deep Drawing, Journal of Solid Mechanics and Materials Engineering **7** 2 (2013) 251-263
7. T. Shimizu, M.Yang, K.I. Manabe, Classification of mesoscopic tribological properties under dry sliding friction for microforming operation, Wear **330-331** (2015) 49-58
8. B. Bhushan, M. Nosonovski, Comprehensive model for scale effects in friction due to adhesion and two- and three-body deformation (plowing), Acta Materialia **52** (2004) 2461-2474
9. E. Brinksmeier, O. Riemer, S. Twardy, Tribological behavior of micro structured surfaces for micro forming tools, International Journal of Machine Tools and Manufacture **50** 4 (2010) 425-430
10. E. Brinksmeier, O. Riemer, S. Twardy, Surface Analysis of Micro Sphere End Milled Cold Work Tool Steels, Proceedings of the International Conference on Nanomanufacturing **2** (2010) 173-177
11. F. Böhmermann, O. Riemer, Tribological Performance of Textured Micro Forming Dies, Dry Metal Forming OAJ FMT **2** (2016) 067-071
12. T. Kayaba, K. Hokkirigawa, K. Kato, Analysis of the abrasive wear mechanism by successive observations of wear processes in a scanning electron microscope, Wear **110** (1986) 419-430

Development and application of methods to characterize micro semi-finished products and micro components

Bernd Köhler*, *Brigitte* Clausen, and *Hans-Werner* Zoch

Leibniz Institute for Materials Engineering IWT, 28359 Bremen, Germany

Abstract. In the production of semi-finished products for the production of microcomponents and the components themselves, the characterization of their physical properties is of particular importance. Due to the often oligocrystalline character of these semi-finished products and components, it is necessary to use a suitable testing technique for static and dynamic investigations, as the mechanical properties are not transferable from the macroscopic point of view. In addition, the micro semi-finished products and components often show inhomogeneities induced by the manufacturing process. On the one hand, these are directly reflected in the microstructure and on the other hand they have an effect on quantities such as hardness or residual stresses, which play a decisive role in the application. Mechanical testing, conventional metallography, scanning electron microscopy (SEM), electron backscatter diffraction (EBSD), ultra-microhardness testing and X-ray residual stress analysis were used as measuring and analysis techniques suitable for the sub-millimeter range. In the following, the possibilities and limitations of two these methods are illustrated using the example of mechanical testing and EBSD. In this paper several examples for possible characterization techniques are given.
Keywords: cold forming, fatigue, micro structure

1 Mechanical testing

1.1 Testing system

The mechanical properties of the manufactured semi-finished products and their post-forming behaviour in the finished product have to be analysed thoroughly. Such analysis necessitate testing systems capable of meeting specific requirements for static and cyclic testing of micro specimens. The system has to allow low forces and displacements to be set and controlled with sufficient accuracy, and provide a method of strain measurement, which takes into account the mechanical sensitivity of the specimens. Specifically with a view to dynamic testing, the machine must allow for precise control of low forces and provide an adequate displacement amplitude at maximum frequency of load cycles, whilst maintaining the preset waveform (e. g. sine wave).

1.1.1 Testing machine

Considering all the above mentioned requirements for static and cyclic testing of micro specimens, an electro-dynamic testing machine type Instron E1000 (Fig. 1) was used to characterize the mechanical properties of thin metal sheets and rotary swagged wires. The test machine is driven by a brushless linear motor and provides a maximum load capacity of ± 710 N for static tests and ± 1000 N for cyclic tests. On the one hand, this load range proved to be sufficient for micro-samples that were strongly work-hardened by forming processes. On the other hand, the load control also allows for loads of just a few Newtons, such as those occurring with sheet thicknesses in the range of 10 µm.
The load was measured with load cells type Instron Dynacell calibrated to ISO 7500-1 with measuring ranges of ± 2 kN and ± 250 N, respectively and automatic inertia compensation.

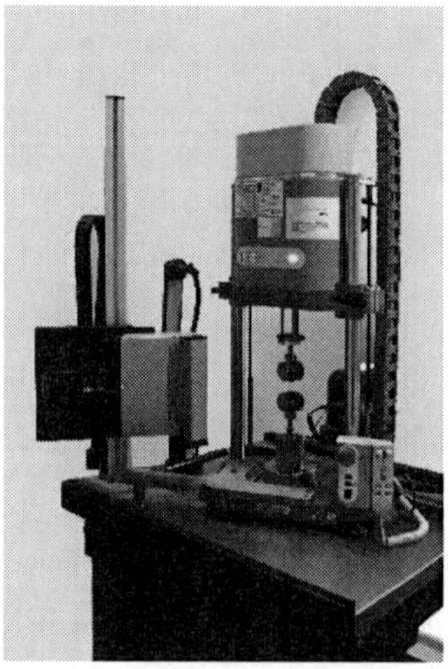

* Corresponding author: koehler@iwt-bremen.de

Fig. 1. Testing machine ElectroPuls E1000.

Special attention was given to the accuracy of load control for small loads and in cyclic operation. Fig. 2 shows the measured variation of the load amplitude for a cyclic tensile test under sinusoidal load at a stress ratio $R = 0.1$ and a frequency $f = 20$ Hz. The specimen was a rotary swaged wire made from steel X5CrNi18-10 (AISI 304) with a diameter of 0.5 mm. The graph shows the feedback command value of $F_a = 40.5$ N. The average load feedback amplitude determined over 2000 load cycles is $F_a = 40.487$ N with a standard deviation of $s = 0.055$ N. The deviation in the applied load is less than 1 %, still the first 1000 cycles were not taken into account in the calculation of the mean value, to ensure that the process of stabilization does not distort the result. Standardization of the standard deviation with respect to the measuring range of the load cell used provides:

$$s_n = s/250 = 0.00022 = 0.022\ \% \qquad (1)$$

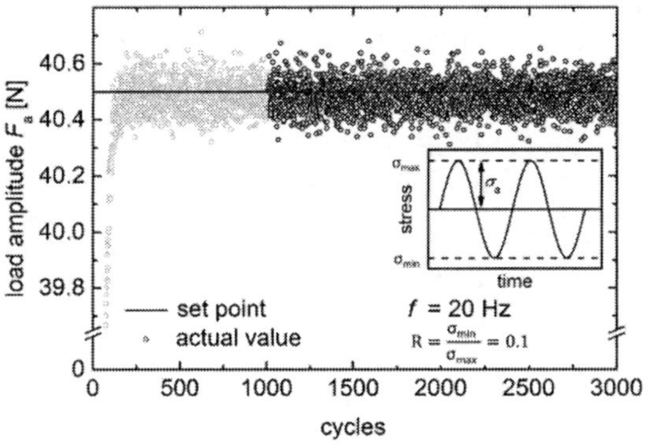

Fig. 2. Fluctuation of load amplitude during cyclic test.

The dynamic performance of the testing system is shown by way of example in Fig. 3 that illustrates the relationship between the testing frequency and the achievable displacement amplitude of the actuator for three different loading conditions. The curve plotted without a specimen installed represents a limiting curve resulting from the maximum achievable acceleration of the moving masses. Looking at the plot with installed specimen at a static mean load of 200 N and a load amplitude of 100 N, it was found that a significant deviation from the limiting curve does not occur until a frequency of 100 Hz has been reached. When the mean load is increased to 500 N, the curve shifts to lower frequencies, i.e. a given actuator rod amplitude will not be achievable under these conditions unless the test frequency is reduced. Basically, the performance diagram shows that testing frequencies of 100 Hz can be achieved with the testing system in cyclic tests, provided that the stiffness of the specimen is adequate. For the samples tested in this project, a sinusoidal loading of the samples could typically be ensured up to about 60 Hz during cyclic testing. At this frequency, fatigue strength tests can be carried out in a reasonable time frame. It takes about 46 hours for a run out to experience 10^7 load cycles.

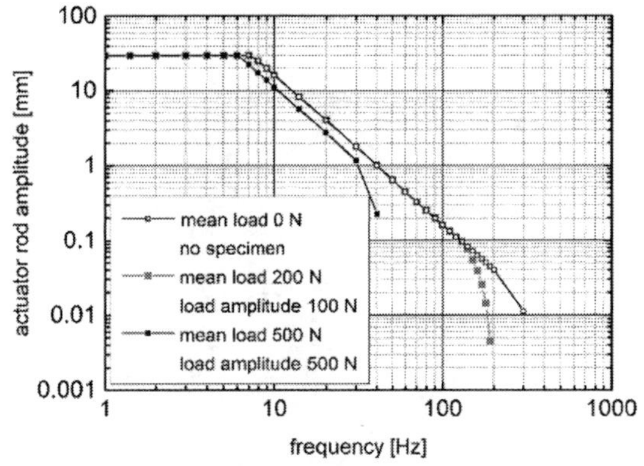

Fig. 3. Dynamic performance plot of the testing machine ElectroPuls E1000

As an example Fig. 4 shows two Woehler diagrams of wires made of steel X5CrNi18-10 (AISI 304) in the initial state as well as after cold forming by rotary swaging to a diameter of 0.5 mm with a feed rate of 1 mm/s. Further results are published by Köhler [8].

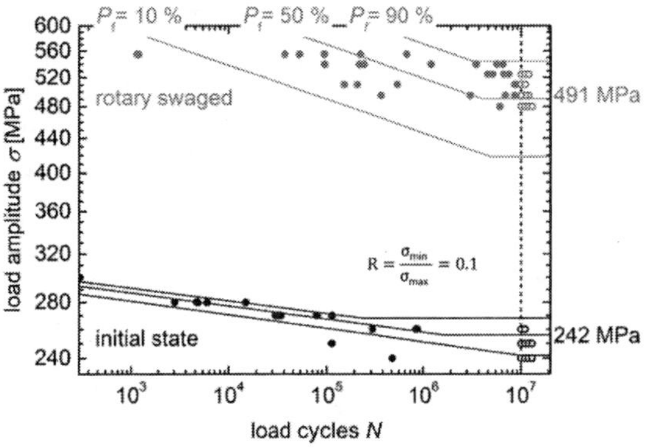

Fig. 4. Woehler diagram of X5CrNi18-10 steel wires before and after cold forming by rotary swaging (● = failure, O = run out)

The lines of the 10, 50 and 90 % failure probability P_f were calculated on the basis of a Weibull distribution [1, 2]. After rotary swaging, the fatigue strength increases from 242 to 491 MPa and the scatter increases. A detailed discussion of the investigation can be found in Köhler and Clausen [3].

1.1.2 Non-contact strain measurement

In view of the micro dimensions of the test specimens (typical sheet thickness ranges between 10 µm and 100 µm), strain measurement using specimen-contacting methods such as strain gauges or clip-on extensometers is not feasible. On the one hand, such methods involve the danger of damaging the specimens during the attachment of the respective strain measurement device, on the other hand, the impact of these measuring methods on the result can no longer be neglected, as it can be with larger specimens.

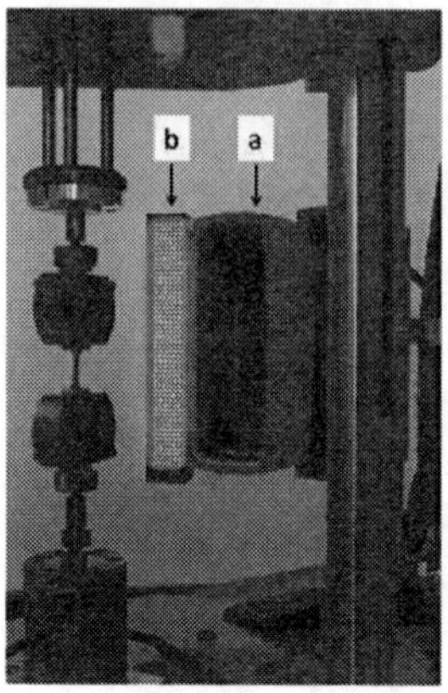

Fig. 5. Video extensometer (a) with integral illumination (b).

For this reason, a non-contacting optical strain measurement in the form of a video extensometer was chosen. The extensometer consists of a high-resolution digital video camera and a LED light source, which illuminates the specimen with pulsed, monochromatic, red polarized light (Fig. 5). In this case, strain measurement is achieved by tracing the axial movement of markings applied on the specimen with the video camera, and calculating strain by means of a real-time image processing system.

Measurement of the original gauge length, which is defined by the markings on the specimen and which is essential for strain measurement, is achieved prior to the test by the calibrated video extensometer with an absolute accuracy of ± 2.5 µm. The markings, which must ensure that they have sufficient contrast to the sample surface, take the form of two spots with a diameter between 0.5 and 3 mm (Fig. 6).

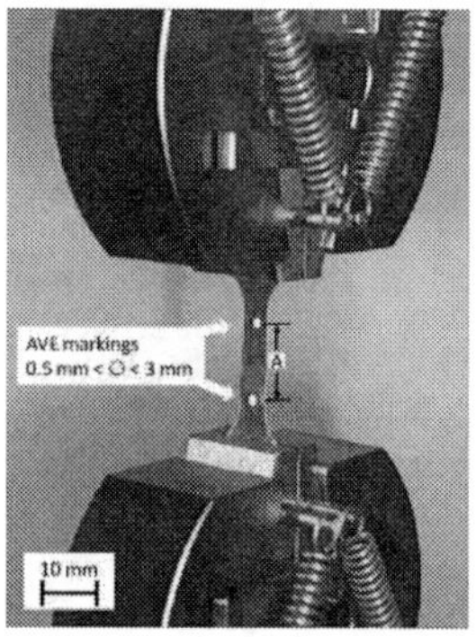

Fig. 6. Gauge marks defining the gauge length A for the AVE applied to a tensile specimen.

Fig. 7 shows the results of static tensile tests with different strain rates conducted on flat specimens. Ultimate tensile strength *UTS*, yield strength σ_y and the elongation at fracture *A* were determined on Al 99.5 micro sheets having a sheet thickness of 100 µm.

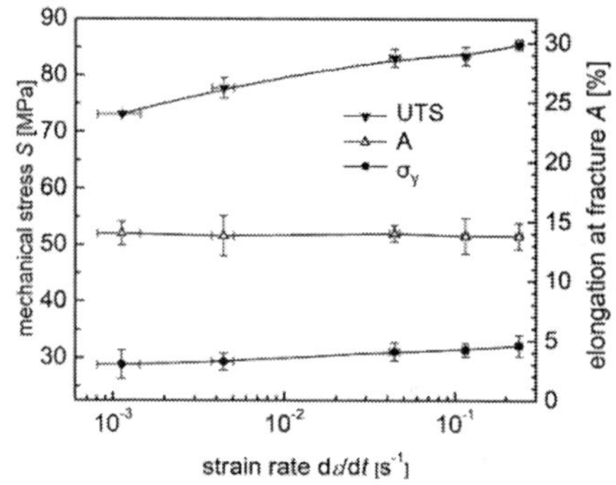

Fig. 7. Mechanical properties of a micro-flat-specimen of Al99.5 (thickness 100 µm) in dependence on the strain rate.

The test system enabled the strain rate to be varied over more than two orders of magnitude between $\dot{\varepsilon} = 10^{-3}$, which is approximately equivalent to that recommended in the standard DIN EN ISO 6892 – 1:2016 for tensile tests at macroscopic samples, till· $\varepsilon = 2 \cdot 10^{-2}\ s^{-1}$, the highest possible strain rate that the extensometer allows with such a specimen. This confirms that the strain rate has a significant influence on the mechanical parameters. Other than known for macro samples the ultimate tensile strength shows a clear increase with increasing strain rate. Only a weak influence is observed on the yield strength, and no influence is discernible on the elongation at break.

1.1.3 Clamping devices

The clamping of flat specimens is relatively easy in the standard wedge grips (Fig. 6), as long as fitted specimens can be used, so that the notching effect at the clamping device cannot have a detrimental effect. However, it is necessary to pay attention to a very exact alignment of the specimens, as otherwise wrinkles will form in the area of the clamps, which can stretch into the test area (Fig. 8). It may be advisable to use suitable auxiliary structures, such as those suggested by Hong and Hoffmann, when installing the specimens [4].

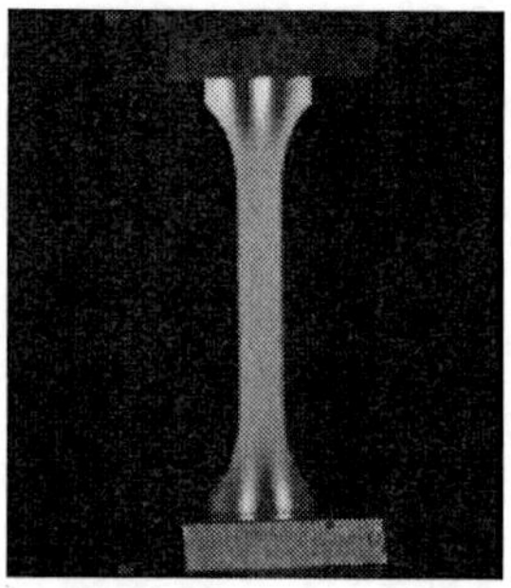

Fig. 8. Wrinkling of thin sheet tensile test specimen (thickness 100 µm) due to eccentric clamping in tensile direction.

The testing of round specimens, on the other hand, poses a greater challenge, as is the case, for example, with the cold forming process of rotary swaging, the principle of which is explained for example by Kuhfuss et al. or Piwek et al. [5, 6]. In this process, a wire shaped semi-finished product with a diameter of 1 mm is rotary swaged to, for example, a diameter of 0.5 mm and cold worked at the same time. The mechanical characterization of the formed state involves thin round specimens, whose tension in a standard clamping device (Fig. 6) with corresponding clamping jaws (Fig. 9) always leads to the problem of the notch effect, which leads to failure at clamping in almost all samples. Thus, in the static tensile test, only the yield strength can be determined for these specimens, but neither tensile strength nor elongation at break. An evaluable cyclical testing is not possible at all.

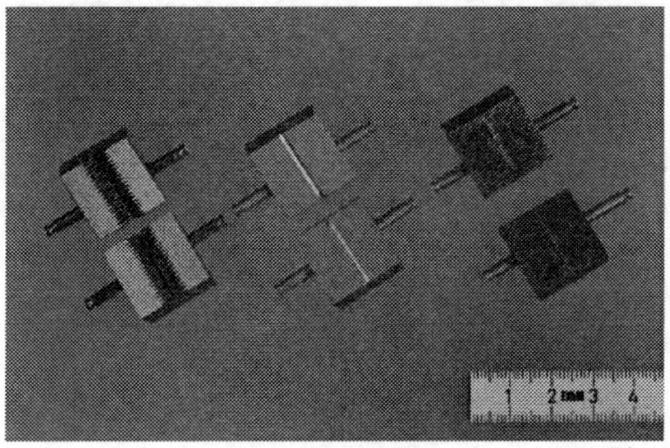

Fig. 9. Wedge jaws for use with standard clamping device (see Fig. 6) for testing round samples of different diameters.

One way of preventing failure in the clamping device is to increase the specimen cross section in the clamping area. This can be done, for example, by gluing a suitable tube onto the sample ends (Fig. 10). In principle, it is possible to ensure that the specimens break in the test area as desired and no longer in the clamping jaws. In order to obtain a sufficiently strong bond, however, it is necessary to cure the adhesive at 180 °C and thus allow the sample to undergo an unwanted heat treatment which can change the material's condition.

Fig. 10. Tensile test specimen of a rotary swaged wire reinforced in the clamping area with a glued-on copper tube.

In order to avoid the need for reinforcement at the clamping ends of the specimens, a concept, known from the testing of fibers, was used: The specimens are guided around a roll at both ends and only mechanically clamped at the end. In this way, the force is dissipated via the friction of the wire on the rollers and the notch effect at the end of the clamping turns out to be harmless.
Ultra-micro-hardness measurements on the rotary swaged wires made of X5CrNi18-10 showed that they undergo a strong work hardening due to the forming process and have a surface hardness up to 550 HV 0.03 [3]. In view of the fact that the clamping fixture should also be used for testing under cyclic tensile stress, it was necessary to adjust the hardness on its surface as high as possible in order to prevent the wire from working into the capstan during the test. For this reason, the nitriding steel 31CrMoV9 was chosen as material for the rollers. After manufacturing, the rolls were quenched and tempered (870 °C 2 h/oil + 550 °C 2 h) and then nitrided (510 °C 24 h) with characteristic number of nitrification $K_n = 3$. This way, a nitriding depth of about 0.3 mm could be achieved and a surface hardness of 900 HV 0.1 or 65 HRC, respectively.

Fig. 11. Clamping device for static and cyclic testing of wires

Fig. 11 shows the manufactured clamping fixture for wires up to a diameter of 1 mm. The specimens are guided on both sides two turns in a groove around the rollers (diameter 80 mm) and are mechanically clamped at the end. In both the static and cyclic tests, failures only occurred in the test area which has a minimum length of 18mm. In addition, unlike the wedge clamping device, even with high-strength specimens, there are no problems with slipping during the test.

2 Electron backscatter diffraction (EBSD)

Electron backscatter diffraction (EBSD) was used as a further characterization method for micro cold formed semi-finished products and components. A Philips XL30 scanning electron microscope with an EBSD detector EDAX DigiView IV was available for this purpose.

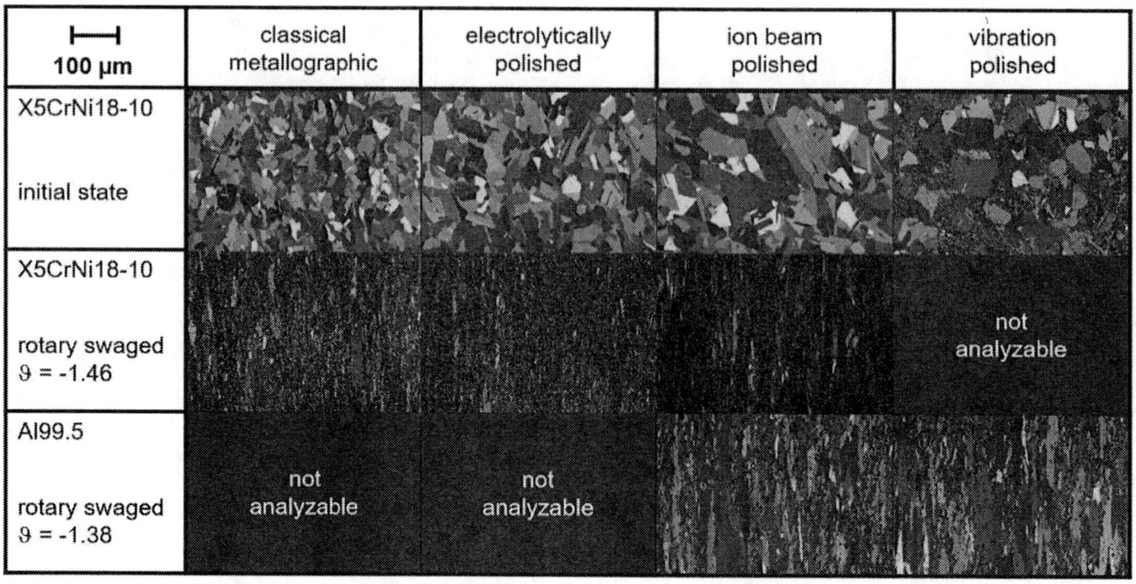

Fig. 13. Grain diagrams for comparison of different preparation methods for EBSD investigations on rotary swaged specimens and material in initial state

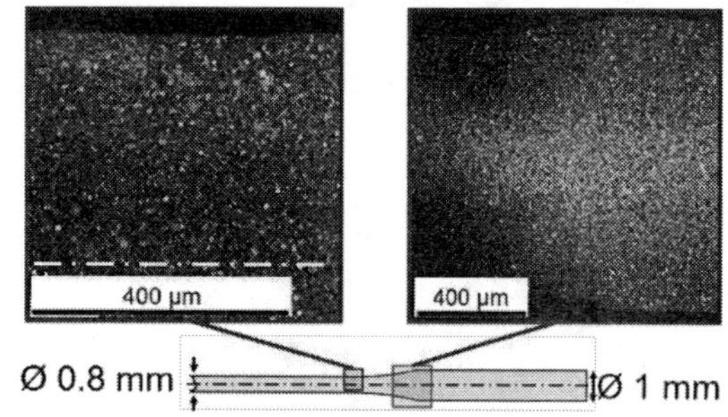

Fig. 12. EBSD grain images from the rotary swaged area (red) and the transition area (blue) of X5CrNi18-10 steel wire.

Fig. 12 shows two EBSD grain images of the centre of a cold formed wire made of steel X5CrNi18-10. The sample was rotary swaged from an initial diameter of 1 mm to a diameter of 0.8 mm. This corresponds to a comparatively low degree of deformation of $\varphi = \ln(A/A_0) = -0.45$, where A is the initial cross section and A_0 is the cross section after forming. One EBSD measurement was taken in the rotary swaged area (Fig. 12, red area), the second measurement in the transition area from the starting material to the formed area (Fig. 12, blue area).

In order to detect the poor indexability of the Kikuchi patterns in these areas, systematic investigations were carried out using various conventional preparation methods on different materials at different degrees of deformation. A selection of the results of these investigations is shown in Fig. 13: The conventional metallographic preparation, with which the sample was prepared in Fig. 12, electrolytic polishing, ion beam polishing and vibration polishing were compared on the basis of three different sample batches, steel X5CrNi18-10 in initial state and in rotary swaged state and Al 99.5, also in rotary swaged condition. At $\varphi = -1.46$ and -1.38 the degree of deformation of the latter two states are significantly larger than the deformation degree of the sample from fig. 12.

It can be seen that the first three procedures lead to comparably good results when the steel is prepared in its initial state. Vibratory polishing provides in principle also large evaluable areas, but has already larger, non-indexable regions. A significantly different picture results from the corresponding strongly cold-formed steel. In this case, the vibration-polished variant could not be analyzed at all and the three other preparation methods each have very large, non indexable areas. Fig. 14 shows an enlarged view of the grain representation before (a) and after forming (b) as well as the corresponding image quality (IQ) diagrams (c and d).

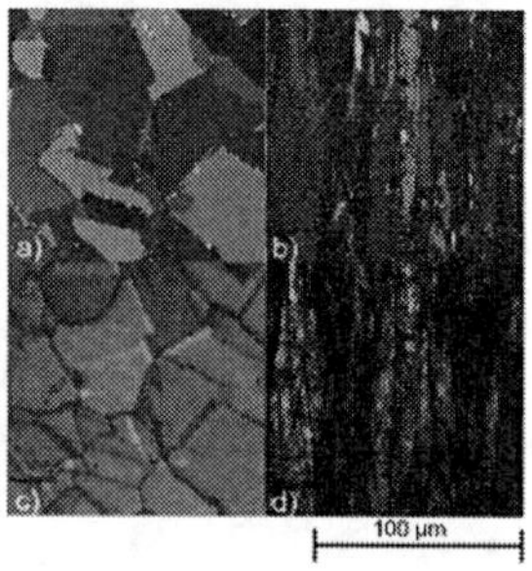

Fig. 14. Detailed depiction of the grain diagrams of X5CrNi18-10 steel wires in the initial state (a) and the rotary swaged state with degree of deformation $\varphi = -1.46$ (b) and the corresponding image quality diagrams (c and d).

* Corresponding author: koehler@iwt-bremen.de

https://doi.org/10.1051/matecconf/201819015012

In the latter case, the darker grey shades in the image denote lower IQ values in the evaluable areas. Any distortion to the crystal lattice within the diffracting volume produce lower quality (more diffuse) diffraction patterns. This enables the parameter to provide a qualitative description of the strain distribution in the microstructure, provided that the underlying structures can still be dissolved in principle [7]. Thus the dark, non-indexable areas are interpreted here as areas of high strain, but possibly also as very fine-grained structures. The evaluable areas of the grain diagrams show a strong grain stretching in the direction of the sample longitudinal axis compared to the initial state in the rotary swaged state.

The same picture can be seen in Fig. 13 for Al99.5 in the rotary swaged condition with approximately the same degree of deformation. In this case, a clear longitudinal stretching of the grains can be seen and, in contrast to steel, measurements on ion-beam and vibration-polished samples can be easily indicated. This could be an indication that the reason for the poor evaluability of the formed steel specimens is due to a fine-grained structure that lies beyond the resolution limit of the measuring system used.

In contrast to steel, the Al99.5 measurements on ion-beam and vibration-polished specimens are quite easy to indicate, whereas the conventional metallographic and electrolytic preparation proves to be unsuitable. This shows that it is not possible to make a general statement about the most suitable preparation method, but that it depends on the material and its condition.

Although, for the above-mentioned reasons, only an incomplete indexing of the Kikuchi pattern is possible in the cold-formed state, at least qualitative findings can be drawn from the EBSD measurements, which cannot be obtained by conventional metallographic methods, as shown in Fig. 15. Four different evaluations of an EBSD mapping on an X5CrNi18-10 steel wire in the initial state and in the rotary swaged state ($\varphi = -1.38$) are shown. The figures show the middle plane of the specimen, with the longitudinal axis of the wires running horizontally.

In contrast to the homogeneous structure of the initial state, a strongly inhomogeneous microstructure can already be seen in the grain depiction after strong cold forming. A similar structure can be seen around the longitudinal axis of the sample as in Figure 14, consisting of relatively large and strongly stretched grains. On the other hand, the edge areas show a clearly finer structure

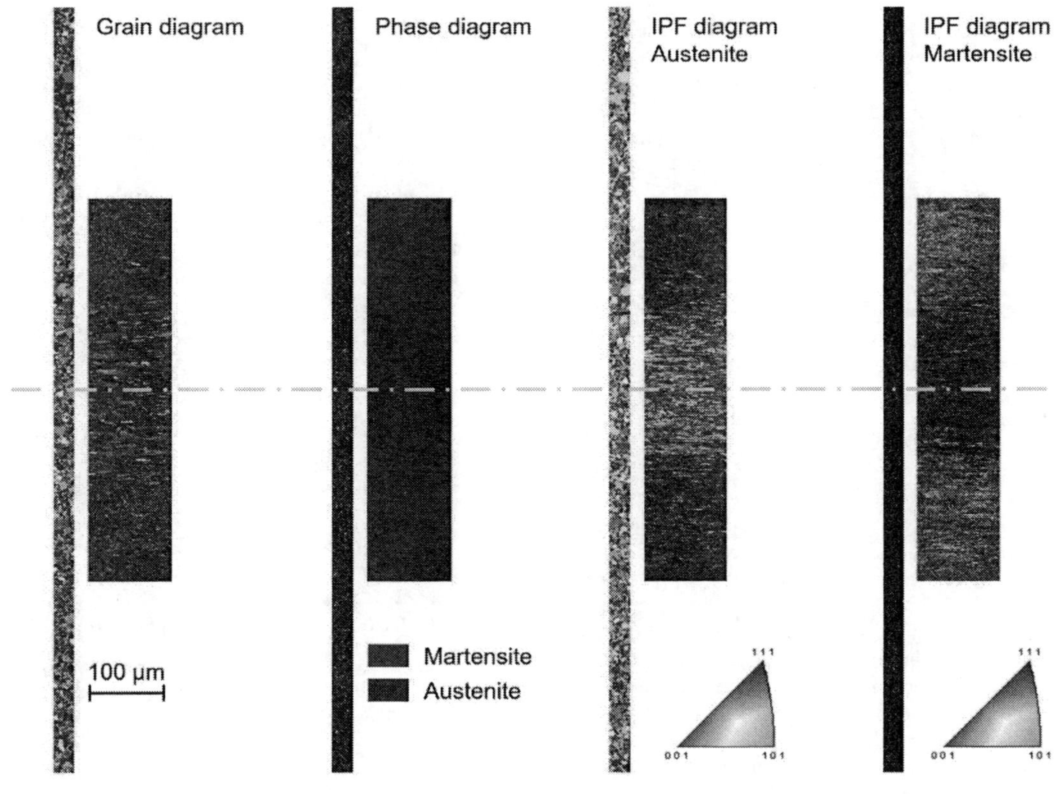

Fig. 15. EBSD mappings of X5CrNi18-10 steel wires: The figure shows a section through the specimen center plane in the initial state, ∅ 1 mm (left) and the rotary swaged state, ∅ 0.5 mm (right), $\varphi = -1.38$, in four different mappings: grain diagram, phase diagram and the inverse pole figure (IPF) diagrams of austenite respectively martensite.

after rotary swaging, in which only very occasionally larger and stretched grains such as in the close to the axis areas can be found. In the phase representation, these peripheral areas are largely identified as martensite, whereas the central part is almost completely austenitic. Obviously, in the originally completely austenitic wire, strong deformation in the periphery areas is accompanied by formation of deformation-induced martensite, which is also confirmed by the distinct ferromagnetic properties of the material after forming.

The inverse pole figure (IPF) diagram for the austenitic phase also shows that the forming process causes texturing. The grains in the austenitic sample center preferably align themselves with the [101]-direction parallel to the sample longitudinal axis. In the predominantly martensitic peripheral regions of the sample, the IPF diagram shows a clearly purple colored portion, which can be assigned to the legend in the standard triangle according to a crystal direction between [001]- and [111]-direction, approximately in the [112]-direction. In addition, there are also smaller green areas where the [101]-direction lies in the direction of the wire axis, as in the austenitic area.

Overall, the analysis using EBSD shows that the workpiece is subjected to an inhomogeneous material state due to the strong cold forming during swaging. This concerns not only the microstructure, but also other variables such as residual stress distribution across the diameter, and naturally has a decisive influence on its mechanical properties. In cyclic testing, for example, failure of the material thus formed is typically based on a very narrowly defined area within the martensitic region below the surface in which a residual stress maximum can be found [3].

7 Summary

The characterization of semi-finished products for the production of micro cold-formed components as well as the components themselves places special demands on the devices and methods used. Mechanical testing of such micro specimens requires testing equipment specifically adapted to their small dimensions.

The mechanical testing of such micro-samples requires a test equipment that allows the reliable measurement and control of static as well as dynamic forces in the area of a few Newtons. Tactile measurement methods, for example in strain measurement, are excluded, so that optical methods are the method of choice. Furthermore, the smaller the specimens to be tested are, the higher the demands on their handling, e. g. for precise alignment in the clamping device of the testing machine. A special challenge arises if the test specimens do not allow to be brought into a conventional sample form similar to that used in macroscopic samples. For this purpose, it may be necessary to use special devices in order to obtain valid results. This is the case, for example, in the case of semi-finished products and components that have no homogeneous properties, so that any mechanical, material-removing processing for sample production would falsify the test result.

In the characterization of microstructures, conventional metallographic methods quickly reach their limits for sample sizes in the distinct sub-millimeter range. Orientation imaging microscopy, which is based on electron backscattering (EBSD), allows statements about phase distribution, grain structure and orientation, is a powerful addition. However, this method requires an extremely complex sample preparation for micro samples. In particular, it was shown that it is necessary to find out the most suitable preparation technique individually according to the material and its condition. Very strongly deformed and/or ultra-fine grain structures could only be investigated to a limited extent. However, there are at least qualitative findings to be gained which cannot be achieved with pure conventional metallography.

8 Acknowledgments

The authors would like to thank the German Research Foundation (DFG) for financial support of the subproject B4 “Component strength” within the collaborative research centre SFB 747 “Micro Cold Forming”.

References

1. E. Macherauch, H.-W. Zoch, *Praktikum in Werkstoffkunde* (Vieweg + Teubner, Wiesbaden, 2011)
2. W. Weibull, *Ingeniörsvetenskapsakademiens Handlingar Nr. 151* (Generalstabens Litografiska Anstalts Förlag, Stockholm, 1939)
3. B. Köhler, B. Clausen, *Fortschritte in der Werkstoffprüfung für Forschung und Praxis, Proceedings Tagung Werkstoffprüfung 2017 Berlin, eds.: H. Frenz, J. B. Langer* (DVM e. V., Berlin, 2017)
4. S. Hong, H. Hoffmann, *Proceedings 1st Colloquium Processscaling* (BIAS Verlag, Bremen, 2003)
5. B. Kuhfuss, E. Moumi, V. Piwek, Microsystem Technologies **14**, 1995 (2008)
6. V. Piwek, B. Kuhfuss, E. Moumi, M. Hork, International Journal of Material Forming **3/1**, 845 (2010)
7. S. T. Wardle, L. S. Lin, A.Cetel, B. L. Adams, *Proceedings 52nd Annual Meeting of the Microscopy Society of America, eds.: G. W. Bailey, A. J. Garrett-Reed* (San Francisco Press, San Francisco, 1994)
8. B. Köhler, B. Clausen, H.W. Zoch, *Tagungsband 8. Kolloquium Mikroproduktion, 27.-28. 11. 17, Bremen, Deutschland. Vollertsen, F. Hopmann, C.; Schulze, V.; Wulfsberg, J. (Hrsg.), 91* (2017)

* Corresponding author: koehler@iwt-bremen.de

Process chain for the fabrication of hardenable aluminium-zirconium micro-components by deep drawing

Anastasiya Toenjes[1,*], *Julien* Kovac[1], *Bernd* Koehler[1], *Axel* von Hehl[1,2], *Andreas* Mehner[1,2], *Brigitte* Clausen[1,2] and *Hans-Werner* Zoch[1,2]

[1]Leibniz Institute for Materials Engineering IWT, 28359 Bremen, Germany
[2]MAPEX Center for Materials and Processing
[1,2]University of Bremen, Faculty of Production Engineering, 28359 Bremen, Germany

Abstract. Today, micro components are used in various industrial sectors such as electronics engineering and medical applications. The final quality of such parts depends on each individual step of the production chain from the manufacturing of semi-finished parts to the post-processing. In this study, magnetron sputtering is used to manufacture thin (15-30 μm) aluminium-zirconium alloy foils for the deep drawing of high strength and hardenable micro cups, which can be, for example, employed as micro valve caps. The development of a novel process chain for the production of these parts includes four different steps, beginning with the production of Al-Zr foils by magnetron sputtering. Secondly, tensile tests are performed with the foils in order to estimate their mechanical properties. Subsequently, micro deep drawing is used to produce the cup's shape, and finally, a heat treatment in a drop-down tube furnace adjusts the cup's hardness during fall. It is shown in particular that Al-Zr foils produced by magnetron sputtering have an attractive cold forming and hardening potential due to a microstructure consisting essentially of an oversaturated solid solution of zirconium in the aluminium matrix. This material state enables adequate formability and simplifies the heat treatment process since no solution annealing is required.

Keywords: Cold forming, Material, Development.

1 Introduction

The trend for micro components in various applications such as medical tools, connectors in mobile phones or fuel injectors requires a decrease of size and an increase in quality over the whole product life cycle. At the same time, the relation between quality of parts, production speed and cost must be balanced. To fulfil these requirements, new production technologies, materials and testing methods are necessary because direct downscaling of processes for macro parts into the micro scale is rarely possible.

In this paper, four steps for the development of a process chain for the production of micro cold-formed cups made of a high-strength aluminium alloy are presented. Among the possible alloying elements for aluminium to increase the hardness, scandium shows the best potential due to the formation of nanoscaled cubic ($L1_2$ Cu_3Au-type) Al_3Sc precipitates. These equilibrium precipitates provide an extraordinarily high lattice misfit without losing the coherence with the aluminium matrix and, hence, result in a significantly higher strength of the alloy compared to semi-coherent precipitates in conventionally age-hardened alloys [1]. However, the limited availability of scandium increases the cost of parts and makes the industrial use of such alloys difficult. Alternatively, other alloying elements such as zirconium in combination with the base material aluminium lead to similar effects based on the formation of zirconium tri-aluminide precipitates (Al_3Zr) [2]. In order to enable high deep drawing ratios, the precipitates should only be formed after the deep drawing process. Indeed, the presence of precipitates in the material before cold forming would result in a limited ductility. Therefore, any post forming heat treatment must be performed on a material that is free of precipitates at this point in the process chain.

Due to high processing temperatures and low solidification and cooling rates during conventional slab casting of Al-Zr alloys it is not possible to obtain more than 0.28 wt. % zirconium in the solid solution with the aluminium. The binary Al-Zr phase diagram shows that the maximum solubility of zirconium in aluminium is only 0.28 wt. % at the melting point (650 °C) [3]. As shown in our previous work, magnetron sputtering enables the fabrication of thin foils with a thickness of 15 - 30 μm containing up to 4 wt. % zirconium in supersaturated solid solution [4-5]. The properties of these foils, i.e. semi-finished products for cold forming, depend on several parameters of the magnetron sputtering process, primarily the target power.

Thus, the mechanical properties of the cold formed components depend on the initial state of the semi-finished products since the forming parameters affect

* Corresponding author: toenjes@iwt-bremen.de

strain hardening of the material during cold forming and the ageing behavior during subsequent heat treatment. By means of the heat treatment process, the resulting mechanical properties have to be adjusted to the application specific requirements.

The objective of this study is the investigation of the influence of the processing parameters of each step of the process chain on the mechanical properties of the resulting components. The mechanical properties of the semi-finished components were tested by tensile tests and micro hardness measurements. The mechanical properties of deep drawn cup before and after the heat treatment were also investigated by the micro hardness.

2 Fabrication of thin Al-Zr alloy foils by magnetron sputtering

2.1 Working principle of magnetron sputtering

Magnetron sputtering is a frequently used physical vapour deposition (PVD) process for the deposition of functional thin films or coatings. The most common applications of this technology is the deposition of hard coatings for cutting and forming tools, coatings for wear and corrosion protection, low friction coatings as well as conducting and semiconducting coatings for microelectronic devices and photovoltaic cells or special biocompatible coatings for medical applications [6].

Magnetron sputtering is based on the vaporization of a target material by low-pressure plasmas. The vaporized material condensates onto the coated part (substrate). A schematic view of the magnetron sputtering process is presented in Fig. 1.

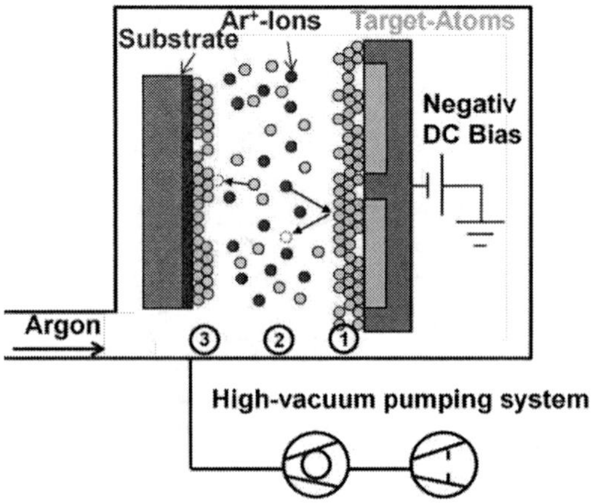

Fig. 1. Sketch of a magnetron sputtering unit. 1) Bombardment of the target with positive argon ions by means of a low-pressure plasma and sputtering of the superficial atoms of the target. 2) Transport of the sputtered atoms through the chamber. 3) Deposition of target atoms on the substrate.

As illustrated in Fig. 1 the magnetron sputtering process is carried out in a vacuum chamber. Chemically inert working gases such as argon or other noble gases are introduced into the vacuum chamber. Typical working gas pressure is about 10^{-3} mbar. The targets are connected to an electric generator (direct current for DC magnetron sputtering) and are negatively biased. The vacuum chamber and the substrate are electrically grounded. At the target, a glow discharge plasma consisting of positively charged argon ions and negatively charged electrons is created by collision ionization in the low pressure noble gas atmosphere. Due to the negative bias, the positive argon ions accelerate to the target and collide with the target atoms. The momentum transfer of the argon ions with the target atoms results in an ejection (sputtering) of the target atoms. The sputtered target atoms move through the chamber and condensate on the substrate and progressively form the coating.

2.2 Magnetron sputtering process for fabrication of thin Al-Zr foils for micro deep drawing

A more unusual aspect of magnetron sputtering is the fabrication of metallic freestanding foils. While cold rolling is generally used to produce 15 to 30 µm thick foils made of conventional materials such as pure or low alloyed aluminium, copper or stainless steel [7], the ability to roll and form high strength materials such as aluminium-zirconium alloys in micro deep drawing is very limited. This limitation is due to the presence of ductility-reducing Al_3Zr precipitates, which form already during continuous casting. Therefore, alternative methods such as magnetron sputtering were considered to manufacture thin foils of aluminium-zirconium alloys. Indeed, magnetron sputtering offers the possibility to produce supersaturated solutions which are more or less free of precipitations. The aim is to improve the ductility of these alloys and to enable a better control of the hardening process of finished parts during artificial aging.

To produce Al-Zr foils by magnetron sputtering, a mirror polished low carbon steel foil with 100 µm thickness was used as substrate. This substrate foil was set to a non-moving substrate holder equipped with a temperature control system. The distance between the target and the substrate was 50 mm and the substrate temperature was maintained at 37 °C during the deposition process. The substrate foils were coated with 20 to 30 µm Al-Zr using a dc magnetron sputtering unit Leybold Z700. The target consisted of a rectangular Al 4 mass % Zr alloy target with dimensions of 500 × 88 × 10 mm³. A flow of 100 sccm of argon was introduced into the chamber for the deposition process, resulting in a pressure of 4×10^{-3} mbar.

Some of the process parameters greatly influence the morphology and the mechanical properties of the final foils [8-9], which affects the whole process chain. In previous investigations, it was shown that the target power is a particularly important process parameter for the mechanical properties of the foils [8]. In this work, the influence of the target power on the whole process chain will be highlighted. Therefore, Al-Zr foils were deposited using a target power of 2 and 3 kW during

https://doi.org/10.1051/matecconf/201819015013

respectively 2 and 1 hour so that similar thicknesses are achieved for each foil variant.

After deposition, the steel substrate was removed by chemical etching in a solution of concentrated nitric acid, resulting in the complete dissolution of the steel foils and leaving the Al-Zr foil undamaged.

2.3 Properties of the produced Al-Zr foils

The main properties of the Al-Zr foils produced with target powers of 2 and 3 kW were already studied in a previous paper establishing the relation between the target power and the hardness of sputtered Al-Zr foils [10]. In both foils studied in this work, 3.9 mass % zirconium was measured, whereas minor traces of iron (0.16 mass %) and silicon were also detected. It was verified by transmission electron microscopy (TEM) that, as expected, the foils consist of a zirconium supersaturated aluminium solution with almost no $AlZr_3$ precipitates. The latter were only observed after aging at 400 °C and 500 °C. Finally, complementary TEM examinations have shown that the density of crystal defects (dislocations) was higher in the foil sputtered at 3 kW. This was attributed to the higher kinetic energy of the sputtered atoms at 3 kW [11].

3 Mechanical testing of the foils

3.1 Tensile test

Single-axis, quasi-static tensile tests were carried out for the PVD produced Al-Zr foils. An electrodynamic testing system of the type Instron ElectroPuls E1000 with an optical strain measurement based on a video extensometer [12]. The specimens were designed as rectangular flat specimens as shown in Fig. 2 (a), with the clamping ends on both sides with glued-on reinforcing steel plates to prevent failure in clamping due to the notch effect. In this way, it was possible to ensure that failure occurs almost exclusively in the test area, as shown in Fig. 2 (b).

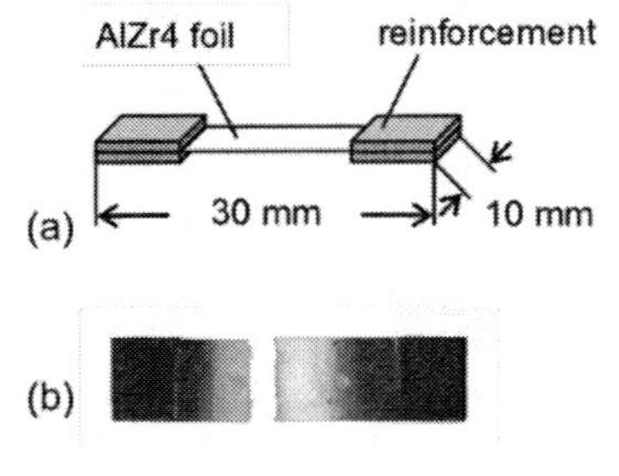

Fig. 2. Sketch of a tensile test sample (a) and tensile test specimen fractured in the test area (b).

The tensile tests were carried out under control of the strain rate which was $\dot{\varepsilon} = 0.00025\ s^{-1}$. Ten tests were carried out for each of the two sample batches. Fig. 3 shows a typical representative stress-strain-diagram for each of the two batches, as an example. In addition, the Hooke's straight line is drawn and parallel to it, the straight lines serving to determine the uniform elongation A_g and the elongation at break A. The PVD foils do not show any constriction prior to failure (see Fig. 2 (b)), so that in this case the strain at the ultimate tensile strength equals A. In addition, the straight line is drawn to determine the 0.2 % yield strength (dashed line) $R_{p0.2}$. It can be seen that this cannot be determined with the 2 kW version, since the failure here already occurs at an elongation of 0.1 %.

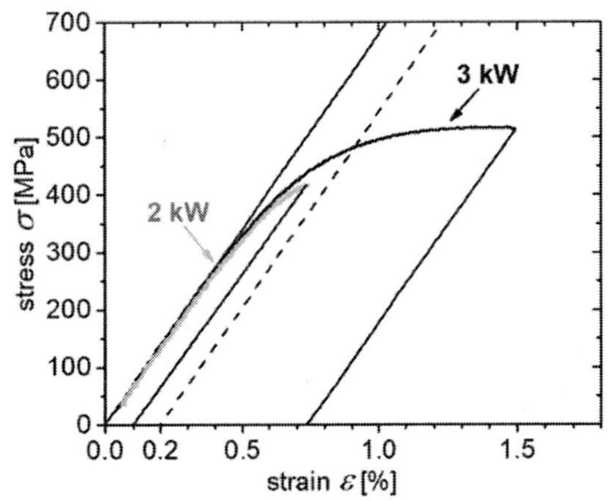

Fig. 3. Exemplary stress-strain diagrams for the two sheet metal foil batches sputtered with 2 kW and 3 kW plasma power respectively.

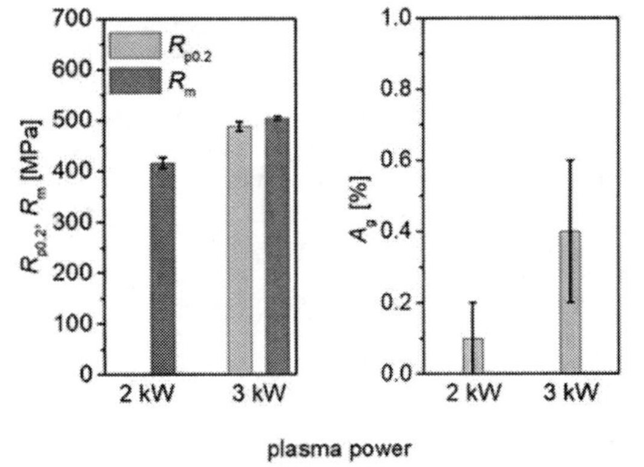

Fig. 4. Mean values of mechanical properties 0.2 % yield strength, $R_{p0.2}$, tensile strength R_m and uniform elongation A_g of all valid tensile tests (as-sputtered condition).

Fig. 4 shows the results of the measured yield strength, tensile strengths and uniform elongations of 10 valid tests for each variant. Plasma power, as the only parameter varied in the manufacturing process leads to different mechanical properties. Here, the batch produced with 3 kW plasma power shows an approximately 88 ± 12 MPa higher tensile strength on average over the ten samples tested. The uniform elongation is also higher compared to the 2 kW sample. However, compared to conventional aluminium alloys the uniform elongation is very low at 0.1 ± 0.1 % and 0.4

± 0.2 % for both variants. Thus, for the 2 kW sample the 0.2 % yield strength could not be determined anymore. With the 3 kW batch, however, their mean value is 488 ± 9 MPa.

The 3 kW variant shows significantly higher tensile strength, which is also indicated in Figure 3. The main reason for this is the increased dislocation density as shown in our previous work induced by the sputtering process at higher plasma power [11]. This observation is in conformity with hardness measurements on both variants, where the 3 kW charge also showed a higher hardness compared to the 2 kW batch [10].

3.2 Hardness of heat treated foils

The different Al-Zr foils (2 kW and 3 kW) were annealed at 300 °C, 400 °C and 500 °C for 10 minutes for 1, 5, 10, 20 and 50 hours in order to investigate the age hardening potential of these alloys. The selected range of temperatures is in the typical range of temperatures for ageing treatment of Al-Zr alloys of 375 to 425 °C [13]. All ageing experiments were carried out in standard air atmosphere muffle furnace.

For the hardness measurements, small pieces of the foils were embedded in a polymer matrix and mechanically polished. Hardness measurements were performed on the polished cross section of each sample using a micro hardness tester (FISCHERSCOPE H100C) with a Vickers indenter. The indentation force was 5 mN and the duration of the penetration, holding and indenter withdrawal phases was 10 seconds for each measurement.

The micro hardness results of the sputtered foils before and after ageing treatment at different temperatures against the ageing time are shown in Figure 5 and 6.

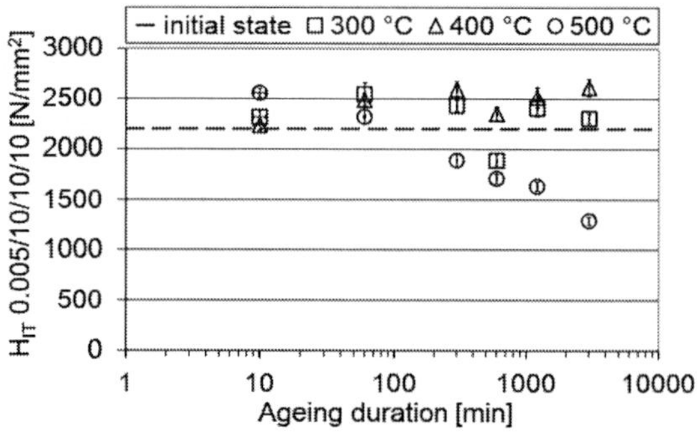

Fig. 5. Micro hardness measurements on the Al-Zr foils produced with 2 kW plasma power before and after several ageing treatments between 300 and 500 °C and 10 minutes to 50 hours.

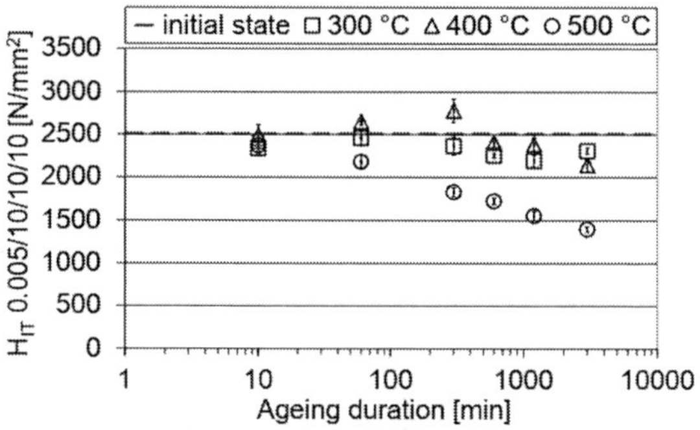

Fig. 6. Micro hardness measurements on the Al-Zr foils produced with 3 kW plasma power before and after several ageing treatments between 300 and 500 °C and 10 minutes to 50 hours.

In the initial state without ageing treatment, the hardness of the 2 kW Al-Zr foils is lower (H_{IT} ≈ 2200 MPa) than the hardness of the Al-Zr foils produced with a plasma power of 3 kW (H_{IT} ≈ 2500 MPa). The hardness of the 2 kW films increases during heat treatment. Thus, the experiments at 500 °C and shortest ageing time of 10 min lead to the maximum hardness of H_{IT} ≈ 2557 MPa. The ageing time of more than 10 hours does not increase the hardness at all tested temperatures. As Fig. 6 shows, a heat treatment does not lead to higher hardness anymore if foils are produced with 3 kW plasma power. An increasing plasma power in the sputtering process induces a higher density of crystal defects (dislocations) in the material. Hence, the hardness of the foils after the ageing treatment is the result of two superimposing mechanisms: recovery and precipitation hardening. The absence of the hardness increase during heat treatment of foils sputtered with 3 kW can be explained by the rapid recovery and the resulting loss of hardness (due to an accelerated heal up of the higher crystal defect density), which could not be compensated by the formation of precipitates within the investigated time and temperature range [10].

4 Micro deep drawing of Al-Zr cups

Micro deep drawing is an efficient cold sheet forming process for the mass production of hollow microscopic parts with complex shape, which are widely used in micro system technologies (MST) and micro electro-mechanical systems (MEMS) such as connector pins, resistor caps, lead chip frames or parts of pressure sensors [14]. The principle of a micro deep drawing process is illustrated in the sketch on Fig. 7. A circular foil portion (blank) is drawn into the cavity of a die by a punch applying a force to the blank body, resulting in a 3-dimensional hollow part which has the shape of a cup. At first, the punch force is transmitted to the workpiece bottom and afterwards to the cup walls and the flanges, where the tangential stress may cause wrinkling of the blank material. To avoid this, the flanges are compressed between a blankholder and the die upper part. A strict control of the blankholder force is necessary for

https://doi.org/10.1051/matecconf/201819015013

successful drawing. Indeed, if this force is too low, wrinkling occurs, whereas if the blankholder force is too high, the blank may be totally clamped by the blankholder and the blank bottom almost immediately fractures due to the rapidly increasing strain hardening caused by the punch force. Another very important process parameter, which defines a suitable process window, is the drawing ratio that is defined as the ratio of the blank radius to the punch radius. If the drawing ratio is too high, the bottom of the workpiece fails due to hardening in the cup wall area. The maximum achievable drawing ratio (without bottom fracture and wrinkling) is called limit drawing ratio and is characteristic of the formability of the foil material. This is however a simplified view and other parameters such as the foils mechanical and tribological properties, the lubrification conditions and the tool geometry must be considered. At this point of the description, micro deep drawing is similar to conventional (macroscopic) deep drawing, the only difference can be seen in the microscopic dimensions of the punch and tool (less than 1 mm) and the smaller thickness of the foil (typically 15 to 50 µm). This small foil thickness results in structural, mechanical and tribological size effects [15], which considerably reduce the formability of micro parts compared to macroscopic parts [16].

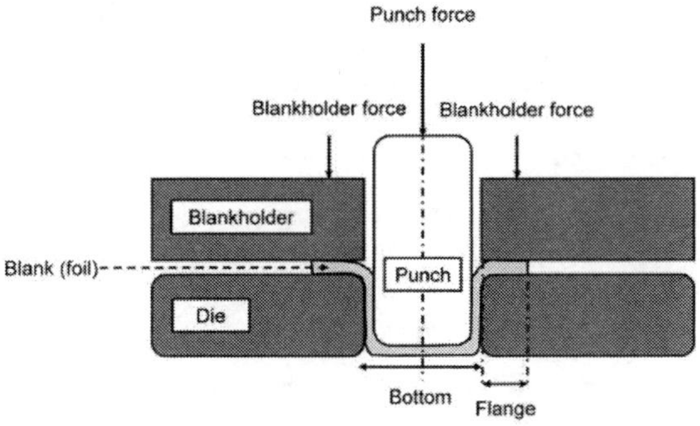

Fig. 7. Schematic of the micro deep drawing process

For the micro deep drawing of Al-Zr foils, the drawing process was carried out on a single axis micro forming press that allows a maximum force of 500 N. The punch force was measured using a piezo load cell of type Kistler 9217A with an accuracy of 0.01 N. The punch velocity was kept constant at 10 mm/s, and HBO 947/11 was used as lubricant. Several circular blanks were cut using a picoseconds pulsed laser with a wavelength of 1030 nm. The blanks were positioned on the die using a pneumatic gripper driven on a linear drive cross table. The position of the blank from the center of the die was measured using a CCD camera of type Allied Vision G 917 B with magnification of 0.75, which enables blank positioning within a radius of maximum 10 µm from the die center. The punch was made of a powder metallurgical tool steel of grade X135CrMoV12, and the blankholder force was always kept constant at 7 N. All relevant tool dimensions were measured using a Keyence VK 9700 laser scanning microscope. The tool dimensions are summarized in Table 1.

Table 1. Dimensions of the tool used for the micro deep drawing of Al-Zr blanks.

Punch diameter [mm]	1.005
Punch radius [mm]	0.12
Drawing die diameter [mm]	1.06
Average drawing die radius [mm]	0.077
Drawing gap [mm]	0.028

Several blanks were cut from the Al-Zr foils sputtered at 2 kW with different diameters so that cups with different drawing ratios between 1.5 and 1.8 could be drawn. Post-drawing observations of the cup revealed that the maximum drawing ratio at which no wrinkling and bottom failure occurs (e.g. limit drawing ratio) is 1.7 as depicted in Fig. 8. This limit drawing ratio is similar to the limit drawing ratio found for commercial Al99.5 foils [17], which shows that the drawability of sputtered aluminium zirconium foils is as high as the formability of pure aluminium.

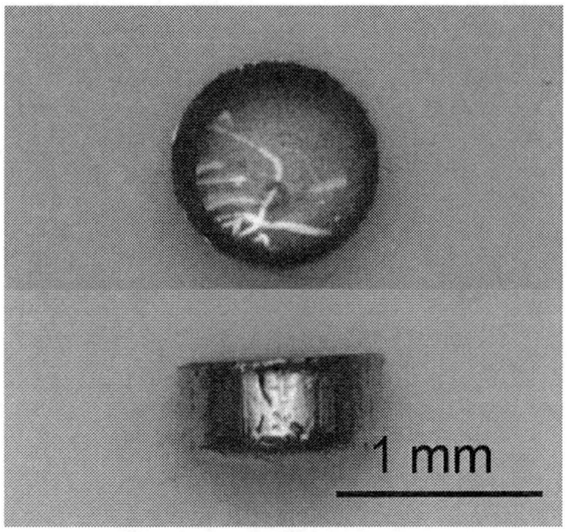

Fig. 8. Overview and side view of a deep drawn cup of aluminium zirconium with a drawing ratio of 1.7 (foils sputtered with 2 kW target power).

5 Heat treatment of the cups

Heat treatment was performed on deep drawn cups (sputtered with 2 kW) with a drawing ratio of 1.5. The experiments were carried out in two furnaces. One part of the cups was fed into the drop-down tube furnace (Fig. 9) at 500 °C and 600 °C. Measurement of the drop time was performed by an area laser light measuring system at the bottom of the furnace [18]. Another part of the cups was aged in a standard air atmosphere muffle furnace at 500 °C and 600 °C for 5 minutes. After ageing, the cups were embedded and micro hardness measurements were performed as described above.

 https://doi.org/10.1051/matecconf/201819015013

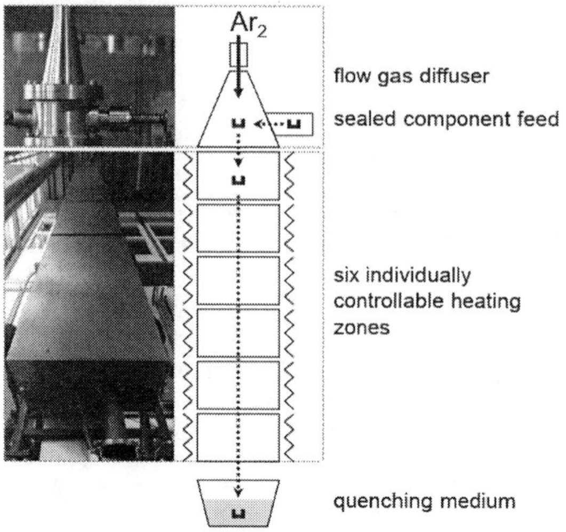

Fig. 9. Drop-down tube furnace.

Fig. 10 shows the results of the micro hardness measurements of the micro cups with a drawing ratio of 1.5 before and after heat treatment. The short-time heat treatment increased the hardness of the parts from HIT ≈ 2290 MPa to HIT ≈ 2720 MPa. Thus, after about 4 seconds, the same hardness could be achieved in the drop-down tube furnace as after 5 minutes in a muffle furnace.

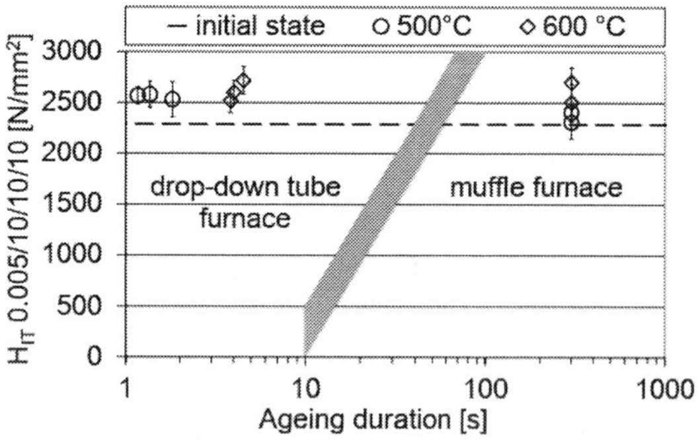

Fig. 10. Micro hardness measurements on deep drawn cup of Al-Zr (sputtered with 2 kW) with a drawing ratio of 1.5 before and after ageing treatments at 500 and 600 °C in drop-down tube furnace and standard air atmosphere muffle furnace.

The variation of the values for the fall duration of the cups is related to the different flow characteristics of the parts depending on their orientation when falling. This leads to the different drag coefficient and thus, to a change in the fall velocity and fall path.

The hardness of the cups after cold forming is about H_{IT} ≈ 2290. This is nearly 100 MPa higher than the hardness of the semi-finished foils produced with 2 kW target power. After short-time heat treatment, the hardness of the deep drawn cups is also higher than the hardness of the aged foils. This increase can be traced back to strain hardening during forming.

6 Recent evolution of the fabrication of hardenable aluminium alloy foils by sputtering

As shown by the previous results, sputtered Al-Zr foils are promising for deep drawing of hardenable micro components. Nevertheless, further improvements in terms of hardness and formability are possible by optimizing the material composition and design and the sputtering technology. Some aspects, which are currently employed to improve the properties of sputtered Al-Zr foils are briefly presented in the subsequent sections. More detailed results will be published later.

6.1 Aluminium-scandium-zirconium alloys

It has been reported by several authors that the addition of zirconium into aluminium-scandium alloys enhanced their mechanical properties [1, 13]. In this context, a higher hardness was reported for Al-Sc-Zr than in both Al-Sc and Al-Zr alloys [13]. Hence, when zirconium is added to aluminium-scandium alloys, a small amount of zirconium is incorporated into Al_3Sc precipitates. The lattice constant of the newly formed Al_3 $(Sc_{1-x}Zr_x)$ precipitates decreases with growing zirconium content (up to x = 0.5) and the misfit between the precipitates and the matrix is lower, which (for a given precipitate size) enhance the coherency of the precipitates [19]. The other interesting effect of the addition of zirconium to Al-Sc alloys is the decrease of the coarsening rate of the precipitates during heat treatment and a higher recrystallization temperature [20], which is related to the slower diffusivity of zirconium in aluminium [21]. Altogether, the lower size of the precipitates, their high dispersivity and their lower misfit with the aluminium matrix explain the higher hardness of Al-Sc-Zr alloys.

6.2 High Power Impulse Magnetron Sputtering (HiPIMS)

Al-Sc-Zr foils were obtained by DC magnetron sputtering (DCMS) using an industrial sputtering unit (CemeCon CC800/9) equipped with 4 targets and a rotating substrate table, which enables the deposition of larger foil areas and an excellent homogeneity of the foil thickness. However, the foils presented a porous columnar structure as shown in Fig. 11a. These porosities are due to the rotating substrate table. Indeed, with this configuration the average kinetic energy of the sputtered atoms is lower than for non-moving substrate since the substrate not always faces toward target during the deposition process. Consequently, the sputtered atoms hit the coating with lower kinetic energy and the impulse driven mobility of the coating atoms is reduced, resulting in a growth process dominated by shadowing effects [22]. In order to get a denser structure, it is necessary to increase the kinetic energy of the sputtered atoms, which can be achieved by either increasing the dc target power or by using High Power Impulse Magnetron Sputtering (HiPIMS). In HiPIMS, the current applied to the target is periodical. Within a period, during the off-cycle the target is not biased and within a short duty-cycle time, extremely high voltage is applied to the target, resulting in a higher degree ionisation than in dc sputtering and higher peak power (10 to a few hundreds of kW). This leads to a higher kinetic energy of the sputtered atoms resulting in a higher mobility of the coating atoms, which reduces the formation of pores [23]. Fig 11 shows the film morphology of Al-Sc-Zr foils deposited by DCMS (Fig. 11a) and by HiPIMS (Fig. 11b). The results in Fig. 11 indicate that the foils deposited by HiPIMS have less and smaller pores compared to the foils deposited by DCMS. In terms of mechanical properties, the higher density of the HiPIMS foils results in foils with much better ductility as shown in the tensile test curves in Fig. 12. Accordingly, a higher limit drawing ratio of 1.8 was also obtained when HiPIMS Al-Sc-Zr foils were tested in micro deep drawing.

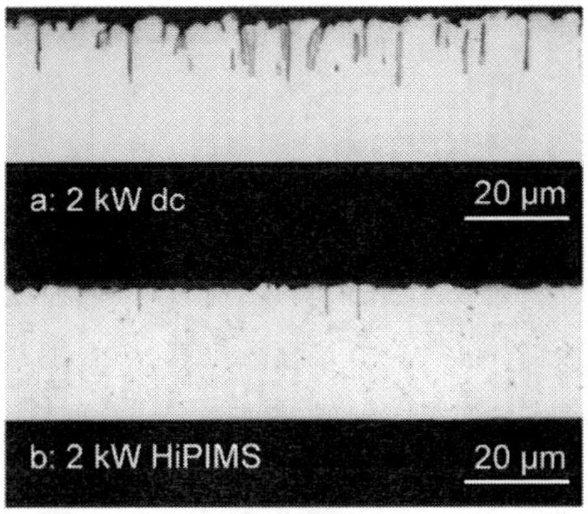

Fig. 11. Optical microscopy pictures of the cross section of two different sputtered Al-Sc-Zr foils. a: 2 kW dc b: 2 kW HiPIMS frequency 1000 Hz duty cycle 20 %.

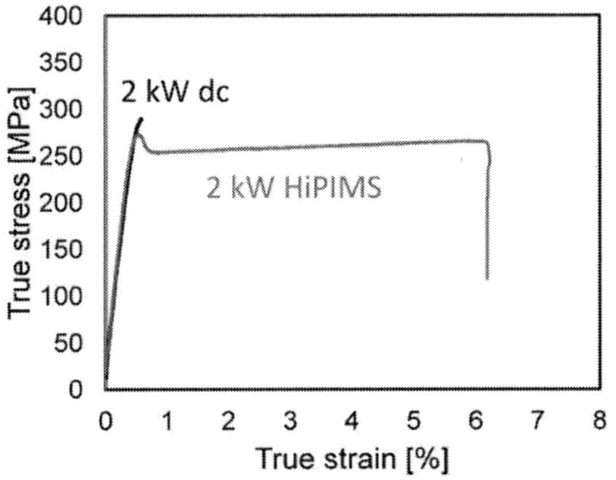

Fig. 12. Tensile test curves of a 2 kW DC sputtered Al 1.8 % Sc 0,3 % Zr foil and of a 2 kW HiPIMS Al 1.8 %Sc 0.3 % Zr.

6.3 Fabrication of bimetallic Al-Sc-Zr / stainless steel foils

Another way to improve the mechanical properties and the drawability of sputtered aluminium alloy foils is the production of bimetallic foils instead of monometallic Al-Sc-Zr foils. Bimetallic sheets consist of a stacking of several layers of two materials, generally a hard material and a soft and ductile material as for example ultra high carbon steel and brass [24]. Similarly, aluminium / stainless steel bimetallic sheets have been produced and investigated by Lee et al. [25]. These authors have shown that the mechanical properties of bimetallic foils in tensile testing are between the mechanical properties of pure aluminium and pure stainless steel and vary proportionally to the volume fraction (i.e. the total thickness) of aluminium in the bimetallic sheets. In every case, the bimetallic aluminium / stainless steel sheets show a better strength and a better ductility than pure aluminium sheets. In deep drawing, investigations on aluminium / stainless steel bimetallic sheets have also shown that limit drawing ratios of bimetallic sheets are higher than the one of pure aluminium and that the limit drawing ratio increases with decreasing aluminium volume fraction [26]. Sputtering was used to manufacture two layers Al-Sc-Zr / X5CrNi18-10 bimetallic foils. Commercial stainless steel foils with thicknesses from 15 to 25 µm were used as substrate foils. These substrate foils were coated with Al-Sc-Zr by magnetron sputtering, so that the total thickness of the bimetallic sheets was always 30 µm. In tensile testing (Fig. 13) these bimetallic foils showed both a better ductility and strength compared to monometallic Al-Sc-Zr foils. The ductility and the tensile strength of the bimetallic sheets increased when the volume fraction of aluminium decreased (the volume fraction of stainless steel increases). In deep drawing as observed for thick sheets [25], the limit drawing ratio depends on the layer stacking sequence of the bimetallic foil with respect to the punch-die system. Hence, when the Al-Sc-Zr layer side was in contact with the die, bimetallic foils only achieved a limit drawing ratio of 1.7 as for pure aluminium. When the stainless steel side of the bimetallic foil faces toward the die, a limit drawing ratio of 1.9 could be achieved which is far better than the 1.7 of the Al-Zr foils presented in this paper.

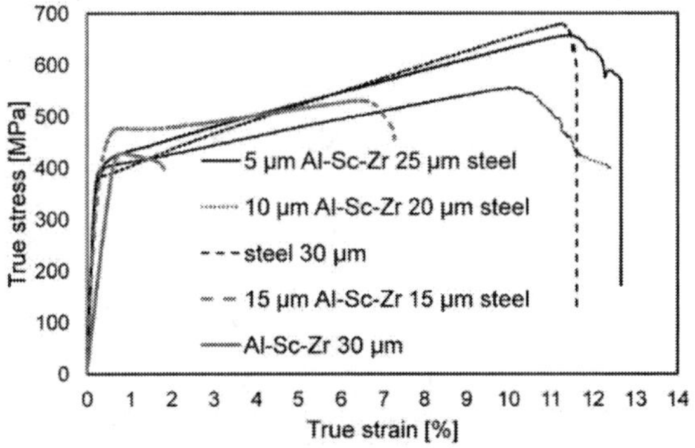

Fig. 13. Tensile test curves of a 30 µm sputtered Al 1.8 % Sc 0,3 % Zr foil and of three Al-Sc-Zr / X5CrNi18-10 bimetallic foils with a total thickness of 30 µm for each foil. The Al-Sc-Zr layer thickness of the bimetallic foils was 5 µm, 10 µm and 15 µm (in all foils the Al-Sc-Zr layer was sputtered at a target power of 4 kW dc).

7 Summary

Four steps for the optimization of the process chain for the production of micro cold formed cups made of a high-strength Al-Zr alloy were presented. The influence of the parameters of the individual process steps on the mechanical properties of the deep drawn cups was investigated. Furthermore, the hardening evolution during ageing treatment of semi-finished foils and deep drawn cups was characterized. The hardening potential of the foils and the micro cups depends on the plasma power during magnetron sputtering. In addition, the maximum hardness of the cups can be increased by short-time heat treatment in a drop-down tube furnace. Improvement in terms of mechanical properties and drawability can be achieved by producing Al-Sc-Zr foils sputtered by High Power Impulse Magnetron Sputtering or by producing bimetallic foils, consisting of a Al-Sc-Zr and a stainless steel layer, instead of monolithic Al-Sc-Zr foils.

8 Acknowledgments

The authors would like to thank the German Research Foundation (DFG) for financial support of the subproject A1 "PVD Sheet Metal", A2 "Heat Treatment" and B4 "Component strength" within the collaborative research centre SFB 747 "Micro Cold Forming".

References

1. J. Røyset, N. Ryum, Inter. Mater. Rev., **50**, 19 (2005)
2. P. Cavaliere, Mater. Character., **57**, 100 (2006)
3. J. Murray, A. Peruzzi, J.P. Abriata, J. Phase Equilib. **13**, 277 (1992)
4. J. Kovac, H.-R. Stock, B. Köhler, H. Bomas, H.-W. Zoch, Surf. Coat. Tech., **215**, 369 (2013)
5. G. Behrens, J. Kovac, B. Köhler, F. Vollersten, H.-R. Stock, T. Nonferr. Metal Soc., **22**, 268 (2012)
6. D. G. Constantin, M. Apreutesei, R. Arvinte, A. Marin, O. C. Andrei, D. Munteanu, RECENT, **12**, 29 (2011).
7. G. Behrens, F. O. Trier, H. Tetzel, F. Vollertsen, J. Mat. Form **9**, 253 (2015)
8. J. Kovac, H.-R. Stock, B. Koehler, H. Bomas, H.-W. Zoch, Surf. Coat. Tech., **215**, 369 (2013)
9. J. Kovac, H.-R. Stock, H.-W. Zoch, Journal of Surface Engineered Materials and Advanced Technology, **2**, 115 (2012)
10. A. Egorova, J. Kovac, A. von Hehl, A. Mehner, H.-W. Zoch, Mat.-Wiss. u Werkstofftech., **47**, 989 (2016)
11. Y. Pauleau, *Residual stresses in physically vapor deposited thin films* In H. S. Nalwa, *Handbook of Thin Films Vol. 1* , Elsevier, Los Angeles (1992)
12. B. Köhler, H. Bomas, H.-W. Zoch, J. Stalkopf, Engineering Integrity, **31**, 18 (2011)
13. K. E. Knipling, D. C. Dunand, D. N. Seidman, Acta Materialia, **56**, 114 (2008)
14. Frank Vollertsen, *Micro Metal Forming* (Springer, Berlin, Heidelberg, 2013)
15. F. Volletsen, Z. Hu, H. Schulze Niehoff, C. Theiler, J. Mat. Proc. Tech., **151**, 1 (2004)
16. F. Vollertsen, Z. Hu, J. Technol. Plast., **32**, 1 (2007)
17. G. Behrens, F. O. Trier, H. Tetzel, F. Vollertsen, Int. J. Mater. Form., **9**, 253 (2015)
18. R. von Bargen, A. von Hehl, H.-W. Zoch, J. Heat Treatm. Mat., **65/2**, 55 (2010)
19. Y. Harada, D. C. Dunand, Mat. Sci. Eng., **A329-331**, 686 (2002)
20. V. I Elagin, V. V. Zakharov, S. G. Pavlenko, T. D. Rostova, Phys. Met. Metall., **60**, 88 (1985)
21. K. Hirano, S. Fujikawa, J. Nuc. Mater., **69-70**, 564 (1978)
22. K.-H. Muller, J. App. Phys., **62**, 1796 (1987)
23. K. Sarakinos, J. Alami, S. Konstantinidis, Surf. Coat. Tech., **204**, 1661 (2010)
24. C. K. Syn, D. R. lesuer, J. Wolfenstine, O. D. Sherby, Met. Trans., **24A**, 1647 (1993)
25. D. N. Lee, Y. K. Kim, J. Mat Sci., **23**, 1436 (1988).
26. M. H. Parsa, K. Yamaguchi, N. Takakura, Int. J. Mech. Sci., **43**, 2331 (2001)

Material flow control in plunge micro rotary swaging

Eric Moumi[1,3*], *Philipp* Wilhelmi[1,3], *Christian* Schenck[1,2,3], *Marius* Herrmann[1,3] and *Bernd* Kuhfuss[1,2,3]

[1]bime, Bremen Institute for Mechanical Engineering, Badgasteiner Str. 1, 28359 Bremen, Germany
[2]MAPEX Center for Materials and Processing
[3]University of Bremen

Abstract. In rotary swaging, the material flow is not fully controlled by closure of the forming dies. This is especially noticeable in plunge rotary swaging of rod, where the workpiece is positioned into the forming zone und processed locally. As result, an uncontrolled elongation of the workpiece in axial direction takes place and an axial position shift of the workpiece relative to the dies occurs. This is a special challenge in production of linked micro parts, where single parts are interconnected in order to enable the handling as a strip and thereby a roll-to-roll production. The axial shift influences not only the subsequent positioning of neighbouring parts, but also the final geometry of the currently processed part. The presented investigation analyses the material flow during plunge micro rotary swaging on basis of in-process measurements of the workpiece shift on both sides of the forming zone as well as with the help of contour measurements of the processed parts. It is shown that the measured shift is strongly influenced by the workpiece clamping and fixation and that it can be controlled by applying low axial forces to the workpiece on one or both sides of the forming zone. Further, the geometry of the workpiece can be affected by these measures.

Keywords: Metal forming, Process control, Geometry

1 Introduction

Rotary swaging is an incremental cold forming process typically applied to produce rotational symmetric parts. There are two main process variants, the infeed and the plunge rotary swaging. They differ mainly in the type and direction of the feed. While the workpiece is fed axially in infeed rotary swaging, the dies are radially fed during plunge rotary swaging [1]. In Figure 1 the principle of the plunge swaging process is presented.

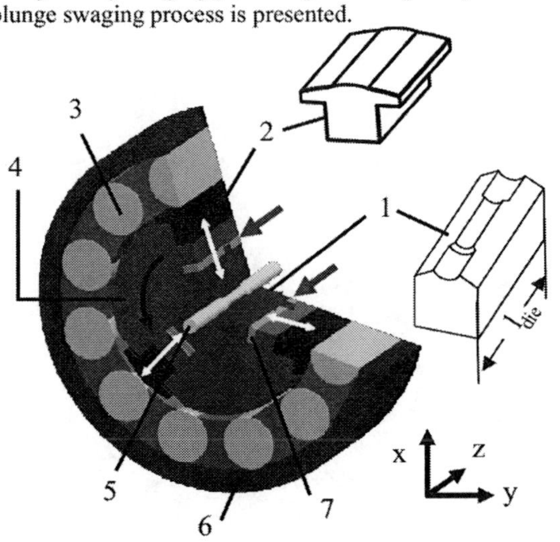

Fig. 1. Principle of plunge rotary swaging; 1 die, 2 base jaw, 3 cylinder roller, 4 drive shaft, 5 workpiece, 6 outer cage.

In plunge rotary swaging the workpiece is positioned in the forming zone between the dies. The swaging head in Figure 1 exhibits three dies, which process the workpiece locally. The forming of the workpiece takes place in the swaging head in small steps by radially oscillating motions and at a superposed radial feed of the dies towards the workpiece. The radial feed is realised with wedges (position 7 in Fig. 1). The radial oscillation of the forming dies (1) is generated by the rotation of the drive shaft (4) and the use of base jaws with cam on the top (2). By means of the roll motion between the base jaws and the cylinder rollers (3), the dies are pushed inwards simultaneously and the cross section of the workpiece (5) is reduced with every overrunning of the cams. An outer cage (6) encloses the described machine components. The workpiece is typically clamped in a collet, which is mounted on a supply system.

As in all forming processes, the understanding of the material flow is essential for a successful process design. Material flow was analysed by the visioplasticity method and the plasticity theory. In the visioplasticity method, the application of a grid in the area of the workpiece to be investigated allows determining the path of the material by a comparison of the grid before and after forming. In that way, statements about the strain and the strain rate could be made [2, 3]. With the improvements in the computational technology, finite element methods (FEM) had become state of the art for analysing forming processes. According to [4] planning of incremental bulk forming processes ideally requires extended process

* Corresponding author: moumi@bimde.de

simulations for gathering information about the material flow and also further product properties. However, the required fine meshes for the incremental forming and the resulting computation times have complicated their application in many cases [5]. Further, due to size effects in micro range [6, 7] advanced input data are required to validate that classical simulation models.

Inadequate material flow may cause several process failures such as flash formation, break and torsion of the workpiece [8]. Furthermore, the geometrical accuracy of the workpiece can also be affected by the material flow as presented in [9, 10]. From investigations on plunge rotary swaging of linked micro parts rupture and misalignment of the workpieces by inadequate material flow was found [11]. Linked parts are suitable in micro manufacturing because the single parts are kept interconnected through a process chain so that handling and transportation issues can be simplified. During the rotary swaging process material flows in axial direction, which causes an elongation of the workpiece. This elongation leads to a relative axial motion of the processed area with respect to the dies although the workpiece is clamped in a collet. This shift can cause geometrical deviations or even a rupture of the workpiece, which must be avoided. Therefore, understanding and controlling the material flow is essential for the design of a robust plunge micro rotary swaging process.

In this work, the influence of applying forces to the workpiece at both sides of the swaging machine in micro plunge rotary swaging was investigated. The evaluation was based on in-process measurements of the workpiece shift on both sides and on contour measurements of the produced parts.

2 Set up and experimental procedure

Forming experiments were conducted with annealed rods made of AISI-304 (X5CrNi 18-10) with an initial diameter $d_0 = 1.5$ mm. The annealing ensured good formability. Before forming, the workpiece was lubricated with "Condocut KNR 22". The same lubricant was also used for continuously lubricating the components in the swaging head close to the dies. The stroke frequency and the radial feed rate were kept constant at 102 Hz and 0.05 mm/s. At the end of the forming, when the radial displacement of the dies was completed they re-opened immediately without any settling time.

The setup for forming experiments consisted of three areas (Figure 2): the areas in the front and in the back of the forming machine, in which sensors for measuring the workpiece shift were implemented, and the machine in which the forming took place. The axial shift of the workpiece in the front (s_{front}) and at the backside (s_{back}) during forming was measured with two laser triangulation sensors. The triangulation sensors provided a resolution of 0.15 µm within a measuring range of 10 mm. The sensors captured the motion of lightweight reflectors, which were fixed on the workpiece and were guided by sliding guides. Adding the absolute values of the shift yielded the total elongation Δl of the workpiece according to Equation 1.

$$\Delta l = |s_{back}| + |s_{front}| \quad (1)$$

Due to the configuration of the forming machine, the distance between the middle of the forming zone of the die and the reflector in the front side was $L_F \approx 48$ mm at the beginning of the reduction and was about 10 times shorter than the distance L_B to the reflector in the backside, which was bridged with a guiding tube.

The geometry of the die and the workpiece are illustrated in Figure 3. The states before (Figure 3 a)) and after forming (Figure 3 b)) are shown. The die featured a length of $l_{die} = 40$ mm and the forming zone $l_f = 20$ mm. The transition zone of the die was designed with an angle $\alpha = 8°$. As the dies reduced the cross section of the workpiece, its diameter changed from d_0 to d_1 and its length increased from l_0 to l_1. In order to ease the handling and to speed up the experiments d_1 was set to 1.2 mm. However, selected measurements for a reduction from $d_0 = 1.0$ mm to $d_1 = 0.8$ mm with another die set delivered the same trends as described below. During this process, an angle occurred at the transition between deformed and the not deformed area of the workpiece.

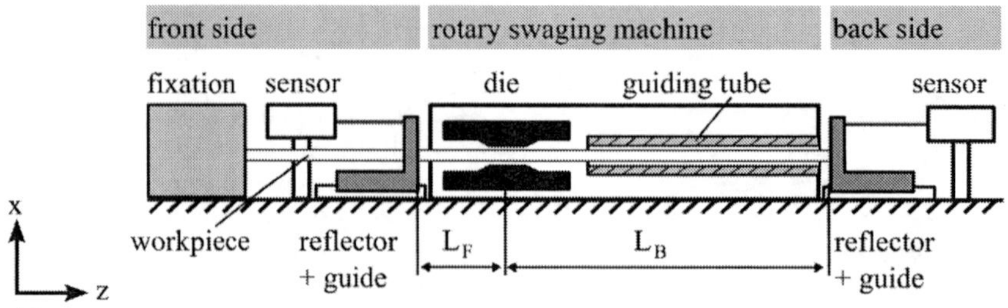

Fig. 2. Experimental setup.

https://doi.org/10.1051/matecconf/201819015014

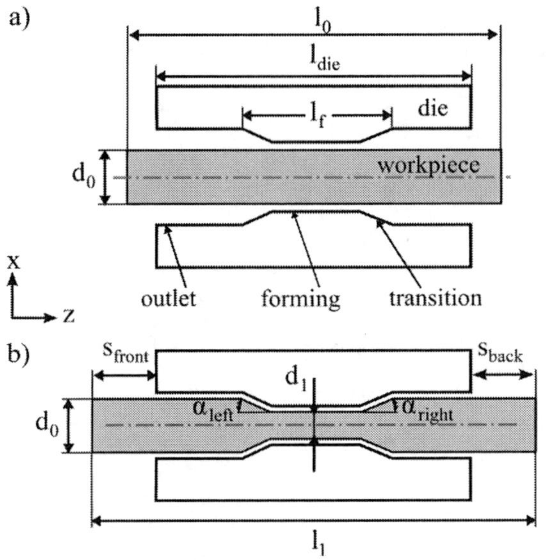

Fig. 3: Geometry of the dies and the workpiece; a) before forming, b) after forming.

To apply forces in axial direction of the workpiece helical springs were applied with four different set-up configurations (Figure 4):

A. Absence of external axial forces, no collet in use
B. Axial force in positive z-direction at the front; resembling a one-sided feed system
C. Axial force in negative z-direction at the front; intermediate state to configuration D
D. Axial forces at both sides away from the swaging head; symmetrical tensioning of the workpiece

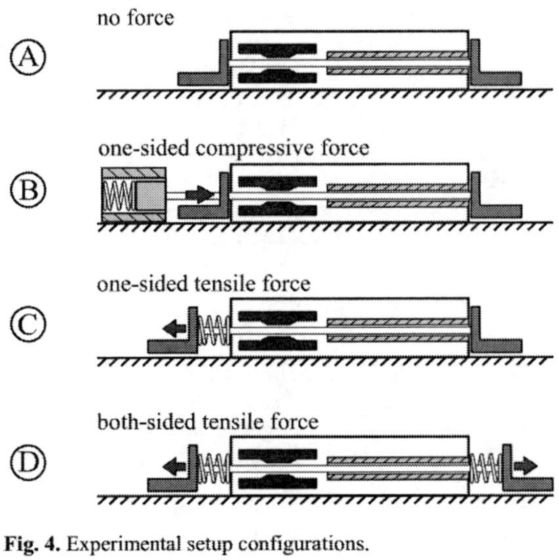

Fig. 4. Experimental setup configurations.

The applied helical springs featured a spring rate of 0.647 N/mm. The external forces were adjusted through appropriate preloading of the springs at the beginning of the forming process, Table 1. By the material flow, the springs were further compressed (configuration B) or stretched (configuration C and D) so that all forces changed almost proportionally with the shift.

Table 1. Axial forces for the four set-up configurations.

Name of configuration	Type of axial forces	Initial force $F_{a\_initial}$ (N)
A	none	-
B	one-sided compressive force	0; 1.7; 3.6
C	one-sided tensile force	7.1
D	both-sided tensile force	7.1

Subsequently to the forming step, the final geometry of the workpiece was measured contactless with an optical micrometer. For that, the diameter of the workpiece was measured every 60° around the z-axis and in 100 µm steps in the same axis. For interpretation, the mean value of the diameter with respect to the z-position of each sample was considered.

3 Results and Discussion

3.1 Axial shift of the workpiece

For all experiments, the measured shifts were always negative in the front side ($s_{front} \leq 0$) and positive in the backside ($s_{back} \geq 0$).

Configuration A
In plunge micro rotary swaging with configuration A (no external axial forces) the material flowed freely in the front and in the back side. An exemplary measurement of the workpiece shift and the elongation during forming are shown in Figure 5.

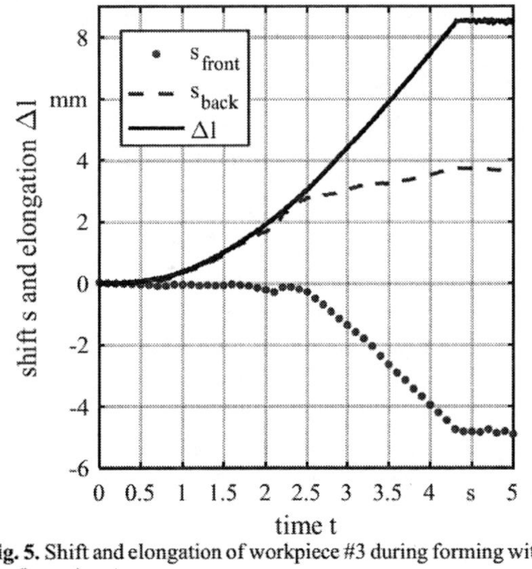

Fig. 5. Shift and elongation of workpiece #3 during forming with configuration A.

* Corresponding author: moumi@bimde.de

Figure 6 shows the absolute values of the maximum shift in the front and at the backside as well as the elongation for five samples formed under same conditions. While the standard deviation of the elongation was less than 90 µm, it was higher than 500 µm for the individually measured shifts. Additionally, the shift s was always higher in the front (Figure 6). As the workpiece was not fixed in axial direction, axial vibrations that were stimulated by the forming strokes changed the contact conditions between the workpiece and the dies. Hence, the workpiece could be shifted between the single strokes. The variation in the shifts could be attributed to the different lengths L_F and L_B. The longer length L_B hindered the material flow to the backside due to friction forces in the guiding tube.

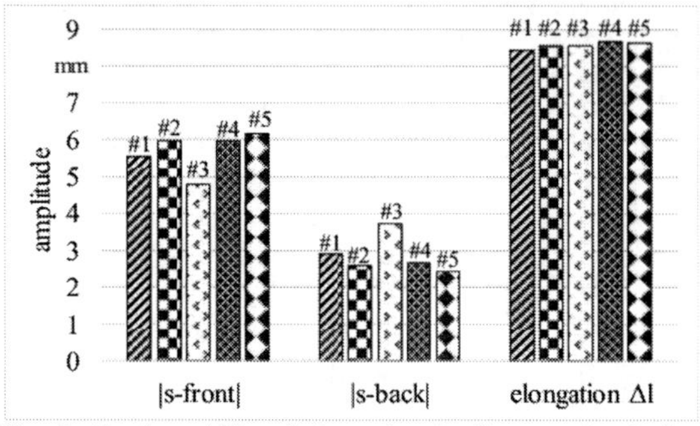

Fig. 6. Absolute shift and elongation for five samples during forming without external axial forces (configuration A).

Configuration B

In Figure 7, the shifts and the elongations for configuration B are presented as a function of the initial compressive force $F_{a\_initial}$.

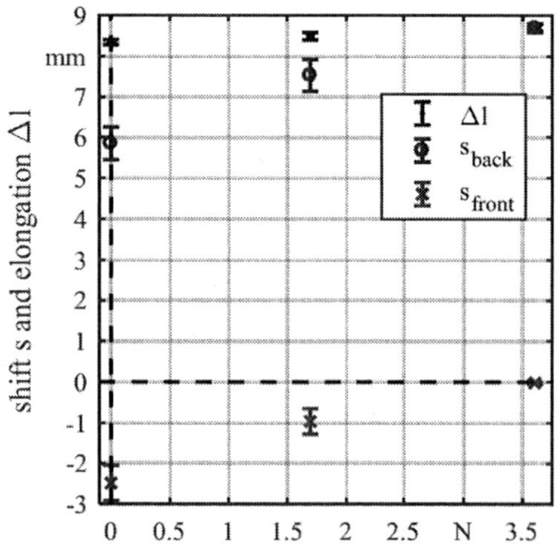

Fig. 7. Shift and elongation for different initial compressive forces (configuration B); mean values and single standard deviations.

Increasing the initial compressive force at the front side led to a reduction of the elongation in the negative z-direction and to a corresponding increase in the positive z-direction. For the investigated initial compressive force of 0 N about 70 % of the material shifted in positive z-direction. Although the axial force rose with the shift, the results accord with configuration A. With an initial force of 3.6 N the material shifted completely in positive z-direction. Additionally, the scatter of the shift values is very low (standard deviation < 100 µm) when an initial force of 3.6 N was applied.

Configuration C and D

In the configuration C and with an initial axial force value of 7.1 N the complete shift took place in the front side (Figure 8). This force value was high enough to direct the material flow completely in negative z-direction.

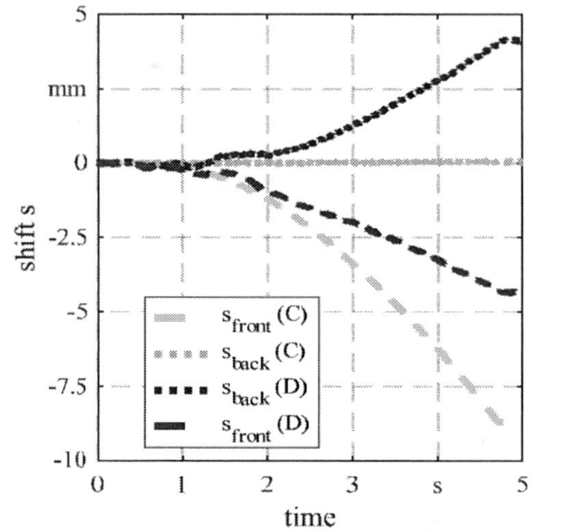

Fig. 8. Shift of the workpiece during forming with external one- and both-sided axial tension force (configurations C and D).

Applying the symmetrical tensioning of the workpiece resulted in an almost equal shift in both directions. The shift at the front side was -4.4 ± 0.09 mm and at the back side 3.9 ± 0.13 mm. The difference between these values could be explained by the different distances from the forming zone to the measuring point of the triangulation sensors (Figure 2) and the friction in the guiding tube, as described above. The standard deviation of the total elongation was less than 100 µm.

3.2 Workpiece Geometry

Exemplary measurements for the resulting geometry are presented in Figure 9 together with the set geometry of the die. The contour of the workpiece changed with the kind and amplitude of the applied forces. It was found that the generated geometry significantly varied when no external force was applied. This coincides with the measured shifts. This low reproducibility can be explained by the uncontrolled motion of the workpiece in axial direction during the complete forming process (configuration A) or at the beginning of the forming (configuration B, $F_{a\_initial}$ = 0 N). For the compressive force of 3.6 N in configuration B (Figure 9b) as well as for the one sided tensile force (Figure 9c) the geometrical deviation

 https://doi.org/10.1051/matecconf/201819015014

between tool and finished workpiece is concentrated to on side.

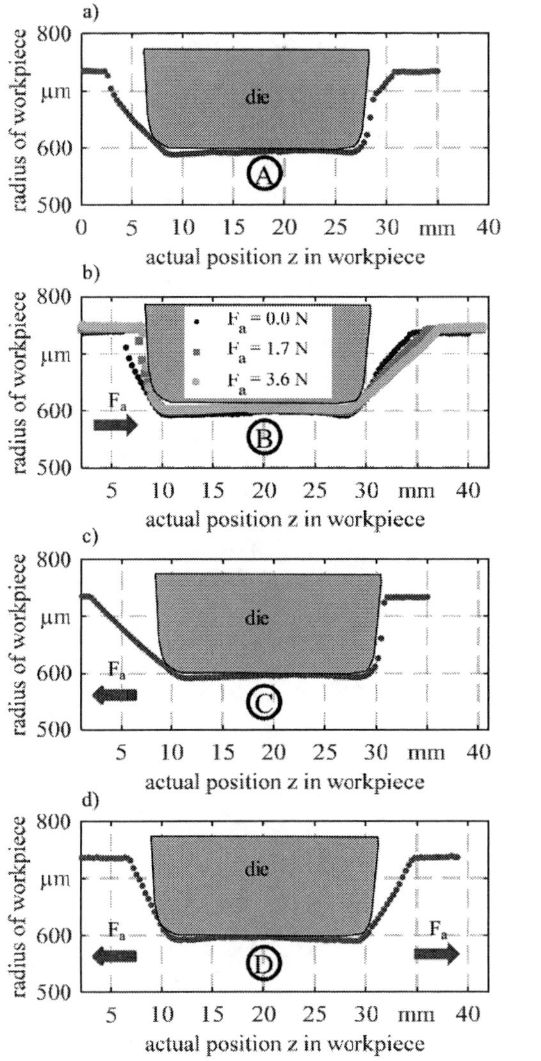

Fig. 9. Workpiece geometry for configuration; a) A, b) B, c) C, d) D.

On one side of the die, the contours of the workpiece and the die fit well and on the other side, contact occurs only in the base and the corner of the die and not anymore on the rising edge. With setting D (both-sides external axial tensile force (Figure 9 d)), this contact between tool and workpiece takes place only in the base of the die during the complete forming process.

To compare the geometry of the workpiece for different settings the symmetry value SYM was defined. This value is based on the relationship of angles on the left and the right side of the workpiece (Equation 2):

$$SYM = \frac{\alpha_{left} - \alpha_{right}}{\alpha_{left} + \alpha_{right}} \quad (2)$$

SYM can take values between -1 and +1. The case where the contour is symmetrical is represented by SYM = 0. This characteristic value does not represent the actual geometry of the transition area between d_0 and d_1 in the workpiece but it shows the direction in which the material flow takes place. In addition, the reproducibility of the forming can be assessed. Figure 10 shows the SYM values of the four configurations and five samples each.

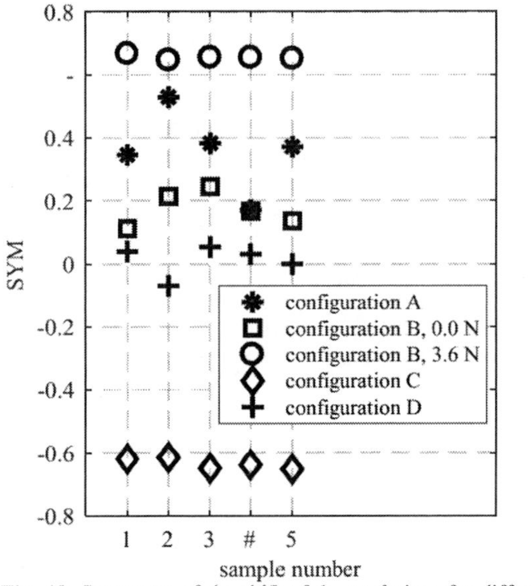

Fig. 10. Symmetry of the shift of the workpiece for different forming settings.

It can be noticed that for configurations A (no external axial forces) and B with a compressive force of 0 N the symmetry and therefore the material flow fluctuates strongly, SYM(A) = 0.36 ± 0.13 and SYM(B0.0 N) = 0.17 ± 0.05. That means manufacturing in those regimes will lead to less reproducibility of the geometry of the workpieces. On the other hand, forming with external axial forces can improve both the reproducibility and the part symmetry, the latter especially with configuration D.

4 Conclusion

In the present study, experimental investigations for a better understanding of the material flow during plunge rotary swaging in micro range were performed. Measured axial shift of the workpiece during forming was correlated with the final workpiece geometry. The main results of the conducted experiments could be drawn:

- The final geometry of the workpiece reacts very sensitively on different external axial forces.
- Low forces at one side can ensure a constant touch between one side of the tool and the workpiece. This contact results in a one-sided imprinting of the tool geometry with high contour accuracy.
- A symmetrical forming is possible if equal tensile axial forces are applied on both sides of the forming zone. Thereby, the workpiece is kept in the same position and following stroke hit the same spot.

The findings from the present investigation are particularly useful for the configuration of the forming in linked parts. As the external axial forces are another effective parameter to control the material flow, the influence on the resulting mechanical properties of the workpieces has also to be investigated. Furthermore, other processing parameters as the calibration time and the feed rate have to be varied, as well as different materials. The calculated standard deviations within this investigation were low, which indicates that size effects do not occur or have minor influence in micro rotary swaging. Thus, numerical process analysis with models that are validated in the macro range could be a possible measure to investigate the material flow in this micro-forming process.

Aknowledgments

The authors gratefully acknowledge the financial support by Deutsche Forschungsgemeinschaft (DFG, German Research Foundation) for the Subprojects A4 "Stoffverdrängen" and C5 “Teileverbunde” within the SFB 747 (Collaborative Research Center) "Mikrokaltumformen - Prozesse, Charakterisierung, Optimierung".

References

1. A. Uhlig, Ph.D. thesis, Technical University of Hannover (1964)
2. L. Chevalier, N. Dahan, J. Mat. Proc. Technol., **31**, 1–2 (1992)
3. R. Paukert, K. Lange, CIRP Ann., **32**, 1, (1983)
4. P. Groche, D. Fritsche, E.A. Tekkaya, J.M. Allwood, G. Hirt, R. Neugebauer, CIRP Ann., **56**, 2, (2007)
5. G. Hirt, R. Kopp, O. Hofmann, M. Franzke, G. Barton, CIRP Ann., **56**, 1, (2007)
6. F. Vollertsen, D. Biermann, H.N. Hansen, I. S. Jawahir, K. Kuzman, CIRP Annals Manuf. Tech. **58**, 566 (2009)
7. X. Lai, L. Peng, P. Hu, S. Lan, J. Ni, Computational Materials Science, **43**, 1003 (2008)
8. B. Kuhfuss, E. Moumi, V. Piwek, Procedings ICOMM2008, 86-91, (2008)
9. P. Wilhelmi, E. Moumi, E. Foremny C. Schenck, B. Kuhfuss, *Proceedings of Newtech Conference*, 75 (2013)
10. P. Wilhelmi, E. Moumi, C. Schenck, B. Kuhfuss, *Proceedings of 5. Kolloquium Mikroproduktion (german)*, (2015)
11. B. Kuhfuss, E. Moumi, K. Tracht, F. Weikert, F. Vollertsen, A. Stephen, *Proceedings 14th ESAFORM*, (2011)

AUTHOR INDEX

AUTHOR INDEX

AUTHOR INDEX